ADVANCED MATHEMATICS

Pure Mathematics

Bob Francis

Devon Further Mathematics Centre

SERIES EDITOR

John Berry

Centre for Teaching Mathematics, University of Plymouth

CONTRIBUTING AUTHORS

Howard Hampson

Sue de Pomerai

Claire Rowland

PROJECT CONTRIBUTORS

Steve Dobbs, Roger Fentem,

Ted Graham, Penny Howe,

Rob Lincoln, Stuart Rowlands,

Stewart Townend, John White

Bob Francis is manager of the Devon Further Mathematics Centre. He has also served as Principal Examiner for MEI Statistics. Bob is currently serving as an advanced practitioner at Devon Further Mathematics Centre.

Howard Hampson was formerly Senior Lecturer and Head of Department in South Devon Further Education College.

Sue de Pomerai works for MEI as Professional Development Leader in the Further Mathematics Support Programme. She also spent many years as a mathematics teacher at The Grammar School, Guernsey.

Claire Rowland was formerly a lecturer at Exeter College.

John Berry is Professor of Mathematics Education at the University of Plymouth and has acted as the Director of their Centre for Teaching Mathematics. As well as being research-active, John is the Mathematics Professor in Residence at Wells Cathedral School in Somerset. John is a consultant to the National Academy of Gifted and Talented Youth delivering workshops to members of The Academy and to primary pupils as part of the SWGate programme. He delivers CPD courses for teachers nationally and internationally and visits schools to run workshops on the teaching and learning of Mathematics. In addition, John also leads a team of teachers delivering a programme of GCSE and A level revision student workshops for TEACHERS FIRST.

William Collins' dream of knowledge for all began with the publication of his first book in 1819. A self-educated mill worker, he not only enriched millions of lives, but also founded a flourishing publishing house. Today, staying true to this spirit, Collins books are packed with inspiration, innovation and practical expertise. They place you at the centre of a world of possibility and give you exactly what you need to explore it.

Collins. Do more

Published by Collins
An imprint of HarperCollins*Publishers*
77-85 Fulham Palace Road
Hammersmith
London
W6 8JB

Browse the complete Collins catalogue at www.collinseducation.com

---

First published in 1996
This edition published in 2011

ISBN 978-0-00-742906-6

10 9 8 7 6 5 4 3 2 1

British Library Cataloguing in Publication Data
A Catalogue record for this publication is available from the British Library

Designed and illustrated by Ken Vail Graphic Design, Cambridge, UK
Project editor Joan Miller
Cover design by Julie Martin
Cover photograph iStock © Fernando Batista
Printed and bound by L.E.G.O. S.p.A., Italy

Contents

Acknowledgements

We are grateful to the following Examination Boards for permission to reproduce questions from their past examination papers and from specimen papers. Full details are given with each question. The Examination Groups accept no responsibility whatsoever for the accuracy or method of working in the answers given, which are solely the responsibility of the author and publishers.

Associated Examining Board (*AEB*)
Northern Examinations and Assessment Board (*NEAB*); also School Mathematics project, 16-19 (*SMP 16–19*)
Oxford and Cambridge Schools Examination Board (*OCSEB*)
Scottish Examination Board (*SEB*)
University of Cambridge Local Examinations Syndicate (*UCLES*)
University of London Examinations and Assessment Council (*ULEAC*)
University of Oxford Delegacy of Local Examinations (*Oxford*); also Nuffield Advanced Mathematics (*Nuffield*)
Welsh Joint Examinations Council (*WJEC*)

We are grateful to the following for permission to reproduce copyright photographs: Dr Tong Brain/Science Photo Library (page 15). Every effort has been made to contact all copyright holders. If any have been inadvertently overlooked the publisher would be pleased to make full acknowledgement at the first opportunity.

Preface

discovering advanced mathematics

Mathematics is not just an important subject in its own right, but also a tool for solving problems. Mathematics at A-level is changing to reflect this: during the A-level course, you must study at least one area of the *application* of mathematics. This is what we mean by 'mathematical modelling'. Of course, mathematicians have been applying mathematics to problems in mechanics and statistics for many years. But now, the process has been formally included throughout A-level maths.

A second innovation is the inclusion of the mathematics of uncertainty – looking at data and probability – in all maths A-levels. In this book, we build on the work done at GCSE and show how important this topic is in modelling. The third innovation is the recognition that numerical methods play an important part in mathematics. This book shows how much easier it is to study this topic now that we have programmable calculators and computers.

Technology is advancing as well. Hand-held calculators that can produce graphs and even do simple algebra are revolutionising the subject. The Common Core for A-level expects you to know how to use appropriate technology in mathematics and be aware that this technology has limitations.

We have written *discovering advanced mathematics* to meet the needs of the new A- and AS-level syllabuses and the Common Core for mathematics. The books provide opportunities to study advanced mathematics while learning about modelling and problem-solving. We show you how to make best use of new technology, including graphics calculators (but you don't need more than a good scientific calculator to work through the book).

In every chapter in this book, you will find:

- an introduction that explains a new idea or technique in a helpful context;
- plenty of worked examples to show you how the techniques are used;
- exercises in two sets, A and B (the B exercises 'mirror' the A set so that you can practise the same work with different questions);
- consolidation exercises that test you in the same way as real exam questions;
- questions from the Exam Boards.

And then, once you have finished the chapter:

- modelling and problem-solving exercises to help you pull together all of the ideas in the chapter.

We hope that you will enjoy advanced mathematics by working through this book.

We thank the many people involved in developing *discovering advanced mathematics*. In particular, we thank the students and staff of Exeter College where much of the material in this book was trialled; Karen Eccles for typing the manuscript, and Jon Reeves for working through the exercises.

Bob Francis, John Berry *January 1996*

PURE MATHS

1 Problem-solving in mathematics

WHAT IS MATHEMATICS?

Let us begin our Advanced Level studies in mathematics by asking the question, *What is mathematics?* Some people call it the analysis of pattern. Some say that it is a language of communication. Others say that mathematics is the art of problem-solving.

In fact, all three definitions are appropriate and, in many cases, all three are applicable at the same time. There are many occasions in this book when we analyse a pattern to generalise a mathematical idea, and we use the language of mathematics to communicate that idea. To begin with, though, we emphasise the art of problem-solving, using mostly the mathematics techniques that are familiar from GCSE studies.

This opening chapter looks at problem-solving in mathematics under five broad headings, each of which features several related problems. These five areas are:

A linear models,
B quadratic models,
C other non-linear models,
D exponential models,
E wave models.

The first problem in each section is laid out on a right-hand page so that you can try solving it on your own. It is then followed by a fully worked solution on the two following pages, which you can use to check your work and clarify any ideas you found tricky. Following this, in each section there are two or three problems for you to tackle without any help. There are answers to these problems at the back of the book.

For each problem you may be asked to:

- complete a table of values,
- draw a graph to illustrate the relationship between two variables,
- solve the problem using your graph,
- formulate, where appropriate, an algebraic relationship between the variables,
- check and extend your work using a graphics calculator or computer package.

1 Problem-solving in mathematics

Much of the mathematics in this book springs from the type of problems set in this chapter. You could work through the whole of this chapter at the beginning of this A Level course, to get a flavour of the types of models to be found in mathematics. Alternatively you could use each section separately as an introduction to one of the main chapters which follow. Wherever appropriate, you will see mathematics in context as well as in theory. Applications of mathematics will appear throughout each topic as it is developed, so you can both use and apply mathematics as well as learn new mathematics for its own sake.

Throughout the book you will be encouraged to use both graphics calculators and computers effectively. In this chapter there are numerous opportunities to confirm and extend your results using a graphics calculator, once you have obtained initial solutions by drawing graphs by hand. Once you have modelled a problem on your calculator, you can very easily change the ground rules and investigate 'What happens if ... ?' Equally, you could solve these problems with the help of a spreadsheet or graph plotter. Use whatever technology you have at hand to bring your mathematics to life.

Example 1.1

Linear models

Problem

> *Yellacabs* taxis charge their customers a fixed cost of £1 plus 40 pence per mile.

1 *Copy and complete the following table of values.*

Distance, d (miles)	0	5	10	15	20
Cost, c (£)	1	3			

2 *Using suitable scales, draw a graph to show the cost of a journey against the number of miles travelled.*

3 *From your graph find:*
a) *the cost of a journey of 8 miles,*
b) *how far you can travel for £6.20.*

> Another taxi-firm, *Maxitaxis*, have no fixed charge, but customers pay 60 pence per mile.

4 *On the same axes, draw a graph of cost against distance for* ***Maxitaxis.***

5 *Use your graphs to:*
a) *find which firm is cheaper, and by how much, for a journey of:*
i) *3 miles,* ***ii)*** *12 miles,*
b) *find which firm offers the greater travelling distance for a cost of £7.50.*

6 *Devise a simple strategy for choosing between* ***Yellacabs*** *and* ***Maxitaxis*** *if you are a potential customer.*

7 *Find formulae, one for* ***Yellacabs*** *and one for* ***Maxitaxis,*** *connecting the cost of a journey (in £) and the distance travelled (in miles).*

CALCULATOR ACTIVITY

Check your work using a graphics calculator

Firstly scale your axes using the range settings:

$x_{min} = 0$; $x_{max} = 20$; $x_{scl} = 5$
$y_{min} = -2$; $y_{max} = 10$; $y_{scl} = 1$.

Using variables x and y (instead of d and c), enter equations:
$y = 0.4x + 1, y = 0.6x$.

Use the TRACE facility to confirm all your results.

Use the ZOOM facility to obtain the desired degree of accuracy.

Problem solution

1 *A taxi fare with* **Yellacabs** *consists of a fixed cost of £1 plus a variable cost of 40 pence for each mile travelled. So:*

a 5-mile journey would cost £1 + 5 × 40 pence = £3

a 10-mile journey would cost £1 + 10 × 40 pence = £5.

The complete table is shown here.

Distance, d (miles)	0	5	10	15	20
Cost, c (£)	1	3	5	7	9

2 *The graph below shows the cost of a journey in a* **Yellacab** *against the number of miles travelled.*

3 *From the graph we can read off:*

a) *an 8-mile journey costs £4.20,*

b) *for £6.20 you can travel a distance of 13 miles.*

Note: *If you use 2 mm graph paper, you will be able to obtain a reasonably accurate result in this case. Sometimes, though, only an estimate is possible, due to the scales chosen.*

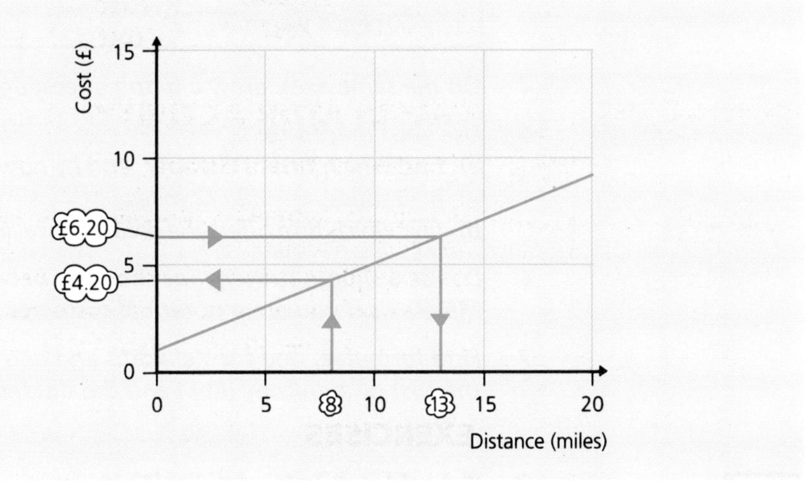

4 ***Maxitaxis*** *charge 60 pence per mile (i.e. no fixed cost, just a variable cost), which is equivalent to £3 for every five miles.*

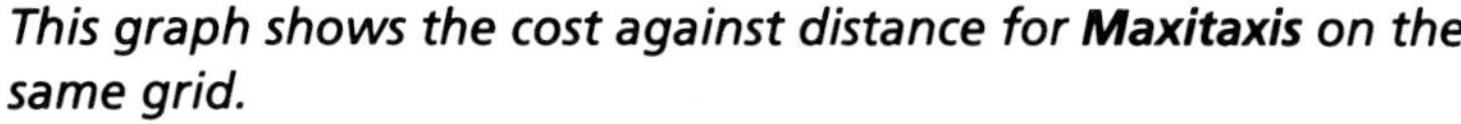

*This graph shows the cost against distance for **Maxitaxis** on the same grid.*

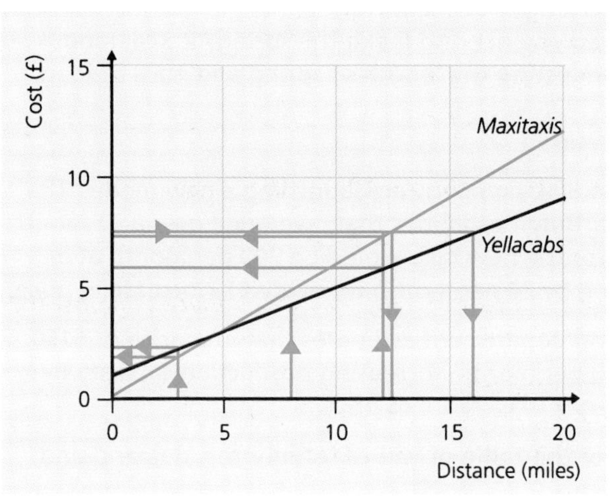

5 *From the graph we can read off :*

*a) i) For a 3-mile journey, **Maxitaxis** are cheaper by 40 pence,*

*ii) For a 12-mile journey, **Yellacabs** are cheaper by £1.40.*

*b) **Yellacabs** offer 16.25 miles compared with **Maxitaxis** who offer 12.5 miles. So:*

6 *For journeys less than five miles **Yellacabs** are cheaper, for journeys greater than five miles **Maxitaxis** are cheaper.*

7 *Let c represent cost and d represent distance, then the formulae relating cost and distance for the two firms are:*

Yellacabs: $c = 0.4d + 1$ ***Maxitaxis:*** $c = 0.6d$

variable cost *fixed cost* *variable cost*

CALCULATOR ACTIVITY

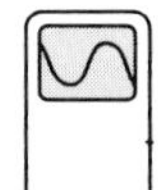

Graphics calculator

You can use a graphics calculator to investigate changes in ways of charging customers.

Find out how your answers to questions **3** to **6** would change if, say:

Yellacabs increased their mileage rate to 50 pence,

Maxitaxis introduced a fixed charge, but reduced their mileage rate.

EXERCISES

1 **Stretching a spring**

A spring stretches by 1 cm for every 100 g of load.

When the spring carries a load of 300 g the length of the spring is 10 cm.

a) Copy and complete the table of values for length of spring in centimetres against mass of a load in grams.

Mass of load (g)	0	100	200	300	400	500
Length of spring (cm)				10		

b) Using suitable scales, draw a graph to show the length of the spring against load.

c) From your graph find:
  i) the length of the spring when the load is 180 g,
  ii) the unstretched length of the spring,
  iii) how large a load you need to make the spring 12.3 cm long.

d) Find a formula connecting the length of spring and mass of load.

2 Marketing a new bicycle

The *Everest Mountain Bike Company* are launching a new model. The market research department estimates that, averaged over a 12-month period, the demand for the new bike would be 200 per month at a price of £150, reducing by 20 per month for every £10 increase in price.

The production control department estimates that at a price of £150 they could supply 100 per month, increasing production capacity by 15 per month for every £10 increase in price.

a) Copy and complete this table of values for monthly demand and supply against price.

Price (£)	150	160	170	180	190	200	210
Demand	200						
Supply	100						

b) On the same axes draw graphs of demand against price and supply against price.

c) What price should be charged for supply to equal demand? How many bikes per month could the company sell at this price?

d) For what range of prices would you think this model to be valid?

e) Find formulae connecting **i)** demand and price, **ii)** supply and price.

3 World record for the mile

The table shows some of the world records (in minutes : seconds) set for the mile from 1913 to 1985.

Athlete	Year	Time taken	Athlete	Year	Time taken
John Paul Jones	1913	4:14.4	Roger Bannister	1954	3:59.4
Paavo Nurmi	1923	4:10.4	Peter Snell	1964	3:54.1
Glen Cunningham	1933	4:06.8	John Walker	1975	3:49.4
Arne Andersson	1943	4:02.6	Steve Cram	1985	3:46.3

a) Using suitable scales plot times taken against year of race.

b) Draw in a line of best fit and find an equation for it.

c) Predict the likely year when the mile will be run in:
  i) 3:40, **ii)** 3:30.

d) Is it reasonable to use the same method to estimate when we are likely to see the first three-minute mile?

Example 1.2

Quadratic models

Problem

Chris is a keen cricketer.

He is practising throwing the ball from the boundary to the wicket-keeper, a distance of 50 m.

An equation to describe the flight-path (trajectory) of the ball is

$y = x - 0.02x^2$

where y represents the height (in metres) of the ball when it has travelled x metres horizontally.

1 *Copy and complete the following table of values.*

x (metres)	0	10	20	30	40	50
y (metres)						

2 *Using suitable scales, draw a graph to show the height of the ball (y metres) against x, the horizontal distance travelled, also in metres.*

3 *From your graph find:*
a) *the height of the ball when $x = 8$,*
b) *how far the ball is, horizontally, from Chris when it is at a height of 10 m,*
c) *the maximum height of the ball during its flight.*

Trevor and Winston also throw the ball from the boundary and aim for the wicket-keeper.

Equations describing the trajectories for their throws are as follows:

Trevor: $y = x - 0.025x^2$ Winston: $y = x - 0.018x^2$

4 *On the same axes, draw graphs to represent these two trajectories.*

5 *Re-work question **3**, for both Trevor and Winston.*

6 *How far from the wicket-keeper will the ball land when:*
a) *Trevor throws the ball,*
b) *Winston throws the ball?*

CALCULATOR ACTIVITY

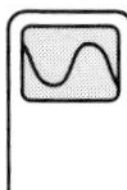

Check your work using a graphics calculator.

Problem solution

1 *The complete table is shown below.*

x (metres)	0	10	20	30	40	50
y (metres)	0	8	12	12	8	0

2 *The graph below shows the height of the ball (y metres) against x, the horizontal distance travelled.*

3 *From the graph we can read off:*

a) *when $x = 8$, height of ball (y) = 6.7 m,*

b) *when the ball is at a height of 10 metres, it is either about 14 m or about 36 m from Chris,*

c) *the maximum height of the ball is 12.5 m, when it is 25 m from Chris, half-way to the wicket-keeper.*

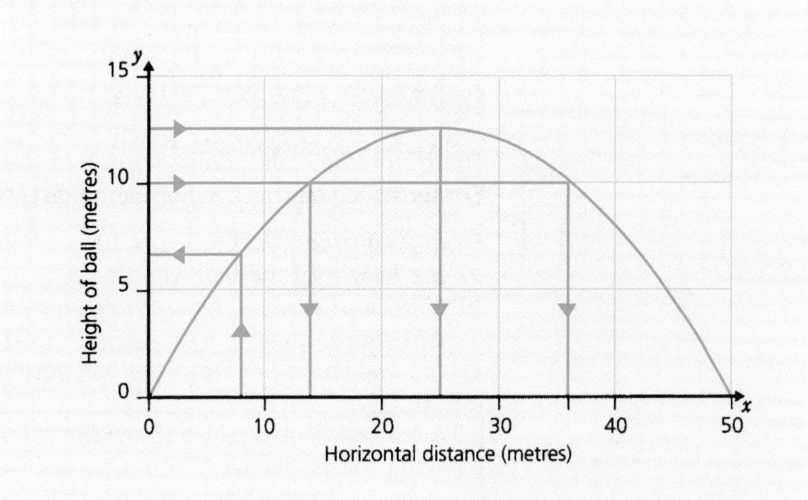

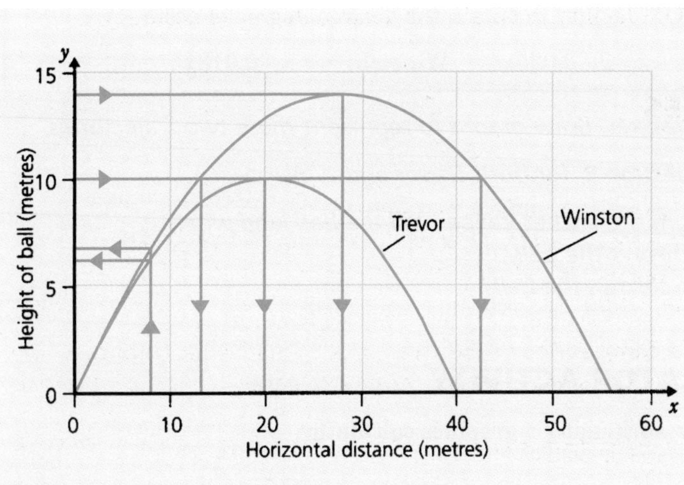

4 *When Trevor and Winston throw the cricket ball from the boundary, both throw the ball at the same angle to the ground, but with different speeds, as shown in the graph (left). Trevor's throw fails to reach the wicket-keeper, whereas Winston's throw sails over the wicket-keeper's head!*

5 *From the graph we can read off the following results.*
Trevor:
a) *When $x = 8$, height of ball (y) = 6.4 m.*
b) *When the ball is at a height of 10 metres, it is 20 m from Trevor.*
c) *The maximum height of the ball is 10 m, when it is 20 m from Trevor, less than half-way to the wicket-keeper.*

Winston:
a) *When $x = 8$ m, height of ball (y) = 6.85 m.*
b) *When the ball is at a height of 10 metres, it is either about 13.5 m or about 42.5 m from Winston.*
c) *The maximum height of the ball is about 13.9 m, when it is 28 m from Winston, more than half-way to the wicket-keeper.*

6 ***a)*** *When Trevor throws the ball, it lands 10 metres short of the wicket-keeper.*
b) *When Winston throws the ball, it lands about 5.5 metres beyond the wicket-keeper.*

CALCULATOR ACTIVITY

Graphics calculator

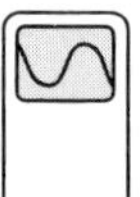

Firstly scale your axes using the range settings:

$x_{min} = 0$; $x_{max} = 70$; $x_{scl} = 10$
$y_{min} = -5$; $y_{max} = 15$; $y_{scl} = 2$.

Enter equations: $y = x - 0.02x^2$ (Chris),
$y = x - 0.025x^2$ (Trevor),
$y = x - 0.018x^2$ (Winston).

Use the TRACE facility to confirm all your results.

Use the ZOOM facility to obtain the desired degree of accuracy.

EXERCISES

1.2

1 **Picture framing**

Ubeen Framed offer a picture-framing service.
The cost of framing square pictures is made up of two parts:
cost of glass: 50 pence per square foot,
cost of frame: £1.50 per foot.

a) Copy and complete the following table of values for total cost (in £) against length of side of frame (in feet).

Length of side of frame (feet)	1	2	3	4	5
Cost (£)					

b) Using suitable scales, draw a graph to show the cost against the length of side.

c) From your graph find:

i) the cost of framing a picture of side 2 feet 6 inches,

ii) how large a picture you could have framed for £25.

d) Find a formula connecting the cost and the length of side of the frame.

e) *Ubeen Framed* also offer a home delivery service, within the local area, for which they charge a fixed fee of £5.

i) On the same axes, draw a graph of total cost (including home delivery) against length of frame.

ii) Re-work part **c)**, assuming the picture is delivered to the customer's home.

iii) How is the formula in **d)** altered?

2 Stopping distance

The highway code contains information about shortest stopping distances for cars on a dry surface, summarised by the following table.

Speed (mph)	Thinking distance (ft)	Braking distance (ft)	Stopping distance (ft)
20	20	20	
30	30		75
40	40	80	
50	50		175
70	70	245	

a) Given that
stopping distance = thinking distance + braking distance,
copy and complete the table.

b) Plot a graph of stopping distance (in feet) against speed (in mph).

c) Use your graph to estimate:

i) the shortest stopping distance for a car travelling at 60 mph,

ii) the speed of a car which stopped in 70 feet.

d) Find a relationship between the overall stopping distance, d, and speed, v, of the form:

$$d = v + kv^2.$$

e) Using your equation from **d)**, check your results to **b)** and **c)** on a graphics calculator.

f) In wet conditions, the minimum recommended *braking* distance is doubled. On the same axes plot a graph of stopping distance (in feet) against speed (in mph) for wet conditions. How are your answers to **c)** modified for wet conditions?

Example 1.3

Other non-linear models

Problem

An open box is to be formed by cutting four squares of side x cm from the corners of a sheet of card 20 cm square.

You are asked to investigate how the volume of the box varies as x varies.

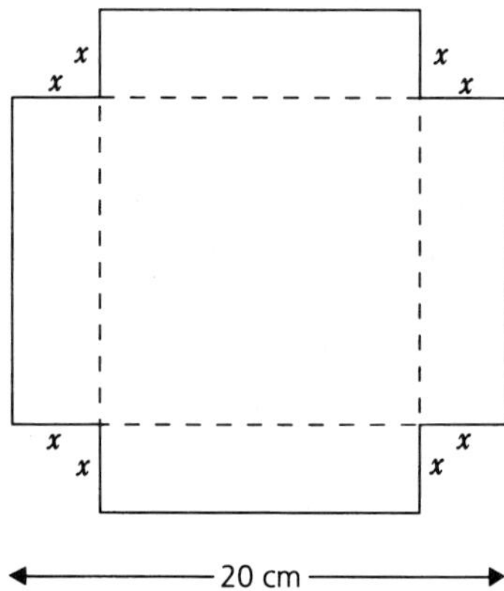

1 *Find a formula connecting the volume of the box, V cm^3, and the length x cm.*

2 *Copy and complete the following table of values.*

x (cm)	0	2	4	6	8	10
V (cm^3)						

3 *Using suitable scales, draw a graph to show the volume, V cm^3, of the box against the length x cm.*

4 *From your graph find:*
a) *the volume for which **i)** $x = 2.5$ cm, **ii)** $x = 6.6$ cm,*
b) *which x-values correspond to a volume of 0.5 litre,*
c) *the value of x for which the box has maximum volume. What is the maximum volume?*

CALCULATOR ACTIVITY

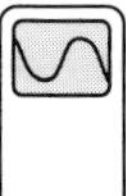

Check your work using a graphics calculator.

Problem solution

1 *The dimensions of the open box are:*

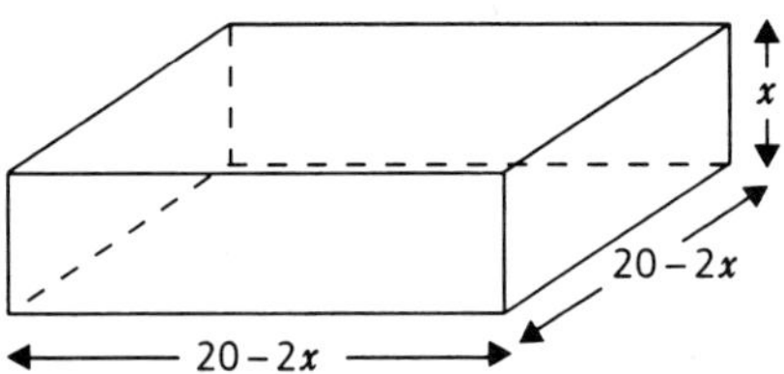

length and width = $20 - 2x$
height = x.
Therefore the volume is given by:
$V = x(20 - 2x)^2$.

2 *Using the formula we can obtain the complete table of values, as shown below.*

x (cm)	0	2	4	6	8	10
V (cm^3)	0	512	576	384	128	0

3 *The graph below shows the volume (V) against the value of* x.

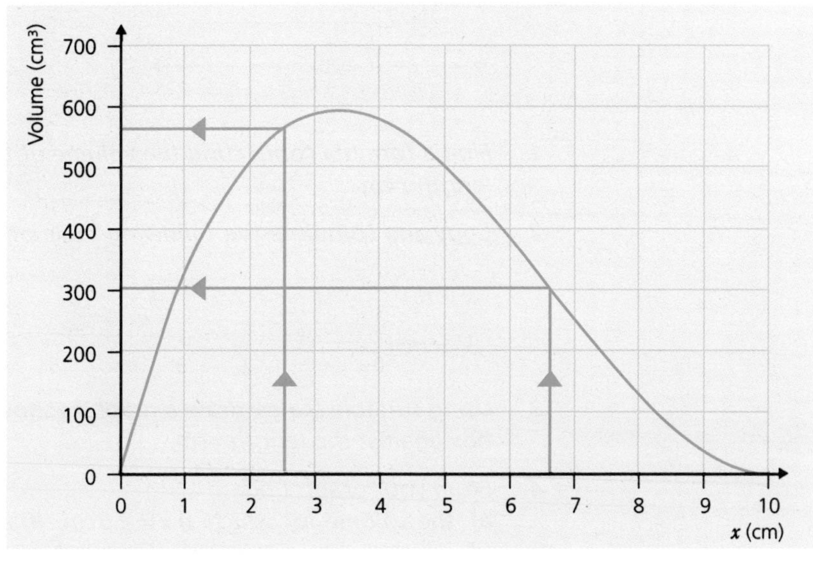

4 *From the graphs (above and opposite) we can read off:*

a) ***i)*** *when* $x = 2.5$, $V = 562.5$,
ii) *when* $x = 6.6$, $V = 305$
(see the diagram above),

b) *when* $V = 0.5$ litre $= 500\,\text{cm}^3$, $x = 1.9$ *or* 5
(see the diagram opposite, top)

c) *maximum volume is* $V = 593$, *when* $x = 3.33$ *(see the diagram opposite, beneath).*

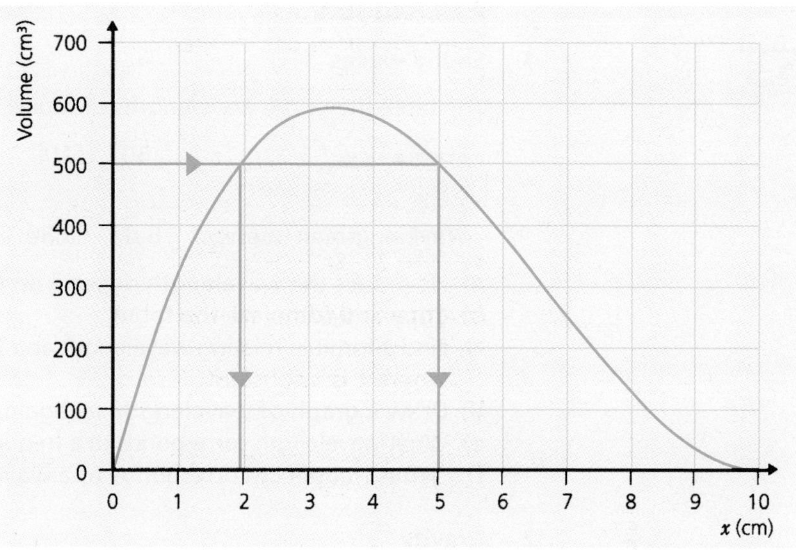

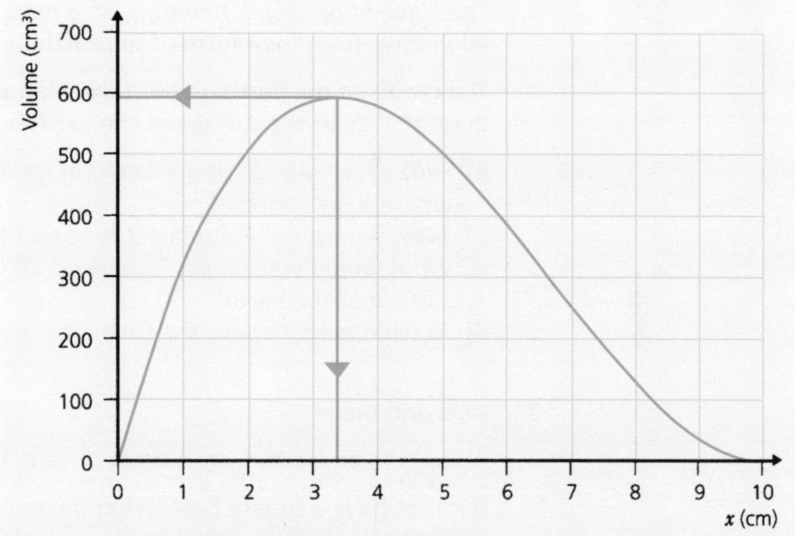

CALCULATOR ACTIVITY

Graphics calculator

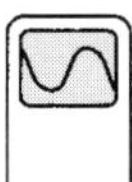

Firstly scale your axes using the range settings:

$x_{min} = 0$; $x_{max} = 10$; $x_{scl} = 1$
$y_{min} = -100$; $y_{max} = 700$; $y_{scl} = 100$.

Using variables x and y (instead of x and V), enter the equation:

$$y = x(20 - 2x)^2.$$

Use the TRACE facility to confirm all your results.
Use the ZOOM facility to obtain the desired degree of accuracy.

EXERCISES

1.3

1 Sound waves

The table shows the wavelengths of sound waves of various frequencies.

Frequency, f (cycles per second)	50	100	150	200		300
Wavelength, w (metres)	6.60	3.30	2.20		1.32	

a) How does the wavelength depend on the frequency?
b) Copy and complete the table.
c) Find a formula relating wavelength and frequency in the form: $w = \frac{k}{f}$, where k is a constant.
d) Draw a graph of wavelength, w, against frequency, f.
e) What wavelength corresponds to a frequency of 400 cycles per second?
f) What frequency corresponds to a wavelength of 1.5 metres?

2 Gravity

The force of gravity, F newtons, on a mass of 1 kg, at a distance d kilometres from the centre of the Earth, is inversely proportional to d^2.

The radius of the Earth is about 6400 km and the force of gravity on a mass of 1 kg, at the surface of the Earth is about 10 newtons.

a) Find a formula relating F and d in the form: $F = \frac{k}{d^2}$, where k is a constant.
b) Draw a graph of F against d for suitable values of d.
c) What would you expect F to be on an object 9000 m from the surface of the Earth?
d) At what height above the Earth's surface would F fall to 8 newtons?

3 Nets and boxes

A box is to be made, from a sheet of card, to have a volume of 500 cm^3.

If it is to have a square base, what dimensions would you choose to minimise the surface area for:

a) an open box (i.e. a box without a lid)
i) Let the length of side of base be x cm and its height be h cm. Explain why $h = \frac{500}{x^2}$.
ii) Show that the formula for the surface area, A, in terms of x, for the open box is $A = x^2 + \frac{2000}{x}$.
iii) Copy and complete the table for the open box.

x	2	4	6	8	10	12	14	16
A	1004	516						

iv) Plot a graph of surface area, A, against length of base, x.
v) What dimensions would you choose to minimise the total surface area?
b) a closed box.

Example 1.4

Exponential models

Problem

A colony of bacteria has a population of 1 million at midday on Sunday.

During the next week, the colony's population **doubles** every 24 hours.

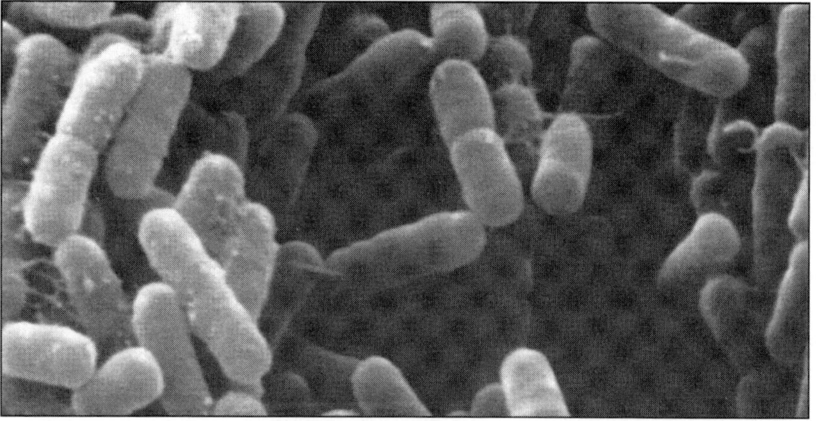

1 *Copy and complete the following table of population values, taken at midday each day (day 0 is Sunday, day 1 is Monday, etc.).*

Day	0	1	2	3	4	5	6	7
Population (millions)	1	2	4					

2 *Using suitable scales, draw a graph to show how the population grows during the week.*

3 *From your graph find:*

a) *the population at midnight on Thursday,*
b) *the day and time by which the population has reached 70 million,*
c) *the rate at which the population is growing on:*
i) *Wednesday (midday),* ***ii)*** *Friday (midnight).*

4 *Find a formula that gives population in millions (P) after t days.*

CALCULATOR ACTIVITY

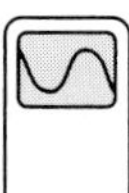

Check your work using a graphics calculator.

Problem solution

1 *Since the population is 1 million on day 0 and doubles every 24 hours, the populations on subsequent days are all powers of 2.*

The complete table is shown below.

Day	0	1	2	3	4	5	6	7
Population (millions)	1	2	4	8	16	32	64	128

2 *The graph below shows the population (in millions) against the number of days.*

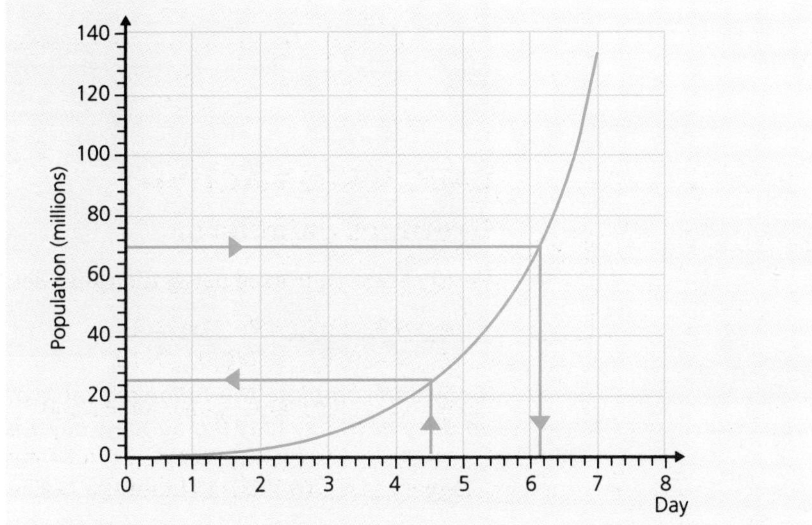

3 *From the graph we can read off the following results.*

a) The population at midnight on Thursday is about 23 million.

b) The population reaches 70 million by about 1330 on Saturday.

c) The rate at which the population is growing at any one time is estimated by finding the gradient of the tangent to the graph. To do this, we draw in the tangent and choose two suitable points on the tangent from which a right-angled triangle may be constructed. The gradient of the line is the difference in population values divided by the difference in times. The graph on the next page shows the method to estimate the rate at which the population is growing on Wednesday (midday), day 3 and Friday (midnight), day 5.5.

d) The rates at which the population is growing are given by:

i) on Wednesday (midday): $\frac{11}{2}$ *= 5.5 million per day,*

ii) on Friday (midnight): $\frac{90}{3}$ *= 30 million per day.*

4 *Let P represent population and t represent time, then the formula relating population (in millions) and time (in days) is:*

$P = 2^t$.

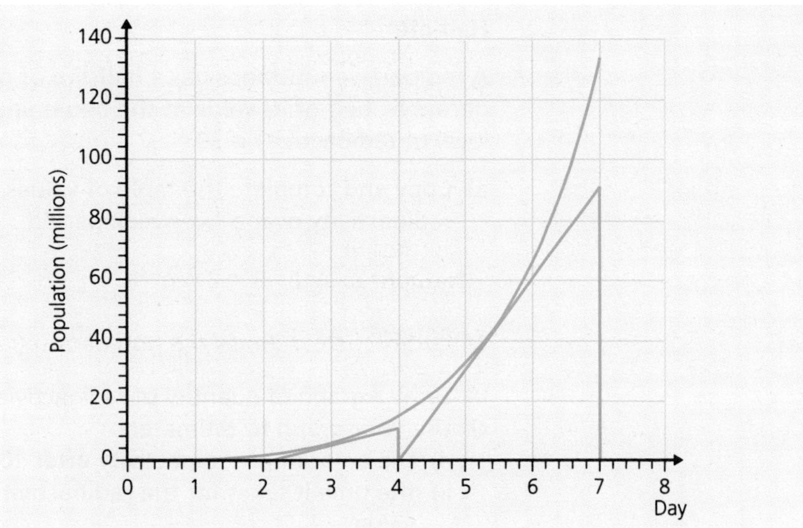

CALCULATOR ACTIVITY

Graphics calculator

Firstly scale your axes using the range settings:

$x_{\min} = 0;\quad x_{\max} = 7;\quad x_{\text{scl}} = 1$
$y_{\min} = -20;\ y_{\max} = 140;\ y_{\text{scl}} = 10.$

Using variables x and y (instead of t and P), enter the equation: $y = 2^x$.

Use the TRACE facility to confirm all your results.
Use the ZOOM facility to obtain the desired degree of accuracy.

EXERCISES

1 **Growing trees**

The quantity of timber in a fast growing forest is estimated to increase by 15% per annum. After planting, the quantity of timber is 10 units.

a) Copy and complete the table of values, showing quantities of timber over a five-year period.

Time, t (years)	0	1	2	3	4	5
Quantity of timber, q	10	11.5				

b) Draw a graph of quantity of timber, q, against time in years, t.
c) Use your graph to estimate:
  i) the quantity of timber after $2\frac{1}{2}$ years,
  ii) the time it takes for the quantity of timber to double its initial size.
d) Find a formula that gives q in terms of t.
e) Check your results to **a), b)** and **c)** using a graphics calculator.

2 Half-life

A radioactive substance has a half-life of one week, i.e. every week it decays by half of its value at the beginning of the week. Its initial level of radioactivity is 20.

a) Copy and complete the table of values, showing levels of radioactivity over a five-week period.

Time, t (weeks)	0	1	2	3	4	5
Radioactivity, r	20	10				

b) Draw a graph of quantity of radioactivity, r, against time in weeks, t.
c) Use your graph to estimate:
  i) the amount of radioactivity after 10 days,
  ii) the time it takes for the radioactivity to reach 10% of its initial value.
d) Find a formula that gives r in terms of t.
e) Check your results to **a)**, **b)** and **c)** using a graphics calculator.

3 Cooling

A flask of water is at a temperature of 100°C. After five minutes it has cooled to 80°C.

a) Assuming that the fall in temperature in any five-minute period is proportional to the temperature at the beginning of the period, copy and complete the following table of values.

Time, t (minutes)	0	5	10	15	20	25
Temperature, T (°C)	100	80	64			

b) Draw a graph of temperature, T, against time in minutes, t.
c) Use your graph to estimate:
  i) the temperature after 8 minutes,
  ii) the time it takes for the water to cool to 50°C.
d) Find a formula that gives T in terms of t.
e) Check your results to **a)**, **b)** and **c)** using a graphics calculator.

Example 1.5

Wave models

Problem

The height of the tide in Portbury harbour during a period of 12 hours (midnight to midday) is given in the table below.

Time (hours)	0	1	2	3	4	5	6	7	8	9	10	11	12
Height (metres)	6.6	5.5	4.0	2.5	1.4	1.0	1.4	2.5	4.0	5.5	6.6	7.0	6.6

1 *Using suitable scales, draw a graph to show the height of the tide against time.*

2 *From your graph find:*
a) *the height of the tide at:*
i) *02.30,* **ii)** *09.45,*
b) *the times at which the height was:*
i) *4.0 metres,* **ii)** *6.0 metres,*
c) *the time of:*
i) *low tide,* **ii)** *high tide.*

3 *Fishing vessels can safely leave or enter the harbour provided the depth of water is at least two metres. For how long is it not safe for a boat to enter or leave the harbour?*

A formula that gives the height in terms of time is $h = 3\cos\big(30(t+1)\big) + 4$.

Check that the values in the table above may be found by substituting values of t in this equation.

CALCULATOR ACTIVITY

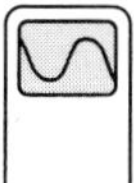

Check your work using a graphics calculator

Firstly make sure that you are in degree MODE.

Secondly scale your axes using the range settings:

$x_{\text{min}} = 0$; $x_{\text{max}} = 12$; $x_{\text{scl}} = 1$
$y_{\text{min}} = 0$; $y_{\text{max}} = 8$; $y_{\text{scl}} = 1$.

Using variables x and y (instead of t and h), enter the equation:

$$y = 3\cos\big(30(x+1)\big) + 4.$$

Use the TRACE facility to confirm all your results.

Use the ZOOM facility to obtain the desired degree of accuracy.

Problem solution

1 *The graph below shows the height of tide against time for the period of time midnight to midday.*

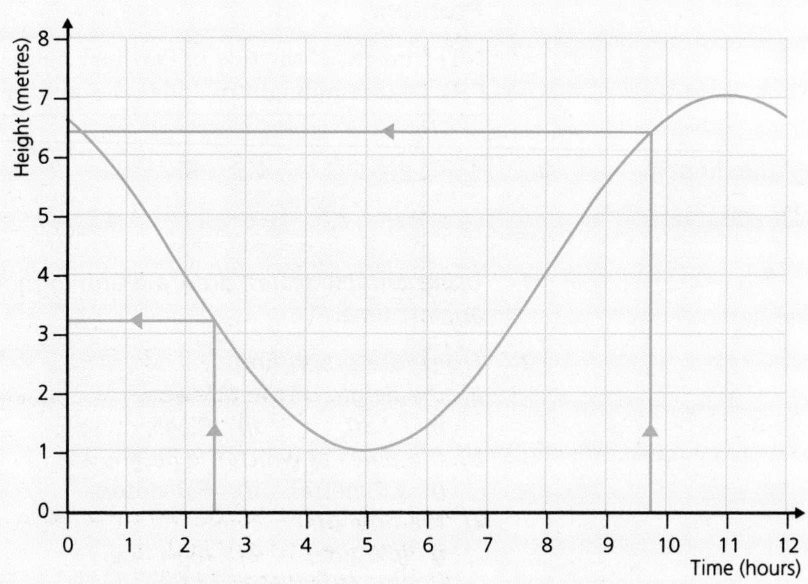

2 ***a)*** *From the graph above we can read off the height of tide:*

i) *at 02.30 = 3.2 metres,*

ii) *at 09.45 = 6.4 metres.*

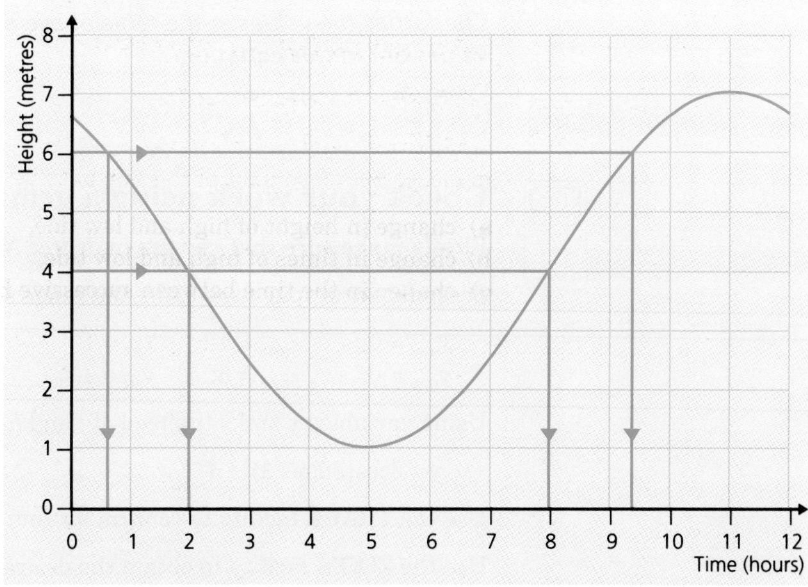

b) *From the graph above we can read off that the height of tide was:*

i) *4.0 metres at 02.00 and 08.00,*

ii) *6.0 metres at 00.36 and 09.24.*

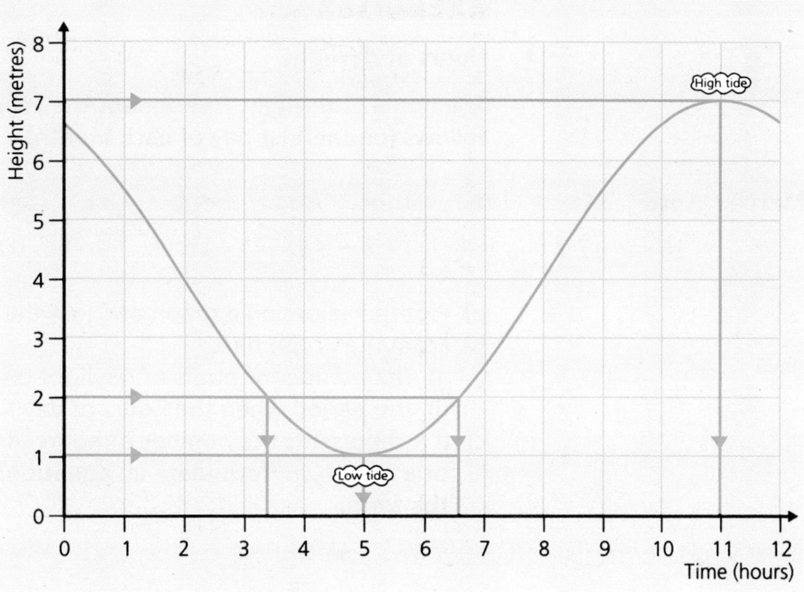

c) From the graph above, we can read off:

i) low tide was at 05.00,

ii) high tide was at 11.00.

3 *Also from the graph above, we can deduce that the tide is two metres in depth at about 03.25 and 06.35, so it is not safe for a boat to enter or leave harbour for a period of just over three hours.*

CALCULATOR ACTIVITY

Graphics calculator

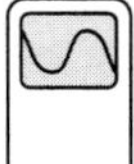

Modify the equation you were given, i.e. $h = 3\cos\big(30(t+1)\big) + 4$, to model the movement of the tide to take account of, say:

a) change in height of high and low tide,

b) change in times of high and low tide,

c) change in the time between successive high tides.

EXERCISES

1.5

1 Hours of daylight

The hours of daylight throughout the year for Casterton are as follows (on the 21st day of each month).

March	April	May	June	July	Aug.	Sept.	Oct.	Nov.	Dec.	Jan.	Feb.
12	15	17.2	18	17.2	15	12	9	6.8	6	6.8	9

a) Plot the values on a graph and join the points with a smooth curve.
b) From the graph find:
i) the number of hours of daylight on 1 October,
ii) the period when the hours of day light are less than 7.5 hours.
c) If h represents the number of hours of daylight on the twenty-first of a month, m, formulate an equation, to give h in terms of m, of the form:

$$h = c + a\sin(bm)$$

where $m = 0$ for March, $m = 1$ for April, ..., $m = 11$ for February, and a, b and c are constants to be found.
d) Use your equation from **c)** to plot the graph on a graphical calculator and check your results from **a)** and **b)**.

2 The big wheel

A Ferris wheel at a fair ground has radius 8 m.

When in motion it completes one revolution in 36 seconds.

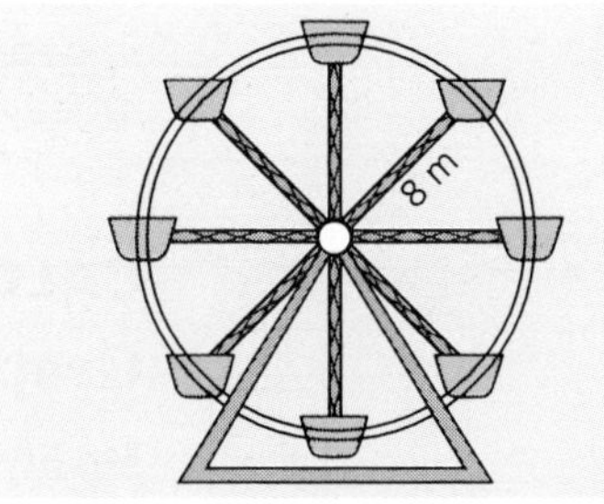

a) Copy and complete the table of values, showing height above the ground against time.

Time (seconds)	0	3	6	9	12	15	18	21	24	27	30	33	36
Height (metres)	1	2.1	5	9	13	15.9	17						1

b) Using suitable scales, draw a graph of height against time.
c) From the graph find:
i) the height of a Ferris wheel car after 25 seconds,
ii) the period of time during which the car is more than 15 m above the ground.
d) If h represents the height of a Ferris wheel car t seconds after the wheel starts to move, find an equation, to give h in terms of t, of the form $h = c - a\cos(bt)$.
e) Use your equation from **d)** to plot the graph on a graphical calculator and check yours results from **a)**, **b)** and **c)**.

Summary

In each of these sections you have been solving real problems using mathematics.

This is called ***mathematical modelling****. Most of this book is about the* ***mathematical methods*** *that are needed to solve problems and to develop further techniques. For example, you will need* ***algebra*** *to be able to understand rates of change. However, it is important to be able to use your mathematics to solve real problems. So the theme of mathematical modelling will run through the book.*

An important part of solving a real problem is called the ***mathematical model****. For the problems in this chapter the mathematical models are described by* ***graphs*** *or* ***equations****. The following table illustrates this using two of the problems in this chapter.*

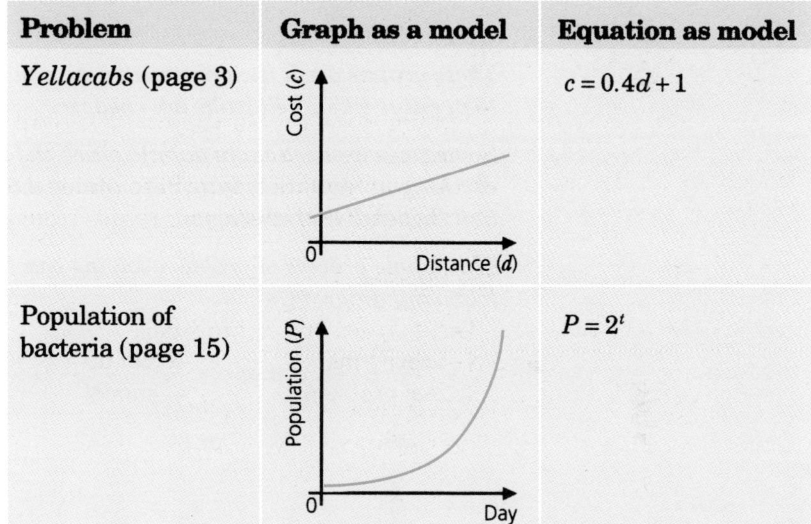

Problem	Graph as a model	Equation as model
Yellacabs (page 3)		$c = 0.4d + 1$
Population of bacteria (page 15)		$P = 2^t$

Here is a list of skills that may be required to solve a real problem:

- *understand the problem,*
- *choose suitable variables (and units),*
- *label the variables,*
- *find the relationship between the variables.*

For example, in the ***Yellacabs*** *problem it was appropriate to choose the cost (£) and distance (miles) as variables. We labelled these variables c and d. The relationship between c and d is then* $c = 0.4d + 1$.

Having found a mathematical model for a situation, the model is then used in two ways:

- *to describe the situation,*
- *to make predictions about the future.*

For example, in the taxis problem we had graphical models for ***Yellacabs*** *and* ***Maxitaxis****.*

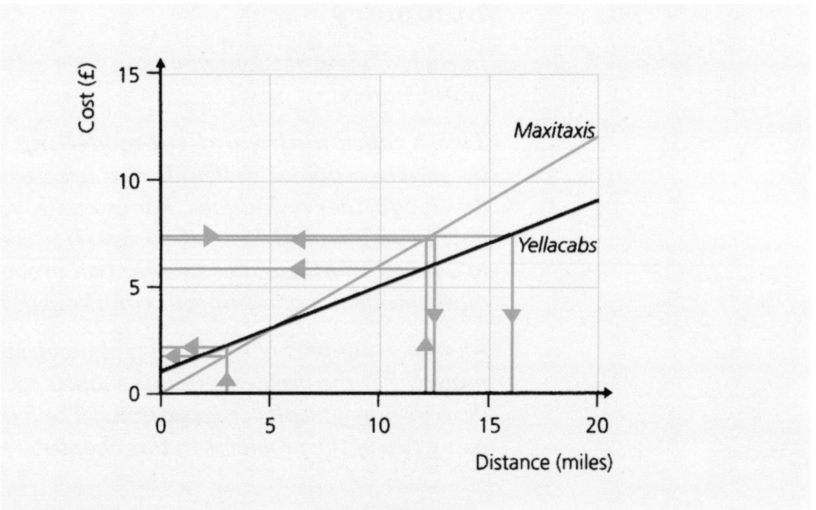

*These graphs are a **description** of the situation. The graphs were used to **predict** when Yellacabs are cheaper.*

Sometimes we need more data to check the model and in the light of the checking process we may refine our model to give a better description and (hopefully) predictions.

This whole process of problem solving can be represented in the following diagram.

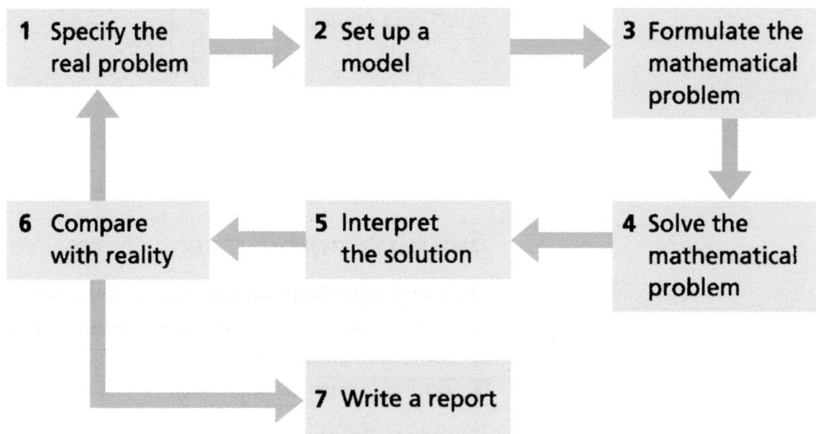

This diagram will appear regularly in the margin as we use mathematics in problem-solving. You do not need to remember the diagram. It is there to help you identify the problem-solving skills that you will be developing as part of your advanced mathematics course.

As you work through this book, you will learn new mathematical methods, which will give you better tools to describe and predict.

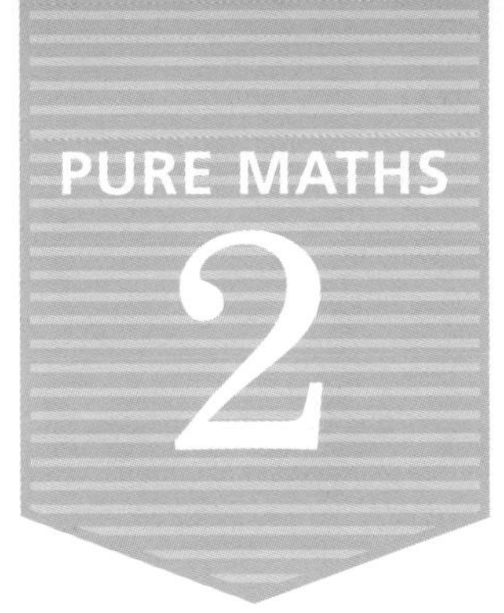

Basic algebra

In this chapter we:

- *review the basic techniques of solving linear equations and inequalities*
- *manipulate expressions involving brackets*
- *practise making a variable the subject of an equation.*

SIMPLE LINEAR EQUATIONS

Solving for x

An **equation** is a mathematical statement that two quantities are equal. It can thought of as a balance. To maintain the balance we have to do the same to both sides (e.g. subtract 5, divide by 9, multiply by 4, etc.).

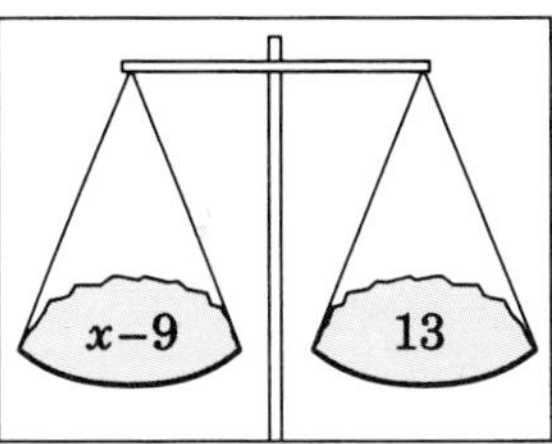

Exploration 2.1

Balancing equations

For each equation, find a value for x that makes it balance.

1 $x + 7 = 12$ **2** $5x = 45$

3 $\frac{x}{4} = 7$ **4** $3x - 4 = 23$

5 $25 - 4x = 9$ **6** $2x - 7 = 8 - 3x$

In each case describe what you did to solve the equation.

Solving equations

For any equation we may be able to find the solution without writing anything down. In solving equations we use the fact that addition is the inverse of subtraction (and vice-versa) and multiplication is the inverse of division (and vice-versa). However, a more formal approach is essential when there is more manipulation to do. We need a strategy for solving equations.

The following examples illustrate how to solve equations by 'doing the same to both sides', in each case to find the value of x (or the unknown given).

Example 2.1

Solve $x - 9 = 13$.

Solution

Add 9 to both sides $\Rightarrow \quad x - 9 + 9 = 13 + 9$

$\Rightarrow \quad x = 22$

Example 2.2

Solve $30 = 8x$.

Solution

Divide both sides by 8 $\Rightarrow \dfrac{30}{8} = x$

$\Rightarrow x = 3.75$

Example 2.3

Solve $5t + 18 = 3$.

Solution

Subtract 18 from both sides $\Rightarrow 5t + 18 - 18 = 3 - 18$

$\Rightarrow 5t = -15$

Divide both sides by 5 $\Rightarrow \dfrac{5}{5}t = \dfrac{-15}{5}$

$\Rightarrow t = -3$

Example 2.4

Solve $10 - 2x = x - 12$.

Solution

Add 2x to both sides $\Rightarrow 10 - 2x + 2x = x - 12 + 2x$

Collect like terms $\Rightarrow 10 = 3x - 12$

Add 12 to both sides $\Rightarrow 10 + 12 = 3x - 12 + 12$

$\Rightarrow 22 = 3x$

Divide both sides by 3 $\Rightarrow \dfrac{22}{3} = \dfrac{3}{3}x$

$\Rightarrow x = 7\frac{1}{3}$

Example 2.5

Solve $\dfrac{3y+2}{4} = 2y$.

Solution

Multiply both sides by 4 $\Rightarrow 4 \times \dfrac{3y+2}{4} = 4 \times 2y$

$\Rightarrow 3y + 2 = 8y$

Subtract 3y from both sides $\Rightarrow 3y + 2 - 3y = 8y - 3y$

$\Rightarrow 2 = 5y$

Divide both sides by 5 $\Rightarrow \dfrac{2}{5} = \dfrac{5}{5}y$

$\Rightarrow y = 0.4$

For each example, check that the value obtained does solve the equation by ensuring that, with this value for the unknown, the equation balances.

e.g. for example 2.3: $t = -3$

$\text{LHS} = 5 \times (-3) + 18$

$= -15 + 18 = 3 = \text{RHS}$

EXERCISES

2.1A

Solve the following equations, in each case checking that the value obtained does balance the equation.

1 $x + 11 = 30$
2 $7 = x - 2.5$
3 $5x = 75$
4 $\frac{1}{3}x = 13$
5 $3t = 16$
6 $2x - 5 = 11$
7 $3x - 7 = 20$
8 $7 = 5x + 22$
9 $30 - x = 11$
10 $7 = 57 - 5m$
11 $5 + 6y = 7$
12 $20 = 8 - 3x$
13 $5x - 10 = 3x + 2$
14 $5x - 3 = 3 - x$
15 $25 - m = 6m + 11$
16 $3x - 22 = 8x + 18$
17 $7 - 2x = 2x - 7$
18 $4x + 5 = 7 - 5x$
19 $\dfrac{7x - 5}{5} = 3x$
20 $\dfrac{20 - 3r}{5} = r$

EXERCISES

2.1B

Solve the following equations, in each case checking that the value obtained does balance the equation.

1 $x - 15 = 68$
2 $8 = 7.7 + m$
3 $14 = 5x$
4 $\frac{1}{8}x = 3.5$
5 $12 = 4x - 12$
6 $11 + 8x = 81$
7 $41 = \dfrac{x}{6} + 12$
8 $0 = 8t + 28$
9 $93 - x = 143$
10 $4 = 5 - 3x$
11 $1 - 9z = -17$
12 $z = 2 + 11z$
13 $5x + 9 = 8x + 7$
14 $3 - 6x = x - 18$
15 $3x - 1 = 6 - x$
16 $8 - 5t = 7 - 9t$
17 $12 + 8x = 15x$
18 $6x + 8 = 4 - 4x$
19 $\dfrac{9x + 7}{5} = 8$
20 $\dfrac{119 - x}{11} = 3x$

Exploration 2.2

Brackets

We often need to find the area of something.

We can find the areas of the following shapes in two different ways.

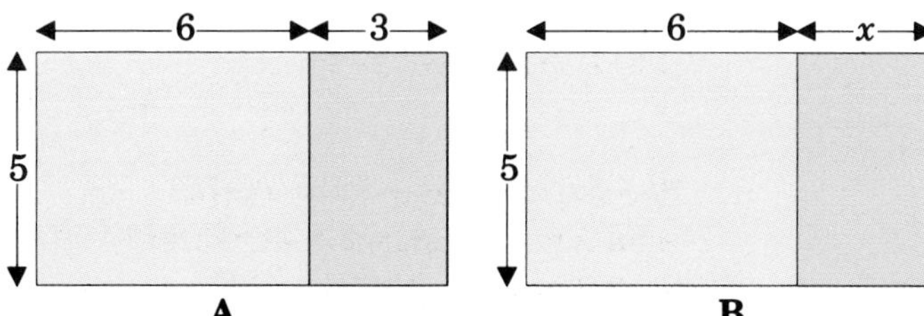

The area of rectangle A may be found as

$5 \times (6 + 3)$ or $5 \times 6 + 5 \times 3$

Either way the area is 45 square units.

The area of rectangle B may be thought of as
$5 \times (6 + x)$ or $5 \times 6 + 5 \times x$

This leads to the algebraic idea of multiplying out or **expanding** brackets.

Multiply the 5 by the 6 *and* by the x.
$5(6 + x) = 30 + 5x$.

Exploration 2.3

Diagrams to represent equations

Draw a diagram to illustrate $3(10 - x) = 30 - 3x$.

We can take the approach used in the last exploration, but in reverse.

area PUTS = area PQRS – area UQRT
→ $3(10 - x) = 30 - 3x$

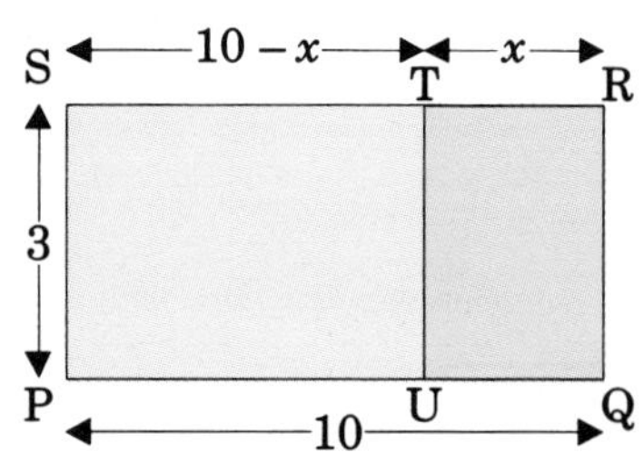

The following examples use similar techniques.

Example 2.6

Multiply out these brackets.

a) $7(x - 10)$ ***b)*** $a(12 - 5x)$
c) $2x(x + 3)$ ***d)*** $-3a(2s - 7t)$

Solution

a) $7(x - 10) = 7(x + (-10))$
$= 7x + (-70)$
$= 7x - 70$

Treating 'subtracting a positive' as 'adding a negative' helps to get the signs right.

b) $a(12 - 5x) = a \times 12 - a \times 5x$
$= 12a - 5ax$

Rearrange the products, writing them with the number first, followed by variables in alphabetical order.

c) $2x(x + 3) = 2x \times x + 2x \times 3$
$= 2x^2 + 6x$

Simplifying products as before, $x \times x$ becomes x^2.

d) $-3a(2s - 7t) = -3a(2s + (-7t))$
$= -3a \times 2s + (-3a) \times (-7t)$
$= -6as + 21at$

Treat 'subtracting a positive' as 'adding a negative' and remember that rules for multiplying directed numbers apply.

Example 2.7

Simplify the following expressions.

a) $3(x - 2y) - 5(x + 3y)$ ***b)*** $(10a - 3)b - 2a(5b - 1)$
c) $3x(2x + 3) - (7 + 4x)$

Solution

a) $3(x - 2y) - 5(x + 3y) = 3(x + (-2y)) + (-5)(x + 3y)$
$= 3x + (-6y) + (-5x) + (-15y)$
$= 3x - 5x - 6y - 15y$
$= -2x - 21y$

b) $(10a - 3)b - 2a(5b - 1) = (10a + (-3))b + (-2a)(5b + (-1))$
$= 10ab + (-3b) + (-10ab) + 2a$
$= 10ab - 3b - 10ab + 2a$
$= 2a - 3b$

c) $3x(2x + 3) - (7 + 4x) = 3x(2x + 3) + (-1)(7 + 4x)$
$= 6x^2 + 9x + (-7) + (-4x)$
$= 6x^2 + 9x - 7 - 4x$
$= 6x^2 + 5x - 7$

Shortcuts

We can put together the ideas from this section to form a set of rules that may be applied to expanding brackets, to give some short-cuts to the calculations.

Rule	Example	
$a(b + c) = ab + ac$	$3(x + 5) = 3x + 15$	(1)
$a(b - c) = ab - ac$	$5(p - q) = 5p - 5q$	(2)
$-a(b + c) = -ab - ac$	$-r(s + t) = -rs - rt$	(3)
$-a(b - c) = -ab + ac$	$-x(x - 4) = -x^2 + 4x$ or $4x - x^2$	(4)

Notice the pattern of the operation signs in each case.

The reverse of expanding brackets is called **factorising**.

Rules (1) and (2) in reverse give $ab + ac = a(b + c)$ and $ab - ac = a(b - c)$

In both cases a is the **common factor**. In practice we try to make a the highest common factor (HCF), which is often obtained by inspection.

Example 2.8

Factorise the following.

a) $3r + 12s$ ***b)*** $25x - 15$
c) $6x^2 - 21x$ ***d)*** $12e + 4f - 20$

Solution

a) *HCF is 3:* $3r + 12s = 3 \times r + 3 \times 4s = 3(r + 4s)$
b) *HCF is 5:* $25x - 15 = 5 \times 5x - 5 \times 3 = 5(5x - 3)$
c) *HCF is 3x:* $6x^2 - 21x = 3x \times 2x - 3x \times 7 = 3x(2x - 7)$
d) *HCF is 4:* $12e + 4f - 20 = 4 \times 3e + 4 \times f - 4 \times 5 = 4(3e + f - 5)$

Example 2.9

Simplify and factorise the following.

a) $(8x + 2y) - (3x - 8y)$
b) $7x(1 - 3x) + x(5 + x)$

Solution

a) $(8x + 2y) - (3x - 8y) = 8x + 2y - 3x + 8y$
$= 5x + 10y = 5 \times x + 5 \times 2y$
$= 5(x + 2y)$

b) $7x(1 - 3x) + x(5 + x) = 7x - 21x^2 + 5x + x^2$
$= 12x - 20x^2 = 4x \times 3 - 4x \times 5x$
$= 4x(3 - 5x)$

EXERCISES

2.2A

1 Multiply out the brackets.

a) $3(x + 4)$ **b)** $-3(x - 4)$
c) $-4(x + 5)$ **d)** $-4(x - 5)$
e) $7(2p + 3q)$ **f)** $7(2p - 3q)$
g) $2x(x - 3)$ **h)** $2x(x + 3)$
i) $-3p(q - 2r)$ **j)** $-3p(q + 2r)$
k) $5a(a - b + c)$ **l)** $5a(a + b - c)$
m) $-2r(-3r + 5s - t)$ **n)** $-2r(3r - 5s + t)$

2 Simplify the following.

a) $3(x + 1) + 2(x + 4)$ **b)** $3(x + 1) - 2(x + 4)$
c) $5(2a - 4) + 3(5a + 2)$ **d)** $5(2a - 4) - 3(5a + 2)$
e) $4(1 - 2x) + 3(3x - 4)$ **f)** $4(1 - 2x) - 3(3x - 4)$
g) $x(2x - 1) + 2x(x + 4)$ **h)** $x(2x - 1) - 2x(x + 4)$
i) $4(x + 5) - 3(x + 2) + 5(x - 1)$ **j)** $a(b + c) + a(b - c) - a(b + c)$
k) $2p - 3q - (5p + 2q)$ **l)** $8(x - y) - (3x - 4y)$

3 Factorise each of these fully.

a) $10a - 5b$ **b)** $7x + 21$
c) $12y + 3$ **d)** $12xy - 3x$
e) $7x^2 - 5x$ **f)** $7t^2 - 35t$
g) $a^2b - ab^2$ **h)** $15cd + 20de$
i) $5r - 20s + 15t$ **j)** $30xy + 6x^2 - 15x$

EXERCISES

1 Multiply out the brackets.

a) $6(3x + 2)$ **b)** $-8(9x + 8)$
c) $2(5 - 7x)$ **d)** $-(4x - 3)$
e) $8(8a - 2b)$ **f)** $-4(9b - 6a)$
g) $4m(7m + 6)$ **h)** $-x(9 + x)$
i) $-4x(5x - 1)$ **j)** $5t(-5 - 2t)$
k) $x(x + y - z)$ **l)** $-y(x - y - z)$
m) $-3a(8a - 8b + c)$ **n)** $4a(2a - 3b - 1)$

2 Simplify the following.

a) $9(x + 7) + 6(x + 7)$ b) $5 - (x - 3)$
c) $9(x - 7) - 6(x + 1)$ d) $7(6x - 1) + 3(5 - 4x)$
e) $5(4 + 2t) - (14 - 2t)$ f) $-(7x + 1) - 3(9 + 8x)$
g) $9x(7x - 6) + 2x(8 + 2x)$ h) $5x(5x + 2) - 4(3x - 7)$
i) $(6x - 7) - 2(x + 1) + 3(3 - x)$ j) $x(y + z) - x(y - z) + 2x(y + z)$
k) $7(6x - 5) - x + 7 - 7(5 + 3x)$ l) $8(a - 5b) - (4a + b) + (a - 6b)$

3 Factorise each of these fully.

a) $28x + 24$ b) $42 - 35a$ c) $3 - 3x$ d) $4m^2 + 10m$
e) $8y^2 - 3y$ f) $2a^2 + 18a$ g) $8x^2y - 7xy^2$ h) $3ab + 6a^2d$
i) $x + 7xy - x^2$ j) $24t^2 - 30t - 42tu^2$

LINEAR EQUATIONS WITH BRACKETS

We can now use the ideas built up in the section above to solve equations with brackets.

Example 2.10

Solve $4(x - 5) = 19$.

Solution

Multiply out $\Rightarrow$ $4x - 20 = 19$
Add 20 to both sides $\Rightarrow$ $4x - 20 + 20 = 19 + 20$
$\Rightarrow$ $4x = 39$
Divide both sides by 4 $\Rightarrow$ $\dfrac{4x}{4} = \dfrac{39}{4}$
$\Rightarrow$ $x = 9.75$

Example 2.11

Solve $3(x + 4) - 5(x - 1) = 23$.

Multiply out $\Rightarrow$ $3x + 12 - 5x + 5 = 23$
Collect like terms $\Rightarrow$ $-2x + 17 = 23$
Subtract 17 from both sides $\Rightarrow$ $-2x + 17 - 17 = 23 - 17$
$\Rightarrow$ $-2x = 6$
Divide both sides by −2 $\Rightarrow$ $\dfrac{-2x}{-2} = \dfrac{6}{-2}$
$\Rightarrow$ $x = -3$

Example 2.12

Solve $5(2x - 3) = 22 - (7x + 3)$.

Solution

Multiply out $\Rightarrow$ $10x - 15 = 22 - 7x - 3$
Collect like terms $\Rightarrow$ $10x - 15 = 19 - 7x$
Add 7x to both sides $\Rightarrow$ $10x - 15 + 7x = 19 - 7x + 7x$
Collect like terms $\Rightarrow$ $17x - 15 = 19$
Add 15 to both sides $\Rightarrow$ $17x - 15 + 15 = 19 + 15$
$\Rightarrow$ $17x = 34$
Divide both sides by 17 $\Rightarrow$ $\dfrac{17}{17}x = \dfrac{34}{17}$
$\Rightarrow$ $x = 2$

EXERCISES

2.3 A

Solve the following equations.

1 $2(x + 1) = 9$
2 $5(m - 2) = 17$
3 $2x = 4(x + 1)$
4 $5(8 - x) = 3x$
5 $3(x - 1) - 4(2x + 3) = 14$
6 $5(y + 2) - 3(y - 5) = 29$
7 $4(x - 5) = 7 - 5(3 - 2x)$
8 $7x - (2 - x) = 0$
9 $3(n + 1) = 10 - (2n + 3)$
10 $3y + 7 + 3(y - 1) = 4(y + 3)$
11 $4(r - 1) + 3(r + 2) = 5(r - 4)$
12 $7 - 2(x - 1) = 3(2x - 1) + 2$

EXERCISES

2.3 B

Solve the following equations.

1 $8(3x - 8) = 26$
2 $2(5 - 3x) = 70$
3 $7(x + 2) = 21x$
4 $2x - 8 = 4(x + 6)$
5 $8(x - 2) - (7x - 9) = -4$
6 $5(4 - 5t) + 8(t - 7) = 32$
7 $7(4x + 3) = 8 - (4x + 3)$
8 $2x - 5(x - 7) = 0$
9 $8(x + 3) = 8 - (x + 3)$
10 $4n - 7 + 6(n + 1) = 8(n - 4)$
11 $9(4x + 1) - 6(8 - 7x) = 5(5x - 1)$
12 $4x - 6 - (2x + 3) = 2 - 3(8x - 2)$

LINEAR EQUATIONS WITH FRACTIONS

Exploration 2.4

Solving equations involving fractions

What operations do we need to carry out, and in what order, to solve this equation?

$$\frac{7x+5}{4} = 2x - 3 \qquad (1)$$

Refine the method to solve

$$\frac{5x-3}{7} = \frac{2x+1}{3} \qquad (2)$$

and $$\frac{2x}{3} - \frac{x}{5} = \frac{x+3}{6} \qquad (3)$$

In each case the first operation is to multiply both sides by the same number to eliminate the fractions.
For equation (1) we multiply both sides by 4.
For equation (2) we multiply both sides by 21.
For equation (3) we multiply both sides by 30.

Whenever we need to solve equations with fractions, first of all we multiply both sides by the **lowest common denominator**.

The following examples illustrate how this method can be applied in a variety of cases, including when a variable (such as x) appears in the denominator.

Example 2.13

Solve $\dfrac{5x-3}{7}=\dfrac{2x+1}{3}$.

Solution

Multiply both sides by 21	$\Rightarrow$	$\dfrac{21(5x-3)}{7}=\dfrac{21(2x+1)}{3}$
Simplify numerical fractions	$\Rightarrow$	$3(5x-3)=7(2x+1)$
Multiply out	$\Rightarrow$	$15x-9=14x+7$
Subtract 14x from both sides	$\Rightarrow$	$15x-9-14x=14x+7-14x$
	$\Rightarrow$	$x-9=7$
Add 9 to both sides	$\Rightarrow$	$x-9+9=7+9$
	$\Rightarrow$	$x=16$

Example 2.14

Solve $\dfrac{2x}{3}-\dfrac{x}{5}=\dfrac{x+3}{6}$.

Solution

Multiply both sides by 30	$\Rightarrow$	$30\left(\dfrac{2x}{3}-\dfrac{x}{5}\right)=30\left(\dfrac{x+3}{6}\right)$
	$\Rightarrow$	$\dfrac{30\times 2x}{3}-\dfrac{30\times x}{5}=\dfrac{30(x+3)}{6}$
Simplify numerical fractions	$\Rightarrow$	$20x-6x=5(x+3)$
Simplify and multiply out	$\Rightarrow$	$14x=5x+15$
Subtract 5x from both sides	$\Rightarrow$	$14x-5x=5x+15-5x$
	$\Rightarrow$	$9x=15$
Divide both sides by 9	$\Rightarrow$	$\dfrac{9x}{9}=\dfrac{15}{9}$
	$\Rightarrow$	$x=1\frac{2}{3}$

Example 2.15

Solve $\dfrac{t-4}{5}+\dfrac{2t-3}{10}=2.5$.

Solution

Multiply both sides by 10	$\Rightarrow$	$10\left(\dfrac{t-4}{5}+\dfrac{2t-3}{10}\right)=10\times 2.5$
	$\Rightarrow$	$\dfrac{10(t-4)}{5}+\dfrac{10(2t-3)}{10}=25$
	$\Rightarrow$	$2(t-4)+(2t-3)=25$
	$\Rightarrow$	$2t-8+2t-3=25$
	$\Rightarrow$	$4t-11=25$
Add 11 to both sides	$\Rightarrow$	$4t-11+11=25+11$
	$\Rightarrow$	$4t=36$
Divide both sides by 4	$\Rightarrow$	$\dfrac{4t}{4}=\dfrac{36}{4}$
	$\Rightarrow$	$t=9$

Example 2.16

Solve $\frac{2y-3}{y}=\frac{5}{4}$

Solution

Multiply both sides by 4y $\Rightarrow \frac{4y(2y-3)}{y}=\frac{5\times 4y}{4}$

$\Rightarrow 4(2y-3)=5y$

$\Rightarrow 8y-12=5y$

Add 12 to both sides $\Rightarrow 8y-12+12=5y+12$

Subtract 5y from both sides $\Rightarrow 8y-5y=5y+12-5y$

$\Rightarrow 3y=12$

Divide both sides by 3 $\Rightarrow \frac{3}{3}y=\frac{12}{3}$

$\Rightarrow y=4$

EXERCISES

2.4A

Solve the following equations.

1 $\frac{x}{2}+7=12$

2 $\frac{x}{3}+10=2x$

3 $\frac{5x}{6}=\frac{1}{4}$

4 $\frac{4x}{3}=\frac{x}{2}+5$

5 $\frac{m}{5}-\frac{m}{6}=0.5$

6 $\frac{t}{5}-\frac{t}{4}=2$

7 $\frac{2s+1}{3}=\frac{4s-1}{5}$

8 $\frac{r}{4}=\frac{36-r}{8}$

9 $\frac{y+3}{4}-\frac{y-3}{5}=2$

10 $\frac{3u+5}{4}=2-\frac{10+u}{6}$

11 $\frac{2x}{15}-\frac{x-6}{12}-\frac{3x}{20}=\frac{3}{2}$

12 $\frac{6x-1}{2}+\frac{8-x}{5}=\frac{17+7x}{4}$

13 $\frac{12}{x}=-4$

14 $\frac{x+2}{x}=3$

15 $\frac{8-x}{x}=\frac{3}{5}$

16 $\frac{3x}{x-2}=4$

17 $\frac{3x-1}{2x-1}=2$

18 $\frac{15}{2y}+\frac{10}{3y}=\frac{5}{6}$

19 $\frac{30}{2x+1}=\frac{14}{x}$

20 $\frac{9}{x+1}-\frac{15}{3x-1}=0$

EXERCISES

2.4B

Solve the following equations.

1 $\frac{x}{4}-1=5$

2 $\frac{2t}{3}+5=9t$

3 $\frac{6x}{5}=\frac{4}{3}$

4 $\frac{x}{9}=\frac{5x}{6}+13$

5 $\frac{x}{8}+\frac{x}{12}=1$

6 $\frac{n}{4}-7=\frac{6n}{7}$

7 $\frac{9x+3}{7}=\frac{1-8x}{2}$

8 $\frac{2x}{21}=\frac{x-15}{33}$

9 $\frac{6n+7}{9}+\frac{8n}{3}=3$

10 $\frac{5-9x}{8}-2=\frac{x}{4}$

11 $\frac{x+4}{14}+\frac{4x-5}{35}-\frac{9}{10}=\frac{x}{5}$

12 $\frac{9x+2}{6}-\frac{1-3x}{7}=\frac{x+6}{3}$

13 $\frac{31}{x}=10$

14 $\frac{8x+1}{x}=2$

15 $\frac{5x-1}{x}=\frac{8}{9}$

16 $\frac{x}{x+3}=6$

17 $\frac{x-4}{9-3x}=\frac{5}{6}$

18 $\frac{3}{4u}-\frac{u+1}{5u}=2$

19 $\frac{6}{2x-1}=\frac{5}{4x+3}$

20 $\frac{45}{7x+4}+\frac{53}{2-5x}=0$

CHANGING THE SUBJECT OF A FORMULA

28° in the shade!

What does 28° mean? Weather reports seem to switch randomly from the Celsius scale to the Fahrenheit. The formula $f = 32 + 1.8c$ is used to convert temperatures in Celsius (°C) into temperatures in Fahrenheit (°F).

When $c = 20$, $f = 32 + 1.8 \times 20 = 68$

We can rearrange the formula to convert temperatures in °F to temperatures in °C.

Rearranging a formula is called **changing the subject**.

In the example above, we began with f as the subject of a formula and wanted to rearrange it to make c the subject.

Treat $f = 32 + 1.8c$ as an equation which can be solved for c. The steps we might take are:

Subtract 32 from both sides $\Rightarrow$ $f - 32 = 32 + 1.8c - 32$

$\Rightarrow$ $f - 32 = 1.8c$

Divide both sides by 1.8 $\Rightarrow$ $\frac{f-32}{1.8}=\frac{1.8c}{1.8}$

$\Rightarrow$ $c=\frac{f-32}{1.8}$

Given any temperature in Fahrenheit, this rearranged formula may be used to give the temperature in Celsius.

When $f = 98.6$, $c = \frac{98.6-32}{1.8} = 37$

We often find that a formula may be rearranged to change the subject, in this way. We simply treat it as an equation and solve for the variable we want to make the subject.

The following examples illustrate a variety of cases.

Example 2.17

In mechanics or physics, a common equation of motion is $v = u + at$. This describes how the velocity of an object in uniform motion changes in relation to the starting velocity, the (constant) acceleration and the time of travel. Suppose we need to find the value of t. We need to make t the subject of $v = u + at$.

Solution

Treat the formula as an equation in t.

$$v = u + at$$

Subtract u from both sides $\Rightarrow v - u = u + at - u$

$\Rightarrow v - u = at$

Divide both sides by a $\Rightarrow \dfrac{v-u}{a} = \dfrac{at}{a}$

$\Rightarrow t = \dfrac{v-u}{a}$

Example 2.18

Make a the subject of $P = 2(a + b)$.

Solution

$$P = 2(a + b)$$

Multiply out the brackets $\Rightarrow P = 2a + 2b$

Subtract $2b$ from both sides $\Rightarrow P - 2b = 2a + 2b - 2b$

$\Rightarrow P - 2b = 2a$

Divide both sides by 2 $\Rightarrow \dfrac{P-2b}{2} = \dfrac{2a}{2}$

$\Rightarrow a = \dfrac{P-2b}{2}$ (1)

Alternatively, we could try a different approach, to give an equivalent formula.

$$P = 2(a + b)$$

Divide both sides by 2 $\Rightarrow \dfrac{P}{2} = \dfrac{2(a+b)}{2}$

$\Rightarrow \dfrac{P}{2} = a + b$

Subtract b from both sides $\Rightarrow \dfrac{P}{2} - b = a + b - b$

$\Rightarrow a = \dfrac{P}{2} - b$ (2)

(1) and (2) are equivalent to each other – try substituting numerical values for b and P and check that you get the same value for a whichever rearrangement you use.

Example 2.19

Make p the subject of $\frac{q-ap}{m}=l$.

Solution

Treat the formula as an equation in p. $\frac{q-ap}{m}=l$

Multiply both sides by m $\Rightarrow$ $\frac{m(q-ap)}{m}=ml$

$\Rightarrow$ $q-ap=ml$

Add ap to both sides $\Rightarrow$ $q-ap+ap=ml+ap$

$\Rightarrow$ $q=ml+ap$

Subtract ml from both sides $\Rightarrow$ $q-ml=ml+ap-ml$

$\Rightarrow$ $q-ml=ap$

Divide both sides by a $\Rightarrow$ $\frac{q-ml}{a}=\frac{ap}{a}$

$\Rightarrow$ $p=\frac{q-ml}{a}$

Example 2.20

Rewrite $\frac{a}{b}+c=d$ in the form $b=\ldots$.

Solution

Treat the formula as an equation in b. $\frac{a}{b}+c=d$

Subtract c from both sides $\Rightarrow$ $\frac{a}{b}+c-c=d-c$

$\Rightarrow$ $\frac{a}{b}=d-c$

Multiply both sides by b $\Rightarrow$ $\frac{a}{b}\times b=b(d-c)$

$\Rightarrow$ $a=b(d-c)$

Divide both sides by $(d-c)$ $\Rightarrow$ $\frac{a}{d-c}=\frac{b(d-c)}{d-c}$

$\Rightarrow$ $b=\frac{a}{d-c}$

Example 2.21

Make s the subject of the formula $3s+t=7r-as$.

Treat the formula as an equation in s. $3s+t=7r-as$

Add as to both sides $\Rightarrow$ $3s+t+as=7r-as+as$

Collect like terms $\Rightarrow$ $3s+as+t=7r$

Subtract t from both sides $\Rightarrow$ $3s+as+t-t=7r-t$

$\Rightarrow$ $3s+as=7r-t$

Take out s as a common factor $\Rightarrow$ $s(3+a)=7r-t$

Divide both sides by $(3+a)$ $\Rightarrow$ $\frac{s(3+a)}{3+a}=\frac{7r-t}{3+a}$

$\Rightarrow$ $s=\frac{7r-t}{3+a}$

Example 2.22

Make x the subject of the formula $y = \frac{x+5}{3-x}$ *and find x when $y = 3$.*

$$y = \frac{x+5}{3-x}$$

Multiply both sides by $(3-x)$ $\Rightarrow$ $(3-x)y = \frac{x+5}{3-x}(3-x)$

$\Rightarrow$ $(3-x)y = x + 5$

Multiply out the brackets $\Rightarrow$ $3y - xy = x + 5$

Add xy to both sides $\Rightarrow$ $3y - xy + xy = x + 5 + xy$

Collect like terms $\Rightarrow$ $3y = x + xy + 5$

Subtract 5 from both sides $\Rightarrow$ $3y - 5 = x + xy + 5 - 5$

$\Rightarrow$ $3y - 5 = x + xy$

Take x out as a common factor $\Rightarrow$ $3y - 5 = x(1 + y)$

Divide both sides by $(1 + y)$ $\Rightarrow$ $\frac{3y-5}{1+y} = \frac{x(1+y)}{1+y}$

$\Rightarrow$ $x = \frac{3y-5}{1+y}$

EXERCISES

2.5 A

For each of the following formulae, make the variable shown in brackets the subject.

1	$v = u + at$	(u)	**2**	$v = u + at$	(a)
3	$A = \frac{1}{2}h(a + b)$	(h)	**4**	$A = \frac{1}{2}h(a + b)$	(a)
5	$y = mx + c$	(c)	**6**	$y = mx + c$	(x)
7	$u = a + (n - 1)d$	(d)	**8**	$u = a + (n - 1)d$	(n)
9	$5x - y = 12$	(y)	**10**	$5x - y = 12$	(x)
11	$4x + 5y = 20$	(y)	**12**	$4x + 5y = 20$	(x)
13	$\frac{3r+5}{s} = 8$	(s)	**14**	$\frac{3r+5}{s} = 8$	(r)
15	$\frac{5p-2q}{r} = w$	(r)	**16**	$\frac{5p-2q}{r} = w$	(p)
17	$\frac{5p-2q}{r} = w$	(q)	**18**	$\frac{a(b-c)}{d} = e$	(a)
19	$\frac{a(b-c)}{d} = e$	(d)	**20**	$\frac{a(b-c)}{d} = e$	(b)
21	$\frac{3e}{g} = f - 7$	(f)	**22**	$\frac{3e}{g} = f - 7$	(e)
23	$\frac{3e}{g} = f - 7$	(g)	**24**	$3a + b = 5a - b$	(a)
25	$px + q = rx + s$	(x)	**26**	$5m - n = 4(3m + n)$	(m)
27	$5m + n = n(3m - 4)$	(m)	**28**	$2(x - 1) = a(b - x)$	(x)
29	$x - 2 = \frac{x+a}{t}$	(x)	**30**	$x(a + b) = d(x + b)$	(x)
31	$y = \frac{x+1}{x-1}$	(x)	**32**	$y = \frac{5-x}{2x-2}$	(x)

EXERCISES

2.5 B

For each of the following formulae, make the variable shown in brackets the subject.

1	$a = bx + c$	(x)	**2**	$v^2 = u^2 + 2as$	(s)
3	$s = \frac{1}{2}(u + v)t$	(t)	**4**	$s = \frac{1}{2}(u + v)t$	(v)
5	$y - y_1 = m(x - x_1)$	(m)	**6**	$c = 2\pi r$	(r)
7	$p = \frac{w}{g}f$	(w)	**8**	$A = 2\pi r^2 + 2\pi rh$	(h)
9	$8x - 7y = 3$	(x)	**10**	$9x + 2y = 6$	(y)
11	$7x - 9y = 3$	(y)	**12**	$I = \frac{Cnp}{100}$	(p)
13	$g = \frac{x+3y}{a}$	(x)	**14**	$2m = \frac{3x-7y}{n}$	(n)
15	$\frac{3a-2b}{5} = m$	(b)	**16**	$\frac{3a}{2} = 9x - 1$	(a)
17	$\frac{tu}{x} = 4 + p$	(x)	**18**	$m = \frac{y-b}{x-a}$	(y)
19	$2x + a = 8x + b$	(x)	**20**	$7x + 4y = 5(x - 2y)$	(x)
21	$8p - 7q = T(5p + q)$	(q)	**22**	$6a(x + y) = 4(x - 7y)$	(x)
23	$m(x + y) = n(x - y)$	(y)	**24**	$x = \frac{y+3}{2-y}$	(y)
25	$m = \frac{y-p}{2y-x}$	(y)	**26**	$f = \frac{8x-5f}{m}$	(f)
27	$n = \frac{B(x-a)}{5(2x+b)}$	(x)	**28**	$\frac{3x+b}{a-ux} = n + 2$	(x)
29	$y = \frac{m}{x} + c$	(x)	**30**	$\frac{a}{u} = \frac{b}{u} + 1$	(u)
31	$\frac{1}{a} + b = c$	(a)	**32**	$\frac{1}{u} + \frac{1}{v} = F$	(v)

LINEAR INEQUALITIES

Exploration 2.5

What is an inequality?

- What do you understand by these expressions?
  $x < 7$ $\quad x \geq 10$ $\quad 0 \geq x$ $\quad -2 \leq x \leq 2$ $\quad 2 < x < 7$
- Illustrate these inequalities, using the real number line.
- Simplify **1** $3x + 1 < 13$ **2** $12 - 5x \leq 20$
  3 $3x - 5 < 6x + 10$ **4** $\frac{10-x}{3} > x + 5$
  5 $3 < 2x - 5 \leq 11$
- If $x > 5$ does it follow that $-x > -5$?
- What do you understand by $|x| > 3$?

Interpreting inequalities

An inequality such as $x < 7$ means that x can take any value less than 7. If x is a real number then the set of values satisfying the inequality may be illustrated by an **open** interval on the real number line.

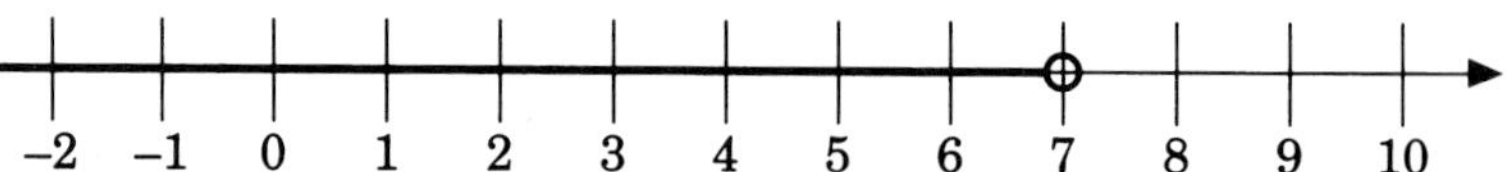

An inequality such as $-2 \leq x \leq 2$ means that x can take any value between −2 and 2 inclusive, illustrated by a closed interval on the real number line.

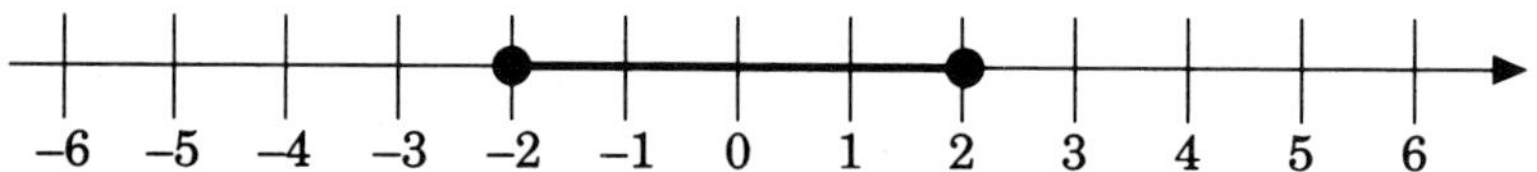

Inequalities such as $12 - 5x \leq 20$ may be simplified by manipulating the inequality, in a similar way to an equation.

$$12 - 5x \leq 20$$

Add $5x$ to both sides $\Rightarrow 12 - 5x + 5x \leq 20 + 5x$

$\Rightarrow 12 \leq 20 + 5x$

Subtract 20 from both sides $\Rightarrow 12 - 20 \leq 20 + 5x - 20$

$\Rightarrow -8 \leq 5x$

Divide both sides by 5 $\Rightarrow -\frac{8}{5} \leq \frac{5x}{5}$

$\Rightarrow -1.6 \leq x$

This is equivalent to $x \geq -1.6$, reading the inequality from right to left. The **solution** to the inequality represents the set of values of x which are *greater than or equal* to −1.6.

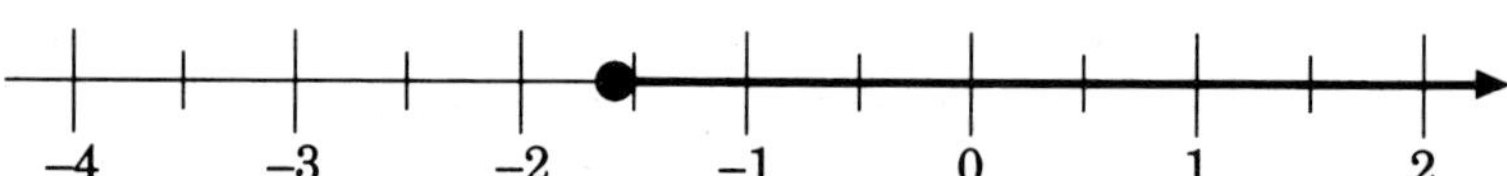

We could take an alternative approach to simplifying $12 - 5x \leq 20$.

$$12 - 5x \leq 20$$

Subtract 12 from both sides $\Rightarrow 12 - 5x - 12 \leq 20 - 12$

$\Rightarrow -5x \leq 8$

Divide both sides by −5 and *reverse* the inequality sign $\Rightarrow \frac{-5x}{-5} \geq \frac{8}{-5}$

$\Rightarrow x \geq -1.6$

Note

When solving an inequality, if you multiply or divide by a negative number you *must* reverse the inequality.

For example:

$12 > 5 \Rightarrow -12 < -5$

or $x > 5 \Rightarrow -x < -5$

The modulus sign

The **modulus** of a number is its **absolute** value, i.e. its **magnitude**, ignoring its sign, and is denoted by enclosing it by a pair of vertical lines.

e.g. $|8.3| = 8.3$
$|-2| = 2$

Algebraic equations and inequalities may also involve the modulus sign.
e.g. $|x| = 5 \Rightarrow x = -5$ or $x = 5$
$|x| \leq 5 \Rightarrow -5 \leq x \leq 5$
$|x| > 3 \Rightarrow x < -3$ or $x > 3$

Note how the last two inequalities may be illustrated.
$|x| \leq 5 \Rightarrow -5 \leq x \leq 5$ which is a closed interval.

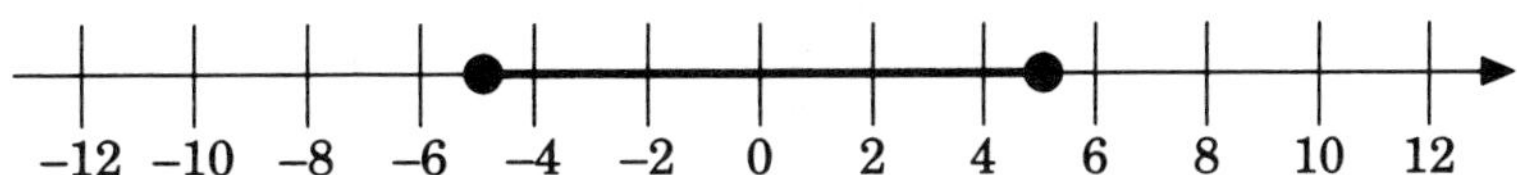

$|x| > 3 \Rightarrow x < -3$ or $x > 3$ which is two open intervals.

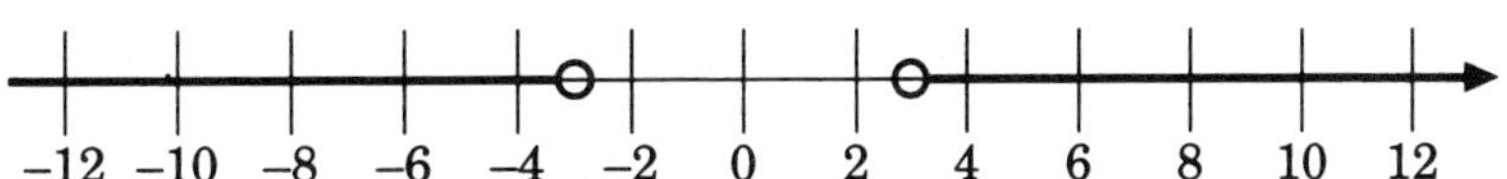

The following examples illustrate the simplification or solution of inequalities and how the solution set may be represented on the real number line.

Example 2.23

Simplify the inequality $3x - 5 < 6x + 10$.

Solution

$$3x - 5 < 6x + 10$$

Subtract 10 from both sides $\Rightarrow 3x - 5 - 10 < 6x + 10 - 10$

$\Rightarrow 3x - 15 < 6x$

Subtract 3x from both sides $\Rightarrow 3x - 15 - 3x < 6x - 3x$

$\Rightarrow -15 < 3x$

Divide both sides by 3 $\Rightarrow \dfrac{-15}{3} < \dfrac{3x}{3}$

$\Rightarrow -5 < x$

or $\Rightarrow x > -5$

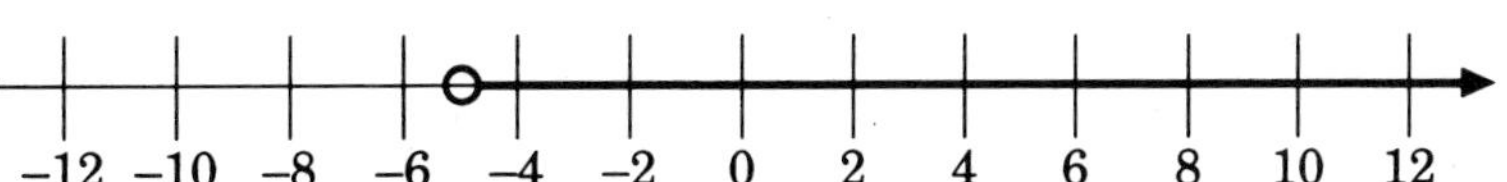

Example 2.24

Solve the inequality $\frac{8-x}{3} \geq x+5$.

Solution

$$\frac{8-x}{3} \geq x+5$$

Multiply both sides by 3 $\Rightarrow \frac{3(x-8)}{3} \geq 3(x+5)$

Multiply out the brackets $\Rightarrow 8 - x \geq 3x + 15$

Add x to both sides $\Rightarrow 8 - x + x \geq 3x + 15 + x$

$\Rightarrow 8 \geq 4x + 15$

Subtract 15 from both sides $\Rightarrow 8 - 15 \geq 4x + 15 - 15$

$\Rightarrow -7 \geq 4x$

Divide both sides by 4 $\Rightarrow \frac{-7}{4} \geq \frac{4x}{4}$

$\Rightarrow -1.75 \geq x$

or $\Rightarrow x \leq -1.75$

Example 2.25

Find the values of x which satisfy $3 < 2x - 5 \leq 11$.

Solution

Treat the double inequality as two separate inequalities.

i.e. ***a)*** $3 < 2x - 5$ ***b)*** $2x - 5 \leq 11$

a) $3 < 2x - 5$

Add 5 to both sides $\Rightarrow 3 + 5 < 2x - 5 + 5$

$\Rightarrow 8 < 2x$

Divide both sides by 2 $\Rightarrow \frac{8}{2} < \frac{2}{2}x$

$\Rightarrow 4 < x \quad (1)$

or $\Rightarrow x > 4$

b) $2x - 5 \leq 11$

Add 5 to both sides $\Rightarrow 2x - 5 + 5 \leq 11 + 5$

$\Rightarrow 2x \leq 16$

Divide both sides by 2 $\Rightarrow \frac{2}{2}x \leq \frac{16}{2}$

$\Rightarrow x \leq 8 \quad (2)$

Combining inequalities (1) and (2) gives the solution set.

$4 < x \leq 8$

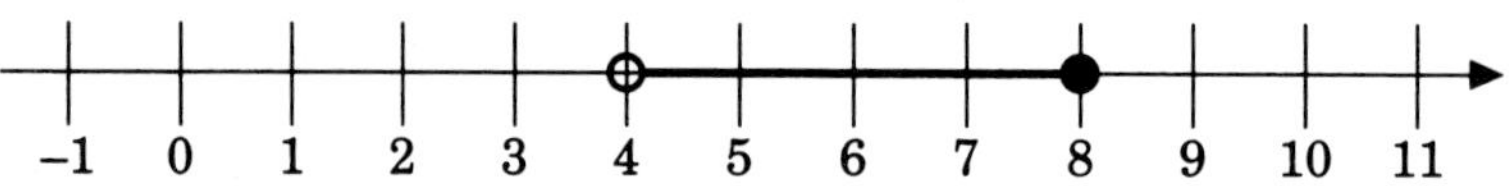

Example 2.26

Solve the inequality $|5x - 8| \geq 12$.

Solution

$|5x - 8| \geq 12$ *means* ***a)*** $5x - 8 \leq -12$ ***b)*** $5x - 8 \geq 12$

a)

$$5x - 8 \leq -12$$

Add 8 to both sides $\Rightarrow 5x - 8 + 8 \leq -12 + 8$

$\Rightarrow 5x \leq -4$

Divide both sides by 5 $\Rightarrow \frac{5x}{5} \leq \frac{-4}{5}$

$\Rightarrow x \leq -0.8$ (1)

b)

$$5x - 8 \geq 12$$

Add 8 to both sides $\Rightarrow 5x - 8 + 8 \geq 12 + 8$

$\Rightarrow 5x \geq 20$

Divide both sides by 5 $\Rightarrow \frac{5x}{5} \geq \frac{20}{5}$

$\Rightarrow x \geq 4$ (2)

Combining inequalities (1) and (2) gives the solution set.

$x \leq -0.8$ *or* $x \geq 4$

Note

Inequalities (1) and (2) cannot be combined into a single inequality because they represent two **disjoint** solution sets on the number line, as shown below.

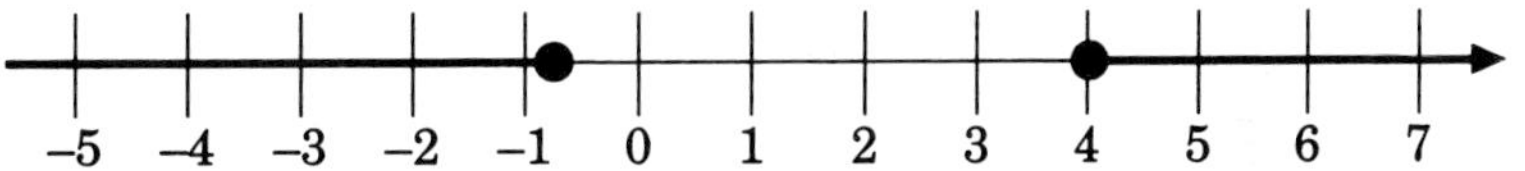

EXERCISES

2.6 A

Solve the following inequalities and illustrate the result on the real number line.

1 $3x + 2 > 8$

2 $5x - 3 \geq 17$

3 $4 + 2x \leq 13$

4 $4(x - 3) < 16$

5 $3x - 2 \geq 4x + 3$

6 $14 - 3x < 5x + 6$

7 $15 + 4x > 3x + 12$

8 $3(x + 1) \geq 5(4 - x)$

9 $\frac{3x + 5}{2} < 4x - 3$

10 $\frac{5 - 2x}{4} \geq \frac{7 - x}{5}$

11 $7 \leq 3x + 1 \leq 16$

12 $15 < 9 - 2x < 31$

13 $2x - 3 < 5$ or $2x - 3 > 17$

14 $-14 \geq 3x - 5$ or $2x + 1 > 5$

15 $3x - 4 \leq 14 < 5x - 6$

16 $12 - 5x < 2 \leq 12 + 5x$

17 $|2x + 3| \leq 7$

18 $|10 - 3x| > 5$

19 $|10x - 4| \geq 16$

20 $|-12 - 5x| < 23$

EXERCISES

2.6 B

Solve the following inequalities and illustrate the result on the real number line.

1 $5x - 1 < 3$

2 $6 \geq x + 1$

3 $5(x + 4) \geq 9$

4 $\dfrac{8x+7}{7} < 2$

5 $8x + 3 > 5x - 21$

6 $4x \geq 7 - 5x$

7 $3(x - 5) < 9 - 7x$

8 $4(9 - 8x) \geq 5(3x + 1)$

9 $\dfrac{3x+2}{4} < 4x + 7$

10 $\dfrac{6x+7}{2} > \dfrac{8-7x}{3}$

11 $-10 < 6x + 2 < 68$

12 $14 \geq 19 - 3x \geq 7$

13 $9x - 2 < 7$ or $9x - 2 > 43$

14 $8 - 7x \leq 22$ or $2x + 8 \leq -5$

15 $9x + 2 < 6 \leq x + 9$

16 $3 + 4x > 2x - 5 > 1 - x$

17 $|x - 8| \leq 3$

18 $|5 - 7x| > 9$

19 $25 \leq |5x - 8|$

20 $|4 - m| < 25$

CONSOLIDATION EXERCISES FOR CHAPTER 2

1 Solve these equations.

a) $3 - 2x = 5$

b) $12x + 8 = 12 + 3x$

c) $\dfrac{9-x}{5} = 7x$

2 Multiply out the brackets.

a) $5(6x - 7)$

b) $-3p(5 + 8p)$

3 Simplify these expressions.

a) $4x(7x + 3) - x(5 - x)$

b) $2 - (5x + 2)$

4 Factorise these expressions.

a) $9a^2b + 6ab^2$

b) $20xy + 5x - 20x^2$

5 Solve these equations.

a) $7(3x - 9) = 42$

b) $2(8x + 7) - (x - 1) = 6$

c) $2x - 8 + 6(1 - 2x) = 9(x + 4)$

6 Solve these equations.

a) $\dfrac{x}{7} + 2 = 3x$

b) $\dfrac{x+7}{6} = \dfrac{6x-1}{9}$

c) $\dfrac{7x}{3} - 6 = \dfrac{4-3x}{8}$

d) $\dfrac{5x+1}{x} = 9$

e) $\dfrac{8}{x+5}+\dfrac{6}{2x-1}=0$

f) $\dfrac{2-2(2-x)}{3}+6=\dfrac{5(8-7x)+7}{2}$

7 a) Make k the subject of $m + n = 2k + fm$.

b) Make c the subject of $\dfrac{ax-c}{a+b}=m$.

c) Make g the subject of $g = 4ag - 7b$.

d) Make y the subject of $x = \dfrac{2y-1}{3-y}$.

e) Make x the subject of $8(3x-4) = 8 - \dfrac{(9+x)}{m}$.

8 Solve these inequalities.

a) $8x + 1 < 9$

b) $2x - 7 \geq 5 - 7x$

c) $7 < 4 - 3x \leq 12$

d) $x + 2 < 9 < 2 - 7x$

e) $|3x - 5| \geq 8$

Summary

This chapter was a review of basic algebraic techniques.

- Solving linear equations
- Solving linear inequalities
- Manipulating expressions involving brackets
- Making a variable the subject of an equation
- The modulus of a number is its absolute value, i.e. its magnitude, ignoring the sign. The magnitude (modulus) of x is denoted by $|x|$.

PURE MATHS

3

Linear functions

In this chapter we:

- *explore the equation of a straight line*
- *discover algebraic relations between pairs of parallel and perpendicular lines*
- *solve simultaneous equations.*

CARTESIAN COORDINATES

Exploration 3.1

Map references

The map shows part of south-west England, together with the national grid references of five places, given as coordinates.

Isles of Scilly	(92,10)
Newquay	(186,64)
Penzance	(148,32)
Plymouth	(250,60)
Truro	(182,44)

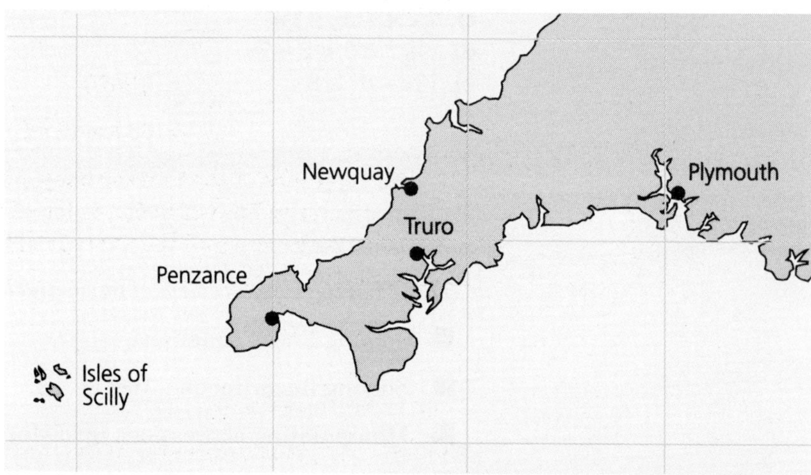

A helicopter flies directly from Penzance to the Isles of Scilly. On the way, it flies directly over a marker buoy which is equidistant between the two places.

- How far is it, as the helicopter flies, from Penzance to the Isles of Scilly?
- What is the national grid reference of the buoy?

Maps and coordinates

If we superimpose a right-angled triangle over the map, with Penzance and the Isles of Scilly at the ends of the hypotenuse and the other two sides running north–south and east–west, we can use Pythagoras' theorem to calculate the distance between the two places.

The lengths of the two perpendicular sides can be calculated by finding the difference in the east–west references and the difference in the north–south references.

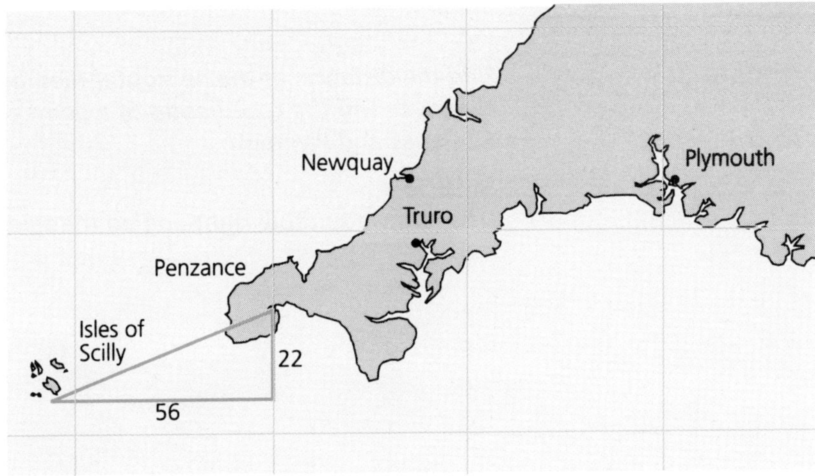

Therefore the distance from Penzance to the Isles of Scilly is given by:

$$\sqrt{(148-92)^2+(32-10)^2} = \sqrt{56^2+22^2}$$

$$= \sqrt{3620}$$

$$= 60 \text{ km (2 s.f.)}$$

The grid reference of the marker buoy, being halfway between Penzance and the Isles of Scilly, is found by taking the average of the east–west references and the average of the north–south references for the two places.

east–west reference $\Rightarrow \dfrac{092+148}{2} = 120$

north–south reference $\Rightarrow \dfrac{010+032}{2} = 021$

The national grid coordinates of the marker buoy are (120, 021).

The general result

The results of the exploration can be generalised for any grid, given the coordinates of two points.

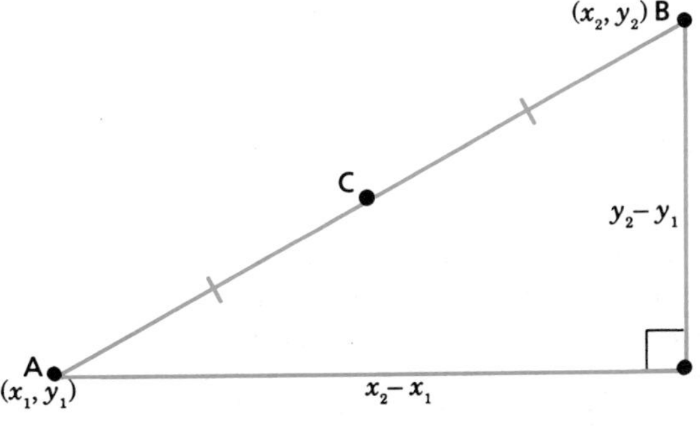

Let point A have coordinates (x_1, y_1) and point B have coordinates (x_2, y_2). Point C is the **midpoint** of the line AB.

$$AB = \sqrt{(x_2 - x_1)^2 + (y_2 - y_1)^2}$$

Coordinates of C are $\left(\frac{1}{2}(x_1 + x_2), \frac{1}{2}(y_1 + y_2)\right)$

Example 3.1

Find the distance, as the helicopter flies, between Newquay and Plymouth and the coordinates of a point exactly halfway between Newquay and Plymouth.

Solution

First, draw a suitable right-angled triangle.

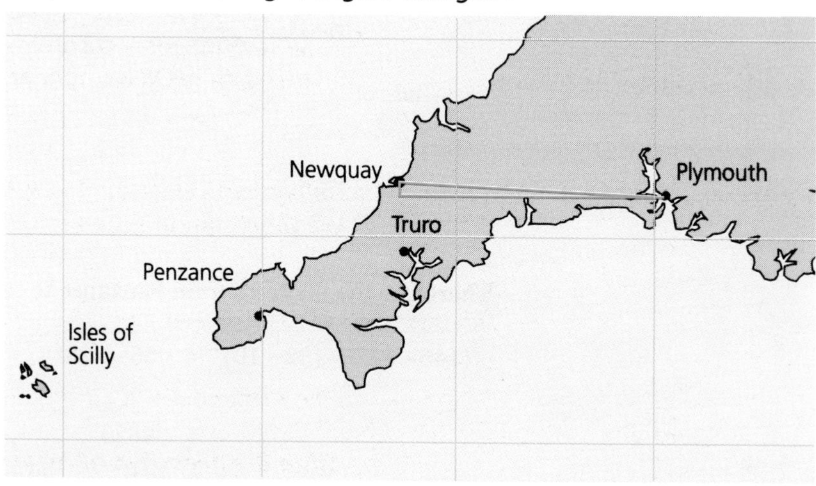

The distance from Newquay to Plymouth is:

$$\sqrt{(250-186)^2+(60-64)^2} = \sqrt{(64^2+(-4)^2)}$$

$$= \sqrt{4112}$$

$$= 64 \text{ km (2 s.f.)}$$

The coordinates of the halfway point are

$\left(\frac{1}{2}(186+250), \frac{1}{2}(64+60)\right)$ or $(218, 62)$.

Note: It would not make sense to quote the distance between Newquay and Plymouth to more than 2 significant figures (i.e. to the nearest kilometre), since the grid references are given to 2 s.f. Note also that the distance is roughly the difference in the east–west references, since the difference in the north–south references is relatively small.

Example 3.2

In ΔPQR, points P, Q and R have coordinates (–7, 6), (10, 2) and (–2, –4) respectively.

a) *Find the distance PQ.*
b) *Find the coordinates of M, the midpoint of PQ.*
c) *Show that $PQ^2 = PR^2 + QR^2$. What does this tell you about ΔPQR?*

Solution

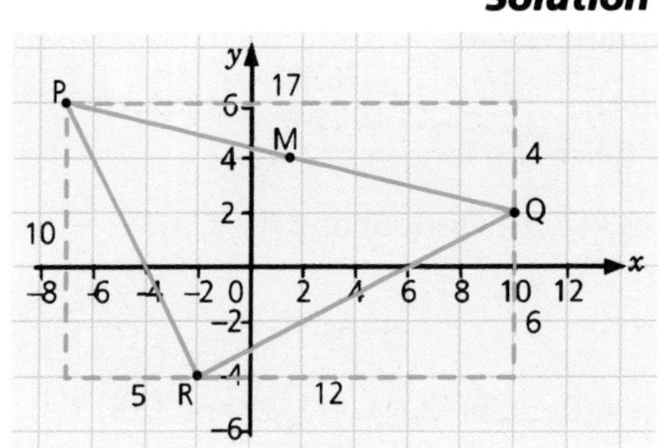

a) $PQ^2 = 17^2 + 4^2 = 305 \Rightarrow PQ = \sqrt{305} = 17.5$ (3 s. f.)

b) *Coordinates of M are* $\left(\frac{1}{2}(-7+10), \frac{1}{2}(6+2)\right)$ *or* $(1.5, 4)$

c) $PR^2 = 10^2 + 5^2 = 125$
$QR^2 = 6^2 + 12^2 = 180$
$\Rightarrow PR^2 + QR^2 = 125 + 180 = 305 = PQ^2$, *as required.*
Since Pythagoras' theorem is only true for right-angled triangles, ΔPQR is a right-angled triangle with $\angle R = 90°$.

Example 3.3

A parallelogram has vertices A(–5, 4), B(1, 6), C(9, –2) and D(3,–4). Show that its diagonals bisect each other.

Solution

The midpoint of AC is $\left(\frac{1}{2}(-5+9), \frac{1}{2}(4+(-2))\right) = (2, 1)$

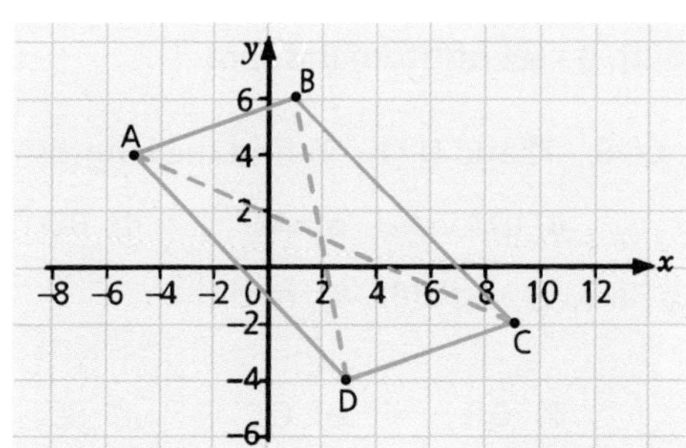

The midpoint of BD is $\left(\frac{1}{2}(1+3), \frac{1}{2}(6+(-4))\right) = (2, 1)$

Since the midpoints of AC and BD have the same coordinates, the diagonals must bisect each other.

EXERCISES

1 Using the map from Exploration 3.1, find the shortest distance between Penzance and Truro. Troon is exactly halfway between Penzance and Truro; find its grid reference.

2 Find the lengths and coordinates of the midpoints of the following lines, illustrating your answers on a grid.

a) AB where A is (2, 1), B is (8, 4)
b) PQ where P is (3, 0), Q is (–5, 5)
c) RS where R is (–1, 5), S is (–4, –3)
d) XY where X is (4, –3), Y is (–2, 8)

3 A triangle XYZ has vertices X(6, 2), Y(2, 5) and Z(–1, 1).

a) Find the lengths of the sides of the triangle.
b) What type of triangle is ΔXYZ?
c) Find the coordinates of point M, the midpoint of XZ.
d) Calculate the area of ΔXYZ.

4 A triangle PQR has vertices at P(4, –2), Q(10, 2) and R(–2, 6). M is the midpoint of PQ and N is the midpoint of QR.

a) Find the coordinates of M and N.
b) Show that the length of PR is twice the length of MN.

5 A parallelogram PQRS has vertices P(2, 1), Q(–1, 5) and R(–3, 3).

a) Find the midpoint of PR.
b) Deduce the coordinates of S.

6 A quadrilateral ABCD has vertices A(0, 5), B(–2, 1), C(1, 2) and D(6, 3). Points E, F, G and H are the midpoints of AB, BC, CD and DA respectively.

a) Find the coordinates of E, F, G and H.
b) Show that EFGH is a parallelogram.

EXERCISES

3.1 B

1 Using the map from Exploration 3.1, find the grid reference of the point midway between Plymouth and Penzance, and the shortest distance I would have to fly to get to Plymouth if I set off from this spot.

2 Using the points A(5, 2), B(–1, 2), C(–6, –7) and D (1, –9) find the lengths:

a) AB b) BC c) CD d) DA e) AC f) BD.

3 Using the points P(7, 4), Q(5, –6), R(–3, 5) and S(0, –8), find the midpoints of:

a) PQ b) PR c) PS d) QR e) QS f) RS.

4 A triangle ABC has vertices A(1, 9), B(–6, 2) and C(6, –3).

a) Find the lengths of the sides of the triangle.
b) Use Pythagoras' theorem to show that ABC is not a right-angled triangle.
c) Find the coordinates of point M, the midpoint of AC, and N, the midpoint of AB.
d) Explain why the area ABC is equal to $\frac{1}{2} \times AB \times CN$, but not equal to $\frac{1}{2} \times AC \times BM$.

5 A quadrilateral PQRS has vertices at P(–4, 8), Q(9, 6), R(11, –7) and S(–2, –5).

a) Find the lengths of the four sides of PQRS.
b) Find the midpoints of the diagonals PR and QS.
c) Without drawing a diagram, prove that PQRS is not square.

6 A quadrilateral ABCD has vertices at A(–1, 7), B(8, 7), C(9, –3) and D(–6, 0). Find the distance between the midpoints of the diagonals, giving your answer correct to 4 s.f.

GRADIENTS

Exploration 3.2

Is there a rule?

The firm 'Have van will travel' produces a table of charges.

Time (hours)	2	4	6	8
Charge (£)	30	40	50	60

What is the fixed charge and what is the variable charge for the service?

Finding a relationship

By plotting the data on a graph we can see more clearly the linear relationship between the cost and time.

From the graph, we can read:

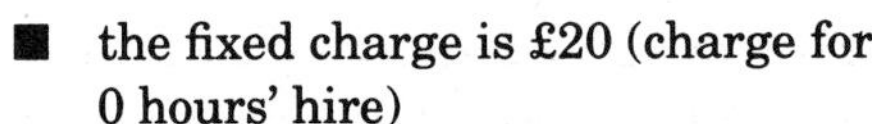

- the fixed charge is £20 (charge for 0 hours' hire)
- the variable charge is £5 per hour.

The variable charge is found from the **slope** or **gradient** of the line through the points, which is found by taking any two points on the line.

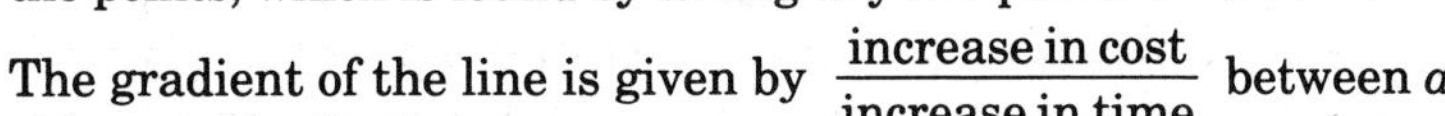

The gradient of the line is given by $\frac{\text{increase in cost}}{\text{increase in time}}$ between *any* two points on the line.

$$\text{Gradient of AB} = \frac{10}{2} = 5$$

$$\text{Gradient of BD} = \frac{20}{4} = 5$$

$$\text{Gradient of AD} = \frac{30}{6} = 5$$

The gradient of the straight line is constant and represents the **rate of change** of cost with respect to time, i.e. the hourly rate.

For any straight line on any grid with variables x and y, the gradient is defined in a similar way.

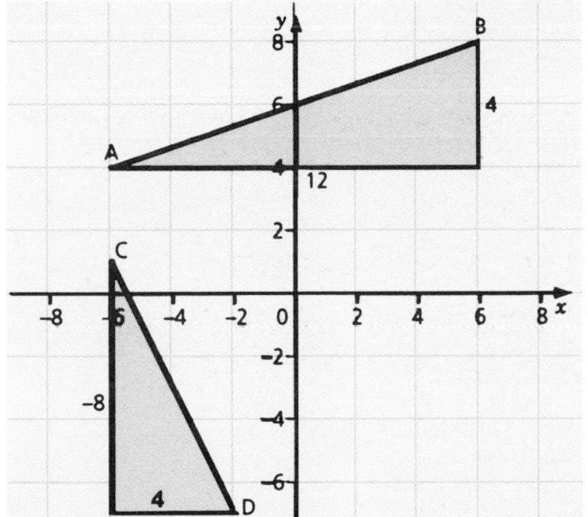

$$\text{gradient} = \frac{\text{change in } y}{\text{change in } x}$$

For line AB: gradient $= \frac{4}{12} = \frac{1}{3}$ *y increases* by 1 for every increase of 3 in x

Line AB has a **positive** gradient.

For line CD: gradient $= \frac{-8}{4} = -2$ *y decreases* by 2 for every increase of 1 in x

Line CD has a **negative** gradient.

The general result

These examples lead to a general definition for the gradient between two points $P(x_1, y_1)$ and point $Q(x_2, y_2)$.

$$\text{gradient of PQ} = \frac{\text{change in } y}{\text{change in } x} = \frac{y_2 - y_1}{x_2 - x_1}$$

A line with a **positive** gradient slopes **upwards** from left to right.

A line with a **negative** gradient slopes **downward** from left to right.

Example 3.4

Three points A, B and C have coordinates (–3, –1), (2, 1) and (12, 5) respectively.

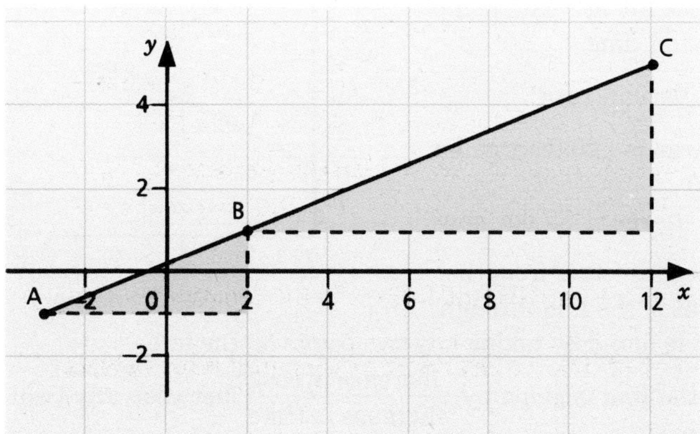

Show that these three points are collinear (lie on a straight line).

Solution

Gradient of AB $= \frac{1-(-1)}{2-(-3)} = \frac{2}{5} = 0.4$

Gradient of BC $= \frac{5-1}{12-2} = \frac{4}{10} = 0.4$

Point B is common to both lines AB and BC. This is sufficient to show that A, B and C are collinear.

Example 3.5

A quadrilateral has vertices P(0, 4), Q(–2, 1), R(0, –2) and S(4, –2). Find one pair of parallel sides and hence show that PQRS is a trapezium.

Solution

Start by plotting quadrilateral PQRS on a grid. From the graph, it seems that PS is parallel to QR. We can confirm this by showing that PS and QR have the same gradient.

Gradient of PS $= \frac{-6}{4} = -1.5$

Gradient of QR $= \frac{-3}{2} = -1.5$

Since lines PS and QR have the same gradient they must be parallel.

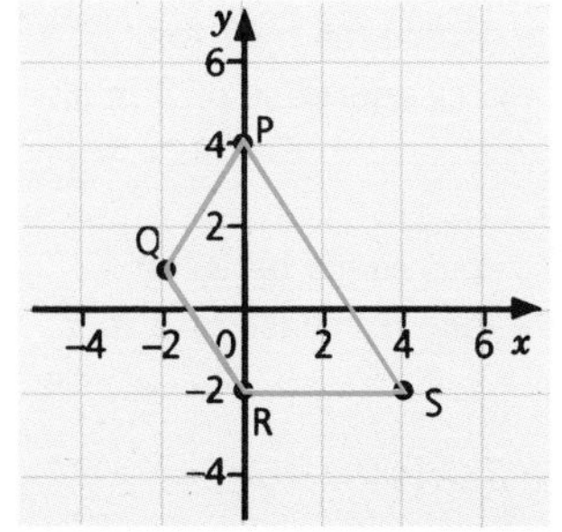

Parallel and perpendicular lines

We have seen that when two lines are **parallel** they have *equal* gradients. What happens when two lines are **perpendicular** to each other (intersect at right angles)?

Exploration 3.3 *Perpendicular lines*

Using the same scales on both axes, plot the points A(2, 10), B(10, 8) and C(6, 2) and join them to form a triangle.

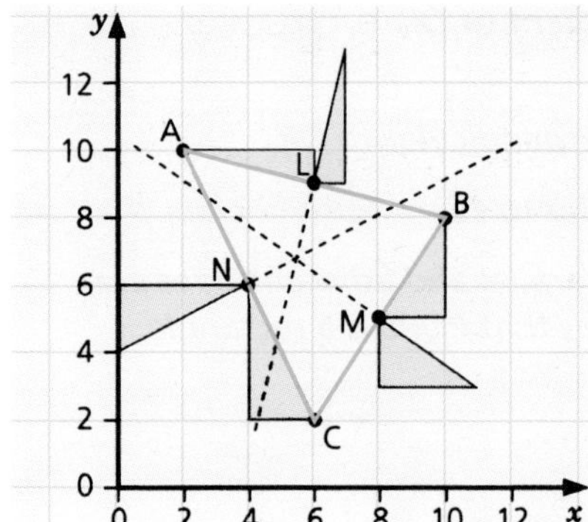

Construct the perpendicular bisectors of AB, BC and CA and comment on what you find.

The diagram shows triangle ABC. To construct a perpendicular bisector you need to find the midpoint and gradients.

For line AB, the midpoint is L(6, 9). The gradient of AB is $-\frac{1}{4}$, i.e. for every 4 units *to the right* you move 1 unit *down*.

For the perpendicular bisector, for every 1 unit to the *right* you move 4 units *up*. So the gradient of the perpendicular bisector is 4.

Working similarly with BC and CA, verify the following summary.

Line	Gradient of line	Gradient of perpendicular bisector
AB	$-\frac{1}{4}$	4
BC	$\frac{3}{2}$	$-\frac{2}{3}$
CA	–2	$\frac{1}{2}$

For each line, find the product of its gradient and the gradient of its perpendicular bisector. What can you say?

The general result

For each line in the exploration, the product of the two gradients is –1.

$$-\frac{1}{4}\times 4=-1 \qquad \frac{3}{2}\times-\frac{2}{3}=-1 \qquad -2\times\frac{1}{2}=-1$$

If two straight lines are perpendicular, the product of their gradients is –1.

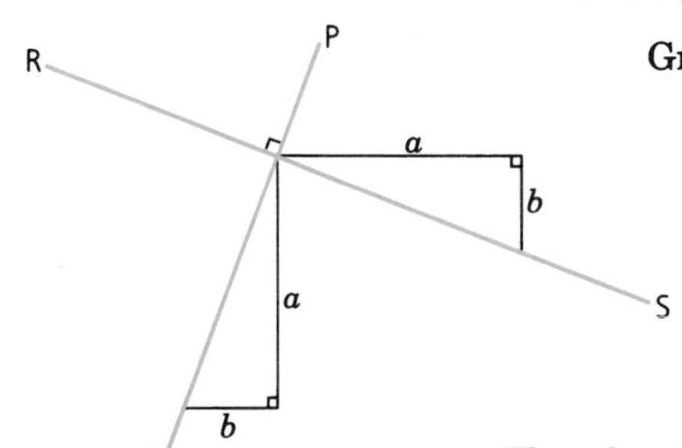

Gradient of PQ is $\frac{a}{b}$ Gradient of RS is $-\frac{b}{a}$

$$\frac{a}{b}\times-\frac{b}{a}=-1$$

Let $m=\frac{a}{b}$, then $-\frac{b}{a}=\frac{-1}{a/b}=-\frac{1}{m}$

Therefore, if a line has gradient m, a perpendicular line has gradient $-\frac{1}{m}$.

Example 3.6

A quadrilateral has vertices P(–6, 6), Q(0, 10), R(8, –2) and S(2, –6). By working out gradients, show that PQRS is a rectangle.

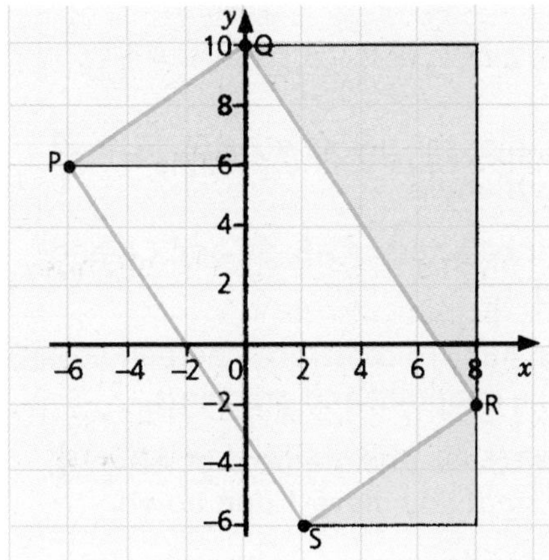

Solution

Gradient of PQ $= \frac{4}{6} = \frac{2}{3}$ *Gradient of RS* $= \frac{4}{6} = \frac{2}{3}$

Since their gradients are the same, PQ and RS are parallel. Similarly, QR and PS are parallel.

Gradient of PQ $= \frac{2}{3}$ *Gradient of QR* $= \frac{-12}{8} = -\frac{3}{2}$

Since $\frac{2}{3} \times -\frac{3}{2} = -1$, *PQ is perpendicular to QR. Similarly, PQ is perpendicular to PS.*

So the opposite sides are parallel and adjacent sides are perpendicular, which means that PQRS is a rectangle.

EXERCISES

3.2A

Eight points have coordinates as follows.
A(3, 5) B(6, 10) C(8, 3) D(6, –2) E(0, –5) F(–4, –7) G(–1, –4) H(–4, 2)

1 Plot all eight points on a grid, using the same scale for both axes.

2 Find the gradients of the lines AF, HD, AD and CH.

3 Show that the points D, E and F are collinear.

4 Show that AC is parallel to DH.

5 Find another pair of parallel lines, giving a reason why you think they are parallel.

6 Show that BC is perpendicular to DG.

7 Find another pair of perpendicular lines, giving a reason why you think they are perpendicular.

8 A point I is such that ACDI is a square. Find the coordinates of I.

9 Using gradients show that ΔEFG is right-angled.

10 A point J is such that DGHJ is a parallelogram. Find the coordinates of J.

EXERCISES

3.2B

Six points have coordinates as follows.
L(8, 1) M(5, 2) N(0, 8) P(–5, 3) Q(–12, –2) R(2, –4)

1 Plot all six points on a grid, using the same scale for both axes.

2 Find the gradient of the lines MR, PM and QL.

3 Show that NM is perpendicular to LR.

4 Show that PN is not parallel to LR.

5 Find a point which is collinear to Q and R, and prove that it is collinear.

6 What is the gradient of a line that is perpendicular to PR? Which line is perpendicular to PR?

7 If S is a point such that PQRS is a parallelogram, find the coordinates of S.

8 If T and U are points such that RLTU is a square, find possible coordinates of T and U.

9 V is a point such that MRV is a right-angled triangle with hypotenuse MV, which has a zero gradient. FInd the coordinates of V.

10 A circle is drawn with MV (as defined in question 9) as its diameter. Find the coordinates (x, y) of as many points as possible that lie on the circumference of this circle (x and y should only be whole numbers).

EQUATION OF A STRAIGHT LINE

Exploration 3.4

Straight-line relationships

Take another look at the 'Have van will travel' problem on page 51 at the beginning of the last section. From the graph we found that:

- the fixed charge is £20 (charge for 0 hours' hire)
- the variable charge is £5 per hour.

Find an equation that connects cost (c) with time (t).

Finding the equation

The cost of hiring a van (c) consists of a fixed charge of £20 plus a variable charge of £$5t$ (£5 × number of hours of hire). Symbolically this is expressed as:

$$c = 20 + 5t$$

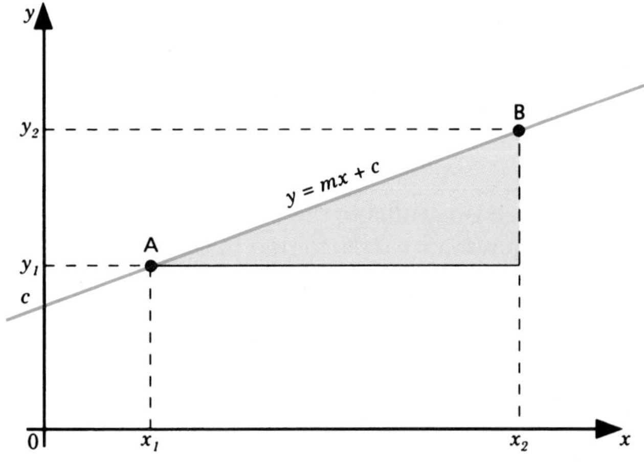

From the graph on page 51 we can see that 20 is the charge for 0 hours hire (the **intercept** on the cost axis) and the **gradient** of the straight line is 5. This principle can be applied generally to any straight line graph on xy-axes with y-intercept $(0, c)$ and gradient m. The equation of the straight line is $y = mx + c$.

Note
When $x = 0$, $y = m \times 0 + c$, which equals c. So, the graph of $y = mx + c$ passes through $(0, c)$.

If A and B are two points on the line with coordinates (x_1, y_1) and (x_2, y_2) then in particular:

$$y_1 = mx_1 + c \qquad (1)$$
$$y_2 = mx_2 + c \qquad (2)$$

Subtracting equation (1) from equation (2):

$$\begin{aligned}(2) - (1) \quad \Rightarrow \quad y_2 - y_1 &= (mx_2 + c) - (mx_1 + c) \\ &= mx_2 + c - mx_1 - c \\ &= mx_2 - mx_1 \\ &= m(x_2 - x_1)\end{aligned}$$

$$\Rightarrow \quad m = \frac{y_2 - y_1}{x_2 - x_1}$$

This is the gradient of the line joining two points

This confirms that m represents the gradient of the line passing through A and B.

CALCULATOR ACTIVITY

Exploration 3.5

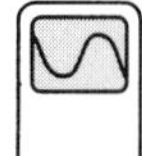

You will need a graphics calculator or graph plotter. Set x and y limits so that the origin is at the centre of the screen and both axes have the same scale.

Test your settings: the graph of $y = x$ should make a 45° angle with both x and y axes.

- Plot graphs of $y = x + c$ by choosing different values for c (e.g. –5, –2, 0, 1, 3, etc.).
- Plot graphs of $y = mx$ by choosing different values for m (e.g. –2, –1, 0.5, 0, 0.5, 1, 2, etc.).
- For the straight line $y = mx + c$, describe the effect of:

a) varying c whilst keeping m fixed
b) varying m whilst keeping c fixed
c) varying both m and c.

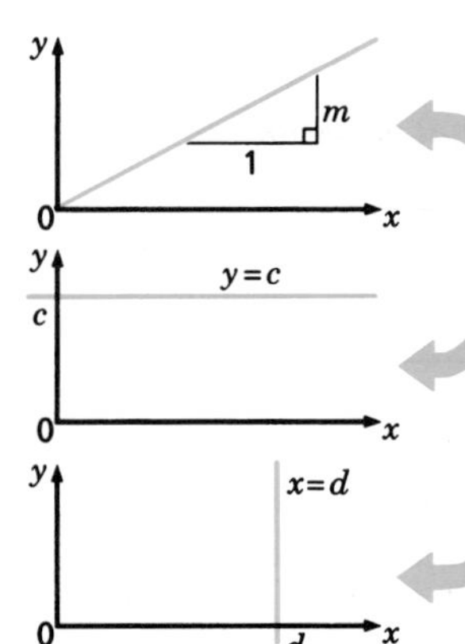

Nearly all straight lines have equations in the form $y = mx + c$. There are some special cases worth noting.

(1) If $c = 0$ then $y = mx$ which passes through (0, 0).

(2) If $m = 0$ then $y = c$, a line parallel to the x-axis ($y = 0$) passing through (0, c).

The problem line is one which is parallel to the y-axis. If it passes through (d, 0) then it has equation $x = d$. Note that in this case you cannot use $y = mx + c$ since m, the gradient, is not defined.

The following examples show how problems involving straight-line graphs and their equations may be tackled.

Example 3.7

Find the equation of the line with gradient 3 which passes through (0, 12). Where does it intersect the x-axis?

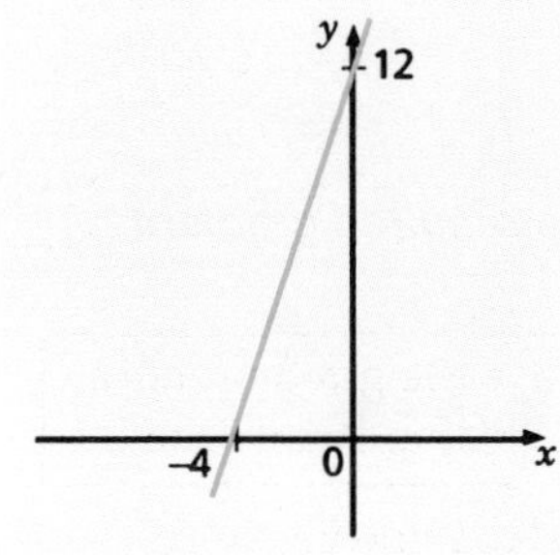

Solution

Gradient = $3 \Rightarrow m = 3$ and y-intercept is (0, 12) $\Rightarrow c = 12$
Therefore the equation of the line is $y = 3x + 12$.
Let the point of intersection with the x-axis be $(d, 0)$, i.e. when $y = 0, x = d$. Then:

$$\begin{aligned} 0 &= 3d + 12 \\ \Rightarrow -12 &= 3d \\ \Rightarrow d &= -4 \end{aligned}$$

The line intersects with the x-axis at (–4, 0).

These results are confirmed on the sketch graph.

Example 3.8

Find the gradient and y-intercept for the following equations.

a) $y = 10 - 4x$ **b)** $3y = 2x - 9$ **c)** $5x + 2y = 10$

In each case, draw a sketch graph, clearly showing where the line intersects with the x and y axes.

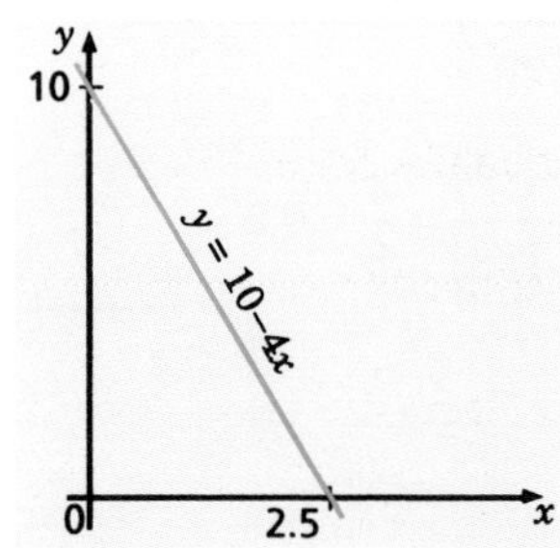

Solution

a) $y = 10 - 4x$ may be written $y = -4x + 10$.

Gradient = -4 and y-intercept is (0, 10)
Let the x-intercept be $(d, 0)$, i.e. when $y = 0, x = d$. Then:

$$\begin{aligned} 0 &= -4d + 10 \\ \Rightarrow 4d &= 10 \\ \Rightarrow d &= 2.5 \end{aligned}$$

The line intersects with the x-axis at (2.5, 0).

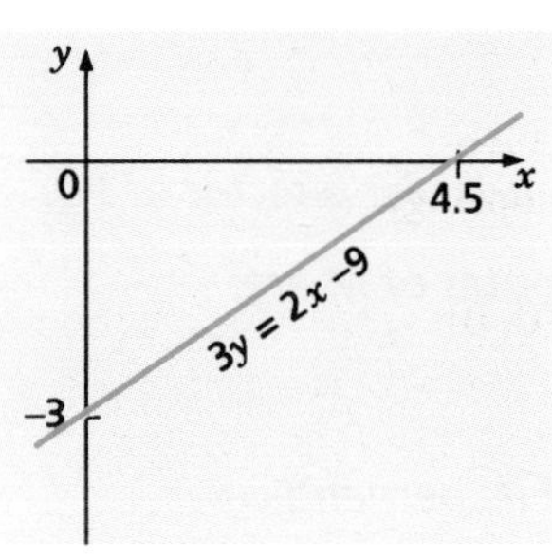

b) $3y = 2x - 9$ may be written $y = \frac{2}{3}x - 3$

Gradient = $\frac{2}{3}$ and y-intercept is (0, –3)

Let the x-intercept be $(d, 0)$. Then from the original equation:

$$\begin{aligned} 3 \times 0 &= 2d - 9 \\ \Rightarrow 9 &= 2d \\ \Rightarrow d &= 4.5 \end{aligned}$$

The line intersects with the x-axis at (4.5, 0).

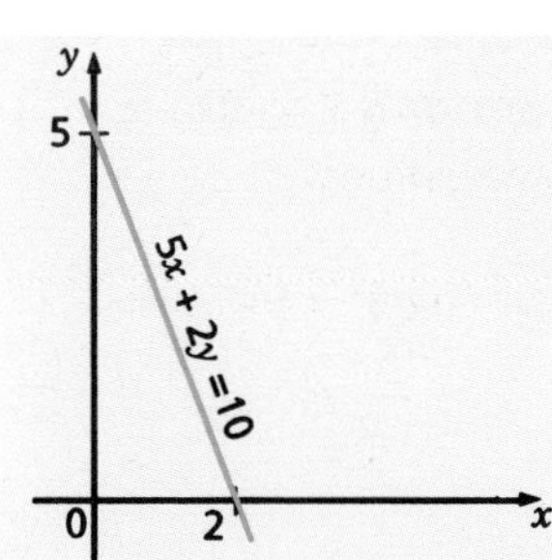

c) $5x + 2y = 10$ may be written $y = -2.5x + 5$.

Gradient = -2.5 and y-intercept is (0, 5)
Let the x-intercept be $(d, 0)$. Then from the original equation:

$$\begin{aligned} 5d + 2 \times 0 &= 10 \\ \Rightarrow 5d &= 10 \\ \Rightarrow d &= 2 \end{aligned}$$

The line intersects with the x-axis at (2, 0).

Example 3.9

Find the equation of the straight line parallel to $y = 12 - 5x$ passing through (2, –4).

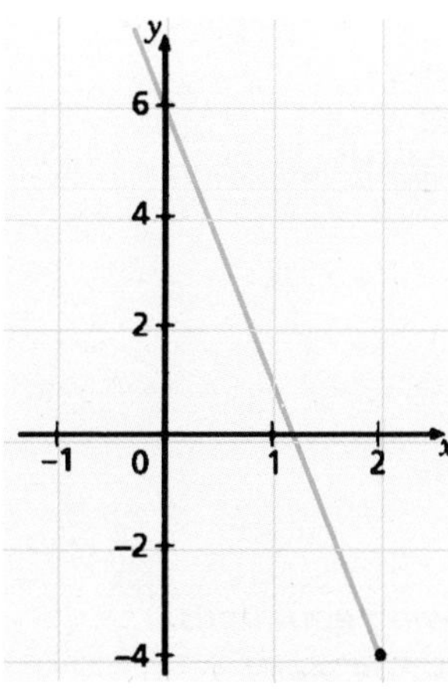

Solution

$y = 12 - 5x$ *may be written* $y = -5x + 12$.
Gradient $= -5$

Since parallel lines have the same gradient, the required gradient is $m = -5$.
Let the equation be $y = -5x + c$.
Since the line passes through (2, –4), when $x = 2, y = -4$.

$$-4 = -5 \times 2 + c$$
$$\Rightarrow \quad c = 10 - 4 = 6$$

The equation of the line is $y = -5x + 6$
or $y = 6 - 5x$
or $5x + y = 6$.

Example 3.10

Find the equation of the straight line passing through P(–3, –1) and Q(5, 3) and show that the equation of the perpendicular bisector of PQ is $2x + y = 3$. Illustrate your answer with a sketch.

Solution

The gradient of the line through P and Q is $\dfrac{3-(-1)}{5-(-3)} = \dfrac{4}{8} = 0.5$

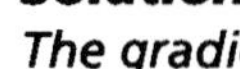

Let the equation be $y = 0.5x + c$.
Since the line passes through (5, 3):

$$3 = 0.5 \times 5 + c$$
$$\Rightarrow \quad c = 3 - 2.5 = 0.5$$

The equation of the line is $y = 0.5x + 0.5$
or $2y = x + 1$

Now we need to find the equation of the perpendicular bisector of PQ.

Let M be the midpoint of PQ. Then the coordinates of M are:

$\left(\dfrac{-3+5}{2}, \dfrac{-1+3}{2}\right)$ *or (1, 1).*

Since the gradient of PQ is 0.5, the gradient of its perpendicular bisector is $\dfrac{-1}{0.5} = -2$.
Therefore the equation of the perpendicular bisector is $y = -2x + c$, *and* $y = 1$ *when* $x = 1$.

$$\Rightarrow \quad 1 = -2 \times 1 + c$$
$$c = 1 + 2 = 3$$

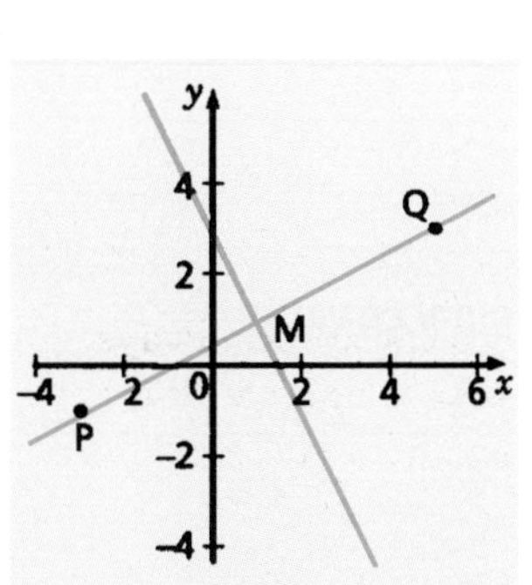

The equation of the perpendicular bisector is:

$$y = -2x + 3$$

or $2x + y = 3$.

The last two worked examples demonstrate two general methods for finding equations of straight lines, given certain initial information:

■ **Finding the equation of a line with gradient m passing through (x_1, y_1)**

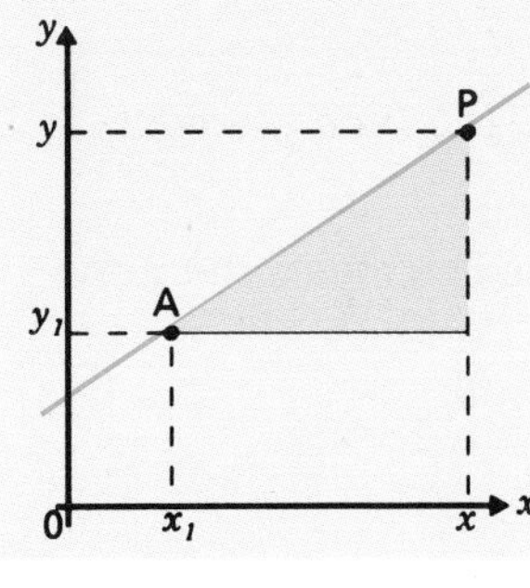

Let $A(x_1, y_1)$ be the fixed point and $P(x, y)$ be a variable point on the line with gradient m. Then:

$$\frac{y - y_1}{x - x_1} = m$$

$$\Rightarrow \quad y - y_1 = m(x - x_1)$$

So the equation of the line with gradient –1.5 passing through (3, –0.5) is

$$y - (-0.5) = -1.5(x - 3)$$

$$\Rightarrow \quad y + 0.5 = -1.5x + 4.5$$

$$\Rightarrow \quad y = 4 - 1.5x$$

or $2y + 3x = 8$.

■ **Finding the equation of the line passing through (x_1, y_1) and (x_2, y_2)**

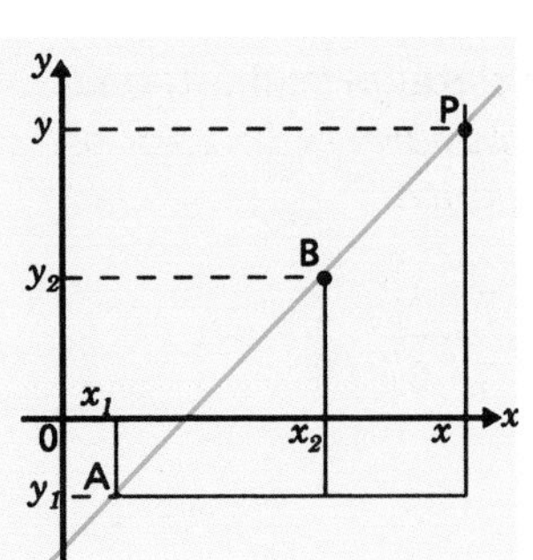

Let $A(x_1, y_1)$ and $B(x_2, y_2)$ be the fixed points and $P(x, y)$ be the variable point on the line.

Since points A, B and P are collinear:

gradient AP = gradient AB

$$\Rightarrow \quad \frac{y - y_1}{x - x_1} = \frac{y_2 - y_1}{x_2 - x_1}$$

$$\Rightarrow \quad \frac{y - y_1}{y_2 - y_1} = \frac{x - x_1}{x_2 - x_1}$$

So the equation of the line passing through (2, 7) and (5, 1) is given by:

$$\frac{y - 7}{1 - 7} = \frac{x - 2}{5 - 2}$$

$$\Rightarrow \quad \frac{y - 7}{-6} = \frac{x - 2}{3}$$

$$\Rightarrow \quad \frac{-6(y - 7)}{-6} = \frac{-6(x - 2)}{3}$$

$$\Rightarrow \quad y - 7 = -2(x - 2)$$

$$\Rightarrow \quad y - 7 = -2x + 4$$

$$\Rightarrow \quad y = -2x + 11$$

or $2x + y = 11$.

You may find both methods useful but in both cases you can always work from first principles, i.e. by finding values for m and c in the equation $y = mx + c$.

EXERCISES

1 Find the gradients and y-intercepts for lines with the following equations.

a) $y = 2x - 5$ **b)** $y = 10 - 3x$
c) $5y = x - 12$ **d)** $x + y = 7$
e) $2y - 3x = 8$ **f)** $4x + 5y = 30$
g) $2x - 6y + 9 = 0$ **h)** $\frac{1}{3}x + \frac{1}{4}y = 1$

2 For each of the following equations:

a) find the coordinates of the points where the line meets the x and y axes, i.e. $(d, 0)$ and $(0, c)$ respectively
b) sketch a graph, showing clearly $(0, c)$ and $(d, 0)$
c) confirm that the gradient $m = -\frac{c}{d}$.

i) $y = 2x + 8$ **ii)** $y = 12 - 4x$
iii) $y = \frac{1}{2}x + 3$ **iv)** $y = 4 - \frac{2}{3}x$
v) $x + y = 10$ **vi)** $2x + y = 10$
vii) $3x - 4y = 18$ **viii)** $\frac{3}{5}y + \frac{1}{2}x = \frac{3}{2}$

3 Find the equations of the lines with the following gradients and y-intercepts, giving your answers in an appropriate form. Illustrate each answer with a sketch.

a) gradient = 3 y-intercept –5
b) gradient = –2 y-intercept 7
c) gradient = $\frac{3}{4}$ y-intercept $2\frac{1}{2}$
d) gradient = $-1\frac{1}{2}$ y-intercept 12
e) gradient = $\frac{5}{3}$ y-intercept 0
f) gradient = 0 y-intercept 8
g) gradient = –0.4 y-intercept –2.4
h) gradient = $2\frac{1}{2}$ y-intercept $3\frac{1}{2}$

4 Find the equations of the lines with the following gradients, passing through the points given. Give each equation in an appropriate form.

a) gradient = 4, passing through (5, –1)
b) gradient = –3, passing through (–1, 6)
c) gradient = $\frac{1}{5}$, passing through (1, 4)
d) gradient = $-\frac{2}{3}$, passing through (6, –3)

5 Find the equations of the lines which pass through the following pairs of points.

a) (2, 0) and (7, 1) **b)** (–3, 7) and (5, 3)
c) (–4, –1) and (5, 2) **d)** (–2, 8) and (3, –5)

6 Points R and S have coordinates (6, 1) and (2, 5).

a) Find the coordinates of the midpoint of RS.
b) Find the gradient of RS.
c) Find the equation of the perpendicular bisector of RS.

7 A triangle has vertices A(5, 1), B(8, 4) and C(2, 5).

a) Find the equation of the line AB.
b) Find the equation of the line passing through C that is perpendicular to AB and meets AB at point D.
c) Show that point D has coordinates (5.5, 1.5).
d) Find the lengths of AB and CD and hence deduce the area of the triangle.

8 A triangle has vertices X(2, 5), Y(–2, –2) and Z(6, 0).

a) Find the coordinates of L, M, and N, which are the midpoints of XY, YZ and ZX respectively.
b) Find the equations of the lines LZ, MX and NY.
c) Show that the lines LZ, MX and NY meet in a single point with coordinates (2, 1).

9 The line l has equation $3x + 4y = 21$. Find the equation of a line that is perpendicular to l and intersects l at a point on the line $y = x$.

EXERCISES

1 Find the gradients and y-intercepts for the lines with the following equations.

a) $y = x + 2$
b) $y = 2 - 3x$
c) $8y = x - 4$
d) $2x + y = 0$
e) $6x - 3y = 5$
f) $7x + 3y = 9$
g) $x + 2y + 7 = 0$
h) $\frac{1}{8}x + \frac{5}{4}y = 6$
i) $y = 7$

2 For each of the following equations, sketch a graph marking where the line crosses each of the axes, and state its gradient.

a) $y = 7x + 6$
b) $y = 6 - 4x$
c) $y = \frac{3}{4}x + 2$
d) $y = \frac{2}{9} - \frac{1}{3}x$
e) $x - y = 5$
f) $7x + 5y = 9$
g) $3x = 3y + 5$
h) $\frac{3}{4}x + \frac{5}{6}y + \frac{1}{2} = 0$
i) a line which passes through the x-axis at $\frac{-2}{3}$ and through the y-axis at $\frac{4}{9}$
j) a horizontal line which passes through the point (6.2, –7.1)

3 Find the equations of the lines with the following gradients and y-intercepts, giving your answers in an appropriate form. Illustrate each answer with a sketch.

a) gradient = 7 y-intercept 4
b) gradient = –9 y-intercept $\frac{2}{3}$
c) gradient = $\frac{4}{3}$ y-intercept $\frac{1}{4}$
d) gradient = $4\frac{1}{2}$ y-intercept $-2\frac{3}{4}$

4 Find the equations of the lines with the following gradients, passing through the given points. Give each equation in an appropriate form.

a) gradient = 3, passing through (8, 1)
b) gradient = –5, passing through (–1, 8)

c) gradient = $\frac{2}{3}$, passing through (4, –5)
d) gradient = $-1\frac{1}{2}$, passing through (7, 0)

5 Find the equations of the lines which pass through the following pairs of points.

a) (2, 2) and (8, 7)
b) (–7, 5) and (–2, –8)
c) ($\frac{1}{2}$, –1) and (4, $-2\frac{1}{2}$)
d) (4.2, 3.8) and (–3.6, 6.2)

6 Find the equations of the lines and give your answers in the form $ax + by + c = 0$.

a) a line of gradient 3 and y-intercept 6
b) a line of gradient –2 which passes through the y-axis at (0, 3)
c) a line of gradient $\frac{3}{7}$ which crosses the y-axis at –9
d) a line of gradient 4 which crosses the x-axis at (8, 0)
e) a line of gradient $-\frac{2}{3}$ which crosses the x-axis at (–3, 0)
f) a line of gradient 9 which passes through the point (4, 6)
g) a line of gradient –4 which passes through the point (–8, 5)
h) a line which passes through the points (4, –9) and (–2, 6)

7 Points A and B have coordinates (–3, 1) and (7, 4)

a) Find the coordinates of the midpoint of AB.
b) Find the gradient of AB.
c) Find the equation of the perpendicular bisector of AB.

8 A parallelogram has vertices P(–20, 10), Q(–6, 16), R(24, 4) and S(10, –2).

a) Find the equation of the line PS.
b) Find the equation of the line passing through Q that is perpendicular to PS.
c) Show that the point T (–10, 6) lies on both of the lines found in parts (a) and (b).
d) Find the lengths of PS and QT and hence find the area of the parallelogram.

9 The line $2x - 3y + 7 = 0$ passes through the x-axis at point P. Find the equation of the line which is perpendicular to $6x + 3y + 4 = 0$ and which passes through point P.

FITTING LINES TO DATA

Exploration 3.6

Line of best fit

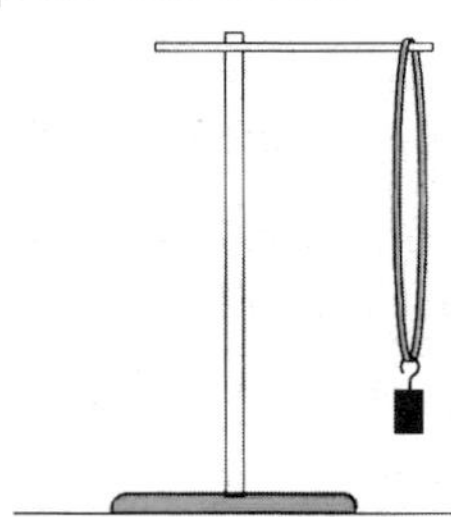

In an experiment to measure the elasticity of a rubber band, a student hung various masses on one end of the rubber band and measured its stretched length. The results were as follows.

Mass in g x	100	200	300	400	500	600
Length in mm y	227	237	253	272	288	298

- Plot this data on graph paper.
- Draw in a straight line which best fits the data.
- Find the equation of the line of best fit.
- Use your equation to predict the length of the rubber band when a mass of 540 g is attached.
- Would it be sensible to use this model to predict the length when a mass of 1 kg is attached?

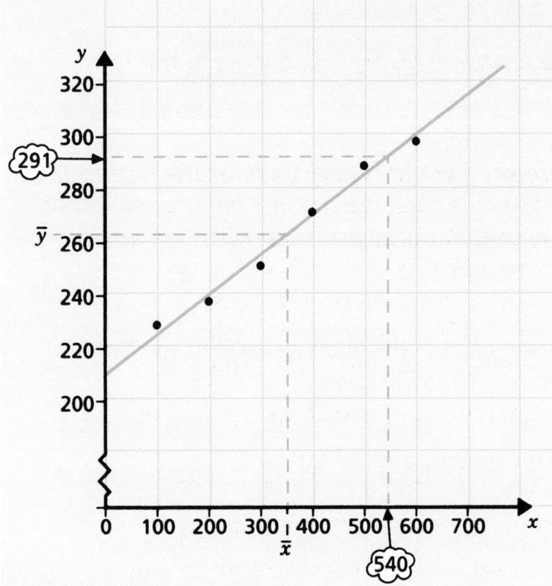

The graph shows the six data points joined with a line of best fit.

A line of best fit should pass through $(\bar{x}, \bar{y})$ where $\bar{x}$ is the mean of the x-values and $\bar{y}$ is the mean of the y-values.

$$\bar{x} = \frac{100 + 200 + 300 + 400 + 500 + 600}{6} = 350$$

$$\bar{y} = \frac{227 + 237 + 253 + 272 + 288 + 298}{6} = 262.5$$

The point $(\bar{x}, \bar{y}) = (350, 262.5)$ is called the **centroid**.

Now draw in the straight line passing through $(\bar{x}, \bar{y})$ which fits the data best.

Interpreting the results

From the graph, the gradient is about 0.15 (for every 100 g added, the rubber band stretches about 15 mm) and the y-intercept is at (0, 210) mm (the unstretched length is 210 mm).

So $m = 0.15$ and $c = 210$.

Therefore the equation of the line of best fit is:
$y = 0.15x + 210$

For a mass of 540 grams, substitute $x = 540$:
$y = 0.15 \times 540 + 210$
$\Rightarrow \; y = 291$

The model predicts that a mass of 540 grams will stretch the band to a length of 291 mm.

Predicting a y-value using an x-value within the range of x-values given is called **interpolation**. Using an x-value from outside the range of x-values is called **extrapolation**. There are dangers in extrapolating. Attaching a mass of 1 kg may break the rubber band!

An alternative way of forming the equation of the line of best fit is to use its gradient (m) and the fact that it must pass through $(\bar{x}, \bar{y})$.

This is especially useful if it is inconvenient to include $x = 0$ on your graph. The following example illustrates the method.

Example 3.11

In athletics, the times set by various world record holders for the mile, together with the date they broke the record are:

Roger Bannister	*6–5–54*	*3 minutes 59.4 seconds*
Derek Ibbotson	*19–7–57*	*3 minutes 57.2 seconds*
Peter Snell	*27–1–62*	*3 minutes 54.4 seconds*
Michel Jazy	*9–6–65*	*3 minutes 53.6 seconds*
Filbert Bayi	*17–5–75*	*3 minutes 51.0 seconds*
Seb Coe	*17–7–79*	*3 minutes 49.0 seconds*
Steve Cram	*27–7–85*	*3 minutes 46.3 seconds*

a) *Using suitable variables x and y, model the data using a straight-line equation.*

b) *Use your model to predict a world record time for the mile in 1995.*

c) *When is it likely that the mile could be run in 3 minutes 30 seconds?*

d) *Is it reasonable to use this model to estimate when we are likely to see the first 3-minute mile?*

Solution

a) *Let x represent the year since 1900, and y represent the time in seconds above 3 minutes. Then we have:*

x	54.3	57.5	62.1	65.4	75.4	79.5	85.5
y	59.4	57.2	54.4	53.6	51.0	49.0	46.3

$\bar{x} = 68.5$ and $\bar{y} = 53.0$, so the line of best fit passes through (68.5, 53.0).

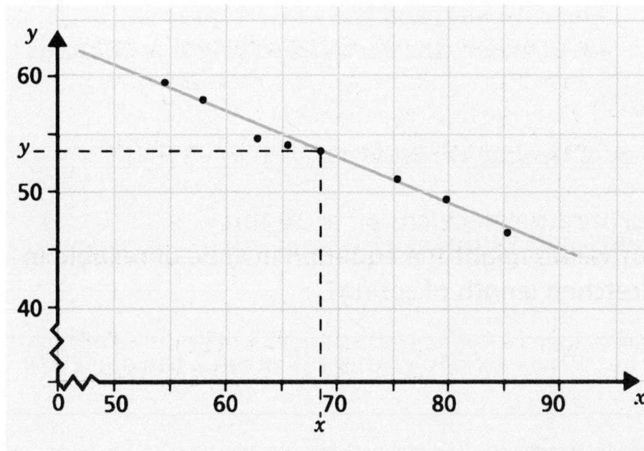

From the graph the gradient $m = -0.38$, therefore the equation of the line of best fit is

$$y - \bar{y} = m(x - \bar{x})$$

$$\Rightarrow \; y - 53.0 = -0.38(x - 68.5)$$

$$\Rightarrow \; y - 53.0 = -0.38x + 26.03$$

$$\Rightarrow \; y = -0.38x + 79.03$$

b) *To predict the world record time for 1995, let $x = 95.5$ (representing a date mid-way through 1995).*

$$\Rightarrow \; y = -0.38 \times 95.5 + 79.03$$

$$y = 42.74$$

So we might expect a time of 3 minutes 42.7 seconds during 1995.

c) *To predict when the mile might be run in 3 minutes 30 seconds, let $y = 30$.*

$$\Rightarrow \; 30 = -0.38x + 79.03$$

$$\Rightarrow \; 0.38x = 79.03 - 30 = 49.03$$

$$\Rightarrow \; x = 49.03/0.38 = 129.0$$

So the model predicts that we shall have to wait until the year 2029 until this record is set!

d) In c) we used the model to extrapolate. The reliability of the model for a prediction so far ahead is questionable. The model would give an even more unreliable prediction of the year in which we are likely to see a 3-minute mile.
The model cannot be valid indefinitely since this would mean that eventually the mile would be run in no time at all!

EXERCISES

3.4 A

1 The table shows some equivalent temperatures in °C and °F.

Temperature (in °C)	20	37	55	70	100
Temperature (in °F)	68	98.6	131	158	212

a) Plot points to represent these pairs of temperatures.
b) Find the equation of the line that fits the data exactly.
c) What temperature in °F corresponds to 0°C? What is the significance of this?
d) What temperature is the same in both scales?

2 In a science experiment a student attached various masses to the end of a spring and measured its stretched length. The results were as follows.

Mass, m (in g)	100	200	300	400	500
Length, l (in cm)	24.2	27.9	31.8	36.1	39.8

a) Plot points to represent lengths of spring for different masses.
b) Calculate the mean mass and mean length for the data given.
c) Find the equation of the line that fits the data best.
d) What would you expect the natural (unstretched) length of the spring to be?
e) What mass would produce an *extension* of 10 cm?
f) For what range of values might the equation in **c)** be unreliable in predicting the stretched length of spring?

3 A particular type of fish (North Seas haddock) has a lifetime of about 10 years. The table shows average masses of haddock at different times in their life.

Age (years)	2	4	6	8	10
Mass (grams)	250	500	820	1110	1300

a) Plot points to represent the mass of a fish at different ages.
b) Calculate the mean age and mean mass for the data given.
c) Find the equation of the line that fits the data best.
d) Use your equation to predict the mass at age 3 years.
e) At what age would you expect a haddock to have a mass of 1 kg?

4 A student conducted an experiment to discover the relationship between the resistance of a piece of wire and its temperature.

At five different temperatures her results were as shown in the table.

Temperature, T (in °C)	20	40	60	80	100
Resistance, R (in ohms)	18.1	21.0	23.8	26.9	30.0

a) Plot the data on a suitable graph.
b) Find the equation of the line of best fit in the form $R = aT + b$.
c) Interpret what a and b represent for the relationship.
d) Use your equation in **b)** to predict:
  i) the resistance in the wire at 130°C,
  ii) the temperature range for which the resistance is less than 10 ohms,
  iii) the temperature for which the resistance is zero – does this make sense?

EXERCISES

1 In an experiment a student found the speed (u) with which a ball hit the floor, and the speed (v) with which it bounced back up off the floor. The results were as follows.

Speed before bounce, u (in m s^{-1})	3.13	3.83	4.43	4.95	5.42	5.86	6.26
Speed after bounce, v (in m s^{-1})	2.29	2.79	3.32	3.60	3.82	4.06	4.34

a) Plot points to represent this data.
b) Calculate the mean speed before bounce and the mean speed after bounce for the given data.
c) Find the equation of the line that fits the data best.
d) If the ball hit the floor with a speed of 4 m s^{-1}, at what speed would you expect it to bounce off the floor?
e) If a ball was placed at rest on the floor, so it touched the floor with a speed of 0 m s^{-1}, would you expect it to bounce off the floor? According to this model, would it bounce off the floor? Comment on your answer.

2 In an attempt to find the thickness of the paper in a telephone book, the thickness of multiple pages was measured and the following data were noted.

Number of pages	506	220	423	104	0
Thickness (in mm)	33	14	27	7	0

a) Plot points to represent this data.
b) Find the equation of the line that you think best fits the data.
c) Estimate the thickness of a single sheet of telephone book paper.

3 The population of Scotland for certain years is given below.

Year	Total population
1821	2 091 521
1831	2 364 386
1841	2 620 184
1851	2 888 742
1881	3 735 573
1891	4 025 647
1921	4 882 497

a) Let x be the number of years after 1800 and y be the population of Scotland in millions. Plot points to represent this data in terms of x and y.
For the first piece of information $x = 21$ and $y = 2.09$ (3 s.f.).

b) Does this data seem to lie on a straight line? If so, find the equation of the line which best fits the data.

c) What do you think the population of Scotland was in 1931?

d) According to this model, what was the population of Scotland in 1700 ($x = -100$)?

e) Some more data on Scotland's population is given below.

Year	Total population
1891	4 025 647
1901	4 472 103
1911	4 760 904
1921	4 882 497
1931	4 842 554

Plot this data, and comment on what you have done through this question.

4 The times taken by the winners of the Women's Olympic 400 m freestyle swimming between 1956 and 1980 are given in the table below.

year	Winner	Time (minutes : seconds)
1956	*Lorraine Gapp*	*4:54.6*
1960	*Chris von Saltza*	*4:50.6*
1964	*Virginnia Duenkel*	*4:43.3*
1968	*Debbie Meyer*	*4:31.8*
1972	*Shane Gould*	*4:19.0*
1976	*Petra Thumer*	*4:09.9*
1980	*Ines Diers*	*4:08.8*

a) Let x be the number of years after 1950 and let y be the time taken (in seconds). Plot points to represent this data as (x, y) coordinates.

b) Draw in a line of best fit and find its equation.

c) Use the equation to predict:

i) the time taken by the 1984 winner,

ii) the time taken by the winner in the year 2000.

d) Comment on the validity of your predictions in **c)**.

SIMULTANEOUS EQUATIONS

Exploration 3.7

Best combinations

Imagine you are a landscape gardener and have been given the job of creating a floral display in the local park. You are given the following brief.

- Use daffodil and/or tulip bulbs only.
- The display should contain no more than 400 bulbs in total.
- Tulip bulbs cost 10 pence and daffodil bulbs cost 6 pence each.
- Total budget for bulbs = £30.

Find a combination of tulip and daffodil bulbs which total 400 and use up all the money.

Finding the combination

The two variables in this problem are the number of tulips and the number of daffodils. Let d represent the number daffodil bulbs and t the number of tulip bulbs purchased for the display.

The two constraints lead to two **inequalities**:

Maximum of 400 bulbs $\Rightarrow$ $d + t \leq 400$

Budget of £30 $\Rightarrow$ $6d + 10t \leq 3000$

cost of d daffodil bulbs

cost of t tulip bulbs

money available in pence

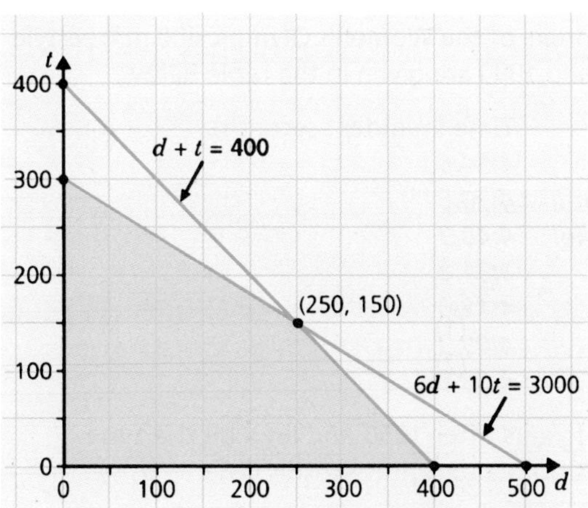

If 400 bulbs are used and a total of £30 is spent on them, the inequalities become equations:

$d + t = 400$ (1)

$6d + 10t = 3000$ (2)

The simultaneous solution of these two equations may be found graphically by finding the point of intersection of the two straight line graphs.

To use 400 bulbs and spend all the money the gardener needs to buy 250 daffodil bulbs and 150 tulip bulbs. Given that the number of daffodil or tulip bulbs cannot be negative, i.e. $d \geq 0$, $t \geq 0$, the shaded area in the diagram represents the set of combinations that satisfy all the constraints.

Simultaneous equations – algebraic methods

The simultaneous solution of equations (1) and (2) above may be found algebraically, rather than graphically. Two methods are in common use.

Method A

$d + t = 400$ (1)

$6d + 10t = 3000$ (2)

Make the coefficient of t the same in both equations:

$(1) \times 10 \Rightarrow \quad 10d + 10t = 4000 \quad (3)$

$(2) \Rightarrow \quad 6d + 10t = 3000 \quad (4)$

Eliminate the variable t:

$(3) - (4) \Rightarrow \quad 4d = 1000 \quad (5)$

$(5) \div 4 \Rightarrow \quad d = \frac{1000}{4} = 250$

Substitute for d in (1):

$\Rightarrow \quad 250 + t = 400$

$\Rightarrow \quad t = 400 - 250 = 150$

Method B

$d + t = 400 \quad (1)$

$6d + 10t = 3000 \quad (2)$

Make d the subject of (1):

$d = 400 - t$

Substitute for d in (2):

$$6(400-t) + 10t = 3000$$
$$2400-6t + 10t = 3000$$
$$4t = 600$$
$$t = 150$$

Substitute for t in (1):

$$d + 150 = 400$$
$$d = 400 - 150 = 250$$

Example 3.12

Solve this pair of simultaneous equations.

$3x + 4y = 29$

$2x - 5y = 4$

Illustrate your solution graphically.

Solution

Method A

$3x + 4y = 29 \quad (1)$

$2x - 5y = 4 \quad (2)$

Make the coefficient of x the same in both equations:

$(1) \times 2 \Rightarrow \quad 6x + 8y = 58 \quad (3)$

$(2) \times 3 \Rightarrow \quad 6x - 15y = 12 \quad (4)$

$(3) - (4) \Rightarrow \quad 23y = 46 \quad (5)$

$(5) \div 2 \Rightarrow \quad y = \frac{46}{23} = 2$

Substitute for y in (2):

$$2x - 5 \times 2 = 4$$
$$\Rightarrow \quad 2x = 14$$
$$\Rightarrow \quad x = 7$$

Method B

$3x + 4y = 29$ (1)

$2x - 5y = 4$ (2)

Make x the subject of (2):

$$2x = 4 + 5y$$

$$x = \frac{4+5y}{2}$$

Substitute for x in (1):

$$3 \times \frac{4+5y}{2} + 4y = 29$$

$$\Rightarrow \quad 3(4 + 5y) + 8y = 58$$

$$\Rightarrow \quad 12 + 15y + 8y = 58$$

$$\Rightarrow \quad 23y = 46$$

$$\Rightarrow \quad y = \frac{46}{23} = 2$$

Substitute for y in (1):

$$3x + 4 \times 2 = 29$$

$$\Rightarrow \quad 3x = 21$$

$$\Rightarrow \quad x = 7$$

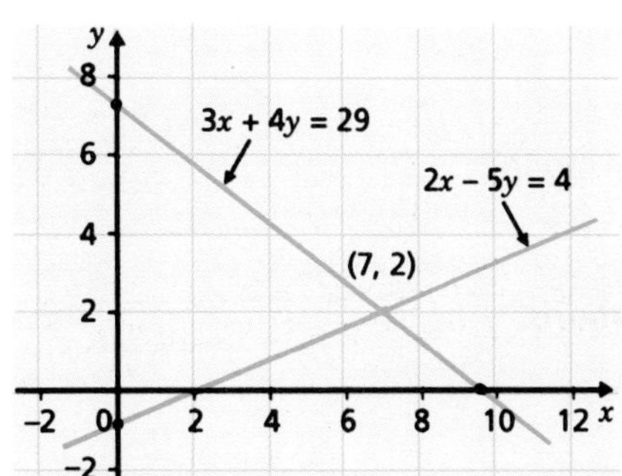

Thus $x = 7$, $y = 2$ satisfy both equations simultaneously. Graphically the point (7, 2) represents the intersection of the lines $3x + 4y = 29$ and $2x - 5y = 4$.

The line $3x + 4y = 29$ passes through $(0, 7\frac{1}{4})$ and $(9\frac{2}{3}, 0)$.

The line $2x - 5y = 4$ passes through $(0, -\frac{4}{5})$ and (2, 0).

The two lines intersect at (7, 2).

Example 3.13

A quadrilateral has vertices A(0, 3), B(3, 4), C(5, –2) and D(–1, 0).

a) *Find the equations of lines AC and BD.*
b) *Show that AC bisects BD at right angles.*

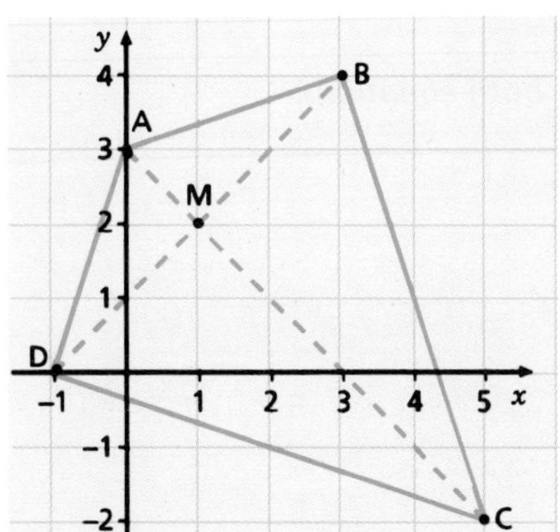

Solution

Start by plotting points A, B, C and D and joining A to C and B to D. From the graph it is evident that M(1, 2) is the midpoint of BD, and AC and BD are perpendicular, i.e. AC bisects BD at right angles.

The following solution provides an algebraic proof.

a) *The equation of AC is found from:*

$$\frac{y-3}{-2-3} = \frac{x-0}{5-0}$$

$$\Rightarrow \quad \frac{y-3}{-5} = \frac{x}{5}$$

$$\Rightarrow \quad y = 3 - x$$

The equation of BD is found from:

$$\frac{y-0}{4-0} = \frac{x-(-1)}{3-(-1)}$$

$$\Rightarrow \quad \frac{y}{4} = \frac{x+1}{4}$$

$$\Rightarrow \quad y = x + 1$$

b) *Solve equations* (1) *and* (2) *simultaneously.*

$$y = 3 - x \qquad (1)$$

$$y = x + 1 \qquad (2)$$

From (1) $y = 3 - x$

Substitute for y *in* (2):

$$3 - x = x + 1$$

$$\Rightarrow \quad 2 = 2x$$

$$\Rightarrow \quad x = 1$$

Substitute for x *in* (1):

$$y = 3 - 1 = 2$$

So the coordinates of M are (1, 2)
But the midpoint of BD is $(\frac{1}{2}(-1 + 3), \frac{1}{2}(0 + 4))$ *or* $(1, 2)$.
The gradient of AC is –1 and the gradient of BD is 1, hence the product of gradients is $(-1) \times 1 = -1$.
Therefore AC bisects BD at right angles.

EXERCISES

3.5 A

For questions **1** to **6**, solve the pairs of simultaneous equations, illustrating your answers graphically.

1 $2x + 3y = 30$
$x + y = 12$

2 $13x + 9y = 36$
$2x + 3y = 6$

3 $3s + 2t = 12$
$2s + 5t = 19$

4 $5p + 3r = 41$
$p - 2r = 3$

5 $2x + 3y = 16$
$5x - 2y = 21$

6 $-3x + 8y = 31.1$
$5x + 3y = 3.7$

7 Find the point of intersection of the lines $y = 3x - 2$ and $5x + 3y = 15$.

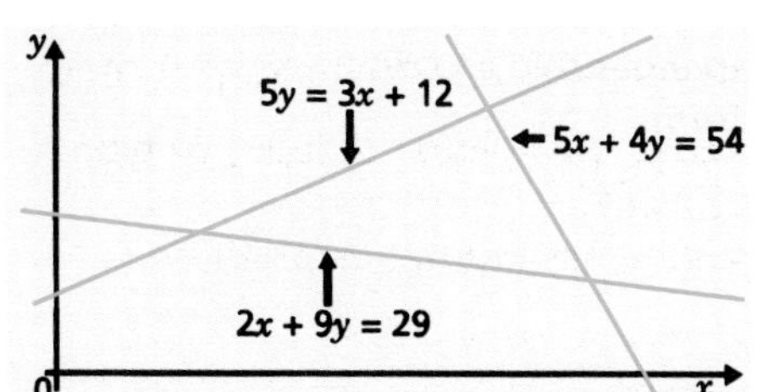

8 The three lines

$5x + 4y = 54 \qquad 2x + 9y = 29 \qquad 5y = 3x + 12$

make a triangle.

Find the coordinates of its vertices.

9 Show that the three lines

$2x + 7y = 42 \qquad y = x - 3 \qquad 3x + y = 25$

are collinear (they meet at a single point).

10 Try to find solutions for these pairs of simultaneous equations.

a) $10x - 2y = 6$
$y = 5x - 3$

b) $6x + 3y = 20$
$y = 7 - 2x$

Give a graphical explanation for your answers.

11 Triangle EFG has vertices E(2, 5), F(–2, –2) and G(6, 0).

Show that the medians meet in a single point and find its coordinates.

12 A triangle has vertices at P(8, 0), Q(1, 1) and R(3, 5).

a) Show that the triangle is isosceles.
b) Find the equation of the straight line through P and R.
c) Find the equation of the perpendicular from Q to PR.
d) Find the coordinates of the foot of the perpendicular from Q to PR.
e) Find the area of ΔPQR in two different ways.

13 A landscape designer has £240 to spend on planting trees and shrubs to landscape an area of 1000 m^2. For a tree he plans to allow 25 m^2 and for a shrub 10 m^2. Planting a tree will cost £2 and planting a shrub will cost £4.

Let t represent the number of trees planted and s the number of shrubs planted.

Assuming he spends all the money and all the area is used up,

a) Explain why $t + 2s = 120$
and $5t + 2s = 200$
b) Solve the equations in **a)** to find how many trees and how many shrubs he should buy.

14 A company manufactures two products, X and Y. Each of these products requires a certain amount of time on the assembly line and a further amount of time in the finishing shop, as indicated below.

	X	Y	Total available per week
No. hours for assembly	5	3	105
No. hours for finishing	2	4	70

Assume that all the time available for assembly and finishing is taken up.

a) Formulate two equations in x and y, the number of items of type X and Y respectively produced per week.
b) Solve the equations to find the number of each type the firm should produce

15 A manufacturer supplies 200 pairs of shoes per month at a price of £30 per pair. However, the demand for the shoes is 2800 pairs per month. At a price of £35 per pair, 400 more pairs can be supplied each month. But, at this increased price, the demand reduces by 100 pairs.

a) Assuming linear relationships, determine the supply and demand relations in the form of a pair of simultaneous equations.
b) Solve the equations to find the equilibrium price and quantity, i.e. the price and quantity when supply equals demand.

EXERCISES

3.5 B

For questions **1** to **6**, solve the pairs of simultaneous equations, illustrating your answers graphically.

1	$5x + 4y = 9$ $6x + 2y = 2$	**2**	$6x - 3y = 8$ $5x + y = 4$
3	$x = 2y$ $6x + 7y = 38$	**4**	$8x - y - 5 = 0$ $6y - 9x + 7 = 0$
5	$3(2x + y) = 4$ $4x + 8(x - y) = 2(9x - 1)$	**6**	$\frac{9}{8}x + \frac{7}{4}y = 3$ $\frac{4}{x} - \frac{2}{y} = 0$

7 Find the point of intersection of the lines $y = 7x - 4$ and $3x + 5y + 1 = 0$.

8 Two of the following lines are parallel. Find which two, and find the coordinates of the points where the third line cuts the parallel lines.
$8x + 12y + 9 = 0 \quad 6y = 3 - 9x \quad 4x + 6y = 3$

9 Show that the three lines
$7y + 68 = 2x \quad 9x + y = 46 \quad 9x - 8y = 118$
all pass through a single point, and find the coordinates of that point.

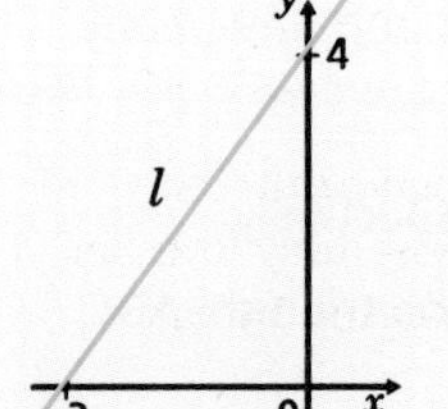

10 Line l is shown on the graph on the left.

Which of the following lines is

a) parallel to **b)** equal to **c)** intersecting l?
i) $2x - 1.5y + 6 = 0$
ii) $3y = 4x + 4$
iii) $y = \frac{3}{4}x - 3$
d) For the intersecting lines, find the point of intersection.

11 Find the point where the diagonals of the quadrilateral A(2, 5), B(0, –9), C(–5, –6), D(–3, 6) cut each other. Show that this is not the midpoint of either of the diagonals.

12 A triangle has vertices L(4, 0), M(–2, 7) and N(3, –4).

a) Find the gradient of LM.
b) Find the equation of l_1, the line through N which is perpendicular to LM.
c) Find the equation of l_2, the line through LM.
d) Find where l_1 crosses l_2. Call this point P.
e) Find the lengths LM and NP and hence find the area of triangle LMN.

13 Use a method similar to that in question 12 to find the area of the triangle with vertices at (3, 6), (–4, 8) and (–1, –5).

14 First and second class postage was introduced in 1968. First class letters weighing less than 4 oz cost 5d to post, and second class letters 4d. If 50 invitations to a party needed to be sent out at these prices, and £1 was available for postage, find how many should be sent first class and how many should be sent second class. Use all the money, so that as many can go first class as possible. £1 = 240d. Assume all the letters weighed less than 4 oz.

15 Towards the end of a holiday spent sailing across the Atlantic, two friends were horrified to discover that the only food left on their boat consisted of brown rice and tins of baked beans. In a book on nutrition they discovered that women of their ages needed 45 g protein and 1940 kcal of energy per day, from their food. According to the labels, the baked beans had 4.7 g of protein and 75 kcal of energy per 100 g, and the brown rice had 2.7 g of protein and 135 kcal of energy per 100 g. Find how many grams of each food they should eat each, per day, to meet the minimum requirements stated in the nutrition book.

CONSOLIDATION EXERCISES FOR CHAPTER 3

1 The points A and B have coordinates (8, 7) and (–2, 2) respectively. A straight line l passes through A and B and meets the coordinate axes at the points C and D.

a) Find, in the form $y = mx + c$, the equation of l.
b) Find the length CD, giving your answer in the form $p\sqrt{q}$, where p and q are integers and q is prime.

(ULEAC Specimen Paper, 1994)

2 A, B and C are the points (0, 2), (4, 6) and (10, 0) respectively.

a) Find the lengths AB, BC and CA of the sides of the triangle ABC, and show that:
$AB^2 + BC^2 = CA^2$
Deduce the size of angle ABC.
b) Find the gradients of the lines AB and BC and show these can be used to confirm your answer in part a) for the size of angle ABC.
c) M is the midpoint of line CA. Show that MA = MB.

(MEI Specimen Paper, 1994)

3 The points P, Q and R have coordinates (2, 4), (7, –2) and (6, 2) respectively.

Find the equation of the straight line l which is perpendicular to the line PQ and which passes through the midpoint of PR.

(AEB Specimen Paper, 1994)

4 Find the equation of the straight line that passes through the points (3, –1) and (–2, 2), giving your answer in the form $ax + by + c = 0$.

Hence find the coordinates of the point of intersection of the line with the x-axis.

(UCLES Specimen Paper, 1994)

5 The line l has equation $2x - y - 1 = 0$. The line m passes through the point A (0, 4) and is perpendicular to the line l.

a) Find an equation of m and show that the lines l and m intersect at the point P(2, 3).

The line n passes through the point B(3, 0) and is parallel to the line m.

b) Find an equation of n and hence find the coordinates of the point Q where the lines l and n intersect.

c) Show that AP = BQ = PQ.

(ULEAC Specimen Paper, 1994)

6 The equations of the sides of a triangle ABC are:

AB $x - 2y + 11 = 0$

BC $y = 7$

AC $2x + y + 7 = 0$

a) Find the coordinates of the points A, B and C.

b) Calculate the size of each of the angles in the triangle ABC.

7 The speed of a car (in mph) accelerating away from traffic lights is noted from the speedometer at one-second time intervals. The results were as follows.

Time from start (t)	1	2	3	4	5	6	7	8
Speed of car (v)	5	11	16	21	27	30	39	45

a) Plot points to represent this data.

b) Find the equation of the line that you think best fits the data.

c) Estimate the acceleration of the car stating your units carefully.

8 The following set of data is taken from a study of primates.

Adult body weight (in grams)	6583	733	582	288	2490	7362	4173	3384
New born baby weight (in grams)	480	97	50	27	234	425	314	107

Source: Unit 5, Edate pack, originally from Ross C (1988) in Journal of Zoology 214: pp. 199–219

a) Rewrite this table in more usual form with the data in each row increasing.
b) Plot points to represent this data.
c) Find the equation of the line that you think best fits the data.
d) How would your line of best fit change if the primate with weight 3384 grams had not been included?
e) What would be your prediction of the birth weight of a gorilla with adult body weight 117 500 grams? The actual baby weight is 2122 grams. Comment on your answer.

9 A farmer has 20 hectares available for growing barley and swedes. He has to decide how much of each to grow. The cost per hectare for barley is £30 and for swedes is £20. The farmer has budgeted £480.

Barley requires one worker-day per hectare and swedes require two worker-days per hectare. There are 36 worker-days available.

a) Formulate two equations in b and s, the area planted in hectares of barley and swedes respectively.
b) Solve the equations to find the area of each crop grown.

10 A camp site for caravans and tents has an area of 1800 m^2 and is subject to the following regulations:

The number of caravans must not exceed six.

Reckoning on four persons per caravan and three persons per tent, the total number of persons must not exceed 48.

At least 200 m^2 must be available for each caravan and 90 m^2 for each tent.

a) Formulate two equations in c and t, the number of caravans and tents respectively.
b) Solve the equations to find the number of each type that the camp site can accommodate.

Summary

Straight lines

- The straight line through the points $P(x_1, y_1)$ and $Q(x_2, y_2)$ has the following properties:

  gradient $\dfrac{y_2 - y_1}{x_2 - x_1}$

  equation $\dfrac{y - y_1}{y_2 - y_1} = \dfrac{x - x_1}{x_2 - x_1}$.

- The midpoint of PQ has coordinates $\left(\frac{1}{2}(x_1 + x_2), \frac{1}{2}(y_1 + y_2)\right)$.

- The length of the line PQ is $\sqrt{(x_2 - x_1)^2 + (y_2 - y_1)^2}$.

Equation of a straight line

- The general equation of a straight line is $y = mx + c$
  where m is the gradient or slope of the line and
  c is called the intercept because the point $(0, c)$ is the intercept of the line with the y-axis.

Parallel and perpendicular lines

- If two straight lines are perpendicular then the product of their gradients is –1.

- The two lines $y = m_1x + c_1$ and $y = m_2x + c_2$ are
  parallel if $m_1 = m_2$
  perpendicular if $m_1 m_2 = -1$.

Quadratic functions

In this chapter we:

- *explore the properties of quadratic functions and their graphs*
- *investigate factorisation of quadratic functions*
- *explore the transformation of quadratic functions and their graphs*
- are *introduced to the method of 'completing the square'.*

PARABOLAE

Exploration 4.1 *Fixed perimeter*

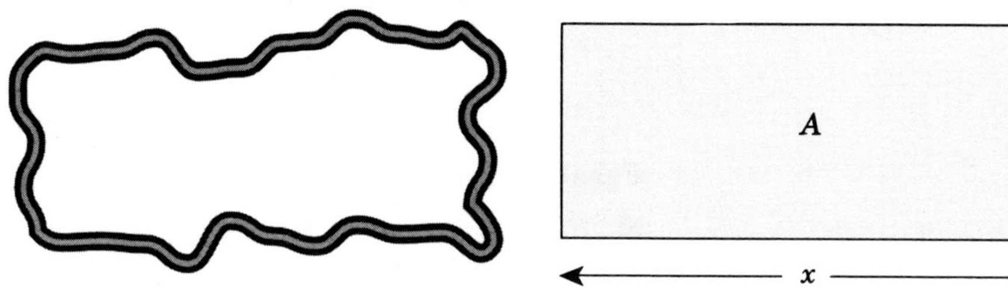

Make a loop of string 40cm long and lay it out in the form of a rectangle.

Investigate the area enclosed by the string.

1 How does the area (A) vary as the length of rectangle (x) varies?

2 What will the value of A be when $x = 7$?

3 What value of x will give $A = 75$?

4 What value of x will maximise the value of A?
What shape is the rectangle then?

5 Find a formula connecting A with x. Use it to illustrate your answers to 2, 3 and 4.

Describe any geometrical properties of the curve.

1 Since the rectangle is formed from a piece of string 40 cm long, the perimeter of the rectangle must be constant at 40 cm. The perimeter of a rectangle is twice the sum of adjacent sides, so the sum of the length and width of a rectangle is always half the perimeter, i.e.

$$\begin{aligned} &\text{length + width} &&= 20 \\ \Rightarrow\ &\text{width} &&= 20 - \text{length} \\ & &&= 20 - x \end{aligned}$$

Area = length × width
$\Rightarrow\ A = x(20 - x)$

2 In particular, when $x = 7$:
$A = 7 \times 13 = 91\ \text{cm}^2$

3 If $A = 75$, then $75 = x(20 - x)$
Looking carefully at this, we can see that $x = 5$ or 15 will satisfy the equation.

4 The area will be greatest if we form a square from the string.
When $x = 10$:
$A = 10 \times 10 = 100\ \text{cm}^2$

5 The solutions to 2, 3 and 4 can be read from the graph of
$A = x(20 - x)$.

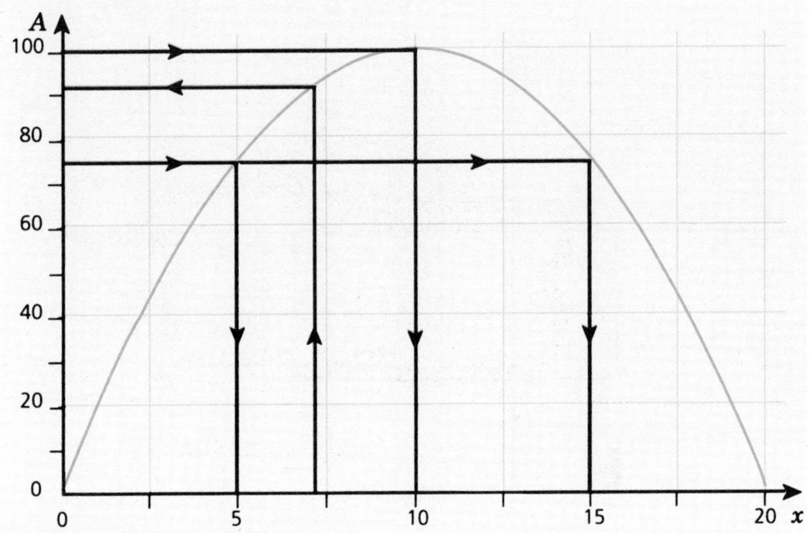

Parabolae as functions

The curve above is an example of a **parabola**. Every parabola is symmetrical and has a line of symmetry passing through the **vertex**. In our example the equation of the line of symmetry is $x = 10$ and the vertex has coordinates (10, 100).

All parabolae can be described by **quadratic functions**. The expression for the area of the rectangle may be expanded by multiplying out the brackets.

$$A = x(20 - x)$$
$$\Rightarrow A = 20x - x^2$$

Any expression with a term in x^2, with or without a term in x and with or without a constant, is a quadratic function. The following are all quadratic functions.

$$y = x^2$$
$$y = 2x^2 - 5$$
$$y = x^2 + 5x - 6$$
$$s = 12t - 5t^2$$

Example 4.1

Plot a graph of $y = x^2 - 2x - 15$ *for* $-4 \le x \le 6$.

From your graph find:

a) *the values of* x *when* $y = 0$,

b) *the values of* x *when* $y = -10$,

c) *the equation of the line of symmetry and the coordinates of the vertex.*

Solution

A table of values for $y = x^2 - 2x - 15$ *gives:*

x	–4	–3	–2	–1	0	1	2	3	4	5	6
y	9	0	–7	–12	–15	–16	–15	–12	–7	0	9

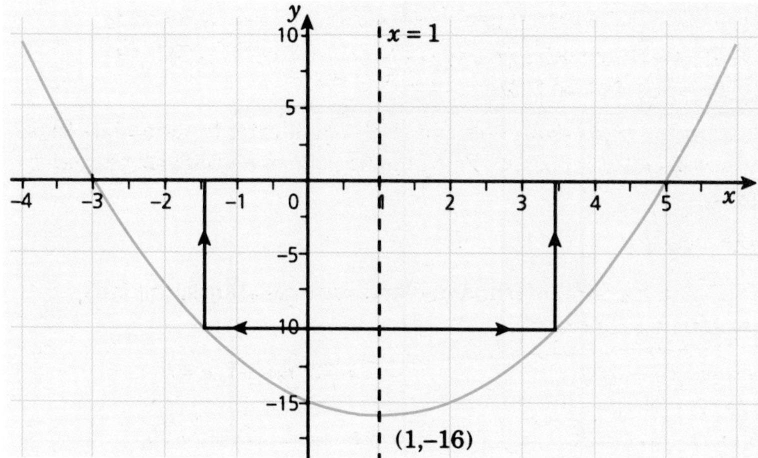

From the graph:

a) *when* $y = 0$, $x = -3$ *or* 5

b) *when* $y = -10$, $x = -1.45$ *or* 3.45

c) *the line of symmetry has equation* $x = 1$ *and the vertex has coordinates (1, –16).*

Example 4.2

An athlete is a shot-putter. On one occasion the flight path (or trajectory) of the shot is modelled by the equation:

$$y = -0.1x^2 + x + 1.5$$

where x *represents the horizontal distance of the shot and* y *represents its height.*
Plot a graph of y *against* x *for* $0 \le x \le 12$ *and use it to find:*

a) *how far the shot travelled horizontally before hitting the ground,*

b) *the maximum height of the shot and the horizontal distance from the shot-putter when this occurred.*

Solution

A suitable table of values might be as shown here.

x	0	2	4	6	8	10	12
y	1.5	3.1	3.9	3.9	3.1	1.5	–0.9

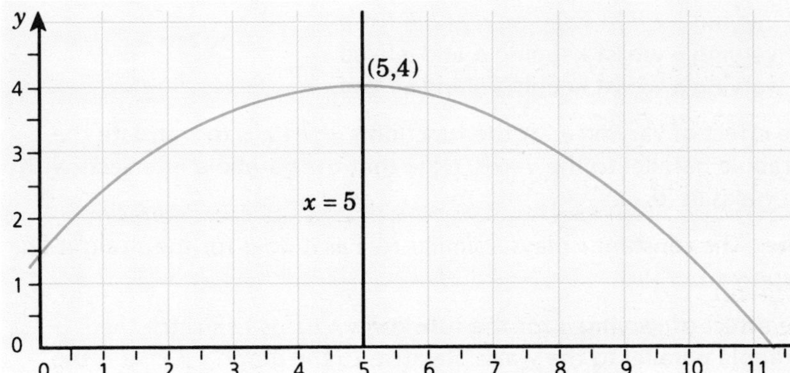

Looking at the table, notice the symmetry in the y-values. The highest y-value will occur between $x = 4$ and $x = 6$, when $x = 5$ and this point should be plotted as well.

a) *From the graph, when* $y = 0$, $x = 11.3$ m.
$\Rightarrow$ horizontal distance travelled by shot = 11.3 m

b) *The curve has its line of symmetry through* $x = 5$*, where the corresponding* y*-value on the curve is* $y = 4$*, which gives the maximum height of the shot, i.e. 4 metres.*

In each of the worked examples so far, the quadratic function has had terms in x^2, x and a constant. The general form of a quadratic function is given by:

$f(x) = ax^2 + bx + c$

where $f(x)$ is a function of x and a, b and c are constants, the only restriction being that $a \neq 0$.

e.g. $f(x) = 2x^2 - x + 5$ $\Rightarrow$ $a = 2, b = -1, c = 5$
$\Rightarrow$ $f(3) = 2 \times 3^2 - 3 + 5 = 20$
e.g. $f(x) = x^2 - 16$ $\Rightarrow$ $a = 1, b = 0, c = -16$
$\Rightarrow$ $f(-4) = (-4)^2 - 16 = 0$
e.g. $f(x) = 8 + 5x - x^2$ $\Rightarrow$ $a = -1, b = 5, c = 8$
$\Rightarrow$ $f(-2) = 8 + 5(-2) - (-2)^2 = -6$

The following exploration examines the effect that varying a, b or c has on the graph of the function.

CALCULATOR ACTIVITY

Exploration 4.2

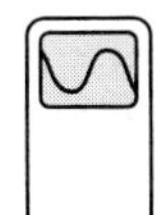

You will need a graphics calculator or graph plotter. Set x and y limits so that the origin is at the centre of the screen and both axes have the same scale.

To test your settings, the graph of $y = x$ should make an angle of 45° with both axes.

- Plot graphs of $y = x^2 + c$ by choosing different values for c (e.g. –5, –2, 0, 1, 3 etc.).
- Plot graphs of $y = ax^2$ by choosing different values for a (e.g. –5, –2, –0.2, 0.3, 1, 3, 4 etc.).
- Plot graphs of $y = x^2 + bx$ by choosing different values for b (e.g. –5, –2, 0, 1, 3 etc.).
- For the parabola $y = ax^2 + bx + c$, describe the effect of:

1 varying c whilst keeping a and b fixed
2 varying b whilst keeping a and c fixed
3 varying a whilst keeping b and c fixed.

The effect of varying c for the function $y = x^2 + c$ is to translate the parabola parallel to the y-axis. Note that the parabola intersects with the y-axis at (0, c).

Note: The constant c plays a similar role as it does for the straight-line graph $y = mx + c$.

The effect of varying a for the function $y = ax^2$ is to stretch the parabola parallel to the y-axis. Relative to the graph $y = x^2$, a is the **stretch factor** which can be fractional or negative.

Note: A negative value for a may be regarded as a positive stretch followed by a reflection in $y = 0$, e.g. consider the graph of $y = -3x^2$ as a stretch factor 3 of the graph $y = x^2$ followed by a reflection in the x-axis.

The effect of varying b for the function $y = x^2 + bx$ is less easy to interpret. As with varying c for the function $y = x^2 + c$, the effect is a translation, but this time in two directions: a translation through $\frac{1}{2}b$ units parallel to the x-axis and a translation through $-(\frac{1}{2}b)^2$ units parallel to the y-axis. To examine the amounts by which the curve $y = x^2$ is translated, consider what happens to the vertex.

e.g. for $y = x^2 + 5x$, the vertex is at (–2.5, –6.25).

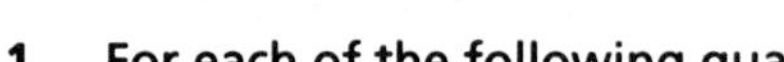

EXERCISES

4.1 A

1 For each of the following quadratic functions:

a) plot a graph of the function,
b) find the coordinates of the y-intercept,
c) find the coordinates of the points where the graph intersects the x-axis,
d) draw in and give the equation of the line of symmetry,
e) write down the coordinates of the vertex.

i) $y = x^2 - 6x + 5$ ii) $y = -x^2 - 2x + 3$
iii) $y = x^2 - 3x + 2.25$ iv) $y = 4x^2 + 8x - 5$
v) $y = 25 + 5x - 2x^2$ vi) $y = x^2 - 16$
vii) $y = x^2 + 2x + 5$ viii) $y = 25 - 4x^2$
ix) $y = 3x^2 - 12x$ x) $y = 8x - 20 - x^2$
xi) $y = 6x - 2x^2$ xii) $y = -(x^2 + 6x + 9)$

2 Find the equation of the line of symmetry of a parabola which intersects the x-axis at (1.2, 0) and (7.8, 0).

3 Find the coordinates of the vertex of the parabola $y = 9 - 7x - 2x^2$, given that it intersects the x-axis at (–4.5, 0) and (1, 0).

4 The St Louis arch in the USA is in the shape of a parabola. The height, y metres, of a point on the arch, in terms of its distance, x metres, from one side, can be modelled by the function $y = \frac{1}{48}x(192 - x)$. Draw a graph to represent the arch, and from your graph find:

a) the width of the arch at ground level,
b) the height of the arch,
c) the equation of its line of symmetry.

5 The spinal compression, H, of an athlete taking prolonged exercise, after t minutes, can be modelled by the function $H = 8.6 + 0.21t - 0.01t^2$.
Draw a graph of H against t for $0 \leq t \leq 50$ and from your graph find:

a) the time at which $H = 8$,
b) the time at which $H = 0$,
c) the time at which H reaches its greatest value and the value it takes at this time.

EXERCISES

1 Without plotting any points, state the coordinates where the following parabolae cut the y-axis.

a) $y = x^2 + 3x + 4$ b) $y = 8x^2 - 3$
c) $y = 3x^2 - 3x + 7$ d) $y = 5x^2$
e) $y = 8 - 8x + 9x^2$ f) $y = 2x - 6x^2$
g) $y = 4x - 7 - 5x^2$ h) $2y = 8x^2 + 4x - 7$

2 Find the equation of the line of symmetry of a parabola which passes through the x-axis at (–2, 0) and (3, 0).

3 Find the coordinates of the vertex of the parabola $y = 2x^2 - 25x + 63$, given that it crosses the x-axis at (3.5, 0) and (9, 0).

4 A ball is thrown up in the air. After t seconds the ball is h metres above the ground where $h = 1.4 + 2.3t - 4.9t^2$. Find:

a) how long it is before the ball hits the ground,
b) the time at which the ball is at its greatest height above the ground, and how high it is at that point.

5 The cross-section of a reflector of a torch is modelled by the part of $y = 6 - 0.24x^2$ which lies above the x-axis, where x and y are both measured in cm. Draw this curve and find:

a) the depth of the reflector,
b) the diameter of the mouth of the reflector.

BRACKETS

Sometimes, in some mathematical expressions, we can separate or group terms which have something in common, or which have to be considered as one item. We have already done this in earlier chapters, when we took out common factors.

Expanding

Exploration 4.3

Area of a rectangle

Find four different expressions for the area of this rectangle.

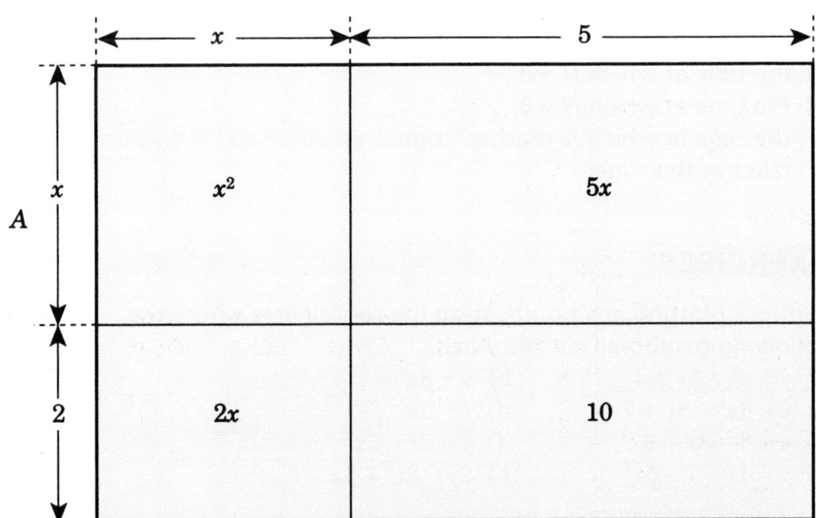

Dividing the area into four smaller rectangles gives:

area $= x^2 + 5x + 2x + 10$
$= x^2 + 7x + 10$

We can divide the rectangle into smaller parts, in different ways, to give these expressions.

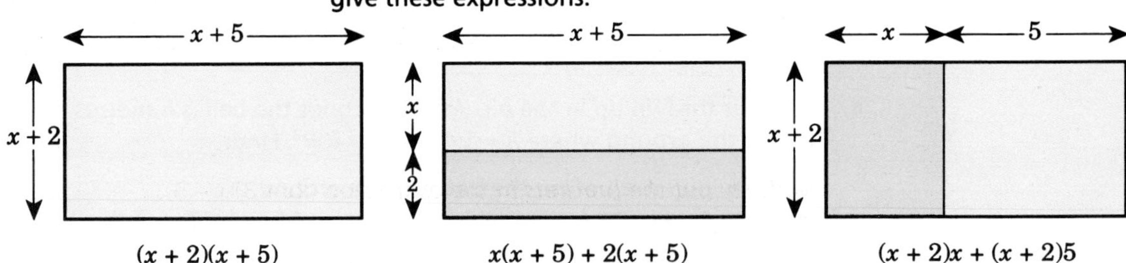

$(x + 2)(x + 5)$ $\quad$ $x(x + 5) + 2(x + 5)$ $\quad$ $(x + 2)x + (x + 2)5$

The four expressions are equivalent and illustrate ways in which pairs of brackets may be expanded.

$(x + 2)(x + 5)$ → $x(x + 5) + 2(x + 5)$ / $(x + 2)x + (x + 2)5$ → $x^2 + 5x + 2x + 10$ → $x^2 + 7x + 10$

The diagram shows how the expansion takes place in two stages. First we expand to give two sets of single brackets. Then we expand these and collect like terms.

With practice, we can cut out the middle stage:

$$(x+2)(x+5) = x^2 + 5x + 2x + 10$$

where **F** means multiply the two first terms
O means multiply the two outside terms
I means multiply the two inside terms
L means multiply the two last terms.

Exploration 4.4

Expanding brackets

Use diagrams to illustrate that
$(x+3)(x-2) = x^2 + x - 6$ and
$(x-3)(x-2) = x^2 - 5x + 6$

Solution

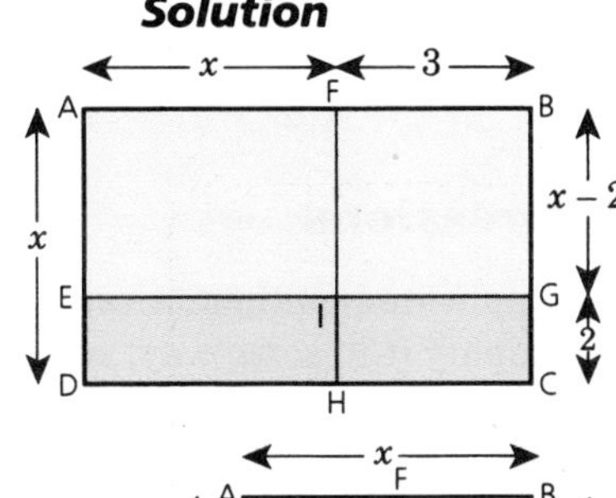

Area ABGE = area AFHD + area BCHF − area DEIH − area CHIG

$(x+3)(x-2) = x^2 + 3x - 2x - 6 = x^2 + x - 6$

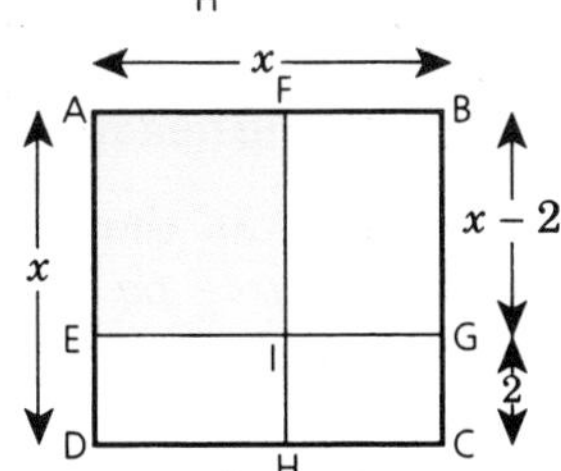

Area AFIE = area ABCD − area BCHF − area CDEG + area CHIG

$(x-3)(x-2) = x^2 - 3x - 2x + 6 = x^2 - x + 6$

Example 4.3

Multiply out the brackets in the expression $(2x+3)(x-5)$.

Solution

$$\begin{aligned}(2x+3)(x-5) &= 2x(x-5) + 3(x-5)\\ &= 2x^2 - 10x + 3x - 15\\ &= 2x^2 - 7x - 15\end{aligned}$$

Example 4.4

Expand the brackets in the expression $(5x-2)(8-3x)$.

Solution

$$\begin{aligned}(5x-2)(8-3x) &= 5x(8-3x) - 2(8-3x)\\ &= 40x - 15x^2 - 16 + 6x\\ &= -15x^2 + 46x - 16\end{aligned}$$

Example 4.5

Multiply out the brackets and simplify this expression.
$(x - 7)(3x + 4) + (2x + 3)(2x - 3)$

Solution

$$\begin{aligned}(x-7)(3x+4)+(2x+3)(2x-3) &= x(3x+4)-7(3x+4)+2x(2x-3)+3(2x-3)\\ &= 3x^2+4x-21x-28+4x^2-6x+6x-9\\ &= 3x^2-17x-28+4x^2-9\\ &= 7x^2-17x-37\end{aligned}$$

With practice, it is possible to find the final x-term by adding the two products which combine to give it. This is illustrated in the next example.

Example 4.6

Expand the brackets and simplify this expression.
$(3x - 1)(5 + 2x) - (4x - 3)^2$

Solution

$$\begin{aligned}(3x-1)(5+2x)-(4x-3)^2 &= (3x-1)(5+2x)-(4x-3)(4x-3)\\ &= 6x^2+13x-5-\{16x^2-24x+9\}\\ &= -10x^2+37x-14\end{aligned}$$

Factorising expressions

When faced with a complicated mathematical expression, it is often useful to be able to simplify it in some way, such as taking out common factors. This is the reverse of expanding brackets and is called **factorising**, as we saw in Chapter 2, *Basic algebra*. We shall now look at ways of factorising quadratic functions (where possible).

If p and q are any two constants, then:

$$\begin{aligned}(x+p)(x+q) &= x^2+(p+q)x+pq\\ \text{e.g. } (x+5)(x+2) &= x^2+(5+2)x+5\times 2\\ &= x^2+7x+10\end{aligned}$$

The constant term in the expansion is the **product** of the constants p and q and the coefficient of x is the **sum** of p and q. This result is very useful when factorising, since:

$$x^2 + (p + q)x + pq = (x + p)(x + q)$$

Example 4.7

Factorise the expression $x^2 + 5x + 6$.

Solution

Let $x^2 + 5x + 6 = x^2 + (p + q)x + pq$
We need to find two constants p *and* q *such that:* $pq = 6$ *and* $p + q = 5$
Possible values of p *and* q *to satisfy* $pq = 6$ *are* $p = 1, q = 6; p = 6, q = 1; p = 2, q = 3; p = 3, q = 2$.

By inspection, either $p = 2, q = 3$ or $p = 3, q = 2$ *will satisfy both conditions.*
$\Rightarrow \; x^2 + 5x + 6 = (x + 2)(x + 3)$ *or* $x^2 + 5x + 6 = (x + 3)(x + 2)$

This method is used again in the following examples.

Example 4.8

Factorise the expression $x^2 - x - 12$.

Solution

Let $x^2 - x - 12 = x^2 + (p + q)x + pq$
Then $pq = -12$ *and* $p + q = -1$.

Possible combinations of p *and* q *to give a product of –12 are:*

p	1	–1	2	–2	3	–3	4	–4	6	–6	12	–12
q	–12	12	–6	6	–4	4	–3	3	–2	2	–1	1

but only $p = 3, q = -4$ *or* $p = -4, q = 3$ *give a sum of –1.*
$\Rightarrow \quad x^2 - x - 12 = (x + 3)(x - 4)$ *or* $x^2 - x - 12 = (x - 4)(x + 3)$

Note: As before, it does not matter which of the two factorised forms we take, since the same factors appear in both expressions, although in a different order.

Example 4.9

Factorise the expression $x^2 - 10x + 21$.

Solution

$x^2 - 10x + 21 = x^2 + (p + q)x + pq$
$\Rightarrow \quad p + q = -10$ *and* $pq = 21$

For a positive product and negative sum, both p *and* q *have to be negative numbers; by inspection:*

$p = -3$ *and* $q = -7$ *or* $p = -7$ *and* $q = -3$

since $-3 + (-7) = -10$ *and* $-7 + (-3) = -10$
$\Rightarrow \quad x^2 - 10x + 21 = (x - 3)(x - 7)$ *or* $x^2 - 10x + 21 = (x - 7)(x - 3)$

The examples so far suggest some useful clues.
$x^2 + 5x + 6 = (x + 2)(x + 3)$

positive product, positive sum $\Rightarrow$ both signs positive

$x^2 - x - 12 = (x + 3)\,x - 4)$
$x^2 + x - 12 = (x - 3)(x + 4)$

negative product $\Rightarrow$ different signs

$x^2 - 10x + 21 = (x - 3)(x - 7)$

positive product, negative sum $\Rightarrow$ both signs negative

Factorisation of $ax^2 + bx + c$

So far, the expressions we have been factorising have been of the form $x^2 + bx + c$. More complex factorisations occur, when the coefficient of x^2 is not 1. We now have to consider the product of the two x terms as well as the product of the two constants. The most effective approach is to try the possibilities systematically until we find the correct factorisation.

Example 4.10

Factorise $2x^2 + 7x + 3$.

Solution

Both signs must be positive; possible combinations which give the correct products are:

$(x + 1)(2x + 3) = 2x^2 + 5x + 3$ ✗
$(x + 3)(2x + 1) = 2x^2 + 7x + 3$ ✓
$(2x + 1)(x + 3) = 2x^2 + 7x + 3$ ✓
$(2x + 3)(x + 1) = 2x^2 + 5x + 3$ ✗

Either of the ✓ lines give the correct factorisation. We really need only try the first two, since the second two are equivalent to the first two.

Example 4.11

Factorise $6x^2 + 13x - 5$.

Solution

The signs must be different; possible combinations which give the correct products are:

$(x + 1)(6x - 5) = 6x^2 + x - 5$ ✗
$(x + 5)(6x - 1) = 6x^2 + 29x - 5$ ✗
$(2x + 1)(3x - 5) = 6x^2 - 7x - 5$ ✗
$(2x + 5)(3x - 1) = 6x^2 + 13x - 5$ ✓

Having found a correct factorisation there is no need to go further.

We could have tried the combination:

$(3x + 1)(2x - 5) = 6x^2 - 13x - 5$

which is almost correct, but the sign for the x term is wrong. This means that changing both signs in the factors will give the correct answer.

To factorise $15x^2 + x - 6$:
$(3x - 2)(5x + 3) = 15x^2 - x - 6$

Change signs Change sign

$(3x + 2)(5x - 3) = 15x^2 + x - 6$

Difference of two squares

A special case of factorising involves the expansion of $(x - p)(x + p)$.

$$(x - p)(x + p) = x^2 - px + px - p^2$$
$$= x^2 - p^2$$
$$\Rightarrow \quad x^2 - p^2 = (x - p)(x + p)$$

Example 4.12

Factorise these expressions.

a) $x^2 - 9$ ***b)*** $16t^2 - 1$ ***c)*** $a^2 - b^2$
d) $10x^2 - 250$ ***e)*** $3y^3 - 3y$

Solution

Each expression is or can be factorised to give a difference of two squares.

a) $x^2 - 9 = x^2 - 3^2 = (x - 3)(x + 3)$

b) $16t^2 - 1 = (4t)^2 - 1^2$
$= (4t - 1)(4t + 1)$

c) $a^2 - b^2 = (a - b)(a + b)$

d) $10x^2 - 250 = 10(x^2 - 25)$
$= 10(x^2 - 5^2)$
$= 10(x - 5)(x + 5)$

e) $3y^3 - 3y = 3y(y^2 - 1^2)$
$= 3y(y - 1)(y + 1)$

EXERCISES

4.2 A

1 Multiply out the brackets.

a) $(x + 5)(x + 8)$ **b)** $(x + 5)(x - 8)$
c) $(x - 5)(x + 8)$ **d)** $(x - 5)(x - 8)$
e) $(2x + 1)(x + 7)$ **f)** $(2x - 1)(x + 9)$
g) $(5t + 3)(t - 1)$ **h)** $(4y - 3)(y - 2)$
i) $(3x + 4)(5x + 1)$ **j)** $(7u - 1)(2u - 1)$
k) $(2t - 3)(2t + 5)$ **l)** $(3a + b)(a - 3b)$
m) $(4x + 1)^2$ **n)** $(2x - 3)^2$
o) $(4z + 1)(4z - 1)$ **p)** $(c - 7)(c + 7)$
q) $(3x + 4)(5 - x)$ **r)** $(2a + b)(b - a)$
s) $5(3x - 1)(x + 3)$ **t)** $3(2x - 5y)^2$

2 Expand the brackets and simplify each expression.

a) $(x - 7)(x + 3) + (x + 2)(x - 4)$
b) $(x + 1)(x - 9) + (x - 1)(x + 9)$
c) $(x + 3)(x + 10) - (x - 3)(x - 10)$
d) $(2x - 1)(x + 5) - (x - 4)^2$
e) $(3x - 2)(3x + 2) + (4x + 5)(x - 7)$
f) $(4 - x)^2 - (3 - x)(5 - x)$
g) $(3a + b)(a - 5b) - (2a + b)^2$
h) $(2x - y)(3x - y) + (4x + 3y)(4x - 3y)$
i) $9(e - 2f)(3e + f) - 5(2e + f)(2e - 3f)$
j) $7(2x - 5)^2 - 3(4x + 3)(4x - 3)$

3 Factorise each of these expressions fully.

a) $x^2 + 9x + 20$ **b)** $x^2 - 9x + 20$
c) $x^2 - x - 20$ **d)** $x^2 + x - 20$
e) $x^2 + 11x + 28$ **f)** $x^2 - 12x + 35$
g) $t^2 + 6t - 91$ **h)** $y^2 - 2y - 63$
i) $x^2 + 6x + 9$ **j)** $x^2 - 14x + 49$
k) $2x^2 + 3x + 1$ **l)** $3x^2 + 17x + 20$
m) $3p^2 + 2p - 8$ **n)** $10p^2 - 13p + 4$
o) $2x^2 + 5xy - 3y^2$ **p)** $16a^2 - 40ab + 25b^2$
q) $60 + 95x + 20x^2$ **r)** $24 + 23z - 12z^2$
s) $12x^2 + 60x + 75$ **t)** $2x^3 - x^2 - 21x$

4 Factorise each of these expressions fully.

a) $x^2 - 49$ **b)** $100 - t^2$
c) $p^2 - q^2$ **d)** $64s^2 - 25$
e) $9 - 16y^2$ **f)** $121t^2 - 1$
g) $36a^2 - b^2$ **h)** $9e^2 - 49f^2$
i) $10x^2 - 40$ **j)** $45 - 20x^2$
k) $7a^2 - 7b^2$ **l)** $75p^2 - 147q^2$
m) $9x - 49x^3$ **n)** $320x^2 - 125$
o) $a^3b - ab^3$

EXERCISES

4.2 B

1 Multiply out the brackets.

a) $(x + 1)(x + 4)$ **b)** $(x + 1)(x - 4)$
c) $(x - 1)(x + 4)$ **d)** $(x - 1)(x - 4)$
e) $(4x + 7)(x + 6)$ **f)** $(x - 1)(6x + 7)$
g) $(y + 5)(4y - 9)$ **h)** $(t - 8)(2t - 7)$
i) $(9x + 7)(3x + 4)$ **j)** $(7x - 1)(6x + 1)$
k) $(4x - 7)(8x - 9)$ **l)** $(2x + y)(x - 2y)$
m) $(5x + 3)^2$ **n)** $(6x - 5)^2$
o) $(x - 8)(x + 8)$ **p)** $(7u + 2)(7u - 2)$
q) $(3x + 7)(3 - x)$ **r)** $(8a + b)(b - 2a)$
s) $8(9x + 4)(3x - 2)$ **t)** $9xy(3x + 2y)^2$

2 Expand the brackets and simplify each expression.

a) $(x - 4)(x + 5) + (x + 8)(x - 3)$
b) $(x + 2)(x - 5) - (x - 2)(x + 5)$
c) $(x - 3)(x - 9) + (x + 3)(x + 9)$
d) $(3x - 4)(x + 8) + (x - 6)^2$
e) $(2x + 7)(5x + 3) - (4x - 3)(2 - x)$
f) $(6 - x)^2 - (8 - x)(5 - x)$
g) $(8a - b)(5a + 3b) - (a - 9b)^2$
h) $(x - 2y)(x + 2y) - (x + 2y)^2$
i) $4(a - 3b)(3b - a) - 3(2a + 2b)(a - 6b)$
j) $5(2x + 9)(9x + 2) - 3(2x + 11)^2$

3 Factorise each of these expressions fully.

a) $x^2 + 7x + 10$ **b)** $x^2 + 3x - 10$
c) $x^2 - 3x - 10$ **d)** $x^2 - 7x + 10$
e) $x^2 + 12x + 32$ **f)** $-x^2 + 12x - 32$
g) $u^2 - 8u - 84$ **h)** $t^2 - 6t - 16$
i) $x^2 - 10x + 25$ **j)** $x^2 + 22x + 121$
k) $3x^2 + 2x - 1$ **l)** $3x^2 - 16x + 5$
m) $6x^2 - 17x - 3$ **n)** $5x^2 + 28x - 12$
o) $7a^2 + 13ab - 2b^2$ **p)** $81x^2 + 36xy + 4y^2$
q) $12x^2 + 14x - 40$ **r)** $82x - 24x^2 - 70$
s) $200x^2 - 560x + 392$ **t)** $25x^3 - 195x^2 - 84x$

4 Factorise each of these expressions fully.

a) $x^2 - 16$ **b)** $49 - u^2$
c) $a^2 - 4b^2$ **d)** $36 - 16x^2$
e) $25n^2 - 9m^2$ **f)** $9x^2 - 1$
g) $4a^2 - 36b^2$ **h)** $64x^2 - 81y^2$
i) $45x^2 - 20$ **j)** $9a^2 - 9b^2$
k) $243p^2 - 48q^2$ **l)** $20a^3 - 5ab^2$
m) $637x^2 - 1573$ **n)** $81t - 16t^3$
o) $36x^3y - 100xy^3$

TRANSFORMING QUADRATIC FUNCTIONS

The parabola $y = x^2$ can be transformed in various ways. In this section, we explore some of those ways. For each transformation the equation of the function will change. Start with the following explorations.

You may find a graphics calculator or computer package useful. Record your findings on sketch graphs.

Exploration 4.5

Transforming $y = x^2$

Describe geometrically the transformation of the parabola $y = x^2$ onto each of the following curves.

1 $y = x^2 - 5$ **2** $y = 3x^2$
3 $y = (x - 4)^2$ **4** $y = -x^2$
5 $y = 2x^2 + 3$ **6** $y = 4(x + 3)^2$
7 $y = (x - 1)^2 + 4$ **8** $y = (x + 2)^2 - 7$

Exploration 4.6

Multiple transformations

Now look at the transformation of $y = x^2$, beginning with geometrical descriptions. In each case, what is the equation of the transformed curve?

1 translation through 7 units parallel to the y-axis

2 translation through 7 units parallel to the x-axis

3 stretch, factor 2 parallel to the y-axis

4 reflection in the x-axis *followed by* translation through 5 units parallel to the y-axis

5 translation through –3 units parallel to the x-axis *followed by* translation through 4 units parallel to the y-axis

6 translation through 5 units parallel to the y-axis *followed by* a stretch, factor $\frac{1}{2}$ parallel to the y-axis.

The geometrical transformations used in the explorations may be summarised as:

	Geometrical transformation	Equation of transformed curve
1	translation through b units parallel to the x-axis	$y = (x - b)^2$
2	stretch, factor a parallel to the y-axis	$y = ax^2$
3	reflection in the x-axis	$y = -x^2$
4	translation through c units parallel to the y-axis	$y = x^2 + c$

When two or more transformations are combined the equation of the transformed curve becomes more complex, for example:

translation through b units parallel to the x-axis
followed by
translation through c units parallel to the y-axis.

The study of these transformations is important, since *any* parabola may be obtained from $y = x^2$ by a combination of one or more of the four transformations **1**, **2**, **3** and **4**, usually in that order.

Example 4.13

The parabola $y = x^2$ undergoes the following transformations:
translation through –5 units parallel to the y-axis
followed by
stretch, factor 2 parallel to the y-axis

a) *Write down the equation of the transformed curve.*
b) *If the order of the transformations was reversed, which one would have to be modified to represent the same curve? Explain your reasoning.*

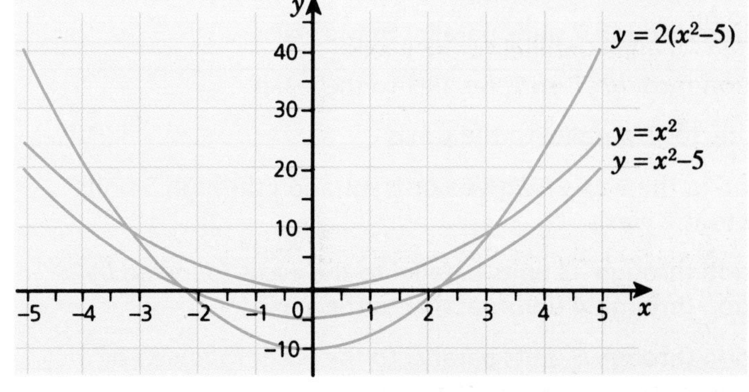

Solution

a) *The equation of the transformed curve is $y = 2(x^2 - 5)$ indicating a translation through –5 units followed by a stretch, factor 2, both parallel to the y-axis.*

b) *Multiplying out the brackets*
$\Rightarrow \quad y = 2(x^2 - 5) = 2x^2 - 10.$
Expressed in this way, we can see that the curve can be obtained by a stretch, factor 2 followed by a translation through –10 units, both parallel to the y-axis. You can see that the translation has to be modified. Try transforming $y = x^2$ using both combinations.

EXERCISES

4.3 A

For each of these questions, begin with the graph of $y = x^2$ and transform it using one or more transformations.

1 Write down the single geometrical transformation to obtain the following curves.

a) $y = x^2 + 3$
b) $y = x^2 - 7$
c) $y = (x - 2)^2$
d) $y = (x + 5)^2$
e) $y = 4x^2$
f) $y = \frac{1}{3}x^2$
g) $y = -x^2$
h) $y = x^2 + 6x + 9$

2 Find the equation of the curve obtained by applying the following transformations to $y = x^2$.

a) translation –5 units parallel to the y-axis
b) translation 3.5 units parallel to the y-axis
c) translation 4 units parallel to the x-axis
d) translation – 7 units parallel to the x-axis
e) stretch, factor 3 parallel to the y-axis
f) stretch, factor $\frac{1}{3}$ parallel to the y-axis.

3 Find, in a suitable order, the combination of geometrical transformations to obtain the following curves.

a) $y = 2x^2 + 6$
b) $y = 2(x^2 + 3)$
c) $y = \frac{1}{5}(x^2 - 5)$
d) $y = \frac{1}{5}x^2 - 1$
e) $y = -(x^2 - 4)$
f) $y = 4 - x^2$
g) $y = 3(x + 4)^2$
h) $y = 5(x - 1)^2$
i) $y = (x - 2)^2 + 3$
j) $y = (x + 5)^2 - 7$
k) $y = 2(x + 3)^2 + 5$
l) $y = -(x + 4)^2$
m) $y = 7 - (x + 4)^2$
n) $y = 0.4(x + 6)^2 - 4.4$
o) $y = 10 - 2(x - 3)^2$
p) $y = -(5(x + 2)^2 + 3)$

4 Find the equation of the curve obtained by applying the following transformations.

a) stretch, factor 2 parallel to the x-axis
followed by
translation –5 units parallel to the y-axis
b) stretch, factor 3 parallel to the y-axis
followed by
reflection in the x-axis
c) translation 3 units parallel to the x-axis
followed by
stretch, factor $\frac{1}{2}$ parallel to the y-axis
d) translation –5 units parallel to the x-axis
followed by
reflection in the x-axis
e) translation 2 units parallel to the x-axis
followed by
translation 5 units parallel to the y-axis

f) reflection in the x-axis
followed by
translation 9 units parallel to the y-axis

g) translation 5 units parallel to the y-axis
followed by
reflection in the x-axis

h) translation –4 units parallel to the x-axis
followed by
stretch, factor 3 parallel to the y-axis
followed by
translation –7 units parallel to the y-axis

i) translation 6 units parallel to the x-axis
followed by
reflection in the x-axis
followed by
translation 4 units parallel to the y-axis

j) translation 3 units parallel to the x-axis
followed by
stretch, factor 5 parallel to the x-axis
followed by
translation –12 units parallel to the y-axis

EXERCISES

1 In each case make a rough sketch of $y = x^2$ and, on the same axes, make a rough sketch of the curve given to show its relationship to $y = x^2$.

a) $y = x^2 + 8$ **b)** $y = 2x^2$ **c)** $y = (x - 3)^2$
d) $y = -x^2$ **e)** $y = x^2 - 2$ **f)** $y = \frac{1}{4}x^2$
g) $y = (x + 4)^2$ **h)** $y = -3x^2$

2 Find the equation of the curve obtained by applying the following transformations to $y = x^2$.

a) stretch, factor 6 parallel to the y-axis
b) translation 3 units parallel to the x-axis
c) translation 3 units parallel to the y-axis
d) translation $-\frac{1}{2}$ units parallel to the x-axis
e) stretch, factor 0.2 parallel to the y-axis
f) translation –8 units parallel to the y-axis

3 Find the equation of the curve obtained by applying the following transfomations to $y = x^2$.

a) translation 5 units parallel to the x-axis
followed by
translation 5 units parallel to the y-axis

b) stretch, factor 7 parallel to the y-axis
followed by
translation –1 unit parallel to the y-axis

c) stretch, factor 4 parallel to the y-axis
followed by
reflection in the x-axis

d) translation 3 units parallel to the y-axis
followed by
stretch, factor $\frac{1}{2}$ parallel to the y-axis

e) reflection in the x-axis
followed by
translation –6 units parallel to the y-axis

f) stretch, factor 3 parallel to the y-axis
followed by
reflection in the x-axis
followed by
translation 2 units parallel to the y-axis

g) translation 5 units parallel to the x-axis
followed by
stretch, factor 0.7 parallel to the y-axis
followed by
reflection in the x-axis

h) translation –2 units parallel to the x-axis
followed by
stretch, factor 7 parallel to the y-axis
followed by
translation 3 units parallel to the y-axis

i) translation 9 units parallel to the x-axis
followed by
stretch, factor 2 parallel to the y-axis
followed by
translation –3 units parallel to the y-axis

j) translation 4 units parallel to the x-axis
followed by
stretch, factor 9 parallel to the y-axis
followed by
reflection in the x-axis
followed by
translation 1 unit parallel to the y-axis

COMPLETING THE SQUARE

In the last section we found that we can transform the quadratic function $y = x^2$ onto another quadratic function by a combination of transformations, for example:

translation 3 units parallel to the x-axis $\quad y = (x - 3)^2$
followed by
stretch, factor 2 parallel to the y-axis $\quad y = 2(x - 3)^2$
followed by
translation –5 units parallel to the y-axis $\quad y = 2(x - 3)^2 - 5$

The transformed quadratic function may be simplified to give:

$$\begin{aligned} y &= 2(x-3)^2 - 5 \\ &= 2(x-3)(x-3) - 5 \\ &= 2(x^2 - 6x + 9) - 5 \\ \Rightarrow y &= 2x^2 - 12x + 13 \end{aligned}$$

Notice that the coefficient of x^2, i.e. 2, is the same as the stretch factor.

Any quadratic function $y = ax^2 + bx + c$ may be obtained by a combination of transformations of $y = x^2$ that gives:

$$y = a(x+p)^2 + q$$

where p and q are constants that depend on a, b and c. The process of expressing $y = ax^2 + bx + c$ in the form $y = a(x+p)^2 + q$ is called **completing the square** and is illustrated in the following examples.

Example 4.14

Complete the square for $y = x^2 + 10x + 8$.

Solution

Here $a = 1$, so:

$$\begin{aligned} & x^2 + 10x + 8 \equiv (x+p)^2 + q \\ \Rightarrow\ & x^2 + 10x + 8 \equiv (x+p)(x+p) + q \\ \Rightarrow\ & x^2 + 10x + 8 \equiv x^2 + 2px + p^2 + q \end{aligned}$$

Comparing the coefficients of x:

$$\begin{aligned} & 10 = 2p \\ \Rightarrow\ & p = 5 \end{aligned}$$

Comparing constants:

$$\begin{aligned} & 8 = p^2 + q \\ \Rightarrow\ & 8 = 25 + q \\ \Rightarrow\ & q = -17 \\ \therefore\ & y = x^2 + 10x + 8 = (x+5)^2 - 17 \end{aligned}$$

Example 4.15

Complete the square for $y = x^2 - 7x + 3$.

Solution

Here $a = 1$, so:

$$\begin{aligned} & x^2 - 7x + 13 \equiv (x+p)^2 + q \\ \Rightarrow\ & x^2 - 7x + 13 \equiv x^2 + 2px + p^2 + q \end{aligned}$$

Comparing the coefficients of x:

$$\begin{aligned} & -7 = 2p \\ \Rightarrow\ & p = -\tfrac{7}{2} \end{aligned}$$

Comparing constants:

$$\begin{aligned} & 13 = p^2 + q \\ \Rightarrow\ & 13 = \tfrac{49}{4} + q \\ \Rightarrow\ & q = 13 - \tfrac{49}{4} = \tfrac{3}{4} \\ \therefore\ & y = x^2 - 7x + 13 = (x - \tfrac{7}{2})^2 + \tfrac{3}{4} \end{aligned}$$

Example 4.16

Complete the square for $s = 3t^2 + 12t - 10$.

Solution

Here $a = 3$, *so:*

$3t^2 + 12t - 10 \equiv 3(t + p)^2 + q$

$\Rightarrow 3t^2 + 12t - 10 \equiv 3(t^2 + 2pt + p^2) + q$

$\Rightarrow 3t^2 + 12t - 10 \equiv 3t^2 + 6pt + 3pt^2 + q$

Comparing coefficients of t:

$12 = 6p$

$\Rightarrow p = 2$

Comparing constants:

$-10 = 3p^2 + q$

$\Rightarrow -10 = 3 \times 4 + q$

$\Rightarrow q = -10 - 12$

$\Rightarrow q = -22$

$\therefore\ s = 3t^2 + 12t - 10 = 3(t + 2)^2 - 22$

Example 4.17

Express $h = 2 + 9t - 5t^2$ *in the form* $q - 5(t + p)^2$.

Solution

$2 + 9t - 5t^2 \equiv q - 5(t + p)^2$

$\Rightarrow 2 + 9t - 5t^2 \equiv q - 5(t^2 + 2pt + p^2)$

$\Rightarrow 2 + 9t - 5t^2 \equiv q - 5p^2 - 10pt - 5t^2$

Comparing coefficients of t:

$9 = -\,10p$

$\Rightarrow p = -\,0.9$

Comparing constants:

$2 = q - 5p^2$

$\Rightarrow 2 = q - 5 \times 0.81$

$\Rightarrow 2 = q - 4.05$

$\Rightarrow q = 6.05$

$\therefore\ h = 2 + 9t - 5t^2 = 6.05 - 5(t - 0.9)^2$

Exploration 4.7

Transformations and symmetry

The parabola $y = x^2$ has a line of symmetry $x = 0$ and vertex (turning point) at (0,0).

Describe what happens to the line of symmetry and vertex of $y = x^2$ under successive transformations to give $y = a(x + p)^2 + q$.

Successive tranformations

1 translation $-p$ units parallel to the x-axis $y = (x + p)^2$
followed by

2 stretch, factor a parallel to the y-axis $y = a(x + p)^2$
followed by

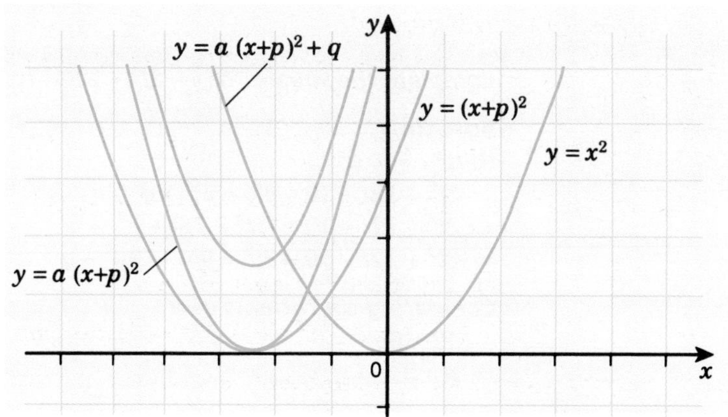

3 translation q units parallel to the y-axis $\quad y = a(x + p)^2 + q$

Transformation	*Vertex*	*Line of symmetry*
1	$(-p, 0)$	$x = -p$
2	$(-p, 0)$	$x = -p$
3	$(-p, q)$	$x = -p$

As a result of successive transformations the vertex (0, 0) and line of symmetry $x = 0$ move as follows.

By completing the square for any quadratic function we can now read off the coordinates of the vertex and the line of symmetry, enabling us to sketch the curve. Using the results from previous worked examples:

4.4 $y = x^2 + 10x + 8 = (x + 5)^2 - 17$
$\Rightarrow \quad p = 5, q = -17$
Vertex: (–5, –17)
Line of symmetry: $x = -5$

4.7 $h = 2 + 9t - 5t^2 = 6.05 - 5(t - 0.9)^2$
$\Rightarrow \quad p = -0.9, a = 6.05$
Vertex: (0.9, 6.05)
Line of symmetry: $t = 0.9$

Example 4.18

A farmer wishes to create six pens, in the arrangement shown here, using 144 metres of fencing.

	y	y	y
x			
x			

a) *Show that the area enclosed is given by* $A = 96x - \frac{16}{3}x^2$.
b) *By completing the square find the maximum area that can be enclosed and the corresponding dimensions of each pen.*

Solution

a) *The framework for the pens requires 8 x-lengths + 9 y-lengths.*

$\Rightarrow \; 8x + 9y = 144$

$\Rightarrow \; 9y = 144 - 8x$

$\Rightarrow \; y = 16 - \frac{8}{9}x$

Area $A = 2x \times 3y = 6xy = 6x(16 - \frac{8}{9}x)$

$\Rightarrow \; A = 96x - \frac{16}{3}x^2$

b) *Let* $96x - \frac{16}{3}x^2 \equiv q - \frac{16}{3}(x + p)^2$

$\Rightarrow \; 96x - \frac{16}{3}x^2 \equiv q - \frac{16}{3}x^2 - \frac{32}{3}px - \frac{16}{3}p^2$

Comparing coefficients of x:

$96 = -\frac{32}{3}p$

$\Rightarrow \; p = 96 \times (-\frac{3}{32}) = -9$

Comparing constants:

$0 = q - \frac{16}{3}p^2$

$\Rightarrow \; 0 = q - \frac{16}{3} \times 81$

$\Rightarrow \; q = 432$

$\Rightarrow \; A = 432 - \frac{16}{3}(x - 9)^2$

From this form of the equation we can see that A will be maximised when x = 9, and its value is 432.

Substituting for x gives:

$y = 16 - \frac{8}{9}x$

$= 16 - \frac{8}{9} \times 9 = 8$

The final framework is as shown here.

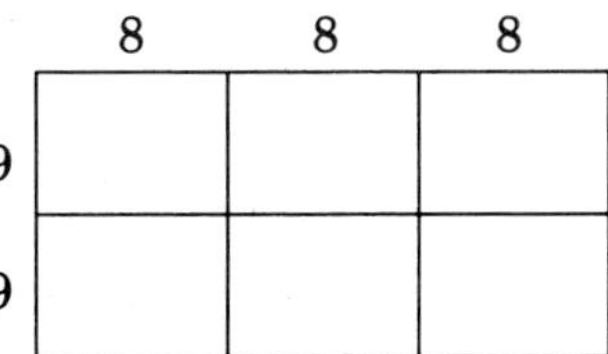

EXERCISES

1 Complete the square for each of these quadratic functions.

a) $y = x^2 + 4x + 11$ **b)** $y = x^2 - 6x + 5$

c) $y = x^2 - 5x - 3$ **d)** $y = x^2 + 3x - 8$

e) $y = x^2 + x + 1$ **f)** $y = x^2 - 7x$

For each of your answers, what is the quick way of finding the value of p?

2 Complete the square for each of these quadratic functions.

a) $y = 2x^2 - 8x + 5$ **b)** $y = 3x^2 + 6x - 7$

c) $y = 2x^2 + 4x + 9$ **d)** $y = 5x^2 - 5x + 8$

e) $s = 3t^2 - t + 5$ **f)** $d = 5t^2 + 2t - 12$

g) $y = 8 + 2x - x^2$ **h)** $y = 10 - 3x - x^2$

i) $y = 7 - 10x - 3x^2$ **j)** $y = 24 + 23z - 12z^2$

3 For each of the following quadratic functions, complete the square, write down the equation of the parabola's line of symmetry and the coordinates of its vertex, and illustrate your answer on a sketch graph.

a) $y = x^2 - 2x - 3$ **b)** $y = x^2 + 5x + 8$
c) $y = x^2 - 2.5x + 7$ **d)** $y = 4x^2 - 8x + 11$
e) $y = 3x^2 + 5x - 8$ **f)** $y = 5x^2 - 3x$
g) $y = 10x - x^2$ **h)** $y = 12 - x - x^2$
i) $h = 15 + 3t - 5t^2$ **j)** $s = 4 - 9t^2$

4 The equation of the curve taken on by a cable of a suspension bridge is $y = 0.001x^2 - x + 28$.

a) By completing the square:
  i) sketch the graph,
  ii) find the line of symmetry of the parabola,
  iii) find the coordinates of the vertex of the parabola.
b) The horizontal distance between the ends of the cable is 100 metres. Find the difference between the maximum and minimum heights of the cable.

5 A farmer wishes to create twelve pens using the arrangement shown, using 240 metres of fencing.

	y	y	y	y
x				
x				
x				

a) Show that the area enclosed is given by $A = 180x - \frac{45}{4}x^2$.
b) By completing the square:
  i) sketch the graph of A against x,
  ii) find the maximum area that can be enclosed,
  iii) find the values of x and y to give the maximum area.

EXERCISES

4.4 B

1 Complete the square for each of these quadratic functions.

a) $x^2 - 10x + 28$ **b)** $x^2 - 3x + 3$
c) $x^2 + 12x + 27$ **d)** $x^2 - 5x + 0.5$
e) $x^2 + 7x$ **f)** $x^2 + 7$

2 Complete the square for each of these quadratic functions.

a) $y = 2x^2 + 12x + 7$ **b)** $y = 3x^2 - 36x + 109$
c) $y = 8x^2 - 80 + 105$ **d)** $s = 3t^2 - 15t - 5$
e) $y = 8x^2 - 2x + 9$ **f)** $y = -x^2 + 14x - 45$
g) $v = 1 - 8t - 2t^2$ **h)** $y = 2x - 2x^2$
i) $y = 1 - 2x + 3x^2$ **j)** $y = 6x - 3 - 5x^2$

3 Give a rough sketch of each of the following parabolae, showing clearly where it crosses the y-axis and the coordinates of its vertex.

a) $y = x^2 - 4x + 7$
b) $y = 8 - x^2$
c) $y = x^2 + 2x - 4$
d) $y = x^2 + 5x - 2$
e) $s = t^2 + 5t + 7$
f) $y = 4x^2 - 32x + 57$
g) $y = 1 - x - x^2$
h) $y = 6x^2 + 8x - 5$
i) $y = 7x - 3x^2$
j) $h = 9t - 5t^2 - 23$

4 The Jodrell Bank radio telescope has a giant circular dish. When it is in a horizontal position, the reflecting surface has a cross-section which follows the curve given by part of the parabola

$$y = 0.003\,968x^2 - 0.992x + 225$$

where x is the distance from the edge of the dish in feet and y is the height above ground level in feet.

a) By completing the square, sketch the cross-section of the dish.
b) How high above the ground level is the rim of the dish?
c) How deep in the dish?
d) What is the radius of the dish?

5 A lawn is a rectangle 10 m by 15 m. It is to have two flower beds cut out of it. One will be a square flower bed and the other will be in the shape of a rectangle, twice as long as it is wide. The gardener has 13.6 m of lawn edging left to put round the flower beds to stop the grass growing into them. He wants to use up all the lawn edging, but requires to keep the lawn as large as possible in area after the flower beds have been cut out.

a) If the square bed has a side of length y metres and the rectangular bed has width x metres, show that the area of the lawn will be $A = 138.44 + 10.2x - 4.25x^2$.
b) By completing the square, find the dimensions of the flower beds so that the area of the lawn is as large as possible.

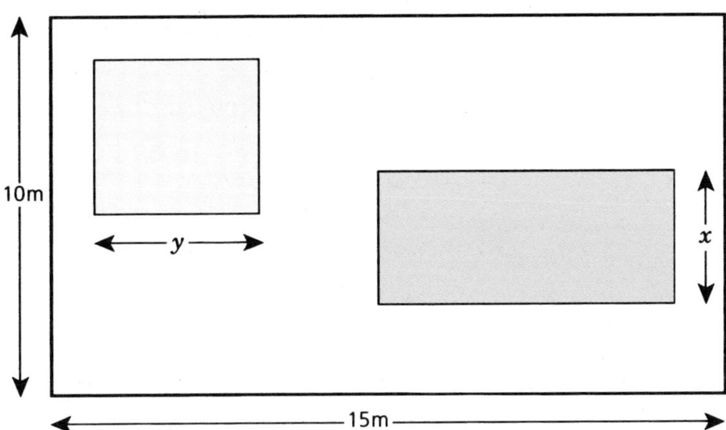

MATHEMATICAL MODELLING ACTIVITY

1 → 2 → 3 ↓ 6 ← 5 ← 4
Specify the real problem

Problem statement

A fleet hire company leases cars to firms on a yearly basis, with a percentage discount in the leasing charge proportional to the number of cars leased.

1 How many cars must the fleet hire company lease to a single firm to maximise income?

2 How would your decision be affected if the fleet hire company wish to maximise profit?

1 → 2 → 3 ↓ 6 ← 5 ← 4
Set up a model

Set up a model

The first stage in setting up a model is to identify the important variables:

- the size of a firm's fleet,
- the income for a contract with a given fleet size,
- the profit for a contract with a given fleet size.

Before formulating the problem mathematically, you will also need to make certain assumptions:

- the income I and profit P depend only on the fleet size S,
- the nominal lease charge is £2000 per car per year,
- the fleet hire company has a special offer: the lease charge per car is discounted by 1 per cent for each car in the fleet, e.g. a fleet size of 20 cars earns a 20 per cent discount,
- each car depreciates by £1000 each year.

1 → 2 → 3 ↓ 6 ← 5 ← 4
Formulate the mathematical problem

Mathematical problem

The total income is given by:

$$\begin{aligned} I &= 2000S - S \times 0.01 \times 2000S \\ &= 2000S - 20S^2 \end{aligned}$$

income before special offer

special offer discount

Assuming that each car depreciates by £1000 each year, the profit is given by:

$$\begin{aligned} P &= 2000S - 20S^2 - 1000S \\ &= 1000S - 20S^2 \end{aligned}$$

1 → 2 → 3
6 ← 5 ← 4

Solve the mathematical problem

Mathematical solution

$I = 2000S - 20S^2$ is a quadratic function in S.

Graphically you can read off the maximum income from the vertex of the parabola.

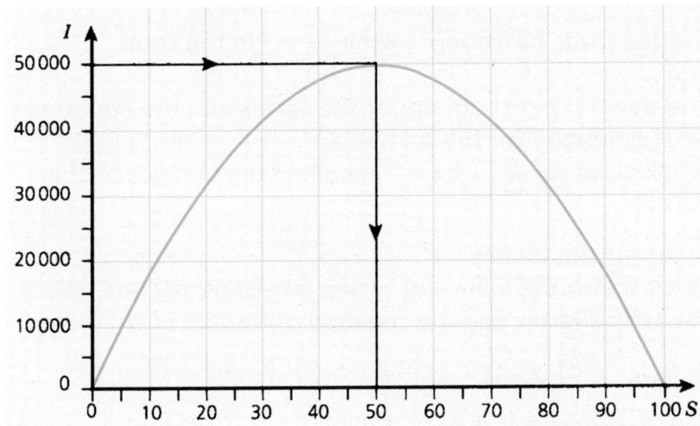

The maximum income of £50 000 is obtained from a contract for a fleet size of 50. Confirm these findings by plotting the graph using a graphics calculator.

The solution may also be found algebraically by completing the square:

$$\begin{aligned} I &= 2000S - 20S^2 \\ &= -20\{S^2 - 100S\} \\ &= -20\{(S - 50)^2 - 2500\} \\ &= 50\,000 - 20(S - 50)^2 \end{aligned}$$

You can now deduce that I has a maximum value of 50 000 when $S = 50$.

By repeating the analysis for the profit function, show that P has a maximum of 12 500 when $S = 25$.

1 → 2 → 3
6 ← 5 ← 4

Interpret the solution

Interpretation

To maximise income, a fleet size of 50 cars produces an income of £50 000 but to maximise profit, the fleet size should be 25 cars, realising a profit of £12 500.

Note that the fleet size for maximum income, $S = 50$, means that the company just breaks even (zero profit) and for $S > 50$ the company makes a loss.

Refinement of the model

Investigate the effect of changes in

1 the annual lease charge
2 the percentage discount per car
3 the annual depreciation.

Introduce a threshold for the discounting process to start.

CONSOLIDATION EXERCISES FOR CHAPTER 4

1 Multiply out the brackets and simplify each expression.

a) $(x + 9)(x - 4)$ **b)** $(2x - 3)(3x + 2)$
c) $(5 - 2x)^2$ **d)** $(3x + 4)(2x - 5) + (x - 7)(x + 7)$
e) $(2x + 9)^2 - (4x + 5)(x - 3)$

2 Factorise each expression fully.

a) $x^2 - 20x + 91$ **b)** $2x^2 + 21x - 11$
c) $12x^2 + 38x + 20$ **d)** $20x^2 - 120x + 36$
e) $245x^2 - 5$

3 Complete the square for the following quadratic functions.

a) $f(x) = x^2 - 20x + 91$ **b)** $f(x) = x^2 + 3x + 10$
c) $f(x) = 2x^2 + 20x + 35$ **d)** $f(x) = 8 + x - x^2$
e) $f(x) = (2x - 1)(2x - 7)$

4 **a)** Express the quadratic function $y = 4x^2 + 6x + 7$ in the form $a(x + p)^2 + q$.
b) Hence write down the coordinates of the vertex and the equation of the line of symmetry for the parabola.
c) Sketch the graph of $y = 4x^2 + 6x + 7$, labelling the vertex and the line of symmetry.

5 Sketch the curve with equation $y = x^2$.
On separate axes, sketch the following curves, labelling each one, giving the coordinates of the vertex and the equation of the axis of symmetry.

a) $y = (x - 2)^2 + 1$ **b)** $y = (x + 1)^2 - 3$ **c)** $y = 3(x - 1)^2 + 5$

6 **a)** Show that $x^2 + 10x + 7 \equiv (x + 5)^2 + a$, where a is to be found.
b) Sketch the graph of $y = x^2 + 4x + 7$, giving the equation of the axis of symmetry and the coordinates of its vertex.

7 **a)** Write $x^2 + 4x + 13$ in the form $(x + p)^2 + q$, where p and q are integers to be found.
b) Find the minimum value of $x^2 + 4x + 13$ and state the value of x for which this minimum occurs.
c) Sketch the graph of $y = x^2 + 4x + 13$

8 **a)** Sketch the curve with equation $y = x^2$.
b) On the same axes sketch curves which represent $y = x^2$ transformed by
i) translation through –5 units parallel to the x-axis,
ii) stretch, factor 3 parallel to the y-axis,
iii) translation through –2 units parallel to the y-axis.
In each case write down the equation of the transformed curve.
c) The transformations in (b) are combined in the following ways:
i) *followed by* ii) *followed by* iii),
ii) *followed by* iii) *followed by* i),
iii) *followed by* ii) *followed by* i).
For each combination, write down the equation of the transformed curve and sketch it.

9 Complete the square for the function $f(x) = ax^2 + bx + c$.

Assuming that a is positive, and by considering the coordinates of the vertex, deduce conditions for which the graph of $f(x)$:

a) crosses the x-axis in two distinct places,
b) touches the x-axis at one point only,
c) lies entirely above the x-axis.

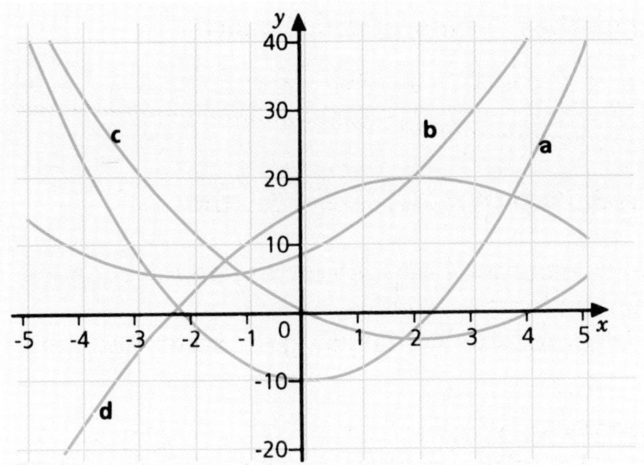

10 The diagram shows four parabolae.

By inspecting the points where they intersect with the axes and the coordinates of the vertex, write down the equation for each one in the form $y = ax^2 + bx + c$.

Summary

Quadratic functions

- A quadratic function is of the general form $f(x) = ax^2 + bx + c$.
- The graph of a quadratic function is called a **parabola**.

Factorising

- The quadratic expression $ax^2 + bx + c$ may sometimes be factorised into the product of linear factors by inspection.
  For example:
  $x^2 - 2x - 15 = (x - 5)(x + 3)$
  $2x^2 + 9x + 4 = (2x + 1)(x + 4)$.
- Special cases of factorising are:
  a perfect square $(x + p)^2 = x^2 + 2px + p^2$
  the difference of two squares $(x - p)(x + p) = x^2 - p^2$.

Completing the square

- Any quadratic function $f(x) = ax^2 + bx + c$ can be written in the form $f(x) = a(x + p)^2 + q$
  This is called completing the square.
- For the parabola $y = ax^2 + bx + c$:
  the **line of symmetry** has equation $x = -p$
  the coordinates of the **vertex** are $(-p, q)$.

PURE MATHS 5

Quadratic equations and inequalities

In this chapter we:

- *solve quadratic equations by factorising, completing the square and using 'the quadratic formula'*
- *see how to solve a pair of simultaneous equations consisting of a quadratic equation and a linear equation*
- *see how to solve quadratic inequalities.*

QUADRATIC EQUATIONS BY FACTORS

The general form of a quadratic function is $y = ax^2 + bx + c$, where a, b and c are constants and $a \neq 0$. A **family** of quadratics is a group which have something in common. The equations all take the same form and the curves are all the same general shape, but they may have slightly different properties.

Exploration 5.1

The family $y = k(x^2 - 9)$

Look carefully at these graphs. They all belong to the family $y = k(x^2 - 9)$, for different values of k.

- Identify the values of k.
- Factorise $k(x^2 - 9)$.
- What do the four graphs all have in common?

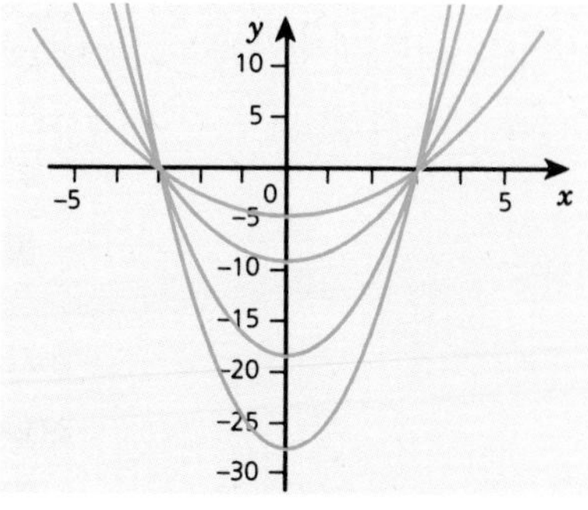

Exploration 5.2

The family $y = k(5x - x^2)$

Now examine carefully another set of graphs. They all belong to the family $y = k(5x - x^2)$, for different values of k.

- Identify the values of k.
- Factorise $k(5x - x^2)$.
- What do the four graphs all have in common?

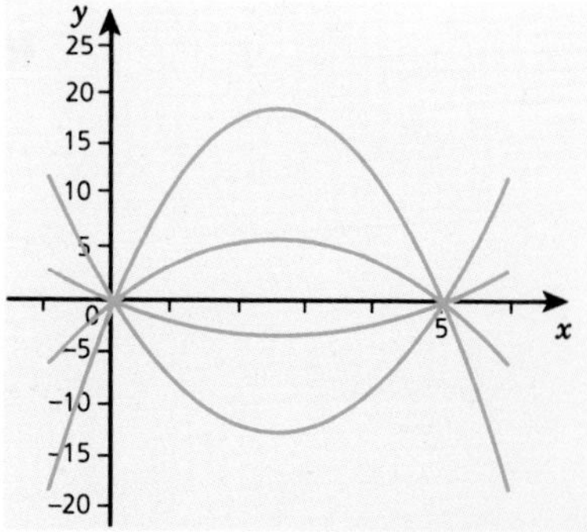

Exploration 5.3

The family $y = k(x^2 + 5x - 14)$

Finally study a third set of graphs. They all belong to the family $y = k(x^2 + 5x - 14)$, for different values of k.

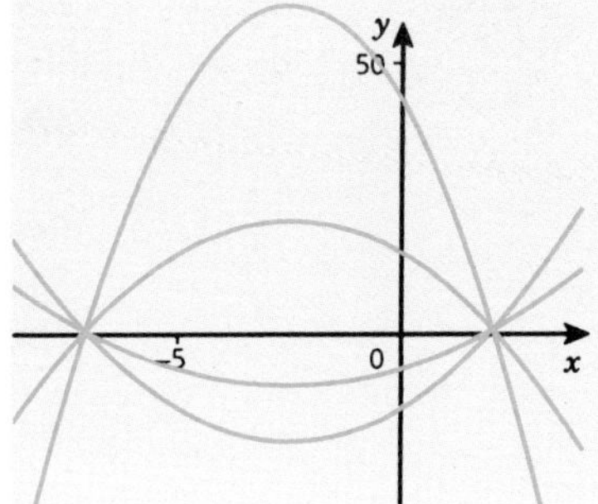

- Identify the values of k.
- Factorise $k(x^2 + 5x - 14)$.
- What do the four graphs all have in common?

Interpreting the results

Each graph represents a quadratic function f(x), which intersects with the x-axis in two places, i.e. the x-values for which f(x) = 0.

For varying values of k your findings might be summarised as:

Exploration	Quadratic function	Factorised form	x-values when $y = 0$
5.1	$k(x^2 - 9)$	$k(x + 3)(x - 3)$	–3, 3
5.2	$k(5x - x^2)$	$kx(5 - x)$	0, 5
5.3	$k(x^2 + 5x - 14)$	$k(x + 7)(x - 2)$	–7, 2

So by factorising the **quadratic function**, the x-values which solve the quadratic equation f(x) = 0, may be fairly easily identified. These x-values are called the **roots** of the quadratic equation. Factorising gives a product of factors which is zero. This means that one or other of the two factors (other than k) must be zero, which in turn gives the roots.

$k(x + 3)(x - 3) = 0 \Rightarrow x + 3 = 0$ or $x - 3 = 0 \Rightarrow x = -3$ or $x = 3$

$kx(5 - x) = 0 \Rightarrow x = 0$ or $5 - x = 0 \Rightarrow x = 0$ or $x = 5$

$k(x + 7)(x - 2) = 0 \Rightarrow x + 7 = 0$ or $x - 2 = 0 \Rightarrow x = -7$ or $x = 2$

The value of the constant k does not affect the solution of the quadratic equation. Provided the quadratic function can be factorised, solutions may be found, as in the following examples.

Example 5.1

Solve the equation $7x^2 - 28 = 0$.

Solution

$7x^2 - 28 = 0$

Take out a constant factor $\Rightarrow 7(x^2 - 4) = 0$

This will factorise further $\Rightarrow 7(x - 2)(x + 2) = 0$

$\Rightarrow x = 2$ *or* $x = -2$

Example 5.2

Solve the equation $16x^2 = 100$.

Solution

$$16x^2 = 100$$

Rearrange in the form $f(x) = 0 \Rightarrow 16x^2 - 100 = 0$

Take out a constant factor $\Rightarrow 4(4x^2 - 25) = 0$

This will factorise further $\Rightarrow 4(2x - 5)(2x + 5) = 0$

$\Rightarrow 2x - 5 = 0$ *or* $2x + 5 = 0$

$\Rightarrow x = 2.5$ *or* $x = -2.5$

Note: For the last two examples we may use an alternative strategy:

$7x^2 - 28 = 0 \Rightarrow 7x^2 = 28 \Rightarrow x^2 = 4 \Rightarrow x = \pm\sqrt{4} = \pm 2$

$16x^2 = 100 \Rightarrow x^2 = 6.25 \Rightarrow x = \pm\sqrt{6.25} = \pm 2.5$

This works well provided the quadratic equation does not contain a term in x. This idea occurs again later.

Example 5.3

Solve the equation $15x^2 - 25x = 0$.

Solution

$$15x^2 - 25x = 0$$

Take out a constant factor $\Rightarrow 5(3x^2 - 5x) = 0$

This will factorise further $\Rightarrow 5x(3x - 5) = 0$

$\Rightarrow x = 0$ *or* $3x - 5 = 0$

$\Rightarrow x = 0$ *or* $x = 1\frac{2}{3}$

Example 5.4

Solve the equation $3x^2 - 24x + 45 = 0$.

Solution

$$3x^2 - 24x + 45 = 0$$

Take out a constant factor $\Rightarrow 3(x^2 - 8x + 15) = 0$

This will factorise further $\Rightarrow 3(x - 3)(x - 5) = 0$

$\Rightarrow x - 3 = 0$ *or* $x - 5 = 0$

$\Rightarrow x = 3$ *or* $x = 5$

Example 5.5

Solve the equation $3x^2 + 11x - 4 = 0$.

Solution

$$3x^2 + 11x - 4 = 0$$

Which will factorise $\Rightarrow (3x - 1)(x + 4) = 0$

$\Rightarrow 3x - 1 = 0$ *or* $x + 4 = 0$

$\Rightarrow x = \frac{1}{3}$ *or* $x = -4$

Some equations may not appear to be quadratic at first glance, but using some algebraic manipulation we can change them into a suitable form. The next two examples illustrate this.

Example 5.6

Solve the equation $x - \dfrac{8}{x} = 2.$

Solution

$$x - \frac{8}{x} = 2$$

Multiply both sides by x	$\Rightarrow$	$x^2 - 8 = 2x$
Rearrange	$\Rightarrow$	$x^2 - 2x - 8 = 0$
This will factorise	$\Rightarrow$	$(x - 4)(x + 2) = 0$
	$\Rightarrow$	$x = 4$ *or* $x = -2$

Example 5.7

Solve the equation $x^4 - 13x^2 + 36 = 0$.

Solution

$$x^4 - 13x^2 + 36 = 0$$

Change to a quadratic equation in x^2	$\Rightarrow$	$(x^2)^2 - 13(x^2) + 36 = 0$
This will factorise	$\Rightarrow$	$(x^2 - 9)(x^2 - 4) = 0$
	$\Rightarrow$	$x^2 = 9 \Rightarrow x = \pm 3$
	or	$x^2 = 4 \Rightarrow x = \pm 2$

Solving problems

As we discovered in Chapter 4, *Quadratic functions,* quadratic equations can sometimes be useful in solving problems. The next worked example extends a problem we met there and uses a quadratic equation to solve it.

Example 5.8

The height, y metres, of a point on the St Louis arch, x metres from one side, can be modelled by the function $y = \dfrac{x(192 - x)}{48}$.

How far apart are points on the arch which are 165 m above the ground?

Solution

For a height of 165 m, let $y = 165$ *in the quadratic function.*

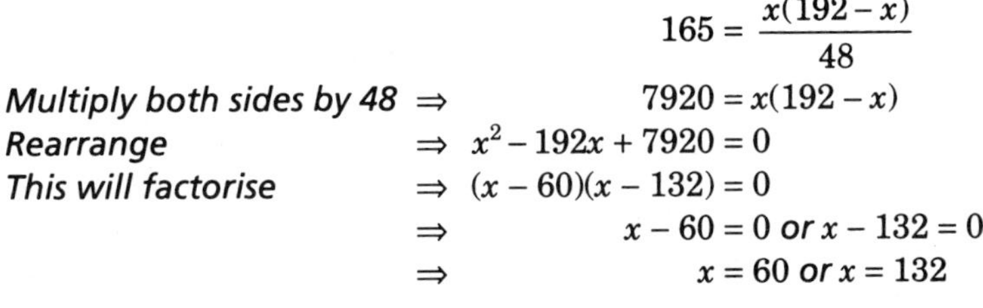

$$165 = \frac{x(192 - x)}{48}$$

Multiply both sides by 48	$\Rightarrow$	$7920 = x(192 - x)$
Rearrange	$\Rightarrow$	$x^2 - 192x + 7920 = 0$
This will factorise	$\Rightarrow$	$(x - 60)(x - 132) = 0$
	$\Rightarrow$	$x - 60 = 0$ *or* $x - 132 = 0$
	$\Rightarrow$	$x = 60$ *or* $x = 132$

This means that two points which are 165 m above the ground have corresponding x-values 60 and 132, i.e. they are 72 metres apart.

Notes

- If you found the factorising in the last example very difficult, there are alternative strategies which appear in the next section.
- Sometimes problems lead to quadratic equations where one of the roots is meaningless (e.g. a negative value for a physical quantity). In such cases the unwanted root is discarded.

A word of warning

From the explorations it is clear that the solution of a quadratic equation in the form $f(x) = 0$ occurs where the graph of $y = f(x)$ intersects with the x-axis. The roots occur at the points where the parabola intersects the x-axis. However, not all parabolae actually intersect with the x-axis. To put it another way, not all quadratic equations have roots!

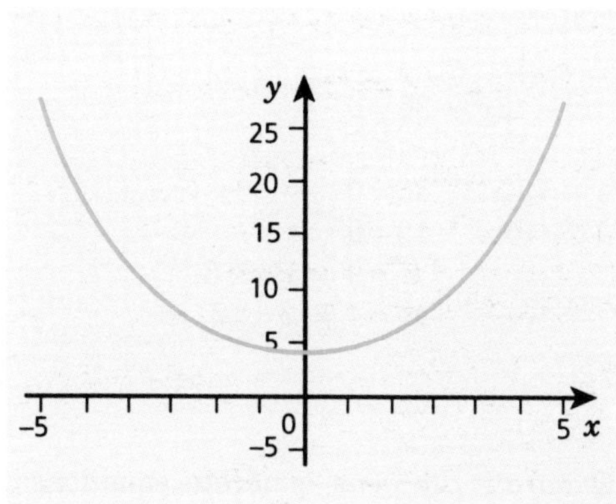

Consider the graph of $y = x^2 + 4$, i.e. $f(x) = x^2 + 4$.

Since the graph lies entirely above the x-axis, $f(x) = 0$ has no solutions.

Trying to solve the equation leads to:

$$x^2 + 4 = 0 \quad \Rightarrow \quad x^2 = -4 \quad \Rightarrow \quad x = \pm\sqrt{-4}$$

which is not possible, at least within the system of real numbers (try finding the square root of -4 on your calculator). There is no real number which has a square of -4. Whenever attempting to solve a quadratic equation leads to taking the square root of a negative number, the roots are said to be **imaginary**.

EXERCISES

For each of the quadratic functions, solve the equation $f(x) = 0$. In each case, sketch the curve $y = f(x)$ and mark on your solutions.

1	$f(x) = x^2 - 7x$	**2**	$f(x) = x^2 - 12x + 32$
3	$f(x) = x^2 - 6x + 9$	**4**	$f(x) = x^2 - x - 12$
5	$f(x) = 8 + 2x - x^2$	**6**	$f(x) = 8x - x^2 - 16$
7	$f(x) = 5x - 2x^2$	**8**	$f(x) = x^2 - 25$

Solve the following equations wherever possible. If the roots are imaginary, say so.

9	$x^2 - 17x = 0$	**10**	$3x^2 - 6x = 0$	**11**	$x^2 = -4x$
12	$x^2 = 49$	**13**	$x^2 - 25 = 0$	**14**	$3x^2 - 147 = 0$
15	$4x^2 - 9 = 0$	**16**	$4x^2 - 25 = 0$	**17**	$x^2 = -9$
18	$p^2 + 51 = 0$	**19**	$-81 + q^2 = 0$	**20**	$81 + r^2 = 0$

Solve the following quadratic equations by factorising.

21 $x^2 + 5x + 6 = 0$ **22** $x^2 + 6x + 8 = 0$
23 $x^2 + 15x + 54 = 0$ **24** $x^2 - 4x + 3 = 0$
25 $t^2 - 3t - 40 = 0$ **26** $x^2 - 17x - 60 = 0$
27 $x^2 + 5x + 4 = 0$ **28** $x^2 + 3x - 4 = 0$
29 $x^2 - x - 12 = 0$ **30** $t^2 - 2t + 1 = 0$
31 $2x^2 - 11x + 5 = 0$ **32** $3x^2 - x - 14 = 0$
33 $6x^2 + 5x + 1 = 0$ **34** $6x^2 + 19x + 10 = 0$
35 $5y^2 + 33y - 14 = 0$ **36** $6x^2 + 11x + 3 = 0$
37 $13x - 14 - 3x^2 = 0$ **38** $7u - 6u^2 + 20 = 0$
39 $9x^2 + 12x + 4 = 0$ **40** $5x^2 + 40x + 80 = 0$

Rearrange the following as quadratic equations and solve by factorising:

41 $x + 1 = \dfrac{6}{x}$ **42** $2x - \dfrac{12}{x} = 5$
43 $x^4 - 13x^2 + 36 = 0$ **44** $x^4 + 5x^2 - 36 = 0$
45 $(x + 1)^2 + 3(x + 1) + 2 = 0$

46 A ball is thrown vertically upwards. Its height, h metres, after t seconds, is given by $h = 16t - 5t^2$.

At what times will the ball be at a height of 12 m?

47 A rectangular sports pitch is 25 metres longer than it is wide and has an area of $3150\,m^2$.

a) If x represents the width of the pitch, express both the length and the area in terms of x.
b) Solve a quadratic equation to find the dimensions of the pitch.

48 The diagram shows a right-angled triangle in which the shortest side is x cm.

a) Use Pythagoras' theorem to formulate a quadratic equation in x.
b) Solve the quadratic equation to find the sides of the triangle.

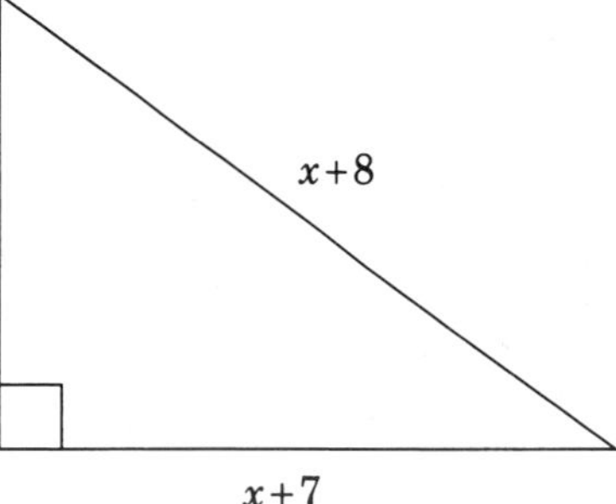

49 A farmer has 31 metres of fencing to enclose three sides of a rectangular pen, the fourth side being a wall. The area of the pen is to be $120\,m^2$. Find the dimensions of the pen.

EXERCISES

For each of the quadratic functions, solve the equation $f(x) = 0$, wherever possible.
In each case, sketch the curve $y = f(x)$ and mark on any solutions.

1 $f(x) = x^2 + 2x$
2 $f(x) = x^2 - 9$
3 $f(x) = 2x^2 + 5x - 3$
4 $f(x) = -9x^2 - 6x - 1$
5 $f(x) = 24 + 6x - 3x^2$
6 $f(x) = 6x^2 - 13x + 6$
7 $f(x) = -5x - 2x^2$
8 $f(x) = x^2 + 10$

Solve the following equations wherever possible.

9 $x^2 + 4x = 0$
10 $x^2 = 3x$
11 $3x^2 - 5x = 0$
12 $x^2 = 17$
13 $x^2 - 26 = 0$
14 $x^2 - 10 = 0$
15 $4x^2 - 64 = 0$
16 $100 - 9x^2 = 0$
17 $x^2 + 16 = 0$
18 $7s^2 + 19 = 0$
19 $7t^2 + 35t = 0$
20 $7u^2 = 17u$

Solve the following quadratic equations by factorising.

21 $x^2 + 4x + 3 = 0$
22 $x^2 + 10x + 24 = 0$
23 $x^2 + 20x + 51 = 0$
24 $x^2 - 12x + 32 = 0$
25 $x^2 - 7x + 10 = 0$
26 $y^2 + 5y - 6 = 0$
27 $3y^2 - 6y - 9 = 0$
28 $2y^2 - 12y + 18 = 0$
29 $5x^2 - 5x - 30 = 0$
30 $x^2 - 2x + 1 = 0$
31 $1 + 6y - 7y^2 = 0$
32 $y^2 + 5y - 14 = 0$
33 $3x^2 + 10x + 3 = 0$
34 $3x^2 + 22x + 35 = 0$
35 $6x^2 + 7x - 3 = 0$
36 $20x^2 + x - 1 = 0$
37 $7x - 6 - 2x^2 = 0$
38 $15 - 4x - 4x^2 = 0$
39 $9x^2 + 6x + 1 = 0$
40 $-x^2 + 6x - 9 = 0$

Rearrange the following as quadratic equations and solve by factorising.

41 $6x + \frac{12}{x} = 17$
42 $\frac{x}{4} + \frac{3}{x} = 2$
43 $(x - 1)(x - 2) = 20$
44 $\frac{1}{(x-3)} - \frac{1}{(x-7)} = \frac{4}{3}$
45 $x^4 + 7x^2 = 8$

46 The diagram shows a right-angled triangle.

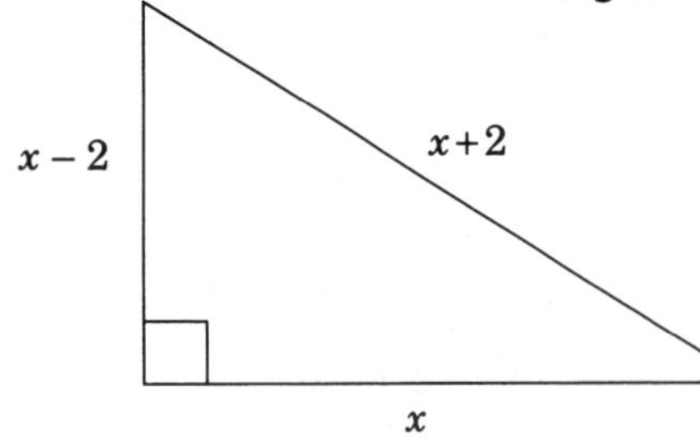

a) Use Pythagoras' theorem to formulate a quadratic equation in x.
b) Solve the quadratic equation to find the sides of the triangle.
c) Try to predict the dimensions of a right-angled triangle with sides of lengths $x - d$, x and $x + d$, in terms of a constant d.

47 **a)** The product of two consecutive odd numbers is 783. What are the numbers?
b) The product of two integers, which differ by 9, is 1540. Find the two integers.

48 **a)** Explain why a polygon, with n sides, has d diagonals where $d = \frac{n(n-3)}{2}$, $n \geq 0$.

b) A polygon has 44 diagonals. How many sides does it have?

QUADRATIC EQUATIONS

Completing the square

In the last section we saw how to solve quadratic equations such as $x^2 = 49$ and $4x^2 - 9 = 0$ using an alternative method.

$x^2 = 49 \Rightarrow x = \pm\sqrt{49} = \pm 7$
$4x^2 - 9 = 0 \Rightarrow 4x^2 = 9 \Rightarrow x^2 = 2.25 \Rightarrow x = \pm\sqrt{2.25} = \pm 1.5$

These are **rational** roots, since they can be expressed precisely as a whole number or a fraction.

This may be extended to equations which have roots that are not exact, e.g.

$5x^2 = 100 \Rightarrow x^2 = 20 \Rightarrow x = \pm\sqrt{20} = \pm 4.47$ (3 s.f.)
$x^2 - 93 = 0 \Rightarrow x^2 = 93 \Rightarrow x = \pm\sqrt{93} = \pm 9.64$ (3 s.f.)

These are **irrational** roots, since they cannot be expressed precisely as a whole number or a fraction. Accuracy is given correct to three significant figures (3 s.f.).

Going one stage further, we can solve quadratic equations such as those in the following examples.

Example 5.9

Solve the equation $(x - 3)^2 = 15$.

Solution

$(x - 3)^2 = 15$

Take the square root of both sides $\Rightarrow$ $x - 3 = \pm\sqrt{15}$
Rearrange $\Rightarrow$ $x = 3 \pm\sqrt{15}$
$\Rightarrow$ $x = -0.873$ *or* $x = 6.87$ (3 s.f.)

Example 5.10

Solve the equation $4(x + 5)^2 - 59 = 0$.

Solution

$4(x + 5)^2 - 59 = 0$

Add 59 to both sides $\Rightarrow$ $4(x + 5)^2 = 59$
Divide both sides by 4 $\Rightarrow$ $(x + 5)^2 = 14.75$
Take the square root of both sides $\Rightarrow$ $x + 5 = \pm\sqrt{14.75}$
Rearrange $\Rightarrow$ $x = -5 \pm\sqrt{14.75}$
$\Rightarrow$ $x = -8.84$ *or* $x = -1.16$ (3 s.f.)

The last two examples pave the way for a method which may be used on any quadratic equation which has a solution, but may not factorise. Firstly, complete the square for the function $f(x)$, then solve the equation $f(x) = 0$ as above. Study the following examples carefully.

Example 5.11

Solve the equation $x^2 - 8x - 11 = 0$.

Solution

		$x^2 - 8x - 11 = 0$
Complete the square	$\Rightarrow$	$(x-4)^2 - 27 = 0$
Add 27 to both sides	$\Rightarrow$	$(x-4)^2 = 27$
Take the square root of both sides	$\Rightarrow$	$x - 4 = \pm\sqrt{27}$
Rearrange	$\Rightarrow$	$x = 4 \pm\sqrt{27}$
	$\Rightarrow$	$x = -1.20$ *or* $x = -9.20$ (3 s.f.)

Example 5.12

Solve the equation $2x^2 + 8x + 3 = 0$.

Solution

		$2x^2 + 8x + 3 = 0$
Complete the square	$\Rightarrow$	$2(x+2)^2 - 5 = 0$
Add 5 to both sides	$\Rightarrow$	$2(x+2)^2 = 5$
Divide both sides by 2	$\Rightarrow$	$(x+2)^2 = 2.5$
Take the square root of both sides	$\Rightarrow$	$x + 2 = \pm\sqrt{2.5}$
Rearrange	$\Rightarrow$	$x = -2 \pm\sqrt{2.5}$
	$\Rightarrow$	$x = -3.58$ *or* $x = -0.419$ (3 s.f.)

Using the formula

The process of solving a quadratic equation by completing the square may be quite time consuming. The structure of the solution is roughly the same each time, only the details are different. This leads to the idea of finding a general purpose formula for solving quadratic equations of the type:
$ax^2 + bx + c = 0$
where a, b and c are constants and $a \neq 0$. It is essential that $a \neq 0$, since in developing the formula we shall need to divide by a.

To solve the equation $ax^2 + bx + c = 0$

		$ax^2 + bx + c = 0$
Complete the square	$\Rightarrow$	$a\left(x + \frac{b}{2a}\right)^2 - \frac{b^2}{4a} + c = 0$
Rearrange	$\Rightarrow$	$a\left(x + \frac{b}{2a}\right)^2 = \frac{b^2 - 4ac}{4a}$
Divide both sides by a	$\Rightarrow$	$\left(x + \frac{b}{2a}\right)^2 = \frac{b^2 - 4ac}{4a^2}$

Take the square root of both sides $\Rightarrow$ $x + \dfrac{b}{2a} = \pm\sqrt{\dfrac{b^2 - 4ac}{4a^2}} = \pm\dfrac{\sqrt{b^2 - 4ac}}{2a}$

Rearrange $\Rightarrow$ $x = \dfrac{-b \pm \sqrt{b^2 - 4ac}}{2a}$

This gives a general formula which can be used to solve any quadratic equation with real roots. Looking again at the formula, we can see that this will be the case provided $a \neq 0$ and $b^2 - 4ac$ is **not** negative. The expression $b^2 - 4ac$ is called the **discriminant** of the expression $ax^2 + bx + c$.

Exploration 5.4

Applying the formula

Apply the formula to the following quadratic equations and see what happens.

a) $x^2 - 2x - 3 = 0$
b) $x^2 - 2x + 1 = 0$
c) $x^2 - 2x - 7 = 0$
d) $x^2 - 2x + 5 = 0$

a) $x^2 - 2x - 3 = 0 \Rightarrow a = 1, b = -2, c = -3$

$$\Rightarrow x = \frac{-(-2) \pm \sqrt{(-2)^2 - 4 \times 1 \times (-3)}}{2 \times 1} = \frac{2 \pm \sqrt{16}}{2}$$

$\Rightarrow x = -1$ or $x = 3$

b) $x^2 - 2x + 1 = 0 \Rightarrow a = 1, b = -2, c = 1$

$$\Rightarrow x = \frac{-(-2) \pm \sqrt{(-2)^2 - 4 \times 1 \times 1}}{2 \times 1} = \frac{2 \pm \sqrt{0}}{2}$$

$\Rightarrow x = 1$

c) $x^2 - 2x - 7 = 0 \Rightarrow a = 1, b = -2, c = -7$

$$\Rightarrow x = \frac{-(-2) \pm \sqrt{(-2)^2 - 4 \times 1 \times (-7)}}{2 \times 1} = \frac{2 \pm \sqrt{32}}{2}$$

$\Rightarrow x = -1.83$ or $x = 3.83$ (3 s.f.)

d) $x^2 - 2x + 5 = 0 \Rightarrow a = 1, b = -2, c = 5$

$$\Rightarrow x = \frac{-(-2) \pm \sqrt{(-2)^2 - 4 \times 1 \times 5}}{2 \times 1} = \frac{2 \pm \sqrt{-16}}{2}$$

$\Rightarrow$ There are no real roots.

In **a)** the discriminant is a perfect square, which means that the equation will factorise to give **rational** roots.

In **b)** the discriminant is zero, which means that the equation will factorise to give **repeated rational** roots.

In **c)** the discriminant is not a perfect square, which means that when completing the square (using the formula) the roots are **irrational**.

In **d)** the discriminant is negative, which means that there are no real roots to the equation, since you cannot take the square root of a negative number.

Exploration 5.5 Solutions and the x-axis

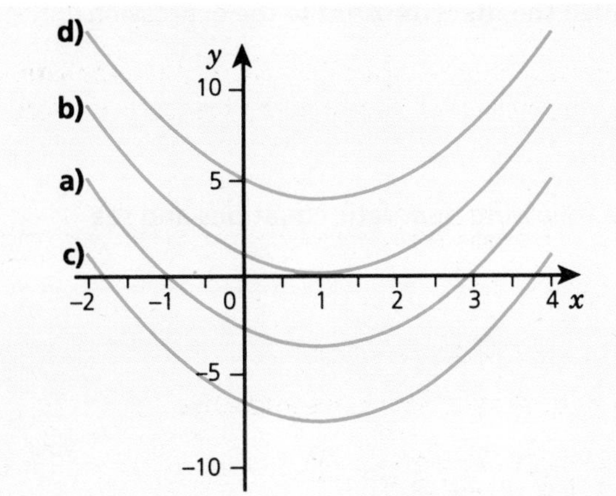

Sketch the graphs of $y = f(x)$ for each function in Exploration 5.4 and relate the solutions of the quadratic equations to where the curves intersect with the x-axis.

Taking the three graphs in turn:

a) $y = f(x)$ intersects with the x-axis at $x = -1$ and $x = 3$.
b) $y = f(x)$ touches with the x-axis at $x = 1$.
c) $y = f(x)$ intersects with the x-axis at $x = -1.83$ and $x = 3.83$.
d) $y = f(x)$ does **not** intersect with the x-axis.

Manipulating equations

As we found earlier, there are equations which may not look at first like quadratic equations, but a little algebraic manipulation can turn them into quadratic equations. The next example illustrates such a situation, where the quadratic function does not factorise and the formula is applied.

Example 5.13

Solve the equation $x = \dfrac{(x+2)}{(x+5)}$

Solution

$$x = \frac{(x+2)}{(x+5)}$$

Multiply both sides by $(x+5)$ $\Rightarrow$ $x(x+5) = x+2$
Multiply out the brackets $\Rightarrow$ $x^2 + 5x = x + 2$
Rearrange $\Rightarrow$ $x^2 + 4x - 2 = 0$

Use the quadratic formula $\Rightarrow x = \dfrac{-4 \pm \sqrt{16 - 4(1)(-2)}}{2} = \dfrac{-4 \pm \sqrt{24}}{2}$

$$\Rightarrow x = \frac{-4 \pm 2\sqrt{6}}{2} = -2 \pm \sqrt{6}$$

$\Rightarrow x = 0.449$ *or* $x = -4.45$ (3 s.f.)

EXERCISES

5.2 A

Solve the following equations, wherever possible, by completing the square, giving your answers correct to 3 s.f. where necessary. If there are no real roots, say so.

1	$x^2 - 8x + 12 = 0$	**2**	$x^2 + 17x - 18 = 0$
3	$x^2 + 8x = 84$	**4**	$x^2 + 12x = 253$
5	$3x^2 + 10x + 3 = 0$	**6**	$9x^2 + 13x = 10$
7	$11x^2 - x = 12$	**8**	$x^2 - 5x + 2 = 0$
9	$3x^2 + 4x = 5$	**10**	$7x^2 + 8x + 2 = 0$

Solve the following equations, wherever possible, using the quadratic formula, giving your answers correct to 3 s.f. where necessary. If there are no real roots, say so.

11	$x^2 + 3x - 4 = 0$	**12**	$3x^2 - 11x + 6 = 0$
13	$x^2 - 3x + 1 = 0$	**14**	$3x^2 - 13x = 10$
15	$8x^2 - 14x - 15 = 0$	**16**	$3x^2 - 5x = 7$
17	$10x^2 + 7x = 12$	**18**	$5x^2 + 4 = 7x$
19	$x^2 + 111x = 3400$	**20**	$12x^2 = x + 1740$

Rearrange the following as quadratic equations and solve.

21	$x + \frac{1}{2x} = 3$	**22**	$\frac{2x+3}{x+4} = \frac{x-5}{x+1}$
23	$\frac{x-2}{2x+1} = \frac{x-3}{x-1}$	**24**	$\frac{12}{x+2} - \frac{1}{x} = 2$
25	$\frac{1}{x+1} - \frac{1}{x-1} = 3$		

26 The diagram shows a regular pentagon, with sides that are each 1 unit long. Each diagonal is of length d units.

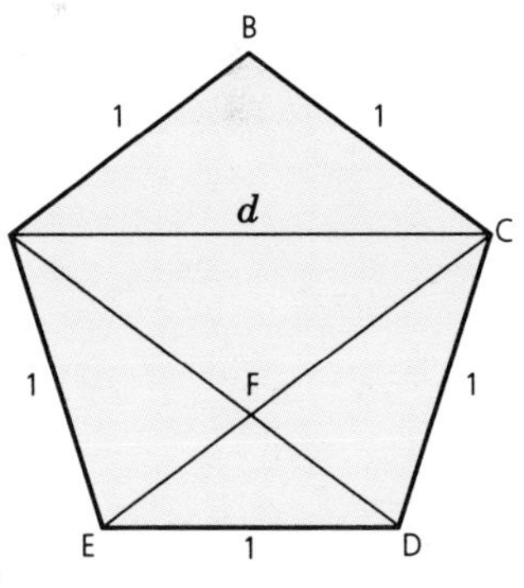

a) Explain why quadrilateral ABCF is a parallelogram.

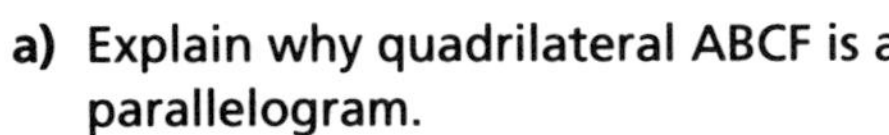

b) Show that the length of DF $= d - 1$.

c) Using similar triangles, formulate an equation involving d and show that it can be rearranged to give $d^2 - d - 1 = 0$.

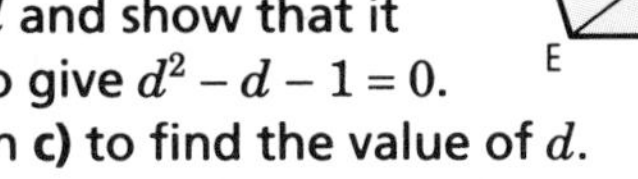

d) Solve the equation in **c)** to find the value of d.

EXERCISES

5.2 B

Solve the following equations, wherever possible, by completing the square, giving your answers correct to 3 s.f. where necessary. If there are no real roots, say so.

1	$x^2 + 12x + 20 = 0$	**2**	$x^2 - x - 20 = 0$
3	$x^2 - 11x = 180$	**4**	$14x - x^2 = 45$
5	$7x^2 - 8x = 18$	**6**	$8x^2 + 38x = -35$
7	$x^2 - 6x + 3 = 0$	**8**	$x^2 + 9x + 4 = 0$
9	$3x^2 - 7x = 1$	**10**	$4x^2 + 3x - 4 = 0$

Solve the following equations, wherever possible, using the quadratic formula, giving your answers correct to 3 s.f. where necessary. If there are no real roots, say so.

11 $x^2 + 8x + 7 = 0$ **12** $2x^2 + x - 4 = 0$

13 $3x^2 - 2x - 2 = 0$ **14** $5x^2 - 12x + 4 = 0$

15 $6x^2 + 11x = 10$ **16** $4x^2 + 5x + 2 = 0$

17 $7x^2 - 9x = 0$ **18** $12x^2 - 25x + 12 = 0$

19 $7x^2 - 26x = 1008$ **20** $x^2 = 5x + 6000$

Rearrange the following as quadratic equations and solve.

21 $3x = \frac{1}{x+1} + 2$ **22** $x + \frac{2x}{x+1} = 3$

23 $\frac{3}{x-6} - \frac{2}{x-5} = 1$ **24** $\frac{4}{x-1} - \frac{3}{x+7} = \frac{1}{18}$

25 $\frac{x-3}{x+2} = \frac{2x+1}{x-3}$

QUADRATIC INEQUALITIES

Exploration 5.6

Ranges of values

Sketch the graph of $y = x^2 + 3x - 4$.
Find the values of x for which:

a) $x^2 + 3x - 4 = 0$ **b)** $x^2 + 3x - 4 < 0$ **c)** $x^2 + 3x - 4 > 0$

Finding the values

a) Since $x^2 + 3x - 4 = (x + 4)(x - 1)$, solve the equation by factorising:

$$x^2 + 3x - 4 = 0$$
$$\Rightarrow \quad (x + 4)(x - 1) = 0$$
$$\Rightarrow \quad x = -4 \text{ or } x = 1$$

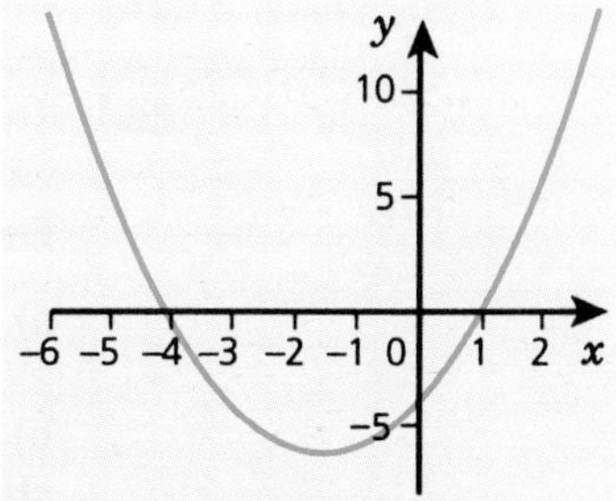

These values correspond to the x-coordinates of the points where the graph crosses the x-axis, i.e. where $y = 0$.

b) From the graph, $y < 0$ over the interval $-4 < x < 1$, so the solution to the inequality $x^2 + 3x - 4 < 0$ is a set of x-values.
The solution set is formally written $\{x: -4 < x < 1\}$.

c) Finally, notice that $y > 0$ on the graph when $x < -4$ or $x > 1$, so the solution to the inequality $x^2 + 3x - 4 > 0$ is the union of two disjoint sets of x-values.
The solution set is formally written

$$\{x: x < -4\} \cup \{x: x > 1\} \int \{x: x < -4 \text{ or } x > 1\}$$

Since $x^2 + 3x - 4 \equiv (x + 4)(x - 1)$ we can see that

$x^2 + 3x - 4 = 0$ when either factor is zero

$x^2 + 3x - 4 < 0$ when the factors have different signs

$x^2 + 3x - 4 > 0$ when the factors have the same signs.

This can be summarised as:

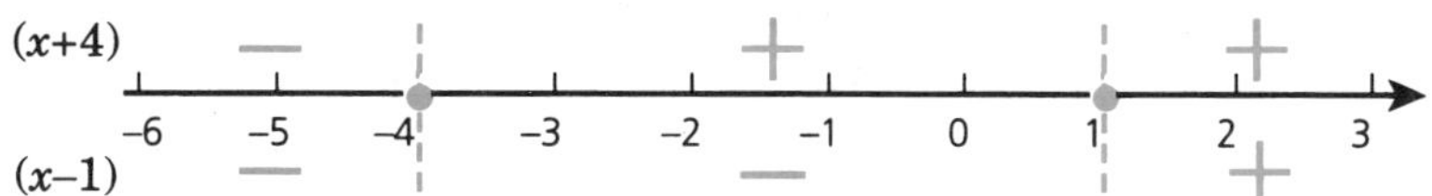

The signs are **different** for the set $\{x: -4 < x < 1\}$.

The signs are the **same** for the set $\{x: x < -4 \text{ or } x > 1\}$.

Example 5.14

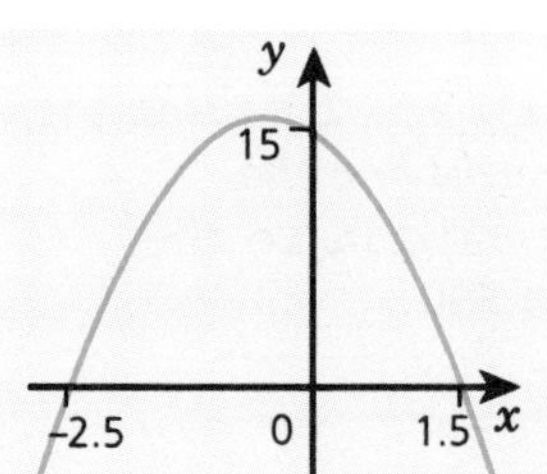

Solve $15 - 4x - 4x^2 \geq 0$.

Solution

Firstly factorise the quadratic function:

$15 - 4x - 4x^2 = (5 + 2x)(3 - 2x)$

$\Rightarrow 15 - 4x - 4x^2 = 0$ *when* $x = -2.5$ or $x = 1.5$

A sketch of the graph $y = 15 - 4x - 4x^2$ *leads to the conclusion:*

$15 - 4x - 4x^2 \geq 0 \Rightarrow -2.5 \leq x \leq 1.5$

Now $15 - 4x - 4x^2 > 0 \Rightarrow$ *the factors must have the **same** sign.*

*The signs are the **same** for the set* $\{x: -2.5 < x < 1.5\}$*, which means that the solution set for the inequality is* $\{x: -2.5 \leq x \leq 1.5\}$.

Functions which do not factorise

For quadratic functions which do *not* factorise, we can try solving the corresponding quadratic equation (giving the **boundary** values), then solving the inequality with reference to a sketch graph. Alternatively, a formal approach is possible, which may require completing the square.

Example 5.15

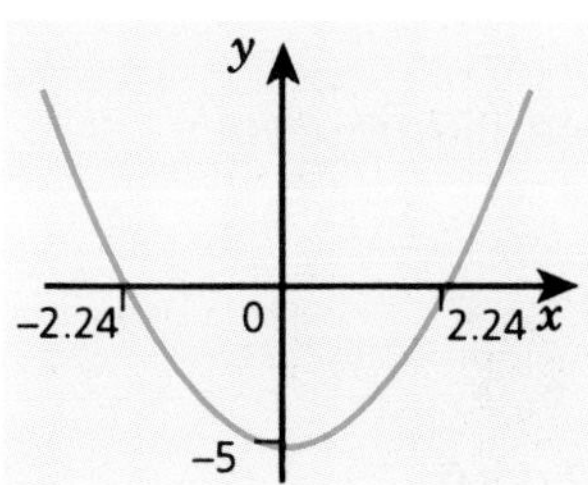

Solve the inequality $x^2 > 5$.

Solution

The equation $x^2 = 5 \Rightarrow x = \pm\sqrt{5}$.

Now $x^2 > 5 \Rightarrow x^2 - 5 > 0$ *and a sketch of the graph of* $y = x^2 - 5$ *leads to the conclusion that the* x*-values satisfying the inequality* $x^2 > 5$ *lie below* $-\sqrt{5}$ *or above* $+\sqrt{5}$.

The solution set is given by $\{x: x < -\sqrt{5} \text{ or } x > +\sqrt{5}\}$*, which is equivalent to* $\{x: |x| > \sqrt{5}\}$.

Example 5.16

Find the set of x-values satisfying $x^2 < 5x - 2$.

Solution

To solve the corresponding equation, use either the method of completing the square or the formula.

$$x^2 = 5x - 2$$
$$\Rightarrow \quad x^2 - 5x + 2 = 0$$
$$\Rightarrow \quad x = \frac{-(-5) \pm \sqrt{(-5)^2 - 4 \times 1 \times 2}}{2 \times 1} = \frac{5 \pm \sqrt{17}}{2}$$
$$\Rightarrow \quad x = 0.438 \text{ or } x = 4.56 \text{ (3 s.f.)}$$

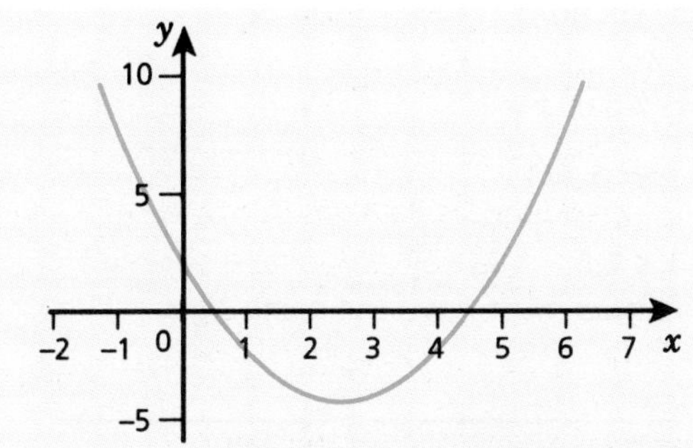

A sketch of the graph of $y = x^2 - 5x + 2$ *leads to the conclusion that the x-values satisfying the inequality lie between 0.438 and 4.56, i.e. the solution set is* $\{x: 0.438 < x < 4.56\}$.

A formal treatment following completion of the square gives:

$$x^2 - 5x + 2 < 0$$

Complete the square $\Rightarrow \left(x - \frac{5}{2}\right)^2 - \frac{17}{4} < 0$

Rearrange $\Rightarrow \left(x - \frac{5}{2}\right)^2 < \frac{17}{4}$

Take the square root of both sides $\Rightarrow \left|x - \frac{5}{2}\right| < \frac{\sqrt{17}}{2}$

$\Rightarrow x - \frac{5}{2} > -\frac{\sqrt{17}}{2}$ *and* $x - \frac{5}{2} < \frac{\sqrt{17}}{2}$

$\Rightarrow x > \frac{5-\sqrt{17}}{2}$ *and* $x < \frac{5+\sqrt{17}}{2}$

$\Rightarrow x > 0.438$ *and* $x < 4.56$ *(3 s.f.)*

EXERCISES

5.3 A

Solve the following inequalities, where possible. Illustrate your answer graphically.

1 $x^2 < \frac{1}{9}$

2 $x^2 \geq \frac{1}{9}$

3 $x^2 + \frac{1}{16} > 0$

4 $x^2 + \frac{1}{16} \leq 0$

5 $(x + 3)(x - 7) > 0$

6 $3x(3 - 2x) > 0$

7 $x^2 < 2x + 8$

8 $x^2 + 10 > 7x$

9 $x^2 - 6x - 7 < 0$

10 $x^2 - 6x + 11 \geq 0$

11 $2x^2 + 5x - 3 > 0$

12 $3x^2 - 4x - 4 < 0$

13 $x^2 + 4x + 4 > 0$

14 $x^2 - 4x + 4 > 0$

15 $x^2 + 4x - 4 > 0$

16 $x^2 - 4x - 4 > 0$

EXERCISES

5.3 B

Solve the following inequalities, where possible. Illustrate your answer graphically.

1 $x^2 - 7 < 0$

2 $x^2 + 7 < 0$

3 $x^2 - 13 > 0$

4 $x^2 + 13 > 0$

5 $(x - 2)(x + 5) \leq 0$

6 $(x + 1)(x + 4) \leq 0$

7 $x^2 > 2x$

8 $x^2 - 5x + 4 > 0$

9 $x^2 - 4x \geq -2$

10 $x^2 - 4x \leq 2$

11 $x^2 + x \leq 1$

12 $x^2 + 5x < 6$

13 $2x^2 + 5x - 4 < 0$

14 $3x^2 - 6x - 2 > 0$

15 $x^2 - 4x - 3 < 0$

16 $8x^2 - 12x - 1 \leq 0$

SIMULTANEOUS EQUATIONS

In Chapter 3, *Linear functions* we saw how to find the point of intersection of two straight line graphs by solving a pair of simultaneous equations. In this section we shall discover how to find the points of intersection of a curve, such as a parabola, and a straight line.

Exploration 5.7

Lines and curves

Draw the graph of $y = x^2 - x - 2$.

Find where the straight lines

a) $y = 3x - 2$ **b)** $y = 3x - 6$ **c)** $y = 3x - 10$

intersect with the parabola.

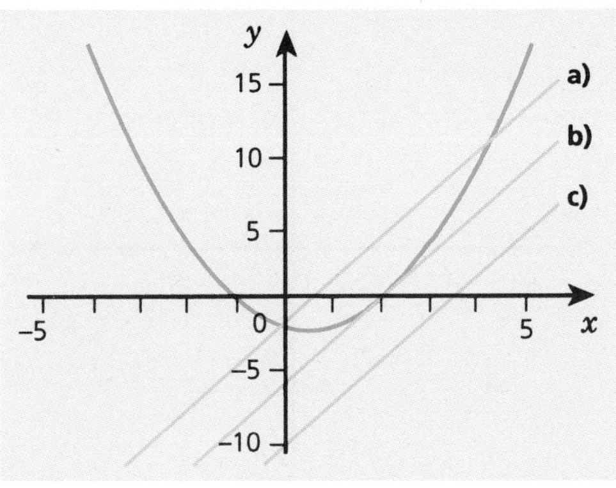

From the graphs:

a) $y = 3x - 2$ **crosses** the parabola at (0, –2) and (4, 10).
b) $y = 3x - 6$ **touches** the parabola at (2, 0).
c) $y = 3x - 10$ **does not** intersect with the parabola at all.

In **a)** the points of intersection represent the pairs of (x, y) coordinates which satisfy the equations $y = x^2 - x - 2$ and $y = 3x - 2$ simultaneously.

To find these pairs algebraically, proceed as follows:

$y = x^2 - x - 2$ (1)
$y = 3x - 2$ (2)

From (2) substitute for y in (1):

$3x - 2 = x^2 - x - 2$

Rearrange $\Rightarrow$ $x^2 - 4x = 0$
Factorise $\Rightarrow$ $x(x - 4) = 0$
$\Rightarrow$ $x = 0$ or $x = 4$
Substitute back in (2) $\Rightarrow$ $y = -2$ or $y = 10$

The line intersects the parabola at (0, –2) and (4, 10).

In **b)** the points of intersection represent the pairs of (x, y) coordinates which satisfy the equations $y = x^2 - x - 2$ and $y = 3x - 6$ simultaneously. By substitution this leads to the quadratic equation:

$x^2 - 4x + 4 = 0$

Factorise $\Rightarrow$ $(x - 2)^2 = 0$
$\Rightarrow$ $x = 2$
Substitute back in (2) $\Rightarrow$ $y = 0$

The line touches the parabola at (2, 0).

In **c)** the points of intersection represent the pairs of (x, y) coordinates which satisfy the equations $y = x^2 - x - 2$ and $y = 3x - 10$ simultaneously. By substitution this leads to the quadratic equation:

$$x^2 - 4x + 8 = 0$$

The quadratic function will not factorise. Using the discriminant from the formula, $b^2 - 4ac = (-4)^2 - 4 \times 1 \times 8 = -16$.

Therefore the quadratic equation has no real roots, which means that there are no points of intersection between the line and the parabola.

Example 5.17

A circle, centre (0, 0) and radius 5, is described by the equation $x^2 + y^2 = 25$.

Find the coordinates of the points where the line $y = 5 - 2x$ meets the circle.

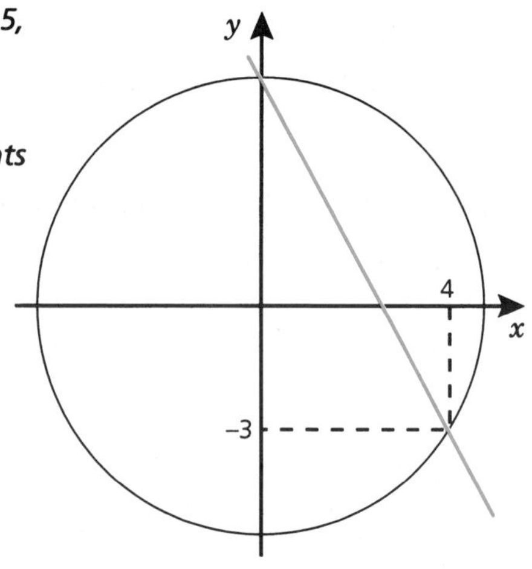

Solution

Solving two simultaneous equations:

$x^2 + y^2 = 25$ (1)
$y = 5 - 2x$ (2)

From (2) substitute for y in (1):

$$x^2 + (5 - 2x)^2 = 25$$

Multiply out the brackets	$\Rightarrow$	$x^2 + 25 - 20x + 4x^2 = 25$
Rearrange	$\Rightarrow$	$5x^2 - 20x = 0$
Factorise	$\Rightarrow$	$5x(x - 4) = 0$
	$\Rightarrow$	$x = 0$ *or* $x = 4$
Substitute back in (2)	$\Rightarrow$	$y = 5$ *or* $y = -3$

The line intersects the circle at (0, 5) and (4, –3).

EXERCISES

5.4 A

For the following pairs, one curve and one line, sketch them on the same axes and find the coordinates of their point(s) of intersection.

1 $y = x^2 - x - 6$
$y = x + 2$

2 $y = x^2 - 4x + 5$
$y = 2x - 4$

3 $y = 4x^2 + 3x + 1$
$5x + y = 6$

4 $y = 0.5x^2 + 3x - 10$
$x + 2y = 10$

Solve the following simultaneous equations, where possible.

5 $2x + 3y = 14$
$xy = 14$

6 $2x + 9y = 14$
$y(x + 1) = -2$

7 $4y = 3x + 7$
$8y = 3x^2 + 5$

8 $x^2 + y^2 = 125$
$x - y = 9$

9 $x + 2y = 7$
$x^2 + 2y^2 = 17$

10 $x + 4y = 14$
$x^2 - y^2 = 56$

EXERCISES

5.4 B

For the following pairs, one curve and one line, sketch them on the same axes and find, where possible, the coordinates of their point(s) of intersection.

1 $y = 2x^2 - 3x - 1$
$y = 3x + 7$

2 $y = (x + 2)^2$
$x + y = 4$

3 $y = x^2 + 3x - 1$
$2x + 3y = 20$

4 $y = x^2 + 3x - 1$
$2x + 3y = -20$

Solve the following simultaneous equations, where possible.

5 $3x + 4y = 15$
$2xy = 9$

6 $y = 2x + 3$
$y(5 - x) = 20$

7 $y = 2x - 2$
$x^2 + y^2 = 8$

8 $x + y = 6$
$4(x - y) = xy$

9 $9x^2 - 4y^2 = 576$
$3x - 2y = 12$

10 $x - y = 3$
$x^2 - y^2 = 69$

What is the significance of your answers to questions 9 and 10?

CONSOLIDATION EXERCISES FOR CHAPTER 5

1 Solve for x the equation $4x(x - 1) = 3$.
Hence, or otherwise, solve for y the equation $4(y + 3)(y + 2) = 3$.

2 Solve the equation $\frac{1}{x} + \frac{1}{3x - 2} = 2$.

3 Find the set of values of x for which $(2x + 1)^2 < 9(4 - x)$.

4 **a)** Factorise the quadratic function f(x), where $f(x) = 5x^2 - 7x - 6$.
b) Find the set of values of x for which $f(x) > 0$.

5 Find the range of values for which:
a) $x > x^2 - 12$
b) $x(2x - 3) < x^2 - 2$
c) $2x(x - 1) < 3 - x$
d) $x < x^2 - 6$
e) $x(8 - x) < 15$
f) $8x + 3 < 3x^2$
g) $2x^2 + 2 > 5x$

6 Eliminate x from the equations
$x - 3y = 5$, $5xy + 3x - 4y = 0$
to form a quadratic equation in y.

Hence show that the two given equations are satisfied simultaneously for only **one** pair of values (x, y).

7 **a)** Factorise the quadratic function $f(x) = 6x^2 - 17x + 7$.
b) Find the set of values for which $f(x) > 0$.
c) Solve the simultaneous equations
$y = 6x^2 - 17x + 7$ and
$y = 7 - x$.
d) Illustrate your answers to **b)** and **c)** graphically.

8 Given that $x + y = 1$ and $16x^2 + y^2 = 65$, calculate the two possible values of x.

9 Solve the simultaneous equations
$3x - y = 2$ and $3xy - 7x^2 = 20$.

10 Solve the simultaneous equations
$y = x + 2$ and $2x^2 - y^2 + 2x + 1 = 0$.

11 Solve the simultaneous equations
$3x + y = 7$ and $xy = -6$.

12 Solve the simultaneous equations
$x + y = 5$ and $\frac{1}{x} - \frac{1}{y} = \frac{1}{6}$.

Summary

Quadratic equations

- In general a quadratic equation $ax^2 + bx + c = 0$ may have
  i) two real roots
  ii) one (repeated) real root
  iii) no real roots.

  Two real roots occur if the parabola $y = ax^2 + bx + c$ intersects the x-axis twice.

  One real root occurs if the parabola $y = ax^2 + bx + c$ touches the x-axis,
  i.e. the x-axis is a tangent to the parabola.

  The parabola may not intersect or touch the x-axis, in which case there are no real roots.

- The **formula** for solving a quadratic equation $ax^2 + bx + c = 0$ is

  $$x = \frac{-b \pm \sqrt{b^2 - 4ac}}{2a}$$

  Real roots occur if the **discriminant** $b^2 - 4ac$ is positive or zero.

Quadratic inequalities

- To solve a quadratic inequality such as $ax^2 + bx + c > 0$:
  i) solve the quadratic equation $ax^2 + bx + c = 0$
  ii) from a sketch graph of $y = ax^2 + bx + c$ deduce the interval(s) for which the inequality holds.

PURE MATHS

6 Polynomial functions

In this chapter we:

- *explore the properties of polynomial functions and their graphs*
- *consider the addition, subtraction and multiplication of polynomials*
- *investigate the factorisation of polynomial functions using the factor theorem*
- *solve polynomial equations and inequalities by factorisation and graph sketching.*

EXPLORING POLYNOMIALS

CALCULATOR ACTIVITY

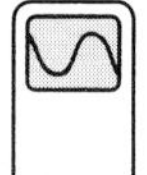

What is a polynomial?

You will need a graphics calculator. By setting the range of x- and y-values appropriately, sketch the following curves.

A $y = x^3$ B $y = x^3 - 10$
C $y = 10 - x^3$ D $y = x^3 - 6x^2 + 12x - 8$
E $y = 10 - 3x - 6x^2 - x^3$ F $y = 2x^3 - 2x^2 - 12x$
G $y = x^3 + x^2 - 8x - 12$ H $y = 12 + 8x - x^2 - x^3$
I $y = x^3 + 2x^2 + 3x - 2$ J $y = 1 + 2x^2 - x^3$

- Investigate the relationship between the graphs, where possible.
- Find where the graphs intersect the axes.
- Find the coordinates of any maximum or minimum points.
- Each graph has rotational symmetry about a certain point. Find the coordinates of that point.

Having explored various properties of the ten graphs you will have noticed several things.

- The highest power of x is x^3, called 'x-cubed', makes each curve the graph of a cubic function.
- The graphs intersect with the y-axis once, but intersect with the x-axis once, twice or three times.
- Some graphs have maximum and minimum points, some neither.

- Graphs with a positive x^3 term have the shape.

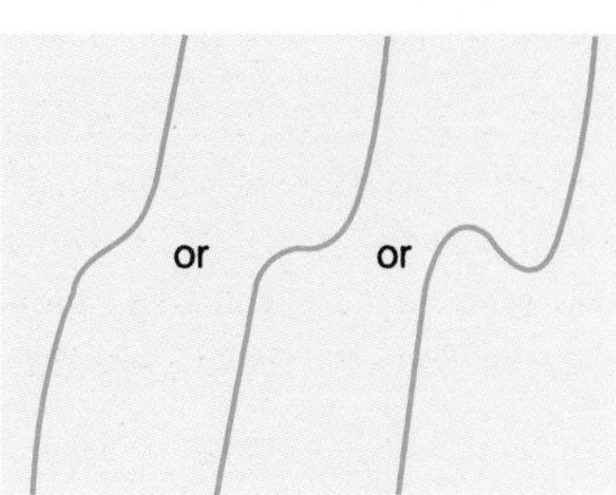

- Graphs with a negative x^3 term look like this curve.

The general form of a cubic function is given by:

$$f(x) = ax^3 + bx^2 + cx + d$$

Both **quadratic** and **cubic** functions are examples of **polynomial** functions, where each term is a multiple of a power of x. This multiple is called a **coefficient**.

Consider the cubic function:

$$f(x) = 4x^3 - 1.5x + 7$$

In its fullest form it would be written: $f(x) = 4x^3 + 0x^2 - 1.5x^1 + 7x^0$

- the coefficient of x^3 is 4
- the coefficient of x^2 is 0
- the coefficient of x^1, or x, is -1.5
- the constant term 7 may be regarded as the coefficient of x^0, which itself is 1 (you will meet this idea in Chapter 12, *Indices*).

Other polynomial functions which you will meet include:

quartic functions, e.g. $f(x) = x^4 - 3x^2 + x - 7$

quintic functions, e.g. $f(x) = 3x^5 + 2x^4 - x^2 + 8$

The **degree** of a polynomial function is the index of the highest power of x with a non-zero coefficient e.g. the cubic function $f(x) = x^3 - 2x^2 + 5x - 7$ is a polynomial of degree 3.

There is no upper limit to the degree of a polynomial. In further study you may meet polynomials with an infinite degree, but restrict your interest to terms in relatively small powers of x.

Example 6.1

Sketch graphs of the cubic functions: $y = -x^3 + 6x^2 + 15x + d$ where:

a) $d = -121$ ***b)*** $d = -100$ ***c)*** $d = -50$
d) $d = 8$ ***e)*** $d = 30$

In each case find:

i) *the coordinate of the maximum and minimum points,*

ii) *the x-value(s) where the curve intersects with the x-axis, i.e. where $f(x) = 0$.*

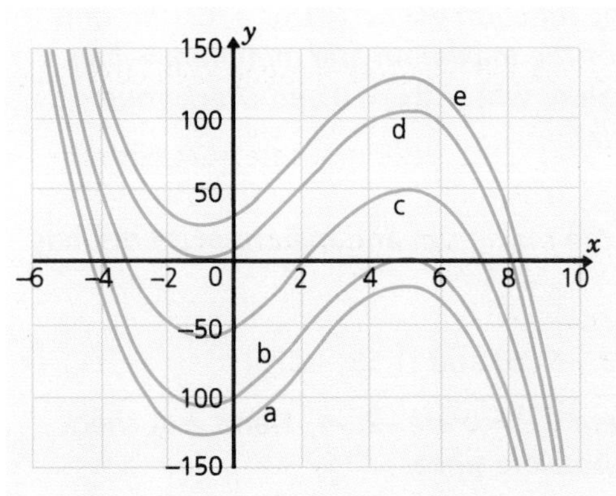

Solution

Drawing all five graphs on one grid gives this result.

All the minimum points occur at $x = -1$ and maximum points at $x = 5$. The x-values where the curves intersect with the x-axis are either exact or can be estimated by using trace and/or zoom on a calculator. Where necessary the results are given to 3 s.f.

Curve	Minimum point	Maximum point	x-values where $f(x) = 0$
$y = -x^3 + 6x^2 + 15x - 121$	(–1, –129)	(5, –21)	–4.25
$y = -x^3 + 6x^2 + 15x - 100$	(–1, –108)	(5, 0)	–4, 5
$y = -x^3 + 6x^2 + 15x - 46$	(–1, –54)	(5, 54)	–3.20, 2, 7.20
$y = -x^3 + 6x^2 + 15x + 8$	(–1, 0)	(5, 108)	–1, 8
$y = -x^3 + 6x^2 + 15x + 30$	(–1, 22)	(5, 130)	8.26

Example 6.2

Sketch graphs of the functions $y = x^3 + cx$ where:

a) $c = -9$ ***b)*** $c = -4$ ***c)*** $c = 0$ ***d)*** $c = 4$ ***e)*** $c = 9$.

In each case find:

i) *the coordinates of the maximum and minimum points,*

ii) *the x-value(s) where $f(x) = 0$.*

Solution

Drawing all five graphs on one grid gives this result.

The x-values where $f(x) = 0$ are exact and can be read off easily. The coordinates of the maximum and minimum points, where they exist, may be estimated by using trace and/or zoom. Where necessary results are given correct to 3 s.f.

Curve	Minimum point	Maximum point	x-values where $f(x) = 0$
$y = x^3 - 9x$	(–1.73, 9.98)	(1.73, –9.98)	–3, 0, 3
$y = x^3 - 4x$	(–1.15, 3.08)	(1.15, –3.08)	–2, 0, 2
$y = x^3$	–	–	0
$y = x^3 + 4x$	–	–	0
$y = x^3 + 9x$	–	–	0

Example 6.3

Sketch the graph of the quartic function $y = 3x^4 - 15x^2 + 12$ and find the coordinates of the maximum and minimum values. Also find the x-values where $f(x) = 0$ and check your answers by calculation.

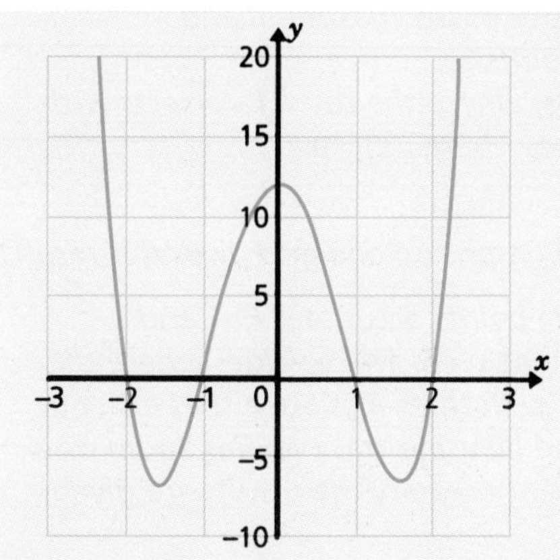

Solution

From the graph the maximum and minimum points may be read off:

maximum : (0, 12)

minimum : (–1.58, –6.75) and (1.58, –6.75)

The x-values where $f(x) = 0$ are –2, –1, 1 and 2. A check by calculation is possible since:

$$f(x) = 0 \Rightarrow 3x^4 - 15x^2 + 12 = 0$$

which is a quadratic equation in x^2.

Factorising: $3(x^4 - 5x^2 + 4) = 0$

$\Rightarrow \quad 3(x^2 - 1)(x^2 - 4) = 0$

$\Rightarrow \quad x^2 - 1 = 0$ *or* $x^2 - 4 = 0$

Now $x^2 - 1 = 0 \Rightarrow (x - 1)(x + 1) = 0 \Rightarrow x = \pm 1$

$x^2 - 4 = 0 \Rightarrow (x - 2)(x + 2) = 0 \Rightarrow x = \pm 2$

CALCULATOR ACTIVITY

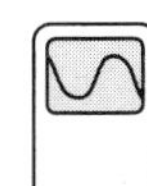

In the previous Calculator activity, the graph of the cubic function $f(x) = x^3 + 2x^2 + 3x - 2$ looks like this.

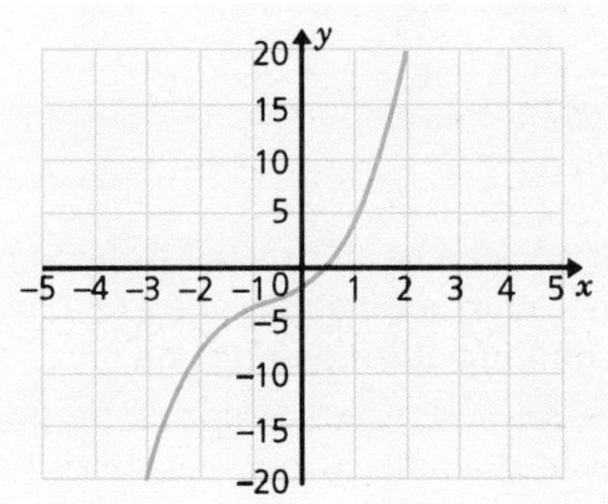

Since the graph is continuous, $f(0) = -2$ and $f(1) = 4$, there is a number a such that $0 < a < 1$ and $f(a) = 0$. From the sketch it looks as though a is about 0.5.

Plot the graph of $y = x^3 + 2x^2 + 3x - 2$ and zoom in to find the value of a correct to 3 d.p. Trace the curve to show that there is a change of sign of the function either side of your computed value. You should find that $a = 0.478$ (3 d.p.), i.e. $0.4775 < a < 0.4785$, since $f(0.4775) < 0$ and $f(0.4785) > 0$.

Using a calculator with a table facility or a spreadsheet will also enable you to apply a **decimal search** to find the value of a to the required degree of accuracy.

- Evaluate the function $f(x)$ between $x = 0$ and $x = 1$, in steps of 0.1: a change of sign in $f(x)$ occurs between $x = 0.4$ and $x = 0.5$.
- Evaluate the function $f(x)$ between $x = 0.4$ and $x = 0.5$, in steps of 0.01: a change of sign in $f(x)$ occurs between $x = 0.47$ and $x = 0.48$.
- Evaluate the function $f(x)$ between $x = 0.47$ and $x = 0.48$, in steps of 0.001: a change of sign in $f(x)$ occurs between $x = 0.477$ and $x = 0.478$.
- Evaluate the function $f(x)$ between $x = 0.477$ and $x = 0.478$, in steps of 0.0001: a change of sign in $f(x)$ occurs between $x = 0.4779$ and $x = 0.4780$.

Take the function $f(x) = 1 + 2x^2 - x^3$ from Calculator activity 6.1 and show that there is a number a such that $2 < a < 3$ and $f(a) = 0$. Use a graphical and/or tabular method to find the value of a correct to 3 d.p.

EXERCISES

For questions **1** to **10**:

a) Sketch the graph of the curve.
b) Find the coordinates of any maximum or minimum points.
c) Find x-values where $f(x) = 0$.

1 $y = x^3 - 7x + 6$

2 $y = x^3 + 4x + 1$

3 $y = 4x^2 - x^3$

4 $y = x^3 - 4x^2$

5 $y = 10 + 17x + 7x^2 - 2x^3$

6 $y = 7 - 3x + 4x^2 - x^3$

7 $y = x^4 - 3x^3 - 11x^2 + 3x + 10$

8 $y = x^4 - x^3 - 2x - 5$

9 $y = 3 - 2x^3 - x^4$

10 $y = 20 + 43x + 20x^2 - 4x^3 - x^4$

EXERCISES

6.1 B

For questions **1** to **10**:

a) Sketch the graph of the curve.
b) Find the coordinates of any maximum or minimum points.
c) Find x-values where $f(x) = 0$.

1	$y = x^3 - 9x$	**2**	$y = x^3 + 4x - 5$
3	$y = 5x^2 - 2x^3$	**4**	$y = 2x^3 - 5x^2$
5	$y = 2 - 11x + 7x^2 - x^3$	**6**	$y = 2 - 8x - 7x^2 - x^3$
7	$y = x^4 - 16x^2$	**8**	$y = x^4 - 6x^2 + 6x + 2$
9	$y = 60 - 89x + 38x^2 + 4x^3 - 3x^4$	**10**	$y = 1 - 6x - 2x^2 + 3x^3 - x^4$

ARITHMETIC OF POLYNOMIALS

Polynomials can be added together, or one can be subtracted from another. They can be multiplied or divided by a constant, or by another polynomial. We start this section by thinking about adding and subtracting polynomial functions.

Exploration 6.1 *Investigating addition and subtraction*

Two polynomial functions are given by $f(x) = x^3 - 3x + 7$ and $g(x) = x^2 + 4x - 3$.

- What does $f(x) + g(x)$ mean?
- What does $f(x) - g(x)$ mean?
- How can you interpret your answers graphically?

Adding and subtracting polynomials

The polynomials $f(x)$ and $g(x)$ may be added or subtracted by collecting like terms.

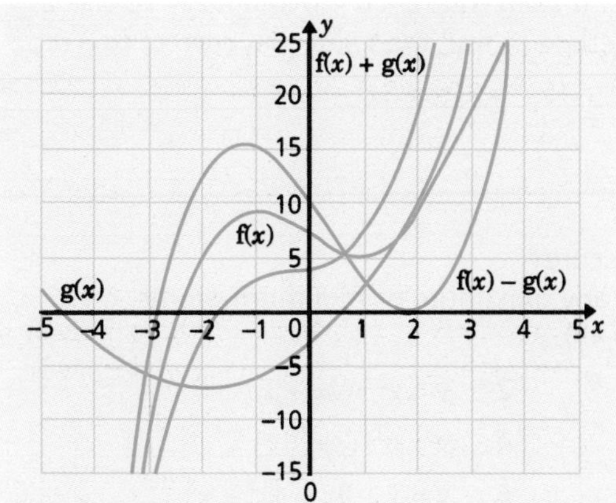

$$f(x) + g(x) = x^3 - 3x + 7 + x^2 + 4x - 3 = x^3 + x^2 + x + 4$$

$$f(x) - g(x) = x^3 - 3x + 7 - (x^2 + 4x - 3)$$
$$= x^3 - 3x + 7 - x^2 - 4x + 3 = x^3 - x^2 - 7x + 10$$

Note: The result of adding or subtracting a cubic function and a quadratic function is a cubic function. The results of the addition and subtraction may be illustrated graphically.

Note what happens when $f(x) + g(x) = 0$ and $f(x) - g(x) = 0$.

In general, if $f(x)$ is a polynomial of degree m and $g(x)$ is a polynomial of degree n, then $f(x) + g(x)$ and $f(x) - g(x)$ will both be polynomials. The degree of the resulting polynomial will be m or n, whichever

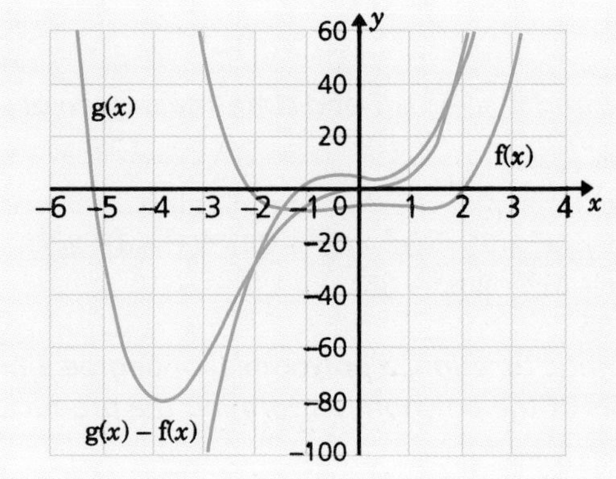

is higher. The only exception is when both f(x) and g(x) have the same degree, n, and the coefficient of x^n is zero in the result, e.g.

$$f(x) = x^4 - 3x^2 + x - 5$$

$$g(x) = x^4 + 5x^3 - x^2$$

$$\Rightarrow \quad g(x) - f(x) = 5x^3 + 2x^2 - x + 5$$

Here the result of subtracting two quartic functions is a cubic function.

Multiplying polynomials

Exploration 6.2 *Investigating multiplication*

How do you multiply two polynimials? This chart illustrates the multiplication of $5x + 3$ and $x^2 + 3x + 2$. Note that every term in one polynomial is multiplied by every term in the other.

	x^2	$3x$	2
$5x$	$5x^3$	$15x^2$	$10x$
3	$3x^2$	$9x$	6

- Write out $(5x + 3)(x^2 + 3x + 2)$ as a cubic polynomial.
- What different ways are there to find this product *without* the aid of a diagram?
- How is the degree of the resulting polynomial found from the degree of $5x + 3$ and $x^2 + 3x+ 2$?

Multiplication without a diagram

Summing the terms in the boxes gives:

$$(5x + 3)(x^2 + 3x + 2) = 5x^3 + 15x^2 + 3x^2 + 10x + 9x + 6 = 5x^3 + 18x^2 + 19x + 6$$

Alternatively, taking the two rows separately:

$$\begin{aligned}(5x + 3)(x^2 + 3x + 2) &= 5x(x^2 + 3x + 2) + 3(x^2 + 3x + 2)\\ &= 5x^3 + 15x^2 + 10x + 3x^2 + 9x + 6\\ &= 5x^3 + 18x^2 + 19x + 6\end{aligned}$$

You should find this a convenient way of multiplying out brackets for any pair of polynomials.

Note: The degree of the product is the sum of the degrees of the parts. In the example above, the **linear** function is of degree 1, the **quadratic** function is of degree 2 and the resulting **cubic** function is of degree 3.

Polynomials of degree greater than 3

The following examples illustrate how these ideas may be extended to multiplication of higher degree polynomials. Some of them include negative coeeficients, too, so watch out for the minus signs.

Example 6.4

Find the product of $x^2 - x + 5$ *and* $x^3 - 4x^2 + 7$ *and state the degree of the resulting polynomial.*

Solution

$(x^2 - x + 5)(x^3 - 4x^2 + 7) = x^2(x^3 - 4x^2 + 7) - x(x^3 - 4x^2 + 7) + 5(x^3 - 4x^2 + 7)$
$= x^5 - 4x^4 + 7x^2 - x^4 + 4x^3 - 7x + 5x^3 - 20x^2 + 35$
$= x^5 - 5x^4 + 9x^3 - 13x^2 - 7x + 35$

The resulting product is a quintic function, a polynomial of degree 5 (= 2 + 3), the sum of the degrees of the polynomials forming the product.

Example 6.5

Simplify $(2x - 1)(2x^3 + 5x^2 - x + 3) - (x^2 - x + 3)(4x^2 - 9)$

Solution

Firstly expand each pair of brackets:

$$\begin{aligned}(2x - 1)(2x^3 + 5x^2 - x + 3) &= 2x(2x^3 + 5x^2 - x + 3) - (2x^3 + 5x^2 - x + 3)\\ &= 4x^4 + 10x^3 - 2x^2 + 6x - 2x^3 - 5x^2 + x - 3\\ &= 4x^4 + 8x^3 - 7x^2 + 7x - 3\end{aligned}$$

$$\begin{aligned}(x^2 - x + 3)(4x^2 - 9) &= x^2(4x^2 - 9) - x(4x^2 - 9) + 3(4x^2 - 9)\\ &= 4x^4 - 9x^2 - 4x^3 + 9x + 12x^2 - 27\\ &= 4x^4 - 4x^3 + 3x^2 + 9x - 27\end{aligned}$$

Then subtract: $(4x^4 + 8x^3 - 7x^2 + 7x - 3) - (4x^4 - 4x^3 + 3x^2 + 9x - 27)$

$$\begin{aligned}&= 4x^4 + 8x^3 - 7x^2 + 7x - 3 - 4x^4 + 4x^3 - 3x^2 - 9x + 27\\ &= 12x^3 - 10x^2 - 2x + 24\\ &= 2(6x^3 - 5x^2 - x + 12)\end{aligned}$$

Note: Although each separate product gives a quartic function, the resulting polynomial is a cubic, since the coefficient of x^4 is 0.

When multiplying out brackets it is sometimes necessary to consider the product of three or more polynomials. In such cases a step-by-step approach is appropriate, as illustrated in the next two examples.

Example 6.6

Multiply out the brackets in $(t - 3)(2t^2 - t + 5)(t^3 + 7t^2 - 3t - 4)$.

Solution

Firstly expand the first pair of brackets.

$$\begin{aligned}(t - 3)(2t^2 - t + 5) &= t(2t^2 - t + 5) - 3(2t^2 - t + 5)\\ &= 2t^3 - t^2 + 5t - 6t^2 + 3t - 15\\ &= 2t^3 - 7t^2 + 8t - 15\end{aligned}$$

Then multiply this result by the remaining polynomial.

$$\begin{aligned}&(2t^3 - 7t^2 + 8t - 15)(t^3 + 7t^2 - 3t - 4)\\ &= 2t^3(t^3 + 7t^2 - 3t - 4) - 7t^2(t^3 + 7t^2 - 3t - 4) + 8t(t^3 + 7t^2 - 3t - 4)\\ &\quad - 15(t^3 + 7t^2 - 3t - 4)\\ &= 2t^6 + 14t^5 - 6t^4 - 8t^3 - 7t^5 - 49t^4 + 21t^3 + 28t^2 + 8t^4 + 56t^3 - 24t^2\\ &\quad - 32t - 15t^3 - 105t^2 + 45t + 60\\ &= 2t^6 + 7t^5 - 47t^4 + 54t^3 - 101t^2 + 13t + 60\end{aligned}$$

Note: The degree of the resulting polynomial is $6 = 1 + 2 + 3$, the sum of the degrees of the polynomials in the original product.

Example 6.7

Expand $(a + bx)^4$, where a and b are constants.

Solution

A useful strategy is to build up the product piecewise:

$$
\begin{aligned}
(a+bx)^2 &= (a+bx)(a+bx) = a(a+bx) + bx(a+bx)\\
&= a^2 + abx + abx + b^2x^2\\
&= a^2 + 2abx + b^2x^2\\
(a+bx)^3 &= (a+bx)(a+bx)^2 = (a+bx)(a^2+2abx+b^2x^2)\\
&= a(a^2+2abx+b^2x^2) + bx(a^2+2abx+b^2x^2)\\
&= a^3 + 2a^2bx + ab^2x^2 + a^2bx + 2ab^2x^2 + b^3x^3\\
&= a^3 + 3a^2bx + 3ab^2x^2 + b^3x^3\\
(a+bx)^4 &= (a+bx)(a+bx)^3 = (a+bx)(a^3+3a^2bx+3ab^2x^2+b^3x^3)\\
&= a(a^3+3a^2bx+3ab^2x^2+b^3x^3) + bx(a^3+3a^2bx+3ab^2x^2+b^3x^3)\\
&= a^4 + 3a^3bx + 3a^2b^2x^2 + ab^3x^3 + a^3bx + 3a^2b^2x^2 + 3ab^3x^3 + b^4x^4\\
&= a^4 + 4a^3bx + 6a^2b^2x^2 + 4ab^3x^3 + b^4x^4
\end{aligned}
$$

Note: This method could be extended to expand $(a + bx)^n$ for $n = 5, 6, 7, \ldots$.

An alternative approach in this example is possible by considering $(a + bx)^4$ as $[(a + bx)^2]^2$:

$$[(a+bx)^2]^2 = (a^2 + 2abx + b^2x^2)(a^2 + 2abx + b^2x^2)$$

which gives the identical result when expanded.

EXERCISES

1 For each pair of polynomials $f(x)$ and $g(x)$:

i) simplify $f(x) + g(x)$ and $f(x) - g(x)$,

ii) illustrate your answers graphically.

a) $f(x) = x^2 + 5x - 8$ $g(x) = x^2 - x + 5$

b) $f(x) = x^3 - 2x^2 + 3x - 1$ $g(x) = 2x^2 + 7x + 3$

c) $f(x) = 2x^3 + x^2 - 4x + 9$ $g(x) = 15 - x + 3x^2 - 2x^3$

d) $f(x) = 3x^4 + 5x^2 - 8$ $g(x) = x^3 - 9x$

e) $f(x) = 5x^3 - x + 8$ $g(x) = x^5 - x^3 + x^2 - 3$

2 Expand the following, in each case stating the degree of the resulting polynomial.

a) $(3x + 2)(x^2 + x + 5)$ **b)** $(5x - 1)(2x^2 - 3x + 1)$

c) $(x^2 + 3x + 5)(x^2 + 2x + 8)$ **d)** $(x^2 + 7x - 3)(2x^2 - x + 10)$

e) $(t - 5)(2t^3 + 4t^2 - t + 6)$ **f)** $(t^2 + 3t - 5)(t^3 - 5t^2 + 2t - 4)$

g) $(2x^3 - 5x^2 + 3x + 9)(5x^3 + 7x^2 - 2x - 5)$ **h)** $(2x - 1)(x^2 - x + 7)(x^2 + 5x + 3)$

3 Simplify these.

a) $(2x - 5)(3x^2 - 2x + 7) + (x^2 + x + 8)(7x - 1)$

b) $(3t^2 + 4t - 2)(4t^2 - 5t + 1)(2t^2 - 5)(6t^2 - 3t + 5)$

c) $(7 - 2x)(3 - 5x + 2x^2 - 4x^3) + (2 + 3x)^3$

d) $(x + 2)(4x - 3)(2x + 7) - (2x - 5)^2(3x + 2)$

e) $(2t - 3)(2t + 3)(t^2 + 5t - 6) + (3t - 1)(t - 4)^3$

4 Functions f, g and h are defined by:

$f(x) = 2x + 3 \qquad g(x) = 10 - x^2 \qquad h(x) = x^3 + 5x^2 - 3$

Simplify the following, and in each case:

i) state the degree of the resulting polynomial,

ii) give the coefficient of x^3.

a) $h(x) - g(x) + f(x)$ **b)** $3f(x) + 2g(x)$ **c)** $f(x) \times g(x)$

d) $g(x) \times h(x)$ **e)** $f(x) \times g(x) \times h(x)$ **f)** $f(x) \times h(x) - [g(x)]^2$

g) $[h(x)]^2 + [g(x)]^3$ **h)** $[f(x)]^4$

EXERCISES

6.2 B

1 For each pair of polynomials $f(x)$ and $g(x)$:

i) simplify $f(x) + g(x)$ and $f(x) - g(x)$,

ii) illustrate your answers graphically.

a) $f(x) = 2x^2 + 3x \qquad g(x) = 2x^2 - 4$

b) $f(x) = x^3 - 3x^2 \qquad g(x) = x^2 + x - 6$

c) $f(x) = 4 - 6x + 2x^3 \qquad g(x) = x^3 + x^2 - 4$

d) $f(x) = x^4 - 3x^3 + 6 \qquad g(x) = x^3 - 2x^2 + 4$

e) $f(x) = x^2 + 4x + 4 \qquad g(x) = x^5 - 3x^2 - 4x + 4$

2 Expand the following, in each case stating the degree of the resulting polynomial.

a) $(x^2 - 2)(x^2 + 1)$ **b)** $(x - 3)(2x^2 - 4x - 3)$

c) $(x^2 - 4x + 3)(2x^2 - x + 2)$ **d)** $(2 - 4x - 3x^3)(6x^2 - 3x + 2)$

e) $(3x^3 - 3x + 2)(4x^2 - 5)$ **f)** $(x^5 + 2x - 1)(x^3 - 3x^2 + 4x)$

g) $(3x^5 - 2x^2 + 3x + 1)(6x^3 - 5x^2 + 6)$ **h)** $(2x + 5)(x^3 - 3x^2 + 4)(2x^2 - x - 2)$

3 Simplify these.

a) $(2x - 1)(x^2 + 4) + (x - 1)(x^2 + 1)$

b) $(2x - 1)(x^2 + 4) - [x(2x^2 - x) + 4(x - 1)]$

c) $(x^2 + 3)(x - 2) + (x + 1)(x + 2)(x - 3)$

d) $(x^2 + 6)^2 - (x + 3)(x^2 + 2) - (x - 1)$

4 Functions f, g and h are defined by:

$f(x) = 3x - 2 \qquad g(x) = x^2 + 9 \qquad h(x) = x^3 + 3x^2$

Simplify the following, and in each case:

i) state the degree of the resulting polynomial,

ii) give the coefficient of x^2.

a) $h(x) + g(x)$ **b)** $f(x) - g(x)$ **c)** $3h(x) - 2f(x)$

d) $f(x) \times g(x)$ **e)** $g(x) \times 2h(x)$ **f)** $[h(x)]^2 - [g(x)]^3$

FACTORISING AND THE FACTOR THEOREM

Exploration 6.3

Finding where the graph crosses the x-axis

Consider the function $f(x) = 2x^2 - 11x + 12$

Sketch the graph of $y = f(x)$ by considering the intersection of the curve with the axes.

Roots of an equation

The graph crosses the y-axis when $x = 0 \Rightarrow y = 12$
$\Rightarrow$ point of intersection is (0, 12).

The graph crosses the x-axis when $y = 0$, i.e. $f(x) = 0$

$2x^2 - 11x + 12 = 0$
$(2x - 3)(x - 4) = 0$ (factorising)
$x = \frac{3}{2} \Rightarrow f(\frac{3}{2}) = 0$

or $x = 4 \Rightarrow f(4) = 0$

By symmetry we deduce that the line of symmetry is given by

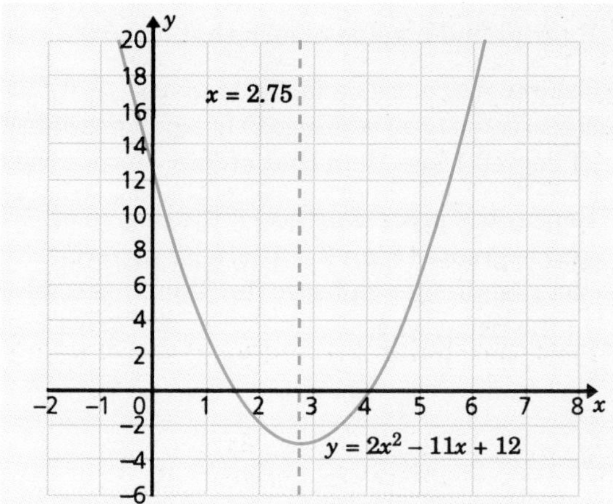

$x = 2.75$, since $\frac{1.5 + 4}{2} = 2.75$.

We now have sufficient information to sketch the graph of $y = f(x)$.

Reversing the above argument for finding the x-values where $f(x) = 0$:

$f(4) = 0 \Rightarrow x - 4$ is a factor of $f(x)$ and
$f(\frac{3}{2}) = 0 \Rightarrow 2x - 3$ is a factor of $f(x)$.

Looked at in this way, by locating where $f(x) = 0$ we can deduce a factor of the function $f(x)$.

This result is expressed formally as the factor theorem, which follows.

The factor theorem

For any polynomial function $f(x)$: $f(a) = 0 \Leftrightarrow x - a$ is a factor
or, more generally: $f(\frac{a}{b}) = 0 \Leftrightarrow bx - a$ is a factor

Exploration 6.4

The factor theorem

This property is especially useful when factorising polynomials of degree 3 or higher. In Exploration 6.2 we discovered that $(5x + 3)(x^2 + 3x + 2) = 5x^3 + 18x^2 + 19x + 6 = f(x)$.

- Since $5x+3$ is a factor of the cubic function, what is the value of $f(-\frac{3}{5})$?
- Can you factorise the quadratic function $x^2 + 3x + 2$?
- Find two further values of a such that $f(a) = 0$.
- Sketch the graph of $y = f(x)$.

Using the factor theorem

$f(-\frac{3}{5}) = f(-0.6) = 5 \times (-0.6)^3 + 18\,(-0.6)^2 + 19\,(-0.6) + 6 = 0$

i.e. $f(-\frac{3}{5}) = 0 \Leftrightarrow 5x + 3$ is a factor.

Factorising $x^2 + 3x + 2$ gives $(x + 1)(x + 2)$, from which we may deduce $f(-1)$ and $f(-2) = 0$

i.e. $f(-1) = 0 \Leftrightarrow x + 1$ is a factor
$f(-2) = 0 \Leftrightarrow x + 2$ is a factor.

A full factorisation of the cubic function as a product of linear factors gives:

$$f(x) = 5x^3 + 18x^2 + 19x + 6 = (5x + 3)(x + 1)(x + 2)$$

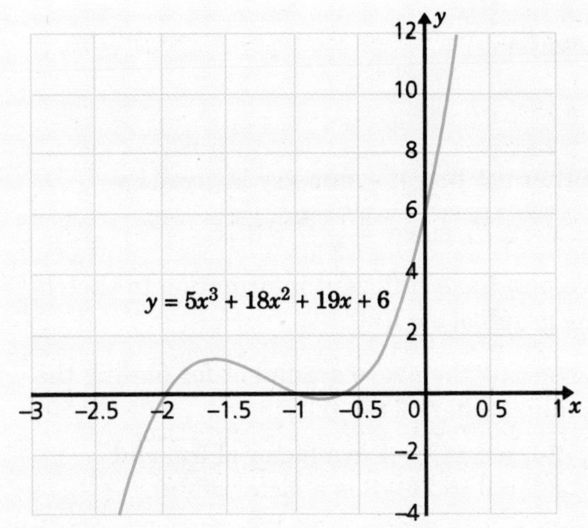

The graph of $y = f(x)$ crosses the y-axis at (0, 6) and the x-axis at (–2, 0), (–1, 0) and (–0.6, 0). This should give you enough information to sketch the graph.

When attempting to factorise a polynomial of degree 3 or higher a useful strategy is:

1 By trial find a value a such that $f(a) = 0$.

2 Take $(x - a)$ out as a factor to give $f(x) = (x - a)g(x)$ where $g(x)$ is a polynomial of degree 1 less than that of $f(x)$.

3 Repeat the process on $g(x)$ if the degree of $g(x)$ is greater than 2. Otherwise factorise the remaining quadratic function, if possible, in the usual way.

Example 6.8

Factorise as fully as possible $f(x) = x^3 + 6x^2 + 3x - 10$.

Solution

By inspection $f(1) = 1^3 + 6 \times 1^2 + 3 \times 1 - 10 = 0 \Rightarrow x - 1$ *is a factor.*
Thus $f(x) = (x - 1)g(x)$ *where* $g(x)$ *is a quadratic function to be found.*

$$x^3 + 6x^2 + 3x - 10 \equiv (x - 1)(px^2 + qx + r)$$

From our experience of multiplying out polynomials we have discovered that there is only one way to produce the term in x^3.

$$x^3 + 6x^2 + 3x - 10 \equiv (x - 1)(px^2 + qx + r)$$

i.e. $x^3 \equiv x \times px^2 = px^3 \Rightarrow p = 1$

Similarly there is just one way to produce the constant term:

$$x^3 + 6x^2 + 3x - 10 \equiv (x - 1)(px^2 + qx + r)$$

i.e. $-10 \equiv -1 \times r \Rightarrow r = 10$

So far we have:

$$x^3 + 6x^2 + 3x - 10 \equiv (x - 1)(x^2 + qx + 10)$$

The term in, say, $6x^2$ *may be produced from a pair of products as shown.*

i.e. $6x^2 \equiv x \times qx - 1 \times x^2$
$6x^2 \equiv qx^2 - x^2$
$q = 7$

This completes the picture.

$$\overbrace{x^3 + 6x^2 + 3x - 10}^{f(x)} \equiv (x - 1)\overbrace{(x^2 + 7x + 10)}^{g(x)}$$

The function $g(x)$ *may be factorised in the usual way, hence:*

$$x^3 + 6x^2 + 3x - 10 \equiv (x - 1)(x + 2)(x + 5)$$

Finally, check the two linear factors found from $g(x)$ *by using the factor theorem.*

$f(-2) = 0 \Leftrightarrow x + 2$ *is a factor*
$f(-5) = 0 \Leftrightarrow x + 5$ *is a factor*
$f(-2) = (-2)^3 + 6 \times (-2)^2 + 3 \times (-2) -10 = -8 + 24 - 6 - 10 = 0$
$f(-5) = (-5)^3 + 6 \times (-5)^2 + 3 \times (-5) - 10 = -125 + 150 - 15 - 10 = 0$

Example 6.9

For the function $f(x) = x^4 - 2x^3 - 5x^2 + 6x + 8$*:*

a) *find* $f(2)$ *and so deduce a linear factor,*
b) *express* $f(x)$ *as a product of a linear and a cubic function,*
c) *factorise the cubic function as a product a linear and a quadratic function,*
d) *deduce that the quadratic function cannot be factorised and so factorise* $f(x)$ *as fully as you can.*

Solution

a) $f(2) = 2^4 - 2 \times 2^3 - 5 \times 2^2 + 6 \times 2 + 8 = 16 - 16 - 20 + 12 + 8 = 0$
$\Rightarrow$ $(x - 2)$ *is a factor of* $f(x)$
$\Rightarrow$ $f(x) \equiv (x - 2)\, g(x)$

b) *To find the exact form of* $g(x)$*, let:*
$x^4 - 2x^3 - 5x^2 + 6x + 8 \equiv (x - 2)(px^3 + qx^2 + rx + s)$
Comparing terms in x^4*:* $x^4 \equiv x \times px^3 = px^4 \Rightarrow p = 1.$
Comparing constants: $8 = (-2) \times s \Rightarrow s = -4$
Thus so far: $x^4 - 2x^3 -5x^2 + 6x + 8 \equiv (x - 2)(x^3 + qx^2 + rx - 4)$
Comparing terms in x^3*:* $-2x^3 \equiv -2x^3 + qx^3 \Rightarrow q = 0$
Comparing terms in x*:* $6x \equiv -2rx - 4x \Rightarrow r = -5$
Therefore: $x^4 - 2x^3 - 5x^2 + 6x + 8 \equiv (x - 2)(x^3 - 5x - 4)$

c) *By trial,* $g(-1) = 0 \Rightarrow x + 1$ *is a factor of* $g(x)$
$\Rightarrow$ $g(x) \equiv (x + 1)h(x)$
To find the exact form of $h(x)$*, let:*
$x^3 - 5x - 4 \equiv (x + 1)(px^2 + qx + r)$
Comparing terms in x^3*:* $x^3 \equiv x \times px^2 \Rightarrow p = 1$
Comparing constants: $-4 \equiv 1 \times r \Rightarrow r = -4$
Thus, so far: $x^3 - 5x - 4 \equiv (x + 1)(x^2 + qx - 4)$
Comparing terms in x*:* $-5x \equiv 1 \times qx - 4x \Rightarrow q = -1$
Therefore: $(x^3 -5x - 4) \equiv (x + 1)(x^2 - x - 4)$

d) *The function* $h(x) \equiv x^2 - x^2 - 4$ *cannot be factorised, hence in fully factorised form:*
$x^4 -2x^3 - 5x^2 + 6x + 8 \equiv (x - 2)(x + 1)(x^2 - x - 4)$

EXERCISES

Factorise the following functions as fully as possible.

1 $x^3 - 2x^2 - 11x + 12$

2 $2x^3 - 7x^2 - 17x + 10$

3 $x^3 - 7x + 6$

4 $x^3 - 3x^2 + 4$

5 $3x^3 + 2x^2 - 19x + 6$

6 $4x^3 + 8x^2 - 11x - 15$

7 $x^3 - 7x^2 + 11x - 2$

8 $x^4 + 2x^3 - 13x^2 - 14x + 24$

9 $4x^4 + 4x^3 - 43x^2 - 22x + 21$

10 $2x^4 - 7x^3 - 6x^2 + 44x - 40$

11 $x^4 + 3x^3 + 3x^2 + 5x - 12$

12 $5x^4 + 13x^3 - 51x^2 - 82x + 40$

EXERCISES

Factorise the following functions as fully as possible.

1 $x^3 - 6x^2 + 11x - 6$

2 $x^3 - 3x^2 + 5x - 6$

3 $5x^3 - 10x^2 - 5x + 10$

4 $2x^3 - x^2 + 2x - 1$

5 $4x^3 + 12x^2 - x - 3$

6 $x^3 + 4x - 5$

7 $x^4 - x^3 - 13x^2 + x + 12$

8 $x^4 + x^3 + x + 1$

9 $3x^4 + 7x^3 - 8x^2 - 14x + 4$

10 $3x^4 - 8x^3 + 9x^2 - 32x - 12$

POLYNOMIAL EQUATIONS AND INEQUALITIES

Exploration 6.5

Cubic functions

A cubic function is given by:

$$f(x) = 2x^3 - 5x^2 - 4x + 3$$

- Without using a graph plotter, find the coordinates of the points of intersection of the curve $x = f(x)$ and the axes and hence **sketch** the curve.

The factor theorem and cubic functions – and above

Identifying where the curve intersects with the y-axis is straightforward:

$f(0) = 3 \Rightarrow$ curve crosses y-axis at $(0, 3)$

To find where the curve intersects with the x-axis we need to find x-values such that $f(x) = 0$, i.e. solve the equation:

$$2x^3 - 5x^2 - 4x + 3 = 0$$

The first step in solving the equation is to see if $f(x)$ will factorise. By trial we see that:

$$f(-1) = 2(-1)^3 - 5(-1)^2 - 4(-1) + 3 = -2 - 5 + 4 + 3 = 0$$

Using the factor theorem, $(x + 1)$ must be a factor of $f(x)$, i.e. $f(x) = (x + 1)\,g(x)$ where $g(x)$ is a quadratic function.

Setting this out as before, let $g(x) = px^2 + qx + r$, then:

$2x^3 - 5x^2 - 4x + 3 \equiv (x + 1)(px^2 + qx + r)$

comparing terms in x^3: $2x^3 \equiv x \times px^2 \Rightarrow p = 2$

comparing constants: $3 \equiv 1 \times r \Rightarrow r = 3$

Substituting values for p and r gives the equivalence:

$$2x^3 - 5x^2 - 4x + 3 \equiv (x + 1)(2x^2 + qx + 3)$$

comparing terms in, say, x:

$$-4x \equiv 3x + qx$$

$\Rightarrow \quad -4 = 3 + q$

$\Rightarrow \quad q = -7$

This gives:

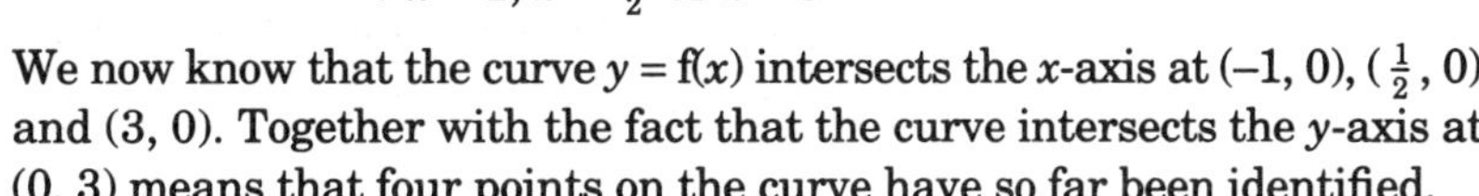

$$\overbrace{2x^3 - 5x^2 - 4x + 3}^{f(x)} \equiv (x + 1)\overbrace{(2x^2 - 7x + 3)}^{g(x)}$$

Factorising $g(x)$ gives the complete factorisation:

$$2x^3 - 5x^2 - 4x + 3 \equiv (x + 1)(2x - 1)(x - 3)$$

Thus $\quad f(x) = 0 \Rightarrow (x + 1)(2x - 1)(x - 3) = 0$

$\qquad \Rightarrow x = 1, x = \frac{1}{2}$ or $x = 3$

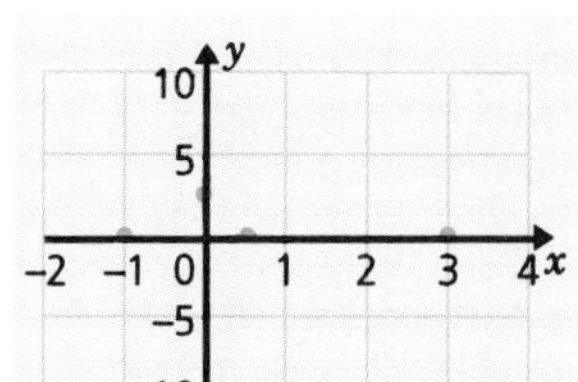

We now know that the curve $y = f(x)$ intersects the x-axis at $(-1, 0)$, $(\frac{1}{2}, 0)$ and $(3, 0)$. Together with the fact that the curve intersects the y-axis at $(0, 3)$ means that four points on the curve have so far been identified.

To enable a sketch to be made it just remains to find out how the curve behaves between $(-1, 0)$ and $(\frac{1}{2}, 0)$ and between $(\frac{1}{2}, 0)$ and $(3, 0)$, and then sketch a typical cubic curve to fit the data.

We already know that the graph intersects the y-axis at $(0, 3)$. By substituting in, say, $x = 1$ we can find the behaviour of the curve between $(\frac{1}{2}, 0)$ and $(3, 0)$:

$$f(1) = 2(1)^3 - 5(1)^2 - 4(1) + 3 = -4$$

which means that the curve also passes through $(1, -4)$.

A complete picture is obtained by continuing the graph below $x = -1$ and above $x = 3$. We know roughly where the maximum and minimum points occur and that there are no more.

From the graph we may also deduce that:
$f(x) > 0$ provided $-1 < x < 0.5$ or $x > 3$ and
$f(x) < 0$ provided $x < -1$ or $0.5 < x < 3$.

By a combination of factorising and graph sketching we can now solve both equations and inequalities for suitable polynomial functions.

Example 6.10

A quartic function is given as $f(x) = x^4 + 2x^3 - 13x^2 - 14x + 24$.

a) *Solve the equation* $f(x) = 0$.
b) *Sketch the graph of* $y = f(x)$.
c) *Solve the inequality* $f(x) \leq 0$.

Solution

a) *By inspection and use of the factor theorem we may deduce that:*

$$x^4 + 2x^3 - 13x^2 - 14x + 24 \equiv (x - 1)(x + 2)(x - 3)(x + 4)$$

since $f(1) = 0$, $f(-2) = 0$, $f(3) = 0$ *and* $f(-4) = 0$.

We may find any one of the linear factors and proceed as before. Alternatively, if we spot two factors then a short-cut is possible. Suppose we find that $f(1) = 0$ *and* $f(-2) = 0$, *then not only are* $(x - 1)$ *and* $(x + 2)$ *linear factors, but also* $(x - 1)(x + 2) = x^2 + x - 2$ *is a quadratic factor, i.e.*

$$x^4 + 2x^3 - 13x^2 - 14x + 24 \equiv (x^2 + x - 2)(px^2 + qx + r)$$

Comparing terms in x^4*:* $\quad x^4 \equiv x^2 \times px^2 \Rightarrow p = 1$

Comparing constants: $\quad 24 \equiv -2 \times r \Rightarrow r = -12$

Thus so far we have: $\quad x^4 + 2x^3 - 13x^2 - 14x + 24 \equiv (x^2 + x - 2)(x^2 + qx - 12)$

Comparing terms in x: $\quad -14x \equiv -12x - 2qx$

$$\Rightarrow \quad -14 = -12 - 2q$$
$$\Rightarrow \quad q = 1$$
$$\Rightarrow \quad x^4 + 2x^3 - 13x^2 - 14x + 24 \equiv (x^2 + x - 2)(x^2 + x - 12)$$

Since $x^2 + x - 12$ *factorises to give* $(x - 3)(x + 4)$, *in fully factorised form:*

$$f(x) = (x - 1)(x + 2)(x - 3)(x + 4)$$

Thus $f(x) = 0$ *when* $x = 1$, $x = -2$, $x = 3$ *or* $x = -4$.

b) *We now know that the curve* $y = f(x)$ *crosses the axes at (0, 24) and(–4, 0), (–2, 0), (1, 0), (3, 0). We can test some intermediate values of* $f(x)$ *to see how the graph behaves.*

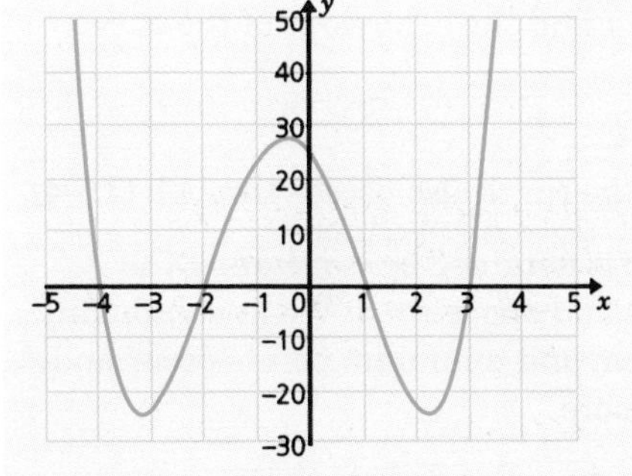

$$\begin{aligned} f(-3) &= (-3)^4 + 2(-3)^3 - 13(-3)^2 - 14(-3) + 24 \\ &= 81 - 54 - 117 + 42 + 24 \\ &= -24 \\ f(0) &= 24 \text{ (by inspection)} \\ f(2) &= 2^4 + 2(2)^3 - 13(2)^2 - 14(2) + 24 \\ &= 16 + 16 - 52 - 28 + 24 \\ &= -24 \end{aligned}$$

We now have sufficient information to complete a sketch.

c) *The curve lies below the* x*-axis for* $-4 < x < -2$ *and* $1 < x < 3$, *hence the solution of the inequality* $f(x) \le 0$ *is* $\{x : -4 \le x \le -2 \text{ or } 1 \le x \le 3\}$.

Example 6.11

The diagram shows the sketch of a cubic function $f(x)$, *which can be expressed as:*

$f(x) = k(x + l)^2(mx + n)$ *where* k, l, m *and* n *are constants.*

a) *By considering the coordinates of the points where the graph meets the axes, find the values of the constants.*

b) *Express* $f(x)$ *in the form*
$$a + bx + cx^2 + dx^3.$$

c) *Solve the inequality* $f(x) > 0$.

Solution

a) *Since the curve just touches the* x*-axis at* $x = -2$ *and crosses it at* $x = 2.5$, *the equation* $f(x) = 0$ *has a* ***repeated*** *root at* $x = -2$ *and a single root at* $x = 2.5$.

From the factor theorem:
$f(-2) = 0 \Rightarrow x + 2$ *is a factor*
$f(2.5) = 0 \Rightarrow 2x - 5$ *is a factor*
The repeated root, $x = -2$*, means that* $(x + 2)$ *is a repeated factor, therefore we can see that in the expression:*

$$f(x) = k(x + l)^2 (mx - n)$$

$l = 2$*,* $m = 2$ *and* $n = 5$*, thus:* $f(x) = k(x + 2)^2(2x - 5)$
Since the curve crosses the y*-axis at (0, 10) this fact may be used to determine the value of* k*.*

$$\begin{aligned} f(0) = k(0 + 2)^2(2 \times 0 - 5) &= 10 \\ \Rightarrow k \times 4 \times (-5) &= 10 \\ \Rightarrow \quad -20k &= 10 \\ \Rightarrow \quad k &= -0.5 \end{aligned}$$

b) *Expanding brackets:*

$$\begin{aligned} f(x) &= -0.5(x + 2)^2(2x - 5) \\ &= -0.5(x^2 + 4x + 4)(2x - 5) \\ &= -0.5(2x^3 + 3x^2 - 12x - 20) \\ &= 10 + 6x - 1.5x^2 - x^3 \end{aligned}$$

c) *The curve lies on or above the* x*-axis for* $x \le 2.5$*, so* $f(x) \ge 0$ *for* $x \le 2.5$*. However* $f(x) = 0$ *for* $x = -2$ *and* $x = 2.5$*, therefore the solution set for the inequality* $f(x) > 0$ *is* $\{x : x < 2.25; x \ne -2\}$*.*

Example 6.12

The function $f(x)$ *is given by:*

$$f(x) = x^3 + (3 - k)x^2 + (4 - 3k)x + 12.$$

a) *Evaluate* $f(-3)$ *and hence write down a linear factor of* $f(x)$*.*
b) *Show that the corresponding quadratic factor is* $x^2 - kx + 4$*.*
c) *Solve the equation* $f(x) = 0$ *for* ***i)*** $k = 5$*,* ***ii)*** $k = 4$*,* ***iii)*** $k = 6$*,* ***iv)*** $k = 3$*.*

Solution

a)
$$\begin{aligned} f(-3) &= (-3)^3 + (3 - k)(-3)^2 + (4 - 3k)(-3) + 12 \\ &= -27 + 9(3 - k) - 3(4 - 3k) + 12 \\ &= -27 + 27 - 9k - 12 + 9k + 12 \\ &= 0 \end{aligned}$$

$\Rightarrow (x + 3)$ *is a factor of* $f(x)$*.*

b) *Let* $x^3 + (3 - k)x^2 + (4 - 3k)x + 12 \equiv (x + 3)(px^2 + qx + r)$
Comparing terms in x^3*:* $x^3 \equiv x \times px^2 \Rightarrow p = 1$
Comparing constants: $12 \equiv 3 \times r \Rightarrow r = 4$
Thus so far we have: $x^3 + (3 - k)x^2 + (4 - 3k)x + 12 \equiv (x + 3)(x^2 + qx + 4)$
Comparing terms in x^2*:* $(3 - k)x^2 \equiv 3x^2 + qx^2$

$$\begin{aligned} &\Rightarrow 3 - k = 3 + q \\ &\Rightarrow q = -k \end{aligned}$$

$\therefore$ *The quadratic factor is* $x^2 - kx + 4$*.*

c) $f(x) = 0 \Rightarrow (x + 3)(x^2 - kx + 4) = 0 \Rightarrow x + 3 = 0$ *or* $x^2 - kx + 4 = 0$
Thus $x = -3$ *is a root, whichever value of* k *is chosen.*

i) $k = 5 \Rightarrow x^2 - 5x + 4 = 0 \Rightarrow (x - 1)(x - 4) = 0$
$\Rightarrow x = 1$ *or* $x = 4$

ii) $k = 4 \Rightarrow x^2 - 4x + 4 = 0 \Rightarrow (x - 2)^2 = 0$
$\Rightarrow x = 2$

iii) $k = 6 \Rightarrow x^2 - 6x + 4 = 0$ *which cannot be solved by factorising.*

Using the formula: $x = \dfrac{6 \pm \sqrt{36 - 4 \times 1 \times 4}}{2} = \dfrac{6 \pm \sqrt{20}}{2}$

$\Rightarrow x = 0.76$ *or* $x = 5.24$

iv) $k = 3 \Rightarrow x^2 - 3x + 4 = 0$, *which cannot be solved by factorising.*

Using the formula: $x = \dfrac{3 \pm \sqrt{9 - 4 \times 1 \times 4}}{2} = \dfrac{3 \pm \sqrt{-7}}{2}$ *which does not give a real root.*

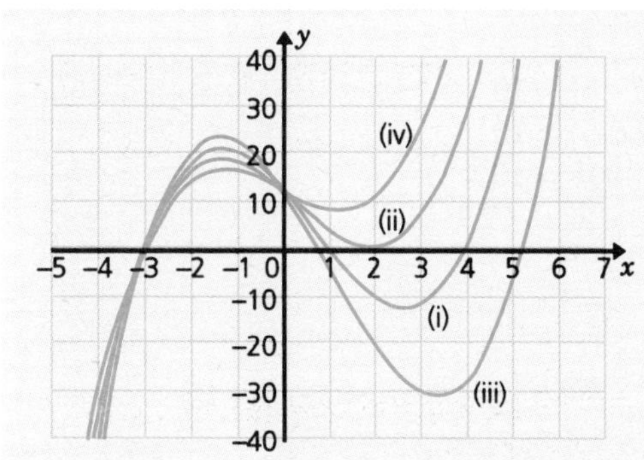

Note: Each value of k chosen above gives a different type of solution set for $f(x) = 0$:

i) $k = 5 \Rightarrow x = -3$, 1 or 4 (three distinct rational roots)
ii) $k = 4 \Rightarrow x = -3$ or 2 (three rational roots; two repeated; one distinct)
iii) $k = 6 \Rightarrow x = -3$, 0.76 or 5.24 (one rational root; two irrational roots)
iv) $k = 3 \Rightarrow x = -3$ (one rational root)

Inspecting where the various curves $y = f(x)$ intersect with the x-axis should help you appreciate the various possibilities.

EXERCISES

For each polynomial $f(x)$:

a) solve the equation $f(x) = 0$, **b)** sketch the graph of $y = f(x)$,
c) solve the inequalities $f(x) > 0$ and $f(x) \le 0$.

1 $f(x) = (x + 2)(2x - 1)(x - 5)$

2 $f(x) = (2x + 7)(x + 1)(2x - 1)(x - 3)$

3 $f(x) = (x - 2)^3(2x + 5)$

4 $f(x) = x^3 + x^2 - 10x + 8$

5 $f(x) = 2x^3 - 7x^2 + 9$

6 $f(x) = x^3 + x^2 - 8x - 12$

7 $f(x) = 5x^3 - 53x^2 + 155x - 75$

8 $f(x) = x^3 - 3x^2 - 7x + 6$

9 $f(x) = 2x^3 - x^2 - 11x - 12$

10 $f(x) = x^4 - 3x^3 - 11x^2 + 3x + 10$

11 $f(x) = x^4 + 4x^3 - 20x^2 - 43x - 20$

12 $f(x) = 4x^4 - 24x^3 + 29x^2 + 21x - 36$

13 $f(x) = 3x^4 - 4x^3 - 38x^2 + 89x - 60$

14 $f(x) = x^5 + 5x^4 + 2x^3 - 14x^2 - 3x + 9$

Find polynomial functions for which graphs are sketched below.

15

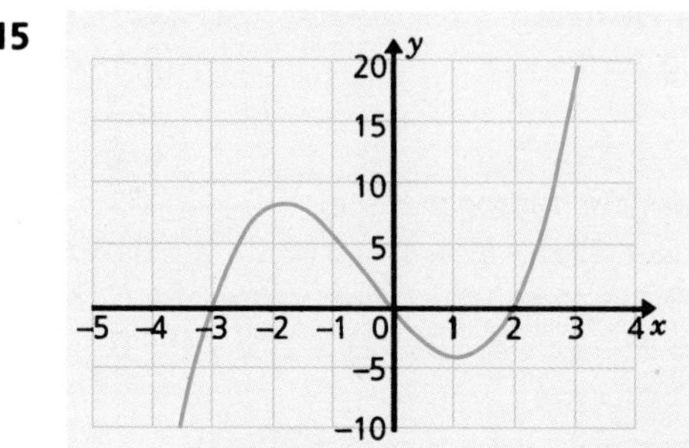

16

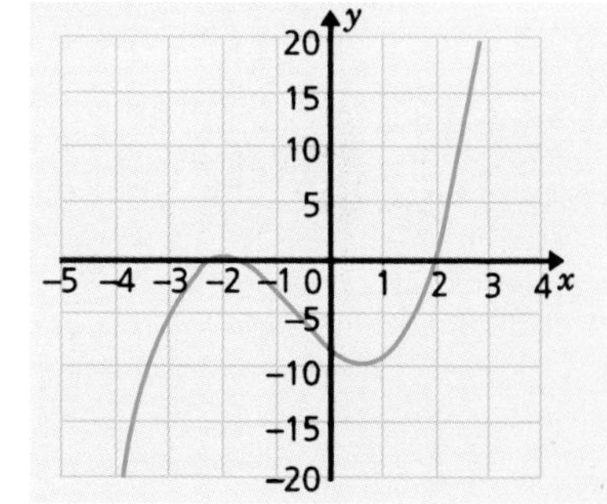

17

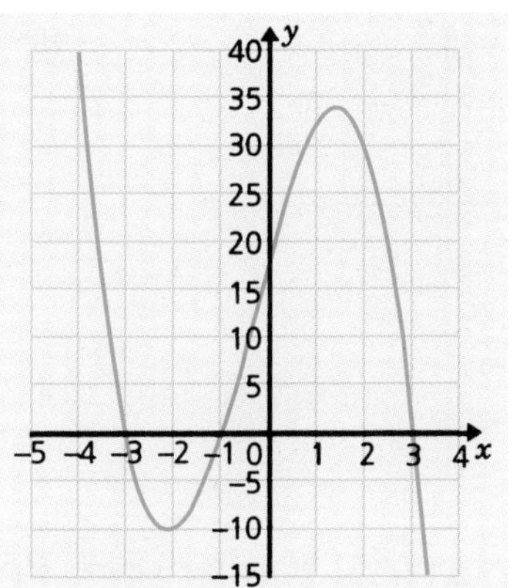

18

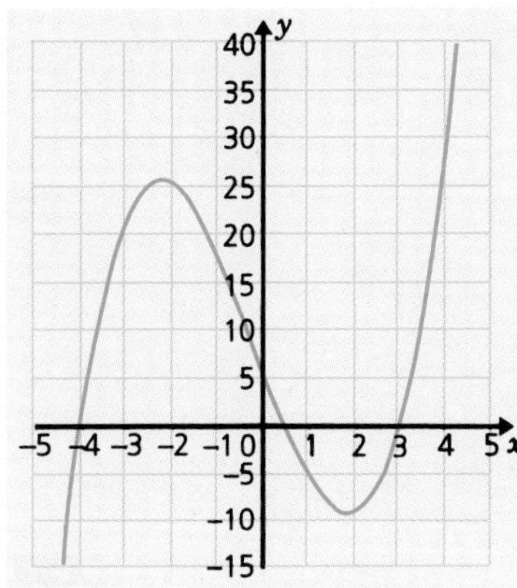

EXERCISES

6.4 B

For each polynomial $f(x)$:

a) solve the equation $f(x) = 0$, **b)** sketch the graph of $y = f(x)$,

c) solve the inequalities $f(x) < 0$ and $f(x) \geq 0$.

1 $f(x) = (x + 3)(2x + 3)(x - 2)$

2 $f(x) = (x - 1)(2x - 3)(2x + 5)$

3 $f(x) = (x + 4)(x^2 - 3x + 1)$

4 $f(x) = 3x^3 - 5x - 2$

5 $f(x) = 2x^3 - 5x - 6$

6 $f(x) = (x + 1)(2x - 3)^3$

7 $f(x) = x^4 + 10x^3 + 35x^2 + 50x + 24$

8 $f(x) = x^4 + 2x^3 - 10x^2 - 17x + 6$

9 $f(x) = 2x^4 + x^3 - 7x^2 - x + 5$

10 $f(x) = 6x + 6x^2 - 2x^3 - x^4 - 9$

11 $f(x) = x^5 + 3x^4 - 5x^3 - 15x^2 + 4x + 12$

12 $f(x) = 2x^5 + 5x^4 - 6x^3 - 19x^2 - 10x$

13 $f(x) = 2x^4 - 35x^3 + 225x^2 - 625x + 625$

14 $f(x) = 3 - 7x + 5x^2 - x^3$

Find polynomial functions for which graphs are sketched below.

15

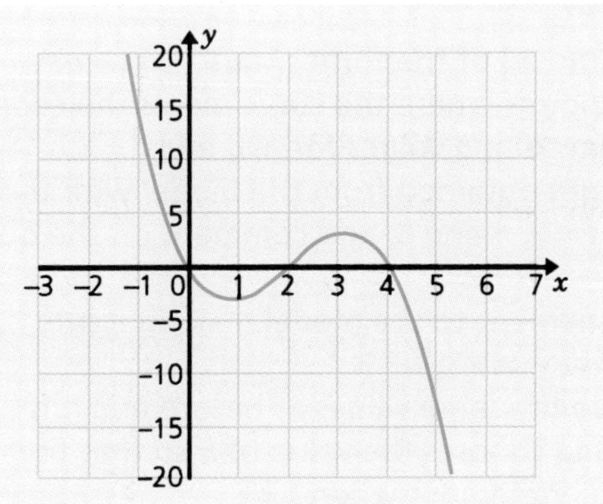

16

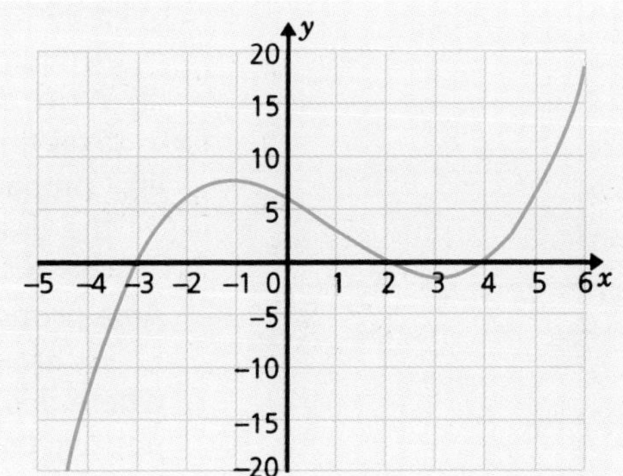

17

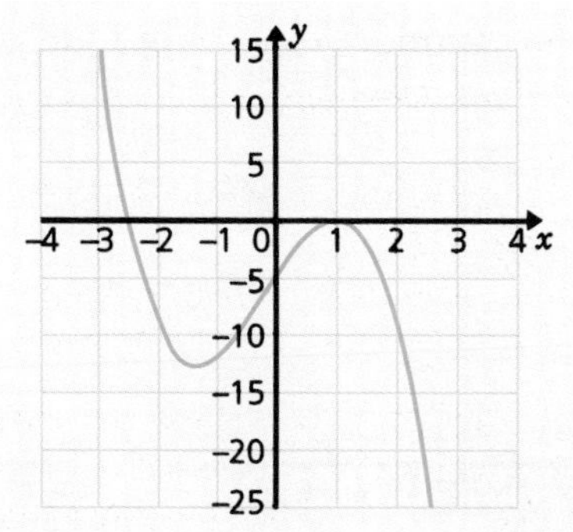

18

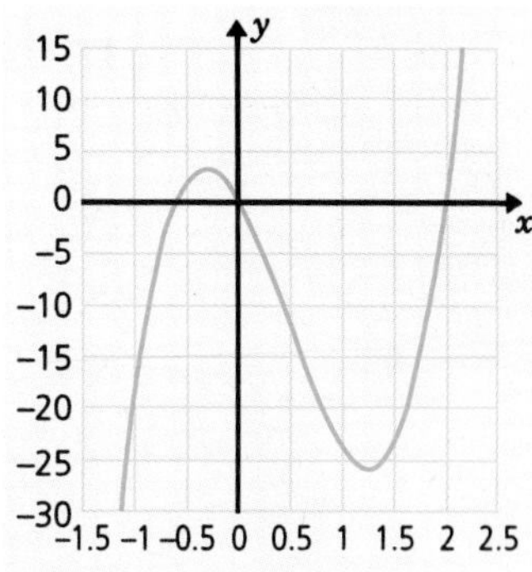

CONSOLIDATION EXERCISES FOR CHAPTER 6

1 $f(x) = x^4 + 3x^3 - 8x^2 - 22x - 24$

a) Verify that $x = 3$ and $x = -4$ are solutions of $f(x) = 0$.
b) Show that they are the only real solutions of $f(x) = 0$ and explain why.
c) Sketch the graph of $f(x)$.

2 Find a cubic polynomial, $f(x)$, such that $f(x) = 0$ at $x = 1$, $x = 2$ and $x = -1$ and the y-intercept is 10.

3 **a)** Given that $x = \frac{1}{2}$ is a root of $f(x) = 8x^3 + 4x^2 + kx + 15$, find the value of k.
b) Find all the roots of $f(x) = 0$.
c) Sketch a graph of $y = f(x)$, clearly indicating where the graph intersects the axes.

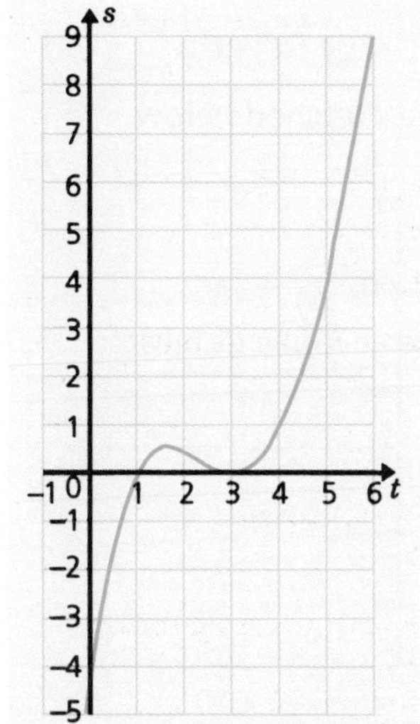

4 **a)** Write down an example of a polynomial in x of order 4.
b) In an experiment, Ama measures the value of s at different times, t. Her results are shown as the curve on the graph to the left.
Ama believes that it is possible to model s as a polynomial in t.
i) Explain why it is reasonable to think that the order of such a polynomial might be 3.
Ama proposes a model of the form $s = a(t - p)(t - q)^2$.
ii) Write down the points where the curve meets the coordinate axes and use them to find values for p, q and a.
iii) Compare the values obtained from the model with those on the graph when $t = 2$, 4 and 5, and comment on the quality of the model.
Ama proposes a refinement to the model making it into
$s = a(t - p)(t - q)^2\,(1 - ht)$.
where a, p and t have the same values as before and h is a small positive constant. Ama chooses the value of h so that the model and the graph are in agreement when $t = 5$.
iv) Find the value of h.

(MEI Specimen Paper Pure 1, 1994)

5 The function f is given by $f(x) = x^3 - 3x^2 - 2x + 6$.
a) Use the factor theorem to show that $(x - 3)$ is a factor of $f(x)$.
b) Write $f(x)$ in the form $(x - 3)(ax^2 + bx + c)$, giving the values a, b and c.
c) Hence solve $f(x) = 0$.

d) Using your solutions to $f(x) = 0$, write down the solutions of the equation $f(x + 1) = 0$.

(SMP 16–19 Specimen Paper 1, 1994)

6 Given that $(x - 2)$ and $(x + 2)$ are each factors of $x^3 + ax^2 + bx - 4$, find the values of a and b.
For these values of a and b, find the other linear factor of $x^3 + ax^2 + bx - 4$.

(UCLES (Modular) Specimen Paper, Question 5 1994)

7 Show that $(x - 2)$ is a factor of $x^3 - 9x^2 + 26x - 24$.
Find the set of values of x for which $x^3 - 9x^2 + 26x - 24 < 0$.

(AEB Specimen Paper 1, Question 7 1994)

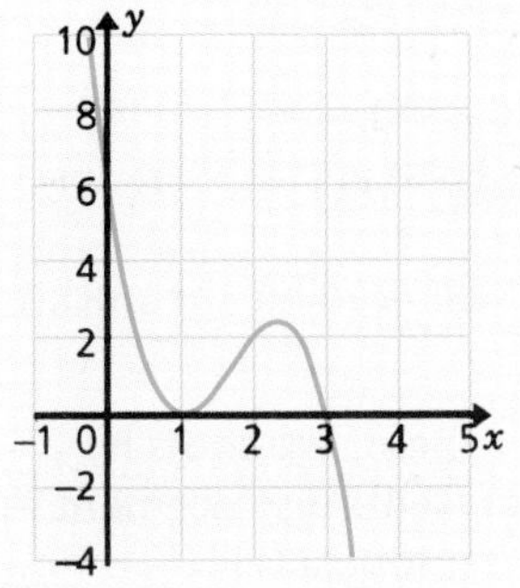

8 The graph shown left shows a curve drawn through a set of points which represents the results of several experiments. It is thought that the relationship between y and x may be modelled by a simple polynomial equation.

a) Explain, with reference to the number of turning points, why the polynomial is not a quadratic.

b) A possible form of the equation is $y = a(x - b)^2(x - c)$.
Write down the values of b and c. Substituting $x = 0$, calculate the value of a.

c) Find the values of the integers p, q, r and s when the equation is written in the form $y = px^3 + qx^2 + rx + s$.

d) A further experiment produces the result that $y = 1.9$ when $x = 2$.
State, with reasons, whether you think this supports the model.

(MEI Pure 1, January 1995)

Summary

Polynomial functions

- The **degree** of a polynomial function $f(x)$ is the index of the highest power of x with non-zero coefficient.
- A polynomial of degree 3 is a **cubic** function.
- The graph of a polynomial function of degree n has at most $n - 1$ maximum or minimum points.

Factorising polynomials

- Polynomial functions may be factorised using the factor theorem, which states:

  for any polynomial function $f(x)$, $f(a) = 0 \Leftrightarrow x - a$ is a factor or, more generally, $f(\frac{a}{b}) = 0 \Leftrightarrow bx - a$ is a factor.

Solving polynomial equations and inequalities

- Polynomial equations may be solved by factorising or decimal search.
- Polynomial inequalities may be solved by considering the intersection of the graph of $y = f(x)$ with the x-axis.

Differentiation 1

- *The rate of change of a function is an important concept in mathematics. The size of a rate of change tells us how quickly a quantity is changing.*

In this chapter we:

- *study differentiation, which is the mathematical representation of rate of change,*
- *use differentiation to find stationary values of functions and solve optimisation problems.*

GRADIENTS AND GRADIENT FUNCTIONS

Exploration 7.1

Alan and Bill were practising shots in ice-hockey. Alan hit the puck to Bill and Bill hit it back again.

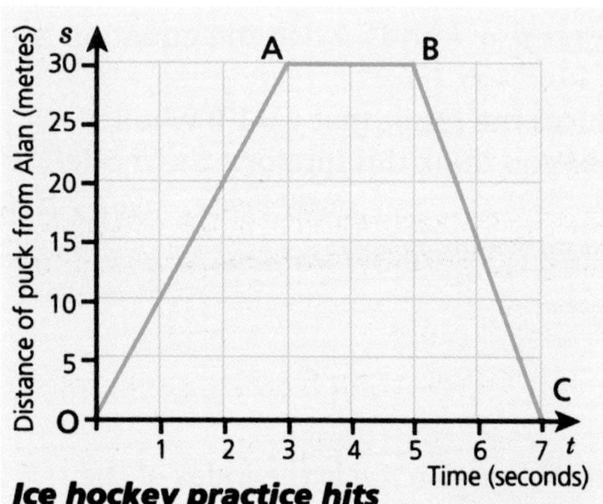

Ice hockey practice hits

A displacement–time graph for these two shots is shown on the left (where s represents displacement in metres and t represents time in seconds).

- Who hit the puck with the greater velocity?
- How do you find the velocity of the puck for each hit?
- What happens to the puck between $t = 3$ and $t = 5$?
- What difference does the direction of slope of the lines make to:
  a) the velocity, **b)** the speed?
- What are the displacement–time equations for the three lines?

The velocity of the puck at any instant in time is given by the **gradient** of the displacement–time graph. Alan hit the puck to Bill with a velocity of $10\,\text{m}\,\text{s}^{-1}$ and Bill hit it back again with velocity $15\,\text{m}\,\text{s}^{-1}$. The gradient of the line for Alan's hit is **positive**, but the gradient of the line for Bill's hit is **negative**, indicating that the shots are hit in opposite directions. The gradient of the line between $t = 3$ and $t = 5$ is **zero**; this means that the puck was at rest. It had a velocity of $0\,\text{m}\,\text{s}^{-1}$.

The gradient of a line measures **rate of change**. **Velocity** is the rate of change of **displacement** with respect to **time**. In this example all three velocities are **constant**, indicated by the straight-line graphs.

The displacement–time equations for the three lines are:

OA: $s = 10t$	AB: $s = 30$	BC: $s = 105 - 15t$
gradient $= 10$	gradient $= 0$	gradient $= -15$

You are more likely to meet a situation like the one in the next exploration.

Exploration 7.2

Find the velocity

Chris is a keen cricketer. He practises his fielding by throwing a cricket ball vertically upwards into the air and then catching it.

The displacement–time equation for one practice throw is given by $h = 20t - 5t^2$. By substituting for various values of t between $t = 0$ and $t = 4$ we can draw up the displacement–time graph shown below.

- What was the average velocity between $t = 1$ and $t = 2$?
- What was the average velocity between $t = 1$ and $t = 1.1$?
- What was the average velocity between $t = 1$ and $t = 1.01$?
- What do you think the instantaneous velocity of the ball is at $t = 1$?

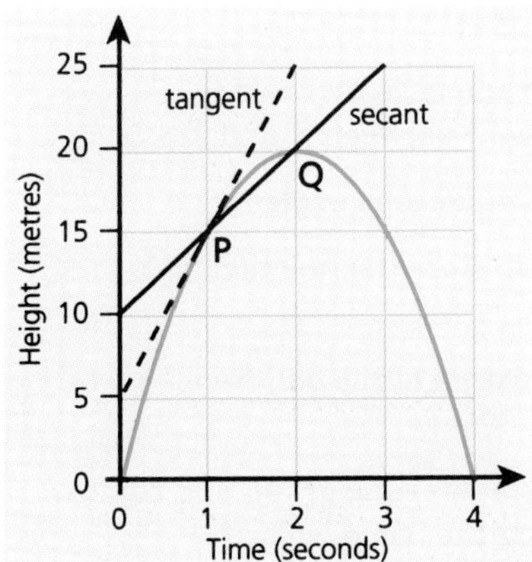

Cricket ball throw

Instantaneous velocity

From the graph we can see that the average velocity of the cricket ball between $t = 1$ and $t = 2$ is given by the gradient of the **secant** passing through points P and Q: i.e. $\frac{20-15}{2-1} = 5\ \text{m s}^{-1}$.

Similarly the average velocity of the cricket ball between $t = 1$ and $t = 1.1$ is given by the gradient: $\frac{15.95-15}{1.1-1} = 9.5\ \text{m s}^{-1}$

and the average velocity of the cricket ball between $t = 1$ and $t = 1.01$ is given by the gradient: $\frac{15.0995-15}{1.01-1} = 9.95\ \text{m s}^{-1}$.

If we continue to reduce the time interval by a factor of 10 each time we find the sequence of average velocities:

between $t = 1$ and $t = 2$	average velocity $= 5\,\text{m s}^{-1}$
between $t = 1$ and $t = 1.1$	average velocity $= 9.5\,\text{m s}^{-1}$
between $t = 1$ and $t = 1.01$	average velocity $= 9.95\,\text{m s}^{-1}$
between $t = 1$ and $t = 1.001$	average velocity $= 9.995\,\text{m s}^{-1}$
between $t = 1$ and $t = 1.0001$	average velocity $= 9.9995\,\text{m s}^{-1}$
between $t = 1$ and $t = 1.000\,01$	average velocity $= 9.999\,95\,\text{m s}^{-1}$.

The **instantaneous velocity** (or just the **velocity**) of the ball at $t = 1$ is defined as the gradient of the **tangent** to the displacement–time curve at the point (1, 15). Numerically, this is the limit of the sequence of average velocities between $t = 1$ and $t = 1 + h$ as $h \to 0$.

By inspection, the limit of this sequence would seem to be 10, i.e. the velocity of the ball at $t = 1$ is $10\,\mathrm{m\,s^{-1}}$.

By repeating this limiting process for various values of t we can build up a set of velocities, each corresponding to a different value of t.

CALCULATOR ACTIVITY

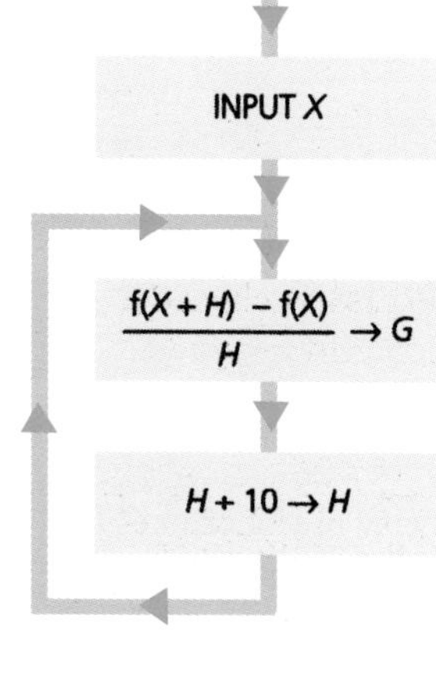

The flowchart outlines a program to find the gradient of a function as the limit of the gradient of secants.

Store X contains the value where the gradient is required.
Store G contains the gradient of the secant between X and $X + H$.

Record values of G until a limit of the sequence is evident.

Before using the program you will need to store the function $20X - 5X^2$ as function f in the function memory.

Use the computer program to find the velocity of the ball from $t = 0$ to $t = 4$ in steps of 0.2.

Using suitable scales, copy the displacement–time graph and superimpose a velocity–time graph by plotting the (time, velocity) coordinates you have found.

You should obtain a graph like this.

The dotted line represents the velocity–time graph. By inspection we can find its equation. The graph is a straight line passing through (0, 20) with gradient −10. From this we deduce two important results.

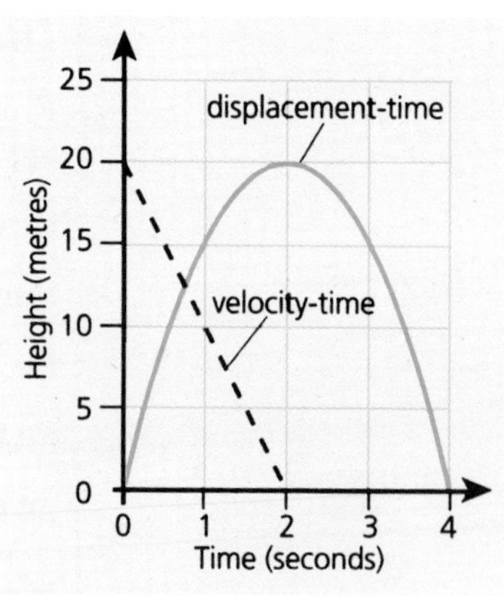

Cricket ball throw

Displacement–time equation

$$h = 20t - 5t^2$$

Velocity–time equation

$$v = 20 - 10t$$

So, given a displacement–time equation, we can find the corresponding velocity–time equation by plotting gradients systematically throughout a time interval.

More generally, given a function $y = \mathrm{f}(x)$, by plotting gradients systematically over a suitable interval of x-values, we can deduce the **gradient function**, denoted by $\mathrm{f}'(x)$, which measures the rate of change of y with respect to x.

CALCULATOR ACTIVITY

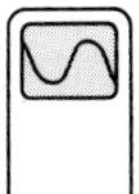

Use the previous Calculator activity to find the gradient of the function $y = x^3$ at each of the x-values –5, –4, –3, –2, –1, 0, 1, 2, 3, 4, 5.

Firstly set function f to be x^3. Use the program repeatedly to obtain the following values.

x	$y = f(x)$	Gradient = $f'(x)$
–5	–125	75
–4	–64	48
–3	–27	27
–2	–8	12
–1	–1	3
0	0	0
1	1	3
2	8	12
3	27	27
4	64	48
5	125	75

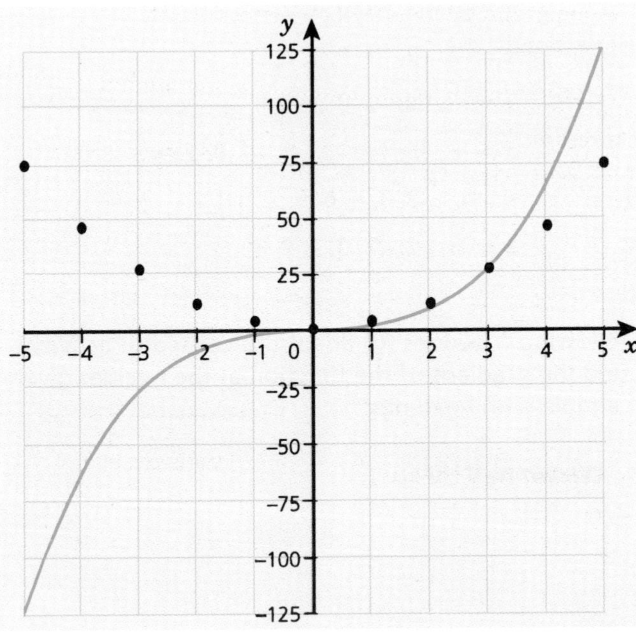

On a graphical calculator, choose a suitable range for x- and y-values.

x_{min}: –5 x_{max}: 5 x_{scl}: 1
y_{min}: –125 y_{max}: 125 y_{scl}: 25

Draw the graph of $y = x^3$ and plot values of the gradient at each of the x-values.

The plotted points suggest that the gradient function is a parabola through (0, 0) of the form

$$f'(x) = kx^2$$

for some constant k. By inspecting gradient values over the interval $-5 \leq x \leq 5$, deduce that $k = 3$, i.e. the gradient function for $y = x^3$ is given by $f'(x) = 3x^2$.

EXERCISES

7.1 A

For each of the following functions $f(x)$, adapt the Calculator activity on page 148 to find the gradient of the function at the x-values given. Hence construct a table with headings:

x	$y = f(x)$	gradient $= f'(x)$

Complete it for the suggested values of x.

Sketch graphs of $f(x)$ and $f'(x)$ and so deduce an appropriate formula for $f'(x)$.

1. $f(x) = x^2$ $\quad x = -3, -2, -1, 0, 1, 2, 3$
2. $f(x) = 5x^2$ $\quad x = -3, -2, -1, 0, 1, 2, 3$
3. $f(x) = 2x^3$ $\quad x = -3, -2, -1, 0, 1, 2, 3$
4. $f(x) = -2x^3$ $\quad x = -3, -2, -1, 0, 1, 2, 3$
5. $f(x) = x^4$ $\quad x = -3, -2, -1, 0, 1, 2, 3$
6. $f(x) = 3x^4$ $\quad x = -3, -2, -1, 0, 1, 2, 3$
7. $f(x) = x^2 + x$ $\quad x = -3, -2, -1, 0, 1, 2, 3$
8. $f(x) = x^2 + 5x$ $\quad x = -6, -5, -4, -3, -2, -1, 0, 1$
9. $f(x) = x^2 - 7x + 10$ $\quad x = 0, 1, 2, 3, 4, 5, 6, 7$
10. $f(x) = x^3 - 3x + 2$ $\quad x = -3, -2, -1, 0, 1, 2, 3$

7.1 B

For each of the following functions $f(t)$ adapt the Calculator activity on page 148 to find the gradient of the function at the t-values given. Hence construct a table with headings:

t	$s = f(t)$	gradient $= f'(t)$

Complete it for the suggested values of t.

1. $f(t) = t$ $\quad t = -3, -2, -1, 0, 1, 2, 3$
2. $f(t) = -5t$ $\quad t = -3, -2, -1, 0, 1, 2, 3$
3. $f(t) = 4t^2$ $\quad t = -3, -2, -1, 0, 1, 2, 3$
4. $f(t) = -3t^2$ $\quad t = -3, -2, -1, 0, 1, 2, 3$
5. $f(t) = 3t^3$ $\quad t = -3, -2, -1, 0, 1, 2, 3$

6 $f(t) = t^5$ $\quad t = -3, -2, -1, 0, 1, 2, 3$

7 $f(t) = t^2 - 2t + 1$ $\quad t = -2, -1, 0, 1, 2, 3, 4$

8 $f(t) = t^2 - 3t$ $\quad t = 0, 1, 2, 3, 4, 5, 6$

9 $f(t) = t^3 - 6t^2 + 12t - 8$ $\quad t = -1, 0, 1, 2, 3, 4, 5$

10 $f(t) = t^2 - 8t$ $\quad t = 5, 6, 7, 8, 9, 10, 11$

DIFFERENTIATION

The results from the previous section gave us several **gradient functions** (or **derivatives**), for example:

$f(x)$	x^2	x^3	x^4	$x^2 + 5x$	$x^2 - 7x + 10$
$f'(x)$	$2x$	$3x^2$	$4x^3$	$2x + 5$	$2x - 7$

Using these and similar results, we can see some important patterns which we can use to find a gradient function. This is called **differentiation**.

1 $f(x) = x^n \Rightarrow f'(x) = nx^{n-1}$, $n = 0, 1, 2, 3, 4, \ldots$

e.g. $f(x) = x^5 \Rightarrow f'(x) = 5x^4$

2 $f(x) = ax^n \Rightarrow f'(x) = anx^{n-1}$ where a is a constant

e.g. $f(x) = 5x^3 \Rightarrow f'(x) = 5 \times 3x^2 = 15x^2$

3 The derivative of a sum (or difference) is the sum (or difference) of the derivatives.

e.g. $f(x) = x^3 - 5x + 10 \Rightarrow f'(x) = 3x^2 - 5$

Definition

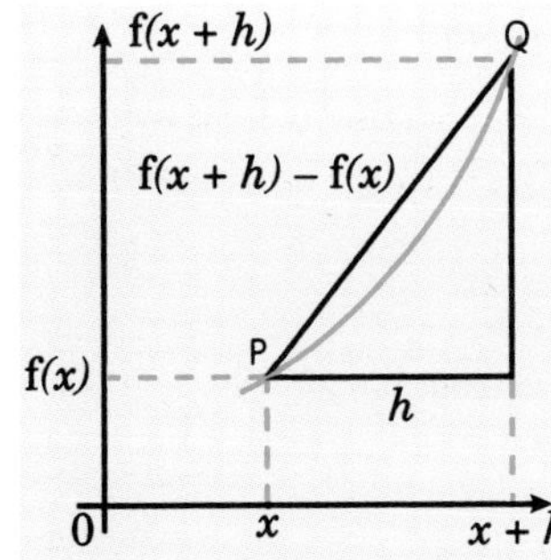

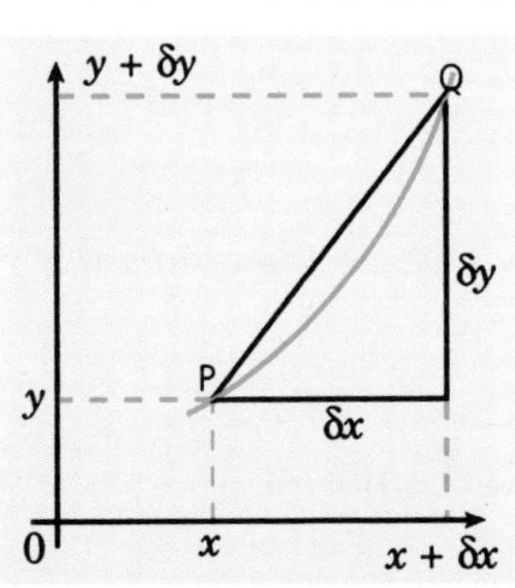

A formal definition of the gradient function $f'(x)$ is given by:

$f'(x)$ is the limit of $\dfrac{f(x+h)-f(x)}{h}$ as $h \to 0$.

Until now we have used function notation:

$f(x)$ for the original function
$f'(x)$ for the gradient function.

The gradient function (or derivative) is also represented by $\dfrac{dy}{dx}$, which is not a fraction, but a single quantity.

$$y = f(x) \Rightarrow \frac{dy}{dx} = f'(x)$$

The notation $\dfrac{dy}{dx}$ comes from the idea that the gradient is the limit of $\dfrac{\delta y}{\delta x}$ as $\delta x \to 0$.

The process is called **differentiation with respect to x**.

Example 7.1

For the function $f(x) = (x+1)(x-4)$, *find:*

a) $f'(1)$ ***b)*** *values of* x *such that* $f'(x) = 0$.

Solution

Multiplying out the brackets:

$(x+1)(x-4) \equiv x^2 - 3x - 4$

Differentiating: $f(x) = x^2 - 3x - 4$

$\Rightarrow f'(x) = 2x - 3$

a) $f'(1) = 2 \times 1 - 3 = -1$

b) $f'(x) = 0 \Rightarrow 2x - 3 = 0 \Rightarrow x = 1.5$

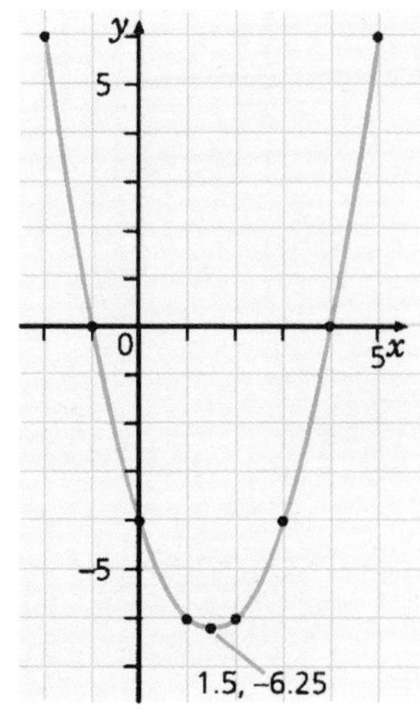

Differentiating w. r. t. any variable

The work so far has involved functions of the independent variable x. The same rule of differentiation works for other variables, for example: $f(t) = at^n \Rightarrow f'(t) = ant^{n-1}$.

This is called **differentiation with respect to *t*.**

Example 7.2

For the function $f(t) = 7 + 5t - t^3$, *find:*

a) $f'(1.5)$ ***b)*** *values of t such that* $f'(t) = 2$.

Solution

a) $f'(t) = 5 - 3t^2 \Rightarrow f'(1.5) = 5 - 3 \times (1.5)^2 = -1.75$

b) $f'(t) = 2 \Rightarrow 5 - 3t^2 = 2 \Rightarrow 3t^2 = 3 \Rightarrow t^2 = 1 \Rightarrow t = \pm 1$

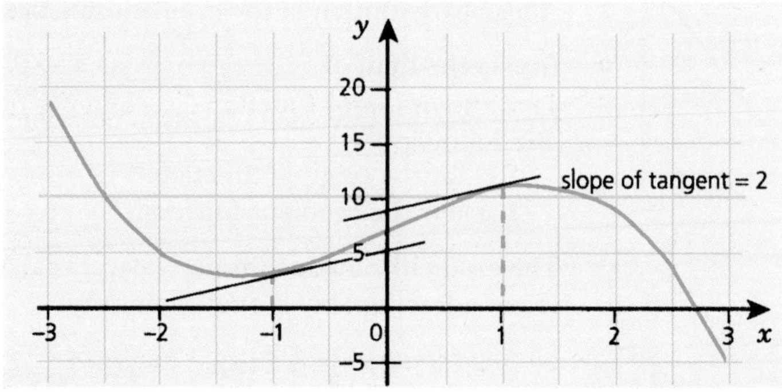

Example 7.3

Find the points on the curve $y = x^3 - 5x^2 + 5x - 3$ *where the gradient is parallel to the straight line* $y = 2x - 7$.

Solution

Differentiating: $y = x^3 - 5x^2 + 5x - 3 \Rightarrow \dfrac{dy}{dx} = 3x^2 - 10x + 5$

Since the gradient of the straight line is 2, then we look for the values of x such that $\frac{dy}{dx} = 2$.

Now: $3x^2 - 10x + 5 = 2 \Rightarrow 3x^2 - 10x + 3 = 0$
Solving the quadratic equation gives:

$(3x-1)(x-3) = 0 \Rightarrow x = \frac{1}{3}$ *or* $x = 3$.

Substituting for these values of x:

$$x = \tfrac{1}{3} \Rightarrow y = \left(\tfrac{1}{3}\right)^3 - 5\left(\tfrac{1}{3}\right)^2 + 5\left(\tfrac{1}{3}\right) - 3 = -\tfrac{50}{27} = -1\tfrac{23}{27}$$

$$x = 3 \Rightarrow y = 3^3 - 5(3)^2 + 5(3) - 3 = -6$$

Therefore the points A and B on the curve where the gradient is 2 *are* $(\frac{1}{3}, -1\frac{23}{27})$ *and* (3, –6).

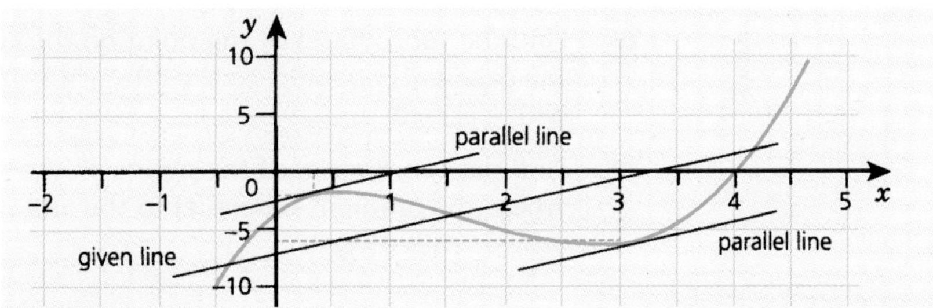

EXERCISES

7.2 A

Differentiate these expressions with respect to x.

1 $10x^2$

2 $-7x^3$

3 $12 - 5x$

4 $x^3 - 2x^2 + 10$

5 $3x^7 - 4x^5 + 2x^3$

6 $\frac{1}{2}x^2 + \frac{1}{3}x - \frac{1}{4}$

7 $3x(x^2 - 4)$

8 $(x-5)(x-3)$

9 $(2x-5)(x+3)$

10 $x(x-5)(x-3)$

Find the gradient function $f'(x)$ and the gradient values for each of the following.

11 $f(x) = 2x^2 - 10,\ f'(5)$

12 $f(x) = 2x - x^3,\ f'(0.5)$

13 $f(x) = 3x(x-7),\ f'(3.5)$

14 $f(x) = 5x^2(x-2),\ f'(-1)$

15 $f(x) = (5+2x)(4-x),\ f'(2)$

Find $\frac{dy}{dx}$ and the derivatives at the points shown, for each of the following.

16 $y = 2x^3 - 5x^2 + x - 8$ at (0, –8)

17 $y = (3x+2)(5x-4)$ at (–2, 56)

18 $y = x(x^2 - 3x + 10)$ at (3, 30)

Find $\frac{ds}{dt}$ and the derivatives at the points shown.

19 $s = 3 + 2t - 4t^2$ at (1, 1)

20 $s = t(3t+2)(5t-4)$ at (–1, –9)

Find point(s) on the following curves where the gradient is as given.

21 $y = 5x^2 - 3x + 5, \frac{dy}{dx} = 7$

22 $y = x^3 - 10x + 5, \frac{dy}{dx} = 17$

23 $u = x^4 - 2x^3, \frac{du}{dx} = 0$

24 $y = u^3 - u^2, \frac{dy}{du} = 1$

25 $y = 3x^4 - 4x^3 - 6x^2 + 12x - 12, \frac{dy}{dx} = 0$

26 Find the gradient of the curve $y = x^3 + 7x^2 - x + 3$ where the curve crosses the y-axis.

27 Find the gradient of the curve $y = (3x + 1)(x - 2)$ where the curve crosses the x-axis.

28 Find the gradient of the curve $s = t(2t - 1)(t + 1)$ where the curve crosses the t-axis.

29 Find the coordinates of the point on the curve $y = 7 - 6x - 2x^2$ where the gradient is parallel to the line $y = 3 - 2x$. Illustrate your answer with a sketch.

30 Find the coordinates of the points on the curve $y = 4x^3 - 5x^2 - 9x + 7$ where the gradient is parallel to the line $y = 3x + 5$. Illustrate your answer with a sketch.

7.2B

Differentiate these expressions.

1 $5x^3$

2 $x^4 + 3x - 7.5$

3 $10(1 - x + x^2 - x^3)$

4 $5t$

5 13

6 $0.31(x^4 - x^2 + 1)$

7 $u^3 - u^2 + 5u - 6$

8 $14x^7 - 10x^5$

9 $(t - 4)(3t + 1)$

10 $5(t - 5)(3t + 1)$

Find the gradient function $f'(x)$ and the gradient values for each of the following.

11 $f(x) = 7, f'(2)$

12 $f(x) = 7 - 5x, f'(2)$

13 $f(x) = 2x^2 - 10, f'(1)$

14 $f(x) = (2x + 1)(2x - 1), f'(0)$

15 $f(x) = \frac{1}{5}(7 - 5x)(3 + x), f'(2)$

Find $\frac{dh}{dt}$ and the derivatives at the points shown, for each of the following.

16 $h = 2t^3 + t^2 - 5t + 3$ at (–2, 1)

17 $h = t^4 - 2t^3 + 0.5t^2 - t + 1$ at (0, 1)

18 $h = \frac{1}{2}(2t - 5)^2$ at (4, 4.5)

19 $h = (t^2 - 1)(t + 2)$ at (1, 0)

20 $h = 3(t + 1)(2t - 5)^2$ at (2, 9)

Find point(s) on the following curves where the gradient is as given.

21 $y = x^4 - x^2, \frac{dy}{dx} = 0$

22 $y = (3x + 4)^2, \frac{dy}{dx} = 2$

23 $s = 3(t+1)(2t-5), \frac{ds}{dt} = 15$

24 $s = t(t+1)(2t-5), \frac{ds}{dt} = 7$

25 $y = (x-1)(x-2)(x-3), \frac{dy}{dx} = 0$

26 The equation of a curve is $y = (3x-1)(x+2)$.

a) Find the gradient of the curve:
- **i)** at the point (1, 6),
- **ii)** at the point where the curve crosses the y-axis,
- **iii)** at each point where the curve crosses the x-axis.

b) Find the coordinates of the point where $\frac{dy}{dx} = 0$.

27 The equation of a curve is $y = x^3 + 5x^2 - 4x - 20$.

a) Find the gradient of the curve:
- **i)** at the point where the curve crosses the y-axis,
- **ii)** at each point where the curve crosses the x-axis.

b) Find the coordinates of the points where $\frac{dy}{dx} = 0$.

28 **a)** Find the gradient of the tangent to the curve $y = x^2 - 9x - 3$ when $x = 5$.

b) Find the equation of this tangent.

c) Illustrate your answer with a sketch.

29 Find the coordinates of the points on the curve $y = 2x^4 - 3x^2 + x - 7$ where the gradient is parallel to the line $y = 3x$.

30 For a certain curve, the gradient function is $\frac{dy}{dx} = 3x^2 - 1$.

Which of the following could be the equation of the curve?

a) $y = 3x^2 - x$
b) $y = x^3 - x$
c) $y = 3x^3 - x + 4$
d) $y = x^3 - x + 4$

RATES OF CHANGE

Modelling a throw of a cricket ball, in the first section of this chapter, we found that the gradient function represented the rate of change of distance with respect to time, i.e. instantaneous **velocity**.

Using mathematical notation, this can be summarised as:

$h = f(t)$ represents height h as a function of time t

$v = \frac{dh}{dt} = f'(t)$ represents the rate of change of h with respect to t.

Similarly:

$a = \frac{dv}{dt} = f''(t)$ represents the rate of change of v with respect to t.

This rate of change of velocity with respect to time gives us the instantaneous **acceleration**. The notation $f''(t)$ means that the original function $f(t)$ has been differentiated **twice**.

Example 7.4

A particle is propelled vertically upwards from a point 6 metres above the ground. Its height, h metres, after time t seconds, is given by:
$h = 6 + 13t - 5t^2$.

a) *Find its velocity after one second.*
b) *Find the time at which it reaches its greatest height, and the height above the ground at this time.*
c) *Find the time when it hits the ground and the speed with which it hits the ground.*
d) *Show that the acceleration is constant.*

Solution

a) Velocity $v = \frac{dh}{dt} = 13 - 10t$.

When $t = 1$, $v = 13 - 10 = 3\,\mathrm{m\,s^{-1}}$.

b) Maximum height occurs when $v = \frac{dh}{dt} = 0 \Rightarrow 13 - 10t = 0$

$$\Rightarrow \quad 10t = 13$$
$$\Rightarrow \quad t = 1.3 \text{ seconds}$$

c) When the particle hits the ground,

$$h = 0 \Rightarrow \quad 6 + 13t - 5t^2 = 0$$
$$\Rightarrow \quad (3 - t)(2 + 5t) = 0$$
$$\Rightarrow \quad t = 3 \text{ or } t = -0.4 \text{ seconds}$$

Velocity of particle after 3 seconds is $v = 13 - 10 \times 3 = -17\,\mathrm{m\,s^{-1}}$.
The speed of the particle when it hits the ground is $17\,\mathrm{m\,s^{-1}}$ and the negative sign in the velocity indicates that it is travelling downwards.

d) The acceleration is given by $a = \frac{dv}{dt} = -10\,\mathrm{m\,s^{-2}}$, which means that throughout its motion, the acceleration is $10\,\mathrm{m\,s^{-2}}$ downwards.

In general, if $y = f(x)$, then $\frac{dy}{dx} = f'(x)$ represents the rate of change of y with respect to x.
The rate of change is often with respect to time, but other situations arise.

Example 7.5

The volume, V, of a sphere in terms of its radius, r, is given by $V = \frac{4}{3}\pi r^3$.

a) *Show that the surface area is equivalent to the rate of change of volume with respect to radius.*
b) *Find the rate of change of volume with respect to radius when $r = 3$.*
c) *What is the volume of the sphere when $\frac{dV}{dr} = 9\pi$?*

Solution

a) $V = \frac{4}{3}\pi r^3 \quad \Rightarrow \quad \frac{dV}{dr} = \frac{4}{3}\pi \times 3r^2 = 4\pi r^2$

where $\frac{dV}{dr}$ represents the rate of change of volume with respect to the radius, and the surface area of a sphere is given by $4\pi r^2$.

b) When $r = 3$, $\frac{dV}{dr} = 4\pi \times 3^2 = 36\pi$

c) $\frac{dV}{dr} = 9\pi \Rightarrow 4\pi r^2 = 9\pi \Rightarrow r^2 = 2.25 \Rightarrow r = \sqrt{2.25} = 1.5$

$$\Rightarrow V = \tfrac{4}{3}\pi r^3 = \tfrac{4}{3}\pi \times 1.5^3 = 4.5\pi$$

EXERCISES

1 A stone is thrown upwards. Its height above ground, h metres, after t seconds, is given by $h = 2 + 9t - 5t^2$.

a) Find the speed of the stone after 1.5 seconds.

b) Find the maximum height of the stone and the time at which this occurs.

c) At what speed is the stone travelling when it hits the ground?

2 During the boost stage of a rocket launch, the height, h metres, after t seconds, is given by $h = 2t^4 - 3t^3 + 45t^2$.

a) Find the velocity v of the rocket at $t = 0$, 1, 2, 3 and 4 seconds.

b) Find the acceleration a of the rocket at $t = 0$, 1, 2, 3 and 4 seconds.

c) Sketch graphs of:
i) h against t, **ii)** v against t, **iii)** a against t,
over the interval $0 \le t \le 4$.

3 Billie is blowing up a spherical balloon.

a) Find a formula for $\frac{dV}{dr}$, the rate of change of volume V with respect to radius r.

b) Find $\frac{dV}{dr}$ when $r = 5$ cm.

c) For what volume does $\frac{dV}{dr} = 1000$?

4 A body, which is initially at rest, is projected in a straight line from a point O. Its distance after t seconds is s metres, where $s = 10t^2 - 2t^3$.

a) Calculate the distance travelled by the body during the third second.

b) Find an expression for the velocity v in terms of t.

c) Sketch a graph of v against t for $0 \le t \le 4$.

d) Calculate the acceleration of the body when $t = 2$.

e) Find the value of t when the body is next momentarily at rest.

5 For a certain production process, the productivity p, when x machines are in use, is given by $p = 500(x + 1)^2 - 500$.

a) Determine the marginal productivity $\frac{dp}{dx}$, when $x = 3$.

b) How many machines are required to raise the marginal productivity to a level of 6000?

7.3 B

1 Fiona throws a ball straight up into the air. The height of the ball, h metres, after t seconds, is given by $h = 1.6 + 10t - 5t^2$.

a) Find the speed of the ball after one second and explain your answer in terms of the motion of the ball.
b) If Fiona catches the ball at the same height as she threw it, find the speed of the ball just before she catches it.
c) Find the acceleration of the ball.
d) For the interval $0 \leq t \leq 2$, sketch graphs of:
i) height against time, ii) velocity against time,
iii) acceleration against time.

2 Alarna and Paul are experimenting to find the formula for the speed of a stone t seconds after it is dropped. They measured the distance fallen by the stone, h metres, at 0.1 second intervals and collect these data.

t (seconds)	0.00	0.10	0.20	0.30	0.40	0.50	0.60
h (metres)	0.00	0.05	0.20	0.45	0.80	1.25	1.70

a) Using the table of values plot the graph of h against t.
b) By trial and improvement, find a function $f(t)$ that is a good fit to the data points.
c) Use differentiation to find a formula for the speed of the ball as a function of t.

3 A yo-yo travels along a straight line, down and up. Sean models the distance, s metres, from his hand at time t seconds by $s = 24t - 3t^2$, $0 \leq t \leq 8$.

a) Find the speed of the yo-yo, v metres per second, at time t seconds.
b) Explain the restriction $0 \leq t \leq 8$.
c) What is the length of the string of the yo-yo?
d) Sketch the graphs of s and v against t.

Describe the motion of the yo-yo, giving its position, speed and direction of motion at $t = 1$, 2, 5 and 7 seconds.

4 The area of a circle, radius r, is given by $A = \pi r^2$.
A stone is dropped vertically into a still pond, causing the surface to ripple with circles of increasing diameter.

a) Show that the circumference is equivalent to the rate of change of area with respect to radius.
b) Find the rate of change of area with respect to radius when $r = 3$.
c) What is the area enclosed by a circular ripple with circumference 8?

5 The number of people, P, newly infected on day t of a flu epidemic is given by $P = 13t^2 - t^3$, $t \leq 13$.

a) Sketch a graph of P against t.
b) Find the rate of change of P with respect to t,
i) on day 5, ii) on day 10. Interpret your answers.
c) Calculate the time t for which $\frac{dP}{dt} = 0$, the value of P at this time, and interpret your results.

STATIONARY VALUES

CALCULATOR ACTIVITY

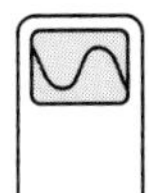

Consider the following four curves.

A $y = x^2 - 7x + 10$ B $y = 10 - x^2$
C $y = x^3$ D $y = x^3 - 12x + 3$

For each equation:

- sketch a graph of the curve,
- find the coordinates of the point(s) where the tangent to the curve is parallel to the x-axis,
- find the sign of the gradient of the curve either side of the point(s) you have found,
- describe what happens to the curve around the point(s) you have found.

Stationary points

When exploring the graphs of functions similar to those in the above Calculator activity, we find one or two points on each curve where the gradient is 0 (i.e. the tangent to the curve is parallel to the x-axis). Such points are called **stationary points**, where the function has a **stationary value**.

For each stationary point, the zero gradient means that $f'(x) = 0$. This means that the coordinates of a stationary point may be found by first solving the equation $f'(x) = 0$.

Once we have found the stationary point, we can classify it as a **maximum**, **minimum** or **point of inflexion** by looking at the behaviour of $f'(x)$ either side of the stationary point.

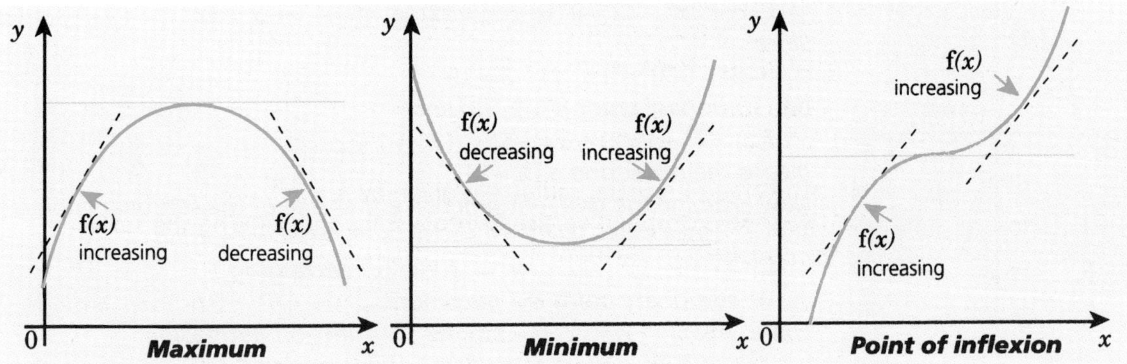

Example 7.6

Find the coordinates of the stationary point for the curve $y = x^2 - 7x + 10$ and state which type it is.

Solution

For a stationary value:

$$f'(x) = 2x - 7 = 0$$
$$\Rightarrow 2x = 7$$

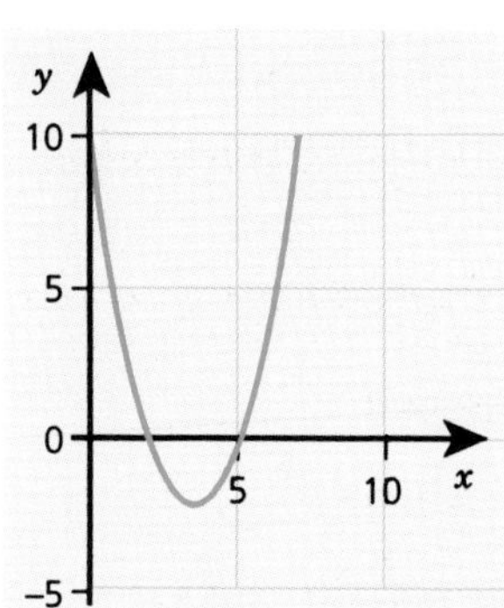

$\Rightarrow x = 3.5$

Since $f(3.5) = -2.25$, *the* ***only*** *stationary point is* $(3.5, -2.25)$.
Now examine the gradient function either side of $x = 3.5$, *for example:*

$f'(3) = -1$ *and* $f'(4) = 1$.

A ***negative*** *gradient at* $x = 3$ *means that the function is* ***decreasing.***
A ***positive*** *gradient at* $x = 4$ *means that the function is* ***increasing.***
The conclusion is that $(3.5, -2.25)$ *is a* ***minimum*** *point.*
This should confirm the result of the exploration of curve A, in Calculator activity 7.3.

Note

For **quadratic** functions only, we can also find the stationary point by completing the square:

$f(x) = x^2 - 7x + 10 = (x - 3.5)^2 - 12.25 + 10 = (x - 3.5)^2 - 2.25$

from which we can deduce the minimum point $(3.5, -2.25)$.

For cubic and other polynomial functions solving the equation $f'(x) = 0$ is the best way to find stationary points for a function, as shown in the next two examples.

Example 7.7

Find the coordinates of any stationary point for the curve $y = x^3 - 12x + 3$ *and classify them.*

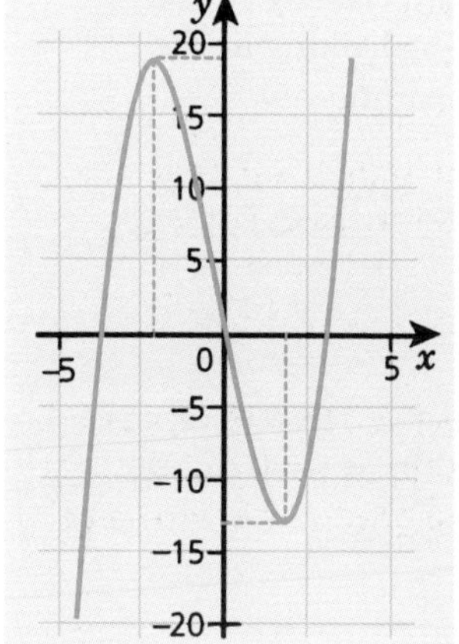

Solution

For a stationary value:

$f'(x) = 3x^2 - 12 = 0 \Rightarrow 3x^2 = 12 \Rightarrow x^2 = 4 \Rightarrow x = \pm 2$

Since

$f(-2) = (-2)^3 - 12(-2) + 3 = 19$

one stationary point is $(-2, 19)$ *and*

$f(2) = 2^3 - 12(2) + 3 = -13$

means the other one is $(2, -13)$.

Now examine the gradient function either side of the stationary points.

For $(-2, 19)$: $f'(-3) = 3(-3)^2 - 12 = 15 \Rightarrow$ *function* ***increasing***

$f'(-1) = 3(-1)^2 - 12 = -9 \Rightarrow$ *function* ***decreasing***

$\Rightarrow$ *stationary point is a* ***maximum.***

For $(2, -13)$: $f'(1) = 3(1)^2 - 12 = -9 \Rightarrow$ *function* ***decreasing***

$f'(3) = 3(3)^2 - 12 = 15 \Rightarrow$ *function* ***increasing***

$\Rightarrow$ *stationary point is a* ***minimum.***

These conclusions should confirm the exploration of curve D.

The nature of the stationary points also gives valuable information when sketching the graph by hand.

Example 7.8

Show that the curve $y = \dfrac{x^3(x-4)}{2}$ *has a point of inflexion at the origin.*

Find and classify the other stationary point.
Sketch the curve, showing clearly the stationary values and where the graph crosses the axes.

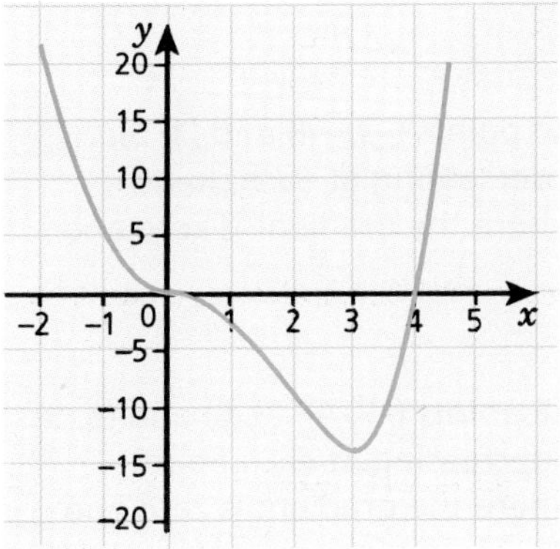

Solution

Firstly simplify the function for differentiating.

$$f(x) = \frac{x^3(x-4)}{2} = \frac{x^4 - 4x^3}{2} = 0.5x^4 - 2x^3$$

For stationary values:

$f'(x) = 2x^3 - 6x^2 = 0 \Rightarrow 2x^2(x-3) = 0 \Rightarrow x = 0 \text{ or } x = 3.$

Since $f(0) = \dfrac{0^3(0-4)}{2} = 0$, *one stationary point is* $(0, 0)$

and $f(3) = \dfrac{3^3(3-4)}{2} = -13.5$ *means the other one is* $(3, -13.5)$.

Now examine the gradient function either side of the stationary points.

For $(0, 0)$: $f'(-1) = 2(-1)^3 - 6(-1)^2 = -8$ $\Rightarrow$ *function* ***decreasing***
$f'(1) = 2(1)^3 - 6(1)^2 = -4$ $\Rightarrow$ *function* ***decreasing***
$\Rightarrow$ *stationary point is a* ***point of inflexion.***

For $(3, -13.5)$: $f'(2) = 2(2)^3 - 6(2)^2 = -8$ $\Rightarrow$ *function* ***decreasing***
$f'(4) = 2(4)^3 - 6(4)^2 = 32$ $\Rightarrow$ *function* ***increasing***
$\Rightarrow$ *stationary point is a* ***minimum.***

EXERCISES

7.4A

1 Confirm the results of the rest of the Calculator activity on page 159 by finding and classifying the stationary points for these curves.
B $y = 10 - 6x^2$ C $y = x^3$.

2 A curve has equation $y = 1 + x - 2x^2$.

a) Find the gradient of the curve at the points (–2, –9) and (2, –5). Decide if the graph is increasing or decreasing at each point.
b) Find the coordinates of the point where the gradient is zero and sketch the curve.

3 A curve has equation $y = x^3 - 6x + 4$.

a) Find the gradient of the curve at the points (–3, –5), (0, 4) and (2, 0). Decide if the graph is increasing or decreasing at each point.
b) Find the coordinates of the two points where the gradient is zero and sketch the curve.

For each of the curves in questions **4** to **14**, find the stationary points and classify them as maximum, minimum or point of inflexion. Sketch a graph of the curve, clearly indicating the stationary point(s).

Check your answers on a graphics calculator.

4 $y = 14 + 5x - x^2$
5 $y = 3(x + 2)(x - 5)$
6 $y = (2x - 5)^2$
7 $y = x^3 - 4x^2 + 9$
8 $y = 10 + 12x - x^2 - \frac{2}{3}x^3$
9 $y = x(3 - x)^2$
10 $y = (2x - 5)^3$
11 $y = x^3 + x^2 - x - 1$
12 $y = x^3(4 - x)$
13 $y = 3 + 8x^2 - x^4$
14 $y = 3x^4 - 16x^3 + 24x^2 - 10$

1 A curve has equation $y = x^2 + 3x - 4$.

a) Find the gradient of the curve at the points (–3, –4) and (1, 0). Decide if the graph is increasing or decreasing at each point.
b) Find the coordinates of the point where the gradient is zero and sketch the curve.

2 A curve has equation $y = 12x + 3x^2 - 2x^3$.

a) Find the gradient of the curve at the points (–2, 4), (0, 0) and (3, 9). Decide if the graph is increasing or decreasing at each point.
b) Find the coordinates of the point where the gradient is zero and sketch the curve.

3 A curve has equation $s = t^2(t - 1)^2$.

a) Find the gradient of the curve at the points (–1, 4), (0, 0) and (2, 4). Decide if the graph is increasing or decreasing at each point.
b) Find the coordinates of the point(s) where the gradient is zero and sketch the curve.

For each of the curves in questions **4** to **14**, find the stationary points and classify them as maximum, minimum or point of inflexion. Sketch a graph of the curve, clearly indicating the stationary point(s).

Check your answers on a graphics calculator.

4 $y = (2 + x)(3 + x)$
5 $s = 40 + 15t - 5t^2$
6 $y = -2(x + 5)^2$
7 $y = x^3 - 3.5x^2 + 2x - 1$
8 $y = x^4 - 3x^2 - 1$
9 $s = t^3 + 6t^2 + 12t + 8$
10 $y = x^4 - 4x^3 - 2x^2 + 12x + 5$
11 $y = \frac{1}{4}x^4 - x^3 + x^2 - 6x + 2$
12 $s = t^3 - t^2 + t - 1$
13 $y = x - x^2 + x^3 - x^4$
14 $y = 3x^4 - 20x^3 + 36x^2 - 5$

OPTIMISATION

In the previous section, we saw that the stationary points often represent maximum or minimum values.

For any function, stationary values can be found by setting

$$\frac{dy}{dx} = f'(x) = 0$$

and solving the resulting equation.

The method is often applied in the area of **optimisation**, where the aim is to maximise or minimise a function.

Example 7.9

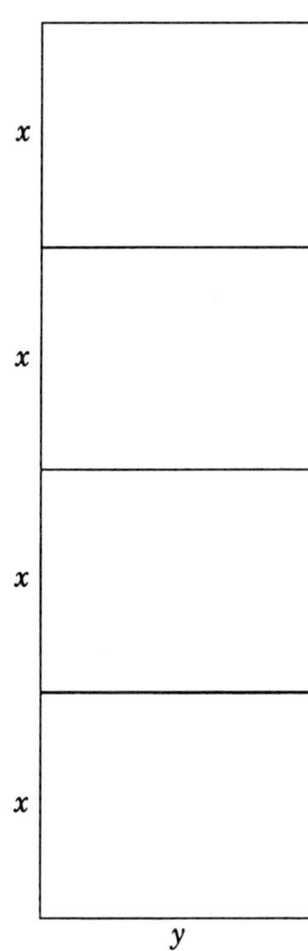

A farmer wishes to create four pens with 200 metres of fencing.

a) *Assuming each pen measures x m by y m, as shown in the diagram, express the area of the pens, A, in terms of the length, x, only.*

b) *Find the values of x and y which gives a maximum area for the pens. What is the maximum value of A?*

Solution

a) The framework for the pens uses eight x-lengths plus five y-lengths. With 200 metres of fencing available, this means:

$8x + 5y = 200 \quad \Rightarrow 5y = 200 - 8x$

$\Rightarrow y = \dfrac{200 - 8x}{5} = 40 - 1.6x$

Area $A = 4x \times y = 4xy = 4x(40 - 1.6x) \Rightarrow A = 160x - 6.4x^2$

b) The maximum area occurs when $\dfrac{dA}{dx} = 0$

$\Rightarrow \dfrac{dA}{dx} = 160 - 12.8x = 0$

$\Rightarrow 12.8x = 160$

$\Rightarrow x = 160 \div 12.8 = 12.5$

$\Rightarrow y = \dfrac{200 - 8 \times 12.5}{5} = 20$

i.e. the maximum area of pens occurs when $x = 12.5$ and $y = 20$, which gives a maximum value of A as:

$4 \times 12.5 \times 20 = 1000.$

The conclusion is that the maximum area enclosed is 1000 m^2.

Example 7.10

An open box is to be formed by cutting four squares of side x cm from the corners of a sheet of card 20 cm square.

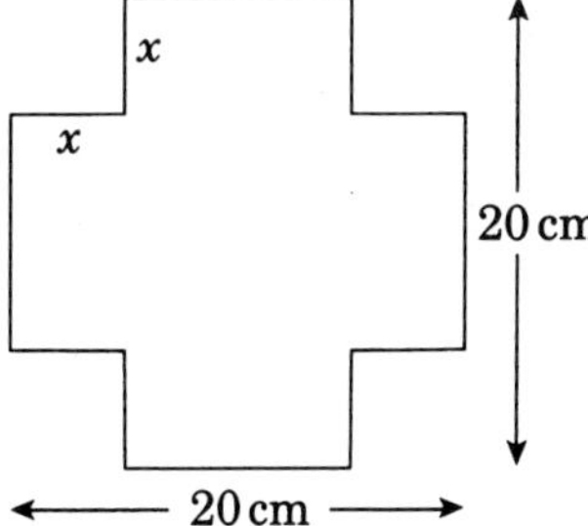

a) *Express the volume of the box, V cm^3, in terms of the length, x cm.*

b) *Find the value of x which gives a maximum volume. What is the maximum volume?*

Solution

a) Height of box $= x$ cm $\Rightarrow$ length and width $= 20 - 2x$ cm.

Volume $V = x(20 - 2x)^2 = x(400 - 80x + 4x^2)$

$\Rightarrow V = 400x - 80x^2 + 4x^3$

b) Maximum volume occurs when $\dfrac{dV}{dx} = 0$

$\Rightarrow \dfrac{dV}{dx} = 400 - 160x + 12x^2 = 0$

$\Rightarrow 3x^2 - 40x + 100 = 0$

$\Rightarrow (3x - 10)(x - 10) = 0$

$$x = 10 \;\Rightarrow\; V = 10 \times (20 - 2 \times 10)^2 = 0$$
$$x = 3\tfrac{1}{3} \;\Rightarrow\; V = \tfrac{10}{3} \times (20 - 2 \times \tfrac{10}{3})^2 = 593 \quad \textit{(3 s.f.)}$$

Conclusion: the maximum volume of $593\,cm^3$ *occurs when* $x = 3\frac{1}{3}$.

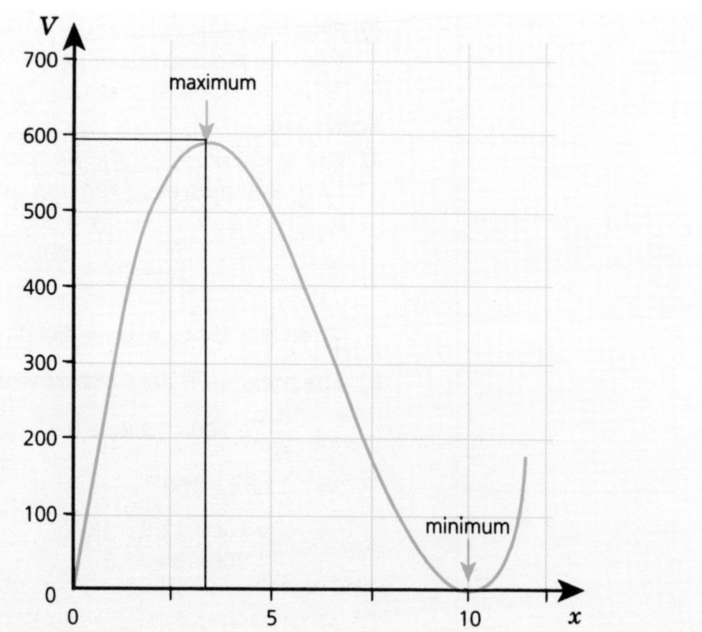

EXERCISES

7.5A

1 A farmer wishes to enclose a rectangular area with a 60-metre length of fencing.

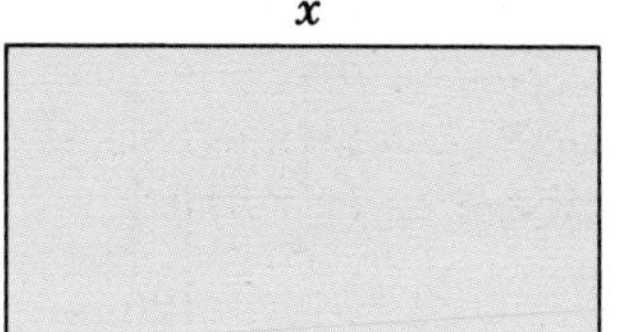

a) Explain why $2x + 2y = 60$ and rearrange this formula in the form $y = \ldots$.
b) Write down a formula for the area, A, in terms of x only.
c) Find the value of x that maximises the value of A.
What is the maximum area?
What particular shape is formed in this case?
d) Sketch a graph of A against x for $0 \le x \le 30$.

2 The farmer in question **1** has alternative sites for the enclosure, one involving the use of a single existing wall (figure A) and the other involving the use of three walls (figure B).

Figure A

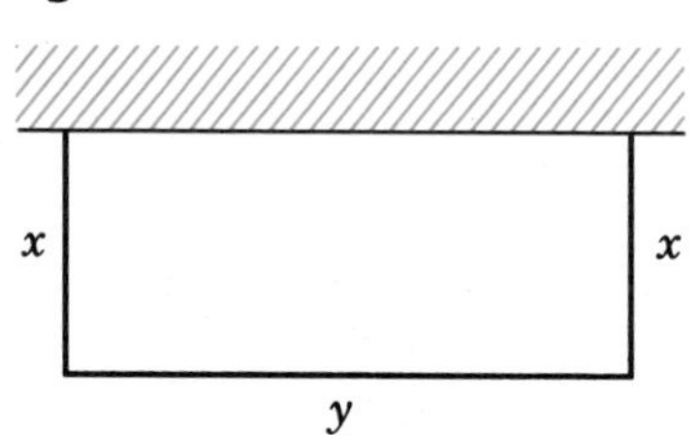

Figure B

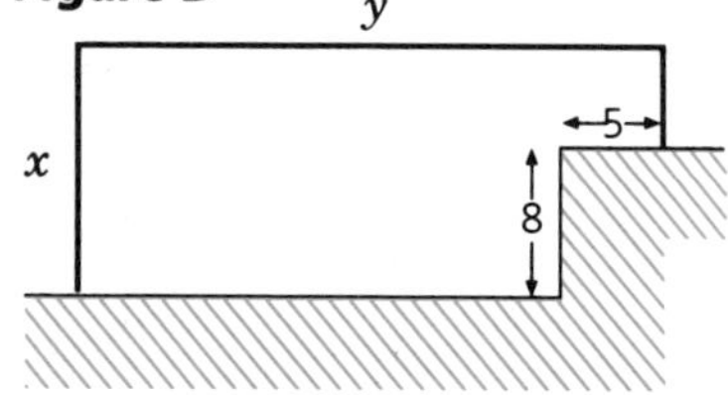

He still has only 60 metres of fencing to use. For each site:

a) write down a formula connecting x and y,
b) write down a formula for the area, A, in terms of x only,
c) find the maximum area that he can enclose, giving the dimensions of the enclosure,
d) sketch graphs of A against x for suitable values of x.

3 A farmer wishes to create six pens with 144 metres of fencing.

a) Assuming each pen measures x m by y m, as shown in the diagram, show that the area of the pens, A, in terms of the length, x, only, is given by:

$A = 96x - \frac{16}{3}x^2$

b) Find the values of x and y which give a maximum area for the pens. What is the maximum value of A?

4 Two industrial plants, 30 miles apart, are polluting a large lake. The pollution level, P parts per million, is given by:

$P = 2x^2 - 48x + 400,\ 0 \leq x \leq 30$

where x measures the distance from plant A towards plant B, in miles.

a) Find where, between A and B, the pollution is the least.
b) Sketch a graph of P against x for $0 \leq x \leq 30$ and deduce which factory pollutes the water more.

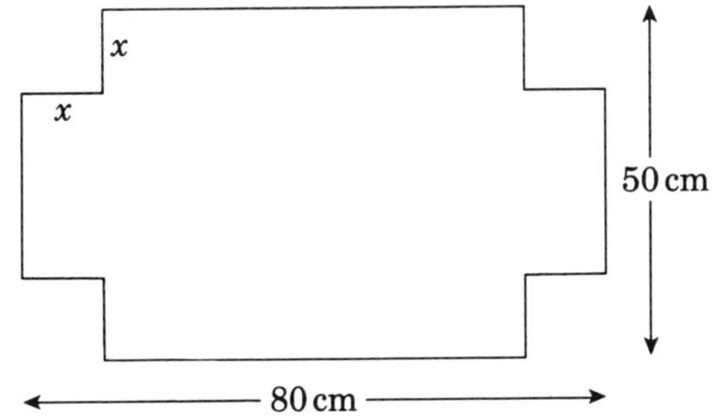

5 An open box is to be made from a sheet of card measuring 80 cm by 50 cm, by cutting a square of side x cm from each corner.

a) Express the volume of the box, $V\text{cm}^3$, in terms of the length, x cm.
b) Find the value of x which gives a maximum volume. What is the maximum volume?

6 A cylinder is to be fitted into a sphere of radius 20 cm. The cylinder has height h cm and base radius r cm.

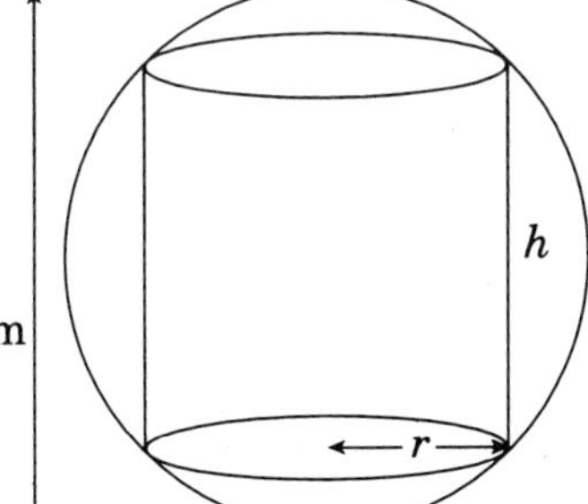

a) Use Pythagoras' theorem to show that

$$r^2 + \frac{h^2}{4} = 400$$

b) Express the volume of the cylinder, $V\text{cm}^3$, in terms of h only.
c) Find the value of h (and r) that maximises V. What is the maximum volume?

7 An athletics track consists of a rectangle with a semi-circle at each end, as in the diagram.

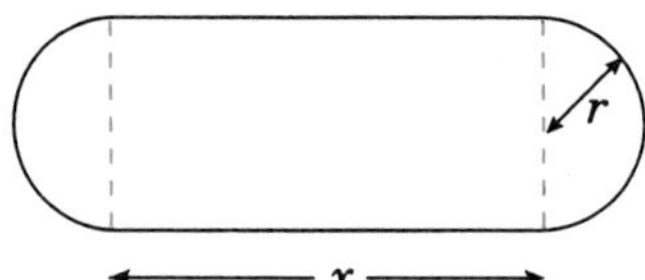

If the perimeter is to be exactly 400 metres, find the dimensions (x and r) that maximise the area of the rectangle.

7.5 B

1 An open box is made from a square sheet of metal with sides 1 metre long, by cutting out a square from each corner, folding up the sides and welding the edges.

Find the length of the box that gives the maximum volume.

2 A farmer wants to make three identical rectangular enclosures, side by side, using an existing wall as one boundary.

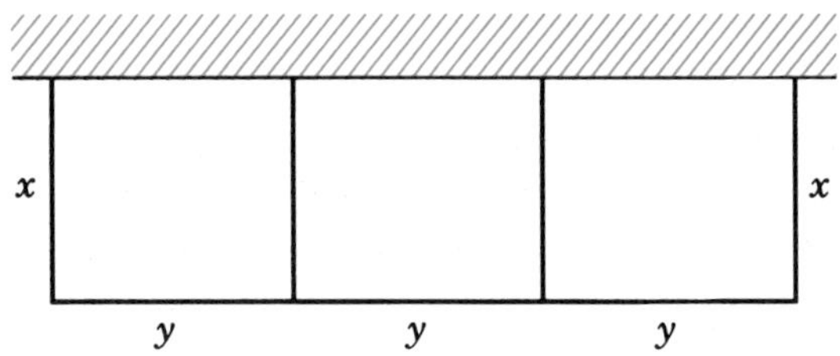

If he has 600 metres of fencing, what should be the dimensions of each enclosure if the total area is to be maximised?

3 The fuel economy E of a car, in miles per gallon, is given by:

$$E = 35 + 0.02v^2 - 3.85 \times 10^{-6}v^4, 5 \leq v \leq 70$$

where v is the speed in miles per hour.

What is the most economical speed to drive so that fuel consumption is a minimum?

4 A building contractor has 150 metres of security fencing and wishes to enclose a rectangular area. The area is divided into two parts with fencing. The areas A_1 and A_2 are such that $A_1 = 2A_2$.

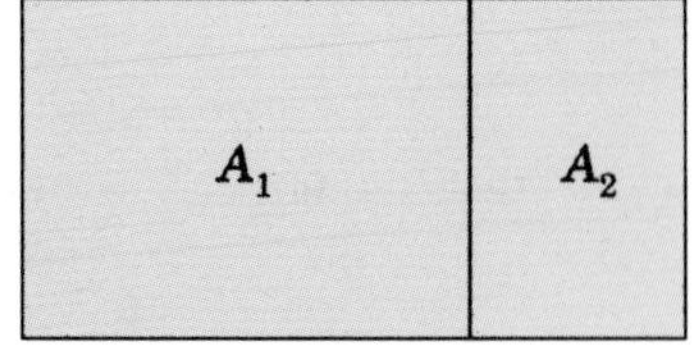

What are the dimensions of the enclosure so that the total enclosed area is a maximum?

5 A Norman window has a semi-circular arch above a rectangular section. Its perimeter is to be 6 m.

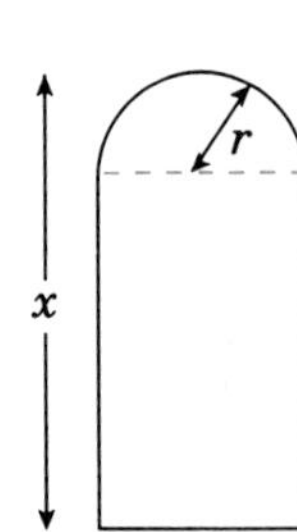

a) Show that $x = 3 - \frac{1}{2}\pi r$.

b) Show that the area of the window is maximised when $r = \dfrac{6}{\pi + 4}$.

c) Find the maximum area of the window.

6 The proportion of the river's energy, E, that can be obtained from an undershot water-wheel is given by:

$$E = 2v^3 - 4v^2 + 2v,\ 0 \le v \le 1$$

where v is the speed of the water-wheel relative to the speed of the river.

Show that only about 30 per cent of the river's energy can be captured, and that this occurs when the speed of the wheel is about one-third of the speed of the river.

7 A cylindrical can without a lid is made from aluminium sheet. If A is the surface area of the sheet used and V is the volume of the can, show that $V = \frac{1}{2}(Ar - \pi r^3)$.

Show that if A is given, the volume is a maximum when the diameter of the can is twice the height of the can.

MATHEMATICAL MODELLING ACTIVITY

1 → 2 → 3 → 4 → 5 → 6

Specify the real problem

Problem statement

A manufacturer of portable telephones has designed a new model. The market research department suggests that the number sold in the first year of production will depend on the selling price in roughly the following way.

Selling price (£)	50	75	100
Number sold	10 000	8000	6000

The company would have fixed costs of £25 000 and the manufacturing cost per item is £25.

- What selling price would you recommend in order that the company:
  a) maximises its revenue, **b)** maximises its profits?

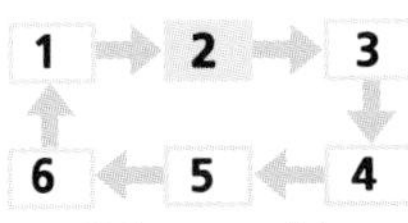

Set up a model

Set up a model

Firstly you need to identify the important variables:

- the selling price for each telephone,
- the number of telephones sold per annum,
- the revenue (receipts from sales) per annum,
- the costs involved (fixed and variable),
- the expected profit.

You will also need to make some assumptions before formulating the problem mathematically, some of which are simplistic, but necessary to tackle the problem initially:

- the relationship between number sold and price is linear,
- all telephones produced in the first year are sold.

Further refinements may be built in later.

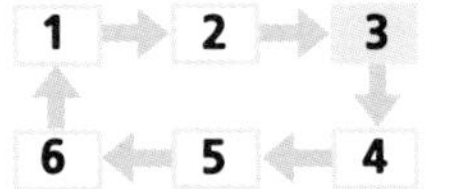

Formulate the mathematical problem

Mathematical problem

Firstly, establish the relationship between the number of telephones the company expects to sell (n) in terms of selling price (s).

The market research figures would seem to indicate a linear relationship, since for each reduction in price of £25, the company would expect to sell 20 000 more.

You should confirm that the equation connecting n and s is: $n = 14\,000 - 80s$. The revenue, R, is number sold × price = ns.

The cost of production, C, is fixed cost + variable cost = 25 000 + 25n.

The profit is the difference between revenue and cost = $R - C$.

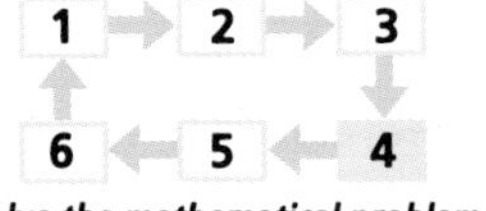

Solve the mathematical problem

Mathematical solution

Given that you need to find a selling price to:

a) maximise revenue or
b) maximise profit during the first year, you must first express both revenue, R, and profit, P, in terms of selling price, s.

Revenue function

$R = ns$ but $n = 14\,000 - 80s$
$\Rightarrow R = (14\,000 - 80s)s$
$\Rightarrow R = 14\,000s - 80s^2$

Profit function

$C = 25\,000 + 25n = 25\,000 + 25(14\,000 - 80s)$
$\quad = 25\,000 + 350\,000 - 2000s$
$\Rightarrow C = 375\,000 - 2000s$

and $P = R - C$
$\Rightarrow P = 14\,000s - 80s^2 - (375\,000 - 2000s)$
$\Rightarrow P = 16\,000s - 80s^2 - 375\,000$

Graphs of R and P in terms of s are shown below.

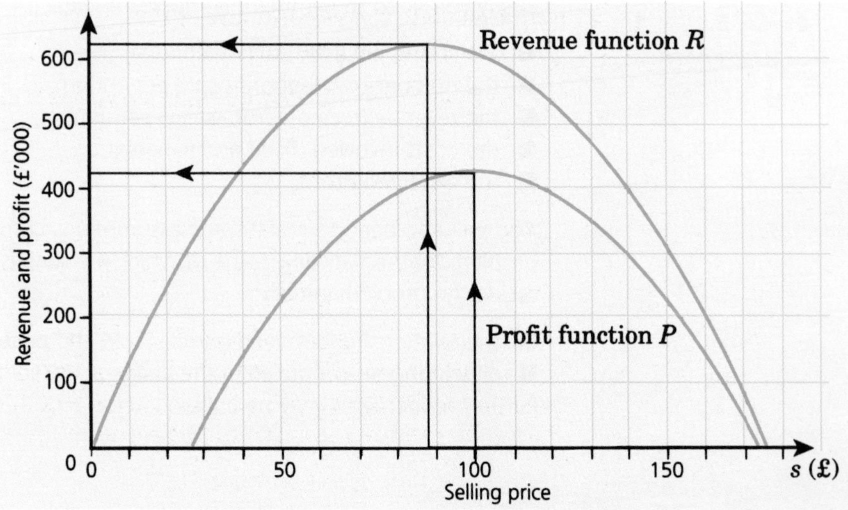

To find where R reaches a maximum:

$$\frac{dR}{ds} = 14\,000 - 160s = 0 \text{ for stationary values}$$

$$\Rightarrow \quad s = \frac{14000}{160} = 87.5$$

To find where P reaches a maximum:

$$\frac{dP}{ds} = 16\,000 - 160s = 0 \text{ for stationary values}$$

$$\Rightarrow \quad s = \frac{16000}{160} = 100$$

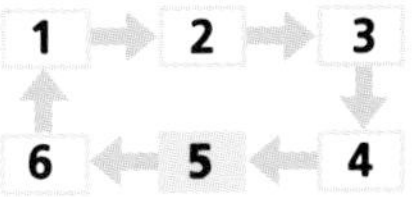

Interpret the solution

Interpretation

To maximise revenue the company should set the selling price at £87.50, but to maximise profit the price should be set at £100.

At a selling price of £87.50, the annual revenue would be £612 500, but costs of £200 000 reduce the profit to £412 500.

At a selling price of £100, the annual revenue would be reduced to £600 000, but reduced costs of £175 000, since fewer telephones are sold at a higher price, realise a profit of £425 000.

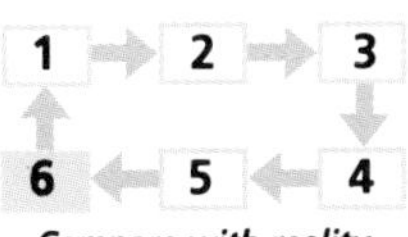

Compare with reality

Refinement of the model

Investigate the effect of:

- a change in the fixed costs,
- a change in the manufacturing cost per item,
- a different relationship between expected number sold and selling price (is a linear relationship realistic?).

CONSOLIDATION EXERCISES FOR CHAPTER 7

1 Given that $y = x^3 - 4x^2 + 5x - 2$, find $\frac{dy}{dx}$.

P is the point on the curve where $x = 3$.

a) Calculate the y-coordinate of P.
b) Calculate the gradient at P.

Find the values of x for which the curve has a gradient of 5.

(MEI, 1992 (part))

2 Given the function $y = 3x^4 + 4x^3$:

a) find $\frac{dy}{dx}$,
b) show that the graph of the function y has stationary points at $x = 0$ and $x = -1$ and find their coordinates,
c) determine whether each of the stationary points is a maximum, minimum or point of inflexion, giving reasons for your answers,
d) sketch the graph of the function y, giving the coordinates of the points where the curve cuts the axes.

(MEI, June 1992)

3 **a)** Differentiate $y = -x^2 + 2x + 3$.
b) Find the maximum value of y.

(MEI, January 1993 (part))

4 **a)** Given that $y = x^3 - x + 6$, find $\frac{dy}{dx}$.
On the curve representing y, P is the point where $x = -1$.
b) Calculate the y-coordinate of the point P.
c) Calculate the value of $\frac{dy}{dx}$ at P.
d) Find the equation of the tangent at P.
The tangent at the point Q is parallel to the tangent at P.
e) Find the coordinates of Q.

(MEI, June 1993 (part))

5 **a)** Given that $y = 5x^3 - 2x^2 + 1$, find $\frac{dy}{dx}$.

b) Hence find the exact values of x at which the graph of $y = 5x^3 - 2x^2 + 1$ has stationary points.

(SMP, Specimen Paper, 1994)

6 A particle moves along a line so that its distance, x metres, from O, after t seconds, is given by $x = t^3 - 9t^2 + 24t$.

a) Find the velocity v in terms of t, and the values of t for which the particle is at rest.
b) Calculate the distance between the two points where the particle is instantaneously at rest.
c) Calculate the acceleration of the particle at the times when it is instantaneously at rest.
d) Find the distance of the particle from O when its acceleration is zero.

(Oxford & Cambridge, Specimen Paper 1, 1994)

7 A curve has equation $y = x^3 + 3x^2 + 4x + 5$.
a) Find $\frac{dy}{dx}$.
b) Prove that the curve has no stationary points.
c) Show that the gradient function has a minimum value of 1 at $x = -1$.
d) Sketch the curve and explain why $x^3 + 3x^2 + 4x + 5 = 0$ has just one real root.

8 A curve has equation $y = 2x^3 - 9x^2 + 12x - 4$.
a) Find $\frac{dy}{dx}$.
b) Find the coordinates of the stationary points, showing that one lies on the x-axis.
c) Determine the values of x for which $y = 0$.
d) Sketch the curve.

9 Functions f and g are defined by:

$$f(x) = -\frac{x^3}{6} + x \text{ and } g(x) = \frac{x^5}{120} - \frac{x^3}{6} + x.$$

a) Evaluate f(0) and g(0).
b) Find $f'(x)$ and $g'(x)$ and evaluate $f'(0)$ and $g'(0)$.
c) Find the coordinates of the stationary points for:
i) $y = f(x)$, **ii)** $y = g(x)$.
d) Determine the values of x for which:
i) $f(x) = 0$, **ii)** $g(x) = 0$.
e) Sketch the curves $y = f(x)$ and $y = g(x)$ on the same axes.

10 An importer and distributor of computers has found an exclusive source of lap-top microcomputers. They will cost him £250 per machine. In addition, he will incur a cost of £5000 to adapt his distribution system to sell them, no matter how many machines he buys. The total cost of adapting his distribution system and buying n machines is £c. Express c in terms of n.

Experience suggests that the number, n, of machines is related to the selling price per machine, £s, by an equation

$$n = a + bs$$

where a and b are constants. The importer has been informed by his market research department that if he fixes the selling price at £400 per machine he is likely to sell about 5500 machines and, if he fixes it at £500, this will fall to about 3500 machines. Find a and b based on the information supplied by the market research department.

Show that the total profit, £p, the importer will make from selling all these machines is given by
$p = 18\,500s - 20s^2 - 3\,380\,000$.

Find the selling price per machine which will maximise the importer's total profit and hence find the number of machines he should purchase and his total profit on selling all the machines.

(NEAB, Specimen Paper 1)

Summary

Differentiation

Notation: $y = \mathrm{f}(x) \Rightarrow \dfrac{\mathrm{d}y}{\mathrm{d}x} = \mathrm{f}'(x)$

1 $\mathrm{f}(x) = x^n \Rightarrow \mathrm{f}'(x) = nx^{n-1}$ for $n = 1, 2, 3, \ldots$

2 $\mathrm{f}(x) = ax^n \Rightarrow \mathrm{f}'(x) = anx^{n-1}$ where a is a constant.

3 The derivative of a sum (or difference) is the sum (or difference) of the derivatives.

Rates of change

For any function $y = \mathrm{f}(x)$,
the **rate of change** of y with respect to x is given by $\dfrac{\mathrm{d}y}{\mathrm{d}x} = \mathrm{f}'(x)$.

In particular:
$s = \mathrm{f}(t)$ gives displacement s as a function of time t
$\Rightarrow v = \mathrm{f}'(t)$ gives velocity v as a function of time t
$\Rightarrow a = \mathrm{f}''(t)$ gives acceleration a as a function of time t.

Stationary values

For any function $y = \mathrm{f}(x)$, **stationary values** occur where $\mathrm{f}'(x) = 0$. A stationary value may be a **maximum**, **minimum** or **point of inflexion**.

Optimisation

Optimum values for problems may be found by formulating a function $y = \mathrm{f}(x)$ and finding value(s) of x such that $\mathrm{f}'(x) = 0$.

Trigonometry 1

In this chapter we:

- *investigate the sine, cosine and tangent of angles greater than 90° and explore the graphs of sine, cosine and tangent functions*
- *meet the idea of modelling periodic functions using sine and cosine*
- *solve simple trigonometric equations.*

TRIGONOMETRIC FUNCTIONS

Exploration 8.1

The position of an object moving in a circle

A big wheel at a fairground has 12 cars, equally spaced around the circumference. The radius of the wheel is 8 metres and the hub of the wheel is 9 metres above the ground.

- How high above the ground is each car?
- How far from the vertical support is each car?

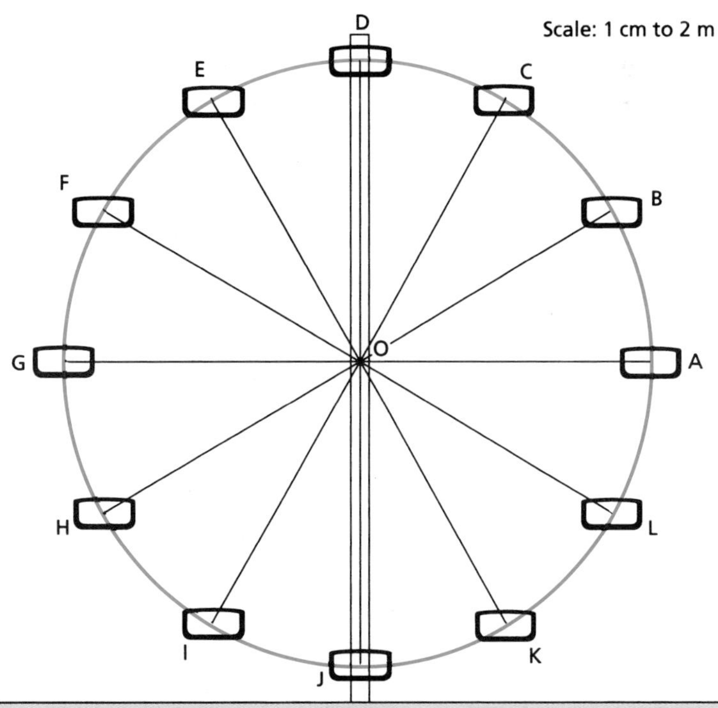

Modelling the big wheel

To model this situation, we can express the coordinates of a car, relative to O as origin, in terms of r, the radius of the wheel, and the angle AOP = θ, where P(x, y) is the position of the car.

In our 'big wheel' example, $r = 8$.

For car B $\theta = 30° \Rightarrow x = 8\cos 30°$ and $y = 8\sin 30°$

from which we can deduce:
  height of car B = 9 + 8sin 30° = 13 m
  distance of car B from support = 8cos 30° = 6.93 m.

For car C $\theta = 60° \Rightarrow x = 8\sin 30°$ and $y = 8\cos 30°$

from which we can deduce:
  height of car C = 9 + 8sin 60° = 15.93 m
  distance of car C from support = 8cos 60° = 4 m.

The positions of all the other cars, relative to O, may be deduced using similar methods, and by consideration of symmetry.

Sines and cosines of any angle

If the radius of the 'wheel' is 1 unit then $x = \cos\theta$ and $y = \sin\theta$, i.e. $\cos\theta$ is the x-coordinate of P and $\sin\theta$ is the y-coordinate of P. These definitions may be used for any angle θ, measured anticlockwise from the positive x-axis. The values for cosine and sine may be found using a calculator.

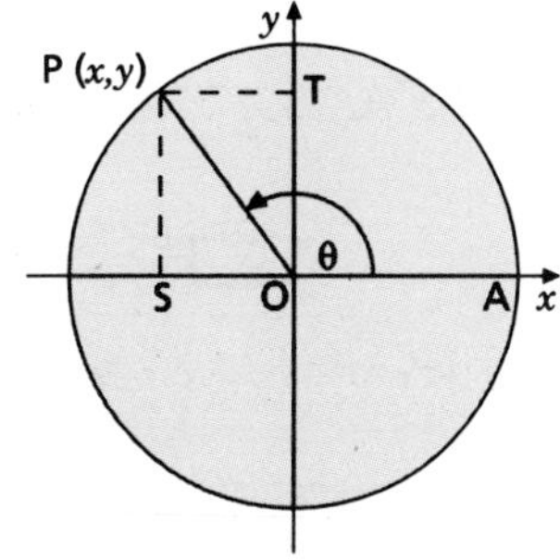

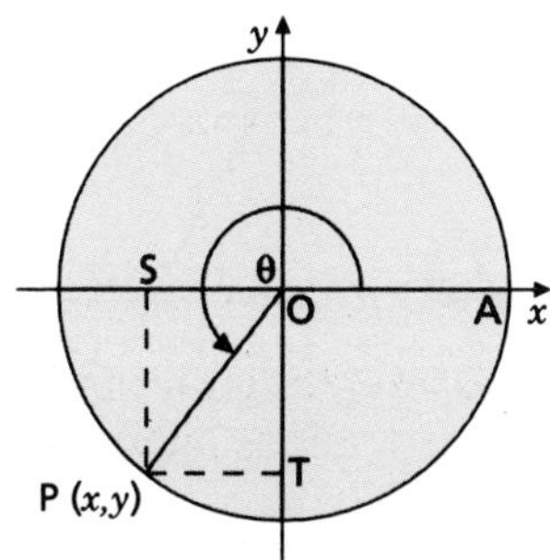

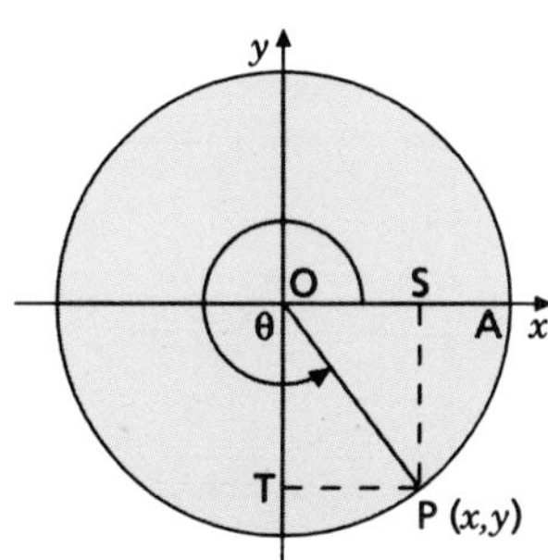

Exploration 8.2

Graphs of the sine and cosine functions

Refer again to the diagram in Exploration 8.1. Check your values for the coordinates of cars D to L relative to O using

$x = 8\cos\theta$ and $y = 8\sin\theta$.

As θ varies from 0° to 360°, sketch graphs of both $x = \cos\theta$ and $y = \sin\theta$, with values of $\cos\theta$ and $\sin\theta$ found from your calculator.

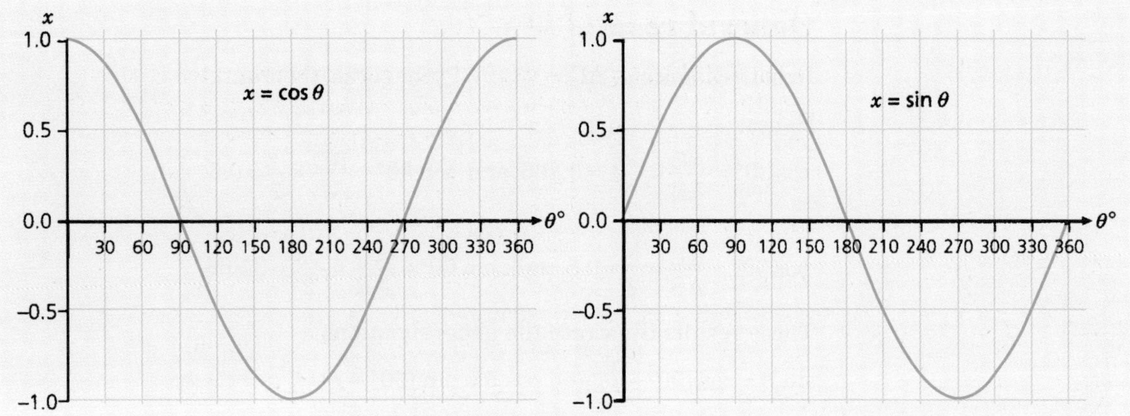

Circular functions

The definitions of x and y as $x = \cos\theta$ and $y = \sin\theta$ are derived from the coordinates of point P, as it moves round the circumference of the unit circle. Therefore they are called **circular functions**.

For either graph, plotting points at intervals of 30° and then joining them with a smooth curve is a good way to sketch it.

Remember these significant values for cosines and sines.

θ	0°	90°	180°	270°	360°
$\cos\theta$	1	0	−1	0	1
$\sin\theta$	0	1	0	−1	0

It is also useful to remember cosines and sines for angles 30° and 60°

Exploration 8.3

Expressions for sin θ and cos θ

Draw two equilateral triangles of side 2 units, one with $\theta = 30°$ and the other with $\theta = 60°$ as indicated.

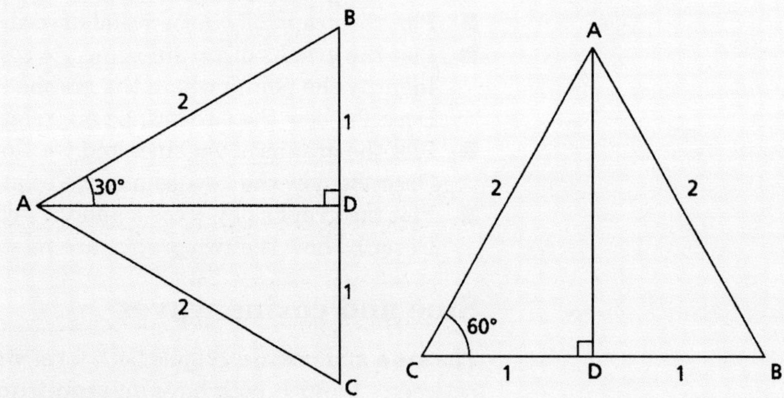

In each case, find expressions for $\cos\theta$ and $\sin\theta$.

General results

In both diagrams, AD = √3 (by Pythagoras' theorem).

Hence:

$$\cos 30° = \frac{AD}{AB} = \frac{\sqrt{3}}{2} \approx 0.866 \text{ and } \sin 30° = \frac{BD}{AB} = \frac{1}{2} = 0.5$$

$$\cos 60° = \frac{BD}{AB} = \frac{1}{2} = 0.5 \text{ and } \sin 60° = \frac{AD}{AB} = \frac{\sqrt{3}}{2} \approx 0.866$$

These results illustrate the generalisations:

$\sin\theta = \cos(90° - \theta)$ and $\cos\theta = \sin(90° - \theta)$

By the symmetry of the cosine and sine graphs, we can also deduce:

$\cos 30° = \cos 330° = \frac{\sqrt{3}}{2} \quad \Rightarrow \quad \cos 150° = \cos 210° = -\frac{\sqrt{3}}{2}$

$\sin 30° = \sin 150° = \frac{1}{2} \quad \Rightarrow \quad \sin 210° = \sin 330° = -\frac{1}{2}$

$\cos 60° = \cos 300° = \frac{1}{2} \quad \Rightarrow \quad \cos 120° = \sin 240° = -\frac{1}{2}$

$\sin 60° = \sin 120° = \frac{\sqrt{3}}{2} \quad \Rightarrow \quad \sin 240° = \cos 300° = -\frac{\sqrt{3}}{2}$

These results illustrate some useful generalisations:

$\sin\theta = \sin(180° - \theta)$ and $\cos\theta = -\cos(180° - \theta)$ $0° \le \theta \le 180°$

$\cos\theta = \cos(360° - \theta)$ and $\sin\theta = -\sin(360° - \theta)$ $0° \le \theta \le 360°$

CALCULATOR ACTIVITY

You will need a graphics calculator.

Firstly scale your axes:

$x_{min} = -360$ $x_{max} = 360$ $x_{scl} = 30$

$y_{min} = -1.5$ $y_{max} = 1.5$ $y_{scl} = 0.5$

and make sure you are in degree mode.

- Plot the graph of $y = \sin x$ and describe its properties.
- Plot the graph of $y = \cos x$ and describe its properties.
- Plot the graphs of $y = \sin x$ and $y = \cos x$.
  Identify the points where the graphs intersect.
  Describe how the two graphs are related.
- Plot the graphs of $y = \sin x$ and $y = -\sin x$.
  Describe how the two graphs are related.
- Plot the graphs of $y = \cos x$ and $y = -\cos x$.
  Describe how the two graphs are related.

Sine and cosine waves

The **sine** and **cosine** graphs both form **waves**. The waves **oscillate** between –1 and 1, both have an **amplitude** of 1 unit. You will note that both wave forms repeat themselves every 360°. This is the **period** of the sine and cosine graphs. Any function with a graph which exhibits a repeating pattern is **periodic**.

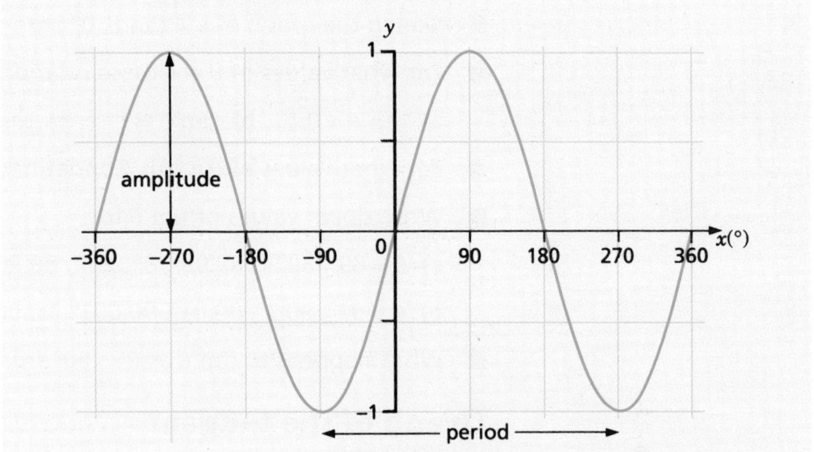

The size of angle is not restricted to the interval $-360° \leq \theta \leq 360°$. Indeed the wave form may be extended indefinitely for any angle, positive or negative.

Try some other intervals on your calculator.

Both wave forms have reflectional *and* rotational symmetry. There are an infinite number of mirror lines and centres of rotational symmetry for the infinite wave forms.

However, taking each wave over the interval $-360° \leq \theta \leq 360°$, we can see that:

- the sine curve has rotational symmetry about the origin, which makes it an **odd** function, and
- the cosine curve has reflectional symmetry in the y-axis, which makes it an **even** function.

Tangents of any angle

Following on from Explorations 8.1 and 8.2, the definitions of cosine and sine gave the coordinates (x, y) of point P as:

$x = r\cos\theta$ and $y = r\sin\theta$.

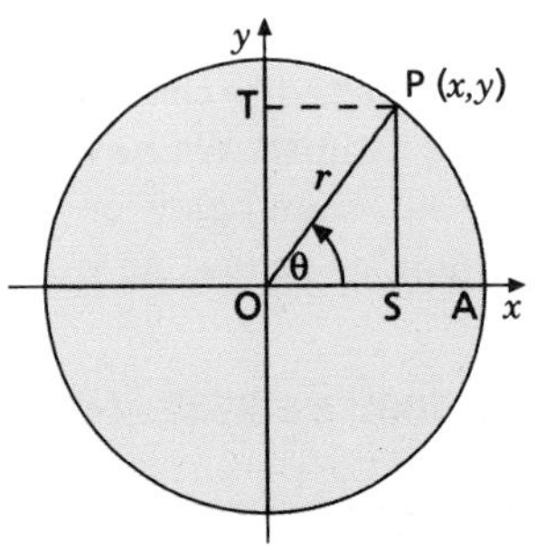

From the diagram:

$$\tan\theta = \frac{PS}{OS} = \frac{y}{x} = \frac{r\sin\theta}{r\cos\theta}$$

$$\Rightarrow \quad \tan\theta = \frac{\sin\theta}{\cos\theta}$$

This definition is extended to any angle θ, positive or negative.

Exploration 8.4

The graph of tan θ

- Sketch the graph of $z = \tan \theta$, $0° \le \theta \le 360°$.
- For what values of θ are these relationships true?

  a) $\tan \theta = 0$ **b)** $\tan \theta = 1$ **c)** $\tan \theta = -1$
- For what values of θ is $\tan \theta$ undefined ? Why?
- Write down values of $\tan \theta$ for:

  a) $\theta = 89°, 89.9°, 89.99°, 89.999°, 89.9999°$, etc.

  b) $\theta = 91°, 90.1°, 90.01°, 90.001°, 90.0001°$, etc.
- What happens to $\tan \theta$ as $\theta \Rightarrow 90°$ from either side of 90°?

Graph of the tangent

The **tangent graph** does not form a wave, but it does repeat itself every 180°, so it is also a periodic graph. The graph behaves in this way whatever interval for θ we take.

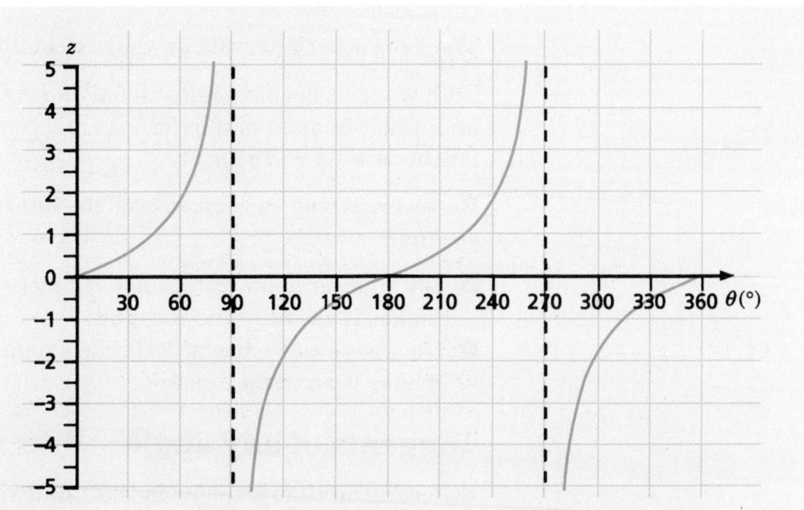

Like the sine graph, the tangent graph has rotational symmetry about the origin, which makes it an **odd** function.

The **range** of values the tangent can take is not restricted to the range −1 to 1. In fact, the range of the tangent function is infinite. We have found that by taking angles as close as we like to 90°, from below or above, we can make the value of the tangent as high, or as low, as we like.

Since the tangent of an angle is not defined for $\theta = 90°$, $\theta = 270°$, etc. the lines $\theta = 90°$, $\theta = 270°$, etc. are **asymptotes**.

EXERCISES

1 Using a scale of 1 cm to 30° on the x-axis and 1 cm to 0.2 units on the y-axis, plot the graphs of $y = \sin x$ and $y = \cos x$ on the same axes, over the interval $-180° \leq x \leq 180°$.

a) Write down the coordinates of the maximum and minimum points for each graph.
b) Write down the coordinates of the points where the graphs intersect; use your calculator to check your approximation.

2 Use an isosceles right-angled triangle to find values for cos 45°, sin 45° and tan 45°. Check the decimal equivalents on your calculator.

a) What angles between –360° and 360° have cosines the same as:
i) cos 45° **ii)** –cos 45°?
b) What angles between –360° and 360° have sines the same as:
i) sin 45° **ii)** –sin 45°?
c) What angles between –360° and 360° have tangents the same as:
i) tan 45° **ii)** –tan 45°?

3 Find values for tan 30° and tan 60° from the diagram on page 175 and the graphs on page 178. Check the decimal equivalents on your calculator.

What angles between –360° and 360° have tangents the same as:
a) tan 30° **b)** tan 60° **c)** –tan 30° **d)** –tan 60°?

4 Using a scale of 1 cm to 30° on the x-axis and 1 cm to 1 unit on the y-axis, plot the graphs of $y = \tan x$ and $y = -\tan x$ on the same axes, over the interval $-180° \leq x \leq 180°$.
a) Write down the values of x for which the functions are not defined.
b) Write down the coordinates of the points where the graphs intersect.

5 Using your calculator and the appropriate graph, find all values of θ, where $0° \leq \theta \leq 360°$, such that:
a) $\sin \theta = 0.7$ **b)** $\sin \theta = 0.25$ **c)** $\sin \theta = -0.4$ **d)** $\sin \theta = -0.866$
e) $\cos \theta = 0.7$ **f)** $\cos \theta = 0.25$ **g)** $\cos \theta = -0.4$ **h)** $\cos \theta = -0.866$
i) $\tan \theta = 0.7$ **j)** $\tan \theta = 2.5$ **k)** $\tan \theta = -0.4$ **l)** $\tan \theta = -1.732$.

6 Using your calculator and the appropriate graph, find all values of θ (where possible), such that $-180° \leq \theta \leq 180°$ and:
a) $\cos \theta = 0.15$ **b)** $\sin \theta = 1.25$ **c)** $\tan \theta = -1$ **d)** $\cos \theta = -0.5$
e) $\sin \theta = -0.33$ **f)** $\tan \theta = 10$ **g)** $\cos \theta = -2.4$ **h)** $\sin \theta = 0.866$
i) $\tan \theta = -0.75$ **j)** $\tan \theta = 2.5$ **k)** $\tan \theta = 0.577$ **l)** $\tan \theta = -4$.

7 Copy and complete the following table and comment on the results (give values of $\sin \theta$ and $\tan \theta$ to five decimal places).

θ	5°	4°	3°	2°	1°	0.5°	0°
$\sin \theta$							
$\tan \theta$							

8.1 B

1 Using a scale of 1 cm to 60° on the x-axis and 1 cm to 0.5 units on the y-axis, plot the graphs of $y = \sin x$, $y = -\sin x$ and $y = \sin(-x)$ on the same axes over the interval $-360° \le x \le 360°$. Comment on your graphs.

2 Using a scale of 1 cm to 60° on the x-axis and 1 cm to 0.5 units on the y-axis, plot the graphs of $y = \cos x$, $y = -\cos x$ and $y = \cos(-x)$ on the same axes over the interval $-360° \le x \le 360°$. Comment on your graphs.

3 Using a scale of 1 cm to 30° on the x-axis and 1 cm to 0.5 units on the y-axis, plot the graphs of $y = \cos x$ and $y = \tan x$ on the same axes over the interval $-90° \le x \le 90°$. State the coordinates of the points of intersection.

4 Using the diagrams on page 175 and the appropriate graphs, write down the sine, cosine and tangent of each of the following angles. Check the decimal equivalents using a calculator.

a) 120° **b)** 150° **c)** 210° **d)** 300°

5 Using your calculator and the appropriate graph, find all values of θ, where $0° \le \theta \le 360°$, such that:

a) $\sin\theta = 0.6$ **b)** $\sin\theta = 0.1$ **c)** $\sin\theta = -0.75$ **d)** $\sin\theta = -0.3$
e) $\cos\theta = 0.6$ **f)** $\cos\theta = 0.1$ **g)** $\cos\theta = -0.75$ **h)** $\cos\theta = -0.3$
i) $\tan\theta = 0.6$ **j)** $\tan\theta = 0.1$ **k)** $\tan\theta = -0.75$ **l)** $\tan\theta = -0.3$.

6 Using your calculator and the appropriate graph, find where possible values of θ, where $-180° \le \theta \le 180°$, such that:

a) $\sin\theta = -0.8$ **b)** $\cos\theta = -0.35$ **c)** $\tan\theta = 50$ **d)** $\cos\theta = 50$
e) $\sin\theta = 0.02$ **f)** $\tan\theta = -0.4$ **g)** $\sin\theta = -1.01$ **h)** $\cos\theta = 0.94$
i) $\tan\theta = -100$ **j)** $\sin\theta = 0.5$ **k)** $\cos\theta = -0.45$ **l)** $\tan\theta = 0.2$.

7 Choose any angle θ between 0 and 90°. Use your calculator to evaluate $(\cos\theta)^2 + (\sin\theta)^2$. Repeat for other values of θ, both inside and outside the interval $0° \le \theta \le 90°$. Comment on your findings.

TRANSFORMATION OF TRIGONOMETRIC FUNCTIONS

The three basic trigonometric graphs, in terms of x and y:

$y = \cos x$, $y = \sin x$ and $y = \tan x$

may all be subject to simple transformations and combination of transformations.

Exploration 8.5

Transforming $y = \sin x$

Beginning with the graph of $y = \sin x$, examine the effect of each of the following transformations.

- reflection in the x-axis
- stretch parallel to the y-axis
- translation parallel to the y-axis
- stretch parallel to the x-axis
- translation parallel to the x-axis

In each case, what happens to the equation for the transformed graph?

Now try again, starting with $y = \cos x$ and then $y = \tan x$.

Combine the transformations in different ways and investigate whether the order of transformations matters.

CALCULATOR ACTIVITY

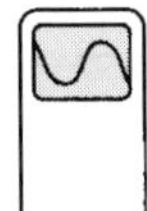

Transformations

The transformation of the trigonometric functions may be illustrated using a suitable graph plotter or on a graphics calculator. Set up the axes using suitable scales, and experiment. Check the solutions of the following worked examples, but always attempt a sketch yourself before using the computer or calculator.

Example 8.1

Sketch three copies of the graph of $y = \sin x$, $0° \le x \le 360°$, $-3 \le y \le 3$.

On each sketch respectively superimpose the graphs of:
a) $y = 3\sin x$ ***b)*** $y = \sin 2x$ ***c)*** $y = 3\sin 2x$.
In each case describe the transformation from $y = \sin x$ to a), b) or c) geometrically.

Solution

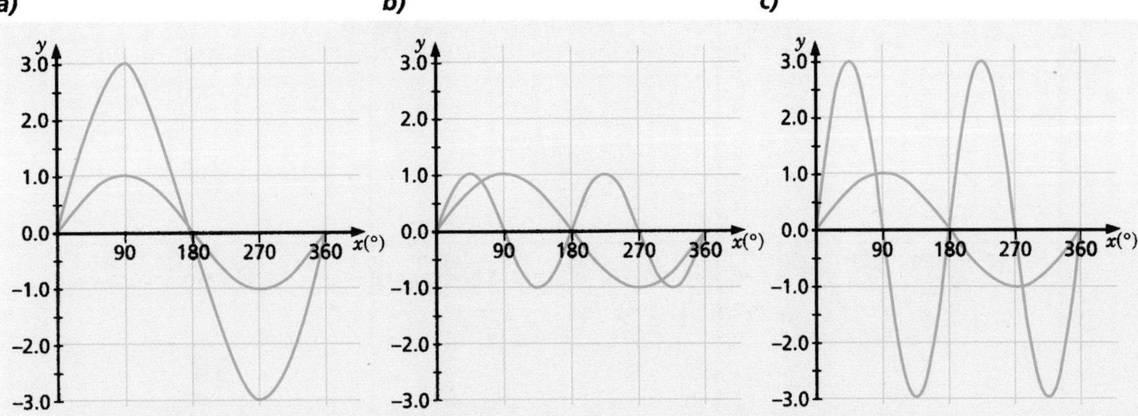

Stretch factor 3, parallel to y-axis Stretch factor 0.5 parallel to x-axis b) followed by a)

Example 8.2

Sketch three copies of the graph of $y = \cos x$, $0° \le x \le 360°$, $-1 \le y \le 3$.

On each sketch respectively show the result of transforming $y = \cos x$ *by:*

a) *translation through +2 units parallel to the* y-axis,

b) *translation through +90° parallel to the* x-axis,

c) ***b)** followed by **a)**.*

In each case give the equation of the transformed graph and state the coordinates of its stationary points.

Solution

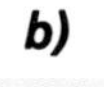

a) **b)** **c)**

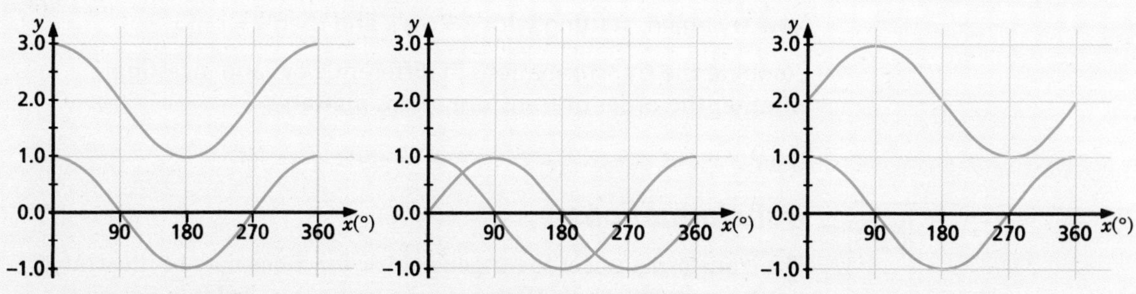

a) $y = \cos x + 2$
(0, 3), (180, 1), (360, 3)

b) $y = \cos(x - 90°)$
(90, 1), (270, −1)

c) $y = \cos(x - 90°) + 2$
(90, 3), (270, 1)

Example 8.3

Sketch three copies of the graph of $y = \tan x$, $0° \le x \le 360°$, $-5 \le y \le 5$.

On each sketch respectively show the result of transforming $y = \tan x$ *by:*

a) *translation through −60° parallel to the* x*-axis,*

b) *stretch, factor 2 parallel to the* y*-axis,*

c) *a) followed by b).*

In each case give the equation of the transformed graph.

Solution

a) **b)** **c)**

a) $y = \tan(x + 60°)$

b) $y = 2\tan x$

c) $y = 2\tan(x + 60°)$

EXERCISES

8.2 A

1 For each of the following trigonometric functions:

a) sketch the graph of the function for $0° \le x \le 360°$,
b) label the points of intersection with the axes,
c) state the range of the function in the form $a \le y \le b$.

i) $y = 3 + \sin x$ ii) $y = -\sin x$ iii) $y = 0.5 \sin x$
iv) $y = 2\cos x$ v) $y = 2 - \cos x$ vi) $y = -3\cos x$
vii) $y = \cos 3x$ viii) $y = \cos 0.5x$ ix) $y = \tan 2x$
x) $y = \sin (x + 90°)$ xi) $y = \cos (x - 60°)$ xii) $y = \tan (x + 45°)$.

2 For each of the following trigonometric functions:

a) sketch graphs of $y = \sin x$ and the given function, for $0° \le x \le 360°$,
b) describe the geometrical transformation(s) that map(s) $y = \sin x$ onto the given function.

i) $y = \frac{1}{2}\sin x$ ii) $y = \sin x - 1$
iii) $y = \sin 3x$ iv) $y = \sin (x + 60°)$
v) $y = -\sin 3x$ vi) $y = \frac{1}{2}\sin 3x$
vii) $y = \frac{1}{2}\sin x + 3$ viii) $y = \sin \frac{1}{2}x$
ix) $y = 5 \sin \frac{1}{2}x$ x) $y = 3\sin (x + 60°)$
xi) $y = \sin (x + 60°) - 5$ xii) $y = 3\sin (x + 60°) - 5$.

3 For each geometrical transformation:

a) sketch graphs of $y = \cos x$, for $0° \le x \le 360°$, and the transformation of $y = \cos x$,
b) write down the equation of the transformed graph,
c) find the coordinates of any stationary points.

i) stretch, factor 4 parallel to the y-axis
ii) translation –3 units parallel to the y-axis
iii) translation –90° parallel to the x-axis
iv) stretch, factor 2.5 parallel to the x-axis
v) stretch, factor $\frac{1}{3}$ parallel to the x-axis *followed by* translation –5 units parallel to the y-axis
vi) translation –90° parallel to the x-axis *followed by* stretch, factor 4 parallel to the y-axis
vii) translation +120° parallel to the x-axis *followed by* reflection in the x-axis
viii) translation –60° parallel to the x-axis *followed by* stretch, factor 0.5 parallel to the y-axis
ix) translation –45° parallel to the x-axis *followed by* stretch, factor 3 parallel to the y-axis *followed by* translation +3 units parallel to the y-axis
x) stretch, factor 2 parallel to the x-axis *followed by* stretch, factor 1.5 parallel to the y-axis *followed by* translation –3.5 units parallel to the y-axis

8.2B

1 For each of the following trigonometric functions:

a) sketch the graph of the function for $0° \le x \le 360°$,
b) label the points of intersection with the axes,
c) state the range of the function in the form $a \le y \le b$.

i) $y = 4 - \cos x$ **ii)** $y = 2 + \sin x$
iii) $y = 3\cos x$ **iv)** $y = -2.5\sin x$
v) $y = 2\cos x$ **vi)** $y = \tan 3x$
vii) $y = \tan (90° - x)$ **viii)** $y = \sin (30° + x)$
ix) $y = \sin 3x - 1$ **x)** $y = \cos (30° + x) + 2$

2 For each of the following trigonometric functions:

a) sketch graphs of $y = \cos x$ and the given function, for $0° \le x \le 360°$,
b) describe the geometrical transformation(s) that map(s) $y = \cos x$ onto the given function.

i) $y = \frac{1}{3}\cos x$ **ii)** $y = 4 + \cos x$
iii) $y = \cos 6x$ **iv)** $y = \cos (90° - x)$
v) $y = -\cos 6x$ **vi)** $y = \frac{1}{3}\cos 2x$
vii) $y = \frac{1}{3}\cos x + 1$ **viii)** $y = \cos \frac{1}{3}x$
ix) $y = 6\cos \frac{1}{3}x$ **x)** $y = \cos \frac{1}{2}x + 2$

3 For each geometrical transformation:

a) sketch the graph of $y = \sin x$ and the transformation of $y = \sin x$,
b) write down the equation of the transformed graph.

i) stretch, factor $\frac{1}{2}$ parallel to the y-axis
ii) stretch, factor $\frac{1}{4}$ parallel to the x-axis
iii) translation –2 units parallel to the y-axis
iv) translation +60° parallel to the x-axis
v) stretch, factor 2 parallel to the y-axis *followed by* translation +1 unit parallel to the y-axis
vi) stretch, factor $\frac{1}{2}$ parallel to the x-axis *followed by* reflection in the x-axis
vii) translation –30° parallel to the x-axis *followed by* translation –2 units parallel to the y-axis
viii) reflection in the x-axis *followed by* translation +3 units parallel to the y-axis
ix) translation +90° parallel to the x-axis *followed by* stretch, factor 2.5 units parallel to the y-axis *followed by* reflection in the x-axis
x) stretch, factor 2 parallel to the x-axis *followed by* stretch, factor –1 parallel to the y-axis *followed by* stretch, factor 0.25 parallel to the y-axis.

TRIGONOMETRIC EQUATIONS

A **trigonometric equation** is an equation which contains a trigonometric function such as sine, cosine or tangent. To **solve** a trigonometric equation, we need to find values of the angle which satisfy the equation.

When using a calculator with the inverse trigonometric functions $\mathbf{sin^{-1}}$, $\mathbf{cos^{-1}}$, $\mathbf{tan^{-1}}$, the value on the display is called the **principal value** and is usually in the range $-180° \le \theta \le 180°$. For example,

$\cos^{-1}(-0.5) = 120°$, $\sin^{-1}(-0.5) = -30°$, $\tan^{-1}(-1) = -45°$.

If the principal value for the angle is not in the required interval, adding multiples of the period of the function (360° for sine and cosine, 180° for tangent) will give a suitable solution. Other solutions may be obtained by using symmetry properties of the graphs, etc.

Example 8.4

Find values of θ *such that* $0° \le \theta \le 360°$ *where* $4\cos\theta = 3$.

Solution

$4\cos\theta = 3 \Rightarrow \cos\theta = 0.75$

$\Rightarrow \theta = \cos^{-1}(0.75) = 41.4°$ (to 1 d.p.) (*from calculator*)

but $0° \le \theta \le 360° \Rightarrow \theta = 41.4°$ or $\theta = 360° - 41.4° = 318.6°$

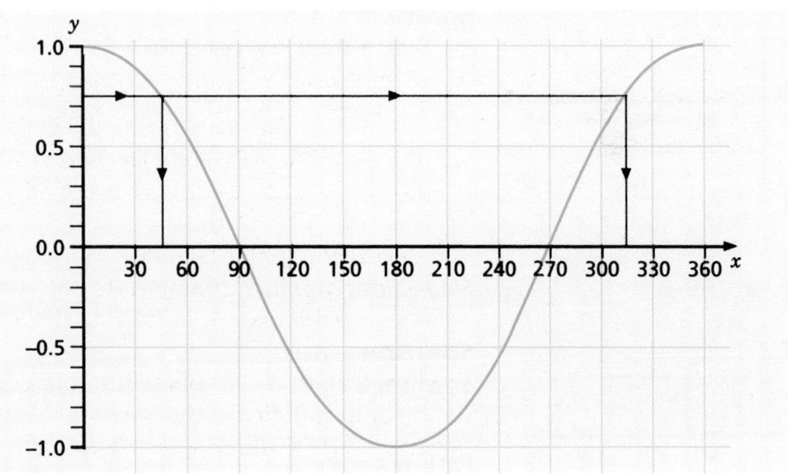

Example 8.5

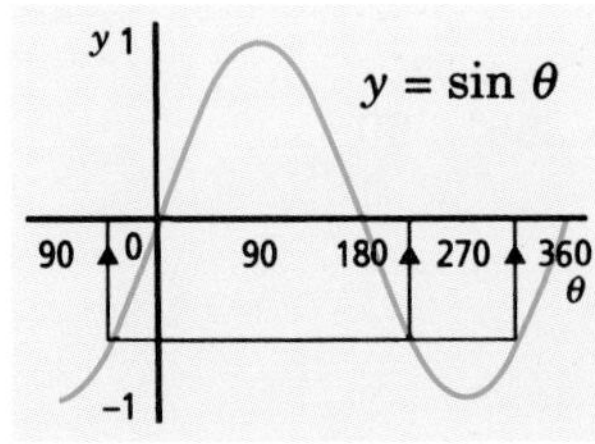

Find values of θ *such that* $0° \le \theta \le 360°$ *where* $5\sin\theta = -3$.

Solution

$5\sin\theta = -3 \Rightarrow \sin\theta = -0.6$

$\Rightarrow \theta = \sin^{-1}(-0.6) = -36.9°$ (to 1 d.p.) (*from calculator*)

but $0° \le \theta \le 360°$ *and* $\sin\theta$ has period 360°:

$\Rightarrow \theta = 180° + 36.9°$ or $\theta = 360° - 36.9°$

$\Rightarrow \theta = 216.9°$ or $\theta = 323.1°$

Example 8.6

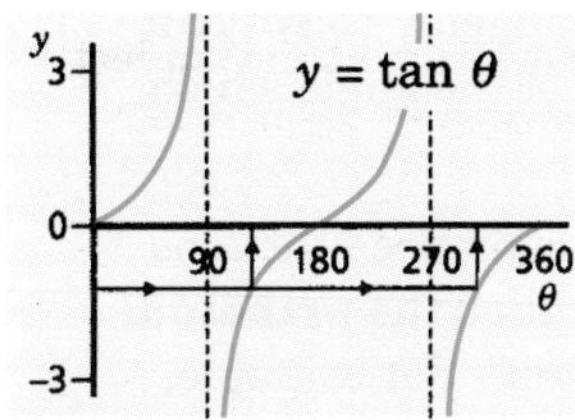

Find values of θ such that $0° \le \theta \le 360°$ *where* $10\tan\theta + 7 = 3$.

Solution

$10\tan\theta + 7 = 3 \Rightarrow \tan\theta = -0.4$

$\Rightarrow \theta = \tan^{-1}(-0.4) = -21.8°$ (to 1 d.p.) (*from calculator*)

but $0° \le \theta \le 360°$ *and* $\tan\theta$ has period 180°:

$\Rightarrow \theta = -21.8° + 180°$ or $\theta = -21.8° + 360°$

$\Rightarrow \theta = 158.2°$ *or* $\theta = 338.2°$

Example 8.7

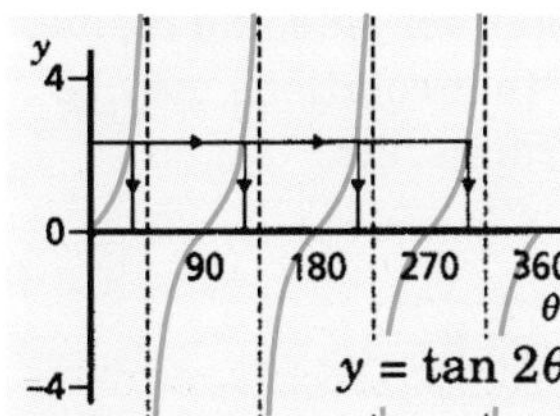

Find values of θ such that $0° \le \theta \le 360°$ *where* $2\tan 2\theta = 5$.

Solution

$2\tan 2\theta = 5 \Rightarrow \tan 2\theta = 2.5$

$\Rightarrow 2\theta = \tan^{-1}(2.5) = 68.2°$ (to 1 d.p.) (*from calculator*)

$\Rightarrow \theta = 68.2° \div 2 = 34.1°$ (to 1 d.p.)

but $0° \le \theta \le 360°$ *and* $\tan 2\theta$ has period 90°:

$\Rightarrow \theta = 34.1°, 34.1° + 90°, 34.1° + 180°$ or $34.1° + 270°$

$\Rightarrow \theta = 34.1, 124.1, 214.1$ or 304.1

Example 8.8

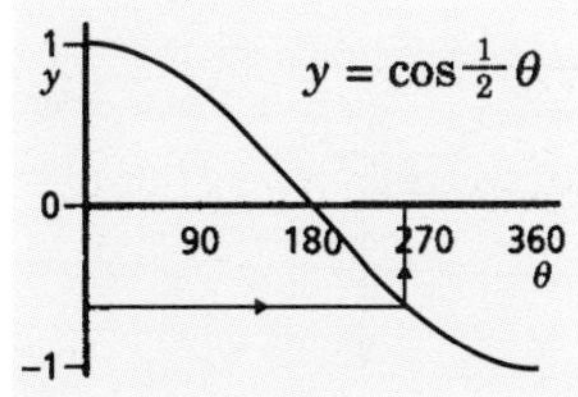

Find values of θ such that $0° \le \theta \le 360°$ *where* $2 + 3\cos\frac{1}{2}\theta = 0$.

Solution

$2 + 3\cos\frac{1}{2}\theta = 0 \Rightarrow \cos\frac{1}{2}\theta = -\frac{2}{3}$

$\Rightarrow \frac{1}{2}\theta = \cos^{-1}(-\frac{2}{3}) = 131.8°$ (to 1 d.p.) (*from calculator*)

$\Rightarrow \theta = 131.8° \times 2 = 263.6°$ (to 1 d.p.)

but $0° \le \theta \le 360°$ *and* $\cos\frac{1}{2}\theta$ has period 720°:

$\Rightarrow \theta = 263.6°$ is the only solution.

Example 8.9

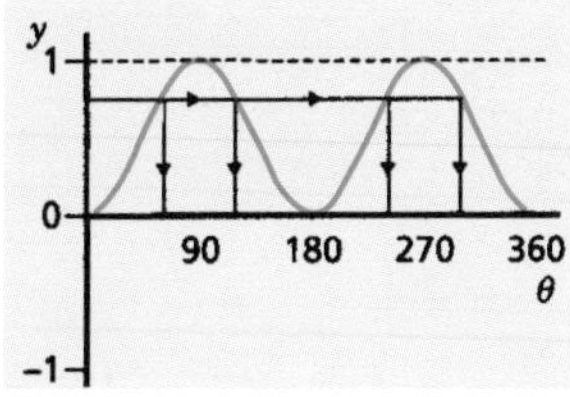

Find values of θ such that $0° \le \theta \le 360°$ *where* $4\sin^2\theta = 3$.

Solution

$4\sin^2\theta = 3 \quad \Rightarrow \sin\theta = \pm\sqrt{\frac{3}{4}} = \pm\frac{\sqrt{3}}{2}$

Either $\sin\theta = +\frac{\sqrt{3}}{2} \quad \Rightarrow \theta = \sin^{-1}(+\frac{\sqrt{3}}{2}) = 60°$

but $0° \le \theta \le 360°$ *and* $\sin\theta$ has period 360°:

$\Rightarrow \theta = 60°$ or $\theta = 180° - 60° = 120°$

or $\sin\theta = -\frac{\sqrt{3}}{2} \quad \Rightarrow \theta = \sin^{-1}(-\frac{\sqrt{3}}{2}) = -60°$

but $0° \le \theta \le 360°$ *and* $\sin\theta$ has period 360°:

$\Rightarrow \theta = 180° + 60°$ *or* $\theta = 360° - 60°$

$\Rightarrow \theta = 240°$ *or* $\theta = 300°$

Solving trigonometric equations

The steps in solving trigonometric equations are:

1 Rearrange the equation to make sine, cosine or tangent the subject.
2 Use a calculator, where necessary, to find a principal value of the angle.
3 Using a graph, or otherwise, find other solutions to the equation in the given interval.

EXERCISES

1 Using your calculator and the appropriate graph, find all values of θ such that $0° \le \theta \le 360°$, where:

a) $5\cos\theta = 4$ **b)** $13\sin\theta = 5$
c) $4\tan\theta = 7$ **d)** $10\cos\theta = -3$
e) $8\sin\theta + 3 = 0$ **f)** $5\tan\theta = -12$
g) $8\cos\theta - 17 = 0$ **h)** $6 - 7\tan\theta = 30$.

2 Using your calculator and the appropriate graph, find all values of θ such that $0° \le \theta \le 360°$, where:

a) $5\sin 2\theta = 4$ **b)** $13\cos 2\theta = 5$
c) $9\tan 2\theta = 7$ **d)** $10\cos\frac{1}{2}\theta = -7$
e) $8\sin\frac{1}{2}\theta + 1 = 0$ **f)** $8\tan\frac{1}{2}\theta = -15$
g) $25\cos 3\theta - 7 = 0$ **h)** $6 - 7\tan\frac{1}{3}\theta = 30$.

3 Using your calculator and the appropriate graph, find all values of θ such that $0° \le \theta \le 360°$, where:

a) $5\cos^2\theta = 4$ **b)** $13\sin^2\theta = 5$
c) $4\tan^2\theta = 7$ **d)** $16\cos^2\theta + 6 = 15$
e) $8\sin^2\theta - 3 = 0$ **f)** $15 - 5\tan^2\theta = 0$.

4 Find (if possible) all values of θ such that $-180° \le \theta \le 180°$, where:

a) $15\cos 2\theta = 7$ **b)** $5\sin\theta = 13$
c) $25\tan^2\theta = 9$ **d)** $14\cos(\theta + 10°) = -7$
e) $8\sin^2\theta + 3 = 0$ **f)** $21 - 7\tan^2 2\theta = 0$
g) $16\cos^3\theta = 2$ **h)** $10 - 8\sin 5\theta = 3$
i) $5 + 2\tan(\theta - 30°) = 8$ **j)** $8 - 4\cos 3\theta = 3$.

1 Using your calculator and the appropriate graph, find all values of θ such that $0° \le \theta \le 360°$, where:

a) $10\sin\theta = 7$ **b)** $5\cos\theta = 2$
c) $3\tan\theta = 8$ **d)** $12\sin\theta = -5$
e) $8\cos\theta = -2$ **f)** $4 - 3\tan\theta = 11$
g) $5\sin\theta + 4 = 1$ **h)** $9\cos\theta - 6 = -14$.

2 Using your calculator and the appropriate graph, find (where possible) all values of θ such that $0° \leq \theta \leq 360°$, where:

a) $10\cos 2\theta = 7$ **b)** $5\sin 2\theta = 2$
c) $7\tan 3\theta = 8$ **d)** $7\cos \frac{1}{2}\theta = -4$
e) $3\tan \frac{1}{4}\theta = -4$ **f)** $5\sin 2\theta + 3 = 6$
g) $11 + 7\cos \frac{1}{2}\theta = 8$ **h)** $9 + 2\tan \frac{1}{2}\theta = 14$.

3 Using your calculator and the appropriate graph, find all values of θ such that $0° \leq \theta \leq 360°$, where:

a) $9\cos^2 \theta = 4$ **b)** $5\sin^2 \theta = 2$
c) $6\tan^2 \theta = 5$ **d)** $12\cos^2 \theta - 2 = 3$
e) $8\sin^2 \theta - 2 = 1$ **f)** $14 - 4\tan^2 \theta = 0$

4 Find (if possible) all values of θ such that $-180° \leq \theta \leq 180°$, where:

a) $17\sin 2\theta = 8$ **b)** $3\cos 3\theta = 4$
c) $9\tan^2 \theta = 36$ **d)** $15\sin (\theta + 14°) = -4$
e) $27\sin^3 \theta = 8$ **f)** $3\tan^2 \theta + 11 = 0$
g) $12 - 15\sin^2 2\theta = 0$ **h)** $1 + 6\cos \frac{1}{2}\theta = -2$
i) $13 - 8\cos^2 \frac{1}{2}\theta = 7$ **j)** $15 - \tan (2\theta + 15°) = 0$

WAVE MODELS

Back to the big wheel!

At the beginning of this chapter, we were looking at the motion of a big wheel and how the position of a car, relative to the hub, can be given in terms of the sine and cosine of the anti-clockwise angle made with the horizontal passing through the hub.

Exploration 8.6

Motion of the cars

Suppose that the wheel is now rotating at two revolutions per minute. Assume that the wheel is in the position shown in Exploration 8.1 when $t = 0$ and that it is rotating anticlockwise.

- Find an equation for the height h metres of car A after t seconds.
- Sketch a graph of h against t for $0 \leq t \leq 30$.
- Find the time interval for which car A is:
  - **a)** less than 5 metres above the ground,
  - **b)** at least 15 metres above the ground.

Repeat for car D, plotting the graph of h against t on the same axes as for car A.

When are cars A and D at the same height as each other?

Repeat for car K, plotting the graph of h against t on the same axes as for car A.

Describe the relationship between the three graphs.

Relating *h* and *t*

The wheel rotates through 360° in 30 seconds, which means that it turns through 12° in one second, or $12t°$ in t seconds, i.e. $\theta = 12t$.

For car A, relative to O, $y = 8\sin\theta \Rightarrow y = 8\sin 12t°$.

Therefore, relative to the ground, height h in terms of time t is $y = 9 + 8\sin 12t°$.

A sketch graph of h against t for car A looks like this.

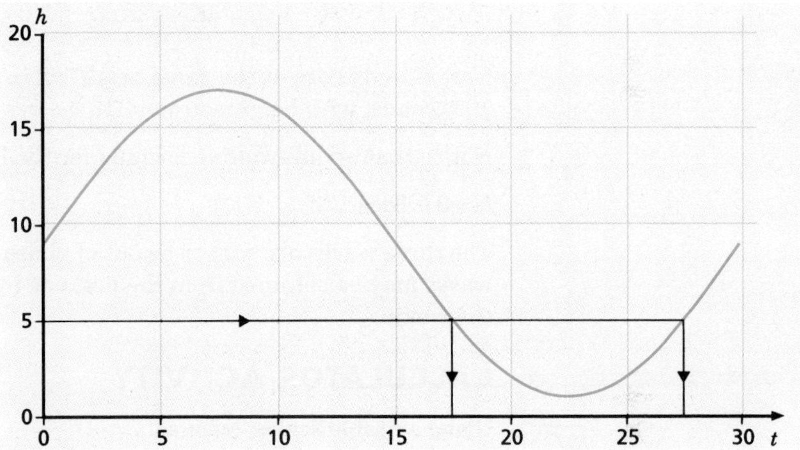

From the graph it looks as though car A is less than 5 metres above the ground from $t = 17$ to $t = 28$, i.e. for about 11 seconds.

Using the techniques of the last section, the corresponding times may be found precisely by solving the equation:

$$5 = 9 + 8\sin 12t°$$
$$\Rightarrow \quad 8\sin 12t° = -4$$
$$\Rightarrow \quad \sin 12t° = -0.5$$
$$\Rightarrow \quad 12t = 210 \text{ or } 12t = 330$$
$$\Rightarrow \quad t = 17.5 \text{ or } t = 27.5 \text{ seconds}$$

Cars D and K are 'ahead' of car A, in terms of anti-clockwise position from A.

For car D $\theta = 90° + 12t°$

For car K $\theta = 210° + 12t°$.

The corresponding height functions are:

For car D $h = 9 + 8\sin(90 + 12t)°$

For car K $h = 9 + 8\sin(210 + 12t)°$

The graphs of h against t for all three cars are shown below.

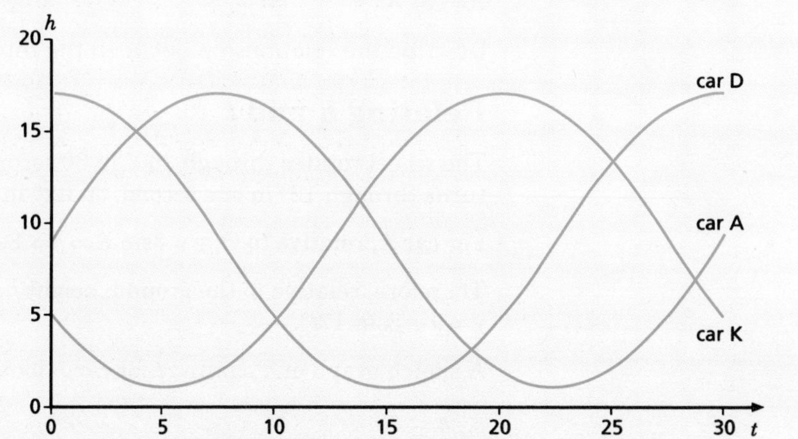

Cars A and D are at the same height after roughly 4 seconds and 19 seconds, with heights around 15 metres and 3 metres respectively.

Notice that an alternative formula for the height of car D is given by:

$h = 9 + 8\cos 12t°$

The three waves are said to be out of phase since the second and third waves may be obtained from the first one by a translation parallel to the x-axis.

CALCULATOR ACTIVITY

Using suitable scales, such as:

$x_{min} = 0;\quad x_{max} = 30;\quad x_{scl} = 2$

$y_{min} = 0;\quad y_{max} = 20;\quad y_{scl} = 2.$

plot the graphs for cars A, D and K.

Check all the answers to the exploration using the trace facility.

Think up other questions and use your calculator to solve them.

Example 8.10

The hours of daylight throughout the year for Casterton are as follows (on 21st of each month).

March	April	May	June	July	Aug.	Sept.	Oct.	Nov.	Dec.	Jan.	Feb.
12	15	17.2	18	17.2	15	12	9	6.8	6	6.8	9

a) *Plot the values on a graph and join the points with a smooth curve.*
b) *From the graph estimate:*
i) the number of hours of daylight on 1st October,
ii) the period when the hours of daylight are less than 7.5 hours.

c) *If h represents the number of hours of daylight on the 21st of a month, m, formulate an equation, to give h in terms of m, of the form:*
$h = a\sin bm° + c$

where m = 0 for March, m = 1 for April, ..., m = 11 for February, and a , b and c are constants to be found.

d) *Use your equation from (c) to plot the graph on a graphical calculator and check yours results from (a) and (b).*

Solution

a)

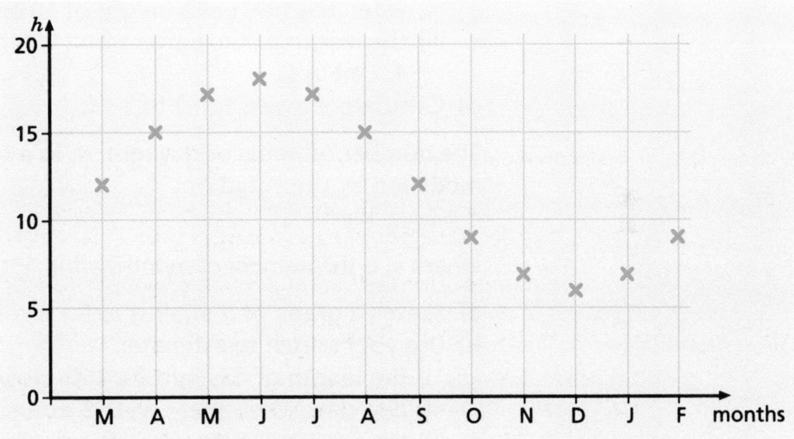

b) i) *Number of hours of daylight on 1st October ≈ 10 hours.*

ii) *There were less than 7.5 hours of daylight between early November and late January.*

c) *The range of values for h is given by $6 \le h \le 18$*
$\Rightarrow a = \frac{1}{2}(18 - 6) = 6.$

The period of the function is 12 months $\Rightarrow b = \frac{360}{12} = 30.$

The value of the constant c is given by $6 + \frac{12}{2} = 12.$
Putting these results together gives the function:
$h = 6\sin 30m° + 12$

d) *Scale the axes using:*
$x_{min} = 0; \quad x_{max} = 30; \quad x_{scl} = 2$

$y_{min} = 0; \quad y_{max} = 20; \quad y_{scl} = 2.$

*and trace the curve to check your solutions to **a)** and **b)**.*
*Part **b) ii)** may be answered by solving the equation:*

$$\begin{aligned} 7.5 &= 6\sin 30m° + 12 \\ \Rightarrow \quad 6\sin 30m° &= -4.5 \\ \Rightarrow \quad \sin 30m° &= -0.75 \\ \Rightarrow \quad 30m &= 228.6 \text{ or } 311.4 \\ \Rightarrow \quad m = 7.62 \text{ or } m &= 10.38 \end{aligned}$$

Therefore the model predicts that there will be less than 7.5 hours daylight between 7.62 and 10.38 months after March 21st, i.e. between November 9th and February 1st (roughly).

EXERCISES

1 The approximate depth of water in a harbour, y metres, is given by:

$y = 4\sin 30t° + 7$

where t is the number of hours after midnight on a certain day.

a) Draw a graph of this function over a period of 24 hours.
b) At what times do the high and low tides occur?
c) Use your graph to estimate:
  i) the height of the tide at 07:30,
  ii) when the tide has a height of 10 metres and is falling,
  iii) the length of time after noon when the depth is less than 4.5 metres.
d) Calculate answers to (c) to 3 s.f.

2 The number of hours of daylight, d, in a certain country can be modelled by the equation:
$d = 5\cos 30m° + 12$

where m is the number of months after 15th June.

a) Sketch a graph of d against m for a complete year (June to June).
b) Use your sketch to estimate:
  i) the length of daylight on 15th November,
  ii) two days when there were 9 hours of daylight,
  iii) the length of time when there were more than 14 hours of daylight.
c) Check your answers to **b)** by calculation.

3 During a period of 13 consecutive days, referred to as days 0, 1, ..., 11, 12, the temperatures in a factory are measured, in °C, and the results are shown as plotted on this diagram.

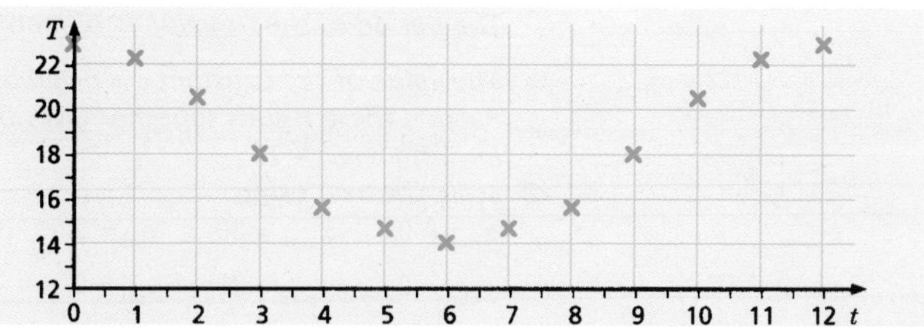

a) Given that the temperature, T°C, can be modelled as a function of time, t days, using a function of the form:
$T = a\cos bt° + c$
deduce the values of the constants a, b and c.
b) From the graph find:
  i) the temperature on day 10,
  ii) the days on which the temperature was below 17.5°C.
c) Plot a graph of T against t using the definition in part a) and check that the curve agrees with the plotted points. Check your answers to part b).

4 A mass, on the end of a spring which is hanging vertically, is pulled down and then let go. The mass begins to oscillate between 1 metre and 2 metres above the floor and completes 20 complete oscillations in one minute.

The height of the mass, h metres, above the floor after t seconds from being let go can be modelled by a function of the form:
$h = a\cos bt° + c$.

a) Find values for a and c, and explain why $b = 120$.
b) Sketch a graph of h against t for $0 \leq t \leq 6$.
c) Find the times, during the first minute, that the mass is:
  i) 1.25 metres above the floor,
  ii) within 15 cm of the equilibrium height (1.5 metres).

5 The height of tides can be modelled fairly accurately by a function of the form:

$h = a\sin bt° + c$

where h is the height of the water in metres and t is time in hours after midnight. Find values for a, b and c for the following tide tables.

a)

	Time	Height
High tide	03:00	2.0 m
Low tide	09:00	0.0 m
High tide	15:00	2.0 m
Low tide	21:00	0.0 m

b)

	Time	Height
High tide	03:00	5.0 m
Low tide	09:00	1.0 m
High tide	15:00	5.0 m
Low tide	21:00	1.0 m

c)

	Time	Height
High tide	03:06	2.0 m
Low tide	09:18	0.0 m
High tide	15:30	2.0 m
Low tide	21:42	0.0 m

8.4 B

1 A mass on the end of a spring which is hanging vertically is pulled down and let go. It oscillates between 1.5 m and 2 m above the floor and completes 32 oscillations in one minute. The height, h metres, of the mass above the floor after t seconds can be modelled by a function of the form $h = a\cos bt° + c$.

a) Find values for a and b and explain why $c = 1.75$ m.
b) Sketch the graph of h against t for $0 \leq t \leq 5$.
c) Find the times during the first 5 seconds, when the mass is within 0.1 m of its equilibrium height.
d) Calculate the height of the mass above the ground after:
i) 25 seconds, **ii)** 47 seconds.

2 The graph shows how an anchored buoy bobs on the waves. The difference between its highest and lowest positions is 2 metres and it returns to the same position every 15 seconds.

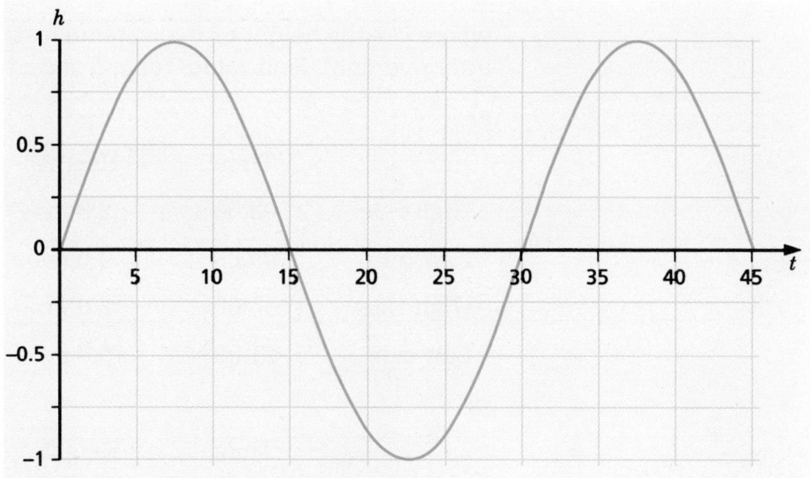

a) Write the equation for its motion.
b) If the sea-bed is 4 m below the lowest position of the buoy, write the equation for its height above the sea-bed at any time.
c) Use the equation to find the height above the sea-bed of the buoy after:
i) 10 seconds, **ii)** 20 seconds, **iii)** 59 seconds.

3 Vibrations, such as those created by plucking a violin string, cause sound waves which have an equation in the form:

$y = A\sin \omega t$

where A is the amplitude of the wave, which is related to the loudness of the sound. Middle C is struck on a piano and the equation of the resulting sound wave is given by $y = 2\sin 95\,040t°$.

a) Draw the graph and the function for $0 \leq t \leq 0.01$.
b) **i)** What is the amplitude of the wave?

ii) If frequency = $\frac{\omega}{360}$ cycles per second, find the frequency for this function.

c) The frequency of a note one octave below middle C is 132 cycles per second. Find the equation of a sound wave struck at that frequency with an amplitude of 5.

4 The graphs below represent two sound wave patterns. Find the equation for each sound wave.

a)

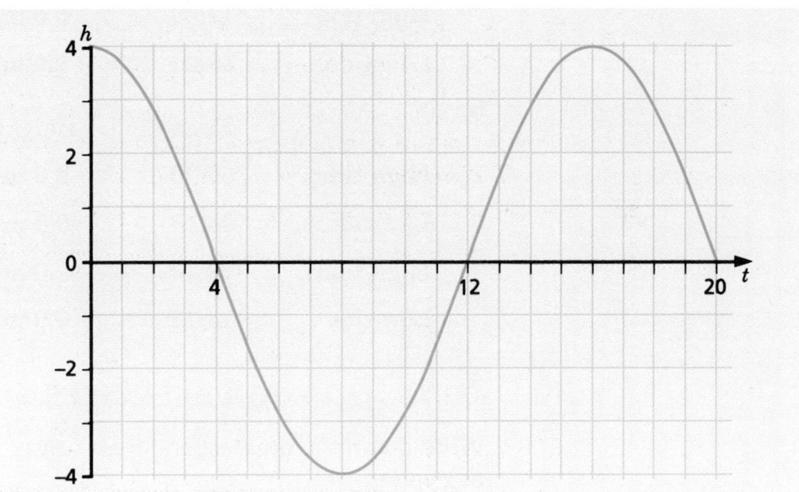

b)

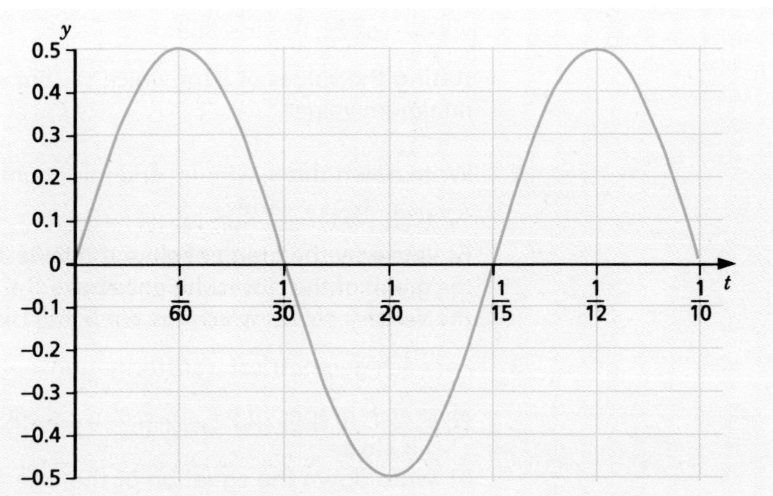

5 The height of tides can be modelled fairly accurately by a function of the form $h = a \cos (bt)° + c$ where h is the height of the water in metres and t is the time in hours after midnight. Find values for a, b and c from the following tide tables.

a)

	Time	Height
High tide	00:00	5.0 m
Low tide	06:00	1.0 m
High tide	12:00	5.0 m
Low tide	18:00	1.0 m

b)

	Time	Height
High tide	00:06	2.0 m
Low tide	06:18	0.0 m
High tide	12:30	2.0 m
Low tide	18:42	0.0 m

CONSOLIDATION EXERCISES FOR CHAPTER 8

1 Write down the greatest and least values of the expression $3 - \cos 2x$ as x varies.

Sketch the graph of the curve with equation:

$y = 3 - \cos 2x,\ 0° \leq x \leq 360°$,

stating the values of x for which the curve has maximum and minimum values.

(Oxford)

2 Write down the maximum and minimum values of the expression $4 - 2\sin 3x$ as x varies.

Explain why the graph of $y = 4 - 2\sin 3x$ does not cut the x-axis. Sketch the graph of the curve with equation $y = 4 - 2\sin 3x,\ 0° \leq x \leq 360°$, stating the values of x for which the curve has maximum and minimum values.

3 For each geometrical transformation:

a) sketch graphs of $y = \sin x,\ 0° \leq x \leq 360°$, and the transformation of $y = \sin x$,
b) write down the equation of the transformed graph,
c) find the coordinates of any stationary points.
  i) stretch factor 2 parallel to the y-axis
  ii) translation –5 units parallel to the y-axis
  iii) translation +60° parallel to the x-axis
  iv) translation –120° parallel to the x-axis *followed by* reflection in the x-axis.

4 For the geometrical transformation stretch factor 2 parallel to the x-axis *followed by* stretch factor 2.5 parallel to the y-axis *followed by* translation –3 units parallel to the y-axis:

a) sketch graphs of $y = \cos x$, $0° \le x \le 360°$, and the transformation of $y = \cos x$,
b) write down the equation of the transformed graph,
c) find the coordinates of any stationary points.

5 **a)** Sketch the graph of $y = \cos x$.
b) Superimpose the graph of $y = \cos 2x$.
c) Use your graphs in parts **a)** and **b)** to sketch the graph of $y = \cos x + \cos 2x$.
Where does your graph cut the x-axis?

What are the greatest and least values of the expression $\cos x + \cos 2x$?

6 State the greatest and least values of $1 + 2\cos 2\theta$ for all values of θ.

Solve the equation $1 + 2\cos 2\theta = 0$, giving all solutions in the interval $0° \le \theta \le 360°$.

Sketch the graph of the curve with equation $y = 1 + 2\cos 2\theta$, $0° \le \theta \le 360°$. *(Oxford)*

7 Using your calculator and the appropriate graph, find all values of x, such that $0° \le x \le 360°$, where:
$12\sin^2 x + 5 = 8$.

8 The depth of water at the entrance to a harbour is y metres at time t hours after low tide. The value of y is given by:

$y = 10 - 3\cos kt$

where k is a positive constant.

Write down, or obtain, the depth of water in the harbour:

a) at low tide, **b)** at high tide.
Show by means of a sketch graph how y varies with t between two successive low tides.

Given that the time interval between low tide and the next high tide is 6.20 hours, calculate, correct to two decimal places, the value of k.
(NEAB Specimen Paper)

9 At a particular point on the earth's surface, the number n of hours of daylight x days after the start of the year is given approximately by the model:
$n = 12 - 6\cos x$.

a) Calculate the length of the day on May 1st, which is day 121.
b) According to the model find the dates in the year when the length of the day is longest and shortest.
c) Explain why the model is not very good and suggest an improvement.

10 A tuning fork vibrates with a frequency of 256 cycles per second. The tip of the tuning fork has an amplitude of 2 mm.

a) Denoting by d (in mm) the displacement of the tip at time t seconds after the start of the vibrations, write down a model describing the vibrations of the tuning fork.

b) Sketch a graph of your model, using appropriate scales on the axes, to show two complete cycles.

Summary

Trigonometric functions

- The functions $\cos\theta$ and $\sin\theta$ are called **circular** functions. It is important to recognise their graphs:

i) $\cos\theta$ ii) $\sin\theta$

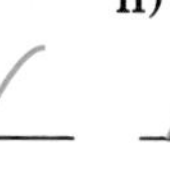

The graphs form waves, oscillating between –1 and 1, with **amplitude** 1 and **period** 360°. Any function with a graph which repeats itself in this way is said to be **periodic**.

Special cases

- Significant values for cosines, sines and tangents are:

θ	0°	30°	45°	60°	90°	180°	270°	360°
$\cos\theta$	1	$\frac{\sqrt{3}}{2}$	$\frac{1}{\sqrt{2}}$	$\frac{1}{2}$	0	–1	0	1
$\sin\theta$	0	$\frac{1}{2}$	$\frac{1}{\sqrt{2}}$	$\frac{\sqrt{3}}{2}$	1	0	–1	0
$\tan\theta$	0	$\frac{1}{\sqrt{3}}$	1	$\sqrt{3}$	∞	0	$-\infty$	0

Trigonometric identities

- Useful identities include:

$\sin\theta = \cos(90° - \theta)$ $\cos\theta = \sin(90° - \theta)$

$\sin\theta = \sin(180° - \theta)$ $\cos\theta = \cos(180° - \theta)$

PURE MATHS

9 Integration 1

In this chapter we:

- *introduce the concept of integration*
- *use integration to find the area under a graph*
- *approximate areas using the trapezium rule*
- *use integration to find velocity from acceleration and displacement from velocity.*

AREAS AND AREA FUNCTIONS

A velocity–time graph can be used to display a journey made by an object, but what other information can it show?

Exploration 9.1

Velocity and distance travelled

Heidi is testing the acceleration and braking power of her new car, on an open stretch of a straight road. A velocity–time graph for the motion of the car is shown below.

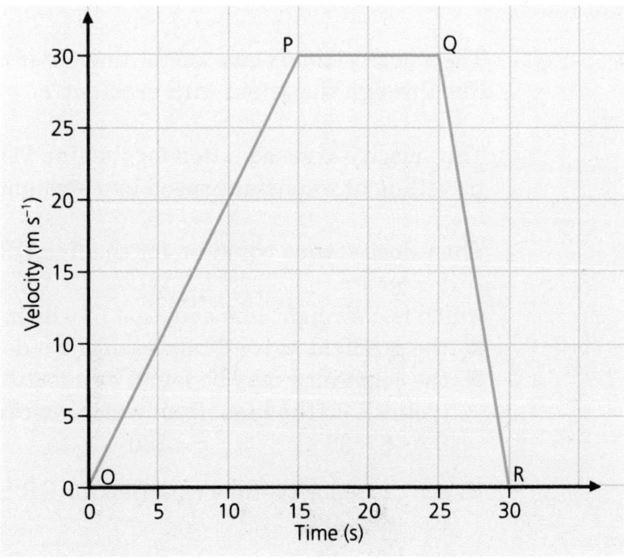

- What was the car's acceleration during the first 15 seconds?
- Describe the motion of the car during the last five seconds.
- How far did the car travel between $t = 15$ and $t = 25$?
- How far did the car travel altogether?
- What are the velocity–time equations for the three lines?

The area under a velocity–time graph

The **acceleration** of the car is given by the gradient of the velocity–time graph. The car accelerates at $2\,\text{m s}^{-2}$ for the first 15 seconds, then travels at a constant velocity of $30\,\text{m s}^{-1}$ for 10 seconds before application of the brakes produces a **deceleration** (negative acceleration) of $6\,\text{m s}^{-2}$.

For the time interval $15 \leq t \leq 25$ the car is travelling at a constant speed, therefore the distance travelled between $t = 15$ and $t = 25$ is $30 \times 10 = 300$ metres.

For the time interval $0 \leq t \leq 15$ the car is accelerating, and for $25 \leq t \leq 30$ the car is decelerating, both at constant rates. To find the distance travelled between $t = 0$ and $t = 15$ we need to multiply the average velocity ($15\,\text{m s}^{-1}$) by the time of travel (15 seconds) to give a distance of $15 \times 15 = 225$ metres.

Similarly the distance travelled between $t = 25$ and $t = 30$ is $15 \times 5 = 75$ metres. Hence the total distance Heidi travels in the car is $225 + 300 + 75 = 600$ metres.

Notice that the three distances correspond to the areas of the rectangle (300 metres) and the two triangles (225 metres and 75 metres). This example illustrates an important principle:

distance travelled = area under a velocity–time graph

The velocity–time equation for line OP is $v = 2t$, since OP is a straight line through the origin with gradient 2.

The velocity–time equation for the line PQ is $v = 30$, since the car is travelling at a constant speed in a straight line.

The velocity–time equation for the line QR is of the form:

$v = mt + c$

which is a straight line equation in which:

- the gradient m is -6 (representing a deceleration of $6\,\text{m s}^{-2}$)
- the constant c may be found by substituting, say, $t = 30$ and $v = 0$ (point R on the line), from which we can deduce:
  $0 = -6 \times 30 + c \quad \Rightarrow \quad c = 180$

Hence the velocity–time equation for QR is:

$v = -6t + 180$

or $v = 180 - 6t$.

Exploration 9.2

Finding the distance

Look more closely at the motion of the car in Exploration 9.1, during the first 15 seconds.

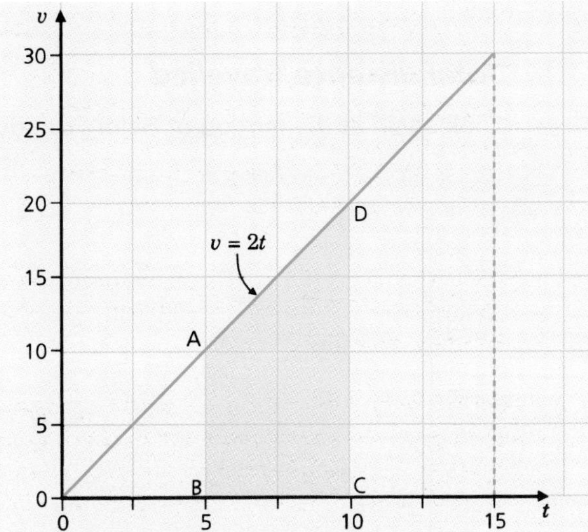

- How far does the car travel between $t = 5$ and $t = 10$?
- How far does the car travel between $t = a$ and $t = b$, where a and b are times in seconds such that $0 \le a < b \le 15$?

The area function

The distance travelled be the car between $t = 5$ and $t = 10$ is given by the area of the trapezium ABCD.

$$\tfrac{1}{2}(10 + 20) \times 5 = 75 \text{ metres}$$

In this case, this area is equivalent to the difference in area of triangles OCD and OAB, i.e.

$$\begin{aligned}\text{area OCD} - \text{area OAB} &= \tfrac{1}{2} \times 10 \times 20 - \tfrac{1}{2} \times 5 \times 10 \\ &= 100 - 25 \\ &= 75 \text{ metres}\end{aligned}$$

Using this idea to find the area between $t = a$ and $t = b$ a very useful generalisation emerges.

$$\begin{aligned}\text{Area of trapezium} &= \tfrac{1}{2}b \times 2b - \tfrac{1}{2}a \times 2a \\ &= b^2 - a^2\end{aligned}$$

This is written as $\left[t^2\right]_a^b$.

$$\left[t^2\right]_a^b = (\text{area of } \Delta \text{ when } t = b) - (\text{area of } \Delta \text{ when } t = a)$$

The function $A(t) = t^2$ is called the **area function**, since:

$$\left[A(t)\right]_a^b = b^2 - a^2 = A(b) - A(a)$$

It follows that $\left[A(t)\right]_0^t = t^2 = A(t)$

which represents the displacement of the car in the first t seconds.

In general:

$f(t) = 2t$	$\Rightarrow$	$A(t) = t^2$
velocity–time equation		*displacement–time equation*

Exploration 9.3

Total distance travelled

Look again at the motion of Heidi's car, during the last five seconds.

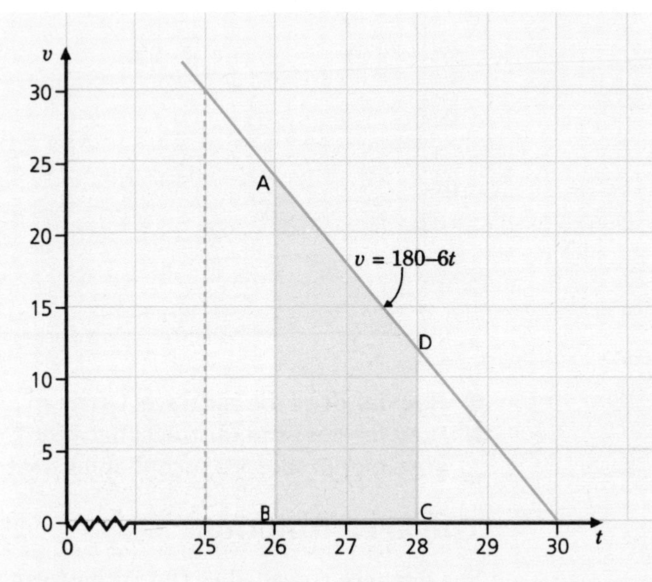

- How far does the car travel between $t = 26$ and $t = 28$?
- How far does the car travel between $t = a$ and $t = b$, where a and b are time in seconds such that $25 \le a < b \le 30$?

Using the area under the graph

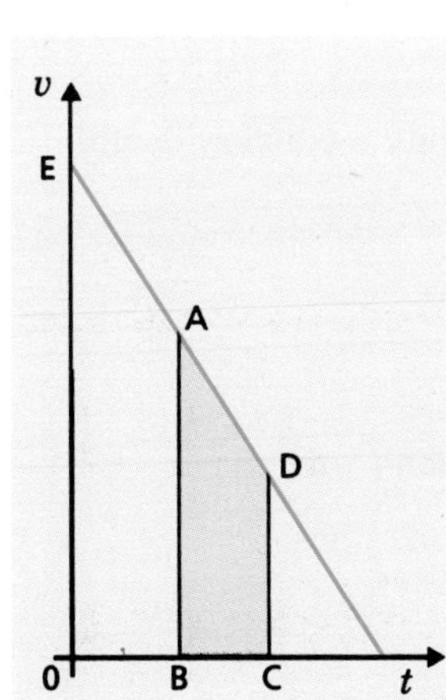

The distance travelled by the car between $t = 26$ and $t = 28$ is given by the area of the trapezium ABCD.

$$\tfrac{1}{2}(24 + 12) \times 2 = 36 \text{ metres}$$

In this case, this is equivalent to the difference in area of the trapezia OEAB and OEDC, i.e.

$$\begin{aligned}\text{area OEDC} - \text{area OEAB} &= \tfrac{1}{2}(180 + 12) \times 28 - \tfrac{1}{2}(180 + 24) \times 26 \\ &= 2688 - 2652 = 36 \text{ metres}\end{aligned}$$

Using this idea to find the areas between $t = a$ and $t = b$, again a useful generalisation emerges.

$$\begin{aligned}\text{area OEDC} - \text{area OEAB} &= \tfrac{1}{2}(180 + 180 - 6b) \times b - \tfrac{1}{2}(180 + 180 - 6a) \times a \\ &= \left(180b - 3b^2\right) - \left(180a - 3a^2\right)\end{aligned}$$

This is written as $\left[180t - 3t^2\right]_a^b$

The function $\mathrm{A}(t) = 180t - 3t^2$ represents the area under the graph $v = 180 - 6t$ between 0 and t.

Therefore it follows that:

$$\mathrm{f}(t) = 180 - 6t \quad \Rightarrow \quad \mathrm{A}(t) = 180t - 3t^2$$

Exploration 9.4

Expressions for the area under a graph

The area under the graph of $y = mx + c$ between $x = a$ and $x = b$ is illustrated in this diagram.

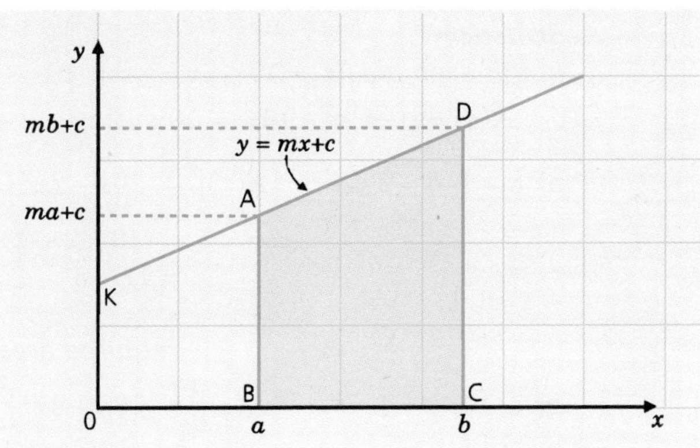

- Find an expression for the area of OBAK.
- Find an expression for the area of OCDK.
- Deduce an expression for the area of ABCD.

The integral function

The area of trapezium OBAK is $\frac{1}{2}(c+(ma+c))a$

The area of trapezium OCDK is $\frac{1}{2}(c+(mb+c))b$

Subtracting, the area of ABCD is:

$$\frac{1}{2}(c+(ma+c))a \;-\; \frac{1}{2}(c+(mb+c))b$$

$$= \frac{1}{2}(mb+2c)b \;-\; \frac{1}{2}(ma+2c)a$$

$$= \frac{1}{2}mb^2+cb \;-\; \left(\frac{1}{2}ma^2+ca\right)$$

This may be written as $\left[\frac{1}{2}mx^2+cx\right]_a^b$.

Therefore, using f(x) for the original function and A(x) for the area function:

$$f(x) = mx + c \Rightarrow A(x) = \frac{1}{2}mx^2 + cx$$

Notation

The area function is also called the **integral function.**
The process of calculating the area enclosed by the graph of $y = f(x)$, the x-axis and the lines $x = a$ and $x = b$ as:

$$[A(x)]_a^b = A(b) - A(a)$$

is called **definite integration**.
We shall meet the technique again later in this chapter.
Areas *above* the x-axis result in a *positive* integral.
Areas *below* the x-axis result in a *negative* integral.

Example 9.1

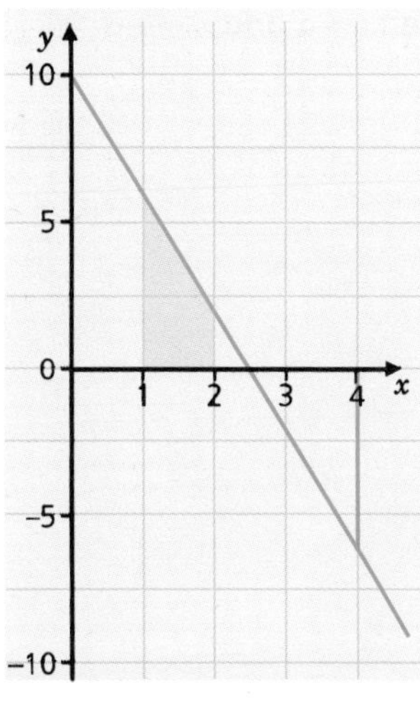

Sketch the graph of $y = 10 - 4x$ *for* $0 \leq x \leq 5$ *and then:*

a) *find the area under the graph for* $a = 1$, $b = 2$,

b) *explain why* $\left[\mathrm{A}(x)\right]_1^4 = 0$.

Solution

In this example $\mathrm{f}(x) = 10 - 4x = -4x + 10$

$$\mathrm{A}(x) = \tfrac{1}{2}(-4)x^2 + 10x \;=\; 10x - 2x^2$$

a) $a = 1, b = 2 \;\Rightarrow\; \left[\mathrm{A}(x)\right]_1^2 = \left[10x - 2x^2\right]_1^2$

$$= (10 \times 2 - 2 \times 2^2) - (10 \times 1 - 2 \times 1^2)$$
$$= 12 - 8$$
$$= 4$$

$\Rightarrow$ *required area = 4 units*^{2}.

b) $a = 1, b = 4 \;\Rightarrow\; \left[\mathrm{A}(x)\right]_1^4 = \left[10x - 2x^2\right]_1^4$

$$= (10 \times 4 - 2 \times 4^2) - (10 \times 1 - 2 \times 1^2)$$
$$= 8 - 8$$
$$= 0$$

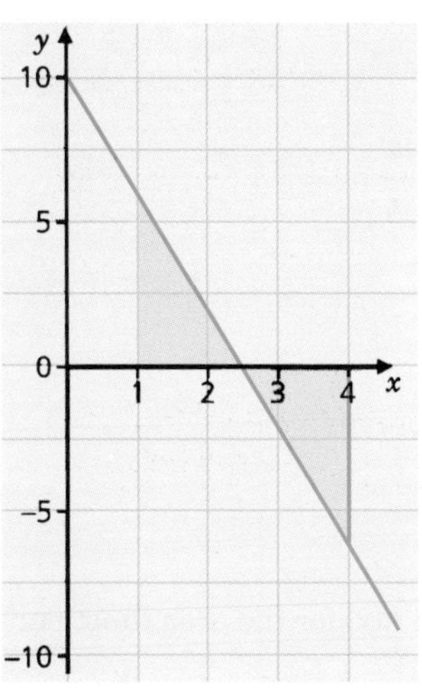

Over the interval $1 \leq t \leq 4$, *the graph is both positive* $(1 \leq t < 2.5)$ *and negative* $(2.5 < x \leq 4)$; *the area* ***above*** *the* x*-axis is* ***positive*** *whereas the area* ***below*** *the* x*-axis is* ***negative***.

The two areas are equal in magnitude, but opposite in sign, which is why their sum is zero.

To find the actual area enclosed by the graph and the x*-axis over the interval* $1 \leq t \leq 4$ *it is necessary to carry out two separate integrations.*

$a = 1, b = 2.5 \Rightarrow \left[\mathrm{A}(x)\right]_1^{2.5} = \left[10x - 2x^2\right]_1^{2.5}$

$$= (10 \times 2.5 - 2 \times 2.5^2) - (10 \times 1 - 2 \times 1^2)$$
$$= 12.5 - 8$$
$$= 4.5$$

$a = 2.5, b = 4 \Rightarrow \left[\mathrm{A}(x)\right]_{2.5}^{4} = \left[10x - 2x^2\right]_{2.5}^{4}$

$$= (10 \times 4 - 2 \times 4^2) - (10 \times 2.5 - 2 \times 2.5^2)$$
$$= 8 - 12.5$$
$$= -4.5$$

Therefore the actual area is $4.5 + |-4.5| = 9$

EXERCISES

1 For each of the following functions, find the area function $\mathrm{A}(x)$ and so calculate the area of the shaded region.

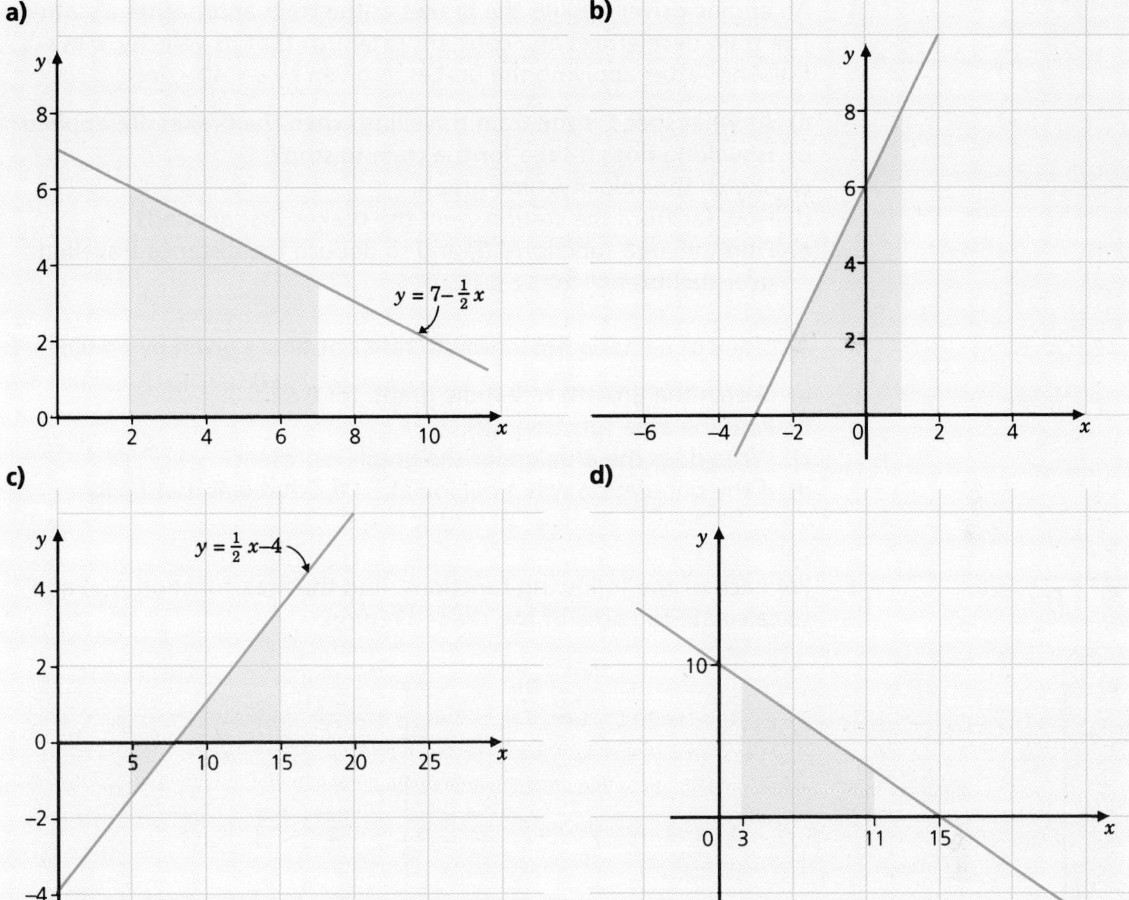

2 For each of the following functions f(x), find the area function A(x) and use it to find the area under the graph of $y = f(x)$ between $x = a$ and $x = b$ as given.
Illustrate your answer with a sketch.

a) $f(x) = 5x - 2$ $\quad a = 1, b = 4$
b) $f(x) = \frac{3}{2}x + 5$ $\quad a = 1, b = 7$
c) $f(x) = 10 - 3x$ $\quad a = -2, b = 3$
d) $f(x) = \dfrac{8 - 2x}{5}$ $\quad a = 2.8, b = 3.2$

3 For each of the following functions, sketch the graph of the function and find the x-value of the point where it crosses the x-axis. Hence find the total area between the graph and the x-axis for the given values of a and b.

a) $y = 4 - 2x$ $\quad a = 1, b = 3$
b) $y = 3x + 3$ $\quad a = -2, b = 2$
c) $y = x - 6$ $\quad a = 0, b = 8$
d) $y = -\frac{1}{4}x - 1$ $\quad a = -4, b = -2$

4 An engine driver applies the brakes as the train approaches a station. The train decelerates at a constant rate such that its velocity, v m s^{-1}, t seconds after applying the brakes, is given by $v = 40 - 2t$.

a) At what speed is the train travelling when the brakes are applied?
b) How long does it take for the train to stop?
c) Sketch the velocity–time graph.
d) How far from the station were the brakes first applied?
e) Find the area function $\mathrm{A}(t)$ and so deduce the distance travelled over the interval $10 \le t \le 15$.

5 A colony of bacteria has a growth rate r at time t given by $r = 0.8t + 1$.

a) Sketch the growth rate–time graph for $0 \le t \le 3$.
b) Find the area function $\mathrm{A}(t)$.
c) What does the area under the graph represent?
d) If the population was 1 million at $t = 0$, what was it at $t = 3$?

9.1 B

1 For each of the following functions, find the area function $\mathrm{A}(x)$ and so calculate the area of the shaded region.

a)

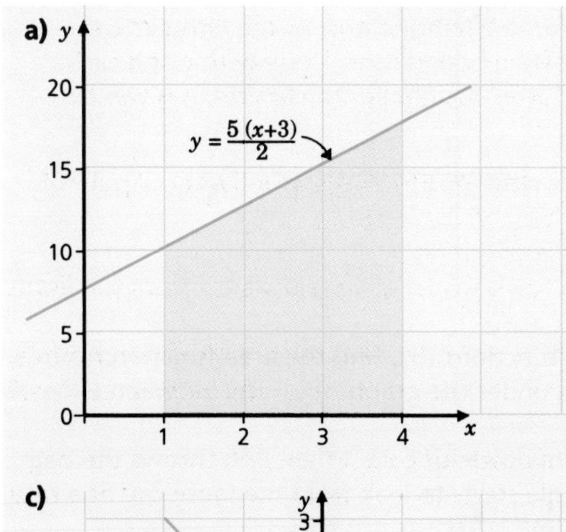

b)

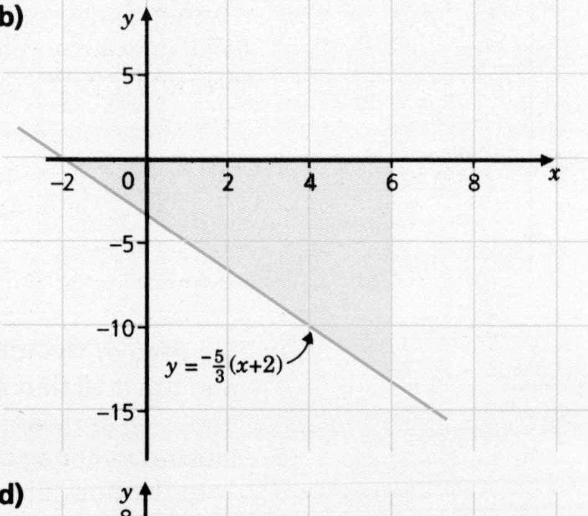

c)

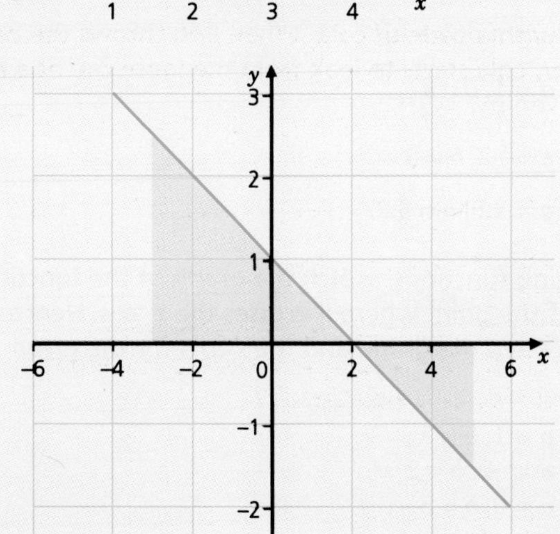

d)

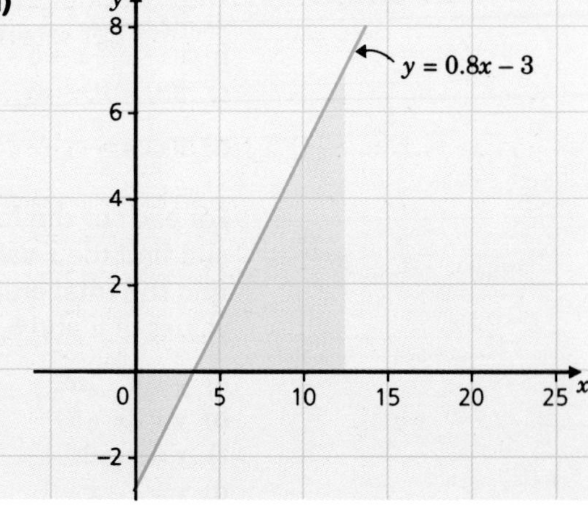

2 For each of the following functions $f(x)$, find the area function $A(x)$ and use it to find the area under the graph of $y = f(x)$ between $x = a$ and $x = b$ as given.

Illustrate your answer with a sketch.

a) $y = 4 - 2x$ $\quad a = -1, b = 1$
b) $y = 3x + 3$ $\quad a = -2, b = 1$
c) $y = x - 6$ $\quad a = 3, b = 4$
d) $y = -\frac{1}{3}x - 1$ $\quad a = -2, b = 3$

3 For each of the following functions, sketch the graph of the function and find the x-value of the point where it crosses the x-axis. Hence find the total area between the graph and the x-axis for the given values of a and b.

a) $f(x) = 3 - \frac{1}{2}x$ $\quad a = 5, b = 7$
b) $f(x) = 2x - 5$ $\quad a = 0, b = 3$
c) $f(x) = 4x + 5$ $\quad a = -1.5, b = -1$
d) $f(x) = -\frac{1}{4}x - 2$ $\quad a = -10, b = 0$

4 A driver is travelling at constant speed on a stretch of dual carriageway in a built-up area when she notices the lights change to red 50 m away. She applies the brakes in such a way that the car's velocity, v m s^{-1}, t seconds after she applies the brakes, is given by $v = 24 - 6t$.

a) At what speed is the car travelling when the brakes are applied?
b) How long does it take for the car to stop?
c) Sketch the velocity–time graph.
d) Find the area function $A(t)$ and deduce the distance from the lights when the car stops.
e) After noticing the lights, what is the maximum time delay possible, before applying the brakes, in order *not* to go through a red light?

5 Bob's sports bag contains a bottle of cola. When Bob throws the bag into the boot of his car, cola starts to leak from the loose cap at a rate R litres per second, given by:

$$R = \frac{1}{175} + \frac{t}{1750}$$

until after 50 seconds the bottle is completely empty.

a) Sketch a graph of R against t for $0 \leq t \leq 50$.
b) Find the area function $A(t)$.
c) What does the area under the graph actually represent?
d) How much cola was in the bottle before it started leaking?
e) After how long was the bottle half empty?

THE TRAPEZIUM RULE

In the first part of this chapter, we found areas under straight line graphs by finding the area of a trapezium. Here we look at how trapezia may be used to estimate the area under a curve.

Exploration 9.5

Curves and trapezia

The diagram shows part of the graph of $y = \frac{12}{x}$.

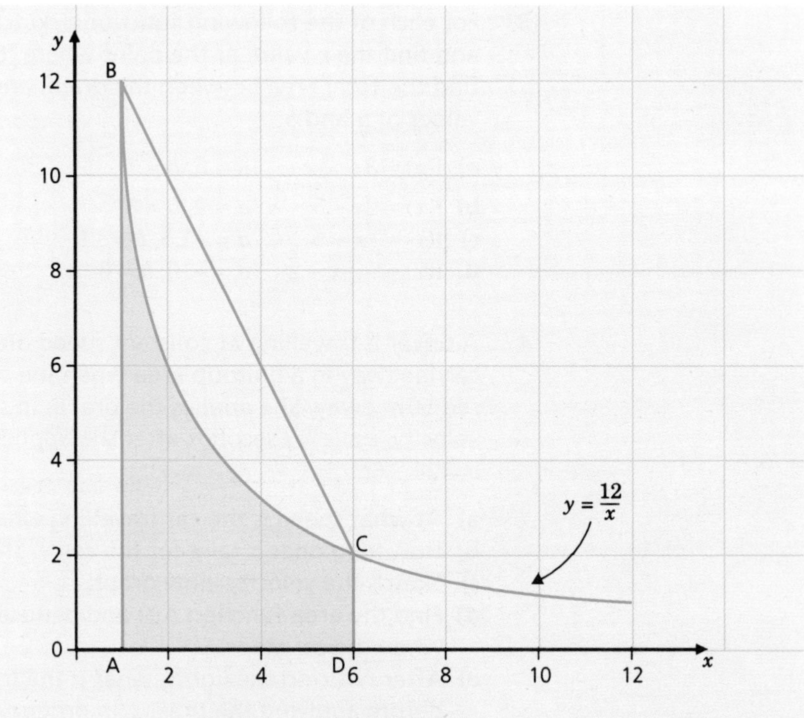

- Estimate the area of the shaded region.
- Find the area of trapezium ABCD.
- Does this over- or under-estimate the shaded area?
- Instead of using one trapezium of width 5 units, use five trapezia of width 1 unit to estimate the area of the shaded region.
- Is this a better estimate? Is it good enough? How can you tell? How could you improve it further?

Fitting trapezia to the curve

The area of trapezium ABCD = $\frac{1}{2} \times (12 + 2) \times 5 = 35$ units^2.
This is an *over-estimate* of the shaded area, with the unshaded area between the graph and the trapezium representing the error.

By using more but narrower trapezia the error may be reduced. Five trapezia, each of width 1 unit, are illustrated on the next diagram.

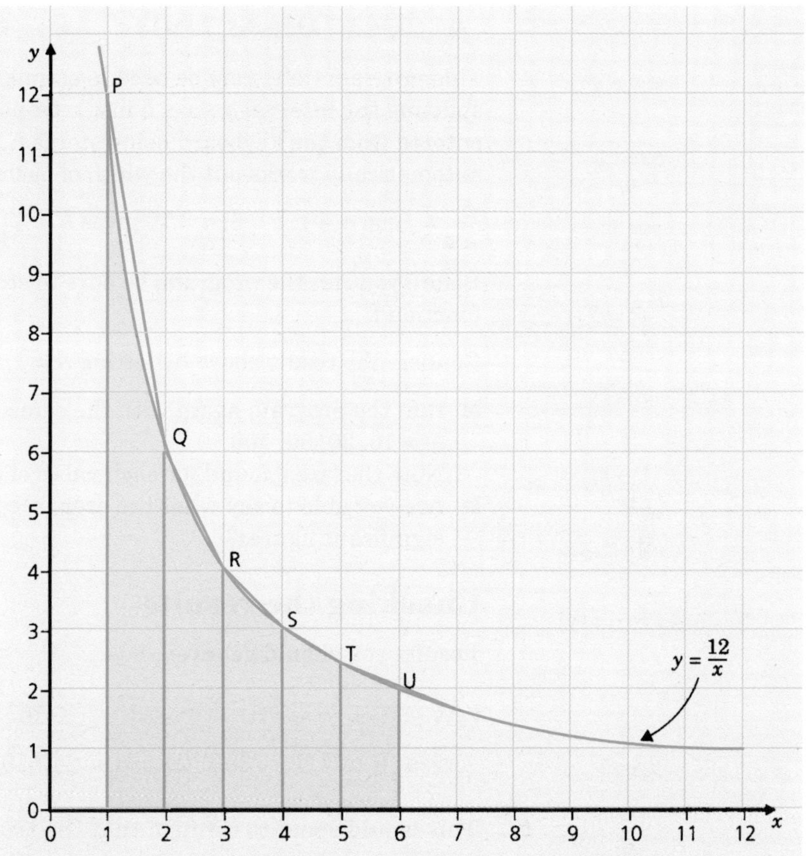

The chords PQ, QR, RS, ST, TU are a much closer fit to the curve than the chord BC. Areas of the five trapezia are:

$\frac{1}{2}(12+6)\times 1 \quad = \quad 9.0$

$\frac{1}{2}(6+4)\times 1 \quad = \quad 5.0$

$\frac{1}{2}(4+3)\times 1 \quad = \quad 3.5$

$\frac{1}{2}(3+2.4)\times 1 \quad = \quad 2.7$

$\frac{1}{2}(2.4+2)\times 1 \quad = \quad 2.2$

Total $= 22.4$

The value of 22.4 is still an over-estimate of the shaded area, but looks a much better approximation than 35 units^2. The error could be further reduced by considering even more, but even narrower, trapezia, e.g. ten trapezia, each of width 0.5 unit. The process of estimating the area under a graph using trapezia is called the **trapezium rule**. Like most numerical methods it may be automated, as illustrated in the following calculator activity.

CALCULATOR ACTIVITY

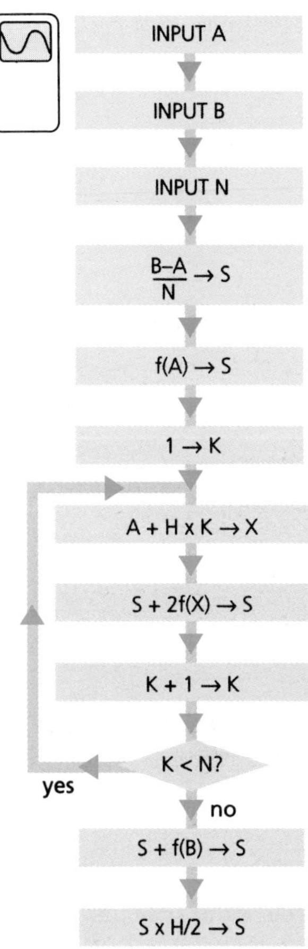

This program (left) may be used to estimate the area under a graph by splitting the interval $a \le x \le b$ into n trapezia. Values for a, b and n are entered from the keyboard using stores A, B and N. The calculator automatically works out the width of each trapezium, h as $\frac{b-a}{n}$, e.g. $a = 1, b = 6, n = 10$ gives $h = \frac{6-1}{10} = 0.5$.

Before you run the program, be sure to store the function $\frac{12}{x}$ as function f.

Confirm the result above by letting $A = 1$, $B = 6$ and $N = 5$.

- Run the program again with the same values for A and B, but $N = 10, 20, 50, 100$.
  Note the 'area' found for each value of N.
- Are you able to say what the area is, correct to three significant figures?

Checking the results

Results you should get are:

N	10	20	50	100
Area (6 s.f.)	21.7385	21.5615	21.5108	21.5035

This would seem to confirm that the true area = 21.5 correct to 3 s.f. The true area, correct to 6 s.f., is 21.5011. Check the following percentages errors in the approximations for different values of N.

N	10	20	50	100
% error	1.104	0.281	0.045	0.011

Example 9.2

Use the trapezium rule to estimate the area under the graph of $y = \frac{4}{1+x^2}$ *over the interval* $0 \le x \le 1$ *using five trapezia.*

Use the program to investigate what happens to the area as you increase the number of trapezia.

Solution

If $a = 0$, $b = 1$ *and* $n = 5$, *then* $h = \frac{b-a}{n} = 0.2$.

A table of function values gives:

x	0	0.2	0.4	0.6	0.8	1.0
y	4	3.8462	3.4483	2.9412	2.4390	2

Areas of the five trapezia are:

$\frac{1}{2}(4+3.8462)\times 0.2 = 0.78462$

$\frac{1}{2}(3.8462+3.4483)\times 0.2 = 0.72945$

$\frac{1}{2}(3.4483+2.9412)\times 0.2 = 0.63895$

$\frac{1}{2}(2.9412+2.4390)\times 0.2 = 0.53802$

$\frac{1}{2}(2.4390+2)\times 0.2 = 0.44390$

Total $= 3.13494$

The approximate area, using five trapezia, is 3.135 (3 d.p.)
Using the program with five trapezia will verify this result.
Increasing the number of trapezia gives, for example:

N	10	20	50	100
Area (6 s.f.)	3.13993	3.14118	3.14153	3.14158

The exact area is π (3.141 59...), so you can see that with many, narrow trapezia a very good approximation is possible.

Example 9.3

Use the trapezium rule to estimate the area under a curve over the interval $3 \le x \le 11$, with x and y values given in the table.

x	3	5	7	9	11
y	8	13.5	10	3.5	0

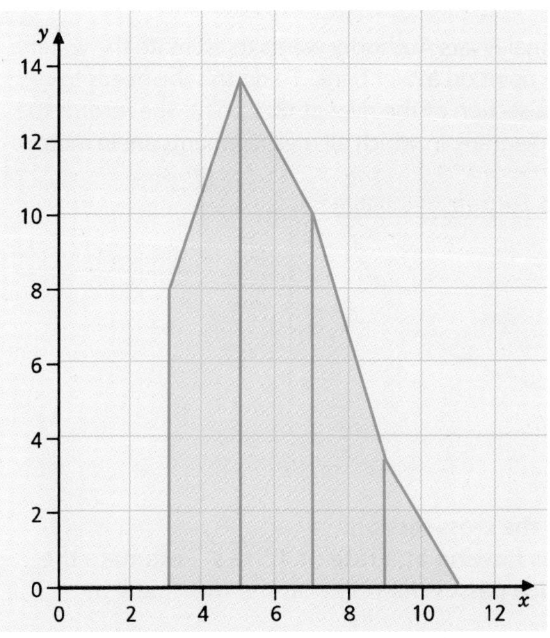

Solution

In this example there is no equation of the form $y = f(x)$ to work with, so we use four trapezia, each of width 2.

Areas of the four trapezia are:

$\frac{1}{2}(8+13.5)\times 2 = 21.5$

$\frac{1}{2}(13.5+10)\times 2 = 23.5$

$\frac{1}{2}(10+3.5)\times 2 = 13.5$

$\frac{1}{2}(3.5+0)\times 2 = 3.5$

So the estimated area under the curve using the trapezium rule with four trapezia is 62.0 square units.

EXERCISES

9.2A

1 Use the trapezium rule, with the number of trapezia indicated in brackets, to estimate the following areas. Illustrate each answer with a sketch.

a) Area under graph of $y = x^2$ $\quad 0 \le x \le 3 \quad$ (6)

b) Area under graph of $y = \dfrac{1}{1+x}$ $\quad 0 \le x \le 1 \quad$ (5)

c) Area under graph of $y = 2^x$ $\quad -1 \le x \le 1 \quad$ (4)

d) Area under graph of $y = x^2 - 9$ $\quad -3 \le x \le 3 \quad$ (6)

2 A circle is drawn, radius 2 units, centre (0,0).
A point P is chosen at random on the circumference. It has coordinates (x, y).

a) Use Pythagoras' theroem to show that $x^2 + y^2 = 4$.
b) Rearrange the equation in **a)** in the form $y = \ldots$ and sketch its graph over the interval $-2 \le x \le 2$.
c) Use the trapezium rule, with four trapezia, to estimate the area under the curve in **b)** over the interval $0 \le x \le 2$.
d) Use the program to see what happens as you increase the number of trapezia (e.g. 10, 50, or even 100).
e) What is the significance of your result in **d)**?

3 Use the trapezium rule to estimate the area under a curve over the interval $0 \le x \le 25$, with x and y values given in this table.

x	0	5	10	15	20	25
y	0	2.2361	3.1623	3.8730	4.4772	5

4 A surveyor from the National Rivers Authority wants to estimate the volume of water passing a certain point on a river bank. To do this she needs to know the area of the cross-section of the river at that point. She records the results of her survey on a diagram, in which all measurements are in metres.

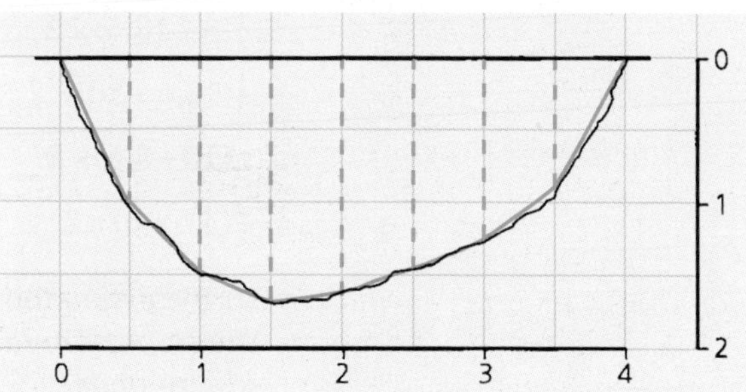

a) Estimate the area of the cross-section.
b) Given that the river is flowing at a rate of $1.5\,\text{m s}^{-1}$, estimate the volume of water which passes the point on the river bank in one minute.

9.2 B

1 Use the trapezium rule, with the number of trapezia indicated in brackets, to estimate the following areas. Illustrate each answer with a sketch.

a) Area under graph of $y = x^3$ $\quad 0 \le x \le 4 \quad (4)$

b) Area under graph of $y = \dfrac{x}{x^2+1}$ $\quad 0 \le x \le 1 \quad (4)$

c) Area under graph of $y = 10^x$ $\quad 0 \le x \le 1 \quad (4)$

d) Area under graph of $y = (x-1)(x-2)$ $\quad -3 \le x \le 3 \quad (6)$

2 The velocity of a decelerating car is recorded each second in the following table.

t	0	1	2	3	4	5	6	7	8	9	10
v	20	12.13	7.36	4.46	2.71	1.64	1.00	0.60	0.37	0.22	0.13

Estimate the distance travelled in the 10 seconds, using the trapezium rule.

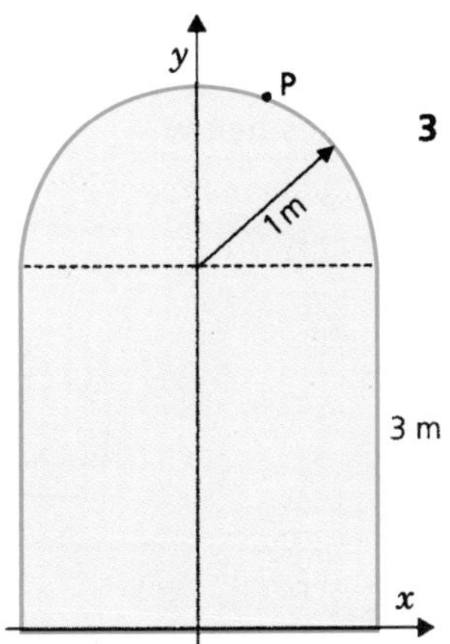

3 A Norman arched window is in the form of a semi-circle of radius 1 m, above a rectangle of height 3 m, as shown in the diagram.

a) Write down the exact area of the window.

b) Use your calculator to find this area correct to 3 s.f.

c) Taking the origin and axes as shown, use Pythagoras' theorem to show show that, for points such as P:

$$y = 3 + \sqrt{1 - x^2}$$

d) Use the trapezium rule with four trapezia to estimate the area of the window.

e) Use the computer program (page 210) to investigate how many trapezia would be necessary to give accuracy to 3 s.f.

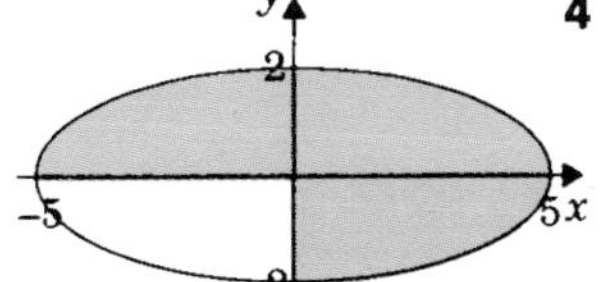

4 The diagram is a sketch of an ellipse, a curve for which the equation may be written:

$$\frac{x^2}{25} + \frac{y^2}{4} = 1$$

a) Show that the equation may be re-arranged in the form:

$$y = \pm 0.4\sqrt{25 - x^2}$$

b) Use the trapezium rule with five trapezia to estimate the shaded area; hence estimate the area of the ellipse.

c) Repeat part **b)** with 10 trapezia, each of width 0.5.

d) The true area of the ellipse is 10π. Calculate the percentage error in your estimates from **b)** and **c)**. Comment on your result.

INTEGRATION: THE PROCESS

The main result from the first part of this chapter was:
$f(x) = mx + c \Rightarrow A(x) = \frac{1}{2}mx^2 + cx$

e.g. $f(x) = 5x - 2 \quad \Rightarrow \quad A(x) = \frac{5}{2}x^2 - 2x$

$f(x) = 1.5x + 5 \quad \Rightarrow \quad A(x) = 0.75x^2 + 5x$

$f(x) = 10 - 4x \quad \Rightarrow \quad A(x) = 10x - 2x^2$

Since the area expression $[A(x)]_0^x = A(x)$, then A(x) represents the area under a curve between 0 and x.
For any function f(x), the process of finding the area function A(x) is called **integration**. A(x) is the **integral** of f(x). What is more, the inverse process, starting with A(x) and finding f(x) is **differentiation**, i.e. $A'(x) = f(x)$.

e.g. $A(x) = 10x - 2x^2 \Rightarrow \quad A'(x) = 10 - 4x$

This works well as long as $f(x) = mx + c$, but what of other functions f(x)? Is it always true that $A'(x) = f(x)$?

Exploration 9.6

Functions and areas

The graph of the function $y = 3x^2$ is drawn on the axes below.

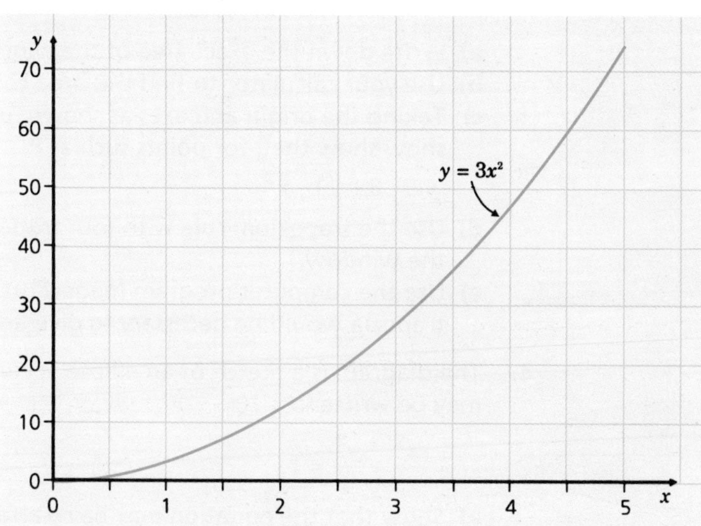

- Using a suitable numerical method, copy and complete this table.
- Find a formula for A(x).
- Describe the relationship between A(x) and f(x).

x	$f(x) = 3x^2$	A(x)
0	0	0
1	3	
2	12	
3	27	
4	48	
5	75	

The inverse of differentiation

The result of the last exploration was that $A(x) = x^3$ and so once again $A'(x) = f(x)$. This is just another illustration of the general result that the process of integration is the *inverse* process to differentiation.

Notation

The integral of a function is denoted by an elongated S, i.e. $\int$.
In short:

$$A(x) = \int f(x)\,dx$$

Integral of f(x) with respect to x
Since the derivative of a constant is 0, the integral of a function f(x) *may* differ by a constant, e.g. all these statements are true:

$$\int 3x^2 dx = x^3$$

$$\int 3x^2 dx = x^3 + 5$$

$$\int 3x^2 dx = x^3 - 2 \text{ etc.}$$

In general:

$$\int 3x^2 dx = x^3 + c \quad \text{where } c \text{ is the constant of integration.}$$

The inclusion of a constant c other than 0 in the integral alters the left-hand boundary value a for the area function.
i.e. $A(x) = [A(x)]_0^x$ only if the constant of integration $c = 0$.

Since the process of integration is the inverse of differentiation, rules can easily be formulated, as in differentiation.

1 $f(x) = x^n \Rightarrow \int f(x)dx = \frac{1}{n+1}x^{n+1} + c$, $n = 0, 1, 2, 3, \ldots$

e.g. $f(x) = x^5 \Rightarrow \int f(x)dx = \frac{1}{6}x^6 + c$

2 $f(x) = ax^n \Rightarrow \int ax^n dx = a\frac{1}{n+1}x^{n+1} + c$, $n = 0, 1, 2, 3, \ldots$

e.g. $f(x) = 5x^3 \Rightarrow \int 5x^3 dx = 5 \times \frac{1}{4}x^4 + c = \frac{5}{4}x^4 + c$

3 The integral of a sum (or difference) is the sum (or difference) of the integrals, e.g.

$f(x) = x^3 - 5x + 10 \Rightarrow \int f(x)dx = \frac{1}{4}x^4 + \frac{5}{2}x^2 + 10x + c$

Exploration 9.7

Areas and integrals

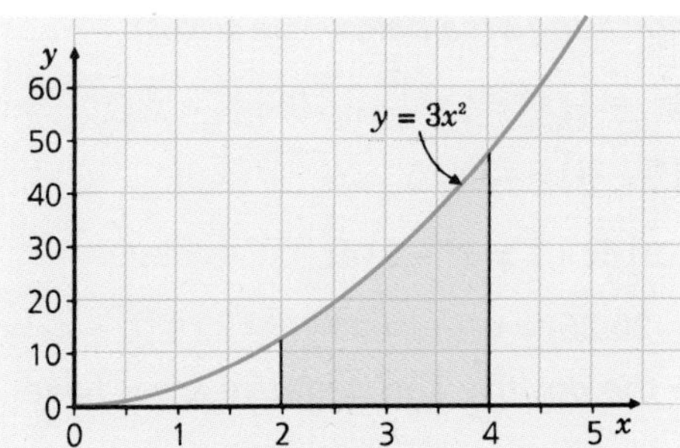

The diagram shows the graph of $f(x) = 3x^2$.

- Use the area function $A(x) = x^3$ developed in the last exploration to find the area of the shaded region.

Definite integrals

The area between $x = 0$ and $x = 4$ is given by A(4).
The area between $x = 0$ and $x = 2$ is given by A(2).
Area under graph between $x = 2$ and $x = 4$ is given by:
$A(4) - A(2) = 4^3 - 2^3 = 64 - 8 = 56$.

Notation

The required area is denoted by the **definite integral**:

upper limit

$$\int_2^4 3x^2\,dx = \left[x^3\right]_2^4 = 4^3 - 2^3 = 56$$

lower limit

Areas can lie above the x-axis, below the x-axis or a mixture of the two.

Example 9.4

The graph shows a sketch of the graph $y = x(x^2 - 9)$.

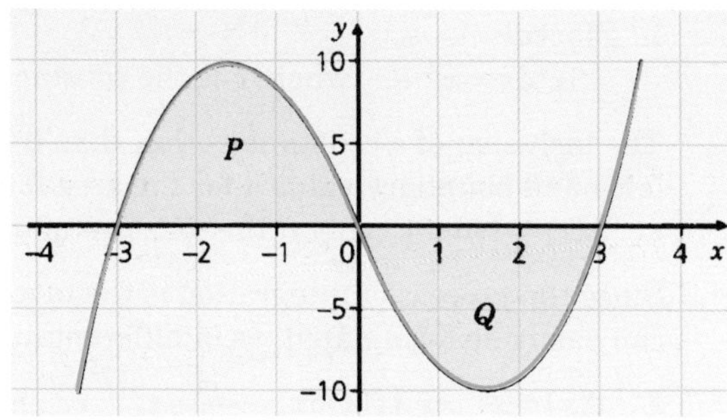

Use a definite integral to find areas P and Q.

Explain why $\int_{-3}^{3} x(x^2 - 9)\,dx = 0$.

Solution

Area P is given by $\int_{-3}^{0} x(x^2 - 9)\,dx = \int_{-3}^{0} (x^3 - 9x)\,dx$

$$= \left[\tfrac{1}{4}x^4 - \tfrac{9}{2}x^2\right]_{-3}^{0} = (0 - 0) - \left(\tfrac{1}{4}(-3)^4 - \tfrac{9}{2}(-3)^2\right)$$

$$= -\left(\tfrac{81}{4} - \tfrac{81}{2}\right) = 20.25$$

Since the function is odd, the graph has rotational symmetry of order 2 about the origin, so area Q = 20.25, which is confirmed by integration (the negative sign indicating area ***below*** *the x-axis).*

$$\int_0^3 x(x^2 - 9)\,dx = \int_0^3 (x^3 - 9x)\,dx = \left[\tfrac{1}{4}x^4 - \tfrac{9}{2}x^2\right]_0^3$$

$$= \left(\tfrac{1}{4}(3)^4 - \tfrac{9}{2}(3)^2\right) - (0 - 0) = \left(\tfrac{81}{4} - \tfrac{81}{2}\right) = -20.25$$

Finally $\int_{-3}^{3} x(x^2 - 9)\,dx = \int_{-3}^{0} x\left(x^2 - 9\right)dx + \int_0^3 x\left(x^2 - 9\right)dx$

$$= 20.25 + (-20.25) = 0$$

Here the zero result occurs since the positive and negative areas have 'cancelled each other out'.

Example 9.5

Find the area enclosed by the graphs of $y = 3x(5 - x)$ *and* $y = 15 - 3x$.

Solution

Sketch the graphs and shade the required area.

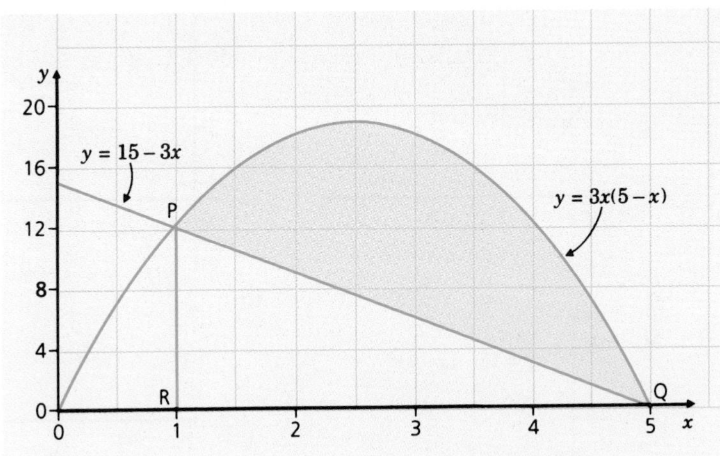

Notice that the graphs intersect at (1, 12) and (5, 0).

The required area is given by $\int_1^5 3x(5-x)\,dx - \int_1^5 (15-3x)\,dx$

shaded area + ΔPQR ***area ΔPQR***

There are two ways of obtaining the required answer.

Method A

Area $= \int_1^5 (15x - 3x^2)\,dx - \int_1^5 (15 - 3x)\,dx$

$= \left[7.5x^2 - x^3\right]_1^5 - \left[15x - 1.5x^2\right]_1^5$

$= \{(7.5 \times 25 - 125) - (7.5 \times 1 - 1)\} - \{(15 \times 5 - 15 \times 25) - (15 \times 1 - 15 \times 1)\}$

$= \{62.5 - 6.5\} - \{37.5 - 13.5\} = 32$

Method B

Area $= \int_1^5 \{(15x - 3x^2) - (15 - 3x)\}\,dx$

$= \int_1^5 (18x - 3x^2 - 15)\,dx$

$= \left[9x^2 - x^3 - 15x\right]_1^5$

$= (9 \times 25 - 125 - 15 \times 5) - (9 \times 1 - 1 - 15 \times 1) = 32$

EXERCISES

9.3 A

1 Find by definite integration the areas of the following shaded regions.

a)

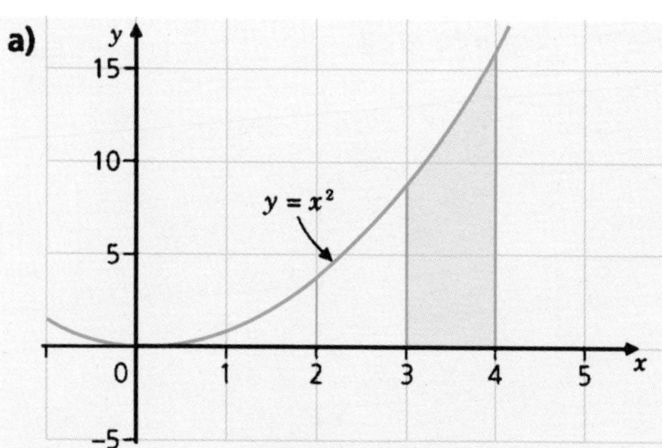

b)

$y = x^2 - 5x + 10$

c)

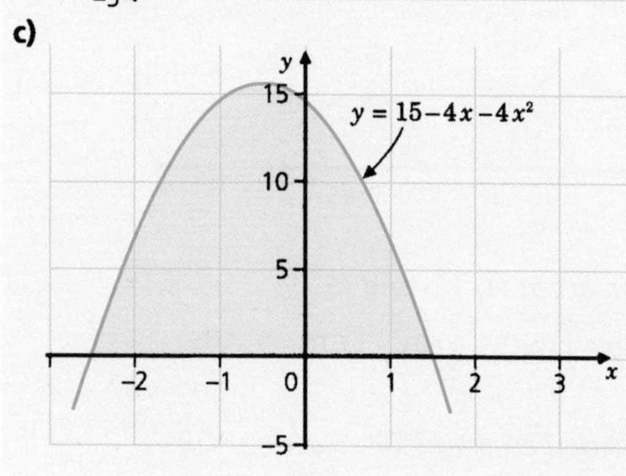

d)

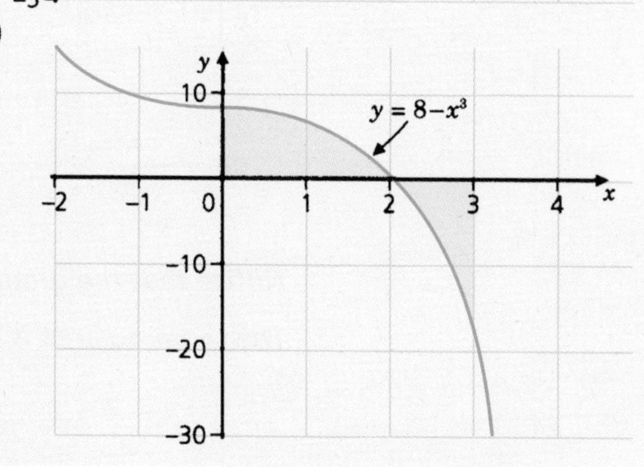

2 Find the areas represented by the following integrals. Illustrate each of your answers with a sketch.

a) $\int_1^2 x^3\,dx$ **b)** $\int_{-1}^0 (x^2 + x)\,dx$ **c)** $\int_{-2}^2 x^4\,dx$ **d)** $\int_{-3}^3 (18 - 2x^2)\,dx$

For each of questions **3 to 11** sketch a graph of the given curve(s) and find the area described.

3 The area between the curve $y = x(3 - x)$ and the x-axis from $x = 0$ to $x = 4$.

4 The area enclosed by the curve $y = (2 + x)(5 - x)$ and the x-axis.

5 The area enclosed by the curve $y = x(x - 4)^2$ and the x-axis.

6 The area bounded by the curve $y = 3x(4 - x^2)$ and the x-axis from $x = -2$ to $x = 2$.

7 The area bounded by the curve $y = (x - 3)(x - 4)$ and the x- and y-axis.

8 The area enclosed by the curve $y = (x - 2)^2$ and the line with equation $y = 2x - 1$.

9 The area enclosed by the curve $y = (x - 1)(x - 4)$ and the line with equation $x + y = 4$.

10 The area enclosed by the curve $y = 0.25x^3$ and the curve $y = 4x(5 - x)$.

11 The area enclosed by the curve $y = 9 - x^2$ and the curve $y = x^2 - 2x - 3$.

9.3 B

1 Find by definite integration the areas of the following shaded regions.

a)

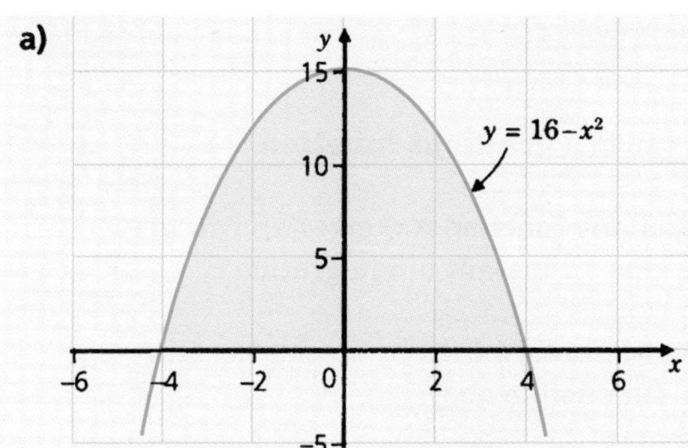

b)

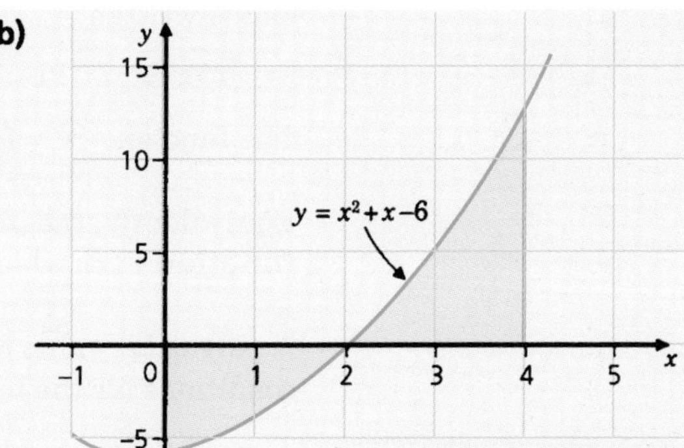

c)

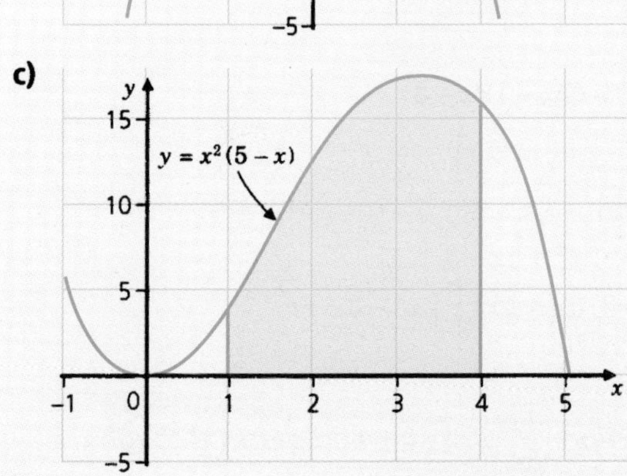

d)

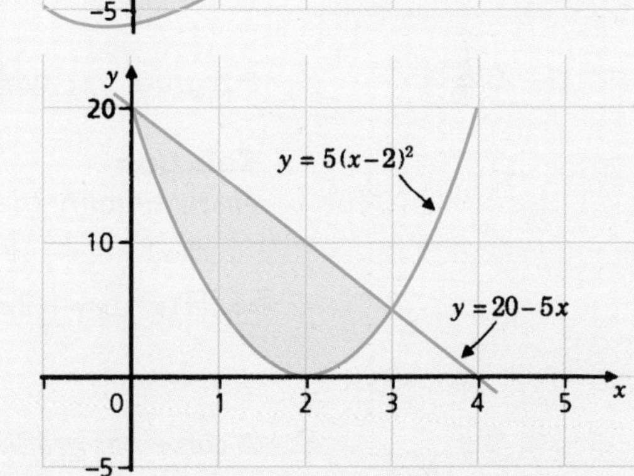

2 Find the areas represented by the following integrals. Illustrate each of your answers with a sketch.

a) $\int_1^2 3x^4 \mathrm{d}x$ **b)** $\int_{-1}^{0.5}(x^2 + x - 1)\,\mathrm{d}x$ **c)** $\int_0^4 x(4 - x)\,\mathrm{d}x$ **d)** $\int_{-4}^{4}(3x^2 - 48)\,\mathrm{d}x$

For each of questions **3** to **11** sketch a graph of the given curve(s) and find the area described.

3 The area between the curve $y = x(4 - x)$ and the x-axis from $x = 0$ to $x = 6$.

4 The area enclosed by the curve $y = (2x - 1)(2x + 1)$ and the x-axis.

5 The area enclosed by the curve $y = x(9x^2 - 1)$ and the x-axis.

6 The area enclosed by the curve $y = x(x - 1)(x - 2)$ and the x-axis.

7 The area enclosed by the curve $y = x(2 - x)$ and the line $y = \frac{1}{2}x$.

8 The area enclosed between the curve $y = x^2 - 3x - 4$ and the line $y = x + 1$.

9 The area enclosed between the curves $y = x^2 - 1$ and $y = 1 - x^2$.

10 The area enclosed between the curves $y = 4 - x^2$ and $y = x^2 - 2x + 4$.

11 The area enclosed between the curves $y = -x^3 + 6x^2 + 2x - 3$ and $y = (x - 3)^2$.

INTEGRATION: REVERSING DIFFERENTIATION

The technique of reversing differentiation, i.e. given $\frac{dy}{dx} = f'(x)$ finding $y = f(x)$, is known as **indefinite integration**, and is written:

$$\int f'(x)dx = f(x)$$

The function $\frac{dy}{dx} = f'(x)$ is called a **gradient function.**

Equivalently, the integral of any function $f(x)$ gives an **integral function** $F(x)$, which contains a constant of integration c.

A particular value for c may be determined by stating certain conditions for the integral function to obey.

Example 9.6

Integrate $f(x)$*, where* $f(x) = (2x-1)(x+3)$.

Solution

Firstly simplify the function.

$$f(x) = (2x-1)(x+3) = 2x^2 + 5x - 3$$

$$\Rightarrow \int f(x)\,dx = \int (2x^2 + 5x - 3)\,dx = \tfrac{2}{3}x^3 + \tfrac{5}{2}x^2 - 3x + c$$

Example 9.7

A curve has gradient function $\frac{dy}{dx} = 4x - 1$ *and passes through the point (1, 5).*

Find its equation in the form $y = f(x)$ *and illustrate your answer with a sketch.*

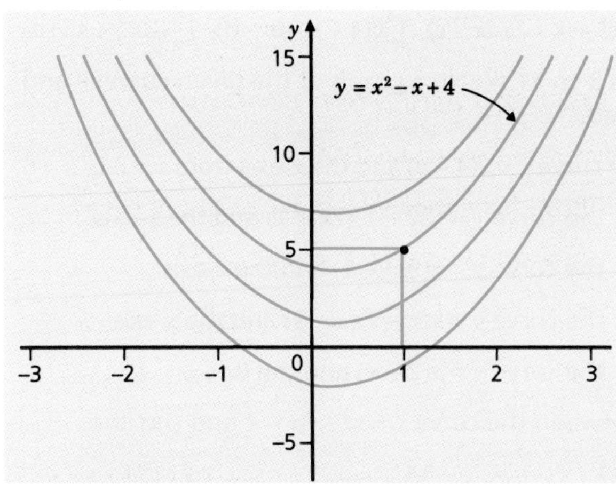

Solution

$\frac{dy}{dx} = 4x - 1 \Rightarrow y = \int (4x-1)\,dx = 2x^2 - x + c$

But $y = 5$ *when* $x = 1$*, since the curve passes through* $(1, 5)$.

$\Rightarrow 5 = 2 \times 1^2 - 1 + c$

$\Rightarrow c = 4$

Therefore the equation of the curve is $y = 2x^2 - x + 4$.

Note

The function $2x^2 - x + c$ represents a family of curves, of which only one curve passes through $(1, 5)$.

$y = 2x^2 - x + 4$.

Example 9.8

For a function $y = f(x)$, $f'(x) = 3x^2 - 2x$ and $f(1) = 3$, find $f(x)$.

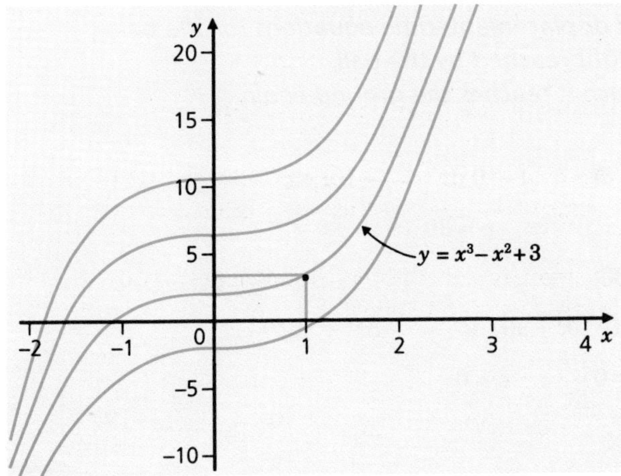

Solution

$f(x) = \int (3x^2 - 2x)\,dx = x^3 - x^2 + c$

But $f(1) = 3 \Rightarrow 1^3 - 1^2 + c = 3$

$\Rightarrow c = 3$

$y = f(x) = x^3 - x^2 + 3$

Note

The function $y = x^3 - x^2 + c$ represents a family of curves, of which only one curve passes through (1, 3).

$y = x^3 - x^2 + 3$.

Example 9.9

Show that $\int (x+a)^2\,dx = \frac{1}{3}(x+a)^3 + k$*, where* a *is a constant.*

Solution

Firstly simplify the function.

$(x + a)^2 = (x + a)(x + a) = x^2 + 2ax + a^2$

$$\Rightarrow \int (x^2 + 2ax + a^2)\,dx = \tfrac{1}{3}x^3 + \tfrac{1}{2}\times 2ax^2 + a^2x + c$$

$$= \tfrac{1}{3}\left(x^3 + 3ax^2 + 3a^2x + a^3\right) + k \quad \{\text{where } c = \tfrac{1}{3}a^3 + k\}$$

$$= \tfrac{1}{3}(x+a)^3 + k$$

Displacement, velocity and acceleration

In Chapter 7, *Differentiation 1* we found that:

$s = f(t)$	gives displacement s as a function of time t
$\Rightarrow \quad v = \dfrac{ds}{dt} = f'(t)$	gives velocity v as a function of time t
$\Rightarrow \quad a = \dfrac{dv}{dt} = f''(t)$	gives acceleration a as a function of time t

Since the process of integration is the reverse of differentiation, it follows that:

$v = \int a\,dt$ and $s = \int v\,dt$

The following example illustrates how velocity–time and displacement–time equations may be deduced, by successive integration, beginning with the acceleration–time equation.

Example 9.10

A ball is thrown vertically upwards from ground level with an initial velocity of 30 m s^{-1}. Assuming a constant acceleration of -10 m s^{-2}, find:

a) velocity–time and displacement–time equations for the ball,
b) the maximum height reached by the ball,
c) the time after which it reaches the ground again.

Solution

a) $a = -10 \Rightarrow v = \int a\,dt = \int -10\,dt = -10t + c$

but $v = 30$ *when* $t = 0 \Rightarrow c = 30$

hence $v = -10t + 30$

$\Rightarrow s = \int v\,dt = \int(-10t + 30)\,dt = -5t^2 + 30t + c$

But $s = 0$ *when* $t = 0 \Rightarrow c = 0$

hence $s = 30t - 5t^2$.

b) *The maximum height occurs when* $\frac{ds}{dt} = v = 0$.
$\Rightarrow -10t + 30 = 0 \Rightarrow t = 3$
$\Rightarrow s = 30 \times 3 - 5 \times 3^2 = 90 - 45 = 45$
$\Rightarrow$ *The maximum height is 45 metres.*

c) *On reaching the ground again:*
$s = 0$
$\Rightarrow 30t - 5t^2 = 0 \Rightarrow 5t(6 - t) = 0$
$\Rightarrow t = 0$ *or* $t = 6$
$\Rightarrow$ *The ball reaches the ground again after 6 seconds.*

EXERCISES

1 For each of the following gradient functions find $y = f(x)$.

a) $f'(x) = 7$
b) $f'(x) = 4x^3$
c) $f'(x) = 6x^2$
d) $f'(x) = x^2 - 4x + 5$
e) $f'(x) = x^7$
f) $\frac{dy}{dx} = 3x(x - 2)$
g) $\frac{dy}{dx} = (x + 4)(x - 3)$
h) $\frac{dy}{dx} = 4x^2(x - 1)$
i) $\frac{dy}{dx} = \frac{x^2 - 1}{x + 1}$
j) $\frac{dy}{dx} = (x + 2)^3$

2 Find the values of the following integrals.

a) $\int(4x - 7)\,dx$
b) $\int x^2(x - 1)\,dx$
c) $\int 2x(x^3 + 4)\,dx$
d) $\int 7(2 - x)^2\,dx$
e) $\int(x - 1)(x + 2)\,dx$
f) $\int(2x - 1)^3\,dx$

3 For each of the following gradient functions, find the equation of the curve that passes through the given point and sketch the curve.

a) $f'(x) = 2x$ — $y = f(x)$ passes through (0, 3)
b) $f'(x) = 4x + 5$ — $y = f(x)$ passes through (0, 2)
c) $f'(x) = 10 - 3x^2$ — $y = f(x)$ passes through (1, 1)
d) $f'(x) = x^2(x + 1)$ — $y = f(x)$ passes through (1, 1)
e) $\frac{dy}{dx} = 6x^2 - 3x$ — $y = f(x)$ passes through (2, 0)
f) $\frac{dy}{dx} = 2x(x - 1)$ — $y = f(x)$ passes through (3, 4)
g) $\frac{dy}{dx} = (3x + 5)^2$ — $y = f(x)$ passes through (1, 0)
h) $\frac{dy}{dx} = (x^2 - 1)(3x + 5)$ — $y = f(x)$ passes through (0, –3)

1 For each of the following gradient functions find $y = f(x)$.

a) $f'(x) = -3$ **b)** $f'(x) = 5x^4$
c) $f'(x) = 0.2x^4$ **d)** $f'(x) = 3.5x^6$
e) $f'(x) = x(x + 1)$ **f)** $\frac{dy}{dx} = 5x(3 - 2x)$
g) $\frac{dy}{dx} = (x - 1)(x - 2)$ **h)** $\frac{dy}{dx} = 3x^3(2x + 7)$
i) $\frac{dy}{dx} = \frac{x^2 - 4}{x - 2}$ **j)** $\frac{dy}{dx} = (3x + 1)^3$

2 Find the values of the following integrals.

a) $\int (3x - 4)\,dx$ **b)** $\int t^3(t^2 - 1)\,dt$
c) $\frac{1}{2}\int t^2(t^3 + 1)\,dt$ **d)** $\int \frac{(5 - x)^2}{2}\,dx$
e) $\int x(x + 1)(x + 2)\,dx$ **f)** $\int (3 + u)^3\,du$

3 For each of the following gradient functions, find the equation of the curve that passes through the given point and sketch the curve.

a) $f'(x) = 0.5$ — $y = f(x)$ passes through (1, 1)
b) $f'(x) = 8x$ — $y = f(x)$ passes through (3, 4)
c) $f'(x) = 11 - 2x$ — $y = f(x)$ passes through (2, 0)
d) $f'(x) = 11 - 2x^2$ — $y = f(x)$ passes through (0, 0)
e) $\frac{dy}{dx} = 3x(x + 2)$ — $y = f(x)$ passes through (–1, 1)
f) $\frac{dy}{dx} = x^2(x + 4)$ — $y = f(x)$ passes through (3, 0)
g) $\frac{dy}{dx} = (3 - 2x)^2$ — $y = f(x)$ passes through (–1, 1)
h) $\frac{dy}{dx} = (3x - 2)^3$ — $y = f(x)$ passes through (0, 0)

CONSOLIDATION EXERCISES FOR CHAPTER 9

1 Calculate the area of the region bounded by the curve $y = x^2 - 3x + 2$ and the x-axis.

2 Sketch the curve $y = (x - 1)(6 - x)$. Find the area enclosed between the curve and the x-axis.

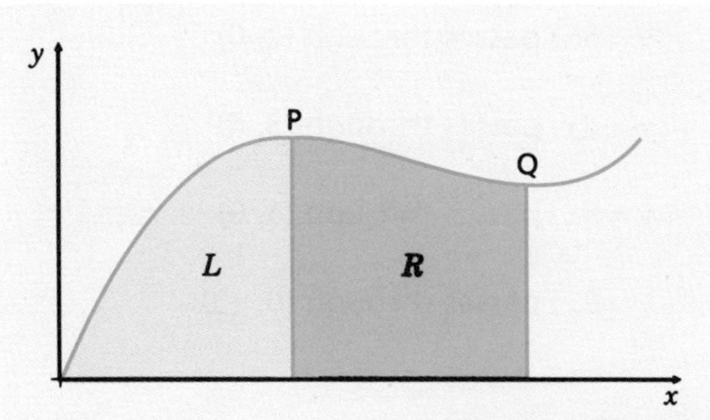

3 The diagram shows a sketch of the curve with equation $y = x^3 - 9x^2 + 24x$.

Maximum and minimum values occur at points P and Q respectively.

a) Calculate the coordinates of P and Q.
b) Show that areas L and R are in the ratio 7:9.

4 Evaluate the following integrals.

a) $\int_1^2 (x^2 + 2)^2 \, dx$ **b)** $\int_{-1}^3 (5x^2 - 2)^2 \, dx$

5 **a)** Show that the line $x + y = 4$ crosses the curve $y = \frac{1}{4}x^2 - x$ at the point A (4, 0) and find the coordinates of B, the other point of intersection.
b) Sketch the line and the curve on the same axes.
c) Evaluate these integrals.

i) $\int_{-4}^0 (\frac{1}{4}x^2 - x) \, dx$ **ii)** $\int_0^4 (\frac{1}{4}x^2 - x) \, dx$

d) Hence find the area of the region bounded by the line AB and the curve $y = \frac{1}{4}x^2 - x$.

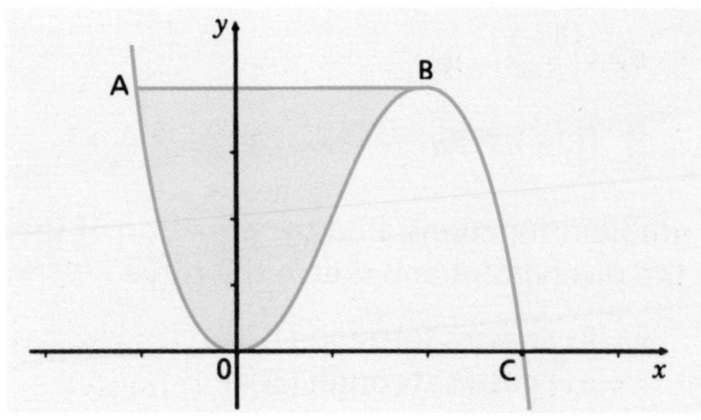

6 The diagram shows a sketch of the curve $y = x^2(3 - x)$, which crosses the x-axis at C and has stationary values at O and B.

a) Find the coordinates of A and B.
b) Calculate the area of the shaded region.
c) Prove that OABC is a parallelogram.

7 A particle P moves in a straight line, starting at rest from the point O. At time t seconds after leaving O, the velocity, v m s^{-1}, of P is given by $v = 4t + 6t^2$
Calculate the distance covered by P over the interval $2 \le t \le 3$.

8 A lift moves vertically without intermediate stops from rest at ground floor level to rest at top floor level. When the lift has been moving for t seconds, where $0 \le t \le 24$, the velocity v ms^{-1} of the lift is given by:

$$v = \frac{t(576 - t^2)}{288}$$

a) Calculate the maximum speed of the lift.
b) Show that the vertical distance between the ground and top floors is 288 m.

9 The velocity v m s^{-1} of a particle P at time t seconds, is given by:
$v = t(t-2)^2 + 3$

Given that P moves in a straight line, calculate:
a) the times at which the acceleration is zero,
b) the distance covered by P over the interval $1 \le t \le 2$.

10 Show that $\int (x+a)^3 \, dx = \frac{1}{4}(x+a)^4 + k$, where a is a constant.

Summary

Integration

- The process of integration is the **inverse** of differentiation.
- The **indefinite integral** of a function f(x) with respect to x is denoted by $\int f(x)dx$.

1 $f(x) = x^n \Rightarrow \int f(x)dx = \frac{1}{n+1}x^n + c$ for $n = 0, 1, 2, 3, \ldots$

2 $f(x) = ax^n \Rightarrow \int f(x)dx = \frac{a}{n+1}x^n + c$ for $n = 0, 1, 2, 3, \ldots$

3 The integral of a sum (or difference) is the sum (or difference) of the integrals.

Areas

- For positive functions, the area function $A(x) = \int f(x)dx$ represents the area under a graph between 0 and x.
- The area under a graph between $x = a$ and $x = b$ is denoted by the **definite integral**

$$\int_a^b f(x)dx = A(b) - A(a)$$

Trapezium rule

- Approximate a region under a graph by a number of trapezia, each with equal width.
- The sum of the areas of the trapezia is an approximation to the required area.
- As the number of trapezia increases, with the width of each one decreasing, the sum of the areas will more closely approximate the required area.

PURE MATHS

10 Trigonometry 2

In this chapter we:

- *introduce the sine rule and cosine rule for triangles*
- *explore the use of trigonometry in problem-solving.*

THE SINE RULE

Trigonometry allows us to deduce general results which are true for any triangle, and then apply them to solve specific triangles. One very important rule relates the sizes of the angles of a triangle to the lengths of the sides.

Exploration 10.1

Scale drawing

A surveyor wants to produce a drawing to include a triangular plot of land. She knows the distance AB and the angles A and B, as shown in this scale drawing.

Scale: 1 cm:10 m

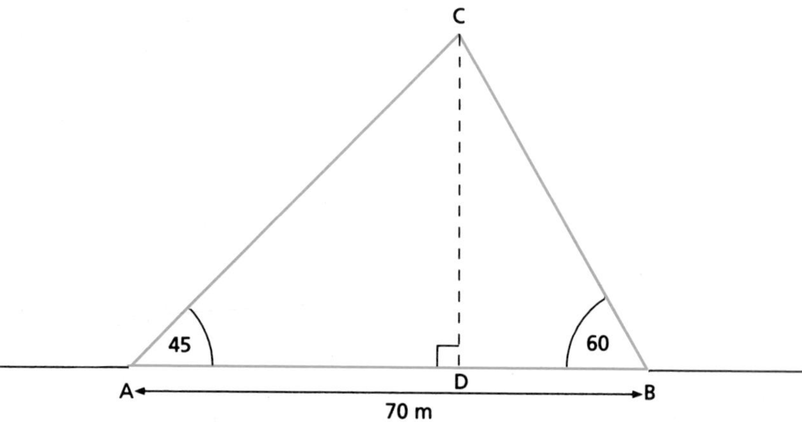

- How can she find the distance of point C from AB?
- How far is C from either A or B?

From the scale drawing, check that
CD ≈ 44 m,
AC ≈ 63 m and
BC ≈ 51 m.

Developing the sine rule

Approximate values for CD, AC and BC may be found by scale drawing, but more precise values may be obtained by calculation, as follows. Label the triangle ABC as in the diagram, with CD = h.

For triangle ACD:

$\frac{h}{b} = \sin A$ or $h = b \sin A$

For triangle BCD:

$\frac{h}{a} = \sin B$ or $h = a \sin B$

$$\therefore \quad a \sin B = b \sin A$$

$$\Rightarrow \quad \frac{a}{\sin A} = \frac{b}{\sin B}$$

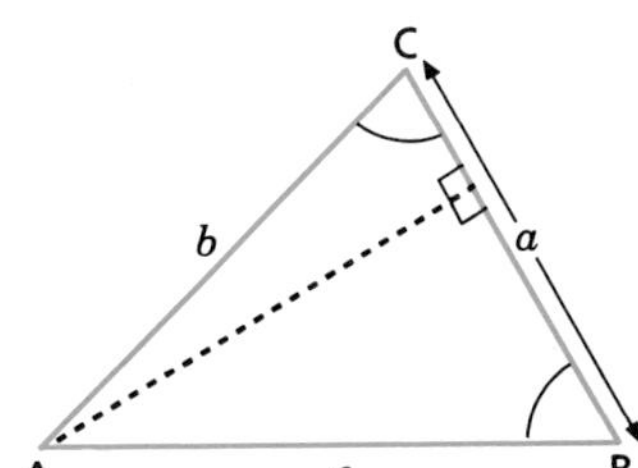

Equally, we can drop a perpendicular from A to BC.
We get the result:

$$\frac{b}{\sin B} = \frac{c}{\sin C}$$

Combining the two results gives the **sine rule** for any triangle ABC.

$$\frac{a}{\sin A} = \frac{b}{\sin B} = \frac{c}{\sin C}$$

Any two of these fractions may be equated to **solve** a triangle, given that we know the values of three quantities and need to find the fourth.

Returning to Exploration 10.1, we know $A = 45°$, $B = 60°$ and $c = 70$ m. Since the three angles of a triangle add up to 180°, we also know that $C = 75°$.

To find a:

$$\frac{a}{\sin A} = \frac{c}{\sin C}$$

$$\Rightarrow \quad \frac{a}{\sin 45°} = \frac{70}{\sin 75°}$$

$$\Rightarrow \quad a = \frac{70 \sin 45°}{\sin 75°} = 51.2 \quad \text{(3 s.f.)}$$

To find b:

$$\frac{b}{\sin B} = \frac{c}{\sin C}$$

$$\Rightarrow \quad \frac{b}{\sin 60°} = \frac{70}{\sin 75°}$$

$$\Rightarrow \quad b = \frac{70 \sin 60°}{\sin 75°} = 62.8 \quad \text{(3 s.f.)}$$

Therefore the lengths of BC and AC are 51.2 m and 62.8 m respectively, correct to 3 s.f. Compare these results with the approximate values obtained from the scale drawing.

Finally:

$h = b \sin A = 62.8 \sin 45° = 44.4$ (3 s.f.)

i.e. point C is 44.4 m from AB.

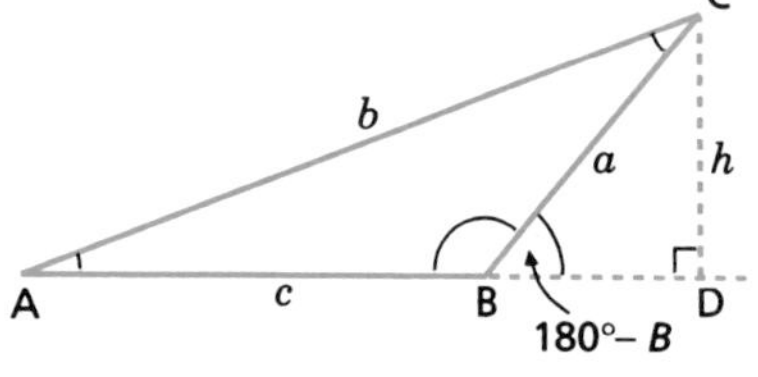

The sine rule also applies to triangles where one of the angles is **obtuse**, i.e. greater than 90° and less than 180°.
Suppose ∠ABC is obtuse, then the basic diagram looks like this.

For triangle ACD:

$$\frac{h}{b} = \sin A \Rightarrow h = b \sin A$$

but for triangle BCD:

$$\frac{h}{a} = \sin(180° - B) \Rightarrow h = a \sin(180° - B)$$

However, since B and $180° - B$ have the same sine:

$$\sin B = \sin(180° - B) \Rightarrow h = a \sin B$$

From then on the argument is the same as before, leading to the same statement of the sine rule.

Example 10.1

Triangle PQR has ∠RPQ = 25°, ∠QRP = 117° and PR = 25 cm. Find ∠PQR and the lengths of PQ and QR.

Solution

$Q = 180° - (25° + 117°) = 38°$ and $q = \text{PR} = 25$ cm.

To find PQ = r:

$$\frac{r}{\sin R} = \frac{q}{\sin Q} \quad \Rightarrow \quad \frac{r}{\sin 117°} = \frac{25}{\sin 38°}$$

$$\Rightarrow \quad r = \frac{25 \sin 117°}{\sin 38°} = 36.2 \text{ (3 s.f.)}$$

To find QR = p:

$$\frac{p}{\sin P} = \frac{q}{\sin Q} \quad \Rightarrow \quad \frac{p}{\sin 25°} = \frac{25}{\sin 38°}$$

$$\Rightarrow \quad p = \frac{25 \sin 25°}{\sin 38°} = 17.2 \text{ (3 s.f.)}$$

The ambiguous case

So far we have used the sine rule to find two lengths, given the length of one side and the angles at either end. The sine rule may also be used to **solve** a triangle (find all unknown quantities) given the size of one angle and the lengths of two sides, one of which is opposite the known angle. However, an ambiguity may arise, as we may recall from work on constructing triangles. The method (and possible ambiguity) is illustrated below.

Example 10.2

In ΔXYZ, X = 40° and XY = 20 cm. Solve the triangle if:

a) *YZ = 25 cm* ***b)*** *YZ = 15 cm.*

Solution

*Sketches of the situations reveal the ambiguity in **b)**.*

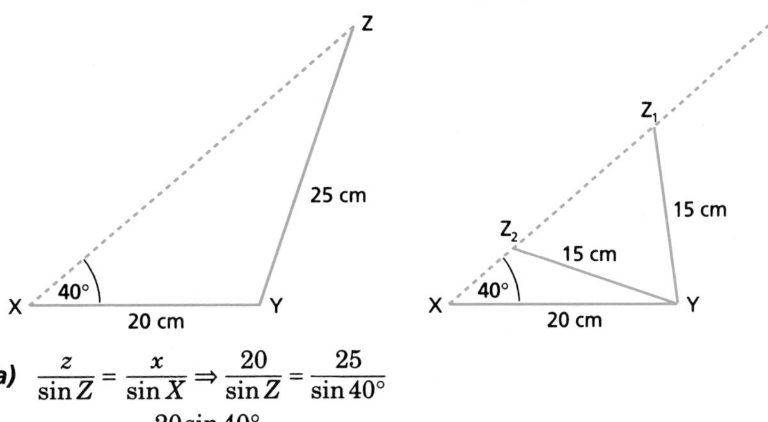

a) $\dfrac{z}{\sin Z} = \dfrac{x}{\sin X} \Rightarrow \dfrac{20}{\sin Z} = \dfrac{25}{\sin 40°}$

$\Rightarrow \sin Z = \dfrac{20 \sin 40°}{25} = 0.514 \ \ (3 \text{ s.f.})$

$\Rightarrow Z = 30.95° \ \ or \ \ 149.05°$

Since $X = 40°$, $40° + 149.05° > 180°$, *so an angle of* $149.05°$ *is impossible, hence* $Z = 30.95°$.

$X = 40° \ \text{ and } \ Z = 30.95° \Rightarrow Y = 180° - (40° + 30.95°)$
$= 109.05°$

Hence to find $XZ = y$*:*

$\dfrac{y}{\sin Y} = \dfrac{x}{\sin X} \ \Rightarrow \ \dfrac{y}{\sin 109.05°} = \dfrac{25}{\sin 40°}$

$\Rightarrow \quad y = \dfrac{25 \sin 109.05°}{\sin 40°} = 36.8 \ (3 \text{ s.f.})$

b) $\dfrac{z}{\sin Z} = \dfrac{x}{\sin X} \Rightarrow \dfrac{20}{\sin Z} = \dfrac{15}{\sin 40°}$

$\Rightarrow \sin Z = \dfrac{20 \sin 40°}{15} = 0.857 \ \ (\ 3 \text{ s.f.})$

$\Rightarrow Z = 59.99° \ \text{ or } \ 121.01°$

since $X = 40°$, $40° + 121.01° < 180°$ *so either value for* Z *is possible.*

Case 1:

$X = 40° \ \text{ and } \ Z = 59.99° \Rightarrow Y = 180° - (40° + 59.99°)$
$= 80.01°$

$\dfrac{y}{\sin Y} = \dfrac{x}{\sin X} \ \Rightarrow \ \dfrac{y}{\sin 80.01°} = \dfrac{15}{\sin 40°}$

$\Rightarrow \quad y = \dfrac{15 \sin 80.01°}{\sin 40°} = 23.0 \ (3 \text{ s.f.})$

Case 2:

$Z = 121.01° \Rightarrow Y = 180° - (40° + 121.01°)$
$= 18.99°$

$\dfrac{y}{\sin Y} = \dfrac{x}{\sin X} \ \Rightarrow \ \dfrac{y}{\sin 18.99^0} = \dfrac{15}{\sin 40°}$

$\Rightarrow \quad y = \dfrac{15 \sin 18.99°}{\sin 40°} = 7.59 \ (3 \text{ s.f.})$

Exploration 10.2

Solving a triangle

Try to solve the triangle XYZ when $X = 50°$, XY = 20 cm and YZ = 15 cm.

What happens and why?

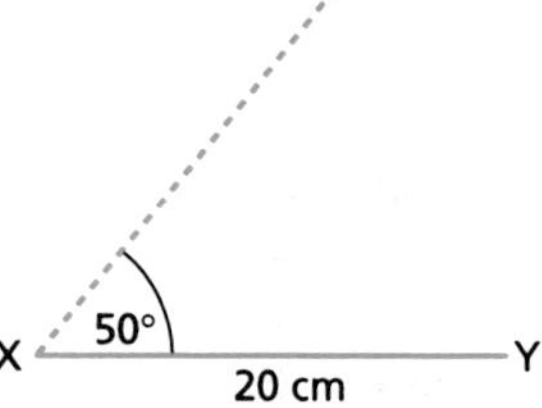

EXERCISES

10.1 A

1 Calculate the lengths of the sides a and b of the triangle in which $c = 12$ cm, $A = 53°$ and $B = 62°$.

2 Find the lengths of the other two sides of these triangles.

a) PQR where $p = 10$ cm, $Q = 85°$, $R = 35°$
b) XYZ where $y = 14$ cm, $X = 42°$, $Z = 38°$

3 A surveyor is unable to gain access to one corner of a triangular field. The side opposite this corner measures 65 metres. If the angles between the 65 m length and the other two sides are 63° and 51° respectively, calculate the lengths of the other two sides.

4 The diagram shows the results of a survey of a plot of land.

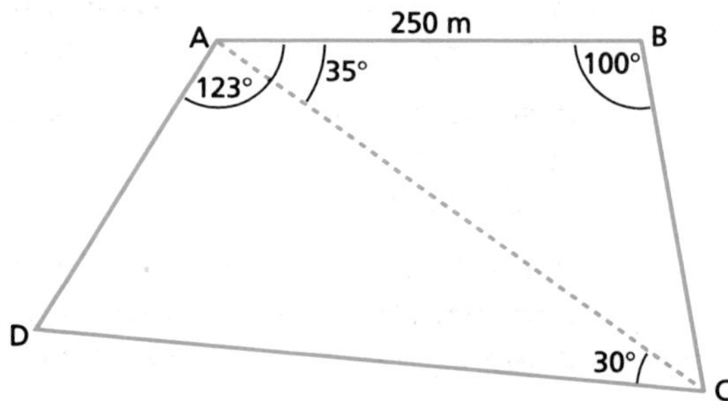

Calculate the length of CD.

5 A vertical radio mast, PQ, is supported by wires PR and PS.

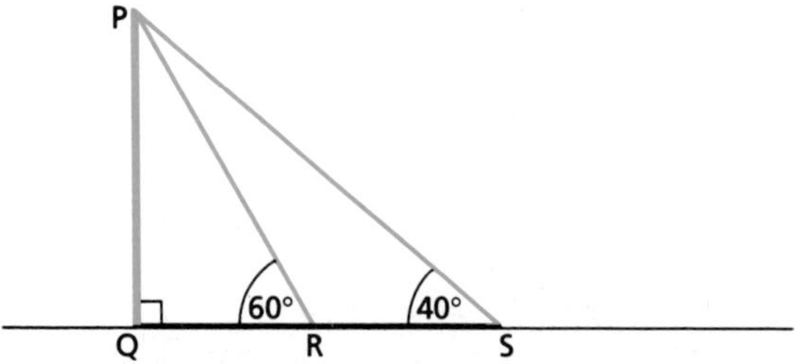

The angles of elevation of the top of the mast from R and S are 60° and 40° respectively. If R and S are 30 m apart, find the length of the wires and the height of the mast.

6 In triangle ABC, $c = 15$ cm, $A = 105°$, $a = 25$ cm.

Find the third side and the remaining angles.

7 In triangle EFG, FG = 30 cm and $F = 52°$. Solve the triangle if:

a) EG = 40 cm **b)** EG = 25 cm.

Illustrate your answers with a sketch.

8 In question **7**, what is the minimum length for EG to make the triangle solvable? What shape will it take in this case?

9 Two speedboats, 200 m apart at P and Q respectively, are both racing towards a buoy at R.

The boat at P travels at 5 m s^{-1} and boat at Q travels at 4 m s^{-1}.

Which boat gets to the buoy first and by what time does it beat the other boat?

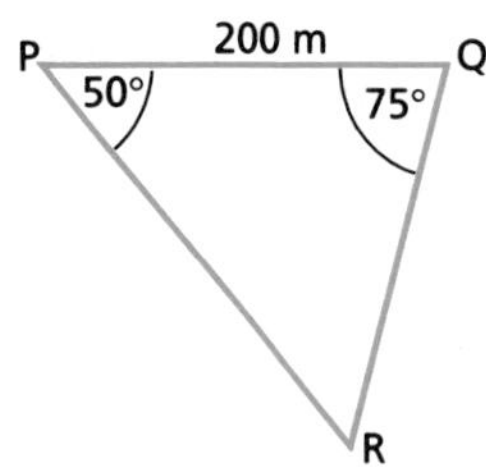

10.1 B

1 Calculate the size of A and the length of the side c for this triangle.

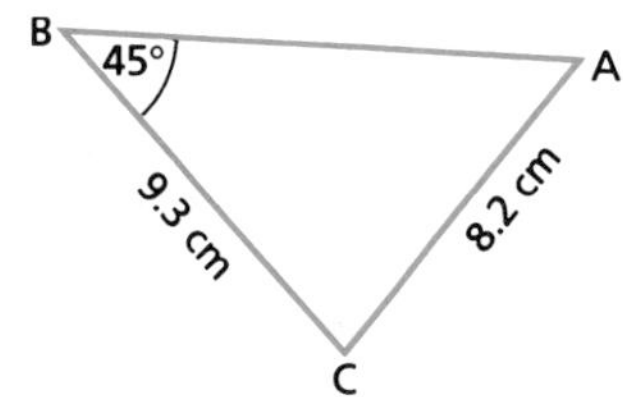

2 In the triangle XYZ, $X = 20°$, XZ = 8 cm, YZ = 5 cm and ∠XYZ is obtuse. Solve the triangle.

3 In triangle PQR, QR = 10 cm, PR = 7 cm and ∠PQR = 30°.

a) Calculate two possible values for P.
b) Draw diagrams to illustrate your answers.

4 A footbridge is to be constructed across a river, from A to B. A surveyor obtains the measurements shown in the diagram. Find the length of the footbridge.

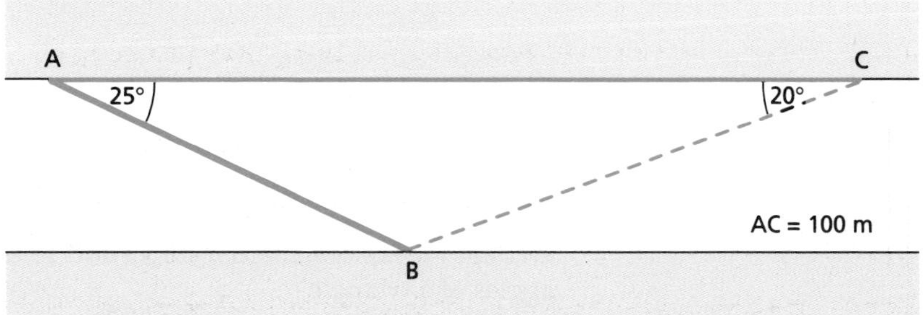

5 A surveyor, wishing to find the height of a building, takes two readings of the angle of elevation of the top of the building, from points A and B, 20 m apart. If the angles are 24° and 50° respectively, what is the height of the building?

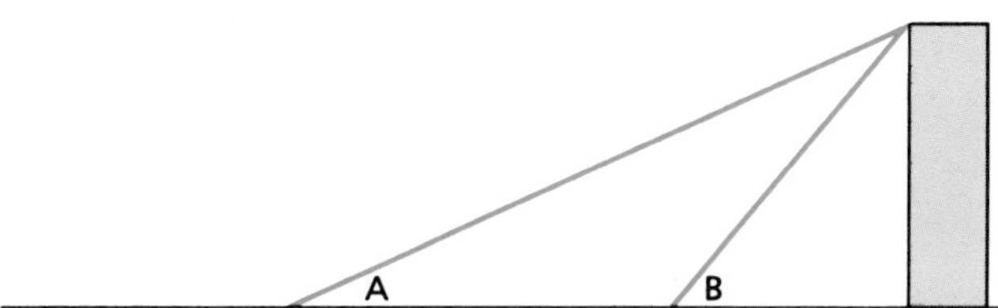

6 A man is walking due north along a footpath. At a point A he sees a church spire on a bearing of 062°. After walking a further 250 m, he is at B and the spire is now on a bearing of 137°.

a) What is the distance of the spire from A?
b) What is the shortest distance of the spire from the footpath?

7 In a motor bike trial, the competitors race around a triangular course, starting and finishing at S. The bearing of T from S is 035° and the bearing of U from S is 075°.

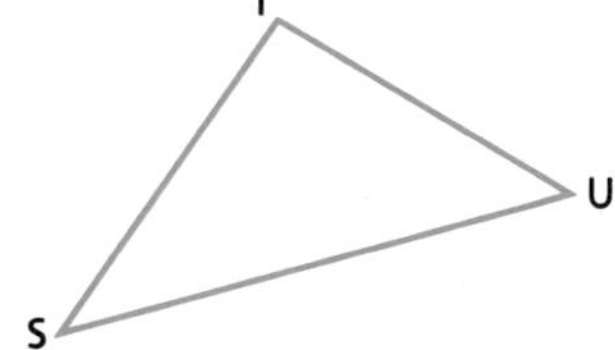

The winning rider is travelling at 15 m s^{-1} and his split times at points T and U are 48 s and 92 s. Find the length of the course, to the nearest metre.

8 In triangle ABC, AB = 50 cm and A = 40°. Solve the triangle if

a) BC = 70 cm, and
b) BC = 35 cm.

Illustrate your answers with a sketch.

9 In question **8**, above, what is the minimum length for BC to make the triangle solvable? Find the area of this triangle.

THE COSINE RULE

Although the sine rule is a valuable aid in solving triangles, it is not always the best rule to use, for example, when we are given three sides, or two sides and the included angle. Can you see why? In this section, we explore another important relation between the sides and angles of a triangle.

Exploration 10.3

Solving a triangle without using the sine rule

A helicopter regularly flies between three rigs in a North Sea oil field. It takes off from rig A and flies directly to rig B on a bearing of 040°, a distance of 50 km. Rig C is 70 km from rig B on a bearing of 110°. For the third leg of its round trip, on what bearing should it fly from C to A, and how long should it take for the journey, assuming an average speed of 200 km per hour?

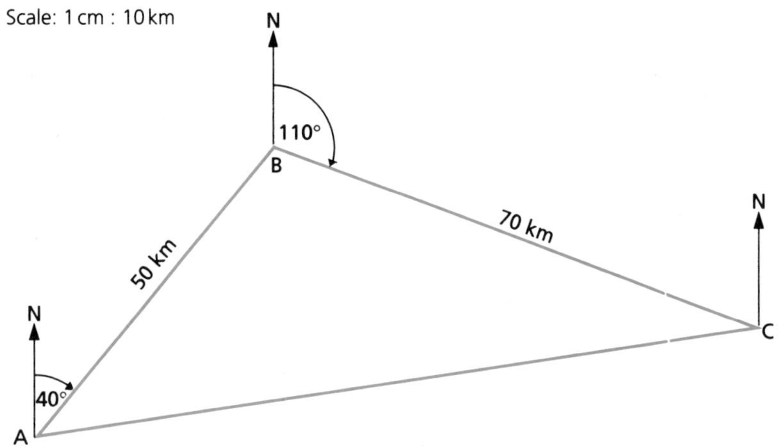

From the scale drawing, check that AC ≈ 99 km and the bearing of A from C is roughly 262°.

Developing the cosine rule

Whilst approximations may be found by scale drawing, a method of calculating the distance AC and the bearing of A from C would be preferable.

Although angle B of triangle ABC may be deduced (it is 110°) the sine rule cannot be applied. (Why not?)

Begin again with a general triangle ABC, with CD = h, AD = x and DB = $c - x$.

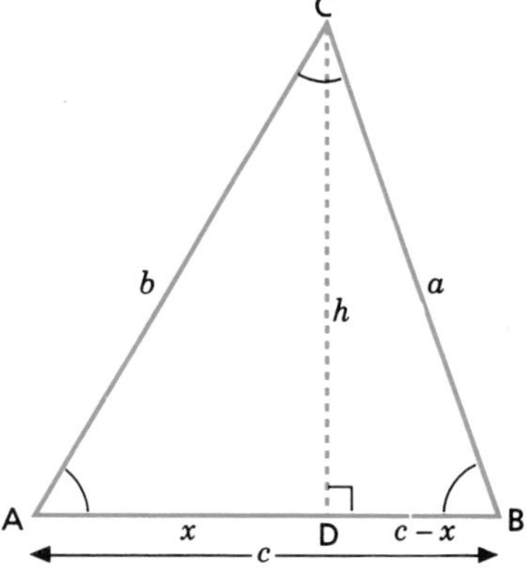

Using Pythagoras' theorem on triangle ACD:

$$b^2 = h^2 + x^2$$

$$\Rightarrow h^2 = b^2 - x^2 \qquad \text{(i)}$$

Similarly for triangle BCD:

$$a^2 = h^2 + (c - x)^2$$

$$\Rightarrow h^2 = a^2 - (c - x)^2 \qquad \text{(ii)}$$

Equating the RHS of equations (i) and (ii):

$$b^2 - x^2 = a^2 - (c-x)^2$$

$$\Rightarrow \quad b^2 - x^2 = a^2 - (c^2 - 2cx + x^2)$$

$$\Rightarrow \quad b^2 - \not{x}^2 = a^2 - c^2 + 2cx - \not{x}^2$$

$$\Rightarrow \quad b^2 = a^2 - c^2 + 2cx$$

$$\Rightarrow \quad a^2 = b^2 + c^2 - 2cx$$

$$\text{but} \quad x = b\cos A$$

$$\Rightarrow a^2 = b^2 + c^2 - 2bc\cos A \qquad (1)$$

This relationship is known as the **cosine rule**. By dropping perpendiculars from A to BC or from B to AC, instead of from C to AB, equivalent versions of the cosine rule may be obtained.

$$b^2 = a^2 + c^2 - 2ac\cos B \qquad (2)$$

$$c^2 = a^2 + b^2 - 2ab\cos C \qquad (3)$$

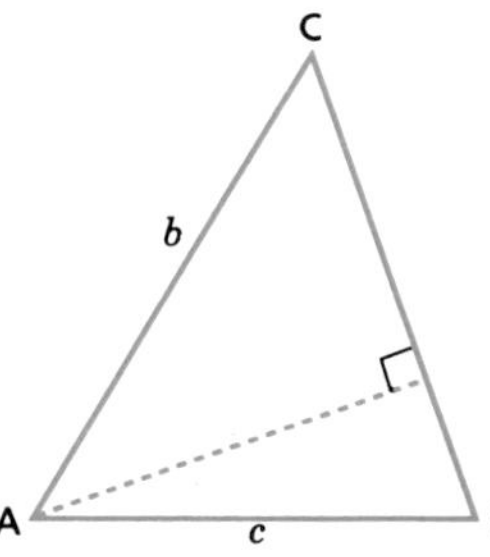

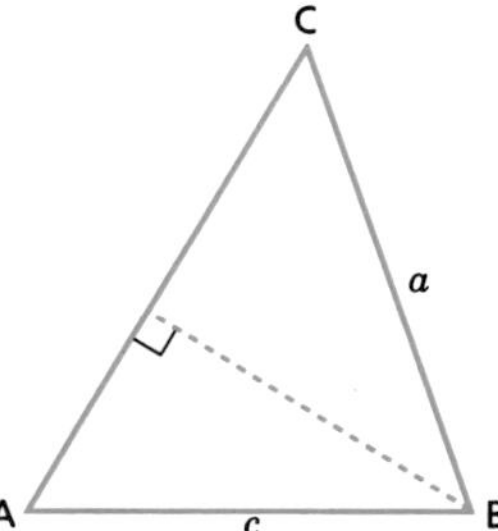

Choose the version that is appropriate for the problem. Returning to the exploration, the distance AC (= b) may be found as follows.

$$b^2 = a^2 + c^2 - 2ac\cos B$$

$$\Rightarrow \quad b^2 = 70^2 + 50^2 - 2 \times 70 \times 50 \times \cos 110°$$

$$\Rightarrow \quad b^2 = 4900 + 2500 - (-2394)$$

$$\Rightarrow \quad b^2 = 9794$$

$$\Rightarrow \quad b = \sqrt{9794} \approx 99 \text{ km}$$

To find the bearing of A from C you need to calculate angle BCA of triangle ABC. When all three sides are known, this may be done by solving version (3) of the cosine rule.

$$c^2 = a^2 + b^2 - 2ab\cos C$$

$$\Rightarrow \quad 50^2 = 70^2 + 9794 - 2 \times 70 \times 99 \times \cos C$$

$$\Rightarrow \quad 2500 = 14\,694 - 13\,860 \times \cos C$$

$$\Rightarrow \quad \cos C = \frac{14\,694 - 2500}{13\,860} = 0.8798 \quad (4 \text{ d.p.})$$

$$\Rightarrow \quad C = 28.4° \quad (3 \text{ s.f.})$$

Since the bearing of B from C is 290°, the bearing of A from C is 290° − 28.4° = 261.6°.

Compare the calculated values with those obtained from the scale drawing.

The cosine rule should be used to solve triangles when you are given:

either two sides and the angle between them

or all three sides.

Example 10.3

In triangle ABC, $a = 5$ cm, $b = 12$ cm. Find the value of c when:

a) $C = 75°$ **b)** $C = 90°$ **c)** $C = 115°$.

Solution

In each case we apply the version of the cosine rule:

$c^2 = a^2 + b^2 - 2ab\cos C$

a)
$$\begin{aligned} c^2 &= 5^2 + 12^2 - 2\times 5\times 12\times \cos 75° \\ &= 25 + 144 - 120\times 0.2588 \\ &= 137.94 \\ \Rightarrow c &= \sqrt{137.94} \approx 11.7 \text{ (3 s.f.)} \end{aligned}$$

b)
$$\begin{aligned} c^2 &= 5^2 + 12^2 - 2\times 5\times 12\times \cos 90° \\ &= 25 + 144 - 120\times 0 \\ &= 169 \\ \Rightarrow c &= \sqrt{169} = 13 \end{aligned}$$

Note: *When $C = 90°$, the cosine rule reduces to Pythagoras' theorem.*

c)
$$\begin{aligned} c^2 &= 5^2 + 12^2 - 2\times 5\times 12\times \cos 115° \\ &= 25 + 144 - 120\times (-0.4226) \\ &= 219.71 \\ \Rightarrow c &= \sqrt{219.71} \approx 14.8 \text{ (3 s.f.)} \end{aligned}$$

Example 10.4

Solve the triangle PQR where $p = 11$, $q = 15$ and $r = 19$.

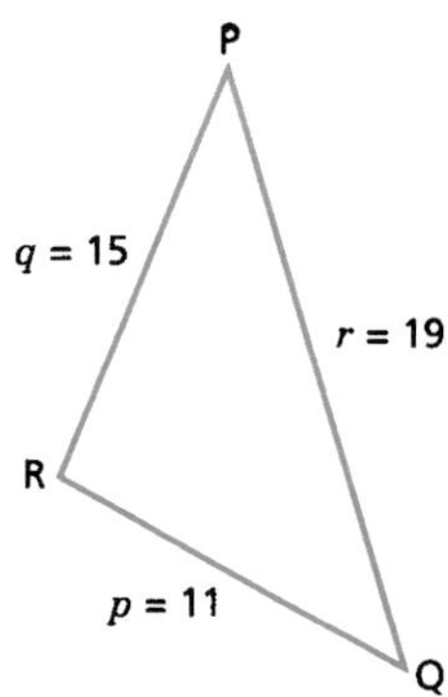

Solution

Firstly, choose an angle to find using the cosine rule, say P. Write out the appropriate version.

$$\begin{aligned} p^2 &= q^2 + r^2 - 2qr\cos P \\ \Rightarrow 11^2 &= 15^2 + 19^2 - 2\times 15\times 19\times \cos P \\ \Rightarrow 121 &= 225 + 361 - 570\times \cos P \\ \Rightarrow \cos P &= \frac{465}{570} = 0.8158 \\ P &= 35.8° \text{ (3 s.f.)} \end{aligned}$$

We could use the cosine rule 'backwards' to find Q or R. Alternatively, now one angle is known, we can use the sine rule.
For example, to find Q:

$$\frac{p}{\sin P} = \frac{q}{\sin Q} \Rightarrow \frac{11}{\sin 35.3°} = \frac{15}{\sin Q}$$

$$\Rightarrow \sin Q = \frac{15\sin 35.3°}{11} = 0.7880 \text{ (4 d.p.)}$$

$$\Rightarrow Q = 52.0° \text{ (3 s.f.)}$$

Finally: $R = 180° - (35.3° + 52.0°) = 92.7°$

EXERCISES

10.2 A

1 In triangle ABC, AB = 8 cm, BC = 12 cm and $B = 65°$.

Find the length of AC and the remaining angles.

2 Solve these triangles.

a) $b = 8$ cm, $c = 12$ cm, $A = 42°$
b) $x = 15$ cm, $y = 22$ cm, $Z = 115°$
c) $p = 7$ cm, $q = 9$ cm, $r = 10$ cm
d) $a = 13$ cm, $b = 17$ cm, $c = 6$ cm

3 Solve the triangle in which $x = 15$ cm, $y = 17$ cm and $z = 8$ cm. What is special about the triangle?

4 Try to solve the triangle in which $p = 20$ m, $q = 12$ m and $r = 7$ m. Make a sketch to explain your result.

5 From a lighthouse A, two ships B and C, lie on bearings of 080° and 220° respectively. If B is 21 km from A and C is 15 km from A, how far apart are the two ships?

6 A ship is sailing on a straight course. At 12.00 it was 20 km due north of a rock R. At 14.00 it was 25 km from R on a bearing of 095°.

a) What was the average speed of the ship?
b) In which direction was the ship sailing?
c) At what time was it closest to the rock?

7 The diagram shows the results of a survey of a plot of land.

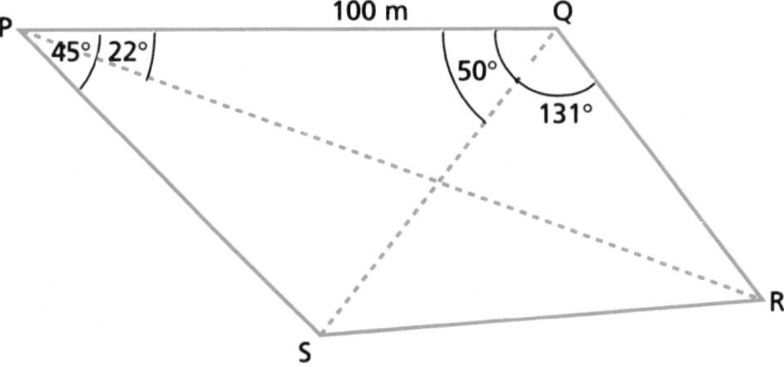

Calculate the length of RS.

8 For any triangle ABC, show that:

$$\cos C = \frac{a^2 + b^2 - c^2}{2ab}$$

Explain why $C < 90° \Leftrightarrow a^2 + b^2 > c^2$

$C = 90° \Leftrightarrow a^2 + b^2 = c^2$

$C > 90° \Leftrightarrow a^2 + b^2 < c^2$

10.2 B

1 For the triangle shown, find the length of PR and the sizes of P and R.

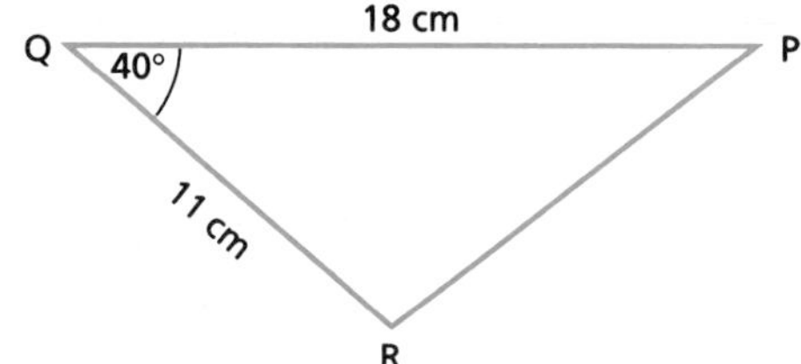

2 In triangle ABC, AB = 12.5 cm, BC = 30 cm and AC = 32.5 cm. Calculate B. Comment on your result. Can you confirm your result by another method?

3 In triangle XYZ, XY = 45 cm, YZ = 23.5 cm and XZ = 20.2 cm. Find Z and explain your result.

4 The circle in the diagram has radius 4 cm and ∠POQ = 110°.

Find the length of PQ.

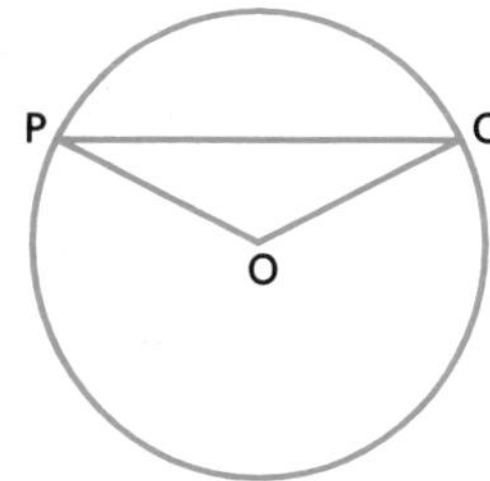

5 A man sets off from town T and walks 4 km on a bearing of 050° to the monument at M. He then walks 2.4 km, on a bearing of 165°, to the pub at P. How far must he walk, and on what bearing, from the pub back to the town?

6 A marker buoy is 12 km from the coastguard station, on a bearing of 030°. The coastguard sees a boat sailing in a straight line towards the buoy. At midday, the boat is 16 km from the coastguard station on a bearing of 142°. At 12.45 pm the boat is 14 km from the coastguard station on a bearing of 082°.

a) Make a sketch showing the positions of the boat relative to the coastguard station.
b) How far is the boat from the buoy at 12.45 pm?
c) Calculate the average speed of the boat.
d) In what direction is the boat sailing?

7 The quadrilateral SABC forms the course for a cycle race. Calculate the length of the course, to the nearest 100 m. The surveyor rounded all the lengths to the nearest 100 m. Calculate, to the nearest metre, the maximum and minimum possible lengths of the course, and the percentage error in each case.

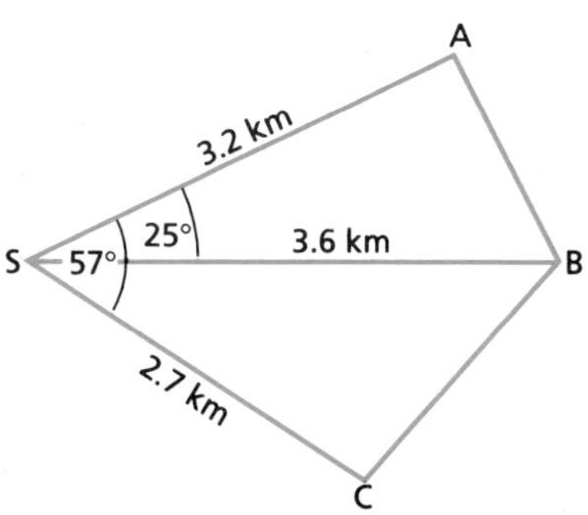

THE AREA OF A TRIANGLE

Exploration 10.4

Finding the area of a triangle

A triangular plot of land has known measurements AC = 120 m, AB = 80 m and $A = 40°$.

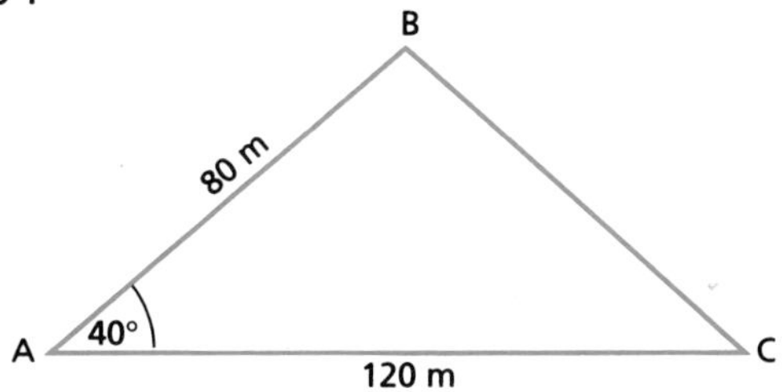

- Find the area of the triangular plot of land.
- Develop a formula for the area of any triangle in terms of two sides and an angle.

The area of a triangle is given by:

$$\frac{\text{base} \times \text{perpendicular height}}{2}$$

Taking AC as base, you need to find the perpendicular height BD. From the right angled triangle ABD:

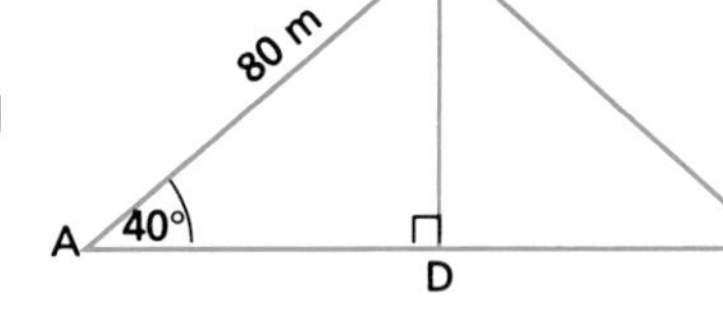

$$\sin A = \frac{BD}{AB} \quad \Rightarrow \quad \sin 40° = \frac{BD}{80}$$

$$\Rightarrow \quad BD = 80 \sin 40° = 51.4 \text{ m}$$

You may now find the area of the plot of land as:

$$\text{area} = \frac{120 \times 51.4}{2} = 3090 \text{ m}^2 \text{ (3 s.f.)}$$

The general case

For any triangle ABC, drop a perpendicular from B to AC. Let BD = h.

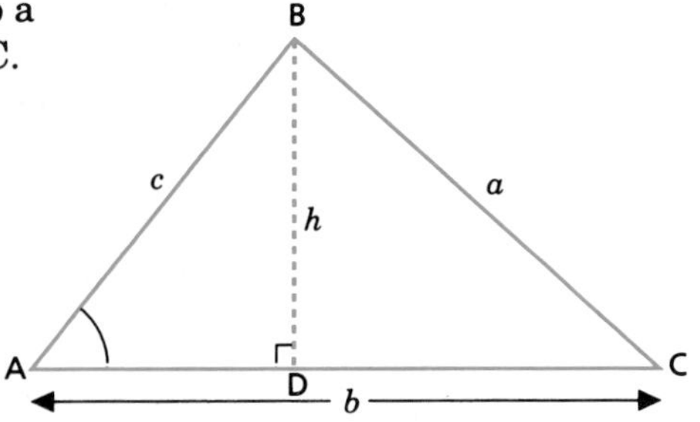

From the diagram:

$$\sin A = \frac{h}{c} \quad \Rightarrow \quad h = c \sin A$$

The area of the triangle is $\frac{b \times h}{2}$ or $\frac{1}{2}bh$ and substituting for h gives:

$$\text{area of triangle} = \tfrac{1}{2}bc \sin A$$

Check that you can use this formula to find the area of the triangular plot of land in the exploration.
The formula also works if angle CAB is obtuse.

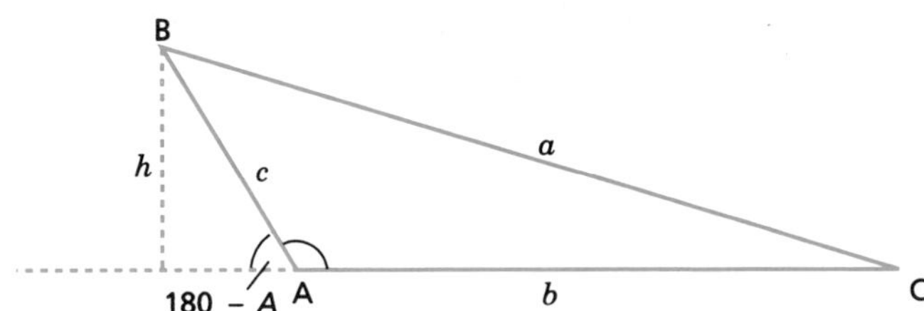

$\sin(180°-A) = \frac{h}{c} \Rightarrow h = c\sin(180°-A)$

But: $\sin(180° - A) = \sin A \Rightarrow h = c \sin A$ and the rest of the proof follows as before.

Example 10.5

Find the area of the triangle ABC with sides 20 cm, 8 cm and 15 cm.

Solution

a) *Let $a = 20$, $b = 8$ and $c = 15$, then use the cosine rule to find A.*

$$a^2 = b^2 + c^2 - 2bc\cos A$$

$$\Rightarrow \cos A = \frac{b^2 + c^2 - a^2}{2bc} = \frac{8^2 + 15^2 - 20^2}{2 \times 8 \times 15}$$

$$\Rightarrow \cos A = -0.4625$$

$$\Rightarrow A = 118° \quad (3 \text{ s.f.})$$

b) *Using the area formula:*

$$\text{area} = \tfrac{1}{2}bc\sin A = \tfrac{1}{2} \times 8 \times 15 \times \sin 118°$$
$$= 53.0 \text{ m}^2 \text{ (3 s.f.)}$$

Note:

By transferring the value of A from the first part of the calculation to the second, corrected to 3 s.f., an error has been introduced. Check the calculations on your own calculator and transfer the value of A by storing it in a memory. You should find that the area is 53.2 m2 (3 s.f.).

Wherever possible, store intermediate results of a calculation to avoid premature approximation, which can result in loss of accuracy.

EXERCISES

10.3 A

1 Find the areas of the following triangles.

a)

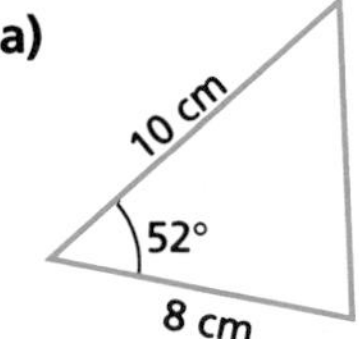

b)

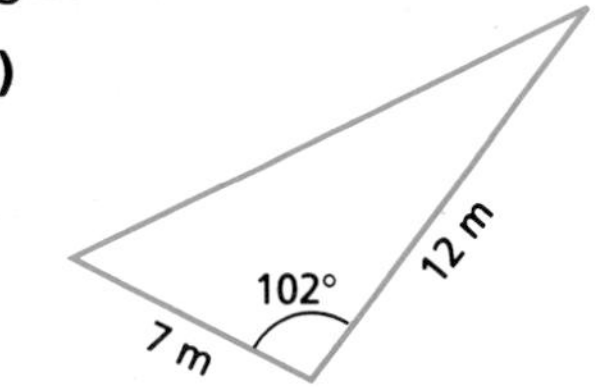

2 Find the areas of the following triangles.

a)

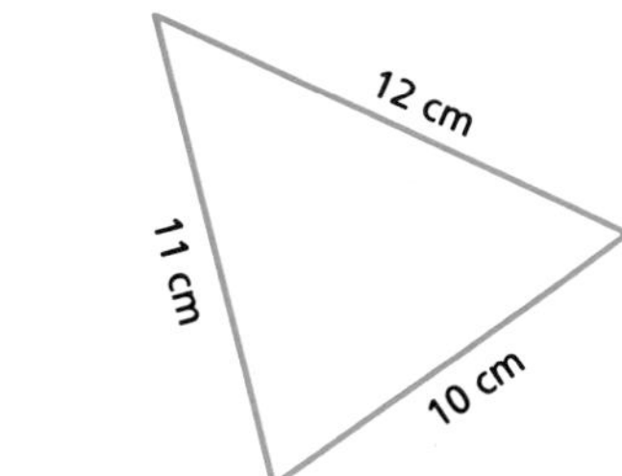

b)

3 Find the areas of the shaded regions.

a)

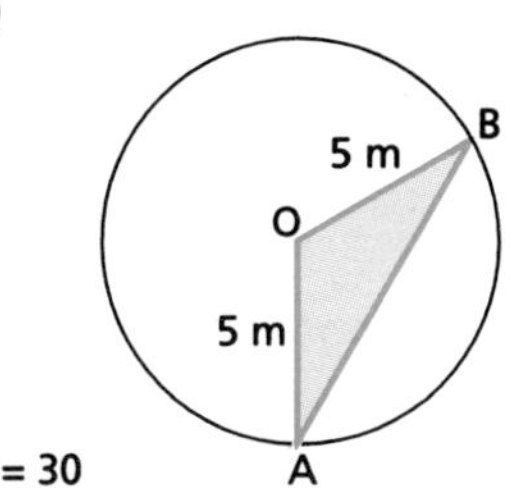

b)

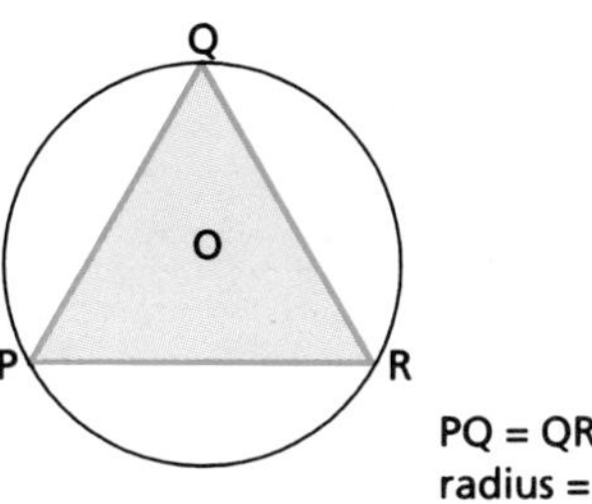

4 Find the area of the triangular plot of land in the diagram.

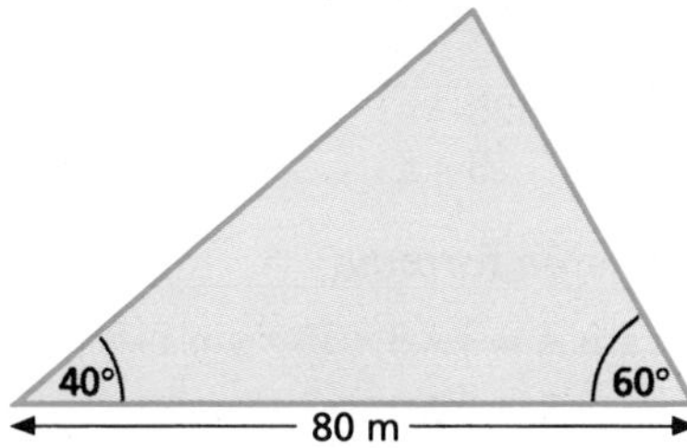

5 An equilateral triangle has side x cm.
Show that the area of the triangle is given by $\frac{\sqrt{3}}{4}x^2$. Use this formula to check your answer to question **3 b)**.

6 A regular hexagon is inscribed in a circle of radius 10 cm.

a) Find the area of the hexagon.

b) Express the area of the hexagon as a percentage of the area of the circle in which it is inscribed.

c) Find a formula for the area of a regular hexagon of side x cm.
Use this formula to check your answer to part **a)**.

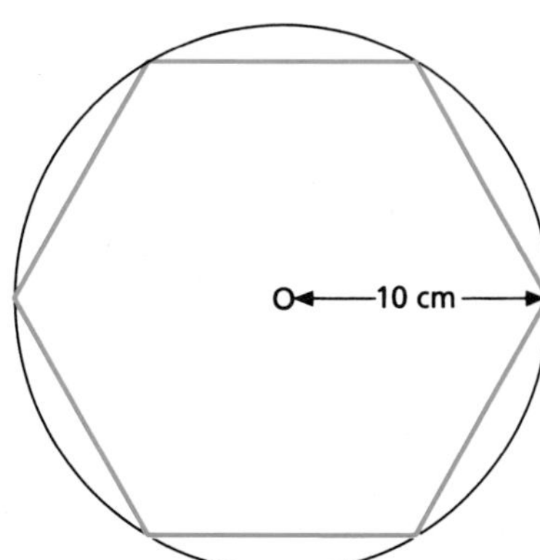

10.3 B

1 Find the areas of these triangles, correct to 3 s.f.

a)

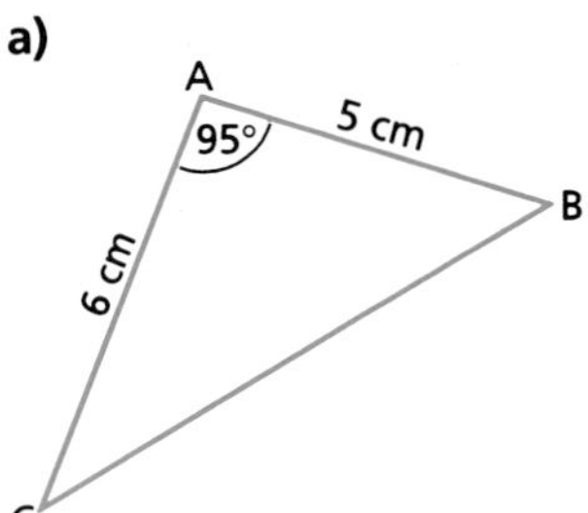

b)

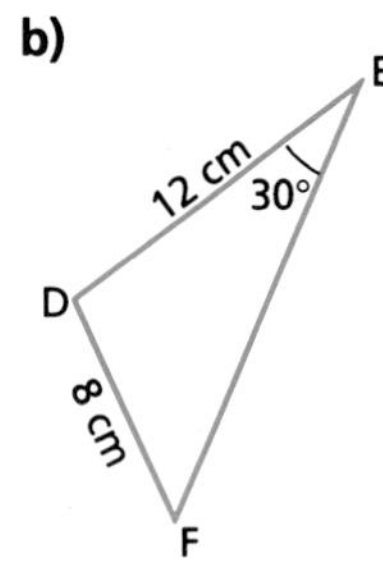

c)

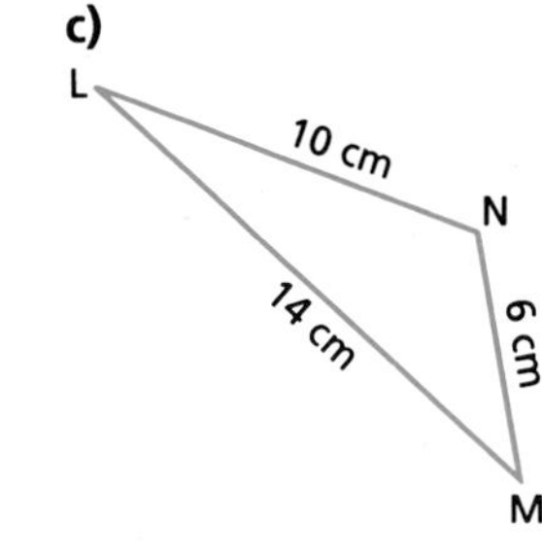

2 The circle in the diagram has radius 7.5 cm.

Calculate the area of the shaded region.

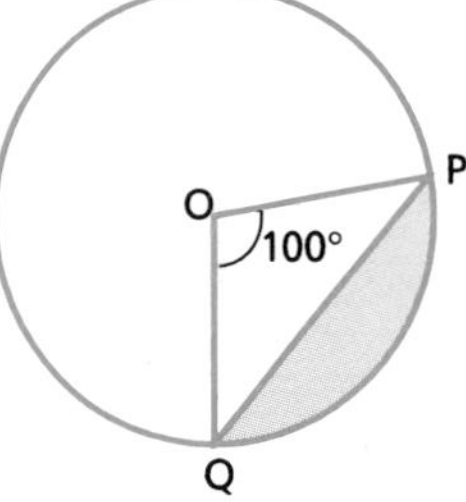

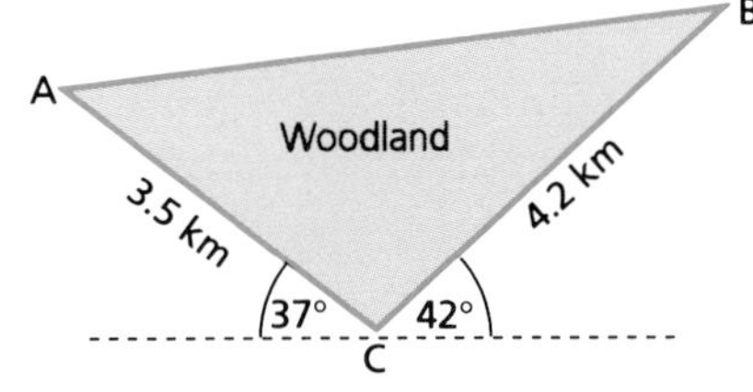

3 A surveyor, wishing to estimate the area of a roughly triangular patch of woodland, takes measurements at C to markers at A and B. Calculate the area of the woodland.

4 ABCDE is a regular pentagon with sides 5 cm.
Calculate its area.

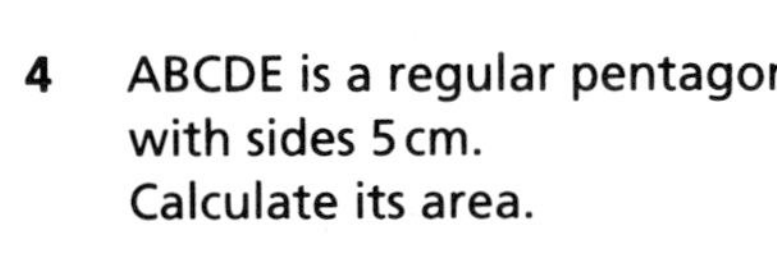

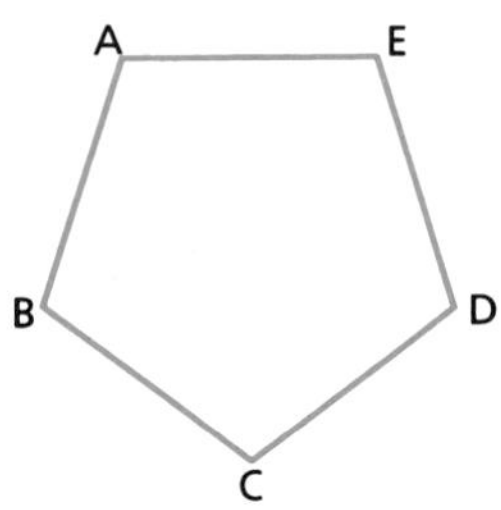

5 **a)** Triangle ABC is isosceles with perimeter 20 cm. If BC = x cm and $B = C = \theta$, show that the area of the triangle is $\frac{1}{8}(20 - x)^2 \sin 2\theta$.

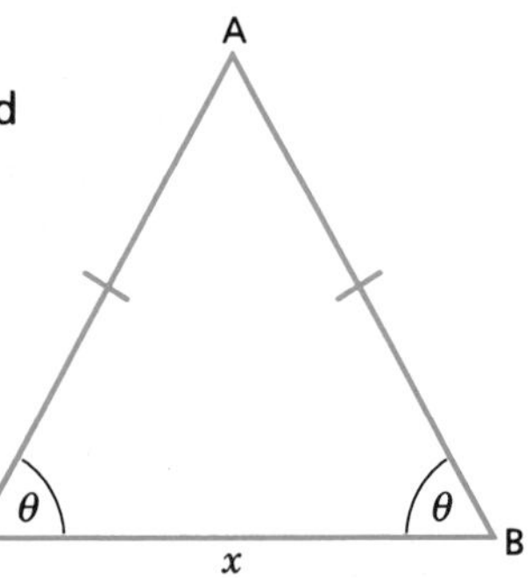

b) Show that $\cos\theta = \dfrac{x}{20 - x}$.

c) Investigate the relationship between x and θ. Which values of x are possible?

d) Investigate the area as x varies.

3D TRIGONOMETRY

Exploration 10.5

Problems with a roof

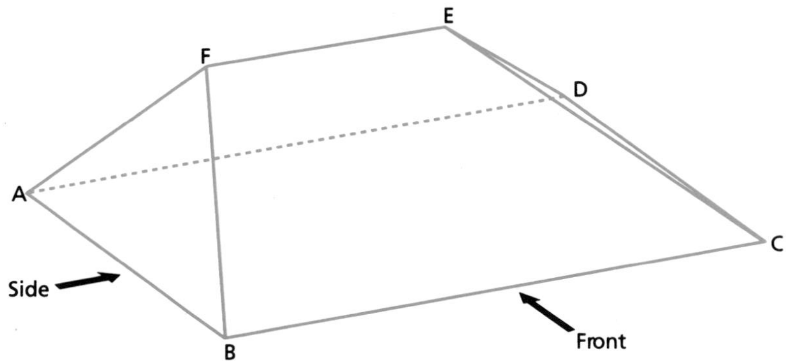

A builder is constructing the roof of a new house. The diagram shows the basic shape of the roof. The architect specifies the following dimensions, which he illustrates on a plan view and side and front elevations.

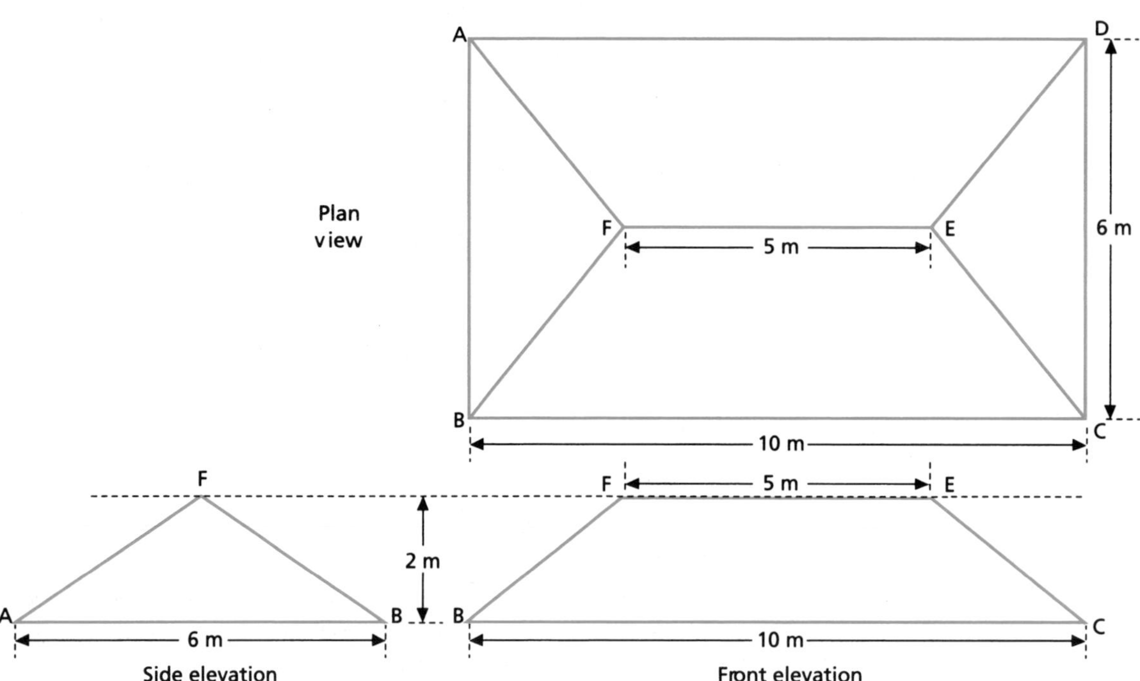

The builder needs further information about the roof structure, as follows:

- pitch of sloping faces (*planning regulations*)
- length of all ridges (*for ridge tiles*)
- surface area of roof (*for roof tiles*).

To find this information about the roof, extract suitable right-angled triangles from the 3D structure.

Pitch of triangular face ABF

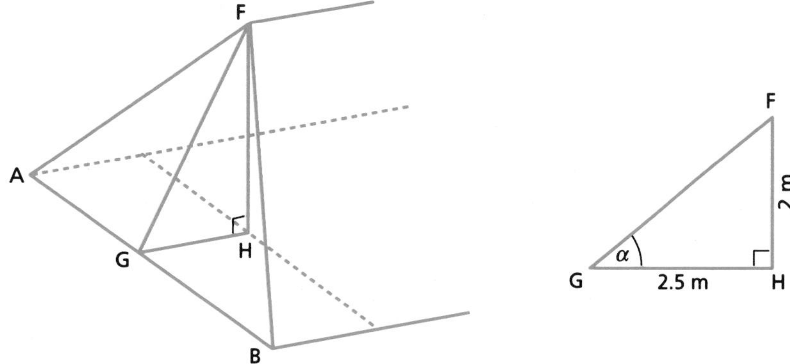

G is the midpoint of AB and H is the point on the rectangular base which is vertically below F. From the plan you can deduce that GH = 2.5 m and from the elevations FH = 2 m. The **pitch** of face ABF is the size of angle α, the angle between FG and GH, both of which are at right-angles to the common line AB (the 'hinge' between the sloping triangle and the horizontal rectangle).

From the right-angled triangle:

$$\tan \alpha = \frac{2}{2.5} = 0.8 \quad \Rightarrow \quad \alpha = 38.7°$$

Pitch of trapezoidal face BCEF

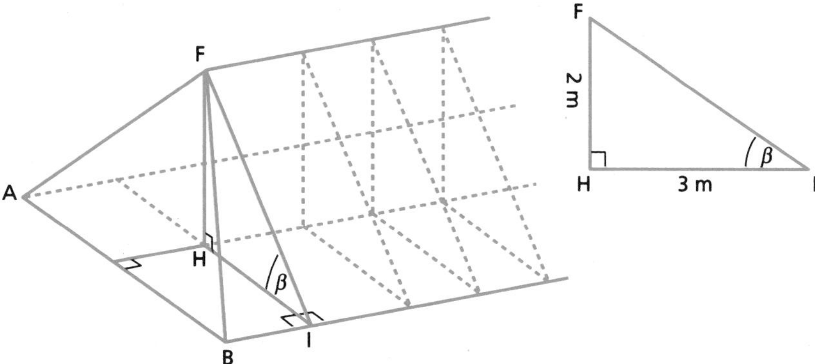

As before, H is vertically below F. To find the pitch of trapezium BCEF you need to find the angle between two lines which are themselves at right angles to the hinge BC. You might have chosen to extract any right-angled triangle parallel to triangle FHI but this one is as good as any.

FH = 2 m and from the plan HI = $\frac{1}{2}$AB = 3 m.

From the right-angled triangle:

$$\tan \beta = \frac{2}{3} = 0.\dot{6} \quad \Rightarrow \quad \beta = 33.4°$$

Since planning regulations stipulate the minimum pitch is 30° for any sloping face, there would seem to be no problem.

Length of ridges

You know that the horizontal ridge EF = 5 m. To find the length of BF, say, it too must form part of a right-angled triangle.

From the diagram, BF is the hypotenuse of triangle FGH, two sides of which are known. Therefore, two applications of Pythagoras' theorem will produce the desired result.

$$\begin{aligned} FG^2 &= FH^2 + GH^2 \\ &= 2^2 + 2.5^2 \\ &= 10.25 \end{aligned}$$

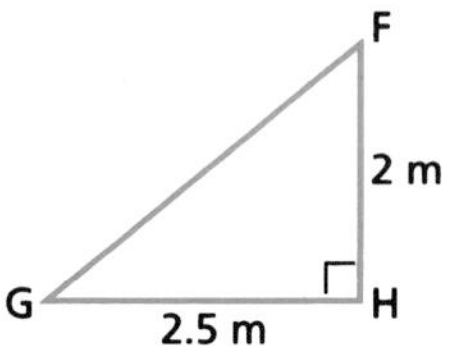

$$\begin{aligned} BF^2 &= BG^2 + FG^2 \\ &= 3^2 + 10.25 \\ &= 19.25 \end{aligned}$$

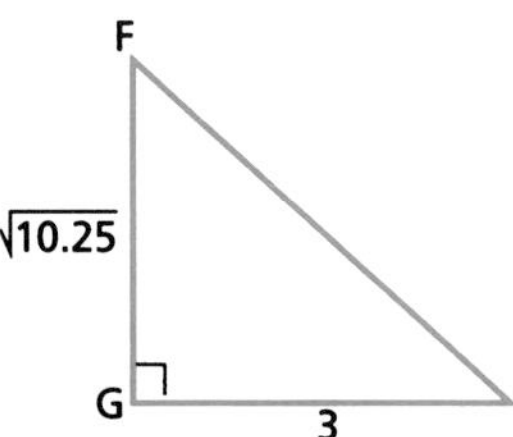

$$\begin{aligned} \Rightarrow BF &= \sqrt{19.25} \\ &= 4.39 \text{ m} \end{aligned}$$

By symmetry, all four sloping ridges are the same length.

Hence the length of all ridges = 5 + 4 × 4.39 = 22.56 m

This is useful for ordering ridge tiles for the roof.

Surface area of roof

For triangle ABF, perpendicular height = FG = $\sqrt{10.25}$.

$$\begin{aligned} \Rightarrow \text{area of triangle} &= \tfrac{1}{2} \times 6 \times \sqrt{10.25} \\ &= 9.60 \text{ m}^2 \end{aligned}$$

For trapezium BCEF, perpendicular height = FI.

From triangle FHI:

$$\begin{aligned} FI^2 &= FH^2 + HI^2 \\ &= 2^2 + 3^2 \\ &= 13 \end{aligned}$$

$$\Rightarrow FI = \sqrt{13} = 3.61 \text{ m}$$

$$\begin{aligned} \Rightarrow \text{area of trapezium} &= \tfrac{1}{2}(10 + 5) \times 3.61 \\ &= 27.1 \text{ m}^2 \end{aligned}$$

$$\begin{aligned} \text{Total surface area of roof} &= 2 \times 9.60 + 2 \times 27.1 \\ &= 73.4 \text{ m}^2 \end{aligned}$$

The builder will now know how many roof tiles are required, based on this calculation.

General procedure

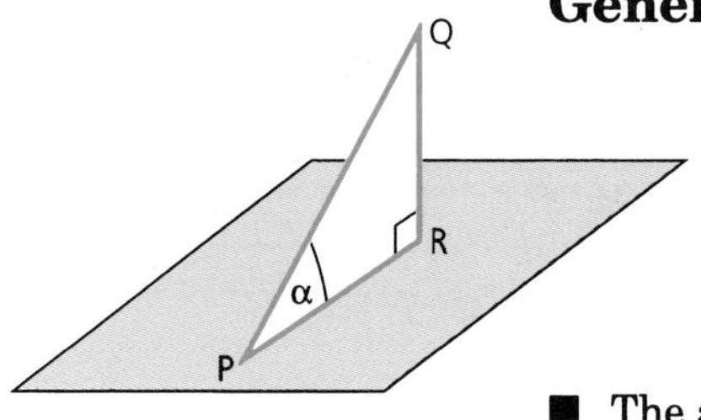

- The angle between a line and a plane is found by dropping a perpendicular from a point on the line to the plane. PQ is a line with P in the plane.
  QR is perpendicular to the plane and PR lies in the plane.
  The angle between PQ and the plane is α.
- The angle between two planes is found by finding the angle between two lines, one in each plane, which are themselves at right angles to the line where the planes meet, the 'hinge'.

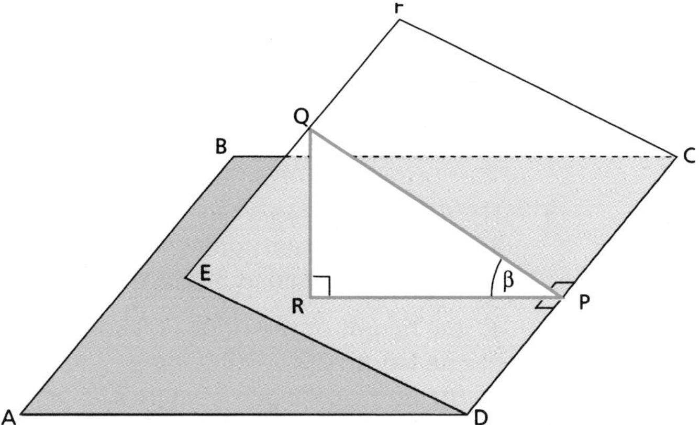

Planes ABCD and CDEF have a common line CD.
PR and PQ lie in these planes and are both at right angles to CD, the 'hinge'.
The angle between the two planes is β.

- For any 3D structure problem, extract the appropriate right-angled triangles and find suitable lengths and angles using trigonometry, Pythagoras' theorem, etc.

EXERCISES

1 The diagram shows a cuboid with AB = 10 cm, BC = 8 cm and CG = 5 cm. Find:

a) the length BD, b) the length BH,
c) the angle BH makes with the base ABCD.

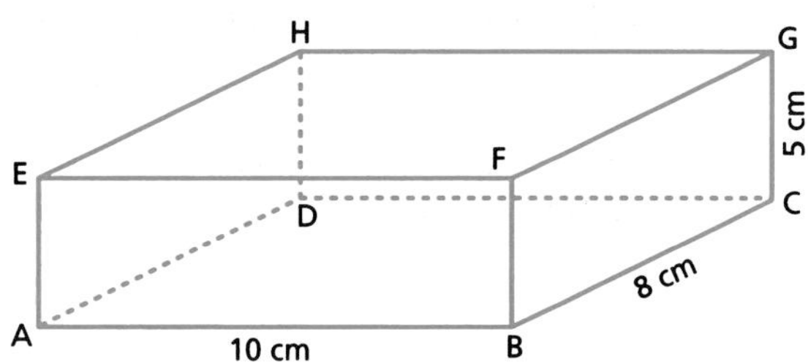

2 In the diagram ABCDE is a square-based pyramid with E vertically above the midpoint of AC. AB = 8 cm, AE = 10 cm. Find:

a) the length AC,
b) the angle AE makes with the base ABCD,
c) the perpendicular height of the pyramid,
d) the angle between triangle BCE and the base ABCD.

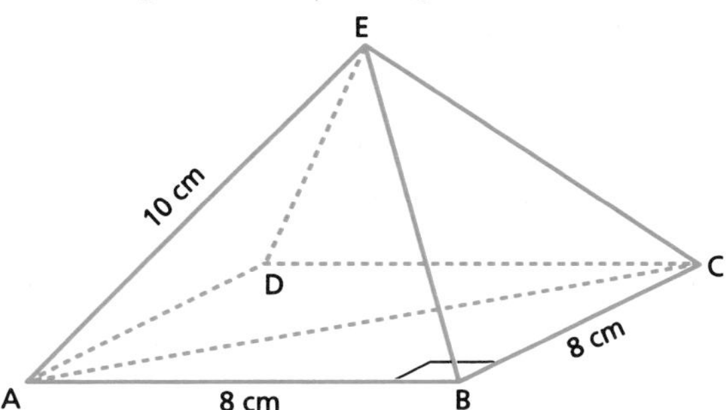

3 The diagram shows a roof with rectangular base ABCD. AB = 6 m, BC = 15 m. The four sloping ridges (AE, BE, CF and DF) are all 5 m in length and inclined at 30° to the base. M is the midpoint of AB. Find:

a) the height of the ridge (EF) above the base,
b) the length EM,
c) the angle between triangle ABE and the base,
d) the length EF,
e) the total surface area of the roof.

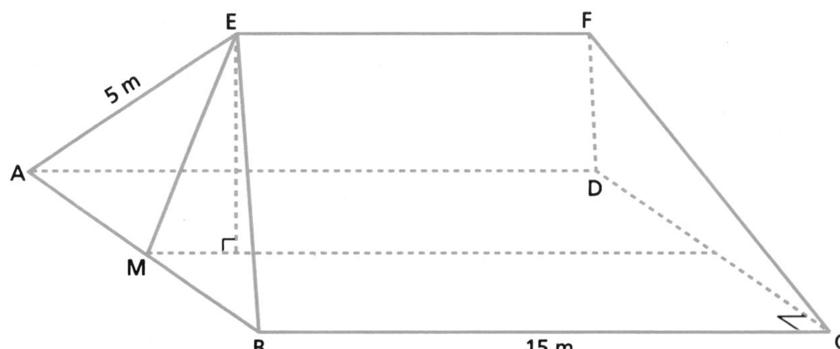

4 The diagram shows a tower PQ. The angle of elevation of the top of the tower (P) from a point R, 100 m due south of Q, is 30°. The angle of elevation of the top of the tower from a point S, north-east of Q, is 25°. Find:

a) the height of the tower,
b) the distance QS,
c) the angle RQS,
d) the distance RS,
e) the bearing of S from R.

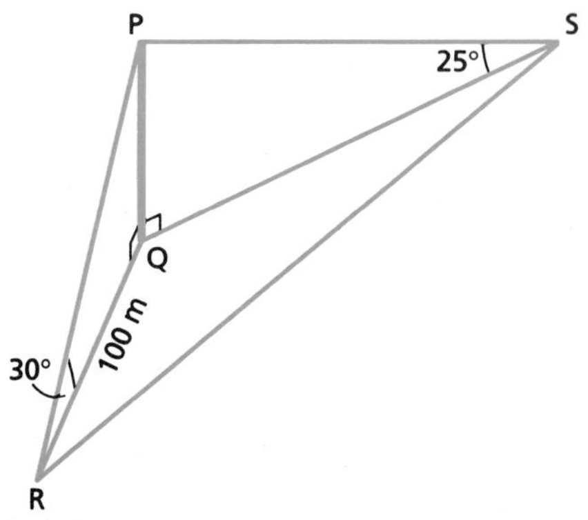

5 An aircraft is flying on a bearing of 120°, at a constant height of 8 km, with a speed of 300 km per hour. At 15.30 it is at A, 50 km due west of a radar station R. At 15.45 it is at B. Find:

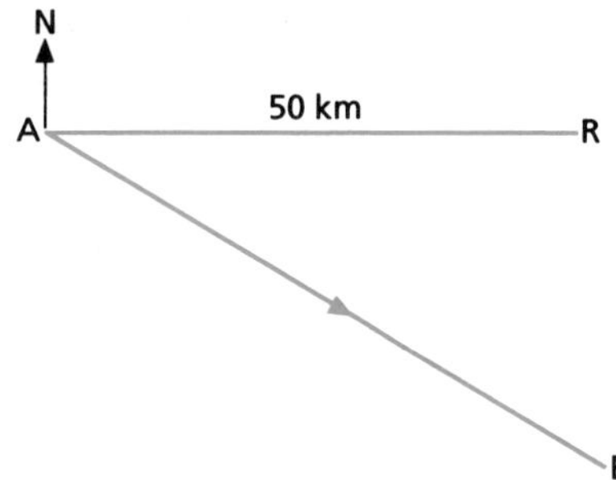

a) the angle of elevation of the aircraft from the radar station at 15.30,
b) the distance AB,
c) the distance and bearing of the aircraft from the radar station at 15.45,
d) the time at which the aircraft is closest to the radar station.

1 The diagram shows a cube with side 8 cm. M is the midpoint of AB. Calculate:

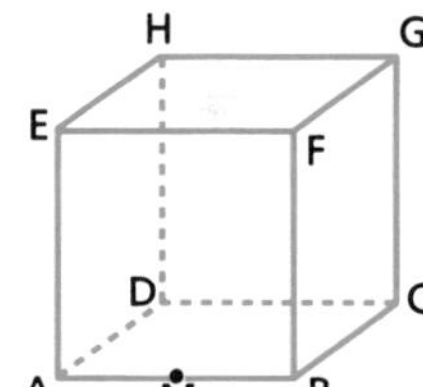

a) the length of MD,
b) the length of MH,
c) the length of EC,
d) the angle EC makes with base ABCD.

2 A kestrel is hovering at a height of 250 m. It is observed from two points, A which is due east of the kestrel and B which is due south of the kestrel.

The angle of elevation of the bird is 40° from A and 30° from B.

What is the distance from A to B?

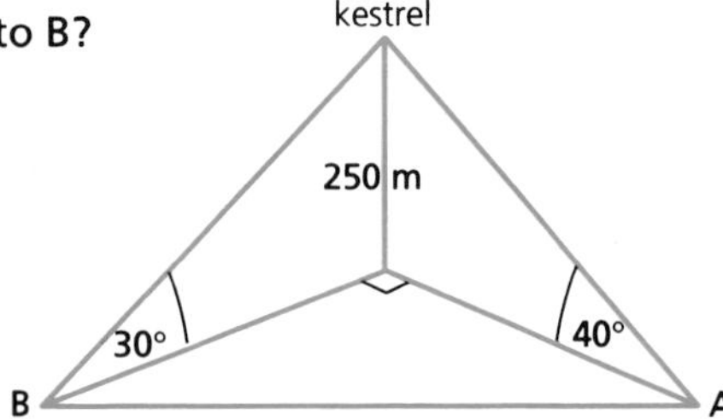

3 The Scout Group's standard issue tent is 2 m wide at the base, 4 m long.

$\angle AEB = \angle DFC = 80°$.

The triangular ends are inclined at an angle of 75° to the base.

M is the midpoint of AB. Find:
a) the length of EM,
b) the height of EF above the base,
c) the length of EF,
d) the total area of fabric, including the sewn-in groundsheet.

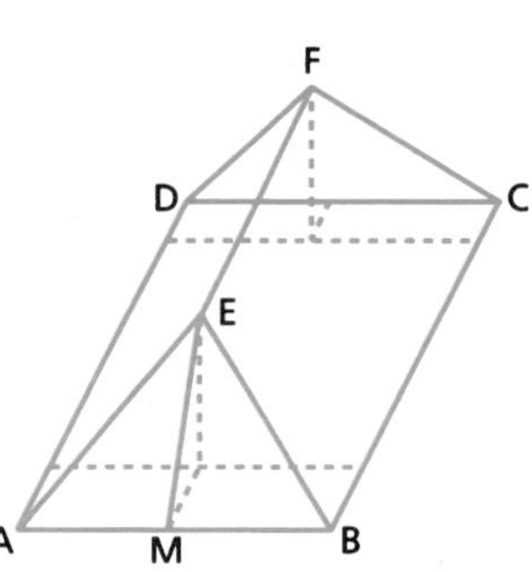

4 A boat is travelling due north. From a point A the angle of elevation of the top of a lighthouse is 5°. Ten minutes later the boat is at point B, where the angle of elevation is 7°.

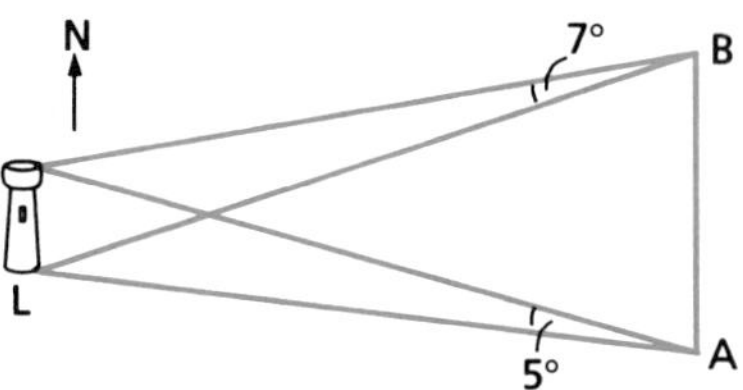

The boat is travelling at an average speed of 12 km per hour and the height of the lighthouse is 120 m. Calculate:

a) the distance AL and the distance BL,
b) the time at which the boat is nearest to the lighthouse, and the distance from the lighthouse at this time.

5 A vertical radio mast has its base at a point C, in the same plane as points A and B, due east and due south of C, respectively. A and B are 100 m apart.

The angle of elevation of the top of the mast from point A is 35°, and from point B, 25°. Find the height of the mast.

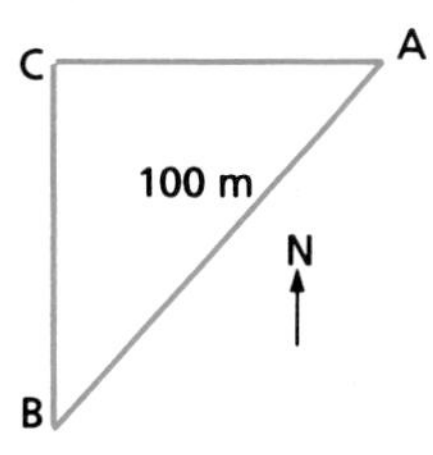

MATHEMATICAL MODELLING ACTIVITY

1 → 2 → 3 → 4 → 5 → 6 → 1
Specify the real problem

Problem statement

Every year power boat races are held in the stretch of water between the islands of Guernsey and Sark, which is known as the *Little Russell*. The sea in this area is very hazardous, with many partly submerged rocks and a large tidal range. This information is shown on the map.

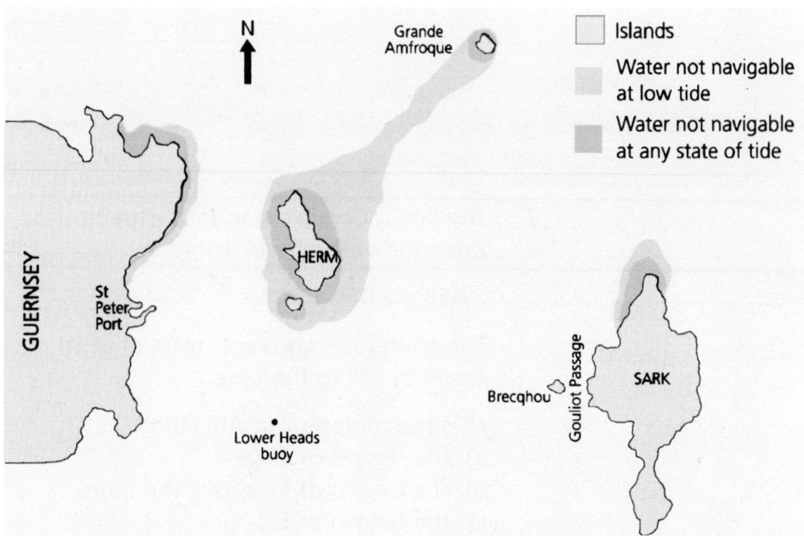

Information

Lower Heads buoy is 2.8 nautical miles from St Peter Port on a bearing of 127°.

Gouliot Passage is 7.1 nautical miles from St Peter Port on a bearing of 115°.

Grande Amfroque is 6.4 nautical miles from St Peter Port on a bearing of 053° and 2.6 nautical miles from the northern edge of the shallows around Herm on a bearing of 059°.

The boats must sail from Brecqhou on a bearing less than 017° and must approach St Peter Port on a bearing not less than 033°, and not more than 042°, to avoid shallows.

Note: 1 nautical mile is approximately 1.3 miles or 2 km.

- Use the information given to design a suitable, safe course of at least 22 nautical miles in length.

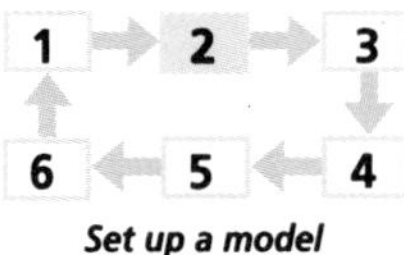

Set up a model

Set up a model

Firstly identify the important variables:

- marker points to give the boats the direction in which to travel
- areas which must be avoided at certain states of the tide.

You will also need to make certain assumptions to simplify the model:

- the course will be set so as to be navigable at all states of the tide
- the boats will travel in straight lines between marker points
- the course will start and end at St Peter Port.

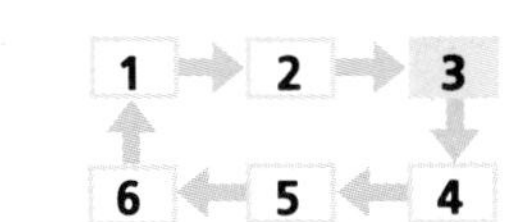

Formulate the mathematical problem

Mathematical problem

By sketching some examples of courses, as below, and using the chart information given, it is possible to consider a problem which can be solved using trigonometry for irregular triangles.

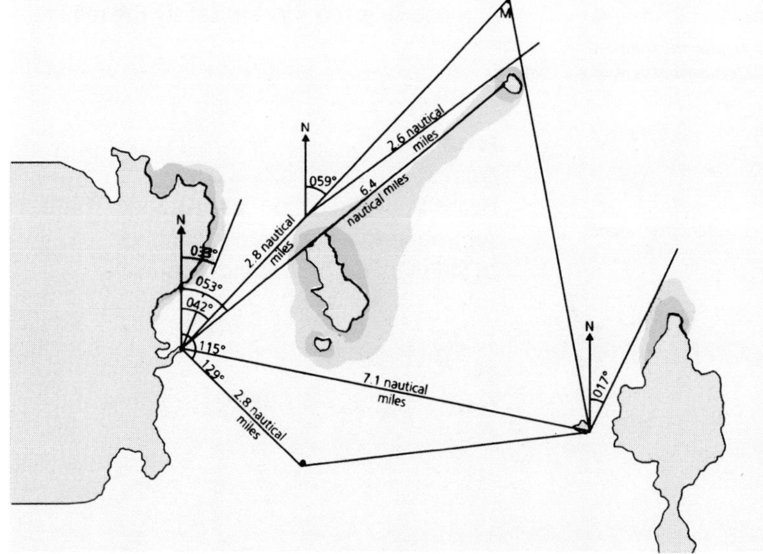

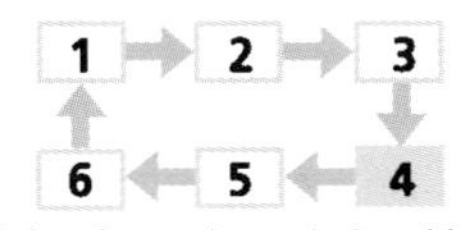

Solve the mathematical problem

Mathematical solution

Idea 1

From St Peter Port, round Brecqhou, Grande Amfroque, past the shallows at the north end of Herm and back to St Peter Port.

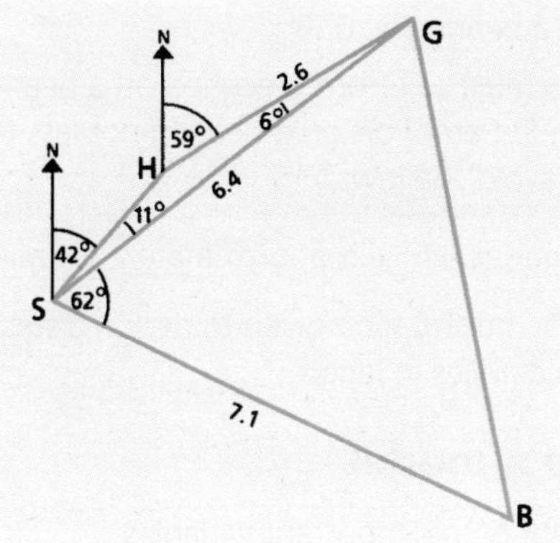

Using ΔSGB:

$$BG^2 = 7.1^2 + 6.4^2 - (2 \times 7.1 \times 6.4 \times \cos 62°)$$

$$\Rightarrow \quad BG = 7.0 \text{ nautical miles (2 s.f.)}$$

Using ΔSHG: $\angle HGS = 180° - (42° + 11° + 121°) = 6°$

$$\Rightarrow \quad SH^2 = 2.6^2 + 6.4^2 - (2 \times 2.6 \times 6.4 \times \cos 6°)$$

$$\Rightarrow \quad SH = 3.8 \text{ nautical miles (2 s.f.)}$$

This gives a total length of course (S→B→G→H→S):

$$7.1 + 7.0 + 2.6 + 3.8 = 20.5 \text{ nautical miles.}$$

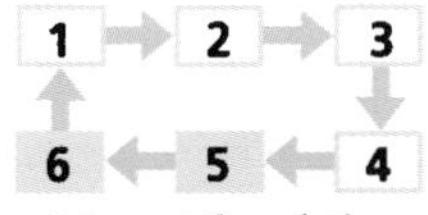

Interpret the solution and compare with reality

Interpretation

This course is too short to satisfy the requirements.

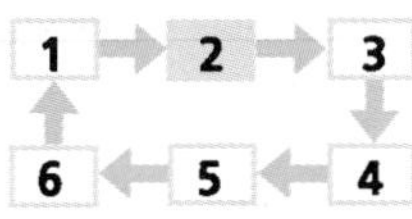

Set up (another) model

Idea 2

From St Peter Port, round Lower Heads buoy, Brecqhou, past Grande Amfroque to a marker, M, and back to St Peter Port on a straight course.

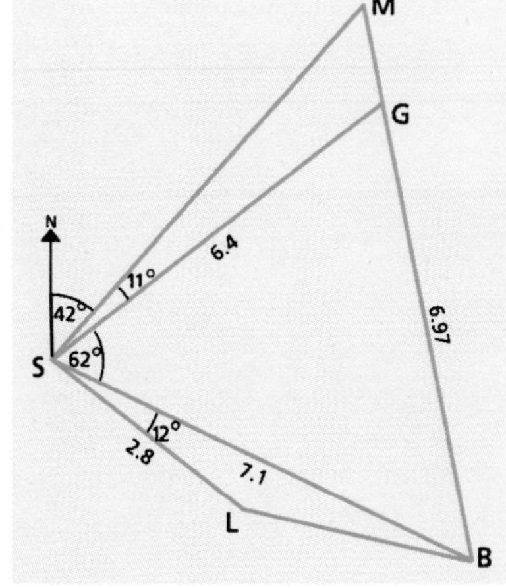

1 → 2 → 3
6 ← 5 ← 4
Formulate and solve the mathematical problem

Using ΔSBL:

$$LB^2 = 7.1^2 + 2.8^2 - (2 \times 7.1 \times 2.8 \times \cos 12°)$$

$$\Rightarrow \quad LB = 4.4 \text{ nautical miles (2 s.f.)}$$

$BG = 6.97$ nautical miles (3 s.f.) *(see Idea 1)*

Using ΔSGB: $\dfrac{6.97}{\sin 62°} = \dfrac{7.1}{\sin \angle SGB}$

$$\Rightarrow \quad \sin \angle SGB = \frac{7.1 \times \sin 62°}{6.97} = 0.899$$

$$\Rightarrow \quad \angle SGB = 64.08° \text{ (4 s.f.)}$$

Thus: $\angle SGM = 180° - 64.08° = 115.92°$

and $\angle GMS = 180° - 115.92° = 53.08°$

Using ΔSGM: $\dfrac{GM}{\sin 11°} = \dfrac{6.4}{\sin 53.08°}$

$$\Rightarrow \quad GM = 1.53 \text{ nautical miles (3 s.f.)}$$

$$SM^2 = 6.4^2 + 1.53^2 - (2 \times 6.4 \times 1.53 \times \cos 115.92°)$$

$$\Rightarrow \quad SM = 7.2 \text{ nautical miles (2 s.f.)}$$

This gives a total length of course (S→L→B→G→M→S):

$$2.8 + 4.4 + 6.97 + 1.53 + 7.2 = 22.9 \text{ nautical miles.}$$

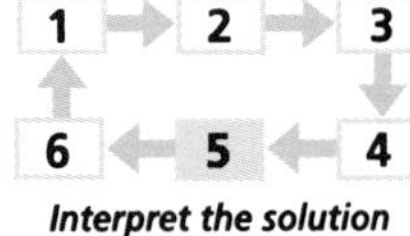

Interpret the solution

Interpretation

This course satisfies the requirements, but involves placing an extra marker buoy in the sea 7.2 nautical miles from St Peter Port on a bearing 042°.

Refinement of the model

Investigate:

- courses of exactly 25 nautical miles
- shorter, simpler courses for smaller boats
- courses which could be set at high tides (see chart for high tide areas).

CONSOLIDATION EXERCISES FOR CHAPTER 10

1 A ship, travelling at a constant speed of 18 km per hour, sails for 2 hours 30 minutes on a bearing of 143°, then for a further 4 hours on a bearing of 267°. How long will it take to return directly to port, and on what bearing must it sail?

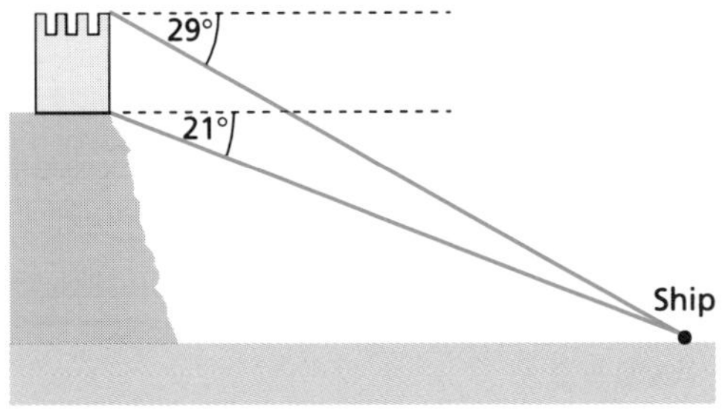

2 A tower, 45 m high, sits on the top of a cliff. The angle of depression from the top of the tower to a nearby ship is 29°, and the angle of depression of the ship from the foot of the tower is 21°.

a) How far is the ship from the base of the cliff?
b) How high is the cliff above sea level?

3 A triangulation point, T, on the top of a hill, is 500 m above sea level. Points A, B and C are in the same horizontal plane. Points A, B, C and T are in the same vertical plane. A surveyor takes the following measurements:

BC = 60 m, ∠ABT = 51°, ∠BCT = 29°.

Find the height of points A, B and C above sea level.

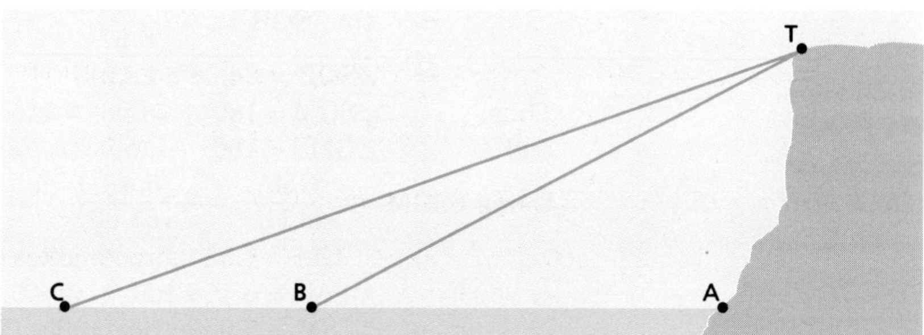

4 The diagram shows a square-based pyamid in which the sides of the square are each 10 cm and each sloping edge is 20 cm. Calculate:

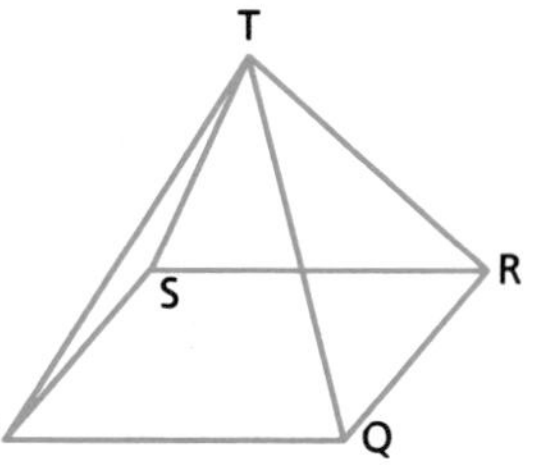

a) the vertical height of the pyramid,
b) the angle between the edge PT and the plane PQRS,
c) the angle between the plane PQT and the plane PQRS.

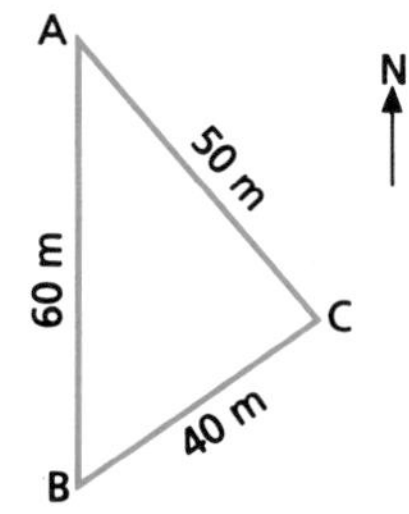

5 The points A, B and C lie in a horizontal plane, where A is 60 m due north of B. The point C, which is east of the line AB, is 50 m from A and 40 m from B.

a) Show that cos∠BAC = 0.75 and find the bearing of C from A.
b) Find the area of triangle ABC.
A vertical pole PM, of height 12 m, is placed at M, the midpoint of AB.
c) Calculate the angle of elevation of P from C.

6 A cuboid ABCDEFGH has a square horizontal base ABCD, where AB is of length $2a$. Each vertical edge is of length a. Calculate:

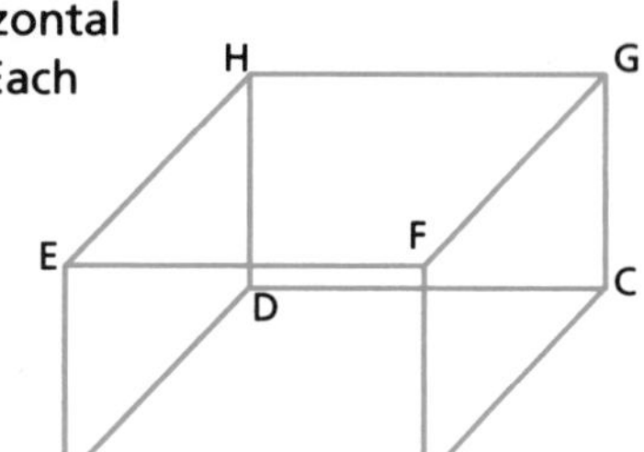

a) the length of GA,
b) the acute angle between GA and the horizontal
c) the acute angle between the plane GBD and the horizontal.

7 In triangle ABC, M is the midpoint of BC.

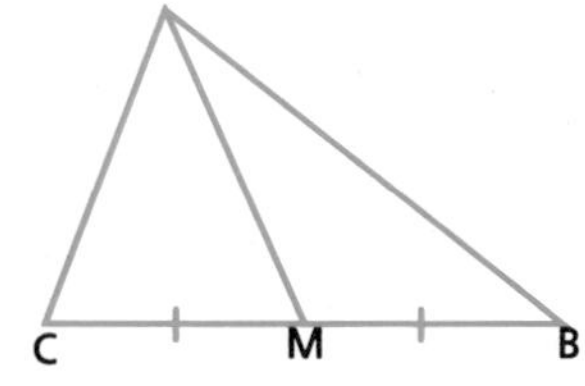

a) Show that cos∠AMC + cos∠AMB = 0.
b) By applying the cosine rule in triangles ABM and ACM, prove that:

$$4AM^2 = 2CA^2 + 2AB^2 - BC^2$$

8 In triangle ABC, BC = 8 cm, AC = 6.5 cm and ∠ABC = 30°. Calculate two possible values of ∠BAC, illustrating your answers diagramatically.

9 In ΔXYZ, XY = 10 cm, YZ = 17 cm and ZX = 21 cm.

a) Show that cos∠YXZ = 0.6.
b) Find the area of ΔXYZ.

10 Triangle ABC has sides a, b and c, as shown in the diagram.

a) Show that area^2 = $\frac{1}{4}b^2c^2\sin^2 A$.

b) Use the cosine rule to show that $\cos^2 A = \dfrac{(b^2+c^2-a^2)^2}{4b^2c^2}$.

c) Deduce that the area of a triangle may be given in terms of a, b and c as:

area = $\frac{1}{4}\sqrt{4b^2c^2-(b^2+c^2-a^2)^2}$

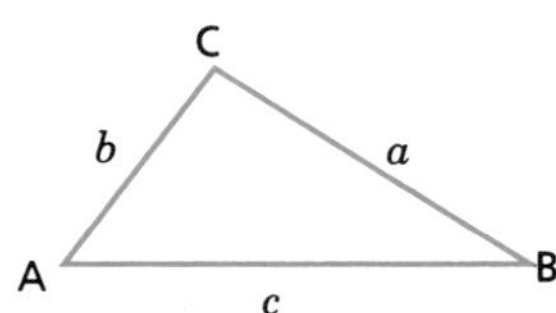

d) If the semi-perimeter, s, is given by $s = \frac{1}{2}(a+b+c)$, show that the area of a triangle may also be written as:

area = $\sqrt{s(s-a)(s-b)(s-c)}$

This is known as Hero's formula and is useful when you need to find the area of a triangle given the values of a, b and c.

Summary

■ For any triangle ABC:

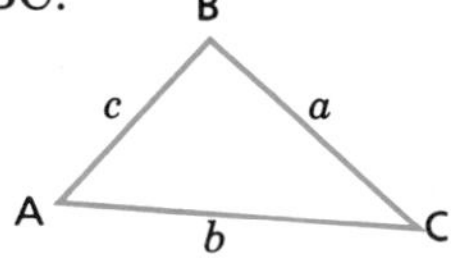

Sine rule

■ $\dfrac{a}{\sin A} = \dfrac{b}{\sin B} = \dfrac{c}{\sin C}$

Cosine rule

■ $c^2 = a^2 + b^2 - 2ab\cos C$

$b^2 = a^2 + c^2 - 2ac\cos B$

$a^2 = b^2 + c^2 - 2ac\cos A$

Area of the triangle

■ Area = $\frac{1}{2}bc\sin A = \frac{1}{2}ac\sin B = \frac{1}{2}ab\sin C$

PURE MATHS

11

Sequences and series 1

In this chapter we:

- *explore the properties of sequences and series*
- *explore particular patterns that give rise to arithmetic and geometric sequences and series*
- *consider the convergence of a geometric series.*

SEQUENCES AND SERIES

Sequences

Exploration 11.1

Looking at sequences

Look at the following sequences of numbers.

a) 5, 10, 15, 20, 25, ...
b) 2, 5, 8, 11, 14, ...
c) 13, 9, 5, 1, –3, ...
d) 1, 2, 4, 8, 16, ...
e) 81, 27, 9, 3, 1, ...
f) 1, –1.1, 1.21, –1.331, 1.461, ...
g) 1, 3, 7, 15, 31, ...
h) 1, 2, 6, 24, 120, ...
i) 1, 4, 9, 16, 25, ...
j) 1, 1, 2, 3, 5, 8, ...

- What are the next three numbers in each sequence?
- For each sequence, can you find a rule for generating the number from the one before it?
- Can you find a formula that will give any number in terms of its position in the sequence?

Notation

Before analysing the results of the exploration, it is worth bringing in some notation. Each number in a sequence is called a **term**, and the terms of any sequence are often written as

$$u_1,\ u_2,\ u_3,\ u_4,\ \ldots,\ u_k,\ \ldots$$

3rd term

kth term

In sequences **a)**, **b)** and **c)** there is a constant difference between consecutive terms, e.g. in **b)**:

$u_1 = 2, u_2 = 5, u_3 = 8, u_4 = 11, u_5 = 14, u_6 = 17, u_7 = 20$, etc.

$\Rightarrow \; u_2 = u_1 + 3$

$\Rightarrow \; u_3 = u_2 + 3$

$\Rightarrow \; u_4 = u_3 + 3$ etc.

and in general: $u_{k+1} = u_k + 3$ for $k = 1, 2, 3, \ldots$

Expressing one term in terms of the term before it, and specifying the first term, is called an **inductive definition**.

A **formula** for each term, u_k, as a function of k, its position in the sequence is easily found for such sequences, e.g. in **b)**:

$u_k = 3k - 1$

In sequences **d)**, **e)** and **f)** there is a constant ratio between consecutive terms, e.g. in **e)**:

the **inductive definition** is given by:

$u_1 = 81$ and $u_{k+1} = \frac{1}{3}u_k$ for $k = 1, 2, 3, \ldots$

and the **formula** for u_k is given by:

$u_k = 81 \times (\frac{1}{3})^{k-1}$ or $243 \times (\frac{1}{3})^k$

Sequences **g)**, **h)**, **i)** and **j)** do not conform to either of the above patterns, but may be defined inductively and have formula definitions.

Sequence	*Inductive definition*	*Formula*
g)	$u_1 = 1$ and $u_{k+1} = 2u_k + 1$	$u_k = 2^k - 1$
h)	$u_1 = 1$ and $u_{k+1} = u_k \times (k + 1)$	$u_k = k!$
i)	$u_1 = 1$ and $u_{k+1} = u_k + (2k + 1)$	$u_k = k^2$
j)	$u_1 = 1, u_2 = 1$ and $u_{k+1} = u_k + u_{k-1}$	?

Note:

The inductive definition for sequence **j)** involves specifying *two* initial terms and the definition of a term involves the *two* previous terms. This particular pattern is called a ***Fibonacci sequence***. There is a formula for the kth term of the sequence, but it is beyond the scope of this work.

Series

When terms of a sequence are added together a **series** is formed.

Exploration 11.2

Find the sum

Look at the following series of numbers, each with a different number of terms.

a) $3 + 8 + 13 + 18 + 23 + 28 + \ldots$

b) $50 + 35 + 20 + 5 - 10 - \ldots$

c) $1 + 2 + 4 + 8 + 16 + 32 + 64 + \ldots$

d) $81 + 27 + 9 + 3 + 1 + \frac{1}{3} + \ldots$

■ Find the sum of each series for the terms shown.

Notation

Before analysing the results of the exploration it is worth bringing in a little more notation.

Sums of consecutive terms are generated according to the pattern:

$S_1 = u_1$
$S_2 = u_1 + u_2$
$S_3 = u_1 + u_2 + u_3$
$S_4 = u_1 + u_2 + u_3 + u_4$
$\vdots$
$S_n = u_1 + u_2 + u_3 + \ldots + u_n$

The terms $S_1, S_2, S_3, S_4, \ldots$ are each **partial sums** of a series.

The sum $S_n = u_1 + u_2 + u_3 + \ldots + u_n$ is often written $\sum_{k=1}^{n} u_k$, which means 'the sum of u_k, for $k = 1$ to n' and Σ is the Greek letter sigma.

For **a)** $\quad u_k = 5k - 2 \Rightarrow S_6 = \sum_{k=1}^{6} u_k = \sum_{k=1}^{6} (5k-2) = 93$

For **b)** $\quad u_k = 65 - 15k \Rightarrow S_5 = \sum_{k=1}^{5} u_k = \sum_{k=1}^{5} (65-15k) = 100$

For **c)** $\quad u_k = 2^{k-1} \Rightarrow S_7 = \sum_{k=1}^{7} u_k = \sum_{k=1}^{7} 2^{k-1} = 127$

For **d)** $\quad u_k = 81\times\left(\frac{1}{3}\right)^{k-1} \Rightarrow S_6 = \sum_{k=1}^{6} u_k = \sum_{k=1}^{6} 81\times\left(\frac{1}{3}\right)^{k-1} = 121\frac{1}{3}$

Example 11.1

A sequence is defined by the formula $u_k = (-1)^{k+1} \times k^2$.

Write down u_1, u_2, u_3, u_4 *and* u_5 *and find* $\sum_{k=1}^{5} u_k$.

Solution

$u_1 = (-1)^2 \times 1^2 = 1$
$u_2 = (-1)^3 \times 2^2 = -4$
$u_3 = (-1)^4 \times 3^2 = 9$
$u_4 = (-1)^5 \times 4^2 = -16$
$u_5 = (-1)^6 \times 5^2 = 25$

$$\sum_{k=1}^{5} u_k = 1 - 4 + 9 - 16 + 25$$
$$= 15$$

Example 11.2

A sequence is defined inductively by $u_{k+1} = \frac{u_k + 3}{2}$, $u_1 = 1$.

Write down u_2, u_3, u_4, u_5 *and find a formula for* u_k.
What do you think happens as $k \to \infty$?

Solution

$$u_2 = \frac{1+3}{2} = 2 \qquad u_3 = \frac{2+3}{2} = 2.5$$

$$u_4 = \frac{2.5+3}{2} = 2.75 \qquad u_5 = \frac{2.75+3}{2} = 2.875$$

A suitable formula is $u_k = 3 - 0.5^{k-2}$. *Check this for yourself.*
As k gets larger, 0.5^{k-2} *gets smaller, which means that as* $k \to \infty$, $0.5^{k-2} \to 0$ *and so* $u_k \to 3$.

EXERCISES

11.1 A

1 For each of the following sequences, find an inductive definition and a formula.

a) 1, 5, 9, 13, … **b)** 1, –2, 4, –8, 16, …
c) 16000, 4000, 1000, 250, … **d)** 1, 8, 27, 64, …

2 Write out each of the following fully and find its value.

a) $\sum_{k=1}^{5} k^2$ **b)** $\sum_{k=0}^{4} (k+1)^2$ **c)** $\sum_{k=3}^{7} (k-2)^2$

d) $\sum_{k=1}^{5} (5k-3)$ **e)** $5\sum_{k=1}^{5} k - 15$ **f)** $\sum_{k=0}^{4} (5k+2)$

3 Use the Σ notation to abbreviate (but do not evaluate) each of the following.

a) 1 + 4 + 9 + 16 + … + 625
b) (1 × 2) + (2 × 3) + (3 × 4) + … + (15 × 16)
c) $1 + \frac{1}{2} + \frac{1}{3} + \ \ldots \ + \frac{1}{100}$

4 For each of the inductive definitions below:

i) write down the values of $u_2, u_3, u_4, u_5, \ldots$
ii) find a formula definition for u_k,
iii) describe what happens to u_k as $k \to \infty$.

a) $u_{k+1} = \frac{u_k + 10}{3}$, $u_1 = 2$ **b)** $u_{k+1} = \frac{2}{u_k + 1}$, $u_1 = 0$

11.1 B

1 For each of the following sequences, find an inductive definition and a formula.

a) 65, 56, 47, 38, … **b)** 1, 0.9, 0.81, 0.729, …
c) 1, 3, 6, 10, 15, … **d)** $1, \frac{1}{2}, \frac{1}{3}, \frac{1}{4}, \ldots$

2 Write out each of the following fully and find its value.

a) $\sum_{k=3}^{8} \frac{1}{k}$ **b)** $\sum_{k=0}^{5} 2^k$ **c)** $\sum_{k=3}^{10} (-1)^k$

d) $\sum_{k=1}^{5} k(k+2)$ **e)** $\sum_{k=1}^{5} k^2 + 2\sum_{k=1}^{5} k$ **f)** $\sum_{k=3}^{7} k(k-2)$

3 Use the Σ notation to abbreviate (but do not evaluate) each of the following.

a) 2 + 16 + 54 + 128 + … + 2000
b) (2 × 5) + (3 × 6) + (4 × 7) + … + (12 × 15)
c) $1 - \frac{1}{2} + \frac{1}{3} - \ \text{L} \ + \frac{1}{21}$

4 For each of the inductive definitions below:

a) write down the values of $u_2, u_3, u_4, u_5, \ldots,$
b) find a formula definition for u_k,
c) describe what happens to u_k as $k \to \infty$.

a) $u_{k+1} = \frac{u_k - 3}{4}, \quad u_1 = 2$

ARITHMETIC SEQUENCES AND SERIES

Exploration 11.3

Regular increases

Karen is beginning work for an engineering company. Her starting salary is £11700. Each year she can expect an increase of £850, rising to a maximum of £18500.

- How much will she earn in her fourth year with the company?
- How long will it take for her salary to rise to the maximum?
- How much will she have earned altogether by the end of the first year that she is on maximum salary?

Interpreting the results

The table shows Karen's salaries until she gets to the top of her range.

Year	Salary £
1	11 700
2	12 550
3	13 400
4	14 250
5	15 100
6	15 950
7	16 800
8	17 650
9	18 500

By inspection we can see that she will earn £14 250 in her fourth year and that it will take eight years for her salary to rise to the maximum.

By summing the second column, we find that she expects to earn a total of £135 900 over a period of nine years.

Without the aid of the table, we could deduce:

In her fourth year she will earn £11700 + 3 × £850 = £14 250

Number of years taken to reach maximum salary:

$$\frac{18\,500 - 11\,700}{850} = 8$$

The salary structure could be illustrated in the following diagram.

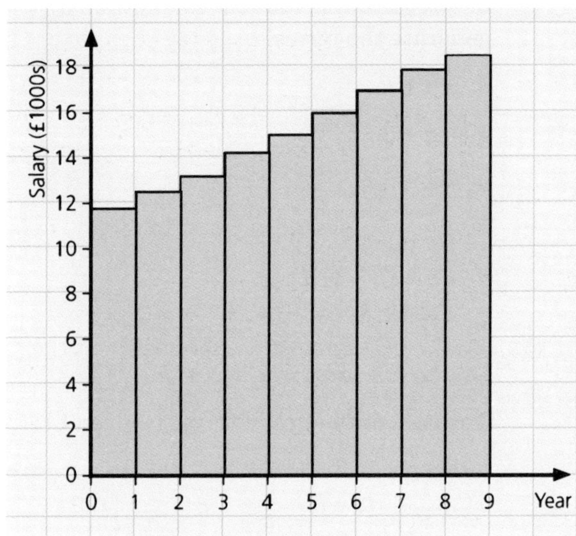

Karen's salary in any one year is represented by the area of a rectangle. Her total earnings are represented by the total shaded area.

Taking a copy of the shaded area, rotating it through a half-turn and placing it above the original diagram gives a rectangle of width 9 (years) and height 11700 + 18500 = 30200. The total earnings is equivalent to **half** the area of the rectangle.

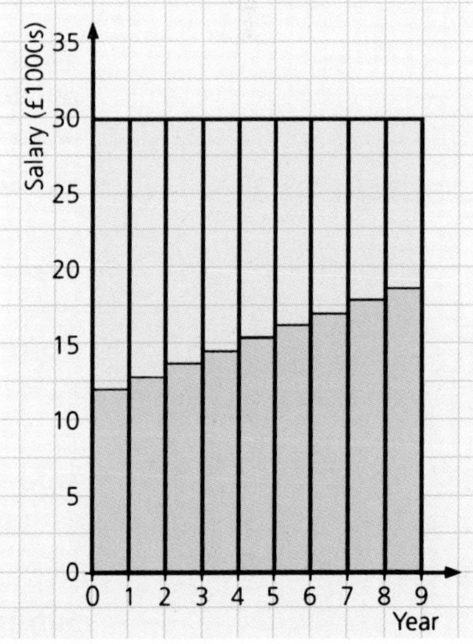

$$\frac{(11\,700 + 18\,500) \times 9}{2}$$
$$= \frac{11\,700 + 18\,500}{2} \times 9$$
$$= 135\,900$$

This gives an easy way of finding the total earnings:

average of first and last year's salaries × **number of years**

Arithmetic sequence

The pattern of salaries in Exploration 11.3: 11700, 12550, 13400, ... is an example of an **arithmetic sequence** with **first term** 11700 and **common difference** 850.

In general, an arithmetic sequence with first term a and common difference d has inductive definition $u_{k+1} = u_k + d$, $u_1 = a$ and will generate the terms:

$u_2 = a + d$

$u_3 = a + 2d$

$u_4 = a + 3d$

$\vdots$

$u_k = a + (k - 1)\,d$

$\vdots$

i.e. the kth term is $u_k = a + (k - 1)\,d$.

The sum of the first n terms is given by:

average of first and last terms × number of terms

$$\Rightarrow \quad S_n = \frac{(a+l)}{2} \times n \equiv \frac{n}{2} \times (a+l)$$

$$\equiv \frac{n}{2}\left[a + (a + (n-1)d)\right]$$

$$\equiv \frac{n}{2}\left[2a + (n-1)d\right]$$

The series generated from an arithmetic sequence is called an **arithmetic progression** (**AP**), so S_n is known as the **sum to *n* terms** of an arithmetic progression.

Example 11.3

The third term of an AP is 17 and the seventh term is 31. Find the common difference d, the first term a, expressions for u_k and S_n, and evaluate u_{10} and S_{10}.

Solution

$u_3 = 17 = a + 2d$ (1)

$u_7 = 31 = a + 6d$ (2)

Subtracting equation (1) from equation (2) gives:

$14 = 4d$

$\Rightarrow d = 3.5$

Substitute for d in (1):

$17 = a + 2 \times 3.5$

$\Rightarrow a = 10$

Hence $u_k = 10 + 3.5(k - 1)$

$$S_n = \frac{n}{2}[2 \times 10 + 3.5(n-1)]$$

In particular:

$$u_{10} = 10 + 3.5 \times 9 = 41.55$$

$$S_{10} = \frac{10}{2}(20 + 3.5 \times 9) = 257.5$$

Example 11.4

For the AP in Example 11.3, find the value of:

a) *k if $u_k = 59$* ***b)*** *n if $S_n = 178$.*

Solution

a)

$$u_k = 10 + 3.5(k-1) = 59$$

$$\Rightarrow 3.5(k-1) = 49$$

$$\Rightarrow k - 1 = 14$$

$$\Rightarrow k = 15$$

b)

$$S_n = \frac{n}{2}[2 \times 10 + 3.5(n-1)] = 178$$

$$\Rightarrow n(20 + 3.5n - 3.5) = 356$$

$$\Rightarrow n(16.5 + 3.5n) = 356$$

$$\Rightarrow 16.5n + 3.5n^2 = 356$$

$$\Rightarrow 7n^2 + 33n - 712 = 0$$

$$\Rightarrow (n-8)(7n+89) = 0$$

Since n has to be a positive integer, the solution must be $n = 8$.

In Example 11.3 the values of a and d were found by solving a pair of simultaneous equations. Another situation that gives rise to simultaneous equations is illustrated in the next example.

Example 11.5

The sum of the first n terms of a series is S_n, where $S_n = 3n^2 + n$.

a) *Show that $u_n = 6n - 2$ and find a and d.*

b) *Find $u_8 + u_9 + u_{10}$ in two different ways.*

Solution

a) *Since* $S_n = 3n^2 + n$, $S_{n-1} = 3(n-1)^2 + (n-1)$

$$\Rightarrow u_n = S_n - S_{n-1}$$

$$= 3n^2 + n - 3(n-1)^2 - (n-1)$$

$$= 3n^2 + n - 3(n^2 - 2n + 1) - n + 1$$

$$= 3n^2 + n - 3n^2 + 6n - 3 - n + 1$$

$$= 6n - 2$$

$a = u_1 = 6 \times 1 - 2 = 4$ *and*
$u_2 = 6 \times 2 - 2 = 10$
$\Rightarrow d = 10 - 4 = 6$

b) *Either*
$u_8 + u_9 + u_{10} = (6 \times 8 - 2) + (6 \times 9 - 2) + (6 \times 10 - 2) = 156$
or
$u_8 + u_9 + u_{10} = S_{10} - S_7 = 3 \times 10^2 + 10 - (3 \times 7^2 + 7) = 156$

EXERCISES

1 Fill in the missing terms in the following arithmetic sequences.

a) $7, -, 15$ **b)** $42, -, -, 18$
c) $12, -, -, -, 42$ **d)** $20, -, -, -, -, -, -22$
e) $2\frac{1}{2}, -, 4, -, -, -, 7$ **f)** $5\frac{2}{3}, -, 1\frac{2}{3}, -, -$

2 How many terms are there in the following arithmetic sequences?

a) $10, 14, 18, \ldots, 46$ **b)** $-16, \ldots, 9, 14, 19$
c) $20, 17.5, \ldots, -22.5, -25$ **d)** $p, p + q, \ldots, p + 10q$

3 For each of the following arithmetic sequences, find expressions for u_k and S_n. Find the tenth term and the sum of the first ten terms.

a) $2, 5, 8, \ldots$ **b)** $150, 143, 136, \ldots$
c) $3.6, 5.5, 7.4, \ldots$ **d)** $3, 2\frac{1}{4}, 1\frac{1}{2}, \ldots$
e) $-13.5, -10.5, -7.5, \ldots$ **f)** $0, x, 2x, \ldots$
g) $p, p + 2q, p + 4q, \ldots$ **h)** $p + q, p, p - q, \ldots$

4 The second term in an arithmetic progression (AP) is 5 and the sixth term is 11. Find an expression for u_k and the value of u_8.

5 In an AP, $u_5 = 13$ and $u_{12} = 27$. Find an expression for S_n and the value of S_8.

6 Find the sum of the following APs.

a) $8 + 10.25 + \ldots + 35$ **b)** $50 + 39 + \ldots - 27$
c) $-14 + \ldots + 31$ (*11 terms*) **d)** $10 + \ldots + 19$ (*13 terms*)

7 In an arithmetic sequence, the eighth term is twice the third term and the terms differ by 12. Show that $u_k = 4.8 + 2.4k$ and find S_8.

8 In an AP, $u_3 = 20$ and $S_5 = 100$. Find an expression for u_k and the value of u_{10}.

9 For the AP $20 + 17 + 14 + \ldots$, $S_n = 65$. Find the value of n.

10 How many consecutive odd numbers (beginning with 1) do I need to add together so that their total exceeds 1000?

11 A skier is skiing down a slope such that distances travelled in successive seconds form an arithmetic sequence. In the first second she travels 4 metres and after 4 seconds has travelled a distance of 25 metres.

a) How far will the skier travel in 8 seconds?
b) How long will it take her to cover a distance of 147 metres?

12 The sum of the first n terms of an arithmetic series is S_n, where $S_n = n(5n - 2)$.

a) Find an expression for u_n.
b) Find the values of a and d.

11.2 B

1 Fill in the missing terms in the following arithmetic sequences.

a) $17, -, 23$
b) $21, -, -, 36$
c) $37, -, -, -, 29$
d) $20, -, -, -, -, -, 5$
e) $\frac{2}{3}, -, \frac{4}{3}, -, 2$
f) $2\frac{2}{3}, -, 1\frac{2}{3}, -, \frac{2}{3}$

2 How many terms are there in the following arithmetic sequences?

a) $11, 14, 17, \ldots, 41$
b) $26, \ldots, 62, 66, 70$
c) $20, 16.5, 13, \ldots, -22$
d) $p + 5q, p + 3q, \ldots, p - 21q$

3 For each of the following arithmetic sequences, find expressions for u_k and S_n. Find the eighth term and the sum of the first eight terms.

a) $2, 4, 6, \ldots$
b) $3, 5, 7, \ldots$
c) $17, 13, 9, \ldots$
d) $\frac{1}{5}, \frac{3}{5}, 1, \ldots$
e) $5, 0, -5, \ldots$
f) $30, 30 - x, 30 - 2x, \ldots$
g) $p, p - 4q, p - 8q, \ldots$
h) $p + q, 2p + 3q, 3p + 5q, \ldots$

4 The third term of an arithmetic progression (AP) is 7 and the seventh term is 15. Find an expression for u_k and the value of u_{11}.

5 In an AP, $u_6 = 21$ and $u_{13} = 56$. Find an expression for S_n and the value of S_{10}.

6 Find the sum of the following APs.

a) $13 + 17 + 21 + \ldots$ (*16 terms*)
b) $37 + 30 + 23 + \ldots$ (*12 terms*)
c) all the odd numbers between 20 and 80
d) all the even numbers between 20 and 80

7 An arithmetic sequence has 13 terms with sum 143. The fifth term is 8. Find the first term.

8 The first term of an AP is 3 and the sum of the first six terms is 4 times the sum of the first three terms. Find the common difference.

9 The positive integers 1, 2, 3, 4, ... form an arithmetic sequence.

a) Using the general formula for S_n, write down and simplify an expression for the sum of the first n positive integers. Hence find, without using a calculator, the sum of the integers:
 i) from 1 to 100
 ii) from 1 to 1000
 iii) from 101 to 200
 iv) from 501 to 1000.

b) Write down and simplify an expression for the sum of the first $2n$ positive integers. Hence find $S_{2n} - S_n$ and use it to check your answers to a) iii) and a) iv).

10 The odd numbers 1, 3, 5, 7, 9, ... form an arithmetic sequence.

a) Without using the formula for S_n, but by adding terms together, write down:
 i) the first odd number,
 ii) the sum of the first two odd numbers,
 iii) the sum of the first three odd numbers, etc.
Continue until you spot a definite pattern in your answers.

b) From your answers to **a)**, write down a formula which gives the sum of the first n odd numbers.
Use the general formula for S_n to verify that your expression for S_n is correct.

11 Use your answers to questions **9** and **10** to deduce a formula for the sum of the first n even numbers in *two* different ways. Verify that your two expressions are equivalent.

12 A major new motorway project is due to last for seven weeks. One hundred men start work on the first Monday. They are paid a flat rate of £360 for a six-day working week with Sundays off. As work progresses, every Monday for the next six weeks an additional sixty men are taken on at the same rate of pay.

How many men are employed by the end of the project and what is the total wage bill?

CALCULATOR ACTIVITY

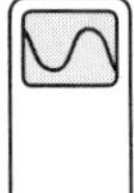

Try the following program (on a calculator or computer) which generates both the terms and the sums of an arithmetic progression. Use it to check your results from Exercises 11.2 and to investigate arithmetic progressions for various values of a and d.

Check the working of Exploration 11.2 by entering 11 700 into store A and 850 into store D. Stop the program when N (representing the year) has reached 9 and U (representing the salary) has reached 18 500.

You should be able to produce a table of values as shown.

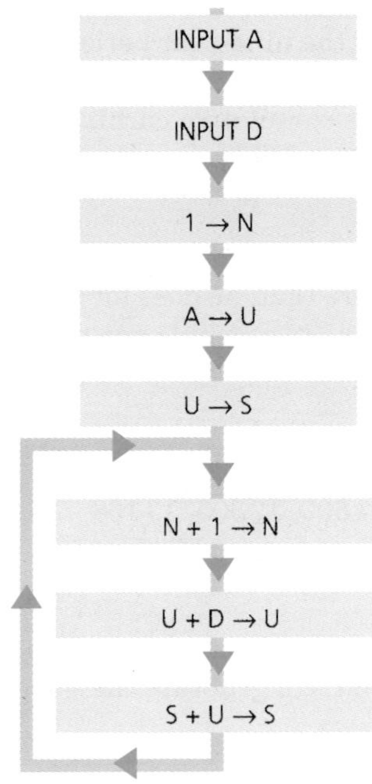

N	U	S
1	11 700	11 700
2	12 550	24 250
3	13 440	37 650
4	14 250	51 900
5	15 100	67 000
6	15 950	82 950
7	16 800	99 750
8	17 650	117 400
9	18 500	135 900

GEOMETRIC SEQUENCES AND SERIES

Exploration 11.4

Percentage increases

Stuart is starting work for a chemical firm. Like Karen (see Exploration 11.3), he is offered a salary scheme with eight annual increments. However, his initial salary is £11 500 and in each subsequent year he will be paid seven per cent more than he was paid the previous year.

- How much will he earn in his second and third years?
- How much will he earn at the top of the scale?
- How much can he be expected to earn altogether over the nine-year period?
- Compare Stuart's salary scheme with Karen's. Who is better off? Why?

Interpreting the results

In his second year with the company Stuart will earn £11 500 × 1.07 = £12 305

In his third year with the company Stuart will earn £12 305 × 1.07 = £11 500 × 1.07^2 = £13 166.35

Stuart will reach the top of the scale in the ninth year, during which he will earn £11 500 × 1.07^8 = £19 759.14

The table shows Stuart's salary structure over a period of nine years. By summing the second column, we can see that he expects to earn a total of £137 745.

Year	Salary (nearest £)
1	11 500
2	12 305
3	13 166
4	14 088
5	15 074
6	16 129
7	17 258
8	18 466
9	19 759

Comparing salaries for Karen and Stuart over the nine-year period, it is clear that for the first five years Karen earns more than Stuart, but from then on Stuart's salary is the greater. His cumulative earnings are £1845 more than Karen's, but his running total only exceeds Karen's for the first time after year 8.

You might argue that Stuart is better off because he will earn more than Karen altogether. However she earns more than he does for the first five years. You might think of other reasons to conclude one way or the other.

Geometric sequence

The pattern of salaries in Exploration 11.4: 11500, 12305, 13166, ... is an example of a **geometric sequence** with **first term** 11500 and **common ratio** 1.07.

In general, a geometric sequence with first term a and common ratio r has inductive definition $u_{k+1} = u_k \times r$, $u_1 = a$ and will generate the terms:

$$\begin{aligned} u_2 &= ar \\ u_3 &= ar^2 \\ u_4 &= ar^3 \\ &\vdots \\ u_k &= ar^{k-1} \\ &\vdots \end{aligned}$$

i.e. the kth term is $u_k = ar^{k-1}$.

The sum of the first n terms, S_n is

$$S_n = u_1 + u_2 + u_3 + \ldots + u_n$$

$$\Rightarrow \quad S_n = a + ar + ar^2 + \ldots + ar^{n-1} \qquad (1)$$

To find a simpler form for S_n we need to carry out some algebraic manipulation.

Multiply both sides of (1) by r:

$$rS_n = ar + ar^2 + \ldots + ar^{n-1} + ar^n \qquad (2)$$

Subtracting equation (2) from equation (1) gives:

$$S_n = a + ar + ar^2 + \ldots + ar^{n-1} \qquad (1)$$

$$rS_n = \quad ar + ar^2 + \ldots + ar^{n-1} + ar^n \qquad (2)$$

$$(1)-(2): \quad S_n - rS_n = a - ar^n$$

$$\Rightarrow \quad S_n(1-r) = a(1-r^n)$$

$$\Rightarrow \quad S_n = \frac{a(1-r^n)}{1-r}$$

Equivalently, subtracting equation (1) from equation (2) gives:

$$rS_n = \quad ar + ar^2 + \ \dots + ar^{n-1} + ar^n \qquad (2)$$

$$S_n = a + ar + ar^2 + \ \dots + ar^{n-1} \qquad (1)$$

$$(2) - (1): \quad rS_n - S_n = ar^n - a$$

$$\Rightarrow \quad S_n(r-1) = a(r^n - 1)$$

$$\Rightarrow \quad S_n = \frac{a(r^n - 1)}{r - 1}$$

Either version of S_n is correct. We usually use whichever one gives a positive denominator. If $r < 1$ use the first one; if $r > 1$ use the second. If $r = 1$ use neither!

Exploration 11.5

Investigating the sequence

In Exploration 11.4, in the analysis of Stuart's salary, the formal approach gives:

$a = 11\,500$, $r = 1.07$, $n = 9$:

$$u_9 = 11500 \times 1.07^8 = 19\,759.14107 \approx 19759$$

$$S_9 = \frac{11500(1.07^9 - 1)}{1.07 - 1} = 137746.8707 \approx 137747$$

How do you explain the difference between S_9 and the sum of the column of salary figures for Stuart?

Note:
The series generated from a geometric sequence is called a **geometric progression (GP)**, so S_n is known as the **sum to *n* terms** of a geometric progression.

Example 11.6

For each of the following GPs, find expressions for u_k and S_n, evaluate u_7 and S_7 and find the least value of n such that $S_n > 1000$.

a) $a = 240$, $r = 0.8$
b) $5 + 7.5 + 11.25 + \dots$
c) $800 + 160 + 32 + \dots$

Solution

a) $u_k = ar^{k-1} = 240 \times 0.8^{k-1}$

$$S_n = \frac{a(1-r^n)}{1-r} = \frac{240(1-0.8^n)}{0.2} = 1200(1-0.8^n)$$

In particular: $u_7 = 240 \times 0.8^6 = 62.9$ (3 s.f.)

$$S_7 = 1200 \times (1 - 0.8^7) = 948 \text{ (3 s.f.)}$$

If $S_n > 1000$, then $1200(1 - 0.8^n) > 1000$

$$\Rightarrow \quad 1 - 0.8^n > \tfrac{5}{6}$$

$$\Rightarrow \quad \tfrac{1}{6} > 0.8^n$$

The required value of n may be found by trial and improvement using a calculator.

$0.8^7 \approx 0.210 > \frac{1}{6}$ $\quad 0.8^8 \approx 0.168 > \frac{1}{6}$ $\quad 0.8^9 \approx 0.134 < \frac{1}{6}$

So the least value of n, such that $S_n > 1000$, is $n = 9$.

b) $a = 5, \quad r = \dfrac{7.5}{5} = 1.5$

$$u_k = 5 \times 1.5^{k-1}$$

$$S_n = \frac{5(1.5^n - 1)}{1.5 - 1} = 10(1.5^n - 1)$$

In particular: $u_7 = 5 \times 1.5^6 = 57.0 \text{ (3 s.f.)}$

$S_7 = 10(1.5^7 - 1) = 161 \text{ (3 s.f.)}$

If $S_n > 1000$, then $10(1.5^n - 1) > 1000$

$\Rightarrow \quad 1.5^n - 1 > 100$

$\Rightarrow \quad 1.5^n > 101$

By trial and improvement:

$1.5^{10} = 57.7$ (3 s.f)

$1.5^{11} = 86.5$ (3 s.f)

$1.5^{12} = 130$ (3 s.f)

So the least value of n, such that $S_n > 1000$, is $n = 12$.

c) $a = 800 \quad r = \dfrac{160}{800} = 0.2$

$\therefore \quad u_k = 800 \times 0.2^{k-1}$

$$S_n = \frac{800(1 - 0.2^n)}{1 - 0.2} = 1000(1 - 0.2^n)$$

In particular: $u_7 = 800 \times 0.2^6 = 0.0512$

$S_7 = 1000(1 - 0.2^7) = 999.9872$

If $S_n > 1000$, then $1000(1 - 0.2^n) > 1000$

$\Rightarrow \quad 1 - 0.2^n > 1$

$\Rightarrow \quad 0.2^n < 0$

Since this is not possible (0.2^n is always positive), there is no value of n such that $S_n > 1000$.

Example 11.7

The second term of a GP is 20 and the fifth term is 2500. Find the common ratio r, the first term a, expressions for u_k and S_n, and evaluate u_{10} and S_{10}.

Solution

$u_2 = 20 = ar^1 \qquad (1)$

$u_5 = 2500 = ar^4 \qquad (2)$

$(2) \div (1) \Rightarrow 125 = r^3 \Rightarrow r = \sqrt[3]{125} = 5$

Substitute for r in (1): $20 = a \times 5 \Rightarrow a = 4$

Hence: $u_k = 4\times5^{k-1}$ and $S_n = \dfrac{4(5^n-1)}{5-1} = 5^n - 1$

In particular: $u_{10} = 45\times5^9 = 7\ 812\ 500$
$S_{10} = 5^{10} - 1 = 9\ 765\ 624$

Example 11.8

Jenny puts £100 into a savings account on 1st January each year. Assuming a constant annual interest rate of eight per cent, how much will her savings be worth after ten years?

Solution

Value of savings after 1 year: $100\times1.08 = S_1$
Value of savings after 2 years: $100\times1.08^2 + 100\times1.08 = S_2$
Value of savings after 3 years: $100\times1.08^3 + 100\times1.08^2 + 100\times1.08 = S_3$
etc.

Rearranging S_3*:*

$$S_3 = 100\times1.08(1+1.08+1.08^2) = 100\times1.08\left(\frac{1.08^3-1}{1.08-1}\right)$$

From which we may deduce, for example, the value of S_{10}.

$$S_{10} = 100\times1.08\left(\frac{1.08^{10}-1}{1.08-1}\right) = 1564.55\ \text{(2 d.p.)}$$

i.e. value of savings after ten years is £1564.55.

EXERCISES

11.3 A

1 Fill in the missing terms in the following geometric sequences.

a) 1, – , 25, 125, – **b)** 27, – , – , (–1)
c) 20, – , 5, – , 1.25 **d)** 1000, – , – , 3375, –
e) 1350, – , – , 400, – **f)** $2\frac{2}{3}$, – , $\frac{3}{8}$, – , –
g) $1, x^2$, – , – , x^8, – **h)** $p^{13}q^7$, – , – , – , – , p^3q^2, pq

2 How many terms are there in the following geometric sequences?

a) 1, 3, 9, …, 19 683 **b)** 1, 1.1, 1.21, …, 1.610 51
c) 32, …, $\frac{1}{8}, -\frac{1}{16}$ **d)** $x^{12}, x^9, \dots, x^{-15}$
e) $\dfrac{p^5}{q^2}, \dots, \dfrac{q^5}{p^2}$ **f)** $\dfrac{32}{9}, \dots, \dfrac{243}{4}$

3 For each of the following geometric sequences, find expressions for u_k and S_n. Find the eighth term and the sum of the first eight terms.

a) 1, 2, 4, … **b)** 500, 250, 125, …
c) 16, –8, 4, –2, 1, … **d)** 1000, 1100, 1210, …
e) $\frac{1}{81}, -\frac{1}{27}, \frac{1}{9}, -\frac{1}{3}, 1, \dots$ **f)** $1, \dfrac{1}{x}, \dfrac{1}{x^2}, \dfrac{1}{x^3}, \dots$

4 In a GP, $u_2 = 3600$ and $u_3 = 4800$. Find an expression for u_k and the value of u_5.

5 The third term of a GP is 90 and the sixth term is 2430. Find an expression for u_k and the value of u_9.

6 In a GP, $u_3 = 3600$ and $u_5 = 1296$. Find two possible values for r and hence two possible expressions for S_n and values for S_5.

7 The second term of a GP is 48 and the sixth term is 243.

a) Find two possible values for r and a.
b) Write down the first eight terms for each pair of values of a and r.
c) Find S_n and the value of S_8 for each GP.
d) How do the two sequences differ?

8 For a GP, $u_k = (-2)^k$.

a) Find values for a and r.
b) Find an expression for S_n and evaluate S_{10}.

9 For the GP 1 + 3 + 9 + ... find the least value of n such that $S_n > 1000$.

10 For the GP 500 + 250 + 125 + ... find the least value of n such that $S_n > 999$.

11 An athlete plans a training schedule which involves running 20 km in the first week of training. In each subsequent week the distance is to be increased by ten per cent over the previous week.

a) Write down an expression for the distance to be covered in the nth week, according to this schedule.
b) Find in which week the athlete would first cover more than 100 km.
c) How far will he have run in his schedule altogether by the end of the week in part b)?

11.3 B

1 For each of the following geometric sequences, find expressions for u_k and S_n. Find the tenth term and the sum of the first ten terms.

a) $2, 6, 18, \ldots$
b) $27, 9, 3, \ldots$
c) $12, -6, 3, -1\frac{1}{2}, \ldots$
d) $1000, 900, 810, \ldots$
e) $1000, 1200, 1440, \ldots$
f) $\frac{\pi}{2}, \frac{\pi^2}{4}, \frac{\pi^3}{8}, \frac{\pi^4}{16}, \ldots$

2 In a GP, $u_3 = 84$ and $u_7 = 1344$. Find an expression for u_k and the value of u_9.

3 The second term of a GP is 6 and the sixth term is $\frac{2}{27}$. Find an expression for u_k and the value of u_{10}.

4 The third term of a GP is $\frac{8}{9}$ and the fifth term is $\frac{32}{81}$.

a) Find two possible values for r and the value of a.
b) Write down the first six terms for each pair of values.
c) Find S_n and the value of S_{10} for each GP.

5 **a)** Take an ordinary sheet of A4 paper and fold it in half, then fold it again and so on. You should find that the first five folds are easy, the sixth is difficult and the seventh is impossible. Assuming that the paper is 0.1 mm thick, write down the thickness of the paper after:
i) 1 fold **ii)** 2 folds **iii)** 3 folds.
b) Deduce an expression for the thickness after n folds.
c) If it were possible, how thick would the paper be after:
i) 7 folds **ii)** 10 folds **iii)** 15 folds?
d) How many folds would, in theory, be necessary to produce a thickness of at least 100 metres?

6 Modern paper sizes are designed on a mathematical principle. The largest is A0, then A1 has half the area of A0, A2 has half the area of A1 and so on.

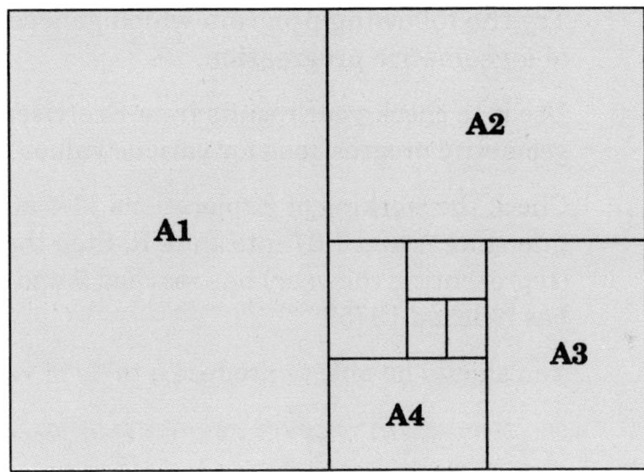

a) Measure the sides of a sheet of A4 paper to the nearest mm and calculate its area in mm^2.
b) Deduce the dimensions of A5, A3, A2, A1 and A0.
c) Deduce the area of each sheet of paper in mm^2.
d) The decreasing areas form a GP. What is its common ratio?
e) The linear dimensions also decrease in geometric progression. Write down the exact value of the common ratio.
f) In an ideal world, what do you think the area of A0 should be? Based on this estimate, estimate the ideal dimensions of a sheet of A4 in mm, correct to 3 d.p.
g) Deduce the theoretical areas and dimensions of A10 and A(–2), if these paper sizes existed.

7 A young mother takes her baby to be weighed once a week at the local clinic. For the first few months the baby gains weight at an average of four per cent per week. After how many weeks will the baby have doubled his birth weight?

8 A rubber ball is dropped onto a hard surface from a height of 2 m and bounces several times. At each bounce it rises to 60 per cent of its previous height. How many bounces will it take before it rises:

a) less than 50 cm, **b)** less than 10 cm?

9 You are offered a job on a short-term contract of 20 weeks and can choose one of the following two salary packages.

a) £500 per week
b) one penny for the first week with salary doubling each week
Which package would you choose, and why?

10 If I place one grain of rice on the first square of a chessboard, two grains on the second square, four grains on the third and so on (doubling the number of grains for each successive square), how many grains will I need altogether?

CALCULATOR ACTIVITY

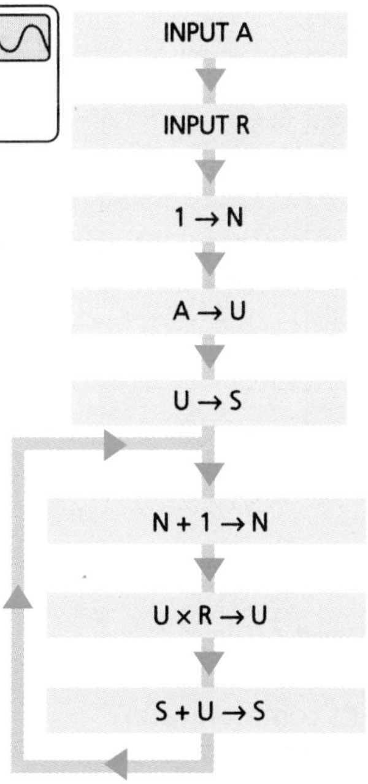

INPUT A
INPUT R
1 → N
A → U
U → S
N + 1 → N
U × R → U
S + U → S

Try the following program which generates the terms and the sums of a geometric progression.

Use it to check your results from Exercises 11.3 and to investigate geometric progressions for various values of a and r.

Check the working of Explorations 11.4 and 11.5 by entering 11 500 into store A and 1.07 into store R. Stop the program when N (representing the year) has reached 9 and U (representing the salary) has reached 19759.

You should be able to produce a table of values like this.

N	U	S
1	11 500	11 500
2	12 305	23 805
3	13 166	36 971
4	14 088	51 059
5	15 074	66 133
6	16 129	82 263
7	17 258	99 521
8	18 466	117 988
9	19 759	137 747

SUM TO INFINITY OF A GEOMETRIC SERIES

Exploration 11.6

Bouncing a ball

The manufacturers of *Superbounce*, a particularly bouncy ball, claim that when the ball is dropped onto a hard surface, it will bounce back to 80 per cent of its original height.

Assuming that when left to bounce repeatedly, the ball always reaches 80 per cent of its previous height and that it is dropped initially from a height of 125 cm, find the pattern of successive heights to which the ball bounces. What happens 'in the long run'?

Consider the first n bounces:

height after the first bounce $= 0.8 \times 125 = 100$ cm

height after the second bounce $= 0.8 \times 100 = 0.8^2 \times 125 = 80$ cm

height after the third bounce $= 0.8 \times 80 = 0.8^3 \times 125 = 64$ cm

etc.

height after the nth bounce $= 0.8^n \times 125$ cm.

As n gets larger, the height of the bounce gets smaller and tends to zero, because as $n \to \infty$, $0.8^n \to 0$.

Successive heights 125, 100, 80, 64, 51.2, ... form a geometric **sequence**.

Successive partial sums 125, 225, 305, 369, 420.2, ... form a geometric **series**.

The first term $a = 125$; the common ratio $r = 0.8$.

CALCULATOR ACTIVITY

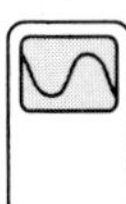

Use the program from the previous Calculator activity to investigate what happens to terms in the sequence and the series for $a = 125$, $r = 0.8$ as $n \to \infty$.

Use the formula for S_n to confirm your investigation.

Repeat this activity by keeping a fixed and varying r.

For what range of values of r does the series converge and for what range does it diverge?

For the ball, $a = 125$ and $r = 0.8$ means that:

$$S_n = \frac{a(1-r^n)}{1-r} = \frac{125(1-0.8^n)}{0.2}$$

Substituting $n = 1, 2, 3, \ldots$ gives $S_n = 125, 225, 305, \ldots$

Running the program until n becomes large reveals that S_n is converging towards a limit, e.g.

$S_{10} = 557.891\,136$
$S_{20} = 617.794\,240\,6$
$S_{30} = 624.226\,287\,5$
$S_{40} = 624.916\,923\,3$
$S_{50} = 624.991\,079\,7$

As $n \to \infty$, $S_n \to 625$, which is confirmed by letting $n \to \infty$ in the formula for S_n.

As $n \to \infty$, $0.8^n \to \infty \quad \Rightarrow \quad S_n \to 125/0.2 = 625$

For the general geometric series, $r^n \to 0$ as $n \to \infty$, provided $-1 < r < 1$, i.e. $|r| < 1$.

This gives the general definition for the **sum to infinity** of a geometric series, S_∞.

$$S_\infty = \lim_{n\to\infty} \frac{a(1-r^n)}{1-r} = \frac{a}{1-r}$$

provided $|r| < 1$.

Note

We usually discount the trivial case $r = 0$, since this would mean that every term of a geometric sequence, after the first one, would be 0. In this case the 'sum to infinity' would just be the first term, a.

Example 11.9

For the geometric sequence 81, 27, 9, ... find expressions for S_n and S_∞. Evaluate S_5, S_{10}, S_{15} and S_∞.

Solution

First term, $a = 81$; common ratio, $r = \frac{1}{3}$.

$$\Rightarrow \quad S_n = \frac{a(1-r^n)}{1-r} = \frac{81(1-(\frac{1}{3})^n)}{1-\frac{1}{3}} = 121.5\times(1-(\tfrac{1}{3})^n)$$

In particular:

$$S_5 = 121.5\times(1-(\tfrac{1}{3})^5) = 121$$

$$S_{10} = 121.5\times(1-(\tfrac{1}{3})^{10}) = 121.4979424...$$

$$S_{15} = 121.5\times(1-(\tfrac{1}{3})^{15}) = 121.4989915...$$

$$S_\infty = \frac{a}{1-r} = \frac{81}{1-\frac{2}{3}} = 121.5$$

Example 11.10

Express the recurring decimal 0.027 027 027... as the sum to infinity of a geometric series and so deduce its fractional equivalent in its simplest form.

Solution

0.027 027 ...= 0.027 + 0.000 027 + 0.000 000 027 + ...
which is the sum to infinity of a geometric series with $a = 0.027$, $r = 0.001$, i.e.

$$S_\infty = \frac{a}{1-r} = \frac{0.027}{1-0.001} = \frac{0.027}{0.999} = \frac{27}{999} = \frac{1}{37}$$

Example 11.11

The sum of the first two terms of a geometric sequence is 15 and the sum to infinity is 27. Find two possible geometric sequences that have this property.

Solution

Let the first term be a and the common ratio r, then:
$a + ar = 15$ (1)

and

$$\frac{a}{1-r} = 27 \qquad (2)$$

From (2): $a = 27(1-r)$

Substitute for a *in* (1).

$\Rightarrow\ 27(1-r) + 27(1-r)r = 15$

$\Rightarrow\ 27 - 27r + 27r - 27r^2 = 15$

$\Rightarrow\ 27 - 27r^2 = 15$

$\Rightarrow\ 27r^2 = 12$

$\Rightarrow\ r^2 = \frac{4}{9}$

$\Rightarrow\ r = \pm\frac{2}{3}$

$r = \frac{2}{3} \quad \Rightarrow \quad a = 27(1-\frac{2}{3}) = 9$

$r = -\frac{2}{3} \quad \Rightarrow \quad a = 27(1+\frac{2}{3}) = 45$

Two possible geometric sequences are:
9, 6, 4, ... and 45, –30, 20, ...

EXERCISES

1 For each of the following geometric series, determine which converge to a limit and in these cases find the sum to infinity.

a) 1 + 1.1 + 1.21 + 1.331 + ... **b)** 1 + 0.9 + 0.81 + 0.729 + ...
c) 112 + 84 + 63 + ... **d)** 112 – 84 + 63 – ...
e) 1 – 4 + 8 – 16 + ... **f)** 20 + 12 + 9.6 + ...

2 Express the following recurring decimals as fractions in their simplest form.

a) $0.6666... = 0.\dot{6}$ **b)** $0.272\,727... = 0.\dot{2}\dot{7}$ **c)** $0.\dot{0}3\dot{7}$

d) $0.\dot{1}4285\dot{7}$ **e)** $0.\dot{3}84\,61\dot{5}$ **f)** $0.\dot{0}09\dot{9}$

3 Find the sum to infinity of a geometric sequence with:

a) second term 40 and fifth term 5
b) second term –40 and fifth term 5.

4 Find the possible sums to infinity of a geometric sequence with:

a) second term –12 and fourth term $-1\frac{1}{3}$
b) third term 32 and fifth term 5.12
c) first term 25 and fifth term 0.04.

5 Write down the first five terms of a geometric sequence with first term 36 and sum to infinity 48.

6 A geometric sequence has third term 21.6 and sum to infinity 150. Find the common ratio r and first term a.

7 A rubber ball is dropped from a height of 5 m and after the first bounce rises to a height of 3 m.

a) Assuming that the heights of bounces form a geometric sequence, find the height after the third bounce.
b) Explain why the total distance travelled by the ball until it hits the ground for the second time is $5 + 2 \times 3$ m.
c) How far does the ball travel altogether before it stops bouncing?

11.4 B

1 Determine which of the following geometric series converge to a limit, and in these cases find the sum to infinity.

a) $4 + 4^2 + 4^3 + \ldots$ **b)** $\frac{1}{4} + (\frac{1}{4})^2 + (\frac{1}{4})^3 + \ldots$
c) $0.6^3 + 0.6^4 + 0.6^5 + \ldots$ **d)** $1 - 1.5 + 1.5^2 - 1.5^3 + \ldots$
e) $8 + 2(1.5) + 0.5(1.5)^2 + 0.125(1.5)^3 + \ldots$
f) $4 - 2(1.5)^2 + 1.5^4 - 0.5(1.5)^6 + \ldots$

2 Express the following recurring decimals as fractions in their simplest form.

a) $0.\dot{7}$ **b)** $1.\dot{7}$ **c)** $0.\dot{1}\dot{2}$

d) $0.\dot{1}2\dot{3}$ **e)** $0.43\dot{1}$ **f)** $0.4\dot{5}\dot{8}$

3 Find the sum to infinity of a geometric sequence with:

a) third term 135 and sixth term 5
b) second term –135 and sixth term 5.

4 A geometric sequence has first term 20 and sum to infinity 80. Find the common ratio and write down the first four terms of the sequence.

5 A geometric sequence has common ratio $\frac{11}{12}$ and its sum to infinity is 24. Find the first term and write down the first six terms of the sequence correct to 2 d.p.

6 The first term of a geometric sequence is 50, the third term is 18 and the common ratio is positive. Calculate:

a) the sum to infinity,
b) the first term of the sequence which is less than 0.01,
c) the least value of n such that the difference between S_∞ and S_n is less than 0.01.

7 **a)** For a certain GP, the sum to infinity is 10 times the first term. Find the common ratio.
b) Is it possible for the sum to infinity of a GP to equal the first term? Give a reason for your answer.
c) Is it possible for the sum to infinity of a GP to be less than the first term? Give a reason for your answer.

MATHEMATICAL MODELLING ACTIVITY

Specify the real problem

Problem statement

When buying a house, purchasers often pay a deposit and take out a **mortgage** (a loan over a fixed term from a bank or building society) for the rest.

- If you took out a mortgage for £40 000, what would you expect to have to repay each year so that you pay off the loan after 25 years?

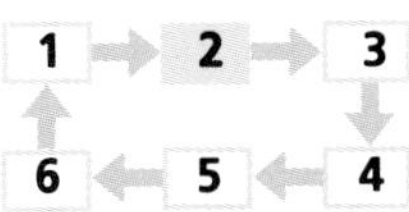

Set up a model

Set up a model

Firstly you need to identify the important variables:

- the rate of interest charged on the loan
- the amount repaid every year
- the amount owed at the beginning of each year.

You will also need to make some assumptions before formulating the problem mathematically, some of which are simplistic, but necessary to tackle the problem initially (Further refinements may be built in later.):

- the rate of interest remains constant
- the interest charged during a year depends on the amount outstanding at the beginning of the year
- a single repayment, £P, is made once a year
- there is no tax relief on interest repaid.

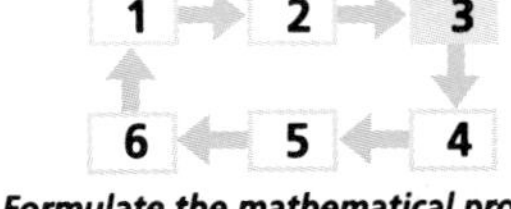

Formulate the mathematical problem

Mathematical problem

Firstly, fix the annual rate of interest at, say, 10 per cent.

Let u_n denote the amount owing to the building society (or bank) at the beginning of year n, then:

$$u_1 = 40\,000$$
$$\text{and } u_{n+1} = u_n \times 1.1 - P \qquad \text{for } n = 1, 2, 3, \ldots$$

The multiplier of 1.1 comes from $1 + \frac{10}{100}$ assuming a rate of interest of 10 per cent.

What value of P should be chosen so that the amount outstanding *after* 25 years, i.e. u_{26}, is 0?

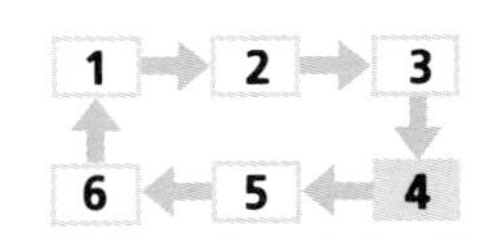

Solve the mathematical problem

Mathematical solution

Since $u_1 = 40\,000$:

$$\begin{aligned}
u_2 &= u_1 \times 1.1 - P = 40\,000 \times 1.1 - P \\
u_3 &= u_2 \times 1.1 - P \\
&= (40\,000 \times 1.1 - P) \times 1.1 - P \\
&= 40\,000 \times 1.1^2 - (P \times 1.1 + P) \\
u_4 &= u_3 \times 1.1 - P \\
&= (40\,000 \times 1.1^2 - P \times 1.1 - P) \times 1.1 - P \\
&= 40\,000 \times 1.1^3 - (P \times 1.1^2 + P \times 1.1 + P)
\end{aligned}$$

A pattern is developing which suggests that:

$$u_{26} = 40\,000 \times 1.1^{25} - (P \times 1.1^{24} + P \times 1.1^{23} + \ldots + P \times 1.1 + P)$$

The sum in brackets, reading from right to left, is the sum of the first 25 terms of a geometric series with first term P and common ratio 1.1, i.e.

$$P + P \times 1.1 + \ldots + P \times 1.1^{23} + P \times 1.1^{24} = \frac{P(1.1^{25} - 1)}{1.1 - 1}$$

Since the mortgage will be repaid after 25 years provided $u^{26} = 0$, you can deduce the value P by solving the equation:

$$0 = 40\,000 \times 1.1^{25} - \frac{P(1.1^{25} - 1)}{1.1 - 1}$$

$$\Rightarrow \quad \frac{P(1.1^{25} - 1)}{1.1 - 1} = 40\,000 \times 1.1^{25}$$

$$\Rightarrow \quad P = \frac{(40000 \times 1.1^{25}) \times 0.1}{1.1^{25} - 1} = 4406.72$$

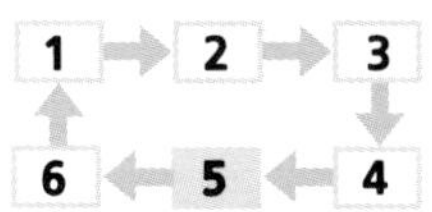

Interpret the solution

Interpretation

If you make an annual payment of just over £4400, at an annual rate of interest of 10 per cent, the mortgage will be paid off over a period of 25 years. This represents total repayments amounting to about £110 000.

Refinement of the model

Investigate the effect of:

a) a change in the rate of interest,
b) monthly repayments instead of annual repayments.

Discovery activity

Modify the flowchart for a geometric sequence, to automate the calculation of the amounts outstanding at the end of each year or each month.

Given an annual rate of interest (APR) calculate the equivalent monthly rate of interest.

CONSOLIDATION EXERCISES FOR CHAPTER 11

1 The ninth term of an arithmetic progression is 52 and the sum of the first twelve terms is 414. Find the first term and the common difference.

(AEB, Specimen Paper 1)

2 A geometric series has first term 3 and common ratio 0.8. Find the sum of the first 24 terms, giving your answer correct to 3 s.f.

(ULEAC, Paper 1, June 1992)

3 The tenth term of an arithmetic progression is 36 and the sum of the first ten terms is 180. Find the first term and the common difference.

(UCLES, Specimen Paper 2)

4 The first term of an AP is −13 and the last term is 99. The sum of all the terms is 1419.

a) Find the number of terms and the common difference.
b) Find also the sum of all the positive terms of the progression.

(AEB, Paper 1, June 1987)

5 The numbers 2, 3 and p are the first, second and third terms respectively of a geometric series. Calculate:

a) the value of p,
b) the sum of the first 21 terms of the series, giving your answer correct to 3 s.f.

(ULEAC, Paper 1, January 1992)

6 **a)** An employer offers the following schemes of salary payments over a five year period.

Scheme X: 60 monthly payments, starting with £1000 and increasing by £6 each month (£1000, £1006, £1012, ...)
Scheme Y: five annual payments, starting with £12 000 and increasing by £d each year [£12 000, £(12 000 + d), ...].

i) Over the complete five year period, find the total salary payable under Scheme X.
ii) Find the value of d which gives the same total salary for both schemes over the complete five-year period.

b) A small ball is dropped from a height of 1 m onto a horizontal floor. Each time the ball strikes the floor it rebounds to 0.6 of the height it has just fallen.

i) Show that, when the ball strikes the floor for the third time, it has travelled a distance 2.92 m.
ii) Show that the total distance travelled by the ball cannot exceed 4 m.

(ULEAC, Paper 1, June 1992)

7 An investment of £2000 is made at the start of a year with a Finance Company. At the end of this year and at the end of each subsequent year the value of the investment is 11 per cent greater than its value at the start of that year.

a) Find, to the nearest £, the value of the investment at the end of:
i) the fifth year,
ii) the tenth year.

A client decides to invest £2000 at the start of each year. Write down a series for which the sum is the total value of this annual investment at the end of 12 years.

b) By finding the sum of your series, determine, to the nearest £, the value of the investment at the end of 12 years.

(London, Paper 1, January 1992)

8 A loan of £50 000 is taken out and repaid over a period of years. At the end of each year the amount still owed is calculated by increasing the amount owed at the start of the year by 10 per cent and then subtracting the total repayments made during the year. Each year the borrower repays £R. The amount still owed after n years is £u_n.

Write down the value of u_0.
Show that $u_{n+1} = 1.1 \times u_n - R$.

Express u_1, u_2 and u_3 in terms of R.
Hence by summing an appropriate geometric series show that

$$u_n = 1.1^n \times 50\,000 - 10R(1.1^n - 1).$$

The loan is to be completely repaid after 25 years. Find the amount which should be repaid each year, giving your answer correct to the nearest penny.

(NEAB, Specimen Paper 1, 1994)

9 **a)** An arithmetic progression begins 25, 39, 53, 67, …
i) Write down an expression for the nth term of the sequence.
ii) Find an expression for the sum of the first n terms of the sequence.

b) A group of schoolchildren decided to collect used aluminium cans to help to raise money for charity. On the first day they collected 25 cans. On each of the next three schooldays they collected 14 cans more than on the previous day (because each day more of their friends became aware that they were collecting the cans). If this pattern continues, how many cans will they collect altogether in the first ten days of their collecting period?

(O&C, Specimen Paper 1, 1994)

10 A pendulum on a faulty clock is set to swing through 10°. Each swing is through an angle which is 0.98 times the previous angle so that the first two swings are through 10° and 0.98 × 10°, whilst further swings continue the terms of this geometric series.

a) Through what angle would the pendulum travel on the fifth swing?
b) What would be the total angle through which the pendulum had swung after eight swings?
c) The clock stops ticking when the pendulum swings through an angle of less than 2°. After how many complete swings does this occur?

(MEI, Pure 1, January 1993)

11 A farmer intends to plant 201 saplings at 1 metre intervals along one edge of a field AB, 200 m long, to form a wind break.

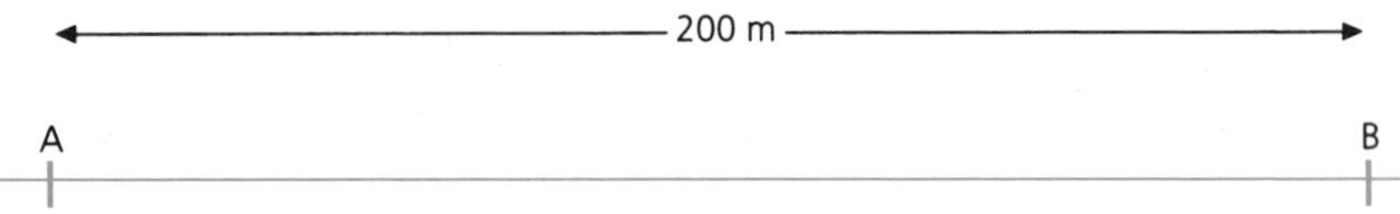

Due to limited access, all the saplings are delivered at A and must be carried to their planting holes one at a time. If the first one is planted at A, how far does the farmer have to walk altogether?

If, instead, the farmer carries two trees at a time, plants the first and then moves on 1 metre to plant the second, before returning to A, how much walking does this save him?

12 All the terms of a certain geometric sequence are positive. The first term is a and the second term is $a^2 - a$. Find the set of values of a for which the series converges. Given that $a = \frac{5}{3}$:

a) find the sum of the first ten terms of the series, giving your answer correct to 2 d.p,
b) show that the sum to infinity of the series is 5,
c) find the least number of terms of the series required to make their sum exceed 4.999.

(AEB, Paper 1, November 1987)

13 A company offers a ten-year contract to an employee. This gives a starting salary of £15 000 a year with an annual increase of eight per cent of the previous year's salary.

a) Show that the amounts of annual salary form a geometric sequence and write down its common ratio.
b) How much does the employee expect to earn in the tenth year?
c) Show that the total amount earned over the ten years is nearly £217 500.

After considering the offer, the employee asks for a different scheme of payment. This has the same starting salary of £15 000 but with a fixed annual pay rise £d.

d) Find d if the total amount paid out over ten years is to be the same under the two schemes.

(MEI, Pure 1, June 1993)

14 **a)** Timber cladding on the end of a shed roof consists of 20 narrow rectangular planks whose longest sides are in arithmetic progression. The shortest plank is 65 cm long, the longest 3.5 m. Find:
i) the difference in length between one plank and the next,
ii) the total length of all the planks.

b) For a demonstration, a 1-metre rule is cut into ten pieces whose lengths are in geometric progression. The length of the longest piece is 7.45 times the length of the shortest piece. The shortest piece is of length a cm and the common ratio is r.
i) Write down the length of the longest piece in terms of a and r.
ii) Find the values of a and r.

(MEI, Pure 1, January 1994)

15 The sum of the first n terms of an arithmetic series is $\frac{1}{2}n(n-20)$.

a) Find the value of n for which the sum equals 400.
b) Find the first term and the common difference.
c) Find in its simplest form the kth term of the series.

(Oxford, 1990)

16 The training program of a pilot requires him to fly 'circuits' of an airfield. Each day he flies three more circuits than the day before. On the fifth day he flew 14 circuits. Calculate how many circuits he flew:

a) on the first day,

b) in total by the end of the fifth day,

c) in total by the end of the nth day,

d) in total from the end of the nth day to the end of the $2n$th day. Simplify your answer.

(MEI, Paper 1, January 1992)

17 As part of a fundraising campaign, I have been given some books of raffle tickets to sell. Each book has the same number of tickets and all the tickets I have been given are numbered in sequence. The number of the ticket on the front of the fifth book is 205 and that on the front of the 19th book is 373.

a) By writing the number of the ticket on the front of the first book as a and the number of tickets in each book as d, write down two equations involving a and d.

b) From these two equations, find how many tickets are in each book and the number on the front of the first book I have been given.

c) The last ticket I have been given is numbered 492. How many books have I been given?

(MEI, Paper 1, June 1992)

Summary

Notation

- u_k represents the kth term of a sequence.
- S_n represents the sum of the first n terms of a series.

Arithmetic sequences and series

- For an arithmetic sequence with **first term** a and **common difference** d:

  Inductive definition: $u_{k+1} = u_k + d, \quad u_1 = a$

  Formula definition: $u_k = a + (k - 1)d$

  Sum to n terms $\quad S_n = \frac{n}{2}[2a + (n-1)d] = \frac{1}{2}(a + l)d$

  where l is the last term

- The series generated from an arithmetic sequence is called an **arithmetic progression** (**AP**).

Geometric sequences and series

- For a geometric sequence with **first term** a and **common ratio** r:

  Inductive definition: $u_{k+1} = ru_k, \quad u_1 = a$

  Formula definition: $u_k = ar^{k-1}$

  Sum to n terms $\quad S_n = \frac{a(1-r^n)}{1-r}$ if $-1 < r < 1$

  $S_n = \frac{a(r^n - 1)}{r - 1}$ if $r < -1$ or $r > 1$

  $S_n = na$ if $r = 1$

  Sum to infinity $\quad S_n = \frac{a}{1-r}$ only if $-1 < r < 1$

- The series generated from an arithmetic sequence is called an **geometric progression** (**GP**).

Indices

In this chapter we explore:

- *the laws of indices for manipulating powers of numbers and variables*
- *the manipulation of surds*
- *the idea and use of exponential functions in modelling.*

LAWS OF INDICES

In this section we shall explore the ideas of indices and powers and show the dramatic effect they have on numbers. To do this, we shall carry out an annotated exploration of a situation that frequently occurs in a biochemistry laboratory.

Exploration 12.1 *Examining indices*

Suppose that the size of a colony of bacteria doubles each day. At midnight (00:00) on a particular day (day 0) it is estimated that there are 1 million bacteria. Some of the figures are given in this table.

Day	–3	–2	–1	0	1	2	3	4	5
Size (in millions)			0.5	1	2				
Size (as a power of 2)									

- Study the patterns made by the figures, then copy and complete the table to show the size of the colony at 00:00 on various days before and after day 0.
- Give the population size (in millions) as a power of 2.

By studying the figures for days 1, 2, 3, ..., we can see that on day n the population size, P, is given by $P = 2^n$. Extending the definition to $n = 0, -1, -2, -3$ gives:

$$2^0 = 1 \qquad 2^{-1} = 0.5 = \frac{1}{2} \qquad 2^{-2} = 0.25 = \frac{1}{2^2} \qquad 2^{-3} = 0.125 = \frac{1}{2^3}$$

Therefore we can generalise to obtain the results:

$$2^0 = 1 \quad \text{and} \quad 2^{-n} = \frac{1}{2^n} \qquad n = 1, 2, 3, \ldots$$

$$a^0 = 1 \quad \text{and} \quad a^{-n} = \frac{1}{a^n} \qquad n = 1, 2, 3, \ldots$$

where a is any positive number.

- Check that your calculator gives equivalent results for various values of a and n:
  e.g. $3^{-2} = \frac{1}{3^2} \qquad 5^0 = 1 \qquad 1.1^{-5} = \frac{1}{1.1^5}$

The summary table of values may be extended like this.

Day	... –7	–6	–5	–4	–3	–2	–1	0	1	2	3	4	5	6	7 ...
Size	... 2^{-7}	2^{-6}	2^{-5}	2^{-4}	2^{-3}	2^{-2}	2^{-1}	2^0	2^1	2^2	2^3	2^4	2^5	2^6	2^7 ...

- By considering the population on day m and the population both n days later and n days earlier, find expressions for $2^m \times 2^n$ and $2^m \div 2^n$.
- At the end of one week, beginning with day 0, the population is 2^7. What will it be at the end of the next week? What will it be five weeks after day 0?

By inspecting the number pattern we see, for example, that:

$$2^2 \times 2^5 = 2^7 \qquad 2^3 \div 2^5 = 2^{-2} \qquad (2^7)^5 = 2^{35}$$

Generalising these results for powers of 2:

$$2^m \times 2^n = 2^{m+n} \qquad 2^m \div 2^n = 2^{m-n} \qquad (2^m)^n = 2^{mn}$$

For any positive number a:

$$a^m \times a^n = a^{m+n} \qquad a^m \div a^n = a^{m-n} \qquad (a^m)^n = a^{mn}$$

e.g. $5^4 \times 5^3 = 5^7 \qquad 1.6^2 \div 1.6^7 = 1.6^{-5} \qquad (3^5)^4 = 3^{20}$

These **rules of indices** also apply when m and/or n are negative:

e.g. $2.1^{-3} \times 2.1^5 = 2.1^2 \qquad 4^2 \div 4^{-5} = 4^7 \qquad (0.8^{-2})^3 = 0.8^{-6}$

- Check these numerical examples on your calculator.

Now let's return to the bacterial colony, and assume growth is continuous.

- What would you expect the population size to be at *midday* (12:00) on day 0, and at midday on day 1?
- What would you expect the population size to be at 08:00 on day 0 and at 16:00 on day 1?

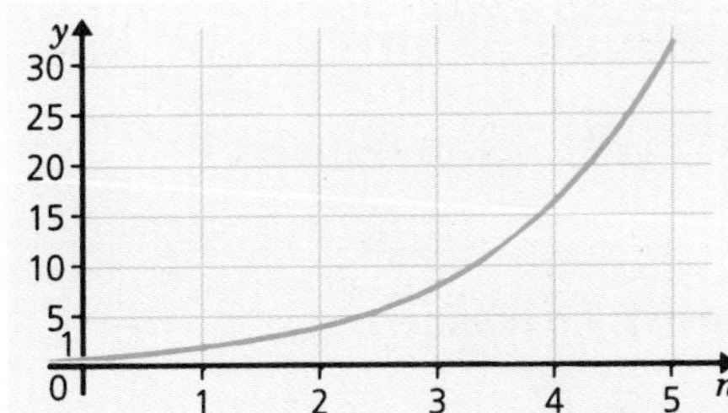

Assuming continuous growth, we can draw a graph of $P = 2^n$, treating n as a continuous variable.

At midday on day 0, $n = \frac{1}{2}$ and, from the graph, $P = 1.4$, correct to two significant figures (2 s.f.) which corresponds to $2^{\frac{1}{2}}$. From a calculator $2^{\frac{1}{2}} = 1.4142$ (5 s.f.).

Assuming the laws of indices still hold true, $(2^{\frac{1}{2}})^2 = 2^1 = 2$, i.e. $2^{\frac{1}{2}} \equiv \sqrt{2}$.

- Use your calculator to check that $2^{\frac{1}{2}}$ and $\sqrt{2}$ give the same decimal expansion.

By similar reasoning, the population at 08:00 on day 0 ($\frac{1}{3}$ of the way through the day) should correspond to $2^{\frac{1}{3}} = 1.26$.
Since $(2^{\frac{1}{3}})^3 = 2$, $2^{\frac{1}{3}} \equiv \sqrt[3]{2}$, the cube root of 2.

The population at 16:00 on day 0 ($\frac{2}{3}$ of the way through the day) will be $2^{\frac{2}{3}} = 1.59$.

Since $2^{\frac{2}{3}} = \left(2^2\right)^{\frac{1}{3}} = \left(2^{\frac{1}{3}}\right)^2$, $\quad 2^{\frac{2}{3}} = \sqrt[3]{4} = \sqrt[3]{2^2} = \left(\sqrt[3]{2}\right)^2$

■ Check these equivalent values give the same decimal expansions.

Generalising this idea for powers of 2:

$$2^{\frac{m}{n}} = \left(2^m\right)^{\frac{1}{n}} \equiv \sqrt[n]{2^m} \quad \text{or} \quad 2^{\frac{m}{n}} = \left(2^{\frac{1}{n}}\right)^m \equiv \left(\sqrt[n]{2}\right)^m$$

For any positive number a:

$$a^{\frac{m}{n}} = \left(a^m\right)^{\frac{1}{n}} \equiv \sqrt[n]{a^m} \quad \text{or} \quad a^{\frac{m}{n}} = \left(a^{\frac{1}{n}}\right)^m \equiv \left(\sqrt[n]{a}\right)^m$$

e.g. $9^{\frac{3}{2}} = \left(9^3\right)^{\frac{1}{2}} = \sqrt{729} = 27$ or $9^{\frac{3}{2}} = \left(9^{\frac{1}{2}}\right)^3 = (3)^3 = 27$

Example 12.1

Simplify the following expressions.

a) $\dfrac{5^7 \times 5^3}{5^8}$ **b)** $5x^2 \times 3x^5$ **c)** $(7a^3b)^2$ **d)** $\dfrac{6p^3q^2r}{2pq^3r^2}$

Solution

a) $\dfrac{5^7 \times 5^3}{5^8} = \dfrac{5^{7+3}}{5^8} = 5^{10-8} = 5^2 = 25$

b) $5x^2 \times 3x^5 = 5 \times 3 \times x^2 \times x^5 = 15x^7$

c) $(7a^3b)^2 = 7a^3b \times 7a^3b = 49a^6b^2$

d) $\dfrac{6p^3q^2r}{2pq^3r^2} = 3p^{3-1}q^{2-3}r^{1-2} = 3p^2q^{-1}r^{-1} = \dfrac{3p^2}{qr}$

Example 12.2

Evaluate the following, without using a calculator.

a) 3^{-2} **b)** $16^{-\frac{1}{2}}$ **c)** $8^{\frac{4}{3}}$

Solution

a) $3^{-2} = \dfrac{1}{3^2} = \dfrac{1}{9}$ **b)** $16^{-\frac{1}{2}} = \dfrac{1}{16^{\frac{1}{2}}} = \dfrac{1}{\sqrt{16}} = \dfrac{1}{4} = 0.25$

c) $8^{\frac{4}{3}} = \sqrt[3]{8^4} = \sqrt[3]{4096} = 16$ *or* $8^{\frac{4}{3}} = \left(\sqrt[3]{8}\right)^4 = 2^4 = 16$

Example 12.3

Without using a calculator, find the missing numbers.

a) $n^3 = -64$ **b)** $n^{\frac{2}{3}} = 25$ **c)** $3^n = \dfrac{1}{243}$

d) $-216^n = -6$ **e)** $25^n = 0.2$

Solution

a) $n^3 = -64 \Rightarrow n = \sqrt[3]{-64} = -4$ *Check:* $(-4)^3 = -64$

b) $n^{\frac{2}{3}} = 25 \Rightarrow \left(\sqrt[3]{n}\right)^2 = 25 \Rightarrow \sqrt[3]{n} = 5 \Rightarrow n = 5^3 \Rightarrow n = 125$

c) *Since* $3^5 = 243, \dfrac{1}{3^5} = 243^{-1} \Rightarrow 3^5 = 243^{-1} \Rightarrow n = -5$

d) *Since* $-6^3 = -216, \sqrt[3]{-216} = -6 \Rightarrow -216^{\frac{1}{3}} = -6 \Rightarrow n = \frac{1}{3}$

e) *Since* $0.2 = \frac{1}{5}$ *and* $\sqrt{25} = 5, \frac{1}{\sqrt{25}} = \frac{1}{5} = 0.2 \Rightarrow n = -\frac{1}{2}$

General results

From Example 12.1 we see that the **square of a product** is the **product of the squares.**

e.g. $(7a^3b)^2 = 49a^6b^2$

Similarly, the **square of a quotient** is the **quotient of the squares.**

e.g. $\left(\frac{5x}{2y^2}\right)^2 = \frac{25x^2}{4y^4}$

Reversing the argument in both cases gives:

- the square root of a product is the product of the square roots,
- the square root of a quotient is the quotient of the square roots.

e.g. $\sqrt{49a^6b^2} = 7a^3b$ and $\sqrt{\frac{25x^2}{4y^4}} = \frac{5x}{2y^2}$

We can extend this to give the general results:

$$(ab)^{\frac{m}{n}} = a^{\frac{m}{n}}b^{\frac{m}{n}} \quad \text{and} \quad \left(\frac{a}{b}\right)^{\frac{m}{n}} = \frac{a^{\frac{m}{n}}}{b^{\frac{m}{n}}}$$

Which can be summarised as:

- the power of a product is the product of the powers,
- the power of a quotient is the quotient of the powers.

Example 12.4

Simplify the following expressions.

a) $\left(\frac{3r}{4st^2}\right)^3$ **b)** $\sqrt[3]{125x^3y^6}$

Solution

a) $\left(\frac{3r}{4st^2}\right)^3 = \frac{(3r)^3}{(4st^2)^3} = \frac{27r^3}{64s^3t^6}$

b) $\sqrt[3]{125x^3y^6} = (125x^3y^6)^{\frac{1}{3}} = 125^{\frac{1}{3}}(x^3)^{\frac{1}{3}}(y^6)^{\frac{1}{3}} = \sqrt[3]{125}\,\sqrt[3]{x^3}\,\sqrt[3]{y^6} = 5xy^2$

Example 12.5

Evaluate the following, without using a calculator.

a) $\sqrt{1\frac{7}{9}}$ **b)** $(2\frac{10}{27})^{\frac{4}{3}}$

Solution

a) $\sqrt{1\frac{7}{9}} = \sqrt{\frac{16}{9}} = \frac{\sqrt{16}}{\sqrt{9}} = \frac{4}{3} = 1\frac{1}{3}$

b) $(2\frac{10}{27})^{\frac{4}{3}} = \left(\frac{64}{27}\right)^{\frac{4}{3}} = \frac{64^{\frac{4}{3}}}{27^{\frac{4}{3}}} = \frac{(\sqrt[3]{64})^4}{(\sqrt[3]{27})^4} = \frac{4^4}{3^4} = \frac{256}{81} = 3\frac{13}{81}$

EXERCISES

12.1 A

1 Simplify the following expressions.

a) $\frac{3^4 \times 3^5}{3^7}$ **b)** $\frac{4^6}{4^3 \times 4^4}$ **c)** $\frac{(7^3)^4}{(7^4)^3}$ **d)** $7x^6 \times 5x^2$

e) $2a^2b^3 \times 4a^3b^2$ **f)** $5p^2qr \times 3pq^2r^3 \times 4p^3qr^2$

g) $(-3r^3s)^2$ **h)** $(2ab)^5$ **i)** $(-4x^3y^2)^3$ **j)** $\frac{7x^2yz^3}{14y^3z}$

k) $\frac{4r^2st \times 6rs^2t}{8rst^2}$ **l)** $\left(\frac{3a^2b^3c}{4abc^3}\right)^3$ **m)** $\sqrt{\frac{9p^6}{49q^2}}$

2 Evaluate the following without using a calculator.

a) 2^{-3} **b)** $49^{\frac{1}{2}}$ **c)** $49^{\frac{3}{2}}$ **d)** $\left(\frac{2}{3}\right)^4$ **e)** $\left(\frac{2}{3}\right)^{-4}$

f) $25^{-\frac{1}{2}}$ **g)** $\sqrt{6\frac{1}{4}}$ **h)** $\sqrt{12.25}$ **i)** $\sqrt[3]{3\frac{3}{8}}$ **j)** $\left(\frac{9}{16}\right)^{-\frac{1}{2}}$

3 Without using a calculator find the value of n.

a) $n^3 = -27$ **b)** $n^5 = 32$ **c)** $n^{-2} = \frac{1}{16}$ **d)** $n^{-3} = \frac{1}{125}$

e) $n^{\frac{1}{3}} = \frac{2}{3}$ **f)** $n^{\frac{2}{3}} = 49$ **g)** $2^n = 1024$ **h)** $-5^n = 625$

i) $-3^n = \frac{1}{81}$ **j)** $225^n = -15$ **k)** $729^n = 9$ **l)** $64^n = \frac{1}{16}$

m) $144^n = \frac{5}{6}$

4 Explain why $\sqrt{a^2b^2} = ab$ but $\sqrt{a^2 + b^2} \neq a + b$.

EXERCISES

12.1 B

1 Simplify the following expressions.

a) $\frac{2^7 \times 2^2}{2^4}$ **b)** $\frac{5^{11}}{5^2 \times 5^7}$ **c)** $\frac{(11^2)^5}{(11^5)^2}$

d) $3y^4 \times 5y^7$ **e)** $4c^2d^5 \times 3c^4d^7$ **f)** $(-5x^4y^3)^2$

g) $2abc \times 5a^3b^2c^3 \times 4a^4b^5c^2$ **h)** $(11a^2b)^9$

i) $(-3m^2n^3p^4)^5$ **j)** $\frac{35p^4q^3r^7}{7p^5qr^5}$ **k)** $\sqrt{\frac{5a^5b^3}{20ab^7}}$

l) $\frac{6x^2y \times 14x^3y^4}{21x^4y^7}$ **m)** $\left(\frac{5r^4s^2t^3}{7r^3s^4t^2}\right)^2$

2 Evaluate the following without using a calculator.

a) 2^{-5} **b)** $64^{\frac{1}{2}}$ **c)** $64^{\frac{3}{2}}$ **d)** $64^{-\frac{1}{2}}$

e) $\left(\frac{3}{4}\right)^3$ **f)** $\left(\frac{3}{4}\right)^{-3}$ **g)** $\left(\frac{125}{8}\right)^{-\frac{1}{3}}$ **h)** $\sqrt{3\frac{1}{16}}$

i) $\sqrt{20.25}$ **j)** $\sqrt[3]{2\frac{10}{27}}$

3 Without using a calculator, find the value of n.

a) $n^5 = 243$ **b)** $n^7 = -128$ **c)** $n^{-3} = \frac{1}{8}$

d) $n^2 = \frac{1}{100}$ **e)** $n^{\frac{1}{3}} = \frac{4}{5}$ **f)** $n^{-\frac{2}{3}} = \frac{1}{36}$

g) $3^n = 729$ **h)** $(-3)^n = 729$ **i)** $625^n = 1$

j) $(-5)^n = -\frac{1}{125}$ **k)** $289^n = -17$ **l)** $1.2^n = \frac{25}{36}$

4 Choose various numbers to show the following results.

$\sqrt{a^2b^2} = ab$ $\sqrt{\frac{b^2}{a^2}} = \frac{b}{a}$ $\sqrt{a^2+b^2} \neq a+b$ $\sqrt{a^2-b^2} \neq a-b$

SURDS

When taking a square root, the answer may be postive or negative. A surd is the square root of a positive number, expressed in a form such as $\sqrt{x}$ or $\sqrt{a^2+b^2}$, without being evaluated numerically. This means that the value of $\sqrt{2}$, for example, is the exact value of the square root of 2, whereas the value given in a table or by a calculator will be an approximation. It is possible to manipulate surds, following the usual rules of operations in arithmetic.

Exploration 12.2

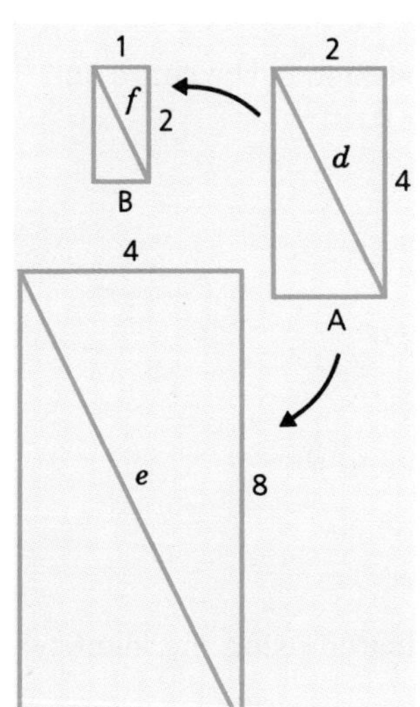

Finding diagonals

Rectangle A measures 2 units by 4 units.

- Express the length of the diagonal, d units, exactly.

Now suppose the rectangle is enlarged in two ways, one with scale factor 2, the other with scale factor $\frac{1}{2}$.

- Find the length of the diagonals of rectangles B and C exactly *in two different ways.*
- Account for any differences in the form of your answers.

Using surds

Using Pythagoras' theorem for rectangle A:

$$d^2 = 2^2 + 4^2 = 20$$

$$\Rightarrow \quad d = \sqrt{20}$$

Since rectangle B is an enlargement, scale factor 2, of rectangle A, the length of its diagonal is $2\sqrt{20}$, since $e = 2d$; similarly $f = \frac{1}{2}d = \frac{1}{2}\sqrt{20}$.

However, applying Pythagoras' theorem for rectangles B and C gives:

$$e^2 = 4^2 + 8^2 = 80 \Rightarrow e = \sqrt{80}$$

and $f^2 = 1^2 + 2^2 = 5 \Rightarrow f = \sqrt{5}$

Comparing the two different ways of finding e and f:

$$2\sqrt{20} \equiv \sqrt{80} \quad \text{and} \quad \tfrac{1}{2}\sqrt{20} \equiv \sqrt{5}$$

The equivalence of the results is justified by rules we discovered in the last section.

$$e = \sqrt{80} = \sqrt{4 \times 20} = \sqrt{4} \times \sqrt{20} = 2\sqrt{20}$$

$$f = \sqrt{5} = \sqrt{\frac{20}{4}} = \frac{\sqrt{20}}{\sqrt{4}} = \tfrac{1}{2}\sqrt{20}$$

The square roots used in these exact expressions are surds. The expression for e may be simplified by reducing the number inside the square root further.

$$e = 2\sqrt{20} = 2\sqrt{4 \times 5} = 2\sqrt{4} \times \sqrt{5} = 4\sqrt{5}$$

Comparing e with d and f gives these results:

$$d = 2\sqrt{5} \quad e = 4\sqrt{5} \quad f = \sqrt{5}$$

Exploration 12.3

Using surds

Function f is defined by $f(x) = x^2 - 4x - 3$.

- Find values for the x-coordinates of the points where the graph of $y = f(x)$ intersects the x-axis using surds:
  a) by completing the square,
  b) using the quadratic formula.
- Investigate the sum and product of the roots of the equation $f(x) = 0$.

Finding the roots

To find the x-coordinates of the points of intersection of the curve and the x-axis, we need to solve the equation $f(x) = 0$.

a) $x^2 - 4x - 3 = 0 \Rightarrow (x-2)^2 - 7 = 0$

$$\Rightarrow (x-2)^2 = 7$$
$$\Rightarrow x - 2 = \pm\sqrt{7}$$
$$\Rightarrow x = 2 \pm \sqrt{7}$$

b) $x^2 - 4x - 3 = 0$

$$\Rightarrow x = \frac{4 \pm \sqrt{(-4)^2 - 4 \times 1 \times (-3)}}{2} = \frac{4 \pm \sqrt{28}}{2}$$
$$= \frac{4 \pm \sqrt{4 \times 7}}{2} = \frac{4 \pm 2\sqrt{7}}{2} = 2 \pm \sqrt{7}$$

The equivalence of the two formats is demonstrated using the identity:

$$\sqrt{28} \equiv \sqrt{4 \times 7} = \sqrt{4} \times \sqrt{7} = 2\sqrt{7} \Rightarrow \tfrac{1}{2}\sqrt{28} = \sqrt{7}$$

Since the roots of the equation are $2 + \sqrt{7}$ and $2 - \sqrt{7}$ their *sum* is given by:

$$2 + \sqrt{7} + (2 - \sqrt{7}) = 4$$

and their *product* is given by:

$$(2 + \sqrt{7})(2 - \sqrt{7}) = 4 - 2\sqrt{7} + 2\sqrt{7} - 7 = 4 - 7 = -3$$

Notice that neither the sum nor the product involves surds. In fact, both are simply linked to the original equation. Can you see how?

Simplest form

It is usual to express surds in their simplest form. This means, for example, that $\sqrt{28}$ would be expressed as $2\sqrt{7}$, so that the number inside the square root sign has no square factors. This is called **minimising** the square root. A surd such as $\sqrt{(21+8)}$ would be expressed as $\sqrt{29}$, by **gathering like terms**.

Any expression involving surds is in its simplest form if:

- numbers inside the square roots are minimised,
- like terms are gathered together,
- fractions do not contain surds in the denominator.

The following examples illustrate how such expressions may be simplified.

Example 12.6

Simplify $7\sqrt{3}-\sqrt{48}+\sqrt{75}$.

Solution

$$7\sqrt{3}-\sqrt{48}+\sqrt{75}=7\sqrt{3}-\sqrt{16\times 3}+\sqrt{25\times 3}=7\sqrt{3}-4\sqrt{3}+5\sqrt{3}=8\sqrt{3}$$

Example 12.7

Simplify $\frac{5}{\sqrt{2}}-\frac{4}{\sqrt{3}}$.

Solution

$$\frac{5}{\sqrt{2}}-\frac{4}{\sqrt{3}}=\frac{5\sqrt{3}-4\sqrt{2}}{\sqrt{6}}=\frac{5\sqrt{3}\sqrt{6}-4\sqrt{2}\sqrt{6}}{6}=\frac{5\sqrt{18}-4\sqrt{12}}{6}$$
$$=\frac{5\sqrt{9\times 2}-4\sqrt{4\times 3}}{6}=\frac{5\times 3\sqrt{2}-4\times 2\sqrt{3}}{6}=\frac{15\sqrt{2}-8\sqrt{3}}{6}$$

Example 12.8

Simplify $\frac{5-\sqrt{2}}{4+\sqrt{3}}$.

Solution

$$\frac{5-\sqrt{2}}{4+\sqrt{3}}=\frac{(5-\sqrt{2})(4-\sqrt{3})}{(4+\sqrt{3})(4-\sqrt{3})}=\frac{20-4\sqrt{2}-5\sqrt{3}+\sqrt{6}}{16+4\sqrt{3}-4\sqrt{3}-3}=\frac{20-4\sqrt{2}-5\sqrt{3}+\sqrt{6}}{13}$$

Rationalising the denominator

In the last example we used a method called **rationalising the denominator**. Wherever a fraction has a denominator of the form $a+b\sqrt{c}$, where a, b and c are integers, multiplying both denominator and numerator by $a-b\sqrt{c}$ will remove the surds from the denominator, since:

$$(a+b\sqrt{c})(a-b\sqrt{c})=a^2+ab\sqrt{c}-ab\sqrt{c}-b^2c=a^2-b^2c$$

EXERCISES

12.2 A

Simplify the following surds.

1 $\sqrt{18}$ **2** $\sqrt{32}$ **3** $\sqrt{28}$ **4** $\sqrt{75}$ **5** $\sqrt{90}$

6 $\sqrt{345}$ 7 $\sqrt{507}$ 8 $\sqrt{60\,000}$ 9 $\sqrt{\frac{3}{4}}$ 10 $\sqrt{\frac{50}{9}}$

Express each of the following as a single surd in its simplest form.

11 $2\sqrt{3}+7\sqrt{3}-\sqrt{3}$ 12 $\sqrt{2}-\sqrt{18}+\sqrt{32}$

13 $\sqrt{45}+4\sqrt{5}-3\sqrt{20}$ 14 $2\sqrt{50}-\sqrt{72}+7\sqrt{8}$

15 $\sqrt{2}(5-\sqrt{2})$ 16 $\sqrt{2^3}-5\sqrt{2}+\sqrt{2^5}$

17 $(3-\sqrt{2})(2+3\sqrt{2})$ 18 $(\sqrt{3}+1)(3-2\sqrt{3})$

19 $(5+\sqrt{5})(5-\sqrt{5})$ 20 $(1-\sqrt{3})(1+\sqrt{3})$

Rationalise the denominators of the following fractions.

21 $\frac{5}{\sqrt{2}}$ 22 $\frac{1}{\sqrt{3}}$ 23 $\frac{7}{\sqrt{7}}$ 24 $\frac{3\sqrt{5}}{\sqrt{6}}$ 25 $\frac{1}{\sqrt{2}}-\frac{1}{\sqrt{3}}$

26 $\frac{5}{\sqrt{7}}+\frac{2}{3}$ 27 $\frac{1}{1+\sqrt{3}}$ 28 $\frac{7}{2+\sqrt{5}}$ 29 $\frac{\sqrt{3}-1}{\sqrt{2}-3}$ 30 $\frac{\sqrt{5}-\sqrt{3}}{\sqrt{5}+\sqrt{3}}$

EXERCISES

12.2 B

Simplify the following surds.

1 $\sqrt{44}$ 2 $\sqrt{125}$ 3 $\sqrt{7\,000\,000}$ 4 $\sqrt{320}$ 5 $\sqrt{261}$

6 $\sqrt{304}$ 7 $\sqrt{496}$ 8 $\sqrt{4205}$ 9 $\sqrt{\frac{32}{25}}$ 10 $\sqrt{\frac{27}{16}}$

Express each of the following as a single surd in its simplest form.

11 $4\sqrt{7}-3\sqrt{7}+6\sqrt{7}$ 12 $4\sqrt{2}-\sqrt{50}+\sqrt{98}$

13 $\sqrt{27}+2\sqrt{243}-\sqrt{75}$ 14 $\sqrt{3^3}-5\sqrt{3}+\sqrt{3^5}$

15 $\sqrt{3}(7+2\sqrt{3})$ 16 $(\sqrt{11}-\sqrt{11})(\sqrt{11}+\sqrt{11})$

17 $(3\sqrt{5}-1)(1+\sqrt{5})$ 18 $(\sqrt{7}-\sqrt{3})(\sqrt{7}+\sqrt{3})$

19 $(\sqrt{13}+1)(\sqrt{13}-13)$ 20 $(\sqrt{x}+\sqrt{y})(\sqrt{x}-\sqrt{y})$

Rationalise the denominators of the following fractions.

21 $\frac{7}{\sqrt{2}}$ 22 $\frac{-4}{\sqrt{5}}$ 23 $\frac{14}{\sqrt{7}}$ 24 $\frac{13}{\sqrt{13}}$ 25 $\frac{2\sqrt{7}}{\sqrt{12}}$

26 $\frac{1}{\sqrt{5}}+\frac{1}{\sqrt{3}}$ 27 $\frac{1}{2-\sqrt{3}}$ 28 $\frac{11}{5+\sqrt{3}}$ 29 $\frac{\sqrt{7}-1}{\sqrt{7}+1}$ 30 $\frac{\sqrt{5}-2}{1+\sqrt{2}}$

FUNCTIONS WITH RATIONAL INDICES

Exploration 12.4

Inverse proportion

A Youth Club is organising an outing by coach. The best quotation they received for the coach hire was £240 for a 59-seater.

■ How much should the Club charge each passenger to cover the cost of the coach hire?

Inverse proportion and hyperbolae

To spread the cost fairly the charge per passenger, £c, could be found by dividing the hire cost of £240 by the number of passengers, n, i.e. $c = \frac{240}{n}$.

So for 30 passengers, the coach fare would be $\frac{240}{30}$ = £8

and for 40 passengers, the coach fare would be $\frac{240}{40}$ = £6

Suppose that no one is willing to pay more than £12 each, then the minimum number of passengers is found by solving the equation:

$$12 = \frac{240}{n} \quad \Rightarrow \quad n = \frac{240}{12} = 20$$

This means the Club needs to attract at least 20 passengers to keep the cost down to £12 per head.

In this example c is **inversely proportional** to n, since $c = \frac{240}{n}$. In this context the formula is only valid for whole-number values of n, which means $n = 1, 2, ..., 59$. Why?

We can draw up a table of values.

n	5	6	8	10	12	15	16	20	24	30	40	48
c	48	40	30	24	20	16	15	12	10	8	6	5

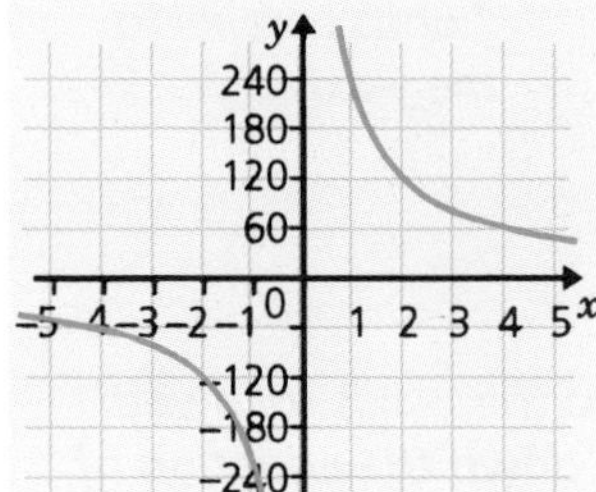

All the coordinates (x, y) lie on the curve $y = \frac{240}{x}$, although not all the points on the curve are solutions to the problem in the example.

When x *can* be any real number, except 0, the curve generated by the equation $y = \frac{240}{x}$ is an example of a **rectangular hyperbola**. As x gets smaller, and therefore closer to 0, from above (e.g. x = 1, 0.1, 0.01, 0.001, ...) y gets larger (e.g. y = 240, 2400, 24 000, 240 000, ...). Similarly as x gets closer to 0 from below (e.g. x = –1. –0.1, –0.01,...) y gets larger in value, although it is also negative (e.g. y = –240, –2400, –24 000, ...). The y-axis is an **asymptote** to the graph. This means the curve approaches the axis but doesn't actually ever touch it.
To summarise:

as $x \to 0$ from above $y \to \infty$
as $x \to 0$ from below $y \to \infty$

The x-axis is also an asymptote, because:

as $x \to \infty$ from above $y \to 0$
as $x \to -\infty$ from below $y \to 0$.

Another way of writing $y = \frac{240}{x}$ is $y = 240x^{-1}$, since $\frac{1}{x} \equiv x^{-1}$ and $\frac{240}{x} = 240 \times \frac{1}{x}$.

Similarly $y = \frac{20}{x^2}$ could be written as $y = 20x^{-2}$, where y is inversely proportional to x^2.

Example 12.9

Sketch graphs of functions with the following equations, in each case stating how you would transform the graph of $y = \frac{12}{x}$ onto it.

a) $y = \frac{12}{x} + 3$ **b)** $y = \frac{12}{x-2}$ **c)** $y = \frac{-12}{x+5}$ **d)** $y = 7 - \frac{6}{x}$

Solution

a)

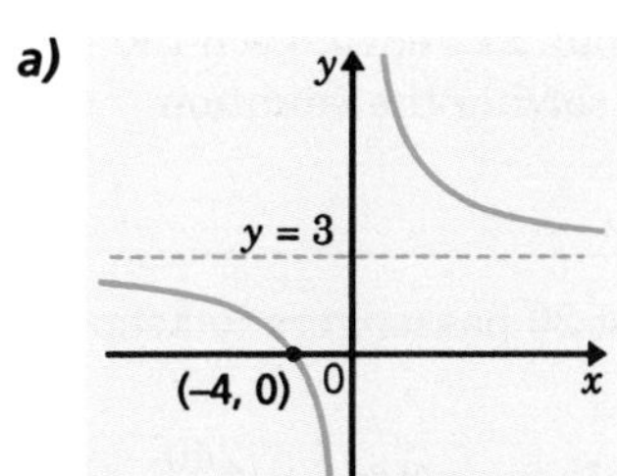

translation through 3 units parallel to the y-axis

b)

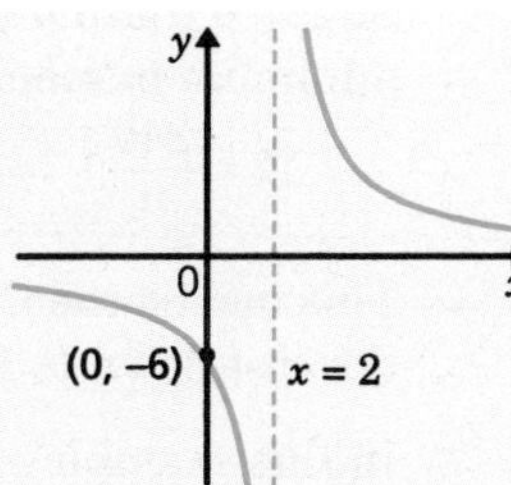

translation through 2 units parallel to the x-axis

c)

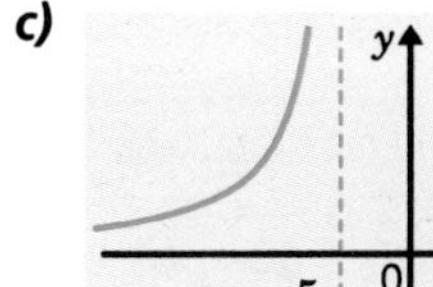

translation through –5 units parallel to the x-axis ***followed by*** *reflection in the x-axis*

d)

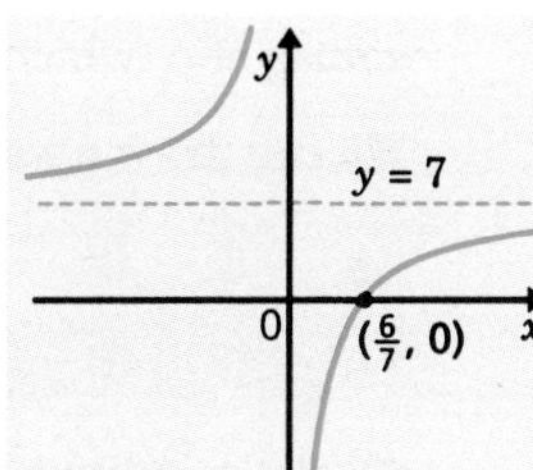

stretch, factor $\frac{1}{2}$, parallel to the y-axis ***followed by*** *reflection in the x-axis followed by translation through 7 units parallel to the y-axis*

Example 12.10

a) *Sketch the graph of $y = \frac{20}{x^2}$.*

b) *Translate the graph in* **a)** *–3 units parallel to the x-axis* ***followed by*** *–8 units parallel to the y-axis, and write down the equation of the transformed graph.*

c) *Find the coordinates of the points where the transformed graph meets the axes.*

Solution

a)

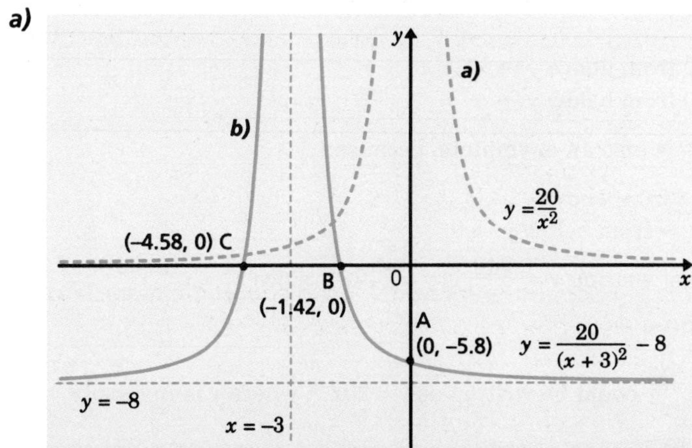

c) At A $x = 0 \Rightarrow \quad y = \frac{20}{3^2} - 8 = -5\frac{7}{9} \quad \Rightarrow$

A is the point $(0, -5\frac{7}{9})$.

At B and C $y = 0$

$\Rightarrow \quad \frac{20}{(x+3)^2} - 8 = 0$

$\Rightarrow \quad 20 = 8(x+3)^2$

$\Rightarrow \quad (x+3)^2 = 2.5$

$\Rightarrow \quad x + 3 = \pm\sqrt{2.5}$

$\Rightarrow \quad x = -4.58 \text{ or } x = -1.42$

So B is (–1.42, 0) and C is (–4.58, 0).

The root function

Another useful function that involves a rational power is the **root function**. The following exploration illustrates how a simple transformation of $y = x^{\frac{1}{2}}$ occurs.

Exploration 12.5

A simple pendulum

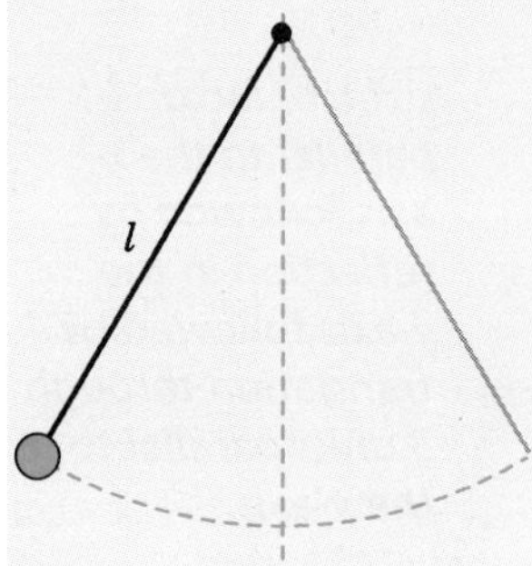

A student is trying to find the connection between the time of swing, t seconds, and length, l metres, for a simple pendulum. She obtains the following experimental results.

Length l (m)	0.5	1.0	1.5	2.0
Time t (s)	1.4	2.0	2.4	2.8

- What is the relationship between time of swing and length?

Analysing the results

When the length is multiplied by 2, the time is multiplied by 1.4 (to 1 d.p.)

l	0.5	1.0
t	1.4	2.0

l	1.0	2.0
t	2.0	2.8

When the length is multiplied by 3, the time is multiplied by 1.7 (to 1 d.p.)

l	0.5	1.5
t	1.4	2.4

When the length is multiplied by 4, the time is multiplied by 2.0

l	0.5	2.0
t	1.4	2.8

This suggests that t is proportional to $\sqrt{l}$.

We can draw up a table to show how t varies with $\sqrt{l}$.

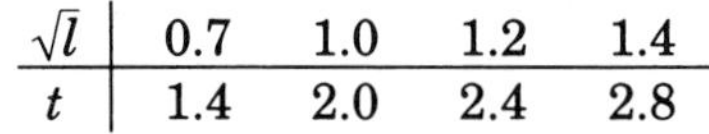

$\sqrt{l}$	0.7	1.0	1.2	1.4
t	1.4	2.0	2.4	2.8

From the table we can see that $t = 2\sqrt{l}$.

The relationship $t = 2\sqrt{l}$ may also be written as $t = 2l^{\frac{1}{2}}$.

Since l represents the length of a pendulum, $l > 0$. More generally, the equation $y = k\sqrt{x}$ is only valid for $x \geq 0$, since it is not possible to take the square root of a negative number.

Just as the equation $y = k\sqrt{x}$ may be written as $y = kx^{\frac{1}{2}}$,

the equation $y = k\sqrt[3]{x}$ may be written as $y = kx^{\frac{1}{3}}$.
This **cube root function** is valid for all real values of x.

Example 12.11

Sketch graphs of functions with the following equations. In each case describe how you would transform the graph of $y = \sqrt[3]{x}$ onto it.

a) $y = \sqrt[3]{x} - 5$ ***b)*** $y = 4\sqrt[3]{x}$

c) $y = \sqrt[3]{x+8}$ ***d)*** $3 - \sqrt[3]{2x}$

Solution

a)

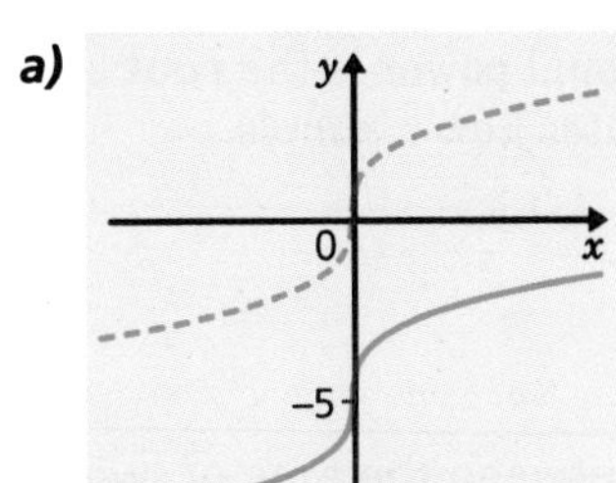

translation through –5 units parallel to the y-axis

b)

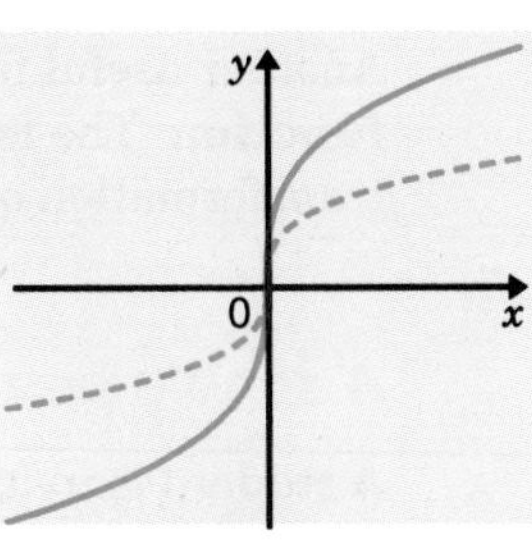

stretch, factor 4, parallel to the y-axis

c)

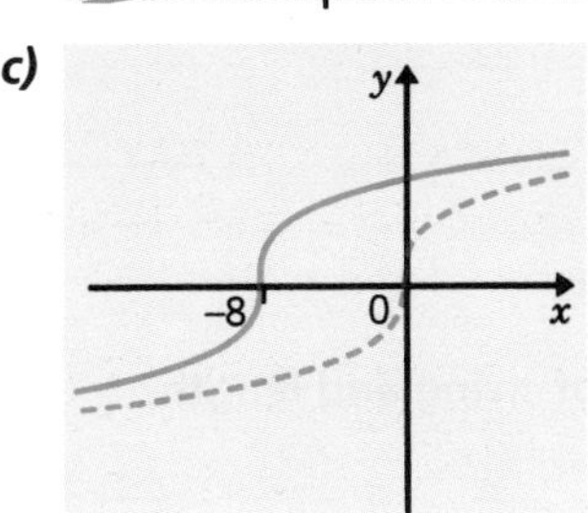

translation through –8 units parallel to the x-axis ***followed by*** *stretch, factor $\frac{1}{2}$, parallel to the y-axis*

d)

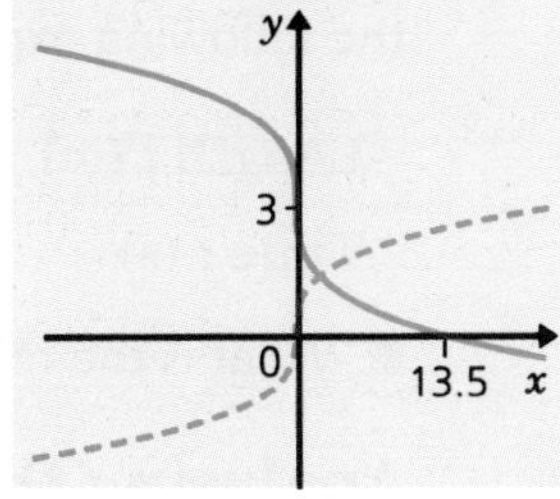

stretch, factor $\frac{1}{2}$, parallel to the x-axis ***followed by*** *reflection in the y-axis followed by translation through 3 units parallel to the y-axis*

EXERCISES

1 Sketch the graph of $y = \frac{4}{x}$.

For each of the following transformations of $y = \frac{4}{x}$, sketch the transformed curve, give its equation and write down the equations of any asymptotes.

a) translation through –3 units parallel to the y-axis
b) translation through –3 units parallel to the x-axis
c) stretch, factor 3, parallel to the y-axis
d) reflection in the x-axis *followed by* translation through 5 units parallel to the y-axis

2 Sketch the graph of $y = -\frac{1}{x^2}$.

For each of the following functions, state how you would transform the graph of $y = -\frac{1}{x^2}$ onto the graph of the given function. In each case sketch the transformed graphs and write down the equations of any asymptotes.

a) $y = -\frac{3}{x^2}$ **b)** $y = -\frac{1}{(x+2)^2}$ **c)** $y = 5 - \frac{1}{x^2}$ **d)** $y = \frac{1}{(x-3)^2}$

3 Each of the following diagrams shows a sketch of the graph of $y = \sqrt{x}$ $(x \geq 0)$, together with a transformation of the graph. Write down the equation of the transformed graph.

a)

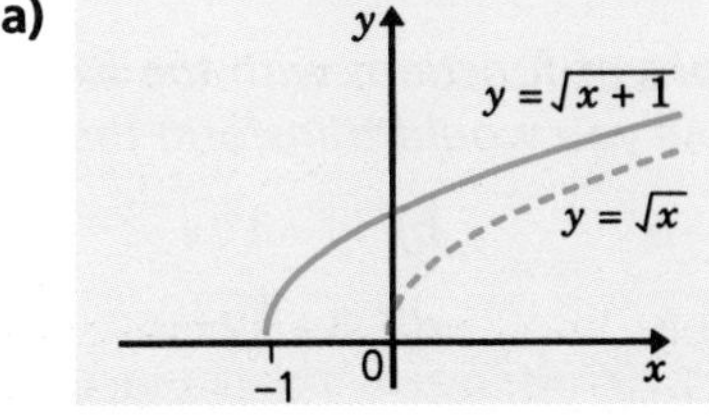

b)

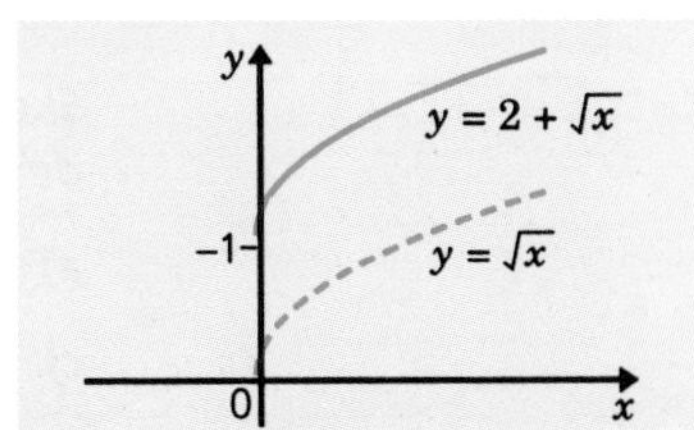

c) 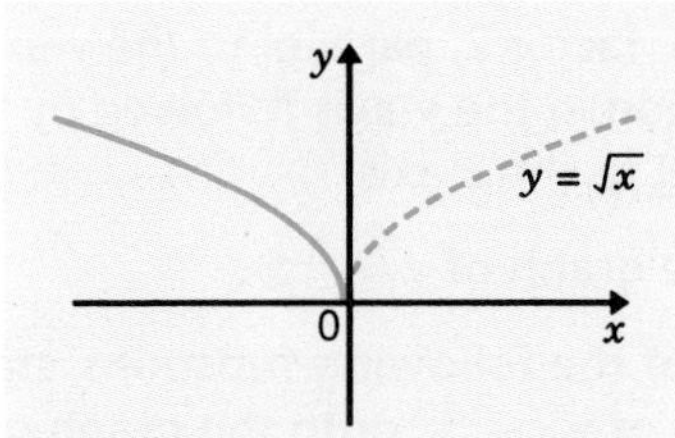

d) 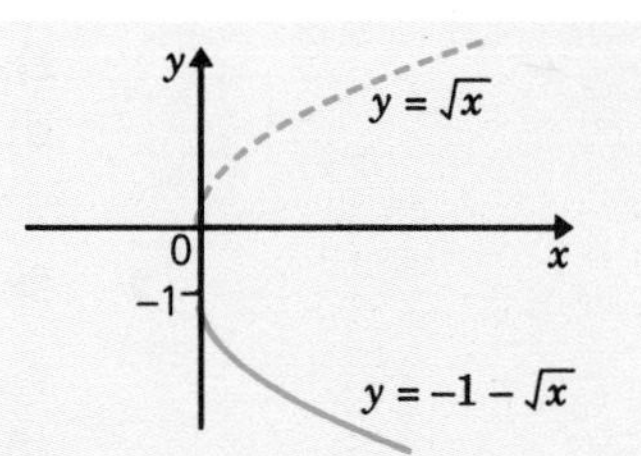

4 Sketch a graph of $y = x + \frac{1}{x}$ by considering what happens to y as:

a) $x \to 0$ from above and below, **b)** $x \to \pm\infty$.

Write down the equations for the two asymptotes. Confirm your answer using a graphics calculator. State the coordinates and nature of the two stationary points.

5 Sketch a graph of $y = x^2$ and a graph of $y = x^2 - \frac{8}{x}$ by considering what happens to y as:

a) $x \to 0$ from above and below, **b)** $x \to \pm\infty$.

Check your sketch using a graphics calculator. Write down the coordinates of the point where the graph intersects the x-axis. Check your answer algebraically. Find the coordinates of the minimum point by trial and improvement.

6 An electrical circuit consists of a battery and a resistance whose value can be changed. In an experiment a student obtained the following set of results for the current in the circuit as a function of the resistance.

Resistance *R* (ohms)	6	9	12	15	20
Current *I* (amps)	2	1.33	1	0.8	0.6

a) Draw a graph showing current against resistance.
b) Deduce the formula between I and R.
c) In a simple circuit of this type it is expected that $IR = E$ where E is the voltage of the battery. Does this relationship hold true in this case? If it does what is the value of E for the battery in this circuit?

7 Two electrical charges attract one another with a force P units which varies inversely as the square of the distance, x units, between them. If $P = 5.4$ when $x = 6$ find P when $x = 9$.

EXERCISES

12.3 B

1 Sketch the graph of $y = \frac{5}{x}$.

For each of the following transformations of $y = \frac{5}{x}$, sketch the transformed curve, give its equation and write down the equations of any asymptotes.

a) translation through +2 units parallel to the y-axis
b) translation through –2 units parallel to the x-axis

c) stretch, factor 2, parallel to the y-axis
d) reflection in the y-axis *followed by* translation through 4 units parallel to the x-axis

2 Sketch the graph of $y = -\frac{1}{x^3}$.

For each of the following functions, state how you would transform the graph of $y = -\frac{1}{x^3}$ onto the graph of the given function. In each case sketch the transformed graphs and write down the equations of any asymptotes.

a) $y = \frac{8}{x^3}$ b) $y = \frac{1}{(x-3)^3}$ c) $y = 3 + \frac{1}{x^3}$ d) $y = \frac{5}{(x+2)^3}$

3 Sketch a graph of $y = x - \frac{1}{x}$ by considering what happens to y as:

a) $x \to 0$ from above and below, b) $x \to \pm\infty$, c) $y \to 0$.

Write down the equations for the two asymptotes. Confirm your answer using a graphics calculator. Are there any stationary points?

4 Sketch the graph of $y = x^3$. On the same diagram sketch a graph of $y = x^3 - \frac{16}{x}$ by considering what happens to y as:

a) $x \to 0$ from above and below, b) $x \to \pm\infty$.

Find where the graph cuts the x-axis. Write down any equations of any asymptotes. Confirm your answer using a graphics calculator.

5 A choir is organising a raffle to raise £300 to buy new music. The total value of the prizes will be £200. The design and print costs of producing n raffle tickets is £$(40 + 0.01n)$.

a) Write down a formula for the selling price of a single raffle ticket, as a function of n, so that the choir raises exactly £300.
b) Sketch a graph of selling price against n.
c) In a survey the choir leader finds out that the local villagers will not pay more than 50p per raffle ticket. How many raffle tickets should the choir sell to raise at least £300?

6 The frequency of an oscillation is defined as the number of complete cycles per second and is measured in hertz. Write down the formula for the frequency, for an oscillation which repeats itself every T seconds, in terms of T. The musical note middle C on a piano has the frequency 256 hertz. Calculate the interval between the oscillations for this note.

7 A student is trying to find the relationship between the depth of water in a leaking tank and the speed of the water leaking out. The following experimental results are obtained.

Depth of water in tank h (m)	0.9	0.8	0.7	0.6	0.5
Speed of water from tank v ($\mathrm{m\,s^{-1}}$)	4.2	3.96	3.7	3.42	3.12

a) Draw a graph of speed against depth of this data.
b) Find a formula relating speed v and depth h.

Theory predicts the formula $\frac{1}{2}v^2 = gh$ where $g = 9.81\,\text{m s}^{-2}$ is a constant (the acceleration due to gravity). How well does the student's experimental data compare with the theory?

EXPONENTIAL FUNCTIONS

When we studied population growth in bacteria at the beginning of this chapter, we met a function of the form $y = a^x$. Functions of this type are **exponential functions**, and their curves always take the same general shape.

Exploration 12.6

The shape of an exponential curve

The general form of the function for bacteria population growth is $y = 2^x$, illustrated in the diagram.

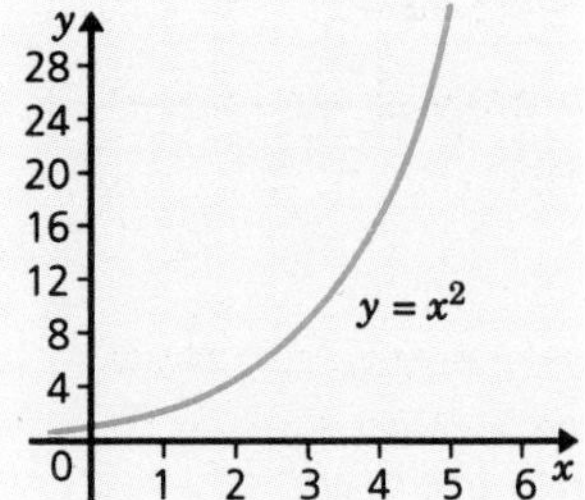

- What happens to y as $x \to \infty$?
- What happens to y as $x \to -\infty$?
- Why is y always positive?
- Estimate the gradient at various points on the curve.
- Describe what happens to the gradient as $x \to \pm\infty$.

The exponential curve

As x gets larger, 2^x gets large (at an increasing rate) so $y \to \infty$ as $x \to \infty$. As x gets smaller, 2^x gets smaller (at a decreasing rate) so $y \to 0$ from above as $x \to -\infty$. [In future work we shall need to use the idea that $2^{-\infty} = 0$]. Since the graph lies entirely above the x-axis y is always positive.

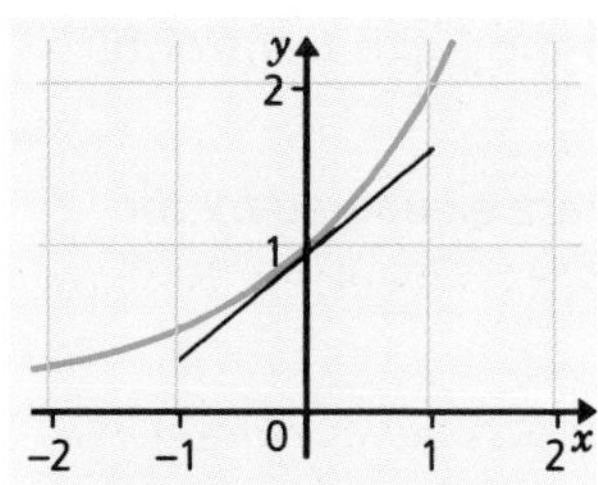

We can find estimates of the gradient at points on the curve, by drawing and measuring the gradient of tangents or by numerical approximation. For example, to estimate the gradient at $x = 0$ we can draw the tangent, as in the diagram. Then we can see that the gradient of the tangent is 0.7.

Numerically, we consider the gradient of the chord PQ, where P and Q have coordinates $(-0.01, 2^{-0.01})$ and $(0.01, 2^{0.01})$ respectively.

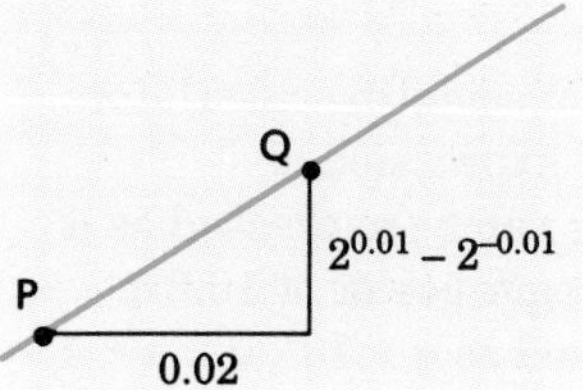

$$\text{Gradient of PQ} = \frac{2^{0.01} - 2^{-0.01}}{0.01-(-0.01)} \approx 0.693$$

Since P and Q are so close together, the gradient of chord PQ is a close approximation to the gradient of the tangent at (0, 1).

Gradients at points around the origin are listed in this table.

x	–3	–2	–1	0	1	2	3
y	0.125	0.25	05	1	2	4	8
gradient	0.087	0.173	0.347	0.693	1.387	2.773	5.545

We can see that as x gets larger, the gradient gets larger (at an increasing rate) and as x gets smaller the gradient gets smaller (at a decreasing rate). The gradient is always positive.

Note that the gradient is always about 0.69 times the y-value, which suggests the following relationships.

$$f(x) = 2^x \Rightarrow f'(x) \approx 0.69 \times 2^x$$

There is a more formal treatment of gradients of exponential functions in Chapter 17, *Exponentials and logarithms*.

Example 12.12

The diagram shows graphs of $y = 2^x$, $y = 2^{x-1}$ and $y = 2^{x+1}$. Identify which is which and describe geometrically the transformation of $y = 2^x$ onto $y = 2^{x-1}$ and $y = 2^x$ onto $y = 2^{x+1}$ in each of two ways.

a)

b)

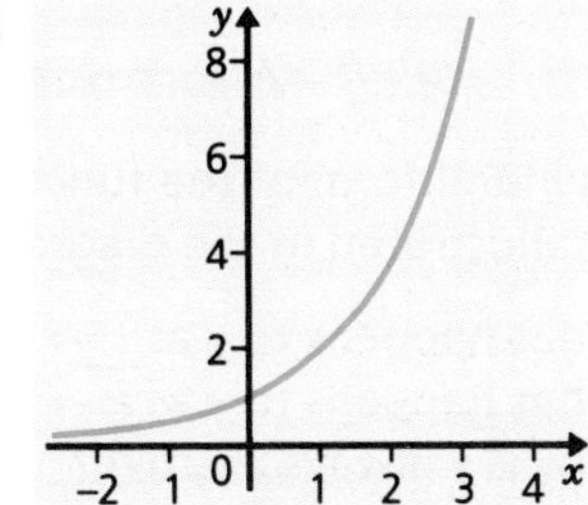

c)

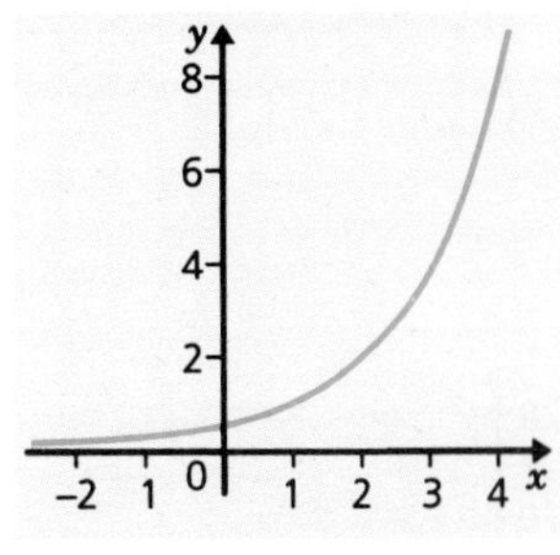

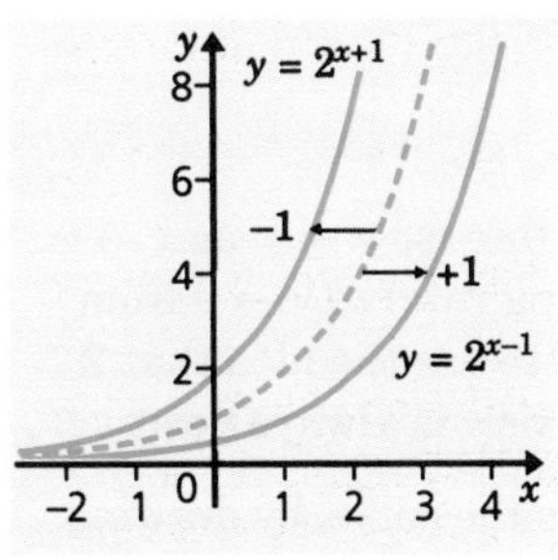

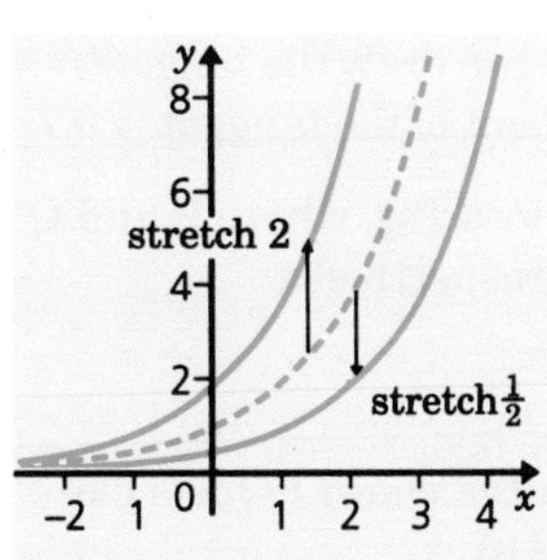

Solution

The graphs illustrated are:

a) $y = 2^{x+1}$ ***b)*** $y = 2^x$ ***c)*** $y = 2^{x-1}$

One way of transforming the graph of $y = 2^x$ onto the graph of $y = 2^{x+1}$ is by a translation of −1 unit parallel to the x-axis. Similarly, a translation of +1 unit parallel to the x-axis transforms $y = 2^x$ onto $y = 2^{x-1}$

Alternatively a stretch, factor 2, parallel to the y-axis transforms $y = 2^x$ onto $y = 2^{x+1}$ and stretch, factor $\frac{1}{2}$, parallel to the y-axis transforms $y = 2^x$ onto $y = 2^{x-1}$.

The reasons for these alternative descriptions are explained by the rules of indices.

$$2^{x+1} = 2^x \times 2^1 = 2 \times 2^x$$
$$2^{x-1} = 2^x \div 2^1 = \tfrac{1}{2} \times 2^x$$

Standard form

A useful exponential function which appears on scientific calculators is 10^x. A common use of the function 10^x is the expression of numbers in standard form, where any number can be expressed as a number between 0 and 1, multiplied by an integer power of 10. In general, a number in standard form is expressed as $a \times 10^b$, $0 < a < 1$, b is an integer.

e.g. $93\,000\,000 = 9.3 \times 10^7$ $0.000\,032\,8 = 3.28 \times 10^{-5}$

When x is a real number, then $f(x) = 10^x$ is a **continuous** function, and values for it may be found using the appropriate key on a calculator. Equations involving 10^x may be solved by a numerical method such as **decimal search**.

Example 12.13

Find values of x, correct to 3 s.f., which satisfy these equations.
a) $10^{-x} = 378$ **b)** $15 + 6x = 10^x$

Solution

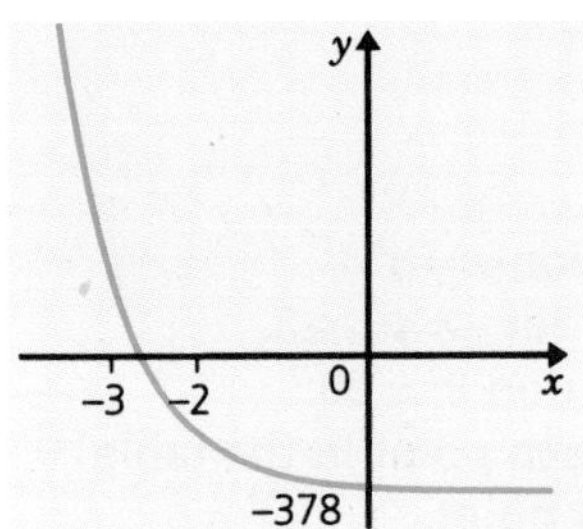

a) $10^{-x} = 378 \Rightarrow 10^{-x} - 378 = 0$
A sketch of $y = 10^{-x} - 378$ shows that the equation has a root between $x = -3$ and $x = -2$, i.e. if α is the root then $-3 < \alpha < -2$.
Using a decimal search with successive refinements, we find that:

$-2.6 < \alpha < -2.5$
$-2.58 < \alpha < -2.57$
$-2.578 < \alpha < -2.577$

This is sufficient to show that $\alpha = -2.58$ (to 3 s.f.). Accuracy to 3 s.f. is confirmed by noting that:

when $x = -2.585$ $y = 6.5918$
when $x = -2.575$ $y = -2.1626$

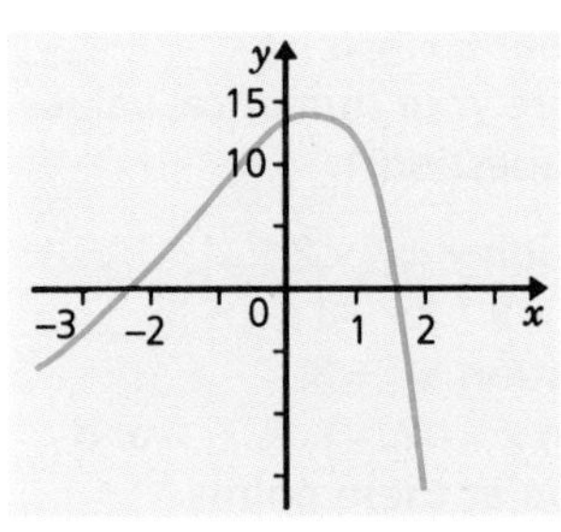

b) $15 + 6x = 10^x \Rightarrow 15 + 6x - 10^x = 0$
A sketch of $y = 15 + 6x - 10^x$ shows that the equation has two roots, α and β, such that:

$-3 < \alpha < -2$ and $1 < \beta < 2$

Using a decimal search with successive refinements, we find that:

$-2.5 < \alpha < -2.4$ $1.3 < \beta < 1,4$
$-2.50 < \alpha < -2.49$ $1.36 < \beta < 1.37$
$-2.500 < \alpha < -2.499$ $1.365 < \beta < 1.366$

This is sufficient to show that $\alpha = -2.50$ and $\beta = 1.37$ (to 3 s.f.).
Note that when $x = -2.5$, $y = 15 + 6(-2.5) - 10^{-2.5}$
$= 15 - 15 - 0.00316 = -0.00316$
Since the term in 10^x is relatively small when $x = -2.5$, then a good approximation to the lower root will be found by solving the equation:

$15 + 6x = 0 \Rightarrow x = -2.5$

and this confirms that $\alpha = -2.50$ (to 3 s.f.).

Modelling problems

The ability to solve equations involving exponential functions is a useful tool in modelling problems, as shown in the next example.

Example 12.14

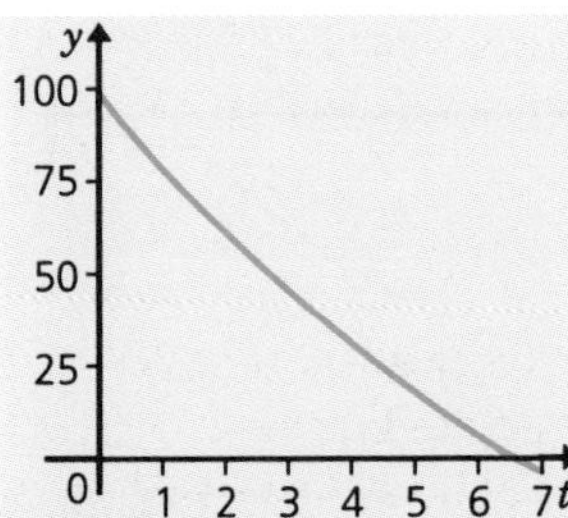

The radioactivity of a substance decays by 10 per cent over a period of a year. Its initial level of radioactivity is 200. Find the time taken for the level to fall to 100, i.e. find its ***half-life***.

Solution

If R is the level of radioactivity after t years, then:

$R = 200 \times 0.9^t$

When $R = 100$:

$100 = 200 \times 0.9^t \Rightarrow 200 \times 0.9^t - 100 = 0$

A sketch graph shows that the equation has a root, α and $6 < \alpha < 7$.

Using a decimal search with successive refinements, we find that:

$$6.5 < \alpha < 6.6$$
$$6.57 < \alpha < 6.58$$
$$6.578 < \alpha < 6.579$$

Therefore $\alpha = 6.58$ (to 3 s.f.), so the half-life of the substance is 6.58 years, which is approximately 6 years 7 months.

EXERCISES

12.4 A

1 a) Copy and complete this table for the function $y = 3^x$.
b) Using suitable scales, plot a graph of $y = 3^x$ for $-3 \leq x \leq 3$.
c) Use your graph to solve the equation $3^x = 20$.
d) Draw tangents to the curve for each point in the table and estimate the gradient at that point.

x	−3	−2	−1	0	1	2	3
y							

2 a) Use a decimal search to solve the equation $3^x - 20 = 0$, giving your answer correct to 3 s.f.
b) Use a numerical method to demonstrate that $f(x) = 3^x \Rightarrow f'(x) \approx 1.1 \times 3^x$ for $x = -2, -1, 0, 1, 2$. Compare your numerical values with the gradient estimates you found in question **1**.

3 a) Copy and complete this table for the function $y = 5 \times 2^{-x}$.
b) Using suitable scales, plot a graph of $y = 5 \times 2^{-x}$ for $-3 \leq x \leq 3$.
c) Use your graph to solve the equation $2^{-x} = 3$.
d) Draw tangents to the curve when $x = -2, -1, 0, 1, 2$ and estimate the gradient of the curve at these points.

x	−3	−2	−1	0	1	2	3
y							

4 a) Use a decimal search to solve the equation $2^{-x} = 3$, giving your answer correct to 3 s.f.
b) Use a numerical method to demonstrate that $f(x) = 5 \times 2^{-x} \Rightarrow f'(x) \approx 3.5 \times 2^{-x}$ for $x = -2, -1, 0, 1, 2$. Compare your numerical values with the gradient estimates you found in question **3**.

5 On the same axes sketch graphs of:

a) $y = 4^x$ **b)** $y = 4^{-x}$ **c)** $y = 4^{\frac{1}{2}-x}$

For parts **b)** and **c)**, describe the geometrical transformations which map the graph of $y = 4^x$ onto the curve.

6 On the same axes sketch graphs of $k \times 10^x$ for $k = -2, -1, -0.5, 0.5, 1, 2$.

7 The diagrams show sketches of two exponential functions, $y = k \times a^x$. In each case find the values of k and a.

a)

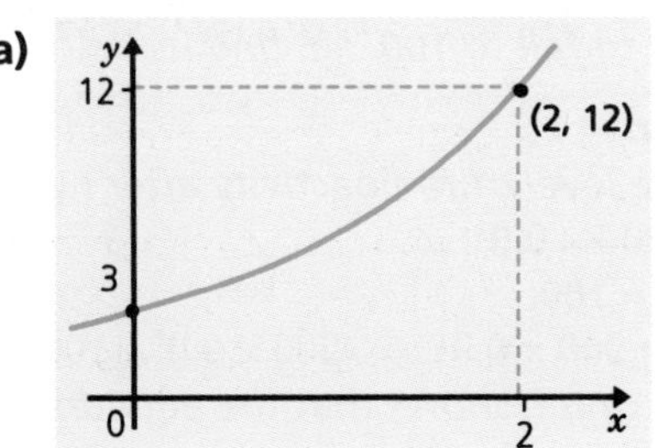

b)

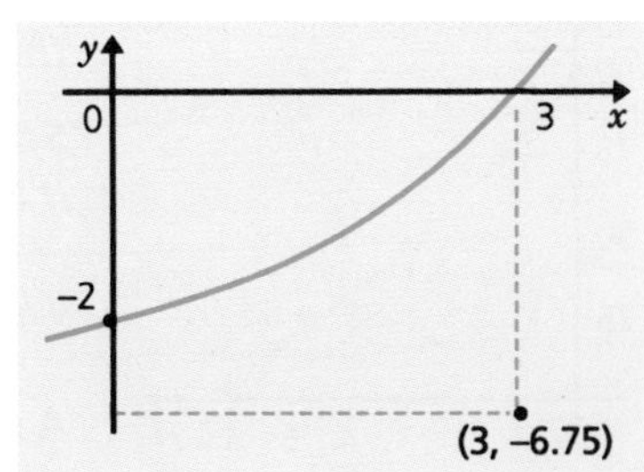

8 a) Sketch a graph of $y = 2^x - 9x + 5$ for $-1 \le x \le 7$.
b) Use your graph to show that the equation $2^x = 9x - 5$ has roots α and β such that $0 < \alpha < 1$ and $5 < \beta < 6$.
c) Use a decimal search to find values for α and β correct to 3 s.f.

9 The population of a country, currently 2 million, is growing at a rate of two per cent per annum.

a) Show that the expected population, P millions, in t years time, is given by $P = 2 \times 1.02^t$.
b) Sketch a graph of P against t for $0 \le t \le 100$.
c) Use your graph to estimate:
i) the size of the population in 35 years' time,
ii) the time taken for P to reach 10 million.

10 The concentration of a drug in a patient's bloodstream is C milligrams per millilitre after t hours, where $C = ka^{-t}$.

For drug A, $k = 2, a = 1.2$ For drug B, $k = 3, a = 1.4$

a) On the same axes, sketch graphs of C against t for both drugs.
b) Which drug will have the greater concentration after:
i) 2 hours, ii) 3 hours?
c) After what period of time would the concentration levels be the same for both drugs?

EXERCISES

12.4 B

1 a) Copy and complete this table for the function $y = 5^x$.
b) Using suitable scales, plot a graph of $y = 5^x$ for $-3 \le x \le 3$.

x	−3	−2	−1	0	1	2	3
y							

c) Use your graph to solve the equation $5^x = 15$.
d) Draw tangents to the curve for each point in the table and estimate the gradient at the solution point of $5^x = 15$.

2 a) Use a decimal search to solve the equation $5^x - 15 = 0$, giving your answer correct to 3 s.f.
b) Use a numerical method to demonstrate that $f(x) = 5^x \Rightarrow f'(x) \approx 1.6 \times 5^x$ for $x = -2, -1, 0, 1, 2$. Compare your numerical values with the gradient estimates you found in question **1**.

3 a) Copy and complete this table for the function $y = 4 \times 3^{-x}$.
b) Using suitable scales, plot a graph of $y = 4 \times 3^{-x}$ for $-3 \le x \le 3$.

x	−3	−2	−1	0	1	2	3
y							

c) Use your graph to solve the equation $3^{-x} = 8$.
d) Draw tangents to the curve when $x = -2, -1, 0, 1, 2$ and estimate the gradient of the curve at these points.

4 a) Use a decimal search to solve the equation $3^{-x} = 8$, giving your answer correct to 3 s.f.
b) Use a numerical method to demonstrate that $f(x) = 4 \times 3^{-x} \Rightarrow f'(x) \approx -4.4 \times 3^{-x}$ for $x = -2, -1, 0, 1, 2$. Compare your numerical values with the gradient estimates you found in question **3**.

5 On the same axes, sketch graphs of:

a) $y = 5^x$ **b)** $y = 5^{-x}$ **c)** $y = 5^{x+1}$ **d)** $y = 5^{x-1}$

For parts **b), c)** and **d)** describe the geometrical transformations which maps the graph of $y = 5^x$ onto the curve.

6 On the same axes, sketch graphs of $y = a^x$ for $a = -2, -1, -\frac{1}{2}, 1, 2$.

7 The diagrams show sketches of two exponential functions, $P = k \times a^t$. In each case find the values of k and a.

a)

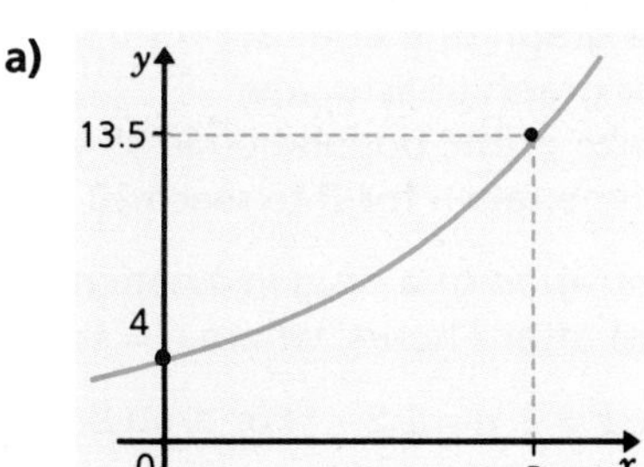

b)

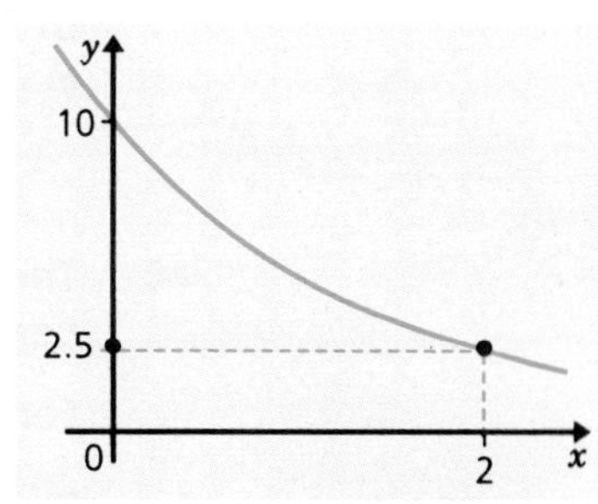

Modelling the change in the population of foxes and rabbits in a park, it is proposed that the numbers of foxes and rabbits change according to exponential laws.

c) Which of the models in graphs **a)** and **b)** could be used to model the population of foxes? Give reasons for your answer.

d) Which of the models in graphs **a)** and **b)** could be used to model the population of rabbits? Give reasons for your answer.

8 Sketch a graph of $y = 2^{-x}$ and $y = x^2 - 1$.

a) Use your graph to show that the equation $2^{-x} = x^2 - 1$ has two solutions, α and β, such that $1 < \alpha < 2$ and β is a negative integer.

b) Use a decimal search to find a value for α correct to 3 s.f.

9 During the first five years after the opening of a theme park in Dorset, the number of visitors increased by 30 per cent per annum. In the first year there were 6000 visitors.

a) Show that the number of visitors, P, after t years is given by:
$P = 6000 \times 1.3^{t-1}\ 1 \leq t \leq 5$.

b) Sketch a graph of P against t for $1 \leq t \leq 5$.

c) If the number of visitors continues to grow at 30 per cent per annum how many people will visit the theme park in the tenth year after opening?

d) The maximum number of visitors per year is to be restricted to 300 000. Use your graph to predict when this will occur.

10 The solar system consists of the sun and the nine planets together with many smaller bodies such as the comets and the meteorites. The nine planets are, in order of distance from the sun, Mercury, Venus, Earth, Mars, Jupiter, Saturn, Uranus, Neptune and Pluto. Between Mars and Jupiter are many very small bodies called the minor planets or asteroids and nearly all of them are too small to be seen by the naked eye from the Earth.

This problem is about Bode's law, developed in the 18th century, that relates the distance of a planet from the sun to a number representing the planet.

Data for Bode's law

The table below gives the average distance of the planets from the sun and the ratio of these distances to the Earth's distance from the sun.

Planet	Distance from the sun R (10^6 km)	Ratio $\frac{R}{R_e}$ where R_e is the Earth's distance from the sun
Mercury	57.9	0.39
Venus	108.2	0.72
Earth	149.6	1.00
Mars	227.9	1.52
Asteroids	433.8	2.9
Jupiter	778.3	5.20
Saturn	1427	9.54
Uranus	2870	19.2
Neptune	4497	30.1
Pluto	5907	38.5

Ignoring Mercury, we assign a number to each planet:

- for Venus, $n = 0$
- for Earth, $n = 1$
- ...
- for Pluto, $n = 8$

Use the data to find a model relating $\frac{R}{R_e}$ and n. What value of n should you give to Mercury so that it, too, fits the model?

CONSOLIDATION EXERCISES FOR CHAPTER 12

1 Simplify the following expressions.

a) $\dfrac{7^2 \times 7^4 \times 7^0}{7^5 \times 7^3}$ **b)** $2a \times 5a^4$ **c)** $3a^2b^2 \times 2a^4b^5$

d) $\left(-2x^7y^{\frac{1}{2}}\right)^4$ **e)** $(-4mnp^3)^2$ **f)** $\sqrt{25x^6y^2z^{-4}}$

g) $\dfrac{6p^2q^3 \times 10pq^5}{15pq}$ **h)** $\sqrt{\dfrac{63r^{11}s^7t^3}{7r^3st^{-1}}}$ **i)** $\left(\dfrac{3x^2yz^3}{2xz}\right)^4$

2 Express each of the following in the form ka^x.

a) $(5a)^2$ **b)** $\dfrac{1}{a^3}$ **c)** $a^2 \times \sqrt{a}$ **d)** $\left(\dfrac{2}{a}\right)^4$

e) $\dfrac{4a}{\sqrt{a}}$ **f)** $\dfrac{\sqrt{a}}{a^2}$ **g)** $\sqrt{9a}$ **h)** $(3a^4)^3$

3 Solve the following equations for x.

a) $x^2 = 0.64$ **b)** $2^x = 128$ **c)** $0.5^x = 8$

d) $2.5^x = 6.25$ **e)** $x^{\frac{1}{4}} = \frac{1}{4}$ **f)** $5^x = 1$

g) $(-2)^x = \frac{1}{64}$ **h)** $(100\,000)^x = 10^{15}$ **i)** $441^x = -21$

4 Simplify the following expressions writing each in the form $x + y\sqrt{b}$ where x and y are integers or fractions.

a) $11\sqrt{3} - 2\sqrt{3} + 5\sqrt{3}$ **b)** $\sqrt{50} - 2\sqrt{2} + \sqrt{18}$ **c)** $\sqrt{5}^3 + 2\sqrt{5} - 4\sqrt{5}^5$

d) $\sqrt{2}\left(3 - 4\sqrt{2}\right)$ **e)** $\dfrac{\sqrt{3} - 2}{\sqrt{3} + 5}$ **f)** $\dfrac{\sqrt{7} + 1}{\sqrt{7} - 1}$

g) $\dfrac{\sqrt{27} - \sqrt{3}}{\sqrt{75} + 1}$ **h)** $\dfrac{\left(\sqrt{5} - 1\right)\left(\sqrt{5} + 2\right)}{\sqrt{5} + 1}$

5 Rationalise the denominators of the following expressions.

a) $\dfrac{25}{\sqrt{5}}$ **b)** $\dfrac{8}{5\sqrt{2}}$ **c)** $\dfrac{10}{3 - \sqrt{7}}$ **d)** $\dfrac{4 - \sqrt{13}}{5 + \sqrt{13}}$

6 Solve the following quadratic equations, giving the answers as surds in their simplest form.

a) $x^2 + 4x + 1 = 0$ **b)** $2x^2 - 5x + 1 = 0$

c) $3x^2 + x - 5 = 0$ **d)** $4x^2 - x - 7 = 0$

7 Sketch the graphs of $y = \dfrac{1}{x^2}$, $y = \dfrac{1}{(x+2)^2}$, $y = \dfrac{4}{x^2}$ and $y = \dfrac{4}{(x-2)^2}$.

State how you could start from a single graph of $y = \dfrac{1}{x^2}$ and transform it to obtain the other graphs.

8 Sketch the graph of $y = \frac{10}{x^2}$. For each of the transformations, sketch the transformed curve, give its equation and write down the equations of the asymptotes.

a) translation through +3 units parallel to the x-axis

b) translation through –3 units parallel to the y-axis

c) stretch, factor 3, parallel to the y-axis

d) reflection in the x-axis *followed by* a translation through 3 units parallel to the y-axis

9 Sketch the graph of $y = x + \dfrac{7}{x^2}$ by considering what happens to y as:

a) $x \to 0$ from above and below,

b) $x \to \pm\infty$.

Find the coordinates of the intersection of the graph with the x-axis. Write down the equations of the asymptotes. Confirm your answer using a graphics calculator. Use a decimal search method to estimate the coordinates of any stationary points, correct to 3 s.f.

10 **a)** Plot the graph of $y = x - 1 + \dfrac{4}{x^2}$ for $-5 \le x \le 5$.

b) Use your graph to find:

i) the coordinates of the stationary point,

ii) the value of x such that $x + \dfrac{4}{x^2} = 1$.

c) Confirm your answer to **b) ii)** by means of a decimal search.

11 By letting $y = 5^x$, solve the equation $5^{2x+1} - 6 \times 5^x + 1 = 0$.

Use a graphics calculator to confirm your answer.

12 Sketch the graph of $y = \sqrt{x}$.

For each of the following equations, describe the geometrical transformation of $y = \sqrt{x}$ and then sketch the graph.

a) $y = \sqrt{x-3}$ **b)** $y = 10 - \sqrt{x}$ **c)** $y = -3\sqrt{x}$ **d)** $y = \sqrt{2x-3}$

13 **a)** Copy and complete this table for the function $y = 1.5 \times 2^x$.

b) Using suitable scales, plot a graph of $y = 1.5 \times 2^x$ for $-3 \le x \le 3$.

x	−3	−2	−1	0	1	2	3
y							

c) Use your graph to solve the equation $2^x = 5$.

d) Draw tangents to the curve for each point in the table and estimate the gradient at that point.

14 **a)** Use a decimal search to solve the equation $2^x - 5 = 0$, giving your answer correct to 3 s.f.

b) Use a numerical method to demonstrate that $f(x) = 1.5 \times 2^x \Rightarrow f'(x) \approx 1.04 \times 2^x$ for $x = -2, -1, 0, 1, 2$. Compare your numerical values with the gradient estimates you found in question **7**.

15 The giant Amazon water lily grows exponentially. Over a period of 25 weeks, its diameter d cm after t weeks is given by $d = 10 \times 2^{\frac{1}{5}t}$.

a) Draw a graph of d against t for $0 \le t \le 25$.

b) Use your graph to find:

i) the diameter of the lily after $12\frac{1}{2}$ weeks,

ii) the time it takes for the diameter to reach 2 metres.

c) Confirm your answer to **b) ii)** by a decimal search.

16 The diagram shows the graphs of $y = 2^x$ and $y = 2^{x+1}$. Describe two different geometrical transformations which map the graph of $y = 2^x$ onto the graph of $y = 2^{x+1}$.

(UCLES Modular Specimen Paper P2, Question 5(Part))

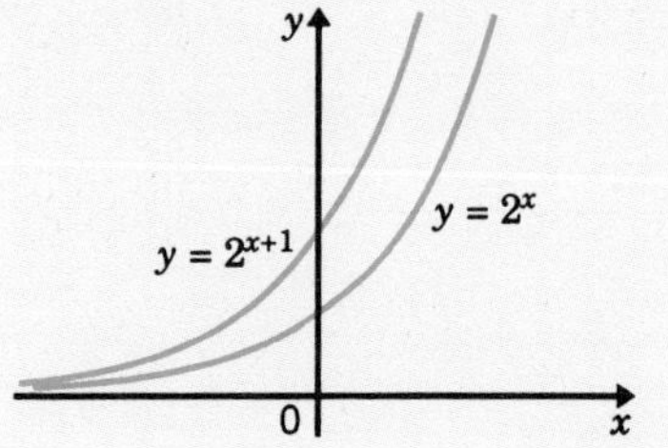

17 **a)** Sketch the graph of $y = \dfrac{x^2}{2^x}$ for $x \ge 0$.

b) The graph has a local maximum. Use graphical methods to find its coordinates.

(Oxford Nuffield, Paper 1, 1995)

Summary

- The laws of indices are

$$a^m \times a^n = a^{m+n}$$

$$\frac{a^m}{a^n} = a^m \div a^n = a^{m-n}$$

$$\left(a^m\right)^n = a^{mn}$$

- A **surd** is an irrational number such as $\sqrt{2}$ or $\sqrt{3}$. Surds can be manipulated following the usual rules of arithmetic.

- $\left(a+\sqrt{b}\right)\left(a-\sqrt{b}\right) = a^2 - b$

- The graph of $y = \frac{a}{x}$, where a is a constant, is called a **rectangular hyperbola**. We say that y is **inversely proportional** to x.

- Functions of the type $f(x) = ax$ are called **exponential functions**, where the constant a is called the **base**.

PURE MATHS 13

Functions

In this chapter we discuss:

- *the formal concept of a function,*
- *the idea of composite functions,*
- *the inverse of a function,*
- *how to transform functions from the same family.*

WHAT IS A FUNCTION?

Exploration 13.1

Different functions

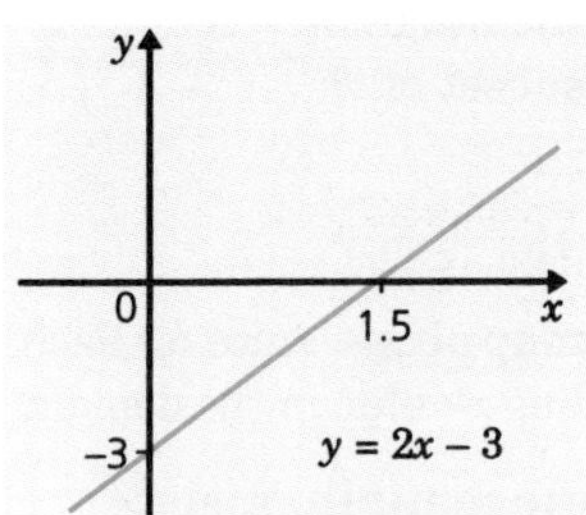

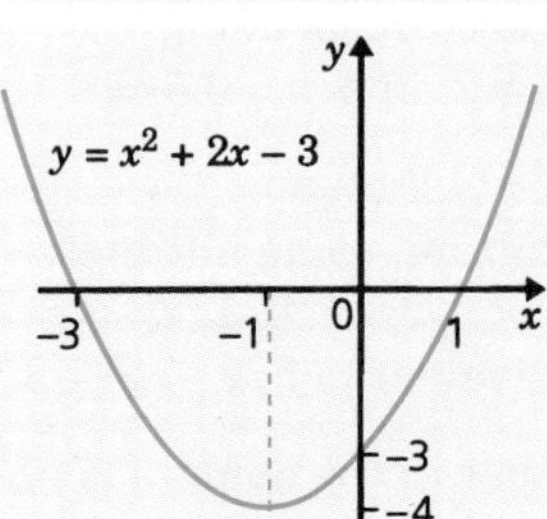

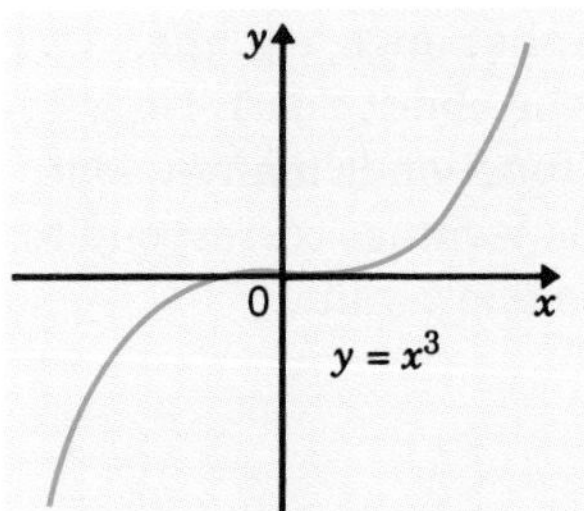

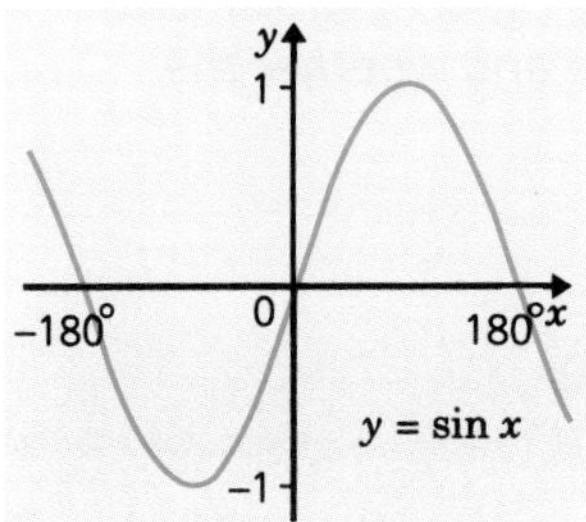

You have already met many different functions such as:

1	$f(x) = 2x - 3$	linear function
2	$f(x) = x^2 + 2x - 3$	quadratic function
3	$f(x) = x^3$	cubic function
4	$f(x) = \sin x$	trigonometric function

- Sketch graphs of $y = f(x)$ for each function.
- What similarities and differences are there between each of the graphs?
- For each function solve the equations $f(x) = 0$, $f(x) = 5$, $f(x) = -5$.

Looking at functions

Sketch graphs of the four functions listed above might look like these.

Each function may be defined for any real number x. The sketch graph can be thought of as a window through which you can see part of the graph. Plot the graphs on a graphics calculator and use the zoom facility to view the graph through different windows.

For all four functions there is a unique y-value corresponding to each x-value, found by putting $y = f(x)$. However, the reverse is only true for functions **1** (linear) and **3** (cubic). The results of solving the equations are summarised as follows.

Function	$f(x) = 0$	$f(x) = 5$	$f(x) = -5$
1 $f(x) = 2x - 3$	1.5	4	–1
2 $f(x) = x^2 + 2x - 3$	–3, 1	–4, 2	–
3 $f(x) = x^3$	0	1.71	–1.71
4 $f(x) = \sin x$	..., –180°, 0°, –180°, ...	–	–

For **2**: when $y > -4$ there are two x-values corresponding to each y-value,
when $y = -4$, $x = -1$ (vertex of parabola),
when $y < -4$ there are no x-values corresponding to each y-value.

For **4**: when $-1 \le y \le 1$ there is an infinite number of x-values corresponding to each y-value,
when $y < -1$ or $y > 1$ there are no x-values corresponding to each y-value.

We can now turn these observations into formal ideas by looking closely at the four functions above.

Domain and range

The set of x-values for which each function is defined is the **domain**. In all four cases the domain could be the set of real numbers, $\Re$, but we could restrict the domain to a subset of the real numbers, e.g. $\{x : -4 \le x \le 2\}$ or $\{x : 0, 1, 2, 3, 4\}$.

The set of y-values, corresponding to every x-value in the domain is called the **range**. Assuming the domain is $\Re$, the range in functions **1** and **3** is also $\Re$, but for **2** and **4** the range is a subset of $\Re$:

For **2**: $f(x) = x^2 + 2x - 3$ range : $\{y : y \ge -4\}$
For **4**: $f(x) = \sin x$ range : $\{y : -1 \le y \le 1\}$

Functions **1** and **3** are examples of **one-to-one mappings**, since for each y-value in the range there is a *unique* corresponding x-value in the domain.

Functions **2** and **4** are examples of **many-to-one mappings**, since each y-value in the range does not have a unique corresponding x-value in the domain (e.g. in **2** $y = 5 \Rightarrow x = -4$ or 2; in **4** $y = 0.5 \Rightarrow x = \ldots, -330°, -210°, 30°, 150°, 390°, \ldots$).

We can illustrate the idea of a mapping by a diagram, using alternative function notation. Suppose that for the domain $\{x : -2, -1, 0, 1, 2\}$ function f is defined by $f(x) = x^2 - 3$ which is equivalent to $f : x \to x^2 - 3$.

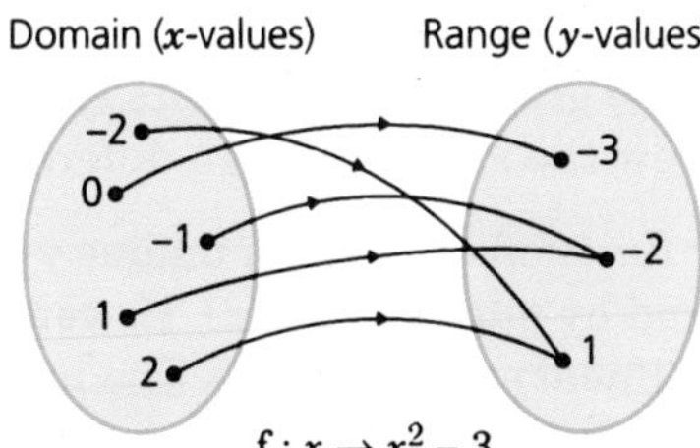

Each number in the domain is an **object** which has a corresponding **image** in the range, e.g. the image of the object 2 is 1. Note that this function is many-to-one since some y-values correspond to more than one x-value.

Example 13.1

Each of the following functions has domain $\{x : 1, 2, 3, 4\}$. For each one draw a mapping diagram, state the range and whether the mapping is one-to-one or many-to-one.

a) $f : x \to 3x - 7$ ***b)*** $f : x \to 5 - x$
c) $f : x \to (x - 2)^2$ ***d)*** $f : x \to \frac{12}{x}$
e) $f : x \to |2x - 6|$

Solution **a)** Range $\{y : -4, -1, 2, 5\}$ one-to-one

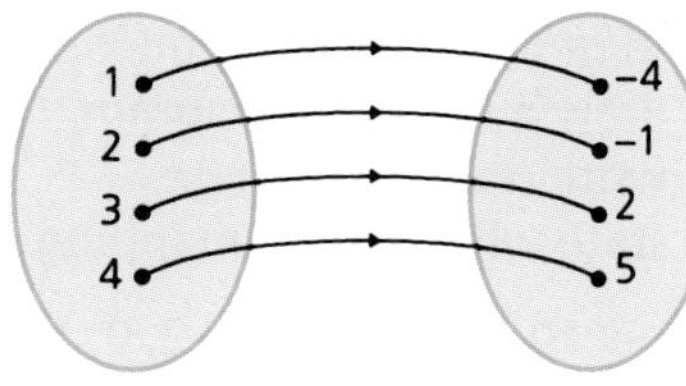

b) Range $\{y : 1, 2, 3, 4\}$ one-to-one

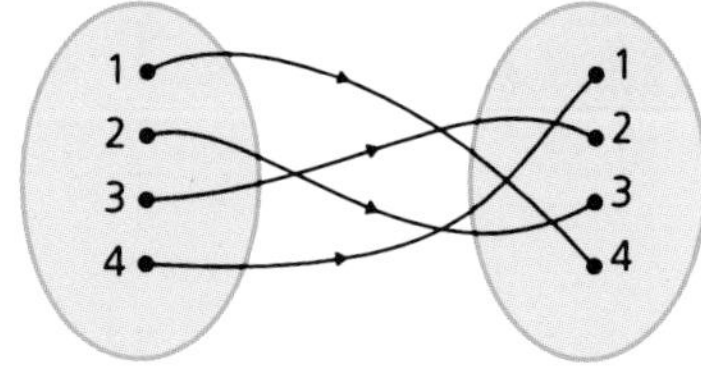

c) Range $\{y : 0, 1, 4\}$ many-to-one

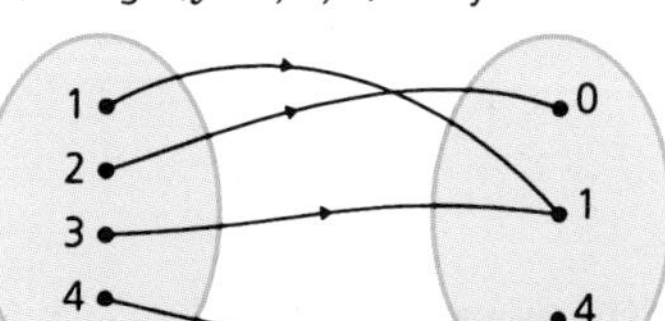

d) Range $\{y : 3, 4, 6, 12\}$ one-to-one

e) Range $\{y : 0, 2, 4\}$ many-to-one

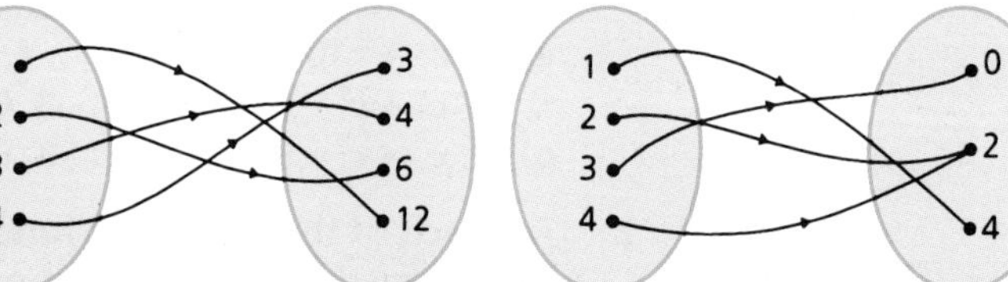

Example 13.2

For each of the following functions, restrict the domain to $\{x : -3 \le x \le 3\}$, sketch the graph of the function, give the range and state which type of mapping it is.

a) $f(x) = 5 - 2x$ ***b)*** $f(x) = 2x - x^2$ ***c)*** $f(x) = |3(x - 1)|$ ***d)*** $f(x) = x(x^2 - 4)$

Solution

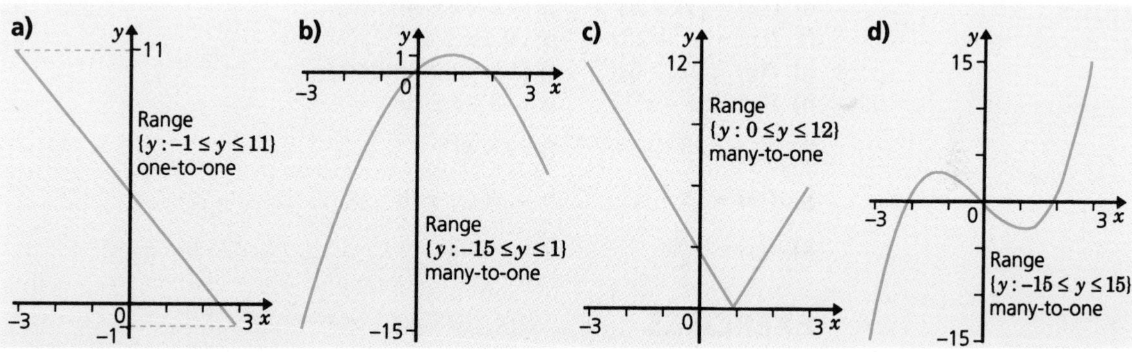

Example 13.3

For each of the following functions:
i) give the largest possible domain, *ii) give the corresponding range,*
iii) sketch a suitable portion of the graph.

a) $f : x \to 9 - x^2$ ***b)*** $f : x \to 5\sqrt{x}$ ***c)*** $f : x \to \frac{4}{x}$

Solution

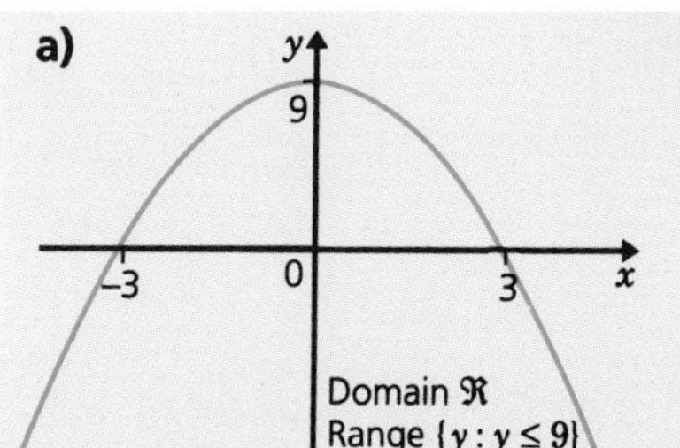

b) y

Domain $\{x : x \ge 0\}$
Range $\{y : y \ge 0\}$

0 x

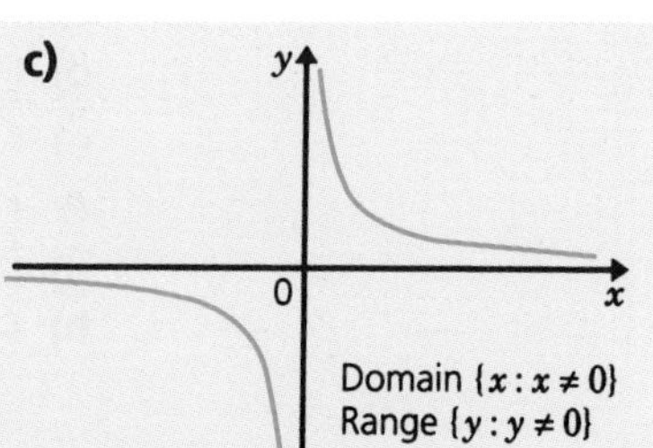

EXERCISES

13.1 A

1 For each of the following functions, with given domain:

i) draw a mapping diagram,
ii) write down the range,
iii) state whether the mapping is one-to-one or many-to-one.

a) $f: x \to 10 - 3x$ $\{x : -2, 0, 1, 4\}$
b) $f: x \to 2x^2$ $\{x : -3, -1, 1, 3\}$
c) $f: x \to 2x^2 - 5$ $\{x : -3, -1, 1, 3\}$
d) $f: x \to \frac{10}{x}$ $\{x : -5, -2, 2, 5\}$
e) $f: x \to |10 - 2x|$ $\{x : 2, 4, 6, 8\}$
f) $f: x \to x^3 - 9x$ $\{x : -3, 0, 3\}$
g) $f: x \to x^3 + 9x$ $\{x : -3, 0, 3\}$

2 For each of the following functions, with given domain:

i) sketch the graph of the function,
ii) write down the range,
iii) state which type of mapping it is.

a) $f(x) = \frac{1}{2}(x + 8)$ $\{x : -4 \le x \le 10\}$
b) $f(x) = 2x^2$ $\{x : -5 \le x \le 5\}$
c) $f(x) = 3x^2$ $\{x : 0 \le x \le 6\}$
d) $f(x) = x^2 - 3x$ $\{x : 0 \le x \le 6\}$
e) $f(x) = |2x - 5|$ $\{x : -2 \le x \le 5\}$
f) $f(x) = 5x^2 - 20x$ $\{x : 0 \le x \le 4\}$
g) $f(x) = 20x - 5x^2$ $\Re$ (all real numbers)
h) $f(x) = 9x - x^3$ $\{x : -3 \le x \le 3\}$
i) $f(x) = \frac{12}{x}$ $\{x : x > 0\}$
j) $f(x) = \frac{5}{x^2}$ $\{x \in \Re : x \ne 0\}$
k) $f(x) = -\frac{5}{x^2}$ $\{x : x > 0\}$

EXERCISES

13.1 B

1 For each of the following functions, with given domain:

i) draw a mapping diagram,
ii) write down the range,
iii) state whether the mapping is one-to-one or many-to-one.

a) $f: x \to 2x + 4$ $\{x : -4, -2, 0, 1, 2\}$
b) $f: x \to 3(5 - 2x)$ $\{x : 0, 2, 4, 6\}$
c) $f: x \to 6x^2$ $\{x : -2, -1, 0, 1, 2\}$
d) $f: x \to 6x^2 - 5$ $\{x : -2, -1, 0, 1, 2\}$
e) $f: x \to \frac{4}{(x + 2)}$ $\{x : -1, 0, 1, 2\}$
f) $f: x \to |6 - 3x|$ $\{x : 0, 2, 4\}$
g) $f: x \to 2x^3$ $\{x : -2, 0, 2\}$
h) $f: x \to x^3 - 4x$ $\{x : -1, 0, 1\}$

2 For each of the following functions, with given domain:
  i) sketch the graph of the function,
  ii) write down the range,
  iii) state which type of mapping it is.

a) $f(x) = 3 - 2x$ $\{x : -2 \le x \le 2\}$

b) $f(x) = \frac{1}{4}(4x + 6)$ $\{x : -2 \le x \le 5\}$

c) $f(x) = \frac{1}{2}x^2$ $\{x : -5 \le x \le 5\}$

d) $f(x) = x^2 + 2x$ $\{x : -5 \le x \le 3\}$

e) $f(x) = 2x^2 - x$ $\{x : -4 \le x \le 5\}$

f) $f(x) = x - 2x^2$ $\{x : -4 \le x \le 5\}$

g) $f(x) = |6 - 3x|$ $\{x : -2 \le x \le 6\}$

h) $f(x) = \frac{3}{x^2}$ $\{x \in \Re : x \ne 0\}$

i) $f(x) = -\frac{4}{x^2 + 1}$ $\{x : -3 \le x \le 3\}$

j) $f(x) = 5x - 2x^3$ $\{x : -3 \le x \le 3\}$

COMPOSITE FUNCTIONS

So far we have used f to represent a function, e.g. $f : x \to 3x - 5$, $f(x) = x^2$, $y = f(x)$, etc. When we are dealing with more than one function we shall also use g and h to represent functions, e.g. $g : x \to 2x$, $h(x) = 2x^3$, etc.

Functions we have met so far may be combined to produce **composite** functions or they may themselves be expressed as a composite of simpler functions.

Exploration 13.2

Using function machines

Three functions f, g and h are defined by:

$f(x) = x - 5$ $g(x) = 3x$ $h(x) = x^2$

all with the domain $\Re$, the set of real numbers.

- Use function machines to illustrate:
  $f(g(x))$ $g(f(x))$
  $g(h(x))$ $h(g(x))$
  $h(f(x))$ $f(h(x))$
- In each case find the range of the composite function.

The meaning of a composite function

To carry out the composite function $f(g(x))$, we first carry out function g on x, then we carry out function f on the result, e.g. 'multiply by 3' then 'subtract 5'.

x → **×3** → $g(x)$ → **−5** → $f(g(x)) = 3x - 5$

The composite function $f(g(x))$ is usually abbreviated to $fg(x)$ and in this case is given by $fg(x) = 3x - 5$.

To carry out the composite function $g(f(x))$, we first carry out function f on x, then we carry out function g on the result, e.g. 'subtract 5' then 'multiply by 3'.

$$x \longrightarrow \boxed{-5} \xrightarrow{f(x)} \boxed{\times 3} \longrightarrow gf(x) = 3(x-5)$$

Notice that the **order** in which the functions are applied is important, since the two composites fg and gf give different functions.

You should be able to check that the other composite results simplify to:

$gh(x) = 3x^2\ (= y)$	range $\{y : y \geq 0\}$
$hg(x) = (3x)^2$	range $\{y : y \geq 0\}$
$hf(x) = (x-5)^2$	range $\{y : y \geq 0\}$
$fh(x) = x^2 - 5$	range $\{y : y \geq -5\}$

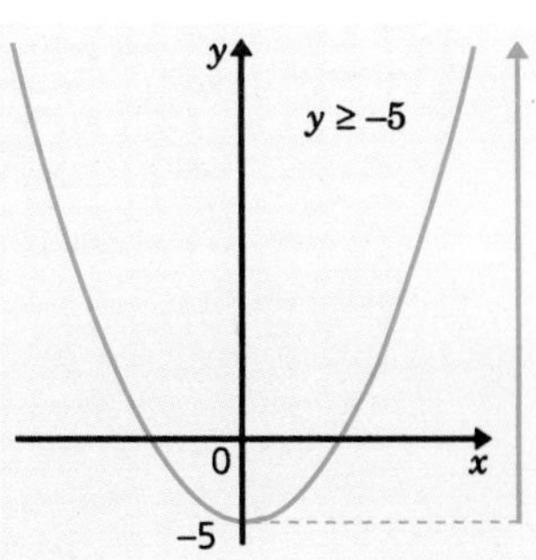

A sketch of the graph of the composite function should help to find the range, e.g. this is the graph of $y = fh(x) = x^2 - 5$.

Example 13.4

Functions f, g, h *are defined by:*

$f : x \rightarrow 8 - x$

$g : x \rightarrow \dfrac{3}{x} \quad (x \neq 0)$

$h : x \rightarrow 2x + 3$

a) *Find expressions for* $gf(x)$, $hg(x)$, $fh(x)$, $hh(x)$.

b) *Show that* $ff(x) = x$ *and* $gg(x) = x$.

c) *Solve the equation* $hgf(x) = x$.

Solution

a) $gf(x) = g(f(x)) = g(8 - x) = \dfrac{3}{(8-x)}$

$hg(x) = h(g(x)) = h\left(\dfrac{3}{x}\right) = 2\dfrac{3}{x} + 3 = \left(\dfrac{6}{x}\right) + 3$

$fh(x) = f(h(x)) = f(2x + 3) = 8 - (2x + 3) = 5 - 2x$

$hh(x) = h(h(x)) = h(2x + 3) = 2(2x + 3) + 3 = 4x + 9$

b) $ff(x) = f(f(x)) = f(8 - x) = 8 - (8 - x) = x$

$gg(x) = g(g(x) = g\left(\dfrac{3}{x}\right) = \dfrac{3}{3/x} = 3 \times \dfrac{x}{3} = x$

c) $hgf(x) = h(g(f(x)))$

$$x \longrightarrow \boxed{\text{Subtract from 8}} \xrightarrow{f(x)} \boxed{\text{Divide into 3}} \xrightarrow{g(f(x))} \boxed{\times 2, +3} \longrightarrow hgf(x)$$

$\Rightarrow hgf(x) = h(g(8 - x)) = h\left(\dfrac{3}{8-x}\right) = 2\dfrac{3}{(8-x)} + 3$

Thus when $\mathrm{hgf}(x) = x$: $2\left(\frac{3}{8-x}\right) + 3 = x$

$\Rightarrow \frac{6}{(8-x)} = x - 3$

$\Rightarrow 6 = (8-x)(x-3)$

$\Rightarrow x^2 - 11x + 30 = 0$

$\Rightarrow (x-5)(x-6) = 0$

$\Rightarrow x = 5$ *or* $x = 6$

EXERCISES

13.2 A

1 Three functions f, g and h are defined by:

$f : x \to 2x \qquad g : x \to 5 - x \qquad h : x \to x^2$

a) Find expressions for **i)** $\mathrm{fg}(x)$, **ii)** $\mathrm{gh}(x)$, **iii)** $\mathrm{hgf}(x)$.

b) Evaluate: **i)** $\mathrm{fg}(-2)$ **ii)** $\mathrm{gh}(-2)$ **iii)** $\mathrm{hgf}(-2)$.

c) Solve the equation $\mathrm{hgf}(x) = 9$.

2 Three functions f, g and h are defined by:

$f(x) = 2x - 1 \qquad g(x) = 3 - 2x \qquad h(x) = \frac{1}{4}(5 - x)$

a) Find a formula for $k(x)$ where $k(x) = f(g(x))$.

b) Find a formula for $h(k(x))$.

c) What is the connection between functions h and k?

(Scottish Higher Paper 1, 1993)

3 Functions f and g are defined by:

$f(x) = x^3 - 2x^2 - 5x + 6 \qquad g(x) = x - 1$

Show that $\mathrm{fg}(x) = x^3 - 5x^2 + 2x + 8$.

(Scottish Higher Paper 2, 1990)

4 Functions f and g are defined by:

$f(x) = \frac{4}{x+3} \qquad g(x) = \frac{4}{x} - 3$

Find both $\mathrm{gf}(x)$ and $\mathrm{fg}(x)$ in their simplest form.

5 Function $f(x)$ is defined by $f(x) = \frac{x}{1+x}$ $(x \neq -1)$.

Find an expression for $\mathrm{ff}(x)$.

6 Three functions f, g and h are defined by:

$f : x \to x - 3 \qquad g : x \to x^2 \qquad h : x \to \frac{12}{x}$

for suitable domains. Find each of the following functions as composites using some of f, g and h.

a) $x^2 - 3$ **b)** $\frac{12}{x-3}$ **c)** $\frac{144}{x^2}$ **d)** $x^2 - 6x + 9$ **e)** $\frac{3(4-x)}{x}$ **f)** x

7 The diagram illustrates three functions f, g and h.

P Q R f g h p q r

The functions f and g are defined by:

$f(x) = 2x + 5 \qquad g(x) = x^2 - 3$

and the function h is defined so that whenever $f(p) = q$ and $g(q) = r$, then $h(p) = r$.

a) If $q = 7$, find the values of p and r.

b) Find a formula for $h(x)$, in terms of x.

(Scottish Higher Paper 1, 1991)

8 Three functions f, g and h are such that:

$$\text{hgf}(x) = \sqrt{25 - x^2}$$

a) State a suitable domain and range so that the graph of $\text{hgf}(x)$ is a semi-circle.

b) Find expressions for functions f, g and h, stating appropriate domains and ranges for each one.

EXERCISES

13.2 B

1 Three functions f, g and h are defined by:

$f : x \to 3x - 4 \qquad g : x \to x^3$

a) Find expressions for $\text{fg}(x)$ and $\text{gf}(x)$.
b) Evaluate: **i)** $\text{fg}(2)$ **ii)** $\text{gf}(2)$.
c) Solve the equation $\text{fg}(x) = 77$.

2 Three functions f, g and h are defined by:

$f(x) = 2x - 5 \qquad g(x) = 3x + 1 \qquad h(x) = 4x$

a) Find expressions for **i)** $\text{fg}(x)$ **ii)** $\text{gh}(x)$ **iii)** $\text{hgf}(x)$.
b) Evaluate **i)** $\text{fg}(-3)$ **ii)** $\text{gh}(-3)$ **iii)** $\text{hgf}(-3)$.
c) Solve the equation $\text{hgf}(x) = 104$.

3 Let $f(x) = 2x + 4$ and $g(x) = \frac{1}{2}x - 2$.

Find: **i)** $\text{fg}(x)$ **ii)** $\text{gf}(x)$.
Comment on your answer.

4 Functions f and g are defined by:

$f(x) = x^2 + 1 \qquad g(x) = 3x^2 + 2x - 4$

Show that $\text{gf}(x) = 3x^4 + 8x^2 + 1$.

5 Let $f(x) = \sqrt{1+x}$ and $g(x) = 3 - x^2$.

Determine $\text{fg}(x)$ and $\text{gf}(x)$.

6 Let $f(x) = 2x + 1$. Find a function $g(x)$ such that $\text{fg}(x) = x^3$.

7 Functions f and g are defined by:

$f(x) = \dfrac{1}{x+3} \qquad g(x) = 5 - \dfrac{4}{x^2}$

Find $\text{fg}(x)$ and $\text{gf}(x)$ in their simplest forms.

8 Let $k(x) = \left(1 - \sqrt{x}\right)^{\frac{2}{3}}$. Find the domain of k. Find three functions $f(x)$, $g(x)$ and $h(x)$ such that $\text{fgh}(x) = k(x)$.

INVERSE FUNCTIONS

For most mathematical operations there is a corresponding operation that has the opposite or reverse effect. For example the opposite of 'add 5' is 'subtract 5'. In this case the operation of 'subtract 5' is the **inverse** operation of 'add 5', and vice-versa.

Exploration 13.3 *Inverses*

For each of the following operations:

- find the inverse operation,
- using a suitable domain, express the operation and its inverse as functions,
- sketch both functions on the same axes.

  a) add three **b)** subtract seven **c)** multiply by five

  d) divide by two **e)** subtract from ten **f)** divide into twenty
- What do all the sketches have in common?

Finding the inverses

The inverse of 'add 3' is 'subtract 3'.
Formally , the function $\mathrm{f}: x \rightarrow x + 3$ has inverse function $\mathrm{f}^{-1}: x \rightarrow x - 3$, both valid for all values of x.

The inverse of 'subtract 7' is 'add 7'.
The inverse of $\mathrm{f}: x \rightarrow x - 7$ is $\mathrm{f}^{-1}: x \rightarrow x + 7$.

The inverse of 'multiply by 5' is 'divide by 5'.
The inverse of $\mathrm{f}: x \rightarrow 5x$ is $\mathrm{f}^{-1}: x \rightarrow \frac{x}{5}$.

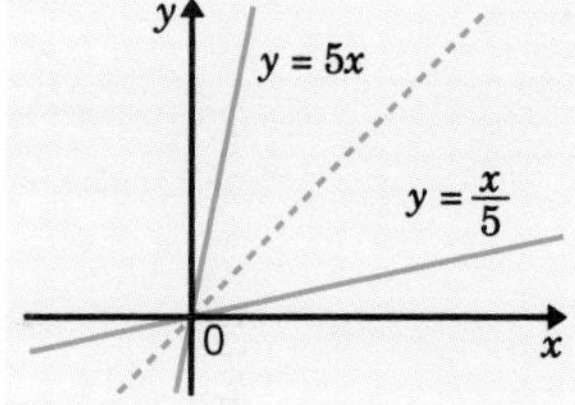

The inverse of 'divide by 2' is 'multiply by 2'.
The inverse of $\mathrm{f}: x \rightarrow \frac{x}{2}$ is $\mathrm{f}^{-1}: x \rightarrow 2x$.

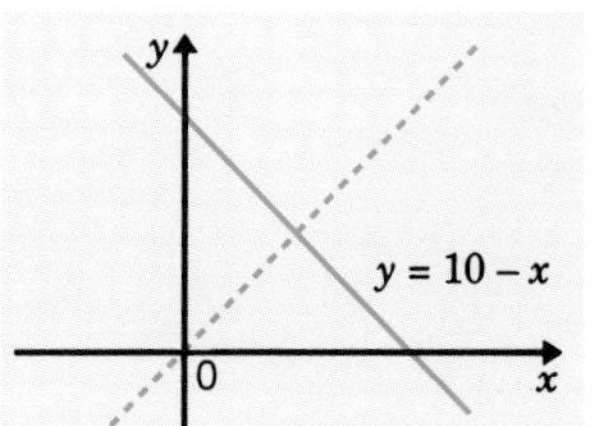

The inverse of 'subtract from 10' is 'subtract from 10'.
The inverse of $\mathrm{f}: x \rightarrow 10 - x$ is $\mathrm{f}^{-1}: x \rightarrow 10 - x$.
This function is **self-inverse**.

The inverse of 'divide into 20' is 'divide into 20'. The inverse of $\mathrm{f}: x \rightarrow \dfrac{20}{x}$ is $\mathrm{f}^{-1}: x \rightarrow \dfrac{20}{x}$.
This is another self inverse function, valid for all values of x except 0.

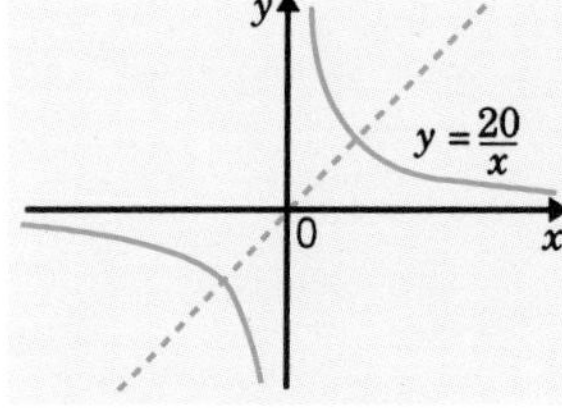

For each sketch the graph of the inverse function $y = \mathrm{f}^{-1}(x)$ is the reflection of $y = \mathrm{f}(x)$ in the line $y = x$. This property is true for any function. When the graph of a function is symmetrical about $y = x$, the function is self-inverse; operations **e)** and **f)** are examples of the two self-inverse functions:

$$\mathrm{f}: x \rightarrow a - x \quad \text{and} \quad \mathrm{f}: x \rightarrow \frac{a}{x}$$

where a is constant.

An important idea is that only functions which are one-to-one mappings have inverse functions. This is brought out in the following example.

Example 13.5

The function $f(x) = x^2$ has domain $\Re$. Show that the inverse relationship is not a function unless the domain of f is restricted.

Solution

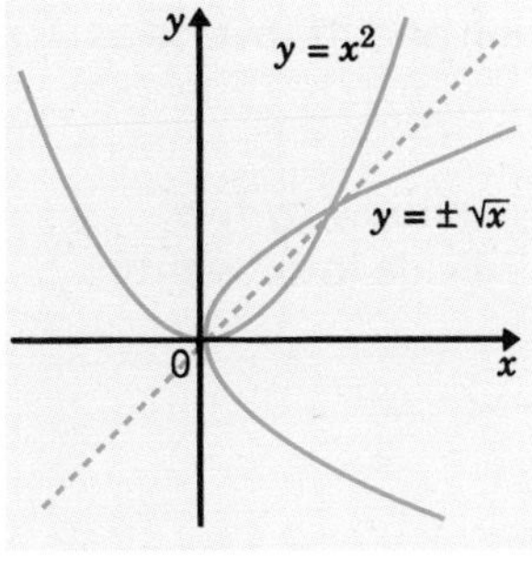

The diagram shows $y = x^2$ and its reflection in the line $y = x$. The inverse relationship may be labelled $y = \pm\sqrt{x}$. Since $f(x) = x^2$ is a many-to-one mapping, the inverse is a one-to-many mapping, which is not a function, e.g. $x = 4 \Rightarrow y = \pm 2$.

By restricting the domain of f to $\{x : x \geq 0\}$ then the reflection of $y = x^2$ is $y = \sqrt{x}$ (the positive square root) which is a function since f is now a one-to-one mapping.

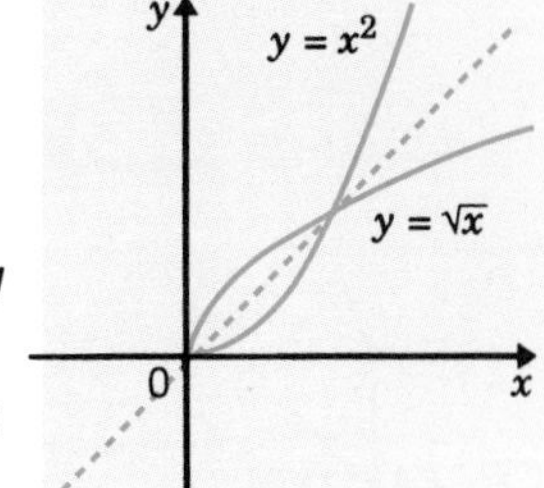

For composite functions, inverses may be found by reversing a function machine diagram or by rearranging a formula. The following examples illustrate when the methods might apply.

Example 13.6

The function $f : x \to 10 - 3x$ is defined for all real numbers. Find the inverse function f^{-1}.

Solution

Method A

Decompose the function f.

$x \longrightarrow$ [$\times 3$] $\xrightarrow{3x}$ [Subtract from 10] $\longrightarrow 10 - 3x = f(x)$

$f^{-1}(x) = \dfrac{10-x}{3} \longleftarrow$ [$\div 3$] $\xleftarrow{10-x}$ [Subtract from 10] $\longleftarrow x$

$\Rightarrow$ *the inverse function is* $f^{-1} : x \to \dfrac{10-x}{3}$

Method B

Let $y = 10 - 3x$ and rearrange to make x the subject.

$$y = 10 - 3x \quad \Rightarrow \quad 3x = 10 - y \quad \Rightarrow \quad x = \frac{10-y}{3}$$

Now reverse the roles of x and y to get $f^{-1} : x \to \dfrac{10-x}{3}$

Example 13.7

By suitably restricting the domain, show that the function f : $x \to x^2 + 2x - 3$ has an inverse function f^{-1}. Sketch graphs to illustrate your answer.

Solution

First complete the square : $x^2 + 2x - 3 = (x + 1)^2 - 4$

Method A

$x \longrightarrow$ [$+1$] $\longrightarrow$ [Square] $\longrightarrow$ [-4] $\longrightarrow (x + 1)^2 - 4 = f(x)$

$f^{-1}(x) = \sqrt{x+4} - 1 \longleftarrow$ [-1] $\longleftarrow$ [Take square root] $\longleftarrow$ [$+4$] $\longleftarrow x$

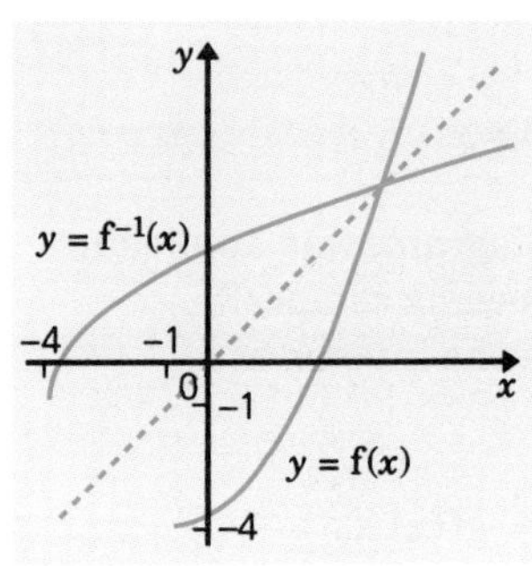

Method B

Let $y = (x+1)^2 - 4$ and rearrange to make y the subject.

$$y = (x+1)^2 - 4$$
$$y + 4 = (x+1)^2$$
$$\sqrt{y+4} = x + 1$$
$$\sqrt{y+4} - 1 = x$$

Now reverse the roles of x and y to get $f^{-1}: x \rightarrow \sqrt{x+4} - 1$.
***Note**: By taking the positive square root, the domain of f must be restricted to $\{x : x \geq -1\}$, which becomes the range of f^{-1}: the range of f is $\{y : y \geq -4\}$, which becomes the domain for f^{-1}. A sketch of the graphs of both functions is shown in the diagram.*

Example 13.8

Function f is defined by $f(x) = \dfrac{ax+1}{x-2}$ for the domain $\{x : x \in \Re, x \neq 2\}$.

***a)** Find the inverse function, f^{-1}.*
***b)** Find a value of a such that the function f is self-inverse, i.e. $f(x) = f^{-1}(x)$.*
***c)** Sketch a graph of $y = f(x)$ for the value of a found in **b)**. What do you notice?*

Solution

***a)** Since x appears more than once in the definition of the function, you need to use Method B, rearrangement of the formula.*

Let $y = \dfrac{ax+1}{x-2}$

$\Rightarrow y(x-2) = ax + 1 \quad \Rightarrow xy - 2y = ax + 1 \quad \Rightarrow xy - ax = 2y + 1$

$\Rightarrow x(y-a) = 2y + 1 \quad \Rightarrow \quad x = \dfrac{2y+1}{y-a}$

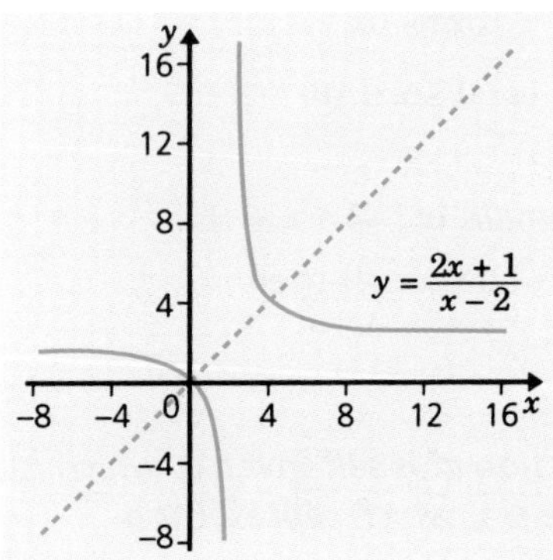

Reversing the roles of x and y : $f^{-1}(x) = \dfrac{2x+1}{x-a}$.

***b)** Since $f(x) = \dfrac{ax+1}{x-2}$ and $f^{-1}(x) = \dfrac{2x+1}{x-a}$, substitution of $a = 2$ will make $f(x) = f^{-1}(x)$.*

$\Rightarrow f(x) = \dfrac{2x+1}{x-2}$ *is a self inverse function.*

***c)** A sketch of the graph of $y = \dfrac{2x+1}{x-2}$ should reveal that it is symmetrical about $y = x$, which gives geometrical support to the idea that the function f is self-inverse.*

Since the action of an inverse function f^{-1} is to reverse the action of the function f, and vice versa, it is generally true that $f^{-1}f(x) = x$ and $ff^{-1}(x) = x$. For example:

let $f(x) = 2x - 3$, then $f^{-1}(x) = \dfrac{x+3}{2}$

$$\Rightarrow \quad f^{-1}f(x) = f^{-1}(2x-3) = \frac{(2x-3)+3}{2} = \frac{2x}{2} = x$$

and $ff^{-1}(x) = f\left(\frac{x+3}{2}\right) = 2\left(\frac{x+3}{2}\right) - 3 = x + 3 - 3 = x$

When Method A, reversing the function flow diagram, was applied in Examples 13.6 and 13.7 the functions f were seen as composite functions. The same principle is used to derive the inverse of a composite function gf.

$x \longrightarrow$ **f** $\xrightarrow{f(x)}$ **g** $\longrightarrow gf(x)$

$f^{-1}g^{-1}(x) \longleftarrow$ **f^{-1}** $\xleftarrow{g^{-1}(x)}$ **g^{-1}** $\longleftarrow x$

provided f^{-1} and g^{-1} exist for suitable domains:

$$(gf)^{-1} \equiv f^{-1}g^{-1}$$

Example 13.9

Functions f and g are defined by:

$f(x) = 16 - x^2 \quad x \in \Re \qquad\qquad g(x) = \sqrt{x} \quad \{x : x \geq 0\}$

a) *Find* $f^{-1}(x)$ *and* $g^{-1}(x)$.

b) *For a suitable domain, find* $fg(x)$ *and* $(gf)^{-1}(x)$.

Solution

a) *First we find* $f^{-1}(x)$.

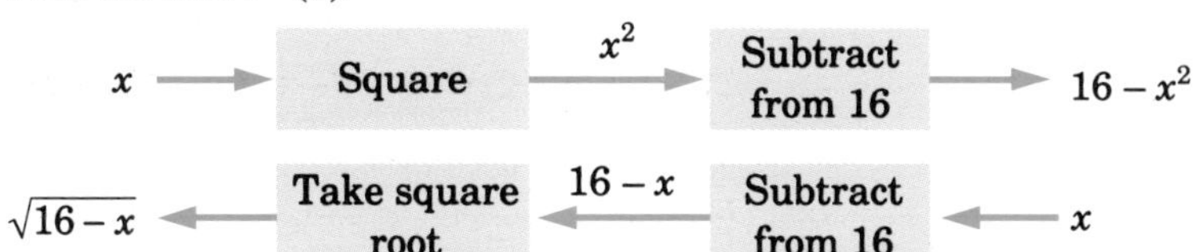

Hence $f^{-1}(x) = \sqrt{16-x}$ *with domain* $\{x : 0 \leq x \leq 16\}$

$g^{-1}(x) = x^2$ *with domain* $\{x : x \geq 0\}$.

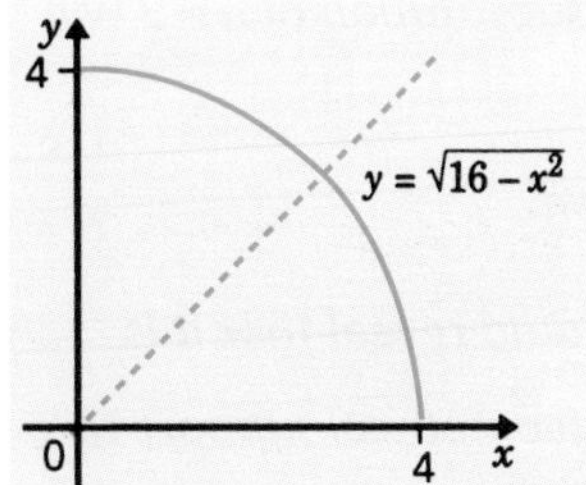

b) $gf(x) = g(16 - x^2) = \sqrt{16 - x^2}$ *with domain* $\{x : -4 \leq x \leq 4\}$.

Provided the domain is restricted to $\{x : 0 \leq x \leq 4\}$ *then:*

$(gf)^{-1}(x) = f^{-1}(g^{-1}(x)) = f^{-1}(x^2) = \sqrt{16 - x^2}$

also with domain $\{x : 0 \leq x \leq 4\}$.

Note*: For the restricted domain the function gf is self-inverse, which is evident from the graph of* $y = gf(x)$, $0 \leq x \leq 4$, *which represents a quarter-circle, centre (0, 0), radius 4 units.*

EXERCISES

13.3 A

In questions **1–10** find the inverse function f^{-1}. Sketch graphs of $y = f(x)$ and $y = f^{-1}(x)$ on the same axes.

1 $f(x) = x - 4, x \in \Re$

2 $f(x) = 3x, x \in \Re$

3 $f(x) = 3x - 4, x \in \Re$

4 $f(x) = \frac{x^2}{5}, x \in \Re, x \geq 0$

5 $f(x) = x^3, x \in \Re$

6 $f(x) = 7 - 2x, x \in \Re$

7 $f(x) = \frac{12}{x+1}, x \in \Re, x \neq -1$

8 $f(x) = \sqrt[3]{x-3}, x \in \Re$

9 $f(x) = 2x^2 + 5, x \in \Re, x \geq 0$

10 $f(x) = (x+2)^2, x \in \Re, x \geq 0$

11 For each of the following functions, the domain is $\{x : x \geq k\}$.

i) Find the least value of k such that f is a one-to-one mapping.

ii) For the value of k found in **i)** find the range of f.

iii) Find the inverse function f^{-1}, assuming the value of k found in **i)**.

iv) Sketch a graph showing both $y = f(x)$ and $y = f^{-1}(x)$.

a) $f : x \to x^2 + 2$

b) $f : x \to (x-3)^2$

c) $f : x \to 5 - (x+1)^2$

d) $f : x \to x^2 - 4x + 3$

e) $f : x \to 12 - 6x - x^2$

12 For each of the following functions, the domain is $\{x : x \neq k\}$.

i) State the value of k.

ii) Find the inverse function f^{-1}.

iii) Write down the domain of f^{-1} and hence the range of f.

a) $f(x) = \dfrac{20}{5-x}$

b) $f(x) = \dfrac{5+x}{x-2}$

c) $f(x) = \dfrac{3x-2}{2x+5}$

13 The function f is defined by $f(x) = \dfrac{ax+b}{bx-a}$, where a and b are constants.

a) Simplify $f(x)$ when **i)** $a = 1, b = 0$, **ii)** $a = 0, b = 1$.
In both cases sketch a graph of $y = f(x)$.

b) Which value of x must be excluded from the domain when $b \neq 0$?

c) Find $f^{-1}(x)$ when $a = -1, b = 4$.

d) Show that the function f is self-inverse for all values of a and b.

14 The function f, with domain $x \geq 0$, is defined by $f(x) = x^2 - 3$.

a) State the range of f.

b) Explain how you know the inverse function f^{-1} exists. Find $f^{-1}(x)$ and state the domain and range of f^{-1}.

c) Sketch, on the same axes, the graphs of $y = f(x)$, $y = f^{-1}(x)$ and $y = x$.

d) Show that $f(x) = f^{-1}(x) \Rightarrow x^2 - x - 3 = 0$.

e) Deduce the value of x for which $f(x) = f^{-1}(x)$.

EXERCISES

13.3 B

In questions **1–10** find the inverse function f^{-1}. Sketch graphs of $y = f(x)$ and $y = f^{-1}(x)$ on the same axes.

1 $f(x) = x - 3, x \in \Re$

2 $f(x) = 6x, x \in \Re$

3 $f(x) = 2x + 3, x \in \Re$

4 $f(x) = x^4, x \in \Re, x \geq 0$

5 $f(x) = \dfrac{x^3}{3}, x \in \Re$

6 $f(x) = \dfrac{x^2-5}{2}, x \in \Re, x \geq -2.5$

7 $f(x) = \dfrac{\sqrt{x+6}}{2}, x \in \Re, x \geq 0$

8 $f(x) = \dfrac{4}{x+2}, x \in \Re, x \geq -2$

9 $f(x) = 3x^2 - 1, x \in \Re, x \geq 0$

10 $f(x) = \dfrac{3}{2+x^2}, x \in \Re, 0 \leq x \leq \sqrt{2}$

11 For each of the following functions, the domain is $\{x : x \geq k\}$.

i) Find the least value of k such that f is a one-to-one mapping.

ii) For the value of k found in **i)** find the range of f.

iii) Find the inverse function f^{-1}, assuming the value of k found in **i)**.

iv) Sketch graphs of $y = f(x)$ and $y = f^{-1}(x)$ on the same axes.

a) $f : x \to 3 - x^2$

b) $f : x \to (x + 2)^2 - 5$

c) $f : x \to x^2 - 5x + 6$

d) $f : x \to 7 - (4 + x)^2$

12 For each of the following functions, the domain is $\{x : x \neq k\}$.

i) State the value of k.

ii) Find the inverse function f^{-1}.

iii) Write down the domain of f^{-1} and hence the range of f.

a) $f(x) = \dfrac{2}{x+5}$ **b)** $f(x) = \dfrac{2x}{5-x}$ **c)** $f(x) = \dfrac{2x+3}{x-4}$

13 A function $f : x \to px + q$ is used to convert temperatures in degrees Celsius to temperatures in degrees Fahrenheit.

a) State the values of p and q.

b) Find the inverse function f^{-1} and interpret what it does.

c) Sketch graphs of $y = f(x)$ and $y = f^{-1}(x)$ on the same axes.

d) Solve the equation $f(x) = f^{-1}(x)$. Interpret your solution.

14 The function g with domain $x \geq 1$ is defined by $g(x) = x^2 - 2x - 5$.

a) State the range of g.

b) Explain how you know the inverse function g^{-1} exists. Find $g^{-1}(x)$ and state the domain and range of g^{-1}.

c) Sketch, on the same axes, the graphs of $y = g(x)$ and $y = g^{-1}(x)$.

d) Find the value of x for which $g(x) = g^{-1}(x)$.

TRANSFORMATION OF GRAPHS

In Chapter 4, *Quadratic functions*, and Chapter 8, *Trigonometry 1*, we explored the effects on graphs of translations, stretches (scaling) and reflections. We now formalise these transformations for graphs of any functions.

CALCULATOR ACTIVITY

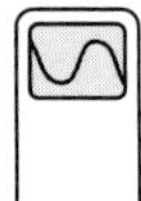

Graphs of transformations

You will need a graphics calculator or graphics package.

Let $f(x) = x^2 - 4x$.

- Using suitable scales plot the graphs of:

  1 $y = f(x)$ and $y = f(x) + a = x^2 - 4x + a$

  2 $y = f(x)$ and $y = f(x + a) = (x + a)^2 - 4(x + a)$

  3 $y = f(x)$ and $y = af(x) = a(x^2 - 4x)$

  4 $y = f(x)$ and $y = f(ax) = (ax)^2 - 4ax$

  for various values of a (e.g. 5, 2, 0.5, –1, –3).

- In each case discuss the effect of a on how the graph was transformed.
- What special effect did taking $a = -1$ have in **3** and **4**?

Geometrical transformations

From your exploration of various transformations of $y = x^2 - 4x$ you may have discovered up to six geometrical transformations of the graph. The conclusions which you might have drawn are summarised below.

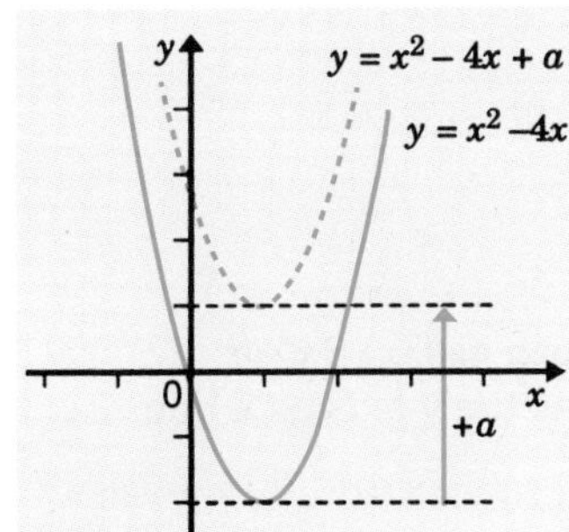

1 $y = \mathrm{f}(x) + a$ translation of $y = \mathrm{f}(x)$ through a units parallel to the y-axis

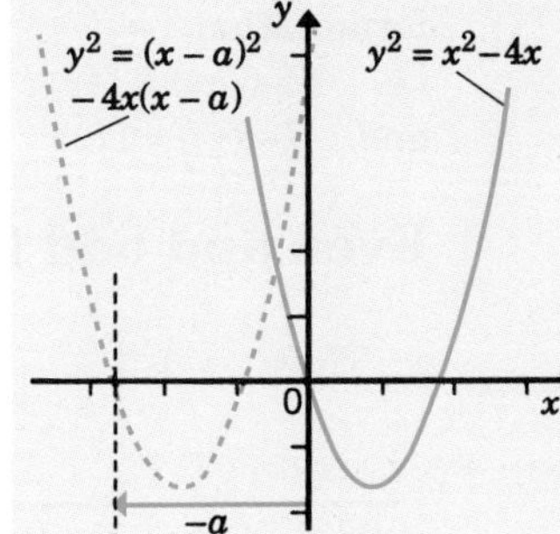

2 $y = \mathrm{f}(x + a)$ translation of $y = \mathrm{f}(x)$ through $-a$ units parallel to the x-axis

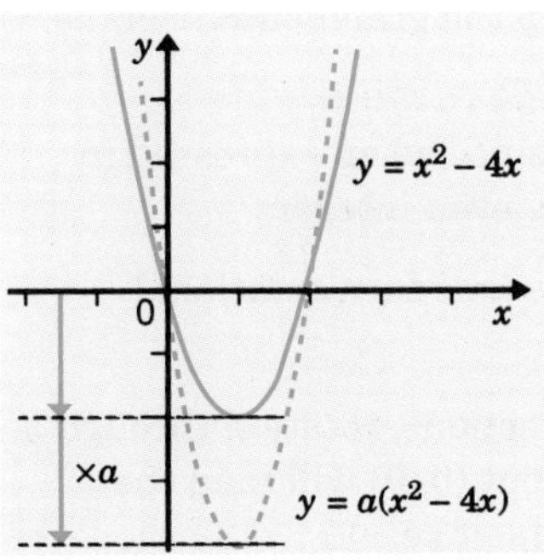

3 $y = a\mathrm{f}(x)$ stretch (scale) of $y = \mathrm{f}(x)$, factor a, parallel to the y-axis

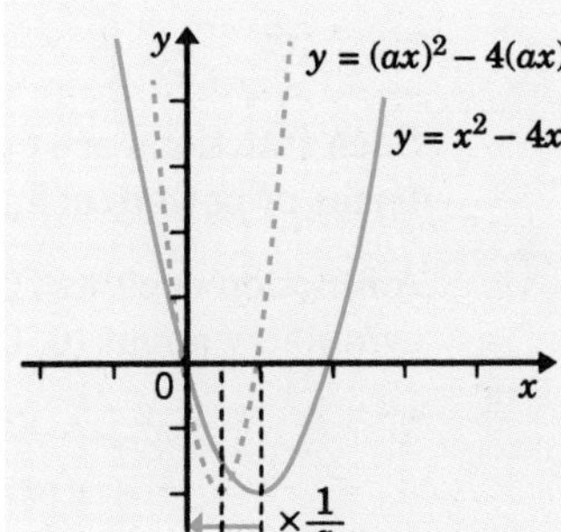

4 $y = \mathrm{f}(ax)$ stretch (scale) of $y = \mathrm{f}(x)$, factor $\frac{1}{a}$, parallel to the x-axis

The special cases of taking $a = -1$ in **3** and **4** leads to two more geometrical transformations that are important:

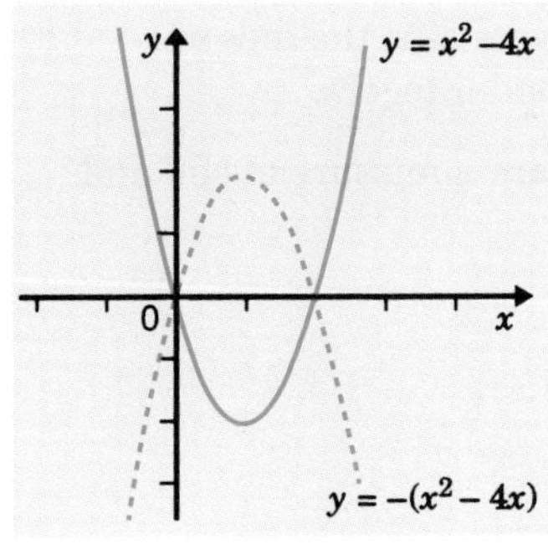

5 $y = -\mathrm{f}(x)$ reflection of $y = \mathrm{f}(x)$ in the x-axis

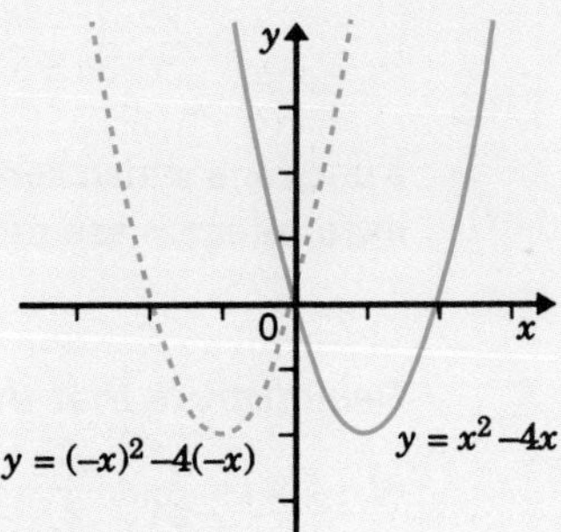

6 $y = \mathrm{f}(-x)$ reflection of $y = \mathrm{f}(x)$ in the y-axis

Combining transformations

The transformations we have met so far may be combined as before to produce graphs of composite functions, for example:

$y = b\mathrm{f}(x + a)$ represents a translation through $-a$ units parallel to the x-axis *followed by* a stretch, factor b, parallel to the y-axis

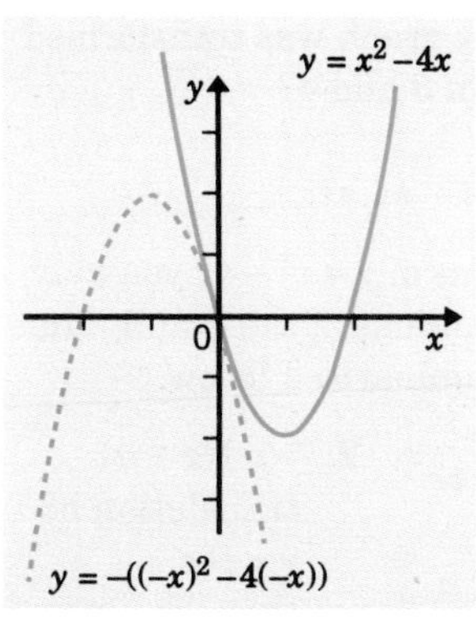

$y = -f(-x)$ represents a reflection in the y-axis *followed by* a reflection in the x-axis; for $f(x) = x^2 - 4x$ this produces a graph as shown.

From the diagram we see that the combined transformation may be described geometrically as a rotation through a half-turn (180°), centre (0, 0).

Note that the transformations *reflect in the x-axis* and *reflect in the y-axis* are both self-inverse, which is evident from the algebraic formulation:

i.e. $f(-(-x)) = f(x)$

and $-(-f(x)) = f(x)$

Even and odd functions

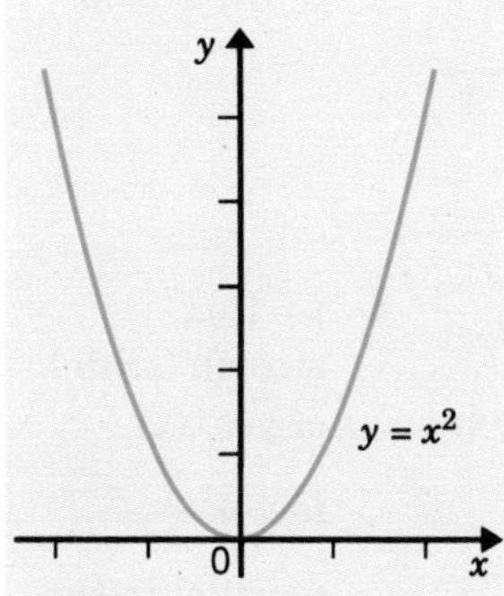

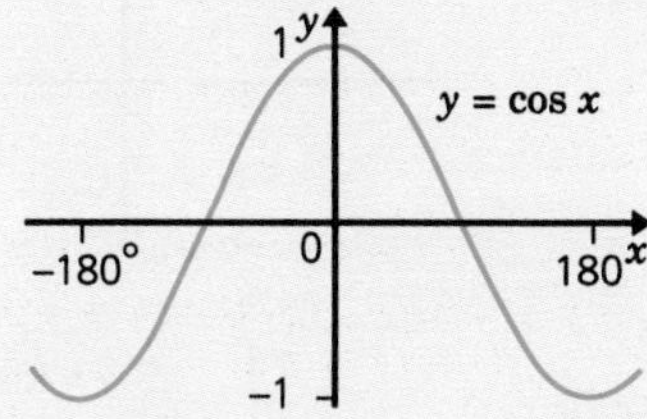

A function such as $f(x) = x^2$ or $f(x) = \cos x$ has the y-axis as a line of symmetry.

For these functions, reflection in the y-axis will map the graph onto itself i.e.

$$f(-x) = f(x)$$

Such functions are called **even functions**, since when expressed in terms of powers of x, the index or indices are even integers.

A function such as $f(x) = x^3$ or $f(x) = \sin x$ has half-turn rotational symmetry about (0, 0).

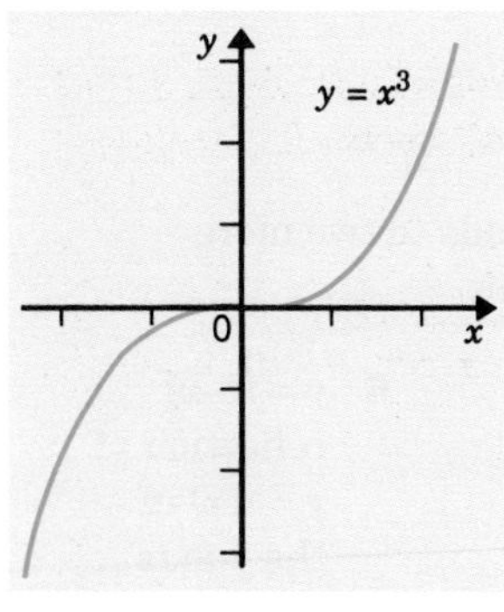

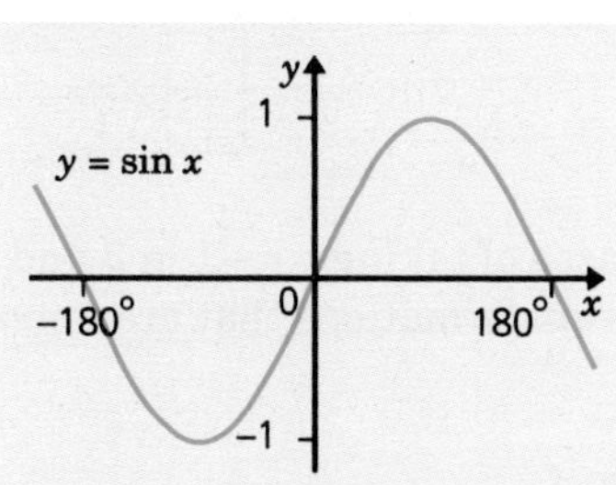

For these functions, rotation through a half-turn about (0, 0) will map the graph onto itself i.e.

$$-f(-x) = f(x)$$

Such functions are called **odd functions**, since when expressed in terms of powers of x, the index or indices are odd integers.

Functions which contain powers of x which are a mixture of odd and even integers are neither odd nor even.

Example 13.10

Demonstrate that the functions:

a) $f(x) = \frac{x^4}{24} - \frac{x^2}{2} + 1$

b) $g(x) = \frac{x^5}{120} - \frac{x^3}{6} + x$

are even and odd respectively. Confirm your results on sketch graphs.

Solution

a) $f(-x) = \frac{(-x)^4}{24} - \frac{(-x)^2}{2} + 1 = \frac{x^4}{24} - \frac{x^2}{2} + 1 = f(x)$

Hence $f(x)$ *is even.*

b) $g(-x) = \frac{(-x)^5}{120} - \frac{(-x)^3}{6} + (-x) = -\frac{x^5}{120} + \frac{x^3}{6} - x$

$$\Rightarrow -g(-x) = -\left(\frac{(-x)^5}{120} - \frac{(-x)^3}{6} - x\right) = \frac{x^5}{120} - \frac{x^3}{6} + x = g(x)$$

Hence $g(x)$ *is odd.*

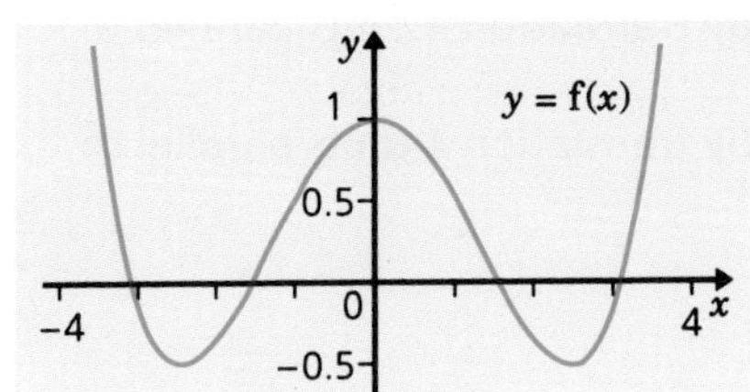

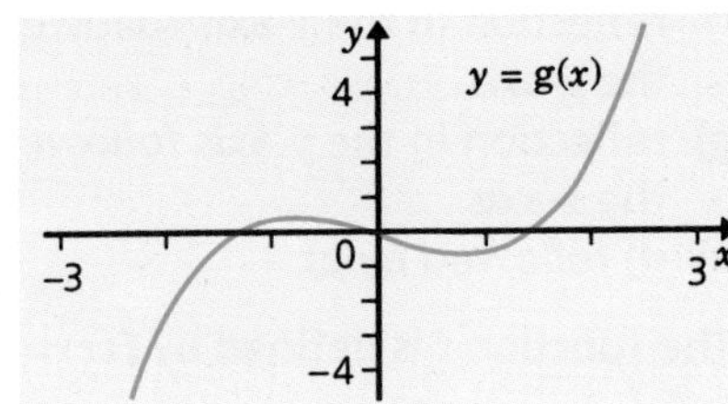

The symmetries of $f(x)$ *and* $g(x)$ *are evident from their graphs.*

Example 13.11

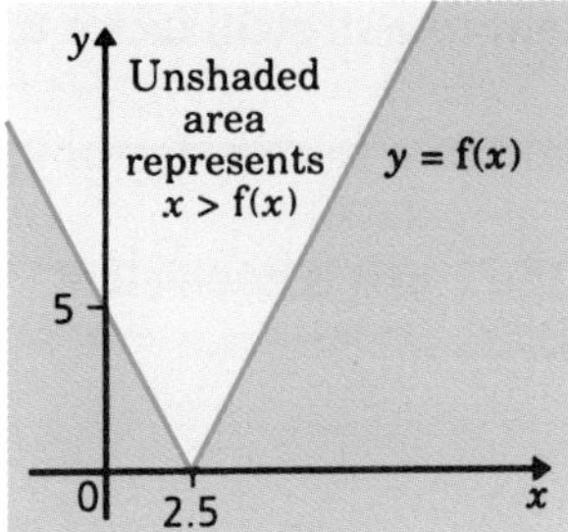

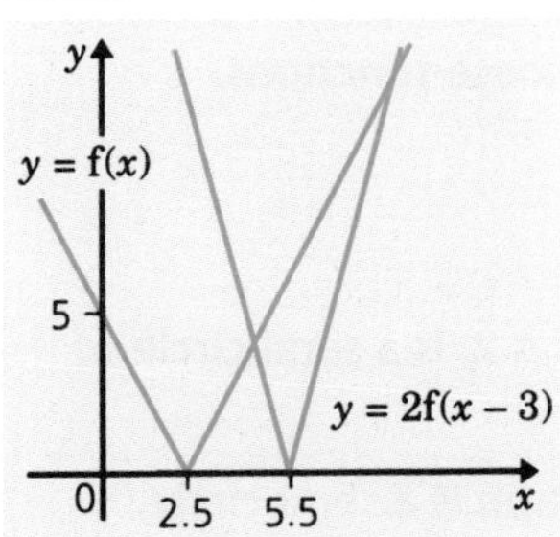

The function f *is defined by* $f : x \to |2x - 5|$.

a) *Sketch a graph of* $y = f(x)$ *and use it to solve the inequality* $f(x) > x$.
b) *Sketch graphs of and simplify:* ***i)*** $y = 2f(x - 3)$ ***ii)*** $y = -f(\frac{1}{2}x)$.

Solution

a) *The graph of* $y = f(x)$ *and* $y = x$ *intersect where:*

$2x - 5 = x \Rightarrow x = 5$

and $-(2x - 5) = x \Rightarrow 3x = 5 \Rightarrow x = 1\frac{2}{3}$

So the points of intersection are $(1\frac{2}{3}, 1\frac{2}{3})$ *and (5, 5).*

From the graph $f(x) > x$ *when* $x < 1\frac{2}{3}$ *or* $x > 5$.

b) ***i)*** *Geometrically* $2f(x - 3)$ *represents a translation through 3 units parallel to the* x*-axis* followed by *a stretch, factor 2, parallel to the* y*-axis, which gives the equation:*

$y = 2f(x - 3) = 2|2(x - 3) - 5| = 2|2x - 11| = |4x - 22|$

ii) *Geometrically* $f(\frac{1}{2}x)$ *represents a stretch, factor 2, parallel to the* x*-axis* followed by *a reflection in the* x*-axis, which gives the equation:*

$y = -f(\frac{1}{2}x) = -|2 \times \frac{1}{2}x - 5| = -|x - 5|$

EXERCISES

13.4A

1 The function f is defined by $f : x \to x^3$. For each of the following transformed functions:

i) describe the transformation geometrically,
ii) sketch graphs of $y = f(x)$ and its image under the transformation.

a) $y = f(x + 3)$ **b)** $y = 2f(x)$ **c)** $y = f(\frac{1}{2}x)$ **d)** $y = \frac{1}{2}f(x) - 4$
e) $y = -f(x - 2)$ **f)** $y = 2f(\frac{1}{2}x)$ **g)** $y = f(5 - x)$ **h)** $y = 3f(2x - 1)$

2 The function g is defined by $g(x) = \frac{12}{x}, x \neq 0$. For each of the following transformations:

i) give the transformed function,
ii) sketch graphs of $y = g(x)$ and its image under the transformation.

a) stretch, factor $\frac{1}{3}$, parallel to the y-axis
b) stretch, factor $\frac{1}{3}$, parallel to the x-axis
c) translation –3 units parallel to the y-axis
d) translation –3 units parallel to the x-axis
e) stretch, factor 2, parallel to the x-axis *followed by* reflection in the x-axis
f) reflection in the y-axis *followed by* translation 4 units parallel to the y-axis
g) reflection in the y-axis *followed by* translation 4 units parallel to the x-axis
h) d) *followed by* e)

3 The function f is defined by $f(x) = 2^x$.

a) Write down the transformed function following:
 i) translation –1 unit parallel to the x-axis,
 ii) stretch, factor 2, parallel to the y-axis.
b) Show that the transformations in a) i) and ii) are equivalent.
c) Find a transformation of function f equivalent to a stretch, factor 8, parallel to the y-axis.
d) Find a transformation of function f equivalent to a stretch, factor 8, parallel to the x-axis.
e) Show that a translation k units parallel to the x-axis is equivalent to a stretch, factor $\frac{1}{2}$, parallel to the y-axis.

4 The function f is defined by $f : x \to \frac{3}{x^2}, x > 0$.

a) Sketch the graph of f.
b) From the graph of f sketch the graphs of these functions.

$g : x \to \frac{3}{(x-1)^2}$ $\quad h : x \to \frac{3}{(x+2)^2}$

c) State the domain and range of g and h.

5 The graph of function $g : x \to \sqrt{4 - x^2}, -2 \le x \le 2$, is a semi-circle, radius 2 units, centre (0, 0).

a) Write down a function h for which the graph is a semi-circle, radius 5 units, centre (0, 0).
b) Which two transformations, when combined, map the graph of g into the graph of h?
c) Prove your assertion in b) by transforming function f into function g.

6 State whether the following functions are even, odd or neither. Sketch a graph of the function, in each case, to support your answer.

a) $y = -\frac{1}{x}$ b) $y = 1 + x^2 + x^4$ c) $y = \frac{10}{x^2}$ d) $y = \tan x$

e) $y = x^3 + 8$ f) $y = \frac{40}{x^3} - \frac{12}{x}$ g) $y = \frac{40}{x^2} - 12$ h) $y = (x + 2)^3$

i) $y = (x^2 - 3)^2$ j) $y = (x^2 - 3)^3$

EXERCISES

13.4 B

1 The function f is defined by $f : x \to 2x^2$. For each of the following transformed functions:

i) describe the transformation geometrically,

ii) sketch graphs of $y = f(x)$ and its range under the transformation.

a) $y = f(x+1)$ **b)** $y = 3f(x)$ **c)** $y = f(2x)$ **d)** $y = \frac{1}{2}f(x) + 2$
e) $y = f(x-3)$ **f)** $y = 2f(x-3)$ **g)** $y = f(4-x)$ **h)** $y = 3f(2x-1) + 4$

2 The function g is defined by $g(x) = \dfrac{3}{2x}, x \neq 0$.

For each of the following transformed functions:

i) state the transformed function,

ii) sketch the graph of $y = g(x)$ and its image under the transformation.

a) translation 3 units parallel to the x-axis
b) translation 3 units parallel to the y-axis
c) stretch, factor 2, parallel to the y-axis
d) stretch, factor $\frac{1}{2}$, parallel to the x-axis
e) stretch, factor $\frac{1}{3}$, parallel to the x-axis *followed by* translation 4 units parallel to the y-axis
f) reflection in the y-axis *followed by* translation –2 units parallel to the y-axis
g) **a)** *followed by* **c)**
h) **d)** *followed by* **b)**

3 The function f is defined by $f(x) = \sin x$.

a) i) Write down the transformed function following a reflection in the y-axis.
ii) Sketch $f(x)$and its image under the transformation.

b) i) Write down the transformed function following a reflection in the x-axis.
ii) Sketch $f(x)$ and its image under the transformation.

c) Comment on your results to **a)** and **b)**.

d) Repeat parts **a)** and **b)** for $g(x) = \cos x$ and $h(x) = \tan x$ and comment.

4 The function f is defined by $f : x \to \dfrac{2}{x^2}, x \neq 0$.

a) Sketch the graph of f.

b) If the domain of f is restricted to $x > 0$, describe the transformation which would give the portion of the curve where $x < 0$

5 The graph of $f : x \to \sqrt{9 - x^2}, -3 \leq x \leq 3$ is a semi-circle of radius 3, centre (0, 0).

a) Describe the transformation which would map f onto a semi-circle radius 3, centre (2, 0).

b) Find the transformed function.

6 State whether the following functions are even, odd or neither. Use sketches to support your answers.

a) $f(x) = x^2 + 5$ **b)** $f(x) = x^3 + x^2 + 2$ **c)** $f(x) = x^4 - x^2$

d) $f(x)=-\frac{3}{2x}$ **e)** $f(x)=\frac{1}{x+1}$ **f)** $f(x)=\frac{1}{x^2+1}$ **g)** $f(x)=\frac{x}{x^3+1}$

h) $f(x)=\frac{1}{4x^3}$ **i)** $f(x)=(x^2+1)^3$ **j)** $f(x)=(x^3+1)^2$

CONSOLIDATION EXERCISES FOR CHAPTER 13

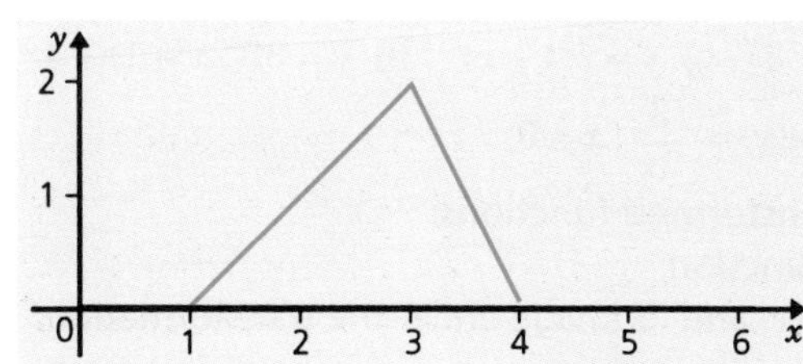

1 The diagram shows a sketch of the curve with equation $y = f(x)$, where $f(x) = 0$ for $x \le 1$ and $x \ge 4$.

Sketch, on separate axes, the graph of the curves with equations:

a) $y = f(x - 1)$ **b)** $y = f(\frac{1}{2}x)$.

(AEB Question 3, Specimen Paper 1, 1994)

2 The functions f and g are defined by:

$f: x \to 3x - 1, x \in \Re$ $g: x \to x^2 + 1, x \in \Re$.

a) Find the range of g.
b) Determine the value of x for which $gf(x) = fg(x)$.
c) Determine the values of x for which $|f(x)| = 8$.

The function $h: x \to x^2 + 3x, x \in \Re, x \ge q$, is one-to-one.

d) Find the least value of q and sketch the graph of this function.

(ULEAC Question 9, Specimen Paper 1, 1994)

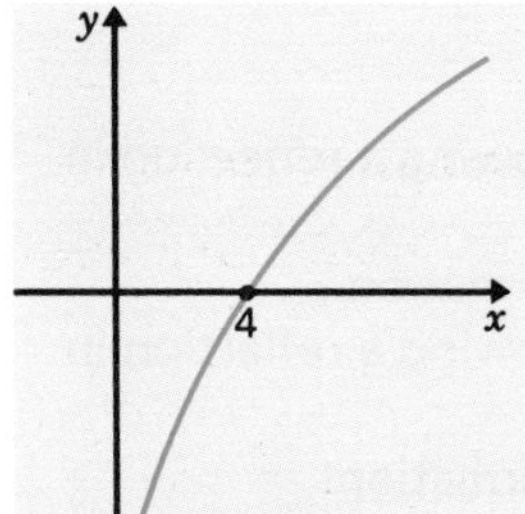

3 The function $f(x)$ is defined for all values of x except $x = 0$ and is an odd function, i.e. $f(-x) = -f(x)$.

a) Part of the graph of $y = f(x)$ is shown in the diagram. Copy and complete the sketch.
b) Draw a separate sketch to illustrate the graph of $y = f(x + 3)$, showing clearly where the graph will intercept the x-axis.

(SMP 16–19 Question 2, Specimen Paper, 1994)

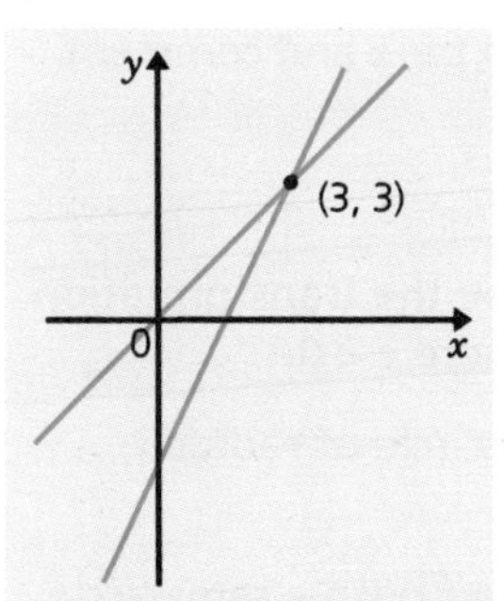

4 The diagram shows the graphs of lines $y = x$ and $y = 2x - 3$, and their point of intersection (3, 3). Sketch on a single diagram the graphs of $y = |x|$ and $y = |2x - 3|$, and give the points of intersection of these graphs.

(UCLES Linear Question 4, Specimen Paper, 1994)

5 The function f has as its domain the set of all non-zero real numbers, and is given by $f(x)=\frac{1}{x}$ for all x in this set. On a single diagram, sketch each of the following graphs, and indicate the geometrical relationships between them.

a) $y = f(x)$ **b)** $y = f(x + 1)$ **c)** $y = f(x + 1) + 2$

Deduce, explaining your reasoning, the coordinates of the point about which the graph of $y=\frac{2x+3}{x+1}$ is symmetrical.

(UCLES Linear Question 12, Specimen Paper, 1994)

6 The diagram shows the graph of $y = x^2(3 - x)$. The coordinates of the points A and B on the graph are (2, 4) and (3, 0) respectively.

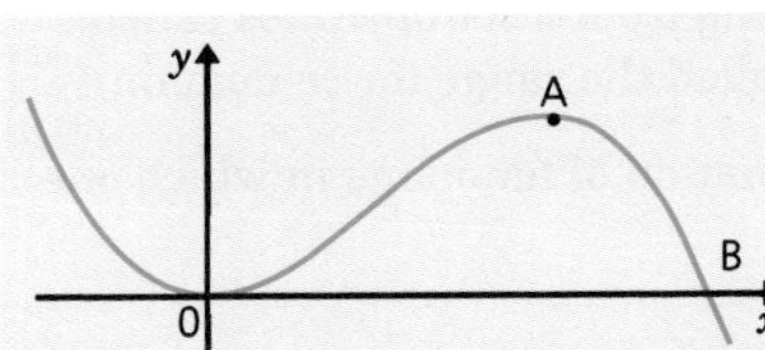

a) Write down the solution set of the inequality $x^2(3 - x) \geq 0$.

b) The equation $3x^2 - x^3 = k$ has three real solutions for x. Write down the set of possible values for k.

c) Functions f and g are defined as follows.

$$f : x \to x^2(3 - x),\ 0 \leq x \leq 2 \qquad g : x \to x^2(3 - x),\ 0 \leq x \leq 3$$

Explain why f has an inverse while g does not.

d) State the domain and the range of f^{-1}, and sketch the graph of f^{-1}.

(UCLES Modular Question 9, Specimen Paper 1, 1994)

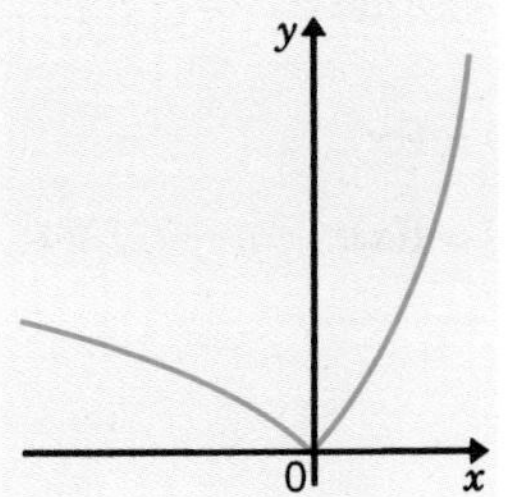

7 The diagram shows the graph of $y = |f(x)|$, for a certain function f with domain $\Re$. Sketch, on separate diagrams, two possibilities for the graph of $y = f(x)$. *(UCLES Modular Question 1, Specimen Paper 2, 1994)*

8 The functions f and g are defined by:

$$f : x \to x^2 + 3,\ x \in \Re \qquad g : x \to 2x + 1,\ x \in \Re.$$

a) Find, in a similar form, the function fg.

b) Find the range of the function fg.

c) Solve the equation $f(x) = 12g^{-1}(x)$.

(ULEAC Question 3, Paper 1, January 1995)

9 The function f is defined by:

$$f : x \to \frac{x+3}{x-1},\ x \in \Re,\ x \neq 1$$

a) Find the set of values of x for which $f(x) < x$.

b) Find, in terms of x, the inverse function of f^{-1} and comment on your result. *(ULEAC Question 3, Paper 2, May 1994)*

10 Functions f and g are defined by:

$$f : x \to \frac{4}{x-2},\ x \in \Re,\ x \neq 2$$
$$g : x \to x + 2,\ x \in \Re$$

a) Show that $gf : x \to \frac{2x}{x-2},\ x \in \Re,\ x \neq 2$

b) Find $(gf)^{-1}$ in the form $(gf)^{-1} : x \to \ldots$

(ULEAC Question 4, Paper 1, January 1994)

11 The functions $f(x)$ and $g(x)$ are defined by $f(x) = x^2 + 1$ and $g(x) = x - 2$.

a) i) Calculate the value of $g(f(2))$.

ii) The function $h(x)$ is defined by $h(x) = g(f(x))$. Find an expression in terms of x for $h(x)$.

b) The domain and range of each of the functions $f(x)$ and $g(x)$ is the set of real numbers. Explain what is meant by this statement.

c) Only one of the functions $f(x)$ and $g(x)$ has an inverse. Write down an expression in terms of x for the inverse function.

d) The function which has no inverse does have an inverse if its domain and range are suitably restricted. Suggest suitable restrictions for the domain and range.

(Oxford Nuffield Question 5, Paper 1, 1995)

Summary

- A function is a one-to-one mapping from a set of values called the **domain** to a set of values called the range (or co-domain).
- A composite function is a combination of functions in which one function is followed by another

  e.g. $fg(x) \equiv f(g(x))$
- Given a function $f(x)$ then the inverse function $f^{-1}(x)$ is such that $f^{-1}f(x) = ff^{-1}(x) = x$.
- An **even** function $f(x)$ has the property $f(-x) = f(x)$.
- An **odd** function $f(x)$ has the property $-f(-x) = f(x)$.

PURE MATHS 14

Sequences and series 2: Binomial expansions

In this chapter we introduce:

- *the binomial theorem expansion of $(a + b)^n$ for integer n*
- *the binomial series expansion of $(1 + x)^n$ for integer, negative and fractional values of n.*

THE BINOMIAL THEOREM

In this section we shall develop the ideas introduced in Chapter 11, *Sequences and series 1*, and examine the binomial theorem and series.

Exploration 14.1

Pascal's triangle

This pattern of numbers is called **Pascal's triangle**.

```
1
1   1
1   2   1
1   3   3   1
1   4   6   4   1
1   5   10  10  5   1
•   •   •   •   •   •   •
•   •   •   •   •   •   •   •
etc.
```

- Continue the pattern for the next four rows.
- What properties can you find?

Patterns in Pascal's triangle

There are several interesting features of Pascal's triangle, for example:

- for any row, the sum of two adjacent numbers gives the number in the next row underneath the second number.

```
e.g.   1   2        6   4        5   10
           3            10           15
```

- the sum of the numbers in each row is a power of 2.
  e.g. $1 + 4 + 6 + 4 + 1 = 16 = 2^4$
- the third column gives the triangle numbers.
  1, 3, 6, 10, 15, ...

Notation

We can refer to any number in Pascal's triangle by specifying its row (n) and column (r), denoting the number by $\binom{n}{r}$.

Both n and r begin at 0, e.g. $\binom{5}{3} = 10$, since 10 is the entry in row 5, column 3.

$n \backslash r$	0	1	2	3	4	5	6	7
0	1							
1	1	1						
2	1	2	1					
3	1	3	3	1				
4	1	4	6	4	1			
5	1	5	10	10	5	1		
6	1	6	15	20	15	6	1	
7	1	7	21	35	35	21	7	1

We have seen that, for example,

$$\binom{6}{4} + \binom{6}{5} = \binom{7}{5} \qquad \text{i.e. } 15 + 6 = 21$$

In general, this gives rise to:

$$\binom{n}{r} + \binom{n}{r+1} = \binom{n+1}{r+1}$$

This is an **inductive** definition, giving one number in the triangle as the sum of two others in the previous row. This is fine for small values of n and r but what of $\binom{15}{4}$? Do we need to extend the triangle to row 15 before we can find its value?

Exploration 14.2

Continuing patterns

Continue the pattern in these expansions.

$$\begin{aligned}
(a+b)^1 &= a + b \\
(a+b)^2 &= a^2 + 2ab + b^2 \\
(a+b)^3 &= a^3 + 3a^2b + 3ab^2 + b^3 \\
(a+b)^4 &= a^4 + 4a^3b + \dots \\
\dots &= \dots \\
\dots &= \dots
\end{aligned}$$

- What patterns do the coefficients form?
- What patterns do the powers of a and b form?
- How can you generate the expansion of $(a + b)^n$?

Coefficients and Pascal's triangle

The coefficients in the expansion of $(a + b)^n$ form the nth row of Pascal's triangle:

$$(a+b)^4 = \mathbf{1}a^4 + \mathbf{4}a^3b + \mathbf{6}a^2b^2 + \mathbf{4}ab^3 + \mathbf{1}b^4$$

Just as the fourth row of the triangle can be generated from the third row, so $(a + b)^4$ can be generated from $(a + b)^3$.

$$\begin{aligned}(a+b)^4 &= (a+b)^3(a+b)\\ &= (a^3 + 3a^2b + 3ab^2 + b^3)(a+b)\\ &= a^4 + 4a^3b + 6a^2b^2 + 4ab^3 + b^4\end{aligned}$$

Note that in the expansion of $(a + b)^4$, reading from left to right, the powers of a descend from 4 to 0, the powers of b ascend from 0 to 4 and for any term, the sum of the indices is 4. This can be extended to $(a + b)^n$.

A formula for $\binom{n}{r}$ arises from a consideration of the expansion of $(a + b)^n$ from first principles.
Look again at $(a + b)^4$.

$$(a+b)^4 = (a+b)(a+b)(a+b)(a+b)$$

The term in a^3b is formed by combining a from three brackets and b from one bracket. This may be done in four ways.

$$\left.\begin{aligned} aaab &= a^3b \quad + \\ aaba &= a^3b \quad + \\ abaa &= a^3b \quad + \\ baaa &= a^3b \end{aligned}\right\} = 4a^3b$$

These four ways are related to the number of ways of choosing one bracket from four to generate the b, the rest generating an a. The number of ways of choosing one object from four is:

$${}^4C_1 = \frac{4!}{1!3!} = 4 \text{ thus } \binom{4}{1} = {}^4C_1$$

Similarly, the term in a^2b^2 is formed by choosing b from two brackets and a from the rest, which can be done in six ways.

$$\binom{4}{2} = {}^4C_2 = \frac{4!}{2!2!} = 6$$

In the expansion of $(a + b)^n$, the term in, say, $a^{n-3}b^3$ is formed by choosing b from three brackets and a from the rest, which gives the coefficient:

$$\binom{n}{3} = {}^nC_3 = \frac{n!}{3!(n-3)!}$$

In general, then, the coefficient of $a^{n-r}b^r$ is formed by choosing b from r brackets and a from the rest, which can be done in

$$\binom{n}{r} = {}^nC_r = \frac{n!}{r!(n-r)!} \quad \text{ways.}$$

This leads to the general statement of the **binomial theorem**.

$$(a+b)^n = \binom{n}{0}a^n + \binom{n}{1}a^{n-1}b + \binom{n}{2}a^{n-2}b^2 + \ldots + \binom{n}{r}a^{n-r}b^r + \ldots + \binom{n}{n-1}ab^{n-1} + \binom{n}{n}b^n$$

Example 14.1

For the expression $(a+b)^{10}$:

a) *find the coefficient of a^3b^7,*
b) *write down the first four terms in descending powers of a.*

Solution

a) *The coefficient of a^3b^7 is* $\binom{10}{7} = {}^{10}C_7 = \frac{10!}{7!3!} = 120$

b)
$$(a+b)^{10} = \binom{10}{0}a^{10} + \binom{10}{1}a^9b + \binom{10}{2}a^8b^2 + \binom{10}{3}a^7b^3 + \ldots$$
$$= a^{10} + 10a^9b + 45a^8b^2 + 120a^7b^3 + \ldots$$

Example 14.2

For the expression $(2p-q)^7$:

a) *find the term in p^2,*
b) *write down the first three terms in ascending powers of q.*

Solution

Firstly treat $(2p-q)^7$ as $(2p+(-q))^7$, i.e.
$a = 2p$, $b = -q$ *and* $n = 7$.

a) *The term in $p^2 \equiv$ term in q^5, which is:*
$${}^7C_5(2p)^2(-q)^5 = 21 \times 4p^2 \times (-q^5) = -84p^2q^5$$

b)
$$(2p-q)^7 = \binom{7}{0}(2p)^7 + \binom{7}{1}(2p)^6(-q) + \binom{7}{2}(2p)^5(-q)^2 + \ldots$$
$$= 2^7p^7 + 7 \times 2^6p^6(-q) + 21 \times 2^5p^5(-q)^2 + \ldots$$
$$= 128p^7 - 448p^6q + 672p^5q^2 + \ldots$$

Example 14.3

Find the term in x^3 in the expansion of $(1-x)(2+x)^5$.

Solution

$$(2+x)^5 = \binom{5}{0}2^5 + \binom{5}{1}2^4x + \binom{5}{2}2^3x^2 + \binom{5}{3}2^2x^3 + \binom{5}{4}2x^4 + \binom{5}{5}x^5$$
$$= 32 + 80x + 80x^2 + 40x^3 + 10x^4 + x^5$$

$\Rightarrow$ *the term in x^3 for*

$$(1-x)(2+x)^5 \equiv (1-x)(32 + 80x + 80x^2 + 40x^3 + 10x^4 + x^5)$$

is $1 \times 40x^3 - x \times 80x^2 = 40x^3 - 80x^3 = -40x^3$

EXERCISES

14.1 A

1 Use Pascal's triangle to expand the following.

a) $(a+b)^6$ **b)** $(p-q)^5$ **c)** $(3r+5)^8$

2 For the expression $(a+b)^{12}$:

a) find the coefficient of a^4b^8,
b) write down the first four terms in descending powers of a.

3 Use the binomial theorem to expand the following expressions.

a) $(x+3y)^4$ **b)** $(2f+3g)^5$ **c)** $(3a-b)^6$
d) $(p+\frac{1}{2}q)^5$ **e)** $(3x-2y)^8$

4 Find the coefficients of the stated terms in the following.

a) b^3 in $(a+b)^9$ **b)** z^2 in $(2z+1)^{10}$ **c)** q^5 in $(2p-q)^8$
d) g^7 in $(5f+2g)^{12}$ **e)** x^{13} in $(1-x)^{20}$ **f)** x^r in $(1-x)^{20}$

5 Find the term in x^3 in each of the following expansions.

a) $(2+x)(3+x)^4$ **b)** $(3-x)(1+x)^7$
c) $(1-x^2)(2-x)^6$ **d)** $(2+x)^2(3+x)^{10}$

EXERCISES

14.1 B

1 Use Pascal's triangle to expand the following expressions.

a) $(m+n)^5$ **b)** $(3-a)^7$ **c)** $(2x+3)^4$

2 For the expression $(x-y)^{10}$:

a) find the coefficient of x^5y^5,
b) write down the first four terms in descending powers of x.

3 Use the binomial theorem to expand the following expressions.

a) $(a+2b)^5$ **b)** $(2x+5b)^4$ **c)** $(4a-2b)^7$
d) $(\frac{1}{2}m-2n)^6$ **e)** $(5-3x)^5$

4 Find the coefficients of the stated terms in the following.

a) a^3 in $(a+b)^4$ **b)** x^4 in $(2x+3)^{13}$ **c)** q^2 in $(p-4q)^{12}$
d) a^5 in $(9a+3b)^7$ **e)** x^5 in $(1-2x)^6$ **f)** x^n in $(2+x)^8$

5 Find the term in x^3 in each of the following expansions.

a) $(3+x)(5+x)^5$ **b)** $(1+x)(x-3)^9$
c) $(5-x^2)(7-x)^6$ **d)** $(3+x)^2(4+3x)^8$

BINOMIAL SERIES

A special case of the binomial theorem arises when $a = 1$ and $b = x$. It can be very useful, as we shall see.

$$(1+x)^n = \binom{n}{0} + \binom{n}{1}x + \binom{n}{2}x^2 + \ldots + \binom{n}{r}x^r + \ldots + \binom{n}{n-1}x^{n-1} + \binom{n}{n}x^n$$

But:

$$\binom{n}{0}=1,\ \binom{n}{1}=n,\ \binom{n}{2}=\frac{n(n-1)}{2!},\ \ldots,\ \binom{n}{r}=\frac{n(n-1)\ldots(n-r+1)}{r!},\ \ldots,\ \binom{n}{n-1}=n,\ \binom{n}{n}=1$$

hence:

$$(1+x)^n=1+nx+\frac{n(n-1)}{2!}x^2+\ldots+\frac{n(n-1)(n-2)\ldots(n-r+1)}{r!}x^r+\ldots+nx^{n-1}+x^n$$

This form of expansion is called the **binomial series**, valid for any positive integer n.

Example 14.4

Expand the following in ascending powers of x, up to and including the term in x^4.

a) $(1+2x)^7$ **b)** $(1-x)^8$ **c)** $(1+x^2)^5$

Solution

a) $(1+2x)^7$

$$\equiv 1+7\times 2x+\frac{7\times 6}{2!}\times(2x)^2+\frac{7\times 6\times 5}{3!}\times(2x)^3+\frac{7\times 6\times 5\times 4}{4!}\times(2x)^4+\ldots$$
$$=1+14x+84x^2+280x^3+560x^4+\ldots$$

b) $(1-x)^8$

$$\equiv 1+8(-x)+\frac{8\times 7}{2!}\times(-x)^2+\frac{8\times 7\times 6}{3!}\times(-x)^3+\frac{8\times 7\times 6\times 5}{4!}\times(-x)^4+..$$
$$=1-8x+28x^2-56x^3+70x^4+\ldots$$

c) $(1+x^2)^5\equiv 1+5x^2+\frac{5\times 4}{2!}\times(x^2)^2+\ldots$
$$=1+5x^2+10x^4+\ldots$$

Example 14.5

Use the binomial series to find the first four terms of $(5+x)(1-3x)^7$.

Solution

$$(1-3x)^7\equiv 1+7(-3x)+\frac{7\times 6}{2!}(-3x)^2+\frac{7\times 6\times 5}{3!}(-3x)^3+\ldots$$
$$=1-21x+189x^2-945x^3+\ldots$$
$$\Rightarrow (5+x)(1-3x)^7\equiv(5+x)(1-21x+189x^2-945x^3+\ldots)$$
$$=5-104x+924x^2-4536x^3+\ldots$$

Example 14.6

The coefficient of x in the expansion of $(1-4x)(1+cx)^6$ is 8. Find the coefficient of x^2.

Solution

$$(1+cx)^6\equiv 1+6cx+15c^2x^2+\ldots$$
$$\Rightarrow (1-4x)(1+cx)^6\equiv(1-4x)(1+6cx+15c^2x^2+\ldots)$$
$$\equiv 1+(6c-4)x+(15c^2-24c)x^2+\ldots$$

Coefficient of x is $8\Rightarrow 6c-4=8\Rightarrow 6c=12\Rightarrow c=2$

$\Rightarrow$ *Coefficient of x^2 is* $15c^2-24c=15\times 4-24\times 2=12$

Approximating powers

It is often appropriate to use the binomial series to approximate powers of numbers *without using a calculator.* This is illustrated in the next example.

Example 14.7

Approximate all of the following numbers, correct to three significant figures.

a) 1.01^4 ***b)*** 0.97^{10} ***c)*** 10.5^6

Solution

a) $1.01^4 = (1+0.01)^4$

$$= 1+4\times 0.01+\frac{4\times 3}{2!}\times(0.01)^2+4\times(0.01)^3+(0.01)^4$$

$$= 1+0.04+6\times 0.0001+4\times 0.000001+0.00000001$$

Only the first two terms are required to give an answer correct to 3 s.f., i.e.

$1.01^4 \approx 1.04$

b) $0.97^{10} = (1-0.03)^{10}$

$$= 1+10\times(-0.03)+\frac{10\times 9}{2!}\times(-0.03)^2+\frac{10\times 9\times 8}{3!}\times(-0.03)^3+\ldots$$

$$= 1-0.3+45\times 0.0009-120\times 0.000027+\ldots$$

$$= 1-0.3+0.0405-0.00324+\ldots$$

Only the first four terms are required to give accuracy to 3 s.f., i.e.

$0.97^{10} \approx 0.737$

c) $10.5^6 = (10\times 1.05)^6 = 10^6\times(1.05)^6 = 10^6\times(1+0.05)^6$

$$= 10^6\times\left(1+6\times 0.05+\frac{6\times 5}{2!}\times 0.05^2+\frac{6\times 5\times 4}{3!}\times 0.05^3+\ldots\right)$$

$$= 10^6\times(1+0.3+15\times 0.0025+20\times 0.000125+\ldots)$$

$$= 10^6\times(1+0.3+0.0375+0.0025+\ldots)$$

Only the first four terms are required to give accuracy to 3 s.f., i.e.

$10.5^6 \approx 10^6\times 1.34 = 1340000$

Exploration 14.3

Approximations and significant figures

Part **a)** of Example 14.7 gave:

$(1.01)^4 = (1+0.01)^4 \approx 1+4\times 0.01 = 1.04$

i.e. $(1.01)^4 = 1.04$ to 3 s.f.

Which of the following approximations are correct to three significant figures?

$(1.02)^4 = (1+0.02)^4 \approx 1+4\times 0.02 = 1.08$
$(1.03)^4 = (1+0.03)^4 \approx 1+4\times 0.03 = 1.12$
$(0.99)^4 = (1-0.01)^4 \approx 1-4\times 0.01 = 0.96$
$(0.98)^4 = (1-0.02)^4 \approx 1-4\times 0.02 = 0.92$
$(0.97)^4 = (1-0.03)^4 \approx 1-4\times 0.03 = 0.88$

- For what range of values of x will it be true that $(1+x)^4 = 1+4x$ to 3 s.f?

- For what range of values of x will approximating $(1+x)^4$ by $1+4x$ produce an error of less than 1%?
- Choose different values of n and investigate the ranges of x values for which the error in approximating $(1+x)^n$ by $1+nx$ is less than 1%.

Errors

The error in approximating $(1+x)^4$ by $1+4x$ is less than 1% provided $-0.0382 \le x \le 0.0438$ (to 3 s.f.). We can check this by working out the percentage error for the boundary values. (Try it. You will notice that the smaller $|x|$ is, the smaller the percentage error.) As n gets larger, the range of values of x which give a percentage error less than 1%, when approximating $(1+x)^n$ by $1+nx$, becomes narrower.

A general principle arising from the exploration is:

$$(1+x)^n \approx 1+nx$$

i.e. terms in x^2, x^3 etc. may be neglected provided $|x|$ is small.

Example 14.8

The equation $5x^3+12x-18=0$ has a root close to 1. Use the substitution $x=1+h$ and the approximation $(1+h)^3 \approx 1+3h$ to find the root to 3 s.f.

Solution

$$5x^3+12x-18=0$$

$$\Rightarrow \quad 5(1+h)^3+12(1+h)-18=0$$

from which we deduce the approximation

$$5(1+3h)+12(1+h)-18=0$$

$$\Rightarrow \quad 5+15h+12+12h-18=0$$

$$\Rightarrow \quad 27h=1$$

$$\Rightarrow \quad h=\tfrac{1}{27}\approx 0.037$$

Therefore the root close to 1 is $1+h=1+\frac{1}{27}=1.04$ (3 s.f.)

EXERCISES

14.2A

1 Use the binomial series to expand the following in ascending powers of x, up to and including the term in x^4.

a) $(1+x)^{10}$ **b)** $(1-x)^7$ **c)** $(1+3x)^4$

d) $(1+\frac{1}{2}x)^{12}$ **e)** $(1-\frac{1}{3}x)^6$ **f)** $(1+ax)^8$

2 Find the first four terms, in ascending powers of x, of each of the following expressions.

a) $(1+3x)(1-x)^6$ **b)** $(1-x)(1+2x)^{10}$

c) $(3+x)(1-3x)^5$ **d)** $(5-x)(1+\frac{1}{5}x)^8$

3 The first three terms in the expansion of $(1+ax)^n$ are $1+10x+40x^2$. Find the values of a and n.

4 The expansion of $(1+rx)^n$ in ascending powers of n, begins:

$$1+x+0.45x^2+\ldots$$

Find the values of r and n.

5 The expansion of $(1 + ax)(1 + bx)^8$, in ascending powers of x, begins:
$1 - 4x - 4x^2 + \ldots$
Find the values for a and b assuming:

a) a and b are integers, **b)** a and b are rational.
Hint: Form a quadratic equation in b. It has two roots, one of which is an integer, the other being a rational number.

6 Expand $(1 + x)^7$ in ascending powers of x up to the term in x^3. Hence approximate 1.03^7 to 3 s.f.

7 Expand $(1 - x)^{10}$ in ascending powers of x up to the term in x^3. Hence approximate 0.98^{10} to 3 s.f.

8 Expand $(1 + \frac{1}{2}x)^8$ in ascending powers of x up to the term in x^2. Hence approximate 2.02^8 to 3 s.f.

9 Use binomial expansions to calculate the following to 5 s.f.

a) $(1.001)^{12}$ **b)** $(3.006)^{10}$ **c)** $(3.996)^7$

10 The equation $x^4 + 0.5x - 1.45 = 0$ has a root close to 1. Use the substitution $x = 1 + h$ and a suitable approximation for $(1 + h)^4$ to find the root to 3 s.f.

EXERCISES

14.2 B

1 Use the binomial series to expand the following in ascending powers of x, up to and including the term in x^4.

a) $(1 + x)^{15}$ **b)** $(1 - x)^9$ **c)** $(1 + 5x)^7$
d) $(1 + \frac{1}{3}x)^{12}$ **e)** $(1 - \frac{2}{3}x)^5$ **f)** $(1 - ax)^{10}$

2 Find the first four terms, in ascending powers of x, of each of the following expressions.

a) $(1 + 2x)(1 + x)^7$ **b)** $(1- 5x)(1 - 2x)^6$
c) $(x + 3)(1 - 9x)^8$ **d)** $(8 - x)(1 + \frac{1}{4}x)^{12}$

3 The first three terms in the expansion of $(1 + ax)^n$ are $1 - 54x + 1296x^2$. Find the values of a and n.

4 When $(1+\frac{x}{r})^n$ is expanded in ascending powers of x, the expansion begins $1+\frac{3}{2}x+\frac{33}{32}x^2+\ldots$. Find the values of r and n.

5 The expansion of $(1 + ax)(1 + bx)^6$, in ascending powers of x, begins $1 + 9x + 24x^2 + \ldots$. Find the values for a and b.

6 The expansion of $(1 + ax)^9$, in ascending powers of x, begins $1 + bx + 8bx^2 + \ldots$. Find a, given that a is not zero.

7 When $(1 - x)(1 + \frac{1}{2}x)^n$ is expanded in ascending powers of x, the term in x^2 is $\frac{25}{4}x^2$. Find the value of n, given that it is a positive integer.

8 Expand $(1 - 2x)^{17}$ in ascending powers of x up to the term in x^3, and hence approximate 0.98^{17} to 3 s.f.

9 Use a binomial expansion to calculate $(2.006)^9$ to 5 s.f.

10 Use the binomial theorem to write down the expansion of $(1+x)^4$. Explain why, if δ is sufficiently small, $(1+\delta)^4 \approx 1+4\delta$.

You are given that the equation $x^4 - 0.7x - 0.32 = 0$ has a solution very close to $x = 1$.

By writing this solution in the form $x = 1+\delta$ and substituting in the equation, use the approximation above to obtain a further estimate for the solution.

(MEI Pure 1, Question 4, June 1991)

BINOMIAL SERIES FOR RATIONAL INDICES

In the previous section we found that, provided $|x|$ is small:

$$(1+x)^n \approx 1+nx \qquad n = 1, 2, 3, 4, 5, \ldots$$

We shall verify that both the binomial expansion and this approximation are valid for all rational numbers, including **fractions** and **negative numbers**.

Exploration 14.4

Fractional indices

- Choose several values of x close to 0 (positive and negative), e.g. 0.01, –0.02, etc. and find values for $\sqrt{1+x}$. Compare your results, when rounded to 2 d.p. and derive a suitable approximation without the use of a calculator.
- Repeat the process, this time for $\sqrt[3]{1+x}$. Again compare your results and find a suitable approximation.

Approximations

The approximation $(1+x)^n \approx 1+nx$, for small values of x, works for rational values of n.

$$\sqrt{1+0.02} = (1+0.02)^{\frac{1}{2}} = 1.009950494 \approx 1.01 = 1+\tfrac{1}{2}\times 0.02$$

$$\sqrt[3]{1+0.03} = (1+0.03)^{\frac{1}{3}} = 1.009901634 \approx 1.01 = 1+\tfrac{1}{3}\times 0.03$$

This is called a **first-order approximation**. A second-order approximation of the form:

$$(1+x)^n \approx 1+nx+\frac{n(n-1)}{2}x^2$$

should give an even better approximation.

When $n = \frac{1}{2}$ and $x = 0.02$, $\sqrt{1.02} \approx 1+\frac{1}{2}\times 0.02 - \frac{1}{8}\times 0.02^2 = 1.00995$

When $n = \frac{1}{3}$ and $x = 0.03$, $\sqrt[3]{1.03} \approx 1+\frac{1}{3}\times 0.03 - \frac{1}{9}\times 0.03^2 = 1.0099$

Try first-order and second-order approximations to $\sqrt[4]{1.04}$.

Exploration 14.5

Negative indices

- Choose several values of x close to 0 (positive and negative) and find values for $\dfrac{1}{1+x}$. Compare your results, rounded to 2 d.p. and derive a suitable approximation which does not require a calculator.

- Repeat the process, this time for $\frac{1}{(1+x)^2}$. Again working to 2 d.p. find a suitable approximation.

Approximations

The first-order approximation $(1 + x)^n \approx 1 + nx$, for small values of x, works for negative values of n:

$$\frac{1}{1+0.02} = (1+0.02)^{-1} = 0.980\,392\,1... \approx 0.98 = 1-0.02$$

$$\frac{1}{(1-0.01)^2} = (1-0.01)^{-2} = 1.020\,304\,0... \approx 1.02 = 1+0.02$$

Second-order approximations of the form:

$$(1+x)^n \approx 1+nx+\frac{n(n-1)}{2}x^2$$

give even better approximations.

When $n = -1$, $x = 0.02$: $\frac{1}{1+0.02} \approx 1-0.02+0.0004 = 0.9804$

When $n = -2$, $x = -0.01$: $\frac{1}{(1-0.01)^2} \approx 1+0.02-0.0003 = 1.0197$

For both rational and negative indices n, ${}^nC_r$ has no meaning since the definition:

$${}^nC_r = \frac{n!}{r!(n-r)!}$$

only applies when n and r are **natural numbers** (non-negative integers). However, the alternative version:

$${}^nC_r = \frac{n(n-1)(n-2)...(n-r+1)}{r(r-1)(r-2)...3\times 2\times 1}$$

can be evaluated for rational and negative n (r is always a natural number).

If n is **not** a natural number, then $(n - 1)$, $(n - 2)$, $(n - 3)$, etc. always give a non-zero result. Therefore the binomial expansion of $(1 + x)^n$ has no last term; it produces an **infinite series**.

$$(1+x)^n = 1+nx+\frac{n(n-1)}{2!}x^2+...+\frac{n(n-1)(n-2)...(n-r+1)}{r!}x^r+...$$

CALCULATOR ACTIVITY

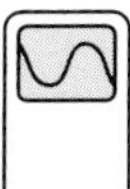

You will need a graphics calculator or graph-drawing package. Scale your axes:

$x_{min} = -3$; $x_{max} = 5$; $x_{scl} = 1$

$y_{min} = -1$; $y_{max} = 5$; $y_{scl} = 1$.

- Plot the graph of $y = \sqrt{1+x}$.
- Superimpose the first-order and second-order polynomial approximations given by the binomial series expansion.
- How closely do the approximations fit the original curve?
- What do all three curves have in common when $x = 0$?
- Repeat for different values of n, e.g. $y = \frac{1}{1+x} \Rightarrow n = -1$.
- Repeat for the expansion of $(1 - x)^n$.

- What happens if n is a positive integer?
- Consider higher order approximations – do they always give a better fit to the original curve?

For the function $y = \sqrt{1+x} = (1+x)^{\frac{1}{2}}$, the first and second order approximations are given by:

$$y = 1 + \tfrac{1}{2}x \quad \text{and} \quad y = 1 + \tfrac{1}{2}x - \tfrac{1}{8}x^2.$$

$y = 1 + \frac{x}{2}$

$y = \sqrt{1+x}$

$y = 1 + \frac{1}{2}x - \frac{1}{8}x^2$

The diagram shows the original curve and the two approximations.

All three curves pass through (0, 1).

The first-order and second-order approximations are a close fit to the original graph provided $|x|$ is small, with the quadratic approximation fitting better around $x = 0$. It is also important to note that all three curves have the same **gradient** when $x = 0$, i.e. the line $y = 1 + \frac{1}{2}x$ is a **tangent** to the original graph at the point (0, 1) and to any higher order approximation.

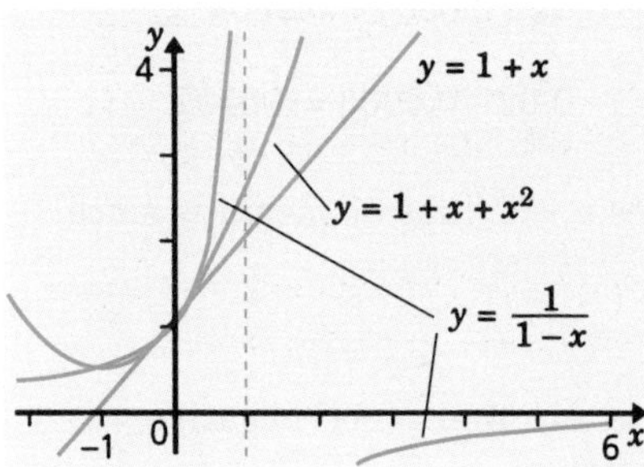

Similar patterns emerge for other values of n and for approximations to $y = (1-x)^n$, e.g. the graph of $y = \dfrac{1}{1-x}$ ($\equiv (1-x)^{-1}$) has first-order and second-order approximations:

$$y = 1 + x \quad \text{and} \quad y = 1 + x + x^2$$

which are illustrated in the diagram.

CALCULATOR ACTIVITY

INPUT N

INPUT X

1 → U

U → S

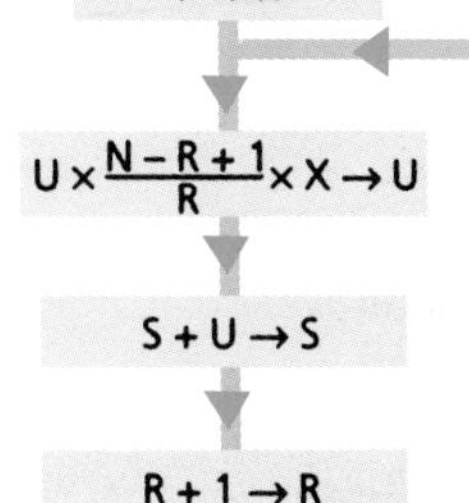

Try the program, outlined by this flowchart, which produces the sum of the first 1, 2, 3, 4, ... terms of a binomial series and will work for any value of n and x.

Run the program with different values of n and x and describe what happens 'in the long run'. Classify the sequence of partial sums as convergent or divergent, oscillating or non-oscillating;

Expected results

You probably chose various values of n and x in the last Calculator activity. Here are some examples showing what you might have found.

1 $n = 0.5$, $x = 0.1$ produces:

1.05, 1.04875, 1.0488125, 1.048808594, 1.048808867, 1.048808847, 1.048808848, ...

This is an **oscillating convergent** sequence, the limit of which is $\sqrt{1+0.1} = 1.048808848\ldots$

2 $n = -2$, $x = 3$ produces:

−5, 22, −86, 319, −1139, 3964, −13532, 45517, −151313, 498226, ...

This is an **oscillating divergent** sequence, which has no limit.

3 $n = -3, x = -2$ produces:

7, 31, 11, 351, 1023, 2815, 7423, 18 943, 47 103, 114 687, 274 431, ...

This is a **non-oscillating divergent** sequence, which has no limit.

4 $n = -0.5, x = -0.2$ produces:

1.1, 1.115, 1.1175, 1.117 937 5, 1.118 016 25, 1.118 030 688, ...

This is a **non-oscillating convergent** sequence, the limit of which is $\frac{1}{\sqrt{1-0.2}} = 1.118\,033\,989...$

Your investigations should confirm the following summary of the behaviour of successive partial sums for the expansion of $(1 + x)^n$.

	$x < -1$	$-1 < x < 0$	$0 < x < 1$	$x > 1$
$n \in N$	oscillating	oscillating	non -oscillating	non -oscillating
$n \notin N$	non-oscillating divergent	non-oscillating convergent	oscillating convergent	oscillating divergent

When $x = 1$, the series for $(1 + x)^n$ converges to $2n$ provided $n > -1$.
When $x = -1$, the series for $(1 + x)^n$ has interesting patterns!

Results

The **binomial expansion** of $(1 + x)^n$ for any rational n, **converges** provided $-1 < x < 1$, i.e. $|x| < 1$.

Similarly, the **binomial expansion** of $(1 + ax)^n$ for any rational n, **converges** provided $-1 < ax < 1$, i.e. $|x| < \frac{1}{a}$.

Example 14.9

Find the first four terms in the expansion of $(1 + 2x)^{-3}$ and state the range of values of x for which the expansion converges. Use the expansion to estimate $\frac{1}{1.02^3}$ to 3 s.f.

Solution

$$(1+2x)^{-3} = 1+(-3)\times 2x + \frac{(-3)\times(-4)}{2!}\times(2x)^2 + \frac{(-3)\times(-4)\times(-5)}{3!}\times(2x)^3 + \ldots$$

$$= 1-6x+24x^2-80x^3+\ldots$$

The series converges provided $-1 < 2x < 1 \Rightarrow -\frac{1}{2} < x < \frac{1}{2}$ *or* $|x| < \frac{1}{2}$.

Substituting $x = 0.01$:

$$\frac{1}{1.02^3} = (1+2\times 0.01)^{-3} = 1-6\times 0.01+24\times 0.01^2-80\times 0.01^3+\ldots$$

$$= 1-0.06+0.0024-0.000080+\ldots$$

$$= 0.942 \quad \text{(3 s.f.)}$$

Example 14.10

Find a quadratic function that approximates $(4+x)^{\frac{1}{2}}$ for small values of x. Use it to approximate $\sqrt{3.97}$ to 6 s.f.

Solution

$$(4+x)^{\frac{1}{2}} = \left(4(1+\tfrac{1}{4}x)\right)^{\frac{1}{2}} = 4^{\frac{1}{2}}(1+\tfrac{1}{4}x)^{\frac{1}{2}} = 2(1+\tfrac{1}{4}x)^{\frac{1}{2}}$$

Now:

$$\begin{aligned}(1+\tfrac{1}{4}x)^{\frac{1}{2}} &= 1+\tfrac{1}{2}\left(\tfrac{1}{4}x\right)+\frac{\left(\frac{1}{2}\right)\left(-\frac{1}{2}\right)}{2!}\left(\tfrac{1}{4}x\right)^2 + \mathrm{L}\\ &= 1+\tfrac{1}{8}x-\tfrac{1}{128}x^2+\mathrm{K}\end{aligned}$$

$$\Rightarrow \quad 2(1+\tfrac{1}{4}x)^{\frac{1}{2}} = 2+\tfrac{1}{4}x-\tfrac{1}{64}x^2+\mathrm{K}$$

which converges provided $-1 < \frac{1}{4}x < 1$

$\Rightarrow -4 < x < 4$ or $|x| < 4$.

$\Rightarrow$ the approximating quadratic function is $2 + \frac{1}{4}x - \frac{1}{64}x^2$.

To approximate $\sqrt{3.97}$ let $x = -0.03$

$$\begin{aligned}\Rightarrow \quad \sqrt{3.97} &\approx 2+\tfrac{1}{4}\times(-0.03)-\tfrac{1}{64}\times(-0.03)^2\\ &= 1.99249 \quad (\text{6 s.f.})\end{aligned}$$

Check the answer using a calculator.

Example 14.11

Expand $\dfrac{(3+x)}{(1-x)^2}$ in ascending powers of x as far as the term in x^4.

Solution

$$\frac{(3+x)}{(1-x)^2} \equiv (3+x)(1-x)^{-2}$$

Now

$$\begin{aligned}(1-x)^{-2} &\equiv 1+(-2)(-x)+\frac{(-2)(-3)}{2!}(-x)^2+\frac{(-2)(-3)(-4)}{3!}(-x)^3\\ &\quad +\frac{(-2)(-3)(-4)(-5)}{4!}(-x)^4+\mathrm{K}\\ &= 1+2x+3x^2+4x^3+5x^4+\mathrm{K}\end{aligned}$$

$$\begin{aligned}\Rightarrow \quad (3+x)(1-x)^{-2} &= (3+x)(1+2x+3x^2+4x^3+5x^4+\mathrm{K}\,)\\ &= 3+(6+1)x+(9+2)x^2+(12+3)x^3+(15+4)x^4+\mathrm{K}\\ &= 3+7x+11x^2+15x^3+19x^4+\mathrm{K}\end{aligned}$$

EXERCISES

1 For each of the following expressions, write down the first five terms of its binomial expansion. In each case state the range of values of x for which the series converges.

a) $(1+x)^{-1}$ **b)** $(1+3x)^{\frac{1}{2}}$ **c)** $(1-\frac{1}{3}x)^{-2}$ **d)** $\sqrt[3]{8+x}$

e) $(1+2x)^{-\frac{1}{2}}$ **f)** $(1+5x)^{\frac{3}{4}}$

2 Use appropriate binomial expansions to approximate each of the following to 3 s.f.

a) $\sqrt{1.04}$ **b)** $\sqrt{96}$ **c)** $\sqrt[3]{8.32}$

3 Expand $\dfrac{(1-x)}{(1+x)}$ in ascending powers of x, up to and including the term in x^3.

4 Show that $\sqrt{\dfrac{(1-x)}{(1+x)}} \equiv (1-x)^{\frac{1}{2}}(1+x)^{-\frac{1}{2}}$.

Using suitable binomial expansions show that:

$$\sqrt{\frac{(1-x)}{(1+x)}} \approx 1 - x + \tfrac{1}{2}x^2 - \tfrac{1}{2}x^3$$

Using a suitable value of x, estimate $\sqrt{\frac{2}{3}}$ to 3 s.f.

5 When $(1 + ax)^{-2}$ is expanded in ascending powers of x, the first four terms are $1 + Px + Qx^2 - \frac{1}{2}x^3$. Find a and then P and Q.

6 The first three terms of the expansion of $(1 + rx)^n$ are $1 - \frac{3}{2}x + \frac{27}{4}x^2$. Find the values of r and n, and the range of values of x for which the series converges.

EXERCISES

14.3 B

1 For each of the following expressions, write down the first four terms of its binomial expansion. In each case state the range of values of x for which the series converges.

a) $(1 + x)^{-3}$ **b)** $(1 + 2x)^{-3}$ **c)** $\left(1-\dfrac{x}{5}\right)^{-1}$ **d)** $\sqrt[3]{27+x}$

e) $(1 + 7x)^{\frac{1}{2}}$ **f)** $(1 - 3x)^{\frac{5}{4}}$

2 Use appropriate binomial expansions to approximate each of the following to 3 s.f.

a) $\sqrt{0.95}$ **b)** $\sqrt{140}$ **c)** $(16.12)^{\frac{1}{4}}$

3 Expand $\dfrac{(1+x)}{\sqrt{1-x}}$ in ascending powers of x, up to and including the term in x^4.

4 The first three terms in the expansion of $(a - 3x)^{\frac{1}{2}}$, in ascending powers of x, are $2 - \frac{3}{4}x + bx^2$ where a and b are constants. Find the values of a and b.

5 Expand $(1 + x + x^2)^{\frac{1}{2}}$ as a series in ascending powers of x up to and including the term in x^4. Find the values of x for which the series is valid.

6 Expand $\left(1-\frac{8}{x}\right)^{\frac{1}{3}}$ as a series in ascending powers of $\dfrac{1}{x}$ up to and including the term in $\frac{1}{x^3}$. State the set of values of x for which the expansion is valid.

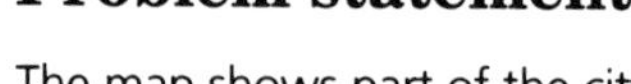

MATHEMATICAL MODELLING ACTIVITY

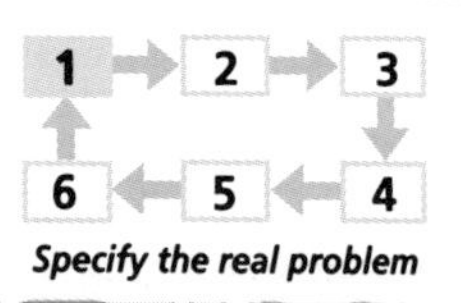

Specify the real problem

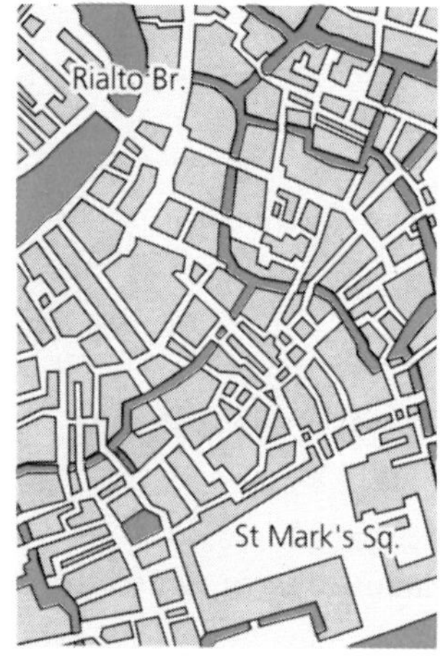

Problem statement

The map shows part of the city of Venice in Northern Italy. It's a beautiful city and each street has old buildings worth viewing. As the map shows the streets and canals are laid out in a roughly rectangular grid. (This layout for roads is typical of many cities around the world especially in America.) With so much to see, there are many different routes between two specific points such as St Mark's Square and the Rialto Bridge. Formulate a model for finding the number of routes between two points in a city made up of rectangular grids.

Set up a model

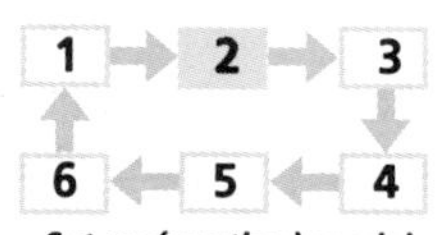

Set up (another) model

The streets in Venice are of different lengths and do not form an exact grid system so we must start with some simplifying assumptions.

- Assume that the city streets form a rectangular grid orientated with horizontal and vertical lines.
- Assume that the distance between adjacent junctions horizontally is a constant h.
- Assume that the distance between adjacent junctions vertically is a constant v.

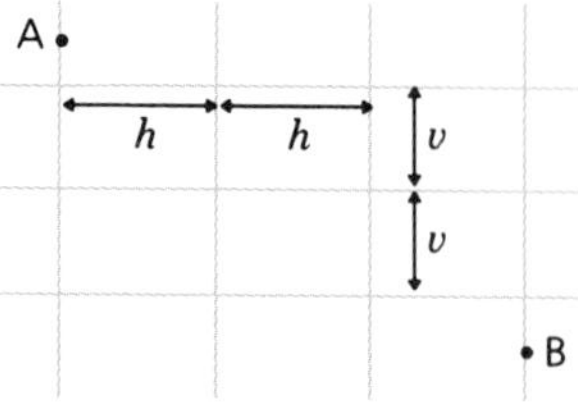

The problem is: how many 'routes of shortest distance' are there between two junctions A and B?

Consider some simple examples. How many routes are there in the following diagrams? (One possible route is shown for each case.)

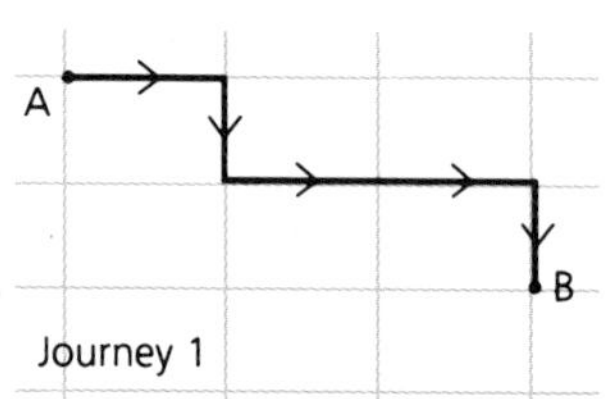

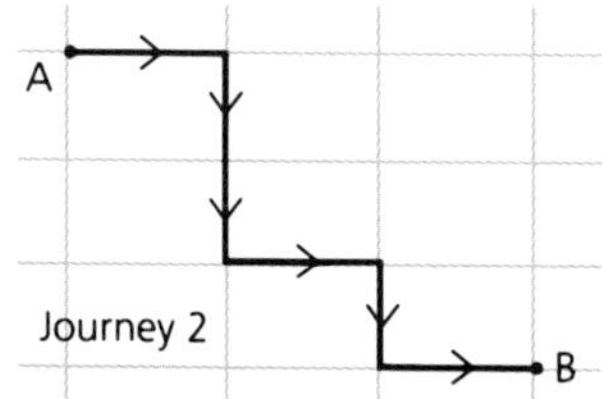

For Journey 1 the shortest distance is $3h + 2v$ and you should have found ten different routes between A and B. Each possible path can be written in the form of a table.
The ten possible routes of shortest distance are shown below.

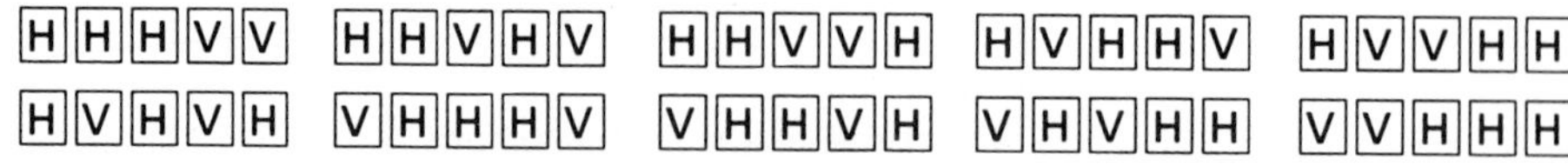

H H H V V   H H V H V   H H V V H   H V H H V   H V V H H
H V H V H   V H H H V   V H H V H   V H V H H   V V H H H

The total number of routes is the same as the number of ways of putting three Hs and two Vs in the five boxes in the table. We can show that this is the coefficient of h^3v^2 in the expansion of $(h + v)^5$.

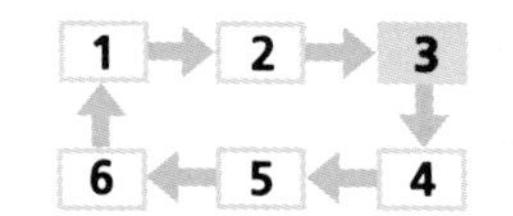

Formulate the mathematical problem

Mathematical problem

For Journey 2 the shortest distance is $3h + 3v$ and the number of different routes is 20. This is the coefficient of h^3v^3 in $(h + v)^6$.

Confirm these results by expanding $(h + v)^5$ and $(h + v)^6$.

We can propose a model. Let $nh + mv$ be the shortest distance between two junctions made up of n horizontal steps and m vertical steps. Then the number of different routes is the coefficient of $h^n v^m$ in the binomial expansion of $(h + v)^{n+m}$.

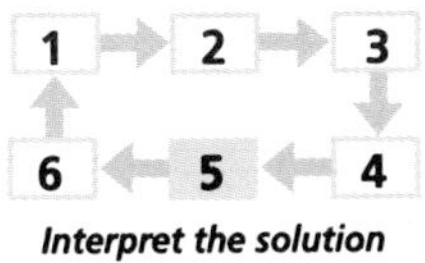

Interpret the solution

Validation of the model

Choose journeys between different junctions and check that the model works for each of your journeys.

Discuss the appropriateness of the model for the streets of Venice.

1 2 3
6 5 4
Compare with reality

Refinement of the model

Investigate the effects of:

- applying the model to Venice,
- applying the model to other cities with grid road systems,
- revising the model for cities whose roads almost form exact rectangular grids.

CONSOLIDATION EXERCISES FOR CHAPTER 14

1 Write down the expansion of $(1 + x)^5$.

Hence, by letting $x = z + z^2$, find the coefficient of z^3 in the expansion of $(1 + z + z^2)^5$ in powers of z.

(UCLES Linear Question 6, Specimen Paper 1, 1994)

2 Find the term independent of x in the expansion of $\left(x^2 - \frac{2}{x}\right)^6$.

(UCLES Modular Question 1, Specimen Paper P3, 1994)

3 **a)** Obtain the first four non-zero terms of the binomial expansion in ascending powers of x of $(1 - x^2)^{-\frac{1}{2}}$, given that $|x| < 1$.

b) Show that, when $x = \frac{1}{3}$, $(1 - x^2)^{-\frac{1}{2}} = \frac{3}{4}\sqrt{2}$.

c) Substitute $x = \frac{1}{3}$ into your expansion and hence obtain an approximation to $\sqrt{2}$, giving your answer to five decimal places.

(ULEAC Question 4, Paper 2, January 1995)

4 In triangle ABC, AB = 3 cm, AC = 4 cm. BC = d cm and $\cos A = x$.

a) Show that $d = 5\sqrt{1 - 0.96x}$.

b) Given that x^3 and higher powers of x may be neglected, use a binomial expansion to express d in the form $p + qx + rx^2$, where p, q and r are constants whose values are to be found.

c) Show that, when $x = 0.005$, the approximation found in part **b)** is correct to 3 decimal places.

(ULEAC Question 7,Paper 2, ...June 1992)

5 Given that $(1 + kx)^8 = 1 + 12x + px^2 + qx^3 + \ldots$ for all $x \in \Re$:

a) find the value of k, the value of p and the value of q.

b) Using your values of k, p and q find the numerical coefficient of the x^3 term in the expansion of $(1 - x)(1 + x)^8$.

(ULEAC Question 5, Paper 1, May 1994)

6 **a)** Expand $(2+x)^4$ and $(2-x)^4$ in ascending powers of x.

b) Use your expansions to show that the equation:

$$(2+x)^4 - (2-x)^4 = 80$$

simplifies to the cubic equation $x^3 + 4x - 5 = 0$.

c) Show that 1 is a root of $x^3 + 4x - 5 = 0$.

d) Show that there are no further real roots of this cubic equation.

(ULEAC Question 4, Paper 1, January 1995)

7 A shelf support is made of wood, with its cross-section in the shape below. The top and bottom edges of the support are horizontal.

The shape of the curved edge of the shelf support is modelled by the equation:

$$8y = (2x-3)^3 \text{ for } -1 \le y \le 1.$$

a) Find the coordinates of the point P.

b) Using Pascal's triangle, or otherwise, write down the expansion of $(2x-3)^3$.

c) Write down the integral which should be calculated in order to find the area AQR between the curve and the x-axis.

d) Carry out the integration to find this area.

e) Hence or otherwise calculate the area of the cross-section of the shelf support.

(MEI Question 5, Paper 1, January 1995)

8 **a)** Write down the expansion of $(2-x)^4$.

b) Find the first four terms in the expansion of $(1+2x)^{-3}$ in ascending powers of x. For what range of values of x is this expansion valid?

c) When the expansion is valid:

$$\frac{(2-x)^4}{(1+2x)^3} = 16 + ax + bx^2 + \ldots$$

Find the values of a and b.

(MEI Question 1, Paper 2, January 1995)

9 **a)** Write down the expansion of $(1+x)^3$.

b) Find the first four terms in the expansion of $(1-x)^{-4}$ in ascending powers of x. For what values of x is this expansion valid?

c) When the expansion is valid:

$$\frac{(1+x)^3}{(1-x)^4} = 1 + 7x + ax^2 + bx^3 + \ldots.$$

Find the values of a and b.

(MEI Question 1, Paper 2, January 1994)

10 **a)** Use the formula for solving a quadratic equation to write down the two roots of the equation $x^2 + x + p = 0$, where p is a constant.

b) Find the values of A, B and C for which:

$$1 + Ap + Bp^2 + Cp^3$$

is equal to the first four terms in the binomial expansion of $\sqrt{1-4p}$. State the condition for which the expansion is valid.

c) Use your expansion to find approximations to the two roots of the equation, assuming that terms in p^4 and higher powers of p are so small that they may be neglected.

d) Use your answer to part **c)** to find the approximate values of the two roots of the equation $10x^2 + 10x + 1 = 0$.

(MEI Question 1, Paper 2, June 1993)

Summary

- The binomial theorem is the expansion of $(a + b)^n$ in the form:

$$(a+b)^n = \binom{n}{0}a^n + \binom{n}{1}a^{n-1}b + \binom{n}{2}a^{n-2}b^2 + \ldots + \binom{n}{r}a^{n-r}b^r + \ldots + \binom{n}{n-1}ab^{n-1} + \binom{n}{n}ab^n$$

where n is a positive integer.

- The binomial series is a special case of the binomial theorem when $a = 1$ and $b = x$ and is usually written

$$(1+x)^n = 1 + nx + \frac{n(n-1)}{2!}x^2 + \ldots + \frac{n(n-1)\ldots(n-r+1)}{r!}x^r + \ldots + nx^{n-1} + x^n$$

- For negative and fractional values of n the binomial series gives an infinite number of terms. It converges provided $|x| < 1$.

PURE MATHS 15

Trigonometry 3

In this chapter we examine:

- *an alternative unit for measuring angles called radians*
- *a method of solving equations containing trigonometric functions*
- *trigonometric identities which are useful for manipulating trigonometric equations.*

RADIANS: CIRCULAR MEASURE

Exploration 15.1

Circular motion on the big wheel

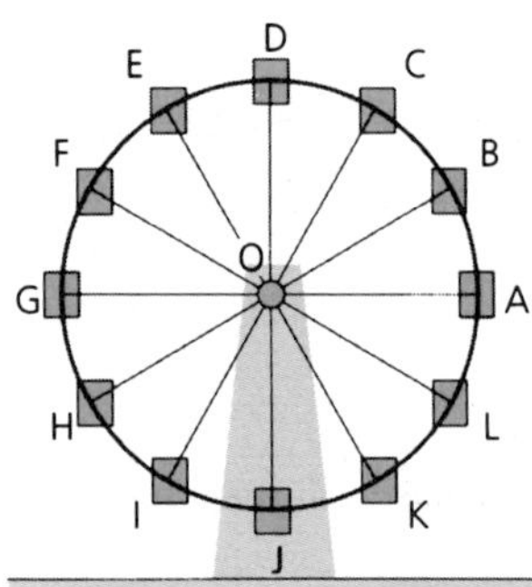

Imagine you are riding on a big wheel at a fairground. The distance of a car from the centre of the wheel is 8 metres. There are 12 cars altogether, equally spaced.

- How far do you travel in one complete revolution of the wheel?
- Assuming passengers are picked up at the bottom of the wheel, how far will you travel between pick-ups?
- Through what angle does the wheel rotate between pick-ups?
- How far will you travel when the wheel rotates through $x°$?
- Through what angle does the wheel rotate when a single car travels 8 metres?

Radians

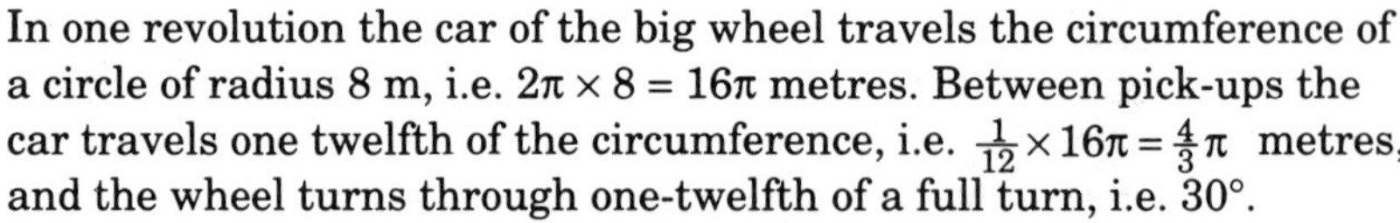

In one revolution the car of the big wheel travels the circumference of a circle of radius 8 m, i.e. $2\pi \times 8 = 16\pi$ metres. Between pick-ups the car travels one twelfth of the circumference, i.e. $\frac{1}{12} \times 16\pi = \frac{4}{3}\pi$ metres, and the wheel turns through one-twelfth of a full turn, i.e. 30°.

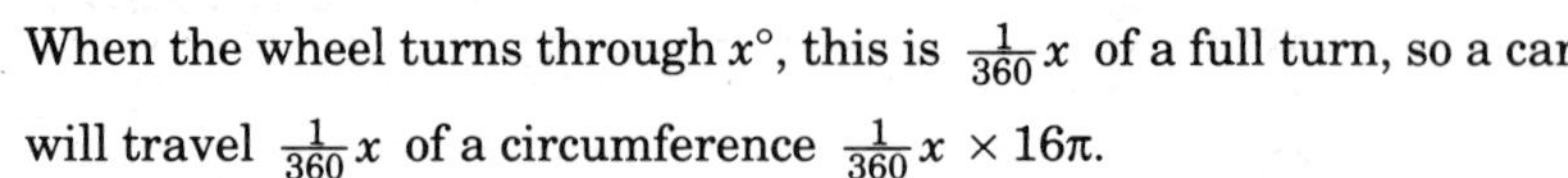

When the wheel turns through $x°$, this is $\frac{1}{360}x$ of a full turn, so a car will travel $\frac{1}{360}x$ of a circumference $\frac{1}{360}x \times 16\pi$.

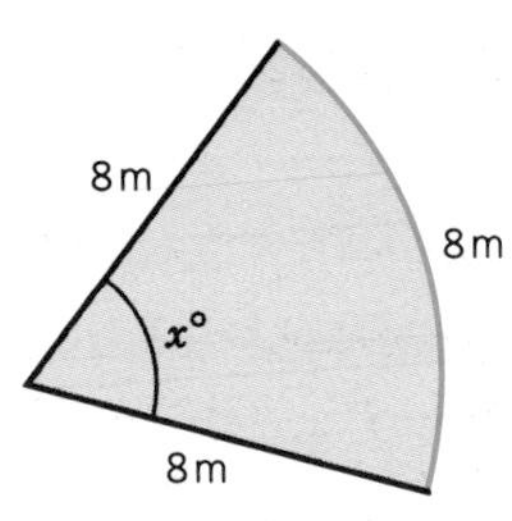

When the car travels 8 metres along an arc of the circle, let the angle the wheel turns through be $x°$, then:

$$\frac{x}{360} \times 16\pi = 8 \Rightarrow x = \frac{8 \times 360}{16\pi} = \frac{180}{\pi}$$

The angle that the wheel turns through when a car travels a distance equal in length to the radius is one **radian**.

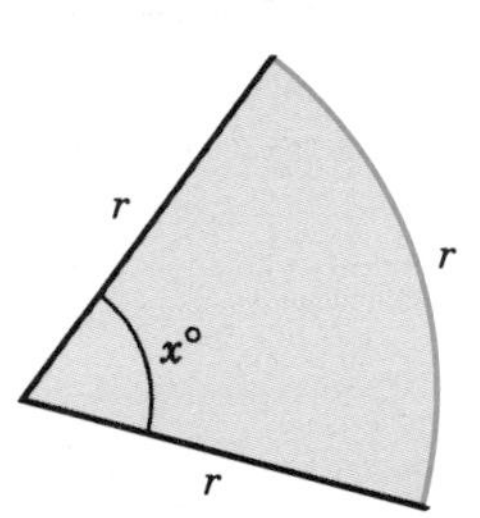

Note: This works for a circle of radius r, for which the circumference is $2\pi r$ and:

$$\frac{x}{360} \times 2\pi r = r \Rightarrow x = \frac{180}{\pi}$$

Geometrically, the angle at the centre of a circle that subtends an arc equal in length to the radius is called a radian and is defined by:

$$1 \text{ radian} \equiv \frac{180°}{\pi} \approx 57.3°$$

Check this on your calculator.

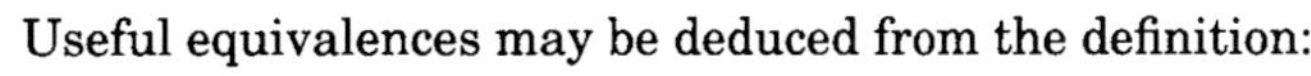

Useful equivalences may be deduced from the definition:

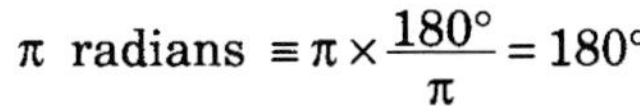

$$\pi \text{ radians} \equiv \pi \times \frac{180°}{\pi} = 180°$$

from which you may deduce:

$$360° \equiv 2\pi \text{ radians}$$

$$90° \equiv \tfrac{1}{2}\pi \text{ or } \frac{\pi}{2} \text{ radians}$$

$$1° \equiv \frac{\pi}{180} \text{ radians} \quad \text{etc.}$$

r
r
1 radian
r

Example 15.1

Convert these angles to radians.

a) 45° ***b)*** 150°

Solution

a) $180° = \pi \text{ radians} \Rightarrow 45° = \frac{\pi}{4} \text{ radians}$

b) $180° = \pi \text{ radians} \Rightarrow 30° = \frac{\pi}{6} \Rightarrow 150° = \frac{5\pi}{6} \text{ radians}$

Example 15.2

Convert these angles to degrees.

a) $\frac{\pi}{6}$ radians ***b)*** $\frac{7\pi}{4}$ radians

Solution

a) $\pi \text{ radians} = 180° \Rightarrow \frac{\pi}{6} \text{ radians} = \frac{180°}{6} = 30°$

b) $\pi \text{ radians} = 180° \Rightarrow \frac{7\pi}{4} \text{ radians} = \frac{7 \times 180°}{4} = 315°$

Note: When an angle is expressed without any units it is assumed to be measured in radians, as in the answers in the next two examples.

Example 15.3

a) *Sketch a graph of* $y = \sin x$*, where* x *is measured in radians, for* $0 \le x \le 2\pi$.

b) *From your graph, find, as multiples of* π*, values of* x *that satisfy the following equations.*

i) $\sin x = 0$ ***ii)*** $\sin x = \pm 1$ ***iii)*** $\sin x = \frac{1}{2}$

iv) $\sin x = -\frac{1}{2}$ ***v)*** $\sin x = \frac{\sqrt{3}}{2}$ ***vi)*** $\sin x = -\frac{\sqrt{3}}{2}$

Solution

a)

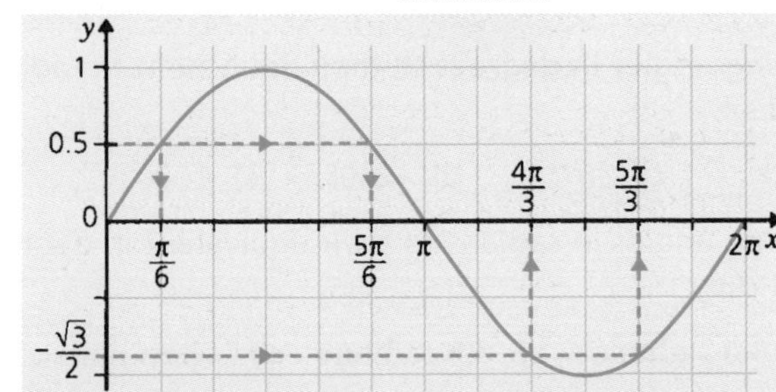

b) ***i)*** $\sin x = 0 \Rightarrow x = 0, \pi$ *or* 2π

ii) $\sin x = \pm 1 \Rightarrow x = \frac{\pi}{2}$ *or* $\frac{3\pi}{2}$

iii) $\sin x = \frac{1}{2} \Rightarrow x = \frac{\pi}{6}$ *or* $\frac{5\pi}{6}$ *(see graph)*

iv) $\sin x = -\frac{1}{2} \Rightarrow x = \frac{7\pi}{6}$ *or* $\frac{11\pi}{6}$

v) $\sin x = \frac{\sqrt{3}}{2} \Rightarrow x = \frac{\pi}{3}$ *or* $\frac{2\pi}{3}$

vi) $\sin x = -\frac{\sqrt{3}}{2} \Rightarrow x = \frac{4\pi}{3}$ *or* $\frac{5\pi}{3}$ *(see graph)*

Check these answers using your calculator in radian mode.

Example 15.4

a) *Sketch a graph of* $y = \cos^2 x$*, where* x *is measured in radians, for* $0 \le x \le 2\pi$. [$\cos^2 x$ *means* $(\cos x)^2$]

b) *From your graph, find, as multiples of* π*, values of* x *that satisfy these equations.*

i) $\cos^2 x = 0$ ***ii)*** $\cos^2 x = 1$ ***iii)*** $\cos^2 x = 0.25$ ***iv)*** $\cos^2 x = 0.75$

Solution

a)

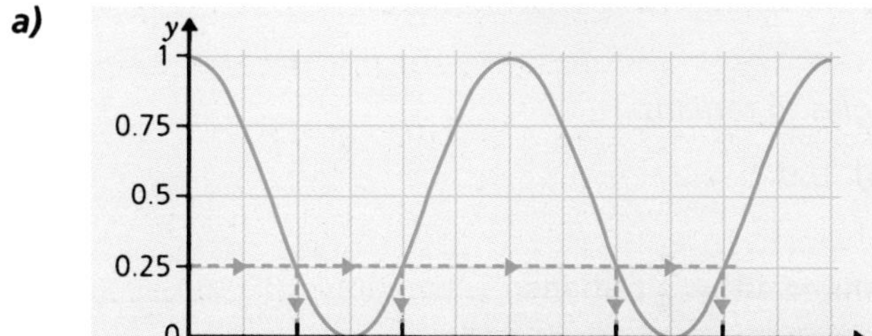

b) i) $\cos^2 x = 0 \Rightarrow x = \frac{\pi}{2}$ *or* $\frac{3\pi}{2}$

ii) $\cos^2 x = 1 \Rightarrow x = 0, \pi$ *or* 2π

iii) $\cos^2 x = \frac{1}{4} \Rightarrow x = \frac{\pi}{3}, \frac{2\pi}{3}, \frac{4\pi}{3}$ *or* $\frac{5\pi}{3}$ *(see graph)*

iv) $\cos^2 x = \frac{3}{4} \Rightarrow x = \frac{\pi}{6}, \frac{5\pi}{6}, \frac{7\pi}{6}$ *or* $\frac{11\pi}{6}$

*Check these answers using your calculator in radian mode. For example, for part **iii)**:*

$$\cos^2 x = 0.25 \Rightarrow \cos x = \pm\sqrt{0.25} = \pm 0.5$$
$$\Rightarrow x = \cos^{-1}(\pm 0.5) \Rightarrow x = \frac{\pi}{3}, \frac{2\pi}{3}, \frac{4\pi}{3} \text{ or } \frac{5\pi}{3}$$

equally: $\cos\frac{\pi}{3} = 0.5 \Rightarrow \cos^2\frac{\pi}{3} = 0.5^2 = 0.25$

$\cos\frac{2\pi}{3} = -0.5 \Rightarrow \cos^2\frac{2\pi}{3} = (-0.5)^2 = 0.25$ *etc.*

EXERCISES

15.1 A

1 Without using a calculator, express the following angles in radians as fractions or multiples of π.

a) 30° **b)** –135° **c)** 720° **d)** –210° **e)** 300

2 Without using a calculator, express the following angles in degrees.

a) $-\frac{2\pi}{3}$ **b)** $\frac{\pi}{8}$ **c)** $-\frac{11\pi}{6}$ **d)** $\frac{5\pi}{12}$ **e)** 1.75π

3 Copy and complete the table of equivalences.

Degrees	0°	30°		90°	120°	150°			240°	270°			360°
Radians	0		$\frac{\pi}{3}$				π	$\frac{7\pi}{6}$			$\frac{5\pi}{3}$	$\frac{11\pi}{6}$	

4 Convert the following angles in degrees to their equivalent in radians, giving answers to 3 s.f.

a) 100° **b)** –29° **c)** 307° **d)** –200° **e)** 1000°

5 Convert the following angles in radians to their equivalent in degrees, giving answers to 1 d.p.

a) 2 radians **b)** 0.5 radians **c)** 6.3 radians **d)** $\sqrt{3}$ radians

6 i) Sketch a graph of $y = \cos x$, where x is measured in radians, for $0 \le x \le 2\pi$.

ii) From your graph, find, in multiples of π, values of x that satisfy the following equations.

a) $\cos x = 0$ b) $\cos x = \pm 1$ c) $\cos x = \frac{1}{2}$ d) $\cos x = -\frac{1}{2}$

e) $\cos x = \frac{\sqrt{3}}{2}$ f) $\cos^2 x = \frac{1}{2}$

Check your results using a calculator in radian mode.

7 a) Sketch a graph of $y = \sin^2 x$, where x is measured in radians, for $0 \le x \le 2\pi$.

b) From your graph, find, in multiples of π, values of x such that:

i) $\sin^2 x = 0$ ii) $\sin^2 x = 1$ iii) $\sin^2 x = 0.25$ iv) $\sin^2 x = 0.5$

Check your results using a calculator in radian mode.

EXERCISES

15.1 B

1 Express the following angles as fractions, or multiples of π, without using a calculator.

a) $60°$ b) $210°$ c) $-45°$ d) $540°$ e) $-120°$

2 Without using a calculator, express the following angles in degrees.

a) $\frac{\pi}{4}$ b) $-\frac{\pi}{6}$ c) $\frac{5\pi}{3}$ d) $-\frac{7\pi}{12}$ e) $\frac{5\pi}{4}$

3 Use a calculator to convert the following angles in degrees to their equivalent in radians, giving answers to 3 s.f.

a) 20° b) –72° c) 400° d) –140° e) 760°

4 Convert the following angles in radians to their equivalent in degrees, giving answers to 1 d.p.

a) 1.5 radians b) 0.4 radians c) 3 radians d) 5 radians e) 7.2 radians

5 a) Sketch a graph of $y = \sin x$, where x is measured in radians, for $0 \le x \le 2\pi$.

b) From your graph, find, in multiples of π, values of x that satisfy the following equations.

i) $\sin x = 0$ ii) $\sin x = \pm 1$ iii) $\sin x = \frac{1}{2}$

iv) $\sin x = -\frac{1}{2}$ v) $\sin x = \frac{\sqrt{3}}{2}$ vi) $\sin^2 x = \frac{1}{2}$

6 a) Sketch a graph of $y = \cos x$, where x is measured in radians, for $-\pi \le x \le \pi$.

b) Sketch a graph of $y = \cos^2 x$, where x is measured in radians, for $-\pi \le x \le \pi$.

c) From your graph, find in multiples of π, values of x that satisfy the following equations.

i) $\cos^2 x = 0$ ii) $\cos^2 x = 1$ iii) $\cos^2 x = 0.25$ iv) $\cos^2 x = 0.5$

7 **a)** Sketch a graph of $y = \tan x$, where x is measured in radians, for $-2\pi \le x \le 2\pi$.

b) Sketch a graph of $y = \tan^2 x$, where x is measured in radians, for $-2\pi \le x \le 2\pi$.

c) From your graph, find, in multiples of π, values of x that satisfy the following equations.

i) $\tan^2 x = 0$ **ii)** $\tan^2 x = 1$ **iii)** $\tan^2 x = 3$

iv) $\tan^2 x = \frac{1}{3}$ **v)** $\tan^2 x$ is undefined

Check your results using a calculator in radian mode.

SECTORS

Arc length

Since an angle of 1 radian subtends an arc of length r, an angle of 2π radians subtends an arc of length $2\pi r$ (the circumference) and in general an angle of θ radians subtends an arc of length

$$\frac{\theta}{2\pi}\times 2\pi r = r\theta$$

Area of a sector

Since the area of a circle, radius r, is πr^2, the area of a sector with angle at the centre θ is given by:

$$\frac{\theta}{2\pi}\times \pi r^2 = \tfrac{1}{2}r^2\theta$$

Example 15.5

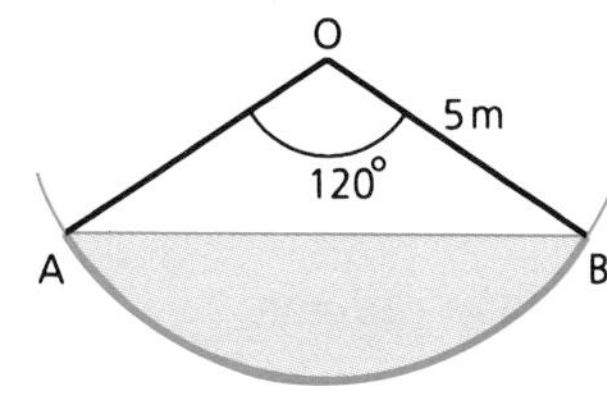

A and B are points on the circumference of a circle, centre O and radius 5 metres, and $\angle\mathrm{AOB} = 120°$. Calculate:

a) *the length of arc AB,*

b) *the area of sector AOB,*

c) *the area of the shaded segment.*

Solution

Firstly convert degrees to radians: $120° \equiv \frac{2\pi}{3}$ radians.

a) *Length of arc* $\mathrm{AB} = r\theta = 5\times\frac{2\pi}{3} \approx 10.5\text{ m}$.

b) *Area of sector* $\mathrm{AOB} = \frac{1}{2}r^2\theta = \frac{1}{2}\times 25\times\frac{2\pi}{3} \approx 26.2\text{ m}^2$

c) *First find the area of* $\Delta\mathrm{AOB}$.

Area of a triangle with adjacent sides b, c and included angle A is given by $\Delta = \frac{1}{2}bc\sin A$.

In this case $b = c = r$ and $A = 120°$, hence:

area of $\Delta\mathrm{AOB}$ *is* $\frac{1}{2}r^2\sin 120° = \frac{1}{2}\times 25\times\frac{\sqrt{3}}{2}$ $(\approx 10.8\text{ m}^2)$

$\Rightarrow$ *area of shaded segment = area of sector – area of triangle*

$$= \tfrac{1}{2}\times 25\times\tfrac{2\pi}{3} - \tfrac{1}{2}\times 25\times\tfrac{\sqrt{3}}{2} = \tfrac{1}{2}\times 25\times \tfrac{2\pi}{3} - \tfrac{\sqrt{3}}{2} = 15.4\text{ m}^2$$

Example 15.6

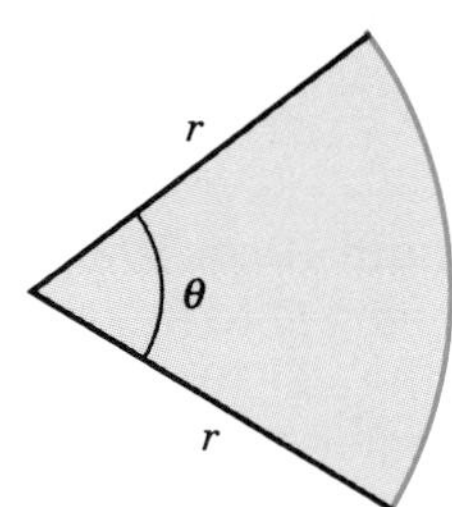

A piece of wire of length 40 cm is bent into the shape of a sector of a circle of radius r cm and angle θ radians.

a) *If $r = 15$, find:*
   i) *the size of angle θ,* **ii)** *the area A of the sector.*

b) *Find a formula for θ in terms of r and show that $A = 20r - r^2$.*

c) *Find the value of r that will give the largest sector area. Write down this area, together with the corresponding value of θ.*

Solution

a) **i)** Perimeter of sector $= 40 \Rightarrow$ arc length $= 40 - 2 \times 15 = 10$
Arc length $= r\theta$ and $r = 15 \Rightarrow 15\theta = 10 \Rightarrow \theta = \frac{2}{3}$
ii) Area $A = \frac{1}{2}r^2\theta = \frac{1}{2} \times 15^2 \times \frac{2}{3} = 75 \text{ cm}^2$

b) Arc length $= r\theta = 40 - 2r \Rightarrow \theta = \dfrac{40 - 2r}{r}$

Area $A = \frac{1}{2}r^2\theta = \frac{1}{2}r^2\dfrac{40-2r}{r} = \frac{1}{2}r^2\dfrac{2(20-r))}{r} = r(20-r)$

$\Rightarrow A = 20r - r^2$

c) For maximum area:
$\dfrac{dA}{dr} = 0 \Rightarrow \dfrac{dA}{dr} = 20 - 2r = 0 \Rightarrow r = 10 \text{ cm}$

When $r = 10$ $A = 20 \times 10 - 10^2 = 100 \text{ cm}^2$

$\theta = \dfrac{40 - 2 \times 10}{10} = 2 \text{ radians}$

EXERCISES

15.2 A

1 Find **i)** the length of the arc PQ, **ii)** the area of sector POQ, **iii)** the area of the shaded segment in these diagrams.

a)

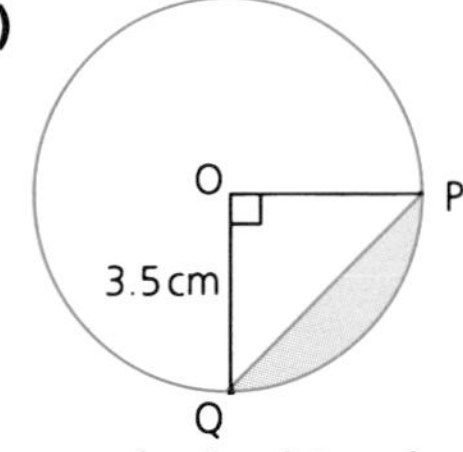

b)

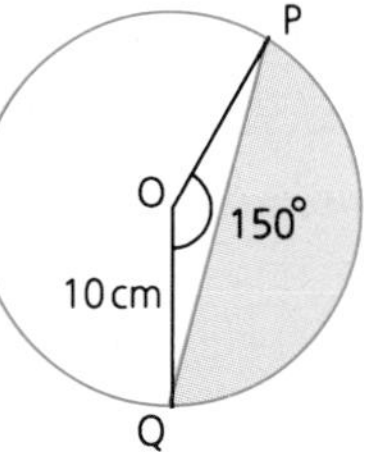

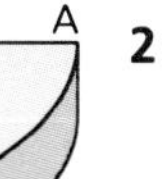

2 An angle θ subtends an arc AB of length 20 cm in a circle, centre O, diameter 30 cm. Find the value of θ and the area of sector AOB.

3 A wedge of cheese, 2 cm thick, has a cross-section which is a sector of a circle. OA = OB = 12 cm. Length of arc AB = 6 cm. Find the volume of the cheese.

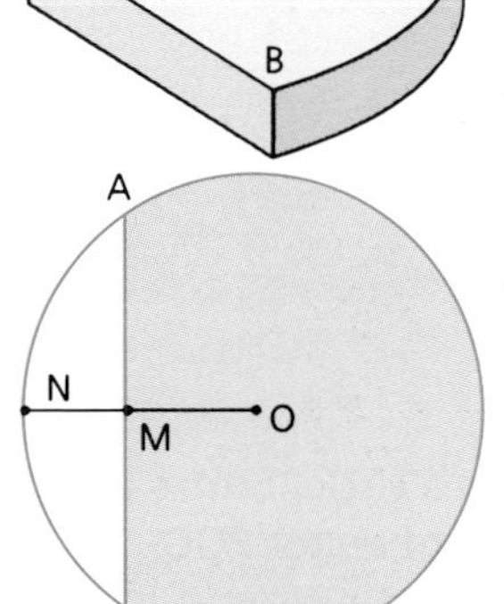

4 An angle θ subtends an arc RS of length 16 cm in a circle of radius r cm. The area of the sector ROS is 50 cm^2. Formulate two equations in r and θ and so find the values of r and θ.

5 The diagram shows a circle of radius 10 cm. M is the midpoint of the chord AB and the length of MN is 2 cm. Find the perimeter and area of the shaded region.

6 A cylindrical pipe, diameter 1 metre, contains water to a depth of 0.3 m. The pipe delivers 50 litres of water per second.

a) Find the cross-sectional area of the water.

b) Find the speed with which the water is flowing through the pipe, in $\mathrm{m\,s^{-1}}$.

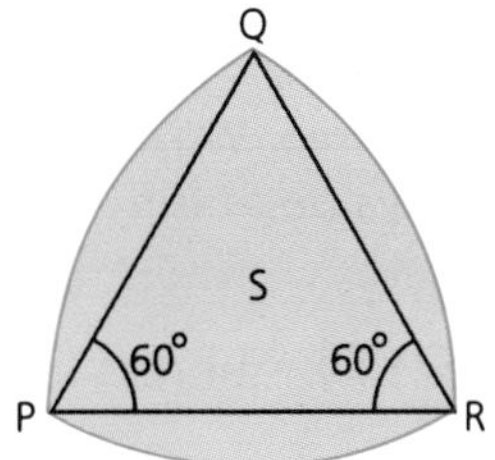

7 The triangle PQR is equilateral with each side of length 10 cm. With centre P and radius 10 cm, a circular arc is drawn joining Q to R. Similar arcs are drawn with centres Q and R and with radii 10 cm, joining R to P and P to Q respectively. The shaded region S is bounded by the three arcs.

Calculate: **a)** the area of ΔPQR, **b)** the area of S.

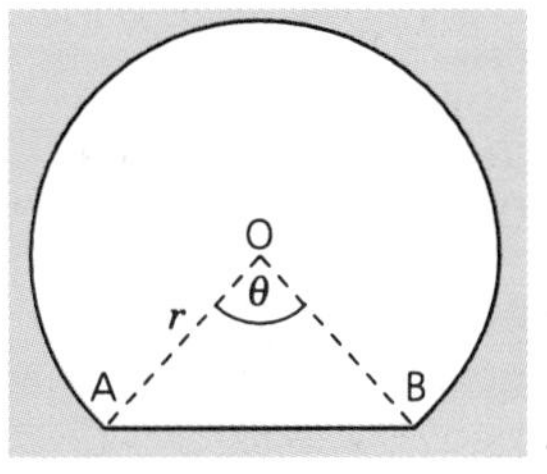

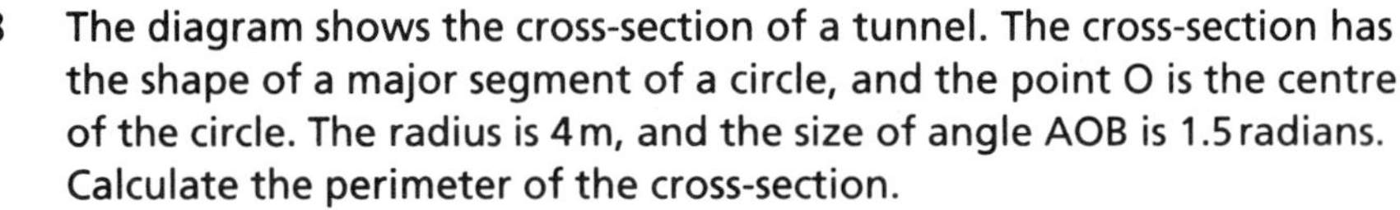

8 The diagram shows the cross-section of a tunnel. The cross-section has the shape of a major segment of a circle, and the point O is the centre of the circle. The radius is 4 m, and the size of angle AOB is 1.5 radians. Calculate the perimeter of the cross-section.

(UCLES (Linear) Specimen Paper 1, 1994)

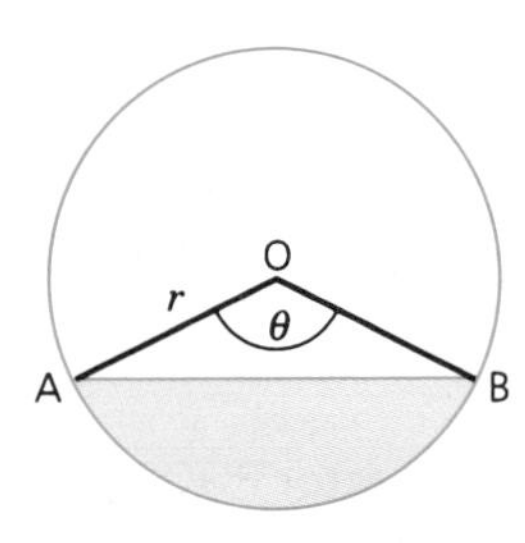

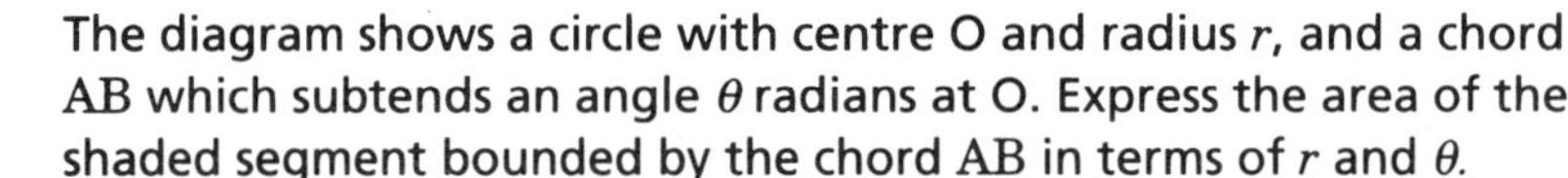

9 The diagram shows a circle with centre O and radius r, and a chord AB which subtends an angle θ radians at O. Express the area of the shaded segment bounded by the chord AB in terms of r and θ.

Given that the area of this segment is one-third of the area of the triangle OAB, show that $3\theta - 4\sin\theta = 0$.

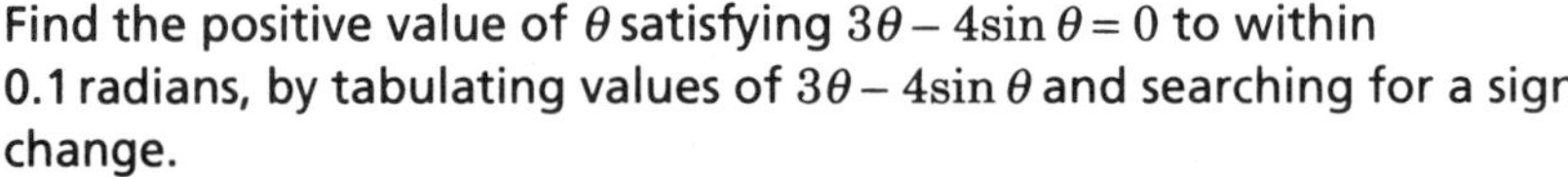

Find the positive value of θ satisfying $3\theta - 4\sin\theta = 0$ to within 0.1 radians, by tabulating values of $3\theta - 4\sin\theta$ and searching for a sign change.

(UCLES (Modular) Specimen Paper 1, 1994)

10 A piece of wire of length l cm is bent into the shape of a sector of a circle of radius r cm and angle θ radians. Find, in terms of l, the value of r that will give the largest sector area.

Write down this area, together with the corresponding value of θ.

EXERCISES

15.2 B

1 Find **i)** the length of the arc PQ, **ii)** the area of sector POQ, **iii)** the area of the shaded segment in these diagrams.

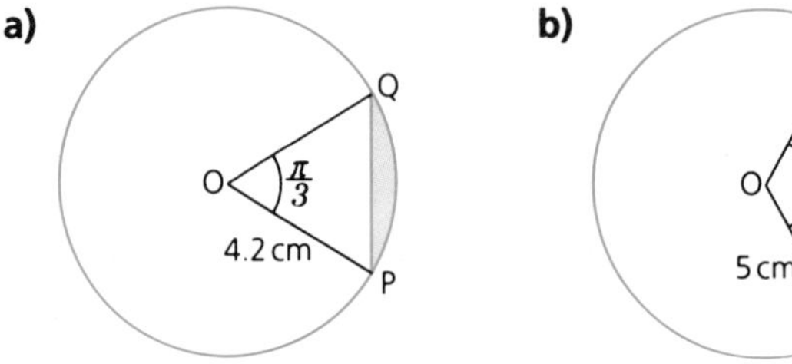

2 An angle θ subtends an arc PQ of length 15 cm in a circle, centre O, diameter 6.5 cm. Find the value of θ and the area of sector POQ.

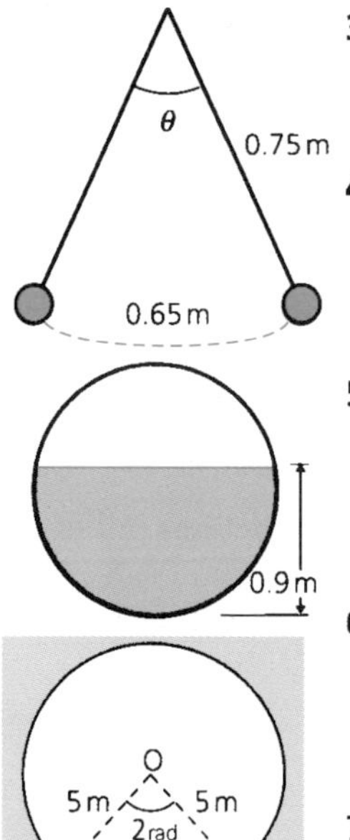

3 An angle θ subtends an arc PQ of length 25 cm in a circle of radius r cm. The area of the sector POQ is 72 cm^2. Formulate two equations in r and θ and so find the values of r and θ.

4 The pendulum of a grandfather clock is 0.75 m long and swings in a circular arc, as shown in the diagram. It takes 1 second to swing from one end of the arc to the other and the 'bob' on the end moves 0.65 m. What angle does it sweep out in 1 second?

5 A cylindrical pipe of diameter 1.5 m, contains water to a depth of 0.9 m.

a) Find the cross-sectional area of water.

b) If the water is flowing at a rate of 60 litres per second, find the speed with which the water is flowing, in m s^{-1}.

6 A railway tunnel has cross-section shaped as a major segment of a circle with diameter 10 m. The angle POQ is 2 radians. If the tunnel is 500 m long, find the area of lining material needed to line the curved surface inside the tunnel.

7 A circular cone made of card has base radius r and slant height l. If the card is unrolled it forms a circular sector. Find the angle of the sector in radians and hence show that the curved surface area of a cone is πrl.

8 The arc PQ of a circle of centre O, radius r subtends an angle α radians ($\alpha < \pi$) at O. Show that the area of triangle OPQ is $\frac{1}{2}r^2 \sin\alpha$ and hence find the area of the minor segment cut off by PQ. What can you deduce about α and $\sin\alpha$ when α is small?

9 **a)** Complete the table for $\sin\alpha$ (where α is in radians).

α	**0.1**	**0.08**	**0.06**	**0.04**	**0.02**
$\sin\alpha$	0.099 833 42				
$\alpha - \sin\alpha$	0.000 166 6				

Comment on the results. Do your comments support your deduction about α and $\sin\alpha$ in question **8**?

10 The diagram shows a triangle ABC in which AB = 5 cm, AC = BC = 3 cm. The circle, centre A, radius 3 cm, cuts AB at X; the circle, centre B, radius 3 cm, cuts AB at Y.

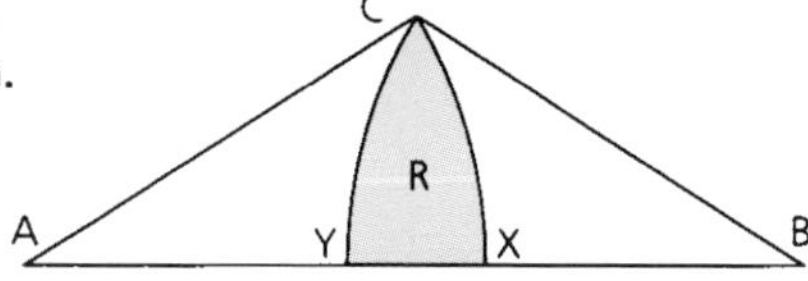

a) Determine the size of angle CAB, giving your answer in radians to four decimal places.

b) The region R, shaded in the diagram, is bounded by the arcs CX, CY and the straight line XY.

Calculate: **i)** the length of the perimeter of R,

ii) the area of the sector ACX,

iii) the area of the region R.

(Oxford, Paper 1, 1991)

RECIPROCAL TRIGONOMETRIC FUNCTIONS

We have already studied the graphs of $\sin x$, $\cos x$ and $\tan x$, all of which can have domain as the set of real numbers except for $\tan x$, which is not valid for $x = \ldots, -270°, -90°, 90°, 270°, \ldots$ (or $x = \ldots, \frac{-3\pi}{2}, \frac{-\pi}{2}, \frac{\pi}{2}, \frac{3\pi}{2}, \ldots$ in radians)

Each one is a periodic function:

- $\sin x$ is an odd function with period 360° (or 2π)
- $\cos x$ is an even function with period 360° (or 2π)
- $\tan x$ is an odd function with period 180° (or π)

We can now introduce three new trigonometrical functions, all of which are reciprocals of the three above.

$$\operatorname{cosec} x = \frac{1}{\sin x} \quad \sec x = \frac{1}{\cos x} \quad \cot x = \frac{1}{\tan x}$$

Exploration 15.2

Graphs of functions

Using a scale of $-360° \le x \le 360°$ [or $-2\pi \le x \le 2\pi$ if working in radian mode] and $-3 \le y \le 3$, use a graphics calculator or computer package to plot the following pairs of graphs separately.

a) $y = \sin x$ and $y = \operatorname{cosec} x$ ($\frac{1}{\sin x}$)

b) $y = \cos x$ and $y = \sec x$ ($\frac{1}{\cos x}$)

c) $y = \tan x$ and $y = \cot x$ ($\frac{1}{\tan x}$)

Study each pair of graphs in turn.

- Describe what happens to one curve as the other curve crosses the x-axis. Why does this happen?
- Describe the nature of any stationary points.
- How are the graphs of $\operatorname{cosec} x$ and $\sec x$ related?
- How are the graphs of $\tan x$ and $\cot x$ related?

These new trigonometrical functions may not be supplied on your calculator but using their definitions, you can find values and solve equations, as shown in the following examples.

Example 15.7

Find: ***a)*** $\operatorname{cosec} 30°$ ***b)*** $\sec^2 135°$ ***c)*** $5\cot(-\frac{2\pi}{3})$

Solution

a) $\operatorname{cosec} 30° = \frac{1}{\sin 30°} = \frac{1}{0.5} = 2$ ***b)*** $\sec^2 135° = \frac{1}{\cos^2 135°} = \frac{1}{0.5} = 2$

c) $5\cot(-\frac{2\pi}{3}) = 5 \times \frac{1}{\tan(-\frac{2\pi}{3})} = 5 \times \frac{1}{1.732} = 2.89$ (3 s.f.)

Example 15.8

Find values of θ such that $0° \le \theta \le 360°$, where:

a) $\operatorname{cosec} \theta = -5$ ***b)*** $\sec 2\theta = 3$ ***c)*** $\cot^2 \theta = 3$

Solution

a) $\text{cosec}\ \theta = -5 \Rightarrow \dfrac{1}{\sin\theta} = -5 \Rightarrow \sin\theta = -0.2$

$\Rightarrow \theta = \sin^{-1}(-0.2) = -11.5°$ *(from calculator)*

but $0° \le \theta \le 360°$*:*
$\Rightarrow \theta = 180° + 11.5°$ *or* $360° - 11.5°$
$\Rightarrow \theta = 191.5°$ *or* $\theta = 348.5°$

b) $\sec 2\theta = 3 \Rightarrow \dfrac{1}{\cos 2\theta} = 3 \Rightarrow \cos 2\theta = \frac{1}{3}$

$\Rightarrow 2\theta = \cos^{-1}\frac{2}{3} = 70.53°$ *(from calculator)*

but $0° \le \theta \le 360° \Rightarrow 0° \le 2\theta \le 720°$
$\Rightarrow 2\theta = 70.53°, 360° - 70.53°, 360° + 70.53°$ *or* $720° - 70.53°$
$\Rightarrow 2\theta = 70.53°, 289.47°, 430.53°$ *or* $649.47°$
$\Rightarrow \theta = 35.3°, 144.7°, 215.3°$ *or* $324.7°$

c) $\cot^2\theta = 3 \Rightarrow \cot\theta = \pm\sqrt{3} \Rightarrow \dfrac{1}{\tan\theta} = \pm\sqrt{3} \Rightarrow \tan\theta = \pm\dfrac{1}{\sqrt{3}}$

Either $\tan\theta = +\dfrac{1}{\sqrt{3}} \Rightarrow \theta = \tan^{-1}(+\dfrac{1}{\sqrt{3}}) = 30°$ *(from calculator)*

but $0° \le \theta \le 360° \Rightarrow \theta = 30°$ *or* $180° + 30° \Rightarrow \theta = 30°$ *or* $210°$

or $\tan\theta = -\dfrac{1}{\sqrt{3}} \Rightarrow \theta = \tan^{-1}(-\dfrac{1}{\sqrt{3}}) = -30°$ *(from calculator)*

but $0° \le \theta \le 360° \Rightarrow \theta = 180° - 30°$ *or* $360° - 30° \Rightarrow \theta = 150°$ *or* $330°$

EXERCISES

15.3 A

1 Find (where possible) the values of the following.

a) $\sec 60°$ **b)** $\text{cosec}\ 100°$ **c)** $\cot -70°$ **d)** $2\sec 32°$
e) $-3\ \text{cosec}\ 30°$ **f)** $5\cot 270°$ **g)** $\sec^2 45°$ **h)** $1 + \tan^2 45°$
i) $\text{cosec}^2\ 60°$ **j)** $\cot^2 60° + 1$

What do you notice about your answers to parts **g)** and **h)** and to parts **i)** and **j)**?

2 Find (where possible) the values of the following.

a) $\sec\frac{5\pi}{3}$ **b)** $\text{cosec}\frac{3\pi}{2}$ **c)** $\cot -\frac{\pi}{4}$
d) $7\sec\frac{\pi}{2}$ **e)** $\frac{1}{2}\text{cosec}\frac{7\pi}{6}$ **f)** $-3\cot\frac{\pi}{10}$

3 Find (if possible) values of θ such that $0° \le \theta \le 360°$ which satisfy these equations.

a) $\sec\theta = 2$ **b)** $\text{cosec}\ \theta = -5$ **c)** $2\cot\theta = 1$ **d)** $\text{cosec}\ 2\theta = 3$
e) $5\sec 2\theta = 8$ **f)** $\cot\frac{1}{2}\theta = -2.5$ **g)** $\sec^2\theta = 2$ **h)** $5\text{cosec}^2\theta = 2$
i) $\cot^2 2\theta = \frac{1}{3}$ **j)** $1 + \cot^2\frac{1}{2}\theta = 5$

4 Find (if possible) values of θ such that $0 \le \theta \le 2\pi$, which satisfy the following equations.

a) $\sec\theta = -1$ **b)** $\text{cosec}\ \theta = 4$ **c)** $2\cot\theta = 5$ **d)** $\text{cosec}\ 2\theta = \frac{2}{\sqrt{3}}$
e) $2\sec 2\theta = 3$ **f)** $\cot 3\theta = -2$ **g)** $4\sec^2\theta = 1$ **h)** $\text{cosec}^2\theta = 2$

EXERCISES

15.3 B

1 Find (where possible) the values of the following.

a) $\sec 20°$ **b)** $\text{cosec} -50°$ **c)** $\cot 170°$ **d)** $\cot 270°$
e) $5 \sec^2 90°$ **f)** $1 + \tan^2 30°$ **g)** $\sec^2 30°$ **h)** $\frac{\cos 40°}{\sin 40°}$
i) $\cot 40°$

What do you notice about your answers to parts **h)** and **i)**? Can you explain this?

2 Find, where possible, the values of the following.

a) $\cot \frac{\pi}{3}$ **b)** $\text{cosec} \frac{\pi}{4}$ **c)** $\sec\left(-\frac{\pi}{6}\right)$
d) $2\sec \frac{3\pi}{2}$ **e)** $\cot^2 \pi$ **f)** $-4\,\text{cosec} \frac{8\pi}{5}$

3 **a)** Complete the table to show the signs of the reciprocal trigonometric functions (their graphs may help you).

	$0 < \theta < \frac{\pi}{2}$	$\frac{\pi}{2} < \theta < \pi$	$\pi < \theta < \frac{3\pi}{2}$	$\frac{3\pi}{2} < \theta < 2\pi$
$\text{cosec}\,\theta$	+			
$\sec\theta$		−		
$\cot\theta$			+	

b) State values of θ for which the following are not defined.
i) $\text{cosec}\,\theta$ **ii)** $\sec\theta$ **iii)** $\cot\theta$

4 Find (if possible) values of θ such that $0 \le \theta \le 2\pi$ which satisfy the following equations.

a) $\cot\theta = \sqrt{3}$ **b)** $\text{cosec}\,\theta = -2$ **c)** $\sec^2\theta = \frac{3}{4}$
d) $\cot 2\theta = -5$ **e)** $4 \sec 3\theta = 6$ **f)** $\text{cosec}\,4\theta = 0.7$
g) $5 \cot 2\theta = -2$ **h)** $1 - \text{cosec}\,2\theta = 0.5$

TRIGONOMETRICAL IDENTITIES

CALCULATOR ACTIVITY

Adding waves

You will need a graphics calculator. Firstly scale the axes:

$x_{min} = 0$; $x_{max} = 360$; $x_{scl} = 30$
$y_{min} = -0.5$; $y_{max} = 1.5$; $y_{scl} = 0.5$.

and make sure you are in degree mode.

- On the same axes:
  plot the graphs of : $y = \sin^2 x$ Enter as $(\sin x)^2$
  $y = \cos^2 x$
  $y = \sin^2 x + \cos^2 x$
- What result does this illustrate?
- Demonstrate the result using a right-angled triangle with base angle θ and hypotenuse 1.

Results

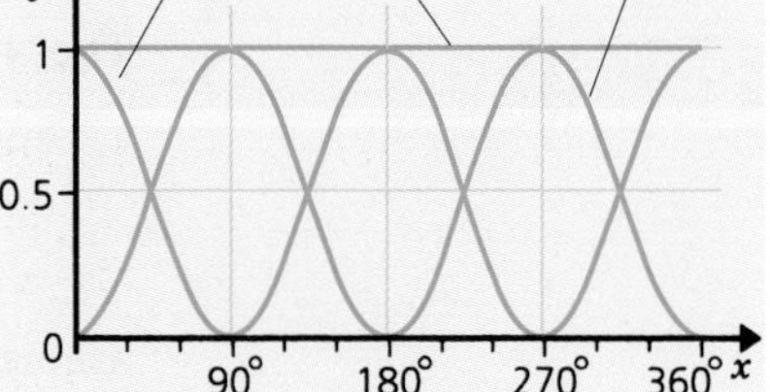

The graphs produced in the Calculator activity should look like those in the diagram.

The result of adding the two waves is always 1, i.e. for any angle:

$$\sin^2\theta + \cos^2\theta = 1$$

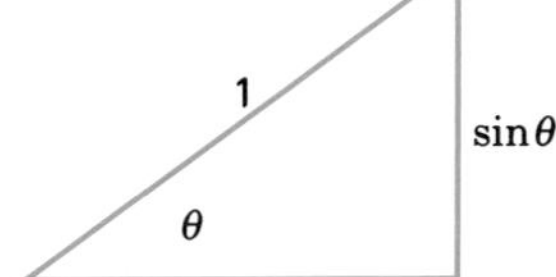

We can obtain the same result from a right-angled triangle.

The height is $\sin\theta$ and base is $\cos\theta$, using the rotating vector definitions. Using Pythagoras' theorem:

$$\sin^2\theta + \cos^2\theta \equiv 1 \qquad (1)$$

This is known as a **Pythagorean identity**.

CALCULATOR ACTIVITY

Adding more waves

Use a graphics calculator. Rescale the axes to:

$x_{min} = 0$ $\quad x_{max} = 360$ $\quad x_{scl} = 30$

$y_{min} = 0$ $\quad y_{max} = 1.5$ $\quad y_{scl} = 1$

and make sure you are in degree mode.

- On the same axes:
  plot the graphs of : $y = \sec^2 x$
  $y = \tan^2 x$
  $y = \sec^2 x - \tan^2 x$
- What result does this illustrate?
- How could you manipulate identity (1) above to get this result?

Results

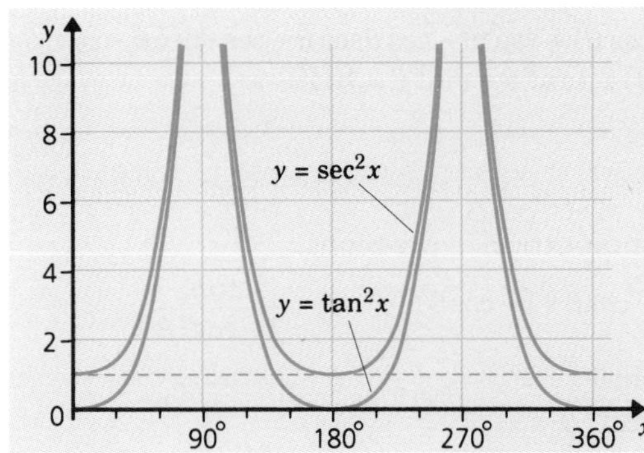

The graphs produced should look like these.

The difference between the two waves is always 1, i.e. for any angle:

$$\sec^2\theta - \tan^2\theta \equiv 1$$

$$\Rightarrow \qquad \sec^2\theta \equiv 1 + \tan^2\theta \qquad (2)$$

We can derive this result from identity (1) by dividing throughout by $\cos^2\theta$.

$$\frac{\sin^2\theta}{\cos^2\theta} + \frac{\cos^2\theta}{\cos^2\theta} \equiv \frac{1}{\cos^2\theta}$$

$$\Rightarrow \tan^2\theta + 1 \equiv \sec^2\theta$$

A third identity may be derived by dividing (1) throughout by $\sin^2\theta$.

$$\frac{\sin^2\theta}{\sin^2\theta} + \frac{\cos^2\theta}{\sin^2\theta} \equiv \frac{1}{\sin^2\theta}$$

$$\Rightarrow 1 + \cot^2\theta \equiv \operatorname{cosec}^2\theta \qquad (3)$$

Using the Pythagorean identities

The three identities:

$$\sin^2\theta + \cos^2\theta \equiv 1$$
$$1 + \tan^2\theta \equiv \sec^2\theta$$
$$1 + \cot^2\theta \equiv \operatorname{cosec}^2\theta$$

are useful for simplifying trigonometrical expressions and establishing further identities, as shown in the following examples.

Example 15.9

Simplify $\dfrac{\sin x}{\sqrt{1-\sin^2 x}}$.

Solution

$$\frac{\sin x}{\sqrt{1-\sin^2 x}} \equiv \frac{\sin x}{\sqrt{\cos^2 x}} \equiv \frac{\sin x}{\cos x} \equiv \tan x$$

Example 15.10

Prove the following identity.

$$\tan\theta + \cot\theta \equiv \sec\theta \operatorname{cosec}\theta$$

Solution

$$\text{LHS} = \tan\theta + \cot\theta = \frac{\sin\theta}{\cos\theta} + \frac{\cos\theta}{\sin\theta} \equiv \frac{\sin^2\theta + \cos^2\theta}{\cos\theta\sin\theta} \equiv \frac{1}{\cos\theta\sin\theta}$$
$$\equiv \frac{1}{\cos\theta} \times \frac{1}{\sin\theta} \equiv \sec\theta \operatorname{cosec}\theta = \text{RHS}$$

Note: Expressing $\tan\theta$ and $\cot\theta$ in terms of $\sin\theta$ and $\cos\theta$, and then simplifying, enables the identity to be proved.

Example 15.11

Prove this identity.

$$(\sec\theta - \cos\theta)(\sec\theta + \cos\theta) \equiv \tan^2\theta + \sin^2\theta$$

Solution

$$\text{LHS} = (\sec\theta - \cos\theta)(\sec\theta + \cos\theta) \equiv \sec^2\theta - \cos\theta\sec\theta + \cos\theta\sec\theta - \cos^2\theta$$
$$\equiv 1 + \tan^2\theta - (1 - \sin^2\theta) \equiv \tan^2\theta + \sin^2\theta = \text{RHS}$$

EXERCISES

1 Simplify the following trigonometrical expressions.

a) $\tan\theta \operatorname{cosec}\theta$ **b)** $\sqrt{(1+\cos\theta)(1-\cos\theta)}$ **c)** $\dfrac{\sin\theta}{1-\cos^2\theta}$

d) $\dfrac{1-\sec^2\theta}{1-\operatorname{cosec}^2\theta}$ **e)** $\sqrt[3]{\dfrac{\sin\theta}{1+\cot^2\theta}}$ **f)** $\sec\theta\sqrt{1-\sin^2\theta}$

2 Prove the following identities.

a) $\operatorname{cosec}\theta - \sin\theta \equiv \cos\theta\cot\theta$ **b)** $(1+\sec\theta)(1-\cos\theta) \equiv \tan\theta\sin\theta$

c) $\dfrac{\operatorname{cosec}\theta}{\operatorname{cosec}\theta - \sin\theta} \equiv \sec^2\theta$ **d)** $(1+\cos\theta+\sin\theta)^2 \equiv 2(1+\cos\theta)(1+\sin\theta)$

e) $(\sin\theta+\cos\theta)^2+(\sin\theta-\cos\theta)^2\equiv 2$

f) $(\sin\theta+\operatorname{cosec}\theta)^2\equiv\cot^2\theta-\cos^2\theta+4$

g) $\dfrac{\sin\theta}{1+\cos\theta}\equiv\dfrac{1-\cos\theta}{\sin\theta}$ Hint: multiply LHS by $\dfrac{1-\cos\theta}{1-\cos\theta}$

EXERCISES

15.4B

1 Simplify the following trigonometrical expressions.

a) $\sin\theta\cot\theta$ b) $\dfrac{\sin\theta}{1+\cot^2\theta}$ c) $\dfrac{1}{\cos\theta\sqrt{(1+\cot^2\theta)}}$ d) $\sqrt{\dfrac{1-\cos^2\theta}{4\sec^2\theta-4}}$

2 Prove the following identities.

a) $\cot\theta+\tan\theta\equiv\sec\theta\operatorname{cosec}\theta$

b) $(\sin\theta+1)^2\equiv 2(1+\sin\theta)-\cos^2\theta$

c) $(1-\cos^2\theta)(1+\cot^2\theta)\equiv 1$

d) $\dfrac{1}{\tan\theta+\cot\theta}\equiv\sin\theta\cos\theta$

e) $(\cos\theta-\sin\theta)^2\equiv 1-2\cos\theta\sin\theta$

f) $\dfrac{2}{\sin\theta+1}-\dfrac{2}{\sin\theta-1}\equiv 4\sec^2\theta$

g) $\sec^2\theta+\operatorname{cosec}^2\theta\equiv\sec^2\theta\operatorname{cosec}^2\theta$

Comment on the last result.

TRIGONOMETRICAL EQUATIONS

Earlier in this chapter, we solved some simple trigonometrical equations that involved a *single* trigonometrical function. Now we can extend the idea to more complex equations, often involving two or more trigonometrical functions, which simplify to a recognisable form (e.g. a quadratic equation).

The examples below illustrate these types of equation.

Example 15.12

Solve the equation $4\sin\theta-3\cos\theta=0$, $0°\le\theta\le 360°$.

Solution

$4\sin\theta-3\cos\theta=0\Rightarrow 4\sin\theta=3\cos\theta$

$\Rightarrow\dfrac{\sin\theta}{\cos\theta}=\dfrac{3}{4}\Rightarrow\tan\theta=0.75$

$\Rightarrow\theta=\tan^{-1}0.75=36.9°$ *(from calculator)*

but $0°\le\theta\le 360°$

$\Rightarrow\ \theta=36.9°$ *or* $180°+36.9°\ \Rightarrow\ \theta=36.9°$ *or* $216.9°$

Example 15.13

Solve the equation $2\cos^2\theta+\cos\theta-1=0$, $0°\le\theta\le 360°$.

Solution

Treat as a quadratic equation in $\cos\theta$.

$2\cos^2\theta+\cos\theta-1=0$

$\Rightarrow(2\cos\theta-1)(\cos\theta+1)=0$

$\Rightarrow 2\cos\theta - 1 = 0$ *or* $\cos\theta + 1 = 0$

$\Rightarrow \cos\theta = 0.5$ *or* $\cos\theta = -1$

Now $\cos\theta = 0.5 \Rightarrow \theta = \cos^{-1}0.5 = 60°$ *(from calculator)*
but $0° \leq \theta \leq 360°$

$\Rightarrow \theta = 60°$ *or* $360° - 60° = 300°$
and $\cos\theta = -1 \Rightarrow \theta = \cos^{-1}(-1) = 180°$

Combining solutions we have:

$\theta = 60°, 180°$ *or* $300°$

Example 15.14

a) *Sketch, on the same grid, the graphs of* $y = 2\cos x$ *and* $y = 3\tan x$*, for* $-\pi \leq x \leq \pi$.

b) *Form and solve a quadratic equation in* $\sin x$ *to find the points of intersection of the two curves in* **a)**.

Solution

a)

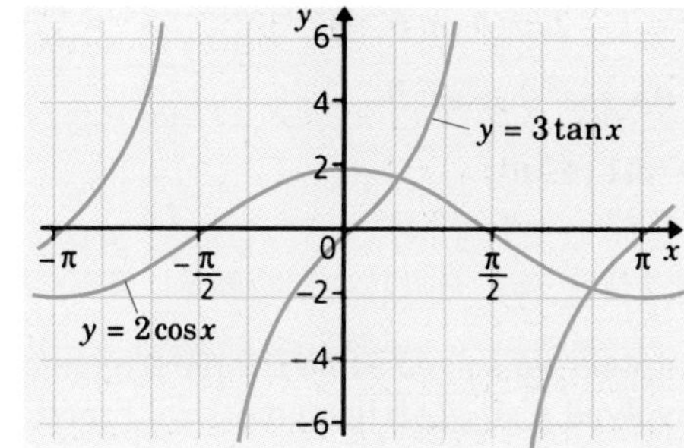

b) *At the points of intersection:*

$$2\cos x = 3\tan x \Rightarrow 2\cos x = 3\frac{\sin x}{\cos x}$$
$$\Rightarrow 2\cos^2 x = 3\sin x$$
$$\Rightarrow 2(1 - \sin^2 x) = 3\sin x$$
$$\Rightarrow 2 - 2\sin^2 x = 3\sin x$$
$$0 = 2\sin^2 x + 3\sin x - 2$$
$$= (2\sin x - 1)(\sin x + 2)$$
$$\Rightarrow 2\sin x - 1 = 0 \quad \text{or} \quad \sin x = -2$$
$$\Rightarrow \sin x = 0.5 \quad \text{or} \quad \sin x = -2$$

Now: $\sin x = 0.5 \Rightarrow x = \sin^{-1}0.5 = \frac{\pi}{6}$ *(from calculator)*

but $-\pi \leq x \leq \pi \Rightarrow x = \frac{\pi}{6}$ *or* $x = \pi - \frac{\pi}{6} = \frac{5\pi}{6}$

Since $\sin x = -2$ *has no solution, the only solutions are* $x = \frac{\pi}{6}$ *or* $x = \frac{5\pi}{6}$.
To find corresponding y*-values, substitute for* x *in, say,* $y = 2\cos x$.

When $x = \frac{\pi}{6}$ $y = 2\cos\frac{\pi}{6} = 2\times\frac{\sqrt{3}}{2} = \sqrt{3}$

When $x = \frac{5\pi}{6}$ $y = 2\cos\frac{5\pi}{6} = 2\times-\frac{\sqrt{3}}{2} = -\sqrt{3}$

Therefore the coordinates of the points of intersection of the two curves are

$\left(\frac{\pi}{6}, \sqrt{3}\right)$ *and* $\left(\frac{5\pi}{6}, -\sqrt{3}\right)$.

EXERCISES

15.5A

1 Transform the following equations into equations in one trigonometrical function and solve them for the domain $0° \leq \theta \leq 360°$.

a) $3\sin\theta + 4\cos\theta = 0$ **b)** $\sin^2\theta - \cos^2\theta = 1$
c) $\tan^2\theta + \sec^2\theta = 9$ **d)** $1 + \cot^2\theta = 8\sin\theta$

2 Solve the following quadratic trigonometrical equations for the domain $0° \le \theta \le 360°$.

a) $4\sin^2\theta + 4\sin\theta + 1 = 0$ b) $4\cos^2\theta - 1 = 0$
c) $3\cos^2\theta - \cos\theta = 2$ d) $4\sec^2\theta - \sec\theta - 5 = 0$

3 Transform the following equations into quadratic equations in one trigonometrical function and solve for the domain $0° \le \theta \le 360°$.

a) $4 + \sin\theta = 6\cos^2\theta$ b) $\sin\theta = \cot\theta$
c) $\cot^2\theta = \operatorname{cosec}\theta + 11$ d) $2\cot\theta + \tan\theta = 2\operatorname{cosec}\theta$

4 Solve the following equations for the domain $-\pi \le x \le \pi$.

a) $\tan^2 x = \tan x + 6$ b) $2\sec^2 x = 5\tan x$
c) $1 + \sin^2 x = 4\sin x$ d) $3\cot x = \operatorname{cosec}^2 x$
e) $\tan x - 2\cot x = 3\sec x$

5 For each of the following pairs of equations:

a) sketch, on the same grid, the graphs of the two functions, for $-\pi \le x \le \pi$,
b) form and solve a quadratic equation to find the points of intersection of the two curves in **a)**.
i) $y = 3\cos x$, $y = 2\sin^2 x$
ii) $y = \tan x$, $y = 5\cos x$
iii) $y = \sin x$, $y = \cot x$

6 a) A function f is defined by: $f(s) = 4s^3 - 4s^2 - s + 1$.
Find $f(1)$ and use the factor theorem to factorise $f(s)$ fully.
b) Solve the equation $4\sin^3 x - 4\sin^2 x - \sin x + 1 = 0$ for the domain $-\pi \le x \le \pi$.

EXERCISES

15.5 B

1 Transform the following equations into equations in one trigonometric function and solve for the domain $0° \le \theta \le 360°$.

a) $\sin\theta - 2\cos\theta = 0$ b) $\cos^2\theta - \sin^2\theta = 0$
c) $\tan^2\theta + \sec^2\theta = 17$ d) $\sec^2\theta - 2\tan\theta = 0$

2 Solve the following quadratic trigonometrical equations for the domain $-180° \le \theta \le 180°$.

a) $6\sin^2\theta - 5\sin\theta + 1 = 0$ b) $9\sin^2\theta - 1 = 0$
c) $\tan^2\theta - 3\tan\theta + 2 = 0$ d) $4\operatorname{cosec}^2\theta - \operatorname{cosec}\theta - 5 = 0$

3 Solve the following equations for the domain $0 \le x \le 2\pi$.

a) $\operatorname{cosec}^2 x - 4 = 0$ b) $8\cos^2 x + 14\sin x = 9$
c) $3\tan x = \sec^2 x$ d) $2\cot x + 3\tan x = 6\cos x$

4 Find the values of x in the range $-\pi \le x \le \pi$ for which $\sec^2 x = 3\tan x - 1$.

5 Find values of x in the range $0 \le x \le 2\pi$ for which $\cot^2 x + \operatorname{cosec} x = 5$.

6 **a)** A function f is defined by $f(s) = 4s^4 + 12s^3 + 7s^2 - 3s - 2$.
Find $f(-1)$ and $f(-2)$ and use the factor theorem to factorise $f(s)$ fully.

b) Solve the equation $4\sin^4\theta + 12\sin^3\theta + 7\sin^2\theta - 3\sin\theta - 2 = 0$ for the domain $0 \leq \theta \leq 2\pi$.

CONSOLIDATION EXERCISES FOR CHAPTER 15

1 The curve with equation $y = 2 + k\sin x$ passes through the point with coordinates $(\frac{\pi}{2}, -2)$. Find:

a) the value of k,
b) the greatest value of y,
c) the values of x in the interval $0 \leq x \leq 2\pi$ for which $y = 2 + 2\sqrt{2}$.

(ULEAC Question 5, Specimen Paper 1, 1994)

2 **a)** Find the values of $\cos x$ for which $6\sin^2 x = 5 + \cos x$.
b) Find all the values of x in the interval $180° < x < 540°$ for which $6\sin^2 x = 5 + \cos x$.

(ULEAC Question 4, Specimen Paper 2, 1994)

3 The depth of water at the entrance to a harbour is y metres at time t hours after low tide. The value of y is given by:

$$y = 10 - 3\cos kt$$

where k is a positive constant. Write down, or obtain, the depth of water in the harbour: **a)** at low tide, **b)** at high tide.

Show by means of a sketch graph how y varies with t between two successive low tides. Given that the time interval between a low tide and the next high tide is 6.20 hours, calculate, correct to two decimal places, the value of k.

(NEAB Question 4, Specimen Paper 1, 1994)

4 A piece of wire of length 4 metres is bent into the shape of a sector of a circle of radius r metres and angle θ radians.

a) State, in terms of θ and r:
i) the length of the arc,
ii) the area A of the sector.
b) Hence show that $A = 2r - r^2$.
c) Find the value of r which will make the area a maximum. Deduce the corresponding value of θ.
d) The figures labelled **A– F** show, all to the same scale, six possible sectors which can be made from the piece of wire. Which of them has the largest area?

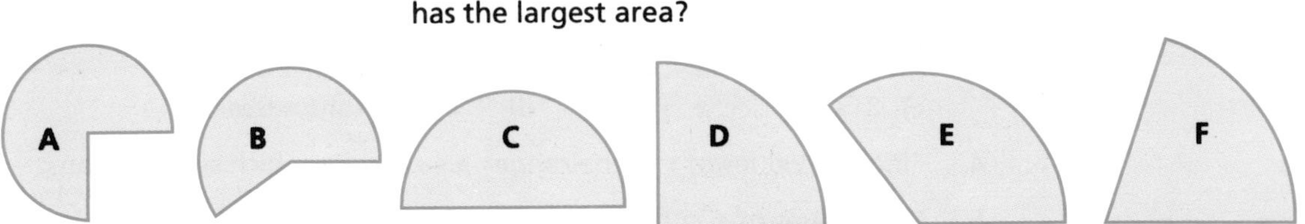

(Oxford and Cambridge Question 9, Specimen Paper 1, 1994)

5 **a)** Solve the equation $\sin 2x = \cos^2 x$, for $0° \le x \le 360°$, giving your answer in degrees to one decimal place where appropriate.

b) Solve the equation $\cos 2x = 2\sin^2 x$ giving the general solution, in radians, as a multiple of π.

(ULEAC Question 7, Paper 2, January 1995)

6 **a)** Given that $(3x + 2)$ is a factor of $3x^3 + Ax^2 - 4x - 4$, show that $A = 5$.

b) Factorise $3x^3 + 5x^2 - 4x - 4$ completely.

c) Given that $0° \le t \le 360°$, find the values of t, to the nearest degree, for which $3\sin^3 t + 5\sin^2 t - 4\sin t - 4 = 0$.

(ULEAC Question 8, Paper 1, May 1994)

7 Find to the nearest 0.1° the angles x between 0° and 180° for which:

a) $\sin 3x = \frac{1}{2}$ **b)** $\sin^2 x = \frac{1}{2}\cos x$.

(ULEAC Question 5, Paper 1, January 1995)

8 The diagram shows an equilateral triangle ABC whose vertices lie on a circle, centre O, of radius r.

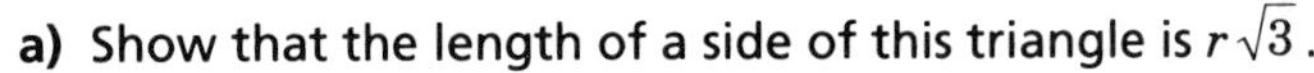

a) Show that the length of a side of this triangle is $r\sqrt{3}$.

b) Show that the ratio of the area of the shaded region to the area of the triangle is $4\pi\sqrt{3} - 9 : 9$.

(ULEAC Question 6, Paper 1, January 1995)

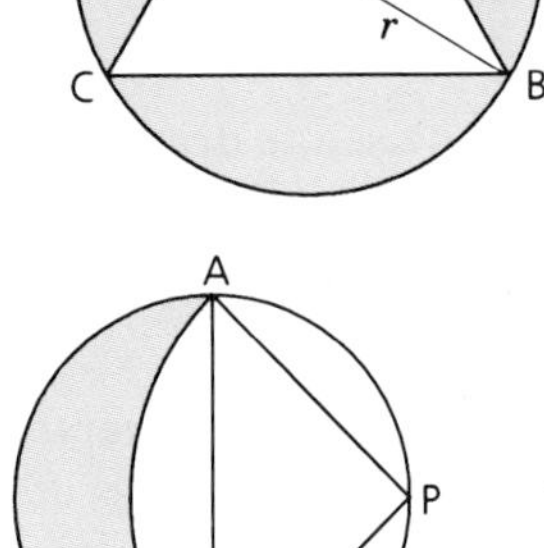

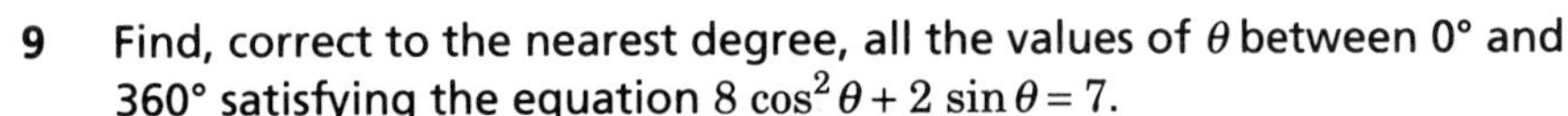

9 Find, correct to the nearest degree, all the values of θ between 0° and 360° satisfying the equation $8\cos^2\theta + 2\sin\theta = 7$.

(WJEC Question 1, Specimen Paper A1, 1994)

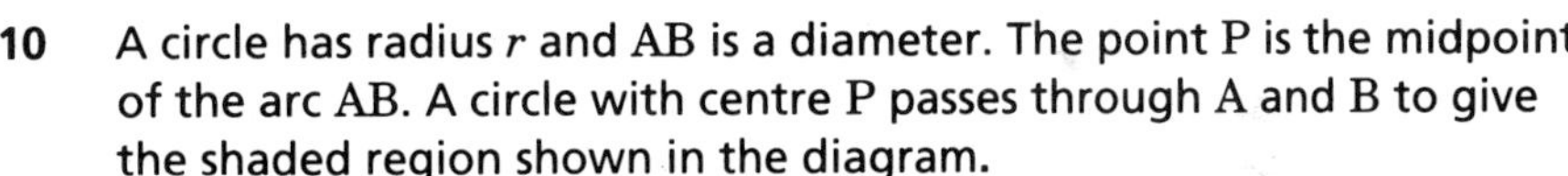

10 A circle has radius r and AB is a diameter. The point P is the midpoint of the arc AB. A circle with centre P passes through A and B to give the shaded region shown in the diagram.

Find the area of the shaded region.

(International Baccalaureate, 1993)

Summary

One **radian** is the angle that gives the same arc length in a sector as the radius of the radius.

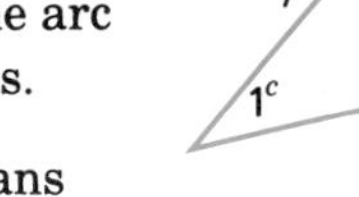

$$1 \text{ radian} = \frac{180°}{\pi} = 57.3° \quad 1° = \frac{\pi}{180} \text{ radians}$$

The arc length in a sector of radius r m and angle θ radians is $r\theta$ m.

The area of the sector is $\frac{1}{2}r^2\theta$ m^2

The reciprocal trigonometric functions are

$$\operatorname{cosec} x = \frac{1}{\sin x} \quad \sec x = \frac{1}{\cos x} \quad \cot x = \frac{1}{\tan x}$$

Three important trigonometric identities are

$$\sin^2\theta + \cos^2\theta = 1$$
$$1 + \tan^2\theta = \sec^2\theta$$
$$1 + \cot^2\theta = \operatorname{cosec}^2\theta$$

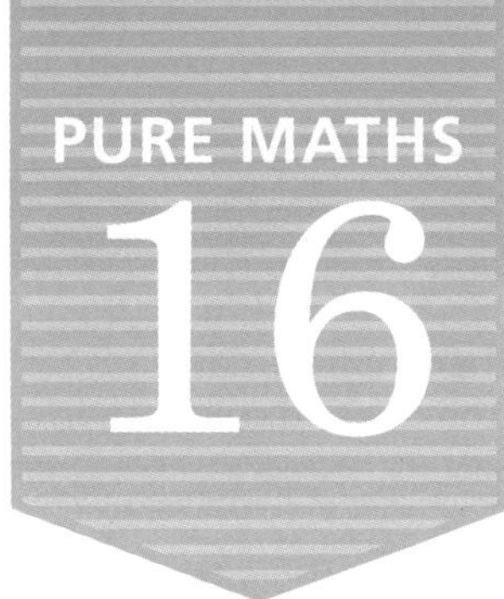

Logarithms

In Chapter 12, Indices, we introduced and developed the idea of exponential functions. In this chapter we introduce the inverse of the exponential functions which are called ***logarithms****.*

DEFINITION AND LAWS OF LOGARITHMS

Exploration 16.1

Drawing graphs of exponential functions

- On graph paper draw graphs of $y = 2^x$ and $y = x$ for $-5 < x < 5$.
- Draw the reflection of the graph of $y = 2^x$ in the line $y = x$.

Repeat the activity for $y = 3^x$ and $y = 5^x$.

Graphs of exponentials

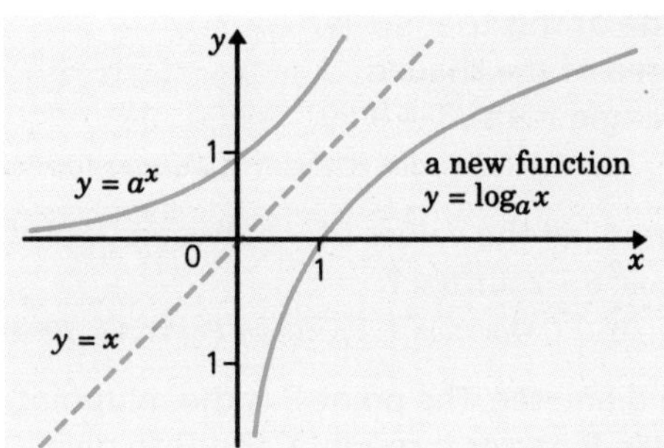

The graphs drawn in Exploration 16.1 had the general shape as shown in the diagram.

In Chapter 13, *Functions*, this idea was used to draw the graphs of inverse functions. For a function $y = f(x)$ the graph of the inverse function $y = f^{-1}(x)$ is a reflection, in the line $y = x$, of the original function $y = f(x)$.

The reflections drawn in Exploration 16.1 are the inverse functions of $y = 2^x$, $y = 3^x$ and $y = 5^x$. They are examples of **logarithmic functions**.

The logarithm of x to base a is defined as the inverse function of a^x in the following way.

If $a^m = b$ then $m = \log_a b$.

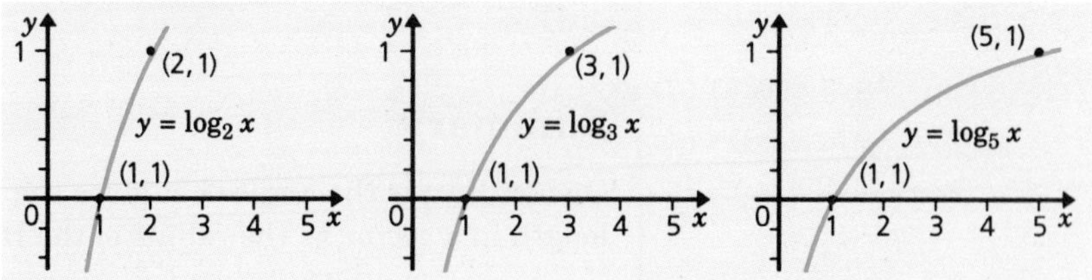

Exploration 16.1 involved graphs of $y = \log_2 x$, $y = \log_3 x$ and $y = \log_5 x$.

In each case the graph passes through the point (1, 0). One important property of logarithms is that:

$\log_a 1 = 0$ for any base a.

This result can be obtained from the definition by letting $b = 1$.

$a^m = 1 \Rightarrow m = 0 \Rightarrow \log_a 1 = 0.$

Example 16.1

Write each of the following in index form.

a) $\log_2 32 = 5$ ***b)*** $\log_3 81 = 4$ ***c)*** $\log_a 5 = b$

Solution

a) *If* $\log_2 32 = 5$ *then the base and index is 2.*

$\log_2 32 = 5 \Rightarrow 32 = 2^5$

b) *Similarly* $\log_3 81 = 4 \Rightarrow 81 = 3^4$

c) *In this case the base and index is a.*

$\log_a 5 = b \Rightarrow 5 = a^b$.

Example 16.2

Write each of the following in logarithmic form.

a) $5^2 = 25$ ***b)*** $x^y = 3$

Solution

a) *The index is 5 and so the base of the logarithm is 5.*

$5^2 = 25 \Rightarrow \log_5 25 = 2$

b) *The index is x and so the base of the logarithm is x.*

$x^y = 3 \Rightarrow \log_x 3 = y$

Common logarithms

Although logarithms may be expressed in terms of any base, there are two that are generally used. One of these is base 10. Logarithms to base 10 are called **common logarithms** and are often written as $\log b$, i.e. without including the base number 10. Find the common logarithm key on your calculator. It will be labelled log (or maybe $\log_{10}$). It is usually on the same key as 10^x, showing that the functions are related.

Use your calculator to check the following values.

$\log_{10} 10 = 1$	$\log_{10} 100 = 2$	$\log_{10} 1000 = 3$
$\log_{10} 20 = 1.30103$	$\log_{10} 0.5 = -0.30103$	$\log_{10} 1 = 0$
$\log_{10} 5.73 = 0.75816$	$\log_{10} 0.12 = -0.92082$	$\log_{10}(-2)$ gives error

Two important properties can be seen from these values.

- If $x < 1$ then $\log_{10} x < 0$.
- If $x < 0$ then $\log_{10} x$ is not defined.

These properties are true for logarithms to any base.

Properties of logarithms

An important property of logarithms is:

$\log_a a = 1$

for any $a \neq 0$. This can be deduced from the definition of logarithms with $b = a$.

If $a^m = a$ then $m = 1 \Rightarrow \log_a a = 1$.

Exploration 16.2

Logarithms of powers of ten

- Write the following numbers as powers of 10.
  100 1000 10 000 100 000 0.1 0.01 0.001
- Use your calculator to find the log of each number to base 10.
- Deduce a law for simplifying $\log_{10} 10^r$.

Developing the properties of logarithms

You will have found that $\log_{10}10^r = r$. This is a special case of one of the properties of logarithms that we will now show.

Consider $x = \log_a b$ and $y = \log_a c$

$\Rightarrow$ $b = a^x$ and $c = a^y$

now $bc = (a^x)(a^y) = a^{x+y}$ *using properties of indices*

$\Rightarrow$ $x + y = \log_a bc$

$\Rightarrow$ $\log_a b + \log_a c = \log_a bc$

The logarithm of a **product** of numbers is the **sum** of the logarithms.

If we divide b by c we have:

$$\frac{b}{c} = \frac{a^x}{a^y} = a^{x-y} \Rightarrow x - y = \log_a \frac{b}{c} \Rightarrow \log_a b - \log_a c = \log_a \frac{b}{c}$$

The logarithm of a **quotient** of numbers is the **difference** between the logarithms.

From this property we deduce that:

$$\log_a \frac{1}{c} = \log_a 1 - \log_a c = -\log_a c$$

The third general law concerns powers of numbers.

$$b^r = (a^x)^r = a^{xr} \Rightarrow xr = \log_a b^r \Rightarrow r\log_a b = \log_a b^r$$

With this last result we can see why the result of Exploration 16.2 works.

$\log_{10} 10^r = r \log_{10} 10 = r$

Since $\log_{10} 10 = 1$.

Example 16.3

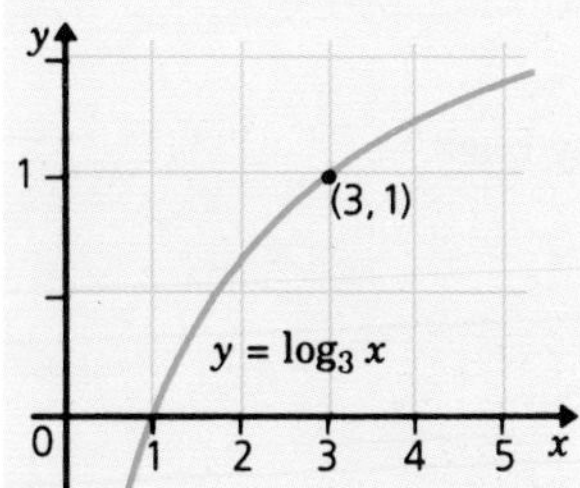

The diagram shows a sketch of the graph of $y = \log_3 x$.

Sketch a graph of $y = \log_3 2x$.

Solution

The rule for the logarithm of a product gives:

$y = \log_3 2x = \log_3 2 + \log_3 x$

The graph of $y = \log_3 2x$ is a translation of $\log_3 x$ of $\log_3 2$ parallel to the y-axis.

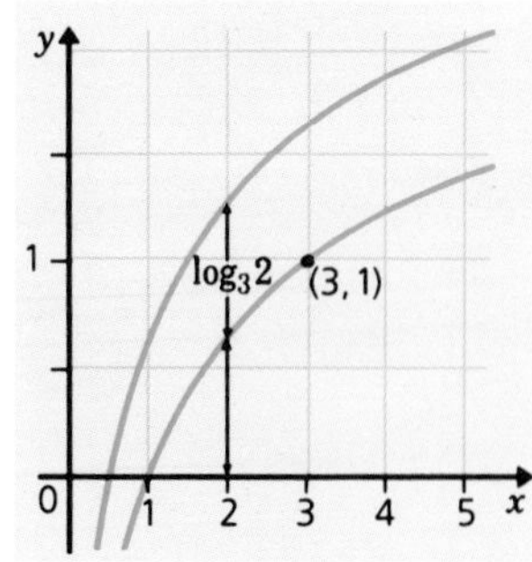

Example 16.4

Write each of the following as a single logarithm.

a) $\log_2 9 - \log_2 16 + \log_2 3$ ***b)*** $3\log_5 2 - 2\log_5 3$

c) $2\log x + 3\log y - \log xy$

Solution

a) $\log_2 9 - \log_2 16 + \log_2 3 = \log_2 \frac{9 \times 3}{16} = \log_2 \frac{27}{16}$

b) $3\log_5 2 - 2\log_5 3 = \log_5 2^3 - \log_5 3^2 = \log_5 \frac{2^3}{3^2} = \log_5 \frac{8}{9}$

c) $2\log x + 3\log y - \log xy = \log x^2 + \log y^3 - \log xy$

$$= \log\left(\frac{x^2y^3}{xy}\right) = \log\left(xy^2\right)$$

Example 16.5

Expand the following in terms of $\log x$, $\log y$ *and* $\log z$.

a) $\log_3 xy$ **b)** $\log_2\left(\frac{x^2y^3}{z^4}\right)$

Solution

a) $\log_3 xy = \log_3 x + \log_3 y$

b) $\log_2\left(\frac{x^2y^3}{z^4}\right) = \log_2 x^2 + \log_2 y^3 - \log_2 z^4 = 2\log_2 x + 3\log_2 y - 4\log_2 z$

Example 16.6

Solve the following equations for x to 4 d.p.

a) $10^x = 13$ **b)** $\log_{10} x = 2.1$

Solution

a) $10^x = 13 \Rightarrow x = \log_{10} 13 = 1.1139$

b) $\log_{10} x = 2.1 \Rightarrow x = 10^{2.1} = 125.8925$

EXERCISES

16.1 A

1 Sketch the graph of $y = \log_{10} x$ for $0 < x \le 10$.

a) Explain why the graph does not intersect the line $y = x$.

b) From your graph, sketch the graphs of $\log_{10}(x-2)$ and $\log_{10} 3x$.

2 Sketch the graph of $y = \log_{10} \frac{1}{x}$ for $0 < x \le 10$.

Use your graph to sketch graphs of $y = \log_{10} \frac{2}{x}$ and $y = \log_{10} \frac{1}{x-2}$

3 Write each of the following in index form.

a) $\log_2 8 = 3$ **b)** $\log_{10} 100\,000 = 5$ **c)** $\log_4 16 = 2$

d) $\log_7 1 = 0$ **e)** $\log_a 3 = b$ **f)** $\log_a x = 5$

g) $\log_a 1 = 0$ **h)** $\log_p q = y$ **i)** $\log_a r = m$

4 Write each of the following in logarithmic form.

a) $3^2 = 9$ **b)** $11^0 = 1$ **c)** $10^4 = 10000$

d) $10^{-3} = 0.001$ **e)** $8^{\frac{1}{3}} = 2$ **f)** $m = n^2$

g) $x^y = 2$ **h)** $10^p = q$ **i)** $u^v = w$

5 Solve the following equations using the log key on your calculator.

a) $10^x = 150$ **b)** $10^{-x} = 0.2$ **c)** $10^{2x} = 72$

d) $10^{2x-1} = 25$ **e)** $10^{0.1x} = 5$ **f)** $10^{0.2x+1} = 18$

6 Solve the following equations using the 10^x key on your calculator.

a) $\log_{10}x = 2.1$ b) $\log_{10}x = 4.2$ c) $\log_{10}3x = 1.4$
d) $\log_{10}5x = 3.5$ e) $\log_{10}0.1x = 2$ f) $\log_{10}(2x-1) = 0.6$
g) $\log_{10}(2x+1) = -0.5$ h) $\log_{10}x - \log_{10}(3x-1) = 1$
i) $\log_{10}x + \log_{10}(x+1) = 0.5$ j) $\log_{10}(x+1) + \log_{10}(x-1) = 1.2$
k) $2\log_{10}x - \log_{10}(2x-1) = 0.1$

7 Write each of the following as a single logarithm in its simplest form.

a) $\log_a 7 + 2\log_a 3 - 3\log_a 5$ b) $3\log 5 + 4\log 2$
c) $\log_2 p + 2\log_2 q + 3$ d) $2\log r + 3\log s - 4\log t - 2$
e) $\frac{1}{2}\log_a x + \frac{1}{3}\log_a y$ f) $\log_a(x^2 - 3x + 2) - \log_a(x-1) + \log_a(x-2)$
g) $\log_2(x^2 - 1) - \log_2(x-1) + \log_2(x+1)$

8 Expand the following in terms of $\log u$, $\log v$ and $\log w$.

a) $\log_2 uv$ b) $\log_3 u^2v^3$ c) $\log \frac{u^a v^b}{w}$ d) $\log_a u^{\frac{1}{2}}v^2w^{\frac{1}{3}}$

e) $\log\sqrt{u^3w^5}$ f) $\log_{10}(100\,v^3)$ g) $\log_a\left(\frac{1}{w^2}\right)$ h) $\log\left(\frac{(u-1)(v+1)}{(w-2)}\right)$

EXERCISES

16.1 B

1 Sketch the graph of $y = \log_{10}x$ for $0 < x \le 10$.

a) Use your graph to show that $\log_{10}10 = 1$ and $\log_{10}1 = 0$.
b) From your graph, sketch the graphs of $\log_{10}(x+2)$ and $\log_{10}5x$.

2 Sketch the graph of $y = \log_{10}\frac{1}{x}$ for $0 < x \le 10$.

Use your graph to sketch graphs of these functions.

a) $y = \log_{10}\frac{3}{x}$ b) $y = \log_{10}\frac{1}{x+2}$

3 Write each of the following in index form.

a) $\log_2 128 = 7$ b) $\log_{10}1\,000\,000 = 6$ c) $\log_5 3125 = 5$
d) $\log_9 1 = 0$ e) $\log_a 7 = b$ f) $\log_a x = 31$
g) $\log_a 1 = 0$ h) $\log_r s = p$ i) $\log_q(2s) = n$

4 Write each of the following in logarithmic form.

a) $7^2 = 49$ b) $130^0 = 1$ c) $16^{\frac{1}{2}} = 4$
d) $10^{-2} = 0.01$ e) $2401^{\frac{1}{4}} = 7$ f) $p = r^3$
g) $x^t = 2.5$ h) $10^{-p} = q$ i) $a^{2b} = 3c$

5 Solve the following equations using the log key on your calculator.

a) $10^x = 169$ b) $10^{-x} = 0.3$ c) $10^{3x} = 90$
d) $10^{2x+1} = 36$ e) $10^{0.1x} = 7$ f) $10^{0.4x-1} = 240$

6 Solve the following equations using the 10^x key on your calculator.

a) $\log_{10}x = 1.1$ b) $\log_{10}x = 2.5$ c) $\log_{10}5x = 1.5$
d) $\log_{10}14x = 0.14$ e) $\log_{10}0.2x = 1$ f) $\log_{10}(3x+1) = 0$
g) $\log_{10}(0.1x - 3) = -0.2$ h) $\log_{10}x - \log_{10}(5x-1) = 0.301\,03$
i) $\log_{10}x + \log_{10}(x-1) = 0.1$ j) $\log_{10}(2x-1) + \log_{10}(2x+1) = 1.8$

7 Write each of the following as a single logarithm in its simplest form.

a) $\log_a 9 + 2\log_a 2 - 3\log_a 3$ **b)** $5\log 3 + 2\log 5$

c) $\log_3 r + 2\log_3 s - 2$ **d)** $\log x + 2\log y - 5\log t + 1$

e) $\frac{1}{4}\log r - \frac{1}{8}\log s$ **f)** $\log_a(x^2 + 5x + 6) - \log_a(x + 2) + \log_a(x + 3)$

g) $\log_5(x^3 - 1) - \log_5(x - 1)$

8 Expand the following in terms of log *r*, log *s* and log *t*.

a) $\log_3 rs$ **b)** $\log_4 s^4 t$ **c)** $\log r^m s^n t^p$

d) $\log_a\left(\frac{\sqrt{r}}{s^3}\right)$ **e)** $\log\sqrt[3]{s^4 t^5}$ **f)** $\log_{10}(0.01\, r^2)$

g) $\log_a\left(\frac{1}{rst}\right)$ **h)** $\log\left(\frac{(2r+1)(3s-2)}{t+3}\right)$

MODELLING WITH LOGARITHMS

The Richter scale

Greece rocked by earthquake

A POWERFUL EARTHQUAKE rocked central and northern Greece, badly damaging the towns of Larissa and Kozani. Fifteen people were hurt. It was the second earthquake to hit Greece this month. It measured 6.6 on the Richter scale.

A sudden rupture in the solid crust of the Earth causes vibrations in the body of the Earth, called **seismic waves**, When these vibrations reach the surface of the Earth they cause the ground to tremble and we call this an **earthquake**. During an earthquake a very large amount of energy is dissipated. For example, the devastating San Francisco earthquake of 1906 had an estimated 10^{17} joules of energy. Physical phenomena which involve such large numbers are usually described in terms of common logarithms to base ten.

Exploration 16.3

The Richter scale

The strength of an earthquake, *M*, is expressed by the **Richter magnitude scale** – a logarithmic scale.

$$M = 0.67\log_{10} E - 2.9$$

where *E* is the energy of the earthquake in joules.

- What was the magnitude of the San Francisco earthquake in 1906?
- How strong, in joules, was the earthquake in Greece in May 1995?
- Approximately how much stronger was the San Francisco earthquake compared with the earthquake in Greece?

Comparing the strength

You should have found that:

- for the San Francisco earthquake $M = 8.5$
- for the Greek earthquake $E = 10^{14}$.

This means that the earthquake in San Francisco had approximately 1000 times more energy than that in Greece. The use of common logarithms allows us to describe physical phenomena with 'sensible numbers' that non-mathematicians can understand.

Noise levels

The intensity of sound is usually measured on a logarithmic scale called the **intensity level**. The formula for measuring sound intensity is:

$$L = 10\log_{10}\frac{I}{I_0}$$

where L is the level of sound in decibels (dB) and I is the observed sound intensity. (The definition of I_0 is $L = 0\,\text{dB}$ when $I_0 = 0.468 \times 10^{-12}\,\text{W m}^{-2}$). The sound intensity for the threshold of our hearing is $1.2 \times 10^{-12}\,\text{W m}^{-2}$ and for the threshold of pain it is $1\,\text{W m}^{-2}$. Notice that for sound levels we are dealing with very small numbers, whereas for earthquakes we have very large numbers.

Example 16.7

Find the level of sound in decibels for the following.

a) *threshold of our hearing*
b) *threshold of pain*
c) *heavy rock music for which $I = 0.15\,\text{W m}^{-2}$*

Find the sound intensity I for a background noise of 50 dB.

Solution

In each case $I_0 = 0.468 \times 10^{-12}$.

a) *For the threshold of hearing* $I = 1.2 \times 10^{-12}$

$$\Rightarrow \quad L = 10\log_{10}\left(\frac{1.2\times 10^{-12}}{0.468\times 10^{-12}}\right) = 4\text{dB}$$

b) *For the threshold of pain* $I = 1$

$$\Rightarrow \quad L = 10\log_{10}\left(\frac{1}{0.468\times 10^{-12}}\right) = 123\text{dB}$$

c) *For heavy rock music* $I = 0.15$

$$\Rightarrow \quad L = 10\log_{10}\left(\frac{0.15}{0.468\times 10^{-12}}\right) = 115\text{ dB}$$

For a background noise of 50 dB we have:

$$50 = 10\log_{10}\left(\frac{I}{0.468\times 10^{-12}}\right)$$

$$\Rightarrow \quad 5 = \log_{10}\left(\frac{I}{0.468\times 10^{-12}}\right)$$

$$\Rightarrow \quad 10^5 = \frac{I}{0.468\times 10^{-12}}$$

$$\Rightarrow I = 10^5 \times 0.468 \times 10^{-12} = 0.468 \times 10^{-7}\,\text{W m}^{-2}$$

We see that heavy rock music produces a sound intensity which is close to the threshold of pain and can cause serious damage to hearing if sustained over a long period of time. On the logarithmic scale a sound intensity 100 times as large as I_0 corresponds to 20 dB; an intensity 1000 times as large corresponds to 30 dB, and so on. This tends to agree with our subjective feeling of sound in that we underestimate very large sounds.

EXERCISES

1 By what factor is the energy in the seismic waves of an earthquake of magnitude 8.0 on the Richter scale larger than in those of an earthquake of magnitude 4.0?

2 The great earthquake in Lisbon, Portugal, in 1755 had an estimated magnitude of 9 on the Richter scale. What was the energy associated with this earthquake?

3 The following table shows the sound intensity level of some common situations. Calculate the sound intensity in $\mathrm{W\,m^{-2}}$ for each case.

Sound	Intensity level
normal breathing	10 dB
normal conversation	60 dB
street traffic in London	70 dB
underground train	100 dB
thunder	110 dB
jet engine at 30 m	120 dB
rupture of eardrum	160 dB

By what factor is normal conversation louder than normal breathing?

4 An office has a general sound level of 60 dB. A new photocopier is introduced into the office that doubles the sound intensity. What is the new sound level in dB?

5 An obvious feature of the stars in the night sky is that they appear to shine with different brightness. Over 2000 years ago the Greek astronomer Hipparcus divided the stars into classes of brightness. The brightest stars were classified as first magnitude and the very faintest that could be seen with the naked eye were classified as sixth magnitude. This classification leads to a logarithmic scale for star brightness:

$$m = -2.5 \log_{10} S$$

where m is the apparent magnitude of a star and S is the apparent brightness of the star. This formula is used for comparing star magnitudes. For this formula the brightest star, Sirius has an apparent magnitude of −1.4.

a) By what factor do the brightness of stars of first ($m = 1$) and sixth ($m = 6$) magnitude differ? This is the brightness ratio between the brightest and faintest stars S_1/S_6.

b) On this scale the sun and moon have apparent magnitudes of –26.8 and –12.6 respectively. What is the brightness ratio for the sun and moon?

c) The faintest observed objects have an apparent magnitude of +28. What is the brightness ratio for this object and the brightest star, Sirius?

EXERCISES

1 What is the energy associated with an earthquake that measures 5.8 on the Richter scale? By what factor is this earthquake stronger than an earthquake that measures 4.8 on the Richter scale?

2 The Richter scale number for the Italian earthquake in November 1980 was 6.8 and for the San Francisco earthquake in 1989 it was 6.1. Show that the Italian earthquake was roughly ten times more intense than the San Francisco earthquake.

3 One noise has a sound intensity of $I_1\,\mathrm{W\,m^{-2}}$ and associated sound level of $L_1\,\mathrm{dB}$. A second noise has a sound intensity of $I_2\,\mathrm{W\,m^{-2}}$ and associated sound level of $L_2\,\mathrm{dB}$. Show that:

$$L_2 - L_2 = 10\log_{10}\left(\frac{I_2}{I_1}\right)$$

Deduce that if the sound intensity increases by a factor 10^s then the intensity level changes by $10s$ dB.

4 As sound waves spread out their intensity falls off according to the law Ir^2 = constant, where I is the sound intensity in W m^{-2} at a distance r from the sound source.

Two people are standing 1 m from heavy motorway traffic where the sound intensity level is 80 dB. How far away from the traffic should they walk so that they can hold a normal conversation?

5 In chemistry, the acidity or alkalinity of a solution is measured by its pH factor defined by

$$\mathrm{pH} = -\log_{10}[\mathrm{H^+}]$$

where $[\mathrm{H^+}]$ is the quantity of hydrogen ions present in the solution. A pH of 7.0 indicates neutral solution for which $[\mathrm{H^+}] = 10^{-7}$ moles per litre. If $\mathrm{pH} < 7.0$ the solution is acidic.

a) What is the pH for a hair shampoo of strength 1.1×10^{-8} moles per litre? Is the shampoo acidic or alkaline?

b) What is the concentration of hydrogen ions for apple juice with a pH value of 5?

SOLUTION OF EQUATIONS OF THE FORM $a^x = b$

Exploration 16.4

Solving an equation

Take the equation $10^{-x} = 378$.

- Draw a graph of $y = 10^{-x} - 378$ and use it to estimate a solution of the equation $10^{-x} = 378$ correct to 1 d.p.
- Use logarithms as the inverse function of 10^{-x} to find x to 4 d.p.

Using common logarithms

The equation in Exploration 16.4 was solved in Example 12.13 on page 301 by a decimal search method, giving $x = -2.58$. From your graph you probably obtained $x = -2.6$ to one decimal place.

From $10^{-x} = 378$ we have $-x = \log_{10}378 = 2.5775 \Rightarrow x = -2.5775$ correct to four decimal places.

This exploration shows how equations of the form $10^x = b$ can be solved using common logarithms. Will it work for other values of a?

Consider $2^x = 20$. The inverse function of 2^x is $\log_2$ so in this case $x = \log_2 20$. Calculators do not use logs to base 2.

Start again but take logs of each side to base 10, since these are available on calculators.

$$2^x = 20 \Rightarrow \log_{10}2^x = \log_{10}20 \Rightarrow x\log_{10}2 = \log_{10}20$$

$$\Rightarrow x = \frac{\log_{10}20}{\log_{10}2} = 4.3219 \text{ to 4 d.p.}$$

This effectively provides a method of solving any exponential equation.

$$a^x = b$$

Take logs to base 10 of both sides.

$$\log_{10}a^x = \log_{10}b \Rightarrow x\log_{10}a = \log_{10}b \Rightarrow x = \frac{\log_{10}b}{\log_{10}a}$$

Example 16.8

Solve these equations.

a) $5^{3x} = 26$ **b)** $3^{-0.1t} = 0.2$

Solution

a) $5^{3x} = 26$

Take logs to base 10.

$$\log_{10}5^{3x} = \log_{10}26 \Rightarrow 3x\log_{10}5 = \log_{10}26 \Rightarrow x = \frac{1}{3}\frac{\log_{10}26}{\log_{10}5} = 0.6748$$

b) $3^{-0.1t} = 0.2$

Take logs to base 10.

$$\log_{10}3^{-0.1t} = \log_{10}0.2 \Rightarrow -0.1t\log_{10}3 = \log_{10}0.2$$

$$\Rightarrow t = -10\frac{\log_{10}0.2}{\log_{10}3} = 14.6497$$

EXERCISES

1 Solve the following equations giving your answers correct to 4 d.p.

a) $10^x = 4$	**b)** $10^x = 132$	**c)** $10^{-x} = 67$	**d)** $10^{2x} = 27$
e) $10^{-3x} = 620$	**f)** $10^{0.1t} = 14.1$	**g)** $2^x = 17$	**h)** $4^{-x} = 0.8$
i) $3^{5x} = 10$	**j)** $5^{2t} = 32$	**k)** $3^{2x-1} = 7$	**l)** $7^{4t+1} = 13$

2 Use logs to base 10 to solve the following equations correct to 4 d.p. Compare your answers with those obtained graphically and by decimal search in Exercise 12.4A.

a) $3^x = 20$ **b)** $2^{-x} = 3$

3 Solve these equations.

a) $5^{2x} - 5^{2x+1} + 6 = 0$ (**Hint:** let $y = 5^x$) b) $3^{2x} = 8^{x+1}$

c) $7^{2x} - 3(7^x) + 2 = 0$ d) $\dfrac{10^{x-1}}{3^x} = 7^{2x+1}$

4 Find the least value of the integer n for which $0.9^n < 0.001$.

EXERCISES

16.3B

1 Solve the following equations giving your answers correct to 4 d.p.

a) $10^x = 1740$ b) $10^{-x} = 5$ c) $10^t = 1.44$ d) $10^{-2t} = 2.6$
e) $10^{-5y} = 0.02$ f) $10^{0.3u} = 16$ g) $7^x = 40$ h) $5^{-x} = 1.5$
i) $4^{0.2x} = 9$ j) $3^{-0.5t} = 10$ k) $2^{3x+5} = 11$ l) $5^{-0.1x+1} = 3$

2 Use logs to base 10 to solve the following equations correct to 4 d.p. Compare your answers with those obtained graphically and by decimal search in Exercise 12.4B.

a) $5^x = 15$ b) $3^{-x} = 8$

3 Solve these equations.

a) $2^{2x} - 7(2^x) + 12 = 0$ b) $5^{3t} = 7^{t-2}$

c) $3^{2p} - 3^{2p+1} + 10 = 0$ d) $\dfrac{10^{x-1}}{9^{x-1}} = 8^x$

4 Find the least value of the integer n for which $1.5^n > 32$.

REDUCTION TO LINEAR FORM

Exploration 16.5

Equations from graphs

- Draw graphs of the following data.
- Find the equation of each curve.

Data set A

x	0.0	0.5	1.0	1.5	2.0	2.5
y	0.0	0.125	1.0	3.975	8.0	15.625

Data set B

x	0.0	0.5	1.0	1.5	2.0	2.5
y	1.0	1.732	3.0	5.196	9.0	15.58

Try to find the equation of the graph of the following data.

Data set C

x	0.0	0.5	1.0	1.5	2.0	2.5
y	0.0	0.4061	1.0	1.694	2.462	3.29

Working with equations

You probably found the equations for data sets A and B quite easily. The curve from set A has equation $y = x^3$ and the curve from set B has equation $y = 3^x$. You may even have spotted the equation from the data before drawing the graph.

For data set C the equation is not so obvious. Its equation is $y = x^{1.3}$ and you should not worry if you did not find it.

Logarithms can be used to help us find equations relating sets of data by transforming the graph of the data from a curve into a straight line.

Case 1: The power law $y = ax^b$

Take the equation $y = ax^b$ where a and b are constants. Take logs of both sides (using $\log_{10}$).

$$\log_{10}y = \log_{10}(ax^b) = \log_{10}a + \log_{10}x^b \Rightarrow \log_{10}y = \log_{10}a + b\log_{10}x$$

Compare this with the equation of a straight line $Y = mX + c$.

If data satisfies a power law $y = ax^b$ then a graph of $Y = \log_{10}y$ against $X = \log_{10}x$ will be a straight line with slope $m = b$ and intercept $c = \log_{10}a$.

Example 16.9

Find the equation of the graph for data set C.

x	0.0	0.5	1.0	1.5	2.0	2.5
y	0.0	0.4061	1	1.694	2.462	3.29

Solution

The graph of y against x is a curve through the origin. Take logs of x and y, avoiding (0, 0).

$\log_{10}x$	−0.30	0	0.18	0.30	0.40
$\log_{10}y$	−0.39	0	0.23	0.39	0.53

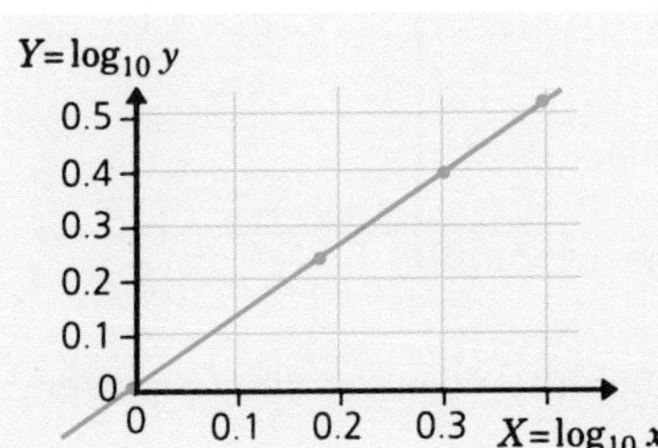

A graph of $\log_{10}y$ against $\log_{10}x$ is a straight line. In this case the points lie very close to the line which passes through the origin.

The slope of the line is 1.3 and the intercept is 0. The equation of the line is:

$$\log_{10}y = 0 + 1.3\log_{10}x$$

we see that $\log_{10}a = 0 \Rightarrow a = 1$ and $b = 1.3$. The law relating x and y is $y = x^{1.3}$.

Case 2: The exponential law $y = ab^x$

Take the equation $y = ab^x$ and take logs of both sides.

$$\log_{10}y = \log_{10}ab^x = \log_{10}a + \log_{10}b^x \Rightarrow \log_{10}y = \log_{10}a + x\log_{10}b$$

Compare this with the equation of a straight line $Y = mX + c$.

If the data satisfies an exponential law $y = ab^x$ then a graph of $Y = \log_{10}y$ against $X = x$ will be a straight line with slope $m = \log_{10}b$ and intercept $c = \log_{10}a$.

Example 16.10

Find the equation of the graph for this data set.

x	0.0	0.5	1.0	1.5	2.0	2.5
y	3.0	3.911	5.1	6.649	8.67	11.30

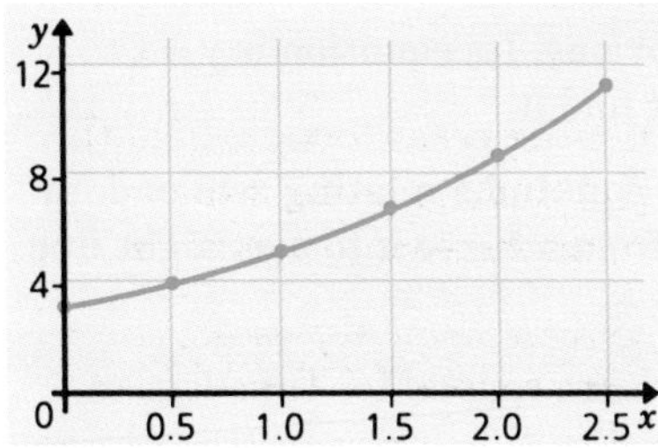

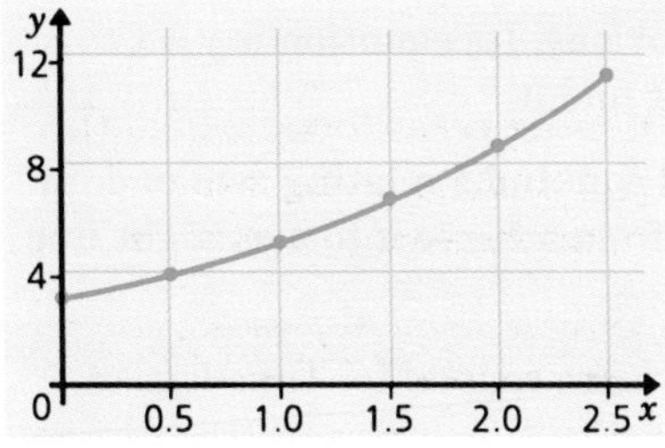

Solution

A graph of y against x is a curve through the point (0, 3).
Take logs of y.

x	0.0	0.5	1.0	1.5	2.0	2.5
$\log_{10}y$	0.48	0.59	0.71	0.82	0.94	1.05

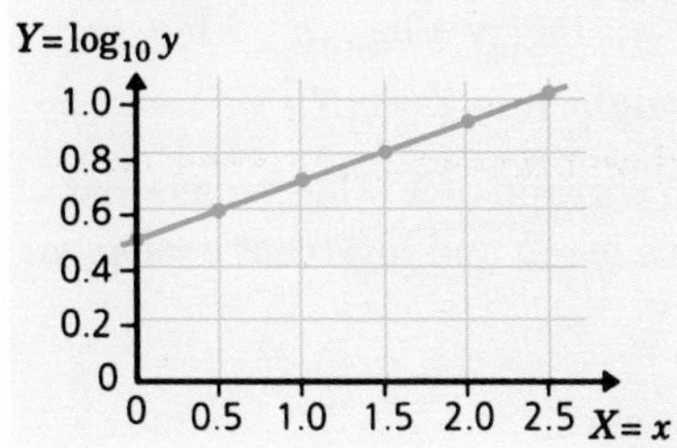

A graph of $\log_{10}y$ *against* x *is a straight line.*
The slope of the line is 0.228 and the intercept is (0, 0.48).
The equation of the line is:

$$\log_{10}y = 0.48 + 0.228x$$

We see that $\log_{10}a = 0.48 \Rightarrow a = 10^{0.48} = 3$ *and* $\log_{10}b = 0.228$
$\Rightarrow b = 10^{0.228} = 1.69$.

The law relating x *and* y *is* $y = 3(1.69)^x$.

Which graph to draw?

When given a set of data the first problem is to decide which graph to draw, **a)** $\log_{10}y$ against $\log_{10}x$ or **b)** $\log_{10}y$ against x. We could draw both graphs; if one of the graphs is a straight line then we should know which model to choose.

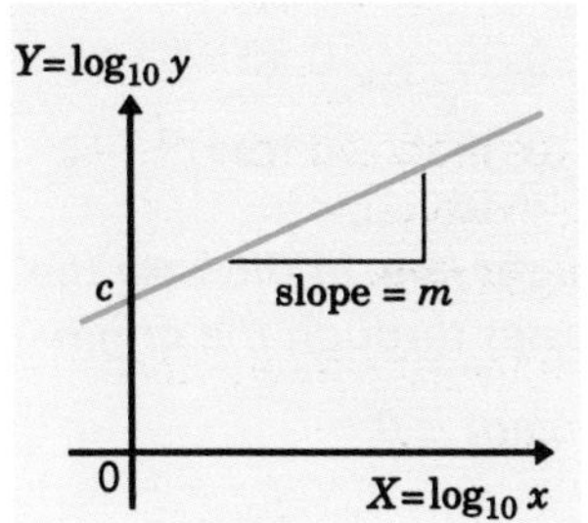

$\Rightarrow y = ax^b$
$b \equiv$ slope of line m
$c \equiv \log_{10}a \Rightarrow a = 10^c$

$\Rightarrow y = ab^x$
$\log_{10}b \equiv$ slope of line m
$\Rightarrow b = 10^m$
$c \equiv \log_{10}a \Rightarrow a = 10^c$

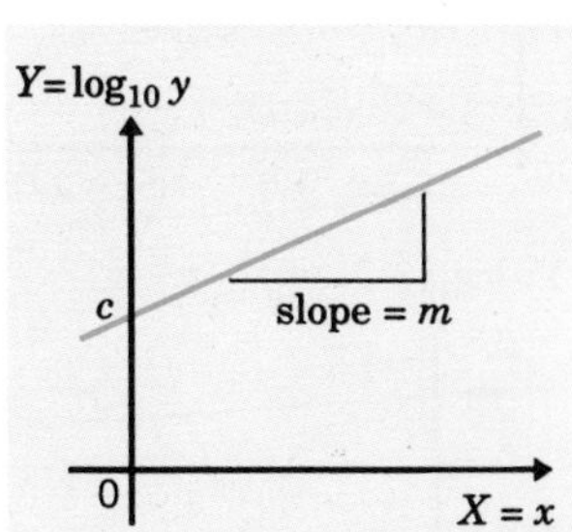

If neither graph is a straight line then the data does not obey a power law or an exponential law.

In some examples we can save time and effort by looking at a graph of the original data. If it passes through the origin, the law is likely to be a power law $y = ax^b$. For example, data set A is a power law because $y = 0$ when $x = 0$.

Exploration 16.6

Finding equations using logs

Take data sets A and B from Exploration 16.5.

- Draw graphs of $\log_{10}y$ against $\log_{10}x$ and $\log_{10}y$ against x for each data set.
- Which data sets give straight lines?
- Use the equations of the straight lines to find the equations relating x and y for each data set.

Confirming the results

You should have confirmed that set A obeys the power law $y = x^3$ with a straight line graph of $\log_{10} y$ against $\log_{10} x$ and set B obeys the exponential law $y = 3^x$ with a straight line graph of $\log_{10} y$ against x.

For examples where reading the $\log_{10} y$ intercept is not a practical proposition, then the rule of thumb, given in Chapter 3, *Linear functions* for lines of best fit, is to fit a straight line through the mean $(\bar{X}, \bar{Y})$ giving the equation:

$$Y - \bar{Y} = m(X - \bar{X})$$

This method is used in the following mathematical modelling activity.

MATHEMATICAL MODELLING ACTIVITY

Problem statement

In the early 17th century the German astronomer, Johannes Kepler formulated three laws for the motion of the planets around the sun. These are known as **Kepler's laws**.

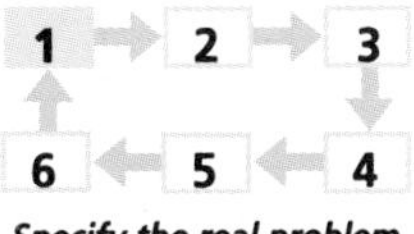

Specify the real problem

1. Each planet moves in an ellipse with the sun at one focus.
2. For each planet, the line from the sun to the planet sweeps out equal areas in equal times.
3. The squares of the orbital periods of the planets vary as the cubes of their mean distance from the sun.

Kepler formulated the first two laws after studying the motion of the planet Mars. It took a further ten years for him to verify these laws and he formulated the third law in 1619 after studying other planets.

The following table shows the average distance from the sun, R, and the period of revolution around the sun, T, for each planet.

Planet	Distance from the sun R ($\times 10^6$ km)	Period of revolution around the sun T (days)
Mercury	57.9	88
Venus	108.2	225
Earth	149.6	365
Mars	227.9	687
Jupiter	778.3	4329
Saturn	1427	10753
Uranus	2870	30660
Neptune	4497	60150
Pluto	5907	90670

Note that the planets outside Saturn were not known to Kepler. Show that Kepler's third law is a good model by fitting an appropriate graph to the data.

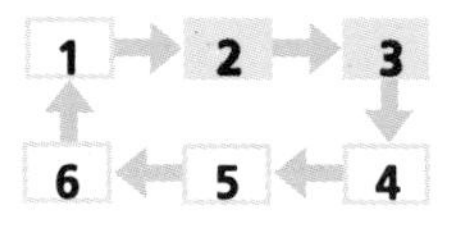

Set up a model

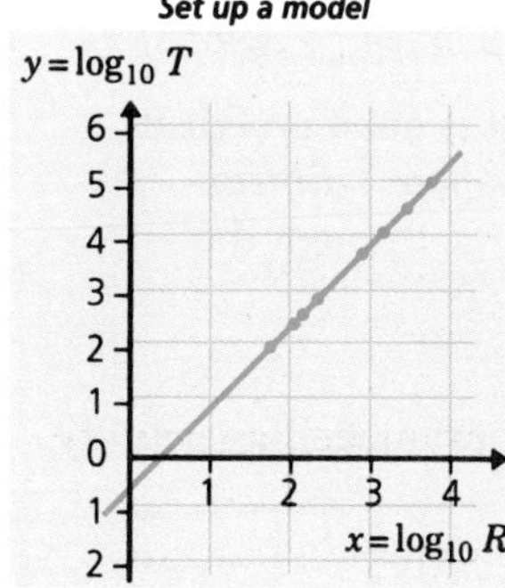

Set up a model

This is an empirical modelling activity obtained from a graph. Kepler's third law suggest a power law of the form:

$$T = KR^{\alpha}$$

where K and α are constants. Taking logs of each side:

$$\log_{10} T = \log_{10} K + \alpha \log_{10} R$$

Thus if Kepler's third law holds true, a plot of $\log_{10} T$ against $\log_{10} R$ should be a straight line. The following table and graph show $\log_{10} T$ and $\log_{10} R$.

$y = \log_{10} T$	$x = \log_{10} R$
1.94	1.76
2.35	2.03
2.56	2.17
2.84	2.36
3.64	2.89
4.03	3.15
4.49	3.46
4.78	3.65
4.96	3.77

1 → 2 → 3 → 4 → 5 → 6

Interpret the solution

Interpret the model

From the graph, the slope is approximately 1.5. The mean values of $\log_{10} R$ and $\log_{10} T$ are (2.80, 3.51) so the equation of the line of best fit is:

$$y - 3.51 = 1.5(x - 2.80)$$
$$\Rightarrow \; y = 1.5x - 0.69$$

Comparing this with the general model:

$$\log_{10} T = \alpha \log_{10} R + \log_{10} K$$

we see that $\alpha = 1.5$ and $\log_{10} K = -0.69 \Rightarrow k = 10^{-0.69} = 0.50$

The empirical model gives:

$$T = 0.5R^{1.5}$$

Squaring both sides gives $T^2 = 0.25R^3$ which agrees well with Kepler's third law.

EXERCISES

16.4A

In problems 1–6 use logs to straighten the curves of the data and hence find the equation for each data set.

1

x	0.0	0.2	0.4	0.6	0.8	1.0
y	0.0	1.78	2.53	3.09	3.58	4.0

2

x	0.0	0.4	0.8	1.2	1.6	2.0
y	5.0	5.88	6.92	8.13	9.57	11.25

3

x	0.5	0.7	0.9	1.1	1.3	1.5
w	16.97	7.32	3.90	2.36	1.56	1.09

4

t	1	2	3	4	5	6
u	2.5	7.58	14.50	22.97	32.83	43.95

5

x	0.2	1.4	2.6	3.8	5.0	6.2
v	13.78	1.10	0.49	0.30	0.21	0.16

6

r	0.1	0.4	0.7	1.0	1.3	1.6
s	2.01	0.606	0.182	0.055	0.017	0.005

7 Experimental values of the pressure P of a given mass of gas corresponding to various values of the volume V are given in a table. According to thermodynamic principles a relationship having the form $P = CV^{-\lambda}$ where λ and C are constants, should exist between the variables.

a) Find the values of λ and C.
b) Write the equation connecting P and V. Estimate P when $V = 100.0$.

Volume V	54.3	61.8	72.4	88.7	118.6	194.0
Pressure P	61.2	49.5	37.6	28.4	19.2	10.1

8 A census of the population of the United States each decade during 1840–1960 is shown in the following table.

Year	1840	1850	1860	1870	1880	1890	1900
Population (in millions)	17.1	23.2	31.4	39.8	50.2	62.9	76.0

Year	1910	1920	1930	1940	1950	1960
Population (in millions)	92.0	105.7	122.8	131.7	151.1	179.3

Formulate a mathematical model to describe the population as a function of time.

9 Burning fossil fuels such as coal and oil adds carbon dioxide to the atmosphere around the Earth. This may be partly removed by biological reactions, but the concentration of carbon dioxide is gradually increasing. This increase leads to a rise in the average temperature of the Earth. The table on the left shows this temperature rise over the 100-year period up to 1980.

Year	**Temperature rise of the Earth above the 1860 figure (°C)**
1880	0.01
1896	0.02
1900	0.03
1910	0.04
1920	0.06
1930	0.08
1940	0.10
1950	0.13
1960	0.18
1970	0.24
1980	0.32

If the average temperature of the Earth rises by about another 6°C from the 1980 value this would have a dramatic effect on the polar ice caps, winter temperatures etc. As the polar ice caps melt, there could be massive floods and a lot of land mass would be submerged. The UK would disappear except for the tops of the mountains!

a) Find a model of the given data and use it to predict when the Earth's temperature will be 7°C above its 1860 value.
b) Discuss how seriously you should take the prediction in **a)**.

10 The table below gives the values of the atmospheric pressure, expressed as a percentage of its sea-level value, at various altitudes.

Altitude h (km)	0	5	10	14	20	24	30
Pressure y (% of sea level value)	100	53	26	14	5.4	2.9	1.2

From this data find a possible law relating pressure and altitude.

EXERCISES

16.4B

In problems **1–6** use logs to straighten the curves of the data and hence find the equation for each data set.

1

x	1	2	3	4	5	6
y	3.4	22.1	66.0	143.6	262.2	429.0

2

t	0	1	2	3	4	5
x	0.3	1.23	5.04	20.67	84.77	347.5

3

t	0	1	2	3	4	5
p	1.5	4.08	11.08	30.12	81.90	222.6

4

t	0.1	0.4	0.7	1.0	1.3	1.6
p	0.01	0.22	0.71	1.50	2.60	4.02

5

u	0.0	0.2	0.4	0.6	0.8	1.0
v	1.70	2.35	3.24	4.47	6.16	8.50

6

t	0.5	1.0	1.5	2.0	2.5	3.0
s	1.225	4.9	11.03	19.6	30.63	44.1

7 In a fishing competition on the River Dart in Devon the following data was collected for the mass of each fish and its length.

Length l (cm)	37	32	44	45	36	41	34
Mass M (kg)	0.77	0.48	1.16	1.39	0.65	1.02	0.58

a) Formulate a mathematical model relating mass and length of the fish for this data.

b) Explain how the model makes sense from theoretical arguments.

8 The viscosity of a liquid is a measure of its resistance to flow. Its value is often found experimentally using a quantity of the liquid between two moving parallel plates and measuring the friction force on one plate. In such an experiment, values of the coefficient of viscosity of lubricating oil against temperature were measured and are shown in

the following table. Show that a power law relation of the form $\mu = AT^m$ exists between the variables, and find values of the constants A and m.

Temperature *T* (°C)	20	40	60	80	100
Viscosity *μ* ($\text{kg m}^{-1}\text{s}^{-1}$)	0.0986	0.0241	0.0110	0.0055	0.0036

9 The heat of combustion H (joule mol^{-1}) for a petroleum hydrocarbon of molecular mass M is given in the following table.

H	213	373	530	688	845	1002	1159
M	16	30	44	58	72	86	100

Show, using logarithms and a suitable graph, that a relationship of the form $H = kM^n$ exists between H and M where k and n are constants. From your graph determine values for k and n and find the value of H when $M = 50$.

10 A census of the population of England and Wales each decade from 1811 to 1911 is shown in the following table.

Year	1811	1821	1831	1841	1851	1861	1871	1881	1891	1901	1911
Population (in millions)	10.2	12.0	13.9	15.9	17.9	20.1	22.7	26.0	29.0	32.5	36.1

a) Formulate a mathematical model to describe the population as a function of time.

b) Use your model to predict the population in 1991. How does it compare with the actual value?

CONSOLIDATION EXERCISES FOR CHAPTER 16

1 The mass of a colony of bacteria, in grams, doubles each day and is given by the formula $P = 7(2^t)$.

a) What is the initial mass of the bacteria?

b) Calculate the number of days after which the total mass of the bacteria is greater than 500 grams.

2 The concentration of a drug in a patient's bloodstream is C milligrams per millilitre after t hours where $C = Ka^{-t}$.

For drug A, $K = 2$ and $a = 1.2$

For drug B, $K = 3$ and $a = 1.4$

Show that the concentration levels are the same for both drugs when

$2(1.2^{-t}) = 3(1.4^{-t})$

and use logs to solve this equation for t.

3 The following table shows three sets of experimental data. One pair of variables satisfies a power law, one pair satisfies an exponential

law and the other pair does not satisfy either of these laws. Use appropriate graphs to find the relationships.

x	0.1	0.4	0.7	1.0	1.3	1.6
y	2.010	0.606	0.182	0.055	0.017	0.005

u	0.1	0.3	0.6	0.9	1.2	1.5
v	1.00	0.96	0.83	0.62	0.36	0.07

t	0.1	0.5	0.9	1.3	1.7	2.1
s	0.084	1.293	3.511	6.561	10.35	14.83

4 A controlled experiment was concerned with estimating the number of microbes, N, present in a culture at time T days after the start of the experiment. Some results from the experiment are shown in the table below.

Time T (days)	3	5	10	15	20
Number of microbes N	900	2000	5000	9000	16 000

By plotting values of $\log N$ against the corresponding values of $\log T$, draw a graph using all the data in the table.

Explain why the graph that you have obtained supports the belief that N and T are related by an equation of the form:

$$N = AT^B$$

where A and B are constants.

Use your graph to find an estimate for A, giving your answer to 1 significant figure, and an estimate for B, giving your answer to 2 significant figures. *(ULEAC Question 6, Paper 2, January 1993)*

5 The table shows experimental values of two quantities x and y which are known to satisfy the equation $yx^n = k$, where n and k are constants.

x	2	3	4	5	6	7
y	39	32	28	25	22.5	20.8

a) Draw a graph of $\log y$ against $\log x$.

b) Use your graph to estimate values for n and k, giving your answers to 2 significant figures. *(ULEAC Question 8, Paper 2, January 1994)*

6 It is believed that the relationship between the variables x and y is of the form $y = Ax^n$

In an experiment the following data are obtained.

x	3	6	10	15	20
y	10.4	29.4	63.2	116.2	178.9

In order to estimate the constants A and n, $\log_{10} y$ is plotted against $\log_{10} x$.

a) Complete the table of values of $\log_{10} x$ and $\log_{10} y$, and draw the graph of $\log_{10} y$ against $\log_{10} x$.
b) Explain and justify how the shape of your graph enables you to decide whether the relationship is indeed of the form $y = Ax^n$.
c) Estimate the values of A and n. *(MEI Question 5, Paper1, June 1993)*

7 An incomplete sketch (not drawn to scale) of the graph of $y = \log_{10}(x + a)$ is shown. Find the value of a.
(Scottish Highers Question 12, Paper 1, 1992)

8 a) On the same diagram, sketch the graphs of $y = \log_{10} x$ and $y = 2 - x$ where $0 < x < 5$.
Write down an approximation for the x-coordinate of the point of intersection.
b) Find the value of this x–coordinate, correct to 2 decimal places.
(Scottish Highers Question 4, Paper 2, 1991)

9 The following table gives the number of people in a particular area who own shares in a public company.

Year	1987	1988	1989	1990	1991
Number of years after 1987, x	0	1	2	3	4
Number of people who own shares, y	12	17	25	37	53

a) Plot the points (x, y) on a the graph. Join the points with a smooth curve.
It is thought that there may be a relationship connecting x and y which is of the form $y = ab^x$.
b) Explain how, by using logarithms (to base 10), the curve given by plotting y against x can be transformed into a straight line.
State the gradient of this straight line and its intercept with the vertical axis, in terms of a and b.
c) Form a table of x and $\log_{10} y$. Plot the points $(x, \log_{10} y)$ on a graph and draw a 'best fit' line through the points and use it to estimate the values of a and b. *(MEI Paper 1, Question 5, June 1994)*

10 a) The variables x, y are thought to satisfy an equation of the form
$$y = ab^x$$
where a, b are unknown constants. To estimate a and b, y is measured when $x = 2.2$ and $x = 3.6$. The respective values of y are 4.21 and 6.22.
Find the values of a and b.
Hence find the value of x when $y = 5$.
b) Six pairs of measurements (x_1, y_1); (x_2, y_2); (x_3, y_3); (x_4, y_4); (x_5, y_5); (x_6, y_6) are now made on x and y. Explain briefly how you would use these values to plot a graph which could be used to investigate whether or not x and y do in fact satisfy the above relationship.
How could you use your graph to estimate a and b?
(WJEC, Specimen Paper A1, Question 11)

Summary

The logarithm is defined from the exponential function a^x in the following way.

$$y = a^x \Rightarrow x = log_a y$$

This is called the logarithm to base a.

Logarithms obey the following rules.

- $\log_a 1 = 0$
- $\log_a a = 1$
- $\log_a(xy) = \log_a x + \log_a y$
- $\log_a \left(\dfrac{x}{y}\right) = \log_a x - \log_a y$
- $\log_a x^r = r \log_a x$

Logarithms to base 10 are found on scientific calculators and are usually written as $\log x$ i.e. omitting the base 10 ($\log_{10} x \equiv \log x$). These logarithms are used in scientific scales and in straightening curves to model data.

- If $y = ax^b$ then a graph of $\log y$ against $\log x$ is a straight line.
- if $y = ab^x$ then a graph of $\log y$ against x is a straight line.

PURE MATHS 17

Exponentials and logarithms

In this chapter we study two special functions for mathematics, science and engineering:

- *the exponential function*
- *the natural logarithmic function.*

EXPONENTIAL GROWTH

CALCULATOR ACTIVITY

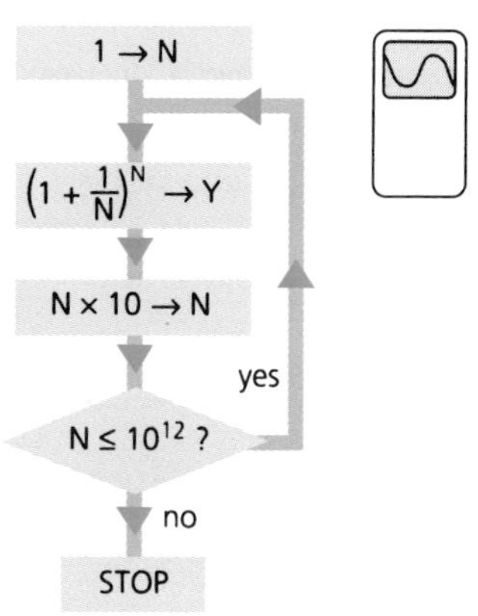

- Use this flowchart to create a program that allows you to complete the table.
- Describe what happens to the sequence of values in the right-hand column as n gets larger.

n	$\left(1+\frac{1}{n}\right)^n$
1	2
10	2.593 742 46
100	

Limits of the function

As n gets larger, the value of $\left(1+\frac{1}{n}\right)^n$ converges towards a limit 2.718 28... which is an irrational number known as e. It has such an important role to play in mathematics that the values of the **exponential function** e^x may be obtained directly from your calculator (show that $e^1 = 2.718\,28...$).

Exploration 17.1

APR

Imagine you have £100 to invest for one year at a nominal annual rate of interest of 8%.

- How much would your investment be worth after one year if interest is compounded:
  a) annually, **b)** quarterly, **c)** monthly, **d)** weekly, **e)** daily?
- What are the corresponding **annual percentage rates** (APRs)?
- What is the APR which corresponds to continuous compounding?

Summarising the results

The results of each computation are summarised in the table overleaf.

The quarterly, monthly, etc. rates of interest are found by dividing the nominal annual rate by 4, 12 etc. The APR, the rate that would provide the same growth if compounded once per year, may be deduced from the corresponding investment values. These values form a convergent sequence as the number of compoundings increases.

	Number of compoundings	Value of investment after 1 year	APR
a)	1	$100 \times (1+0.08) = £108$	8.00%
b)	4	$100 \times \left(1+\frac{0.08}{4}\right)^4 = 100 \times 1.02^4 = £108.24$	8.24%
c)	12	$100 \times \left(1+\frac{0.08}{12}\right)^{12} = 100 \times 1.0067^{12} = £108.30$	8.30%
d)	52	$100 \times \left(1+\frac{0.08}{52}\right)^{52} = 100 \times 1.0015^{52} = £108.32$	8.32%
e)	365	$100 \times \left(1+\frac{0.08}{365}\right)^{365} = 100 \times 1.0002^{365} = £108.33$	8.33%

Even hourly compounding (8760 times a year) has no effect on the investment value, to the nearest penny, also giving an APR of 8.33%.

Continuous compounding, corresponding to continuous growth, occurs as the number of compoundings per year **tends to infinity**.

Suppose there are n compoundings per year, then the value of the investment of £100 after one year is given by:

$$A = 100 \times \left(1+\frac{0.08}{n}\right)^n$$

Let $k = \frac{1}{0.08}n$, then $n = 0.08k$, which gives:

$$A = 100 \times \left(1+\frac{1}{k}\right)^{0.08k} = 100 \times \left[\left(1+\frac{1}{k}\right)^k\right]^{0.08}$$

As $n \to \infty$, $k \to \infty$, and from the last Calculator activity you will have discovered that as $k \to \infty$, $(1+\frac{1}{k})^k \to$ e (2.718 28), which means that the limiting value of A is $100 \times e^{0.08} = 108.33$ (2 d.p.), which also gives an APR of 8.33%.

The results of Exploration 17.1 may be generalised as follows.

An amount £P is invested at a nominal annual rate of interest R%, with n compoundings per year. The value of the investment £A after one year is given by:

$$A = P \times \left(1+\frac{R}{100n}\right)^n$$

Let $k = \dfrac{100n}{R}$, then $n = \dfrac{Rk}{100}$, which gives:

$$A = P \times \left[\left(1+\frac{1}{k}\right)^k\right]^{\frac{R}{100}}$$

As $n \to \infty$, $k \to \infty \Rightarrow A = Pe^{\frac{R}{100}}$.

The APR corresponding to continuous compounding is given by:

$$\frac{A-P}{P} \times 100 = \frac{Pe^{\frac{R}{100}} - P}{P} \times 100 = 100\left(e^{\frac{R}{100}} - 1\right)$$

The corresponding formulae for the value of the investment after t years are:

$$A = P \times \left(1 + \frac{1}{k}\right)^{kt} \text{ and } A = P\,e^{\frac{Rt}{100}}$$

CALCULATOR ACTIVITY

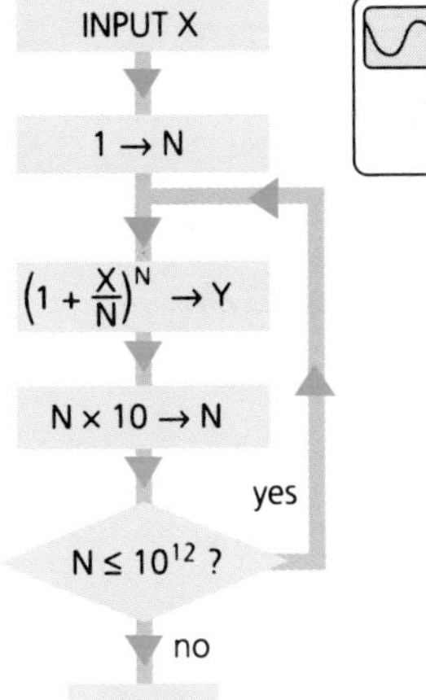

- Use the flowchart to create a program to find the limit of the sequence $\left(1+\frac{x}{n}\right)^n$ as $n \to \infty$ for various values of x.
- In each case compare the limiting value with e^x from your calculator.

The limit of the sequence

You will have seen that the limit of the sequence as $n \to \infty$ is equivalent to e^x, i.e.

$$\underset{n \to \infty}{\text{limit}} \left(1+\frac{x}{n}\right)^n = e^x$$

Using the binomial theorem:

$$\left(1+\frac{x}{n}\right)^n = 1 + \frac{nx}{n} + \frac{n(n-1)}{2!}\frac{x^2}{n^2} + \frac{n(n-1)(n-2)}{3!}\frac{x^3}{n^3} + \ldots + \ldots + \frac{x^n}{n^n}$$

As $n \to \infty$ $\frac{n-1}{n}$, $\frac{n-2}{n}$, etc. $\to 1$, in which case:

$$\underset{n \to \infty}{\text{limit}} \left(1+\frac{x}{n}\right)^n = 1 + x + \frac{x^2}{2!} + \frac{x^3}{3!} + \frac{x^4}{4!} + \ldots\ldots$$

You have found two ways of expressing the limit of $\left(1+\frac{x}{n}\right)^n$ as $n \to \infty$; equating them gives:

$$e^x \equiv 1 + x + \frac{x^2}{2!} + \frac{x^3}{3!} + \frac{x^4}{4!} + \ldots$$

The series expansion of e^x is valid for all values of x since it is derived from a binomial expansion with a positive integer power (n). Since we are using the limiting case as $n \to \infty$, the expansion has an infinite number of terms but is convergent for any real value of x.

Example 17.1

Expand the series to find e correct to 3 d.p.

Solution

$$e^x = 1 + x + \frac{x^2}{2!} + \frac{x^3}{3!} + \frac{x^4}{4!} + \ldots$$

$$\Rightarrow \; e^1 = 1 + 1 + \frac{1}{2!} + \frac{1}{3!} + \frac{1}{4!} + \ldots$$

$$= 1 + 1 + 0.5 + 0.166\,666\,666\,7 + 0.041\,666\,666\,7 + 0.008\,333\,333\,3 + 0.000\,198\,412\,7 + \ldots$$

Accuracy to 3 d.p. is achieved after eight terms, the sum of which is 2.718 253 968 (10 s.f.), giving e = *2.718 (3 d.p.).*

CALCULATOR ACTIVITY

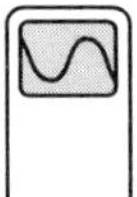

The process of finding the series expansion for e^x, up to any number of terms, may be programmed using this flowchart.

Check that the series is convergent for a variety of x-values. For which range of values does the series oscillate?

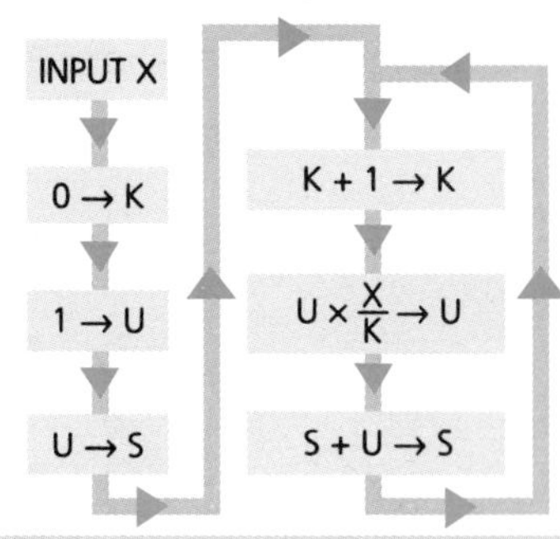

EXERCISES

1 A sum of £2000 is invested at a nominal interest rate of 12%. Calculate its value:

a) after one year if compounding is quarterly,
b) after one year if compounding is monthly,
c) after four years if compounding occurs every six months,
d) after six years with quarterly compounding.

2 Find the APR that is equivalent to:

a) 6% nominal rate compounded semi annually,
b) 8% nominal rate compounded quarterly,
c) 12% nominal rate compounded monthly,
d) 10% nominal rate compounded weekly.

3 Find the value of each of the following investments, each compounded continuously.

a) £5000 for three years at a nominal annual rate of 6%
b) £2000 for five years at a nominal annual rate of 8%
c) £1000 for six years at a nominal annual rate of 10%
d) £3000 for four years at a nominal annual rate of 5%

4 An investment is compounded continuously at a nominal rate of 8% per annum. How long does it take the investment to:

a) double in value, **b)** triple in value?

5 Use the series expansion for e^x to find each of the following to 3 d.p. Check your answers using your calculator function.

a) e^3 **b)** $\sqrt[3]{e}$ **c)** $\frac{1}{e^2}$ **d)** $\sqrt[3]{e^2}$

6 Bank A pays 8.1% compounded quarterly on its savings accounts. Bank B offers 7.8% but compounded daily. Where would you invest your money?

7 You have £500 to invest and you are attracted by two five-year savings schemes. The first offers 12.5% p.a. nominal interest compounded twice yearly and the second just 12.25% nominal interest rate but compounded monthly.

a) Calculate the APR in each case, correct to 2 d.p.
b) Which is worth more after five years and by how much?

EXERCISES

17.1 B

1 £3500 is invested at a nominal interest rate of 12.5%. Calculate its value to the nearest penny:

a) after one year if compounding is monthly,
b) after one year if compounding is weekly,
c) after five years with monthly compounding,
d) after five years with weekly compounding.

2 Find the APR that is equivalent to:

a) 5% nominal rate compounded quarterly,
b) 11% nominal rate compounded monthly,
c) 7% nominal rate compounded weekly,
d) 9% nominal rate compounded daily (in a non-leap-year).

3 Find the value of the following continuously compounded investments.

a) £1500 for four years at a nominal annual rate of 5%
b) £850 for seven years at a nominal annual rate of 7.5%
c) £2400 for ten years at a nominal annual rate of 12%
d) £10 000 for 20 years at a nominal annual rate of 11.25%

4 An investment is compounded continuously at a nominal rate of 12.5% per annum. How long does it take for the investment to

a) double in value, **b)** triple in value, **c)** quadruple in value?

Give your answers to the nearest month.

5 Suppose that in the year 1800, one of your wise old ancestors invested £1 in a bank at 5% interest compounded quarterly. You will inherit this investment in the year 2000. How much will you inherit? How much was it worth in the year 1900?

6 Building Society A pays 5.75% compounded quarterly on its savings accounts. Society B offers just 5.5% but compounded daily. Where would you invest your money?

7 Use the series expansion for e^x to find each of the following correct to 3 d.p. Check your answers using your calculator function.

a) e^2 **b)** $\frac{1}{e}$ **c)** $\sqrt{e}$ **d)** $\frac{1}{\sqrt{e}}$

In each case describe the series as oscillating or non-oscillating.

THE EXPONENTIAL FUNCTION

Exploration 17.2

Exponential graphs

On the same axes draw accurately graphs of:

a) $y = e^x$ **b)** $y = -e^x$ **c)** $y = e^{-x}$ **d)** $y = -e^{-x}$

for the domain $\{x: -2 \le x \le 2\}$.

■ Describe geometrically how each graph is related to each of the others.

- Estimate the gradient of the curve $y = e^x$ at:
  i) $x = 0$ **ii)** $x = 1$ **iii)** $x = -1$.
  What do you notice?
- Use these gradient values to estimate the gradients at the corresponding x-values on the other three curves.

Interpreting the graphs

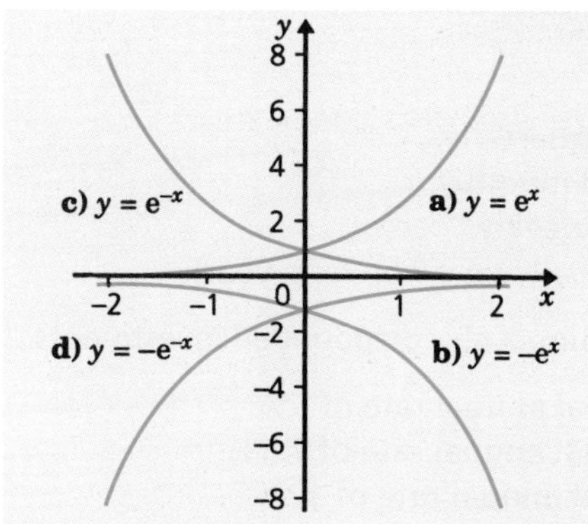

You should obtain the graphs shown below.

Graph **b)** is a reflection of graph **a)** in the x-axis.
Graph **c)** is a reflection of graph **a)** in the y-axis.
Graph **d)** is a rotation through 180° of graph **a)**, centre (0, 0).

Similar relationships exist between other pairs of graphs.

For the curve $y = e^x$, you should have found the following gradients.

i) at (0, 1) gradient = 1
ii) at (1, 2.718) gradient ≈ 2.7
iii) at (−1, 0.368) gradient ≈ 0.4.

It looks as though the gradient equals the function value on each occasion. This is true for any x-value, a property that is brought out in Chapter 18, *Differentiation 2*.

Example 17.2

On separate grids sketch a graph of $y = e^x$ and each of the following equations. In each case describe the geometrical transformation(s) which map(s) $y = e^x$ onto the given equation.

a) $y = 3e^x$ ***b)*** $y = e^{2x}$ ***c)*** $y = 10 - e^x$ ***d)*** $y = 0.5e^{-x}$

Solution

Stretch, factor 3, parallel to the y-axis

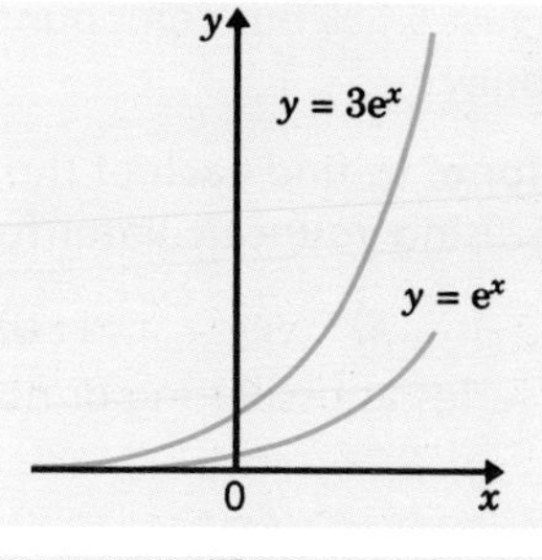

Stretch, factor 0.5, parallel to the x-axis

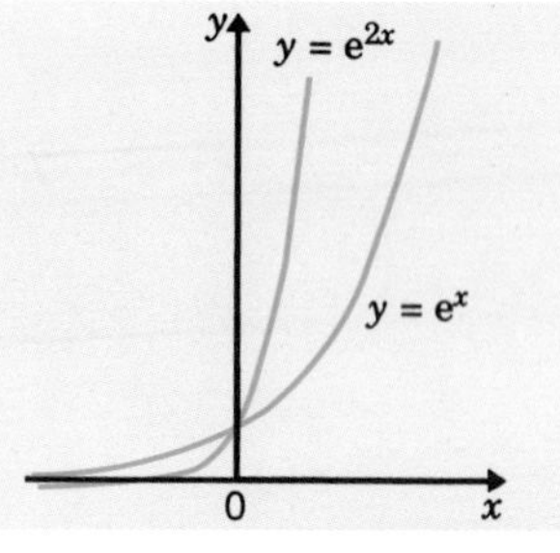

Reflection in the x-axis

followed by

translation 10 units parallel to y-axis

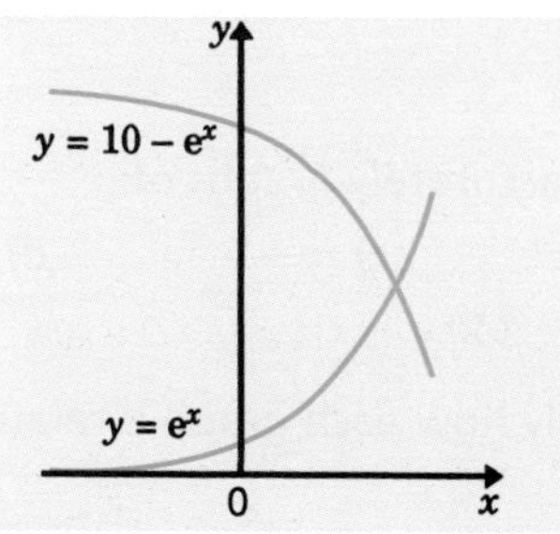

Reflection in the y-axis

followed by

stretch factor 0.5 parallel to the y-axis

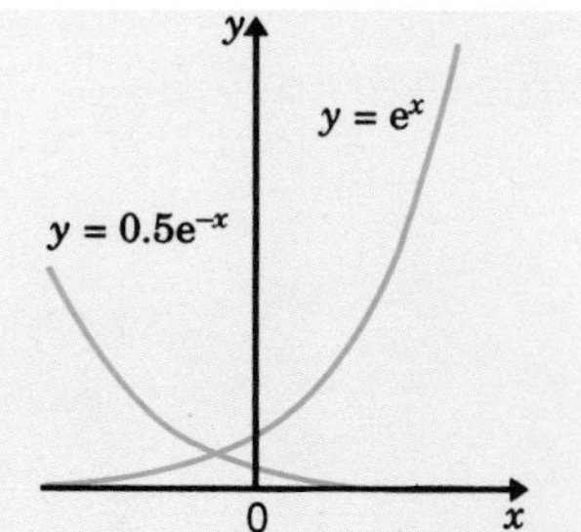

Example 17.3

The population, P, of a certain country is found to be given in millions by the formula:

$P = 10e^{0.03t}$ *where t is time in years since 1980.*

a) *What was the population in 1980?*

b) *Predict the population in the year 2000.*

c) *Draw a graph of P against t for $0 \le t \le 30$.*

d) *Use your graph to estimate when the population will be double its size in 1980.*

e) *Show that the annual percentage growth rate is constant. State its value.*

Solution

a) *In 1980* $t = 0 \Rightarrow P = 10e^0 = 10$ million.

c)

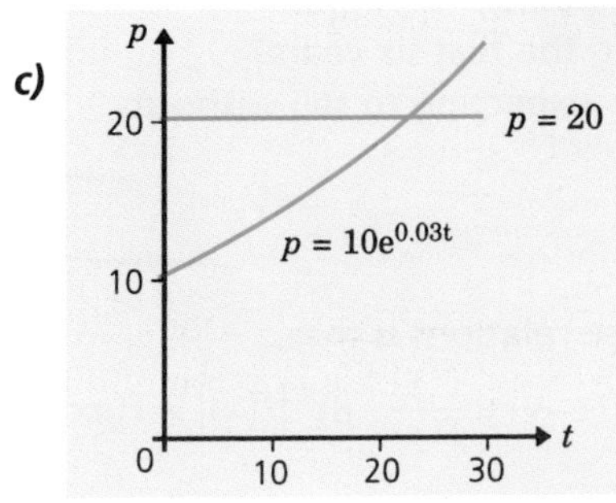

b) *In 2000* $t = 20 \Rightarrow P = 10e^{0.03 \times 20} = 10e^{0.6} = 18.2$ million *(3 s.f.)*

d) *From the graph, when* $P = 20$, $t \approx 23$, *i.e. population should double by the year 2003.*

e) *When* $t = n$, $P = 10e^{0.03n}$ *and when* $t = n + 1$, $P = 10e^{0.03(n+1)}$.

The increase in population between $t = n$ *and* $t = n + 1$ *is:*

$$10e^{0.03(n+1)} - 10e^{0.03n} = 10e^{0.03n}(e^{0.03} - 1)$$

$$\Rightarrow \text{percentage increase} = \frac{10e^{0.03n}\left(e^{0.03} - 1\right)}{10e^{0.03n}} \times 100$$

$$= 100(e^{0.03} - 1) = 3.05\%\ \textit{(3 s.f.)}$$

EXERCISES

17.2 A

1 Given $f(x) = e^x$, show that each of these relations is true.

a) $f(0) = 1$ **b)** $f(x + y) = f(x)f(y)$ **c)** $\dfrac{f(x)}{f(y)} = f(x-y)$ **d)** $\left[f(x)\right]^n = f(nx)$

2 Draw accurate graphs for the following functions for $-3 \le x \le 3$.
Use your graph to estimate:

i) the x-value of the given y-value,

ii) the gradient of the curve when $x = 1$.

a) $y = 0.3e^x;\ y = 2$ **b)** $y = e^{x-2};\ y = 4$ **c)** $y = 2e^{-x};\ y = 3.5$

d) $y = 5 + 1.5e^x;\ y = 5.3$ **e)** $y = 8 - e^{0.5x};\ y = -2$ **f)** $y = e^{-x^2};\ y = 0.9$

3 On separate grids sketch a graph of $y = e^x$ and each of the following equations. In each case describe the geometrical transformation(s) which map(s) $y = e^x$ onto the given equation.

a) $y = -2e^x$ **b)** $y = 10 - e^{-x}$ **c)** $y = e^{\frac{1}{2}x}$ **d)** $y = 0.5e^{x+1}$

e) $y = e^{2x} - 5$ **f)** $y = 12 - 5e^x$

4 The population, P, of a certain city at time t (measured in years) is given by the formula:

$P = 50\,000\, e^{0.05t}$

Draw an accurate graph of P against t for the domain $0 \le t \le 20$.
Use your graph to estimate:

a) the time it takes for the population to double,

b) the rate of growth of the population at $t = 10$.

5 A machine is purchased for £10 000 and depreciates continuously from the date of purchase. Its value, V, after t years is given by the formula:

$$V = 10\ 000\ e^{-0.2t}$$

a) Draw a graph of V against t for the domain $0 \le t \le 10$.
b) From your graph estimate:
   i) the value of the machine after eight years,
   ii) the time it takes for its value to fall by £6000.
c) Calculate the percentage decline in value each year.

6 The sales of a new textbook in hundreds of book(s), S, after it has been on the market for t years is given by the formula:

$$S = 60(1 - e^{kt}).$$

a) Find the value of k if 2000 books are sold in the first year.
b) Draw a graph of S against t for the domain $0 \le t \le 6$.
c) How many books will be sold during the first six years?
d) How many books are the publishers expecting to sell without a reprint?

EXERCISES

17.2 B

1 Given $f(x) = e^{-x}$, show that each of these relations is true.

a) $f(0) = 1$ b) $f(x + y) = f(x)f(y)$ c) $f(x-y) = \dfrac{f(x)}{f(y)}$ d) $[f(x)]^n = f(nx)$

2 Show that the above relationships still hold for $f(x) = e^{ax}$ where a is a constant ($\neq 0$).

3 Draw accurate graphs of the following functions for the domain $-2 \le x \le 2$.
Use your graph to estimate:
   i) the x-value for the given y-value,
   ii) the gradient of the curve when $x = -1$.

a) $y = 2e^{-x}$; $y = 0.75$ b) $y = 3e^{2x}$; $y = 3$ c) $y = 2 - e^{3x}$; $y = -18$
d) $y = e^{x+4}$; $y = 20$ e) $y = 2e^{x^2}$; $y = 4$

4 On separate grids sketch a graph of $y = e^x$ and each of the following equations. In each case describe the geometrical transformation(s) which map(s) $y = e^x$ onto the given equation.

a) $y = \frac{1}{5}e^x$ b) $y = e^{x+1}$ c) $y = e^{2x-3}$ d) $y = 2.5e^{-2x}$ e) $y = 4 - 4e^{-x}$

5 In a simple population model, the number of individual members of the population, N, at time t is given by:

$$N = N_0 e^{(B-D)t}$$

where N_0 is the initial size of the population, B is the number of births per 1000 per year and D is the number of deaths per 1000 per year.

a) What happens if: i) $B > D$, ii) $B < D$, iii) $B = D$?
b) If, in a certain population, $B = 25$, $D = 15$ and $N_0 = 10\,000$ estimate the size of the population in: i) 5 years, ii) 10 years, iii) 50 years.

c) Explain why, in the real world, your answer to **b) iii)** is probably unrealistic.

6 A patient is injected with a certain drug which gradually diffuses from the bloodstream such that the quantity left in the bloodstream (in milligrams) after t hours is given by:

$$Q = Q_0 e^{-0.2t}$$

where Q_0 is the initial amount injected.

a) If the patient is given an initial dose of 10 milligrams, draw a graph of Q against t for the domain $0 \le t \le 12$.

b) From your graph estimate:

i) the amount present after 5 hours,

ii) how long (to the nearest 15 minutes) it takes for 9 milligrams to diffuse,

iii) the rate of diffusion after 5 hours.

THE NATURAL LOGARITHM FUNCTION

In Chapter 16, *Logarithms*, we saw that each exponential function had an inverse function, e.g.

$$f(x) = 10^x \Rightarrow f^{-1}(x) = \log_{10} x$$

$$f(x) = 2^x \Rightarrow f^{-1}(x) = \log_2 x$$

Similarly, if $f(x) = e^x$ then $f^{-1}(x) = \log_e x$.

Logarithms to the base e are so important in mathematics that they are often called the **natural logarithms** and denoted by $\ln x$.

Exploration 17.3

Graphing natural logarithms

Using axes with the same scales, draw graphs of $y = e^x$ and $y = \ln x$.

- Describe the behaviour of each curve geometrically.
- How are the two curves related geometrically?

The graphs of natural logs

The graphs of $y = e^x$ and $y = \ln x$ are shown below.

y
3
2
1
0
-1
-2
-3
-2
-1
1
2
x
$y = e^x$
$y = \ln x$

The graph of $y = e^x$ passes through (0, 1). As $x \to -\infty$, $y \to 0$, i.e. the negative x-axis is an asymptote to the curve; as $x \to +\infty$, $y \to +\infty$. The gradient of the curve is always positive and gets steeper as x increases.

The graph of $y = \ln x$ passes through (1, 0). As $x \to 0$, $y \to -\infty$, i.e. the negative y-axis is an asymptote to the curve; as $x \to +\infty$, $y \to +\infty$. The gradient of the curve is always positive and gets shallower as x increases.

As with all inverse functions, the two graphs are reflections of each other in the line $y = x$.

$$f^{-1}f(x) = x \Rightarrow \ln e^x = x \text{ and } e^{\ln x} = x$$

Exploration 17.4

Domain of the logarithmic function

On the same axes draw accurately graphs of:

a) $y = \ln x$ **b)** $y = \ln(-x)$ **c)** $y = -\ln x$ **d)** $y = -\ln(-x)$

for suitable domains.

- Write down the largest possible domain for each graph.
- Describe how the graph of $y = \ln x$ may be transformed onto each of the other graphs.

Transforming the graphs

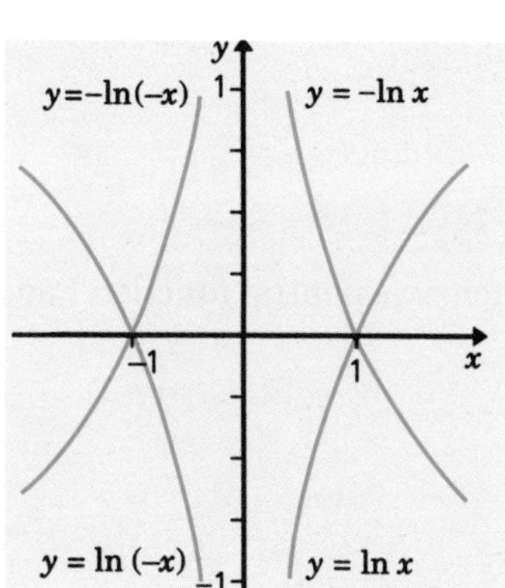

The largest possible domain for $y = \ln x$ and $y = -\ln x$ is $\{x: x > 0\}$ and for $y = \ln(-x)$ and $y = -\ln(-x)$ is $\{x: x < 0\}$.

$y = \ln(-x)$ is a reflection of $y = \ln x$ in the y-axis.
$y = -\ln(x)$ is a reflection of $y = \ln x$ in the x-axis.
$y = -\ln(-x)$ is a rotation of $y = \ln x$ about $(0, 0)$ through $180°$.

Note that $y = -\ln x$ is also the result of composite reflections in two different ways. Describe them for yourself.

The laws of logarithms established in Chapter 16, *Logarithms*, apply to natural logarithms.

$$\ln(ab) = \ln a + \ln b$$
$$\ln\left(\tfrac{a}{b}\right) = \ln a - \ln b$$
$$\ln a^n = n \ln a$$

They are demonstrated in the following example and in the exercises.

Example 17.4

On the same axes sketch the graphs of the following.

a) $\ln x$ ***b)*** $\ln(3x)$ ***c)*** $\ln(3x) - \ln x$

*Use the law of logarithms to explain your graph for **c)** and so describe how the graph for **a)** may be transformed onto the graph for **b)** using a translation.*

Solution

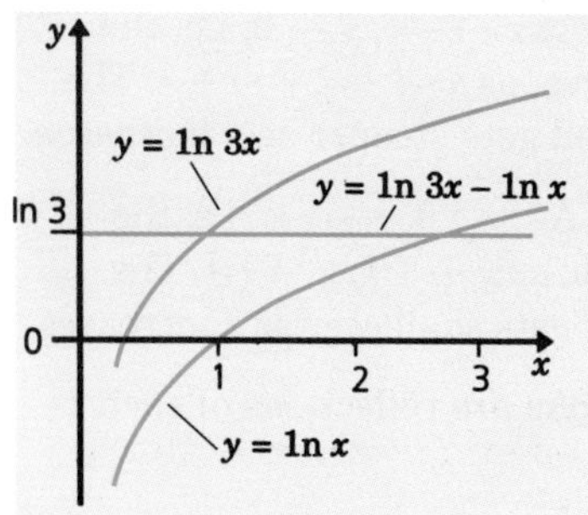

Since $\ln(3x) - \ln x = \ln\left(\frac{3x}{x}\right) = \ln 3$*, the graph of* $y = \ln(3x) - \ln x = \ln 3$ *is a line parallel to the x-axis.*
The graph of $y = \ln(3x)$ *is therefore a translation of* $y = \ln x$ *through* $\ln 3$ *units parallel to the y-axis.*

$$\ln(3x) = \ln x + \ln 3$$

Note: *The graph of* $y = \ln(3x)$ *is also a stretch, factor* $\frac{1}{3}$ *of* $y = \ln x$ *parallel to the x-axis, since for any function* $f(x)$, $f(ax)$ *represents a stretch factor* $\frac{1}{a}$ *of* $f(x)$ *parallel to the x-axis.*

EXERCISES

17.3 A

1 Use the laws of logarithms to express each of these as a single logarithm.

a) $\ln 3 + \ln 7$ **b)** $\ln 15 - \ln 4$ **c)** $2\ln 6 - \ln 12$

d) $\frac{1}{2}\ln 25 + \frac{1}{3}\ln 27$ **e)** $n\ln a + \ln b$

2 Simplify these expressions.

a) $5\ln e^2$ **b)** $\ln \sqrt{e}$ **c)** $3\ln e^3 - 12\ln \sqrt[3]{e}$ **d)** $\ln e^{x2}$ **e)** $\dfrac{\ln e^{4\sin x}}{\ln e^{5\cos x}}$

3 **a)** On the same axes sketch the graphs of $y = \frac{1}{2}e^x$ and $y = \ln 2x$.
b) Explain why $f(x) = \frac{1}{2}e^x \Rightarrow f^{-1}(x) = \ln 2x$.

4 **a)** On the same axes sketch the graphs of $y = \ln x$ and $y = \ln(\frac{1}{5})$.
b) Describe how the graph of $y = \ln x$ may be transformed onto the graph of $y = \ln(\frac{1}{5}x)$ in *two* different ways.

5 **a)** On the same axes sketch the graphs of $y = \ln x$, $y = \ln x^2$ and $y = \ln \sqrt{x}$.
b) Describe how the graph of $y = \ln x$ may be transformed onto:
i) the graph of $y = \ln x^2$,
ii) the graph of $\ln \sqrt{x}$.

c) Find inverse functions for $y = \ln x^2$ and $y = \ln \sqrt{x}$ for suitable domains.

6 **a)** On the same axes sketch graphs of $y = \ln x$ and $y = \ln(\frac{1}{x})$.
b) Describe the geometrical transformation suggested in part **a)**.
c) Use the laws of logarithms to justify your conclusion in part **b)**.
d) Write down the inverse function of $y = \ln(\frac{1}{x})$.

EXERCISES

1 Express each of these as a single logarithm.

a) $\ln 4 + \ln 6$ **b)** $\ln 17 - \ln 3$ **c)** $3\ln 10 - 2\ln 5$

d) $\frac{1}{5}\ln 32 + \frac{1}{4}\ln 81$ **e)** $\frac{1}{3}\ln 125 + \frac{1}{2}\ln 25$

2 Simplify each of these.

a) $\frac{1}{3}\ln e^3$ **b)** $6\ln \sqrt[3]{e}$ **c)** $7\ln e^4 - 6\ln \sqrt{e}$ **d)** $\ln 3e^2$

e) $e^{4\ln x}$ **f)** $e^{\frac{1}{2}\ln x^6}$

3 Given that $\ln a = 5$ and $\ln b = 7$ express the following as simple exponentials.

a) ab **b)** $a + b$ **c)** $\dfrac{a}{b}$

4 **a)** On the same axes sketch the graphs of $y = \frac{1}{3}e^{-x}$ and $y = -\ln 3x$.
b) Explain why $f(x) = \frac{1}{3}e^{-x} \Rightarrow f^{-1}(x) = -\ln 3x$.

5 **a)** On the same axes sketch the graphs of $y = \ln x$ and $y = \ln 5x$.
b) Describe how the graph of $y = \ln x$ may be transformed onto the graph of $y = \ln 5x$ in *two* different ways.

6 **a)** On the same axes sketch the graphs of $y = \ln x$, $y = \ln x^3$ and $y = \ln \sqrt[3]{x}$.

b) Describe how the graph of $y = \ln x$ may be transformed onto:
i) the graph of $y = \ln x^3$, **ii)** the graph of $\ln \sqrt[3]{x}$.

c) Find inverse functions for $y = \ln x^3$ and $y = \ln \sqrt[3]{x}$ for suitable domains.

SOLVING EQUATIONS WITH EXPONENTIALS AND LOGARITHMS

Exploration 17.5

Radioactive decay

A radioactive substance has a **half-life** of five years, i.e. it takes five years for the level of radioactivity to decay to half its original level.

- After how many years will the level of radioactivity be one-quarter of its original level?
- The radioactive substance is deemed to be 'safe' when the level of radioactivity is 1% of its original level. How long will this take?

A suitable model

A suitable model for the decaying process is:

$$A = A_0 e^{-kt}$$

where A represents the amount of radioactivity after t years and A_0 and k are constants to be found. When $t = 0$, $A = A_0 e^0 = A_0$, i.e. A_0 represents the original level of radioactivity.

When $t = 5$, the level of radioactivity has halved. Substituting $t = 5$ and $A = \frac{1}{2}A_0$ gives:

$$\frac{1}{2}A_0 = A_0 e^{-5k}$$

$$\Rightarrow \quad \frac{1}{2} = e^{-5k}$$

Taking the natural logarithm of both sides gives:

$$\ln \frac{1}{2} = \ln e^{-5k}$$

$$\Rightarrow \quad \ln \frac{1}{2} = -5k$$

$$\Rightarrow \quad k = \frac{\ln \frac{1}{2}}{-5} = 0.139 \text{ (3 s.f.)}$$

so the radioactive decay is modelled by the equation:

$$A = A_0 e^{-0.139t}$$

When at one-quarter of its original level $A = \frac{1}{4}A_0$

$$\Rightarrow 0.25A_0 = A_0 e^{-0.139t}$$

which you can solve for t.

$$0.25 = e^{-0.139t} \Rightarrow \ln 0.25 = \ln e^{-0.139t} \Rightarrow \ln 0.25 = -0.139t$$

$$\Rightarrow \quad t = \frac{\ln 0.25}{-0.139} = 10$$

For the safety level $A = 0.01A_0$

$$\Rightarrow \quad 0.01A_0 = A_0 e^{-0.139t} \Rightarrow 0.01 = e^{-0.139t} \Rightarrow \ln 0.01 = -0.139t$$

$$\Rightarrow \quad t = \frac{\ln 0.01}{-0.139} = 33.2 \text{ (3 s.f.)}$$

i.e. the level of radioactivity will have reached a safe level after a little more than 33 years.

The property that the exponential and logarithm functions are inverses of each other can be used in any problem where one of the functions needs to be reversed.

Example 17.5

The growth of a population is modelled by the logistic equation:

$$P(t) = \frac{30}{1+5e^{-0.1t}}$$

where t is time in years and $P(t)$ is the population in millions.

a) *Find the initial population size.*

b) *Describe what happens to $P(t)$ as $t \to \infty$.*

c) *Sketch a graph of the function for $0 \le t \le 100$.*

d) *By what time will the population have grown to 25 million?*

Solution

a) *Initially* $t = 0 \Rightarrow P(0) = \dfrac{30}{1+5e^0} = 5$ million

b) *As* $t \to \infty$ $e^{-0.1t} \to 0 \Rightarrow P(t) \to 30$, *i.e. the limiting population size is 30 million.*

c) *The sketch graph shows a typical logistic curve.*

d) $P(t) = 25 \Rightarrow 25 = \dfrac{30}{1+5e^{-0.1t}}$

$\Rightarrow 25(1+5e^{-0.1t}) = 30 \Rightarrow 5(1+5e^{-0.1t}) = 6$

$\Rightarrow 5 + 25e^{-0.1t} = 6 \Rightarrow 25e^{-0.1t} = 1 \Rightarrow e^{-0.1t} = 0.04$

$\Rightarrow \ln e^{-0.1t} = \ln 0.04 \Rightarrow -0.1t = \ln 0.04$

$\Rightarrow t = \dfrac{\ln 0.04}{-0.1} \Rightarrow t = 32.2$ *(3 s.f.)*

i.e. the population reaches 25 million after 32.2 years.

Example 17.6

Solve this equation.

$$e^{4x} + e^{2x} - 6 = 0$$

Solution

Let $u = e^{2x}$, *then* $e^{4x} = (e^{2x})^2 = u^2$

$\Rightarrow u^2 + u - 6 = 0 \Rightarrow (u-2)(u+3) = 0 \Rightarrow u = 2$ *or* $u = -3$

$u = 2 \Rightarrow e^{2x} = 2 \Rightarrow \ln e^{2x} = \ln 2 \Rightarrow 2x = \ln 2 \Rightarrow x = \dfrac{\ln 2}{2} = 0.347$ *(3 s.f.)*

$u = -3 \Rightarrow e^{2x} = -3$

which is not possible, since $e^{2x} > 0$ *for any real value of* x.

So the only solution is $x = 0.347$ *(3 s.f.).*

Example 17.7

Solve the equation $a^x = e^5$ *given that* $\ln a = 2$.

Solution

Start with $a^x = e^5$ *and take natural logarithms of each side.*

$$\ln a^x = \ln e^5 \quad \Rightarrow \quad x \ln a = 5 \ln e = 5$$

since $\ln e = 1$. *Substituting for* $\ln a = 2$ *gives the solution* $x = \frac{5}{2}$.

EXERCISES

17.4 A

1 Solve these equations.

a) $e^x = 10$ b) $2e^{5t} = 30$ c) $8e^{2x} = 5$ d) $4e^{-x} = 15$ e) $10 = 7e^{-3t}$

2 Solve these equations.

a) $\ln x = 3$ b) $5 + 2\ln x = 0$ c) $5\ln x^2 - 3\ln x = 2.1$
d) $\ln 2x + \ln 3x = 20$ e) $\ln(x^2 - 6) - \ln x = 0$

3 Solve these equations, where possible.

a) $e^{2x} - 7e^x + 12 = 0$ b) $e^{5t} - 2e^{3t} = 0$ c) $2e^{10t} - 9e^{5t} = 5$
d) $3e^{4x} + 10e^{2x} + 3 = 0$ e) $e^x - 3 = \sqrt{e^x}$

4 The population of a certain country is given by:

$$P = 15e^{0.02t}$$

where P is the population in millions and t is the time in years measured from 1980. When will the population reach 25 million, assuming the formula continues to hold?

5 a) The half-life of radium is 1590 years. Calculate its decay constant.
b) If 10 grams of radium is left for 1000 years, how much will remain?
c) How long will it take for a quantity of radium to decay to 10% of its original value?

6 The decay constant for C^{14} (carbon-14) is 1.24×10^{-4}, when t is measured in years.

a) Calculate the percentage of the original specimen that remains after 2000 years and 10 000 years.
b) Find the half-life of C^{14}.

7 When cancer cells are subjected to radiation treatment, the proportion, P, of cells that survive the treatment is given by:

$$P = P_0 e^{-kr}$$

where r is the radiation level and k a constant. It is found that 40% of the cancer cells survive when $r = 500$. What should the radiation level be in order to allow only 1% to survive?

8 A cup of coffee, at temperature T °C cools down according to the formula:

$$T = 70e^{-kt} + 20$$

where t is the time in seconds after the cup of coffee was made.

a) What is the initial temperature of the coffee?
b) The coffee cools by 20°C after the first minute. Find the value of k.
c) How long will it take for the temperature to drop to 50°C?

9 A learning curve is given by:

$$y = A(1 - e^{-kt})$$

where y is a measure of efficiency in terms of time t, and A and k are constants. When $t = 1$, $y = 10$ and when $t = 2$, $y = 15$.

a) Find values of A and k. b) Find the value of y when $t = 3$.
c) How long will it take for y to reach 19?

10 The spread of information through a population is modelled by the logistic equation:

$$p = \frac{1}{1+Ce^{-kt}}$$

where p represents the proportion of the population that is aware of the information after time t, and C and k are constants.

At $t = 0$, 10% of all stockbrokers have heard about the impending financial collapse of a company. Two hours later 25% have heard about it.

a) Show that $C = 9$ and $k = \frac{1}{2}\ln 3$.
b) How long will it be before 75% have heard about it?

EXERCISES

17.4 B

1 Solve these equations.

a) $e^x = 5$ **b)** $3e^{4x} = 7$ **c)** $2e^{-x} = 3$ **d)** $4 - 3e^x = 1$ **e)** $2e^x = 3e^2$

2 Solve these equations.

a) $\ln x = 7$ **b)** $3 - 4\ln x = 0$ **c)** $3\ln x^3 + 4\ln x = 9$
d) $\ln 7x + \ln 5x = 1$ **e)** $\ln(2x - 3) - \ln x = 0$

3 Solve these equations, where possible;

a) $e^{2x} - 7e^x + 10 = 0$ **b)** $e^{2x} - 2e^x - 35 = 0$ **c)** $e^{6x} - e^{3x} - 2 = 0$
d) $e^{4x} + 4e^{2x} + 5 = 0$ **e)** $e^{2x} + 7e^x + 12 = 0$
f) $\frac{e^x + e^{-x}}{2} = 2$ **g)** $\frac{e^x - e^{-x}}{2} = 1$

(**Hint:** in **f)** and **g)** multiply both sides by $2e^x$ and rearrange as quadratics.)

4 The decay of a radioactive isotope is modelled by the law $M = M_0e^{-kt}$ where M is the mass remaining (in g) after t days.

A certain isotope has a half-life of 10 days and after this time 40 g remain.

a) How much of the isotope was present initially? (i.e. what is M_0?)
b) Find the value of the decay constant, k.
c) How much will remain after: **i)** 20 days, **ii)** 50 days?
d) When will 90% have decayed (to nearest day)?

5 The half-life of radium is 1600 years. How long will it take for 20 g of radium to decay to:

a) 15 g **b)** 5 g **c)** 1 g?

6 **a)** 90% of the original amount of a radioactive isotope remains after five days. Find the half-life in days.
b) For a longer lived substance, it takes 1 year for 5% to decay. Find the half-life in years.
c) Certain substances are very short-lived. If a substance loses 99% of its mass in $\frac{1}{100}$ of a second, find the half-life.

7 A culture is grown in a Petri dish in a biology lab under controlled conditions. The number (N) of bacteria grows from 100 to 500 in 24 hours and obeys the law $N = N_0 e^{kt}$ where N_0 is the initial number of bacteria and k is a constant.

a) Find the time taken, to the nearest minute, for the number of bacteria to double.
b) Assuming conditions remain the same, how many bacteria do you predict after 48 hours.
c) Would you expect to be able to predict how many bacteria there would be after one week? Explain your reasoning carefully.

8 Bacteria are growing rapidly according to the law $N = N_0 e^{kt}$ where N is the number of bacteria present at time t hours and N_0 and k are constants.

If 200 000 bacteria grow to 300 000 in five hours, how many bacteria do you predict after ten hours?

9 The population of Great Britain at the time of the Queen's Coronation in 1953 was approximately 52 million. In 1993, this figure had grown to close on 60 million. Assuming a law of the form $N = N_0 e^{kt}$, no major wars, plagues, etc, estimate the population of Great Britain, to the nearest hundred thousand, in the year:

a) 2000 **b)** 2001 **c)** 2050. (**Hint:** take 1953 as year zero)

10 Atmospheric pressure, P, at height x feet above sea level can be modelled by the formula:

$$P = P_0 e^{-kx}$$

The pressure is measured in millibars (mbar) and at sea level the pressure is approximately 1000 mbar on average, at 15°C. At 1000 feet the pressure is approximately 967 mbar.

a) What is the atmospheric pressure:
 i) on top of High Willhays (Dartmoor) at 2038 feet,
 ii) on top of Snowdon (3559 feet),
 iii) on top of Ben Nevis (4406 feet)?
b) At what height would you expect atmospheric pressure to be 1 mbar?

MATHEMATICAL MODELLING ACTIVITY

Carbon dating

Background

The molecular structure of living things is based on the element carbon. There are two types or **isotopes** of carbon, carbon-14 and carbon-12, labelled C^{14} and C^{12} respectively. A molecule of C^{14} is heavier than a molecule of C^{12} and C^{14} is radioactive.

Whilst a plant is alive it takes in carbon dioxide from the air and converts it into useful products. The process of conversion does not distinguish between carbon dioxide containing C^{12} and carbon dioxide containing C^{14}.

Hence the ratio of $C^{14} : C^{12}$ in living tissue is approximately the same as the ratio of $C^{14} : C^{12}$ in the carbon dioxide in the air.

However, when a plant or animal dies it no longer takes in carbon dioxide. The amount of C^{12} remains constant but the radioactive carbon-14, C^{14} decays exponentially with a half-life of 5760 years. By measuring the ratio of $C^{14} : C^{12}$ it is possible to estimate the length of time that has elapsed since the plant or animal died.

This provides a method for archaeologists to estimate the age of fossil remains and other antiquities.

Problem statement

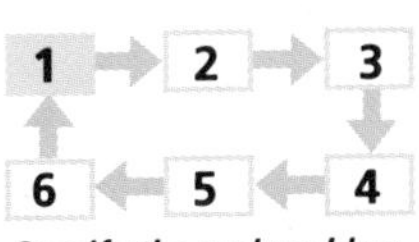
Specify the real problem

In fossilised seeds found recently the residual C^{14} was only 0.5% of the amount present in the air today.

How old were the fossils?

Before the findings, the site was thought to be at least 60 000 years old.

What percentage of C^{14} had been expected prior to finding the seeds?

Set up a model

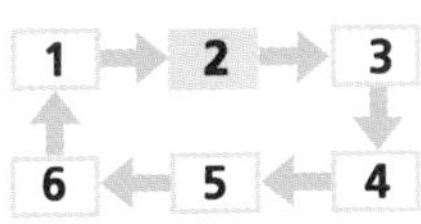
Set up a model

The important variables in this problem are the percentage of C^{14} remaining, $c(t)$ after time t years. In an experiment $c(t)$ could be measured by comparing the amount of C^{14} with the amount of C^{12} present at the present time.

You will need to make some assumptions before formulating the mathematical problem to solve. Here we assume:

- the amount of C^{12} remains constant in dead material,
- when the seeds were living the ratio of $C^{12} : C^{14}$ in the seeds was the same as the ratio at the present day,
- the C^{14} decays exponentially with time,
- environmental conditions have not changed over the last 60 000 years.

Mathematical problem

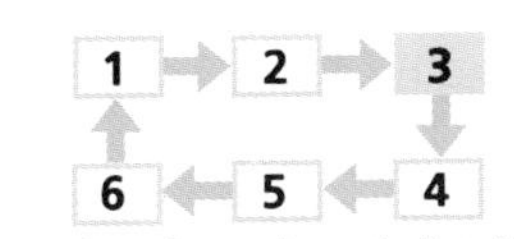
Formulate the mathematical problem

The three assumptions lead to the model:

$$c = c_0 e^{-kt}$$

where c_0 is the percentage of C^{14} initially present in the seeds when they were alive. We take $c_0 = 100$.

The mathematical problem is to use the background information to find a value for k then:

a) to find t when $c = 0.5\%$,
b) to find c when $t = 60\,000$ years.

Mathematical solution

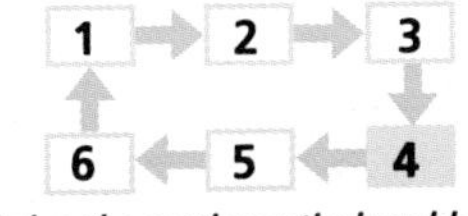
Solve the mathematical problem

The background information gives a half-life for C^{14} as 5760 years.

Hence $c = 50\%$ when $t = 5760 \quad \Rightarrow \quad 50 = 100e^{-5760k}$

$$\Rightarrow\ k = \frac{\ln 2}{5760} = 0.000\,12 \text{ (to 2 s.f.)}$$

The model for carbon dating is $c = 100e^{-0.000\,12t}$

When $c = 0.5,\ t = -\frac{1}{0.000\,12}\ln\left(\frac{0.5}{100}\right) = 44\,000$ (to 2 s.f.)

When $t = 60\,000,\ c = 100e^{-0.000\,12 \times 600\,00} = 0.075$

Interpretation

Interpret the solution

The data for the fossilised seeds suggests that the site where they were found is 44 000 years old (to 2 s.f.). If the site was 60 000 years old then the percentage of C^{14} would be expected to be 0.075% (to 2 s.f.).

Criticism

Compare with reality

There are several sources of error in our modelling, in particular:

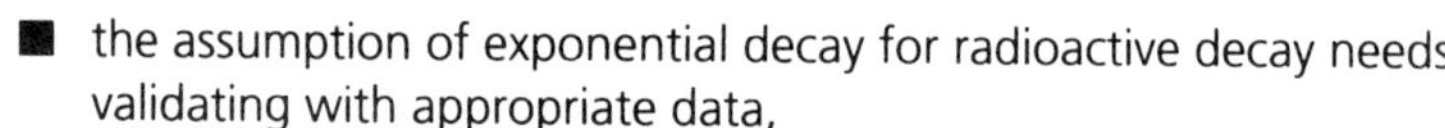

- the assumption of exponential decay for radioactive decay needs validating with appropriate data,

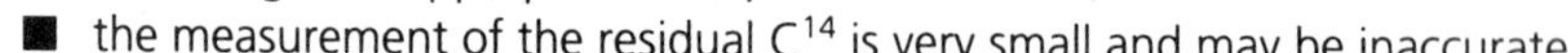

- the measurement of the residual C^{14} is very small and may be inaccurate,
- the assumption about the environment is important; in recent years the C^{14} levels in the atmosphere have increased because of atmospheric testing of nuclear weapons so it will not be possible to use carbon dating for plants and animals that have died in the past fifty years. This will lead to an imbalance of C^{12} and C^{14} in living material.

CONSOLIDATION EXERCISES FOR CHAPTER 17

1 Find the APR (annual percentage rate) that is equivalent to:

a) 10% nominal rate compounded monthly,
b) 8% nominal rate compounded every three months,
c) 12% nominal rate compounded continuously.

2 A metal spoon is being used to stir some food in boiling water which is at a temperature of 100°C. The spoon is placed in cold water at 15°C. After ten seconds the temperature of the spoon is 60°C. The spoon cools so that after t seconds its temperature is given by $T = 15 + Ae^{-kt}$.

a) Find the values of A and k.
b) The spoon can be handled when its temperature is 45°C. After how many seconds will this temperature be reached?

3 The following graphs represent the functions:

A: $y = e^{-x}$ B: $y = e^{-2x}$ C: $y = e^{x}$
D: $y = e^{2x}$ E: $y = 2e^{-x}$ F: $y = 2e^{x}$

Without using a graphics calculator or graph-drawing software, label each graph with its function.

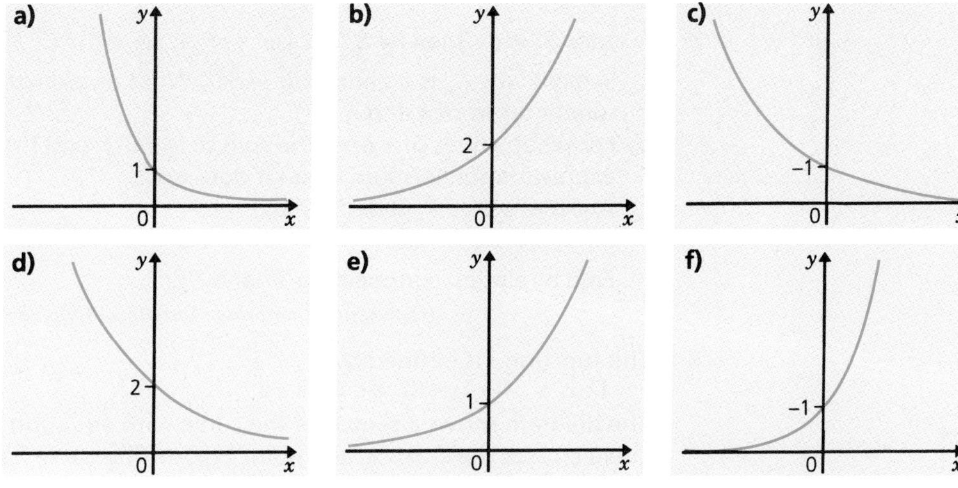

4 In Winchester Guildhall there is a round table that legend says belonged to King Arthur, who was alive in the fifth century AD. When the table was analysed for its carbon content archaeologists found the table to contain 91% of residual C^{14}.

a) Find an approximate age for the round table. Could it have belonged to King Arthur?
b) The method of measuring residual C^{14} is not very accurate. Suppose that there is an error of ± 6% in the measurement. Is it possible that the table could be King Arthur's?

5 A culture of bacteria grows in the laboratory so that its mass M, in mg, after t days satisfies the logistic model:

$$M = \frac{L}{1 + De^{-at}}$$

Initially the amount of culture was 10 mg. After 1 day the mass was 14 mg and after 2 days the mass was 18 mg.

a) Find the values of the constants L, D and a.
b) After how many days is the mass of culture equal to 25 mg?
c) Draw a graph of M against t.
d) What is the limiting amount of mass of culture?

6 A microbiologist measures the population of a certain type of bacterium. He starts the experiment at time $t = 0$. The population, n at time t hours is given by the formula:

$$n = A(1 - e^{-Bt})$$

where A and B are positive constants.

a) Sketch a graph of n against t.

When $t = 2$, $n = 10\,000$ and when $t = 4$, $n = 15\,000$.

b) Show that $2e^{-4B} - 3e^{-2B} + 1 = 0$.
c) Use the substitution $y = e^{-2B}$ to show that $2y^2 - 3y + 1 = 0$.
d) Solve this equation for y and hence show that $B = 0.347$ to 3 significant figures.
e) Determine, to the nearest 100, the maximum population size.

(Oxford Question 9, Specimen Paper 1, 1994)

7 A series S_n is defined by $S_n = 1 + e^x + e^{2x} + \ldots\ldots + e^{(n-1)x}$.

a) Explain why S_n is a geometric series. Write an expression for its sum in terms of x and n.

b) For what values of x does the sum to infinity exist? Write an expression for this sum when it does exist.

c) Another series T_n is defined by:

$$T_n = e + e^{1+x} + e^{1+2x} + e^{1+3x} + \ldots + e^{1+(n-1)x}$$

Find a relation connnecting T_n and S_n.

(Oxford and Cambridge Question 14, Specimen Paper 1, 1994)

8 The function f is defined by:

$\mathrm{f}: x \text{ a } -\ln(x-2),\ x \in \Re,\ x > 2.$

The diagram shows a sketch of the curve with equation $y = \mathrm{f}(x)$. The curve crosses the x-axis at the point $\mathrm{P}(p, 0)$. The curve has an asymptote, shown by a broken line, whose equation is $x = q$.

a) Write down the value of p and the value of q.

b) Find the function f^{-1} and state its domain.

c) Sketch the curve with equation $y = \mathrm{f}^{-1}(x)$ and its asymptote.

Write on your sketch the coordinates of any point where the curve crosses the coordinate axes and the equation of the asymptote.

(London Question 5, Paper 2, January 1993)

9 **a)** For a particular radioactive substance, the mass m (in grams) at time t (in years) is given by $m = m_0e^{-0.02t}$ where m_0 is the original mass. If the original mass is 500 grams, find the mass after ten years.

b) The half-life of any material is the time taken for half of the mass to decay. Find the half-life of this substance.

c) Illustrate all of the above information on a graph.

(Scottish Higher Question 4, Paper 2, 1992)

10 A medical technician obtains this print-out of a wave from generated by an oscilloscope. The technician knows that the equation of the first branch of the graph (for $0 \le x \le 3$) should be of the form $y = ae^{kx}$.

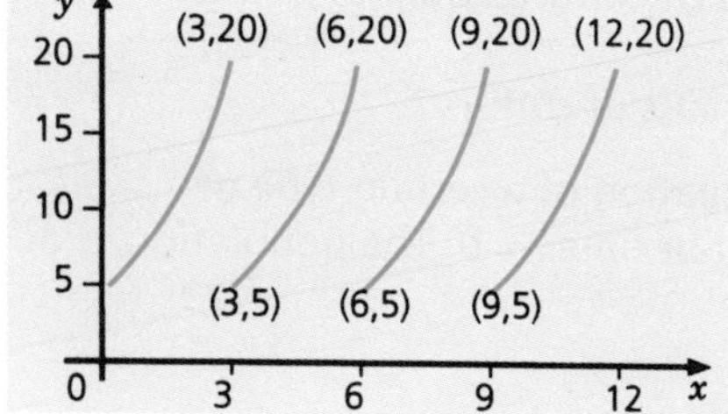

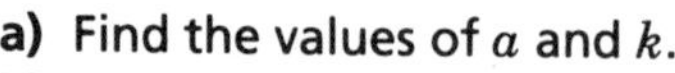

a) Find the values of a and k.

b) Find the equation of the second branch of the curve (i.e. for $3 \le x \le 6$).

(Scottish Highers Question 15, Paper 1, 1993)

11 The function f is defined by $\mathrm{f}(x) = e^{x^2+1}$ with domain $x < 0$.

a) Find an expression for $\mathrm{f}^{-1}(x)$, where f^{-1} denotes the inverse function of f. State the domain of f^{-1}.

b) Show graphically that the equation $e^{x^2+1} = 5 - x^2$ has a single negative root and show that its value lies between –0.8 and –0.7.

c) The function g is defined on the domain $x > 1$ and $\mathrm{fg} = e^x$. Obtain an expression for $\mathrm{g}(x)$ in terms of x. State, with a reason, whether or not the function gf can be formed.

(WJEC Question 14, Specimen Paper A1, 1994)

Summary

- The exponential function e^x is defined in two equivalent ways:

  $$e^x = \lim_{n \to \infty} \left(1 + \frac{x}{n}\right)^n$$

  $$e^x = 1 + x + \frac{x^2}{2!} + \frac{x^3}{3!} + \frac{x^4}{4!} + \ldots$$

- The natural logarithm function $\ln x$ is the inverse function of e^x, that is, if $y = e^x$ then $x = \ln y$.

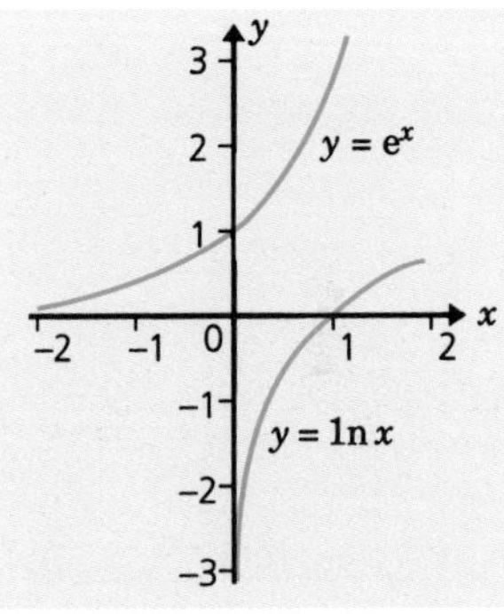

- The function $y = e^{kx}$ models exponential growth if $k > 0$ and exponential decay if $k < 0$.

- Graphs of $y = e^{kx}$ for $k > 0$ and $k < 0$ are shown on the right.

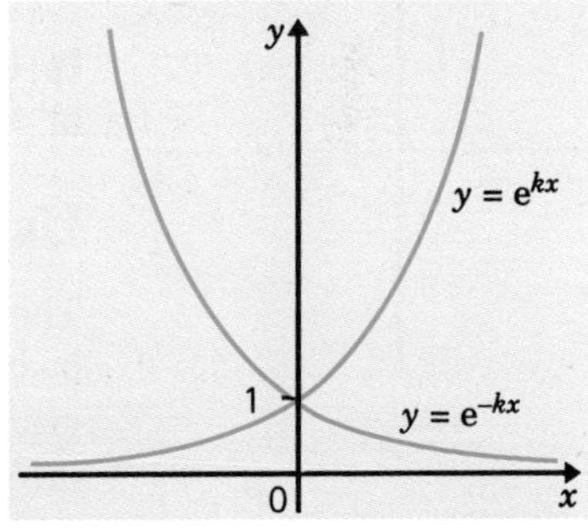

PURE MATHS

18 Differentiation 2

In this chapter we:

- *introduce the derivative of non-integer powers of x, exponential and logarithmic functions*
- *introduce the second derivative and shows how it can be used to classify stationary points.*

FUNCTIONS WITH RATIONAL EXPONENTS

In Chapter 7, *Differentiation 1*, we differentiated polynomial functions, based on the rule for differentiating powers of x:

i.e. $f(x) = x^n \Rightarrow f'(x) = nx^{n-1}$, $n = 0, 1, 2, \ldots$

We can now extend this definition for any rational index n (e.g. $n = -1$, $n = \frac{1}{2}$).

Exploration 18.1

The graph of $y = f(x) = \frac{1}{x}$

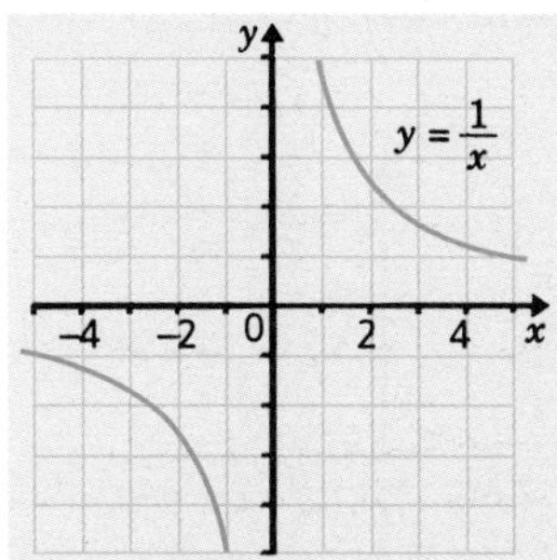

The diagram shows the graph of $y = f(x) = \frac{1}{x}$. Use a numerical method to estimate the gradient for various values of x.

- Use your numerical values to sketch a graph of the gradient function.
- Find a formula for $f'(x)$.

Using the limiting process to find the gradient

The limiting process introduced in Chapter 7, *Differentiation 1*, may be used to find the gradient at any value of x ($x \neq 0$). Consider the gradient at $x = 1$ as the limit of gradients of secants, reducing the x-interval by a factor of 10 each time:

between $x = 1$ and $x = 2$ gradient of secant = –0.5
between $x = 1$ and $x = 1.1$ gradient of secant = –0.909 090
between $x = 1$ and $x = 1.01$ gradient of secant = –0.990 099
between $x = 1$ and $x = 1.001$ gradient of secant = –0.999 000 999

So the limit of the gradient of the secant between $x = 1$ and $x = 1 + h$, as $h \to 0$, will be –1, which suggests $f'(x) = -1$.

Repeating the process for other x-values gives a table of gradient function values.

	–5	–2	–1	–0.5	–0.2	0.2	0.5	1	2	5
$f'(x)$	–0.04	–0.25	–1	–4	–25	–25	–4	–1	–0.25	–0.04

Expressing each decimal as its fraction equivalent gives a clue as to the algebraic formula for $f'(x)$.

x	–5	–2	–1	–0.5	–0.2	0.2	0.5	1	2	5
f′(x)	$-\frac{1}{25}$	$-\frac{1}{4}$	–1	–4	–25	–25	–4	–1	$-\frac{1}{4}$	$-\frac{1}{25}$

We can now spot the pattern for the gradient function:

$$f'(x) = -\frac{1}{x^2}$$

If we express f(x) as a power of x we see that the rule for differentiating powers of x does indeed work for $n = -1$.

$$f(x) = \frac{1}{x} = x^{-1} \Rightarrow f'(x) = -1 \times x^{-2} = -\frac{1}{x^2}$$

In fact the rule can be extended to differentiate any negative power of x,

e.g: $f(x) = \frac{3}{x^2} = 3x^{-2} \Rightarrow f'(x) = 3 \times (-2)x^{-3} = -\frac{6}{x^3}$

Exploration 18.2

The function $y = f(x) = \sqrt{x}$

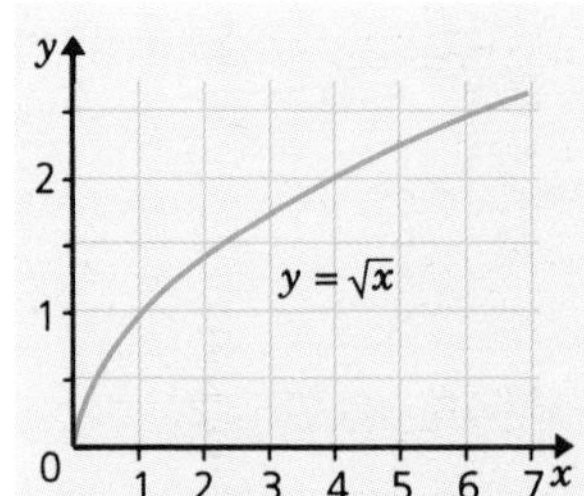

The graph of $y = f(x) = \sqrt{x}$, $x \geq 0$ is shown in the diagram.

- Use a numerical method to estimate the gradient for various values of x.
- Use your numerical values to sketch a graph of the gradient function.
- Find a formula for f′(x).

Finding the gradient

Using a limiting process, we can find the gradient at any x-value ($x > 0$). Consider the gradient at $x = 1$ as the limit of gradients of secants, again reducing the x-interval by a factor of 10 each time:

between $x = 1$ and $x = 2$	gradient of secant = 0.414 213 562 4
between $x = 1$ and $x = 1.1$	gradient of secant = 0.488 088 481 7
between $x = 1$ and $x = 1.01$	gradient of secant = 0.498 756 211 2
between $x = 1$ and $x = 1.001$	gradient of secant = 0.499 875 062 4

By extending this pattern, we find that the limit of the gradient of the secant between $x = 1$ and $x = 1 + h$, as $h \to 0$, will be –0.5.

Repeating this process for other x-values gives a table of gradient function values.

x	0.01	0.04	0.0625	0.25	1	4	16	25	100
f′(x)	5	2.5	2	1	0.5	0.25	0.125	0.1	0.05

Again, by expressing each decimal as an equivalent fraction, we spot a pattern for the gradient function.

x	$\frac{1}{100}$	$\frac{1}{25}$	$\frac{1}{10}$	$\frac{1}{4}$	1	4	16	25	100
f′(x)	5	$2\frac{1}{2}$	2	1	$\frac{1}{2}$	$\frac{1}{4}$	$\frac{1}{8}$	$\frac{1}{10}$	$\frac{1}{20}$

Confirm that the gradient function is given by:

$$f'(x) = \frac{1}{2\sqrt{x}}$$

By expressing $f(x)$ as a power of x, we can see that the rule for differentiating powers of x works for $n = \frac{1}{2}$.

$$f(x) = \sqrt{x} = x^{\frac{1}{2}} \Rightarrow f'(x) = \tfrac{1}{2}x^{-\frac{1}{2}} = \frac{1}{2\sqrt{x}}$$

The rule can be extended to differentiate any functional power of x,

e.g: $f(x) = 12\sqrt[3]{x} = 12x^{\frac{1}{3}} \Rightarrow f\ (x) = 12 \times \tfrac{1}{3}x^{-\frac{2}{3}} = \dfrac{4}{\sqrt[3]{x^2}}$

We have found that the rule for differentiating powers of x works for any rational index. The following examples will show how the technique can be used and applied.

Example 18.1

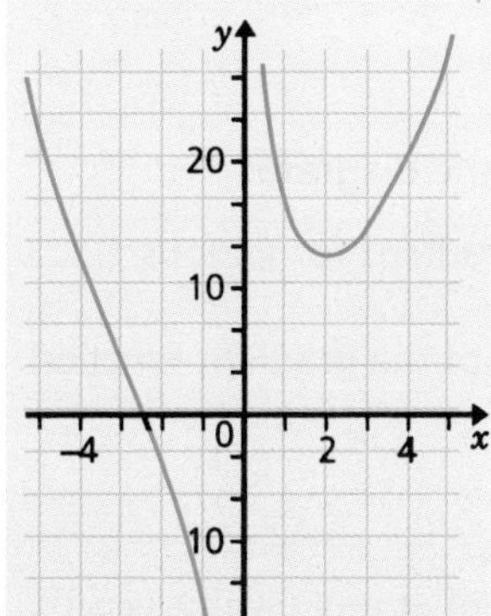

The diagram shows a sketch of $y = f(x) = x^2 + \dfrac{16}{x}$.

a) *Find* $f'(x)$ *and evaluate* $f'(-2)$.

b) *Find the coordinates of the stationary point S.*

Solution

a) $y = f(x) = x^2 + \dfrac{16}{x} = x^2 + 16x^{-1} \Rightarrow f'(x) = 2x - 16x^{-2} = 2x - \dfrac{16}{x^2}$

Hence $f'(-2) = 2 \times (-2) - \dfrac{16}{(-2)^2} = -8$

b) *For a stationary value,*

$$f'(x) = 0 \Rightarrow 2x - \frac{16}{x^2} = 0 \Rightarrow 2x^3 - 16 = 0 \Rightarrow x^3 = 8 \Rightarrow x\sqrt[3]{8} = 2$$

Now $f(2) = 2^2 + \frac{16}{2} = 12$, *which means that the coordinates of S are (2,12).*

Example 18.2

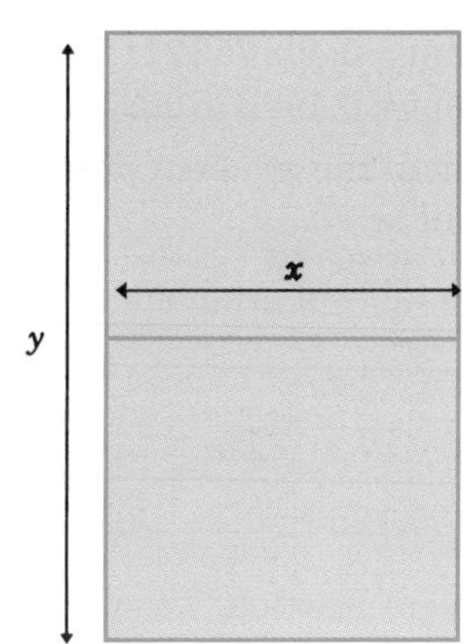

A farmer wishes to enclose a rectangular field of area 16 000 m² and to divide it in half. Fencing to enclose the field costs £3 per metre and the fencing to divide the field costs £2 per metre.

a) *Express the total cost of fencing as a function of x, the width of the field.*

b) *Find the dimensions and the total cost of fencing for the field that is **least** expensive to fence.*

Solution

a) *Given that the area of the field is to be 16 000 m², from the diagram we find:*

$$16\,000 = xy$$

$$y = \frac{16\,000}{x}$$

The cost of fencing the perimeter $= 3(2x + 2y) = 6x + 6y$

The cost of the dividing fence $= 2x$

Therefore the total cost of fencing is given by:

$$C = 8x + 6y = 8x + 6 \times \frac{16\,000}{x} \Rightarrow C = 8x + \frac{96\,000}{x}$$

b) *To minimise C:*

$\dfrac{dC}{dx} = 8 - \dfrac{96\,000}{x^2} = 0$ *for stationary values.*

$\Rightarrow 8x^2 = 96\,000 \Rightarrow x^2 = 12\,000 \Rightarrow x = \sqrt{12\,000} \approx 110$ m

Substituting for x gives:

$y = \dfrac{16\,000}{110} \approx 145$ m

Total cost of fencing $= 8 \times 110 + \dfrac{96\,000}{110} \approx £1750$

Example 18.3

A function is given by $y = 2\sqrt{x}(5-x)$.

a) *Find where the graph of this function intersects with the axes.*

b) *Find the value of* $\frac{dy}{dx}$ *at the points found in* **a)**.

c) *Find the stationary point and determine which type it is.*

d) *Use the information found in* **a), b)** *and* **c)** *to sketch the graph.*

Solution

a) *The graph intersects with the x-axis when* $y = 0$.

$y = 0 \Rightarrow 2\sqrt{x}(5-x) = 0 \Rightarrow x = 0$ *or* $x = 5$

Hence the points of intersection are (0,0) and (5,0).

b) $2\sqrt{x}(5-x) = 10\sqrt{x} - 2x\sqrt{x} = 10x^{\frac{1}{2}} - 2x^{\frac{3}{2}}$

$\Rightarrow \dfrac{dy}{dx} = 10 \times \tfrac{1}{2}x^{-\frac{1}{2}} - 2 \times \tfrac{3}{2}x^{\frac{1}{2}} = 5x^{-\frac{1}{2}} - 3x^{\frac{1}{2}} = \dfrac{5}{\sqrt{x}} - 3\sqrt{x} = \dfrac{5-3x}{\sqrt{x}}$

When $x = 0$, $\frac{dy}{dx}$ *is undefined.*

When $x = 5$, $\dfrac{dy}{dx} = \dfrac{5 - 3 \times 5}{\sqrt{5}} = -4.47$ *(3 s.f.)*

c) *At the stationary point,* $\dfrac{dy}{dx} = 0 \Rightarrow \dfrac{5-3x}{\sqrt{x}} = 0$

$\Rightarrow 5 - 3x = 0 \Rightarrow 3x = 5 \Rightarrow x = 1\tfrac{2}{3} \approx 1.67$

When $x = 1\tfrac{2}{3}$, $y = 2\sqrt{1\tfrac{2}{3}}\left(5 - 1\tfrac{2}{3}\right) = 8.61$ *(3 s.f.)*

To determine which type the stationary point is, inspect the gradient at, say, $x = 1$ *and* $x = 2$.

$x = 1 \Rightarrow \dfrac{dy}{dx} = \dfrac{5 - 3 \times 1}{\sqrt{1}} > 0$

$x = 2 \Rightarrow \dfrac{dy}{dx} = \dfrac{5 - 3 \times 2}{\sqrt{2}} < 0$

Hence (1.67, 8.61) is a maximum point.

d) *Combining information from earlier parts of the question helps to sketch the graph.*

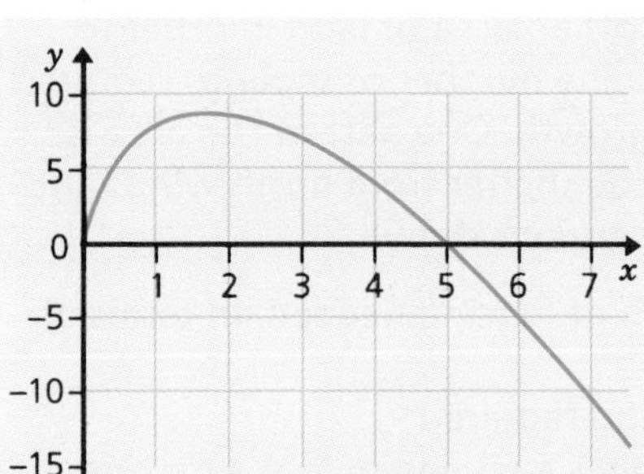

EXERCISES

1 Differentiate each of the following.

a) $\dfrac{7}{x}$ **b)** $\dfrac{-12}{x^2}$ **c)** $5x - \dfrac{16}{x^3}$ **d)** $3x\left(\dfrac{2}{x^2} - 5\right)$

e) $15\sqrt{x}$ **f)** $-\frac{4}{\sqrt{x}}$ **g)** $\sqrt{x}(x-8)$ **h)** $x^2(4+\sqrt{x})$

2 For each of the following functions:
i) find where $f'(x) = 0$, **ii)** identify any stationary points,
iii) sketch a graph of the function.

a) $f(x) = x + \frac{1}{x}$ **b)** $f(x) = 2x - \frac{3}{x^2}$ **c)** $f(x) = \frac{x^2 - 3}{x^3}$

d) $f(x) = x(3 - \sqrt{x})$ **e)** $f(x) = x^2(\sqrt{x} - 3)$ **f)** $f(x) = \frac{50}{x}(3 - \sqrt{x})$

3 A box is to be made with a volume of 500 cm^3. It is to have a square base, of side x cm.

a) Express the height, h, in terms of x.
b) Find a formula for the total surface area, A cm^3, in terms of x for:
i) an open box (i.e. a box without a lid),
ii) a closed box.
c) For each of **i)** and **ii)** in **b)** determine the dimensions to minimise the surface area A, cm^3. State the corresponding values of A.

4 A cylindrical tank, open at the top and of height h m, radius r m, is to hold 2 m^3.

a) Find h in terms of r.
b) Find a formula for the total surface area, A m^2, in terms of r.
c) Find the smallest value of A. How are r and h related in this case?

5 A function is given by $f(x) = \sqrt{x}(x - 12)$.

a) Where does the graph of $y = f(x)$ intersect with the axes?
b) Find $f'(x)$ and evaluate $f'(0)$ and $f'(12)$.
c) Determine the coordinates of the stationary point and which type it is.
d) Sketch the graph of $y = f(x)$.

6 A forestry worker needs to partition off an area of land in order to provide a nursery for young saplings. The number of saplings required will need an area of 338 m^2 for best growth. One part of the area is provided by a boundary wall to provide shelter from north winds and the area will be rectangular with dimensions shown.

a) Find a formula for the total length of fencing needed, in terms of x only.
b) Find the minimum length of fencing required.
c) Use a method which shows *why* this is a minimum, not a maximum.

7 The cost per hour of operating a long-distance lorry is given by £kv^2 where v is the average speed for a journey and k is a constant. The driver works at a fixed hourly rate £C. Show that the most economical speed over a fixed distance, d, is given by:

$v^{\frac{3}{2}} = \frac{2C}{k}$ (independent of d).

EXERCISES

18.1 B

1 Differentiate each of the following , evaluating the gradient at the x-value shown.

a) $-\dfrac{16}{x^2}$ $(x = 2)$ **b)** $\dfrac{25}{2x^2}$ $(x = 5)$ **c)** $3x^4 - \dfrac{3}{x^4}$ $(x = 1)$

d) $\left(4x - \dfrac{1}{x}\right)\left(\dfrac{3}{x^2} + 2\right)$ $(x = -2)$ **e)** $20\sqrt{x} - \dfrac{6}{\sqrt{x}}$ $(x = 9)$

f) $x^2 + x + \dfrac{1}{x} + \dfrac{1}{x^2}$ $(x = -1)$ **g)** $\sqrt{x}\left(x + \sqrt{x}\right)$ $(x = 64)$

h) $\dfrac{3}{\sqrt{x}}\left(2x^2 + \sqrt{x}\right)$ $\left(x = \tfrac{1}{16}\right)$

2 For each of the following functions:
i) find where $f'(x) = 0$, **ii)** identify any stationary points,
iii) sketch a graph of the function.

a) $f(x) = \dfrac{1}{x} - 2x^2$ **b)** $f(x) = \dfrac{1}{x} - \dfrac{1}{x^2}$ **c)** $f(x) = \dfrac{x^2 - 5}{x^3}$

d) $f(x) = x\left(5 - 2\sqrt{x}\right)$ **e)** $f(x) = x^2\left(\sqrt{5} - \sqrt{x}\right)$ **f)** $f(x) = \dfrac{3}{x}\left(5 - \sqrt{x}\right)$

3 The rectangular block shown has a volume of 14 112 cm³.

a) Find an expression for A, the total surface area, in terms of x only.
b) Calculate the value of x that gives A a stationary value.
c) Find this value of A and find whether it is maximum or minimum.

4 The closed cylindrical can shown has a volume 512π cm³.

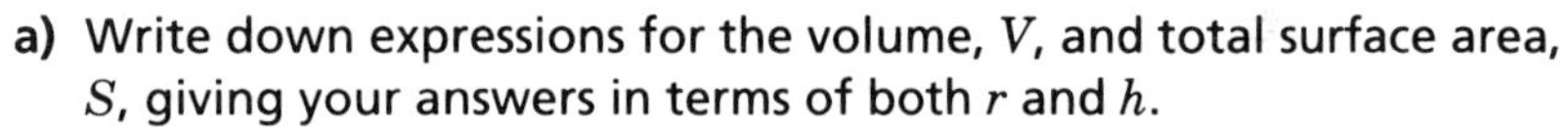

a) Write down expressions for the volume, V, and total surface area, S, giving your answers in terms of both r and h.
b) Eliminate h from your two equations and hence find an equation giving S in terms of r only.
c) Calculate the value of r for which S has a stationary value.
d) Does this value of r make the surface area a maximum or minimum? Give reasons for your answer.
e) Find the values of S and h for this value of r.

5 A function is defined by $f(x) = 10x + \dfrac{27}{x^2}$.

a) Find where the graph crosses the x-axis.
b) Find $f'(3)$ and $f'(9)$.
c) Find the coordinates of the stationary point and whether it is maximum or minimum.
d) Sketch the graph for the domain $0 < x \le 9$.

6 A regulation states that all parcels above a certain volume must be bound by at least three lengths of tape, as shown in the diagram. Given that the parcel has a square base find, to the nearest cm, the minimum length of tape required to secure a parcel of volume 8000 cm³.

7 The velocity of a wave of length x in deep water is given by the formula:
$v = k\sqrt{\frac{x}{c} + \frac{c}{x}}$, where k and c are constants.

a) Let $y = k\sqrt{\frac{x}{c} + \frac{c}{x}}$ and hence find the value of x that minimises y.

b) Deduce the minimum velocity.

DIFFERENTIATION OF EXPONENTIAL AND LOGARITHMIC FUNCTIONS

Exploration 18.3

Differentiating $f(x) = e^{kx}$

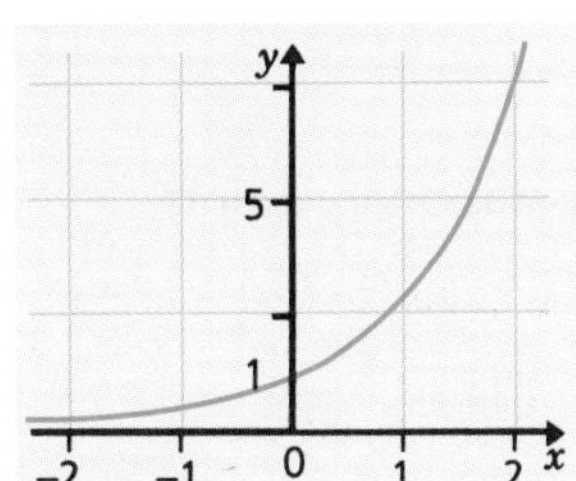

The diagram shows a graph of $y = f(x) = e^x$.

- Use a numerical method to estimate values of the gradient function $f'(x)$ for $x = -2, -1, 0, 1$ and 2.
- What is the connection between $f(x)$ and $f'(x)$?
- Repeat the activities for the functions:

  $f(x) = e^{2x}$ $\quad$ $f(x) = e^{\frac{1}{2}x}$ $\quad$ $f(x) = e^{-x}$

Values of the gradient function

Values of $f(x)$ and $f'(x)$ for the function $f(x) = e^x$, to 3 d.p, are shown in this table.

x	−2	−1	0	1	2
$f(x)$	0.135	0.368	1.000	2.718	7.389
$f'(x)$	0.135	0.368	1.000	2.718	7.389

The gradient function takes the same values as $f(x)$, i.e.

$$f(x) = e^x \quad \Rightarrow \quad f'(x) = e^x$$

or $\quad y = e^x \Rightarrow \frac{dy}{dx} = e^x$

Finding numerical gradients for the other functions reveals different, but consistent patterns.

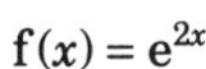

$f(x) = e^{2x}$

x	−2	−1	0	1	2
$f(x)$	0.018	0.135	1.000	7.389	54.598
$f'(x)$	0.037	0.271	2.000	14.778	109.196

Conclusion: $f(x) = e^{2x} \quad \Rightarrow \quad f'(x) = 2e^{2x}$

$f(x) = e^{\frac{1}{2}x}$

x	−2	−1	0	1	2
$f(x)$	0.368	0.607	1.000	1.649	2.718
$f'(x)$	0.184	0.303	0.500	0.824	1.359

Conclusion: $f(x) = e^{\frac{1}{2}x} \quad \Rightarrow \quad f'(x) = \frac{1}{2}e^{\frac{1}{2}x}$

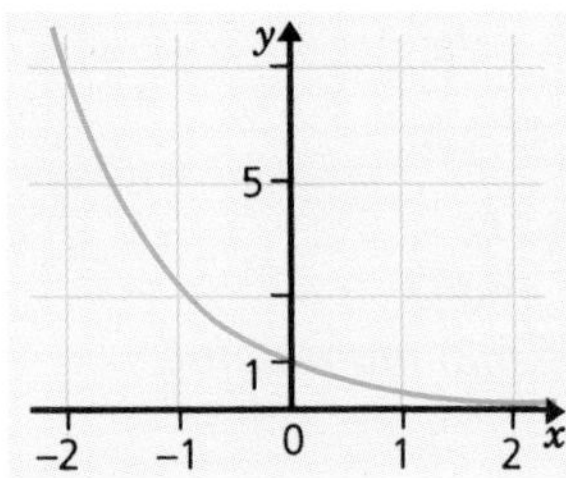

$f(x) = e^{-x}$

x	–2	–1	0	1	2
$f(x)$	7.389	2.718	1	0.368	0.135
$f'(x)$	–7.389	–2.718	–1	–0.368	–0.135

Conclusion: $f(x) = e^{-x} \quad \Rightarrow \quad f'(x) = -e^{-x}$

From the exploration we see that a rule emerges for differentiating exponential functions of the form e^{kx}.

$$f(x) = e^{kx} \quad \Rightarrow \quad f'(x) = ke^{kx}$$

for any rational number k.

Exploration 18.4

An alternative way of differentiating e^{kx}

In Chapter 17, *Exponentials and logarithms*, we introduced the series expansion for e^x:

$$e^x \equiv 1 + x + \frac{x^2}{2!} + \frac{x^3}{3!} + \frac{x^4}{4!} + \ldots$$

which is true for all values of x.

- Differentiate the exponential series to show that $\frac{d}{dx}e^x = e^x$, i.e. the derivative of e^x is e^x.
- Write down the series expansion for e^{2x} and so find $\frac{d}{dx}e^{2x}$.
- Use a similar method to find $\frac{d}{dx}e^{kx}$ for other values of k, e.g. $k = 3$, $k = -1$, $k = \frac{1}{2}$.

Examining the results

$$\frac{d}{dx}e^x = \frac{d}{dx}\left(1 + x + \frac{x^2}{2!} + \frac{x^3}{3!} + \frac{x^4}{4!} + \ldots\right) = 0 + 1 + \frac{2x}{2!} + \frac{3x^2}{3!} + \frac{4x^3}{4!} + \ldots = 1 + x + \frac{x^2}{2!} + \frac{x^3}{3!} + \ldots = e^x$$

Replacing x by $2x$ in the series expansion gives:

$$e^{2x} \equiv 1 + 2x + \frac{(2x)^2}{2!} + \frac{(2x)^3}{3!} + \frac{(2x)^4}{4!} + \ldots \equiv 1 + 2x + \frac{4x^2}{2!} + \frac{8x^3}{3!} + \frac{16x^4}{4!}\ldots$$

Differentiating with respect to x:

$$\frac{d}{dx}\left(1 + 2x + \frac{4x^2}{2!} + \frac{8x^3}{3!} + \frac{16x^4}{4!}\ldots\right) = 0 + 2 + \frac{4(2x)}{2!} + \frac{8(3x^2)}{3!} + \frac{16(4x^3)}{4!} + \ldots$$

$$= 2 + 4x + \frac{8x^2}{2!} + \frac{16x^3}{3!} + \ldots = 2\left(1 + 2x + \frac{4x^2}{2!} + \frac{8x^3}{3!} + \ldots\right)$$

$$\Rightarrow \frac{d}{dx}e^{2x} = 2e^{2x}$$

Replacing x with $3x$, $\frac{1}{2}x$, $-x$ etc. in the series expansion enables us to demonstrate the general result.

$$\frac{d}{dx}e^{kx} = ke^{kx}$$

Example 18.4

The diagram shows a sketch of $y = f(x) = e^{0.5x} - 4x + 5$.

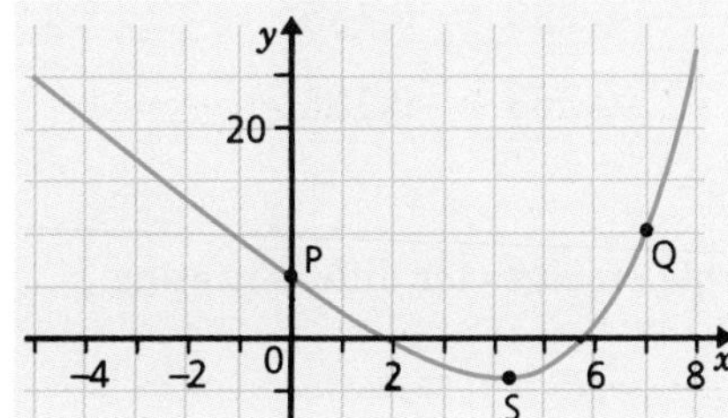

a) *Find the coordinates of points P and Q.*
b) *Find the gradient of the curve at P and Q.*
c) *Find the coordinates of point S, the minimum point.*

Solution

a) *At P,* $x = 0 \Rightarrow f(0) = e^0 - 4 \times 0 + 5 = 6$*, hence point P has coordinates (0, 6).*
At Q, $x = 7 \Rightarrow f(7) = e^{3.5} - 4 \times 7 + 5 = 10.115$*, hence point Q has coordinates (7, 10.115).*

b) $f'(x) = 0.5e^{0.5x} - 4$*, hence:*
at point P $f'(0) = 0.5e^0 - 4 = -3.5$
at point Q $f'(7) = 0.5e^{3.5} - 4 = 12.558$

c) *For a stationary value* $f'(x) = 0$
$\Rightarrow\ 0.5e^{0.5x} - 4 = 0 \ \Rightarrow\ e^{0.5x} = 8 \ \Rightarrow\ \ln e^{0.5x} = \ln 8 \ \Rightarrow\ 0.5x = \ln 8$
$\Rightarrow x = 2\ln 8 = 4.16$ *(3 s.f.)*
Since $f(4.16) = -3.64$ *(3 s.f.), the coordinates of S are (4.16, –3.64).*

Example 18.5

A population is growing according to the formula:
$P = 20e^{0.02t}$
where P is the population in millions and t is the time in years measured from 1990.

a) *When will the population reach 25 million?*
b) *At what rate will the population be growing at that time?*

Solution

a) *When* $P = 25$,

$$25 = 20e^{0.02t} \Rightarrow 1.25 = e^{0.02t} \Rightarrow \ln e^{0.02t} = \ln 1.25 \Rightarrow 0.02t = \ln 1.25$$
$$\Rightarrow t = 50\ln 1.25 = 11.2 \quad (3\ s.f.)$$

i.e. the population will reach 25 million during the year 2001.

b) $P = 20e^{0.02t} \Rightarrow \dfrac{dP}{dt} = 0.4e^{0.02t}$

$t = 11.2 \Rightarrow 0.02t = \ln 1.25$

$$\Rightarrow \frac{dP}{dt} = 0.4e^{\ln 1.25} = 0.4 \times 1.25 = 0.5$$

i.e. when the population reaches 25 million it is increasing at a rate of 0.5 million per year.

Differentiating a constant raised to a power

We have seen that $f(x) = e^x \ \Rightarrow\ f'(x) = e^x$, but can we deduce anything about the derivative of a^x, where a is a constant?

Since the functions e^x and $\ln x$ are inverse functions of each other:

$$a \equiv e^{\ln a}$$
$$\Rightarrow a^x = \left(e^{\ln a}\right)^x = e^{x\ln a}$$

This important equivalence enables us to differentiate a^x:

$$f(x) = a^x = e^{x \ln a}$$
$$\Rightarrow \; f'(x) = \ln a \times e^{x \ln a}$$
$$\text{e.g. } f(x) = 2^x \;\; \Rightarrow \;\; f'(x) = \ln 2 \times 2^x$$

We now know how to compute rates of change for any exponential function, as illustrated in the following example.

Example 18.6

The alcohol level in a person's blood at time t is given by $A = 0.35 \times 0.5^t$, where A is measured in mg per ml and t is the number of hours after consuming four units of alcohol. Find the rate at which the alcohol level is falling after 1 hour.

Solution

$$A = 0.3 \times 0.5^t \Rightarrow \frac{dA}{dt} = 0.3 \times \ln 0.5 \times 0.5^t$$

when $t = 1$, $\dfrac{dA}{dt} = 0.3 \times \ln 0.5 \times 0.5 = -0.104 \quad (3\text{ s.f.})$

i.e. the alcohol level is falling at 0.104 mg/ml per hour.

The derivative of the inverse function

Having explored the derivative of $f(x) = e^x$, we now look at the derivative of the inverse function $f^{-1}(x) = \ln x$.

Exploration 18.5

Differentiating $g(x) = \ln x$

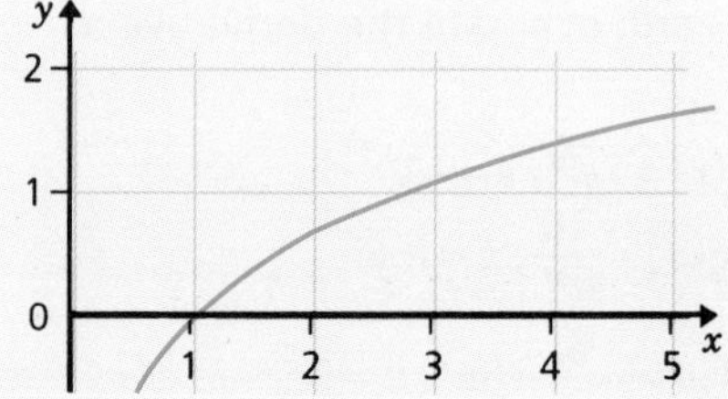

The diagram shows a graph of $y = g(x) = \ln x$.

- Use a numerical method to estimate values of the gradient function $g'(x)$ for $x = 0.5, 1, 2, 3, 4, 5$.
- What is the connection between x and $g'(x)$?
- Repeat these activities for these functions.
  $g(x) = \ln 2x \qquad g(x) = \ln 0.5x$

The gradient function

Values of $g(x)$ and $g'(x)$, to 3 d.p, for the function $g(x) = \ln x$ are given in this table.

x	0.5	1	2	3	4	5
$g(x)$	−0.693	0.000	0.693	1.099	1.386	1.609
$g'(x)$	2.000	1.000	0.500	0.333	0.250	0.200

The gradient function takes the same values as $\frac{1}{x}$.

i.e. $g(x) = \ln x \Rightarrow g'(x) = \dfrac{1}{x}$

or $\quad y = \ln x \Rightarrow \dfrac{dy}{dx} = \dfrac{1}{x}$

Finding numerical gradients for the other functions reveals similar patterns, which leads to the generalisation:

$$g(x) = \ln kx \Rightarrow g'(x) = \frac{1}{x}$$

for any positive constant k.

Example 18.7

A function f is given by $f(x) = x^2 - 8 \ln 0.5x$, $x > 0$.

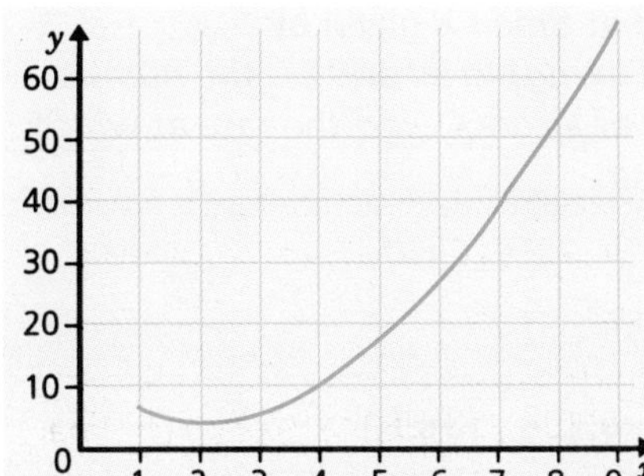

a) *Sketch a graph of* $y = x^2 - 8 \ln 0.5x$.

b) *Show that it has a minimum point at (2,4).*

Solution

a) *A sketch of* $y = f(x)$ *shows the curve's stationary point is a minimum.*

b) $f(x) = x^2 - 8\ln 0.5x \Rightarrow f'(x) = 2x - 8 \times \frac{1}{x}$

For a stationary point:

$$f'(x) = 0 \Rightarrow 2x - \frac{8}{x} = 0 \Rightarrow 2x^2 - 8 = 0 \Rightarrow x^2 = 4$$

$\Rightarrow x = 2$ *(function only defined for* $x > 0$*)*

Since $f(2) = 2^2 - 8 \ln 1 = 4$, *the curve has a stationary point at (2, 4).*

By examining the gradient either side of $x = 2$ *we can show that it is a minimum point.*

Check that $f'(1.9) = -0.411$ *(3 s.f.) and* $f'(2.1) = 0.390$ *(3 s.f.).*

EXERCISES

18.2 A

1 Differentiate the following functions and evaluate the derivative at the value shown.

a) $f(x) = e^{0.5x}$, $x = 2.5$ **b)** $f(x) = 2e^{5x}$, $x = -1$

c) $f(x) = e^{-x}$, $x = -0.3$ **d)** $g(x) = \frac{10}{e^{4x}}$, $x = 0.2$

e) $f(t) = 5e^t + \frac{3}{e^{2t}}$, $t = 1.5$ **f)** $f(u) = \frac{6}{e^u} - 3u^2$, $u = 2$

g) $g(x) = 5^x$, $x = -1.2$ **h)** $g(t) = 0.02 \times 3^t$, $t = 2.1$

i) $h(x) = -2 \times 0.3^{2x}$, $x = 0$ **j)** $f(x) = 5 \ln x$, $x = 8$

k) $g(x) = -\ln 2x$, $x = 5$ **l)** $f(t) = 2 \ln t^3$, $t = 0.8$

m) $f(x) = 10 \ln \sqrt{x} - e^{0.5x}$, $x = 2$ **n)** $g(x) = \ln x e^x$, $x = 1$

2 For each of the following functions, find:

i) $f'(x)$ **ii)** the values of x for which the derivative is given.

a) $f(x) = e^{2x} - 0.5x$, $f'(x) = 5$ **b)** $f(x) = 0.25e^{-0.8x} - 3$, $f'(x) = -10$

c) $f(x) = 10 - 4^{-x}$, $f'(x) = \ln 2$ **d)** $f(x) = 2\ln\frac{7}{x}$, $f'(x) = -0.75$

e) $f(x) = 2e^{-x} - \ln x^2$, $f'(x) = -1$

3 Find and classify the stationary points for the following functions.

a) $f(x) = e^x - 2x$ **b)** $f(x) = 5x - e^{1.2x} + 10$ **c)** $g(x) = 2x - 3\ln x$

d) $g(x) = 12\ln x^2 - x^3$ **e)** $g(x) = 7\ln \frac{1}{x} + 4x$

4 The population of a village in 1980 was 2000. The local authority planned for an annual increase in the population of 4%. On a continuous scale, the population, P, after t years should be given by $P = 2000\,e^{0.04t}$. Find:

a) the time it takes for the population to reach 3000,
b) the rate of growth of the population at this time.

5 A car is purchased for £10 000 and depreciates continuously from the date of purchase. Its value, £V, after t years is given by the formula $V = 10\,000e^{-0.3t}$. Find:

a) the rate of depreciation after two years,
b) the time by which the rate of depreciation has fallen to £600 per yeaı

6 The number of bacteria in a certain culture doubles every three houı After six hours the estimated number of bacteria present is 10 000.

a) Find f(t), the number of bacteria present after t hours (known as the population function).
b) Find f′(t), the rate at which the population is growing after t hour
c) After how many hours will the population reach 15 000?

EXERCISES

18.2 B

1 Differentiate the following functions and evaluate the derivative at the value shown.

a) $f(x) = e^{0.25x}, x = 4$ **b)** $f(x) = 3e^{7x}, x = 2$ **c)** $f(x) = -e^{-x}, x = -6$

d) $h(x) = \frac{7}{e^{0.5x}}, x = 0$ **e)** $g(t) = 3e^{2t} - \frac{3}{e^{2t}}, t = 3$

f) $g(u) = 3e^{-2u} + 7u^3, u = 2$ **g)** $f(x) = 7^x, x = 5$

h) $g(t) = 10\,000 \times 5^{-2t}, t = 0.5$ **i)** $f(p) = -\frac{2}{3} \times 0.1^{\frac{3}{2}p}, p = 1$

j) $f(t) = 7\ln 2t, t = 5$ **k)** $g(z) = z^4 - 3\ln 5z, z = 8$

l) $f(t) = \ln t^3 - \ln t^2, t = 2$ **m)** $h(x) = 2\ln \frac{1}{\sqrt{x}}, x = 9$

n) $h(x) = \ln\left(x^2 e^{3x}\right), x = 3$

2 For each of the following functions, find:
i) f′(x) **ii)** the values of x for which the derivative is given.

a) $f(x) = \frac{1}{2}\left(e^x + e^{-x}\right), f'(x) = 0$ **b)** $f(x) = x - \ln x, f'(x) = \frac{2}{3}$

c) $f(x) = \sqrt{x} - \ln\sqrt{x}, f'(x) = 0$ **d)** $f(x) = \frac{1}{2}\left(e^x - e^{-x}\right), f'(x) = \sqrt{5}$

e) $f(x) = 5\ln \frac{2}{x^2}, f'(x) = 4$

3 Find and classify the stationary points for the following functions.

a) $f(x) = \sqrt{e^x} - x$ **b)** $f(x) = 3x^2 - \frac{1}{2}\ln x$ **c)** $f(x) = e^{2x} + 2e^x - 12x$

d) $f(x) = 4^x - 2^x$ **e)** $f(x) = \ln\left(x^2 e^{-2x}\right)$

4 The concentration of a medical drug injected into the bloodstream can be modelled by the function:

$$c(t) = k\left(e^{-at} - e^{-bt}\right)$$

where a, b and k are all constants (> 0).

a) Show that $c(t) \to 0$ as $t \to \infty$.
b) Find $c'(t)$ and explain its significance.
c) Find the time when $c'(t) = 0$.
d) Does this give a maximum or minimum value of $c(t)$?

5 A bacterial culture starts with 2000 bacteria. After one hour the count is 8000.

a) Find the doubling period.
b) Find the population function.
c) Evaluate the growth rate after one hour and after two hours.

6 **a)** By writing e^{2x+3} as $e^{2x}e^3$, find $\frac{d}{dx}e^{2x+3}$

b) Use a similar method to find $\frac{d}{dx}e^{3x-2}$ and $\frac{d}{dx}e^{5-7x}$

c) Deduce the value of $\frac{d}{dx}e^{ax+b}$

d) Use the above result to calculate, to 2 d.p, the gradient of $y = e^{2x-1}$ at the point where it meets the y-axis.

TANGENTS AND NORMALS

Exploration 18.6

Equation of a tangent

The diagram shows the graph of $y = \frac{1}{2}x(x+2)(x-3)$ with **tangents** to the curve drawn at points A and B.

- Find the equations of the two tangents.
- Find the coordinates of the point C.

Tangents

From the diagram we can see:

- for the tangent at A: gradient = 1 y-intercept = –6
  $\Rightarrow$ equation of tangent is $y = x - 6$
- for the tangent at B: gradient = –0.5 y-intercept = 1.5
  $\Rightarrow$ equation of tangent is $y = -0.5x + 1.5 \quad \Rightarrow \quad x + 2y = 3$
- point C, the intersection of the two tangents, has coordinates (5, –1).

All three deductions have been made from the graph. However, we must be able to complete the task more formally.

Let $y = f(x) = \frac{1}{2}x(x+2)(x-3) = \frac{1}{2}x^3 - \frac{1}{2}x^2 - 3x$

$\Rightarrow \frac{dy}{dx} = f'(x) = \frac{3}{2}x^2 - x - 3$

At A(2, –4), f(2) = 1, which enables us to find the equation of the tangent. Recall that the equation of a line with gradient m, passing through (x_1, y_1) is given by:

$$y - y_1 = m(x - x_1)$$

In this case, $x_1 = 2$, $y_1 = -4$ and $m = 1$

$\Rightarrow y - (-4) = 1(x - 2) \Rightarrow y = x - 6$ (1)

Similarly at B(–1, 2), f′(1) = –0.5

$\Rightarrow y - 2 = -0.5(x - (-1)) \Rightarrow y = -0.5x + 1.5$ or $x + 2y = 3$ (2)

Solving equations (1) and (2) simultaneously enables us to find the coordinates of C.

From (1): $y = x - 6$

Substitute in (2): $x + 2(x - 6) = 3 \Rightarrow 3x - 12 = 3 \Rightarrow x = 5$

Substitute in (1): $y = 5 - 6 = -1$

So the coordinates of C are (5, –1).

The normal

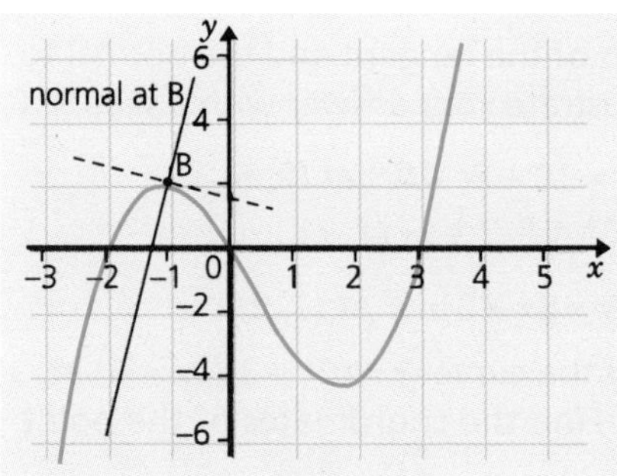

Closely related to the tangent is the **normal**, a straight line which passes through a point on a curve and is perpendicular to the tangent at that point. For perpendicular lines, the product of the gradients is –1, so the gradient of a normal can easily be deduced from the gradient of the tangent.

At B(–1, 2) the gradient of the tangent is –0.5, so the gradient of the normal is $\frac{-1}{-0.5} = 2$.

Therefore the equation of the normal through (–1, 2) is:

$$y - 2 = 2(x - (-1)) \Rightarrow y = 2x + 4$$

Example 18.8

Find the equation of the normal at a point on the curve $y = e^{2x} - 0.5x$ *where the gradient is 5.*

Solution

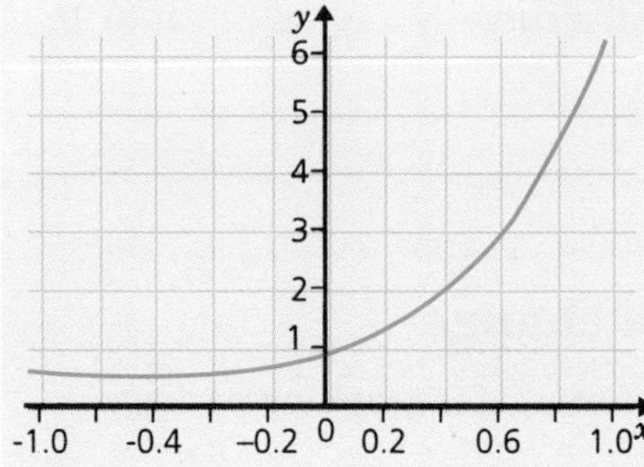

$y = e^{2x} - 0.5x \Rightarrow \frac{dy}{dx} = 2e^{2x} - 0.5$

If the gradient of the curve is 5 then:

$2e^{2x} - 0.5 = 5 \Rightarrow 2e^{2x} = 5.5 \Rightarrow e^{2x} = 2.75$

$\Rightarrow \ln e^{2x} = \ln 2.75 \Rightarrow 2x = 1.0116 \Rightarrow x = 0.5058$

When $x = 0.5058$, $y = e^{1.0116} - 0.5 \times 0.5058 = 2.4971$

Gradient of tangent $= 5 \Rightarrow$ *gradient of normal* $= -\frac{1}{5} = -0.2$.

Therefore the equation of the normal through (0.5058, 2.4971) is:

$y - 2.4971 = -0.2(x - 0.5058) \Rightarrow y = -0.2x + 2.5983$ *or* $x + 5y = 12.99$

EXERCISES

18.3 A

For questions **1** to **6**, find the gradient of the **tangent** and the gradient of the **normal** for each curve at the point given. Deduce the **equations** of the tangent and the normal passing through this point, illustrating your answer with a sketch.

1 $y = x^2 - 5$ at $(2, -1)$

2 $y = 2 - 3x - x^3$ at $(-1, 6)$

3 $y = \frac{12}{x}$ at $(-2, -6)$

4 $y = 10\sqrt{x}$ at $(9, 30)$

5 $s = t(t - 1)(t + 3)$ at $(-1, 4)$

6 $x = 2e^{-t}$ at $(0, 2)$

7 Find the equations of the tangents to the curve $y = 15 + 2x - x^2$ at the points where it meets the x-axis. Find the coordinates of the point where the tangents intersect.

8 Find the equations of the tangents to the curve $y = x^3 - 8x + 5$ which have a gradient 4.

9 The normals to the curve $y = x^2 + x - 12$ at (2,–6) and (p, q) are perpendicular. Find the values of p and q.

10 Find the equation of the tangent to the curve $y = 2x^3 - 3x^2 + 5x - 4$ at the point (0.5, –2). Illustrate your answer with a sketch.

EXERCISES

18.3 B

For questions **1** to **6**, find the equation of the tangent and the normal to the curve through the given point. Illustrate your answer with a sketch.

1 $y = 4(x + 3)^2$, at $(-4, 4)$

2 $s = 10 + 8t - 5t^2$ at $(2, 6)$

3 $y = 5x^3 - 6x^2$ at $(1, -1)$

4 $y = 5x^2 - \frac{1}{x^2}$ at $(1, 4)$

5 $u = 3x\left(5 - \sqrt{x}\right)$ at (4, 36)

6 $y = 2x \pm 3\ln x$ at $(1, 2)$

7 Find the equations of the normals to the curve $y = (2x + 1)(2x - 5)$ at the points where it meets the x-axis. Find the coordinates of the point where the normals intersect.

8 Find the equation of the normals to the curve $y = \frac{x^3}{3} - x^2 + 2$ which have gradient $-\frac{1}{3}$.

9 Find the equations of the tangents to the curve $y = x(x - 1)(x + 3)$ at the point where it intersects with the line $y = 2x + 6$.

10 Find the area enclosed by normal to the curve $y = 10 - e^{-0.5x}$ at (0,9), the line $y = 0$ and the line $x = 0$.

SECOND DERIVATIVES

CALCULATOR ACTIVITY

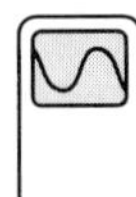

Rates of change of rates of change

You will need a graphics calculator or computer graphing package. Scale the axes as follows.

$x_{min} = -3$; $x_{max} = 3$; $x_{scl} = 1$
$y_{min} = -5$; $y_{max} = 5$; $y_{scl} = 1$.

- Plot the graph of $y = x(x^2 - 3)$.
- Find the coordinates of the stationary points.
- Superimpose the graph of $y = 3(x^2 - 1)$. What value does it take when $x = \pm 1$?
- Superimpose the graph of $y = 6x$. What values does it take when $x = 0$ and $x = \pm 1$?
- How are the three graphs related?

Turning points

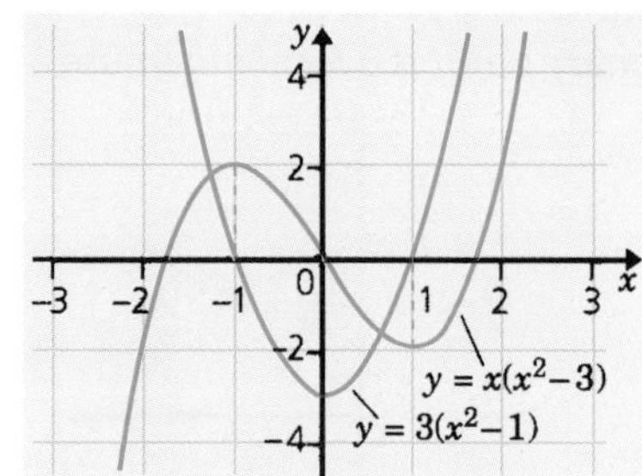

The graphs of $y = x(x^2 - 3)$ and $y = 3(x^2 - 1)$ are shown in the diagram.

Let $y = x(x^2 - 3) = x^3 - 3x$. Then $\frac{dy}{dx} = 3x^2 - 3 = 3\left(x^2 - 1\right)$

i.e. $y = 3\left(x^2 - 1\right)$ is the gradient function for $y = x(x^2 - 3)$.

Let $f(x) = x(x^2 - 3)$. Then $f'(x) = 3\left(x^2 - 1\right)$.

$f'(-1) = f'(1) = 0$, indicating that at $x = \pm 1$ the curve $y = f(x)$ has stationary values (a maximum at $(-1, 2)$ and a minimum at $(1, -2)$).

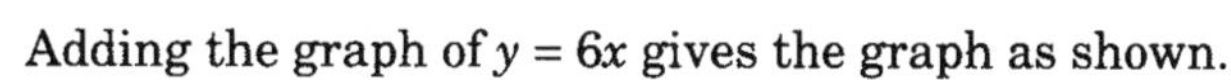

Adding the graph of $y = 6x$ gives the graph as shown.

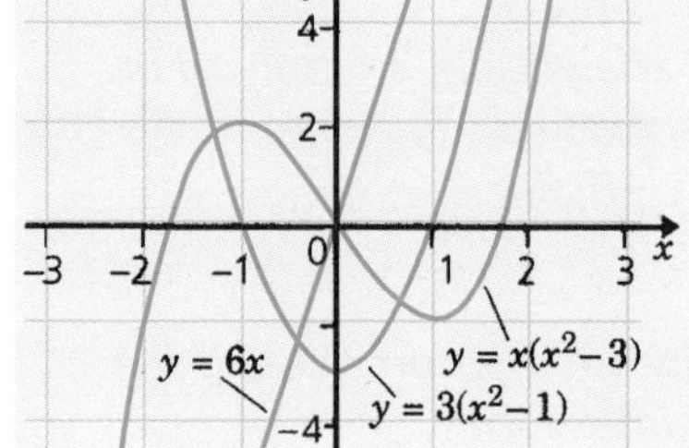

Let $y = 3(x^2 - 1) = 3x^2 - 3$, Then $\frac{dy}{dx} = 6x$

$\Rightarrow$ $y = 6x$ is the gradient function for $y = 3(x^2 - 1)$.

Since $f'(x) = 3(x^2 - 1)$ the function $f''(x) = 6x$ is called the **second derivative** of $f(x)$.

$f''(0) = 0$ indicates that at $x = 0$ the function $y = 3(x^2 - 1)$ has a stationary value and $y = x(x^2 - 3)$ has a **point of inflexion**.

$f''(-1) = -6$, i.e. $f''(-1) < 0$, which indicates that the gradient function $y = f'(x)$ is decreasing at $x = -1$; this means that the stationary point $(-1, 2)$ is a local **maximum**.

$f''(1) = 6$, i.e. $f''(1) > 0$, which indicates that the gradient function $y = f'(x)$ is increasing at $x = 1$; this means that the stationary point at $(1, -2)$ is a local **minimum**.

The results of the exploration provide two useful ideas involving second derivatives.

At stationary points $(a, f(a))$ when $f'(a) = 0$

- if $f''(a) < 0$ then $(a, f(a))$ is a local **maximum**,
- if $f''(a) > 0$ then $(a, f(a))$ is a local **minimum**.

At a point of inflexion $(b, f(b))$, $f''(b) = 0$.

However, the converse is not always true, i.e. $f''(b) = 0$ does not always give a point of inflexion. For example:

$$y = f(x) = x^4 \Rightarrow f'(x) = 4x^3 \Rightarrow f''(x) = 12x^2$$

a local **minimum** point occurs at (0,0) even though $f''(0) = 0$.

A necessary and sufficient condition for a point of inflexion at $(b, f(b))$ is that $f''(b) = 0$ and there is a change of sign in $f''(x)$ either side of $x = b$.

Example 18.9

a) Find and classify the stationary point of the graph of $y = e^x - x$.
b) Show that $y = e^x - x$ *has a point of inflexion, but no stationary points.*
c) Use your calculator and a numerical method to find stationary points and points of inflexion for the graph of $y = e^x - x$.

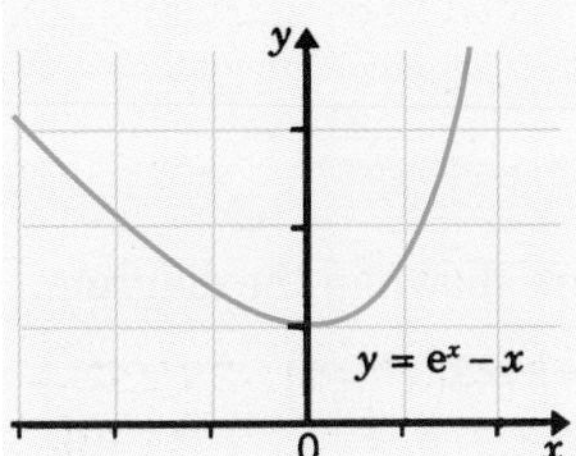

Solution

a) $f(x) = e^x - x \Rightarrow f'(x) = e^x - 1$
For stationary values, $f'(x) = 0 \Rightarrow e^x - 1 = 0 \Rightarrow e^x = 1 \Rightarrow x = \ln 1 = 0$
Since $f(0) = 1 - 0 = 1$, $y = e^x - x$ has a stationary point at (0, 1).
Since $f''(x) = e^x$, $f''(0) = 1 \Rightarrow$ the stationary point is a local minimum.

b) $f(x) = e^x - x^2 \Rightarrow f'(x) = e^x - 2x$
For stationary values, $f'(x) = 0$, but $e^x - 2x > 0$ for all real values of x, since $f'(x)$ has a local minimum at $(\ln 2, 2 - 2\ln 2) \approx (0.693, 0.614)$, where $f''(x) = e^x - 2$.

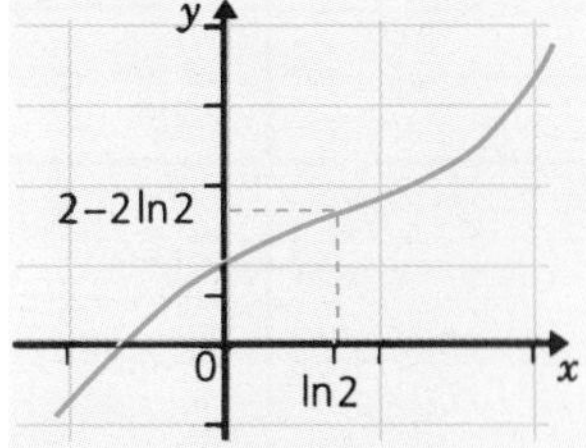

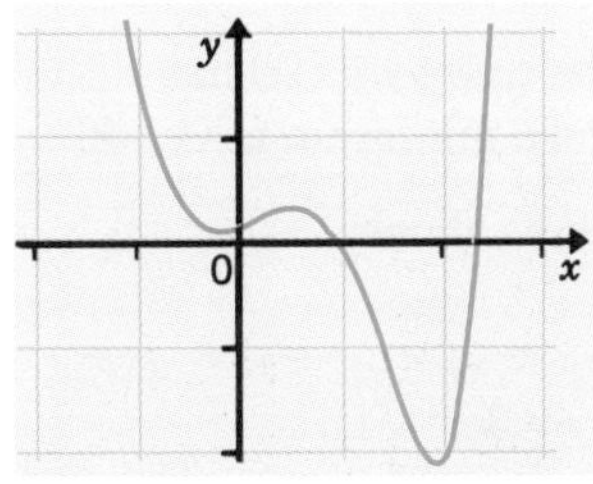

c) $f(x) = e^x - x^3 \Rightarrow f'(x) = e^x - 3x^2$
For stationary values, $f'(x) = 0 \Rightarrow e^x - 3x^2 = 0$
Since this equation cannot be solved analytically, a graphical or decimal search reveals three stationary values (figures to 3 s.f.), at (−0.459, 0.729), (0.910, 1.73) and (3.73, −10.2).

Example 18.10

Show that all cubic functions have a point of inflexion. What is the geometrical significance of the point of inflexion?

Solution

Let a general cubic function $f(x) = ax^3 + bx^2 + cx + d$, where a, b, c, d are constants.

For a point of inflexion we require $f''(x) = 0$ and a change of sign.

$$f(x) = ax^3 + bx^2 + cx + d \Rightarrow f'(x) = 3ax^2 + 2bx + c$$
$$\Rightarrow f''(x) = 6ax + 2b$$

Now $f''(x) = 0 \Rightarrow 6ax + 2b = 0 \Rightarrow x = -\dfrac{b}{3a}$

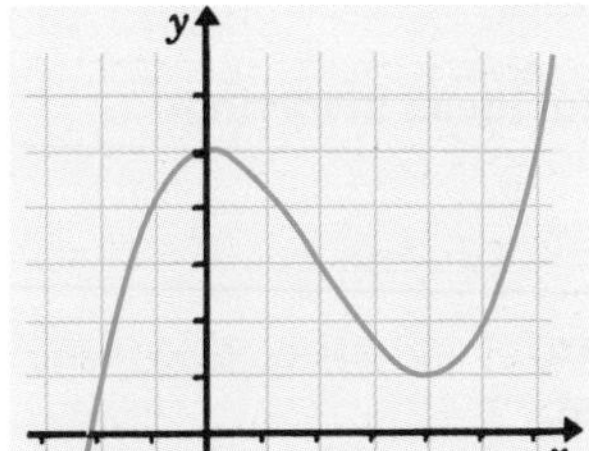

Since $y = 6ax + 2b$ gives a straight line graph, y must change sign either side of $x = -\frac{b}{3a}$, so every cubic has a point of inflexion.

Geometrically, all cubic graphs have rotational symmetry about the point of inflexion, e.g. for $y = x^3 - 3x^2 + 5$ (i.e. $a = 1, b = -3, c = 0, d = 5$) the point of inflexion occurs when $x = -\frac{-3}{3\times 1} = 1$ and $y = 3$.

EXERCISES

18.4A

1 For the following functions, find $f''(x)$ and its value at the x–value given.

a) $f(x) = x^3 - 4x$, $x = 2$
b) $f(x) = 3 + x^2 - x^4$, $x = 3$
c) $f(x) = ax^2 + bx + c$, $x = -1$
d) $f(x) = x^2 - \frac{1}{x^2}$, $x = -1$
e) $f(x) = e^{-x}$, $x = 0$
f) $f(x) = 2\ln x$, $x = 1$
g) $f(x) = \sqrt{x}(4 - x)$, $x = 4$
h) $f(x) = 2^x$, $x = 3$

2 For the following functions, use $f''(x)$ to determine the nature of any stationary points.

a) $f(x)=2x^3+3x^2-12x+5$ b) $f(x)=x+\frac{4}{x^2}$

c) $f(x)=10x-e^{2x}$ d) $f(x)=7+2x^2-x^4$

3 For the following functions, find any points of inflexion.

a) $f(x)=12-x+3x^2-x^3$ b) $f(x)=3x^4-4x^3$

c) $f(x)=5x^3-\frac{15}{x}$ d) $f(x)=x\left(x-4\sqrt{x}\right)$

e) $f(x)=e^{-x}-3x^2$ f) $f(x)=4x^3+3\ln x$

4 In the following expressions, x represents the displacement, in metres, of a particle from an origin O after t seconds, $t \geq 0$. In each case find:

i) the acceleration a at the given time t,

ii) the time at which the given acceleration a occurs.

a) $x=t^2(5-t)$ $t=5\text{s}$ $a=1\,\text{m}\,\text{s}^{-2}$

b) $y=10(1-e^{-t})$ $t=2\text{s}$ $a=-2\,\text{m}\,\text{s}^{-2}$

c) $x=\frac{20}{t}$ $t=2\text{s}$ $a=320\,\text{m}\,\text{s}^{-2}$

EXERCISES

18.4 B

1 For the following functions, find $f''(x)$ and its value at the x-value given.

a) $f(x)=1+x+\dfrac{x^2}{2}+\dfrac{x^3}{6}$, $x=2$ b) $f(x)=\frac{1}{2}x^2(5-x)$, $x=2$

c) $f(x)=2x^3+\frac{3}{x}$, $x=-1$ d) $f(x)=4e^{\frac{1}{2}x}$, $x=-1$

e) $f(x)=9-\ln x^2$, $x=3$ f) $f(x)=2x^2\left(3-\sqrt{x}\right)$, $x=5$

g) $f(x)=5\times 3^{-x}$, $x=0.5$ h) $f(x)=3^x-2^x$, $x=-1$

2 For the following functions, use $f''(x)$ to determine the nature of any stationary points.

a) $f(x)=10+3x^2-x^3$ b) $f(x)=2x^2-\frac{5}{x}$

c) $f(x)=2x^2(3-\sqrt{x})$ d) $f(x)=10+3(x-2)^2-(x-2)^3$

3 For the following functions, find any points of inflexion.

a) $f(x)=x^3-2x^3+3x-4$ b) $f(x)=4x^5-5x^4$ c) $f(x)=2x-\frac{5}{x^2}$

d) $f(x)=(2x-5)^3$ e) $f(x)=10x^2-e^{2x}$ f) $f(x)=x^5+\ln x^2$

4 In the following expressions, x represents the displacement, in metres, of a particle from an origin O after t seconds, $t \geq 0$. In each case find:

i) the acceleration a at the given time t,

ii) the time(s) at which the given acceleration occurs.

a) $x=\frac{1}{10}t^2(6-t)^2$ $t=3\,\text{s}$ $a=1.2\,\text{m}\,\text{s}^{-2}$

b) $x=t-\frac{1}{6}t^3+\frac{1}{120}t^5$ $t=1.5\,\text{s}$ $a=1.5\,\text{m}\,\text{s}^{-2}$

c) $x=3\left(1+4e^{-\frac{1}{2}t}\right)$ $t=0.8\,\text{s}$ $a=1\,\text{m}\,\text{s}^{-2}$

MATHEMATICAL MODELLING ACTIVITY

Problem statement

Specify the real problem

A small company that distributes books and software needs a policy for controlling levels of stock. If the company has too little stock there is a danger of orders to customers being delayed; however if stock is too high money is tied up unnecessarily and the company needs to rent storage space.

After the company has been running for one year, the following data is available for future planning.

a) Ordering costs (administration and postage) are £1.50 per order.
b) Annual cost of holding an item in stock is calculated at 20% of the item price.
c) The current price of a popular software package is £33.
d) The average number of this package sold per month is 35.

What ordering policy should the company adopt?

Set up a model

Set up a model

First you need to identify the important variables:

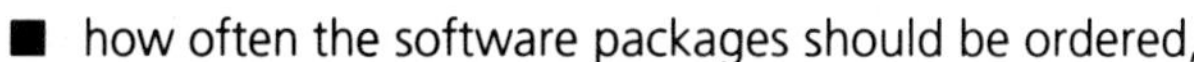

- how often the software packages should be ordered,
- the selling price for each package,
- the total number sold each year,
- the costs involved (fixed and variable),
- the number of packages in stock as a function of time.

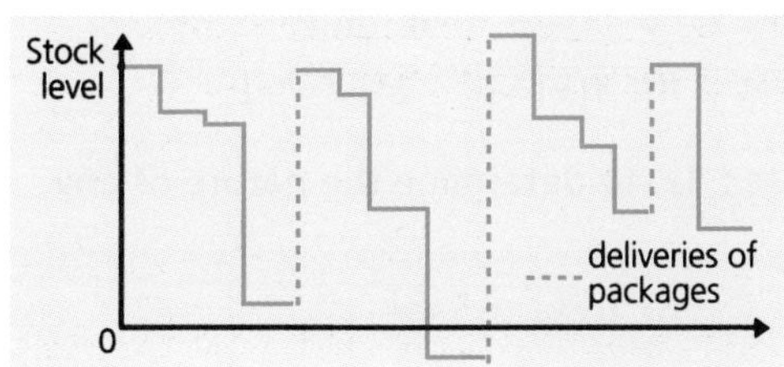

To proceed further we need to make some assumptions about stock level and time. The figure shows what has happened in the first year of running the company.

Notice that for one period there were orders from customers waiting for a delivery.

To formulate a simple model we shall assume:

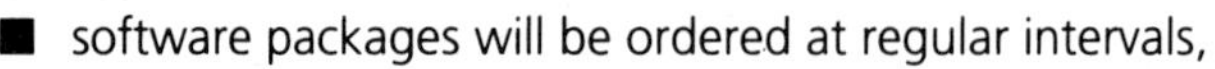

- software packages will be ordered at regular intervals,
- the stock level in store reduces linearly with time,
- each order is constant,
- there are no discounts available for large orders.

Mathematical problem

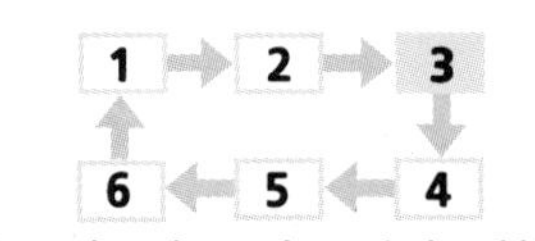

Formulate the mathematical problem

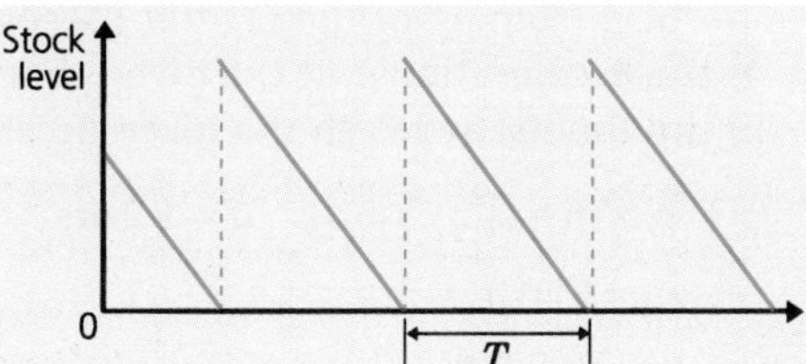

We begin by defining variables.

Let x be the number of packages ordered each T weeks, and C be the total annual costs to the company.

The annual costs are made up of two parts: the ordering costs and the storage costs.

$$\text{ordering costs per year} = \frac{\text{total number sold}}{\text{order size}} \times \text{order costs} = \frac{35 \times 12 \times 1.5}{x} = \frac{630}{x}$$

$$\text{storage costs per year} = 20\% \text{ of average stock held} \times \text{item price} = 0.2 \times \frac{x}{2} \times 33 = 3.3x$$

$$C = \frac{630}{x} + 3.3x \qquad (1)$$

The mathematical problem is to minimise C.

Mathematical solution

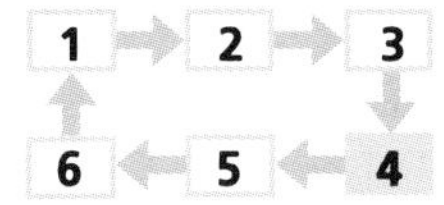

Solve the mathematical problem

Using calculus, we find that the minimum value of C occurs when $\frac{dC}{dx} = 0$ and $\frac{d^2C}{dx^2} > 0$. Differentiating **(1)**:

$$\frac{dC}{dx} = -\frac{630}{x^2} + 3.3$$

and $\frac{dC}{dx} = 0$ when $x^2 = \frac{630}{3.3} = 190.19 \quad \Rightarrow \quad x = 13.82$

Also $\frac{d^2C}{dx^2} = \frac{1260}{x^3} > 0$ for $x = 13.82$. Hence we have a minimum.

Interpretation

Interpret the solution

Clearly we cannot order a fraction of a software package. We suggest that the company orders 14 packages. Since the total number of packages sold is 420 the company will need to make an order 30 times per year.

Refinement of the model

Compare with reality

Discuss the assumptions made and suggest other simplifying assumptions that have not been written down. Investigate the effect of changes to these assumptions.

CONSOLIDATION EXERCISES FOR CHAPTER 18

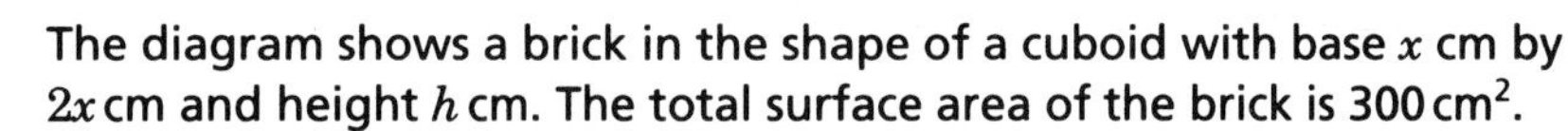

1 The diagram shows a brick in the shape of a cuboid with base x cm by $2x$ cm and height h cm. The total surface area of the brick is 300 cm^2.

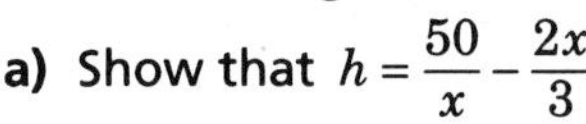

a) Show that $h = \frac{50}{x} - \frac{2x}{3}$

b) The volume of the brick is V cm^3. Express V in terms of x only.

c) Given that x can vary, find the maximum value of V.

d) Explain why the value of V you have found is the maximum.

(ULEAC Question 7 Specimen Paper 1, 1995)

2 The function f is defined by $f(x) = e^x - 5x, x \in \Re$.

a) Determine $f'(x)$.

b) Find the value of x for which $f'(x) = 0$, giving your answer to 2 d.p.

c) Show, by calculation, that there is a root, α, of the equation $f(x) = 0$ such that $0.2 < \alpha < 0.3$,

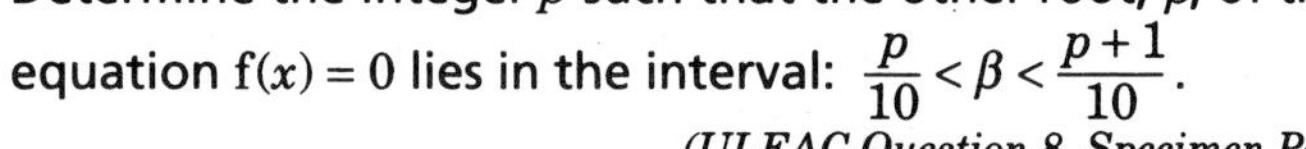

d) Determine the integer p such that the other root, β, of the equation $f(x) = 0$ lies in the interval: $\frac{p}{10} < \beta < \frac{p+1}{10}$.

(ULEAC Question 8, Specimen Paper 1, 1995)

3 The curve with equation $ky = a^x$ passes through the points with coordinates (7, 12) and (12, 7). Find, to 2 significant figures, the values of the constants k and a.

Using your values of k and a, find the value of $\frac{dy}{dx}$ at $x = 20$, giving your answer to 1 decimal place.

(ULEAC Question 6, Specimen Paper 2, 1995)

4 In a medical treatment 500 milligrams of a drug are administered to a patient. At time t hours after the drug is administered X milligrams of the drug remain in the patient. The doctor has a mathematical model which states that:

$$X = 500e^{-\frac{1}{5}t}$$

Find the value of t, correct to two decimal places, when $X = 200$.

Express $\frac{dX}{dt}$ in terms of t. Hence show that when $X = 200$ the rate of decrease of the amount of the drug remaining in the patient is 40 milligrams per hour.

(NEAB Question 5, Specimen Paper 1, 1994)

5 Use differentiation to find the coordinates of the stationary points on the curve:

$$y = x + \frac{4}{x}$$

and determine whether each stationary point is a maximum point or a minimum point. Find the set of values of x for which y increases as x increases.

(UCLES Modular Question 8, Specimen Paper 2, 1994)

6 **a)** Suggest a suitable domain for the function:

$$f : x \to \sqrt{x-5}$$

and draw a sketch of its graph.

b) Show that the x-coordinate of the point of intersection of the line $y = 10 - x$ and the curve $y = \sqrt{x-5}$ lies between 8 and 9.

(Oxford & Cambridge Question 5, Specimen Paper 1, 1994)

7 A forest fire spreads so that the number of hectares burnt after t hours is given by $h = 30(1.65)^t$.

a) By what constant factor is the burnt area multiplied from time $t = N$ to time $t = N + 1$? Express this as a percentage increase.

b) 1.65 can be written as e^K. Find the value of K.

c) Hence show that $\frac{dh}{dt} = 15e^{Kt}$.

d) This shows that $\frac{dh}{dt}$ is proportional to h. Find the constant of proportionality.

(Oxford & Cambridge Question 1, Specimen Paper 2, 1994)

8 Open display boxes are to be 12 cm high and have a base area of 1000 cm^2. Sketches of one of the boxes and its net are shown in the diagrams.

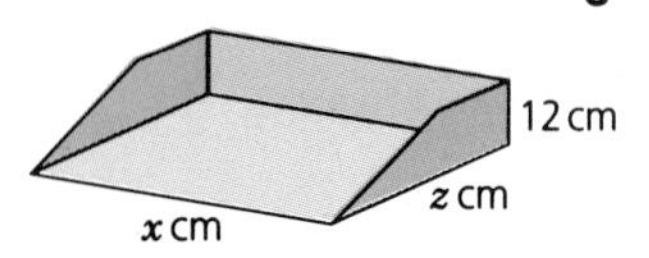

a) The box is x cm wide. The depth from front to back is z cm as shown. Express z in terms of x.

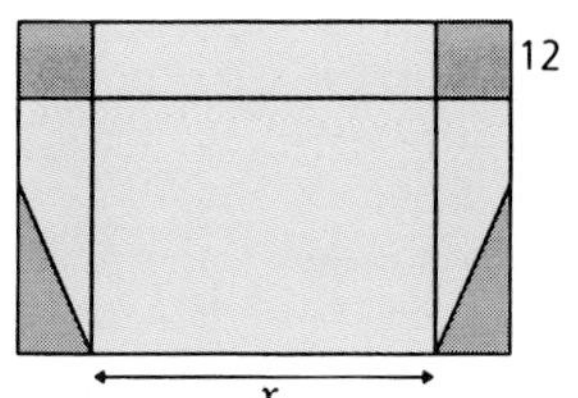

The net of the box is stamped on to a rectangular piece of card, the shaded corners are removed and the card is folded and stuck to form the box.

b) Find, in terms of x, the dimensions of the rectangle of card.

c) Show that the total area A cm² of the rectangle of card is given by:

$$A = 1288 + 12x + \frac{24000}{x}$$

c) Find $\frac{dA}{dx}$ and the least value of A.

(MEI Question 3, Paper 1, June 1994)

9 You are given that $f(x) = 4 + 2x^3 - 3x^4$.

a) Find $f'(x)$ and the values of x for which $f'(x) = 0$. Hence find the coordinates of the stationary points.

b) Find $f''(x)$ and its value at each stationary point.

c) Determine the nature of the stationary point at which $f'(x)$ and $f''(x)$ are both zero, explaining your method and showing your working.

d) State the nature of the other stationary point and, without further calculation, sketch the graph of $y = f(x)$ for values of $x = -1$ and $x = 1$.

(MEI Question 1, Paper 3, January 1993)

10 The diagram shows the flat surface of a tray consisting of a rectangular region WXYZ and a semi-circular region at each end with WX and YZ as diameters. The rectangle WXYZ has area 200 cm².

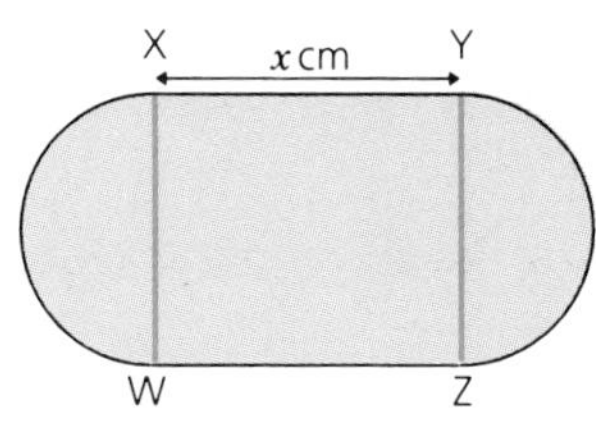

a) Given that XY = x cm, show that the perimeter P cm of the tray is given by the formula:

$$P = 2x + \frac{200\pi}{x}$$

b) Find the minimum value of P as x varies.

(ULEAC Question 7, Paper 1, January 1995)

Summary

- The following table shows the derivative of the functions introduced so far.

$f(x)$	x^n	e^{kx}	a^x	$\ln x$
$f'(x)$	nx^{n-1}	ke^{kx}	$a^x \ln a$	$\frac{1}{x}$

(for any number n)

- The second derivative of a function $f(x)$ is the derivative of the first derivative. $f''(x) = \frac{d^2f}{dx^2} = \frac{d}{dx}\left(\frac{df}{dx}\right)$

- At a stationary point $(a, f(a))$ of a function $f(x)$, $f'(a) = 0$:

  $f''(a) < 0 \Rightarrow (a, f(a))$ is a local maximum,
  $f''(a) > 0 \Rightarrow (a, f(a))$ is a local minimum.

- At a point of inflexion $(b, f(b))$, $f''(b) = 0$.

Integration 2

In this chapter we introduce:

- *the rule for differentiation of functions with rational exponents*
- *the method of finding volumes of revolution.*

INTEGRATING FUNCTIONS WITH RATIONAL EXPONENTS

In Chapter 9, *Integration 1*, we integrated polynomial functions, obtaining indefinite integrals such as:

$$\int x^2dx = \frac{1}{3}x^2 + c$$

and $\frac{ds}{dt} = 15 - 10t \Rightarrow s = 15t - 5t^2 + k$

In Chapter 18, *Differentiation 2*, we found that the rules for differentiating polynomials can be extended to differentiating functions with rational exponents, such as:

$$f(x) = \frac{3}{x} = 3x^{-1} \Rightarrow f'(x) = -3x^{-2} = -\frac{3}{x^2}$$

and $y = 10\sqrt{x} = 10x^{\frac{1}{2}} \Rightarrow \frac{dy}{dx} = 5x^{-\frac{1}{2}} = \frac{5}{\sqrt{x}}$

Now we can extend the rules for integrating polynomials to consider integrating functions with rational exponents.

Exploration 19.1

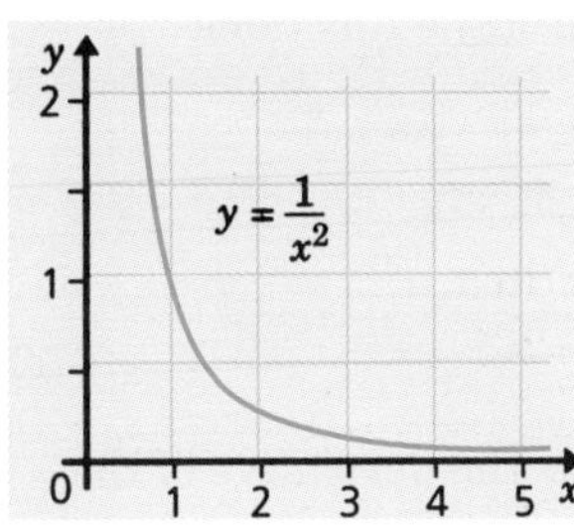

Rational exponents

The diagram shows a sketch of $f(x) = \frac{1^2}{x}$, $x \neq 0$.

- Use a numerical method to estimate areas given by $A = \int_1^a \frac{1}{x^2}dx$ for $a = 2, 3, 4, 5, 10$.
- Repeat for $a = 0.5, 0.2, 0.1$.
- Use the numerical results to find the area function $A(x)$.
- Is it true that $A'(x) = f(x)$?

Numerical results

The numerical calculations should give the following results.

a	$\int_1^a \frac{1}{x^2}dx$	a	$\int_1^a \frac{1}{x^2}dx$
2	0.5	10	0.9
3	0.67	0.5	−1.0
4	0.75	0.2	−4.0
5	0.8	0.1	−9.0

Check that $A(x) = 1 - \frac{1}{x}$ i.e. $\int_1^a \frac{1}{x^2}\,\mathrm{d}x = 1 - \frac{1}{a}$

$$A(x) = 1 - \frac{1}{x} = 1 - x^{-1} \Rightarrow A'(x) = x^{-2} = \frac{1}{x^2} = \mathrm{f}(x)$$

Since $A'(x) = \mathrm{f}(x)$, the process of differentiation may be reversed to integrate $\mathrm{f}(x) = \frac{1}{x^2}$,

i.e. $\mathrm{f}(x) = \frac{1}{x^2} = x^{-2} \Rightarrow \int \mathrm{f}(x)\,\mathrm{d}x = \frac{x^{-1}}{-1} + c = -\frac{1}{x} + c$

Note that in this case the constant of integration $c = 1$, which is determined by the lower bound of the definite integral $\int_1^a \frac{1}{x^2}\,\mathrm{d}x$.

Fractional indices

The process of integration can also be extended to fractional indices, as illustrated in the following example.

Example 19.1

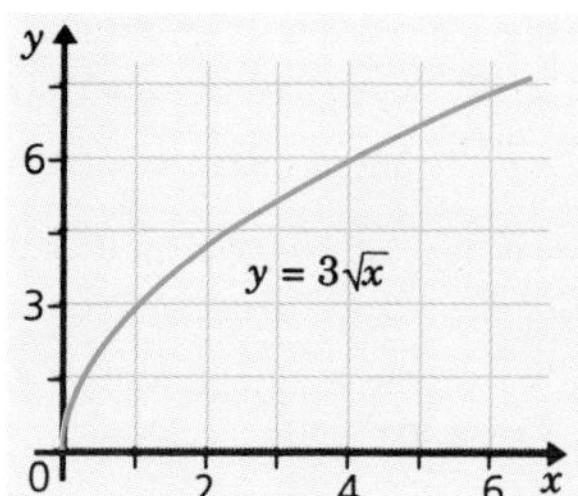

The diagram shows a sketch of $\mathrm{f}(x) = 3\sqrt{x} \quad x \geq 0$.

a) *Use a numerical method to estimate areas given by:* $\int_0^a 3\sqrt{x}\,\mathrm{d}x$ for $a = 1, 4, 9, 16, 25$

b) *Show that* $\int_0^a 3\sqrt{x}\,\mathrm{d}x = \left[2\sqrt{x}^3\right]_0^a$ *is compatible with your numerical results.*

Solution

a) *Results of numerical calculations are shown in this table.*

a	1	4	9	16	25
$\int_0^a 3\sqrt{x}\mathrm{d}x$	2	16	54	128	250

b) *Applying the integration rules to fractional indices:*

$$\int_0^a 3\sqrt{x}\,\mathrm{d}x = \int_0^a 3x^{\frac{1}{2}}\,\mathrm{d}x = \left[\tfrac{2}{3} \times 3x^{\frac{3}{2}}\right]_0^a = \left[2\sqrt{x}^3\right]_0^a$$

for $a = 1$: $\int_0^1 3\sqrt{x}\,\mathrm{d}x = \left[2\sqrt{x}^3\right]_0^1 = 2 - 0 = 2$

for $a = 4$: $\int_0^4 3\sqrt{x}\,\mathrm{d}x = \left[2\sqrt{x}^3\right]_0^4 = 16 - 0 = 16$ *etc.*

Rules for integrating rational powers

To summarise, we have found that the rules for integrating rational powers are the same as for polynomial functions.

$$\int x^n\mathrm{d}x = \frac{1}{n+1}x^{n+1} + c \quad n \neq -1$$

This holds for all rational numbers n, except $n = -1$, since this would lead to division by 0. When $n = -1$, $\mathrm{f}(x) = x^{-1} = \frac{1}{x}$. We shall look at this integration later in the chapter.

We can apply the process of integrating with rational powers to finding areas and reversing the process of differentiation, as illustrated in the next two examples.

Example 19.2

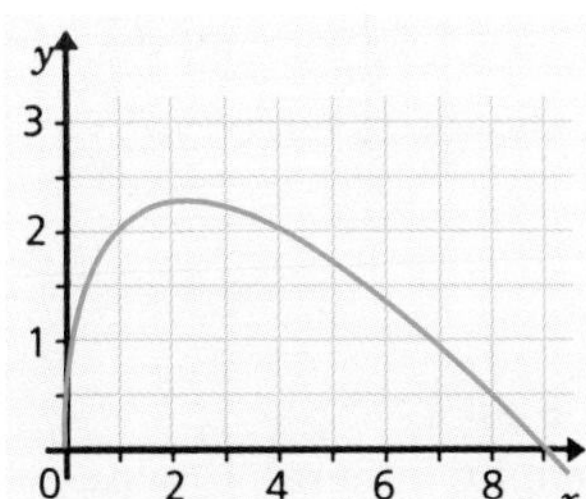

Find the area enclosed by the graph of $y = 3\sqrt{x} - x$ and the x-axis.

Solution

A sketch of $y = 3\sqrt{x} - x$ reveals that it lies above the x-axis for $0 < x < 9$ and cuts it at $x = 0$ and $x = 9$.

The points of intersection of the graph and the x-axis occur where $y = 0$, so may be found algebraically.

$$0 = 3\sqrt{x} - x \Rightarrow x = 3\sqrt{x} \Rightarrow x^2 = 9x$$

$$\Rightarrow x^2 - 9x = 0 \Rightarrow x(x-9) = 0$$

$$\Rightarrow x = 0 \text{ or } x = 9$$

The required area is given by:

$$\int_0^9 \left(3\sqrt{x} - x\right) dx = \int_0^9 \left(3x^{\frac{1}{2}} - x\right) dx = \left[\tfrac{2}{3} \times 3x^{\frac{3}{2}} - \tfrac{1}{2}x^2\right]_0^9 = \left[2\sqrt{x}^3 - 0.5x^2\right]_0^9$$

$$= \left(2 \times \sqrt{9}^3 - 0.5 \times 9^2\right) - \left(2 \times \sqrt{0}^3 - 0.5 \times 0^2\right)$$

$$= (54 - 40.5) - 0 = 13.5$$

Example 19.3

The function $y = \mathrm{f}(x)$ has gradient function:

$$\mathrm{f}'(x) = \sqrt{x}(5-x)$$

and $\mathrm{f}(1) = 3$. Find $\mathrm{f}(4)$.

Solution

$$\mathrm{f}'(x) = \sqrt{x}(5-x) \Rightarrow \mathrm{f}(x) = \int \sqrt{x}(5-x)dx = \int \left(5\sqrt{x} - x\sqrt{x}\right)dx$$

$$= \int \left(5x^{\frac{1}{2}} - x^{\frac{3}{2}}\right)dx = \tfrac{2}{3} \times 5x^{\frac{3}{2}} - \tfrac{2}{5}x^{\frac{5}{2}} + c = \tfrac{10}{3}\sqrt{x}^3 - \tfrac{2}{5}\sqrt{x}^5 + c$$

Now $\mathrm{f}(1) = 3 \Rightarrow 3 = \tfrac{10}{3} - \tfrac{2}{5} + c \Rightarrow c = 3 - \tfrac{10}{3} + \tfrac{2}{5} = \tfrac{1}{15}$

Hence $\mathrm{f}(4) = \tfrac{10}{3}\sqrt{4}^3 - \tfrac{2}{5}\sqrt{4}^5 + \tfrac{1}{15} = \tfrac{10}{3} \times 8 - \tfrac{2}{5} \times 32 + \tfrac{1}{15} = 13\tfrac{14}{15}$

EXERCISES

19.1A

1 Find the following integrals.

a) $\int \frac{7}{x^2} dx$ b) $\int \frac{-12}{x^3} dx$ c) $\int \left(5x - \frac{16}{x^4}\right) dx$ d) $\int 3x\left(\frac{2}{x^3} - 5\right) dx$

e) $\int 15\sqrt{x}\, dx$ f) $\int \frac{-4}{\sqrt{x}} dx$ g) $\int \sqrt{x}(x-8) dx$ h) $\int x^2\left(4 + \sqrt{x}\right) dx$

i) $\int \frac{4}{\sqrt{x}}\left(x^2 - 9\right) dx$ j) $\int \frac{9}{x^2}\left(1 + \sqrt{x}\right) dx$

2 Evaluate the following integrals.

a) $\int_2^5 \frac{10}{x^2} dx$ b) $\int_4^9 6\sqrt{x}\, dx$ c) $\int_1^{2.25} \left(10 - 3\sqrt{x}\right) dx$ d) $\int_1^2 \frac{8}{x^2}\left(1 + x^3\right) dx$

e) $\int_{0.5}^{1.5} (2x+5)\sqrt{x}\, dx$ f) $\int_1^4 \frac{5}{x^4}(x-1)(x+3) dx$

3 Find the areas represented by the following integrals. Illustrate each answer with a sketch.

a) $\int_{-3}^{-1} \frac{15}{x^2}\,dx$ b) $\int_{1}^{4} \frac{-16}{x^3}\,dx$ c) $\int_{2}^{5}\left(5-\frac{8}{x^2}\right)dx$ d) $\int_{1}^{5}\left(x+\frac{5}{x^2}\right)dx$

e) $\int_{0}^{25}\left(x-5\sqrt{x}\right)dx$ f) $\int_{0}^{6}\sqrt{x}(6-x)\,dx$ g) $\int_{0}^{1}\left(\sqrt{x}-\sqrt[3]{x}\right)dx$

h) $\int_{1}^{4}\left(\frac{1}{x^3}-\frac{1}{x^2}\right)dx$

4 For each of the following, find the area of the shaded region.

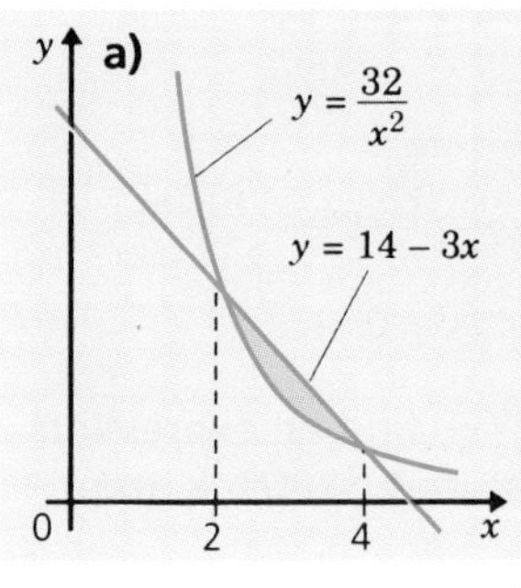

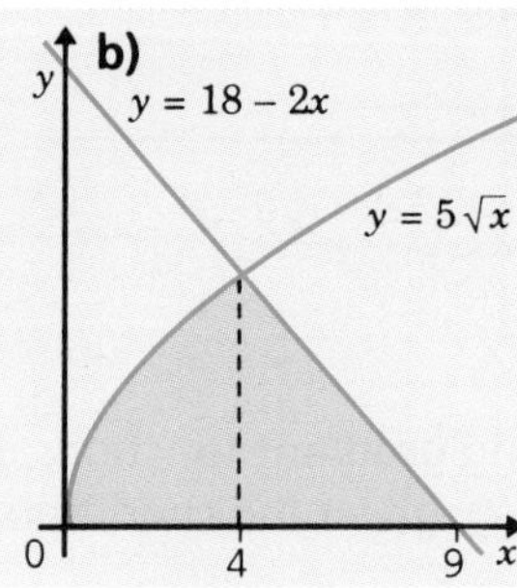

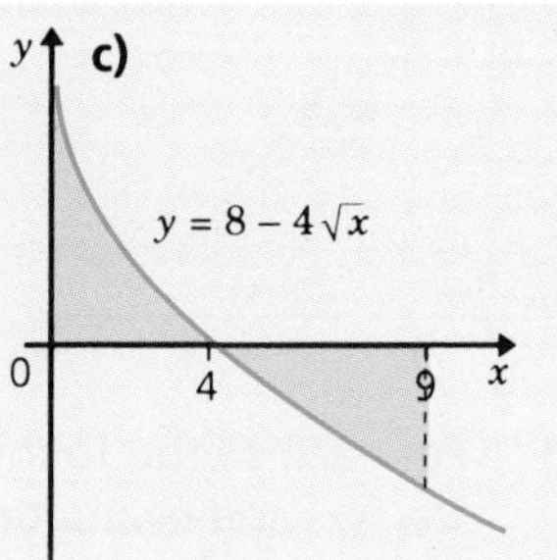

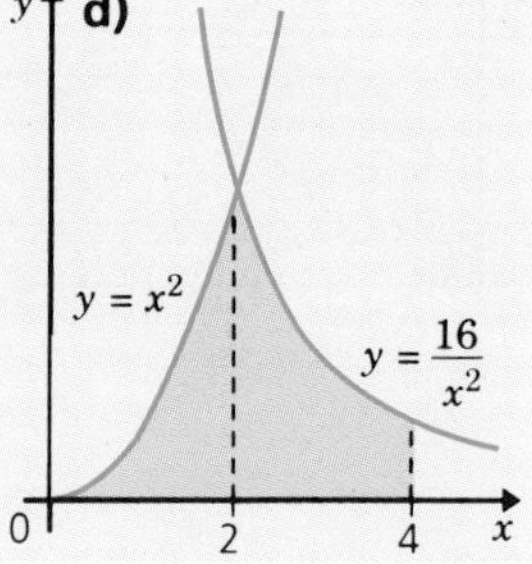

5 For each of the following gradient functions, $f'(x)$, find an expression for $f(x)$ and evaluate $f(4)$, given the condition stated. Sketch a graph of $y = f(x)$.

a) $f'(x) = 5 - \frac{2}{\sqrt{x}}$, $f(0) = -3$ b) $f'(x) = 4 - \frac{3}{x^2}$, $f(1) = 7$

c) $f'(x) = 9\sqrt{x}$, $f(0) = 2$ d) $f'(x) = \frac{2}{x^3} - \frac{3}{x^4}$, $f(1) = 2$

EXERCISES

19.1B

1 Find the following integrals.

a) $\int \frac{dx}{2x^3}$ b) $\int \frac{-5}{x^9}\,dx$ c) $\int\left(3x^2-\frac{3}{x^4}\right)dx$ d) $\int 11x^2\left(7-\frac{3}{x^5}\right)dx$

e) $\int\left(\sqrt{x}+\frac{1}{\sqrt{x}}\right)dx$ f) $\int \frac{x-1}{\sqrt{x}}\,dx$ g) $\frac{1}{2}\int\left(3\sqrt{x}-\frac{1}{\sqrt{x}}\right)dx$

h) $\int x\left(\sqrt{x}-7\right)dx$ i) $\int \frac{\sqrt{x}}{x^3}\,dx$ j) $\int \frac{(x-2)(x+2)}{\sqrt{x}}\,dx$

2 Evaluate the following integrals.

a) $\int_{1}^{3} \frac{5}{x^3}\,dx$ b) $\int_{0}^{4} 3\sqrt{x}\,(1+x)\,dx$ c) $\int_{1}^{2}\left(\frac{5}{\sqrt{x}}+9\sqrt{x}\right)dx$

d) $\int_{2}^{3} \frac{x^4-4}{x^3}\,dx$ e) $\int_{0}^{1}(x-1)(x+1)\sqrt{x}\,dx$ f) $\int_{-2}^{-1} \frac{(x+2)^2}{x^4}\,dx$

3 Find the areas represented by the following integrals. Illustrate each of your answers with a sketch.

a) $\int_{-2}^{-1} \frac{3}{x^2}\,dx$ b) $\int_{2}^{3} \frac{-20}{x^5}\,dx$ c) $\int_{0}^{1}\left(x^2-2\sqrt{x}\right)dx$ d) $\int_{0}^{2}\left(1-\sqrt{x}\right)^2 dx$

e) $\int_1^2 3\left(x^2+\frac{1}{x^2}\right)\mathrm{d}x$ **f)** $\int_1^2 \left(\frac{1}{x}+\sqrt{x}\right)\left(x+\frac{1}{\sqrt{x}}\right)\mathrm{d}x$

g) $\int_0^1\left(\sqrt{x}-\sqrt[3]{x}\right)^2\mathrm{d}x$ **h)** $\int_1^4 \frac{x^2+2x+5}{x^{\frac{3}{2}}}\mathrm{d}x$

4 For each of the following, find the area of the shaded region.

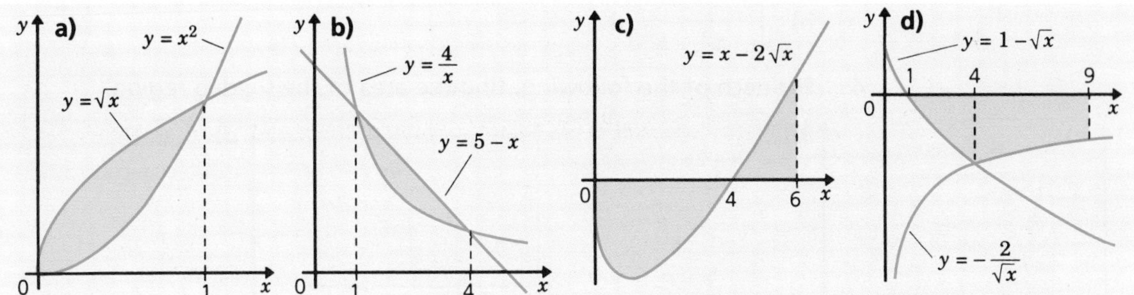

5 For each of the following gradient functions, $f'(x)$, find an expression for $f(x)$ and evaluate $f(2)$, given the condition stated. Sketch a graph of $y = f(x)$.

a) $f'(x)=\frac{3}{\sqrt{x}}-1,\ f(1)=2$ **b)** $f'(x)=3-\frac{7}{x^3},\ f\left(\frac{1}{2}\right)=10$

c) $f'(x)=\frac{1}{\sqrt{x}}-\sqrt{x},\ f(4)=5$ **d)** $f'(x)=\frac{3}{x^4}-\frac{4}{x^5},\ f(1)=2$

INTEGRATING EXPONENTIAL FUNCTIONS AND $\frac{1}{x}$

In Chapter 18, *Differentiation 2*, we found that certain rules applied to differentiating exponential and logarithmic functions.

- $f(x) = e^x \quad \Rightarrow f'(x) = e^x$
- $f(x) = \ln x \quad \Rightarrow f'(x) = \frac{1}{x}$

Reversing the process of differentiation suggests the following indefinite integrals.

- $f(x) = e^x \quad \Rightarrow \int f(x)\mathrm{d}x = e^x + c$
- $f(x) = \frac{1}{x} \quad \Rightarrow \int f(x)\mathrm{d}x = \ln x + c$

How are these results related to finding areas under graphs?

Exploration 19.2

Areas under graphs

The diagram shows a sketch of $f(x) = e^x$.

- Use a numerical method to estimate areas given by $\int_0^a e^x\,\mathrm{d}x$ for $a = 1, 2, 3, 4, 5$.
- Use the numerical results to find the area function $A(x)$.
- Is it true that $A'(x) = f(x)$?

Analysing the results

Results of numerical calculations should give you the following results (given to 3 d.p.).

a	1	2	3	4	5
$\int_0^a e^x \, dx$	1.718	6.389	19.086	53.598	147.413

Check that $A(x) = e^x - 1$, i.e. $\int_0^a e^x \, dx = e^a - 1$

$A(x) = e^x - 1 \Rightarrow A'(x) = e^x = f(x)$

i.e. the process of differentiation may be reversed to integrate e^x.

Exploration 19.3

$f(x) = \frac{1}{x}$

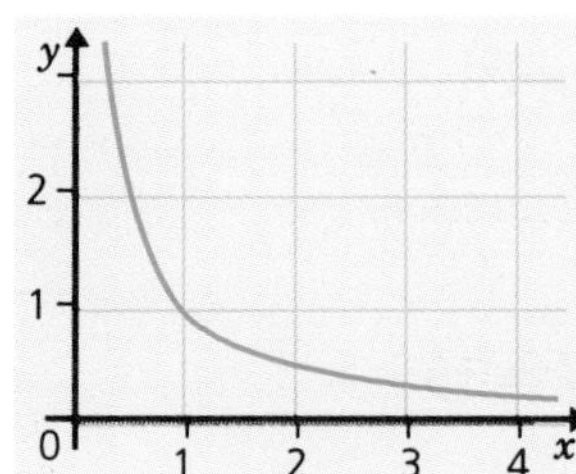

The diagram shows a sketch of $f(x) = \frac{1}{x}$.

- Use a numerical method to estimate areas given by $\int_1^a \frac{1}{x} dx$ for $a = 2, 3, 4, 5, 10$.
- Repeat for $a = 0.5, 0.2, 0.1$.
- Use the numerical results to find the area function $A(x)$.
- Is it true that $A'(x) = f(x)$?

Analysing the results

Results of numerical calculations should give you the following (given to 3 d.p.).

a	2	3	4	5	10	0.5	0.2	0.1
$\int_1^a \frac{1}{x} dx$	0.693	1.099	1.386	1.609	2.303	–0.693	–1.609	–2.303

Check that $A(x) = \ln x$ i.e. $\int_1^a \frac{1}{x} dx = \ln a$

$A(x) = \ln x \Rightarrow A'(x) = \frac{1}{x} = f(x)$

i.e. the process of differentiation may be reversed to integrate $\frac{1}{x}$.

Having seen that $\int e^x \, dx = e^x + c$ and $\int \frac{1}{x} dx = \ln x + c$ we can also reverse associated results for differentiation.

$\int e^{kx} \, dx = \frac{1}{k} e^{kx} + c$ e.g. $\int e^{3x} \, dx = \frac{1}{3} e^{3x} + c$

$\int a^x \, dx = \frac{1}{\ln a} a^x + c$ e.g. $\int 10^x \, dx = \frac{1}{\ln 10} 10^x + c$

$\int a^{kx} \, dx = \frac{1}{k \ln a} a^{kx} + c$ e.g. $\int 2^{5x} \, dx = \frac{1}{5 \ln 2} 2^{5x} + c$

Example 19.4

Evaluate these definite integrals.

a) $\int_0^3 5e^{-x}\,dx$ ***b)*** $\int_0^2 (2^x - 0.5x^3)\,dx$ ***c)*** $\int_1^5 \frac{9-x^2}{x}\,dx$

Solution

a) $\int_0^3 5e^{-x}\,dx = \left[-5e^{-x}\right]_0^3 = (-5e^{-3}) - (-5e^0) = -0.2489 + 5 = 4.75$ *to 3 s.f*

b) $$\int_0^2 (2^x - 0.5x^3)\,dx = \left[\frac{1}{\ln 2}2^x - 0.125x^4\right]_0^2$$
$$= \left(\frac{1}{\ln 2}2^2 - 0.125\times 2^4\right) - \left(\frac{1}{\ln 2}2^0 - 0.125\times 0^4\right)$$
$= 3.7708 - 1.4427 = 2.33$ *to 3 s.f.*

c) $$\int_1^5 \frac{9-x^2}{x}\,dx = \int_1^5 \left(\frac{9}{x} - x\right)dx = \left[9\ln x - 0.5x^2\right]_1^5$$
$$= (9\ln 5 - 0.5\times 5^2) - (9\ln 1 - 0.5\times 1^2)$$
$= 1.9849 - (-0.5) = 2.48$ *to 3 s.f.*

Example 19.5

Find the area enclosed by the graphs $y = \frac{2}{x}$ *and* $2x + y = 5$.

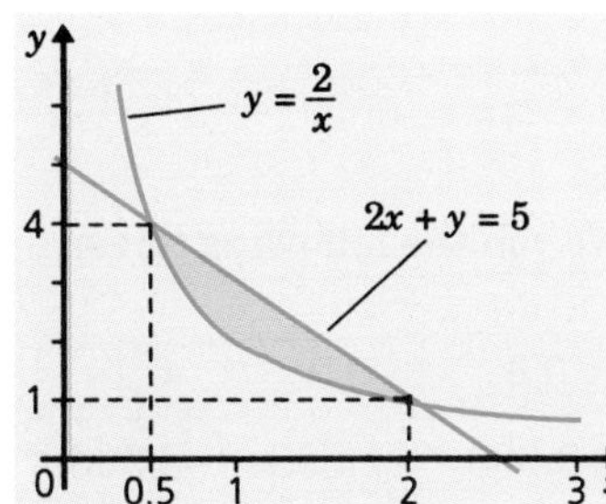

Solution

A sketch graph reveals that $y = \frac{2}{x}$ *and* $2x + y = 5$ *intersect at (0.5, 4) and (2, 1).*

The shaded area is given by:

$$\int_{0.5}^2 (5-2x)\,dx - \int_{0.5}^2 \frac{2}{x}\,dx = \int_{0.5}^2 \left(5 - 2x - \frac{2}{x}\right)dx = \left[5x - x^2 - 2\ln x\right]_{0.5}^2$$
$$= (10 - 4 - 2\ln 2) - (2.5 - 0.25 - 2\ln 0.5)$$
$= 4.6137 - 3.6363 = 0.977$ *to 3 s.f.*

EXERCISES

19.2A

1 Find the following integrals.

a) $\int 4e^x\,dx$ **b)** $\int e^{-x}\,dx$ **c)** $\int 6e^{2x}\,dx$ **d)** $\int 2e^{-3x}\,dx$

e) $\int 2^x\,dx$ **f)** $\int 0.5^{2x}\,dx$ **g)** $\int 3^{-x}\,dx$ **h)** $\int 5^{-2x}\,dx$

2 Find the following integrals.

a) $\int \frac{4}{x}\,dx$ **b)** $\int -\frac{3}{x}\,dx$ **c)** $\int \frac{x^2-9}{x}\,dx$ **d)** $\int \frac{2x-5}{x^2}\,dx$

3 Evaluate the following definite integrals.

a) $\int_0^2 0.5e^x\,dx$ **b)** $\int_{-\infty}^0 4e^x\,dx$ **c)** $\int_{-1}^1 3^x\,dx$ **d)** $\int_1^6 \frac{12}{x}\,dx$

e) $\int_1^3 (10x - e^x)\,dx$ **f)** $\int_0^5 \left(e^{-x} + \frac{x^2}{4}\right)dx$ **g)** $\int_2^6 \left(\frac{x^2-16}{x}\right)dx$ **h)** $\int_{-3}^1 2^{-x}\,dx$

4 Find the areas of the shaded regions.

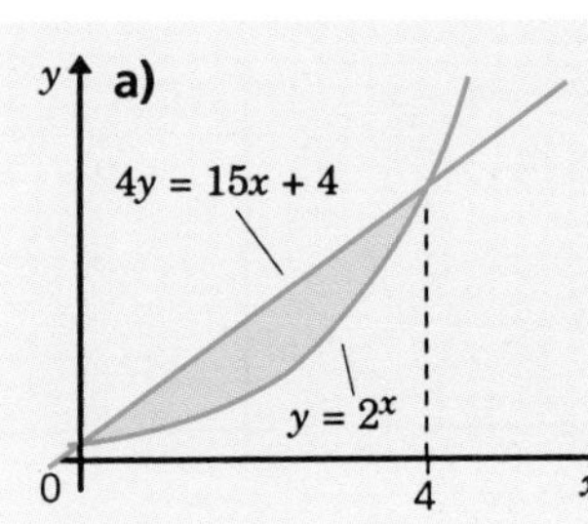

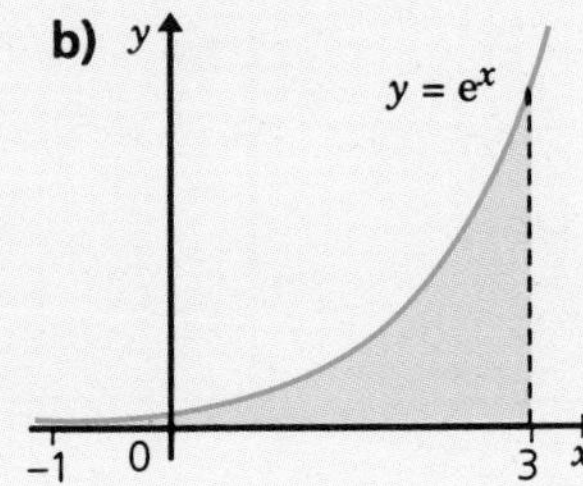

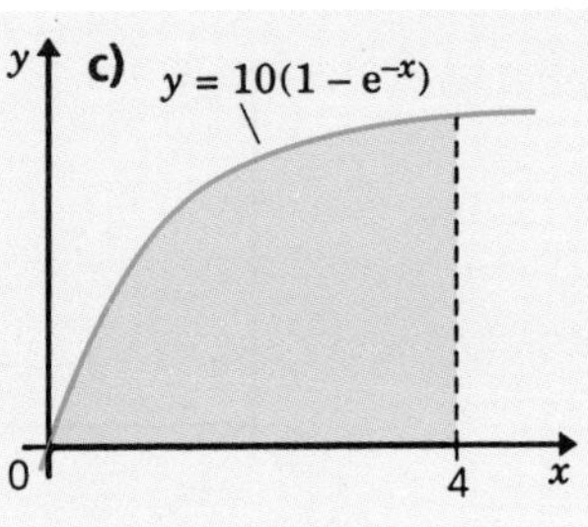

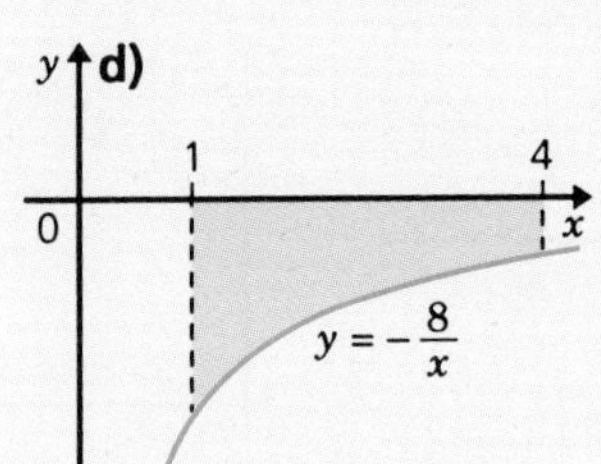

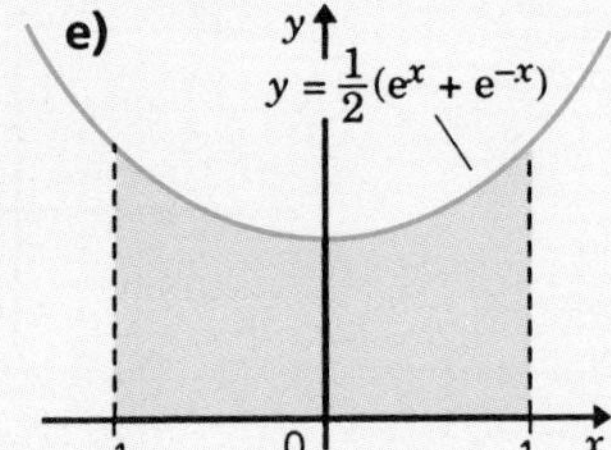

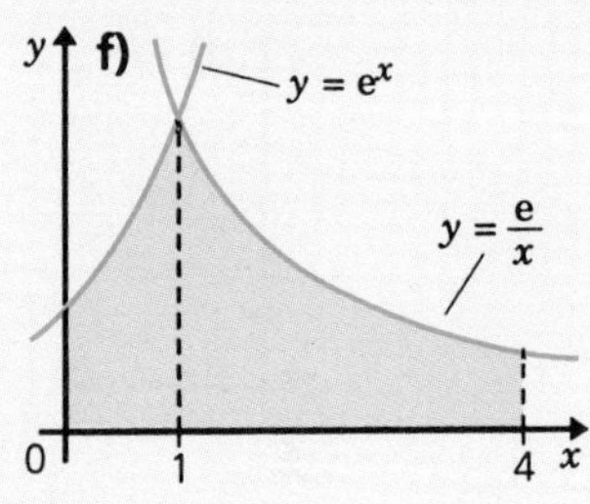

5 **a)** By writing e^{2x+3} as $e^{2x}e^3$, find $\int e^{2x+3}\,dx$.

b) Using a similar method, find $\int e^{4x-7}\,dx$.

c) Deduce $\int e^{ax+b}\,dx$ where a and b are constants.

d) Use your result to evaluate these integrals.

i) $\int_0^{0.5} e^{5x-1}\,dx$ **ii)** $\int_2^4 e^{3-x}\,dx$ **iii)** $\int_{-2}^2 \sqrt{e^x}\,dx$

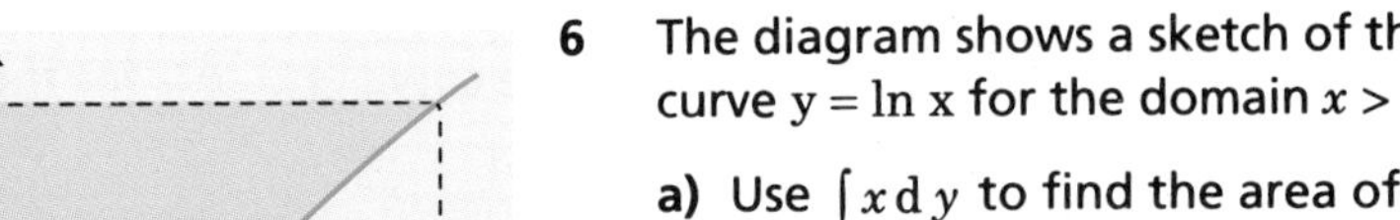

6 The diagram shows a sketch of the curve $y = \ln x$ for the domain $x > 0$.

a) Use $\int x\,dy$ to find the area of the shaded region.

b) Deduce the value of $\int_1^e \ln x\,dx$

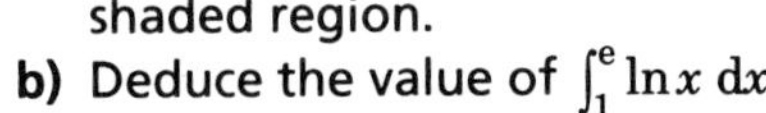

7 Find the area of the shaded region in the diagram on the right.

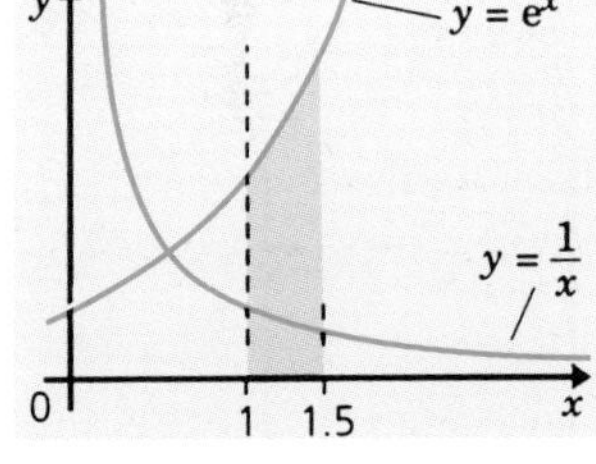

EXERCISES

19.2 B

1 Find the following integrals.

a) $\int 7e^{3x}\,dx$ **b)** $\int e^{-2x}\,dx$ **c)** $\int 12e^{-3x}\,dx$ **d)** $\int (9e^{-3x} + 15e^{-5x})\,dx$

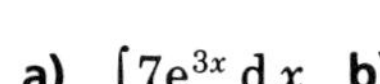

e) $\int 5^x\,dx$ **f)** $\int 0.25^{-x}\,dx$ **g)** $\int 7^{3x}\,dx$ **h)** $\int 8^{-4x}\,dx$

2 Find the following integrals.

a) $\int \frac{11}{x}\,dx$ **b)** $\int -\frac{17}{x}\,dx$ **c)** $\int \frac{13 - x^2}{x}\,dx$ **d)** $\int \frac{7x^2 + 15x - 1}{x^3}\,dx$

3 Evaluate the following definite integrals.

a) $\int_0^{1.5} 1.3e^x\,dx$ **b)** $\int_{-\infty}^0 \frac{e^x}{4}\,dx$ **c)** $\int_{-0.5}^{0.5} 5^x\,dx$ **d)** $\int_1^4 \frac{1}{3x}\,dx$

e) $\int_4^3 (x^2 - 2^x)\,dx$ **f)** $\int_0^5 \left(e^{-3x} - \frac{x^3}{3}\right)dx$ **g)** $\int_4^6 \frac{x^2 - 25}{x}\,dx$ **h)** $\int_{-1}^2 5^{-x}\,dx$

4 Find the areas of the shaded regions.

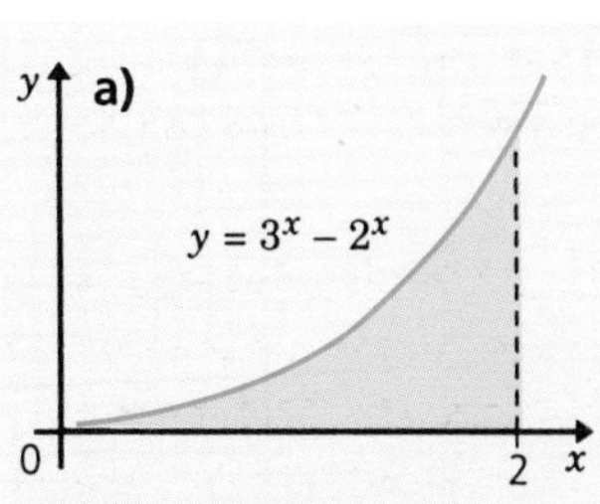

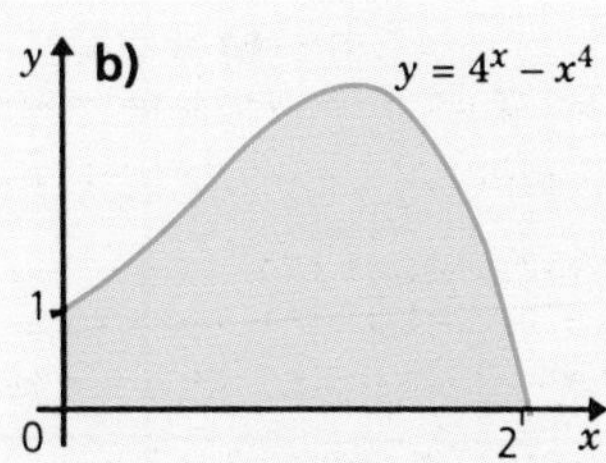

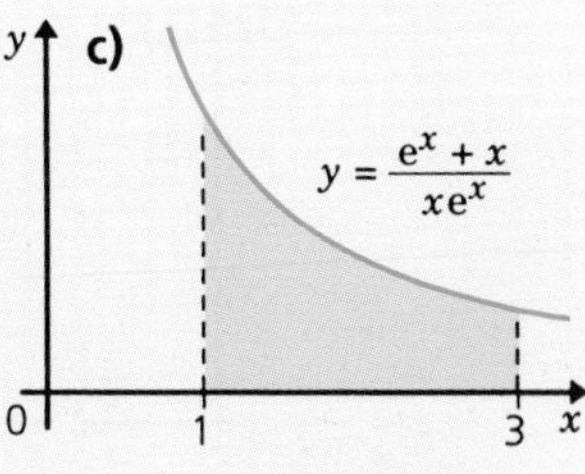

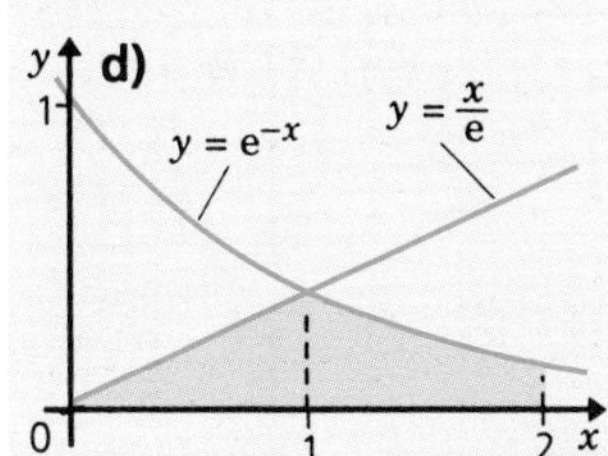

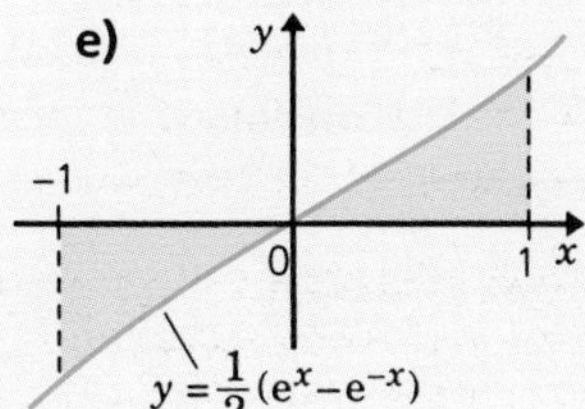

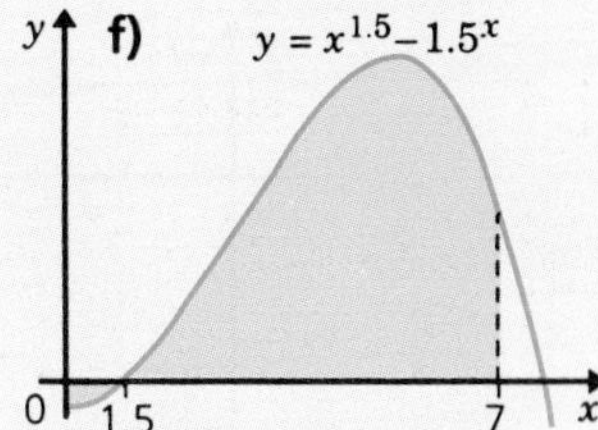

5 **a)** By writing e^{3x+5} as $e^{3x}e^5$, differentiate e^{3x+5} and use your result to find $\int e^{3x+5}\,dx$.

b) Similarly, differentiate e^{4x-11} and use your result to find $\int e^{4x-11}\,dx$.

c) Deduce $\int e^{ax+b}\,dx$ where a and b are constants.

d) Use your general result to evaluate these integrals.

i) $\int_1^2 e^{1-2x}\,dx$ **ii)** $\int_{-2}^0 \frac{1}{e^{x+1}}\,dx$ **iii)** $\int_2^4 \frac{1}{\sqrt{e^x}}\,dx$

6 The diagram on the left shows a sketch of the curve $y = \frac{1}{x+2}$ for the domain $-2 \le x \le 2$.

a) Show that $x = \frac{1}{y} - 2$.

b) Use $\int x\,dy$ to find the area of the shaded region.

c) Deduce the value of $\int_{-1}^0 \frac{1}{x+2}\,dx$

d) Evaluate $\left[\ln(x+2)\right]_{-1}^0$ and compare it with your result in **c)**.

e) Suggest an answer for the integral $\int \frac{1}{x+a}\,dx$.

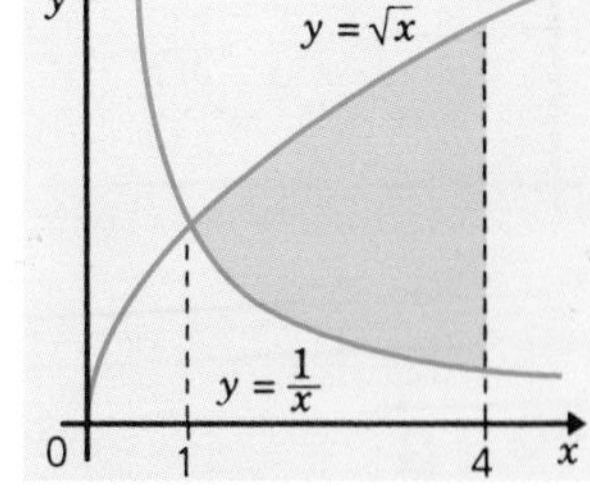

7 Find the area of the shaded region in the diagram on the right.

VOLUMES OF REVOLUTION

Exploration 19.4

Strips and slices

The diagram shows a rectangular strip under a curve $y = f(x)$, with length y and width δx.

- What is the area of the strip?
- If the strip is rotated through 360° about the x-axis, what solid is traced out?
- What is the volume of this shape?

Discs

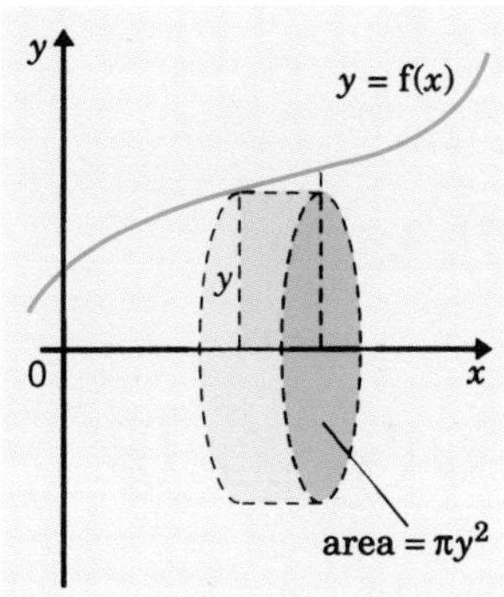

The area of the rectangular strip is $y\delta x$. When it is rotated through 360° about the x-axis a cylindrical disc is traced out with cross-sectional area πy^2 and thickness δx.

If many such strips are considered, then the integral $\int_a^b y\,dx$ can be found as the limit of the sum of such strips as $\delta x \to 0$ and the number of strips tends to infinity. This represents the area under the graph over the interval $a \le x \le b$. Similarly, if many such discs are considered, then the integral $\int_a^b \pi y^2\,dx$ can be found as the limit of the sum of such discs as $\delta x \to 0$ and represents the **volume of the solid of revolution** found by rotating the curve through 360° about the x-axis over the interval $a \le x \le b$.

By considerations of symmetry, when a curve is rotated about the y-axis over the interval $c \le y \le d$, then the volume of solid of revolution is given by $\int_c^d \pi x^2\,dy$.

Example 19.6

Find the volume generated when the curve $y = 2x(x-3)$ is rotated through 360° about the x-axis over the interval $0 \le x \le 3$. Illustrate the volume generated in a sketch.

Solution

The volume generated $= \int_0^3 \pi y^2\,dx$

$$= \int_0^3 \pi(2x(x-3))^2\,dx = 4\pi\int_0^3 x^2(x-3)^2\,dx$$
$$= 4\pi\int_0^3 (x^4 - 6x^3 + 9x^2)\,dx$$
$$= 4\pi\left[\tfrac{1}{5}x^5 - \tfrac{3}{2}x^4 + 3x^3\right]_0^3$$
$$= 4\pi\left\{\left(\tfrac{1}{5}\times 3^5 - \tfrac{3}{2}\times 3^4 + 3\times 3^3\right) - (0 - 0 + 0)\right\}$$
$$= 32.4\pi$$

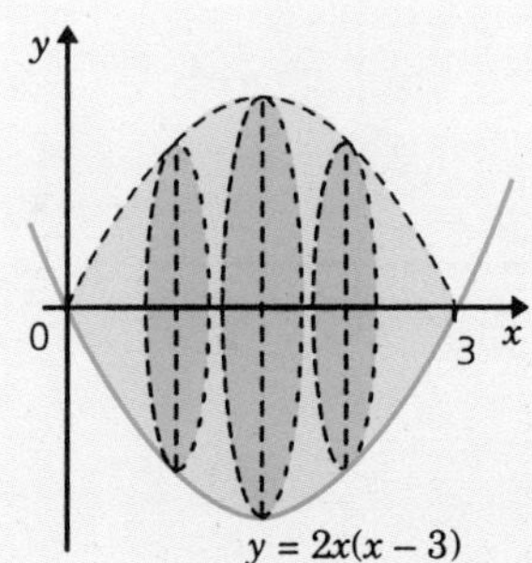

Example 19.7

Find the volume generated when the area enclosed by the graphs of $y = \frac{2}{x}$ and $2x + y = 5$ are rotated:

***a)** about the x-axis,* ***b)** about the y-axis.*

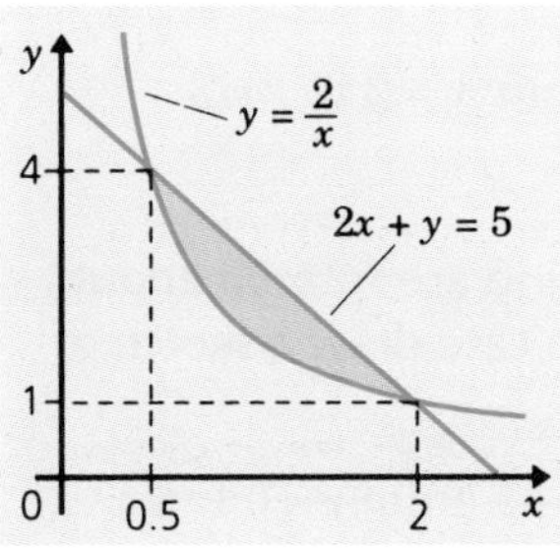

Solution

a) *Volume generated* $= \int_{0.5}^2 \pi(5-2x)^2\,dx - \int_{0.5}^2 \pi\left(\tfrac{2}{x}\right)^2\,dx$

$$= \pi\int_{0.5}^2 \left\{(5-2x)^2 - \tfrac{4}{x^2}\right\}dx = \pi\int_{0.5}^2 \left(25 - 20x + 4x^2 - \tfrac{4}{x^2}\right)dx$$
$$= \pi\left[25x - 10x^2 + \tfrac{4}{3}x^3 + \tfrac{4}{x}\right]_{0.5}^2$$
$$= \pi\left\{\left(25\times 2 - 10\times 2^2 + \tfrac{4}{3}\times 2^3 + \tfrac{4}{2}\right) - \left(25\times 0.5 - 10\times 0.5^2 + \tfrac{4}{3}\times 0.5^3 + \tfrac{4}{0.5}\right)\right\}$$
$$= \pi\left(22\tfrac{2}{3} - 18\tfrac{1}{6}\right) = 4\tfrac{1}{2}\pi$$

b) *For the line* $2x + y = 5$*, rearrange to give* $x = \frac{1}{2}(5 - y)$ *and for the curve* $y = \frac{2}{x}$*, rearrange to give* $x = \frac{2}{y}$*.*

Volume generated = $\int_1^4 \pi\left(\frac{1}{2}(5-y)\right)^2 \mathrm{d}y - \int_1^4 \pi\left(\frac{2}{y}\right)^2 \mathrm{d}y$

$$= \pi\int_1^4 \left\{\left(\tfrac{1}{2}(5-y)\right)^2 - \frac{4}{y^2}\right\}\mathrm{d}y = \pi\int_1^4\left(\tfrac{1}{4}\left(25 - 10y + y^2\right) - 4y^{-2}\right)\mathrm{d}y$$

$$= \pi\left[\tfrac{1}{4}\left(25y - 5y^2 + \tfrac{1}{3}y^3\right) + \tfrac{4}{y}\right]_1^4$$

$$= \pi\left\{\left(\tfrac{1}{4}\left(25\times 4 - 5\times 4^2 + \tfrac{1}{3}\times 4^3\right) + \tfrac{4}{4}\right) - \left(\tfrac{1}{4}\left(25\times 1 - 5\times 1^2 + \tfrac{1}{3}\times 1^3\right) + \tfrac{4}{1}\right)\right\}$$

$$= \pi\left(11\tfrac{1}{3} - 9\tfrac{1}{12}\right) = 2\tfrac{1}{4}\pi$$

CALCULATOR ACTIVITY

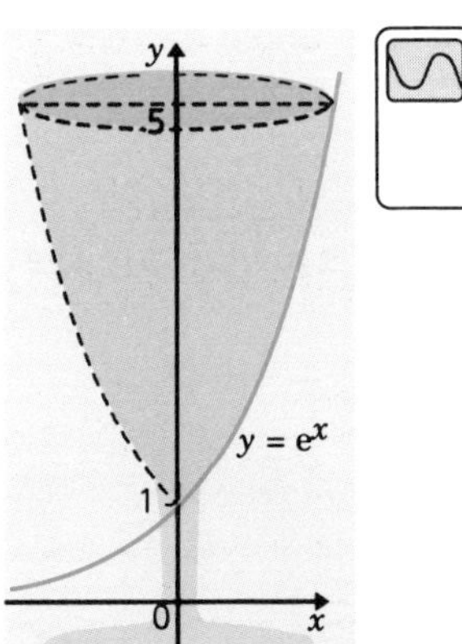

A wine glass is to be modelled by rotating the graph of $y = \mathrm{e}^x$ about the y-axis over the interval $1 \le y \le 5$, where 1 unit represents 1 cm.

- A standard 75 cl bottle of wine is to be shared equally among six people, using glasses of the type proposed in the model. Are the specifications suitable?

The volume generated = $\int_1^5 \pi x^2 \mathrm{d}y$

since $y = \mathrm{e}^x$, $x = \ln y \Rightarrow$ volume = $\int_1^5 \pi(\ln y)^2 \mathrm{d}y$

The integral cannot be found analytically without employing techniques which we have not yet met. Use a numerical method (e.g. the trapezium rule) to evaluate the integral correct to 3 s.f.

Analysing the results

Since each glass needs to hold at least 12.5 cm^3, the volume of the glass is just under 15.3 cm^3, so the specifications are suitable.

EXERCISES

19.3 A

1 Find the volumes generated when the following are rotated about the x-axis, between the bounds given. In each case draw a sketch of the resulting volume of revolution.

a) $y = 2x$ from $x = 0$ to $x = 3$ **b)** $y = 2\sqrt{x}$ from $x = 0$ to $x = 4$

c) $y = x^2$ from $x = 0$ to $x = 2$ **d)** $y = 2^x$ from $x = 2$ to $x = 3$

e) $y = 10(1 - \mathrm{e}^{-x})$ from $x = 0$ to $x = 2$

2 Find the volumes generated when the following are rotated about the y-axis, between the bounds given. In each case draw a sketch of the resulting volume of revolution.

a) $y = 16 - x$ from $y = 0$ to $y = 16$ **b)** $y = \sqrt{x - 2}$ from $y = 0$ to $y = \sqrt{3}$

c) $y = x^{\frac{2}{3}}$ from $y = 0$ to $y = 2$

3 For each of the following curves:

i) find the x-coordinates of the points where the curve cuts the x-axis,

ii) sketch the curve between these values of x,

iii) find the volume generated when the region between the curve and the x-axis is rotated about that axis.

a) $y = x(6 - x)$ b) $y = x^2 - x - 12$

c) $y = \sqrt{x}(3 - x)$ d) $y = \sqrt{x}\left(1 - \sqrt{x}\right)$

4 Find the volumes generated when the shaded areas are rotated about the x-axis.

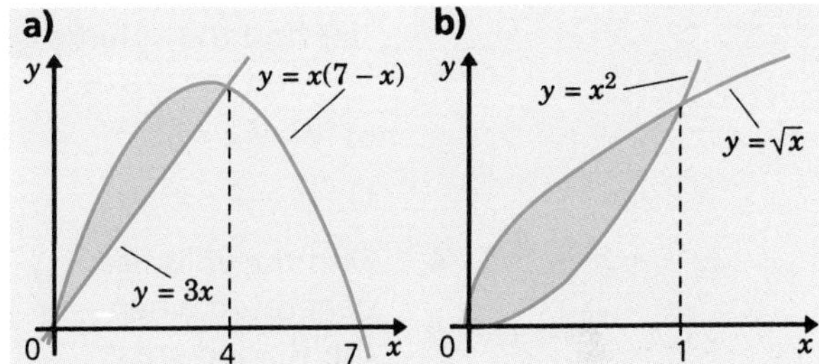

5 Find:

a) the area of the region bounded by the curve $y = \dfrac{6}{x^2}$, the x-axis and the lines $x = 1$ and $x = 2$,

b) the volume generated when this region is rotated through 360° about the x-axis.

6 Find:

a) the x-coordinates of A and B,

b) the area of the shaded region,

c) the volume generated when this area is rotated through 360° about the x-axis.

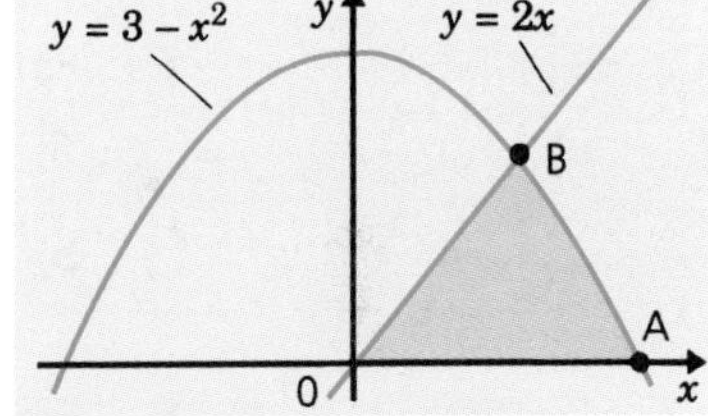

7 When the straight line shown in the diagram is rotated about the x-axis, it forms a cone of base radius a and height h.

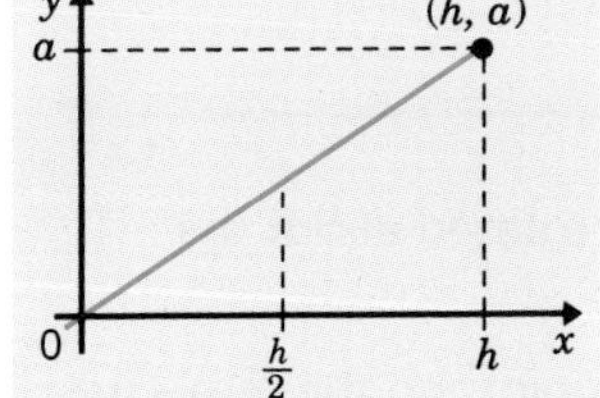

a) Find the equation of the line.

b) Show, by integration, that the volume of a cone is $\frac{1}{3}\pi a^2 h$.

c) If the region between the line, the x-axis and the lines $x = \frac{1}{2}h$ and $x = h$ is rotated about the x-axis, it generates a bucket-shaped object (known as a frustum). Show that the volume of this frustum is $\frac{7}{24}\pi a^2 h$.

d) Repeat part **c)** for a more general vertical line $x = b$.

EXERCISES

19.3 B

1 Find the volumes generated when the following are rotated about the x-axis, between the bounds given. In each case draw a sketch of the resulting volume of revolution.

a) $y = 16 - x$ from $x = -4$ to $x = +4$ b) $y = \sqrt{x - 2}$ from $x = 2$ to $x = 5$

c) $y = x^{\frac{2}{3}}$ from $x = 0$ to $x = 8$ d) $y = 3^x$ from $x = 0$ to $x = 2$

e) $y = 2 - e^x$ from $x = 0$ to $x = \ln 2$

2 Find the volumes generated when the following are rotated about the y-axis, between the bounds given. In each case draw a sketch of the resulting volume of revolution.

a) $y = 2x$ from $y = 0$ to $y = 3$
b) $y = 2\sqrt{x}$ from $y = 0$ to $y = 6$
c) $y = x^2$ from $y = 0$ to $y = 4$

3 For each of the following curves:

i) find the x-coordinates of the points where the curve cuts the x-axis,
ii) sketch the curve between these values of x,
iii) find the volume generated when the region between the curve and the x-axis is rotated about that axis.

a) $y = x(x - 5)$
b) $y = x^2(5 - x)$
c) $4y = 4 - x^2$
d) $y = \sqrt{x}\left(2\sqrt{x} - x\right)$

4 Find the volumes generated when the shaded areas are rotated about the x-axis.

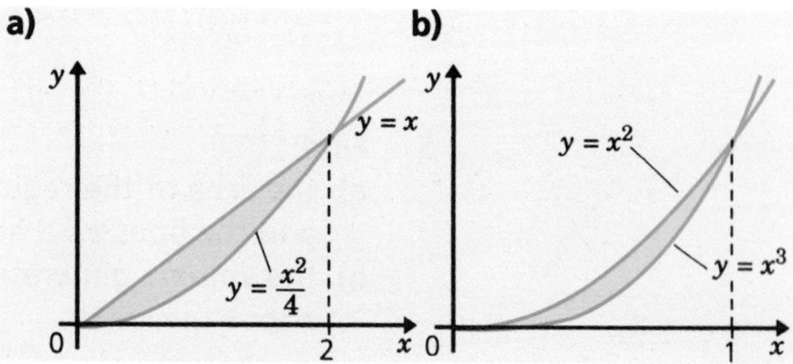

5 Find:

a) the area of the region bounded by the curve $y = 2 + \frac{3}{x}$, the x-axis and the lines $x = 3$ and $x = 6$.
b) the volume generated when this region is rotated through 360° about the x-axis.

6 Find:

a) the x-coordinate of A,
b) the area of the shaded region,
c) the volume generated when this area is rotated through 360° about the x-axis.

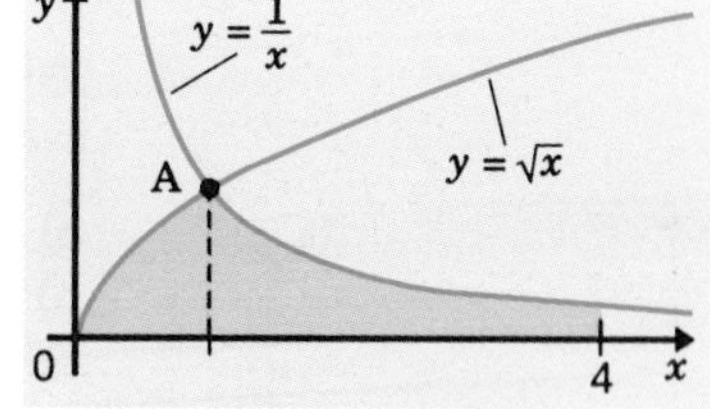

7 When the semi-circle shown in the diagram is rotated about the x-axis, it forms a sphere of radius a.

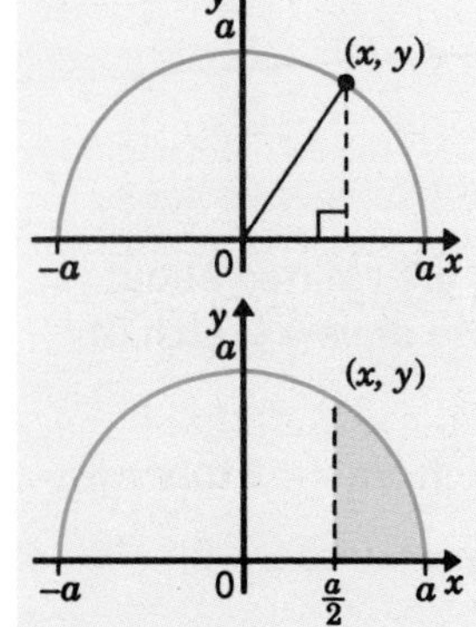

a) Using Pythagoras' theorem, verify that all points on the semi-circle satisfy the equation $y = \sqrt{a^2 - x^2}$.
b) Show, by integration, that the volume of a sphere is $\frac{4}{3}\pi a^3$.
c) If the region between the curve, the x-axis and the line $x = \frac{1}{2}a$ (on the RHS of the line) is rotated about the x-axis, show that it generates a cap-shaped object of volume $\frac{5}{24}\pi a^3$.
d) Repeat part **b)** for a more general vertical line $x = b$.

CONSOLIDATION EXERCISE FOR CHAPTER 19

1 Find the volumes generated when the following regions are rotated about the given axes.

a) $y = 2x + 1$ about the x-axis for $1 \le x \le 4$
b) $y = \frac{6}{x}$ about the x-axis from $x = 1$ to $x = 3$
c) $y = e^x$ about the x-axis from $x = 0$ to $x = 2.5$
d) $y = \frac{1}{2}(x + 4)$ about the y-axis from $y = 2$ to $y = 8$
e) $y = (x - 3)^2$ about the x-axis from $x = 0$ to $x = 3$
f) $y = \frac{8}{\sqrt{x}}$ about the x-axis between $x = 2$ and $x = 16$
g) $y = \ln x$ about the y-axis from $y = 0$ to $y = 1$
h) $y = \sqrt{4x - 1}$ $1 \le x \le 5$, about the x-axis
i) $y = \frac{\sqrt{x+1}}{x^{\frac{3}{2}}}$ $1 \le x \le 2$, about the x-axis

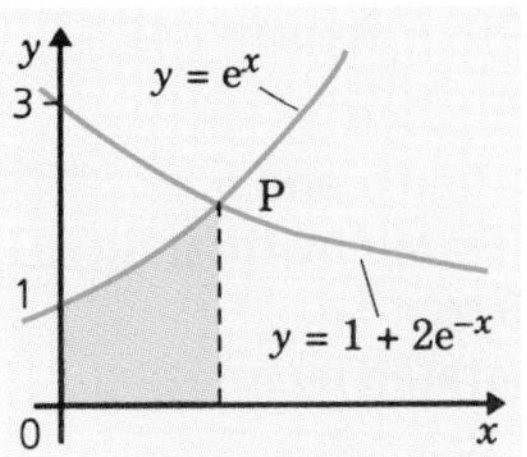

2 The diagram shows part of the curves $y = e^x$ and $y = 1 + 2e^{-x}$ and their point of intersection P.

a) Show that the x-coordinate of P is $\ln 2$.
b) Find the area of the shaded region
c) Hence find the area of the region bounded by the two curves and the y-axis.

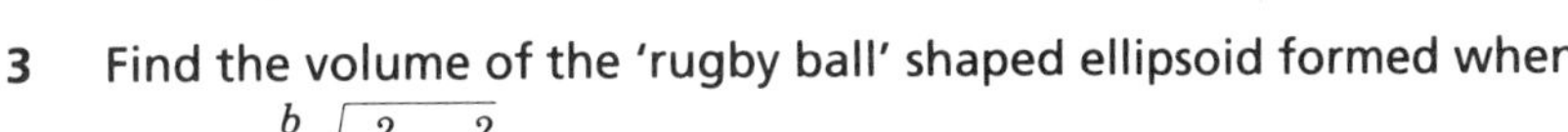

3 Find the volume of the 'rugby ball' shaped ellipsoid formed when

$$y = \frac{b}{a}\sqrt{a^2 - x^2}$$

is rotated about the x-axis. What happens when $a = b$?

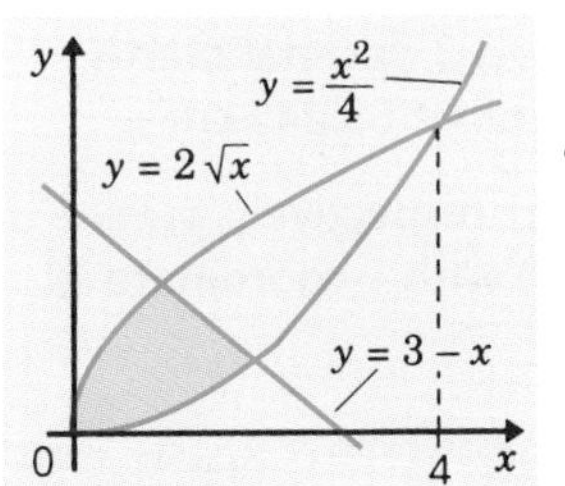

4 **a)** Find the area of the region shaded in the diagram.
b) Find the volume generated when this region is rotated through 360° about the x-axis.

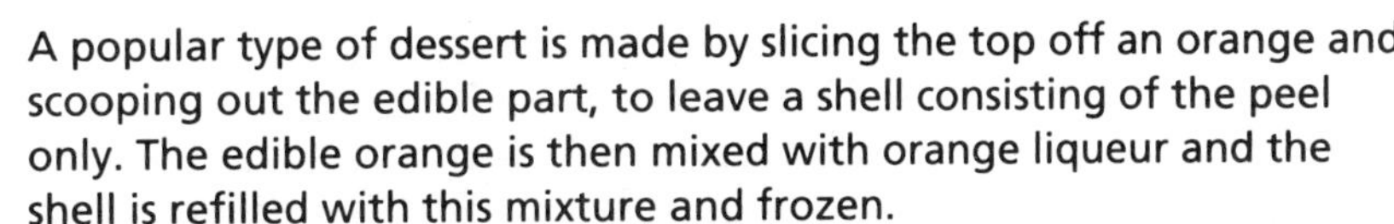

5 A popular type of dessert is made by slicing the top off an orange and scooping out the edible part, to leave a shell consisting of the peel only. The edible orange is then mixed with orange liqueur and the shell is refilled with this mixture and frozen.

Assuming you have a perfectly spherical orange of radius 5 cm and a slice of depth is 1 cm removed, this dessert can be modelled by rotating a portion of the circle shown in the diagram about the y-axis. The surface of the dessert is level with the line $y = 4$.

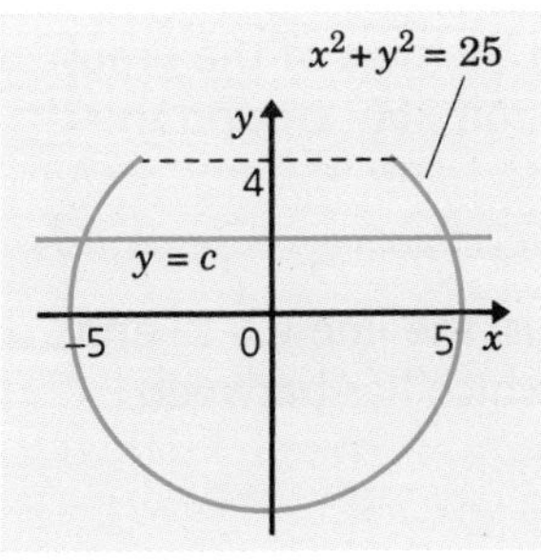

a) Use integration to find the volume of the dessert, leaving your answer in terms of π.
b) Assuming someone eating the dessert tries to keep the surface horizontal, find, in terms of π, the volume remaining when the **depth** of dessert has been halved.
c) When one-third of the dessert has been eaten, the horizontal surface is level with $y = c$ as shown in the diagram. Show that c satisfies the equation:

$$c^3 - 75c + 74 = 0$$

Use the factor theorem to find the only sensible value of c and hence the depth of dessert remaining.

d) Find a similar cubic equation that is satisfied when two-thirds of the dessert have been eaten. Using trial and improvement or an iterative method, show that c is now negative and find the depth of dessert remaining in this case, to 3 d.p.

Extension

e) Consider the same model with a more general orange of radius a where the slice removed leaves the original top of the dessert level with $y = b$.

f) Discuss the limitations of this model and suggest a better one. Investigate your improved model.

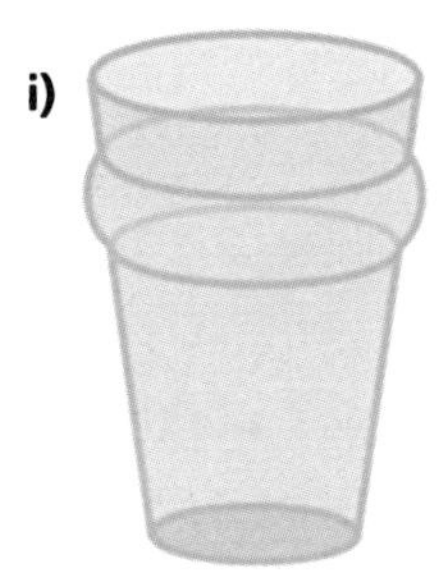

6 **Calculator exercise**: All units are in cm. Work to as many decimal places as you can, in this question.

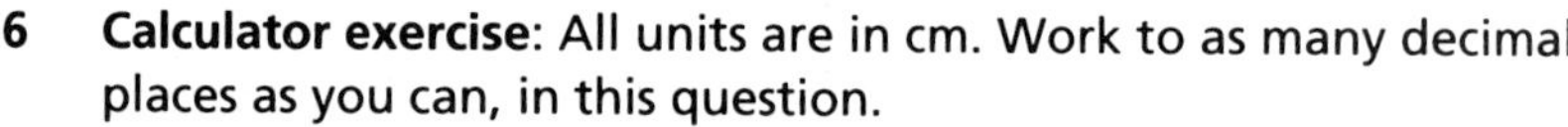

A common type of drinking glass, shown in diagram **i)** can be modelled by rotating the relevant parts of the line $y = \frac{x}{20} + 4$ and the quadratic curve $y = 4.55 - \frac{1}{2}x^2$ through 360° about the x-axis, as shown in **ii)**.

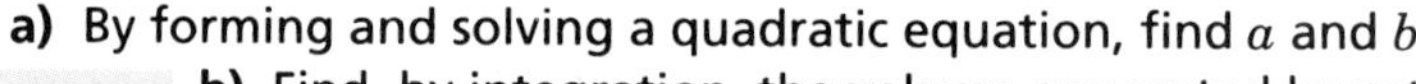

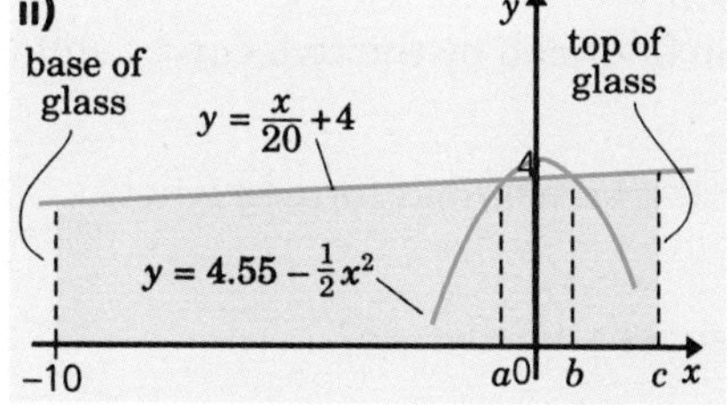

a) By forming and solving a quadratic equation, find a and b.

b) Find, by integration, the volume generated by rotating the line about the x-axis from $x = -10$ to $x = a$.

c) Find, by integration, the volume generated by rotating the curve about the x-axis from $x = a$ to $x = b$.

d) If the glass, filled to the brim, is to hold one pint, use a trial and improvement method to find c, to 3 decimal places.

e) Find the depth of liquid in the glass, to 3 decimal places, when it is holding half a pint (one pint = 568.245 cm^3).

7 Curves C_1 and C_2 have equations $y = \frac{1}{x}$ and $y = kx^2$ respectively, where k is a constant. The curves intersect at the point P, with x-coordinate $\frac{1}{2}$.

a) Determine the value of k.

b) Find the gradient of C_1 at P.

c) Calculate the area of the finite region bounded by C_1, C_2, the x-axis and the line $x = 2$, giving your answer to 2 decimal places.

(ULEAC Question 10, Specimen Paper 1, 1994)

8 **a)** Sketch the area represented by the integral $\int_2^5 x^2\,dx$ and calculate the value of the integral, showing your working.

b) Use your result from **a)** to calculate the value of $\int_4^{25} \sqrt{y}\,dy$.
Note that you will obtain no marks unless you use your result from ***a)***.

c) Calculate the volume obtained by revolving the part of the graph of $y = \sqrt{x+1}$ between $x = 3$ and $x = 5$, through 360° around the x-axis, showing your working.

(Oxford (Nuffield) Question 7, Specimen Paper 2, 1994)

9 The region R is bounded by the x-axis, the y-axis, the line $y = 12$ and the part of the curve with equation $y = x^2 - 4$ which lies between $x = 2$ and $x = 4$.

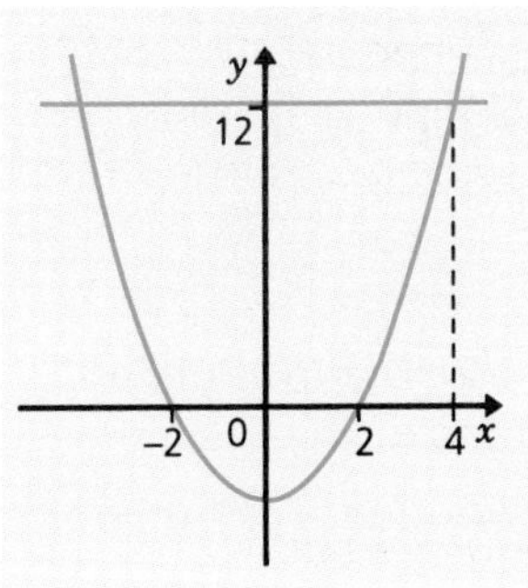

a) Copy the sketch graph and shade the region R.

The inside of a vase is formed by rotating the region R through 360° **about the y-axis**. Each unit of x and y represents 5 cm.

b) Write down an expression for the volume of revolution of the region R **about the y-axis**.
c) Find the capacity of the vase in litres.
d) Show that when the vase is filled to $\frac{5}{6}$ of its internal height it is three-quarters full.

(MEI Question 6, Paper 1, June 1992)

10 The diagram shows two curves with equations $y = x^2$ and $y^2 = x$.

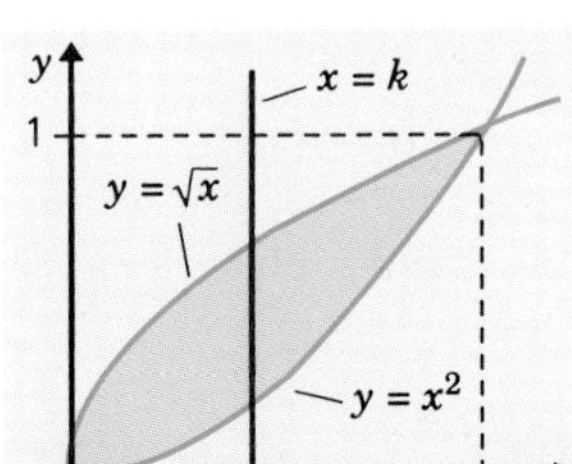

The area completely enclosed between these two curves is divided in half by the line with equation $x = k$.

a) Represent these two equal areas by two separate integrals each involving k.
b) Equate the integrals and show that k is given by the equation:

$2k^3 - 4k^{\frac{3}{2}} + 1 = 0$

c) Use the substitution p^2 for k^3 to find the value of k.

(Scottish Higher Question 7, Paper 2, 1990)

Summary

- $\int x^n \mathrm{d}x = \frac{1}{(n+1)} x^{n+1} + c \quad n \neq -1$

This result holds for all rational numbers except $n = -1$, in this case

- $\int \frac{1}{x} \mathrm{d}x = \ln x + c$
- $\int \mathrm{e}^{kx} \mathrm{d}x = \frac{1}{k} \mathrm{e}^{kx} + c$
- The **volume of revolution** about the x-axis of the area bounded by $y = \mathrm{f}(x)$, $x = a$, $x = b$ and the x-axis is given by $V = \int_a^b \pi y^2 \mathrm{d}x$

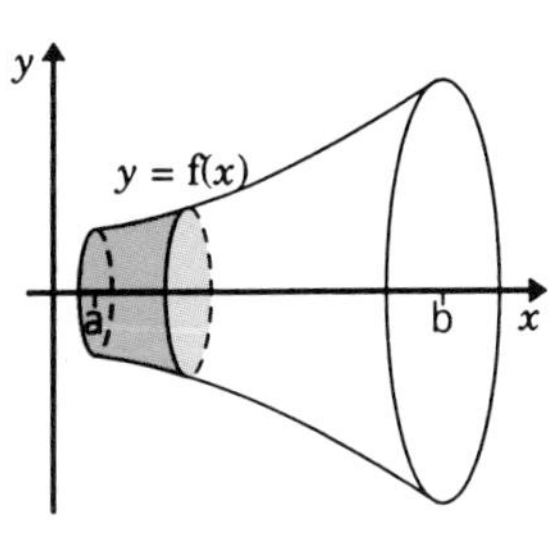

Differentiation 3

In this chapter we introduce:

- *the chain rule for differentiating composite functions (or a function of a function)*
- *rules for differentiating products and quotients of functions*
- *the differentiation of inverse functions.*

THE CHAIN RULE

Exploration 20.1

Differentiating $f(x) = (x^2 + 1)^2$

Look at the following function $f(x) = (x^2 + 1)^2$.

- Expand $f(x)$ to give a polynomial in x.
- Find $f'(x)$.
- Factorise $f'(x)$ showing that $(x^2 + 1)$ is a factor.

Repeat the exploration for $f(x) = (x^2 + 1)^3$ and $f(x) = (x^2 + 1)^4$.

- Can you identify a rule for differentiating $f(x) = (x^2 + 1)^n$?

Composite functions

The functions $(x^2 + 1)^2$, $(x^2 + 1)^3$ and $(x^2 + 1)^4$ are examples of **composite functions** or **functions of a function**. This statement can be justified by letting $y = f(x)$ and noting that $y = u^n$ where $u = x^2 + 1$ so that y is a function of u and u is a function of x.

The answers to Exploration 20.1 are summarised in this table.

$f(x)$	$(x^2+1)^2$	$(x^2+1)^3$	$(x^2+1)^4$
$f'(x)$	$4x(x^2+1)$	$6x(x^2+1)^2$	$8x(x^2+1)^3$

We can generalise y to $f(x) = (x^2 + 1)^n \Rightarrow \frac{dy}{dx} = f'(x) = 2xn(x^2 + 1)^{n-1}$ suggesting the rule:

$$\frac{dy}{dx} = \frac{dy}{du} \times \frac{du}{dx}$$

This is called the **chain rule**.

To use the chain rule we introduce a new variable u.
For example, suppose:

$$y = (4x - 3)^3$$

Then let $u = 4x - 3$ so that $y = u^3$.

Now $\frac{dy}{du} = 3u^2$ and $\frac{du}{dx} = 4$, so using the chain rule:

$$\frac{dy}{du} = (3u^2) \times 4 = 12u^2 = 12(4x - 3)^2$$

Exploration 20.2

Differentiating composite functions

Differentiate each of the following functions:

- by multiplying out the brackets and differentiating term by term,
- by using the chain rule.

a) $y = (3x + 1)^2$ **b)** $y = (1 - 2x)^3$ **c)** $y = (x^3 + 2)^2$

Does the chain rule work in each case?

Example 20.1

For the function $f(x) = e^{-x^2}$ *find:*

a) $f'(x)$ ***b)*** *the stationary points of* $f(x)$.

Hence sketch a graph of $f(x)$.

Solution

a) *Let* $u = x^2$ *then* $y = e^{-u}$.

Differentiating each function:

$$\frac{du}{dx} = 2x \text{ and } \frac{dy}{du} = -e^{-u}.$$

Applying the chain rule:

$$\frac{dy}{dx} = \frac{dy}{du} \times \frac{du}{dx} = (-e^{-u}) \times (2x)$$

Replacing u *by* x^2 *gives:*

$$f'(x) = \frac{dy}{dx} = -2xe^{-x^2}$$

b) *For a stationary value:*

$$f'(x) = -2xe^{-x^2} = 0 \quad \Rightarrow x = 0$$

since $f(0) = e^{-0} = 1$ *the only stationary point is (0, 1).*

Now examine the gradient either side of $x = 0$*; for example:*

$f'(-0.5) = e^{-0.25}$ *and* $f'(0.5) = -e^{-0.25}$.

We deduce that the stationary point is a local maximum, and since $f(x) \to 0$ *as* $x \to -\infty$ *and as* $x \to \infty$ *the graph of* $f(x)$ *is bell-shaped as shown in the diagram.*

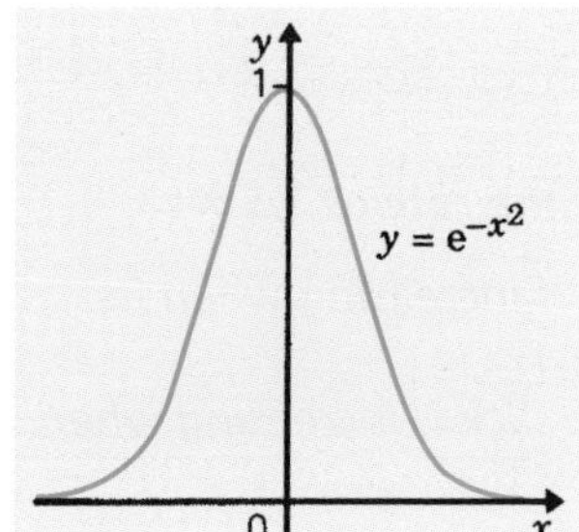

Note: This is an important function in statistics; it is used in defining the normal probability distribution function.

Example 20.2

Find $\frac{dy}{dx}$ *for the function* $y = \frac{1}{(x^3 - 1)^4}$.

Solution

Let $u(x) = x^3 - 1$ *then* $y = \frac{1}{u^4} = u^{-4}$

Differentiating each function:

$$\frac{du}{dx} = 3x^2 \text{ and } \frac{dy}{du} = -4u^{-5}$$

Applying the chain rule:

$$\frac{dy}{dx} = \frac{dy}{du} \times \frac{du}{dx} = (-4u^{-5}) \times (3x^2) = -\frac{12x^2}{u^5}$$

Replacing u by $x^3 - 1$ gives:

$$\frac{dy}{dx} = -\frac{12x^2}{(x^3 - 1)^5}$$

Rates of change

From Chapter 7, *Differentiation 1*, we know that the rate of change of a variable is given by $\frac{dy}{dt}$ where y is a function of time.

Sometimes y is a function of a variable other than t, for example, $y = f(s)$. Then we can use the chain rule to give:

$$\frac{dy}{dt} = \frac{dy}{ds} \times \frac{ds}{dt}$$

Example 20.3

The velocity of an object v m s^{-1} is given by:

$$v = 1 + s^2$$

where s is its displacement from a fixed point, in metres. Find the acceleration of the object when $s = 3$ m.

Solution

Acceleration a is the rate of change of velocity with time.
Applying the chain rule:

$$a = \frac{dv}{dt} = \frac{dv}{ds} \times \frac{ds}{dt} = v\frac{dv}{ds}$$

since $\frac{ds}{dt} = v$, *the velocity, and* $v = 1 + s^2$,

$$\frac{dv}{ds} = 2s \Rightarrow a = (1 + s^2) \times (2s)$$

when $s = 3$, $a = (1 + 9) \times (6) = 60\,\text{m s}^{-2}$.

Example 20.4

The radius of a circular oil slick is increasing at the rate of 1.5 m s^{-1}.

a) *Find the rate at which the area of the slick is increasing when its radius is 20 m.*
b) *Find the rate at which the perimeter of the slick is increasing when its radius is 20 m.*

Solution

a) *Area of circular oil slick of radius r is* $A = \pi r^2$.

Differentiating, $\frac{dA}{dr} = 2\pi r = 40\pi$ *(when* $r = 20\,\text{m}$*)*

The rate at which the area is changing is:

$$\frac{dA}{dt} = \frac{dA}{dr} \times \frac{dr}{dt} \text{ (the chain rule)}$$

and $\frac{dr}{dt} = 1.5\,\text{m s}^{-1}$

So $\frac{dA}{dt} = (40\pi) \times (1.5) = 60\pi\ \text{m}^2\,\text{s}^{-1}$

b) *Perimeter of circular oil slick of radius r is* $C = 2\pi r$.

$$\frac{\mathrm{d}C}{\mathrm{d}t} = \frac{\mathrm{d}C}{\mathrm{d}r} \times \frac{\mathrm{d}r}{\mathrm{d}t} \text{ (the chain rule)}$$

$$\Rightarrow \frac{\mathrm{d}C}{\mathrm{d}r} = (2\pi)\times(1.5) = 3\pi \text{ m s}^{-1}.$$

Example 20.5

A metal rod, of circular cross-section, is being heated and is expanding so that the volume of the rod increases at a rate of 200 cm³ s⁻¹. After t seconds the length of the rod is ten times the radius of the rod. Find the rate of increase of the radius when the rod is 60 cm long.

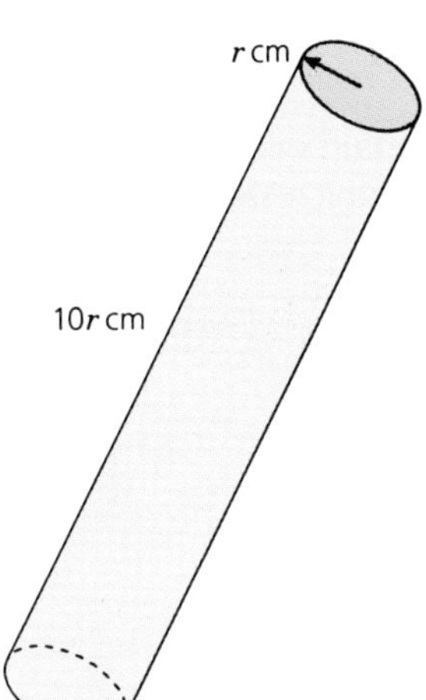

Solution

Let the radius of the rod be r cm at time t, then the length of the rod is $10r$ *cm. The volume of the rod* $V = (\pi r^2) \times 10r = 10\pi r^3$ *cm³.*

$$\frac{\mathrm{d}V}{\mathrm{d}t} = 200 \text{ cm}^3 \text{ s}^{-1} \text{ (given in the question)}$$

with $V = 10\pi r^3$

$$\Rightarrow \frac{\mathrm{d}V}{\mathrm{d}t} = \frac{\mathrm{d}V}{\mathrm{d}r} \times \frac{\mathrm{d}r}{\mathrm{d}t} \text{ (using the chain rule)}$$

$$\Rightarrow \frac{\mathrm{d}V}{\mathrm{d}t} = 30\pi r^2 \times \frac{\mathrm{d}r}{\mathrm{d}t}$$

Hence the rate of increase of the radius is:

$$\frac{\mathrm{d}r}{\mathrm{d}t} = \frac{1}{30\pi r^2} \times \frac{\mathrm{d}V}{\mathrm{d}t}$$

When the length of the rod is 60 cm its radius is 6 cm.

So $\frac{\mathrm{d}r}{\mathrm{d}t} = \frac{1}{30\pi(6)^2} \times 200 = 0.059 \text{ cm s}^{-1}$ *(to 2 s.f.).*

EXERCISES

20.1 A

1 Differentiate these expressions.

a) $(2x-1)^5$ **b)** $(x^2+1)^4$ **c)** $\frac{1}{4x+1}$ **d)** $\frac{1}{x^3 - x^2}$ **e)** $\sqrt{x-2}$

f) $\ln(x^2-1)$ **g)** e^{x^2+2x} **h)** $e^{\sqrt{x}}$ **i)** $(e^x + x)^4$ **j)** $\ln(4x-1)^3$

2 A curve has equation $y = e^{x^2}$. Find the coordinates of the point when the gradient is zero and sketch the curve.

3 A curve has equation $y = \left(x + \frac{1}{x}\right)^2$ for $x > 0$.

a) Find the gradient of the curve at the points (0.5, 6.25) and (2, 6.25). Decide if the graph is increasing or decreasing at each point.

b) Find and classify the stationary point and sketch the curve.

4 The velocity, $v \text{ m s}^{-1}$, of an object is given by:

$$v = \frac{4}{1+2s}$$

where s is its displacement from a fixed point, in metres. Find the acceleration of the object when:

a) $s = 0$ **b)** $s = 1\text{ m}$ **c)** $s = 10\text{ m}$.

5 A spherical balloon is being blown up so that its radius increases at a constant rate of $0.01\,\mathrm{m\,s^{-1}}$.

a) Find the rate of increase of its volume when the radius is 0.2 m.
b) Find the rate of increase of the surface area of the balloon when the radius is 0.2 m.

6 The pressure P ($\mathrm{N\,m^{-2}}$) and volume V ($\mathrm{m^3}$) of a gas obey Boyle's law $P = \frac{c}{V}$ where c is a constant. In an experiment the volume of a quantity of gas is modelled by $V = 0.3\mathrm{e}^{-4t} + 0.2$ at time t. Find the rate of change of the pressure when:

a) $t = 0\,\mathrm{s}$ **b)** $t = 2\,\mathrm{s}$ **c)** $t = 100\,\mathrm{s}$.

7 Wine is spilled onto a carpet forming a circular stain which increases in area at a rate of $150\,\mathrm{mm^2\,s^{-1}}$. Find the rate at which the radius is changing when the area of the stain is $1200\,\mathrm{mm^2}$.

EXERCISES

20.1 B

1 Differentiate these expressions.

a) $(3x + 2)^4$ **b)** $(t^3 - 1)^2$ **c)** $\frac{1}{s-7}$ **d)** $\frac{1}{(2t+1)^3}$

e) $\frac{1}{4x^3 + x^2}$ **f)** $\mathrm{e}^{x^3-2x^2}$ **g)** $\mathrm{e}^{-2\sqrt{x}}$ **h)** $\frac{1}{\sqrt{\mathrm{e}^x - x^2}}$

i) $\ln(2s + 3)^4$ **j)** $\ln(4x^2 + 3x)$

2 A curve has equation $y = \sqrt{x^2 + 3x}$.

a) Find the equation of the tangent to the curve at the point (1, 2). Is the graph increasing or decreasing at this point?
b) Find the coordinates of any stationary points that exist for this curve.
c) For what values of x does the curve exist?
d) Find the slope of the tangent to the curve at the points (0, 0) and (–3, 0).
e) Sketch the curve.

3 Find any stationary points on the following curves.

a) $y = \mathrm{e}^{x^2-1}$ **b)** $y = \ln(x^2 + 1)$ **c)** $y = \left(x - \frac{1}{x}\right)^4$ **d)** $y = \frac{1}{\sqrt{x^2 - 2x + 4}}$

4 The radius of a circular ink blot is increasing at a rate of $0.5\,\mathrm{cm\,s^{-1}}$. Find the rate of increase of the area of the ink blot when its radius is 3 cm.

0.6 m
1.6 m
0.7 m

5 A domestic bath is modelled as a rectangular tank with dimensions as shown in the diagram. The bath is being filled by a mixer tap which delivers water at a rate of $0.07\,\mathrm{m^3\,min^{-1}}$. At time t the depth of water in the bath is h m. How fast is h changing?

6 A current I amps flows through a resistance R ohms. The power developed is given by $P = I^2R$. Find the rate of change of power for a resistance of 20 ohms if:

$I = 5 + 2\mathrm{e}^{-3t}$

where t is the time in seconds.

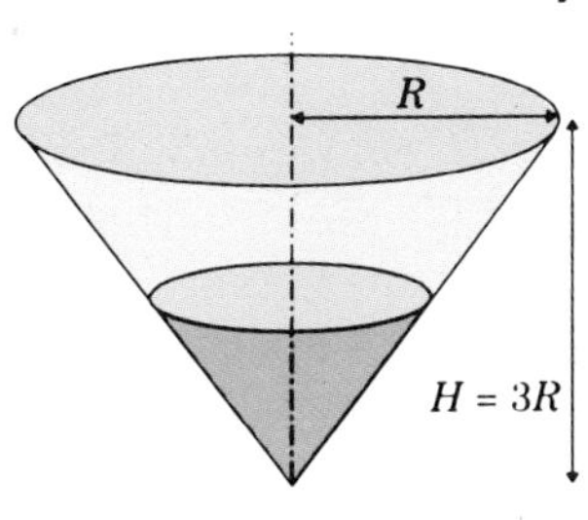

7 A cone has a vertical height, H, which is three times its base radius, R. The cone is supported with its axis of symmetry vertical as shown in the diagram. The cone is being filled with oil at the rate of 0.01 litres per second. At time t the depth of oil in the cone is h cm.

a) Show that the volume of oil in the cone at time t is:

$$V = \frac{\pi}{27}h^3$$

b) Find an expression for the rate of change of h in cm s^{-1}.

THE PRODUCT RULE

Exploration 20.3

Differentiating products of functions

Look at the function $y = \mathrm{f}(x) = x^2(x + 1)^2$ which is a product of the two functions $u = x^2$ and $v = (x + 1)^2$.

- Differentiate $\mathrm{f}(x)$ by multiplying out the brackets and differentiating term by term.
- Is it true that $\frac{dy}{dx} = \frac{du}{dx} \times \frac{dv}{dx}$?
- Can you identify a rule for differentiating a product of two functions?

Repeat these steps for:

a) $\mathrm{f}(x) = x^3(2x - 1)^2$
b) $\mathrm{f}(x) = x^4(x^2 + x - 1)$.

The derivative of a product of two functions

From Exploration 20.3 we find that the derivative of a product of two functions does **not** equal the product of the derivative of the two functions.

To develop a rule, let $u = \mathrm{p}(x)$ and $v = \mathrm{q}(x)$. Consider the product $y = uv$ or $\mathrm{f}(x) = \mathrm{p}(x)\mathrm{q}(x)$ and the basic definition of differentiation given by:

$$\mathrm{f}'(x) \text{ is the limit of } \frac{\mathrm{f}(x+h) - \mathrm{f}(x)}{h} \text{ as } h \to 0.$$

For $\mathrm{f}(x) = \mathrm{p}(x)\mathrm{q}(x)$:

$$\begin{aligned} \mathrm{f}(x+h) - \mathrm{f}(x) &= \mathrm{p}(x+h)\mathrm{q}(x+h) - \mathrm{p}(x)\mathrm{q}(x) \\ &= [\mathrm{p}(x+h) - \mathrm{p}(x)]\mathrm{q}(x+h) + \mathrm{p}(x)[\mathrm{q}(x+h) - \mathrm{q}(x)] \end{aligned}$$

these two extra terms cancel

Dividing both sides by h gives:

$$\frac{\mathrm{f}(x+h) - \mathrm{f}(x)}{h} = \left[\frac{\mathrm{p}(x+h) - \mathrm{p}(x)}{h}\right]\mathrm{q}(x+h) + \mathrm{p}(x)\left[\frac{\mathrm{q}(x+h) - \mathrm{q}(x)}{h}\right]$$

and taking the limit as $h \to 0$ gives the **product rule**:

$$\mathrm{f}'(x) = \mathrm{p}'(x)\,\mathrm{q}(x) + \mathrm{p}(x)\,\mathrm{q}'(x)$$

$$\text{or } \frac{dy}{dx} = \frac{du}{dx}v + u\frac{dv}{dx}$$

Example 20.6

Find $\frac{dy}{dx}$ for the function $y = x^2(x+1)^2$.

Solution

Here $u = x^2$ and $v = (x+1)^2$ so:

$$\frac{du}{dx} = 2x \text{ and } \frac{dv}{dx} = 2(x+1)$$

Apply the product rule:

$$\frac{dy}{dx} = \frac{du}{dx}v + u\frac{dv}{dx} = 2x \times (x+1)^2 + x^2 \times 2(x+1)$$
$$= 2x(x+1)[(x+1)+x] = 2x(x+1)(2x+1)$$

This should agree with the result from Exploration 20.3.

Example 20.7

For the function $f(x) = xe^{-2x}$:

a) *find $f'(x)$,*
b) *find and classify the stationary points of $f(x)$,*
c) *sketch a graph of $y = xe^{-2x}$.*

Solution

a) $u = x$ *and* $v = e^{-2x}$

$$\Rightarrow f'(x) = \frac{du}{dx}v + u\frac{dv}{dx} = 1 \times e^{-2x} + x \times -2e^{-2x} = (1-2x)e^{-2x}$$

b) *For a stationary point:*

$$f'(x) = (1-2x)e^{-2x} = 0$$
$$\Rightarrow 1 - 2x = 0$$
$$\Rightarrow x = \tfrac{1}{2}$$

There is one stationary point at $(\frac{1}{2}, \frac{1}{2}e^{-1})$.
Now investigate the gradient either side of the stationary point.

$$f'(0) = e^0 = 1 \text{ and } f'(1) = -e^{-2}$$

so the stationary point is a local maximum.

c) *The graph of $y = xe^{-2x}$ passes through the origin and $y \to 0$ as $x \to \infty$ and $y \to -\infty$ as $x \to -\infty$.*

The graph of $f(x) = xe^{-2x}$ is shown in the diagram.

EXERCISES

1 Differentiate these expressions.

a) $(x-1)(x^2+1)$ **b)** $\sqrt{x}(1+x)$ **c)** $(x^2-1)(2x+1)^2$ **d)** $x(x-1)^{-1}$

e) $(x+1)(x-1)^3$ **f)** $(t^2+2)(t^3-1)$ **g)** $\sqrt{(s+1)(s-1)^2}$ **h)** u^2e^{-3u}

i) $(x-1)^2e^x$ **j)** $(2x+3)e^{-x^2}$ **k)** $(2x+1)\ln(2x+1)$ **l)** $e^{-x}\ln x$

2 A curve has equation $y = (x^2-2)e^{-2x}$.

a) Find the equation of the tangent at the point (0, –2). Is the curve increasing or decreasing at this point?

b) Find and classify the stationary points of the curve.

3 A curve has equation $y = x^3 e^{x^2}$. Find the equation of the tangent to the curve at the points (0, 0), (1, e^1) and (–1, $-e^1$).

Is the curve increasing or decreasing at each point? Classify the stationary point at the origin.

4 An object moves along the x-axis so that its position at time $t \geq 0$ is modelled by $x = (3 + 2t)e^{-0.25t}$.

a) Find a formula for the velocity of the object.
b) For what values of t is the velocity positive? Explain what a negative velocity means.
c) At what instant of time is the velocity greatest in magnitude? What is the velocity at this time?
d) What happens to the object as t increases?

5 A rectangular beam is to be cut from a log of cylindrical cross-section of radius 0.4 m.

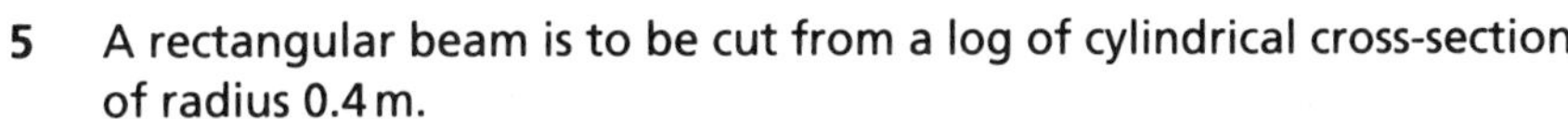

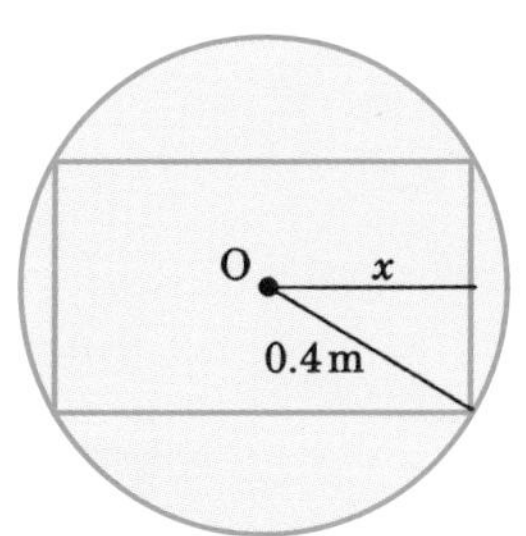

a) If x is half of the width of the beam (see diagram) show that the cross-sectional area of the beam is:

$A = 4x\sqrt{0.16 - x^2}$

b) What is the largest possible cross-sectional area of the beam?

20.2 B

1 Differentiate these expressions.

a) $(2x+1)(x-1)$ **b)** $(3t^2-5)(2t+3)$ **c)** $(s+3)(s^3-2s-4)$ **d)** $x^{-3}(2x+1)$

e) $s\sqrt{s^2+1}$ **f)** $u e^{-u^2}$ **g)** $(1+3v^2)e^{-2v}$ **h)** $(1+x)^{-2}(3x^2+x-1)$

i) $(2-3x)\ln(0.5x-2)$ **j)** $\left(t-\frac{1}{t}\right)e^{t^2}$ **k)** $e^{3x}\ln(2x-1)$ **l)** $(t^2+2t-1)^{\frac{1}{3}}(5t-4)$

2 Find the product rule for the derivative of a function of three functions $f(x) = u(x)v(x)w(x)$.
Apply your rule to differentiate the following expressions.

a) $(x^2-1)(x+1)(2x^3-x^2)$ **b)** $(2x+1)(3x^2+1)e^{-2x}$
c) $(x^2+x)e^{-3x}\ln(x+1)^2$

3 Find and classify the stationary points of the curve with equation $y = x^3(x^2-1)^2$.

4 A curve has equation $y = (x^2-1)\ln(2x+3)$.

a) Find the equation of the tangent to the curve at the points $(0, -\ln 3)$ and $(1.5, 1.25\ln 6)$. Is the curve increasing or decreasing at these points?
b) Find and classify the stationary points of the curve.
c) Sketch the curve for values of x between –2 and 2 for which the function exists.

5 The radius r m of an oil slick on a lake t hours after it was first noticed is given by:

$$r = 0.6 + (1.1 + 2t^2)e^{-0.5t}$$

a) Find the maximum radius of the slick and the time at which this occurs.
b) What are the initial and final radii of the oil slick?

THE QUOTIENT RULE

Exploration 20.4

Quotients of functions

Look at the function $f(x) = \dfrac{x+2}{x-1}$ which is a quotient of two functions $u(x) = x + 2$ and $v(x) = x - 1$.

- Writing the function as $y = (x + 2)(x - 1)^{-1}$ use the product rule to find $\dfrac{dy}{dx}$.
- Is it true that $\dfrac{dy}{dx} = \dfrac{\frac{du}{dx}}{\frac{dv}{dx}}$?

Repeat these steps for: **a)** $y = \dfrac{x}{e^x}$ **b)** $y = \dfrac{e^x}{x}$.

The derivative of a quotient

From Exploration 20.4 we should see that the derivative of a quotient is somewhat different from the ratio of the derivatives of u and v.

To develop a rule consider the quotient $y = \dfrac{u}{v}$ written as a product:

$$y = uv^{-1}$$

Apply the product rule:

$$\frac{dy}{dx} = \frac{du}{dx}v^{-1} + u\frac{d}{dx}\left(v^{-1}\right)$$

To differentiate v^{-1} we apply the chain rule:

$$\frac{d}{dx}\left(v^{-1}\right) = \frac{d}{dv}\left(v^{-1}\right) \times \frac{dv}{dx} = -v^{-2}\frac{dv}{dx}$$

Thus $\dfrac{dy}{dx} = \dfrac{du}{dx}v^{-1} - uv^{-2}\dfrac{dv}{dx}$

$$\Rightarrow \quad \frac{dy}{dx} = \frac{v\frac{du}{dx} - u\frac{dv}{dx}}{v^2}$$

This is the **quotient rule**.

Example 20.8

Find $\dfrac{dy}{dx}$ for the function $y = \dfrac{x+2}{x-1}$.

Solution

Let $u = x + 2$ and $v = x - 1$ then:

$\dfrac{du}{dx} = 1$ *and* $\dfrac{dv}{dx} = 1$

Apply the quotient rule:

$$\frac{dy}{dx} = \frac{v\frac{du}{dx} - u\frac{dv}{dx}}{v^2} = \frac{(x-1)(1) - (x+2)(1)}{(x-1)^2} = \frac{-3}{(x-1)^2}$$

This should agree with the result from Exploration 20.4.

Example 20.9

For the function $y = f(x) = \frac{x}{1+x^2}$:

a) *find* $f'(x)$,

b) *find and classify the stationary points of* $f(x)$,

c) *sketch a graph of* $y = \frac{x}{1+x^2}$.

Solution

a) $u = x$ *and* $v = 1 + x^2 \Rightarrow \frac{du}{dx} = 1$ *and* $\frac{dv}{dx} = 2x$

$$f'(x) = \frac{dy}{dx} = \frac{v\frac{du}{dx} - u\frac{dv}{dx}}{v^2} = \frac{(1+x^2)(1) - x(2x)}{(1+x^2)^2} = \frac{1-x^2}{(1+x^2)^2}$$

b) *For a stationary point:*

$$f'(x) = \frac{1-x^2}{(1+x^2)^2} = 0$$

$\Rightarrow 1 - x^2 = 0 \Rightarrow x = -1$ *or* $x = 1$

There are two stationary points at $(-1, -\frac{1}{2})$ *and* $(1, \frac{1}{2})$.

Now we investigate the gradient either side of each stationary point.

i) *Close to* $(-1, -\frac{1}{2})$: $f'(-2) = -\frac{3}{25}$ *and* $f'(0) = 1$

The point $(-1, -\frac{1}{2})$ *is a local minimum.*

ii) *Close to* $(1, \frac{1}{2})$: $f'(0) = 1$ *and* $f'(2) = -\frac{3}{25}$

The point $(1, \frac{1}{2})$ *is a local maximum.*

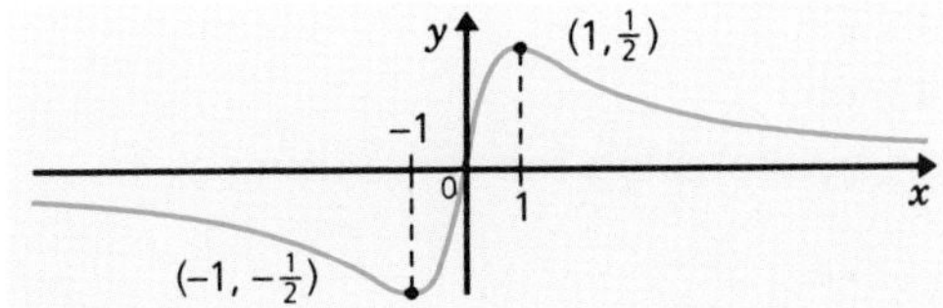

c) *The graph of* $y = \frac{x}{1+x^2}$ *passes through the origin,*

$y \to 0$ *as* $x \to \pm\infty$.

The graph of $f(x) = \frac{x}{1+x^2}$ *is shown in the diagram.*

EXERCISES

1 Differentiate these expressions.

a) $\frac{(x-4)^2}{x}$ **b)** $\frac{t^2}{t+3}$ **c)** $\frac{2-x}{x^2}$ **d)** $\frac{t^{\frac{5}{4}}}{4t-3}$

e) $\frac{x^2+5x+4}{2x^2-1}$ **f)** $\frac{t+t^{-1}}{t-t^{-1}}$ **g)** $\frac{e^{-2x}}{1+x}$ **h)** $\frac{1-t^2}{\ln(t+1)}$

i) $\frac{e^{x^2}}{1-x}$ **j)** $\frac{\ln(2t-3)}{2t-3}$ **k)** $\frac{x^3+x^2+x+1}{x+1}$ **l)** $\frac{\ln t}{t^3-2}$

2 A curve has equation $y = \frac{x^2-1}{2x+4}$.

a) Find and classify the stationary points of the curve.

b) Sketch the curve using a graphical calculator and confirm your answer to **a)**.

3 Find and classify the stationary points of the curve with equation:

$$y = \frac{e^x}{1+x^2}.$$

4 A simple model for the flow rate, $f(v)$ (s^{-1}), of cars along a straight level road is: $$f(v)=\frac{v}{4+0.7v^2}$$

where v is the speed of the cars in m s^{-1}. Find the speed of the cars which gives a maximum flow rate. (Show that this speed does give a maximum value.)

5 A curve has equation $y=\frac{e^{2x}-e^{-2x}}{e^{2x}+e^{-2x}}$. Show that $\frac{dy}{dx}=2(1-y^2)$.

Hence show that the curve has no stationary points.

1 Differentiate these expressions.

a) $\frac{(x+2)^2}{x^3}$ **b)** $\frac{5u}{(1-u)^3}$ **c)** $\frac{2x^2}{x-2}$ **d)** $\frac{s^{\frac{7}{3}}}{3s-1}$

e) $\frac{x^2-3x+2}{x^2+1}$ **f)** $\frac{s+2}{s^2-2s+1}$ **g)** $\frac{e^{4x}}{3-2x}$ **h)** $\frac{1+x+x^2}{\ln x}$

i) $\frac{e^{-2x^2}}{2+3x}$ **j)** $\frac{\ln(x-1)}{x-1}$ **k)** $\frac{e^{-x}}{1+e^{-2x}}$ **l)** $\frac{\sqrt{1-s}}{s+2}$

2 Find and classify the stationary points of the curve with equation

$$y=\frac{x^2}{x^2+4}$$

Sketch the curve, using graphical calculator, and confirm your answer.

3 Find and classify the stationary points of each of the following curves.

a) $y=\frac{x+1}{x^3-2x^2+x}$ **b)** $y=\frac{1+x}{1-x}$

In each case, check your answers by sketching the curve using a graphical calculator.

4 An object moves along the x-axis so that its position at time $t \geq 0$ is given by:

$$x=\frac{4t^2+t+4}{t^2+1}$$

a) Obtain an expression for the velocity of the object and find when the velocity is zero.

b) What is the greatest distance of the object from the origin?

c) What is the maximum speed of the object?

5 The power delivered into the load, x, of a class A amplifier of output resistance R ohms is given by: $P(x)=\frac{V^2x}{(x+R)^2}$

where V is the output voltage. Find the value of x such that P is a maximum.

DIFFERENTIATION OF INVERSE FUNCTIONS

Sometimes a function is given in the form $x = g(y)$ instead of $y = f(x)$. From Chapter 13, *Functions*, g is called the **inverse function** of f. To differentiate an inverse function, we need another important result.

Exploration 20.5

Differentiating an inverse function

Take the function $x = \sqrt{y}$ for which $y = x^2$.

- Find $\frac{dx}{dy}$ and $\frac{dy}{dx}$.
- Try to formulate a simple relation between $\frac{dx}{dy}$ and $\frac{dy}{dx}$.

Verify your result for:

a) $y = x^3, x = y^{\frac{1}{3}}$ **b)** $y = \frac{1}{3}(2x - 1)$, and $x = \frac{1}{2}(3y + 1)$

The rule for differentiating inverse functions

The rule for differentiating inverse functions is: $\frac{dy}{dx} = \frac{1}{\left(\frac{dx}{dy}\right)}$

It is important to note that this rule only applies for **first-order** differentiation. This rule will be particular use in Chapter 27, *Curves 2*, when we need to differentiate the inverse trigonometric functions.

Example 20.10

Find $\frac{dy}{dx}$ for the function $x = y\ln y$.

Solution

Since $x = y\ln y$ is a product of two functions:

$$\frac{dx}{dy} = y\left(\frac{1}{y}\right) + (\ln y) = 1 + \ln y \Rightarrow \frac{dy}{dx} = \left(\frac{1}{\frac{dx}{dy}}\right) = \frac{1}{1 + \ln y}$$

Example 20.11

Find $\frac{dy}{dx}$ for the function $y = a^x$ where a is constant.

Solution

The first step is to take logs of both sides.

$\ln y = x\ln a \quad \Rightarrow x = \frac{\ln y}{\ln a} \Rightarrow \frac{dx}{dy} = \frac{1}{y\ln a}$ *since $\ln a$ is constant.*

Applying the rule for differentiating inverse functions:

$$\frac{dy}{dx} = \frac{1}{\left(\frac{dx}{dy}\right)} = \frac{1}{\left(\frac{1}{y\ln a}\right)} = y\ln a$$

In this case we substitute for y, so that: $\frac{dy}{dx} = a^x \ln a$

This confirms the result we first met in Chapter 18, Differentiation 2.

EXERCISES

In the following expressions find $\frac{dy}{dx}$.

1 $x = ye^{-y}$ **2** $x = \ln(y - 1)$ **3** $y = 2^x$ **4** $x = e^y - e^{-y}$

The following problems contain a mixture of products and quotients of functions and composite functions. In each case identify the type of function and then use the appropriate rule to find its derivative.

5 $x\sqrt{x+2}$ **6** $(2x^3 - 3)^4$ **7** $\frac{x}{\sqrt{x+1}}$ **8** e^{4x^2+x-1}

9 $e^{x^2}(x^2 - 3x + 2)$ **10** $e^{-0.1t}(t + 2)$ **11** $\dfrac{1}{(x^4 + 1)^2}$ **12** $(t - 1)(t - 2)^2$

13 $\sqrt{t^2 - t + 1}$ **14** $\dfrac{e^t}{t + t^{-1}}$

20.4 B

In each of the following expressions find $\dfrac{dy}{dx}$.

1 $x = \ln(2y + 1)$ **2** $y = 4(3)^x$ **3** $x = e^{2y} + e^{-2y}$ **4** $x = y^2 - 2y + 1$

The following problems contain a mixture of products and quotients of functions and composite functions. In each case identify the type of function and then use the appropriate rule to find its derivative.

5 $\dfrac{x}{x^2 - 4}$ **6** $u \ln u$ **7** $(2t - \sqrt{t})(t + \sqrt{t})$ **8** $e^{t^2 - 3t + 4}$

9 $e^{2x-5}(x^2 - 4x - 1)$ **10** $\ln(t^2 + 3)$ **11** $\dfrac{x}{e^x - e^{-x}}$ **12** 10^x

13 $\dfrac{3x + 1}{x - 2}$ **14** $\sqrt{s(2 - s)^3}$

MATHEMATICAL MODELLING ACTIVITY

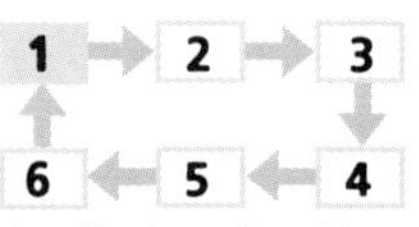

Specify the real problem

Problem statement

Mathematical models of traffic flow can help to show how to avoid traffic jams and to ensure optimum flow conditions in congested situations such as road tunnels, contraflow sections of motorways etc.

- What is the speed of traffic required to obtain maximum traffic flow along a single carriageway road?

1 → 2 → 3, 6 ← 5 ← 4

Set up a model

Set up a model

First you need to identify the important variables:

- the speed of the traffic,
- the length of the vehicles,
- the distance between the vehicles,
- the flow rate defined as the number of vehicles per hour passing a fixed point.

You will also need to make some assumptions before formulating the mathematical problem to solve. Here let's assume:

- all vehicles are identical of length L m,
- all vehicles travel at the same constant speed v m s^{-1},
- the distance between the vehicles is the safe stopping distance given in the Highway Code (see Chapter 1, page 10).

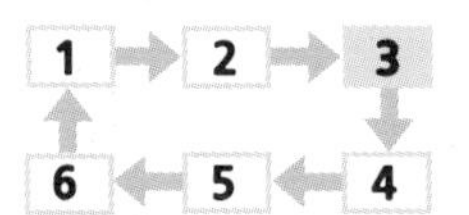

Formulate the mathematical problem

Mathematical problem

Firstly, establish a relationship between the flow rate, f, and the speed of the vehicles, v. If the vehicles pass a given point t seconds apart then in one hour the flow rate (i.e. number of vehicles per hour) is:

$$f = \frac{3600}{t} \quad \text{(vehicles per hour)}$$

If the distance between the front of each pair of vehicles is d m then $d = vt$ giving:

$$f = \frac{3600v}{d} \quad \text{(vehicles per hour)}$$

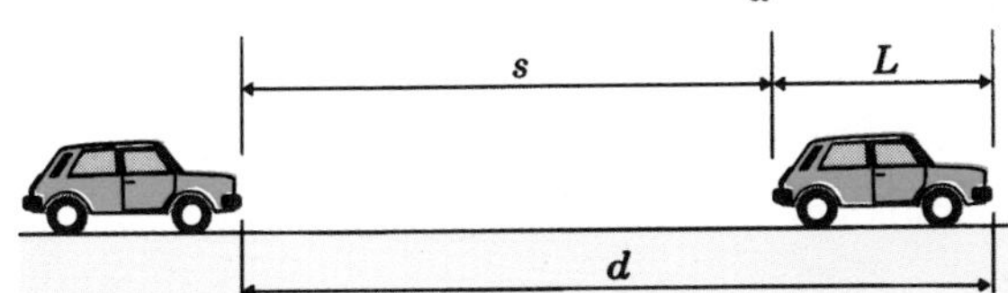

The distance d is made up of the length of a vehicle, L, and the 'safe stopping distance', s.

$$d = s + L$$

The formula for the flow rate then becomes:

$$f = \frac{3600v}{s+L}$$

Secondly, establish a model for the 'safe stopping distance'. Using the data from the *Highway Code* with distances in metres and speeds in m s^{-1}, confirm that the equation relating s and v is:

$$s = 0.682v + 0.076v^2$$

The mathematical problem is to find the value of v that will maximise the function:

$$f = \frac{3600v}{L + 0.682v + 0.076v^2}$$

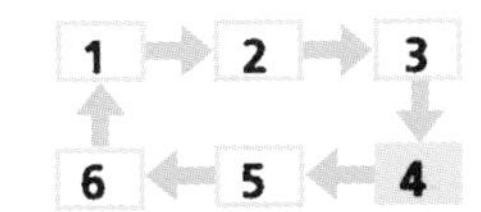

Solve the mathematical problem

Mathematical solution

f(v) is a quotient of two functions. Applying the quotient rule:

$$f'(v) = \frac{3600\left[1\left(L + 0.682v + 0.076v^2\right)\right] - v(0.682 + 0.152v)}{\left(L + 0.682v + 0.076v^2\right)^2}$$

$$= \frac{3600\left[L - 0.076v^2\right]}{\left(L + 0.682v + 0.076v^2\right)^2}$$

For a maximum value:

$$f'(v) = 0 \Rightarrow L - 0.076v^2 = 0 \Rightarrow v = \sqrt{\frac{L}{0.076}} = 3.627\sqrt{L}$$

■ Confirm that this value of v gives a maximum value of f(v).

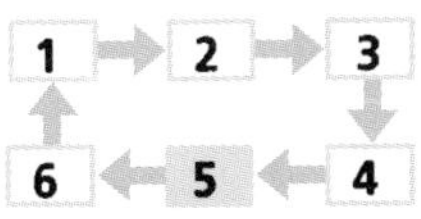

Interpret the solution

Interpretation

Results for small and medium sized cars and for a juggernaut are given in the following table (to the nearest integer).

Vehicle length L (m)	Speed v (m s^{-1})	Speed v (mph)	Separation s (m)	Flow rate f
4	7	16	9.5	2017
5	8	18	10.3	1880
15	14	31	24.4	1278

In simple terms, cars could be advised to travel at about 17 mph with a separation of about two car lengths in order to maximise the flow rate.

1 → 2 → 3
6 ← 5 ← 4
Compare with reality

Criticism

Speeds as low as 17 mph are unlikely to be popular with motorists. Observing vehicles you will notice that vehicles tend to travel closer than the safe stopping distances recommended by the *Highway Code*.

Refinement of the model

Investigate the effect of different models for the distance between the vehicles. For example:

- s = thinking distance from *Highway Code*,
- s = braking distance from *Highway Code*,
- s = thinking distance + fraction of braking distance.

[Reference: **Exploring Mechanics** *by the MEW Group published by Hodder and Stoughton]*

CONSOLIDATION EXERCISES FOR CHAPTER 20

1 Find the coordinates of the points of the curve $y = \dfrac{x^2 - 1}{x}$ at which the gradient of the curve is 5.

2 Differentiate the following expressions.

a) $(2x^2 - 1)(x^3 + 4)^3$ **b)** $\dfrac{x^2 + 1}{\sqrt{x}}$ **c)** $(x^2 + 2)^3$ **d)** $e^x \ln x$

e) $\dfrac{x - 1}{2x - 3}$ **f)** $\left(x + \dfrac{1}{x}\right)^{-1}$

3 A curve has equation $y = \dfrac{x + 3}{\sqrt{1 + x^2}}$. Find and classify the stationary points of the curve.

4 A curve has equation $y = (x - 1)(x - c)^2$ where c is a constant.

a) Investigate the nature of the stationary points in the two cases:
i) $c > 1$ **ii)** $0 < c < 1$.

b) On separate diagrams sketch the curves for the two cases:
i) $c > 1$ **ii)** $0 < c < 1$.

5 The radius of a circular disc is increasing at a rate of 0.01 mm s^{-1}. Find the rate at which the area is increasing when its radius is 25 mm.

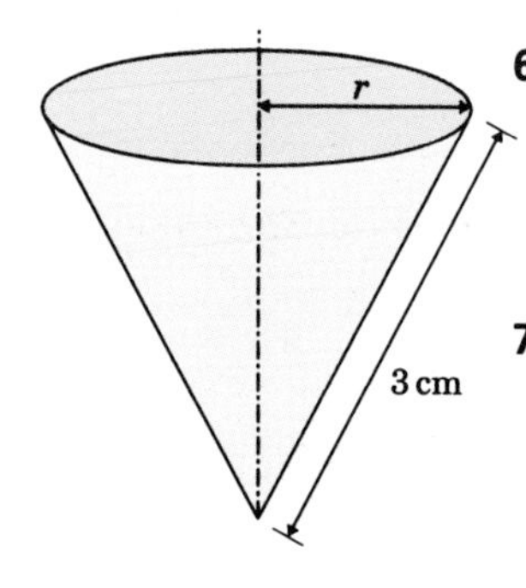

6 The figure shows a conical container with slant height 3 cm and radius of base r cm. It encloses a volume V cm^3 where $V = \frac{1}{3}\pi r^2 \sqrt{9 - r^2}$.

a) Determine the value of r for which $\frac{dV}{dr} = 0$.

b) Show that this value of r gives a maximum value of V.

7 For a lens of constant focal length f cm, the object distance

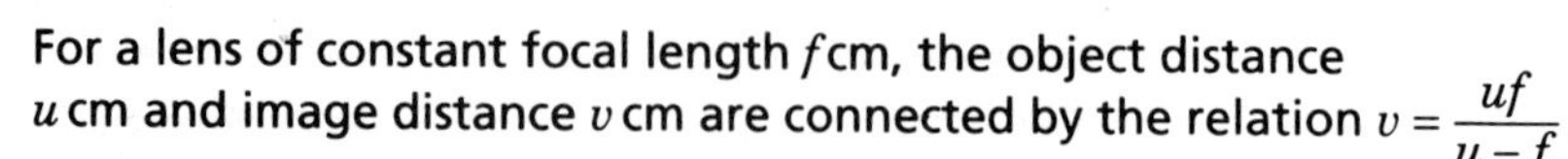

u cm and image distance v cm are connected by the relation $v = \dfrac{uf}{u - f}$.

a) Find the derivative $\frac{dv}{du}$.

b) If u is decreasing at a rate of 1.5 cm s^{-1} and f = 8 cm, calculate the rate at which v is changing when u = 40 cm. Is v increasing or decreasing?

8 A conveyor belt delivers coal onto a stockpile at the rate of 0.3 m^3 s^{-1}. The coal forms a conical heap with height equal to its radius. How fast is the height of the cone increasing when it consists of 20 m^3 of coal?

9 The radius r cm of a circular ink spot, t seconds after it first appears, is given by:

$$r = \frac{1+4t}{2+t}.$$

Calculate:

a) the time taken for the radius to double its initial value,

b) the rate of increase of the radius in cm s^{-1} when $t = 3$,

c) the value to which r tends as t tends to infinity.

(AEB Question 10, Specimen Paper 1)

9√3 km
platform
100 km
x
y
refinery

10 An oil production platform $9\sqrt{3}$ km offshore, is to be connected by a pipeline to a refinery on shore, 100 km down the coast from the platform as shown in the diagram.

The length of underwater pipeline is x km and the length of pipeline on land is y km. It costs £2 million to lay each kilometre of pipeline underwater and £1 million to lay each kilometre of pipeline on land.

a) Show that the total cost of this pipeline is £$C(x)$ million where:

$$C(x) = 2x + 100 - \left(x^2 - 243\right)^{\frac{1}{2}}$$

b) Show that $x = 18$ gives a minimum cost for this pipeline. Find this minimum cost and the corresponding total length of the pipeline.

(Scottish Highers Question 11, Paper 2, 1993)

Summary

- **The chain rule:** If y is a **composite function** of x so that $y = \mathrm{f}(u)$ where $u = u(x)$ then:

$$\frac{\mathrm{d}y}{\mathrm{d}x} = \frac{\mathrm{d}y}{\mathrm{d}u} \times \frac{\mathrm{d}u}{\mathrm{d}x}$$

This is often called the function of a function rule.

- **The product rule:** If y is a **product** of two functions so that $y = uv$ then:

$$\frac{\mathrm{d}y}{\mathrm{d}x} = v\frac{\mathrm{d}u}{\mathrm{d}x} + u\frac{\mathrm{d}v}{\mathrm{d}x}$$

- **The quotient rule:** If y is a **quotient** of two functions so that $y = \frac{u}{v}$ then:

$$\frac{\mathrm{d}y}{\mathrm{d}x} = \frac{v\frac{\mathrm{d}u}{\mathrm{d}x} - u\frac{\mathrm{d}v}{\mathrm{d}x}}{v^2}$$

- **The inverse function rule:** If the relation between x and y is given in the form $x = \mathrm{f}(y)$ then:

$$\frac{\mathrm{d}y}{\mathrm{d}x} = \left(\frac{\mathrm{d}x}{\mathrm{d}y}\right)^{-1}$$

PURE MATHS

21 Numerical solution of equations

In this chapter we shall solve equations numerically using:

- *the method of interval bisection*
- *fixed point iteration*
- *Newton–Raphson iteration*

INTERVAL BISECTION

In earlier chapters we have solved various kinds of equations analytically. This means that there is an analytical method by which equations can be solved by algebraic manipulation. Examples of such analytical methods are:

- **quadratic equations**

  $2x^2 - 5x + 1 = 0 \Rightarrow x = 2.28$ or 0.219 (3 s.f.)

  solved by using the quadratic formula

- **exponential equations**

$$5e^{0.2t} = 18 \Rightarrow e^{0.2t} = 3.6$$
$$\Rightarrow 0.2t = \ln 3.6$$
$$\Rightarrow t = 6.40 \text{ (3 s.f.)}$$

However, there are many equations for which there is no easy analytical method available, so a **numerical** method needs to be developed to find solutions to any desired degree of accuracy.

In the first exploration we use a graphical approach to solving a general cubic equation. We shall then develop three different numerical methods and compare then using this cubic equation.

We have already used the idea of finding a solution by **decimal search** in earlier chapters. The three techniques studied in this chapter may be compared amongst themselves and with the method of decimal search.

Exploration 21.1

Solving equations graphically

Find solutions to these equations.

a) $x^3 - 7x + 6 = 0$

b) $x^3 - 7x + 3 = 0$

Illustrate your solutions graphically.

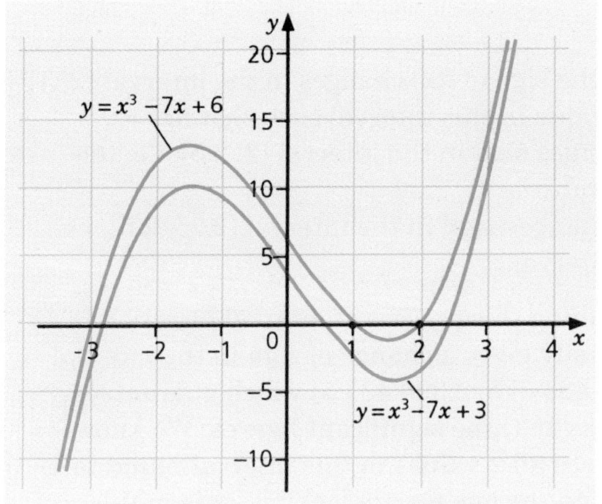

Conclusions

In **a)** the LHS factorises:

$$x^3 - 7x + 6 = 0 \Rightarrow (x + 3)(x - 1)(x - 2) = 0$$
$$\Rightarrow x = -3,\ x = 1 \text{ or } x = 2$$

(three **rational** roots)

In **b)** the LHS will not factorise, but the equation has three **irrational** roots, which can be seen from the graphical illustration for the equations.

Unlike the general quadratic equation it is not easy to develop an **analytical method** for cubic equations which do not factorise. Plotting the graph $y = x^3 - 7x + 3$ and zooming in to where the curve crosses the x-axis gives:

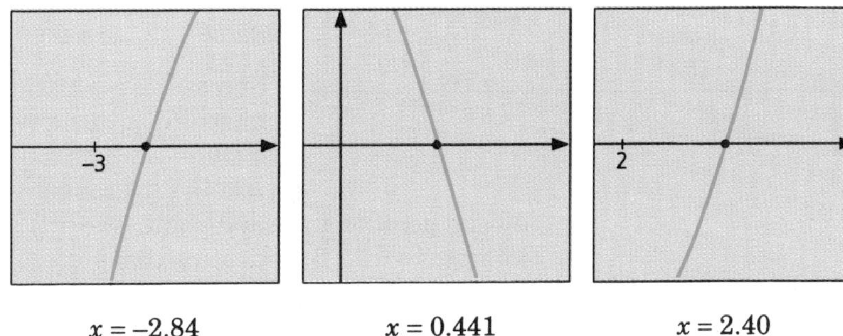

$x = -2.84$ $x = 0.441$ $x = 2.40$

where each root is given correct to three significant figures.

This **graphical method** gives reasonable accuracy after two or three applications of the zoom function on a graphical calculator. However, several **numerical methods** have been devised to enable us systematically to find closer and closer approximations to the roots of an equation. If one approximation can be guaranteed to give a closer approximation than the previous one, then a systematic search will eventually produce an approximation as close to the true value as we want to take it. Such a search method is called an **iteration**.

Exploration 21.2

Interval bisection

The upper root of the equation $x^3 - 7x + 3 = 0$ lies between 2 and 3.

- Find f(2.5), where $f(x) = x^3 - 7x + 3$, and deduce that the upper root lies between $x = 2$ and $x = 2.5$.
- Find f(2.25) and deduce that this root lies between $x = 2.25$ and $x = 2.5$.
- Continue this process of bisecting intervals until you have isolated the root between x-values which differ by less than 0.001.

Conclusions

Since $f(2) = -3$ and $f(3) = 9$, the sign of $f(x)$ changes in the interval [2, 3] which means there is an x-value in this interval such that $f(x) = 0$.
Since $f(2.5) = 1.125$, $f(x)$ changes sign in the interval [2, 2.5], i.e. the upper root lies in this interval.
Since $f(2.25) = -1.359$, $f(x)$ changes sign in the interval [2.25, 2.5], i.e. the upper root lies in this interval.

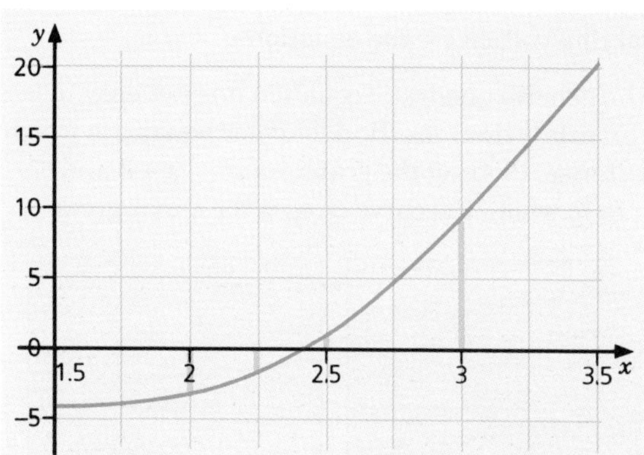

Continuing this process of interval bisection eventually gives a change of sign in the interval [2.396 484 375, 2.398 437 5] which guarantees accuracy to three significant figures. We know this since all x-values in this interval round to 2.40, and the one we are looking for is in this interval, $f(2.396\,484\,375) \approx -0.012\,05$ and $f(2.398\,437\,5) \approx 0.007\,96$.

Successive evaluations of the function, taking the midpoint of the previous interval, leads us to deduce in which half of the current interval the root lies, by comparing values of the function at an end point and the mid-point. The interval is bisected to give a new interval to test. By repeating this process as many times as we wish, we can narrow down the interval within which the root lies indefinitely.

A formal algorithm for interval bisection

Find an interval $[a, b]$ such that $f(a)$ and $f(b)$ have opposite signs.

Let $c = \dfrac{(a+b)}{2}$ and compare $f(a)$ and $f(c)$. If $f(a)$ and $f(c)$ have opposite signs then c becomes the new b, otherwise c becomes the new a.
Repeat this process until the interval $[a, b]$ is narrow enough to ensure the required degree of accuracy.

Example 21.1

Find the middle root of the equation $x^3 - 7x + 3 = 0$ by interval bisection, giving your answer correct to three significant figures.

Solution

Inspecting the graph from the exploration, we see that the middle root lies in the interval $[0, 1]$. *Taking* $a = 0$ *and* $b = 1$, $c = 0.5$.
Since $f(0) = 3$ *and* $f(0.5) = -0.375$, $f(a)$ *and* $f(c)$ *have opposite signs,* c *becomes the new* b.
Taking $a = 0$ *and* $b = 0.5$, $c = 0.25$.
Since $f(0) = 3$ *and* $f(0.25) = 1.265\,625$, $f(a)$ *and* $f(c)$ *have the same sign,* c *becomes the new* a.
Taking $a = 0.25$ *and* $b = 0.5$, $c = 0.375$.
Since $f(0.25) = 1.265\,625$ *and* $f(0.375) = 0.427\,734$, $f(a)$ *and* $f(c)$ *have the same sign,* c *becomes the new* a.

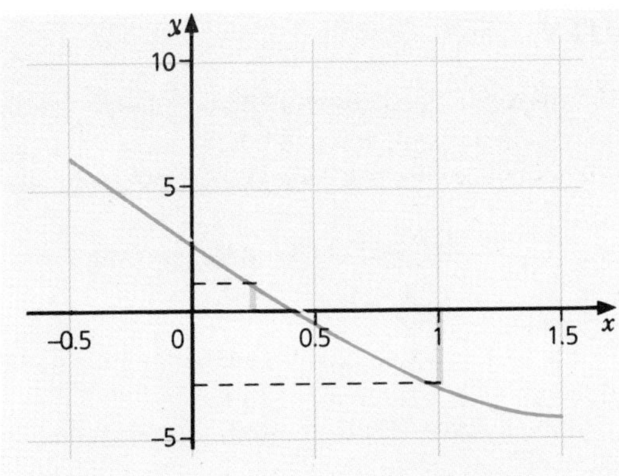

The process continues until the interval $[a, b]$ *within which the root lies is narrow enough to give the desired degree of accuracy. After thirteen steps, we find the root lies in the interval* $[0.440\ 674, 0.440\ 918]$ *which means that the middle root is* **0.441** *correct to three significant figures.*

The table shows a spreadsheet with the values of a, b, c, $f(a)$, $f(b)$ *and* $f(c)$ *for the first thirteen steps.*

n	a	b	c	$f(a)$	$f(b)$	$f(c)$
1	0	1	0.5	3	–3	–0.375
2	0	0.5	0.25	3	–0.375	1.265625
3	0.25	0.5	0.375	1.265625	–0.375	0.427734
4	0.375	0.5	0.4375	0.427734	–0.375	0.02124
5	0.4375	0.5	0.46875	0.02124	–0.375	–0.17825
6	0.4375	0.46875	0.453125	0.02124	–0.17825	–0.07884
7	0.4375	0.453125	0.445313	0.02124	–0.07884	–0.02888
8	0.4375	0.445313	0.441406	0.02124	–0.02888	–0.00384
9	0.4375	0.441406	0.439453	0.02124	–0.00384	0.008695
10	0.439453	0.441406	0.44043	0.008695	–0.00384	0.002426
11	0.44043	0.441406	0.440918	0.002426	–0.00384	–0.00071
12	0.44043	0.440918	0.440674	0.002426	0.00071	0.000859
13	0.440674	0.440918	0.440796	0.000859	–0.00071	0.000076

The way in which the root is trapped in successively narrower intervals is illustrated in the diagram.

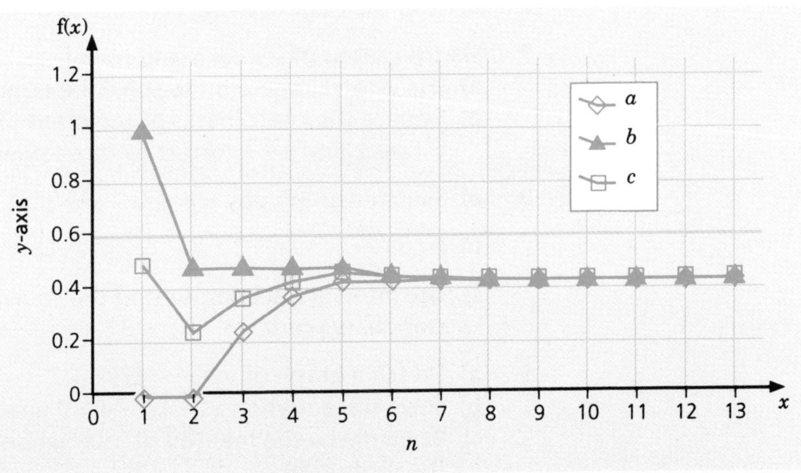

CALCULATOR ACTIVITY

The output on the spreadsheet was produced by automating the algorithm according to this flowchart.

Use the flowchart to write a program for the method of interval bisection and test it by finding all three roots of the equation $x^3 - 7x + 3 = 0$.

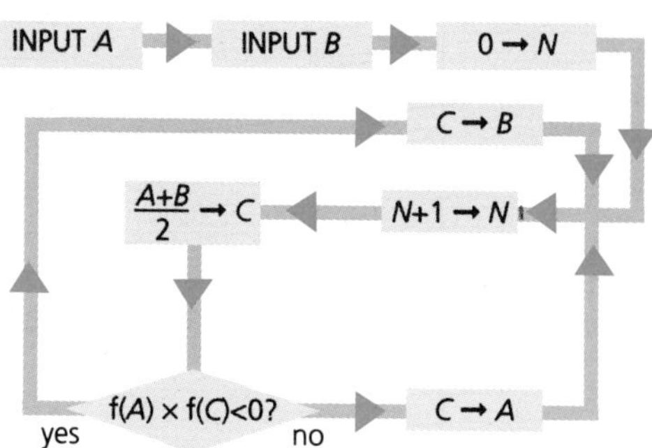

EXERCISES

For questions **1–5**, study the function f(x) and then:

a) sketch the graph of $y = \text{f}(x)$,
b) find the interval(s) $[a, b]$ within which the root(s) of $\text{f}(x) = 0$ lie, where a and b are consecutive integers,
c) use the method of interval bisection to find the root(s) of $\text{f}(x) = 0$, correct to three significant figures.

1 $\text{f}(x) = x^3 + 5x - 9$

2 $\text{f}(x) = x^3 + 5x^2 - 9$

3 $\text{f}(x) = e^x - x - 5$

4 $\text{f}(x) = -x^3 + 5x^2 - x - 4$

5 $\text{f}(x) = x^4 + x^3 - 4x^2 - 3x + 4$

6 The equation $x^4 + x^3 - 4x^2 - 3x + 4 = 0$ has just two roots (see question **5**), but the equation $x^4 + x^3 - 4x^2 - 3x + 3 = 0$ has four roots.
a) Explain, with the aid of a sketch, why the procedure for finding the negative roots needs to be modified.
b) Find the negative roots correct to three significant figures.

7 Sketch graphs of $y = \cos x$ and $y = x$ $(-\pi \le x \le \pi)$.
a) Use interval bisection to solve the equation $\cos x - x = 0$
b) Deduce the coordinates of the point of intersection between $y = \cos x$ and $y = x$ correct to three significant figures.

8 **a)** Sketch a graph of $y = 8 + 2x - x^2 - x^4$.
b) Find $\dfrac{dy}{dx}$.
c) Use interval bisection to find the coordinates of the graph's only stationary point.

9 **a)** Sketch a graph of $y = x^4 - 4x^2 + 4$.
b) Solve the equation $x^4 - 4x^2 + 4 = 0$ analytically.
c) Explain why the method of interval bisection would fail to produce the roots in **b)**.

10 **a)** Show that the equation $3x + 2\ln x = 0$ has just one root, in the interval [2, 3].
b) Use interval bisection to find the root correct to **i)** 1 d.p, **ii)** 2 d.p, **iii)** 3 d.p. noting the number of steps required in each case.
c) What is the least number of steps required to ensure accuracy to 3 d.p., using interval bisection on any equation, beginning with interval $[a, b]$ where a and b are consecutive integers?

EXERCISES

21.1 B

For questions **1–5**, study the function $f(x)$ and then:
a) sketch the graph of $y = f(x)$,
b) find the interval(s) $[a, b]$ within which the root(s) of $f(x) = 0$ lie, where a and b are consecutive integers,
c) use the method of interval bisection to find the root(s) of $f(x) = 0$, correct to three significant figures.

1 $f(x) = 7 - 3x^2 - x^3$

2 $f(x) = 7 - 4x^2 - x^3$

3 $f(x) = 2^x - 5x - 1$

4 $f(x) = x^4 - x^3 - x^2 - x - 1$

5 $f(x) = 2x^2 + x - 2e^x$

6 The equation $2x^2 + x - 2e^x = 0$ has just one root (see question 5), but the equation $2x^2 + 3x - 2e^x = 0$ has more roots.
a) Explain, with the aid of a sketch, why great care needs to be taken in finding the positive roots of the second equation.
b) Use the interval bisection method to find the two positive roots of $2x^2 + 3x - 2e^x = 0$, correct to three significant figures.

7 Sketch graphs of $y = 1 - \tan x$ and $y = 2x$ $(-\pi \le x \le \pi)$.
a) Use interval bisection to solve the equation $2x + \tan x - 1 = 0$.
b) Deduce the coordinates of the point of intersection of $y = 1 - \tan x$ and $y = 2x$ $(-\pi \le x \le \pi)$ correct to three significant figures.

8 **a)** Sketch a graph of $y = x^2 + e^x$. **b)** Find $\frac{dy}{dx}$.
c) Use interval bisection to find the coordinates of the graph's stationary point.

9 **a)** Sketch a graph of $y = x^6 + 6x^3 + 9$.
b) Solve the equation $x^6 + 6x^3 + 9 = 0$ analytically.
c) Explain why the method of interval bisection would fail to produce the roots in **b)**.

10 **a)** Use a graphics calculator to plot a graph of $y = \sin\frac{1}{x}$ $(-\pi \le x \le \pi)$.
b) Use interval bisection to find the highest root of the equation $\sin\left(\frac{1}{x}\right) = 0$
c) How many roots does the equation in **b)** have?

FIXED POINT ITERATION CALCULATOR ACTIVITY

- Using the same scales on both axes plot the graphs of $y = x$ and $y = \dfrac{x^3 + 3}{7}$ for the domain $-5 \le x \le 5$.
- Use a combination of zoom and trace to find the coordinates of the three points of intersection, correct to three decimal places.
- What equation, of the form $f(x) = 0$, are the x-coordinates for the points of intersection the solution?

Conclusions

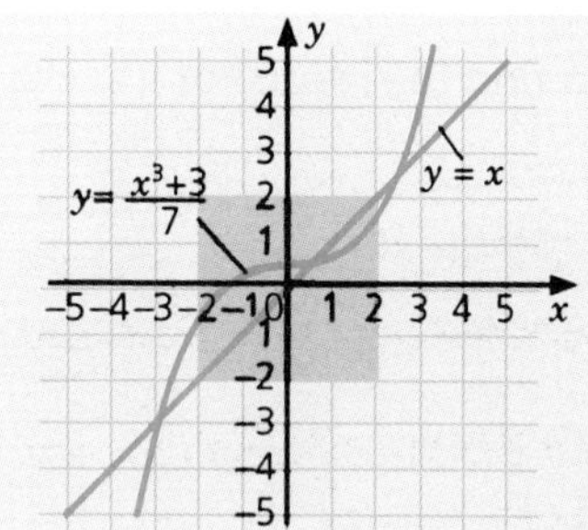

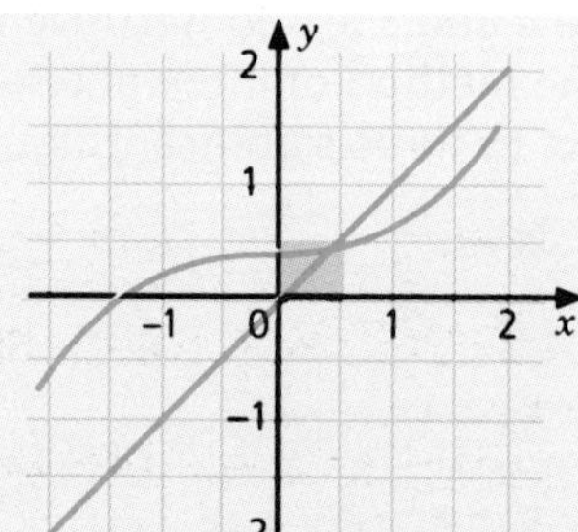

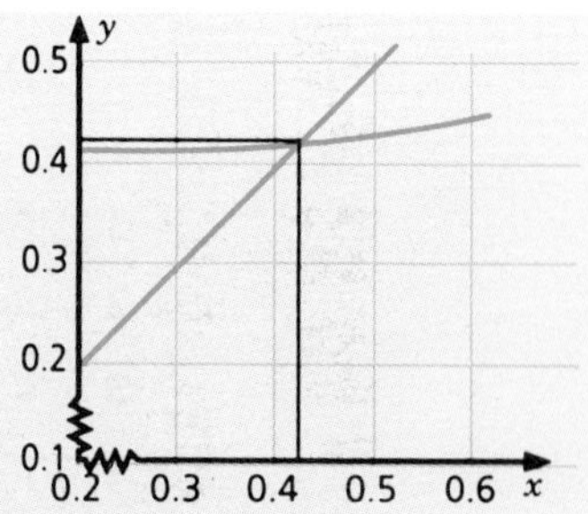

The superimposed graphs show three points of intersection.

Zooming in on the middle point of intersection reveals the coordinates as (0.441, 0.441), correct to three decimal places.

Zoom in similarly to find the lower and upper points of intersections, i.e. (–2.84, –2.84) and (2.40, 2.40).

By comparison with the results of Exploration 21.2 you will see that the x- (and y-) coordinates of the points of intersection are precisely the three roots of the equation $x^3 - 7x + 3 = 0$. This is not surprising since where our two graphs intersect, comparing right-hand sides:

$$x = \frac{x^3 + 3}{7} \Rightarrow 7x = x^3 + 3 \Rightarrow 0 = x^3 - 7x + 3$$

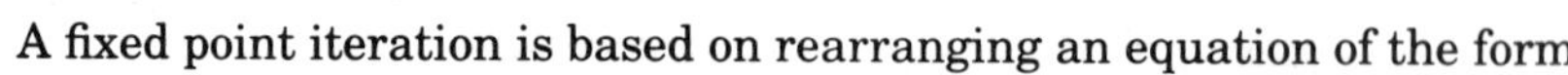

A fixed point iteration is based on rearranging an equation of the form $f(x) = 0$ into the form $x = g(x)$ [in our example $g(x) = \dfrac{x^3 + 3}{7}$] and finding where the two graphs intersect.

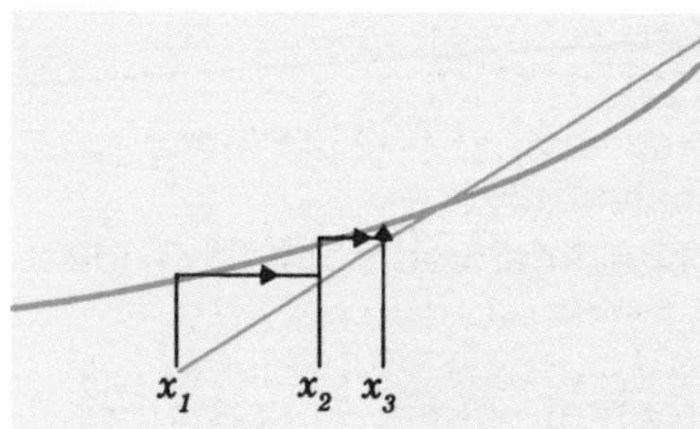

The iterative sequence for the middle root is generated as follows.

Let x_0 be a first approximation to a root of $x = g(x)$. Then a second approximation x_1 is generated by: $x_1 = g(x_0) \quad \Rightarrow \quad x_1 = \dfrac{x_0^3 + 3}{7}$

Likewise: $x_2 = g(x_1) \quad \Rightarrow \quad x_2 = \dfrac{x_1^3 + 3}{7}$

and in general: $x_{n+1} = g(x_n) \quad \Rightarrow \quad x_{n+1} = \dfrac{x_n^3 + 3}{7}$

Let, say $x_0 = 0$, then $x_1 = \dfrac{0^3+3}{7} = 0.428\,57$ (5 d.p.)

$$\Rightarrow \quad x_2 = \frac{0.428\,57^3+3}{7} = 0.439\,82$$

$$\Rightarrow \quad x_3 = \frac{0.439\,82^3+3}{7} = 0.440\,73$$

$$\Rightarrow \quad x_4 = \frac{0.440\,73^3+3}{7} = 0.440\,80 \text{ etc.}$$

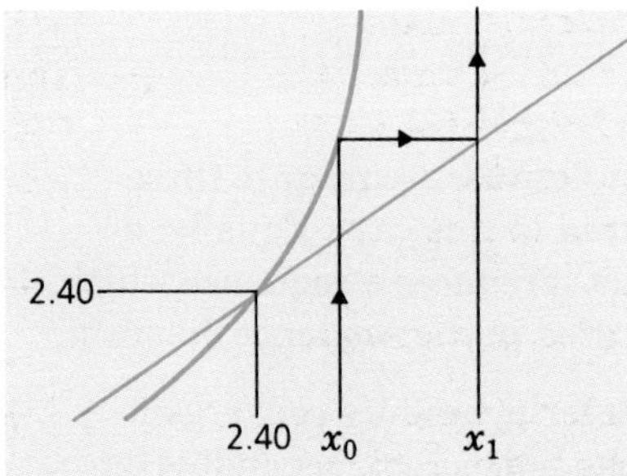

Notice that the sequence $x_0, x_1, x_2, x_3, x_4 \ldots$ converges towards the middle root which is 0.441 (correct to three decimal places).

Try other starting values (x_0) between say, −5 and 5.

Describe what happens to the sequence $x_0, x_1, x_2, x_3, x_4 \ldots$

You should have found that any sequence $x_0, x_1, x_2, x_3, x_4 \ldots$, such that x_0 lies between the lower and upper roots, converges towards the middle root. Any value for x_0 outside this range produces a divergent sequence, e.g. let $x_0 = 3$.

The convergence of the sequence $x_0, x_1, x_2, x_3, x_4, \ldots$ on the middle root for starting values between the lower and upper roots is based on the following idea.

A sequence of the form $x_{n+1} = \text{g}(xn)$ will only converge on a root provided $|\text{g}'(\alpha)| < 1$, and, perhaps only for suitably chosen values for x_0.

In our example $\text{g}(x) = \dfrac{x^3+3}{7} \Rightarrow \text{g}'(x) = \dfrac{3x^2}{7}$ and for the middle root,

$\alpha \approx 0.441$, $\text{g}'(\alpha) = \text{g}'(0.441) = \dfrac{3\times 0.441^2}{7} \approx 0.083$, which satisfies the condition $|\text{g}'(\alpha)| < 1$.

However, for the lower root:

$$\alpha = -2.84 \Rightarrow \text{g}'(\alpha) = \text{g}'(-2.84) = \frac{3\times(-2.84)^2}{7} \approx 3.46$$

and for the upper root:

$$\alpha = 2.40 \Rightarrow \text{g}'(\alpha) = \text{g}'(2.40) = \frac{3\times 2.40^2}{7} \approx 2.47$$

For both lower and upper roots we see that $|\text{g}'(\alpha)| > 1$, which means that the iterative sequence will not converge on either of them. Geometrically this is evident from the sequence diagram above.

To obtain the lower and upper roots of the equation, $\text{f}(x) = 0$, another rearrangement in the form $x = \text{g}(x)$ is required.

$$x^3 - 7x + 3 = 0 \quad \Rightarrow \quad x^3 = 7x - 3 \quad \Rightarrow \quad x = \sqrt[3]{7x-3}$$

which leads to the iteration:

$$x_{n+1} = \sqrt[3]{7x_n - 3}$$

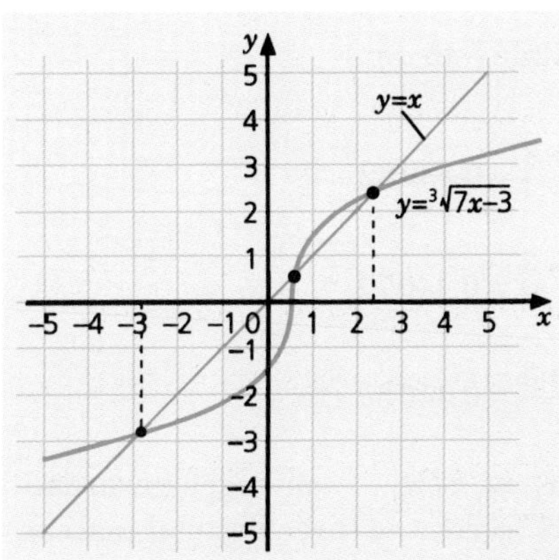

Now we can plot $y = x$ and $y = \sqrt[3]{7x-3}$ on a single graph.

Taking $x_0 = 2$ produces the iteration:

$$x_1 = \sqrt[3]{7\times 2+3} \qquad = 2.223\,98 \qquad (5 \text{ d.p.})$$

$$\Rightarrow \quad x_2 = \sqrt[3]{7\times 2.223\,98+3} \quad = 2.324\,99$$

$$\Rightarrow \quad x_3 = \sqrt[3]{7\times 2.324\,99+3} \quad = 2.367\,79$$

$$\Rightarrow \quad x_4 = \sqrt[3]{7\times 2.367\,79+3} \quad = 2.385\,48$$

The sequence $x_0, x_1, x_2, x_3, x_4, \ldots$ is converging towards the upper root, but accuracy to three significant figures is only obtained by carrying out iterations up to, say, x_9, i.e:

$$x_8 = 2.397\,33 \qquad \text{and} \qquad x_9 = 2.397\,53$$

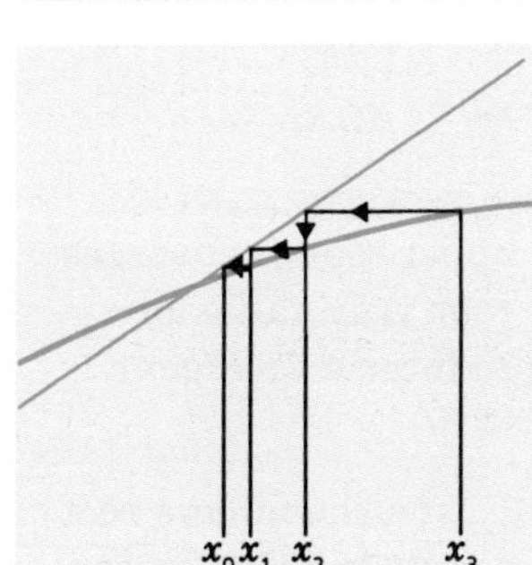

Since x_8 and x_9 agree to four significant figures, accuracy to three significant figures, i.e. $x = 2.40$, is assured in this case. Equally, a sequence with $x_0 = 3$ (just above the root) produces a sequence which converges 'from above', which is illustrated in the sequence diagram.

Using this iteration, all starting values for x_0 produce convergent sequences, either to the lower root or the upper root, depending on which side of the middle root you take for x_0.

We can see from the graphs of $y = x$ and $y = \sqrt[3]{7x-3}$ that for $\alpha = -2.84$ and $\alpha = 2.40$, $|g'(\alpha)| < 1$, since the gradient of the curve $g(x)$ is evidently less than 1 for these values of x, whereas for $\alpha = 0.441$, $|g'(\alpha)| > 1$.

Algebraically:

$$g(x) = \sqrt[3]{7x-3} = (7x-3)^{\frac{1}{3}} \quad \Rightarrow \quad g'(x) = \tfrac{1}{3}(7x-3)^{-\frac{2}{3}} \times 7 \;=\; \frac{7}{3\sqrt[3]{(7x-3)^2}}$$

$g\,(-2.84) = 0.2895 \quad g'(2.40) = 0.4056 \quad g'(0.441) = 11.88$

Both iterative sequences we have met so far produce **staircase** sequence diagrams. A second type, the **cobweb** sequence diagram is illustrated in the following example.

Example 21.2

Find the only root of the equation $x^3 + 4x - 3 = 0$ using fixed point iteration $x_{n+1} = \dfrac{3 - x_n^3}{4}$

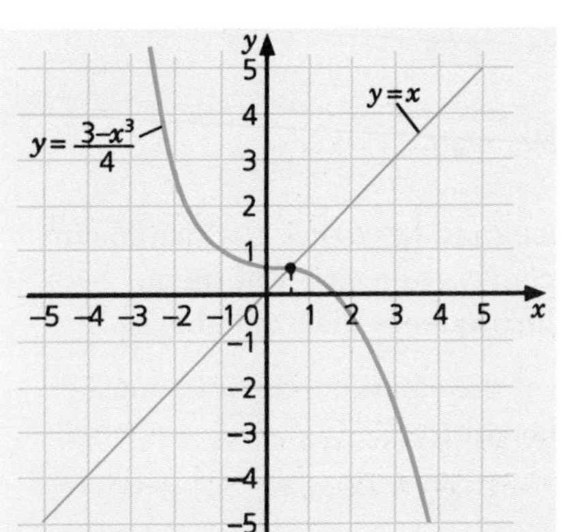

Solution

The graphs of $y = x$ and $y = \dfrac{3 - x^3}{4}$ show that the root of $f(x) = 0$ lies between 0 and 1.

Taking $x_0 = 1$ as our starting value produces the iteration:

$$x_1 = \frac{3 - x_0^3}{4} = 0.5$$

$$x_2 = \frac{3 - 0.5^3}{4} = 0.718\,75$$

$$x_3 = \frac{3 - 0.71875^3}{4} = 0.657\,17 \text{ (5 d.p.)}$$

$$x_4 = \frac{3 - 0.65717^3}{4} = 0.679\,05 \text{ (5 d.p.)}$$

*The sequence $x_0, x_1, x_2, \ldots$ is **oscillating** and converging towards the only root if $x^3 + 4x - 3 = 0$, as illustrated in the sequence diagram.*

Check that the root is 0.674, correct to three decimal places. This is evident after ten iterations.
Taking $x_0 = 2$ also produces a convergent sequence, but taking $x_0 = 3$ produces an oscillating divergent sequence, even though $|g'(0.674)| < 1$:

$$g(x) = \frac{3 - x^3}{4} \Rightarrow \quad g'(x) = -\frac{3}{4}x^2 \quad \Rightarrow \quad g'(0.674) = -\tfrac{3}{4} \times 0.674^2 = -0.341$$

The diagram shows a spreadsheet output of the solution.

Use the flowchart to write a program for the fixed point iteration method and test it by working through the last example.

- Investigate the range of starting values x_0 for which the iteration converges.

- The interval of starting values for which the sequence converges is given by $-1.7814 \leq x \leq 2.1634$, working to five significant figures. Obtain the upper and lower bounds of this interval by decimal search or other simple systematic search. Can you adapt the method of interval bisection?

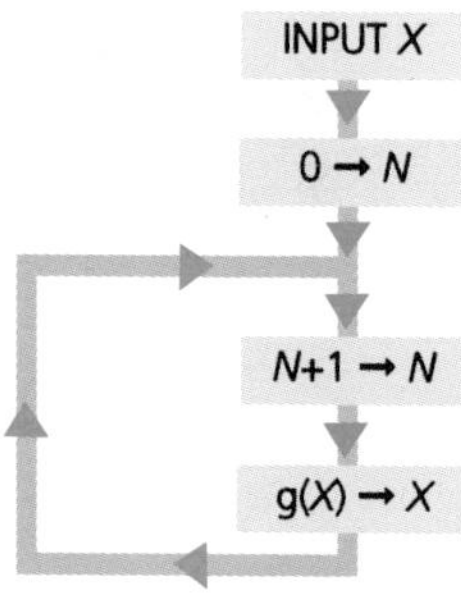

Exploration 21.3

Calculator chaos

The quadratic equation $2x^2 - x - 1 = 0$ has two rational roots:
$2x^2 - x - 1 = 0 \Rightarrow (2x + 1)(x - 1) = 0 \Rightarrow x = -0.5$ or $x = 1$.

Rearranging the equation into the form $x = g(x)$ in one particular way gives $x = 2x^2 - 1$.

Now $g(x) = 2x^2 - 1 \Rightarrow g'(x) = 4x$. In particular:
$g'(-0.5) = -2$, and $g'(1) = 4$

so since $|g'(\alpha)| > 1$ for both roots, we would *not* expect a sequence converging towards either root from the iteration $x_{n+1} = 2x_n^2 - 1$.

- Use your program for fixed point iteration for various starting values x such that:

  a) $|x_0| > 1$, **b)** $|x_0| < 1$, **c)** $x_0 = \pm 0.5$ or ± 1.
- Describe the behaviour of the iteration in all three cases.

Conclusions

For values of x_0 such that $|x_0| > 1$ the sequence diverges, but for values of x_0 such that $|x_0| < 1$, apart from ± 0.5, the sequence is neither convergent nor divergent, in fact it is **chaotic**.

For $x_0 = 0.8$, the sequence continues (figures rounded to three significant figures):
0.28, –0.843, 0.422, –0.644, –0.171, –0.942, ...

For $x_0 = 0.81$ the sequence continues:
0.312, –0.805, 0.296, –0.824, 0.360, –0.742, ...

A slight variation in the starting value produces a completely different chaotic pattern.

- Use the iteration $x_{n+1} = axn^2 - 1$, for various values of a, to produce sequences. Examine their behaviour for various starting values, x_0, $|x_0| < 1$.
- Which range of values of a produce:
  i) divergent sequences,
  ii) convergent sequences,
  iii) chaotic sequences?

EXERCISES

1 **a)** Sketch the graph of $y = 4x^3 - x - 1$ and show that $4x^3 - x - 1 = 0$ has just one root, between 0 and 1.

b) Use the iteration: $x_{n+1} = \sqrt[3]{\dfrac{x_n + 1}{4}}$ to find the root, correct to three significant figures.

c) Illustrate your answer to **b)** using a staircase diagram.

2 **a)** Show that the equation $5x - 2e^{-x} = 0$ can be written in the form $x = 0.4e^{-x}$.
b) Use the fixed point iterative method to find, correct to three significant figures, the solution of the equation.
c) Illustrate your answer to **b)** using a cobweb diagram.

3 The equation $e^{2x} + x - 4 = 0$ has one root, between 0 and 1.
a) Show that the rearrangement $x = 4 - e^{2x}$ produces a sequence that oscillates between 4 and –2976.96, whatever value of x_0 you choose. Explain your answer graphically.
b) Show that the equation can be rearranged into the form $x = 0.5 \ln(4 - x)$ and use the iteration $xn_{+1} = 0.5 \ln(4 - x_n)$ to find the root, correct to three significant figures.

4 By choosing suitable forms for the function g(x), use the fixed point iterative method to find three roots of the equation $x^3 - 3x + 1 = 0$, giving your answer correct to three significant figures.

5 The equation $x + 2 \ln x = 0$ has a root in the interval $[0, 1]$. Show that one and only one of the following iterative forms will converge to that root.
a) $xn_{+1} = -2 \ln xn$ **b)** $x_{n+1} = e^{-\frac{1}{2}x_n}$
Find the root, correct to three significant figures.

EXERCISES

21.2 B

1 **a)** Sketch the graph of $y = 8 - 3x - 2x^2$.
b) Show, by differentiation, that the gradient function is always negative.
c) Use the iteration $x_{n+1} = \sqrt[3]{4 - 1.5x_n}$ to find the only root of the equation, correct to three significant figures.
d) Illustrate your answer to **c)** using a cobweb diagram.

2 **a)** Sketch a graph of $y = 5x - e^x$ and show that $y = 0$ for two different values of x.
b) Show that the equation $5x - e^x = 0$ can be written in the form $x = 0.2e^x$.
c) Use the iteration $x_{n+1} = 0.2e^{x_n}$ to find the lower root of $5x - e^x = 0$ correct to three significant figures.
d) Find another form of g(x) as a basis to find the upper root, correct to three significant figures.

3 **a)** Show that the equation $x^3 - 5x^2 + 7 = 0$ has a root between $x = 1$ and $x = 2$.
b) Use the iteration $x_{n+1} = \dfrac{-7}{x_n(x_n - 5)}$ to find the root, correct to three significant figures.
c) Show that the equation $x^3 - 5x^2 + 7 = 0$ has two other roots. Use suitable rearrangements in the form $x = g(x)$ to find them correct to three significant figures.

4 The function $f(x) = x^4 + x^2 - 3x + 7$ has one stationary point.
a) Using a fixed point iteration to solve a suitable cubic equation, find the coordinates of the stationary point.
b) Determine the nature of the stationary point and so sketch a graph of $y = f(x)$.

5 **a)** Show that the iteration given by:

$$x_{n+1} = \frac{1}{2}\left(x_n + \frac{a}{x_n}\right)$$

will converge towards the square root of a.
b) Use the iteration to find *both* values of $\sqrt{20}$, to three significant figures, illustrating your solutions diagramatically.

NEWTON–RAPHSON ITERATION

The two methods we have used so far produce sequences which converge to the roots of an equation $f(x) = 0$, but both involve quite a lot of repetitive work to achieve a given degree of accuracy. The Newton–Raphson iteration is based on the idea of drawing tangents to a curve in such a way that each new tangent drawn gives a closer approximation to the root.

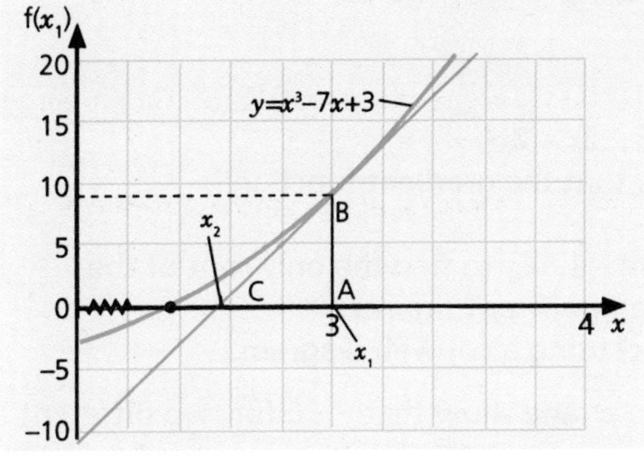

Returning to our example, let us try to find the upper root of the equation:

$f(x) = x^3 - 7x + 3 = 0$

Let $x_1 = 3$ be a first approximation to the root. Now:

$f'(x) = 3x^2 - 7 \quad \Rightarrow \quad f(3) = 3^3 - 7 \times 3 + 3 = 9$

and $\quad f'(3) = 3 \times 3^2 - 7 = 20$

Now we draw in a tangent to the graph of $y = f(x)$ at the point $(3, 9)$ with gradient 20.

The point where the tangent crosses the x-axis is the next approximation to the root, x_2. We can see that x_2 is considerably closer to the upper root than x_1. Now we repeat the process, this time drawing a tangent at $(x_2, f(x_2))$ to generate x_3, and so on.

The algebraic connection between x_1 and x_2 is found by considering the gradient of the tangent in two different ways.

The gradient of the tangent through $(x_1, f(x_1))$ is $f'(x_1)$, but geometrically, part of the tangent BC is the hypotenuse of a right-angled triangle. We can see that the gradient of BC is given by:

$$\frac{AB}{AC} = \frac{f(x_1)}{x_1 - x_2}$$

Equating the two expressions for the gradient of the tangent gives:

$$f'(x_1) = \frac{f(x_1)}{x_1 - x_2}$$

Now we can rearrange this formula to make x_2 the subject.

$$x_1 - x_2 = \frac{f(x_1)}{f'(x_1)} \Rightarrow x_2 = x_1 - \frac{f(x_1)}{f'(x_1)}$$

This leads to the general iteration formula:

$$x_{n+1} = x_n - \frac{f(x_n)}{f'(x_n)}, \quad n = 0, 1, 2, \ldots$$

We can see how the iteration works by following through the successive approximations to the upper root of $x^3 - 7x + 3 = 0$, taking $x_0 = 3$:

$$x_1 = x_0 - \frac{f(x_0)}{f'(x_0)} = 3 - \frac{f(3)}{f'(3)} = 3 - 0.45 = 2.55$$

$$x_2 = x_1 - \frac{f(x_1)}{f'(x_1)} = 2.55 - \frac{f(2.55)}{f'(2.55)} = 2.55 - 0.138\,43 = 2.411\,57$$

$$x_3 = x_2 - \frac{f(x_2)}{f'(x_2)} = 2.411\,57 - \frac{f(2.411\,57)}{f'(2.411\,57)} = 2.411\,57 - 0.013\,78 = 2.397\,80$$

$$x_4 = x_3 - \frac{f(x_3)}{f'(x_3)} = 2.397\,80 - \frac{f(2.397\,80)}{f'(2.397\,80)} = 2.397\,80 - 0.000\,13 = 2.397\,66$$

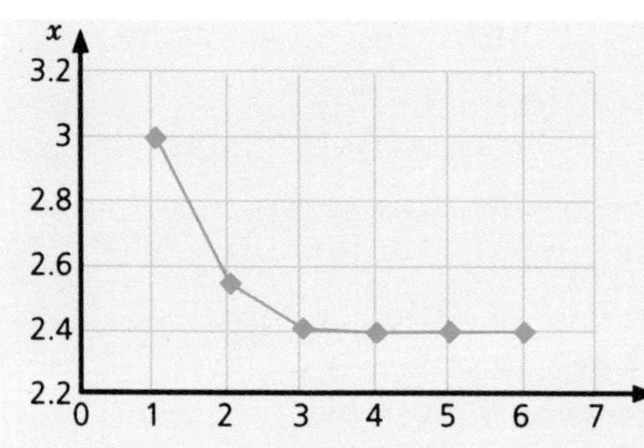

Note that the value of $\frac{f(x_n)}{f'(x_n)}$ is converging rapidly to 0 since f(*xn*) is converging rapidly to zero – which is what we want.

After just four iterations it is evident that the upper root is 2.398 correct to three decimal places.

The diagram shows a spreadsheet output of the solution.

CALCULATOR ACTIVITY

- *Use the flowchart to write a program for the Newton–Raphson iteration and use it to find all three roots of the equation* $x^3 - 7x + 3 = 0$.
- *Does the iteration always produce a convergent sequence, whatever value of* x_0 *you might choose?*
- *What ranges should the initial approximation lie in so that it converges on:*
  - ***a)*** *the lower root,*
  - ***b)*** *the middle root,*
  - ***c)*** *the upper root?*

INPUT *X*
0 → *N*
N+1 → *N*
X–f(*x*)/f′(*x*) → *X*

Conclusions

The iteration will always converge for a continuous function f(x), for any initial approximation x_0 (provided $f'(x_0) \neq 0$), but predicting which root will be found from a particular starting point can be difficult.

If x_0 is taken close to a root then the iteration should converge towards that root. If x_0 is taken further away then the sequence may converge towards any of the roots.

When an equation has, say, a single root but the curve $y = f(x)$ has turning points, problems can arise if an unsuitable value for x_0 is chosen. This is illustrated in the following example.

Example 21.3

A function is given by $f(x) = x^3 + 2x^2 + 3$.

a) *Sketch the graph of* $y = f(x)$ *and show that the equation* $f(x) = 0$ *has just one root in the interval* $[-3, -2]$.

b) *Use the Newton–Raphson iteration to find the root* $f(x) = 0$ *using the initial approximations:*

i) $x_0 = -3$ ***ii)*** $x_0 = -2$ ***iii)*** $x_0 = -1$.

c) *Comment on any differences you find.*

Solution

a)

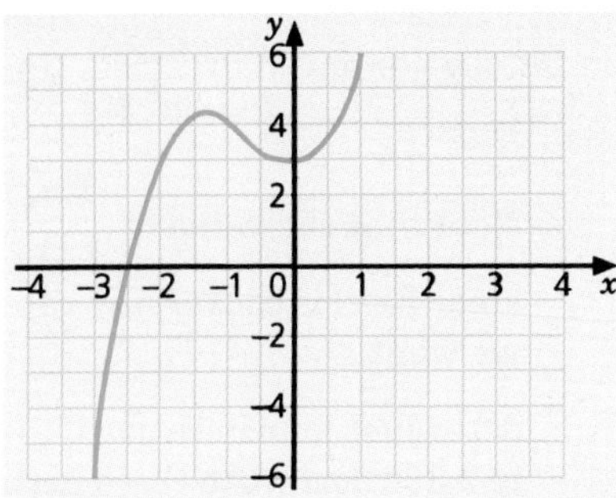

b) $f(x) = x^3 + 2x^2 + 3 \Rightarrow f'(x) = 3x^2 + 4x$

Hence the Newton–Raphson iteration is given by:

$$x_{x+1} = x_n - \frac{f(x_n)}{f'(x_n)} = x_n - \frac{x_n^3 + 2x_n^2 + 3}{3x_n^2 + 4x_n}$$

i) *Taking* $x_0 = -3$:

$x_1 = -2.6$
$x_2 = -2.493\,117$
$x_3 = -2.485\,620$
$x_4 = -2.485\,584$

ii) *Taking* $x_0 = -2$

$x_1 = -2.75$
$x_2 = -2.521\,390$
$x_3 = -2.486\,373$
$x_4 = -2.485\,584$

In both cases the iteration soon converges on the single root, $x = -2.49$ *(correct to three significant figures).*

iii) *Taking* $x_0 = -1$:

$x_1 = 3$
$x_2 = 1.769\,231$
$x_3 = 0.870\,588$
$x_4 = -0.028\,574$
$x_5 = 26.808\,238$
$x_6 = 17.659\,140$

In this case the iteration does not seem to be converging.

c) *The critical feature here is that between the starting point* $x_0 = -1$ *and the root the curve has a turning point. The maximum point seems to act as a barrier to reaching the root. Although the behaviour of the sequence seems chaotic, it does in fact settle down eventually to converge on the root. Picking up the iteration after 47 steps gives:*

$x_{47} = 0.197\,541$
$x_{48} = -3.203\,746$
$x_{49} = -2.683\,345$
$x_{50} = -2.506\,645$
$x_{51} = -2.485\,860$
$x_{52} = -2.485\,584$

Provided a suitable starting point is chosen, the Newton–Raphson iteration is usually very efficient, in that convergence towards a root is quick compared with the general fixed point iteration. The following exploration reveals an interesting idea for the case of a cubic equation with three real roots.

Exploration 21.4 *Newton–Raphson*

- Show that cubic equation $x^3 - 2x^2 - x + 2 = 0$ has three roots, $x = -1$, 1 and 2. Apply the Newton–Raphson iteration with initial value

$$x_0 = \frac{1+2}{2} = 1.5$$

- Repeat for $x_0 = \frac{-1+2}{2} = 0.5$ and $x_0 = \frac{-1+1}{2} = 0$.
- What happens to the iteration in each case?
- Can you generalise the result?
- Does it apply to all cubic equations with three roots?

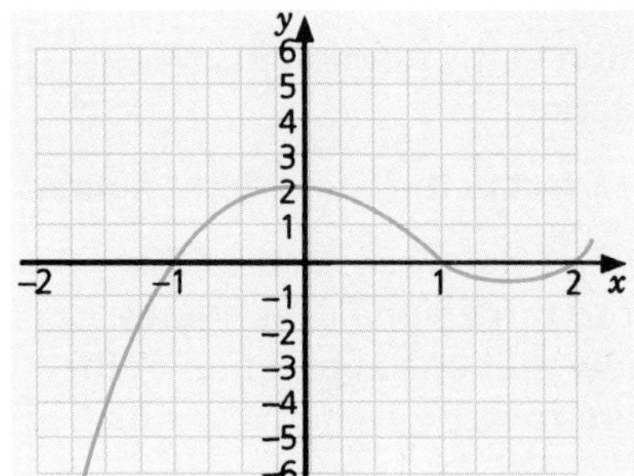

Conclusions

From the graph of $y = x^3 - 2x^2 - x + 2$ we can see that $y = 0$ when $x = -1$, 1 or 2.

Using the Newton–Raphson iteration with $x_0 = 1.5$:

$$x_1 = x_0 - \frac{\left(x_0^3 - 2x_0^2 - x_0 + 2\right)}{3x_0^2 - 4x_0 - 1} = 1.5 - \frac{(-0.625)}{-0.25} = -1$$

After one step of the iteration, $x_1 = -1$, which is the lower root. Using the iteration with $x_0 = 0.5$:

$$x_1 = x_0 - \frac{\left(x_0^3 - 2x_0^2 - x_0 + 2\right)}{3x_0^2 - 4x_0 - 1} = 0.5 - \frac{1.125}{-2.25} = 1$$

Using the iteration with $x_0 = 0$:

$$x_1 = x_0 - \frac{\left(x_0^3 - 2x_0^2 - x_0 + 2\right)}{3x_0^2 - 4x_0 - 1} = 0 - \frac{2}{-1} = 2$$

By starting the iteration at the mid-point of the middle and upper roots, we find the lower root in one step. Similarly, by taking x_1 as the mid-point of the lower and upper roots, we find the middle root in one step, etc.

To generalise, if a and b are any two of three roots, then taking $x_0 = \frac{a+b}{2}$ gives $x_1 = c$.

The method does seem to apply to any cubic equation with three distinct roots.

EXERCISES

For questions **1–5**, study the function $f(x)$ and then:

a) sketch the graph of $y = f(x)$,

b) using suitable initial approximation(s) find the root(s) of $f(x) = 0$, correct to three significant figures,

c) illustrate the method on your graph.

1 $f(x) = 4x^3 - 7x + 5$

2 $f(x) = 4x^3 - 7x + 1$

3 $f(x) = 5x^4 - 2x^3 - 9x^2 - 4x + 1$

4 $f(x) = 5 - x^2 - e^x$

5 $f(x) = \ln x - x^2 + 10$

6 **a)** Sketch the graph of $y = e^x - x - 4$.
b) Explain why the equation $e^x = x + 4$ has two roots.
c) Use the Newton–Raphson iteration to find the roots correct to three significant figures.
d) Why does the Newton–Raphson iteration break down if you take $x_0 = 0$?

7 By applying the Newton–Raphson iteration to the function $f(x) = 1 - \frac{a}{x^2}$ develop an iterative formula for calculating $\sqrt{a}$.

Confirm your iteration by finding $\sqrt{10}$ correct to three significant figures.

8 Use the Newton–Raphson iteration to find the three roots of the equation $2x^3 - 4x^2 - 5x + 1 = 0$, correct to three significant figures.
If a and b are any two of the roots, show that one application of the Newton–Raphson process gives the third root, c, given $x_0 = (a + b)/2$
Illustrate your working graphically.

9 Use the Newton–Raphson iteration to find all stationary points on the graph of $y = x^4 + 2x^3 - 5x^2 - 3x$.
Illustrate your answer with a sketch.

10 Let $f(x) = xe^x - 1 - x$.
a) Sketch a graph of $y = f(x)$.
b) Find the positive root of the equation $f(x) = 0$, correct to three significant figures, taking initial approximation $x_0 = 1$:
i) using the fixed point iteration with $f(x) = \ln\left(1 + \frac{1}{x}\right)$,
ii) using the Newton–Raphson iteration.
c) Compare the efficiency of the two methods in **b)**.

EXERCISES

21.3 B

For questions **1–5**, study the function $f(x)$ and then:
a) sketch the graph of $y = f(x)$,
b) using suitable initial approximation(s) find the root(s) of $f(x) = 0$, correct to three significant figures,
c) illustrate the method on your graph.

1 $f(x) = 10 - 6x + 3x^2 - x^3$

2 $f(x) = 10 + 6x - 3x^2 - x^3$

3 $f(x) = 3x^4 + 3x^3 + 5x - 1$

4 $f(x) = x^5 + 4x^2 - 2$

5 $f(x) = 2^x - 3x - 4$

6 **a)** Sketch the graph of $y = x - 1 + \frac{4}{x^2}$.

b) Use the Newton–Raphson iteration to find the only root of the equation $x + \frac{4}{x^2} = 1$, with $x_0 = -1$.

c) What problems are encountered by taking:

i) $x_0 = 2$

ii) $x_0 > 0, x_0 \neq 2$?

7 By applying the Newton–Raphson iteration to the function $f(x) = 1 - \frac{a}{x^3}$, develop on iterative formula for calculating $\sqrt[3]{a}$.

Confirm your iteration by finding $\sqrt[3]{10}$ correct to three significant figures.

8 Use the Newton–Raphson iteration to find the three roots of the equation $2x^3 - 11x + 1 = 0$, as accurately as your calculator allows. If a and b are any two of the roots, show that one application of the Newton–Raphson process gives the third root, c. Illustrate your working graphically.

9 Use the Newton–Raphson iteration to find all stationary points on the graph of $y = e^x - x^3$. Illustrate your answer with a sketch.

10 Let $f(x) = \frac{2x}{e^x} + 3$.

a) Sketch a graph of $y = f(x)$.

b) Find the negative root of the equation $f(x) = 0$, correct to three significant figures, taking initial approximation $x_0 = -1$:

i) using the fixed point iteration with $g(x) = -1.5e^x$,

ii) using the Newton–Raphson iteration.

c) Compare the efficiency of the two methods in **b)**.

MATHEMATICAL MODELLING ACTIVITY

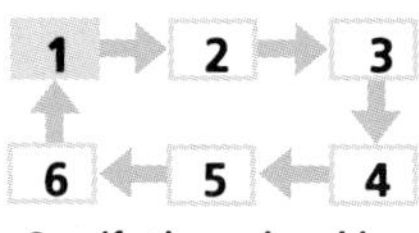

Specify the real problem

Problem statement

An investor is considering a manufacturing project which requires an investment of £10 000 now. She anticipates net returns of £5000 after years 1 and 2, £3000 after year 3 and £1000 after year 4. Should the investor go ahead with the project or not?

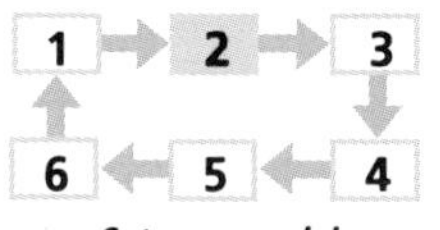

Set up a model

Set up a model

The main purpose of making an investment is to make money! The amount available, £10 000 could be left in a bank or building society to earn interest. This is a 'safe investment'. The essence of investing in a manufacturing project is to obtain a larger return than bank interest. However there is a greater risk!

The important variables are:

- the investment,
- the net returns,
- bank or building society interest rate.

In the problem we will investigate this risk by calculating the loss of the investment and interest and the gain from the anticipated net returns and interest over the five year period.

We make the following assumptions:

- the net returns are guaranteed,
- the net returns are invested in a bank,
- the interest rate over the five year period is constant.

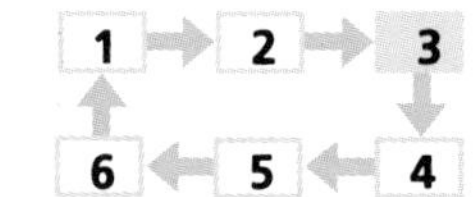

Formulate the mathematical problem

Mathematical problem

The capital available of £10 000 would be worth:

$$L = £10\,000\left(1+\frac{r}{100}\right)^4$$

if invested in a bank at a constant interest rate of r per cent. This is the total amount 'lost' if the investment goes ahead.
You should confirm that the total return plus interest is:

$$I = £5000\left(1+\frac{r}{100}\right)^3 + 5000\left(1+\frac{r}{100}\right)^2 + 3000\left(1+\frac{r}{100}\right) + 1000\,.$$

This is the total income if the investment goes ahead and the net returns are obtained.
The 'profit' on the investment is:

$$P = I - L$$

and if $P > 0$ we would recommend that the person goes ahead with the investment.
The mathematical problem is to find the value of r for $P = 0$. This is the critical case, if r is less than this critical value then $P > 0$.

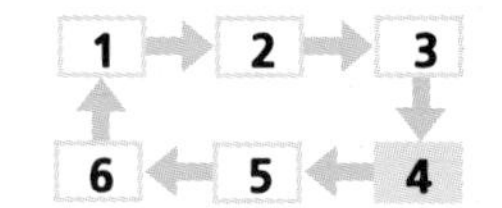

Solve the mathematical problem

Mathematical solution

If we let $1+\frac{r}{100} = x$ and $P = f(x)$ then:
$f(x) = 5000x^3 + 5000x^2 + 3000x + 1000 - 10\,000x^4$
and we must find x such that $f(x) = 0$.
Using the Newton–Raphson method:

$$x_{n+1} = x_n - \frac{f(x_n)}{f'(x_n)}$$

$$= x_n - \frac{\left(5x_n{}^3 + 5x_n{}^2 + 3x_n + 1 - 10x_n{}^4\right)}{\left(15x_n{}^2 + 10x_n + 3 - 40x_n{}^3\right)}$$

choose $x_0 = 1$ (i.e. $r = 0$) for the starting value.

The first five iterations give:

$x_1 = 1.3333$
$x_2 = 1.2202$
$x_3 = 1.1923$
$x_4 = 1.1907$
$x_5 = 1.1907$

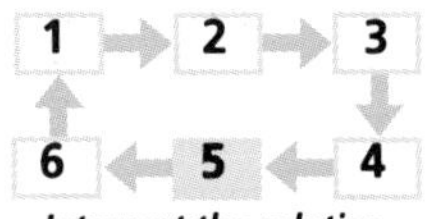

Interpret the solution

Interpretation

When $x = 1.1907$ the profit function is zero. Since $x = 1+\frac{r}{100}$ the critical interest rate is 19.07 per cent.

The graph of P against x shows that if $0 < r < 19.07$ the investment will make a profit. For example, for a constant interest rate of five per cent over the four year period the profit would be £3296.

Since bank interest rates to an investor are likely to be less than 19 per cent, the investment of £10 000 is recommended.

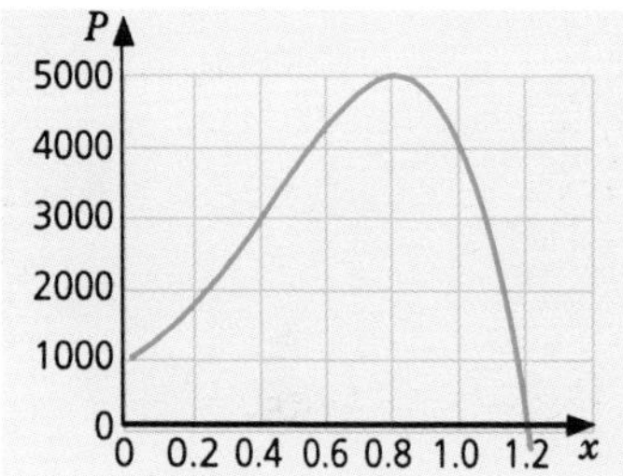

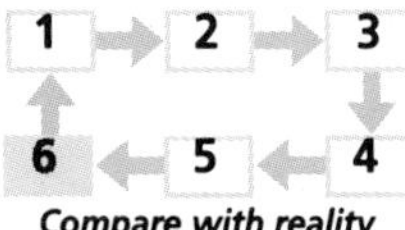

Compare with reality

Refinement of the model

Give a brief criticism of the model. Investigate the effect of changing the assumptions and the amount of the net returns.

CONSOLIDATION EXERCISES FOR CHAPTER 21

1 **a)** Show that the equation $x^3 + 3x^2 - 7 = 0$ may be rearranged into the form $x = \sqrt{\left(\frac{a}{x+b}\right)}$, and state the values of a and b.

b) Hence, using the iteration formula: $x_{n+1} = \sqrt{\left(\frac{a}{x_n+b}\right)}$ with $x_0 = 2$, together with your values of a and b, find the approximate solution x_4 of the equation, giving your answer to an appropriate degree of accuracy. Show your intermediate answers and explain why the degree of accuracy you have given for x_4 is appropriate.

(ULEAC Question 2, Specimen Paper P2, 1994)

2 By sketching the graphs of $y = 1 + \sin x$ and $y = x$, verify that the equation $x = 1 + \sin x$ has exactly one root.
Taking a suitable integer as first approximation, use an iterative method to find the value of this root correct to three decimal places. Explain how you can be confident that your result is accurate to three decimal places.

(Oxford Question 8, Specimen Paper P4, 1994)

3 **a)** Show that if x is a fixed point of iteration $x_{n+1} = \sqrt{3x_n + 2}$ then x satisfies the equation $x^2 - 3x - 2 = 0$.

b) Carry out four iterations for $x_{n+1} = \sqrt{3x_n + 2}$ using $x_0 = 1$ to obtain x_4.

c) On the same axes, sketch the graphs of $y = x$ and $y = \sqrt{3x+2}$.

Use your graphs to illustrate how the iteration $x_{n+1} = \sqrt{3x_n + 2}$ converges using $x_0 = 1$.

d) Suppose now that $x_0 = 6$. Use a sketch graph, as in part **c)**, to decide whether the sequence $x_{n+1} = \sqrt{3x_n + 2}$ converges or diverges. Do not carry out any calculations.

e) Write down the Newton–Raphson recurrence relation for the equation $x^2 - 3x - 2 = 0$ and use six iterations of the Newton-Raphson recurrence relation using $x_0 = 1$.

(Oxford (Nuffield) Question 7, Specimen Paper 3, 1994)

4 Show that the equation $x^3 - x^2 - 2 = 0$ has a root α which lies between 1 and 2.

a) Using 1.5 as a first approximation for α, use the Newton–Raphson method once to obtain a second approximation for α, giving your answer to three decimal places.

b) Show that the equation $x^3 - x^2 - 2 = 0$ can be arranged in the form $x = \sqrt[3]{\text{f}(x)}$ where f(x) is a quadratic function.

Use an iteration of the form $x_{n+1} = \text{g}(x_n)$ based on this rearrangement and with $x_1 = 1.5$ to find x_2 and x_3, giving your answers to three decimal places.

(AEB Question 11, Specimen Paper 1, 1994)

5 The chord AB of a circle subtends an angle θ radians at the centre O of the circle, as shown in the diagram. Find an expression for the shaded area, in terms of r and θ. Given that this shaded area is $\frac{1}{6}$ of the area of the circle, show that θ is given by:

$$\sin\theta = \theta - \frac{\pi}{3}.$$

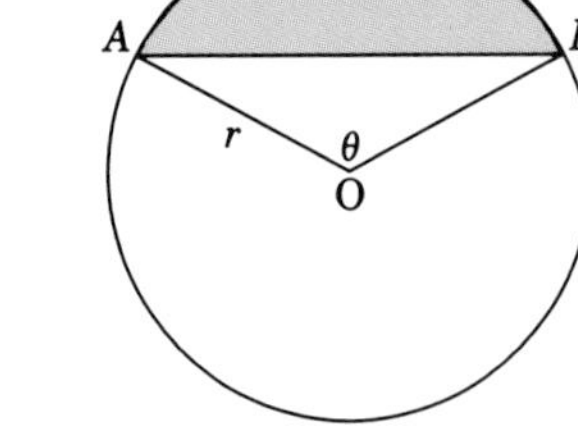

By sketching the graphs of $y = \sin\theta$ and $y = \theta - \frac{\pi}{3}$ on the same diagram, verify that $\theta = 2$ is an approximate solution of the equation $\sin\theta = \theta - \frac{\pi}{3}$. Find a better approximation for θ using one application of Newton's approximation.

(NEAB Question 9, Specimen Paper 2, 1994)

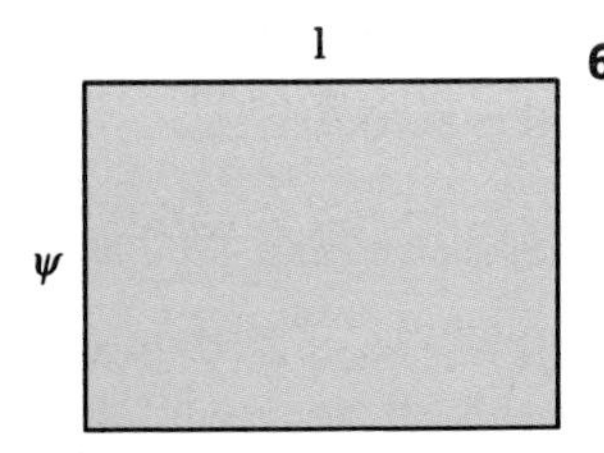

6 A golden rectangle has one side of length 1 unit and a shorter side of length ψ units, where ψ is called the golden section.

ψ can be found using the iterative formula: $x_{n+1} = \sqrt[3]{x_n(1 - x_n)}$.

Choosing a suitable value for x_1 and showing intermediate values, use this iterative formula to obtain the value of ψ to two decimal places.

(SMP 16–19 Question 1, Specimen Paper, 1994)

7 On a single diagram, sketch the graphs of $y = \ln(10x)$ and $y = \frac{6}{x}$, and explain how you can deduce that the equation $\ln(10x) = \frac{6}{x}$ has exactly one real root.
Given that the root is close to 2, use the iteration:

$$x_{n+1} = \frac{6}{\ln(10x_n)}$$

to evaluate the root correct to three decimal places.
The same equation may be written in the form $x\ln(10x) - 6 = 0$. Taking $f(x)$ to be $x\ln(10x) - 6$, find $f'(x)$, and show that the Newton-Raphson iteration for the root of $f(x) = 0$ may be simplified to the form:

$$x_{n+1} = \frac{x_n + 6}{1 + \ln(10x_n)}.$$

Starting with initial approximation 2, it takes nine iterations using:

$$x_{n+1} = \frac{6}{\ln(10x_n)}$$

before successive iterates agree to six decimal places. Find how many iterations are needed to achieve this degree of accuracy using the Newton–Raphson method with the same initial value.

(UCLES Linear Question 15, Specimen Paper 1, 1994)

8 The sequence given by the iteration formula: $x_{n+1} = 2(1 + e^{-x_n})$, with $x_1 = 0$, converges to α. Find α correct to three decimal places, and state an equation of which α is a root.

(UCLES (Modular) Question 3, Specimen Paper 3, 1994)

9 Use the Newton–Raphson method, with initial approximation 14, to find, correct to two decimal places, the positive root of the equation $x = 5\ln(x + 2)$.

(UCLES (Modular) Question 3, Specimen Paper 4, 1994)

10 Mavis wants to buy a computer originally advertised at £400, so she starts to do a paper round. She puts a fixed amount of her earnings each week into a savings account, which with interest produces **total** savings after t weeks of approximately £$2500(e^{0.002t} - 1)$.
Meanwhile the price of the computer drops by an average of £2 a week. Find whether she will have saved enough to buy the computer after 50 weeks.
Now take the formula for her savings to be exact, and assume that the price falls by exactly £2 a week. Find an equation for the number of weeks she will need to work before she has enough money to buy the computer. Show that this equation can be put into the form:
$t = a\ln(b - ct)$
where a, b, c are numbers. By carrying out five steps of the iteration:
$t_{n+1} = a\ln(b - ct_n)$
starting with the first approximation $t_0 = 50$, find the solution of this equation, rounding up your answer to the next highest integer.

(Oxford & Cambridge Question 3, Specimen Paper 2, 1994)

11 A sketch graph of $y = x^2 + px + q$ is shown. The vertex of the graph is at $(2, -3)$.

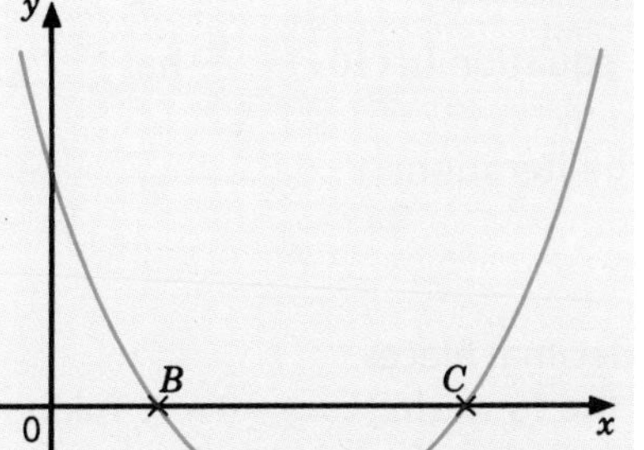

a) Find p and q. Hence calculate the x-coordinates of the points B and C.

b) Draw a sketch graph to show that the equation

$x^2 + px + q = \sqrt{x}$ has two solutions.

c) Use the iterative formula: $x_{n+1} = \frac{1}{4}\left(x_n^2 - \sqrt{x_n} + 1\right)$

and a start value of 0.2 to find the smaller of the two roots correct to two decimal places, showing all intermediate values.

(SMP 16–19 (Pure with Applications) Question 10, June 1994)

12 An iterative sequence is defined by: $x_{n+1} = \dfrac{2x_n^3 + a}{3x_n^2} \ (n = 0, 1, 2, \ldots)$

where $a > 0$.

a) Assuming that x_n tends to a limit L as $n \to \infty$, show that $L = \sqrt[3]{a}$.

b) **Use this result** to find $\sqrt[3]{3}$ correct to three decimal places, showing your working carefully.

(WJEC Question 4, Specimen Paper A1, 1994)

Summary

■ **Interval bisection**

For a continuous function f and interval $[a, b] = a \le x \le b$, if $f(a)$ and $f(b)$ have opposite signs, then $f(x) = 0$ for some x in this interval.

Let $c = \dfrac{(a+b)}{2}$.

If $f(a)$ and $f(c)$ have opposite signs, let $c = b$ otherwise let $c = a$. Repeat until the interval $[a, b]$ containing a root of the equation is narrow enough to give the desired accuracy.

■ **Fixed point iteration**

To solve $f(x) = 0$, rearrange into the form $x = g(x)$.

The iteration: $x_{n+1} = g(x_n)\ n = 0, 1, 2, \ldots$

generates a convergent sequence towards a root α provided $|g'(\alpha)| < 1$ and initial approximation x_0 is close to α.

■ **Newton–Raphson iteration**

To solve $f(x) = 0$, find $f'(x)$ and use the iteration:

$$x_{n+1} = x_n - \frac{f(x_n)}{f'(x_n)} \quad n = 0, 1, 2, \ldots$$

which generates a convergent sequence towards a root α provided the initial approximation x_0 is close to α.

PURE MATHS

22

Trigonometry 4: The calculus of trigonometric functions

- *In Chapter 7, Differentiation 1, we introduced the process of differentiation by finding a gradient function numerically. We extended this idea in finding the gradient function of e^x and $\ln x$ in Chapter 18. We shall use the same idea to discover the gradient functions for sin x, cos x and tan x.*
- *To define the derivatives of trigonometric functions we must measure angles in radians. Make sure that you – and your calculator – are in radian mode throughout this chapter!*

DIFFERENTIATION

Exploration 22.1

Differentiating sin x, cos x and tan x

The diagram shows the graph of $f(x) = \sin x$ for $0 \leq x \leq 2\pi$.

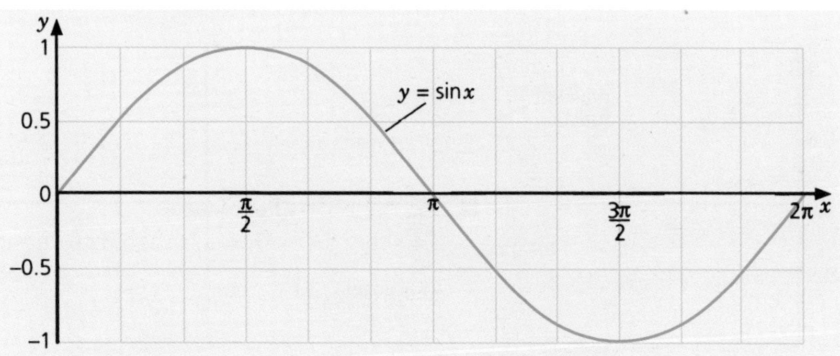

- Use a numerical method to estimate the gradient, $f'(x)$, at intervals of $\frac{\pi}{6}$.
- Copy this table and use your values to complete it.

x	0	$\frac{\pi}{6}$	$\frac{\pi}{3}$	$\frac{\pi}{2}$	$\frac{2\pi}{3}$	$\frac{5\pi}{6}$	π	$\frac{7}{6}\pi$	$\frac{4}{3}\pi$	$\frac{\pi}{2}$	$\frac{5}{3}\pi$	$\frac{11}{6}\pi$	2π
$f(x)$	0	0.5	0.866	1	0.866	0.5	0	–0.5	–0.866	–1	–0.866	–0.5	0
$f'(x)$													

- Sketch the graph of $y = f(x)$ and superimpose on it a graph of the gradient function.
- Deduce the algebraic form for $f'(x)$.
- Repeat the exploration for graphs of:
  $y = \cos x$, $y = \tan x$, $y = \sin ax$, $y = \cos ax$, $y = \sin^2 x$, etc.
- What patterns do you get in finding the gradient functions?

Interpreting the results

The final row in the table should look like this.

$f'(x)$	1	0.866	0.5	0	–0.5	–0.866	–1	–0.866	–0.5	0	0.5	0.866	1

Superimposing the graph of the gradient function on the original graph gives this result.

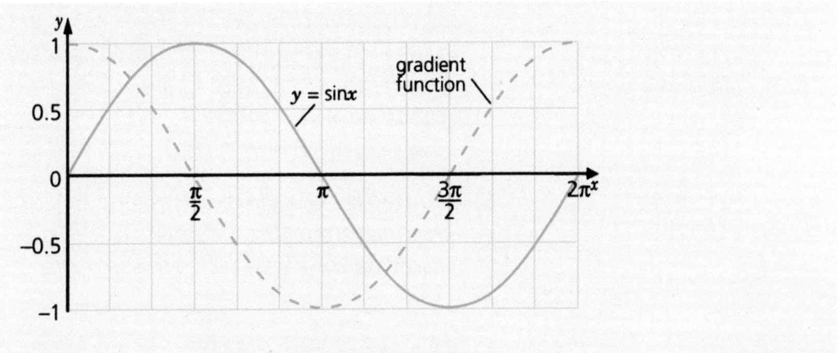

We can see that the gradient function is $f'(x) = \cos x$ i.e. the derivative of $\sin x$ is $\cos x$. Graphical outputs from completing some of the other explorations give these curves.

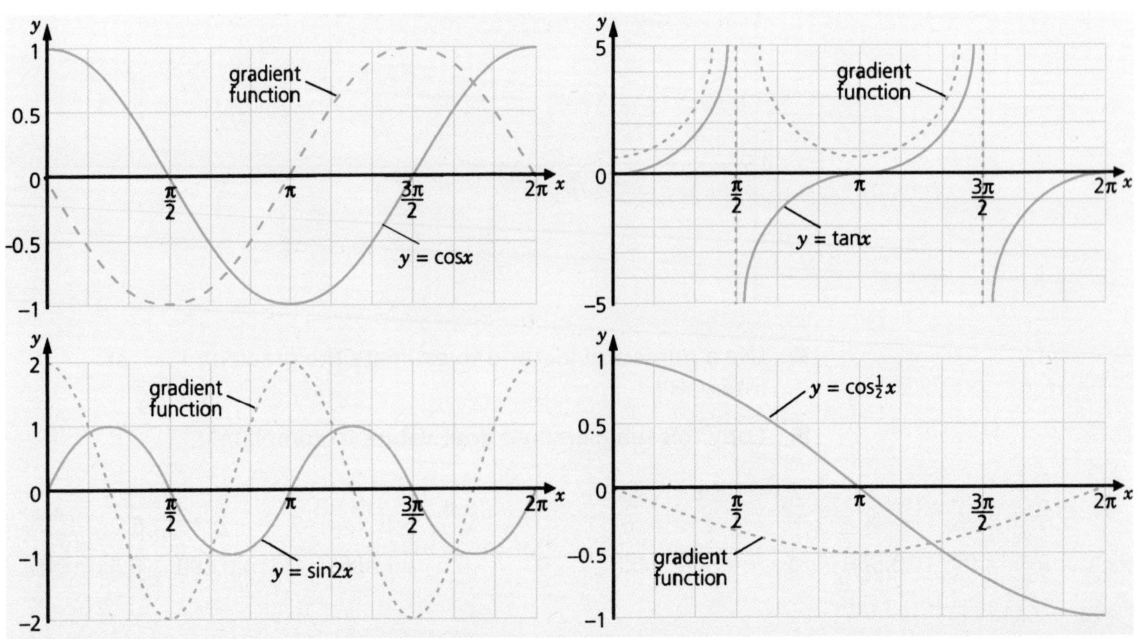

Examining the shape of the gradient function for each graph reveals the results in this table.

f(x)	f′(x)
$\sin x$	$\cos x$
$\cos x$	$-\sin x$
$\tan x$	$\sec^2 x$
$\sin ax$	$a\cos ax$
$\cos ax$	$-a\sin ax$

For $y = \sin^2 x$ and $y = \cos^2 x$, we can obtain the graphical output for the gradient functions as shown in this diagram.

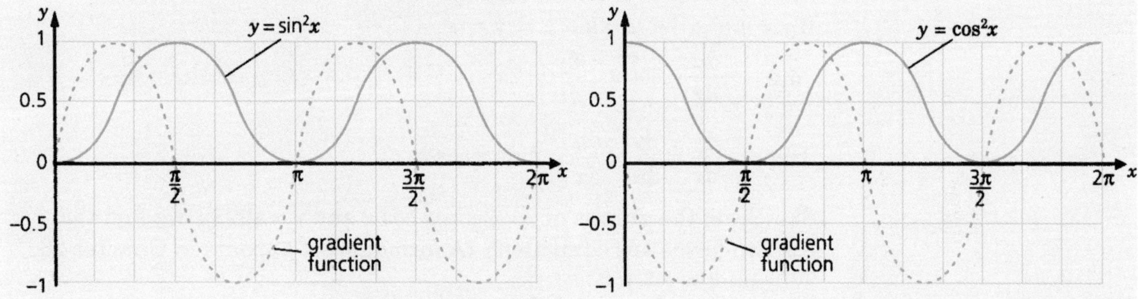

Examining the shape of the gradient function for each graph gives the additional results as in this table.

f(x)	f′(x)
$\sin^2 x$	$\sin 2x$
$\cos^2 x$	$-\sin 2x$

Starting with the two results:

$f(x) = \sin x \Rightarrow f'(x) = \cos x$ and

$f(x) = \cos x \Rightarrow f'(x) = -\sin x$

We can use techniques developed in Chapter 20, *Differentiation 3*, to prove the other results, as illustrated in the following examples.

Example 22.1

Show that the derivative of $\sin ax$ *is* $a\cos ax$.

Solution

Let $y = \sin ax$, *then we require* $\frac{dy}{dx}$.

Using the chain rule:

Let $u = ax \Rightarrow y = \sin u \Rightarrow \frac{dy}{du} = \cos u$ and $\frac{du}{dx} = a$

hence $\frac{dy}{dx} = \frac{dy}{du} \times \frac{du}{dx} = \cos u \times a = a\cos ax$

Example 22.2

Show that the derivative of $\tan x = \sec^2 x$.

Solution

Let $y = \tan x = \frac{\sin x}{\cos x}$, then we require $\frac{dy}{dx}$.

Using the quotient rule:
let $u = \sin x$ *and* $v = \cos x \Rightarrow \quad \frac{du}{dx} = \cos x \quad \frac{dv}{dx} = -\sin x$

Hence $\frac{dy}{dx} = \frac{v\frac{du}{dx} - u\frac{dv}{dx}}{v^2} = \frac{\cos^2 x + \sin^2 x}{\cos^2 x} = \frac{1}{\cos^2 x} = \sec^2 x$

An alternative method

The derivative of $\sin^2 x$ can also be found by applying the chain rule, but a different result is obtained.
If $y = \sin^2 x$, let $u = \sin x \Rightarrow y = u^2$

and $\frac{du}{dx} = \cos x \quad \frac{dy}{du} = 2u$

Hence $\frac{dy}{dx} = \frac{dy}{du} \times \frac{du}{dx} = 2\sin x \cos x$

Sketching the graphs of $y = 2 \sin x \cos x$ and $y = \sin 2x$, we find that the two forms are equivalent. (A formal proof appears in Chapter 26, *Trigonometry 5*.)

Now that we can differentiate trigonometric functions we can find rates of change for various wave models, as illustrated in the following example.

Example 22.3

The depth of water in a harbour, y metres, can be roughly modelled by the equation:

$$y = 4\sin\left(\frac{\pi}{6}t\right) + 7$$

where t is the number of hours after midnight on a certain day. Find the rate of change of height at 10.00 and determine the times in the day when the water level is falling fastest.

Solution
A sketch graph of $y = 4\sin\left(\frac{\pi}{6}t\right) + 7$ *for* $0 \le t \le 12$ *is shown in the diagram.*

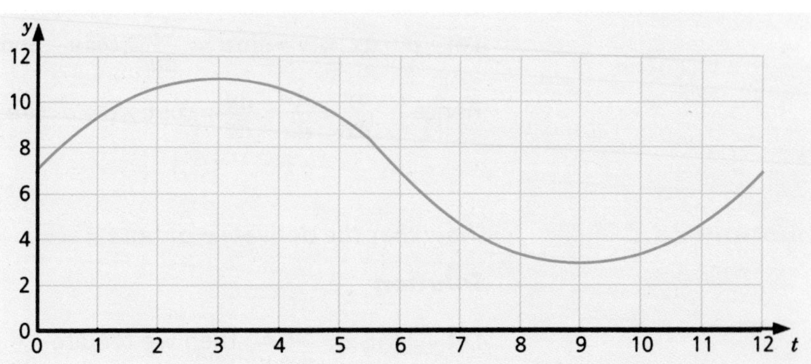

$$y = 4\sin\left(\frac{\pi}{6}t\right) + 7 \Rightarrow \frac{dy}{dt} = 4 \times \frac{\pi}{6}\cos\left(\frac{\pi}{6}t\right) = \frac{2}{3}\pi\cos\left(\frac{\pi}{6}t\right)$$

when $t = 10,\ \frac{dy}{dt} = \frac{2}{3}\pi\cos\left(\frac{\pi}{6} \times 10\right) = \frac{1}{3}\pi \approx 1.05$

i.e. the depth is rising at a rate of just over 1 metre per hour.

To find the times at which the water level is falling fastest we need to minimise $\frac{dy}{dt}$ *by differentiating again and equating to zero.*

$\frac{d^2y}{dt^2} = -\frac{\pi^2}{9}\sin\left(\frac{\pi}{6}t\right) = 0$ *for stationary value*

$\Rightarrow t = 0,\ 6,\ 12,\ 18$ *or* 24

From the shape of the original graph we can see that minimum values for $\frac{dy}{dt}$ *will occur at 06.00 and 18.00.*

EXERCISES

22.1A

1 Differentiate the following with respect to x.

a) $3\sin x$ **b)** $4\cos x$ **c)** $\frac{1}{2}\tan x$ **d)** $\frac{1}{4}\sin x$

e) $3\sin x + 4\cos x$ **f)** $5 - \tan x$ **g)** $x^2 + 8\sin x$ **h)** $e^{3x} + \tan x - 6$

2 Differentiate the following with respect to x.

a) $3\sin 7x$ **b)** $-4\cos 11x$ **c)** $\frac{5}{2}\tan 4x$

d) $3\sin(2x - 5)$ **e)** $4\tan\frac{1}{2}x - 6\sin\frac{1}{3}x$ **f)** $15\cos\frac{1}{5}x - 12\sin\frac{1}{4}x$

g) $\frac{5\ln 6x + 9\cos(5x + 7)}{10}$ **h)** $e^{4x} + 4^x - \sin(1 - 3x)$

3 Use the chain rule to differentiate these.

a) $\sin^2 x$ **b)** $\sin x^2$ **c)** $\cos\left(\frac{1}{x}\right)$

d) $\sqrt{1 + \cos x}$ **e)** $\tan^3 2x$ **f)** $\ln(1 + \sin 2x)$

g) $e^{\sin x}$ **h)** $\sin e^x$ **i)** $\ln\cos x$

j) $\cos\ln x$

4 Differentiate the following products.

a) $x\tan x$ **b)** $\sin x\cos 2x$ **c)** $x^3\cos 2x$

d) $e^x\cos x$ **e)** $e^{-\frac{1}{2}x}\sin 2x$ **f)** $\sqrt{x}\tan x$

g) $\sin x\ln x$ **h)** $xe^{\sin x}$

5 Use the quotient rule to differentiate these.

a) $\dfrac{\sin x}{x}$ **b)** $\dfrac{\sin^4 x}{\cos^3 x}$ **c)** $\dfrac{e^x}{\sin x}$

d) $\dfrac{\sin x}{\sqrt{x}}$ **e)** $\dfrac{x^2}{\sin x + \cos x}$ **f)** $\dfrac{x^2 + x + 1}{\tan x}$

6 By writing $\sec x$ as $\dfrac{1}{\cos x}$ and using the quotient rule, show that:

$\dfrac{d}{dx}(\sec x) = \sec x \tan x$.

Use a similar method to differentiate $\operatorname{cosec} x$ and $\cot x$.

Hence differentiate $\sec(3x - 1)$, $\operatorname{cosec} \frac{1}{2}x$ and $\frac{4}{5}\cot(\frac{3}{4}x - \frac{\pi}{4})$.

7 Show that the tangent to the curve $y = 3\cos x + \ln 4x$ at the point where $x = \pi$, has equation

$y = \dfrac{1}{\pi}x + \ln 4\pi - 4$

and find the equation of the normal at this point.

8 Find the equation of the tangent to the curve $y = 5\cos(4x - \frac{\pi}{6})$ at the point where $x = \dfrac{\pi}{4}$.

9 Find the equations of the tangent and normal to $y = \dfrac{3}{2}\sin x - 2\sin 2x$ when $x = \dfrac{\pi}{2}$.

10 Find the maximum and minimum values of the following in the domain $0 \le x \le \pi$.

a) $\sin x \cos x$ **b)** $\cos x - \sin x$

c) $\sin^3 x \cos x$ **d)** $2\sin x - \cos 2x$

11 Find the stationary points of the following functions in the domain $0 \le x \le \pi$ and distinguish between them. In each case, give a rough sketch of the graph of the function.

a) $f(x) = \cos 3x$ **b)** $f(x) = x - \sin 2x$

c) $f(x) = e^x \sin x$ **d)** $f(x) = \dfrac{4\sin x}{2 - \cos x}$

12 A mass, on the end of a spring, is oscillating in a vertical line between 30 cm and 70 cm above a bench, completing 30 oscillations per minute. The height of the mass, h cm, above the bench after t seconds from being let go from its lowest position can be modelled by a function of the form $h = c - a\cos(bt)$.

a) Find values for a and c and explain why $b = \pi$.

b) Sketch a graph of h against t for $0 \le t \le 2$.

c) Find the velocity of the mass at **i)** $t = 1$, **ii)** $t = 1.5$.

d) Find the acceleration of the mass at **i)** $t = 1$, **ii)** $t = 1.5$.

e) Comment on your answers to **c)** and **d)**.

EXERCISES

22.1 B

The following results, which may be needed to answer some of the questions in this exercise, may be assumed.

$\dfrac{d}{dx}(\sec x) = \sec x \tan x \quad \dfrac{d}{dx}(\operatorname{cosec} x) = -\operatorname{cosec} x \cot x \quad \dfrac{d}{dx}(\cot x) = -\operatorname{cosec}^2 x$

1 Differentiate the following with respect to x.

a) $7\sin x$ **b)** $-\frac{3}{4}\cos x$ **c)** $-11\tan x$

d) $\dfrac{\cos x}{5}$ **e)** $\dfrac{\sin x - \cos x}{10}$ **f)** $\dfrac{x^5 - 5\ln x + 3\cos x}{15}$

g) $3\tan x + \ln 7x$ **h)** $\dfrac{25e^{\frac{x}{5}} - 2\cos x}{20}$

2 Differentiate the following with respect to x.

a) $\dfrac{5\sin 5x}{4}$ **b)** $-\dfrac{\tan \frac{4}{3}x}{4}$ **c)** $\frac{3}{4}\cos 2x$

d) $-\frac{6}{7}\cos 11x + \frac{1}{4}\tan 11x$ **e)** $\frac{5}{2}\sin(4x-3) + \frac{2}{3}\cos(3x-4)$

f) $\sin(\frac{2\pi}{3} - 5x) + \cos(\frac{3\pi}{4} + 6x)$ **g)** $11\ln\dfrac{x}{5} - 5\tan\dfrac{x}{4} + \sqrt{x}$

h) $\dfrac{\sqrt{x}^7}{21} + \cos(\frac{\pi}{9} - 6x) - \dfrac{5}{x}$

3 Use the chain rule to differentiate these.

a) $\cos^3 x$ **b)** $\sin^4 5x$ **c)** $\cos\sqrt{x}$

d) $\sqrt{x + \cos x}$ **e)** $\cos^6 \frac{1}{2}x$ **f)** $\ln(x + \sin x)$

g) $e^{\cos x}$ **h)** $\cos e^x$ **i)** $\ln(\sin 3x + \cos 3x)$

j) $\tan\left(\dfrac{1}{x^2}\right)$

4 Differentiate the following products.

a) $x^2 \sin^2 x$ **b)** $\cos^2 x \sin 4x$ **c)** $\tan 3x \sin 4x$

d) $x \sin 4x$ **e)** $\sin(4x + \frac{\pi}{3})\cos(2x - \frac{\pi}{4})$ **f)** $e^{-x}\sin x$

g) $(1+x)^3\cos^3 x$ **h)** $\sec x \ln x$

5 Use the quotient rule to differentiate these.

a) $\dfrac{\cos x}{x}$ **b)** $\dfrac{1+\sin x}{1-\cos x}$ **c)** $\dfrac{\tan x}{e^x}$

d) $\dfrac{\ln x}{\tan x}$ **e)** $\dfrac{1+\sin x}{1-\sin x}$ **f)** $\dfrac{1+\sec x}{1-\sec x}$

6 By converting $x°$ to radians, find $\frac{d}{dx}(\sin x°)$, $\frac{d}{dx}(\cos x°)$ and $\frac{d}{dx}(\tan x°)$.

Hence differentiate $\sin 5x°$, $\cos\frac{1}{2}x°$ and $\tan(3x - 45)°$.

7 Show that the tangent to the curve $y = \frac{2}{\pi}\sin x - \frac{1}{x}$, at the point where $x = \frac{\pi}{2}$, has the equation $\pi^2 y = 4x - 2\pi$ and find the equation of the normal at this point.

8 Find the equations of the tangent and normal to $y = 5\tan(\frac{1}{2}x - \frac{\pi}{4})$ at the point where $x = \pi$.

9 Find the equations of the tangent and normal to $y = 16\sin x - 4e^{8x}$ when $x = 0$.

10 Find the maximum and minimum values of the following in the domain $0 \le x \le \pi$.

a) $2\sin x + \cos x$ **b)** $3\cos x - 4\sin x$

c) $3\cos x - \cos 3x$ **d)** $\sin x^2$

11 Find the stationary points of the following functions in the domain $0 \le x \le \pi$ and distinguish between them. In each case, give a rough sketch of the graph of the function.

a) $f(x) = \sin 5x$ **b)** $f(x) = 2x - \tan x$

c) $f(x) = \tan^2 x + 2\tan x$ **d)** $f(x) = 4\sin x + \frac{9}{1 + \sin x}$

12 The height of a car's suspension can be modelled by the equation

$h = 10e^{-0.1t}\cos\left(\frac{\pi}{6}t\right) + 15$, where h is height in cm and t is time in seconds.

a) Sketch a graph of h against t for $0 \le t \le 20$.

b) Find the value of:

$v = \frac{dh}{dt}$ and $a = \frac{dv}{dt}$ when

i) $t = 8$, **ii)** $t = 15$.

c) Relate your answers to part **b)** to the shape of the graph from part **a)**.

d) Find the time at which the suspension unit is closest to the ground. Find the clearance between the unit and the ground at this time.

INTEGRATION

In the first part of this chapter we discovered how to differentiate various trigonometric functions. Considering the process of integration as the reverse of differentiation gives the results in this table.

The following examples illustrate the way in which integrals involving trigonometrical functions can be used.

$f(x)$	$\int f(x)dx$
$\sin x$	$-\cos x$
$\cos x$	$\sin x$
$\sec^2 x$	$\tan x$
$\sin ax$	$-\frac{1}{a}\cos ax$
$\cos ax$	$\frac{1}{a}\sin ax$

Example 22.4

Find these integrals.

a) $\int \cos\ 3x\ \mathrm{d}x$ ***b)*** $\int 5\sec^2\tfrac{1}{2}x\ \mathrm{d}x$ ***c)*** $\int \tan^2 x\ \mathrm{d}x$

Solution

a) $\int \cos\ 3x\ \mathrm{d}x = \tfrac{1}{3}\sin\ 3x + c$

b) $\int 5\sec^2\tfrac{1}{2}x\ \mathrm{d}x = 10\tan\tfrac{1}{2}x + c$

c) $\int \tan^2 x\ \mathrm{d}x = \int\left(\sec^2 x - 1\right)\mathrm{d}x = \tan x - x + c$

Example 22.5

Explain, in terms of area, why: $\int_0^{\pi} 2\cos x\ \mathrm{d}x = 0$

and find the smallest positive value of θ *such that* $\int_0^{\theta} 2\cos x\ \mathrm{d}x = 1$.

Solution

The integral represents the area between the graph $y = 2\cos x$ *and the* x*-axis over the interval* $0 \le x \le \pi$.

From the diagram we can see that half the area lies above the x*-axis and half below. The definite integral is:*

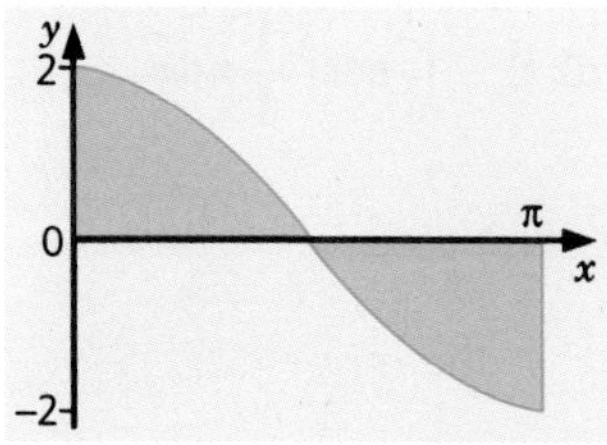

$$\int_0^{\pi} 2\cos x\ \mathrm{d}x = \left[2\sin x\right]_0^{\pi} = 0 - 0 = 0$$

If $\int_0^{\theta} 2\cos x\ \mathrm{d}x = 1$, *then* $\left[2\sin x\right]_0^{\theta} = 1$

$$\Rightarrow \quad 2\sin\theta - 2\sin 0 = 1$$

$$\Rightarrow \quad \sin\theta = \frac{1}{2}$$

$$\Rightarrow \quad \theta = \frac{\pi}{6} \text{ (smallest positive value)}$$

Example 22.6

Show that: $\dfrac{\mathrm{d}}{\mathrm{d}x}\left(\mathrm{e}^{-x}(\cos x + \sin x)\right) = -2\mathrm{e}^{-x}\sin x$

and hence find the value of $\int_0^{\pi} \mathrm{e}^{-x}\sin x$.

Illustrate your answer with a sketch.

Solution

Let $y = \mathrm{e}^{-x}(\cos x + \sin x) = \mathrm{e}^{-x}\cos x + \mathrm{e}^{-x}\sin x$.

Using the product rule to differentiate each product separately:

$$\frac{\mathrm{d}y}{\mathrm{d}x} = \mathrm{e}^{-x}(-\sin x) + \left(-\mathrm{e}^{-x}\right)\cos x + \mathrm{e}^{-x}(\cos x) + \left(-\mathrm{e}^{-x}\right)\sin x$$

$$= -2\mathrm{e}^{-x}\sin x$$

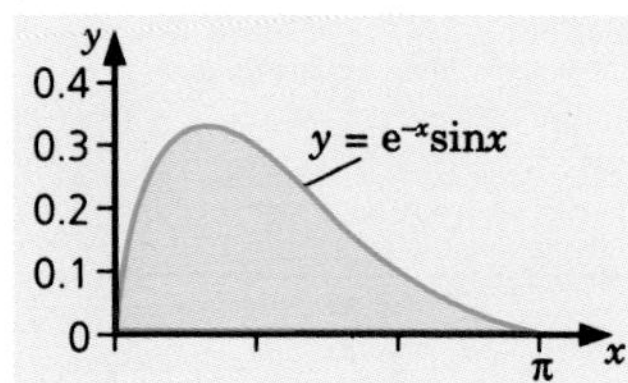

Hence $\displaystyle\int_0^{\pi} \mathrm{e}^{-x}\sin x\ \mathrm{d}x$

$$= -\tfrac{1}{2}\int_0^{\pi} -2\mathrm{e}^{-x}\sin x\ \mathrm{d}x$$

$$= -\tfrac{1}{2}\left[\mathrm{e}^{-x}(\cos x + \sin x)\right]_0^{\pi}$$

$$= -\tfrac{1}{2}\left\{\mathrm{e}^{-\pi}(-1+0) - \mathrm{e}^{0}(1+0)\right\}$$

$$= \tfrac{1}{2}\left(\mathrm{e}^{-\pi} + 1\right) \approx 0.522$$

EXERCISES

22.2 A

1 Find the following indefinite integrals.

a) $\int \sin 3x\,dx$ b) $\int \cos 6x\,dx$ c) $\int \cos \frac{1}{2}x\,dx$

d) $\int \frac{1}{2}\cos x\,dx$ e) $\int \cos \frac{3}{2}x\,dx$ f) $\int \sec^2 4x\,dx$

g) $\int 4\sec^2 x\,dx$ h) $\int (3\sin x - 2\cos x)\,dx$ i) $\int \frac{1}{2}(\sin x + \cos x)\,dx$

j) $\int \sin(2x+4)\,dx$ k) $\int 3\sec^2(\frac{\pi}{3} - x)\,dx$ l) $\int \cos(2x + \frac{\pi}{4})\,dx$

2 Evaluate the following definite integrals.

a) $\int_0^{\pi} \sin x\,dx$ b) $\int_{\frac{\pi}{6}}^{\frac{\pi}{2}} \cos x\,dx$ c) $\int_{-\frac{\pi}{4}}^{\frac{\pi}{4}} \sec^2 x\,dx$

d) $\int_0^{\frac{\pi}{4}} \cos 2x\,dx$ e) $\int_{-\frac{\pi}{2}}^{\frac{\pi}{2}} (1 + 2\sin 2x)\,dx$ f) $\int_{\frac{\pi}{2}}^{\pi} (2\cos x - \sin x)\,dx$

g) $\int_{-\frac{\pi}{2}}^{\frac{\pi}{2}} (\cos x + \sin 2x)\,dx$ h) $\int_0^{\frac{\pi}{9}} (\sin^2 x + \cos^2 x)\,dx$ i) $\int_0^{\pi} \sec^2 \frac{1}{4}x\,dx$

3 Show that $\frac{d}{dx}(\sin x + \frac{1}{2}\sin 2x + \frac{1}{3}\sin 3x) = (1 + 2\cos x)\cos 2x$ and hence find the value of $\int_0^{\frac{\pi}{4}} (1 + 2\cos x)\cos 2x\,dx$.

4 The diagram shows the graphs of $y = \sin x$ and $y = \cos x$ in the domain $-\frac{\pi}{2} \le x \le \pi$.

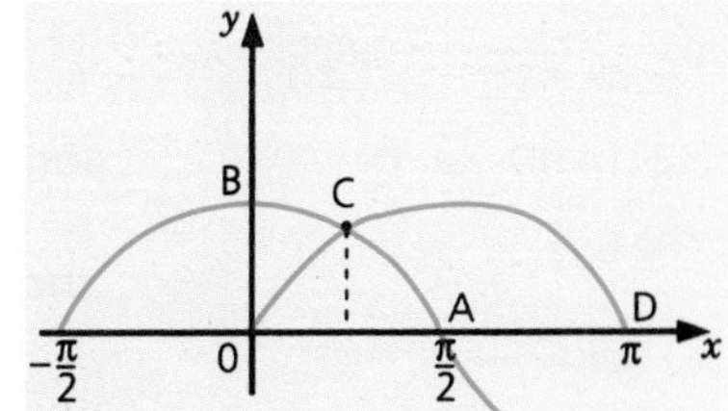

a) Find area OAC.
a) Find area OBC.
b) Find area ACD.

5 The diagram shows the graphs of $y = x$ and $y = \cos x$ in the domain $0 \le x \le \frac{\pi}{2}$.

a) Verify that the graphs intersect very close to $x = 0.7$.
b) Use an iterative method to find the x-coordinate of the point of intersection correct to **nine** d.p.
c) Hence find the area of the darker shaded region correct to nine d.p.
d) Similarly, find the area of the lighter shaded region correct to nine d.p.

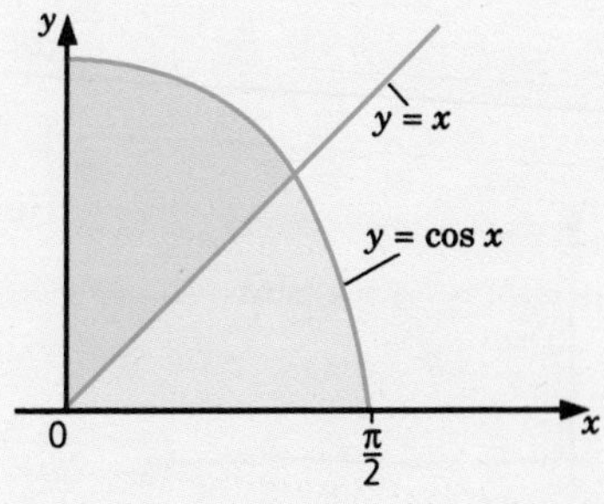

6 a) Differentiate $y = \ln \sin x$ and hence write down $\int \cot x \mathrm{d}x$.

b) Find $\int \cot 5x \mathrm{d}x$, $\int 3\cot \frac{1}{4}x \mathrm{d}x$ and $\int \cot(4x+\pi)\mathrm{d}x$.

EXERCISES

22.2 B

1 Find the following indefinite integrals.

a) $\int 3\cos 4x \mathrm{d}x$ b) $\int 5\sin \frac{1}{3}x \mathrm{d}x$ c) $\int \frac{1}{5}\sin \frac{1}{5}x \, \mathrm{d}x$

d) $\int \frac{\sin x}{5}\mathrm{d}x$ e) $\int \sin \frac{1}{5}x \mathrm{d}x$ f) $\int \sec^2 \frac{3}{2}x \mathrm{d}x$

g) $\int 7\sec^2 \frac{x}{7}\mathrm{d}x$ h) $\int (4\sin 3x - 5\cos 7x)\mathrm{d}x$ i) $\int \frac{\sin \frac{1}{2}x + \cos \frac{1}{2}x}{2}\mathrm{d}x$

j) $\int \sin(\frac{x}{2} - 1)\mathrm{d}x$ k) $\int 2\sec^2(x - \frac{\pi}{2})\mathrm{d}x$ l) $\int \cos(\frac{\pi}{4} - 3x)\mathrm{d}x$

2 Evaluate the following definite integrals.

a) $\int_0^{\frac{\pi}{2}} \cos x \mathrm{d}x$ b) $\int_{\frac{3\pi}{2}}^{2\pi} \sin x \mathrm{d}x$ c) $\int_{-\frac{\pi}{8}}^{\frac{\pi}{8}} \frac{\sin 4x}{2}\mathrm{d}x$

d) $\int_0^{\frac{\pi}{12}} 3\cos 2x \mathrm{d}x$ e) $\int_{-\frac{\pi}{6}}^{\frac{\pi}{6}} (\cos x - \sin x)\, \mathrm{d}x$ f) $\int_0^{\pi} 3\sin(\pi - x)\mathrm{d}x$

g) $\int_{\frac{\pi}{4}}^{\frac{\pi}{2}} (\sin x + \sin 2x)\mathrm{d}x$ h) $\int_0^{\pi} (1 - \cos 2x)\mathrm{d}x$ i) $\int_0^{\frac{\pi}{8}} \sec^2 2x \mathrm{d}x$

3 Differentiate $e^{-x}(\sin x - \cos x)$ and hence find the value of $\int_0^{\frac{\pi}{2}} e^{-x}\cos x \mathrm{d}x$.

Illustrate the area represented by this integral.

4 The diagram shows the graphs of $y = \sin x$ and $3\pi y = 2\pi - 3x$ in the domain $0 \le x \le \pi$.

a) Verify that the graphs intersect at $\left(\frac{\pi}{6}, \frac{1}{2}\right)$ and find the x-coordinate of A.

b) Find the area of OAC.

c) Find the area of OBC.

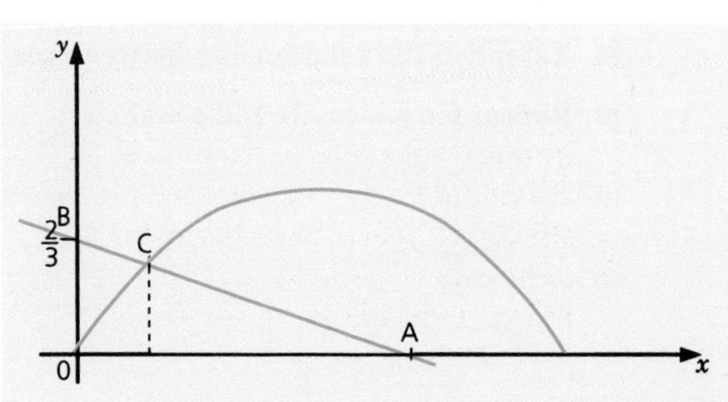

5 The diagram shows the graphs of $y = \sin x$ and $y = e^{-x}$ in the domain $0 \le x \le \pi$.

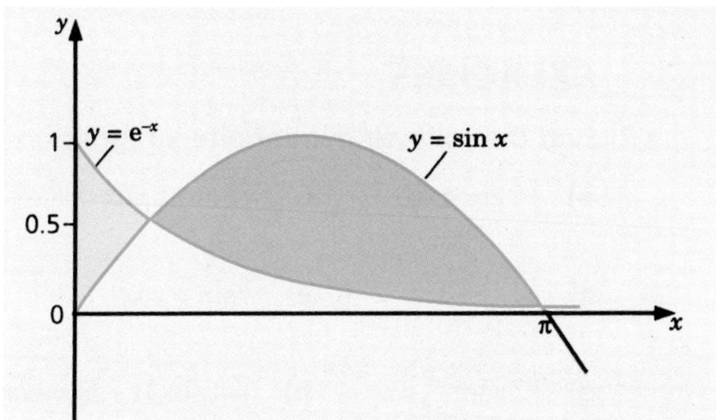

a) Verify that the graphs intersect very close to $x = 0.6$ and $x = 3$.

b) Use an iterative method to find the x-coordinates of the points of intersection correct to **nine** d.p.

c) Hence find the area of the **lighter** shaded region correct to nine d.p.

d) Similarly, find the area of the **darker** shaded region correct to nine d.p.

6 **a)** Differentiate $y = \ln \cos x$ and hence write down $\int \tan x \mathrm{d}x$.

b) Find $\int \tan 3x \mathrm{d}x$, $\int \tan \frac{1}{2}x \mathrm{d}x$ and $\int \tan(\frac{\pi}{4} - 2x)\mathrm{d}x$.

INTEGRATION OF POWERS OF TRIGONOMETRIC FUNCTIONS

CALCULATOR ACTIVITY

You will need a graph plotter.

Make sure you are using radian mode and set the axes as follows.

$x_{\min} = 0$; $x_{\max} = 2\pi$; $x_{\text{scl}} = \dfrac{\pi}{6}$

$y_{\min} = -1$; $y_{\max} = 2$; $y_{\text{scl}} = 0.5$

- Plot the graphs of $y = \cos 2x$ and $y = \cos^2 x$.
- By a sequence of geometrical transformations show how you can transform the graph of $y = \cos^2 x$ onto the graph of $y = \cos 2x$.
- Establish the relationship between $\cos 2\theta$ and $\cos^2\theta$ for any angle θ.
- Repeat for $y = \cos 2x$ and $y = \sin^2 x$.

A stretch, factor 2, parallel to the y-axis followed by translation through -1 unit parallel to the y-axis will transform $y = \cos^2 x$ onto $y = \cos 2x$. From this we deduce the identity:

$$\cos 2\theta \equiv 2\cos^2 \theta - 1$$

A similar result connecting $\cos 2\theta$ and $\sin^2\theta$ follows since:

$$\cos 2\theta \equiv 2(1 - \sin^2\theta) - 1$$

$$\Rightarrow \cos 2\theta \equiv 1 - 2\sin^2\theta$$

This shows that to transform $y = \sin^2 x$ onto $y = \cos 2x$ the operations are stretch, factor 2, followed by reflection in the x-axis followed by a translation through $+1$ unit parallel to the y-axis.

The last calculator activity should also lead to the same conclusion.

The two trigonometrical identities may be used to give both $\cos^2\theta$ and $\sin^2\theta$ in terms of $\cos 2\theta$:

$$\cos 2\theta \equiv 2\cos^2\theta - 1 \Rightarrow \cos^2\theta \equiv \tfrac{1}{2} + \tfrac{1}{2}\cos 2\theta$$

$$\cos 2\theta \equiv 1 - 2\sin^2 \theta \Rightarrow \sin^2 \theta \equiv \tfrac{1}{2} - \tfrac{1}{2}\cos 2\theta$$

These identities are very useful when we are integrating powers of trigonometrical functions. This is illustrated in the following examples.

Example 22.7

The curve $y = 2\sin x$, $0 \le x \le \pi$ is rotated through 2π about the x-axis. Find the volume of the solid of revolution generated.

Solution

The volume swept out is illustrated in the diagram.

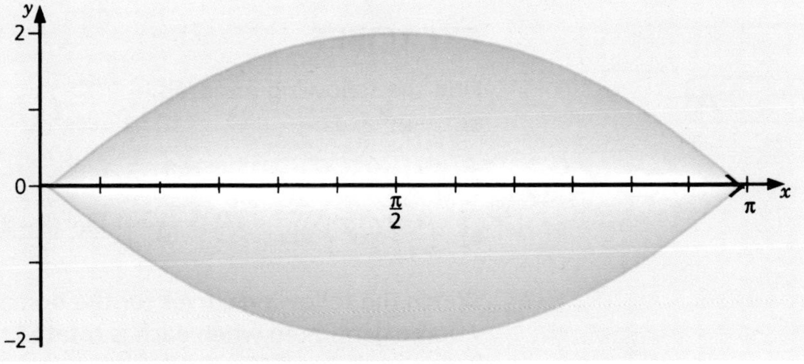

$$\text{Volume} = \int_0^{\pi} \pi y^2 dx = \pi\int_0^{\pi} 4\sin^2 x dx$$
$$= \pi\int_0^{\pi}(2 - 2\cos 2x)dx = \pi[2x - \sin 2x]_0^{\pi}$$
$$= \pi\{(2\pi - 0) - (0 - 0)\} = 2\pi^2$$

Example 22.8

Sketch the graph of $y = \cos^3 x$, $-\pi \le x \le \pi$. Hence find the area enclosed by the curve and the x-axis over the interval $-\frac{\pi}{2} \le x \le \pi$.

Solution

The graph of $y = \cos^3 x$, $-\pi \le x \le \pi$, is shown in the diagram.

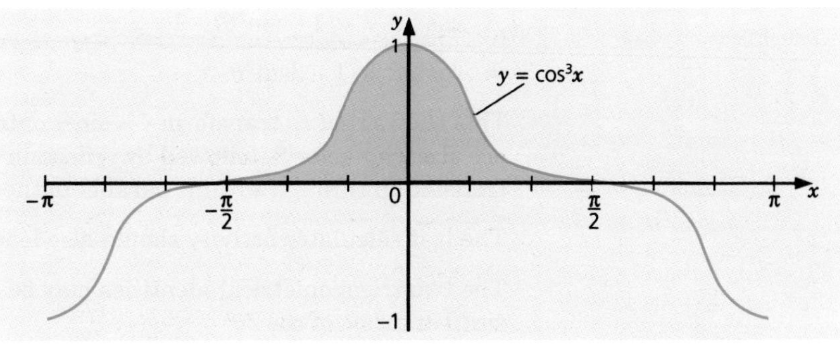

The area of the shaded region is given by:

$$\int_{-\frac{\pi}{2}}^{\frac{\pi}{2}} \cos^3 x\,dx$$

To integrate $\cos^3 x$ we make the following rearrangement.

$$\cos^3 x \equiv \cos^2 x \cos x \equiv \left(1-\sin^2 x\right)\cos x$$

$$\equiv \cos x - \sin^2 x \cos x$$

$$\int_{-\frac{\pi}{2}}^{\frac{\pi}{2}} \cos^3 x\,dx = \int_{-\frac{\pi}{2}}^{\frac{\pi}{2}} \left(\cos x - \sin^2 x \cos x\right)dx$$

$$= \left[\sin x - \frac{1}{3}\sin^3 x\right]_{-\frac{\pi}{2}}^{\frac{\pi}{2}} = \left(1-\frac{1}{3}\right)-\left(-1+\frac{1}{3}\right) = \frac{4}{3}$$

EXERCISES

22.3 A

1 Find the following integrals.

a) $\int \sin^2 2x\,dx$ b) $\int -\cos^2(3x - \frac{\pi}{3})dx$

2 Evaluate these.

a) $\int_{-\pi}^{\pi} 3\cos^2 \frac{1}{2}x\,dx$ b) $\int_0^1 \frac{\sin^2(3-2x)}{5}\,dx$

3 Sketch the following curves for the domain $0 \le x \le 2\pi$ and find the volume generated when each is rotated through 2π about the x-axis, between the x-values given.

a) $y = 1 + \sin x$ from $x = 0$ to $x = \frac{\pi}{2}$

b) $y = \sec x$ from $x = -\frac{\pi}{4}$ to $x = \frac{\pi}{4}$

c) $y = \tan x$ from $x = 0$ to $x = \frac{\pi}{4}$

4 **a)** Differentiate $\sin^3 x$ and hence show that $\int \sin^2 x \cos x \mathrm{d}x = \frac{1}{3}\sin^3 x + c$.

b) By writing $\cos^3 x$ as $\cos^2 x \cos x$, use an appropriate identity and the above result, to find $\int \cos^3 x \mathrm{d}x$.

c) Find the volume generated when the curve $y = \cos^{\frac{3}{2}} x$ is rotated about the x-axis from $x = 0$ to $x = \frac{\pi}{2}$.

5 **a)** Show that $\int \cos^2 2x \mathrm{d}x = \frac{1}{2}x + \frac{1}{8}\sin 4x + c$.

b) By writing $\cos^4 x$ as $(\cos^2 x)^2$, show that $\cos^4 x \equiv \frac{1}{4}(1 + 2\cos 2x + \cos^2 2x)$.

c) Use your result to find $\int \cos^4 x \mathrm{d}x$.

d) Find the volume generated when the curve $y = 3\cos^2 x$ is rotated completely about the x-axis from $x = 0$ to $x = \frac{\pi}{2}$.

6 **a)** Show that $\frac{\mathrm{d}}{\mathrm{d}x}(\sec x) = \sec x \tan x$.

b) Differentiate $\tan^2 x$ and use the result from **a)** to differentiate $\sec^2 x$. Explain why your two answers are the same.

7 **a)** Use the result of question **6a)** to show that $\frac{\mathrm{d}}{\mathrm{d}x}(\ln(\sec x + \tan x)) = \sec x$.

b) Deduce the expression for the indefinite integral $\int \sec x \, \mathrm{d}x$.

c) Use the above result to evaluate $\int_{-\frac{\pi}{3}}^{\frac{\pi}{3}} \sec x \mathrm{d}x$ and $\int_0^{\frac{\pi}{2}} \sec \frac{1}{2}x \mathrm{d}x$.

8 Use an appropriate double angle formula and the result of **7b)** above, to find $\int_{-\frac{\pi}{6}}^{\frac{\pi}{6}} \frac{\cos 2x}{\cos x} \mathrm{d}x$.

EXERCISES

22.3 B

1 Find the following integrals.

a) $\int \cos^2 5x \mathrm{d}x$ **b)** $\int \frac{1}{2}\sin^2 \frac{1}{4}x \mathrm{d}x$

2 Evaluate these.

a) $\int_0^{2\pi} \sin^2 2x \mathrm{d}x$ **b)** $\int_{\frac{\pi}{3}}^{\frac{\pi}{2}} \cos^2(x - \frac{\pi}{6}) \mathrm{d}x$

3 **a)** Use the fact that $\sin 2x = 2\sin x\cos x$ to find $\int \sin x\cos x\,dx$.

b) Hence find the volume generated when $y = \sin x + \cos x$ is rotated through 360° about the x-axis between $x = 0$ and $x = \frac{3\pi}{4}$.

4 **a)** Differentiate $\cos^3 x$ and hence show that $\int \cos^2 x\sin x\,dx = -\frac{1}{3}\cos^3 x + c$.

b) By writing $\sin^3 x$ as $\sin^2 x\sin x$, use an appropriate identity and the above result, to find $\int \sin^3 x\,dx$.

c) Find the area between the curve $y = 3\sin^3 x + \sin x$ and the x-axis from $x = 0$ to $x = \pi$.

5 **a)** Find $\frac{d}{dx}\left(e^x(\sin x - \cos x)\right)$ and $\frac{d}{dx}\left(e^x(\sin x + \cos x)\right)$.

b) Deduce the values of $\int e^x \sin x\,dx$ and $\int e^x \cos x\,dx$.

c) Find the area between the curve $y = e^x \sin x$ and the x-axis from $x = 0$ to $x = \pi$.

d) Find the volume generated when the area under $y = e^{0.5x}\sqrt{\cos x}$ is rotated completely about the x-axis from $-\frac{\pi}{2}$ to $\frac{\pi}{2}$.

6 **a)** Show that $\frac{d}{dx}(\operatorname{cosec} x) = -\operatorname{cosec} x\cot x$ and that $\frac{d}{dx}(\cot x) = -\operatorname{cosec}^2 x$.

b) Use the results from **a)** to differentiate $\operatorname{cosec}^2 x$ and $\cot^2 x$. Explain why your two answers are the same.

7 **a)** Use the results from question **6a)** to show that:
$\frac{d}{dx}(\ln(\operatorname{cosec} x + \cot x)) = -\operatorname{cosec} x$.

b) Deduce the indefinite integral $\int \operatorname{cosec} x\,dx$.

c) Use the above result to evaluate $\int_{\frac{\pi}{4}}^{\frac{3\pi}{4}} \operatorname{cosec} x\,dx$ and $\int_{\frac{\pi}{2}}^{\pi} \operatorname{cosec} \frac{1}{3}x\,dx$.

8 Use an appropriate double angle formula and the result of **7b)** above, to find $\int_{\frac{\pi}{4}}^{\frac{\pi}{2}} \frac{\cos 2x}{\sin x}\,dx$.

SMALL ANGLES

Exploration 22.2

sin x for small values of x

You will need a graph plotter.

- Working in radians, plot the graphs of $y = \sin x$ and $y = x$ on the same axes for $-\pi \le x \le \pi$.
- Tabulate values of $\sin x$ for *small* values of x, e.g. $-0.1 \le x \le 0.1$.

- Describe the relationship between $\sin x$ and x when x is small and measured in radians.
- Repeat for $y = \tan x$ and $y = x$.
- Repeat for $y = \cos x$ and $y = 1 - \frac{1}{2}x^2$.

Interpreting the results

The three pairs of graphs, together with tabulated output are shown in the diagrams.

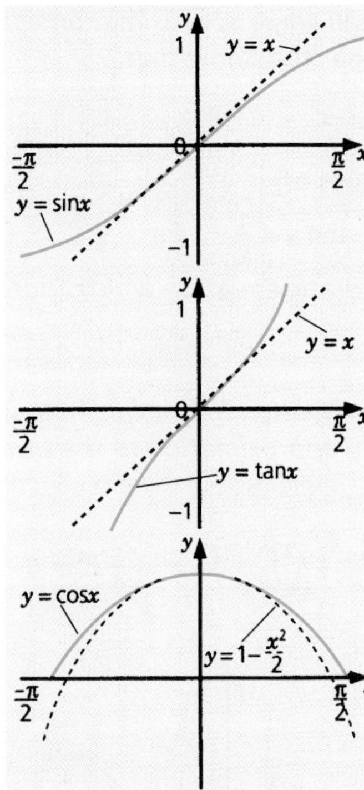

x	$\sin x$
−0.10	−0.099 83
−0.05	−0.049 98
0	0
0.05	0.049 98
0.10	0.099 83

x	$\tan x$
−0.10	−0.100 33
−0.05	−0.050 04
0	0
0.05	0.050 04
0.10	0.100 33

x	$\cos x$	$1-\frac{1}{2}x^2$
−0.10	0.995 00	0.995
−0.05	0.998 75	0.998 75
0	1.000 00	1
0.05	0.998 75	0.998 75
0.10	0.995 00	0.995

From the graphs and tables we see that, provided x is measured in radians and $|x|$ is small, the following **small angle approximations** are valid.

$\sin x \approx x \qquad \tan x \approx x \qquad \cos x \approx 1 - \frac{1}{2}x^2$

These approximate equivalences, for small values of x can be useful in finding approximate solutions to problems.

Example 22.9

Find small angle approximations to the following, where x is measured in radians.

a) $\dfrac{\sin 3x}{5x^2}$ ***b)*** $\dfrac{\sin x + \tan x}{1 - \cos 2x}$

Solution

a) $\dfrac{\sin 3x}{5x^2} \approx \dfrac{3x}{5x^2} = \dfrac{3}{5x}$

b) $\dfrac{\sin x + \tan x}{1 - \cos 2x} \approx \dfrac{x + x}{1 - \left(1 - \dfrac{(2x)^2}{2}\right)} = \dfrac{2x}{2x^2} = \dfrac{1}{x}$

Example 22.10

a) *Using a small angle approximation where terms in x^3 and higher may be neglected, show that:*

$$\frac{\cos^2 x}{x} \approx \frac{1}{x} - x$$

b) *Sketch graphs of:*

$$y = \frac{\cos^2 x}{x} \text{ and } y = \frac{1}{x} - x \text{ for } -\tfrac{\pi}{2} \le x \le \tfrac{\pi}{2}.$$

c) *Find a quadratic equation which approximates the solution of:*

$$\frac{\cos^2 x}{x} = 5$$

d) *Solve your equation in **c)**, explaining why only one of the roots is a very good approximation to the true solution.*

Solution

a) $\dfrac{\cos^2 x}{x} \approx \dfrac{\left(1 - \frac{1}{2}x^2\right)^2}{x} = \dfrac{1 - x^2 + \frac{1}{4}x^4}{x} \approx \dfrac{1}{x} - x$

b)

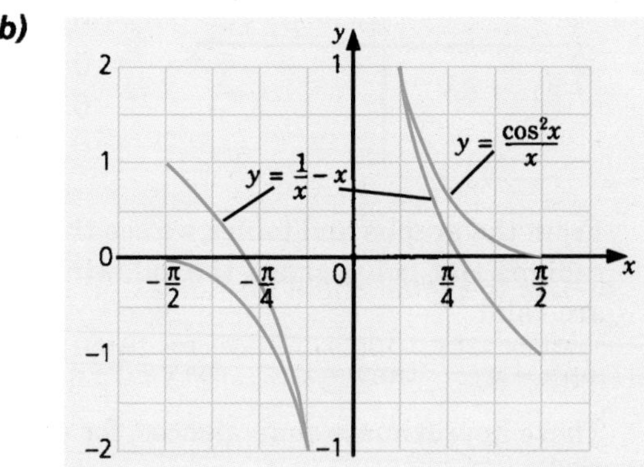

c) $\dfrac{\cos^2 x}{x} = 5 \Rightarrow \dfrac{1}{x} - x = 5$

will give a close approximation to the root of $\dfrac{\cos^2 x}{x} = 5$ *for small* x.

Also: $\dfrac{1}{x} - x = 5 \Rightarrow 1 - x^2 = 5x$

$$\Rightarrow x^2 + 5x - 1 = 0$$

d) *Solutions to the quadratic equation are $x = 0.193$ and $x = -5.19$ (correct to three significant figures). The positive root, 0.193 agrees with the true solution correct to three significant figures, whereas the negative root is totally inadequate.*

EXERCISES

22.4 A

1 Find the approximate values of the following expressions when x is small and measured in radians.

a) $\dfrac{\sin 5x}{\sin 2x}$ **b)** $\dfrac{10x}{\sin 5x}$ **c)** $\dfrac{\cos 2x}{x \sin x}$

d) $\dfrac{3\tan x - x}{\sin 3x}$ **e)** $\dfrac{1-\cos 2x}{\sin x \tan x}$ **f)** $\dfrac{\cos x - 1}{\sin x}$

g) $\sqrt{2+2\cos x}$ **h)** $\dfrac{1-\cos 3x}{1-\cos 4x}$ **i)** $\dfrac{\sin x \cos x}{\sqrt{2}+\sin x}$

2 **a)** Use the small angle approximations to show that the function $f(x) = 1 + \sin^2 x \cos x$ can be approximated by the quadratic function $g(x) = 1 + x^2$ when x is small and measured in radians.

b) Sketch graphs of the two functions for the domain $-1 \le x \le 1$.

c) It can be shown that the integral $\int_0^{\frac{\pi}{6}} f(x)\,dx = 0.565\,265$ correct to six d.p.
Verify that even though $\frac{\pi}{6}$ is not a particularly small angle,

$\int_0^{\frac{\pi}{6}} g(x)dx$ and $\int_0^{\frac{\pi}{6}} f(x)dx$ are the same, correct to two d.p.

d) Show that $f'(x) = 2\sin x - 3\sin^3 x$ and hence write down a cubic approximation to $f'(x)$.

e) Estimate the value of $f'(x)$ when $x = \frac{\pi}{6}$ and compare it with the true value.

3 **a)** Use the small angle formulae and the binomial theorem to find a quadratic approximation for the function $f(x) = \dfrac{3\tan x}{1+\sin x}$ and hence find the approximate value of the integral $\int_0^{0.1} \dfrac{3\tan x}{1+\sin x}dx$.

b) Revise your estimate for the integral by using a cubic approximation for $f(x)$.

EXERCISES

22.4 B

1 **a)** Use the small angle approximations to show that the function $f(x) = \cos x \cos 2x$ can be approximated by the quadratic function $g(x) = 1 - 2.5x^2$ when x is small and measured in radians.

b) Sketch graphs of the two functions for the domain $-1 \le x \le 1$.

c) It can be shown that the integral $\int_{-0.2}^{0.2} f(x)dx$ = 0.386 88 correct to five d.p. Verify that $\int_{-0.2}^{0.2} g(x)dx$ = 0.386 67 correct to five d.p.

d) Find $f'(x)$ and hence find the gradient of $f(x)$ when $x = 0.2$.

e) Write down a cubic approximation to $f'(x)$ and find an approximation for the gradient of $f(x)$ when $x = 0.2$.

f) Comment on the relative accuracy of your integration and differentiation approximations.

2 **a)** By graph sketching, verify that $2\cos(x - \frac{\pi}{3}) \equiv \cos x + \sqrt{3}\sin x$.

b) Use the small angle formulae to show that $2\cos(x - \frac{\pi}{3}) = 1 + \sqrt{3}x - \frac{1}{2}x^2$ when x is small and measured in radians.

c) Find the true value and an approximate value for $\int_{-0.05}^{0.05} \cos(x - \frac{\pi}{3})\,dx$ and compare them.

d) If x is so small that powers greater than 2 can safely be ignored, show further that $4\cos^2(x - \frac{\pi}{3}) \approx 1 + 2\sqrt{3}x + 2x^2$.

e) Use result **d)** to find the approximate value of $\int_{-0.05}^{0.05} \cos^2(x - \frac{\pi}{3})\,dx$.

3 **a)** Use the small angle approximations and the binomial theorem to find a cubic approximation to $f(x) = \dfrac{\sin 4x + \tan x}{5\cos 2x}$ when x is small and measured in radians.

b) Sketch graphs of $f(x)$ and your cubic function for the domain $-1 \le x \le 1$.

c) Find an approximate value for the integral $\int_0^{0.15} f(x)dx$ correct to six d.p.

d) The true value of the integral is 0.010 748 correct to six d.p. Can you explain why the result of your calculation in **c)** is not particularly accurate?

MATHEMATICAL MODELLING ACTIVITY

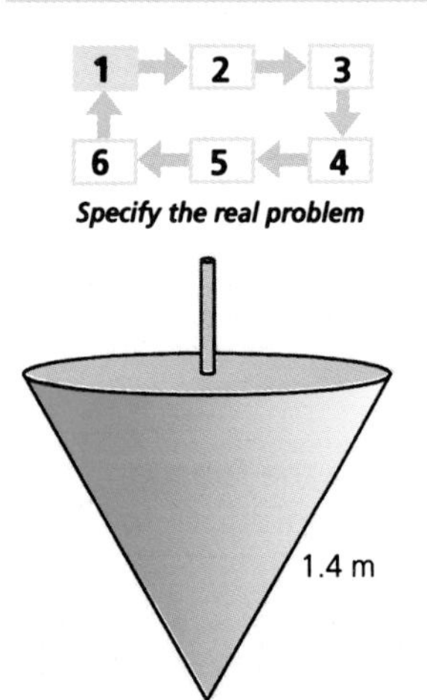

Problem statement

A manufacturer has designed a new concrete mixer in the shape of a cone of side length 1.4 metres. The cone is mounted with its axis vertical. Concrete is made by rotating a mixture of sand, cement and water inside the cone.

A builder wants to make the largest amount of concrete in any mix.

- What angle, between the side and the vertical through the tip of the cone, would you recommend to give the cone the maximum volume?

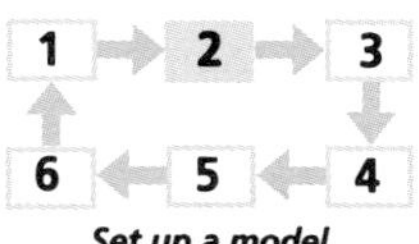

Set up a model

Set up a model

Firstly, identify the important variables. These are:

- the volume of the cone
- the angle between the side of the cone and the vertical axis.

You will also need to assume that the cone is full and does not spill when mixing.

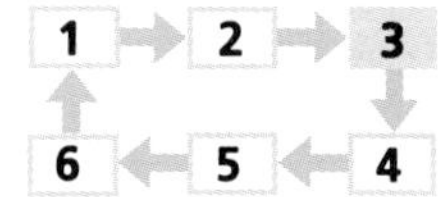

Formulate the mathematical problem

Mathematical problem

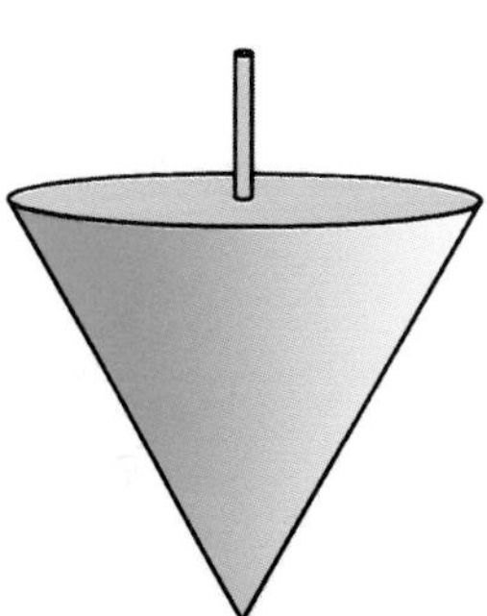

Firstly, establish the relationship between the volume within the cone (V) and the angle (θ).

The volume of a cone is $\frac{1}{3}$ × area of base × height. You should confirm that the equation connecting V and θ is:

$$V = \tfrac{1}{3}\pi(1.4)^3 \sin^2\theta\cos\theta = 2.87\sin^2\theta\cos\theta \text{ m}^3$$

The mathematical problem is to find the value of θ to maximise the volume V.

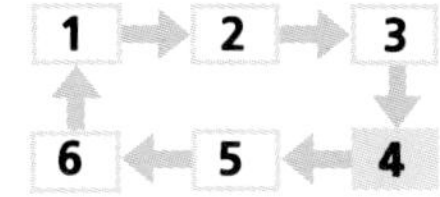

Solve the mathematical problem

Mathematical solution

To find where V reaches a maximum:

$$\frac{dV}{d\theta} = 2.87\left(2\sin\theta\cos^2\theta - \sin^3\theta\right) = 0 \text{ for stationary values}$$

$$\Rightarrow \sin\theta = 0 \text{ or } \cos^2\theta = \tfrac{1}{3}$$

Solving for $\theta \Rightarrow \theta = 0°$ or $\theta = 54.7°$.

To investigate the stationary values consider $\frac{d^2V}{d\theta^2}$.

$$\frac{dV}{d\theta} = 2.87\sin\theta\left(3\cos^2\theta - 1\right)$$

$$\frac{d^2V}{d\theta^2} = 2.87\cos\theta\left(3\cos^2\theta - 1\right) + 2.87\sin\theta(-6\cos\theta\sin\theta)$$

For $\theta = 0$, $\frac{d^2V}{d\theta^2} = 5.74 > 0$ which (clearly) leads to a minimum value of V.

For $\theta = 54.7°$, $\frac{d^2V}{d\theta^2} = -6.63 < 0$ which leads to a maximum value for V.

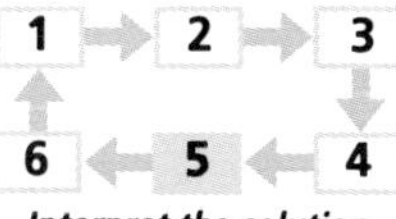

Interpret the solution

Interpreting the solution

To maximise the volume of concrete to be made in the concrete mixer the cone should be designed with an angle (θ) of 54.7°. This may seem to be a large angle! Discuss the effects of such a large angle and consider alternative angles.

CONSOLIDATION EXERCISES FOR CHAPTER 22

1 If $\frac{dy}{dx} = 6e^{3x} + 3\sin 2x$, find an expression for y.

(Oxford Question 3, Specimen Paper 2, 1994)

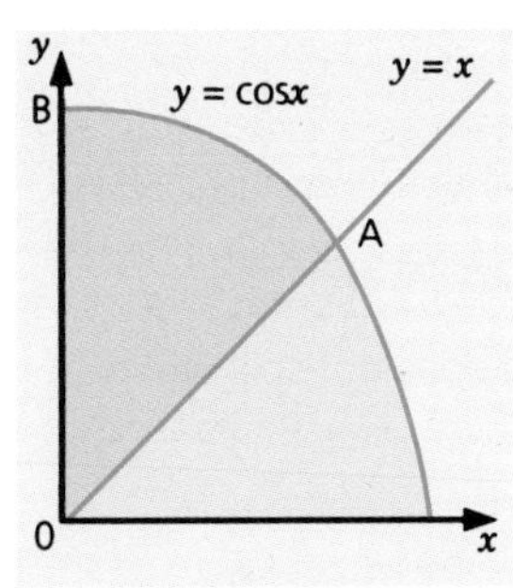

2 **a)** Part of the graphs for $y = x$ and $y = \cos x$ are shown. Taking $x = 0.8$ as the x-coordinate of the point A, calculate the area OAB shaded on the sketch.

b) Given that $x = 0.8$ is an approximate solution to the equation $x = \cos x$, use a single application of the Newton-Raphson method to obtain a better solution.

c) Draw a sketch which illustrates that the equation $x = \cos x$ has only one solution.

(SMP 16–19 Question 14, Specimen Paper, 1994)

3 **a)** Show that there is a root of the equation $8\sin x - x = 0$ lying between 2.7 and 2.8.

b) Taking 2.8 as a first approximation to this root, apply the Newton-Raphson procedure once to $f(x) \equiv 8\sin x - x$ to obtain a second approximation, giving your answer to two decimal places.

c) Explain, with justification, whether or not this second approximation is correct to two decimal places.

d) Evaluate $f\left(\frac{5\pi}{2}\right)$, and hence, by sketching suitable graphs,

determine the number of roots of the equation $8\sin x = x$ in the range $x > 0$.

(ULEAC Question 8, Specimen Paper 3, 1994)

4 Given that x is sufficiently small, use the approximations $\sin x \approx x$ and $\cos x \approx 1 - \frac{1}{2}x^2$ to show that:

$$\frac{\cos x}{1+\sin x} \approx 1 - x + \tfrac{1}{2}x^2.$$

A student estimates the value of $\frac{\cos x}{1+\sin x}$ when $x = 0.1$ by evaluating

the approximation $1 - x + \frac{1}{2}x^2$ when $x = 0.1$. Find, to three decimal places, the percentage error made by the student.

(ULEAC Question 3, Paper 2, June 1993)

5 Differentiate $e^{2x}\cos x$ with respect to x.
The curve C has equation $y = e^{2x}\cos x$.

a) Show that the turning points on C occur where $\tan x = 2$.

b) Find an equation of the tangent to C at the point where $x = 0$.

(ULEAC Question 6, Paper 2, June 1992)

6 Consider the function $y = e^{-x}\sin x$, where $-\pi \le x \le \pi$.

a) Find $\frac{dy}{dx}$.

b Show that, at stationary points, $\tan x = 1$.

c) Determine the coordinates of the stationary points, correct to two significant figures.

d) Explain how you could determine whether your stationary points are maxima or minima. You are not required to do any calculations.

(MEI Question 3, Paper 2, January 1992)

7 **a)** **i)** Show that $(\cos x + \sin x)^2 = 1 + \sin 2x$, for all x.

ii) Hence, or otherwise, find the derivative of $(\cos x + \sin x)^2$.

b) **i)** By expanding $(\cos^2 x + \sin^2 x)^2$, find and simplify an expression for $\cos^4 x + \sin^4 x$ involving $\sin 2x$.

ii) Hence, or otherwise, show that the derivative of $\cos^4 x + \sin^4 x$ is $-\sin 4x$.

(MEI Question 4, Paper 2, June 1992)

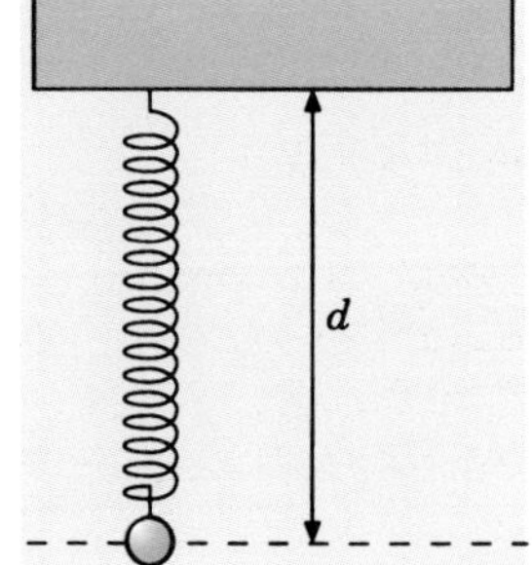

8 A mass is suspended from the end of a spring as shown. The spring is stretched and then released so that it oscillates. The distance, d cm, of the mass from the support t seconds after its release is given by $d = 10 + 3\cos 4t$.

a) State the distance of the mass from the support when the spring is released. Find the period of oscillation of the mass.

b) Calculate the time when d is first 9 cm.

c) Calculate the speed of the mass 2 seconds after its release and state whether it is moving up or down.

(SMP 16–19 (Pure with Applications) Question 9, June 1994)

9 The function f is defined by $f(x) = \sin 2x + 2\cos^2 x$, where x is real and measured in radians.

a) Find the general solution of the equation $f(x) = 0$.

b) Prove that $1 - \sqrt{2} \le f(x) \le 1 + \sqrt{2}$.

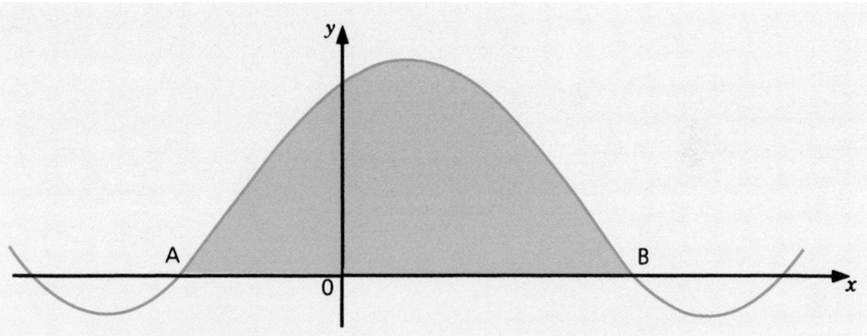

c) The diagram shows part of the curve with equation $y = \sin 2x + 2\cos^2 x$.
Find the area of the shaded region, giving your answer in terms of π.

(AEB Question 12, Paper 1, June 1990)

10 Sketch, for $0 \le x \le 2\pi$, the curve $y = \cos x$.
Hence sketch on different axes, for $0 \le x \le 2\pi$, the curves:

i) $y = \sec x$

ii) $y = \sec x - \cos x$.

Write down an integral equal to the area of the region R enclosed by the curve $y = \sec x - \cos x$, the x-axis and the line $x = \frac{\pi}{4}$. Hence find the area of R.

If R is rotated about the x-axis through four right angles, find the volume generated in terms of π.

(WJEC Question 12, Specimen Paper A1, 1994)

Summary

■ Differentiation of trigonometric functions:

$f(x)$	$f'(x)$
$\sin ax$	$a \cos ax$
$\cos ax$	$-a \sin ax$
$\tan ax$	$a \sec^2 ax$
$\sec (ax)$	$a \sec (ax) \tan (ax)$
$\operatorname{cosec} (ax)$	$-a \operatorname{cosec} (ax) \cot (ax)$
$\cot (ax)$	$-a \operatorname{cosec}^2 (ax)$

where a is constant.

■ Small angle approximations:

when x is sufficiently small

$\sin x \approx x$
$\cos x \approx 1 - \frac{1}{2}x^2$.

PURE MATHS

23

Rational functions

In this chapter we shall study:

- *rational functions and their graphs,*
- *the idea of vertical, horizontal and slant asymptotes,*
- *partial fractions,*
- *a method for developing power series for rational functions.*

THE GRAPHS OF RATIONAL FUNCTIONS

Exploration 23.1

Comparing curves

Draw graphs of the two functions $y = x^3 - x^2 - x$ and $y = \frac{1}{x-1}$ for the domain $\{x : -1 \le x \le 2\}$.

Scale your calculator:

$x_{\min} = -1$; $x_{\max} = 3$; $x_{\text{scl}} = 1$
$y_{\min} = -4$; $y_{\max} = 4$; $y_{\text{scl}} = 1$

- Describe the features of each curve such as stationary points, intersections with the axes.
- What would you say is the essential difference for each curve?

Now add the graphs of the functions $y = 2\left(x^2 - 3x + 2\right)$ and $y = \frac{1}{2\left(x^2 - 3x + 2\right)}$.

- What can you say about these curves?

Interpreting the results

The four graphs are shown in the following diagrams.

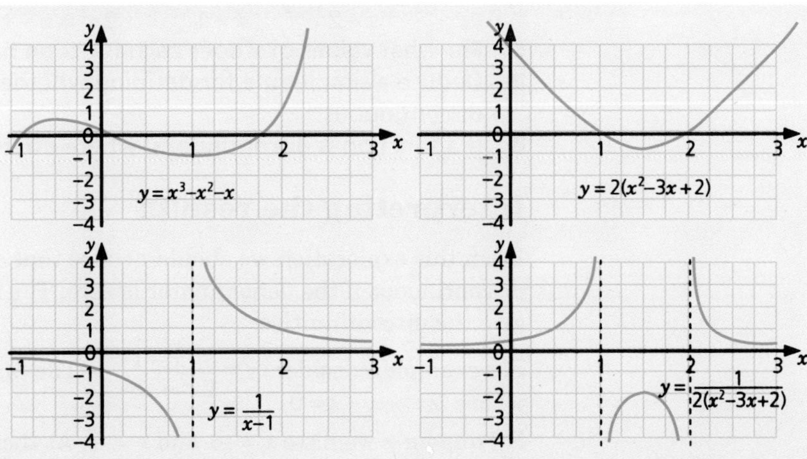

The graphs of the functions $y = x^3 - x^2 - x$ and $y = 2(x^2 - 3x + 2)$ remain finite over the domain $\{x : -1 \le x \le 3\}$.
These are examples of **continuous functions**. We can draw the curves 'without removing the pencil from the paper'.

The graphs of the functions $y = \frac{1}{x-1}$ and $y = \frac{1}{2(x^2 - 3x + 2)}$ have breaks in them. They are examples of **discontinuous functions**. To draw these graphs with pencil and paper we would need to 'remove the pencil' as the curve 'jumps' at various values of x.

Take a closer look at $y = \frac{1}{x-1}$. We can see that y increases without limit as x approaches 1 from above. The function does not exist when $x = 1$. A limit in which a function increases or decreases without bound as x approaches a value c is called an **infinite limit**. It is at this value of $x = c$ that the function is discontinuous.

For the function $y = \frac{1}{2(x^2 - 3x + 2)}$, an infinite limit occurs at two values of x, which are $x = 1$ and $x = 2$.

The functions $y = \frac{1}{x-1}$ and $y = \frac{1}{2(x^2 - 3x + 2)}$ are examples of **rational functions**. A
rational function is a quotient of two functions in which both numerator and denominator are polynomials.

Further examples of rational functions are:

$y = \frac{x-1}{2x^2 + x - 1}$, $y = \frac{4}{x}$, $y = \frac{x}{x-1}$

The functions $y = \frac{\sin x}{x^2}$ and $y = \frac{x}{e^{x^2}}$ are not rational functions.

Exploration 23.2

Discontinuities

Draw the graphs of the following functions.

$y = \frac{4}{x}$, $y = \frac{1}{x-2}$, $y = \frac{x}{x-1}$, $y = \frac{x-1}{2x^2 + x - 1}$

- For what values of x does each function have an infinite limit?
- Deduce a simple rule for deciding whether a function is discontinuous.
- If a function is discontinuous, where do the discontinuities occur?

Interpreting the results

From this exploration we should deduce that a function is discontinuous if the denominator is zero. The roots of the denominator give the discontinuities.

For example, consider $y = \frac{x-1}{2x^2 + x - 1}$. The function is discontinuous where $2x^2 + x - 1 = 0$.

Solving for x we have $x = -1$ and $x = \frac{1}{2}$. At these values of x the function has an infinite limit.

For the other functions in Exploration 23.2:

$y = \dfrac{4}{x}$ is discontinuous at $x = 0$

$y = \dfrac{1}{x-2}$ is discontinuous at $x = 2$

$y = \dfrac{x}{x-1}$ is discontinuous at $x = 1$.

Vertical asymptotes

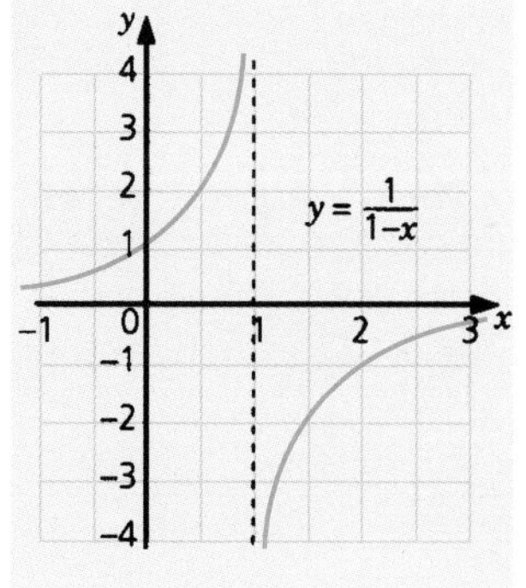

Look again at the graph of $y = \dfrac{1}{1-x}$.

If we could extend the graph towards infinity, it would get closer and closer to the vertical line $x = 1$. This line is called a **vertical asympote** of the graph. For this function the graph approaches the asymptote at large positive values of y on one side, and at large negative values of y on the other side.

For some functions the graph approaches the asymptotes at large positive values of y on both sides or large negative values of y on both sides. These diagrams show examples.

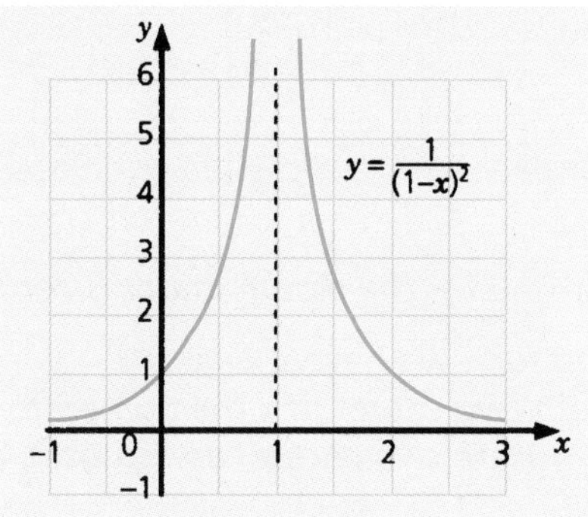

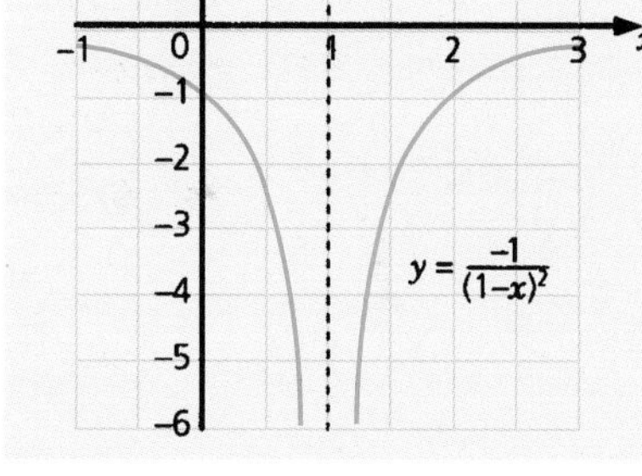

$y = \dfrac{1}{(1-x)^2}$ $\qquad$ $y = \dfrac{-1}{(1-x)^2}$

We are already familiar with discontinuous functions and vertical asymptotes from the work on trigonometric functions, in Chapter 8, *Trigonometry 1*. The function $y = \sin x$ is continuous whereas the function $y = \tan x$ is discontinuous with asymptotes at $x = \pm\frac{\pi}{2}, \pm\frac{3\pi}{2}, \pm\frac{5\pi}{2}, \ldots\ldots$

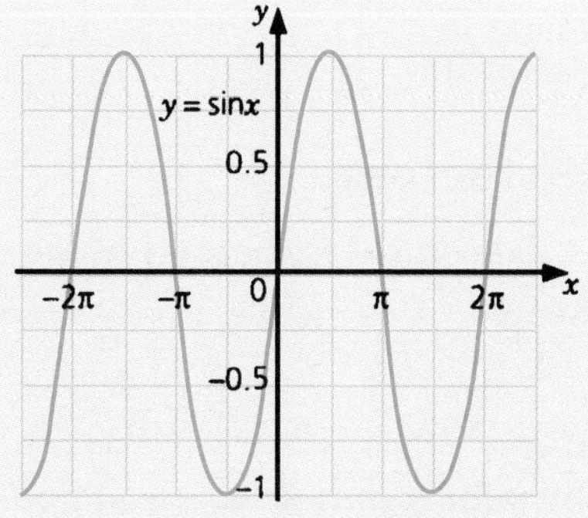

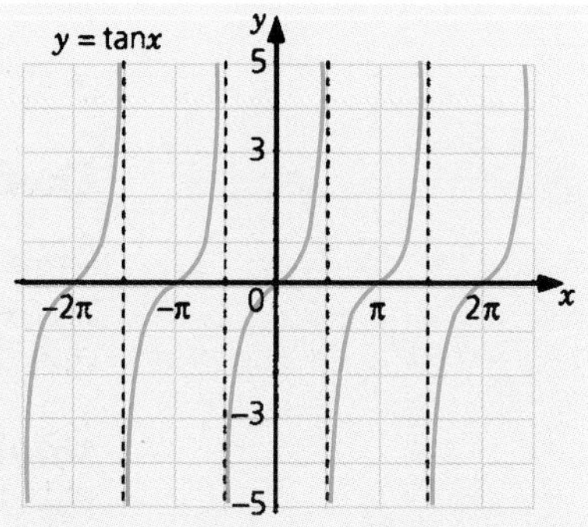

Finding vertical asymptotes

Example 23.1

Determine all the vertical asymptotes of the graphs of the following functions.

a) $y = \dfrac{1}{2(x+3)}$ **b)** $y = \dfrac{x^2+1}{x^2-1}$

Solution

Vertical asymptotes occur at the values of x where the denominator is zero and the numerator is not zero.

a) *For* $y = \dfrac{1}{2(x+3)}$ *the denominator is zero when $x = -3$.*
The numerator is always 1. Hence $x = -3$ is a vertical asymptote, as shown in the diagram.

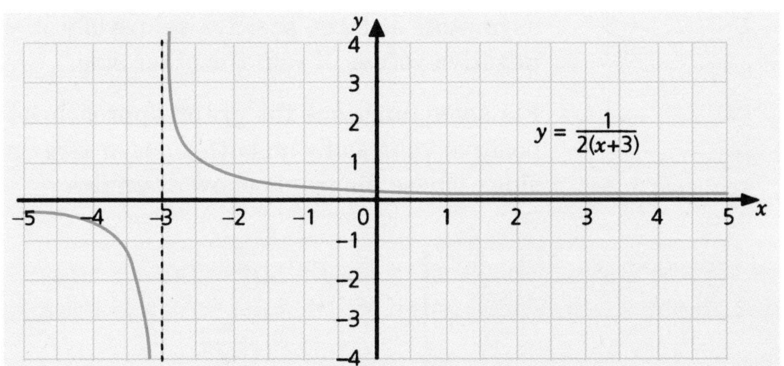

b) *For* $y = \dfrac{x^2+1}{x^2-1}$ *the denominator is zero when* $x^2 - 1 = 0$.
Solving for x we have $x = -1$ or $x = 1$.
At these two values of x the numerator is not zero. Hence $x = -1$ and $x = 1$ are two vertical asymptotes, as shown in the diagram.

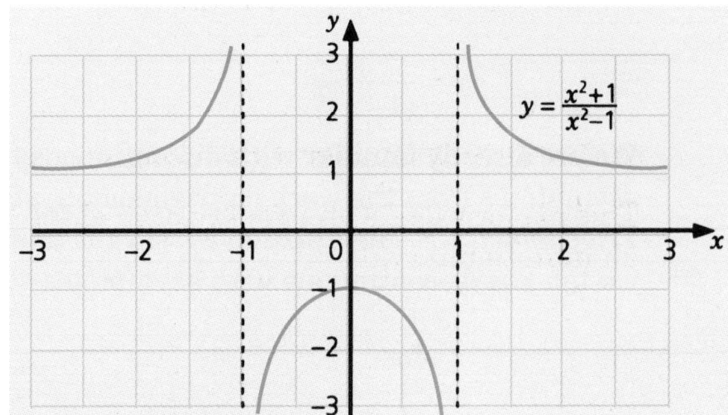

Common factors

Example 23.2

Determine the vertical asymptotes of the graph of the function
$y = \dfrac{x^2+3x-4}{x^2-1}$.

Solution

The denominator is zero when $x = -1$ and $x = 1$.
For $x = -1$ the numerator has value -6 so that $x = -1$ is a vertical

asymptote. For $x = 1$ the numerator also has a value of 0. So $(x - 1)$ is a factor of $x^2 + 3x - 4$. Factorising the numerator and denominator:

$$y = \frac{x^2 + 3x - 4}{x^2 - 1} = \frac{(x + 4)(x - 1)}{(x + 1)(x - 1)}$$

and dividing top and bottom by $(x - 1)$ (provided $x \neq 1$):

$$y = \frac{x + 4}{x + 1} \qquad x \neq -1$$

*The graph of this function shows that $x = -1$ is a vertical asymptote but $x = 1$ is not a vertical asymptote. In fact the graph is undefined at $x = 1$. This point is called a **removable discontinuity**.*

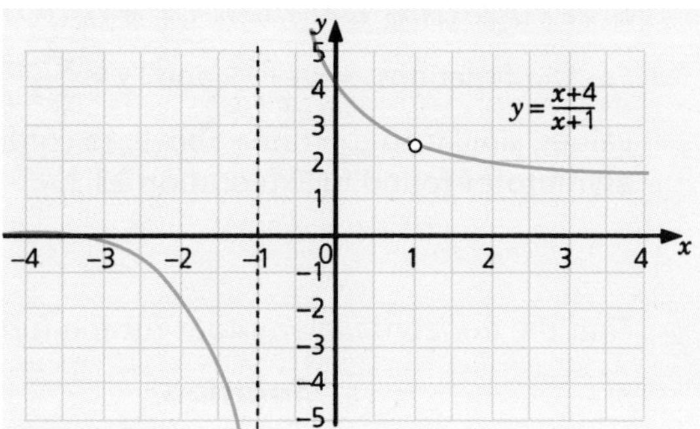

Exploration 23.3

Horizontal asymptotes

Draw the graph of the function $y = \frac{4x^2}{x^2 + 1}$.

- Are there any vertical asymptotes?
- What happens to the graph as x increases to large positive numbers and as x decreases to large negative numbers?

Repeat the exploration for the functions

$y = \frac{x}{x + 1}$ and $y = \frac{x^2 - 3x + 2}{2x^2 + x + 1}$.

Interpreting the results

This exploration is an investigation of the behaviour of a function on an infinite interval.

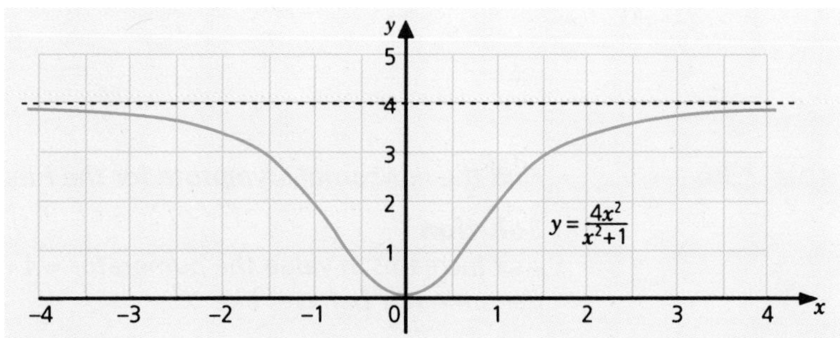

The graph of $y = \frac{4x^2}{x^2 + 1}$ approaches the horizontal line $y = 4$.

The following table of values shows the process.

$\leftarrow x$ decreases $\quad x$ increases $\rightarrow$

x	−1000	−100	−10	−1	0	1	10	100	1000
y	3.999996	3.9996	3.96	2	0	2	3.96	3.9996	3.999996

$\leftarrow y$ approaches 4 $\quad y$ approaches 4 $\rightarrow$

The line $y = 4$ is an example of a **horizontal asymptote**.

Exploration 23.4

Investigating the limit of a function

For the functions $y = \dfrac{x}{x+1}$ and $y = \dfrac{x^2 - 3x + 2}{2x^2 + x + 1}$ calculate tables of values, similar to the table above, to confirm the horizontal asymptotes found in Exploration 23.3.

Example 23.3

Find the horizontal asymptote for the function $y = \dfrac{3x^2 - x + 1}{2x^2 + x - 1}$.

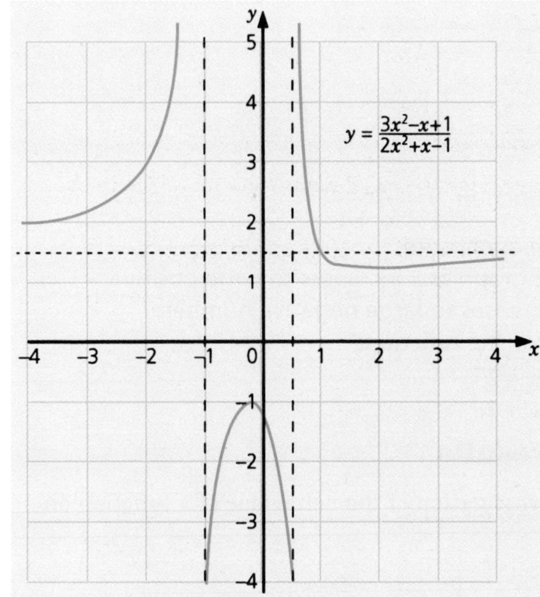

Solution

One method is to draw a graph of the function. Then from the graph we see that the curve tends towards the horizontal line $y = 1.5$.

An algebraic approach is to divide the dominant term in the numerator by the dominant term in the denominator.

As x *increases:* $3x^2 - x + 1$ *behaves like* $3x^2$

As x *increases:* $2x^2 + x - 1$ *behaves like* $2x^2$

$$\Rightarrow \quad y = \frac{3x^2 - x + 1}{2x^2 + x - 1} \rightarrow \frac{3x^2}{2x^2} \rightarrow 1.5$$

The horizontal asymptote is $y = 1.5$.

Example 23.4

Find the horizontal asymptote for the function $y = \dfrac{4}{x^2 + 2}$.

Solution

As x *increases in value the numerator* = 4 *(for all values of* x*) and the denominator behaves like* x^2.

$\Rightarrow y = \dfrac{4}{x^2+2} \to \dfrac{4}{x^2} \to 0$

The x-axis, $y = 0$ is a horizontal asymptote.

The graph shows the functions $y = \dfrac{4}{x^2+2}$ and $y = \dfrac{4}{x^2}$.

For values of x less than −4 or greater than 4 the curves are approximately identical. This confirms that $y = \dfrac{4}{x^2+2}$ approaches $y = \dfrac{4}{x^2}$ as x continuously increases and decreases in value.

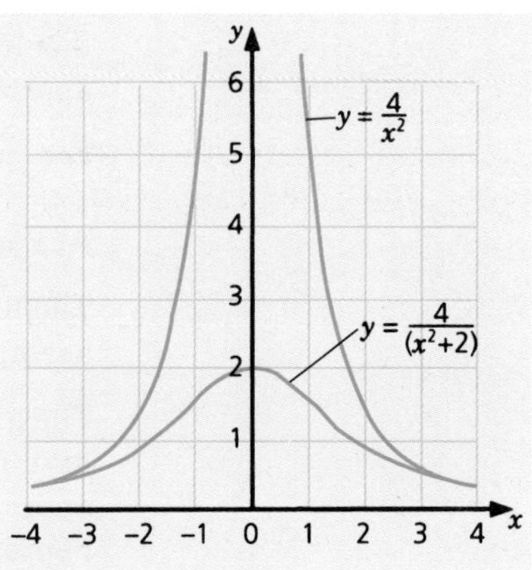

CALCULATOR ACTIVITY

Slant asymptotes

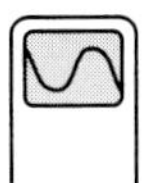

Scale your calculator to:

$x_{min} = -5$; $x_{max} = 5$; $x_{scl} = 1$

$y_{min} = -25$; $y_{max} = 25$; $y_{scl} = 5$

Draw the graph of the function $y = \dfrac{3x^3}{x^2-4}$.

a) Describe the properties of the graph.

b) What happens as x increases to ∞ and decreases to −∞? Add the line $y = 3x$ to your graph.

Rescale your calculator and repeat the activity for the functions $y = \dfrac{x^2-x+4}{x-1}$ and $y = \dfrac{x^2-2x+4}{x-1}$ using the line $y = x$ for part **b)**.

Interpreting the results

For $y = \dfrac{3x^3}{x^2-4}$ the denominator is zero when

$x^2 - 4 = 0 \Rightarrow x = -2$ or $x = 2$.

At $x = -2$ the numerator $3(-2)^3 = -24$.

The line $x = -2$ is a vertical asymptote.

Similarly the line $x = 2$ is a vertical asymptote.

The graph has no horizontal asymptote. What happens as the value of x increases? Using the same approach as in Example 23.4:

$y = \dfrac{3x^3}{x^2-4} \to \dfrac{3x^3}{x^2} \to 3x$

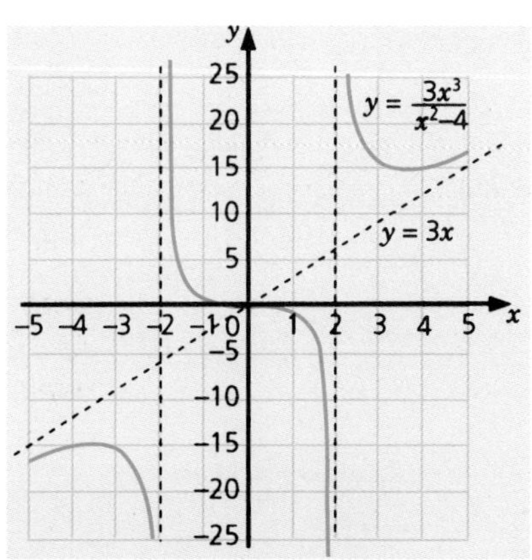

The line $y = 3x$ is called a **slant asymptote**. The graph of the function approaches the slant asymptote $y = 3x$ as x approaches $\pm\infty$.

For $y = \dfrac{x^2 - x + 4}{x - 1}$ the denominator is zero when $x - 1 = 0 \Rightarrow x = 1$.

At $x = 1$ the numerator $= 1^2 - 1 + 4 = 4$.

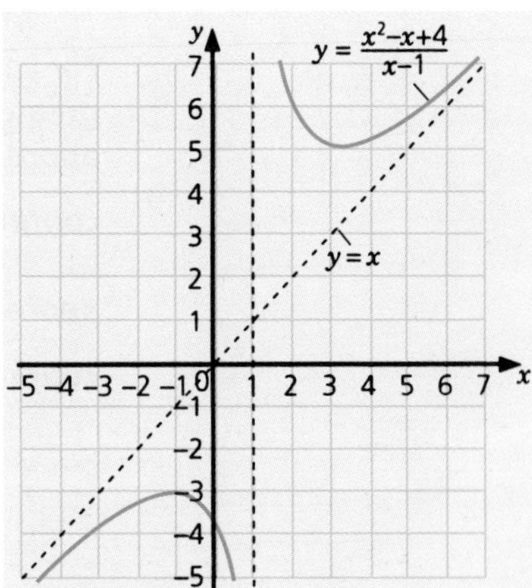

The line $x = 1$ is a vertical asymptote.

The graph has no horizontal asymptote. What happens as x approaches $\pm\infty$? Using the same approach as in Example 23.4, when x takes large positive (or large negative) values:

$$y = \frac{x^2 - x + 4}{x - 1} \to \frac{x^2 - x}{x - 1} \to x$$

The graph of the function approaches the slant asymptote $y = x$ as x approaches $\pm\infty$.

For $y = \dfrac{x^2 - 2x + 4}{x - 1}$, $x = 1$ is a vertical asymptote.

The slant asymptote looks as if it might be $y = x$ again. However the algebra shows that this is not correct. When x takes large positive (or large negative) values:

$$y = \frac{x^2 - 2x + 4}{x - 1} \to \frac{x^2 - 2x}{x - 1} \to x - 1$$

The graph of the function approaches the slant asymptote $y = x - 1$ as x approaches $\pm\infty$.

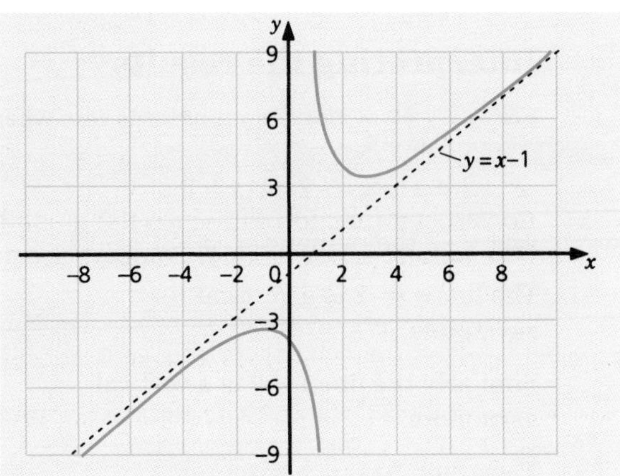

EXERCISES

23.1 A

1 Find the vertical asymptotes (if any) of the following functions.

a) $y = \frac{1}{x^2}$ **b)** $y = \frac{1}{x-3}$ **c)** $y = \frac{1}{x^2+x-2}$

d) $y = \frac{x-1}{x-3}$ **e)** $y = \frac{1}{x^2+x+1}$ **f)** $y = \frac{x^2-1}{2x^3-11x^2+17x-6}$

Use a graphics calculator or a computer graphing program to confirm your results.

2 Find the horizontal asymptotes of the following functions.

a) $y = \frac{x-1}{x-3}$ **b)** $y = \frac{1}{x^2+x-2}$ **c)** $y = \frac{x^2-3x+2}{x^2-1}$

d) $y = \frac{2+x}{1-x}$ **e)** $y = \frac{x^3-x^2+1}{4x^3+x-1}$ **f)** $y = \frac{x^2-1}{2x^3-11x^2+17x-6}$

Use a graphics calculator or a computer graphing program to confirm your results.

3 For the following functions determine whether there is a removable discontinuity at $x = 1$.

a) $y = \frac{x^2-3x+2}{x^2-1}$ **b)** $y = \frac{x^2+3x+2}{x^2-1}$

c) $y = \frac{x^2+6x-7}{x-1}$ **d)** $y = \frac{x^2+1}{x-1}$

4 Sketch a graph of each of the following functions. In each case label the stationary points, intercepts, removable discontinuities and asymptotes. (Use differentiation to find the coordinates of the stationary points.)

a) $y = \frac{2x}{x^2-1}$ **b)** $y = \frac{x+3}{x}$ **c)** $y = \frac{x^3}{x^2-1}$

d) $y = \frac{x^2-4x+3}{x^2+2x-3}$ **e)** $y = \frac{2x^2-5x+5}{x-2}$ **f)** $y = \frac{4(x-1)^2}{x^2-4x+5}$

5 Construct a function for which the graph has the following features. (The answer is not unique. Compare the graph of your function with those found by other students.)

a) vertical asymptote at $x = 2$
horizontal asymptote at $y = 1$

b) vertical asymptote at $x = -1$
slant asymptote at $y = x + 1$

6 Find the vertical and horizontal asymptotes of the function:

$y = \frac{4x}{\sqrt{x^2-4}}$.

What is the domain of the function?
Sketch the graph of the function to confirm your answers.

EXERCISES

23.1B

1 Find the vertical asymptotes (if any) of the following functions.

a) $y = \frac{3}{x}$ **b)** $y = \frac{4}{4+x}$ **c)** $y = \frac{4-x}{5-x}$

d) $y = \frac{x}{4x^2-1}$ **e)** $y = \frac{3x-1}{2x^2+x-1}$ **f)** $y = \frac{x^2+2}{x^3-6x^2+11x-6}$

Use a graphics calculator or a computer graphing program to confirm your results.

2 Find the horizontal asymptotes of the following functions.

a) $y = \frac{4-x}{5-x}$ **b)** $y = \frac{x}{4x^2-1}$

c) $y = \frac{x^2+x+2}{3x^2-x+2}$ **d)** $y = \frac{5x^2+2x-1}{7x^2-x+1}$

Use a graphics calculator or a computer graphing program to confirm your results.

3 For the following functions determine whether there is a removable discontinuity at $x = 2$.

a) $y = \frac{x^2-3x+2}{x^2-2}$ **b)** $y = \frac{x^2+x-1}{x-2}$

c) $y = \frac{3x-6}{x-2}$ **d)** $y = \frac{x^2+3x+2}{x^2-3x+2}$

4 Sketch a graph of each of the following functions. In each case label the stationary points, intercepts, removable discontinuities and asymptotes. (Use differentiation to find the coordinates of the stationary points.)

a) $y = \frac{x^2+1}{x-2}$ **b)** $y = \frac{2x+5}{x}$ **c)** $y = \frac{x^2-6x+12}{x-4}$

d) $y = \frac{2(x+1)^2}{x^2-2x+3}$ **e)** $y = \frac{4(x-1)^2}{3(x+2)^2}$ **f)** $y = \frac{3x-2}{\sqrt{4x^2+1}}$

5 Construct a function for which the graph has the following features. (The answer is not unique. Compare the graph of your function with those found by other students.)

a) vertical asymptote at $x = -3$
horizontal asymptote at $y = 0$

b) vertical asymptote at $x = 0$
slant asymptote at $y = 2x - 1$

6 Consider the function $y = \frac{ax}{x-b}$ where a and b are constants.

a) Investigate the graph of the function if $b \neq 0$ and a is varied. Choose positive and negative values of a.

b) Investigate the graph of the function if $a \neq 0$ and b is varied.

PARTIAL FRACTIONS

Exploration 23.5

Combining fractions

For each of these expressions, combine the two functions to give a single fraction.

- $\dfrac{1}{2x+1}+\dfrac{1}{x-2}$
- $\dfrac{2x}{x^2+1}+\dfrac{1}{x+1}$
- $\dfrac{2}{(x-1)^2}+\dfrac{3}{(x-1)}$

Interpreting the results

In this exploration, two separate rational fractions have been combined to produce a single fraction with a common denominator. For the first problem:

$$\frac{1}{2x+1}+\frac{1}{x-2}\equiv\frac{(x-2)+(2x+1)}{(2x+1)(x-2)}\equiv\frac{3x-1}{(2x+1)(x-2)}$$

For the second and third problems:

$$\frac{2x}{x^2+1}+\frac{1}{x+1}\equiv\frac{3x^2+2x+1}{(x^2+1)(x+1)}$$

$$\frac{2}{(x-1)^2}+\frac{3}{(x-1)}\equiv\frac{3x-1}{(x-1)^2}$$

There are many problems where we *start* with a single fraction and need to *split* it into the sum of two separate fractions. The process of decomposing a fraction is called splitting up a function into **partial fractions**. For example,

$\dfrac{1}{2x+1}$ and $\dfrac{1}{x-2}$ are partial fractions for $\dfrac{3x-1}{(2x+1)(x-2)}$.

Of course in this case we know what the partial fractions are. The following examples show the method of approach when we do not already know what the partial fractions are.

Linear factors in the denominator

Example 23.5

Express $\dfrac{3x+4}{(x-1)(x+6)}$ *in partial fractions.*

Solution

Rewrite the original expression as:

$$\frac{3x+4}{(x-1)(x+6)}\equiv\frac{A}{x-1}+\frac{B}{x+6}$$

where A and B are constants. Combining the fractions on the right-hand side:

$$\frac{3x+4}{(x-1)(x+6)}\equiv\frac{A(x+6)+B(x-1)}{(x-1)(x+6)}$$

$$\Rightarrow\quad 3x+4\equiv A(x+6)+B(x-1)$$

This identity must hold true for all values of x, so we choose any value of x to find A and B.

When $x = 1$: $\quad 7 = 7A + 0B \quad \Rightarrow A = 1$

When $x = -6$: $\quad -14 = 0A + (-7)B \Rightarrow B = 2$

Note that the choice of x-values gives one constant immediately because one of the brackets is zero.

Hence $\dfrac{3x+4}{(x-1)(x+6)} \equiv \dfrac{1}{x-1} + \dfrac{2}{x+6}$

Example 23.6

Express $\dfrac{2}{(2x^2+9x+4)}$ *in partial fractions.*

Solution

The first step is to write the denominator $2x^2+9x+4$ *as a product of linear factors.*

We have $2x^2+9x+4 = (2x+1)(x+4)$

$$\frac{2}{2x^2+9x+4} \equiv \frac{A}{2x+1} + \frac{B}{x+4} \equiv \frac{A(x+4)+B(2x+1)}{(x+4)(2x+1)}$$

$$\Rightarrow \qquad 2 \equiv A(x+4) + B(2x+1)$$

When $x = -4$: $\quad 2 = 0A + (-7)B \Rightarrow B = -\frac{2}{7}$

When $x = -\frac{1}{2}$: $\quad 2 = \frac{7}{2}A + 0B \Rightarrow A = \frac{4}{7}$

Hence $\dfrac{2}{2x^2+9x+4} = \dfrac{4}{7(2x+1)} - \dfrac{2}{7(x+4)}$

Quadratic factors in the denominator

Example 23.7

Express $\dfrac{1-7x}{(x^2+1)(x-3)}$ *as a sum of partial fractions.*

Solution

The partial fractions are of the form $\dfrac{Ax+B}{x^2+1}$ *and* $\dfrac{C}{x-3}$*. Note that if there is a quadratic factor in the denominator it is necessary to start with a linear factor in the numerator.*

$$\frac{1-7x}{(x^2+1)(x-3)} \equiv \frac{Ax+B}{x^2+1} + \frac{C}{x-3}$$

$$\equiv \frac{(Ax+B)(x-3)+C(x^2+1)}{(x^2+1)(x-3)}$$

$$\Rightarrow \qquad 1-7x \equiv (Ax+B)(x-3) + C(x^2+1)$$

When $x = 3$: $\quad -20 = 0(3A+B) + 10C \Rightarrow C = -2$

The values of A and B can be found by substituting any values for x (except $x = 3$ because this has already been used).

When $x = 0$: $\quad 1 = B(-3) + C \quad \Rightarrow B = -1$

When $x = 1$: $\quad -6 = (A+B)(-2) + 2C \Rightarrow A = 2$

Hence $\dfrac{1-7x}{(x^2+1)(x-3)} \equiv \dfrac{2x-1}{x^2+1} - \dfrac{2}{x-3}$

Repeated factors in the denominator

Example 23.8

Express $\frac{2x+1}{x(x+3)^2}$ *as partial fractions.*

Solution

When the denominator includes a repeated factor we must include one fraction for each power of the factors.

$$\frac{2x+1}{x(x+3)^2} \equiv \frac{A}{x} + \frac{B}{x+3} + \frac{C}{(x+3)^2}$$

$$\equiv \frac{A(x+3)^2 + Bx(x+3) + Cx}{x(x+3)^2}$$

$$\Rightarrow \quad 2x+1 \equiv A(x+3)^2 + Bx(x+3) + Cx$$

When $x = 0$: $\quad 1 = 9A + 0B + 0C \Rightarrow A = \frac{1}{9}$

When $x = -3$: $\quad -5 = 0A + 0B - 3C \Rightarrow C = \frac{5}{3}$

To find B *we choose any value of* x *(except* $x = 0$ *and* $x = -3$*).*

Choose $x = 1$: $\quad 3 = 16A + 4B + C \quad \Rightarrow B = -\frac{1}{9}$

Hence $\frac{2x+1}{x(x+3)^2} \equiv \frac{1}{9x} - \frac{1}{9(x+3)} + \frac{5}{3(x+3)^2}$

Improper fractions

In all of the fractions that we have dealt with so far, the highest power in the numerator has been less than the highest power in the denominator. Such a fraction is called a **proper fraction**.

When the highest power in the numerator is greater than or equal to the highest power in the denominator, the fraction is called an **improper fraction**. For example:

$\frac{x}{x^2-1}$ is a proper fraction,

$\frac{x^2}{x^2-1}$ and $\frac{x^3}{x^2-1}$ are improper fractions.

Improper fractions can be rewritten in a form which only includes proper fractions by dividing the denominator into the numerator.

Example 23.9

Express $\frac{x^3-x+1}{x^2-3x+2}$ *as a sum of partial fractions.*

Solution

This is an improper fraction. The first step is to write the fraction in the form $Ax + B + \frac{R}{x^2-3x+2}$ *where* R *is the remainder.*

The fraction $\frac{R}{x^2 - 3x + 2}$ *is now a proper fraction which can be expressed as partial fractions.*

$$\frac{x^3 - x + 1}{x^2 - 3x + 2} \equiv Ax + B + \frac{C}{x-2} + \frac{D}{x-1}$$

$$\equiv \frac{(Ax + B)(x-2)(x-1) + C(x-1) + D(x-2)}{(x-2)(x-1)}$$

$$\Rightarrow x^3 - x + 1 \equiv (Ax + B)(x-2)(x-1) + C(x-1) + D(x-2)$$

When $x = 1$: $1 = 0(A + B) + 0C + (-1)D \quad \Rightarrow D = -1$

When $x = 2$: $7 = 0(A + B) + C + 0D \quad \Rightarrow C = 7$

When $x = 0$: $1 = 2B - C - 2D \quad \Rightarrow B = 3$

When $x = -1$: $1 = 6(B - A) - 2C - 3D \quad \Rightarrow A = 1$

Hence $\frac{x^3 - x + 1}{x^2 - 3x + 2} = x + 3 + \frac{7}{x-2} - \frac{1}{x-1}$

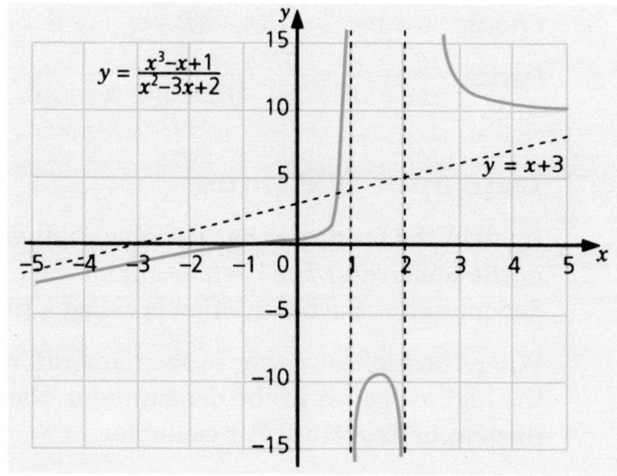

The diagram shows a graph of the function. There are vertical asymptotes $x = 2$ *and* $x = 1$*. The line* $y = x + 3$ *is a slant asymptote.*

EXERCISES

23.2A

1 Write the following in partial fractions.

a) $\frac{2}{(x-1)(x+1)}$ **b)** $\frac{1}{(x-2)(x-1)}$ **c)** $\frac{4}{x(x-1)}$

d) $\frac{(x+1)}{(x-1)(x+3)}$ **e)** $\frac{2x-1}{x^2-3x+2}$ **f)** $\frac{1}{x^2+x-2}$

g) $\frac{x^3}{x^2+x-2}$ **h)** $\frac{5(x+1)}{25-x^2}$ **i)** $\frac{1}{x(x^2+1)}$

j) $\frac{2}{(x-4)(x^2+3)}$ **k)** $\frac{2+x}{(1+x)^2(3x+2)}$ **l)** $\frac{x}{(x-1)(x-2)^2}$

2 Express $\dfrac{13x+16}{(x-3)(3x+2)}$ in partial fractions.

Hence find the value of $\dfrac{d}{dx}\left[\dfrac{13x+16}{(x-3)(3x+2)}\right]$ when $x = 2$.

3 Given that $y = \dfrac{2}{(x-3)(x-1)}$ express y as a sum of partial fractions.

Hence find $\dfrac{dy}{dx}$ and $\dfrac{d^2y}{dx^2}$.

Find and classify the stationary points of y.

EXERCISES

23.2 B

1 Write the following in partial fractions.

a) $\dfrac{1}{x(x-3)}$ b) $\dfrac{1}{(x+1)(x+2)}$ c) $\dfrac{2x-3}{x^2-10x}$

d) $\dfrac{3-x}{(x+1)(2x-1)}$ e) $\dfrac{1}{4x^2-9}$ f) $\dfrac{x+1}{x^2+4x+3}$

g) $\dfrac{2x-3}{(x-1)^2}$ h) $\dfrac{3}{x(3x-1)^2}$ i) $\dfrac{2x-3}{x^3+10x}$

j) $\dfrac{x^2}{(x+1)(x-1)}$ k) $\dfrac{x^3-x+3}{x^2+x-2}$ l) $\dfrac{x+3}{(x^2+2)(x^2+1)}$

2 Express each of the following functions as a sum of partial fractions and hence find the first and second derivatives. Find and classify the stationary points of each function.

a) $y = \dfrac{3x-14}{(x-2)(x+6)}$ b) $y = \dfrac{(x^2-9)}{(x^2-4)}$

POWER SERIES FOR AN ALGEBRAIC FRACTION

The binomial expansion of the partial fractions of an algebraic fraction gives a simple method of giving a power series representation of the function.

Exploration 23.6

The binomial expansion revisited

- Use the binomial expansion to find a series of ascending powers of x up to and including the term in x^2 for the function $\dfrac{1}{1+2x}$.
  State the range of values of x for which the expansion is valid.
- Repeat the exploration for the function $\dfrac{1}{x-2}$.
- Deduce the series expansion for the function $\dfrac{3x-1}{(2x+1)(x-2)}$ which has partial fractions $\dfrac{1}{2x+1}+\dfrac{1}{x-2}$.

Interpreting the results

The binomial expansion of $\dfrac{1}{1+ax} = (1+ax)^{-1}$ is $1 - ax + a^2x^2 - a^3x^3 + \text{K}$ and converges when $|ax| < 1$. For the functions in Exploration 23.6:

$$\frac{1}{1+2x} = 1 - 2x + 4x^2 - 8x^3 \quad \text{for } |x| < \tfrac{1}{2}$$

$$\frac{1}{x-2} = -\frac{1}{2(1-\frac{1}{2}x)} = -\frac{1}{2}\left(1 + \frac{x}{2} + \frac{x^2}{4} + \frac{x^3}{8}\right) = -\frac{1}{2} - \frac{x}{4} - \frac{x^2}{8} - \frac{x^3}{16} - \text{K}$$

for $|x| < 2$

Adding the two series:

$$\frac{1}{1+2x} + \frac{1}{x-2} = \left(1 - 2x + 4x^2 - 8x^3 + \text{K}\right) + \left(-\frac{1}{2} - \frac{x}{4} - \frac{x^2}{8} - \frac{x^3}{16} - \text{K}\right)$$

$$= \frac{1}{2} - \frac{9}{4}x + \frac{31}{8}x^2 - \frac{129}{16}x^3 + \text{K}$$

Looking at the two ranges of values of x for convergence we see that $|x| < \frac{1}{2}$ is contained within $|x| < 2$. The interval of convergence is therefore $|x| < \frac{1}{2}$ or $-\frac{1}{2} < x < \frac{1}{2}$.

Example 23.10

Given that x is small, find a cubic polynomial approximation for

$$f(x) = \frac{3x}{(x+2)(x-1)}.$$

Hence find the slope of the tangent to the graph of the function at $x = 0$, i.e. $f'(0)$.

Solution

Let $\dfrac{3x}{(x+2)(x-1)} = \dfrac{A}{x+2} + \dfrac{B}{x-1} = \dfrac{A(x-1) + B(x+2)}{(x+2)(x-1)}$

$\Rightarrow \quad 3x = A(x-1) + B(x+2)$

When $x = 1$: $\quad 3 = 0A + 3B \quad \Rightarrow B = 1$

when $x = -2$: $\quad -6 = -3A + 0B \quad \Rightarrow A = 2$.

Hence $\dfrac{3x}{(x+2)(x-1)} = \dfrac{2}{x+2} + \dfrac{1}{x-1}$

Applying the binomial expansion to the partial fractions:

$$\frac{2}{x+2} = \frac{1}{1+\frac{1}{2}x} = 1 - \frac{x}{2} + \frac{x^2}{4} - \frac{x^3}{8} + \ldots \quad |x| < 2$$

$$\frac{1}{x-1} = \frac{-1}{1-x} = -1(1 + x + x^2 + x^3) + \ldots \quad |x| < 1$$

Adding the two power series:

$$f(x) = \frac{3x}{(x+2)(x-1)} = -\frac{3x}{2} - \frac{3x^2}{4} - \frac{9x^3}{8} + \ldots \quad \textit{for } |x| < 1$$

$$\Rightarrow f'(x) = -\frac{3}{2} - \frac{3x}{2} - \frac{27x^2}{8} + \ldots$$

In particular, $f'(x) = -\frac{3}{2}$. *The slope of the tangent to the graph of* $y = f(x)$ *at* $x = 0$ *is* $-\frac{3}{2}$.

The figure shows graphs of the function $y = \dfrac{3x}{(x+2)(x-1)}$ *and the polynomials* $p_1 = -\frac{3}{2}x$, $p_2 = -\frac{3}{2}x - \frac{3x^2}{4}$ *and* $p_3 = -\frac{3}{2}x - \frac{3x^2}{4} - \frac{9x^3}{8}$ *drawn on the same axes.*

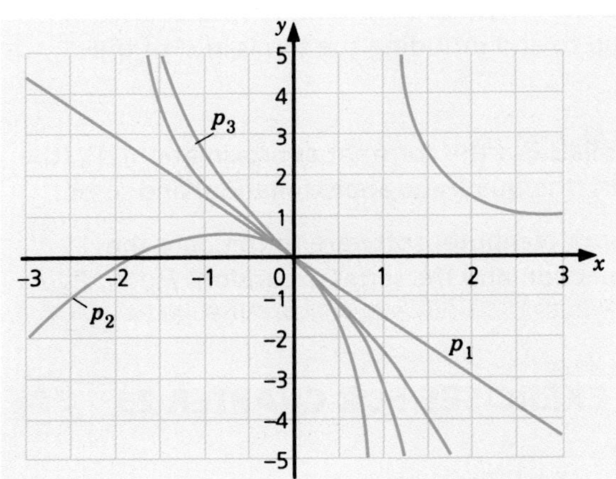

Notice how the fit of the polynomials improves for a larger range of values of x as the degree of the polynomials increases.

EXERCISES

23.3 A

1 Expand the following functions as series of ascending powers of x up to and including the term in x^3. State the range of values of x for which the expansion is valid.

a) $\dfrac{x+2}{1-x}$ b) $\dfrac{x+1}{x+2}$ c) $\dfrac{x^2-1}{x+3}$

d) $\dfrac{\sqrt{1-x}}{1+x}$ e) $\dfrac{1}{(2-x)(1+2x)}$ f) $\dfrac{1}{(1-x)(2-x)}$

g) $\dfrac{5}{(1+2x)(1-3x)}$ h) $\dfrac{x+1}{(1-x)(3+x)}$

2 Use partial fractions and the binomial expansion to find a linear approximation for $\dfrac{1}{(1-2x)(2+x)}$.

3 If x is very small, find a quadratic approximation for $\dfrac{12}{(3+x)(1-x)^2}$.

EXERCISES

23.3 B

1 Expand the following functions as series of ascending powers of x up to and including the term in x^3. State the range of values of x for which the expansion is valid.

a) $\dfrac{x-3}{x+1}$ b) $\dfrac{x}{1-2x}$ c) $\dfrac{2x^2+1}{1-x}$

d) $\frac{1}{(2-x)(1+x)}$ e) $\frac{\sqrt{4-x}}{1-3x}$ f) $\frac{x-1}{(2-3x)(1+x)}$

g) $\frac{x}{(1+x)^2(2+x)}$ h) $\frac{x^2+1}{x^2+3x-4}$

2 If x is very small find a cubic polynomial approximation for $\frac{x^2+1}{x^3-2x^2-x+2}$.

3 Use partial fractions and the binomial expansion to find a series in ascending powers of x up to and including the terms in x^4 of the function $y=\frac{x+2}{1-x}$.

Write down the polynomials P_0 (the constant approximation), P_1 (the linear approximation), P_2 (the quadratic approximation) and so on.

Use a graphics calculator or computer software to compare the graphs of the original function and the series expansions P_0, P_1, P_2, P_3 and P_4.

CONSOLIDATION EXERCISES FOR CHAPTER 23

1 A curve has equation $y=\frac{1}{(x+1)(x-5)}$.

a) Find the asymptotes of the function.
b) Express y as a sum of partial fractions.
c) Find and classify the stationary points of y.
d) Sketch the curve using the information from parts **a)–c)**.

2 A curve has equation $y=\frac{2x^3+x^2-7x+7}{x^2+x-2}$.

a) Find the vertical asymptotes of the function.
b) Express y as a sum of partial fractions.
c) Deduce the equation of the slant asymptote.
d) Use a graphics calculator or computer software to sketch the curve and hence validate your answers.

3 Express $y=\frac{4x}{(1+x)^2(2-x)}$ as a sum of partial fractions.

A curve has equation $y=\frac{4x}{(1+x)^2(2-x)}$.

Determine the equation of the normal to the curve at the point (1, 1).

4 Express $\frac{1+2x}{(1+x)(1-2x)}$ in partial fractions and hence obtain its series expansion in ascending powers of x up to and including the term in x^3.

State the range of values of x for which the expansion is valid.

5 Express $f(x) \equiv \frac{4-x}{(1-x)(2-x)}$ in partial fractions. Hence, or otherwise, for $|x|<1$, obtain the expansion of $f(x)$ in ascending powers of x up to and including the term in x^3, simplifying each coefficient.

(ULEAC Question 4, Paper 2, January 1992)

6 $f(x) \equiv \dfrac{9-3x-12x^2}{(1-x)(1+2x)}$.

a) Given that $f(x) \equiv A + \dfrac{B}{1-x} + \dfrac{C}{1+2x}$, find the values of the constants A, B and C.

b) Given that $|x| < \frac{1}{2}$, expand $f(x)$ in ascending powers of x up to and including the term x^3, simplifying each coefficient.

c) Hence, or otherwise, find the value of $f'(0)$.

(ULEAC Question 8, Paper 2, January 1993)

7 $f(x) \equiv \dfrac{4+x}{(1+2x)(1-x)^2}$.

a) Express $f(x)$ in partial fractions.

b) Given that $|x| < \frac{1}{2}$, expand $f(x)$ in ascending powers of x, up to and including the term in x^3.

(ULEAC Question 5, Paper 2, May 1994)

8 **a)** $\dfrac{14x^2+13x+2}{(x+1)(2x+1)^2} \equiv \dfrac{A}{x+1} + \dfrac{B}{2x+1} + \dfrac{C}{(2x+1)^2}$

Find the values of the constants A, B and C.

b) Given that $y = \dfrac{14x^2+13x+2}{(x+1)(2x+1)^2}$, hence or otherwise find the value of $\dfrac{dy}{dx}$ at $x = 0$.

(ULEAC Question 6, Paper 2, January 1995)

9 The function $f(x)$ is defined by:

$f(x) = \dfrac{2x+7}{x+3}$ $x \varepsilon \Re$, $x \neq -3$.

Find $f'(x)$ and hence show that the gradient of the curve $y = f(x)$ is negative for all values of x in the domain of f.

Sketch the graph of the curve $y = f(x)$. Label its asymptotes and its points of intersection with the x- and y-axes.

Write down the range of f.

Show that:

$f(x) = a + \dfrac{b}{x+3}$

where a and b are constants which are to be determined. Hence describe the geometrical transformation which transforms the graph of $y = f(x)$ into the graph of $y = \frac{1}{x}, x \neq 0$

(NEAB Question 14, Specimen Paper 1, 1994)

10 Express $\dfrac{1-x-x^2}{(1-2x)(1-x)^2}$ as the sum of three partial fractions. Hence, or otherwise, expand this expression in ascending powers of x up to and including the term in x^3. State the range of values of x for which the full expansion is valid.

(NEAB Question 12, Specimen Paper 1, 1994)

11 Find the equations of the asymptotes of the curve $y = \dfrac{x^2+1}{x+1}$.

(UCLES Modular Question 1, Specimen Paper P4, 1994)

Summary

■ **Rational functions**

A rational function is a quotient of two polynomial functions $y = \dfrac{f(x)}{g(x)}$. If there exists a value of $x = \alpha$, say, such that $g(\alpha) = 0$ but $f(\alpha) \neq 0$ then y has a **vertical asymptote** $x = \alpha$ and is said to be discontinuous at $x = \alpha$. If $y \to c$ as $x \to \pm\infty$ then $y = c$ is a **horizontal asymtote**. For some functions $y \to ax + b$ as $x \to \pm\infty$ then the line $y = ax + b$ is called a **slant asymptote**. If $f(\alpha) = 0$ and $g(\alpha) = 0$ at a point $x = \alpha$, then $x = \alpha$ is called a **removable discontinuity**.

■ **Partial fractions**

The process of decomposing a fraction is called 'splitting the function into partial fractions'.

PURE MATHS

24 Integration 3

In this chapter we introduce methods or techniques of integration; in particular:

- *the use of substitutions,*
- *the method of integration by parts,*
- *the use of partial fractions as a means of transforming integrals.*

Some expressions have an integral that is a recognisable function, such as sin x. There are many others for which an exact integral cannot be found. In the latter cases it is necessary to adopt an approximate or numerical approach such as the trapezium rule. In Chapter 9, *Integration 1* and Chapter 19, *Integration 2* we integrated standard functions such as powers of x, polynomial, exponential and logarithmic functions. In this chapter we introduce methods of integration which attempt to transform non-standard functions which are difficult to integrate into standard functions that can be integrated.

USING SUBSTITUTIONS, PART 1

Exploration 24.1

Expanding and integrating

Look at the function $y = f(x) = (x + 1)^3$.

- Expand $f(x)$ to give a polynomial in x.
- Evaluate $\int f(x)\,dx$.
- Factorise your answer into the form $a(x + 1)^4 + b$.

Repeat the exploration for $f(x) = (x + 1)^4$ and $f(x) = (5x + 3)^2$. Can you identify a rule for integrating $f(x) = (ax + b)^n$?

Interpreting the results

The functions $(x + 1)^3$, $(x + 1)^4$ and $(5x + 3)^2$ are examples of **composite functions** or **functions of a function**. For example if we let $u = 5x + 3$ then for $(5x + 3)^2$:
$f(u) = u^2$ and $u = 5x + 3$
so f is a function of u and u is a function of x.

Summarising the results of Exploration 24.1 gives the following results.

$f(x)$	$\int f(x)\,dx$	
$(x + 1)^3$	$\frac{1}{4}x^4 + x^3 + \frac{3}{2}x^2 + x + c$	$\frac{1}{4}(x+1)^4 + \left(c - \frac{1}{4}\right)$
$(x + 1)^4$	$\frac{1}{5}x^5 + x^4 + 2x^3 + 2x^2 + x + c$	$\frac{1}{5}(x+1)^5 + \left(c - \frac{1}{5}\right)$
$(5x + 3)^2$	$\frac{25}{3}x^3 + 15x^2 + 9x + c$	$\frac{1}{15}(5x+3)^3 + \left(c - \frac{27}{15}\right)$

From this exploration we can deduce that:

- expanding the functions and integrating term by term is inefficient and tedious,
- there must be a simple rule to integrate such functions.

The **chain rule for differentiation** (Chapter 20, *Differentiation 3*) provides a useful technique of integration.

The chain rule is:

$$\frac{df}{dx} = \frac{df}{du} \times \frac{du}{dx}$$

To find $\frac{df}{dx}$ if $f = (5x + 3)^3$, let $u = (5x + 3)$ and $f = u^3$

$$\Rightarrow \quad \frac{df}{dx} = \frac{df}{du} \times \frac{du}{dx} = 3u^2 \times 5 = 15u^2 = 15(x + 3)^2$$

When using the chain rule, we choose a new variable u, differentiate the function of u and then substitute back for x.

The same idea can be used as a technique of integration. We can change the variable and use the chain rule.

If $y = \int f(x)\,dx$ then $\frac{dy}{dx} = f(x)$.

Let $x = g(u)$ for some function g and a new variable u, so that y is now a function of u, i.e. $y = f(g(u))$. This is a composite function of u. Using the chain rule:

$$\frac{dy}{du} = \frac{dy}{dx} \times \frac{dx}{du} = f(x) \times \frac{dx}{du} = f(g(u))\frac{dx}{du}$$

and integrating with respect to u:

$$y = \int f(g(u))\,\frac{dx}{du}\,du$$

Compare this with $y = \int f(x)\,dx$ to give a rule for integration.

Integration by substitution

- Choose a new variable u to replace x.
- Replace dx by $\frac{dx}{du}\,du$.
- Integrate the new function of u.

Now let's try this method on the functions in Exploration 24.1.

For $\int (x+1)^3 dx$ let $u = x + 1$; then $\frac{du}{dx} = 1$ so $dx = du$.

Substitute throughout for all the xs to give:

$$\int (x+1)^3 dx = \int u^3 du = \tfrac{1}{4}u^4 + c$$

↑

standard function

For $\int (5x+3)^2 dx$ let $u = 5x + 3$; then $\frac{du}{dx} = 5$ so that $dx = \frac{1}{5} du$.

Substitute through for all the xs to give:

$$\int (5x+3)^2 dx = \int u^2 \times \tfrac{1}{5}\, du = \tfrac{1}{5}\int u^2 du = \tfrac{1}{5}\left(\tfrac{1}{3}u^3\right) + c = \tfrac{1}{15}(5x+3)^3 + c$$

Note that in finding dx in terms of du we use the rule:

$$\frac{dx}{du} = \frac{1}{\frac{du}{dx}}$$

Exploration 24.2

Integrating functions

Integrate each of the following functions:

- by multiplying out the brackets and integrating term by term,
- by using the given substitution.

a) $f(x) = (3x + 7)^2$ let $u = 3x + 7$

b) $f(x) = x^2 (x^3 - 3)$ let $u = x^3 - 3$

c) $f(x) = x^2(x^3 - 3)^2$ let $u = x^3 - 3$

Interpreting the results

The functions $x^2(x^3 - 3)$ and $x^2(x^3 - 3)^2$ are the first examples of products of functions that can be integrated. For $x^2(x^3 - 3)^2$ letting $u = x^3 - 3$ gives $\frac{du}{dx} = 3x^2$ and $dx = \frac{1}{3x^2} du$. Then:

$$\int x^2(x^3-3)^2 dx = \int x^2(u^2)\frac{1}{3x^2}\, du = \tfrac{1}{3}\int u^2 du = \tfrac{1}{9}u^3 + c = \tfrac{1}{9}(x^3-3)^3 + c$$

↑

no xs allowed in here

The substitution $u = x^3 - 3$ works because the derivative of u is also part of the integral. All terms in x must disappear, leaving the new integral in terms only of u. This is how we know that the substitution is a good choice.

Example 24.1

Evaluate $y = \int (3x^2 + x - 4)^3 (6x+1)\, dx$.

Solution

Let $u = 3x^2 + x - 4 \Rightarrow \frac{du}{dx} = 6x + 1$.

Then:

$$dx = \frac{dx}{du}\, du = \frac{1}{(6x+1)}\, du$$

Replacing $(3x^2 + x - 4)^3$ *by* u^3 *and* dx *by* $\frac{1}{(6x+1)}\, du$:

$$y = \int \left(3x^2 + x - 4\right)^3 (6x+1)\,\mathrm{d}x = \int u^3 (6x+1) \frac{1}{(6x+1)}\,\mathrm{d}u = \int u^3 \mathrm{d}u$$

$$\Rightarrow y = \tfrac{1}{4}u^4 + c = \tfrac{1}{4}\left(3x^2 + x - 4\right)^4 + c$$

Example 24.2

Evaluate $y = \int_0^1 x\left(5 + 2x^2\right)^4 \mathrm{d}x$.

Solution

In this example we need to change all the xs *in the function, and also the limits.*

Let $u = 5 + 2x^2 \Rightarrow \dfrac{\mathrm{d}u}{\mathrm{d}x} = 4x$

Then:

$$\mathrm{d}x = \frac{\mathrm{d}x}{\mathrm{d}u}\,\mathrm{d}u = \frac{1}{4x}\,\mathrm{d}u$$

The x *integral will be turned into a* u *integral. The limits on* u *are found from* $u = 5 + 2x^2$ *and the* x *limits* 0 *and* 1.

$x = 0 \Rightarrow u = 5$ *and* $x = 1 \Rightarrow u = 7$

Replacing $(5 + 2x^2)^4$ *by* u^4, $\mathrm{d}x$ *by* $\dfrac{1}{4x}\,\mathrm{d}u$ *and changing the limits gives:*

$$y = \int_5^7 x\left(u^4\right)\frac{1}{4x}\,\mathrm{d}u = \tfrac{1}{4}\int_5^7 u^4 \mathrm{d}u = \tfrac{1}{4}\left[\frac{u^5}{5}\right]_5^7 = \frac{7^5}{20} - \frac{5^5}{20} = 684.1$$

Example 24.3

Find the total area enclosed by the curve $y = xe^{1-2x^2}$, *the* x*-axis and the lines* $x = 1$ *and* $x = -0.5$.

Solution

For problems of this type it is very important to sketch the function. A graphics calculator or computer software is valuable to provide a quick sketch. The graph of the curve $y = xe^{1-2x^2}$ *and the total area required is as shown.*

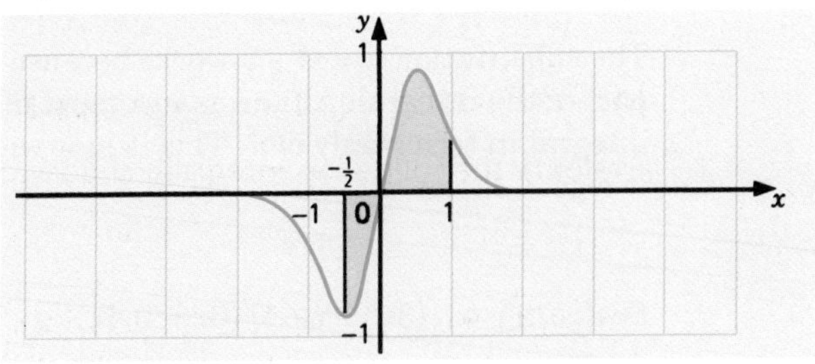

The area between $x = -\frac{1}{2}$ *and* $x = 0$ *is*

$$A_1 = -\int_{-\frac{1}{2}}^{0} xe^{1-2x^2}\,\mathrm{d}x$$

and the area between $x = 0$ *and* $x = 1$ *is*

$$A_2 = \int_0^1 xe^{1-2x^2}\,\mathrm{d}x.$$

The total area is $A_1 + A_2$.

Let $u = 1 - 2x^2 \Rightarrow \dfrac{du}{dx} = -4x$

Then:

$dx = \dfrac{dx}{du}du = -\dfrac{1}{4x}du$

For the limits $x = -\frac{1}{2}$ *and* $x = 0$ *we have* $u = \frac{1}{2}$ *and* $u = 1$.

For the limits $x = 0$ *and* $x = 1$ *we have* $u = 1$ *and* $u = -1$.
Subsituting for all the xs and the limits:

$$A_1 = -\int_{\frac{1}{2}}^{1} -\tfrac{1}{4}e^u du = \left[\tfrac{1}{4}e^u\right]_{\frac{1}{2}}^{1} = \tfrac{1}{4}\left(e^1 - e^{\frac{1}{2}}\right) = 0.2674 \text{ (to 4 d.p.)}$$

$$A_2 = \int_{1}^{-1} -\tfrac{1}{4}e^u du = \left[-\tfrac{1}{4}e^u\right]_1^{-1} = -\tfrac{1}{4}\left(e^{-1} - e^1\right) = 0.5876$$

The total area is $A_1 + A_2 = 0.8550$ *(correct to four d.p.).*

We must be careful when interpreting integrals by areas. It would be easy to write the area – wrongly – as:

$$\int_{-\frac{1}{2}}^{1} xe^{1-2x^2}dx = \int_{\frac{1}{2}}^{-1} -\tfrac{1}{4}e^u du = -\tfrac{1}{4}\left[e^u\right]_{\frac{1}{2}}^{-1} = -\tfrac{1}{4}\left(e^{-1} - e^{\frac{1}{2}}\right) = 0.3202$$

This is **not** the total area required. To avoid making the mistake, always sketch the graph of the function.

EXERCISES

24.1 A

1 By using the given substitution evaluate the integrals.

a) $\int (x+5)^4 dx \quad u = x + 5$

b) $\int (2x-3)^5 dx \quad u = 2x - 3$

c) $\int_0^3 \sqrt{x+1}\, dx \quad u = x + 1$

d) $\int x(3x^2+2)^{\frac{5}{4}} dx \quad u = 3x^2 + 2$

e) $\int_1^2 e^{4x-1} dx \quad u = 4x - 1$

f) $\int_0^1 8xe^{4x^2} dx \quad u = 4x^2$

g) $\int (2x+3)(x^2+3x+7)dx \quad u = x^2 + 3x + 7$

h) $\int_1^2 (3x^2-1)\left(x^3-x+1\right)^2 dx \quad u = x^3 - x + 1$

2 Evaluate the following integrals, choosing an appropriate substitution.

a) $\int (x-5)^6 dx$

b) $\int_0^1 (2x+1)^3 dx$

c) $\int \sqrt{3x-1}\, dx$

d) $\int x(x^2-1)^2 dx$

e) $\int (2x-1)(x^2-x+3)^{-\frac{1}{2}} dx$

f) $\int_{-1}^{1} (4x+1)(2x^2+x+3)^5 dx$

g) $\int e^{2x+5} dx$

h) $\int (2x-3)e^{x^2-3x+1} dx$

3 Find the total area enclosed by the curve $y = 16 - (x-2)^4$ and the x-axis.

4 Find the total area enclosed by the curve $y = xe^{2-3x^2}$, the x-axis and the lines $x = 1$ and $x = -1$.

5 Find the volume generated when the area between the lines $y = 4 - x$, the y-axis and the x-axis is rotated completely about the x-axis.

6 Find the volume generated when the area between the curve $y = \sqrt{x}\left(2 - x^2\right)$ and the x-axis is rotated about the x-axis.

7 In mechanics we define the work done by a force F(x) on an object that moves along the x-axis from $x = a$ to $x = b$ to be $W = \int \mathrm{F}(x)\,\mathrm{d}x$. Calculate the work done by an object that stretches a non-linear spring from $x = 0.2\,\text{m}$ to $x = 0.4\,\text{m}$. The force exerted on the object by the spring is $20(x - 0.1)^{\frac{3}{2}}$, for $x > 0.1$.

8 The length of the curve $y = \mathrm{f}(x)$ between the points on the curve $(a, \mathrm{f}(a))$ and $(b, \mathrm{f}(b))$ is given by $s = \int_a^b \sqrt{1 + \mathrm{f}'(x)^2}\,\mathrm{d}x$.

Find the length of the curve $y = x^{\frac{3}{2}}$ between the points on the curve (1, 1) and (4, 8).

EXERCISES

24.1B

1 By using the given substitution evaluate the integrals.

a) $\int (3x+4)^3\mathrm{d}x \quad u = 3x+4$ b) $\int 2(x-3)^{-4}\mathrm{d}x \quad u = x-3$

c) $\int \sqrt{2x-1}\,\mathrm{d}x \quad u = 2x-1$ d) $\int_0^2 (2x+1)^{\frac{5}{2}}\,\mathrm{d}x \quad u = 2x+1$

e) $\int_2^3 \frac{2}{(3x-4)^2}\mathrm{d}x \quad u = 3x-4$ f) $\int \mathrm{e}^{x-1}\mathrm{d}x \quad u = x-1$

g) $\int 3x^2(x^3-3)\mathrm{d}x \quad u = x^3-3$ h) $\int (4x+1)\mathrm{e}^{2x^2+x-2}\,\mathrm{d}x \quad u = 2x^2+x-2$

2 Evaluate the following integrals choosing an appropriate substitution.

a) $\int (x+3)^4\mathrm{d}x$ b) $\int_1^2 (4x+3)^2\mathrm{d}x$

c) $\int \sqrt{2x+3}\,\mathrm{d}x$ d) $\int_{-1}^1 (x-1)\sqrt{x^2-2x+3}\,\mathrm{d}x$

e) $\int \frac{1}{(4x+2)^3}\mathrm{d}x$ f) $\int \mathrm{e}^{3x-1}\mathrm{d}x$

g) $\int (2x+1)\mathrm{e}^{x^2+x+4}\mathrm{d}x$ h) $\int_0^2 x\mathrm{e}^{-x^2}\mathrm{d}x$

3 Find the total area enclosed by the curve $y = x(2-x^2)^2$ and the x-axis.

4 Find the total area enclosed by the curve $y = (2x-1)\mathrm{e}^{x^2-x}$, the y-axis and the line $x = 1$.

5 Find the volume generated when the area between the lines $y = 2 - x$, the x-axis and the y-axis is rotated completely about the x-axis.

6 Find the volume generated when the area between the curve $y = (1 - 0.5x)^2$ and the x-axis is rotated completely about the x-axis.

7 For a straight thin rod lying along the axis between $x = a$ and $x = b$, the position of the centre of mass (or centroid) is given by:

$$\bar{x} = \frac{\int_a^b x\rho(x)\,\mathrm{d}x}{\int_a^b \rho(x)\,\mathrm{d}x}$$

where $\rho(x)$ is the density of the rod.

Find the position of the centre of mass of a rod of length 1 metre having density $\rho(s) = (3 + s^2)^2$ where s is the distance along the rod from one end.

8 If a square plate of side length a stands vertically in a liquid then the force exerted on one side of the plate by the pressure in the liquid is given by $\int_{h_1}^{h_2} ah\rho\,\mathrm{d}h$ where ρ is the density of the liquid, h_1 and h_2 are the depths of the top and bottom of the plate respectively.

Find the force on a square plate of side length 0.5 metres in a liquid of density $\rho = 2\mathrm{e}^{-h^2}$ when one edge of the plate lies in the surface of the liquid.

USING SUBSTITUTIONS PART 2

Any function of the form $\frac{\mathrm{f}'(x)}{\mathrm{f}(x)}$ can be integrated using the substitution $u = \mathrm{f}(x)$.

Exploration 24.3

Use substitution

Make the substitution $u = ax + b$ to find $\int \frac{1}{ax+b}\,\mathrm{d}x$.

Use your answer to evaluate:

a) $\int \frac{1}{x+1}\,\mathrm{d}x$

b) $\int \frac{1}{3x+2}\,\mathrm{d}x$

c) $\int \frac{1}{1-4x}\,\mathrm{d}x$.

Interpreting the results

The integral $\int \frac{1}{ax+b}\,\mathrm{d}x = \frac{1}{a}\ln|ax+b|$ is an important result for the evaluation of $\int \frac{\mathrm{f}'(x)}{\mathrm{f}(x)}\,\mathrm{d}x$.

Example 24.4

Evaluate $y = \int \frac{x+1}{x^2+2x-1}\,\mathrm{d}x$.

Solution

Let $u = x^2 + 2x - 1 \Rightarrow \frac{\mathrm{d}u}{\mathrm{d}x} = 2x + 2 = 2(x+1)$

Then:

$$dx = \frac{dx}{du}du = \frac{1}{2(x+1)}du$$

Replacing x^2+2x-1 *by* u *and* dx *by* $\frac{1}{2(x+1)}du$:

$$y = \int \frac{(x+1)}{u}\cdot\frac{1}{2(x+1)}du = \frac{1}{2}\int \frac{1}{u}du = \frac{1}{2}\ln|u| + c = \frac{1}{2}\ln\left|x^2+2x-1\right| + c$$

Example 24.5

Evaluate $y = \int_1^2 \frac{x^2+1}{x^3+3x}dx$.

Solution

Let $u = x^3+3x \Rightarrow \frac{du}{dx} = 3x^2+3 = 3(x^2+1)$

Then:

$$dx = \frac{dx}{du}du = \frac{1}{3(x^2+1)}du$$

For the limits $x = 1 \Rightarrow u = 4$ and $x = 2 \Rightarrow u = 14$.

Replacing (x^3+3x) *by* u, dx *by* $\frac{1}{3(x^2+1)}du$ *and changing the limits gives:*

$$y = \int_1^2 \frac{x^2+1}{x^3+3x}dx = \int_4^{14}\frac{(x^2+1)}{u}\cdot\frac{1}{3(x^2+1)}du = \tfrac{1}{3}\int_4^{14}\frac{1}{u}du = \tfrac{1}{3}\ln[u]_4^{14}$$

$$= \tfrac{1}{3}\ln 14 - \tfrac{1}{3}\ln 4 = 0.4176 \text{ (to 4 d.p.)}$$

Example 24.6

Evaluate $y = \int_0^1 \frac{8x+3}{\sqrt{1+3x+4x^2}}dx$.

Solution

Not all quotients will lead to natural logarithms.

Let $u = 1+3x+4x^2 \Rightarrow \frac{du}{dx} = 3+8x$

Then:

$$dx = \frac{dx}{du}du = \frac{1}{(3+8x)}du$$

For the limits $x = 0 \Rightarrow u = 1$ and $x = 1 \Rightarrow u = 8$.

Making the substitutions:

$$y = \int_1^8 \frac{(8x+3)}{\sqrt{u}}\cdot\frac{1}{(8x+3)}du = \int_1^8 u^{-\frac{1}{2}}du = \left[2u^{\frac{1}{2}}\right]_1^8 = 2\sqrt{8} - 2\sqrt{1} = 3.657$$

Example 24.7

Find the volume generated when the area between the curve $y = \sqrt{\frac{x-1}{x^2-2x-1}}$, *the* x*-axis and* y*-axis is rotated completely about the* x*-axis.*

Solution

The figure shows a graph of $y=\sqrt{\dfrac{x-1}{x^2-2x-1}}$, *the area to be rotated and the volume of the region.*

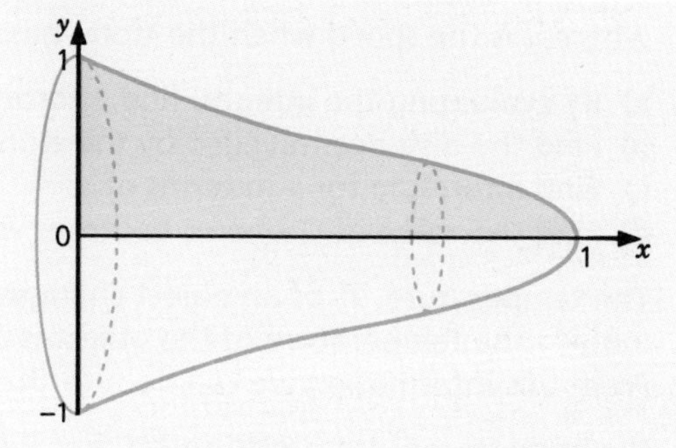

The volume of the region is $V=\int_0^1 \pi y^2 \mathrm{d}x$ *(see Chapter 19, Integration 2).*

Substituting for $y \Rightarrow V=\int_0^1 \dfrac{\pi(x-1)}{(x^2-2x-1)}\mathrm{d}x$

Let $u=x^2-2x-1$ *then* $\mathrm{d}x=\dfrac{1}{2(x-1)}\mathrm{d}u$ *and the new limits are* $u=-1$ *and* $u=-2$. *Making the subtitutions:*

$$V=\int_{-1}^{-2}\pi\frac{(x-1)}{u}\cdot\frac{1}{2(x-1)}\mathrm{d}u=\frac{\pi}{2}\int_{-1}^{-2}\frac{1}{u}\mathrm{d}u=\frac{\pi}{2}\ln|-2|-\ln|-1|=\frac{\pi}{2}\ln 2=1.089$$

EXERCISES

24.2 A

1 By using the given substitution evaluate the integrals.

a) $\int \dfrac{3}{1+2x}\mathrm{d}x \quad u=1+2x$ **b)** $\int \dfrac{6}{2+3x}\mathrm{d}x \quad u=2+3x$

c) $\int_1^3 \dfrac{3x}{2+x^2}\mathrm{d}x \quad u=2+x^2$ **d)** $\int \dfrac{3x^2-2}{1-2x+x^3}\mathrm{d}x \quad u=1-2x+x^3$

e) $\int_0^1 \dfrac{x^2}{\sqrt{x^3+5}}\mathrm{d}x \quad u=x^3+5$ **f)** $\int \dfrac{\mathrm{e}^{-2x}}{1-\mathrm{e}^{-2x}}\mathrm{d}x \quad u=1-\mathrm{e}^{-2x}$

2 Evaluate the following integrals choosing an appropriate substitution.

a) $\int \dfrac{1}{2-3x}\mathrm{d}x$ **b)** $\int_0^2 \dfrac{2}{2t+1}\mathrm{d}t$ **c)** $\int \dfrac{x+1}{\sqrt{x^2+2x+2}}\mathrm{d}x$

d) $\int \dfrac{1}{x\ln x}\mathrm{d}x$ **e)** $\int \dfrac{\mathrm{e}^x-\mathrm{e}^{-x}}{\mathrm{e}^x+\mathrm{e}^{-x}}\mathrm{d}x$ **f)** $\int \dfrac{x^2+2x}{x^3+3x^2+1}\mathrm{d}x$

g) $\int_1^2 \dfrac{2t^3-t}{t^4-t^2+1}\mathrm{d}t$ **h)** $\int \dfrac{\mathrm{e}^x(1+x)}{x\mathrm{e}^x-1}\mathrm{d}x$

3 Find the total area enclosed by the curve $y=\dfrac{10}{2x+5}$, the y-axis and the line $x=2$.

4 Find the total area enclosed by the curve $y=\dfrac{2x^2}{x^3+1}-1$ and the x-axis.

(The curve cuts the x-axis at $x=\dfrac{(1-\sqrt{5})}{2}$, $x=1$ and $x=\dfrac{(1+\sqrt{5})}{2}$.)

5 The distance, s, in metres, moved by a stone of mass 1 kg dropped, from rest, down a well satisfies the equation:

$$s=\int_0^u \frac{v\,\mathrm{d}v}{10-0.4v^2}$$

where u is the speed when the stone has dropped through s metres.

a) By evaluating the integral find a formula for s as a function of u.
b) Find the distance travelled by the stone when its speed is 4 m s^{-1}.
c) Find a formula for u in terms of s.
d) Find the value of u when s becomes very large.

6 The temperature, T, of an object changes so that $\frac{\mathrm{d}T}{\mathrm{d}t}=-0.5(T-40)$. Initially the temperature of the object is 80°C.
From this information we can deduce that the time (in minutes) for the body to cool from 80°C to X°C is $t=\int_{80}^{X}\frac{-2}{T-40}\,\mathrm{d}T$.

a) Evaluate the integral to find t as a function of X.
b) How long does it take for the temperature to fall below 60°C?
c) Find a formula for X in terms of t.
d) Find the value of X when t becomes very large. What does this mean physically?

EXERCISES

24.2 B

1 By using the given substitution evaluate these integrals.

a) $\int \frac{5}{1+4x}\,\mathrm{d}x \quad u=1+4x$ **b)** $\int \frac{3}{0.5-2x}\,\mathrm{d}x \quad u=0.5-2x$

c) $\int_0^1 \frac{x}{3+2x^2}\,\mathrm{d}x \quad u=3+2x^2$ **d)** $\int_{-1}^{2} \frac{2x-3}{x^2-3x+5}\,\mathrm{d}x \quad u=x^2-3x+5$

e) $\int \frac{4t^3-3}{\sqrt{t^4-3t}}\,\mathrm{d}t \quad u=t^4-3t$ **f)** $\int \frac{5\mathrm{e}^{3x}}{2+\mathrm{e}^{3x}}\,\mathrm{d}x \quad u=2+\mathrm{e}^{3x}$

2 Evaluate the following integrals choosing an appropriate substitution.

a) $\int \frac{1}{2+7x}\,\mathrm{d}x$ **b)** $\int_0^1 \frac{t}{4-t^2}\,\mathrm{d}t$ **c)** $\int \frac{v+0.5}{\sqrt{v^2+v-1}}\,\mathrm{d}v$

d) $\int_0^2 \frac{t+1}{\sqrt{t^2+2t+1}}\,\mathrm{d}t$ **e)** $\int \frac{x^2-x+1}{2x^3-3x^2+6x+1}\,\mathrm{d}x$ **f)** $\int_2^5 \frac{t}{(t^2-3)^{\frac{3}{2}}}\,\mathrm{d}t$

g) $\int \frac{2\mathrm{e}^{2x}-1}{\mathrm{e}^{2x}-x}\,\mathrm{d}x$ **h)** $\int_1^5 \frac{\mathrm{e}^x}{(1-\mathrm{e}^x)^2}\,\mathrm{d}x$

3 Find the volume generated when the area between the curve $y=\frac{\sqrt{x}}{(1+3x^2)}$, the x-axis and the line $x = 1$ is rotated completely about the x-axis.

4 Find the total area enclosed by the curve $y=\frac{5}{1+2x}$, the y-axis and the line $x = 1$.

5 The distance, s, in metres, moved by a body of mass 1 kg thrown straight upwards when its speed is U m s^{-1} satifies the equation:

$$s = \int_{10}^{U} -\frac{v}{10 + 0.2v^2}\,dv$$

where the speed of the body is 10 m s^{-1} when released at $s = 0$. (The term $0.2v^2$ is the air resistance on the body.)

a) By evaluating the integral find a formula for s as a function of U.

b) Find the maximum height reached by the body.

6 A sky diver falls so that her speed, U m s^{-1}, and time of drop, t s, are related by:

$$t = \int_{0}^{U} \frac{1}{10 - 0.3v}\,dv$$

where the sky diver jumps from a balloon which is at rest.

a) By evaluating the integral find a formula for t as a function of U.

b) Find a formula for U as a function of t.

c) Find the value of U when t becomes very large. What does this mean for the sky diver?

INTEGRATION BY PARTS

This method of integration works for some products of functions.

Exploration 24.4

Integrate by parts

Look at the function $f(x) = (x + 1)e^x$.

- Differentiate $f(x)$ using the product rule.
- Hence deduce the value of $\int xe^x dx$.
- Is it true that $\int uv dx = \int u dx \int v dx$?

Interpreting the results

From this exploration, we see that the integral of a product of two functions does not equal the product of the integrals of the two functions.

To develop a rule, we start with the product rule for differentiation:

$$\frac{d(uv)}{dx} = \frac{du}{dx}v + u\frac{dv}{dx}$$

Integrating both sides with respect to x:

$$uv = \int v\frac{du}{dx}dx + \int u\frac{dv}{dx}dx$$

$$\Rightarrow \int u\frac{dv}{dx}dx = uv - \int v\frac{du}{dx}dx$$

This is the formula for **integration by parts**. It is the integral of a product of functions written as u and $\frac{dv}{dx}$. The formula tells us that one function $\frac{dv}{dx}$ is being integrated and the other function, u, is being differentiated. The following example shows how the method works.

Example 24.8

Evaluate $y = \int_0^1 xe^{4x}dx$.

Solution

Let $u = x \Rightarrow \frac{du}{dx} = 1$

Let $\frac{dv}{dx} = e^{4x} \Rightarrow v = \frac{1}{4}e^{4x}$

$$\int_0^1 xe^{4x}\,dx = \left[x\left(\tfrac{1}{4}e^{4x}\right)\right]_0^1 - \int_0^1\left(\tfrac{1}{4}e^{4x}\right)\cdot 1\,dx$$

$$= \left(\tfrac{1}{4}e^4 - 0\right) - \left[\tfrac{1}{16}e^{4x}\right]_0^1 = \tfrac{1}{4}e^4 - \left(\tfrac{1}{16}e^4 - \tfrac{1}{16}\right) = \tfrac{3}{16}e^4 + \tfrac{1}{16}$$

Example 24.9

Evaluate $y = \int_0^{\frac{\pi}{4}} x\cos 2x\,dx$.

Solution

Let $u = x \Rightarrow \frac{du}{dx} = 1$

Let $\frac{dv}{dx} = \cos 2x \Rightarrow v = \frac{1}{2}\sin 2x$

$$\int_0^{\frac{\pi}{4}} x\cos 2x\,dx = x\left(\tfrac{1}{2}\sin 2x\right)_0^{\frac{\pi}{4}} - \int_0^{\frac{\pi}{4}}\left(\tfrac{1}{2}\sin 2x\right)\cdot 1\,dx$$

$$= \tfrac{\pi}{4}\left(\tfrac{1}{2}\sin\tfrac{\pi}{2}\right) - 0 - \left[-\tfrac{1}{4}\cos 2x\right]_0^{\frac{\pi}{4}}$$

$$= \tfrac{\pi}{8} + \left(\tfrac{1}{4}\cos\tfrac{\pi}{2} - \tfrac{1}{4}\cos 0\right) = \tfrac{\pi}{8} - \tfrac{1}{4}$$

Exploration 24.5

Choosing u wisely

- What happens if, in Example 26.8, you let $u = e^{4x}$ and $\frac{dv}{dx} = x$?
- Suggest a rule for choosing u in an integral of the form $\int x^n f(x)\,dx$ where n is a positive integer.

Interpreting the results

You will have found that using $u = e^{4x}$ with the method of integration by parts does not make the problem easier. We get:

$$\int xe^{4x}dx = \frac{x^2}{2}e^{4x} - 2\int x^2e^{4x}dx$$

and this is more difficult than the original integral.

A general rule which works in cases where we can integrate $f(x)$ is to let $u = x^n$. The second integral then contains nx^{n-1} i.e. the power of x reduces.

Example 24.10

Evaluate $y = \int e^{2x} \sin x \, dx$.

Solution

Let $u = e^{2x} \Rightarrow \frac{du}{dx} = 2e^{2x}$

Let $\frac{dv}{dx} = \sin x \Rightarrow v = -\cos x$

$$\int e^{2x} \sin x \, dx = e^{2x}(-\cos x) - \int (2e^{2x})(-\cos x)\, dx + c$$

$$= -e^{2x}\cos x + 2\int e^{2x} \cos x dx + c \qquad (1)$$

where c is a constant of integration.

Now we have a second integral to evaluate using integration by parts.

It is important to continue to let $u = e^{2x}$ and $\frac{du}{dx} = 2e^{2x}$.

Let $\frac{dv}{dx} = \cos x$ so $v = \sin x$

$\int e^{2x} \cos x \, dx = e^{2x} \sin x - \int (2e^{2x})(\sin x)\, dx$

Substituting into equation (1) gives:

$$\int e^{2x} \sin x \, dx = -e^{2x}\cos x + 2\left(e^{2x}\sin x - 2\int e^{2x}\sin x \, dx\right)$$

$$= -e^{2x}\cos x + 2e^{2x}\sin x - 4\int e^{2x}\sin x \, dx + c$$

It appears that we are back where we started! However we can proceed by adding $4\int e^{2x}\sin x \, dx$ to each side. Then:

$$5\int e^{2x}\sin x \, dx = -e^{2x}\cos x + 2e^{2x}\sin x + c$$

$$\Rightarrow \int e^{2x}\sin x \, dx = -\tfrac{1}{5}e^{2x}\cos x + \tfrac{2}{5}e^{2x}\sin x + A$$

where $A = \frac{1}{5}c$ is a constant of integration.

Example 24.11

Evaluate $y = \int x^2 e^x \, dx$.

Solution

Let $u = x^2 \Rightarrow \frac{du}{dx} = 2x$

Let $\frac{dv}{dx} = e^x \Rightarrow v = e^x$

$$\int x^2 e^x \, dx = x^2 e^x - \int 2xe^x dx = x^2 e^x - 2\int xe^x dx + c$$

Now we have a second integral to evaluate using integration by parts.

Let $u = x \Rightarrow \frac{du}{dx} = 1$

Let $\frac{dv}{dx} = e^x \Rightarrow v = e^x$

$$\int xe^x \, dx = xe^x - \int 1e^x \, dx = xe^x - \int e^x \, dx = xe^x - e^x$$

$$\Rightarrow \int x^2 e^x \, dx = x^2 e^x - 2\left(xe^x - e^x\right) + c = \left(x^2 - 2x + 1\right)e^x + c$$

The message in the example is to keep going and not give up.

Example 24.12

Evaluate $\int x^2 \ln x \, \mathrm{d}x$.

Solution

Suppose that we follow our rule and let $u = x^2$ *and* $\frac{\mathrm{d}v}{\mathrm{d}x} = \ln x$.

Then $v = \int \ln x \, \mathrm{d}x$ *is not a simple integral. The method of integration by parts will not make the integration easier.*

Instead of following the rule, in this case let $u = \ln x \Rightarrow \frac{\mathrm{d}u}{\mathrm{d}x} = \frac{1}{x}$ *and let* $\frac{\mathrm{d}v}{\mathrm{d}x} = x^2 \Rightarrow v = \frac{1}{3}x^3$.

$\int x^2 \ln x \, \mathrm{d}x = \frac{1}{3}x^3 \cdot \ln x - \int \frac{1}{3}x^3 \cdot \frac{1}{x} \, \mathrm{d}x = \frac{1}{3}x^3 \ln x - \frac{1}{3}\int x^2 \, \mathrm{d}x$

Now the second integral is straightforward.

$\int x^2 \, \mathrm{d}x = \frac{1}{3}x^3 \Rightarrow \int x^2 \ln x \, \mathrm{d}x = \frac{1}{3}x^3 \ln x - \frac{1}{9}x^3 + c$

For the integration of products of functions $\int fg \, \mathrm{d}x$ try $u = f$ and $\frac{\mathrm{d}v}{\mathrm{d}x} = g$ and see if the integration by parts helps. If not then try $u = g$ and $\frac{\mathrm{d}v}{\mathrm{d}x} = f$.

This method may not always be successful for integrating products of functions. With experience, it is possible to sense when this method will work. For some problems, as in the next example, the method of integration by parts may give a sequence from which we can make some progress.

Example 24.13

The integral I_n *is defined by* $I_n = \int x^n e^x \, \mathrm{d}x$ *for positive integers.*

a) *Show that after one integration,* I_n *satisfies the sequence:* $I_n = x^n e^x - nI_{n-1}$

b) *Evaluate* I_0 *directly.*

c) *Use* I_0 *and the sequence to evaluate* $\int x^2 e^x \, \mathrm{d}x$ and $\int x^4 e^x \, \mathrm{d}x$.

Solution

a) *Let* $u = x^n \Rightarrow \frac{\mathrm{d}u}{\mathrm{d}x} = nx^{n-1}$ *and let* $\frac{\mathrm{d}v}{\mathrm{d}x} = e^x \Rightarrow v = e^x$.

$I_n = \int x^n e^x \, \mathrm{d}x = x^n e^x - \int nx^{n-1} \cdot e^x \, \mathrm{d}x = x^n e^x - n\int x^{n-1} e^x \, \mathrm{d}x$

$\Rightarrow I_n = x^n e^x - nI_{n-1}$

b) *For* $n = 0$, $I_0 = \int x^0 e^x \, \mathrm{d}x = \int e^x \, \mathrm{d}x = e^x$

c) *Using the sequence in part* **a)**:

$I_1 = xe^x - I_0 = xe^x - e^x$

$I_2 = x^2e^x - 2I_1 = x^2e^x - 2(xe^x - e^x) = (x^2 - 2x + 2)e^x$

$I_3 = x^3e^x - 3I_2 = x^3e^x - 3(x^2 - 2x + 2)e^x = (x^3 - 3x^2 + 6x - 6)e^x$

$$I_4 = x^4e^x - 4I_3$$
$$= x^4e^x - 4\left(x^3 - 3x^2 + 6x - 6\right)e^x = \left(x^4 - 4x^3 + 12x^2 - 24x + 24\right)e^x$$
$$\Rightarrow \int x^2e^x \, dx = I_2 + c = \left(x^2 - 2x + 2\right)e^x + c$$
$$\Rightarrow \int x^4e^x \, dx = I_4 + c = \left(x^4 - 4x^3 + 12x^2 - 24x + 24\right)e^x + c$$

EXERCISES

24.3 A

1 Use the method of integration by parts to evaluate the following integrals.

a) $\int xe^{-3x} \, dx$ **b)** $\int_0^1 xe^{2x} \, dx$ **c)** $\int x^4 \ln x \, dx$

d) $\int_1^2 x^2e^{-3x} \, dx$ **e)** $\int e^x(x+3) \, dx$ **f)** $\int x\sin 3x \, dx$

g) $\int_0^{\pi/2} x^2 \cos x \, dx$ **h)** $\int_1^4 \ln x \, dx$ **i)** $\int \ln(2x+3) \, dx$

j) $\int \left(x^2 + x + 2\right)e^x \, dx$ **k)** $\int_1^2 x^n \ln x \, dx \quad n \neq -1$ **l)** $\int x(x-1)^4 \, dx$

m) $\int_0^1 3x^2(2x+1)^5 \, dx$ **n)** $\int e^{2x} \sin 3x \, dx$

2 Let n be any positive integer and define I_n to be the integral $I_n = \int_1^2 (\ln x)^n \, dx$.

a) Show that $\int_1^2 \ln x \, dx = 2\ln 2 - 1$.

b) Show that after integration, I_n satisfies the sequence $I_n = 2(\ln 2)^n - nI_{n-1}$.

c) Use the sequence to find the value of $\int_1^2 (\ln x)^4 \, dx$.

3 The integral I_n is defined by $I_n = \int x^n e^{-x^2} \, dx$ where n is a positive even integer.

a) Show that after one integration, I_n satisfies the sequence
$I_n = -\frac{1}{2}x^{n-1}e^{-x^2} + \frac{(n-1)}{2}I_{n-2}$.

b) Use the sequence to express $\int x^2e^{-x^2} \, dx$, $\int x^4e^{-x^2} \, dx$ and $\int x^6e^{-x^2} \, dx$ in terms of I_0.

4 The integral I_n is defined by $I_n = \int x^n \cos x \, dx$ where n is a positive integer.

a) Show that after one integration I_n satisfies the sequence
$I_n = x^n \sin x - n\int x^{n-1} \sin x \, dx$.

b) Establish the sequence for I_n.

$I_n = x^n \sin x + nx^{n-1} \cos x - n(n-1)I_{n-2}$.

c) Evaluate I_0 and use the sequence to evaluate $\int x^4 \cos x \, dx$ and $\int x^6 \cos x \, dx$.

EXERCISES

24.3 B

1 Use the method of integration by parts to evaluate the following integrals.

a) $\int xe^{5x}\,dx$ b) $\int_0^1 xe^{-0.5x}\,dx$ c) $\int x^2e^{2x}\,dx$

d) $\int_1^2 x^5\ln x\,dx$ e) $\int_0^1 x^2(1-x)^9\,dx$ f) $\int (x+1)\ln(x+1)\,dx$

g) $\int \left(2x^2-3x+1\right)e^{-x}\,dx$ h) $\int x\sin 4x\,dx$ i) $\int_0^{\frac{\pi}{4}} x^3\cos 2x\,dx$

j) $\int x^3e^{x^2}\,dx$ k) $\int_{-1}^2 (x+1)^2e^{-3x}\,dx$ l) $\int_1^3 \left(x^2-x\right)(3-x)^7\,dx$

m) $\int \dfrac{x}{\sqrt{1+x}}\,dx$ n) $\int e^{-x}\sin 2x\,dx$

2 The integral I_n is defined by $I_n = \int x^n\ln x\,dx$.

a) Evaluate the integral for I_n when $n \neq -1$.

b) Show that when $n=-1$, $I_{-1} = \frac{1}{2}(\ln x)^2$.

3 The integral I_n is defined by $I_n = \int_0^1 \left(1-x^2\right)^{n+\frac{1}{2}}\,dx$ where n is a non-negative integer.

a) Show that after one integration I_n satisfies the sequence
$I_n = \dfrac{(2n+1)}{2(n+1)} I_{n-1}$.

b) Given that $I_0 = \frac{\pi}{4}$ use the sequence to find $\int_0^1 \left(1-x^2\right)^{\frac{5}{2}}\,dx$.

4 Establish a sequence for the integral:
$I_n = \int x^n\sin x\,dx$
where n is a positive integer.

Evaluate I_1 and use your sequence to evaluate $\int x^5\sin x\,dx$.

INTEGRATION BY PARTIAL FRACTIONS

Exploration 24.5

Integrate using partial fractions

Use the function:

$$f(x) = \frac{2}{(x-1)(x-3)}.$$

- Write $f(x)$ as partial fractions.
- By integrating each term evaluate:

$$\int \frac{2}{(x-1)(x-3)}\,dx.$$

Interpreting the results

Using partial fractions gives a useful technique for evaluating some integrals involving fractions of polynomials. In the exploration the partial fractions for $\mathrm{f}(x)$ are:

$$\mathrm{f}(x) = \frac{1}{x-3} - \frac{1}{x-1}$$

Now each term can be integrated to give:

$$\int \frac{2}{(x-1)(x-3)}\,\mathrm{d}x = \int \left(\frac{1}{x-3} - \frac{1}{x-1}\right)\mathrm{d}x = \ln|x-3| - \ln|x-1| + c = \ln\left|\frac{x-3}{x-1}\right| + c$$

It is convenient in these problems to write the constant of integration as $\ln A$. The solution can then be written as $\ln A\left|\frac{x-3}{x-1}\right|$.

Before using partial fractions, look very carefully at the fraction to be integrated. Substitution might be a quicker method of approach.

Example 24.14

Evaluate the integral $\int \frac{x-2}{x^2-4x+3}\,\mathrm{d}x$

a) *using a substitution,*

b) *using partial fractions.*

Solution

a) *With a fraction of polynomials it is always worth investigating if it is in the form* $\frac{\mathrm{f}'(x)}{\mathrm{f}(x)}$.

Let $u = x^2 - 4x + 3$ *so* $\frac{\mathrm{d}u}{\mathrm{d}x} = 2x - 4 = 2(x-2)$

and $\mathrm{d}x = \frac{\mathrm{d}x}{\mathrm{d}u}\mathrm{d}u = \frac{1}{2(x-2)}\mathrm{d}u$

Substituting for $x^2 - 4x + 3 = u$ *and for* $\mathrm{d}x$:

$$\int \frac{x-2}{x^2-4x+3}\,\mathrm{d}x = \int \frac{(x-2)}{u}\cdot\frac{1}{2(x-2)}\,\mathrm{d}u = \tfrac{1}{2}\int \frac{1}{u}\,\mathrm{d}u = \tfrac{1}{2}\ln|u| + c$$

Replacing u *by* $x^2 - 4x + 3$:

$$\int \frac{x-2}{x^2-4x+3}\,\mathrm{d}x = \tfrac{1}{2}\ln\left|x^2 - 4x + 3\right| + c$$

b) *To use partial fractions factorise* $x^2 - 4x + 3$ *to give* $(x-3)(x-1)$. *Now write the function as partial fractions.*

$$\frac{x-2}{x^2-4x+3} = \frac{x-2}{(x-3)(x-1)} = \frac{\frac{1}{2}}{x-3} + \frac{\frac{1}{2}}{x-1}$$

Integrating each side:

$$\int \frac{x-2}{x^2-4x+3}\,\mathrm{d}x = \int \frac{\frac{1}{2}}{x-3} + \frac{\frac{1}{2}}{x-1}\,\mathrm{d}x = \tfrac{1}{2}\ln|x-3| + \tfrac{1}{2}\ln|x-1| + c$$

$$= \tfrac{1}{2}\ln|x-3||x-1| + c = \tfrac{1}{2}\ln\left|x^2 - 4x + 3\right| + c$$

The answers are exactly of the same form, however the substitution method is often more straightforward.

Example 24.15

Evaluate $\int_2^5 \frac{x^2-2}{x^2-1}\,dx$.

Solution

This improper fraction must be divided out into the form $A+\frac{B}{x^2-1}$.

$$\frac{x^2-2}{x^2-1}=\frac{(x^2-1)-1}{(x^2-1)}=1-\frac{1}{x^2-1}=1-\left(\frac{\frac{1}{2}}{(x-1)}-\frac{\frac{1}{2}}{x+1}\right)$$

Then:

$$\int_2^5 \frac{x^2-2}{x^2-1}\,dx=\int_2^5\left(1-\frac{\frac{1}{2}}{(x-1)}+\frac{\frac{1}{2}}{(x+1)}\right)dx=\left[x-\tfrac{1}{2}\ln|x-1|+\tfrac{1}{2}\ln|x+1|\right]_2^5$$

$$=\left(5-\tfrac{1}{2}\ln 4+\tfrac{1}{2}\ln 6\right)-\left(2-\tfrac{1}{2}\ln 1+\tfrac{1}{2}\ln 3\right)$$

$$=3-\tfrac{1}{2}\ln 2=2.6534$$

Example 24.16

When modelling the growth of a bacterial colony in a laboratory experiment using the logistic equation, it is found that the area of a shallow dish occupied by the bacteria and the time are related by:

$$t=\int_1^A \frac{16}{x(12-x)}\,dx$$

where the area of the dish is A cm^2 and the time is t days.
At the start of the experiment the area of culture is 1 cm^2.

a) *Evaluate the integral to form t as a function of A.*
b) *After how many days will the area of culture be 8 cm^2?*
c) *Rearrange the formula in part **a)** to give a formula for A in terms of t. Sketch a graph of A against t.*
d) *According to the model, what is the maximum area of shallow dish occupied by the bacteria?*

Solution

a) *Using partial fractions* $\frac{16}{x(12-x)}=\frac{4}{3}\left(\frac{1}{x}+\frac{1}{12-x}\right)$

Then $t=\int_1^A \frac{16}{x(12-x)}\,dx=\int_1^A \frac{4}{3}\left(\frac{1}{x}+\frac{1}{12-x}\right)dx$

$$=\left[\frac{4}{3}\left(\ln|x|-\ln|12-x|\right)\right]_1^A$$

$$=\frac{4}{3}\left(\ln\frac{11A}{12-A}\right)$$

b) *For* $A=8$, $t=\frac{4}{3}\ln\frac{88}{4}=\frac{4}{3}\ln 22=4.12$.

The area of culture is 8 cm^2 after 4.12 days.

c) $t=\frac{4}{3}\ln\frac{11A}{12-A}\Rightarrow \ln\frac{11A}{12-A}=0.75t$

Taking exponentials of each side: $\frac{11A}{12-A}=e^{0.75t}$

Rearranging for A gives:

$$A = \frac{12e^{0.75t}}{11+e^{0.75t}}$$

A graph of A against t shows the logistic model that is common in population modelling.

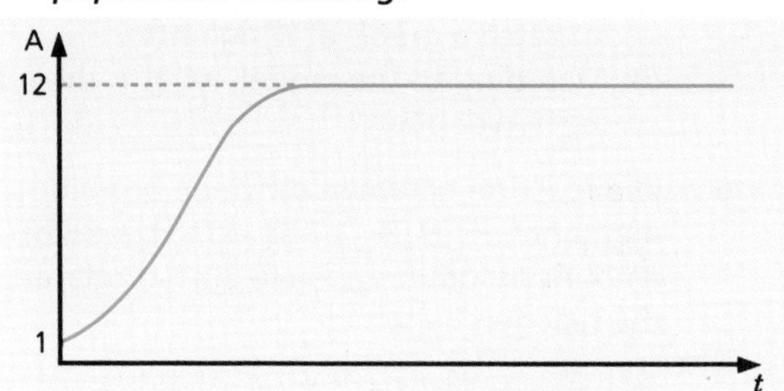

d) *As t increases the graph shows that the area occupied by the bacteria tends to a maximum value 12 cm². This can be confirmed algebraically in the following way.*

$$t \to \infty \Rightarrow A \to \frac{12e^{0.75t}}{e^{0.75t}} = 12$$

EXERCISES

24.4 A

1 In each problem express the given function in partial fractions and hence evaluate the integral.

a) $\int \frac{1}{x(2+x)}\,dx$ **b)** $\int_0^1 \frac{1}{(x+1)(x+2)}\,dx$ **c)** $\int \frac{2}{(2-x)(2+x)}\,dx$

d) $\int \frac{x}{(x-1)(x+1)}\,dx$ **e)** $\int_1^2 \frac{x}{(x+1)(x+2)}\,dx$ **f)** $\int_1^2 \frac{1}{t^2(1+t)}\,dt$

g) $\int \frac{5}{x^2+x-6}\,dx$ **h)** $\int \frac{3}{x^2-1}\,dx$ **i)** $\int \frac{x+2}{(x-2)(x+3)}\,dx$

j) $\int_2^4 \frac{1}{x(x-1)(x+1)}\,dx$ **k)** $\int \frac{x^2-4}{x(x-1)}\,dx$ **l)** $\int \frac{x^3+2}{x(x+1)}\,dx$

2 Find the total area enclosed by the curve $y = \frac{x^2+7x+8}{x^2+3x+2}\,dx$, the y-axis, the x-axis and the line $x = 4$.

3 Find the total area enclosed by the curve $y = \frac{x}{4-x^2}$, the x-axis and the lines $x = -1$ and $x = 1$.

4 Find the volume generated when the area between the curve

$y = \frac{1}{\sqrt{x(5-x)}}$, the x-axis and the lines $x = 1$ and $x = 4$ is rotated

completely about the x-axis.

5 In modelling the growth of a sunflower plant it is found that the height of a sunflower planet, H m, is related to the time, t days, by

$t = \int_{0.2}^{H} \frac{25}{h(2.5-h)}\,dh.$

At the start of the experiment the height of the plant is 0.2 m.

a) Evaluate the integral to find t as a function of H.

b) After how many days is the sunflower plant **i)** 1 metre high, **ii)** 2 metres high?

c) Rearrange the formula in part **a)** to give a formula for H in terms of t. Sketch a graph of H against t.

d) According to the model, what is the maximum height of the sunflower plant?

6 In a reaction between ethylene bromide and potassium iodide in 99% methanol, $C_2H_4Br_2 + 3KI \to C_2H_4 + 2KBr + KI_3$ it is found that the amount of iodine I_2, x mol dm^{-3} is related to the time, t minutes, after the reaction began by:

$$kt = \int_0^x \frac{\mathrm{d}c}{(a-c)(a-3c)}$$

where $k = 0.3$ dm^3 mol^{-1} min^{-1} is the reaction rate constant and a is the initial concentration of the chemicals.

a) Evaluate the integral to form t as a function of x.

b) Rearrange the formula in **a)** to give a formula for x in terms of t.

c) Find the value of x when t becomes very large.

7 The slope of the tangent to a curve at any point is given by:

$$\frac{\mathrm{d}y}{\mathrm{d}x} = \frac{1}{x^2+4x+3}$$

The curve passes through the point (0, 0). Find the equation of the curve.

EXERCISES

24.4 B

1 In each problem express the given function in partial fractions and hence evaluate the integral.

a) $\int \frac{1}{x(x+1)}\,\mathrm{d}x$ **b)** $\int_3^5 \frac{1}{(x-1)(x-2)}\,\mathrm{d}x$ **c)** $\int \frac{3}{(x+2)(x-3)}\,\mathrm{d}x$

d) $\int \frac{x}{(x-1)(x-2)}\,\mathrm{d}x$ **e)** $\int \frac{x+1}{(x-1)(x+2)}\,\mathrm{d}x$ **f)** $\int \frac{2x+3}{(x-1)(x+1)}\,\mathrm{d}x$

g) $\int \frac{1}{t^2-25}\,\mathrm{d}t$ **h)** $\int_0^4 \frac{v}{25-v^2}\,\mathrm{d}v$ **i)** $\int \frac{1}{(x-1)(x+1)(x+2)}\,\mathrm{d}x$

j) $\int \frac{8-x}{x^3-4x}\,\mathrm{d}x$ **k)** $\int_2^3 \frac{t^2-t+1}{t(t^2-1)}\,\mathrm{d}t$ **l)** $\int_0^1 \frac{x}{(1+x)^2}\,\mathrm{d}x$

2 Find the total area enclosed by the curve $y = \frac{5x}{x+4}$, the x-axis and the line $x = 4$.

3 Find the total area enclosed by the curve $y = \frac{7(2-x)}{(x+2)^2}$, the y-axis, the x-axis and the line $x = 4$.

4 Find the volume generated when the area between the curve $y = \dfrac{1}{\sqrt{(x+1)(x+4)}}$, the y-axis and the line $x = 2$ is rotated completely about the x-axis.

5 In a certain autocatalytic reaction, the reaction time t is related to the amount of product x, by the formula:

$$kt = \int_{\frac{1}{2}M}^{x} \frac{\mathrm{d}c}{c(M-c)}$$

where k and M are constants, and initially $x = \frac{1}{2}M$ when $t = 0$.

a) Evaluate the integral.

b) Rearrange the equation connecting t and x from part **a)** giving x in terms of t.

c) Find the value of x when t becomes very large.

6 The speed, U m s^{-1}, of a body moving in a fluid at time t seconds satisfies the equation:

$$t = \int_{10}^{U} \frac{-2\mathrm{d}v}{v(1+0.2v)}$$

where initially the speed of the body is 10 m s^{-1}.

a) Evaluate the integral.

b) Find the time taken for the body to reach a speed of 5 m s^{-1}.

c) Rearrange the formula in part **a)** to obtain a formula for U in terms of t.

d) Find the value of U when t becomes very large.

7 The speed of a parachutist U m s^{-1} at time t seconds satisfies the equation:

$$t = \int_{0}^{U} \frac{\mathrm{d}v}{10-0.2v^2}.$$

a) Evaluate the integral giving t as a function of U.

b) Rearrange the formula in part **a)** to obtain a formula for U in terms of t.
Sketch the graph of U against t.

c) Find the values of U when t becomes very large. Explain what this means for the parachutist.

SUMMARY OF INTEGRATION TECHNIQUES

We now have several methods of integration. When we have to evaluate an integral it is important to run through the various methods to see if one will work. The following flowchart (overleaf) gives a systematic way to approach the evaluation of an integral. There are other methods but we do not need to cover them at this level.

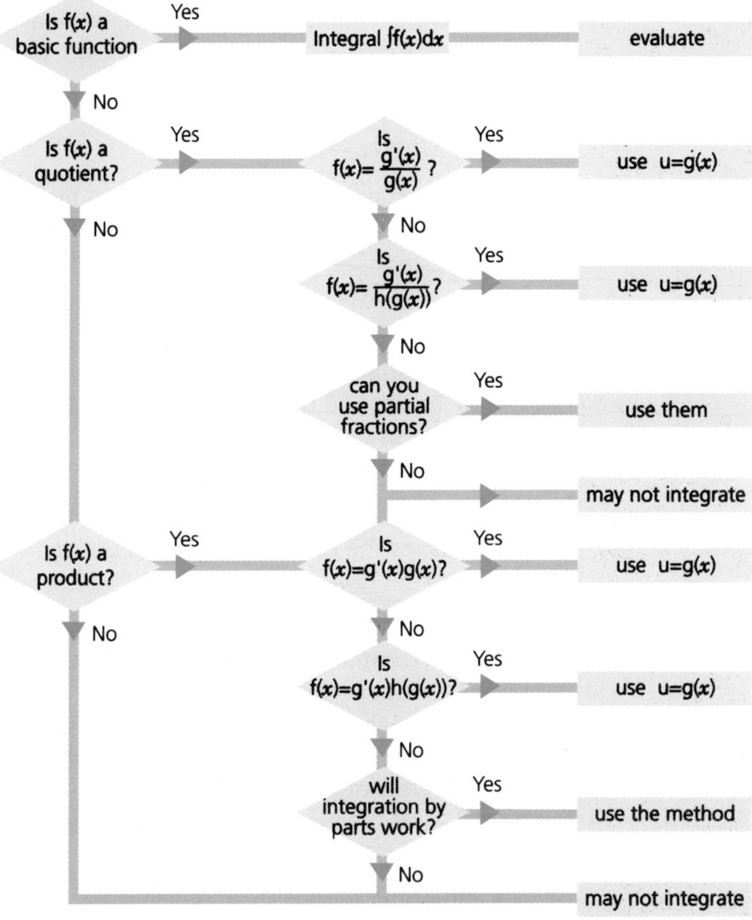

CONSOLIDATION EXERCISES FOR CHAPTER 24

1 Use the flowchart to classify each of the following integrals. Hence evaluate the integral using the appropriate method.

a) $\int_2^3 \frac{x+2}{(x+5)(x-1)}\,\mathrm{d}x$ **b)** $\int xe^{-2x}\,\mathrm{d}x$ **c)** $\int xe^{-3x^2}\,\mathrm{d}x$

d) $\int_0^1 \frac{3x^2+1}{x^3+x+4}\,\mathrm{d}x$ **e)** $\int_0^3 \frac{1}{\sqrt{1+t}}\,\mathrm{d}t$ **f)** $\int t\sqrt{7+t^2}\,\mathrm{d}t$

g) $\int x^2 \ln x\,\mathrm{d}x$ **h)** $\int_2^5 t\sqrt{2t^2-5}\,\mathrm{d}t$ **i)** $\int \frac{1}{3x+5}\,\mathrm{d}x$

j) $\int \frac{t^2}{\left(t^3+9\right)^5}\,\mathrm{d}t$ **k)** $\int x^2 \sin 4x\,\mathrm{d}x$ **l)** $\int e^{-2x}\cos 3x\,\mathrm{d}x$

2 In a physics experiment it is found that the temperature, T°C, at time t s is given by:

$$T = 90 - \int \frac{80}{(2t+1)^2}\,dt\,.$$

a) Find a formula for the temperature as function of time.
b) Find the value of T when t becomes very large.

3 The length of a curve $y = f(x)$ between the points on the curve $(a, f(a))$ and $(b, f(b))$ is given by:

$$s = \int_a^b \sqrt{1 + f'(x)^2}\,dx$$

Find the length of the curve $y = 3x^{\frac{3}{2}}$ between the points on the curve (0, 0) and (1, 3).

4 Find the volume generated when the area between the curve $y = \dfrac{5}{\sqrt{(x+1)(x+2)}}$, the y-axis and the line $x = 2$ is rotated completely about the x-axis.

5 The concentration, x, of one of the products in an nth order chemical reaction is related to the reaction time t by the formula:

$$kt = \int_0^x \frac{dc}{(a-c)^n}$$

where a and k are constants. Evaluate this integral in the two cases:
a) $n = 1$ and **b)** $n \neq 1$.
If T denotes the value of t when $x = \frac{1}{2}a$ find the value of T in each case.

6 In a particular third-order chemical reaction the reaction time t is related to the amount of product, x, by the formula:

$$kt = \int_0^x \frac{dc}{(2-c)(3-c)(1-c)}$$

where k is called the velocity constant.

a) Evaluate the integral.
b) Rearrange the equation connecting x and t from part **a)** to form a quadratic equation in x. Hence write x as a function of t.

7 The speed, U m s^{-1}, of a body moving in a fluid at time t seconds satisfies the equation:

$$t = \int_{100}^{U} \frac{-dv}{v(1 + 0.1v)}$$

where the speed of the body at time $t = 0$ is 100 m s^{-1}.

a) By evaluating the integral find a formula that gives t as a function of U.
b) Find the time taken for the body to reach 50 m s^{-1}.
c) Find a formula for U in terms of t.
d) Find the value of U when t becomes very large.

8 $\mathrm{f}(x) \equiv \dfrac{x^2+6x+7}{(x+2)(x+3)}$, $x \in \Re$.

Given that $\mathrm{f}(x) \equiv A + \dfrac{B}{x+2} + \dfrac{C}{x+3}$

a) find the values of the constants A, B and C,

b) show that $\int_0^2 \mathrm{f}(x)\ \mathrm{d}x = 2 + \ln\dfrac{25}{18}$.

(ULEAC Question 7, Specimen Paper 2, 1994)

9 A sketch of the curve with equation $y = \dfrac{1+2x}{(1-x)^2}$ is shown below.

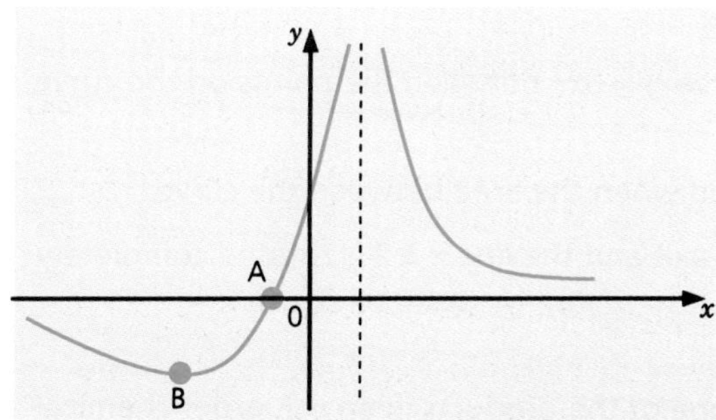

a) Write down the coordinates of A, and state the equation of the asymptote, shown as a broken line.

b) Find the equation of the tangent to the curve with equation:

$$y = \frac{1+2x}{(1-x)^2} \text{ at point (0, 1).}$$

Hence determine the coordinates of the other point at which this tangent intersects the curve.

(Oxford Question 6, Specimen Paper 2, 1994)

10 **a)** Express $\mathrm{f}(x) = \dfrac{x^2+7x+2}{(1+x^2)(2-x)}$ in terms of partial fractions.

b) Hence prove that $\int_0^1 \mathrm{f}(x)\,\mathrm{d}x = \dfrac{11}{2}\ln 2 - \dfrac{\pi}{4}$.

(Oxford Question 6, Specimen Paper 5, 1994)

11 **a)** Differentiate $(1+x^3)^{\frac{1}{2}}$ with respect to x.

b) Use the result from a), or an appropriate substitution, to find the value of:

$$\int_0^2 \frac{x^2}{\sqrt{(1+x^3)}}\ \mathrm{d}x.$$

(AEB Question 8, Specimen Paper 1, 1994)

12 The diagram shows a sketch of the curve defined for $x > 0$ by the equation $y = x^2 \ln x$.

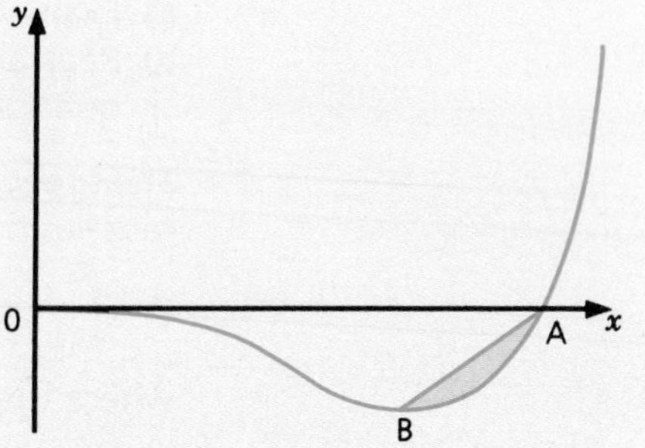

The curve crosses the x-axis at A and has a local minimum at B.

a) State the coordinates of A and calculate the gradient of the curve at A.

b) Calculate the coordinates of B in terms of e and determine the value of $\dfrac{\mathrm{d}^2y}{\mathrm{d}x^2}$ at B.

c) The region bounded by the line segment at AB and an arc of the curve is R, as shaded in the diagram. Show that the area of R is $\left(\dfrac{4-\mathrm{e}^{-\frac{3}{2}}-9\mathrm{e}^{-1}}{36}\right)$.

(AEB Question 13, Specimen Paper 1, 1994)

13 A particle P moves along a straight line which passes through a point O. At time t seconds the velocity of P is v m s^{-1}, its displacement from O is s metres and its acceleration is:

$$-\frac{1}{(1+t)^2}\,\mathrm{m\,s^{-2}}.$$

Initially P is at rest and its displacement from O is 1 metre. Find v in terms of t. Hence show that:

$s = \ln(1 + t) - t + 1$

Show that when P reaches O, t lies between 2 and 2.2.

Obtain this value of t to four significant figures by using the iteration $t_{n+1} = \ln(t_n + 1) + 1$ six times, starting with $t_0 = 2.1$ and listing the result of each iteration obtained on your calculator.

(NEAB Question 8, Specimen Paper 1, 1994)

14 Using the substitution $2x = \sin\theta$, or otherwise, find the exact value of

$$\int_0^{\frac{1}{4}} \frac{1}{\sqrt{(1-4x^2)}}\,\mathrm{d}x.$$

(NEAB Question 2, Specimen Paper 2, 1994)

15 Differentiate $x \ln x$ with respect to x and hence show that the curve $y = x \ln x$ $(x > 0)$ has a minimum point for $x = \frac{1}{e}$. State the corresponding y-coordinate. Give a reason why the curve has no point of inflection. Sketch the curve for $x \geq \frac{1}{e}$ Find the point A where the curve meets the line $y = x$ and find the area enclosed by the line OA, the curve and the x-axis.

(NEAB Question 14, Specimen Paper 2, 1994)

16 Use two trapezia of equal width to estimate the value of $\int_2^3 \ln x\,\mathrm{d}x$, giving your answer correct to two significant figures.

Explain with the aid of a sketch why this estimate is less than the true value of the integral.

Use integration by parts to find the exact value of the integral, giving your answer in terms of logarithms.

(UCLES (Linear) Question 13, Specimen Paper 1, 1994)

17 Use partial fractions to find the exact value of:

$$\int_0^{\frac{1}{2}} \frac{1}{1-x^2}\,\mathrm{d}x$$

giving your answer in a simplified form involving a single logarithm.

(UCLES Question 5, Specimen Paper 4, 1994)

18 **a)** Use the chain rule to differentiate $(3 + 2x)^4$ with respect to x.

b) Hence evaluate $\int_0^1 (3+2x)^3\,\mathrm{d}x$.

c) Expand $(3 + 2x)^3$ in powers of x, and use your answer to check the value of the integral in **b)**.

(Oxford and Cambridge Question 7, Specimen Paper 1, 1994)

19 **a)** Using the substitution $u = x^2 + 4$, or otherwise, find the indefinite integral $\int \frac{x}{x^2+4}\,dx$.

b) You are given that $f(x) = \frac{2}{(2x+1)} - \frac{x}{(x^2+4)}$.

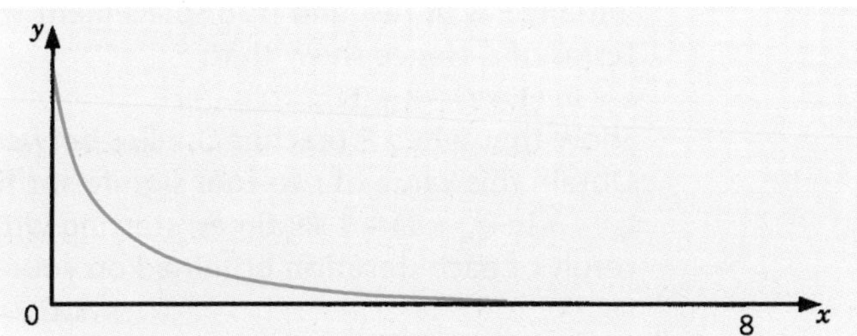

The figure shows the graph of $y = f(x)$ between $x = 0$ and $x = 8$. Show that the area of the shaded region between this graph and the axes is equal to **exactly** $\frac{1}{2}\ln 17$.

c) Find an expression for $f'(x)$. Use this to calculate the gradient of the graph at $x = 0$ and $x = 8$.

d) Write $f(x)$ as a single fraction in its simplest form.

(Oxford & Cambridge Question 6, Specimen Paper 2, 1994)

20 Use integration by parts to show that $\int_2^4 x\ln x\,dx = 7\ln 4 - 3$.

(ULEAC Question 2, Paper 2, January 1995)

21 **a)** By using integration by parts, show that $\int x\ln x\,dx = \frac{x^2}{2}\ln x - \frac{x^2}{4} + C$, where C is a constant.

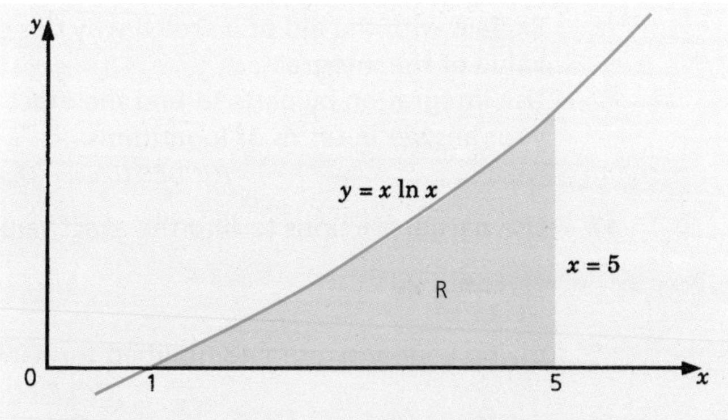

The above diagram shows the shaded region R which is bounded by the curve with equation $y = x\ln x$, the x-axis, and the line with equation $x = 5$.

b) Find the area of R.

Estimate the area of R by tabulating values of $x \ln x$ at $x = 1, 2, 3, 4$ and 5 and applying the trapezium rule, giving the result to two decimal places.

c) Find the percentage error in using the trapezium rule to estimate R.

(ULEAC Question 9, Paper 2, January 1993)

22 **a)** You are given that $f(x) = 2x^3 - x^2 - 7x + 6$.

i) Show that $f(1) = 0$.
Hence find the three factors of $f(x)$.

ii) Solve the inequality $f(x) > 0$.

b) **i)** Given that $\dfrac{x^2+2x+7}{(2x+3)(x^2+4)} \equiv \dfrac{A}{(2x+3)} + \dfrac{Bx+C}{(x^2+4)}$ find the values of the constants A, B and C.

ii) Use your answer to **b) i)** to find $\displaystyle\int \frac{x^2+2x+7}{(2x+3)(x^2+4)}\,dx$.

(MEI Question 1, Paper 3, January 1994)

Summary

- **Integration by substitution**

To evaluate $\int f(x)dx$ when $f(x)$ is not a basic function:

If $f(x) = g'(x)g(x)$	let $u = g(x)$
If $f(x) = \frac{g'(x)}{g(x)}$	let $u = g(x)$
If $f(x) = h(g(x))g'(x)$	let $u = g(x)$.

- **Integration by parts**

To integrate a product of functions try the formula for integration by parts

$\int u \frac{dv}{dx} dx = uv - \int v \frac{du}{dx} dx$

A flowchart showing the methods of integration is shown on page 552.

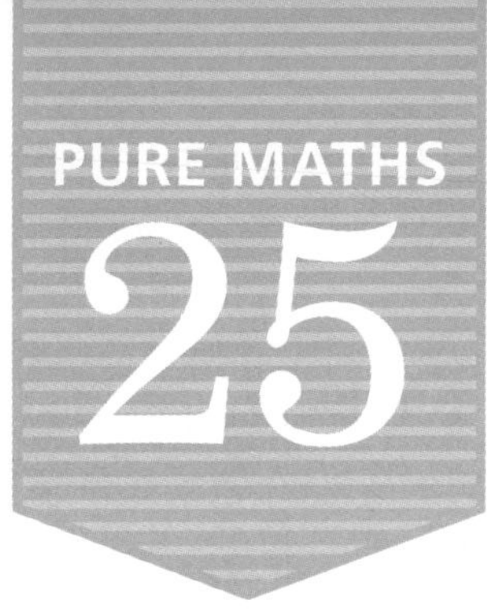

Curves 1

In this chapter we introduce:

- *the basic geometry of the circle,*
- *three other special curves – the ellipse, hyperbola and parabola,*
- *a calculus technique to find the equations of tangents and normals to these curves, called implicit differentiation.*

CARTESIAN EQUATION OF A CIRCLE

CALCULATOR ACTIVITY

Investigating circles

Set the scales on your graphics calculator or computer program to:

$x_{min} = -5$; $x_{max} = 5$; $x_{scl} = 1$
$y_{min} = -5$; $y_{max} = 5$; $y_{scl} = 1$

Use the basic equation $x^2 + y^2 = 1$.

- Rearrange the equation into the forms $y = +\sqrt{f(x)}$ and $y = -\sqrt{f(x)}$.
- Draw graphs of the functions $y = \pm\sqrt{f(x)}$.
- Where does the graph cut the x- and y-axes?
- Describe the graph of the equation $x^2 + y^2 = 1$.

Repeat the activity for the equations $x^2 + y^2 = 4$, $x^2 + y^2 = 9$

and $x^2 + y^2 = 25$.

Deduce the graph of the equation $x^2 + y^2 = r^2$.

Interpreting the results

The following table shows the conclusions that we should expect from this activity.

Equation	Curve
$x^2 + y^2 = 1$	circle of radius 1, centre (0, 0)
$x^2 + y^2 = 4$	circle of radius 2, centre (0, 0)
$x^2 + y^2 = 9$	circle of radius 3, centre (0, 0)
$x^2 + y^2 = 25$	circle of radius 5, centre (0, 0)

We conclude that the equation $x^2 + y^2 = r^2$ is the equation of a circle of radius r and centre (0, 0).

But what about circles that are not centred at (0, 0) but at some other point, say (a, b)?

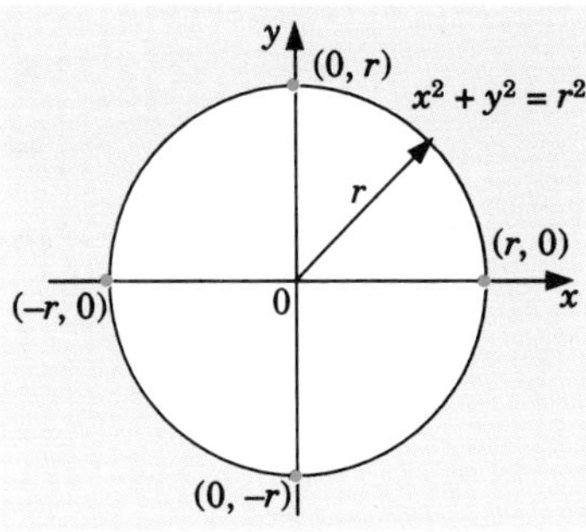

Exploration 25.1

Equations of circles

Take the equation of a circle of radius r and centre (0, 0), $x^2 + y^2 = r^2$.

- Describe the effect on the circle of replacing x by $(x - a)$.
- Describe the effect on the circle of replacing y by $(y - b)$.
- Deduce the equation of a circle of radius r and centre (a, b).

Sketch the circle without using a calculator.

Interpreting the results

The equation $(x - a)^2 + y^2 = r^2$ is a translation of the circle parallel to the x-axis and the equation $x^2 + (y - b)^2 = r^2$ is a translation of the circle parallel to the y-axis.

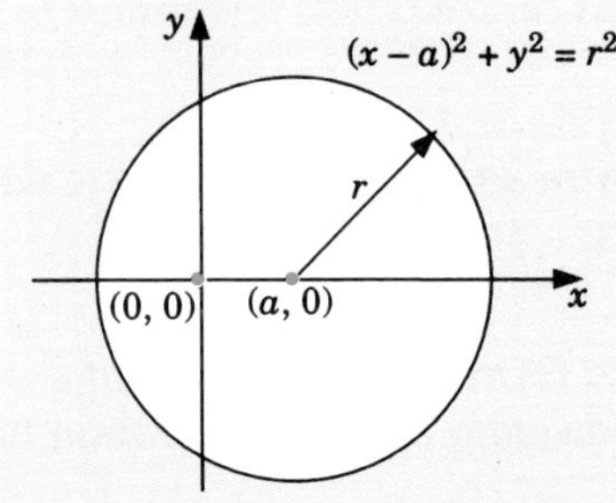

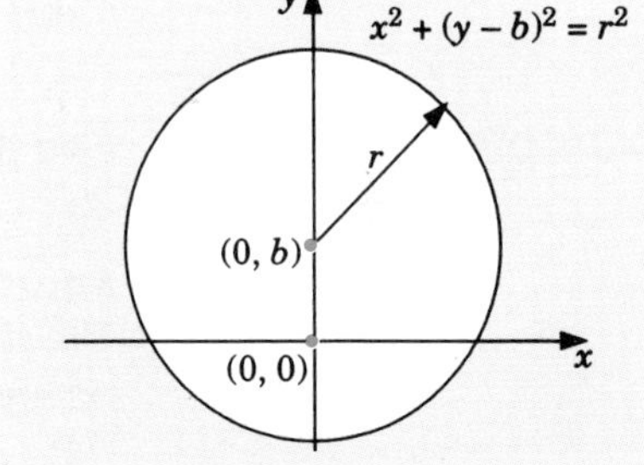

Putting these two translations together we deduce the general result that the equation of a circle of radius r and centre (a, b) is

$$(x - a)^2 + (y - b)^2 = r^2.$$

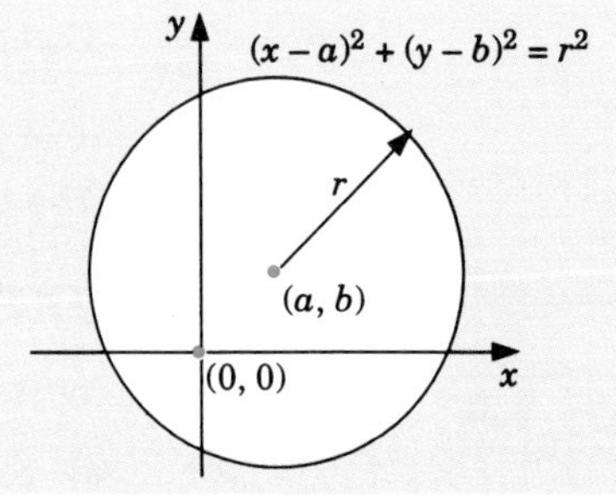

Example 25.1

Find the equation of a circle with centre (1, 3) and radius 2.

Solution

Substituting for $a = 1$, $b = 3$ *and* $r = 2$ *in the general formula for a circle gives:*

$$(x - 1)^2 + (y - 3)^2 = 2^2$$

Multiplying out the brackets:

$$x^2 - 2x + 1 + y^2 - 6y + 9 = 4$$
$$\Rightarrow x^2 + y^2 - 2x - 6y + 6 = 0$$

The diagram shows the circle.

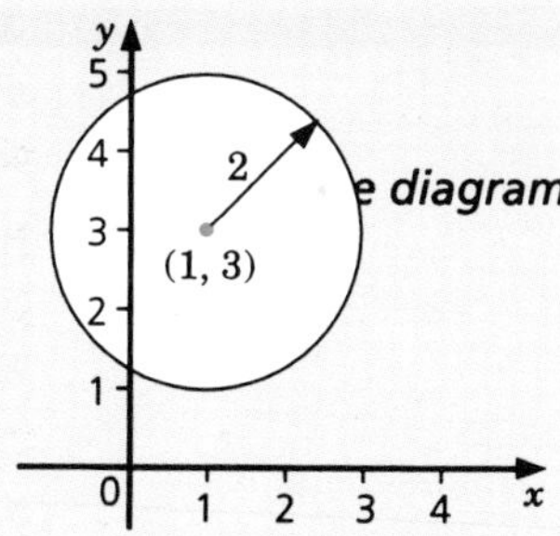

Example 25.2

A circle has the equation $x^2 - 6x + y^2 + 8y - 11 = 0$. *Find the centre and radius.*

Solution

The first step is to write the equation in standard form. To do this complete the square for each pair $(x^2 - 6x)$ *and* $(y^2 + 8y)$.

$x^2 - 6x = (x-3)^2 - 9$
$y^2 + 8y = (y+4)^2 - 16$

The equation of the circle becomes:

$(x-3)^2 - 9 + (y+4)^2 - 16 - 11 = 0$

$\Rightarrow (x-3)^2 + (y+4)^2 = 36 = 6^2$

The circle has centre (3, −4) and radius 6.

Example 25.3

Find the centre and radius of the circle with equation $2x^2 + 2y^2 + 3x - 4y + 1 = 0$.

Solution

The standard form of the equation of a circle always starts with $x^2 + y^2$. *So dividing the given equation by 2:*

$x^2 + y^2 + \frac{3}{2}x - 2y + \frac{1}{2} = 0$

Completing the squares:

$\left(x + \frac{3}{4}\right)^2 - \frac{9}{16} + (y-1)^2 - 1 + \frac{1}{2} = 0 \quad \Rightarrow \left(x + \frac{3}{4}\right)^2 + (y-1)^2 = \frac{17}{16}$

The circle has centre $(-\frac{3}{4}, 1)$ *and radius* $\frac{\sqrt{17}}{4}$.

EXERCISES

25.1 A

1 Write down the equation of each circle.

a) centre (0 , 0), radius 7 **b)** centre (4 , 0), radius 4
c) centre (2 , 1), radius 5 **d)** centre (–3 , 2), radius 3
e) centre (0 , –5), radius 2.5 **f)** centre (–1, –1), radius 6

2 Find the centre and radius of the circles with the following equations.

a) $x^2 + y^2 = 121$ **b)** $x^2 + (y-5)^2 = 81$

c) $(x-3)^2 + (y-4)^2 = 25$ **d)** $(x-2)^2 + (y+3)^2 = 16$

e) $(x+4)^2+y^2-49=0$ f) $x^2+y^2-2y+1=36$

g) $x^2-4x+y^2=60$ h) $x^2+3x+y^2-5y=66$

i) $x^2+y^2-2x-4y-131=0$ j) $4x^2-28x+4y^2-16y=35$

3 Determine which of the following equations represent circles.

a) $x^2+y^2=16$ b) $x^2-y^2=16$ c) $x^2+y^2+8x-9=0$

d) $x^2+2y^2-2x-3=0$ e) $x^2=y^2-4y+1$ f) $x^2+y^2+6x+8y=0$

4 Find the equation of the circle with PQ as diameter where:

a) P is the point (2, 0) and Q is the point (8, 8),
b) P is the point (–1, 1) and Q is the point (0, 2),
c) P is the point (–2, –3) and Q is the point (1, 2).

5 Find the coordinates of the point where a circle with equation $(x-15)^2+(y+5)^2=169$ intersects the x-axis.
Show that the circle does not intersect the y-axis.

6 Find the coordinates of the points of intersection of the circle $x^2+y^2-2x-3y-131=0$ with the x- and y-axes.

EXERCISES

25.1 B

1 Write down the equation of each circle.

a) centre (0, 0), radius 6 b) centre (3, 0), radius 1
c) centre (0, –2), radius $\frac{1}{2}$ d) centre (3, –2), radius 4

e) centre (–4, –5), radius 1.5 f) centre $(\frac{1}{2}, \frac{1}{4})$, radius 2

2 Find the centre and radius of the circles with the following equations.

a) $x^2+y^2=81$ b) $(x-1)^2+y^2=49$

c) $x^2+(y+2)^2=100$ d) $(x-1)^2+(y+2)^2=4$

e) $(x+3)^2+(y-1)^2-16=0$ f) $x^2+y^2-6x+2y+1=0$

g) $x^2+y^2-12x-52=0$ h) $x^2+y^2-5y+0.25=0$

i) $2x^2+2y^2-3x-5y=0$ j) $3x^2+3y^2-6x+12y=17$

3 Determine which of the following equations represent circles.

a) $x^2-y^2=25$ b) $x^2+y^2=25$

c) $x^2+2y^2-2x+3y-5=0$ d) $x^2+y^2-2x-4y+5=0$

e) $x^2+y^2+x+y-1=0$ f) $y^2=x^2+3x-2$

4 Find the coordinates of the points of intersection P and Q of the circle $x^2+y^2-12x-16y+15=0$ with the x- and y-axes. Is PQ a diameter of the circle?

5 Find the equation of the circle with AB as diameter where:
  a) A is the point (0, 2) and B is the point (2, 0),
  b) A is the point (2, 3) and B is the point (–1, –1),
  c) A is the point (0, 0) and B is the point (5, 12).

6 The circle shown has equation $x^2 + y^2 - 4x + 6y - 3 = 0$ and the line AB is a diameter that is parallel to the x-axis. Find the coordinates of A and B.

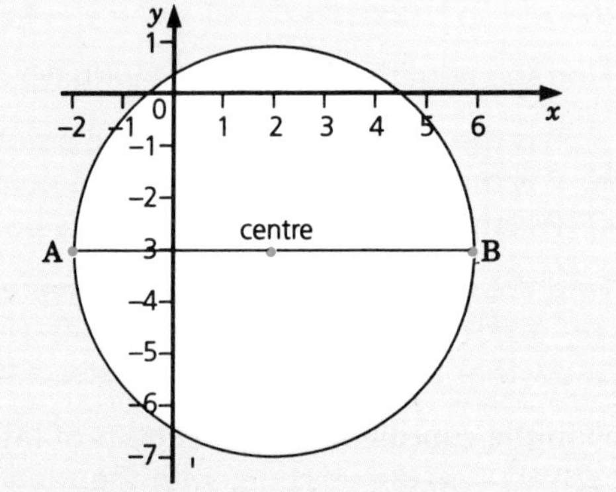

THE ELLIPSE, HYPERBOLA AND PARABOLA CALCULATOR ACTIVITY

Different curves

Set the scales on your graphics calculator or computer program to:

$x_{min} = -2;$ $x_{max} = 2;$ $x_{scl} = 1$
$y_{min} = -2;$ $y_{max} = 2;$ $y_{scl} = 1$

a) Take the equation $\frac{x^2}{4} + \frac{y^2}{4} = 1$.

Rearrange in the form $y = \pm f(x)$.

Draw the graph of the equation and describe the curve.

b) Take the equation $\frac{x^2}{4} + y^2 = 1$.

Rearrange in the form $y = \pm g(x)$.
Draw the graph of the equation and describe the curve.
Where does it cut the x-axis and y-axis?

c) Take the equation $\frac{x^2}{4} - \frac{y^2}{4} = 1$

Rearrange in the form $y = \pm h(x)$.
Draw the graph of the equation and describe the curve.
Where does it cut the x-axis? Does it cut the y-axis?

Interpreting the results

The following figures show the three curves that we should expect to obtain in the calculator activity.

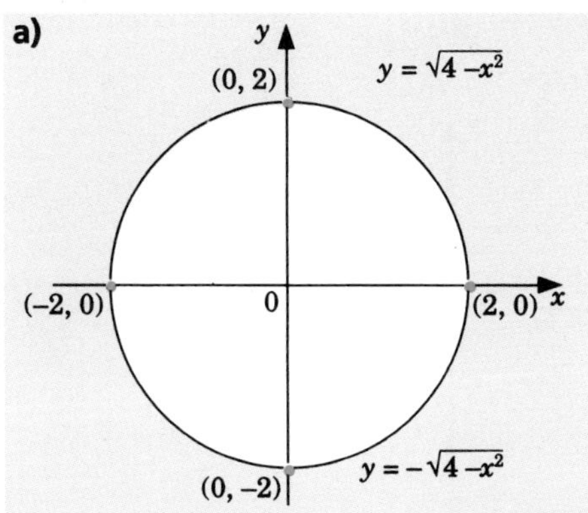

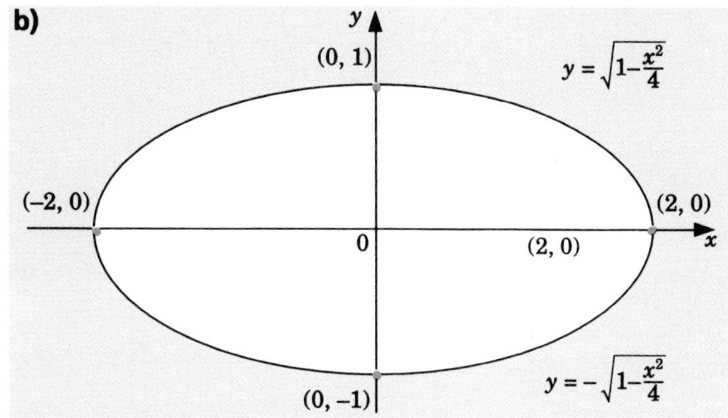

For each curve, we have seen that a change in the sign or value of the coefficient of y^2 in the equation of a circle leads to a different curve. These are other members of a special family of curves called **conic sections**. Curve b) is an example of an **ellipse**. Curve c) is an example of a **hyperbola**. There is a fourth conic section called a **parabola**. An example of a parabola with formula $y^2 = 4x$ is shown in **d)**.

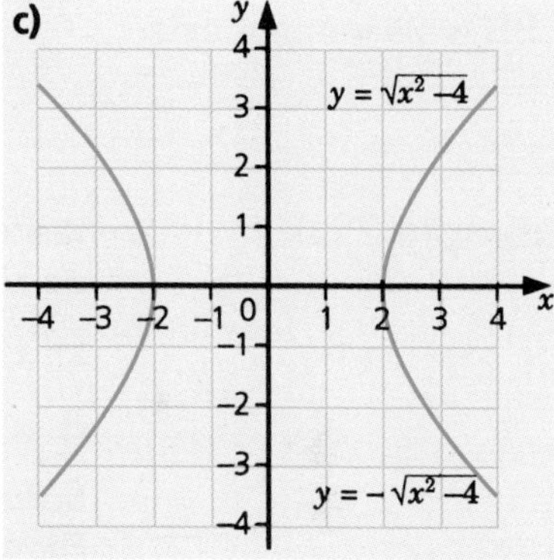

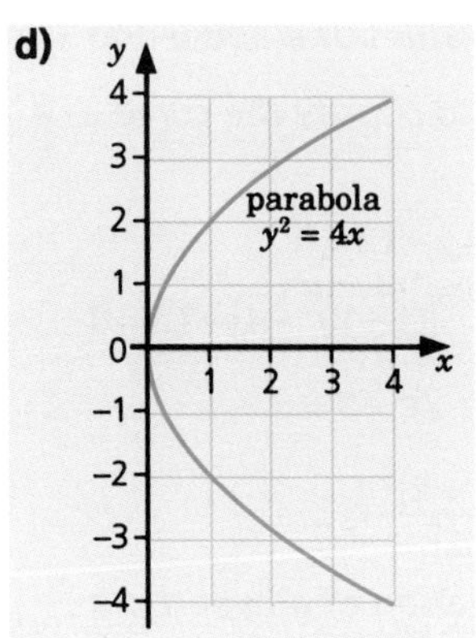

The curves are called conic sections because each one can be obtained by cutting a slice or section through a cone.

We have already seen that the general equation of a circle, with centre (0, 0) and radius r, is $x^2 + y^2 = r^2$ which can be written as:

$$\frac{x^2}{r^2} + \frac{y^2}{r^2} = 1$$

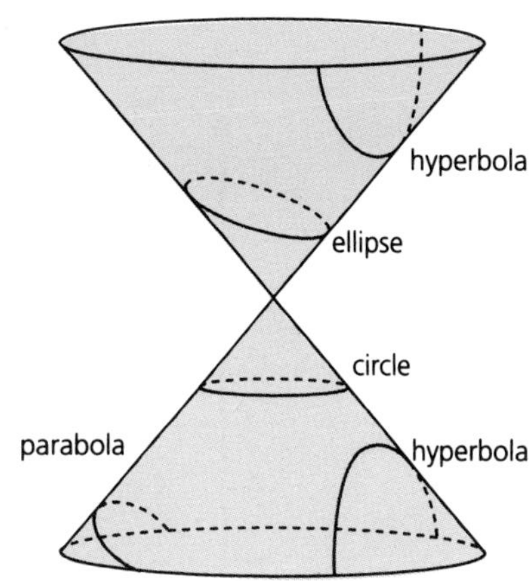

The following table shows the general equations for the other conic sections.

Ellipse	Hyperbola	Parabola
$\frac{x^2}{a^2}+\frac{y^2}{b^2}=1$	$\frac{x^2}{a^2}-\frac{y^2}{b^2}=1$	$y^2=4ax$
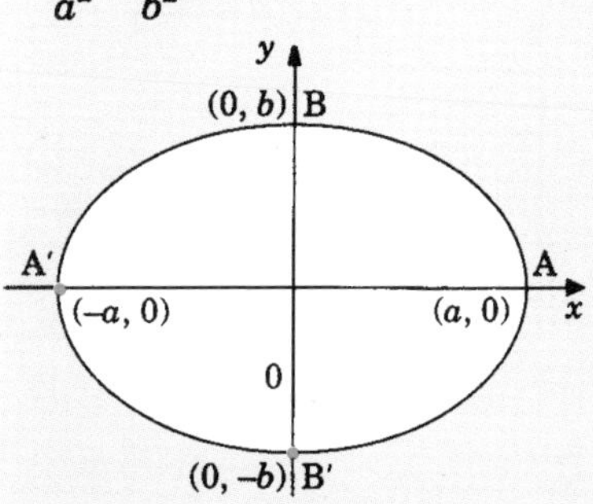	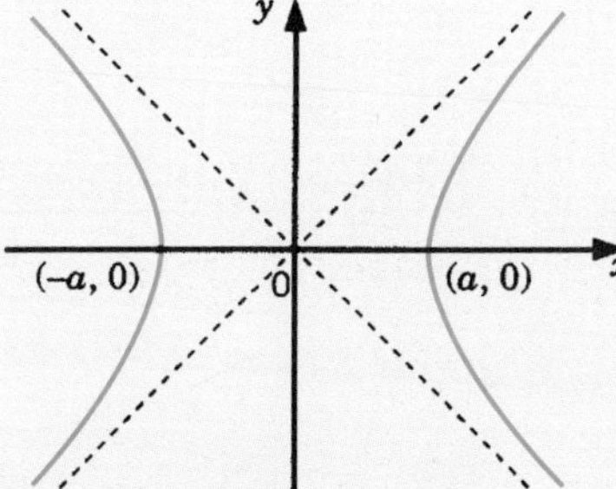	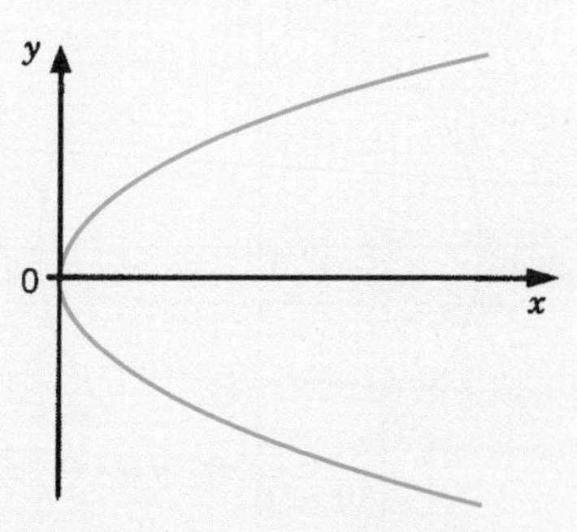
AA′ = 2a = major axis BB′ = 2b = minor axis	slant asymptotes $y=\frac{b}{a}x$ and $y=-\frac{b}{a}x$	

Example 25.4

Write each of the following equations in standard form and hence classify the curves.

a) $x^2-y^2-2x-4y-12=0$ **b)** $2x-y^2-2y-1=0$

c) $x^2+y^2-2x+4y-4=0$ **d)** $x^2+4y^2-2x+16y+8=0$

Solution

For each equation we need to complete the square so that it can be written in the form $\frac{(x-?)^2}{a^2}\pm\frac{(y-?)^2}{b^2}=1$. *The values of* a *and* b *and the occurrence of the + or – sign will allow us to classify the curve as a circle, ellipse, hyperbola or parabola.*

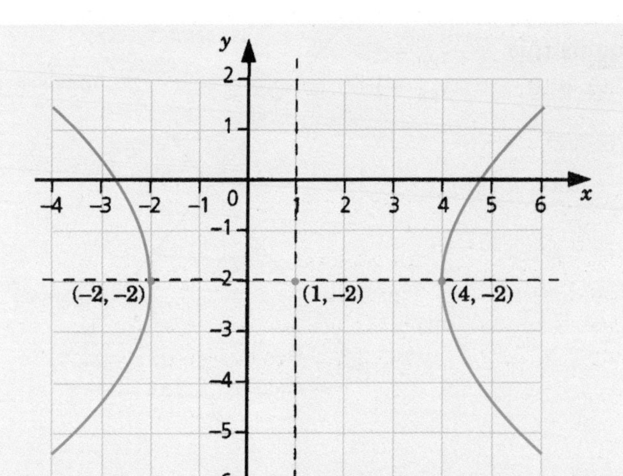

a) $x^2-y^2-2x-4y-12=0$

$$\Rightarrow\left((x-1)^2-1\right)-\left((y+2)^2-4\right)-12=0$$

$$\Rightarrow(x-1)^2-(y+2)^2-9=0$$

$$\Rightarrow\frac{(x-1)^2}{3^2}-\frac{(y+2)^2}{3^2}=1$$

This is the equation of a hyperbola with centre (1, –2).

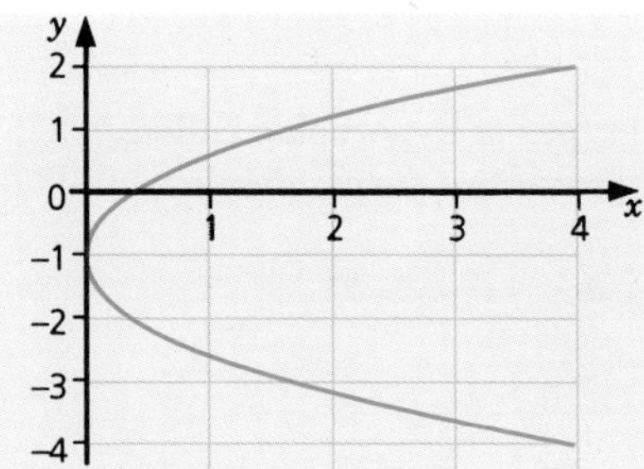

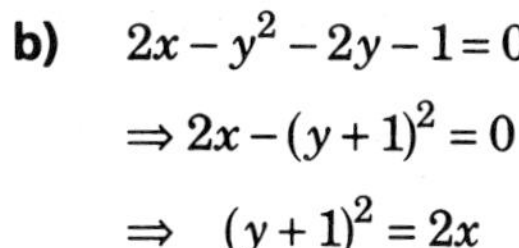

b) $2x - y^2 - 2y - 1 = 0$

$\Rightarrow 2x - (y+1)^2 = 0$

$\Rightarrow \quad (y+1)^2 = 2x$

This is a parabola with centre (0, –1).

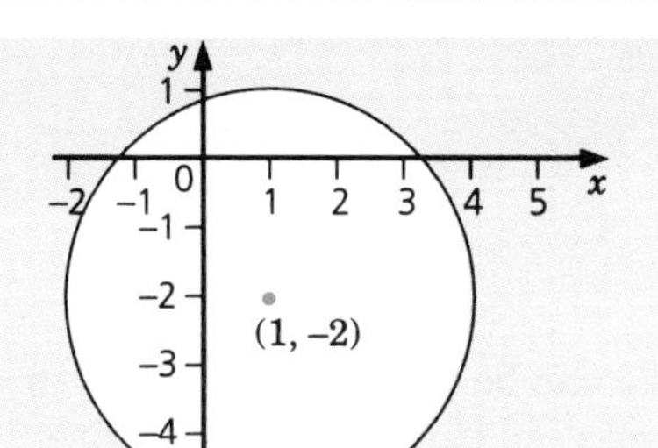

c) $x^2 + y^2 - 2x + 4y - 4 = 0$

$\Rightarrow \left((x-1)^2 - 1\right) + \left((y+2)^2 - 4\right) - 4 = 0$

$\Rightarrow (x-1)^2 + (y+2)^2 - 9 = 0$

$\Rightarrow \dfrac{(x-1)^2}{3^3} + \dfrac{(y+2)^2}{3^3} = 1$

This is the equation of a circle with centre (1, –2) and radius 3.

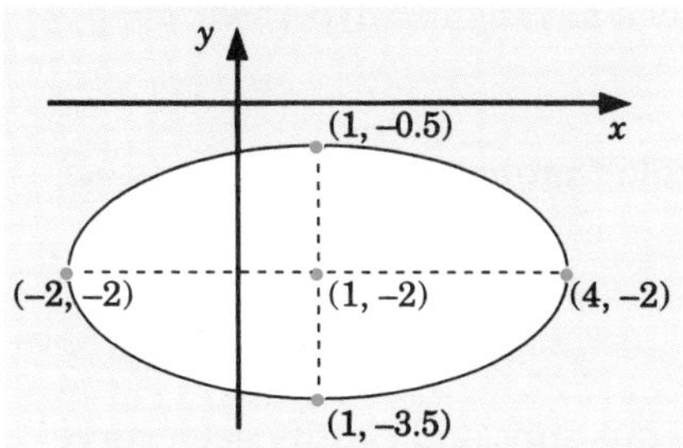

d) $x^2 + 4y^2 - 2x + 16y + 8 = 0$

$\Rightarrow \left((x-1)^2 - 1\right) + 4\left((y+2)^2 - 4\right) + 8 = 0$

$\Rightarrow (x-1)^2 + 4(y+2)^2 - 9 = 0$

$\Rightarrow \dfrac{(x-1)^2}{3^2} + \dfrac{(y+2)^2}{\left(\frac{3}{2}\right)^2} = 1$

This is the equation of an ellipse with centre (1, –2), major axis 6 and minor axis 3.

CALCULATOR ACTIVITY

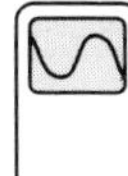

Asymptotes

Set the scales on your graphics calculator or computer program to:

$x_{min} = -10$; $x_{max} = 10$; $x_{scl} = 1$
$y_{min} = -10$; $y_{max} = 10$; $y_{scl} = 1$

Use the equation $x^2 - y^2 = 25$.

- Draw the graph of the equation; which conic section have you drawn?
- Find the equations of the slant asymptotes.
- What is the angle between the asymptotes?
- On the same axes draw the graph of the equation $xy = 25$.
- Describe the transformation between the two curves.

Interpreting the results

The curves in this activity are special types of hyperbolae called **rectangular hyperbolae**. The asymptotes are perpendicular.

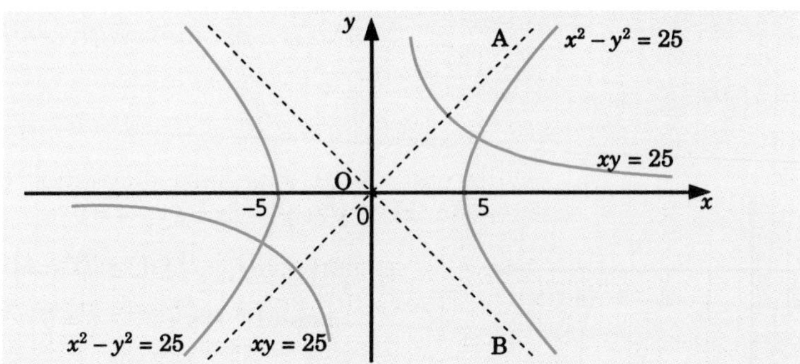

The two branches of the curves $xy = 25$ can be made to coincide with the two branches of $x^2 - y^2 = 25$ by a rotation through 45° clockwise. The x-axis will then lie along OB and the y-axis will lie along OA.

In the standard form of the equation of a rectangular hyperbola, $a = b$ so:

$\frac{x^2}{a^2} - \frac{y^2}{a^2} = 1$ or $x^2 - y^2 = a^2$

A special type of rectangular hyperbola that has the x- and y-axes as asymptotes is $xy = a^2$.

EXERCISES

25.2 A

1 Sketch the graphs of the following equations and classify each curve as one of the conic sections.

a) $\frac{x^2}{9} + \frac{y^2}{25} = 1$ b) $4x^2 + y^2 - 16x - 6y - 24 = 0$

c) $\frac{x^2}{9} - \frac{y^2}{25} = 1$ d) $y^2 - 8x = 0$

e) $xy - 4 = 0$ f) $\frac{y^2}{4} - \frac{x^2}{16} = 1$

2 Write down the equations of the following ellipses.

a) centre (0, 0), major axis 4, minor axis 6
b) centre (1, 2), major axis 2, minor axis 3
c) centre (–2, –1), major axis 5, minor axis 4

3 Write down the equations of the rectangular hyperbolae which pass through these pairs of points.

a) (3, 0) and (–3, 0) b) (4, 0) and (–2, 0)
c) (0, 2) and (0, –2)

4 Write each of the following equations in standard form and hence classify the curve as one of the conic sections.

a) $x^2 + 2y^2 + 2x - 12y - 6 = 0$ b) $x^2 - 2y^2 + 2x - 8 = 0$

c) $x^2 - y^2 - 3x + 5y - 20 = 0$

EXERCISES

25.2 B

1 Sketch the graphs of the following equations and classify each curve as one of the conic sections.

a) $9x^2 + 4y^2 = 36$ b) $9x^2 - 4y^2 = 36$

c) $x^2 - y^2 - 2x - 3y - 2.25 = 0$ d) $x^2 + y^2 - 2x - 3y + 2.25 = 0$

e) $9x^2 + 4y^2 + 36x + 24y + 71 = 0$ f) $y^2 - 4x - 2y - 7 = 0$

2 Write each of the equations in question **1** in standard form and confirm your classification.

3 Write down the equations of the following ellipses.
a) centre (0, 0), major axis 8, minor axis 2
b) centre (–1, 3), major axis 1, minor axis 3
c) centre (0.5, 2.5), major axis 3, minor axis 4

4 Write down the equation of the hyperbola:
a) which passes through the point (4, 0) and has slant asymptotes $y = \pm 2x$,
b) which passes through the point (–1, 0) and has slant asymptotes $y = \pm x$,
c) which passes through the point (2, 0) and the point (4, 5√3).

INTERSECTION OF CURVES AND LINES

Example 25.5

Find the points of intersection of the circle $x^2 + y^2 - 2x - 4y - 20 = 0$ *and the line* $y = 2x + 5$.

Solution

In general a line will:

- *intersect a circle at two distinct points,*
- *touch a circle as a tangent, or*
- *miss the circle completely.*

A graph of the line and circle shows that, in this case, there are two points of intersection.

To find the coordinates of the points of intersection we solve the equations:

$x^2 + y^2 - 2x - 4y - 20 = 0$

$y = 2x + 5$

simultaneously. The x*-values are given by:*

$x^2 + (2x + 5)^2 - 2x - 4(2x + 5) - 20 = 0$

$\Rightarrow 5x^2 + 10x - 15 = 0$

$\Rightarrow 5(x - 1)(x + 3) = 0$

$\Rightarrow x = 1$ *or* $x = -3$

The points of intersection of the line and the circle are (1, 7) and (–3, –1).

Example 25.6

The line through the origin $y = mx$ is a tangent to the circle $x^2 + y^2 - 2x - 4y + 1 = 0$. Find the possible values of m and the corresponding coordinates of the points of contact.

Solution

Substituting $y = mx$ in $x^2 + y^2 - 2x - 4y + 1 = 0$:

$$x^2 + m^2x^2 - 2x - 4mx + 1 = 0$$

$$\Rightarrow \quad (1+m^2)x^2 - (2+4m)x + 1 = 0$$

The line is a tangent if there is only one root to this equation.

$$\Rightarrow \quad (2+4m)^2 = 4(1+m^2)$$

$$\Rightarrow \quad 4 + 16m + 16m^2 = 4 + 4m^2$$

$$\Rightarrow \quad 12m^2 + 16m = 0$$

$$\Rightarrow \quad m = 0 \text{ or } m = -\tfrac{4}{3}$$

There are two tangents to the circle which pass through the origin and these have equations $y = 0$ and $y = -\frac{4}{3}x$.

The points of contact are (1, 0) and $(-\frac{6}{25}, \frac{8}{25})$ = (–0.24, 0.32).

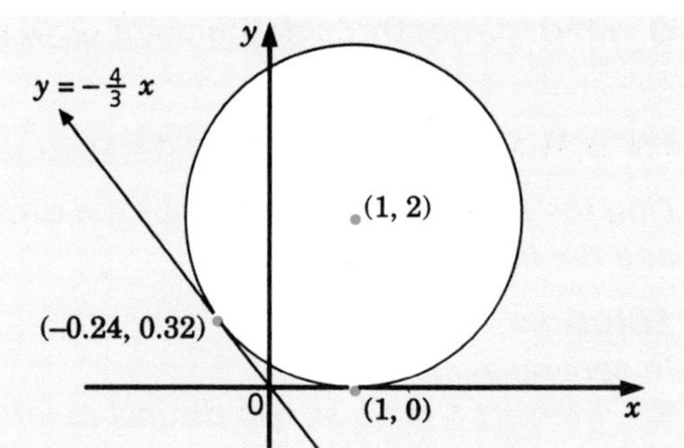

Example 25.7

Find the length of the tangent from the point (2, 6) to the circle $x^2 + y^2 - 2x - 4y + 1 = 0$.

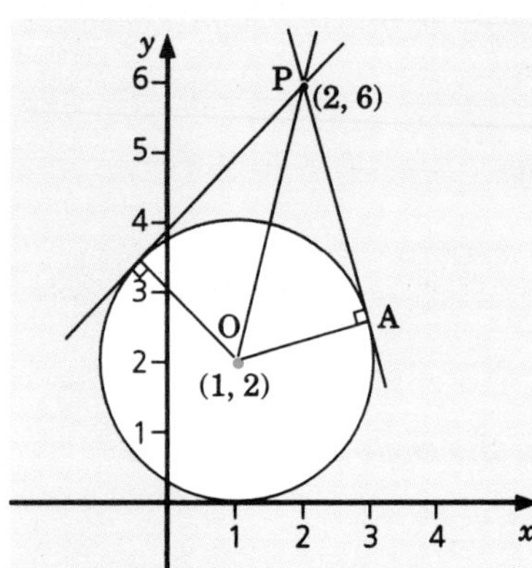

Solution

In standard form the circle has equation:

$$(x-1)^2 + (y-2)^2 = 4$$

The centre is (1, 2) and the radius is 2.

Since each tangent is perpendicular to a radius, Pythagoras' theorem can be used to give:

$$\text{OP} = \sqrt{(6-2)^2 + (2-1)^2} = \sqrt{17}$$

$$\text{AP} = \sqrt{\text{OP}^2 - \text{OA}^2} = \sqrt{17-4} = \sqrt{13}$$

Example 25.8

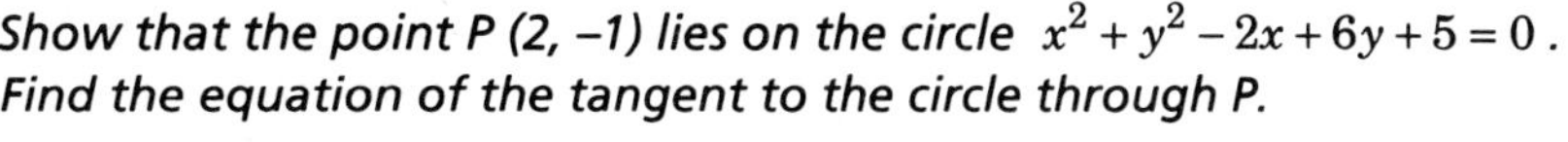

Show that the point P (2, –1) lies on the circle $x^2 + y^2 - 2x + 6y + 5 = 0$. Find the equation of the tangent to the circle through P.

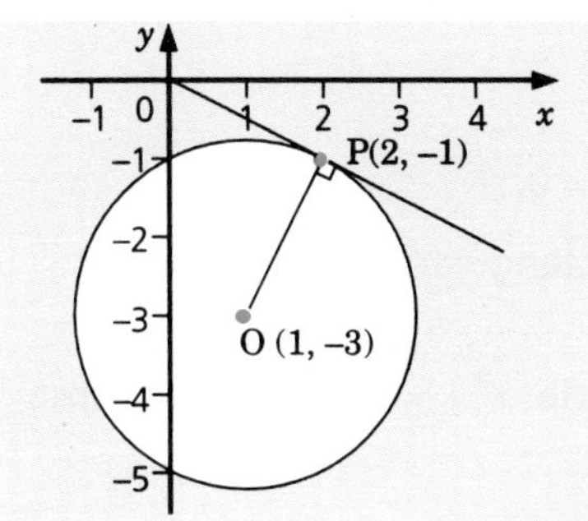

Solution

In standard form the equation of the circle is:

$$(x-1)^2 + (y+3)^2 = 5$$

The circle has centre (1, –3) and radius $\sqrt{5}$.

The point (2, –1) lies on the circle since $(2-1)^2 + (-1+3)^2 = 5$. The radius OP is perpendicular to the tangent. The gradient of OP is:

$$\frac{(-1)-(-3)}{2-1} = 2$$

Hence the gradient of the tangent is $-\frac{1}{2}$.

The equation of the tangent is $\frac{y-(-1)}{x-2} = -\frac{1}{2} \Rightarrow y = -\frac{1}{2}x$.

EXERCISES

25.3 A

1 Find the points of intersection of the given curve and line.

a) circle $x^2 + y^2 = 9$, line $y = x - 3$

b) circle $x^2 + y^2 + 4y = 0$, line $y = 2x - 1$

c) circle $x^2 + y^2 + 8x - 9 = 0$, line $y = 1 - 3x$

d) circle $x^2 + y^2 - 4x + 6y - 12 = 0$, line $y = -x$

e) ellipse $9x^2 + 4y^2 = 36$, line $y = x + 2$

2 The line $y = mx$ is a tangent to the circle $x^2 + y^2 + 2x + 14y + 40 = 0$. Find the possible values of m and the coordinates of the points of contact.

3 The line $y = mx + 1$ is a tangent to the circle $x^2 + y^2 - 10x - 12y + 51 = 0$. Find the possible values of m and the coordinates of the points of contact.

4 Find the length of the tangent from the given point to the given circle.

a) point (0, 0), circle $x^2 - y^2 + 6x - 4y + 12 = 0$

b) point (2, 0), circle $x^2 + y^2 + 2x + 4y + 4 = 0$

c) point (4, 5), circle $x^2 + y^2 = 9$

5 Show that the given point P lies on the given circle. Find the equation of the tangent to the circle through P.

a) P(3, 4), circle $x^2 + y^2 = 25$

b) P(6, 3), circle $x^2 + y^2 - 10x - 12y + 51 = 0$

c) P(7, –2), circle $x^2 + y^2 - 6x - 2y - 15 = 0$

6 The line $y = mx + c$ is a tangent to the circle $x^2 + y^2 = r^2$.
Show that $c^2 = r^2(1 + m^2)$.

7 Find the values of c for which the line $y = 2x + c$:

a) touches,
b) cuts in two points,
c) does not meet the circle $x^2 + y^2 - 2x = 0$.

8 Find the points of intersection of the rectangular hyperbola $xy = 1$ and the circle $x^2 + y^2 + 2x + 2y - 6 = 0$.

9 Find the points of intersection of the circle $x^2 + y^2 = 9$ and the ellipse $\frac{x^2}{16} + \frac{y^2}{2} = 1$.

10 The circle shown in the diagram has equation $(x - 1)^2 + (y - 1)^2 = 5$.

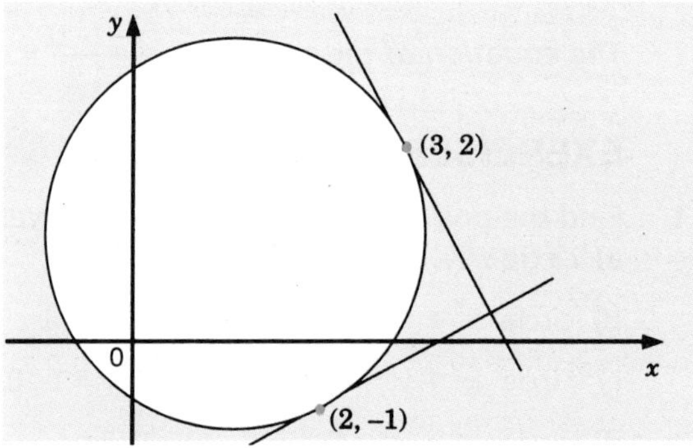

Tangents are drawn at the points (3, 2) and (2, –1).
Write down the coordinates of the centre of the circle and hence show that the tangents are perpendicular to each other.

(Question 5, SEB Paper 1, 1994)

11 An ear-ring is to be made from silver wire and is designed in the shape of two touching circles with two tangents to the outer circle as shown in the top diagram.
The bottom diagram shows a drawing of this ear-ring related to the coordinate axes.
The circles touch at (0, 0).
The equation of the inner circle is $x^2 + y^2 + 3y = 0$.
The outer circle intersects the y-axis at (0, –4).
The tangents meet the y-axis at (0, –6).
Find the total length of silver wire required to make this ear-ring.

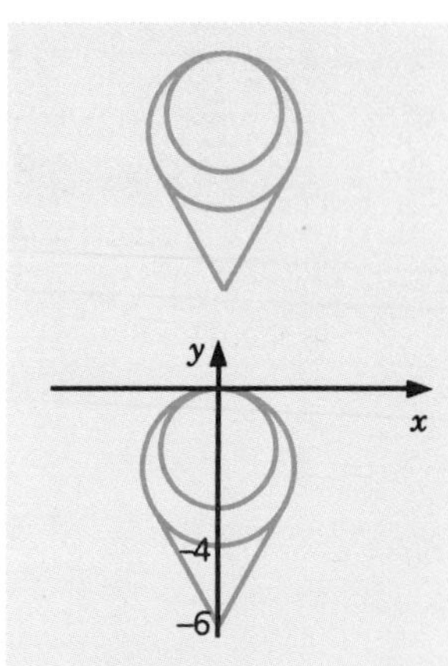

(Question 5, SEB Paper 2, 1989)

EXERCISES

25.3 B

1 Find the points of intersection of the given curve and line.

a) circle $x^2 + y^2 = 2$, line $y = 1 - 2x$

b) circle $x^2 + y^2 - 3x - 2 = 0$, line $y = x + 1$

c) circle $x^2 + y^2 - 3y + 3 = 0$, line $2y + x = 3$

d) hyperbola $xy = 5$, line $y = x - 4$

e) ellipse $4x^2 + y^2 = 4$, line $3y - x = 0$

2 The line $y = mx$ is a tangent to the circle $x^2 + y^2 - 2x + 6y + 2 = 0$. Find the possible values of m and the coordinates of the points of contact.

3 The line $y = mx + 3$ is a tangent to the circle $x^2 + y^2 - 4x - 5 = 0$. Find the possible values of m and the coordinates of the points of contact.

4 Find the values of m for which the line $y = mx + 4$:

a) touches,
b) cuts in two points,
c) does not meet the circle $x^2 + y^2 + 4y - 5 = 0$.

5 The line $y = mx + c$ is a tangent to the ellipse $\frac{x^2}{a^2} + \frac{y^2}{b^2} = 1$.
Show that $c^2 = a^2m^2 + b^2$.

6 Find the length of the tangent from the given point to the given circle.

a) point (0, 0), circle $x^2 + y^2 + 6x + 8y + 16 = 0$

b) point (0, 5), circle $x^2 + y^2 + 4y - 5 = 0$

c) point (–2, –3), circle $x^2 + y^2 - 3x - y + 1 = 0$

7 Show that the points (–2, –4), (3, 1) and (–2, 0) lie on the circle $x^2 + y^2 - 2x + 4y - 8 = 0$. Find the equation of the tangent to the circle through each point.

8 Find the points of intersection of the rectangular hyperbola $xy = 4$ and the hyperbola $x^2 - 2y^2 - 4 = 0$.

9 Find the points of intersection of the parabola $y^2 = 4x$ and the circle $x^2 + y^2 + 2x - 7 = 0$.

10 AB is a tangent at B to the circle with centre C and equation $(x - 2)^2 + (y - 2)^2 = 25$. The point A has coordinates (10, 8). Find the area of triangle ABC.

(SEB Question 16, Paper 1, 1992)

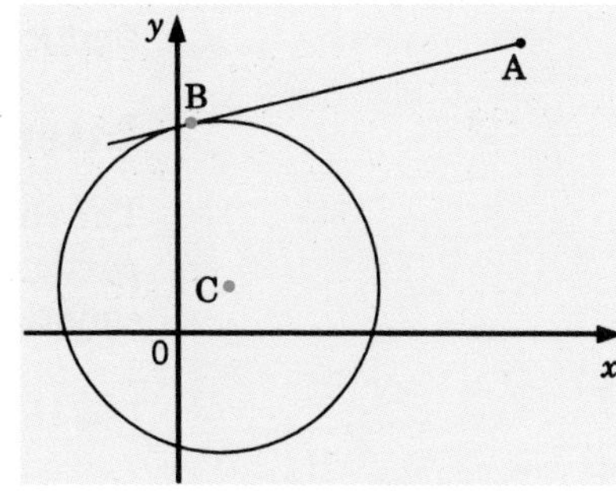

11 A bakery firm makes gingerbread men each 14 cm high with circular heads and bodies.

The equation of the 'body' is $x^2 + y^2 - 10x - 12y + 45 = 0$ and the line of centres is parallel to the y-axis. Find the equation of the 'head'.

(SEB Question 7, Paper 1, 1990)

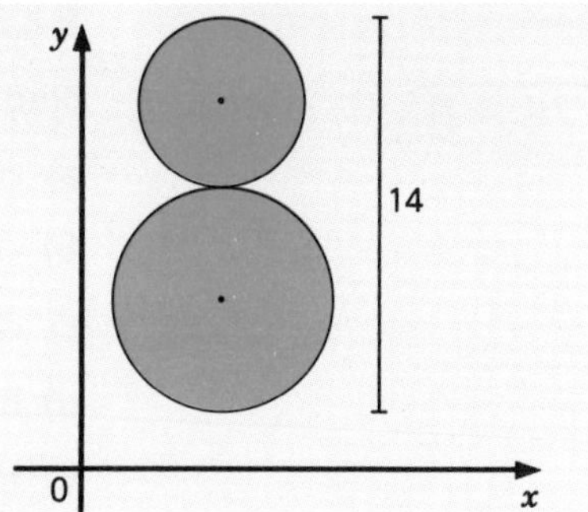

IMPLICIT DIFFERENTIATION

The functions we have met so far in this book have almost all been written in the form $y = \mathrm{f}(x)$. This is called the **explicit form** of a function. The relation $y = x^2 - 2x + 1$ is an example of an explicit function. Some functions are written in an **implied form** from which it is difficult to extract the relation in the form $y = \mathrm{f}(x)$. For example, $x\mathrm{e}^y - y = 0$ is written in such a form. This is called **the implicit form** of a function or an **implicit function**.

To differentiate functions given in implicit form, we need to use a technique called **implicit differentiation**.

Exploration 25.2

Implicit differentiation

Take the equation of a circle $x^2 + y^2 = 1$.

- Rearrange the equation into the form $y = \mathrm{f}(x)$.
- Find $\frac{\mathrm{d}y}{\mathrm{d}x}$.
- Rearrange your formula for $\frac{\mathrm{d}y}{\mathrm{d}x}$ to show that $x + y\frac{\mathrm{d}y}{\mathrm{d}x} = 0$.

Repeat the steps for the following equations.

a) For $2x^2 - 3y^2 - x = 0$ show that $4x - 6y\frac{\mathrm{d}y}{\mathrm{d}x} - 1 = 0$

b) For $\mathrm{e}^y x - x^2 = 0$ show that $x\mathrm{e}^y\frac{\mathrm{d}y}{\mathrm{d}x} + \mathrm{e}^y - 2x = 0$

Deduce a technique for differentiating an implicit function.

Interpreting the results

The chain rule (as well as the other rules) of differentiation is particularly important in implicit differentiation. Consider the equation $x^2 + y^2 = 1$ and differentiate with respect to x term by term.

Use the chain rule.

$$\frac{d(x^2)}{dx} = 2x \text{ and } \frac{d(y^2)}{dx} = \frac{d(y^2)}{dy}\frac{dy}{dx} = 2y\frac{dy}{dx}$$

$$\frac{d(1)}{dx} = 0$$

Adding the first two results gives:

$$2x + 2y\frac{dy}{dx} = 0 \Rightarrow x + y\frac{dy}{dx} = 0$$

This is a more straightforward method of approach than setting up an explicit function first as we did in Exploration 25.2.

Example 25.9

Differentiate the following to find $\frac{dy}{dx}$.

a) $2x^2 - 3y^2 - x = 0$

b) $e^y x - x^2 = 0$

Compare the method with your approach in Exploration 25.2.

Solution

a) $2x^2 - 3y^2 - x = 0$

Differentiating term by term:

$4x - 6y\frac{dy}{dx} - 1 = 0$, *since*

$\frac{d(3y^2)}{dx} = \frac{d(3y^2)}{dy}\frac{dy}{dx} = 6y\frac{dy}{dx}$ *by the chain rule.*

Rearranging to find $\frac{dy}{dx}$*:*

$$\frac{dy}{dx} = \frac{4x-1}{6y}$$

b) $e^y x - x^2 = 0$

Differentiating term by term:

$\left(e^y\frac{dy}{dx}\right)x + e^y - 2x = 0$ *using the product rule and chain rule.*

Rearranging to find $\frac{dy}{dx}$ *we get* $\frac{dy}{dx} = \frac{2x - e^y}{xe^y}$

Example 25.10

Find the value of $\frac{dy}{dx}$ for the equation $2x^2 + 3y^2 - 3x + 2y = 0$ at the point $(1, \frac{1}{3})$. Hence find the equations of the tangent and normal to the curve at $(1, \frac{1}{3})$.

Solution

Differentiating term by term:

$$4x + 6y\frac{dy}{dx} - 3 + 2\frac{dy}{dx} = 0$$

$$\Rightarrow \quad \frac{dy}{dx} = \frac{3-4x}{2+6y}$$

At the point $\left(1, \frac{1}{3}\right)$:

$$\frac{dy}{dx} = \frac{3-4(1)}{2+6\left(\frac{1}{3}\right)} = -\frac{1}{4}$$

The gradient of the tangent at (1, $\frac{1}{3}$*) is* $-\frac{1}{4}$.

The equation of the tangent is:

$$\frac{y-\frac{1}{3}}{x-1} = -\frac{1}{4} \quad \Rightarrow \quad 12y + 3x = 7$$

The gradient of the normal at (1, $\frac{1}{3}$*) is* $-\frac{1}{\left(-\frac{1}{4}\right)} = 4$.

The equation of the normal is $\frac{y-\frac{1}{3}}{x-1} = 4 \Rightarrow 3y - 12x = -11$

EXERCISES

25.4A

In questions **1–10**, find $\frac{dy}{dx}$ using implicit differentiation.

1 $x^2 + y^2 = 5$

2 $x^2 - y^2 = 5$

3 $x^2 + y^2 - 2x + 4y - 1 = 0$

4 $(x-1)^2 - 2(y+1)^2 = 4$

5 $x^3 - xy + y^2 = 6$

6 $\sqrt{xy} = x - 2y$

7 $e^{-2y} \sin 5x = 1$

8 $\sin x + \cos y = 0.5$

9 $y = \cos(xy)$

10 $e^{3y} = x^2 - xy$

In questions **11–18**, find the value of $\frac{dy}{dx}$ at the given point.

11 $x^2 + y^2 = 25$, point (3, 4)

12 $x^2 - y^2 = 3$, point (2, –1)

13 $x^2 + y^2 - 4x + 6y - 13 = 0$, point (1, 2)

14 $x^2 - xy + y^2 = 13$, point (3, –1)

15 $(x+y)^3 = x^3 + y^3$, point (–1, 1)

16 $\sqrt{xy} = x + y - 3$, point (4, 1)

17 $\left(x^2 + y^2\right)^2 = 4x^2 y$, point (1, 1)

18 $x \sin y = 1$, point $\left(2, \frac{\pi}{6}\right)$

19 Show that for the circle with equation $(x-a)^2 + (y-b)^2 = r^2$, $\frac{dy}{dx} = -\frac{(x-a)}{(y-b)}$.

20 If $x^2 + y^2 = 9$ use implicit differentiation to show that $\frac{d^2y}{dx^2} = \frac{-9}{y^3}$.

In questions **21–26** find the equations of the tangent and normal to the conic section at the given point.

21 circle $x^2 + y^2 = 16$, point $\left(2, \sqrt{12}\right)$

22 circle $x^2 + y^2 - 2x + 4y - 20 = 0$, point (5, 1)

23 circle $x^2 + y^2 + 6x - 2y - 3 = 0$, point (–1, –2)

24 ellipse $9x^2 + 4y^2 + 54x - 8y - 95 = 0$, point (1, 4)

25 parabola $y^2 - 2y + 1 - 2x = 0$, point (2, 3)

26 hyperbola $x^2 - y^2 + 2x - 6y - 20 = 0$, point (3, –1)

27 Show that the equation of the tangent to the circle $(x-a)^2 + (y-b)^2 = r^2$ at the point (x_1, y_1) has equation $(y-y_1)(y-b) + (x-x_1)(x-a) = 0$.

EXERCISES

25.4 B

In questions **1–10**, find $\frac{dy}{dx}$ using implicit differentiation.

1 $x^2 + y^2 = 8$

2 $x^2 - y^2 = 8$

3 $2x^2 + 3y^2 = 6$

4 $y^2 - 4y = 3x - 4$

5 $(x-4)^2 + (y+3)^2 = 9$

6 $x^2 - 2y^2 + 3x + 5y - 11 = 0$

7 $x \sin y = y \cos x$

8 $x^2 e^{-y} - xy = 0$

9 $\sin(x+y) = x$

10 $y = \sin(xy)$

In questions **11–18**, find the value of $\frac{dy}{dx}$ at the given point.

11 $x^2 + y^2 = 5$, point (1, 2)

12 $x^2 + 2y^2 = 19$, point (–1, 3)

13 $x^2 + 3xy + y^3 = 11$, point (2, 1)

14 $x^2 + 9y^2 - 4x + 3y - 6 = 0$, point (0, –1)

15 $\left(1 - x^2\right)y = 2x$, point (0, 0)

16 $x^2 \cos y = 1$, point $\left(\sqrt{2}, \frac{\pi}{3}\right)$

17 $e^x y = 4 + xy$, point (0, 4)

18 $y\sqrt{x} - x\sqrt{y} = 6$, point (4, 9)

19 Show that for the rectangular hyperbola with equation $(x-a)^2 - (y-b)^2 = r^2$, $\frac{dy}{dx} = \frac{(x-a)}{(y-b)}$.

20 If $y \sin x = e^x$ show that $\frac{d^2y}{dx^2} + 2\cot x \frac{dy}{dx} - 2y = 0$.

In questions **21–26** find the equations of the tangent and normal to the conic section at the given point.

21 circle $x^2 + y^2 = 36$, point $(-4, \sqrt{20})$

22 circle $x^2 + y^2 + 2y - 1 = 0$, point (1, –2)

23 circle $x^2 + y^2 - 3x + 8y - 20 = 0$, point (3, 2)

24 hyperbola $xy - y + 3x = 11$, point (2, 5)

25 ellipse $4x^2 + y^2 - 4x + 2y - 16 = 0$, point (–1, 2)

26 hyperbola $2x^2 - y^2 + 4x + 4y - 43 = 0$, point (4, –1)

27 Show that the equation of the tangent to the hyperbola

$\frac{(x-a)^2}{A^2} - \frac{(y-b)^2}{B^2} = 1$ at the point (x_1, y_1) is

$A^2(y - y_1)(y - b) - B^2(x - x_1)(x - a) = 0$.

CONSOLIDATION EXERCISES FOR CHAPTER 25

In questions **1–4**, classify the conic sections. For each circle find the radius and centre and for each ellipse find the axis-length and centre. In each case sketch the graph of the function without using a graphics calculator.

1 $x^2 - y^2 = 16$

2 $x^2 + y^2 - 4x + 6y - 87 = 0$

3 $2x^2 + y^2 - 8x - 6y - 8 = 0$

4 $y^2 - 2y - 4x = 7$

5 Find the equation of the tangent and normal to the curve with equation $x^2 + y^2 - 10x - 12y + 51 = 0$ at the point (6, 3).

6 Find the range of values of m so that the line $y = mx$ and the circle $x^2 + y^2 - 6x - 8y + 24 = 0$

a) cut in two points,
b) touch at two points
c) do not intersect.

7 Show that an equation of the normal to the hyperbola with equation $xy = c^2$, at the point $\left(cp, \frac{c}{p}\right)$ is $y - \frac{c}{p} = xp^2 - cp^3$.

(ULEAC Question 3 (part), Paper 2, January 1994)

8 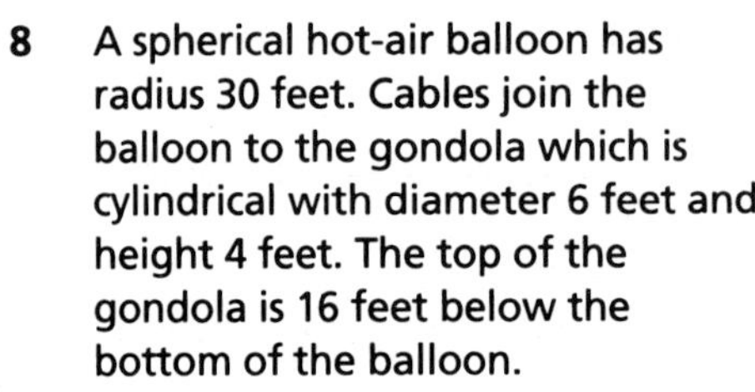
A spherical hot-air balloon has radius 30 feet. Cables join the balloon to the gondola which is cylindrical with diameter 6 feet and height 4 feet. The top of the gondola is 16 feet below the bottom of the balloon.

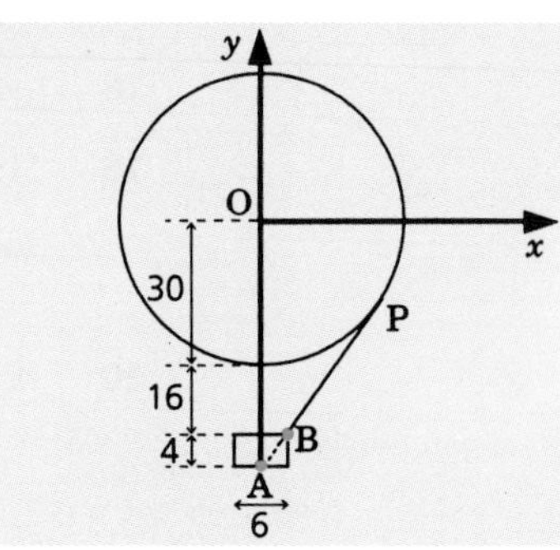

Coordinate axes are chosen as shown in the diagram. One of the cables is represented by PB and PBA is a straight line.

a) Find the equation of the cable PB.

b) State the equation of the circle representing the balloon.

c) Prove that this cable is a tangent to the balloon and find the coordinates of the point P.

(SEB Question 9, Paper II 1992)

9 A penny-farthing bicycle on display in a museum is supported by a stand at points A and C. A and C lie on the front wheel.

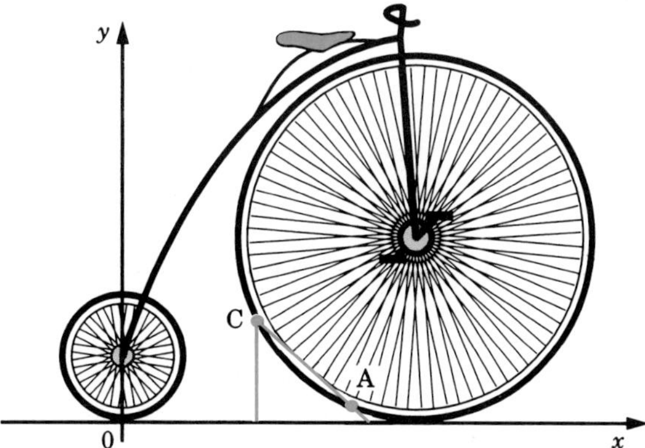

With coordinate axes as shown and 1 unit = 5 cm, the equation of the rear wheel (the small wheel) is $x^2 + y^2 - 6y = 0$ and the equation of the front wheel is $x^2 + y^2 - 28x - 20y + 196 = 0$.

a) **i)** Find the distance between the centres of the two wheels.

ii) Hence calculate the clearance, i.e. the smallest gap, between the front and rear wheels. Give your answer to the nearest millimetre.

b) B(7, 3) is half-way between A and C, and P is the centre of the front wheel.

i) Find the gradient of PB.

ii) Hence find the equation of AC and the coordinates of A and C.

(SEB Question 4, Paper II, 1994)

10 The points (5, 5) and (–3, –1) are the ends of a diameter of the circle C with centre A. Write down the coordinates of A and show that the equation of C is $x^2 + y^2 - 2x - 4y - 20 = 0$.

The line L with equation $y = 3x - 16$ meets C at the points P and Q. Show that the x-coordinates of P and Q satisfy the equation

$x^2 - 11x + 30 = 0$.

Hence find the coordinates of P and Q.

(WJEC Question 3, Specimen Paper A1, 1994)

Summary

The **conic sections** are the following curves:

- circle, centre (a, b), radius r, $(x - a)^2 + (y - b)^2 = r^2$
- ellipse, centre (0, 0), major axis $2a$, minor axis $2b$, $\dfrac{x^2}{a^2} + \dfrac{y^2}{b^2} = 1$
- hyperbola, centre (0, 0), slant asymptotes $y = \pm\dfrac{b}{a}x$, $\dfrac{x^2}{a^2} - \dfrac{y^2}{b^2} = 1$
- parabola, through (0, 0), $y^2 = 4ax$

The **implicit form** of a function is a relation of the form $f(x, y) = 0$. To differentiate functions given in implicit form we use **implicit differentiation**.

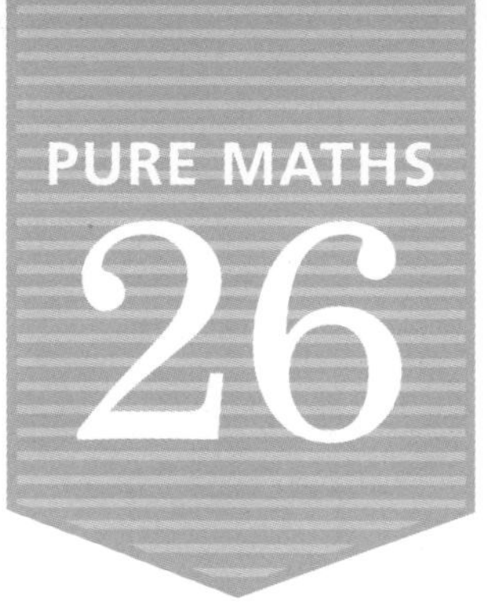

Trigonometry 5

In this chapter we shall introduce:

- *compound-angle formulae, or expansions for expressions such as* $\sin(\theta + \varphi)$, $\cos(\theta + \varphi)$,
- *double-angle expansions,*
- *the use of these expansions to model waves.*

COMPOUND-ANGLE FORMULAE

In studying trigonometry we have discovered various relationships between trigonometric functions such as:

$$\tan\theta \equiv \frac{\sin\theta}{\cos\theta} \quad \cos(90° - \theta) \equiv \sin\theta \quad \cos^2\theta + \sin^2\theta \equiv 1$$

In each case the expressions contain trigonometric functions of a single variable, θ. In this chapter we shall find other useful relationships involving trigonometric functions of two variables, say θ and φ. We begin with an exploration.

Exploration 26.1

Coordinates and angles

The diagram shows three points, O, P and Q where O is the origin and P and Q are in the first quadrant such that OP = OQ = 1, OP makes an angle φ with the x-axis, OQ makes an angle θ with the x-axis and $\theta > \varphi$.

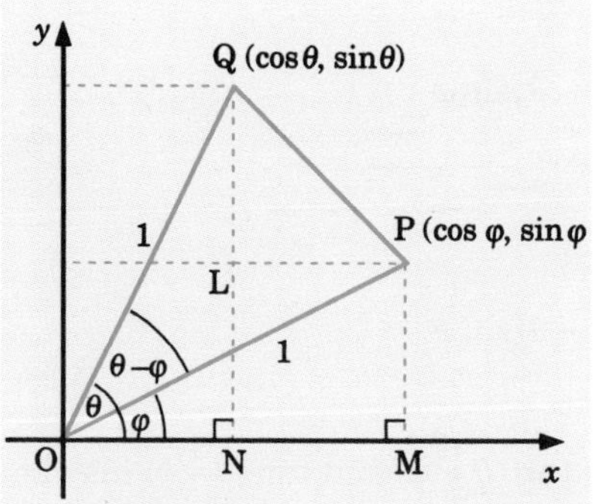

- Explain why the coordinates of P and Q are $(\cos\varphi, \sin\varphi)$ and $(\cos\theta, \sin\theta)$ respectively.
- Find an expression for PQ^2.
- Use the cosine rule on ΔOPQ to derive a formula for $\cos(\theta - \varphi)$.

Interpreting the results

The lines OP and OQ are hypotenuses of right-angled triangles OPM and OQN respectively. Since OP = 1, OM = $\cos\varphi$ and PM = $\sin\varphi$. Similarly we can see that ON = $\cos\theta$ and QN = $\sin\theta$.

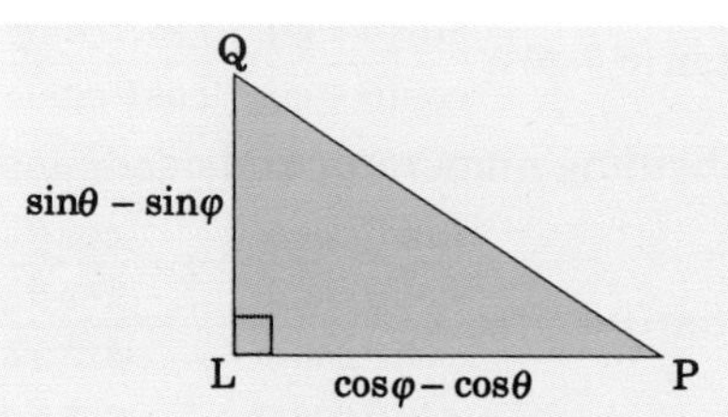

Treating PQ as the hypotenuse of ΔPQL, by Pythagoras' theorem:

$$\begin{aligned}PQ^2 &= PL^2 + QL^2\\ &= (\cos\varphi - \cos\theta)^2 + (\sin\theta - \sin\varphi)^2\\ &= \left(\cos^2\varphi - 2\cos\varphi\cos\theta + \cos^2\theta\right) + \left(\sin^2\theta - 2\sin\theta\sin\varphi + \sin^2\varphi\right)\\ &= \left(\cos^2\theta + \sin^2\theta\right) + \left(\cos^2\varphi + \sin^2\varphi\right) - 2\cos\theta\cos\varphi - 2\sin\theta\sin\varphi\\ &= 2 - 2(\cos\theta\cos\varphi + \sin\theta\sin\varphi)\end{aligned}$$

Using the cosine rule on ΔOPQ:

$$\begin{aligned}PQ^2 &\equiv OP^2 + OQ^2 - 2\times OP\times OQ\times\cos(\theta-\varphi)\\ &\equiv 1 + 1 - 2\times 1\times 1\times\cos(\theta-\varphi)\end{aligned}$$

$$\Rightarrow 2 - 2(\cos\theta\cos\varphi + \sin\theta\sin\varphi) \equiv 2 - 2\cos(\theta-\varphi)$$

$$\Rightarrow \cos(\theta-\varphi) = \cos\theta\cos\varphi + \sin\theta\sin\varphi$$

Check the equivalence by inserting values for θ and φ and showing the two sides balance, e.g. with $\theta = 70°$ and $\varphi = 10°$, show that both sides equal 0.5, i.e:

cos 60° = cos 70° cos 10° + sin 70° sin 10°

We have used a geometrical argument to develop our first compound-angle formula. Now we can use various properties of trigonometrical functions to develop five more formulae.

Replacing φ by $-\theta$ gives:

$$\cos(\theta - (-\varphi)) \equiv \cos\theta\cos(-\varphi) + \sin\theta\sin(-\varphi)$$
$$\Rightarrow \cos(\theta+\varphi) \equiv \cos\theta\cos\varphi - \sin\theta\sin\varphi$$

Compound-angle formulae for $\sin(\theta + \varphi)$ and $\sin(\theta - \varphi)$ can be developed from the respective formulae for $\cos(\theta + \varphi)$ and $\cos(\theta - \varphi)$:

$$\begin{aligned}\sin(\theta+\varphi) &\equiv \cos(90° - (\theta+\varphi))\\ &\equiv \cos((90° - \theta) - \varphi)\\ &\equiv \cos(90° - \theta)\cos\varphi + \sin(90° - \theta)\sin\varphi\end{aligned}$$

$$\Rightarrow \sin(\theta+\varphi) \equiv \sin\theta\cos\varphi + \cos\theta\sin\varphi$$

$$\begin{aligned}\sin(\theta-\varphi) &\equiv \cos(90° - (\theta-\varphi))\\ &\equiv \cos((90° - \theta) + \varphi)\\ &\equiv \cos(90° - \theta)\cos\varphi - \sin(90° - \theta)\sin\varphi\end{aligned}$$

$$\Rightarrow \sin(\theta+\varphi) \equiv \sin\theta\cos\varphi + \cos\theta\sin\varphi$$

Finally, compound-angle formulae for $\tan(\theta + \varphi)$ and $\tan(\theta - \varphi)$ may be deduced from the definition of tangent in terms of sine and cosine:

$$\tan(\theta+\varphi) \equiv \frac{\sin(\theta+\varphi)}{\cos(\theta+\varphi)} \equiv \frac{\sin\theta\cos\varphi + \cos\theta\sin\varphi}{\cos\theta\cos\varphi - \sin\theta\sin\varphi}$$

Dividing numerator and denominator by $\cos\theta\cos\varphi$ gives:

$$\tan(\theta+\varphi) \equiv \frac{\dfrac{\sin\theta\cos\varphi}{\cos\theta\cos\varphi} + \dfrac{\cos\theta\sin\varphi}{\cos\theta\cos\varphi}}{\dfrac{\cos\theta\cos\varphi}{\cos\theta\cos\varphi} - \dfrac{\sin\theta\sin\varphi}{\cos\theta\cos\varphi}}$$

$$\equiv \frac{\dfrac{\sin\theta}{\cos\theta} + \dfrac{\sin\varphi}{\cos\varphi}}{1 - \dfrac{\sin\theta}{\cos\theta} \times \dfrac{\sin\varphi}{\cos\varphi}}$$

$$\Rightarrow \tan(\theta + \varphi) \equiv \frac{\tan\theta + \tan\varphi}{1 - \tan\theta\tan\varphi}$$

Similar reasoning, or replacing φ by $-\varphi$ gives:

$$\tan(\theta + (-\varphi)) \equiv \frac{\tan\theta + \tan(-\varphi)}{1 - \tan\theta\tan(-\varphi)}$$

$$\Rightarrow \tan(\theta - \varphi) \equiv \frac{\tan\theta - \tan\varphi}{1 + \tan\theta\tan\varphi}$$

The following examples illustrate some of the ways in which the compound-angle formula may be used.

Example 26.1

Without using a calculator, find the value of:

a) *sin 15°* ***b)*** *tan 105°.*

Solution

a) $\sin 15° = \sin(45° - 30°) = \sin 45° \cos 30° - \cos 45° \sin 30°$

$$= \tfrac{1}{\sqrt{2}} \times \tfrac{\sqrt{3}}{2} - \tfrac{1}{\sqrt{2}} \times \tfrac{1}{2} = \frac{\sqrt{3} - 1}{2\sqrt{2}} = \frac{\sqrt{2}(\sqrt{3} - 1)}{4}$$

b) $\tan 105° = \tan(60° + 45°) = \dfrac{\tan 60° + \tan 45°}{1 - \tan 60° \tan 45°}$

$$= \frac{\sqrt{3} + 1}{1 - \sqrt{3}}$$

$$= \frac{-(1 + \sqrt{3})^2}{2}$$

Example 26.2

Without using a calculator, find the exact value of:

a) $\cos(\alpha + \beta)$ ***b)*** $\tan(\alpha - \beta)$

given that α *and* β *and are both acute angles and* $\tan\alpha = \frac{4}{3}$ *and* $\sin\beta = \frac{5}{13}$.

Solution

First, deduce values for $\sin\alpha$, $\cos\alpha$, $\cos\beta$, $\tan\beta$ *from right-angled triangles.*

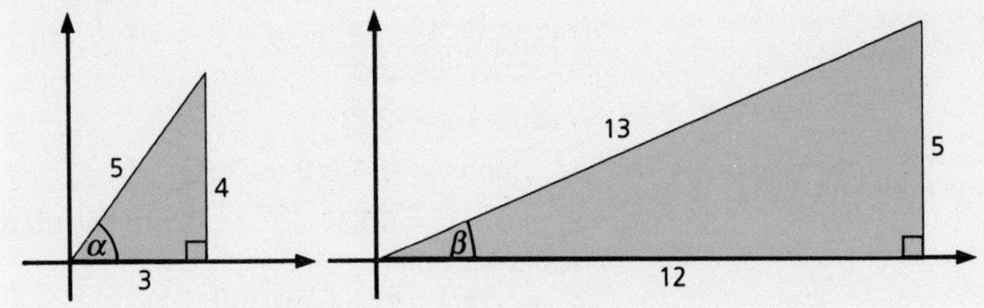

$\sin\alpha = \frac{4}{5}, \cos\alpha = \frac{3}{5}$ *and* $\cos\beta = \frac{12}{13}, \tan\beta = \frac{5}{12}$

a) $\cos(\alpha+\beta) = \cos\alpha\cos\beta - \sin\alpha\sin\beta = \frac{3}{5}\times\frac{12}{13} - \frac{4}{5}\times\frac{5}{13}$
$= \dfrac{36-20}{65} = \dfrac{16}{65}$

b) $\tan(\alpha-\beta) = \dfrac{\tan\alpha - \tan\beta}{1+\tan\alpha\tan\beta} = \dfrac{\frac{4}{3}-\frac{5}{12}}{1+\frac{4}{3}\times\frac{5}{12}} = \dfrac{\frac{11}{12}}{\frac{14}{9}} = \dfrac{11}{56}$

Example 26.3

Show that $\sin(A+B)\sin(A-B) \equiv (\sin A + \sin B)(\sin A - \sin B)$.

Solution

$$\begin{aligned}
LHS &\equiv \sin(A+B)\sin(A-B)\\
&\equiv (\sin A\cos B + \cos A\sin B)(\sin A\cos B - \cos A\sin B)\\
&\equiv \sin^2 A\cos^2 B + \cos A\sin B\sin A\cos B - \cos A\sin B\sin A\cos B\\
&\quad - \cos^2 A\sin^2 B\\
&\equiv \sin^2 A\left(1-\sin^2 B\right) - \left(1-\sin^2 A\right)\sin^2 B\\
&\equiv \sin^2 A - \sin^2 A\sin^2 B - \sin^2 B + \sin^2 A\sin^2 B\\
&\equiv \sin^2 A - \sin^2 B\\
&\equiv (\sin A + \sin B)(\sin A - \sin B) \equiv RHS
\end{aligned}$$

Example 26.4

The diagram shows graphs of $y = -2\cos x$ *and* $y = \sin(x+45°)$, $0 \le x \le 360°$. *Find the coordinates of points A and B, where the graphs intersect.*

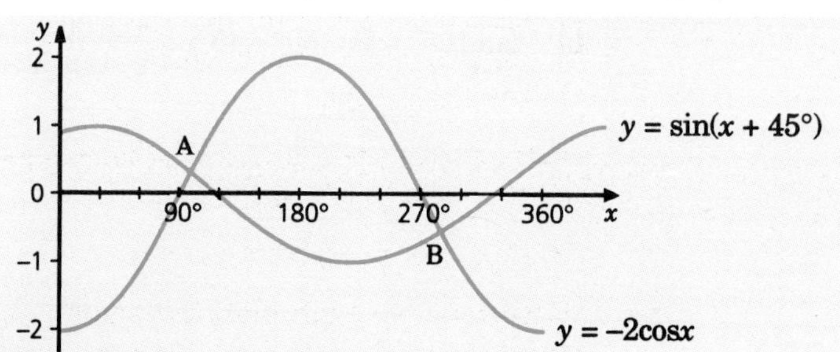

Solution

At the point of intersection:

$\sin(x+45°) = -2\cos x$

$\Rightarrow \sin x\ \cos 45° + \cos x\ \sin 45° = -2\cos x$

$\Rightarrow \dfrac{\sin x}{\cos x}\cos 45° + \dfrac{\cos x}{\cos x}\sin 45° = -\dfrac{2\cos x}{\cos x}$

$\Rightarrow \tan x \times \dfrac{1}{\sqrt{2}} + \dfrac{1}{\sqrt{2}} = -2$

$\Rightarrow \tan x + 1 = -2\sqrt{2}$

$\Rightarrow \quad \tan x = -1 - 2\sqrt{2} \approx -3.83$

$\Rightarrow \quad x = -75.4°$ *(from calculator)*

But $0 \le x \le 360°$

$\Rightarrow\ x = 180° - 75.4° = 104.6°$
or $x = 360° - 75.4° = 284.6°$

EXERCISES

26.1 A

1 Use compound-angle formulae to find the following.
a) cos 15° **b)** tan 15° **c)** sin 75°
d) tan 75° **e)** cos 105° **f)** sin 165°

2 Without using a calculator, find the values of the following.

a) $\sin 52° \cos 22° - \cos 52° \sin 22°$

b) $\cos 17° \cos 43° - \sin 17° \sin 43°$

c) $\dfrac{\tan 73° - \tan 13°}{1 + \tan 73° \tan 13°}$

3 Acute angles α and β are such that tan $\alpha = \frac{15}{8}$ and sin $\beta = \frac{12}{13}$. Find the exact value of each of the following.
a) sin $(\alpha + \beta)$ **b)** cos $(\alpha - \beta)$ **c)** tan $(\alpha + \beta)$

4 Obtuse angles α and β are such that sin $\alpha = \frac{7}{25}$ and cos $\beta = -\frac{4}{5}$. Find the exact value of each of the following.
a) sin $(\alpha - \beta)$ **b)** cos $(\alpha + \beta)$ **c)** tan $(\alpha - \beta)$

5 Find algebraically the exact value of $\sin\theta° + \sin(\theta + 120)° + \cos(\theta + 150)°$.

6 If A and B are acute angles such that tan $A = \frac{1}{2}$ and tan $B = \frac{1}{3}$ show that $A + B = 45°$.

7 Triangle PQR has vertices P(3, 7), Q(1, 2), R(–4, –8). Find:
a) the gradient of the line PQ,
b) the gradient of the line QR,
c) the angle PQR.

8 Lines l and m have gradients $\frac{2}{3}$ and $\frac{3}{2}$ respectively. Find the value of tan θ, where θ is the acute angle between lines l and m.

9 Find the acute angle between the lines $y = 2x - 1$ and $2y = x + 1$.

10 Write down an expression for tan (60° + 30°). What can you deduce about tan 90°?

11 If tan $(x + 45°) = 2$, find the smallest positive value of x.

12 If tan$P = \frac{1}{3}$ and tan$Q = \frac{1}{2}$, and $R = P + Q$, find possible values for R such that $0 \le R \le 360°$.

13 Show that:
a) $\sin(A + B) + \sin(A - B) \equiv 2\sin A \cos B$

b) $\cos(A - B)\cos(A + B) \equiv (\cos A - \cos B)(\cos A + \cos B)$.

14 Show that $\tan A + \tan B \equiv \dfrac{\sin(A + B)}{\cos A \cos B}$.

15 If $\sin(x - \alpha) = k\sin(x + \alpha)$,
a) express tan x in terms of k and tan α,
b) determine possible values of x when $k = \frac{1}{2}$, $\alpha = 30°$, such that $0 \le x \le 360°$.

16 Solve the equation $\sin(x+60°) = \cos x$, $0 \le x \le 360°$.

17 Solve the equation $2\cos(x-20°) + \cos x = 0$, $-180° \le x \le 180°$.

18 Sketch the graphs of $y = 2\sin x$ and $y = \sin(x-30°)$, $0 \le x \le 360°$.

Find the coordinates of their points of intersection, P and Q.

19 Sketch, on the same axes, graphs of $y = \cos(2x+30°)$ and $y = 3\sin x$, $-180° \le x \le 180°$. Find the coordinates of the points where the graphs intersect.

20 The equation $\tan\left(x+\frac{\pi}{4}\right) + 2\tan x = 0$ has four solutions for the domain $-\pi \le x \le \pi$. Find the four values of x which solve the equation.

EXERCISES

26.1 B

1 By calculating each side separately, verify the identity $\sin(A+B) \equiv \sin A\cos B + \cos A\sin B$ for the following pairs of angles.

a) $A = 30°, B = 135°$ **b)** $A = 23°, B = 211°$
c) $A = 0.765$ radians, $B = 3.468$ radians

2 Repeat question 1 for the identity $\cos(A+B) \equiv \cos A\cos B - \sin A\sin B$.

3 Repeat question 1 for the identity $\tan(A+B) \equiv \dfrac{\tan A + \tan B}{1 - \tan A\tan B}$.

4 Rewrite the following expressions as single sines or cosines.

a) $\sin 53°\cos 51° + \cos 53°\sin 51°$ **b)** $\sin 24°\cos 20° - \cos 24°\sin 20°$
c) $\cos 57°\cos 17° + \sin 57°\sin 17°$ **d)** $\sin 41°\sin 19° - \cos 41°\cos 19°$
e) $\sin\frac{5\pi}{12}\cos\frac{\pi}{12} - \cos\frac{5\pi}{12}\sin\frac{\pi}{12}$ **f)** $\sin\frac{5\pi}{12}\cos\frac{\pi}{12} + \cos\frac{5\pi}{12}\sin\frac{\pi}{12}$
g) $\sin 2x\cos x - \cos 2x\sin x$ **h)** $\sin 2x\cos x + \cos 2x\sin x$
i) $\cos^2 x - \sin^2 x$ **j)** $\sin(x+y)\cos y - \cos(x+y)\sin y$

5 If $\sin\frac{\pi}{4} = \frac{\sqrt{2}}{2}$, $\sin\frac{\pi}{6} = \frac{1}{2}$ and $\sin\frac{\pi}{3} = \frac{\sqrt{3}}{2}$, evaluate each of the following, without using a calculator.

a) $\sin\left(\frac{\pi}{4}+\frac{\pi}{6}\right)$ **b)** $\cos\left(\frac{\pi}{4}+\frac{\pi}{6}\right)$ **c)** $\tan\left(\frac{\pi}{4}+\frac{\pi}{6}\right)$

6 Use the compound-angle identities to verify the following.

a) $\sin(x+180°) = -\sin x$ **b)** $\cos(x+180°) = -\cos x$
c) $\sin(x+360°) = \sin x$ **d)** $\cos(x+360°) = \cos x$

7 This question should be attempted without using a calculator.

If $\cos A = \frac{3}{5}$, $\cos B = \frac{12}{13}$ and both A and B are acute angles:

a) find $\sin A$, $\tan A$, $\sin B$ and $\tan B$,
b) use the compound-angle formulae to find $\sin(A+B)$, $\sin(A-B)$, $\cos(A+B)$, $\cos(A-B)$, $\tan(A+B)$ and $\tan(A-B)$.

8 If A is the acute angle with sine $\frac{3}{5}$ and B is the obtuse angle with sine $\frac{5}{13}$, find $\cos(A+B)$ and $\tan(A-B)$, without using a calculator.

9 Prove the following identities.
$\sin(X+Y)\sin(X-Y) \equiv \sin^2 X - \sin^2 Y$
$\cos(X+Y)\cos(X-Y) \equiv \cos^2 X - \sin^2 Y$
Hence, without using a calculator, evaluate $\sin 75°\sin 15°$, $\cos 75°\cos 15°$, $\sin 105°\sin 15°$ and $\cos 105°\cos 15°$.

10 Show that $\cos(x+45°) = \frac{\sqrt{2}}{2}(\cos x - \sin x)$ and

$\sin(x+45°) = \frac{\sqrt{2}}{2}(\cos x + \sin x)$.
Hence find the exact values of $\cos 75°$, $\sin 75°$, $\cos 105°$ and $\sin 105°$.

11 Prove the identity $\frac{\sin(x+y)}{\cos x \cos y} \equiv \tan x + \tan y$.

12 The diagram shows two straight lines, gradients m_1 and m_2, which intersect at P. The angle between the lines is θ, as shown.

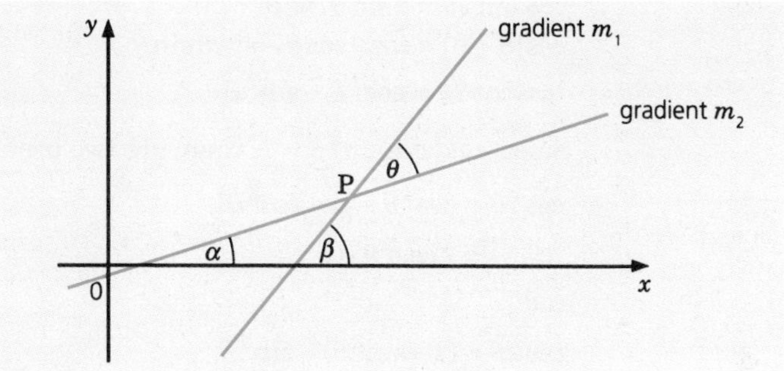

a) Find the relationship between θ, α and β.

b) Hence show that $\tan\theta = \frac{m_1 - m_2}{1 + m_1 m_2}$.

c) Find the angle between the lines $y = 3x + 2$ and $y = 2x + 3$.

13 Use the compound-angle identities to verify the following.
a) $\tan(x+180°) = \tan x$ **b)** $\tan(180°-x) = \tan x$
c) $\sin(90°+x) = \cos x$ **d)** $\cos(90°-x) = \sin x$

14 If A is the obtuse angle with sine $\frac{7}{25}$ and B is the obtuse angle with cosine $-\frac{5}{13}$, find $\sec(A+B)$ and $\cot(A-B)$, without using a calculator.

15 Show that $\tan(x+45°) \equiv \frac{1+\tan x}{1-\tan x}$ and $\tan(x-45°) \equiv \frac{\tan x - 1}{\tan x + 1}$.

16 Prove the identity $\sin(x+y) + \cos(x-y) \equiv (\sin x + \cos x)(\sin y + \cos y)$.

17 Starting with the identity for $\tan(A+B)$, show that

$\cot(A+B) = \frac{\cot A \cot B - 1}{\cot A + \cot B}$.

18 Using a graphics calculator or computer software, sketch graphs of $y = 2\cos x$ and $y = \sin(x - 60°)$ for $-180 \le x \le 180°$.
a) Use your graph to obtain estimates for the coordinates of their points of intersection.

b) Find the exact values of the coordinates by solving the equation $2\cos x = \sin(x - 60°)$.

19 Sketch, on the same axes, graphs of $y = 5\sin(x - 30°)$ and $y = 2\cos(x + 60°)$ for $0 \le x \le 360°$. Find the coordinates of the points where the graphs intersect.

20 Given that $2\sin(\theta + 60°) = \cos(\theta + 60°)$ show that $\tan\theta = \sqrt{3}a + b$ where a and b are integers.

DOUBLE-ANGLE FORMULAE

In the first section of this chapter we discovered six compound-angle formulae. A very useful *particular* case occurs when $\theta = \varphi$ in the trigonometric formulae for the sum of two angles.

Let $\theta = \varphi$, then we have:

$$\sin(\theta + \theta) = \sin\theta\cos\theta + \cos\theta\sin\theta$$
$$\Rightarrow \sin 2\theta = 2\sin\theta\cos\theta$$
$$\cos(\theta + \theta) = \cos\theta\cos\theta - \sin\theta\sin\theta$$
$$\Rightarrow \cos 2\theta = \cos^2\theta - \sin^2\theta$$

Since $\cos^2\theta + \sin^2\theta = 1$, there are two useful alternative forms for $\cos 2\theta$:

$$\cos 2\theta = \cos^2\theta - \left(1 - \cos^2\theta\right)$$
$$= 2\cos^2\theta - 1$$

and:

$$\cos 2\theta = (1 - \sin^2\theta) - \sin^2\theta$$
$$= 1 - 2\sin^2\theta$$

Finally:

$$\tan(\theta + \theta) = \frac{\tan\theta + \tan\theta}{1 - \tan\theta\tan\theta}$$
$$\Rightarrow \tan 2\theta = \frac{2\tan\theta}{1 - \tan^2\theta}$$

Note that letting $\theta = \varphi$ in the trigonometrical formulae for the difference of the two angles gives two obvious and one interesting result.

$$\sin(\theta - \theta) = \sin\theta\cos\theta - \cos\theta\sin\theta$$
$$\Rightarrow \sin 0 = 0$$
$$\tan(\theta - \theta) = \frac{\tan\theta - \tan\theta}{1 + \tan^2\theta}$$
$$\Rightarrow \tan 0 = 0$$
$$\cos(\theta - \theta) = \cos\theta\cos\theta + \sin\theta\sin\theta$$
$$\Rightarrow \quad 1 = \cos^2\theta + \sin^2\theta$$

which is the basic Pythagorean identity.

The following examples illustrate some useful applications of the double-angle formulae.

Example 26.5

Given that $\cos\theta = 0.96$ *and* θ *is acute, find the value of each of the following.*

a) $\sin\theta$ ***b)*** $\sin 2\theta$ ***c)*** $\cos 2\theta$ ***d)*** $\tan 2\theta$

Solution

a) $\sin^2\theta = 1-\cos^2\theta \Rightarrow \sin\theta = \sqrt{1-\cos^2\theta} = \sqrt{1-0.96^2}$
$= \sqrt{0.0784}$
$= 0.28$

b) $\sin 2\theta = 2\sin\theta\cos\theta = 2\times 0.28\times 0.96 = 0.5376$

c) $\cos 2\theta = \cos^2\theta - \sin^2\theta = 0.96^2 - 0.28^2 = 0.8432$

d) $\tan\theta = \dfrac{\sin\theta}{\cos\theta} = \dfrac{0.28}{0.96} = \dfrac{7}{24} \Rightarrow \tan 2\theta = \dfrac{2\tan\theta}{1-\tan^2\theta}$

$= \dfrac{2\times\frac{7}{24}}{1-\left(\frac{7}{24}\right)^2} = \dfrac{336}{527} \approx 0.6376$ (correct to four decimal places)

Example 26.6

Show that $\dfrac{1-\cos 2A+\sin 2A}{1+\cos 2A+\sin 2A} \equiv \tan A$.

Solution

$$\frac{1-\cos 2A+\sin 2A}{1+\cos 2A+\sin 2A} \equiv \frac{2\sin^2 A+2\sin A\cos A}{2\cos^2 A+2\sin A\cos A}$$

$$\equiv \frac{2\sin A(\sin A+\cos A)}{2\cos A(\cos A+\sin A)} \equiv \tan A$$

Example 26.7

The diagram shows the graphs of $y = 5\cos 2x$ *and* $y = 3\cos x - 4$ *for* $0 \le x \le 360°$. *Find the coordinates of the points of intersection.*

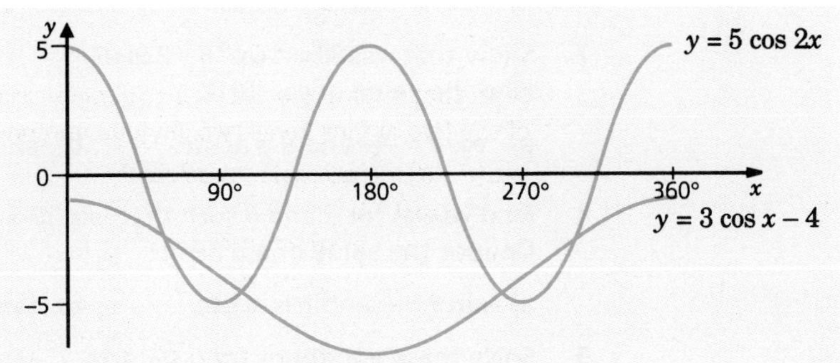

Solution

At the points of intersection:

$5\cos 2x = 3\cos x - 4$

$\Rightarrow 5\left(2\cos^2 x - 1\right) = 3\cos x - 4$

$\Rightarrow 10\cos^2 x - 5 = 3\cos x - 4$

$\Rightarrow 10\cos^2 x - 3\cos x - 1 = 0$

$\Rightarrow (2\cos x - 1)(5\cos x + 1) = 0$

$\Rightarrow 2\cos x - 1 = 0 \quad \Rightarrow \cos x = \frac{1}{2}$

or $5\cos x + 1 = 0 \quad \Rightarrow \cos x = -\frac{1}{5}$

$\cos x = \frac{1}{2} \quad \Rightarrow x = 60° \quad$ or $x = 300°$

$\cos x = -\frac{1}{5} \Rightarrow x = 101.5°$ or $x = 258.5°$

The points of intersection are (60°, –2.5), (101.5°, –4.6), (258.5°, –4.6) and (300°, –2.5).

EXERCISES

1 Given that θ is acute and $\tan\theta = \frac{3}{4}$, find the values of the following.
a) $\sin\theta$ **b)** $\cos\theta$ **c)** $\sin 2\theta$
d) $\cos 2\theta$ **e)** $\tan 2\theta$

2 Given that $\sin\theta = 0.8$ and $\frac{\pi}{2} < \theta < \pi$, find the values of the following.
a) $\sin 2\theta$ **b)** $\cos 2\theta$ **c)** $\sin 4\theta$

3 Given that $25\cos\theta + 7 = 0$ and θ is obtuse, find the exact values of the following.
a) $\cos 2\theta$ **b)** $\sin 2\theta$ **c)** $\tan 2\theta$

4 Show that the following identities are true.
a) $\dfrac{\sin 2\theta}{1+\cos 2\theta} \equiv \tan\theta$ **b)** $\dfrac{1-\cos 2\theta}{\sin 2\theta} \equiv \tan\theta$

5 Show that the following identities are true.
a) $\sin 2\theta \equiv \dfrac{2\tan\theta}{1+\tan^2\theta}$ **b)** $\cos 2\theta \equiv \dfrac{1-\tan^2\theta}{1+\tan^2\theta}$

6 Show that the following identities are true.
a) $\cos^4\theta - \sin^4\theta = \cos 2\theta$ **b)** $\cos 4\theta = 8\cos^4\theta - 8\cos^2\theta + 1$

7 Show that $\cos 3\theta = 4\cos^3\theta - 3\cos\theta$.
Find the value of $\cos 3\theta$ if:
a) $\cos\theta = \frac{2}{\sqrt{5}}$ and θ is acute, **b)** $\sin^2\theta = 0.51$ and θ is obtuse.

8 Find values for a and b such that $\sin 3\theta = a\sin\theta + b\sin^3\theta$.
Deduce the value of $\sin 3\theta$ if:
a) $\sin\theta = \frac{1}{3}$ and θ is acute, **b)** $\tan\theta = \frac{9}{40}$ and θ is acute.

9 Solve these equations for $0 \le x \le 2\pi$.
a) $\sin 2x + \cos x = 0$ **b)** $\cos^2 x - \sin 2x = 0$
c) $5\sin 2x + 3\sin x = 0$

10 Solve these equations for $0 \le x \le 360°$.
a) $\cos x + \cos 2x = 0$ **b)** $\cos 2x + 4\sin x = 3$
c) $3\cos 2x - \sin x + 2 = 0$

11 The diagram shows the graphs of $y = \cos 2x$ and $y = 3\sin x - 1$, $-180° \leq x \leq 180°$. Find the coordinates of P and Q, the points of intersection.

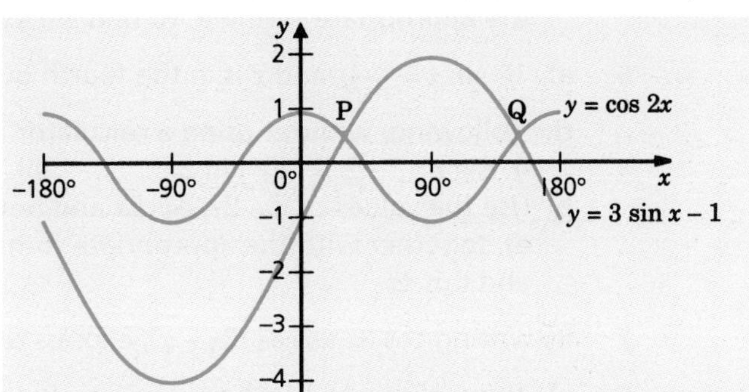

12 The diagram shows the graphs of $y = \sin 2x$ and $y = \tan x$, $0 \leq x \leq 2\pi$. Find the coordinates of the points of intersection.

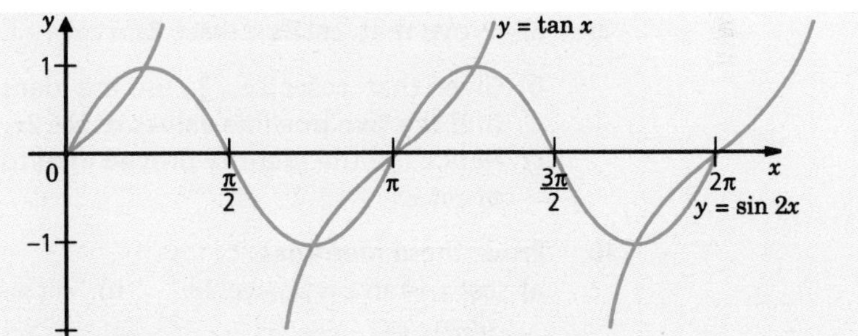

EXERCISES

26.2 B

1 On the same axes, draw the graphs of the functions $y = \cos 2x$ and $y = 2\cos^2 x$ for the domain $-2\pi \leq x \leq 2\pi$. What is the transformation that maps $y = \cos 2x$ onto $y = 2\cos^2 x$? Deduce the relationship between the two functions.

2 By calculating both sides separately, verify the identity $\cos 2x \equiv 2\cos^2 x - 1$ for the following angles.

a) 23° **b)** –511°
c) any angle of your choice in degrees
d) –13.759 radians **e)** $\frac{\pi}{18}$ radians
f) any angle of your choice in radians

3 Repeat question **2** to verify the identity $\sin 2x \equiv 2\sin x \cos x$.

4 Use the appropriate double-angle formula in each case to evaluate the following, without using a calculator.

a) $2\sin 30° \cos 30°$ **b)** $2\cos^2 22.5° - 1$ **c)** $1 - 2\sin^2 15°$

d) $\cos^2 45° - \sin^2 45°$ **e)** $\dfrac{2\tan\frac{\pi}{6}}{1-\tan^2\frac{\pi}{6}}$

5 **a)** If $\cos x = \frac{3}{4}$ and x is in the first quadrant, find the value of the following, without using a calculator.

i) $\sin x$ **ii)** $\sin 2x$ **iii)** $\cos 2x$ **iv)** $\tan 2x$

b) The double-angle formulae work for all angles. Use the values of $\sin 2x$, $\cos 2x$ and $\tan 2x$ that you found in part **a)**, together with the appropriate formula, to find $\sin 4x$, $\cos 4x$ and $\tan 4x$.

6 **a)** If $\sin x = -\frac{1}{\sqrt{5}}$ and x is in the fourth quadrant, find the values of the following, without using a calculator.

i) $\cos x$ **ii)** $\sin 2x$ **iii)** $\cos 2x$ **iv)** $\tan 2x$

b) Use the values of $\sin 2x$, $\cos 2x$ and $\tan 2x$ that you found in part **a)**, together with the appropriate formula, to find $\sin 4x$, $\cos 4x$ and $\tan 4x$.

7 By writing $\cos 3x$ as $\cos(2x + x)$, express $\cos 3x$ in terms of $\cos x$ only.

8 **a)** Prove that $\sin 4x \equiv 4 \sin x \cos x \cos 2x$.

b) Hence, without using a calculator, find the value of $\sin 4x$ when $\sin x = \frac{3}{5}$.

9 **a)** Prove that $\cot 2x + \operatorname{cosec} 2x \equiv \cot x$.

b) Given that $\operatorname{cosec} 2x = \frac{5}{4}$, use the identity $1 + \cot^2 2x \equiv \operatorname{cosec}^2 2x$ to find the two possible values of $\cot 2x$.

c) Hence use the identity proved in **a)** to find the two possible values of $\cot x$.

10 Prove these identities.

a) $\cot x + \tan x \equiv 2\operatorname{cosec} 2x$ **b)** $\cot x - \tan x \equiv 2 \cot 2x$

c) $\dfrac{\sin 2x + (1 - \cos 2x)}{\sin 2x + (1 + \cos 2x)} \equiv \tan x$ (The brackets give a hint where to begin.)

11 Solve the following equations for all values in the range $0 < x < 360°$.

a) $\sin x + \cos 2x = 0$ **b)** $2\cos 2x + 4\cos x - 1 = 0$

c) $4\cos 2x + 2\sin x = 3$

12 Simplify the following expression.

$$\sqrt{\frac{1 - \cos\theta}{1 + \cos\theta}}$$

COMBINING WAVES

Exploration 26.2

Adding waves

You will need a graphics calculator or equivalent computer program. Set the range of values as follows.

$x_{min} = 0$; $x_{max} = 360$; $x_{scl} = 3($

$y_{min} = -5$; $y_{max} = 5$; $y_{scl} = 1$

Make sure you – and your calculator – are working in degree mode.

- Draw the graphs of $y = 3\cos x$, $y = 4\sin x$ and $y = 3\cos x + 4\sin x$ on the same axes.
- What is the result of adding the two waves?
- What is the amplitude and the period of the resultant wave?
- How would you transform:

  a) $y = \cos x$, **b)** $y = \sin x$

  to produce the same graph as $y = 3\cos x + 4\sin x$?
- For what values of x does $3\cos x + 4\sin x = 0$?

Interpreting the results

The diagram shows the three graphs.

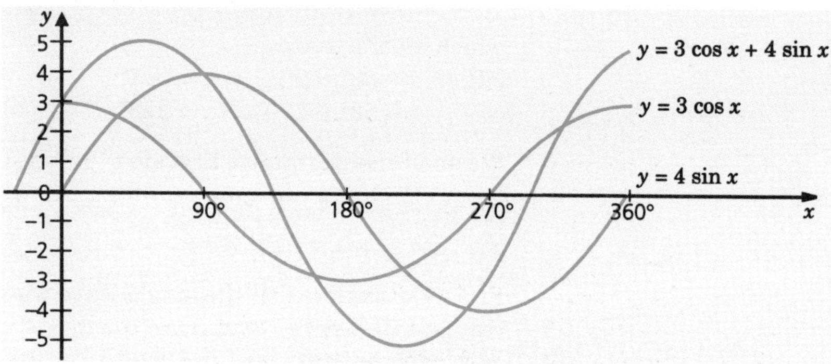

The graph of $y = 3\cos x + 4\sin x$ is a wave with period 360° and amplitude 5. The functions $y = 5\cos x$ and $y = 5\sin x$ both have this property, but need to be translated parallel to the x-axis to fit the graph if $y = 3\cos x + 4\sin x$.

Exploration 26.2

(continued)

Now clear the screen and draw the graph of $y = 3\cos x + 4\sin x$. Try to find a value for α such that $y = 5\sin(x + \alpha)$ gives the same curve.

When you have found a suitable value for α, repeat the exercise, but this time attempt to fit $y = 5\sin(x + \alpha)$ to the graph of $y = 3\cos x + 4\sin x$.

Which values of α fit the functions best?

Interpreting the results

In the first case $y = 5\cos(x - 53°)$ fits the graph of $y = 3\cos x + 4\sin x$ well. Alternatively, $y = 5\sin(x + 37°)$ also provides a good fit.

By zooming in on the points of intersection of $y = 3\cos x + 4\sin x$ and the x-axis, you should find that solutions of the equation $3\cos x + 4\sin x = 0$ are as shown in this diagram.

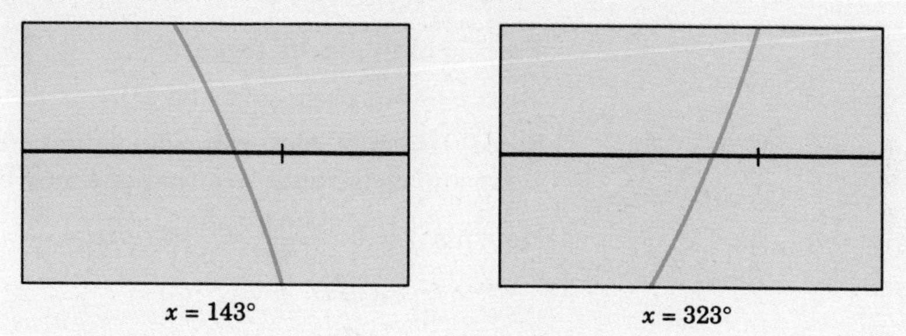

Use the trace function to find each solution correct to the nearest degree.

Knowing alternative expressions for $3\cos x + 4\sin x$ enables us to find the solutions directly, for example:
$3\cos x + 4\sin x = 0 \Rightarrow 5\cos(x - 53°) = 0$

which means that:
either $x - 53° = 90° \Rightarrow x = 143°$
or $x - 53° = 270° \Rightarrow x = 323°$

We shall now formalise this idea.

An alternative form for $a\cos x + b\sin x$ enables us to solve equations of the type $a\cos x + b\sin x = c$.

From our exploration the single wave would seem to have the form $r\cos(x - \alpha)$, where r and α are constants, or $r\sin(x - \alpha)$. We shall deal with each in turn.

Using the compound-angle formula for $\cos(x - \alpha)$:

$$r\cos(x-\alpha) \equiv r(\cos x\cos\alpha + \sin x\sin\alpha)$$
$$\equiv r\cos\alpha\cos x + r\sin\alpha\sin x$$

If this is equivalent to:
$a\cos x + b\sin x$

then:

$$a = r\cos\alpha \quad \text{and} \quad b = r\sin\alpha$$
$$\Rightarrow \quad a^2 = r^2\cos^2\alpha \quad \text{and} \quad b^2 = r^2\sin^2\alpha$$
$$\Rightarrow a^2 + b^2 = r^2\cos^2\alpha + r^2\sin^2\alpha$$
$$= r^2(\cos^2\alpha + \sin^2\alpha)$$
$$= r^2$$

$$r = \sqrt{a^2 + b^2}$$

In our example, $a = 3$, $b = 4 \Rightarrow r = \sqrt{3^2 + 4^2} = 5$

Now substitute for r to find α:

$$a = r\cos\alpha \Rightarrow \cos\alpha = \tfrac{a}{r} \quad \Rightarrow \alpha = \cos^{-1}\left(\tfrac{a}{r}\right)$$
$$b = r\sin\alpha \Rightarrow \sin\alpha = \tfrac{b}{r} \quad \Rightarrow \alpha = \sin^{-1}\left(\tfrac{b}{r}\right)$$

In our example:

$$\alpha = \cos^{-1}\left(\tfrac{3}{5}\right) = 53.1° \text{ or } \alpha = \sin^{-1}\left(\tfrac{4}{5}\right) = 53.1°$$

Putting both results together:

$$3\cos x + 4\sin x \equiv 5\cos(x - 53.1°)$$

Since α is acute, both principal values from the calculator will be the same and consistent, so either one may be used.

Note also that $\dfrac{r\sin\alpha}{r\cos\alpha} = \dfrac{b}{a} \Rightarrow \tan\alpha = \dfrac{b}{a}$

$$\Rightarrow \alpha = \tan^{-1}\left(\frac{b}{a}\right).$$

In our example, $\alpha = \tan^{-1}\left(\tfrac{4}{3}\right) = 53.1°$

For any expression of the type $a\cos x + b\sin x$ we could also use $r\sin(x+\alpha)$ as the equivalent single wave function.

$$r\sin(x+\alpha) \equiv r(\sin x\cos\alpha + \cos x\sin\alpha)$$
$$\equiv r\sin\alpha\cos x + r\cos\alpha\sin x$$
$$\Rightarrow a = r\sin\alpha \quad \text{and} \quad b = r\cos\alpha$$
$$\Rightarrow r = \sqrt{a^2+b^2} \text{ and } \alpha = \tan^{-1}\left(\tfrac{a}{b}\right)$$

For our example $r = 5$ and $\alpha = \tan^{-1}\left(\tfrac{3}{4}\right) = 36.9°$

$$\Rightarrow 3\cos x + 4\sin x \equiv 5\sin(x+36.9)°$$

If either a or b is negative, the same approach may be taken, but care must be taken over the choice of α, since the principal value might not be suitable.

Example 26.8

Express $5\cos x - 12\sin x$ in the form $r\sin(x+\alpha)$.

Solution

$$r\sin(x+\alpha) \equiv r\sin\alpha\cos x + r\cos\alpha\sin x$$
$$\Rightarrow 5 = r\sin\alpha \quad \text{and} \quad -12 = r\cos\alpha$$
$$\Rightarrow r = \sqrt{5^2+(-12)^2} \text{ and } \alpha = \tan^{-1}\left(\tfrac{-5}{12}\right)$$

Now $r = \sqrt{169} = 13$ and the principal value of α is –22.6°.
However:

$$5 = 13\sin\alpha \Rightarrow \alpha = \sin^{-1}\left(\tfrac{5}{13}\right) = 22.6°$$
$$-12 = 13\cos\alpha \Rightarrow \alpha = \cos^{-1}\left(-\tfrac{12}{13}\right) = 157.4°$$

The only value of α compatible with all three requirements is $\alpha = 157.4°$.

$$\Rightarrow 5\cos x - 12\sin x = 13\sin(x+157.4°)$$

Using the wave form $r\cos(x\pm\alpha)$ may prove easier, as in the following example.

Example 26.9

a) *Find a single wave function equivalent to $f(x) = \sqrt{3}\cos x - \sin x$.*

b) *Solve the equation $f(x) = 0, -180° \le x \le 180°$.*

c) *Determine maximum and minimum values of $f(x)$ and the values of x at which they occur for the domain $-180° \le x \le 180°$.*

Solution

a) *A suitable compound formula is $\cos(x+\alpha)$, since:*

$$r\cos(x+\alpha) \equiv r\cos\alpha\cos x - r\sin\alpha\sin x$$

Matching this to $\sqrt{3}\cos x - \sin x$, take $a = \sqrt{3}$ and $b = 1$.

$$\Rightarrow r\cos\alpha = \sqrt{3} \text{ and } r\sin\alpha = 1$$

From this we can deduce:

$r^2 = \sqrt{3}^2 + 1^2 = 4$

$\Rightarrow r = 2$

and:

$\alpha = \cos^{-1}\left(\frac{\sqrt{3}}{2}\right) = 30°$ *or* $\alpha = \sin^{-1}\left(\frac{1}{2}\right) = 30°$.

Therefore $f(x) = \sqrt{3}\cos x - \sin x \equiv 2\cos(x + 30)°$

b) $f(x) = 0 \Rightarrow 2\cos(x + 30)° = 0$

$\Rightarrow x + 30° = 90° \Rightarrow x = 60°$

or $x + 30° = -90° \Rightarrow x = -120°$

c) *The maximum value of* $f(x)$ *is 2 (i.e. when* $\cos x = 1$*), which occurs when* $x + 30° = 0 \Rightarrow x = -30°$.

The minimum value of $f(x)$ *is –2 (i.e. when* $\cos x = -1$*), which occurs when* $x + 30° = 180° \Rightarrow x = 150°$.

The single wave form is useful for solving equations of the form $a\cos x + b\sin x = c$. This is illustrated in the following example.

Example 26.10

Solve the equation $2\sin x - 3\cos x = 1, 0 < x < 2\pi$.

Solution

First rearrange into an appropriate form.

$3\cos x - 2\sin x = -1$

Now express the left-hand side in a suitable form.

$3\cos x - 2\sin x = r\cos(x + \alpha)$

$\Rightarrow r = \sqrt{3^2 + 2^2} = \sqrt{13}$

$\alpha = \tan^{-1}\left(\frac{2}{3}\right) = 0.588$

Hence $\sqrt{13}\cos(x + 0.588) = -1$

$\Rightarrow x + 0.588 = \cos^{-1}\left(-\frac{1}{\sqrt{13}}\right)$

$\Rightarrow \quad x + 0.588 = 1.852$

or $\quad x + 0.588 = 4.431$

$x = 1.264$ *or* $x = 3.843$

EXERCISES

26.3 A

1 On the same axes, plot carefully graphs of the functions $y = 3\cos x + 4\sin x$, $y = 5\sin x$ and $y = 5\cos x$, using sufficient values of x chosen from the domain $-360° \le x \le 360°$.

a) Describe the transformation that maps $y = 3\cos x + 4\sin x$ onto $y = 5\sin x$ and hence write $y = 3\cos x + 4\sin x$ in a form involving just one trigonometric function.

b) Describe the transformation that maps $y = 3\cos x + 4\sin x$ onto $y = 5\sin x$ and hence write $y = 3\cos x + 4\sin x$ in another form involving just one trigonometric function.

2 Express the following in the form given, where R_1 and R_2 are positive and both α and β are in degrees.

a) $5\cos x - 12\sin x$ as $R_1\cos(x-\alpha)$ and $R_2\sin(x+\beta)$

b) $7\cos x - 24\sin x$ as $R_1\cos(x+\alpha)$ and $R_2\sin(x-\beta)$
In each case, verify that $R_1 = R_2$.

3 Rewrite each of the following expressions in the form $R\cos(x-\alpha)$.

a) $\sin x + \cos x$ b) $4\cos x + 3\sin x$

4 Rewrite each of the following expressions in the form $R\cos(x+\alpha)$.

a) $4\sin x - 3\cos x$ b) $12\cos x - 5\sin x$

5 Rewrite each of the following expressions in the form $R\sin(x-\alpha)$.

a) $2\sin x - \cos x$ b) $\sqrt{3}\sin x - 3\cos x$

6 Rewrite each of the following expressions in the form $R\sin(x+\alpha)$.

a) $4\sin x + 5\cos x$ b) $\sqrt{2}\cos x + \sin x$

7 Express $8\sin x - 15\cos x$ in the form $R\sin(x-\alpha)$, where α is an acute angle and hence find the greatest and least values of the following.

a) $8\sin x - 15\cos x$ b) $\dfrac{1}{8\sin x - 15\cos x}$

8 Find the maximum and minimum values of the following functions in the domain $0 \le x \le 360°$. In each case, sketch the graph and find the values of x in the domain $0 \le x \le 360°$ at which the turning points occur.

a) $\sqrt{3}\cos x - \sin x$ b) $8\cos x - 15\sin x + 3$

c) $\dfrac{1}{\cos x + \sin x}$ d) $(5\cos x + 12\sin x)^2$

If you have a graphics calculator, use it to sketch the graphs in each case.

EXERCISES

1 On the same axes, plot carefully drawn graphs of the functions $y = 5\sin x + 12\cos x$, $y = 13\sin x$ and $y = 13\cos x$, using sufficient values of x chosen from the domain $-360° \le x \le 360°$.

a) Describe the transformation that maps $y = 5\sin x + 12\cos x$ onto $y = 13\sin x$ and hence write $y = 5\sin x + 12\cos x$ in a form involving just one trigonometric function.

b) Describe the transformation that maps $y = 5\sin x + 12\cos x$ onto $y = 13\cos x$ and hence write $y = 5\sin x + 12\cos x$ in another form involving just one trigonometric function.

2 Express the following in the form given, where R_1 and R_2 are positive and both α and β are in degrees.

a) $3\cos x - 4\sin x$ as $R_1\cos(x-\alpha)$ and $R_2\sin(x+\beta)$

b) $8\cos x - 15\sin x$ as $R_1\cos(x+\alpha)$ and $R_2\sin(x-\beta)$
In each case, verify that $R_1 = R_2$.

3 Rewrite each of the following expressions in the form $R\cos(x-\alpha)$.

a) $\sqrt{2}\cos x+\sin x$ b) $3\cos x+\sin x$

4 Rewrite each of the following expressions in the form $R\cos(x+\alpha)$.

a) $\cos x-\sqrt{3}\sin x$ b) $4\cos x-3\sin x$

5 Rewrite each of the following expressions in the form $R\sin(x-\alpha)$.

a) $3\sin x-4\cos x$ b) $\cos x-\sin x$

6 Rewrite each of the following expressions in the form $R\sin(x+\alpha)$.

a) $7\cos x+2\sin x$ b) $3\cos x+11\sin x$

7 Express $7\cos x + 24\sin x$ in the form $R\cos(x-\alpha)$, where α is an acute angle and hence find the greatest and least values of the following.

a) $7\cos x+24\sin x$ b) $\dfrac{1}{7\cos x+24\sin x}$

8 Find the maximum and minimum values of the following functions in the domain $0 \le x \le 360°$. In each case, sketch the graph and find the values of x in the domain $0 \le x \le 360°$ at which the turning points occur.

a) $2\sin x-3\cos x$ b) $3\cos x-7\sin x+2$

c) $\dfrac{1}{\sqrt{2}\cos x-\sin x}$ d) $(2\cos x+3\sin x)^2$

If you have a graphics calculator, use it to sketch the graphs in each case.

MATHEMATICAL MODELLING ACTIVITY

1 → 2 → 3 → 4 → 5 → 6

Specify the real problem

Problem statement

When two musical instruments are being played together it is important that they are in tune otherwise the sound is not very good. For example, if you pluck the strings on two guitars which are *almost* in tune (i.e. they almost play the same note), the sound appears to arrive in pulses getting louder then softer, a phenomenon called **beats**. This is, in fact, a useful method of tuning two instruments. By adjusting the note produced by one instrument to a standard note, then tuning the second instrument until the beats disappear, a constant pure sound is heard.

The task is to formulate a model to describe this phenomenon.

1 → 2 → 3 → 4 → 5 → 6

Set up a model

Set up a model

First you need to identify the simplifying assumptions and the important variables to describe the sound waves from the two musical instruments.

- Assume that the sound produced by each instrument is modelled by a sine function.
- Assume that the amplitude of each sound wave is the same.
- Assume that the frequency of each sound wave is slightly different.
- Let the amplitude of each sine function be a.
- Let the frequencies of the two sine functions be ω_1 and ω_2 where $\omega_1 - \omega_2$ is small.

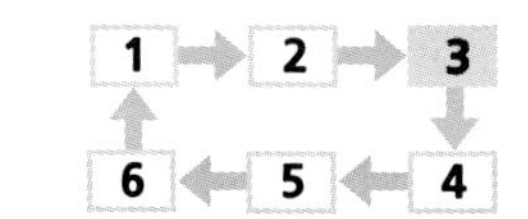

Formulate the mathematical problem

Mathematical problem

Firstly, establish a function to describe the sound wave from each musical instrument. Using the assumptions and variables, confirm that a simple model of each sound wave is $x_1 = a\sin\omega_1 t$ and $x_2 = a\sin\omega_2 t$.

The problem is to combine the sine functions to find the combined sound wave produced when two musical notes are produced simultaneously.

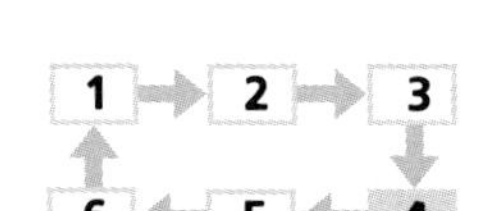

Solve the mathematical problem

Mathematical solution

The resultant sound wave has equation:

$$x = x_1 + x_2 = a\sin\omega_1 t + a\sin\omega_2 t$$

$$= 2a\sin[\tfrac{1}{2}(\omega_1 + \omega_2)t]\cos[\tfrac{1}{2}(\omega_1 - \omega_2)t].$$

Check that this is correct by expanding these formulae.

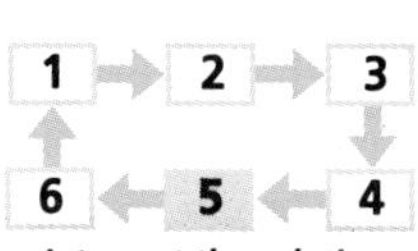

Interpret the solution

Interpret the solution

To interpret the mathematical solution, choose values for ω_1 and ω_2. Suppose that the standard note is chosen as middle C on the piano with frequency 256 hertz. Let $\omega_1 = 256$ and suppose that $\omega_2 = 276$. Then:

$$x = 2a\sin 266t\cos 10t$$

A graph of this resultant wave is shown in the diagram.

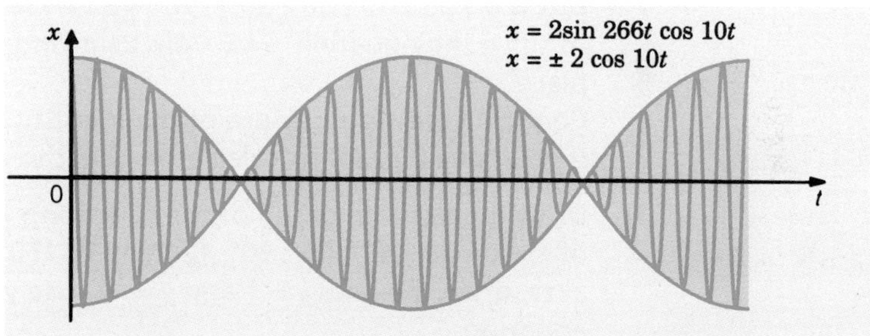

$$x = 2\sin 266t\cos 10t$$

$$x = \pm 2\cos 10t$$

The amplitude of the wave grows to double the original amplitude a and then decays to zero. This process is then repeated. The mathematical model describes the physical phenomenon of beats.

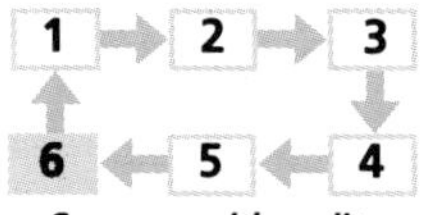

Compare with reality

Refinement of the model

Investigate the effects of:

- changes to ω_2 so that $|\omega_2 - \omega_1|$ increases in size,
- changes to the chosen standard note.

Is it easier to tune notes of high or low frequency?

CONSOLIDATION EXERCISES FOR CHAPTER 26

1 Expand the LHS of the equation $\sin(x+45°)=2\cos x$ using the appropriate addition theorem and hence:
a) show that $\tan x = 2\sqrt{2}-1$,
b) find all values of x from 0 to 360°.

2 Using the addition theorems, show that the equation $\sin\left(x+\frac{\pi}{4}\right)=\cos x+\cos\left(x+\frac{\pi}{4}\right)$ is equivalent to the equation $\tan x=\frac{\sqrt{2}}{2}$ and hence find the general solution of the original equation, giving your answer in radians.

3 Solve the equation $\cos(x+30°)=2\sin(x+60°)$ in the domain $0\le x\le 360°$, by first showing that $\tan x=-\frac{\sqrt{3}}{3}$.

4 Use a method similar to those above to solve the equation $2\cos x=\sin\left(x+\frac{\pi}{6}\right)$, giving the general solution, in radians.

5 Given that $\tan x\tan(x+45°)=3$, by expanding $\tan(x+45°)$ and letting $t=\tan x$, show that $t^2+4t-3=0$. Solve this equation to find the two possible values of t and hence find *all* the values of x that satisfy the original equation.

6 Draw an accurate graph of the function $y=\sin x-\sin 2x$ using values of x = 0°, 30°, 45°, 60°, 90°, 120°, 135°, 150° and 180°. Use your graphs to solve the equation $\sin x=\sin 2x$ giving answers to one decimal place.
Confirm your result by use of the identity $\sin 2x=2\sin x\cos x$.
If you have a graphics calculator, use it to confirm this result.

7 Draw an accurate graph of the function $y=\sin x-\cos 2x$ using values of x = 0°, 30°, 45°, 60°, 90°, 120°, 135°, 150° and 180°. On the same axes, draw the lines $y=1$ and $y=-1$. Use your graphs to solve these equations.
a) $\sin x=\cos 2x$ **b)** $\sin x=\cos 2x+1$ **c)** $\sin x=\cos 2x-1$
Confirm your results by solving the equations using, in each case, an appropriate identity for $\cos 2x$. If you have a graphics calculator, use it to confirm these results.

8 Find the values of x between 0 and 2π radians satisfying $\cos 2x=2\sin x$. Illustrate your answer by sketching the graphs of $\cos 2x$ and $2\sin x$ on the same axes. If you have a graphics calculator, use it to help you sketch the graphs.

9 Solve the following equations for $0\le x\le 360°$.
a) $\sin 2x+\cos x=0$
b) $\sin 2x+3\sin x=0$
c) $\cos 2x=1+2\sin x$
d) $\cos 2x+\sin x-1=0$
e) $3\cos 2x=-\sin^2 x-2\sin x$
f) $\cos 2x+2\sin^2 x=4\cos^2 x$

10 The double-angle formulae can be used for any angle and its double, for example, $\sin 4x \equiv 2\sin 2x \cos 2x$. Write down the three corresponding formulae for $\cos 4x$. Use an appropriate one of these, in each case, to solve the equations:
a) $\cos 4x - 3\cos 2x + 2 = 0$ **b)** $2\cos 4x + 4\sin 2x + 1 = 0$
giving all solutions between 0 and 180°.

11 Solve the equation $\sin 2x \equiv \tan x$ in the domain $0 \leq x \leq 2\pi$.

12 By using the identity for $\tan 2x$ and letting $\tan x = t$, solve the following equations for values of x between 0 and 360°.
a) $\tan 2x = \tan x$ **b)** $\tan 2x = 4\cot x$ **c)** $\tan x \tan 2x = 2$

13 Find a positive number R and an acute angle α such that $8\sin x - 15\cos x = R\sin(x - \alpha)$.
Hence find all solutions of the equation $8\sin x - 15\cos x = 17$ that lie between –360° and 360°.

14 By writing the LHS in the form indicated, where R is a positive number and α is an acute angle, solve the following equations for values of x in the domain $0 \leq x \leq 360°$.
a) $3\sin x + 4\cos x = 2$, $R\sin(x + \alpha)$
b) $2\cos x - \sin x = 0.5$, $R\cos(x + \alpha)$
c) $\cos x + \sin x = 1$, $R\cos(x - \alpha)$
d) $3\sin x - 4\cos x = 4$, $R\sin(x - \alpha)$

15 Find the general solutions of the following equations, in degrees.
a) $5\sin x + 12\cos x = 6.5$ **b)** $\sqrt{2}\sin x + \cos x = 1$
c) $\cos x + \sin x = \sqrt{2}$ **d)** $\cos x - 3\sin x = 1$

16 Find the general solutions of the following equations, in radians.
a) $6\cos x + 8\sin x = 9$ **b)** $4\cos x + \sin x = 3$
c) $5\cos x - 12\sin x = 6.5$ **d)** $\sec x = 1 - \sqrt{3}\tan x$

17 Find the general solution of the equation $7\cos x + 24\sin x = 12.5$ and hence find all solutions in the domain $0 \leq x \leq 360°$. Deduce the solutions of the equations:
a) $7\cos\frac{x}{2} + 24\sin\frac{x}{2} = 12.5$ **b)** $7\cos 2x + 24\sin 2x = 12.5$
lying in the same domain.

18 Rewrite $2\sin x + \cos x$ in the form $R\sin(x + \alpha)$ and hence solve the equation $2\sin x + \cos x = 1$ giving values of x between 0 and 360°. Deduce the solutions, in the same domain, of these equations.
a) $2\sin\frac{x}{2} + \cos\frac{x}{2} = 1$ **b)** $2\sin 3x + \cos 3x = 1$

19 Find the positive constant R and the acute angle A for which $\cos x + \sin x \equiv R\cos(x - A)$.

a) Find the general solution, in radians, of the equation $\cos x + \sin x = 1$.

b) Deduce the greatest value of $\cos x + \sin x$.

(ULEAC Question 6, Specimen Paper 3, 1994)

20 **a)** Find the value of the acute angle α for which $5\cos x - 3\sin x = \sqrt{34}\cos(x + \alpha)$ for all x.
Giving your answers correct to one decimal place.

b) Solve the equation $5\cos x - 3\sin x = 4$ for $0 \leq x \leq 360°$.

c) Solve the equation $5\cos 2x - 3\sin 2x = 4$ for $0 \leq x \leq 360°$.

(MEI Question 3, Specimen Paper 3, 1994)

21 Show that $(\cos x + \sin x)^2 = 1 + \sin 2x$.
Sketch the graph of $y = (\cos x + \sin x)^2$ for $0 \leq x \leq 2\pi$, indicating clearly the value of x for which $y = 0$.
Deduce the exact values of x in this interval for which $(\cos x + \sin x)^2 = 1.5$.

(NEAB Question 6, Specimen Paper 2, 1994)

22 **a)** Express the function $3\cos x + 4\sin x$ in the form $R\sin(x + \alpha)$, stating the values of R and α.

b) Write down the maximum value of $3\cos x + 4\sin x$.

c) Solve the equation $3\cos x + 4\sin x = 0$ for $0 \leq x \leq 360°$.

(SMP 16–19 Question 8, Specimen Paper, 1994)

23 Show that $\tan\theta + \cot\theta = \dfrac{2}{\sin 2\theta}$.
Hence, or otherwise, solve the equation $\tan\theta + \cot\theta = 4$, giving all the values of θ between 0 and 360°.

(UCLES Linear Question 10, Specimen Paper 1, 1994)

24 Prove that $\sin 3\theta = 3\sin\theta - 4\sin^3\theta$.
Hence find all values of θ, for $0 \leq \theta \leq 360°$, which satisfy the equation $\sin 3\theta = 2\sin\theta$.

(UCLES (Modular) Question 7, Specimen Paper P3, 1994)

25 The diagram shows a fixed semicircle, with centre O and radius r, and an inscribed rectangle ABCD. The vertices A and B of the rectangle lie on the circumference of the semicircle, and C and D lie on the diameter. The size of angle BOC is θ radians.

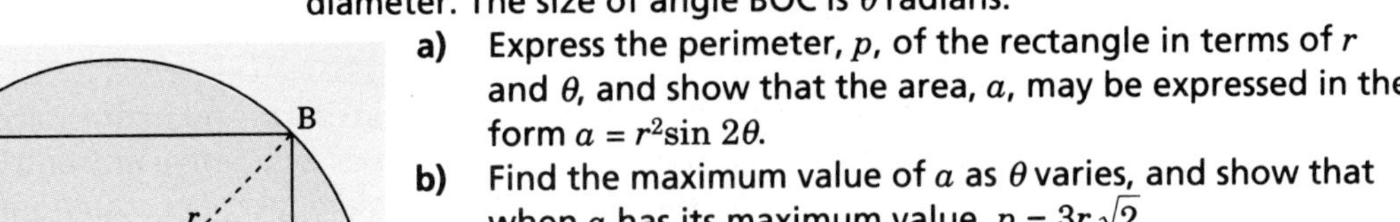

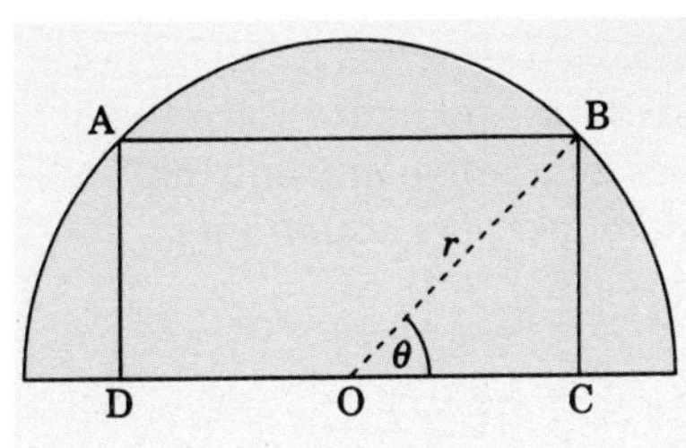

a) Express the perimeter, p, of the rectangle in terms of r and θ, and show that the area, a, may be expressed in the form $a = r^2\sin 2\theta$.

b) Find the maximum value of a as θ varies, and show that when a has its maximum value, $p = 3r\sqrt{2}$.

c) Find, in terms of r, the area of the rectangle which has maximum perimeter.

(UCLES (Modular) Question 8, Specimen Paper P3, 1994)

26 **a)** Write $f(x) = \sin x + 2\sin(x + 60°)$ in the form $f(x) = a\sin x + b\cos x$, where a and b are numbers.

b) Hence write $f(x)$ in the form $r\sin(x + c)$ where r is a positive number and c is an acute angle (in degrees).

c) For what values of x in the domain $0 \le x \le 90°$ does $f(x)$ take:

i) its greatest value, **ii)** its smallest value?

(Oxford and Cambridge Question 2, Specimen Paper 2, 1994)

27 In the diagram, angle QPT = angle SQR = θ, angle QPR = α, PQ = a, QR = b, PR = c, angle QSR = angle QTP = 90°, SR = TU.

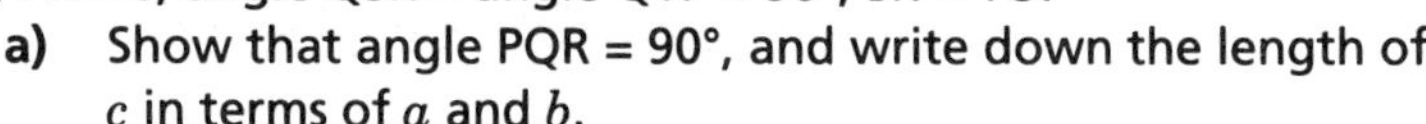

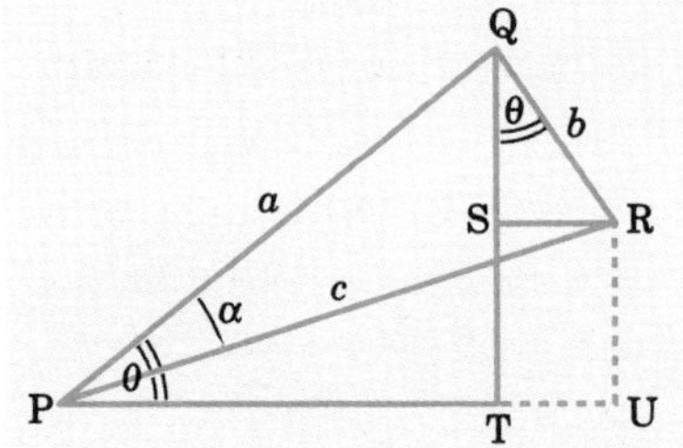

a) Show that angle PQR = 90°, and write down the length of c in terms of a and b.

b) Show that PU may be written as $a\cos\theta + b\sin\theta$ and as $c\cos(\theta - \alpha)$.

Write down the value of $\tan\alpha$ in terms of a and b.

c) In the case when $a = 4$, $b = 3$, find the acute angle α.

d) Solve the equation $4\cos\theta + 3\sin\theta = 2$ for $0 \le \theta \le 360°$.

(MEI Question 3, Paper 2, June 1992)

28 **a)** Use a trigonometrical formula to expand $\cos(x + \alpha)$.

b) Express $y = 2\cos x - 5\sin x$ in the form $r\cos(x + \alpha)$, giving the positive value of r and the smallest positive value of α.

c) State the maximum and minimum values of y and the corresponding values of x for $0 \le x \le 360°$.

d) Solve the equation $2\cos x - 5\sin x = 3$ for $0 \le x \le 360°$.

(MEI Question 2, Paper 2, June 1994)

29 $8\cos x - 15\sin x \equiv R\cos(x + A)$

a) Find the positive constant R.

b) Find the acute angle A, giving your answer in degrees to 1 decimal place.

c) Hence, or otherwise, find the maximum value of $8\cos x - 15\sin x$.

d) Find the value of x, in the range $0 \le x \le 360°$, which will give the maximum value, giving your answer in degrees to 1 decimal place.

(ULEAC Question 6, Paper 2, January 1994)

30 **a)** Show that $7\cos x - 4\sin x$ may be expressed in the form $R\cos(x + \alpha)$, where R is $\sqrt{65}$ and $\tan\alpha = \frac{4}{7}$.

b) Find, in radians to 2 decimal places, the smallest positive value of x for which $7\cos x - 4\sin x$ takes its maximum value.

c) Find, in radians to 2 decimal places, the two smallest positive values of x for which $7\cos x - 4\sin x = 4.88$.

The curve C has equation $y = (7\cos x - 4\sin x + 4)^{\frac{1}{2}}$.

d) Take corresponding values of y at $x = 0, \frac{\pi}{6}, \frac{\pi}{3}$ and $\frac{\pi}{2}$ for the curve C and use the trapezium rule to find an estimate for the area of the finite region bounded by the curve C, the y-axis and the x-axis for $0 \le x \le \frac{\pi}{2}$ giving your answer to 1 decimal place.

(ULEAC Question 9, Paper 2, June 1993)

31 **a)** Express $3\cos x + \sin x$ in the form $R\cos(x - \alpha)$, giving the value of R and the smallest positive value of α.

b) Use your answer to part **a)** to solve the equation $3\cos x + \sin x = 1$, for $0 \le x \le 360°$.

c) Solve the equation $(3\cos x)^2 = (1 - \sin x)^2$ by substituting $\cos^2 x$ in terms of $\sin x$ and solving the resulting quadratic equation in $\sin x$.

d) Explain why the answers to **b)** and **c)** are not the same.

(MEI Question 5, Paper 2, January 1995)

Summary

- **Compound-angle formulae**

$\sin(\theta + \varphi) \equiv \sin\theta\cos\varphi + \cos\theta\sin\varphi$ $\sin(\theta - \varphi) = \sin\theta\cos\varphi - \cos\theta\sin\varphi$

$\cos(\theta + \varphi) \equiv \cos\theta\cos\varphi - \sin\theta\sin\varphi$ $\cos(\theta - \varphi) = \cos\theta\cos\varphi + \sin\theta\sin\varphi$

$\tan(\theta + \varphi) \equiv \dfrac{\tan\theta + \tan\varphi}{1 - \tan\theta\tan\varphi}$ $\tan(\theta - \varphi) = \dfrac{\tan\theta - \tan\varphi}{1 + \tan\theta\tan\varphi}$

- **Double-angle formulae**

$\sin 2\theta \equiv 2\sin\theta\cos\theta$

$\cos 2\theta \equiv \cos^2\theta - \sin^2\theta \equiv 2\cos^2\theta - 1 \equiv 1 - 2\sin^2\theta$

$\tan 2\theta \equiv \dfrac{2\tan\theta}{1 - \tan^2\theta}$

- To solve equations of the form $a\cos x + b\sin x = c$ we first write the left-hand side as $R\cos(x - \alpha)$ or $R\sin(x + \alpha)$ and then find the appropriate values of R and α.

PURE MATHS

27

Curves 2

In this chapter we shall introduce:

- *an alternative method of describing the equation of a curve using a parameter,*
- *the method of parametric differentiation.*

PARAMETRIC EQUATIONS

In Chapter 25, *Curves 1*, the equations of the conic sections were given as direct relationships between x and y. Sometimes it is more convenient to express x and y in terms of a third quantity called a **parameter**. The pair of equations for x and y given in terms of the parameter are called the **parametric** equations of the curve.

Exploration 27.1

Time as the parameter

A tennis player serves a tennis ball from a height of 2 metres above the ground with speed 30 m s^{-1}. The horizontal and vertical displacements of the ball at time t seconds after serving are:

$x = 30t$

$y = 2 - 5t^2$

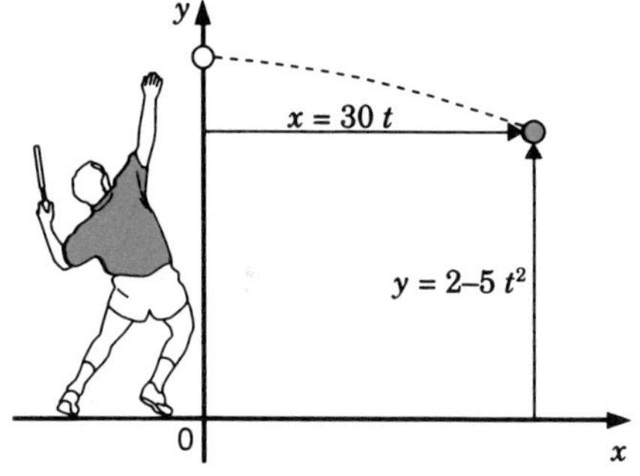

- Eliminate t to find the equation of the path of the tennis ball $y = f(x)$.
- Sketch the graph of $f(x)$.
- Where does the ball first bounce?
- When does the ball first bounce?

Interpreting the results

The equation of the path is $y = 2 - \frac{1}{180}x^2$ which is the equation of a parabola. The ball hits the ground when $y = 0$. Either reading off the x-value from the graph or solving $2 - \frac{1}{180}x^2 = 0$, we find that the ball hits the ground when $x = 19$ metres (correct to two significant figures).

This occurs after $\dfrac{19}{30} = 0.63$ seconds.

In this exploration, time is a parameter. The coordinates of the position of the tennis ball are functions of time. Eliminating time gives the path of the ball. This is a common method used in Mechanics. The equations $x = 30t$ and $y = 2 - 5t^2$ are the parametric equations of the path of the ball.

Exploration 27.2

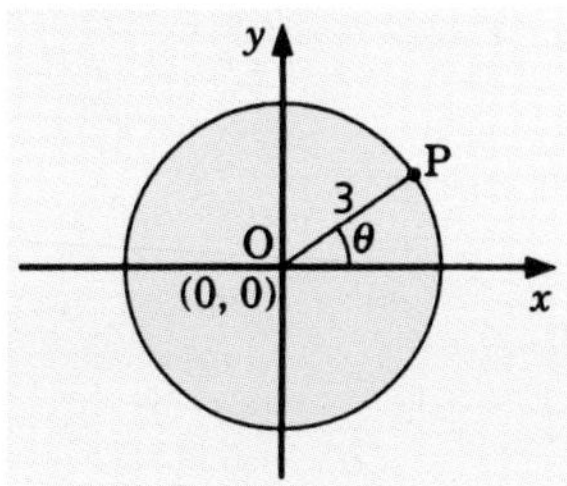

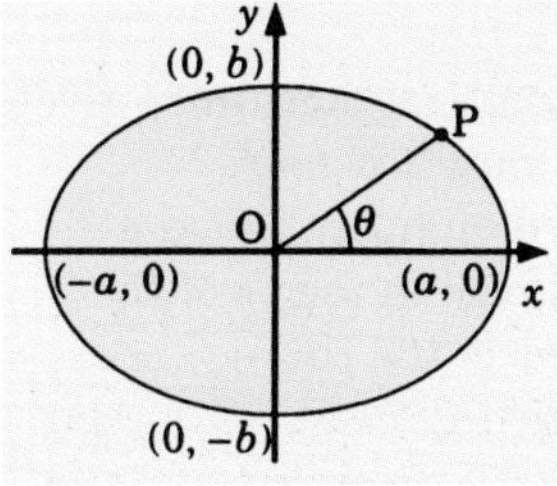

Circles and ellipses

The figure shows the circle centre (0, 0) and radius 3. The angle θ is the angle between OP and the x-axis.

- Write down the coordinates of the point P in terms of θ: $x = f(\theta)$, $y = g(\theta)$.
- Eliminate θ between f and g and show that the equation of the circle is $x^2 + y^2 = 9$.
- Deduce the parametric equations of a circle centre (0, 0) and radius r.
- Investigate the ellipse shown and propose parametric equations for an ellipse of major axis $2a$ and minor axis $2b$.

Interpreting the results

For a circle of radius r, centre (0, 0) the parametric equations are $x = r\cos\theta$ and $y = r\sin\theta$.

When we try to find the parametric equations of an ellipse, a simple geometrical approach does not give the answer as it did for a circle. Begin with the Cartesian equation:

$$\frac{x^2}{a^2} + \frac{y^2}{b^2} = 1$$

and compare it with the trigonometric identity $\cos^2\theta + \sin^2\theta \equiv 1$.

Now if we choose $\frac{x}{a} = \cos\theta$ and $\frac{y}{b} = \sin\theta$ then the parametric equations of the ellipse are $x = a\cos\theta$ and $y = b\sin\theta$.

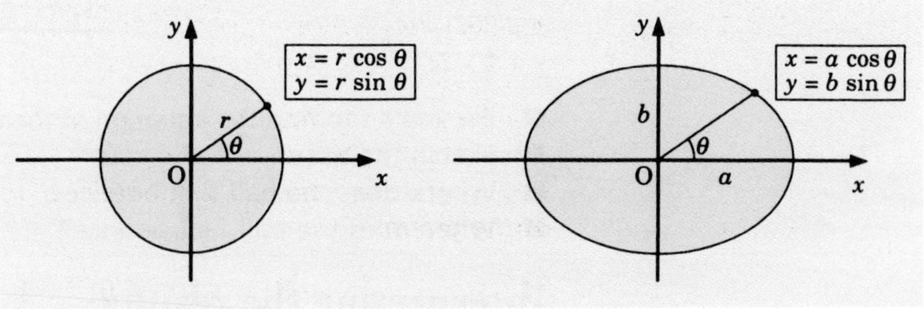

Example 27.1

Sketch the curve given by the parametric equations $x = 2t - 3$ and $y = t + 2$.

Solution

There are two methods of approach.

Using the values of the parameter

Draw up a table of values for t, x and y.

t	−2	0	2	4
x (= $2t - 3$)	−7	−3	1	5
y (= $t + 2$)	0	2	4	6

Plotting a graph of y against x shows that in this case the parametric equations give a straight line.

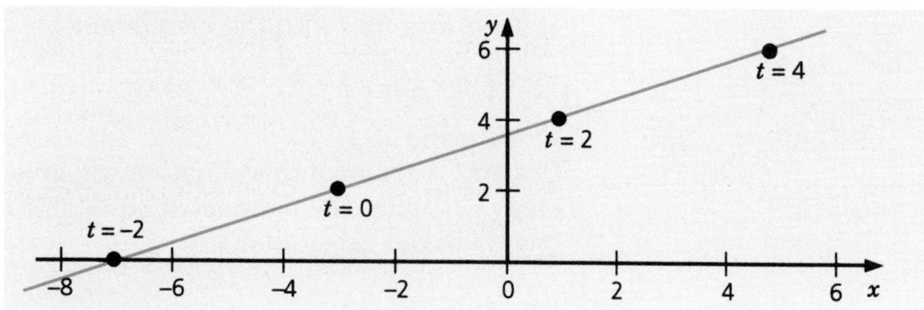

Eliminating the parameter t

The second method is to eliminate the parameter t to find the corresponding Cartesian equation. From $y = t + 2 \Rightarrow t = y - 2$. Substitute for t into the x-equation:

$x = 2(y - 2) - 3 = 2y - 7$

$\Rightarrow \; 2y - x = 7$

This is the equation of a straight line with gradient $\frac{1}{2}$ and intercept (0, 3.5), which is the same as we drew (above) using the first method.

Example 27.2

Sketch the curve given by the parametric equations $y = 2t - 1$ and $x = t^2 - 4$.

Solution

Using the values of the parameter

We can choose a small number of values for the parameters but the problem is always deciding which values to choose. A useful approach is to find where the curve cuts the axes.

$y = 2t - 1 \Rightarrow y = 0$ *when* $t = \frac{1}{2} \Rightarrow x = t^2 - 4 = -3.75$

$x = t^2 - 4 \Rightarrow x = 0$ *when* $t = \pm 2 \Rightarrow y = 3$ *or* $y = -5$

The curve cuts the axes at the points (–3.75, 0), (0, 3) and (0, –5). Choosing other values of t between –3 and + 3 will give a good idea of the graph.

t	–3	–1	0	1	3
$x(= t^2 - 4)$	5	–3	–4	–3	5
$y\ (= 2t - 1)$	–7	–3	–1	1	5

The graph of the function is shown in the figure. It is a parabola.

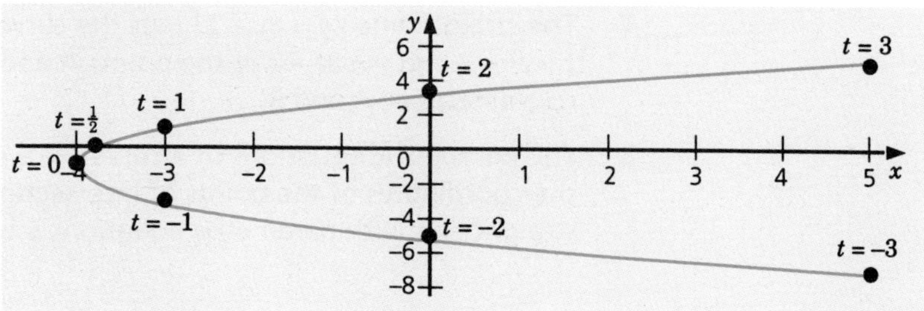

Eliminating the parameter t
From $y = 2t - 1 \Rightarrow t = \frac{1}{2}(y + 1)$.

Substituting for t into the x-equation:

$$x = \left(\tfrac{1}{2}(y+1)\right)^2 - 4 = \tfrac{1}{2}y^2 + y - 3.75$$

$\Rightarrow \ 2x = y^2 + 2y - 7.5$
Drawing a graph of this equation will give a parabola as before. In many examples the parametric equations are easier to deal with and involve easier calculations.

EXERCISES

27.1 A

1 Find the Cartesian equation corresponding to the following pairs of parametric equations and hence classify the curves using the definitions of Chapter 25.

a) $x = 2\cos\theta, y = 2\sin\theta$
b) $x = 4\sin\theta, y = 3\cos\theta$
c) $x = 2t, y = \dfrac{1}{t}$
d) $x = t - 1, y = t^2 + 1$
e) $x = 1 + 3\cos\theta, y = 2 + 3\sin\theta$

2 The parametric equations of a curve are $x = 2\sec t$ and $y = 3\tan t$.
a) Starting from $\sin^2 t + \cos^2 t = 1$ show that $\sec^2 t = 1 + \tan^2 t$.
b) Find the Cartesian equation of the curve by eliminating the parameter t using the result in **a)**.
c) Classify the resulting curve.

3 Find the Cartesian equation of the straight line with parametric equations $x = 2t - 4$ and $y = t - 1$.
Given that the line can also be represented by the parametric equations $x = \dfrac{a}{T-1}$ and $y = \dfrac{T}{T-1}$ where a is a constant, find:
a) the value of a,
b) the value of T at the point where $t = 4$.
(UCLES A/O Question 8 (part), Paper 2, November 1989)

4 Find the Cartesian equation and hence classify the curve with parametric equations $x = 5\sin^2 t$ and $y = \cos t$.

5 Sketch the curve given by the parametric equations $y = 3t^2 + 1$ and $x = 6t$.

6 Sketch the curve given by the parametric equations $x = \dfrac{1}{t}$ and $y = t - \dfrac{1}{t}$.

7 The straight line $2y = 3x - 11$ cuts the curve with parametric equations $x = t^2 + 3$ and $y = 3t - 1$ at the points A and B. Calculate the coordinates of A and B.

8 The parametric equations of a curve are $x = t^2 + 3$ and $y = 2t + 1$. Find the coordinates of the points of intersection of the curve and the straight line with parametric equations $x = T + 5$ and $y = 1 - 2T$.

EXERCISES

27.1 B

1 Find the Cartesian equation corresponding to the following pairs of parametric equations and hence classify the curves using the definitions of Chapter 25.

a) $x = 5\sin\theta, y = 5\cos\theta$ **b)** $x = 5t, y = \frac{2}{t}$

c) $x = 2t^2 - 1, y = 3t + 1$ **d)** $x = 2\cos\theta - 3, y = 5\sin\theta + 1$

e) $x = t + 1, y = t - 1$

2 The parametric equations of a curve are $x = e^t + e^{-t}$ and $y = e^t - e^{-t}$.

a) Eliminate t to find the Cartesian equation of the curve.

b) Classify the curve.

3 Find the Cartesian equation and hence classify the curve with parametric equations $x = 1 + \tan\theta$ and $y = 1 - \sec\theta$.

4 The parametric equations of a curve are $x = 2\cos\theta + \sin\theta$ and $y = \cos\theta - 2\sin\theta$.

a) Write down expressions for $2x + y$ and $x - 2y$ in terms of θ.

b) Hence determine the Cartesian equation of the curve.

c) Classify the curve.

5 Sketch the curve given by the parametric equations $x = t^2 + 2t$ and $y = t^2 - 2t$.

6 Sketch the curve given by the parametric equations $x = 8t^3$ and $y = t^2$.

7 The parametric equations of a curve are $x = t^2 + 2$ and $y = 2t - 1$. Find the coordinates of the points of intersection of the curve and the straight line with Cartesian equation $2x + y = 7$.

8 The parametric equations of a curve are $x = 2t - \frac{1}{t}$ and $y = 2t + \frac{1}{t}$.

a) Sketch the curve.

b) A and B are points on the curve at which the values of t are 1 and $\frac{1}{2}$. Find the equation of the straight line through A and B.

PARAMETRIC DIFFERENTIATION

Exploration 27.3

Differentiating parameters

A curve is given by the parametric equations $x = 30t$ and $y = 2 - 5t^2$.

- By eliminating t find the Cartesian equation of the curve.
- Find the slope of the tangent, $\frac{dy}{dx}$, at any point on the curve.
- Find $\frac{dy}{dt}$ and $\frac{dx}{dt}$ and show that $\frac{dy}{dx} = \frac{\frac{dy}{dt}}{\frac{dx}{dt}}$.

Repeat the tasks for the curves with the following parametric equations.

a) $x = t^2 + 4t, y = 3t - 1$ **b)** $x = 16t^3, y = 2t^2$

Interpreting the results

There are two methods of finding the equation of the tangent or normal to a curve given by parametric equations. We can either eliminate the parameter and differentiate in the normal way or work directly from the parametric equations. Exploration 27.3 should have demonstrated that using the parametric equations is more straightforward.

The chain rule provides the formula in the exploration. Let $y = \mathrm{u}(t)$, $x = \mathrm{v}(t)$ be the parametric equations of a curve and $y = \mathrm{f}(x)$ be the Cartesian equation obtained by eliminating t. Then applying the chain rule:

$$\frac{\mathrm{d}y}{\mathrm{d}t} = \frac{\mathrm{d}y}{\mathrm{d}x} \times \frac{\mathrm{d}x}{\mathrm{d}t} \Rightarrow \mathrm{u}'(t) = \mathrm{f}'(x) \times \mathrm{v}'(t)$$

The slope of the tangent to the curve is $\mathrm{f}'(x) = \dfrac{\mathrm{u}'(t)}{\mathrm{v}'(t)}$.

In terms of x, y and t $\dfrac{\mathrm{d}y}{\mathrm{d}x} = \dfrac{\frac{\mathrm{d}y}{\mathrm{d}t}}{\frac{\mathrm{d}x}{\mathrm{d}t}}$

This is called **parametric differentiation.**

Example 27.3

Find the equations of the tangent and the normal, at the point (2, 2), to the curve with parametric equations $x = t^2 + 1$ and $y = t^2 + t$.

Solution

At the point (2, 2), $t^2 + 1 = 2$ and $t^2 + t = 2 \Rightarrow t = 1$. The slope of the tangent at (2, 2) is the value of $\frac{\mathrm{d}y}{\mathrm{d}x}$ when $t = 1$.

$$\frac{\mathrm{d}y}{\mathrm{d}x} = \frac{\frac{\mathrm{d}y}{\mathrm{d}t}}{\frac{\mathrm{d}x}{\mathrm{d}t}} = \frac{2t+1}{2t}$$

When $t = 1$, $\frac{\mathrm{d}y}{\mathrm{d}x} = \frac{3}{2}$.

The equation of the tangent is $\dfrac{y-2}{x-2} = \dfrac{3}{2} \Rightarrow 2y = 3x - 2$.

The slope of the normal at (2, 2) is $-\frac{2}{3}$.

The equation of the normal is $\dfrac{y-2}{x-2} = -\dfrac{2}{3} \Rightarrow 3y = 10 - 2x$.

Example 27.4

A curve is defined parametrically by the equations $x = t^2 + 2t$ and $y = t^2 - 2t$. Find:

a) *the gradient of the chord joining the points A and B for which $t = 1$ and $t = 4$, respectively,*

b) $\frac{\mathrm{d}y}{\mathrm{d}x}$ *in terms of t,*

c) *the value of t at the point on the curve at which the tangent is parallel to AB. Find the equation of the tangent at this point.*

(Question 8 (part), UCLES (A/O) Paper 2, June 1989)

Solution

a) $t = 1 \Rightarrow x = 3$ *and* $y = -1$

$t = 4 \Rightarrow x = 24$ *and* $y = 8$

Points A and B are (3, –1) and (24, 8) respectively.

The chord joining the points A and B has gradient $\frac{8-(-1)}{24-3} = \frac{9}{21} = \frac{3}{7}$.

b) $\frac{dy}{dx} = \frac{\frac{dy}{dt}}{\frac{dx}{dt}} = \frac{2t-2}{2t+2}$

c) *The tangent to the curve is parallel to the line AB when*

$\frac{2t-2}{2t+2} = \frac{3}{7} \Rightarrow 14t - 14 = 6t + 6 \Rightarrow t = \frac{5}{2}$

At $t = \frac{5}{2}$ *we have* $x = \frac{45}{4}$ *and* $y = \frac{5}{4}$

The equation of the tangent at this point is:

$$\frac{y - \frac{5}{4}}{x - \frac{45}{4}} = \frac{3}{7} \Rightarrow 7y = 3x - 25$$

EXERCISES

27.2A

In questions **1–6** find the equations of the tangent and normal to the curve at the point with the given value of t.

1 $x = t^2 - 1$, $y = t^2 + t$ at $t = 1$

2 $x = 3\cos t + 1$, $y = 3\sin t - 1$ at $t = \frac{\pi}{3}$

3 $x = \frac{4}{t}$, $y = 3t - 1$ at $t = 2$

4 $x = e^t - e^{-t}$, $y = e^t + e^{-t}$ at $t = 0$

5 $x = 1 + \tan t$, $y = 1 - \sec t$ at $t = \frac{\pi}{4}$

6 $x = \frac{t}{t+1}$, $y = \frac{t-1}{t}$ at $t = 2$

7 Find and classify the stationary point of the curve given by the parametric equations $x = 2t^3$, $y = (t + 1)^2$. Sketch the curve.

8 Find and classify the stationary points of the curve given by the parametric equations $x = t$, $y = t^3 - t$. Sketch the curve.

EXERCISES

27.2B

In questions **1–6** find the equations of the tangent and normal to the curve at the point with the given value of t.

1 $x = t^2$, $y = 4t$ at $t = -1$

2 $x = 5\sin t - 1$, $y = 5\cos t + 1$ at $t = \frac{\pi}{6}$

3 $x = 3\cos t - 2$, $y = 2\sin t + 3$ at $t = \frac{4\pi}{3}$

4 $x = \frac{1+t}{1-t}$, $y = \frac{1+2t}{1-2t}$ at $t = 0$

5 $x = e^t$, $y = \sin t$ at $t = 0$

6 $x = t - \cos t$, $y = \sin t$ at $t = \frac{\pi}{4}$

7 A curve is given by the parametric equations $x = \sin t$ and $y = \cos 2t$.
Find $\frac{dy}{dx}$ in terms of x and show that $\frac{d^2y}{dx^2} = -4$.

8 The parametric equations of a curve are $x = 4t^2$ and $y = 2t + \frac{1}{t}$.

a) Find $\frac{dy}{dx}$ when $t = \frac{1}{2}$ and hence find the equations of the tangent and normal at this point.
b) Find if either the tangent or normal intersects the curve again.

CONSOLIDATION EXERCISES FOR CHAPTER 27

1 A particle orbits the origin so that its position at time t is given by the equations $x = 2\cos t$ and $y = 2\sin t$, where t is in radians.
a) Calculate the particle's position and speed when $t = \frac{1}{3}\pi$.
b) Calculate the particle's direction of travel when $t = \frac{1}{3}\pi$.
c) Find the Cartesian equation of the orbit.
d) Confirm your answer to part **b)** by using your Cartesian equation.
(Oxford (Nuffield) Question 5, Specimen Paper 3, 1994)

2 The equation of a curve is given in terms of the parameter t by the equations $x = 2t$ and $y = \frac{2}{t}$, where t takes positive and negative values.
a) Sketch the curve.
At the point on the curve with parameter t,
b) show that the gradient of the curve is $-t^{-2}$,
c) find and simplify the equation of the tangent.
P and Q are the points where a tangent on this curve crosses the x- and y-axes, and O is the origin.
d) Show that the area of the triangle OPQ is independent of t.
(MEI Question 2, Specimen Paper 3, 1994)

3 The parametric equations of a curve are $x = 4t$, $y = \frac{4}{t}$, where the parameter t takes all non-zero values. The points A and B on the curve have parameters t_1 and t_2 respectively.
a) Write down the coordinates of the mid-point of the chord AB.
b) Given that the gradient of AB is –2, show that $t_1t_2 = \frac{1}{2}$.
c) Show that the coordinates of the mid-point of any chord with gradient –2 may be expressed in the form: $x = 2T$, $y = 4T$, where T is a parameter, and hence state the equation of the line on which all such mid-points lie.
d) Find the coordinates of the points on the curve at which the gradient of the tangent is –2.
(UCLES (Modular) Question 9, Specimen Paper 3, 1994)

4 The equation of a curve is given in terms of the parameter t by the equations $x = 4t$ and $y = 2t^2$ where t takes positive and negative values.

a) Sketch the curve.

P is the point on the curve with parameter t.

b) Show that the gradient at P is t.

c) Find and simplify the equation of the tangent at P.

The tangents at two points Q (with parameter t_1) and R (with parameter t_2) meet at S.

d) Find the coordinates of S.

e) In the case when $t_1 + t_2 = 2$ show that S lies on a straight line. Give the equation of the line.

(MEI Question 4, Paper 2, June 1993)

5 The diagram shows a sketch of the curve given parametrically in terms of t by the equations $x = 1 - t^2$, $y = 2t + 1$.

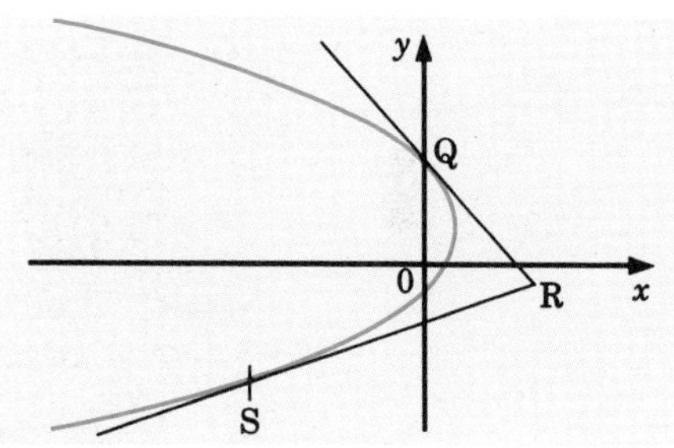

a) Show that the point Q (0, 3) lies on the curve, stating the value of t corresponding to this point.

b) Show that, at the point with parameter t, $\frac{dy}{dx} = -\frac{1}{t}$.

c) Find the equation of the tangent at Q.

d) Verify that the tangent at Q passes through the point R (4, –1).

e) The other tangent from R to the curve touches the curve at the point S and has equation $3y - x + 7 = 0$. Find the coordinates of S.

(MEI Question 4, Paper 2, January 1995,)

6 Show that an equation of the normal to the hyperbola with parametric equations $x = ct$, $y = \frac{c}{t}$, $t \neq 0$, at the point $\left(cp, \frac{c}{p}\right)$ is $y - \frac{c}{p} = xp^2 - cp^3$.

(ULEAC Question 3, Paper 2, January 1994)

7 The curve C shown in the diagram is given by $x = t + \frac{1}{t}$, $y = t - \frac{1}{t}$, $t > 0$, where t is a parameter.

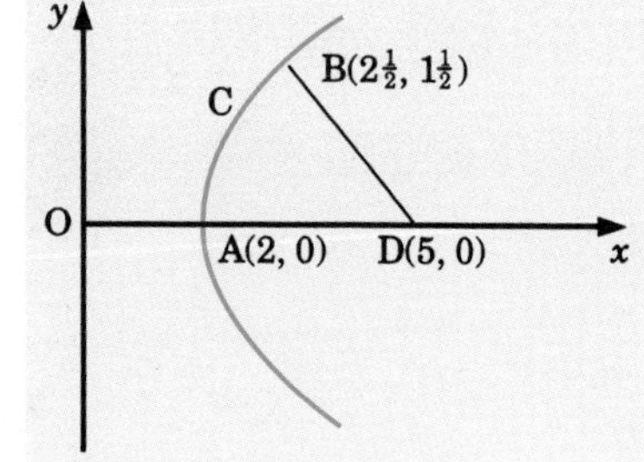

a) Find $\frac{dy}{dx}$ in terms of t and deduce that the tangent to C at the point A(2, 0) has equation $x - 2 = 0$.

b) For every point (x, y) on C, show that $x^2 - y^2 = 4$.

The point B has coordinates $(2\frac{1}{2}, 1\frac{1}{2})$ and the point D has coordinates (5, 0). The region R is bounded by the lines AD and BD and arc AB of the curve C. The region R is rotated through 2π radians about the x-axis to form a solid of revolution.

c) Find the volume of the solid, leaving your answer in terms of π.

(ULEAC Question 9, Paper 2, May 1994)

8 The curve shown in the diagram has parametric equations $x = t^2$, $y = t^3$, where $t \geq 0$ is a parameter. Also shown is part of the normal to the curve at the point where $t = 1$.

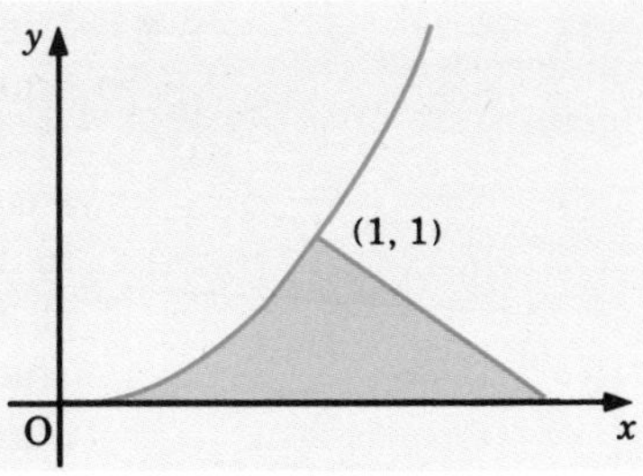

a) Find an equation of this normal.

b) Find the area of the finite region bounded by the curve, the x-axis and this normal.

(ULEAC Question 8, Paper 2, January 1995,

9 A curve is given parametrically by the equations $x = c(1 + \cos t)$, $y = 2c\sin^2 t$, $0 \leq t \leq \pi$, where c is a positive constant.

a) Show that $\dfrac{dy}{dx} = -4\cos t$.

At point P on the curve, $\cos t = \frac{3}{4}$.

b) Show that the normal to the curve at P has equation $24y - 8x - 7c = 0$.

c) Sketch the curve for $0 \leq t \leq \pi$.

(ULEAC Question 10 (part), Paper 2, January 1992)

10 Show that the equation of the normal to the parabola $y^2 = 4x$ at the point $P(p^2, 2p)$ is $y + px = 2p + p^3$.
Given that $p \neq 0$ and that the normal at P cuts the x-axis at the point $B(b, 0)$ obtain an expression for p^2 in terms of b. Deduce that $b > 2$.
The normal at P meets the normal at $Q(q^2, 2q)$ at the point R. Show that the coordinates of R are $(2 + p^2 + pq + q^2, -pq(p + q))$.
Given that $pq = 2$, show that these coordinates can be written in the form $(r^2, 2r)$ where r is a function of p. What does this result tell you about the point R?
The normal at R meets the parabola again at the point $S(s^2, 2s)$. Find s in terms of r.

(WJEC Question 15, Specimen Paper A1, 1994)

Summary

- The parametric equations of a curve are functions $x = u(t)$ and $y = v(t)$ in which the Cartesian coordinates of a point on the curve are given in terms of a third variable t called a parameter.
- The Cartesian equation of the curve is found by eliminating t to give $y = f(x)$.
- The slope of the tangent to the curve is given by parametric differentiation.
- $$\frac{dy}{dx} = \frac{\frac{dy}{dt}}{\frac{dx}{dt}} = \frac{v'(t)}{u'(t)}$$

PURE MATHS

28 Differential equations

In this chapter we shall:

- *introduce the idea of differential equations and give examples of how they are used in problem-solving,*
- *explore the analytical methods of solving simple first-order differential equations,*
- *introduce Euler's numerical method for solving first-order differential equations.*

FORMING DIFFERENTIAL EQUATIONS

Throughout this book it has been clear that one of the reasons for studying mathematics is that it provides a precise language for modelling many of the physical laws and processes in the real world. Many real-life situations involve growth and/or decay. Therefore the mathematical models that we formulate will involve **derivatives** because differentiation describes rates of change. Equations involving derivatives are called **differential equations**. This chapter is about forming and solving equations involving first-order derivatives, and these are called **first-order differential equations**.

Exploration 28.1

Newton's law of cooling

Suppose that you are in a hurry to go out but you want to drink a cup of hot coffee before you go. The initial temperature of the coffee is 90°C and you can start to drink the coffee when its temperature is 45°C. The temperature of the room is 20°C. Formulate a model to find out how long you will have to wait.

What assumptions and simplifications have you made?

Interpreting the results

To formulate a model we need to know something about how a liquid cools. Experimental evidence shows that the rate at which the temperature changes is proportional to the difference in temperature between the liquid and the surrounding air. This is called **Newton's law of cooling**. If T is the temperature of the liquid at time t then in this case:

$$\frac{dT}{dt} = -k(T - 20)$$

where k is the constant of proportionality and the negative sign shows that the temperature is reducing. When the coffee is made its temperature is 90°C. So $T = 90°C$ when $t = 0$.

In formulating this model we have assumed that,

- the temperature throughout the coffee is uniform,
- the temperature of the surrounding air is constant,
- the rate of cooling of a body is proportional to the temperature of the body above that of the surrounding air.

Equation $\frac{dT}{dt} = -k(T - 20)$ is an example of a **first-order differential equation**. The solution for $T(t)$ will give the temperature distribution as a function of time. The statement $T = 90°C$ when $t = 0$ is called **the initial condition.**

Example 28.1

The graph of a function passes through the point (1, 3). The gradient function of the graph is $(x - 1)y^2$. Formulate a problem for finding the equation of the graph.

Solution

The gradient function of a graph is $\frac{dy}{dx}$. So we can write:

$$\frac{dy}{dx} = (x - 1)y^2$$

which is a first-order differential equation. Since the graph passes through the point (1, 3) the initial condition for this problem is $y = 3$ when $x = 1$.

Example 28.2

In the atmosphere the pressure decreases with height at a rate that is proportional to the pressure. At a height of 1 km the pressure is 6.7×10^4 Pa and the rate of change of pressure is –9000 Pa km^{-1}. At ground level the atmospheric pressure is 10^5 Pa.
Formulate a problem consisting of a differential equation and initial condition to model the pressure in the atmosphere. (The Pascal (Pa) is the unit of pressure.)

Solution

Let the pressure in the atmosphere be denoted by P. The quantity $\frac{dP}{dh}$ gives the rate at which the pressure is changing with height h.
As the rate of change is proportional to pressure then:

$$\frac{dP}{dh} \propto P$$

$$\Rightarrow \frac{dP}{dh} = -kP$$

where k is a positive constant of proportionality and the negative sign shows that P is decreasing with height. When the pressure is 6.7×10^4 Pa the pressure change is 9000 Pa km^{-1}, then $k = \frac{9000}{7\times10^4} = 0.13$ (correct to two significant figures).

The differential equation for P is:

$$\frac{dP}{dh} = -0.13P$$

and the initial condition is $P = 10^5$ Pa *when* $h = 0$.

Mixing liquids

Example 28.3

A cylindrical tank A is filled with 800 litres of water containing a concentration of dye. A second tank B is partially filled with 200 litres of pure water. The liquid in tank A feeds into tank B at the rate of 20 litres per minute and is thoroughly mixed with the water in tank B. The mixed solution is drawn off tank B at a rate of 15 litres per minute.

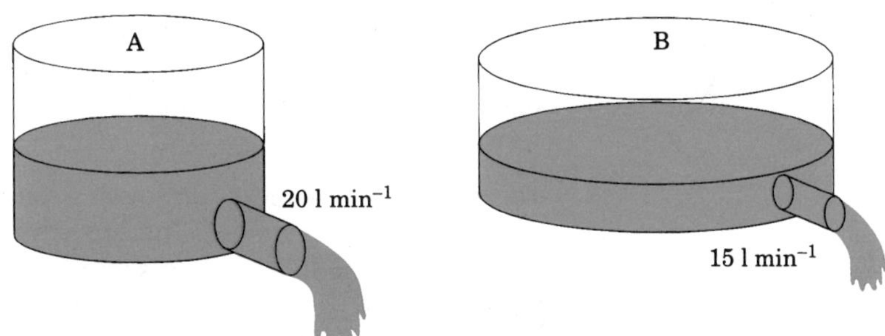

If the concentration of dye in tank A is 1 kg per litre, find an expression for the amount of dye in tank B at any time.

Solution

Let $q(t)$ *be the total amount (i.e. mass) of dye in tank B at time* t. *The rate at which the dye is accumulating in tank B is given by:*

$$\frac{dq}{dt} = \text{amount of dye entering per minute} - \text{amount of dye leaving per minute}$$

The amount of liquid entering tank B from tank A is 20 litres per minute. Since the concentration in tank A is 1 kg per litre:

$$\text{amount of dye entering per minute} = \left(1\,\text{kg}\,\text{l}^{-1}\right)\left(20\,\text{l}\,\text{min}^{-1}\right) = 20\,\text{kg}\,\text{min}^{-1}$$

The amount of liquid leaving tank B is 15 litres per minute so at any time the amount of liquid in tank B is increasing at the rate of $(20 - 15) = 5$ *litres per minute.*

The volume of liquid in tank B after t minutes is $(200 + 2t)$ *litres.*

The concentration of dye in tank B after t *minutes is* $\dfrac{q}{(200+2t)}\,\text{kg}\,\text{l}^{-1}$.

$$\text{amount of dye leaving per minute} = \left(\frac{q}{200+2t}\,\text{kg}\,\text{l}^{-1}\right)\left(15\,\text{l}\,\text{min}^{-1}\right) = \frac{15q}{200+2t}\,\text{kg}\,\text{min}^{-1}$$

$$\Rightarrow \frac{dq}{dt} = 20 - \frac{15q}{(200+2t)}$$

Initially tank B contains pure water so the initial condition is $q = 0$ *when* $t = 0$.
The differential equation for $\frac{\mathrm{d}q}{\mathrm{d}t}$ *with this initial condition can be solved to give the desired equation for the amount of dye in tank B.*

EXERCISES

28.1A

In the following problems *do not attempt to solve* the differential equations.

1 The slope of the tangent to a curve at the point (x, y) is given by $(x + y)$. The curve passes through the point (0, 1). Set up a problem consisting of a differential equation and initial condition to find the equation of the curve.

2 An object moves so that its velocity is proportional to $\sqrt{1-x^2}$ where x is its position at time t. The object starts at the origin. Form a differential equation and initial condition to find the position as a function of time.

3 The rate of growth of a sunflower after germination is initially proportional to its height. The growth rate is 2.5 cm per day when its height is 10 cm. In modelling the growth it is assumed that the initial height is 2 cm.

Formulate a problem consisting of the differential equation and initial condition to find the height of the sunflower at any time.

4 The rate at which the population of a colony of birds is increasing is proportional to its population. Currently the population is 100 000 and is increasing at a rate of 2000 per year. Form a differential equation for the population of the colony.

5 A cup of hot chocolate cools at a rate proportional to the temperature of the chocolate above that of the surrounding air. The chocolate is made from boiling milk and has an initial temperature of 90°C. It cools at the rate of 0.4°C per second in a room where the temperature is 20°C.

Formulate a problem consisting of a differential equation and initial condition to model this situation.

6 The area, A, of a circular oil slick is increasing at a rate proportional to its radius, r.

a) Write down an expression for $\frac{\mathrm{d}A}{\mathrm{d}t}$.

b) Find an expression for $\frac{\mathrm{d}A}{\mathrm{d}r}$ in terms of r.

c) Use the chain rule to find a differential equation involving $\frac{\mathrm{d}r}{\mathrm{d}t}$ and r.

7 Air is escaping from a spherical balloon at a rate that is proportional to its surface area. If the air is escaping at 2.5 $cm^3 s^{-1}$ when its radius is 6 cm, formulate a differential equation for the rate of change of the radius.

8 A tank made of porous material has a square base of side 2 m and vertical sides. The tank contains water which seeps out through the base and sides at a rate proportional to the total area in contact with the water. When the depth is 3 m it is observed that the level of the water is falling at a rate of 0.2 $m h^{-1}$. Formulate a differential equation model to describe the depth of water in the tank as a function of time.

9 When drugs are injected into the bloodstream they lose effectiveness as they pass out of the system. A simple model assumes that the rate of decay in concentration of the drug is proportional to the present concentration. Set up a differential equation model for the drug concentration for a drug at the rate 0.02 $g cm^{-3} s^{-1}$ when the concentration is 0.1 $g cm^{-3}$.

10 A water storage tank has a rectangular cross-section. Its base is 1 m by 1.25 m and its depth is 1 m. Water flows into the tank at a constant rate of 0.005 $m^3 s^{-1}$. Water flows out of the tank at a rate that is proportional to the depth of water in the tank; the constant of proportionality is 0.001 $m^2 s^{-1}$. The tank is initially half full of water. Formulate a differential equation and initial condition to describe how the depth of water changes with time. Deduce from the differential equation whether the tank fills or empties.

EXERCISES

28.1 B

In the following problems *do not attempt to solve* the differential equations.

1 As a radioactive substance decays it loses its mass at a rate proportional to its mass at the present time. Write down a differential equation to model this statement.

2 The population of a colony of rabbits in a park increases at a rate proportional to the population. Initially there are ten rabbits in the park. When the population is 100 rabbits the colony is increasing at a rate of seven rabbits per month. Set up a problem consisting of a differential equation and initial condition to model this situation.

3 A curve has equation $y = f(x)$. The function $f(x)$ is not known but the slope of the tangent to the curve at any point (x, y) is given by $x^2 - y^2$. We know that the curve goes through the point (1, 1). Set up a problem consisting of a differential equation and initial condition to find the function $f(x)$.

4 A cold can of cola is removed from a refrigerator where it had a temperature of 2°C and placed in a warm room which is at a temperature of 20°C. The cola warms so that the rate of increase in temperature is proportional to the difference between its

temperature and the temperature of the surroundings. The rate of increase is 3°C per minute when the temperature is 10°C.
Formulate a problem consisting of a differential equation and initial condition to model this situation.

5 Air is escaping from a spherical balloon at a rate that is proportional to the volume of air left in the balloon. When the radius of the balloon is 15 cm the air is leaving at a rate of 10 cm^3 s^{-1}. The initial radius of the balloon is 20 cm.
   a) Form a differential equation for the volume, V.
   b) Set up a problem consisting of a differential equation and initial condition involving the radius, r, of the balloon.

6 The volume of a balloon is increasing at a constant rate of 20 cm^3 s^{-1} as air is pumped in. Form a differential equation for the radius of the balloon.

7 Living tissue contains 10^{-22} g of carbon-14 in every gram of carbon. A sample of dead wood is estimated to contain 3.5×10^{-23} g of carbon-14 in every gram of carbon. It is assumed that the proportion of carbon-14 in carbon in living tissue was the same when the wood died as it is in living tissue today. The rate of decay of carbon-14 is assumed to be proportional to the amount of carbon-14 present.
Formulate a differential equation and initial condition to model this situation.

8 In a electrical circuit, the current is reducing at a rate which is proportional to the current flowing at that time. When the current is 40 milliamps, it is falling at a rate of 0.5 milliamps per second. Set up a differential equation which models this situation.

9 Water is pumped into a cone shaped tank at a constant rate of 30 cm^3 s^{-1}. The cone has dimensions shown in the diagram.
The water leaks from a hole in the base of the cone at a rate proportional to the depth of water in the tank. The constant of proportionality is 0.5 cm^2 s^{-1}. Initially the cone has depth of 0.5 m.
Formulate a differential equation and initial condition to describe how the depth of water changes with time. Deduce from the differential equation whether the cone initially fills or empties. Is there a constant depth for the water?

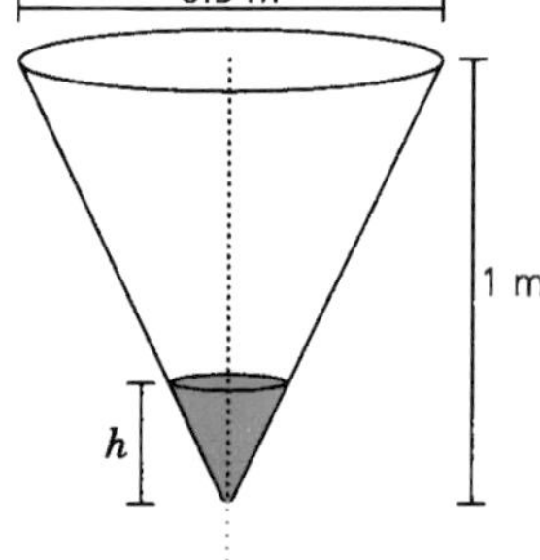

10 An individual in a population of 1500 people working in a company becomes infected with a virus. It is assumed that the rate at which the virus spreads throughout the company is proportional to the number of people infected, P, and to the number of people not infected.
Form a differential equation to model the number of people infected as a function of time.

SOLVING DIFFERENTIAL EQUATIONS

Exploration 28.2

Method of direct integration

We have solved differential equations of the type $\frac{dy}{dx} = f(x)$ in Chapters 9, *Integration 1*, and 19, *Integration 2*. The solution of $\frac{dy}{dx} = f(x)$ is found simply by integrating.

- Solve the differential equation $\frac{dy}{dx} = 3x^2 - 1$ by direct integration.
- Sketch several solution curves.
- Sketch the solution curve that passes through the point (1, 0).

Interpreting the results

Integrating $\frac{dy}{dx} = 3x^2 - 1$ gives $y = x^3 - x + c$ where c is a constant of integration. The diagram shows some of the solution curves for different values of c.

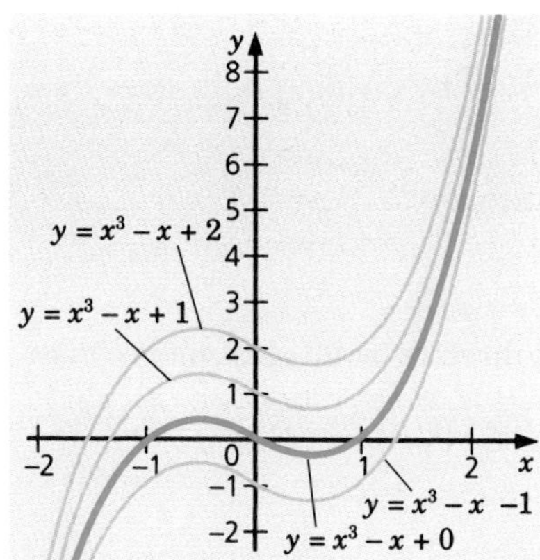

The curve through the point (1, 0) is the bold curve for $c = 0$.

The solution $y = x^3 - x + c$ is called the **general solution** of the differential equation. It contains one unknown constant. When we solve a differential equation we obtain a **family of solutions**. Each value of c gives a different member of the family.

The solution through the point (1, 0) for which $c = 0$ is one member of that family. It is called **the particular solution** of the differential equation.

We usually find the value of the constant for the particular solution using algebra. For example, in this case the general solution is $y = x^3 - x + c$ and $y = 0$ when $x = 1$.

Substituting for $y = 0$ and $x = 1$ into the general solution gives:

$0 = 1 - 1 + c \Rightarrow c = 0$

The particular solution is:
$y = x^3 - x$

Example 28.4

Find the particular solution of the differential equation $\frac{dy}{dx} = \frac{1}{x(x-1)}$ *for which* $y = 3$ *when* $x = 2$.

Solution

By direct integration:

$$y = \int \frac{1}{x(x-1)} dx$$

$$y = \int \left(\frac{1}{x-1} - \frac{1}{x}\right) dx \quad \textit{(using partial fractions)}$$

$$y = \ln(x-1) - \ln x + c$$

This is the general solution. When $x = 2$, $y = 3$

$\Rightarrow 3 = \ln 1 - \ln 2 + c$
$\Rightarrow c = 3 + \ln 2$

So the particular solution is:

$y = \ln(x-1) - \ln x + 3 + \ln 2$

which can be simplified to:

$$y = \ln\left(\frac{2(x-1)}{x}\right) + 3$$

Exploration 28.3

Growth and decay problems

Many quantities, such as population growth and radioactive decay, vary in such a way that the rate of change of the quantity is proportional to the quantity itself. In such models the differential equation is:

$$\frac{dy}{dx} = ky$$

where k is a constant of proportionality.

- Solve the differential equation $\frac{dy}{dx} = 2y$ by dividing both sides by y and integrating.
- Write the general solution in the form $y = \dots$.

Interpreting the results

The function $\frac{dy}{dx} = 2y$ cannot be solved directly by integration because the right-hand side is a function of y. However if we divide both sides by y we get:

$$\frac{1}{y}\frac{dy}{dx} = 2$$

and integrating both sides:

$$\int \frac{1}{y}\,dy = \int 2\,dx$$

Now both sides can be integrated.

$$\ln y = 2x + c$$

Raising both sides to the power e (remember $e^{\ln y} = y$)

$$y = e^{2x+c} = e^{2x}e^{c}$$

This expression can be made simpler by replacing e^{c} with a new constant A, so the general solution is:

$$y = Ae^{2x}$$

Example 28.5

A model for the pressure in the atmosphere P is:

$$\frac{dP}{dh} = -0.13P$$

where h is the height above sea level. When $h = 0$, $P = 10^5$ Pa (see Example 28.2).

Solution

Divide by P and integrate.

$\int \frac{1}{P}\mathrm{d}P = \int -0.13\mathrm{d}h$

$\ln P = -0.13h + c$

Raising each side to the power e:

$P = \mathrm{e}^{-0.13h+c} = \mathrm{e}^{-0.13h}\mathrm{e}^{c} = A\mathrm{e}^{-0.13h}$

where $\mathrm{e}^c = A$.

This is the general solution.

When $h = 0$, $P = 10^5$.

$10^5 = A\mathrm{e}^0 = A$

So the particular solution is $P = 10^5\mathrm{e}^{-0.13h}$.

Method of separation of variables

Exploration 28.3 and Example 28.5 are special cases of **the method of separation of variables**. Consider a general first order differential equation which may be written as:

$$\frac{\mathrm{d}y}{\mathrm{d}x} = \mathrm{u}(x)\mathrm{v}(y)$$

The special feature of this equation is that the right-hand side **separates** into the product of two functions. One is a function of x only and the other is a function of y only. Dividing both sides by $\mathrm{v}(y)$ and multiplying by $\mathrm{d}x$ leads to:

$$\frac{1}{\mathrm{v}(y)}\mathrm{d}y = \mathrm{u}(x)\mathrm{d}x$$

This process is called **separating the variables**. Once the variables are separated, the general solution is given by integrating both sides.

$$\int \frac{1}{\mathrm{v}(y)}\mathrm{d}y = \int \mathrm{u}(x)\mathrm{d}x + c$$

Example 28.6

Find the particular solution of the differential equation $\frac{\mathrm{d}y}{\mathrm{d}x} = -xy^3$ *that satisfies the initial condition* $y = 1$ *when* $x = 1$.

Solution

The differential equation is written in the separated form.

$$\frac{\mathrm{d}y}{\mathrm{d}x} = (\ -x\ )(\ y^3\ )$$

Separating the variables:

$$\frac{1}{y^3}\mathrm{d}y = -x\mathrm{d}x$$

Integrating both sides:

$$\int \frac{1}{y^3}\,dy = \int -x\,dx + c \Rightarrow -\frac{1}{2y^2} = -\frac{1}{2}x^2 + c$$

This is the general solution.

When $x = 1$, $y = 1$ so:

$-\frac{1}{2} = -\frac{1}{2} + c \Rightarrow c = 0$

The particular solution is $y^2 = \frac{1}{x^2}$ *or* $y = \pm\frac{1}{x}$.

Example 28.7

A curve passes through the point $(0, -\frac{1}{2})$ and its gradient function is $y^2 \cos x$. Find the equation of the curve.

Solution

The quantity $\frac{dy}{dx}$ is the gradient so the differential equation for the curve is:

$$\frac{dy}{dx} = y^2 \cos x$$

Separating the variables:

$$\frac{1}{y^2}\,dy = \cos x\,dx$$

Integrating both sides:

$$\int \frac{1}{y^2}\,dy = \int \cos x\,dx + c \Rightarrow -\frac{1}{y} = \sin x + c$$

This is the general solution.

Since the curve passes through the point $(0, -\frac{1}{2})$:

$$-\frac{1}{\left(-\frac{1}{2}\right)} = \sin 0 + c \Rightarrow c = 2$$

The particular solution is $y = \frac{-1}{2 + \sin x}$.

EXERCISES

28.2 A

1 Find the general solution of the following differential equations by integrating.

a) $\frac{dy}{dx} = x^2$ **b)** $\frac{dy}{dx} = \sin x$ **c)** $\frac{dy}{dx} = e^{-2x}$

d) $\frac{dy}{dx} = \frac{1}{x(x+1)}$ **e)** $\frac{dy}{dt} = \sqrt{t}$ **f)** $\frac{ds}{dt} = 4t^{\frac{3}{2}}$

2 Find the general solution of the following differential equations by the method of separation of variables.

a) $\frac{dy}{dx} = \frac{x}{y}$ **b)** $\frac{dy}{dx} = xy^2$ **c)** $\frac{dy}{dx} = \frac{\sin x}{y}$

d) $\frac{dy}{dx} = xe^{-4y}$ **e)** $\frac{dy}{dx} = y\cos x$ **f)** $\frac{dy}{dx} = (1+y)(1-y)$

3 Find the particular solution of the following differential equations

using the given conditions.

a) $\frac{dy}{dx} = x^2 + 1$ and $y = 4$ when $x = 3$

b) $\frac{dy}{dx} = 2y(1+x)$ and $y = 1$ when $x = 0$

c) $\frac{dy}{dx} = y^2 \sin x$ and $y = 1$ when $x = 0$

d) $\frac{dx}{dt} = 2t\left(x^2 - 1\right)$ and $x = 2$ when $t = 1$

e) $\frac{dP}{dt} = 3P(1-P)$ and $P = 0.5$ when $t = 0$

f) $\frac{dx}{dt} = e^{-2x} \sin 3t$ and $x = 0$ when $t = 0$

g) $\frac{dy}{dx} = \frac{(3+x)\left(1+y^2\right)}{\left(1-x^2\right)y}$ and $y = \sqrt{3}$ when $x = 0$

h) $\frac{y}{x}\frac{dy}{dx} = \frac{y^2 - 1}{x^2 - 1}$ and $y = 3$ when $x = 2$

4 A curve passes through the point (0, 2) and its gradient function is $2y - 1$. Find the equation of the curve and use a graphical calculator to sketch the curve.

5 A curve passing through point (–3, 1) has gradient function $\frac{y+1}{x^2 - 1}$. Find the equation of the curve and use a graphics calculator to sketch the curve.

6 The tangent to a curve at any point (x, y) has slope inversely proportional to x^2. The curve passes through the two points (1, 1) and (2, 3). Find the equation of the curve.

7 Solve problems **3–10** from Exercises 28.1A on pages 616–17. Use your solutions to answer each of the following questions

Problem 3 Sketch a graph of the height of a sunflower against time. Is this a good model of sunflower growth? Describe how you would expect the graph to look.

Problem 5 How many seconds does it take for the temperature to reach 30°C?

Problem 8 If the initial depth of the water is 4 m, how long does it take to half empty the tank?

Problem 9 How long does it take for the drug to decay to 25 per cent of the initial dose?

Problem 10 Find how long it takes for the tank to overflow.

EXERCISES

28.2 B

1 Find the general solution of the following differential equations by integration.

a) $\frac{dy}{dx} = x^3$ b) $\frac{dy}{dx} = \cos x - 1$ c) $\frac{dy}{dx} = e^{4x}$

d) $\frac{ds}{dt} = \frac{1}{(t+1)(t+2)}$ **e)** $\frac{dP}{dt} = te^t$ **f)** $\frac{dv}{ds} = -4s^{\frac{5}{4}}$

2 Find the general solution of the following differential equations by the method of separation of variables.

a) $\frac{dy}{dx} = \frac{y}{x^2}$ **b)** $\frac{dP}{dt} = 2P(3-P)$ **c)** $\frac{dy}{dx} = \frac{x+2}{y}$

d) $r\frac{dr}{d\theta} = \sin^2\theta$ **e)** $\frac{dy}{dx} = \frac{(1+y)(3+2y)}{1-x}$ **f)** $v\frac{dv}{ds} = 3-2v^2$

3 Find the particular solution of the following differential equations using the given conditions.

a) $\frac{dy}{dx} = 3x^2 - 2x + 1$ and $y = 2$ when $x = 1$

b) $\frac{dy}{dx} = y\sin x$ and $y = 1$ when $x = 0$

c) $\frac{dy}{dx} = \sqrt{y}e^{-x}$ and $y = 4$ when $x = 0$

d) $(1+x)\frac{dy}{dx} - y(y+1) = 0$ and $y = 1$ when $x = 1$

e) $\frac{dT}{dt} = -2(20 - T)$ and $T = 5$ when $t = 0$

f) $v\frac{dv}{ds} = 10 - 3v$ and $v = 1$ when $s = 0$

g) $\frac{ds}{dt} = e^{-2s}\sin 3t$ and $s = 0$ when $t = 0$

h) $\frac{dv}{dt} = -0.2(v + v^2)$ and $v = 40$ when $t = 0$

4 Water leaks from a tank so that the rate of change of depth of water h m is modelled by:

$$\frac{dh}{dt} = -\frac{3h^2}{4}$$

where t is time in minutes. The initial depth of water in the tank is 1 m.

a) Find the particular solution of the differential equation.

b) How long does it take for the depth of the water in the tank to decrease to 10 cm?

5 A curve passes through the point (3, 2) and its gradient function is $\frac{y+2}{x-2}$. Find the equation of the curve.

6 The tangent to a curve at any point (x, y) has slope proportional to $\sqrt{y}$. The curve passes through the two points (0, 4) and (2, 9). Find the equation of the curve.

7 Solve problems **2, 4–10** from Exercise 28.1B on pages **617–18**. Use your solutions to answer each of the following questions.

Problem 2 Sketch a graph of the size of the colony of rabbits against time. Is this a good model? Describe how you would expect the graph to look.

Problem 4 Andrew likes to drink cola at a temperature of 10°C. How long does he have to wait after removing the cola from the refridgerator to be able to drink the cola?
Problem 5 How long does it take for the radius of the balloon to reduce to 10 cm?
Problem 7 How old is the sample of dead wood?

EULER'S NUMERICAL STEP-BY-STEP METHOD

The methods of the last section provide exact solutions in the form $y = f(x)$. These are called **analytical solutions**. The differential equation $\frac{dy}{dx} = x + y^2$ cannot be solved exactly. In such cases we have to adopt methods for estimating solutions. These are called **numerical methods**. Here we shall introduce one simple method to illustrate the ideas. For a first order differential equation $\frac{dy}{dx} = f(x, y)$, the gradient function $f(x, y)$ gives the slope of the tangent at any point on a solution curve. If we do not know the actual solution curve, then a sequence of straight line segments is used as an approximation. The slopes of these straight line segments are given by the gradient function $f(x, y)$. The diagram illustrates the idea. Suppose the differential equation is $\frac{dy}{dx} = x + y^2$ with initial condition (1, 2).

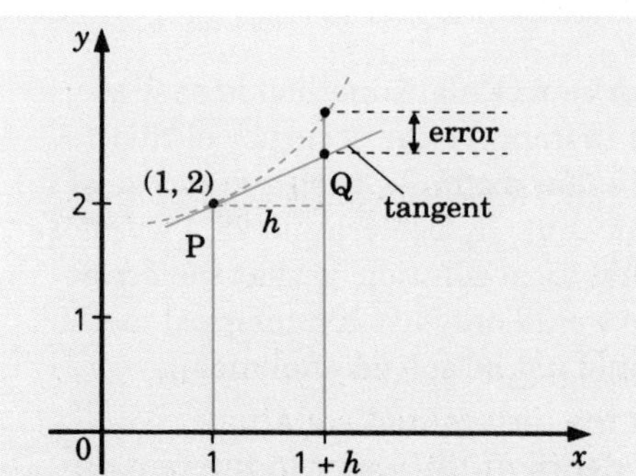

The gradient function tells us in which direction to leave the point (1, 2). When $x = 1$ and $y = 2$ the slope of the tangent is $f(1, 2) = 5$. We choose a step size h so that the tangent arrives at point Q. At Q we form another line segment by recalculating the gradient function with the coordinates of Q. Gradually we build up a sequence of line segments. The method is attributed to a famous 18th-century mathematician Leonhard Euler and is called **Euler's numerical step-by-step method** (or often just 'Euler's method').

Exploration 28.4

A formula for Euler's method

Use graph paper for this exploration, choosing large scales on the x-axis. Look at the differential equation $\frac{dy}{dx} = 3x^2 + 1$ with initial condition (1, 2).

- On your graph paper, draw a sketch of the figure above using gradient function $3x^2 + 1$ with initial condition (1, 2).
- Choosing $h = 0.2$ find the coordinates of point Q using gradient function $3x^2 + 1$ to form the line PQ.
- Using the coordinates of Q, draw in the next line segment. Find the coordinates of the next point R using a step size 0.2.

- Repeat the process forming five line segments.
- Find the exact solution of the problem and draw the exact solution curve between $x = 1$ and $x = 2$.
- Form a table showing the estimated values and exact values for y for each step i.e. $y(1.2)$, $y(1.4)$, $y(1.6)$, … .
- Comment on how you might improve the accuracy of the numerical method.

Interpreting the results

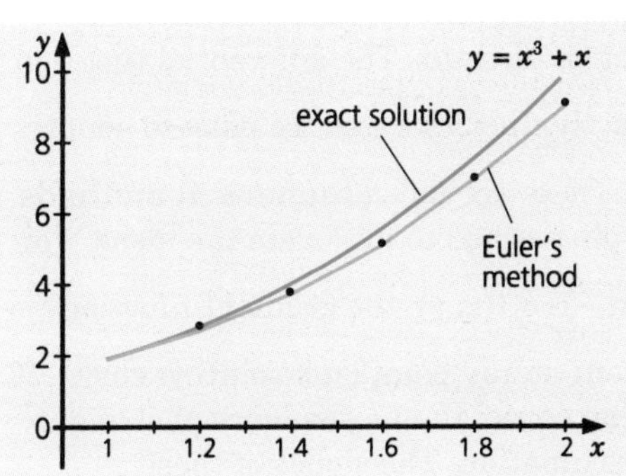

For this problem we can find an exact solution by direct integration. The general solution is $y = x^3 + x + c$. With $y = 2$ when $x = 1$ the constant of integration $c = 0$. The particular solution is $y = x^3 + x$.

You should have drawn the graph on the left, which shows the line segments and exact solution.

The following table compares the answers.

	$y(1.2)$	$y(1.4)$	$y(1.6)$	$y(1.8)$	$y(2.0)$
Analytical solution	2.928	4.144	5.696	7.632	10
Numerical solution	2.8	3.864	5.24	6.976	9.12

The error after five steps is 8.8 per cent. Euler's method is easy to use but may produce errors. One way to improve the accuracy of this method is to reduce the step size h. For example, with a step size of 0.1 the error after ten steps is 4.45 per cent. In Exploration 28.4 we chose a differential equation with an analytical solution so that the error could be identified. In practice we would only use a numerical method for a differential equation that could not be solved analytically.

Exploration 28.5

Improving the estimate

A computer package was used to estimate the value of $y(2.0)$ given by Euler's method when solving the differential equation $\frac{dy}{dx} = \sqrt{x+y}$ with initial condition $y = 1$ when $x = 1$. The following table shows the results for different step sizes.

Step size	Estimate to $y(2.0)$
0.2	2.726 79
0.1	2.770 11
0.05	2.792 12
0.01	2.809 89
0.005	2.812 12

- Plot a graph of these estimates to $y(2.0)$ against the step size h.
- Comment on how good a straight line graph would be.
- Use your graph to estimate the true value of $y(2.0)$ correct to four decimal places.

Interpreting the results

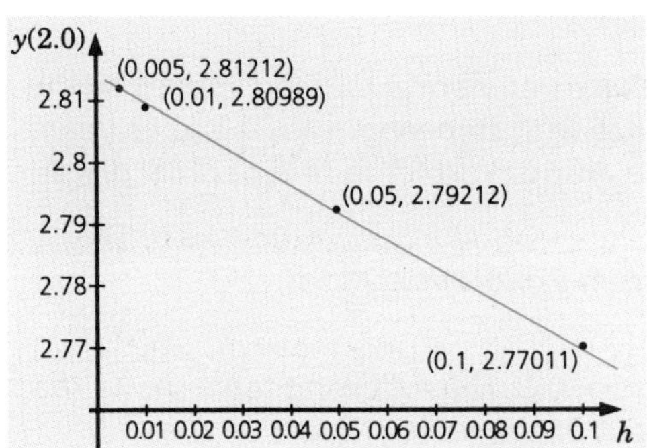

The graph shows that the estimates to $y(2.0)$ obtained using Euler's method depends reasonably linearly on the step size. The graph can be used to estimate the exact value by taking the intercept on the vertical axis of the best straight line through the plotted points. The answer is 2.814 (correct to three decimal places).

A formula for Euler's method

The real power of a numerical method is that it can be used in a calculator or computer program, thus avoiding many long-winded calculations. The first step in writing a program is to find an algebraic formula to define the sequence of estimates.

Consider the initial value problem $\frac{dy}{dx} = f(x, y)$ with initial condition $y(x_0) = y_0$. With a step size h the figure shows the first straight line segment that estimates the value of y at $x = x_0 + h$. The values of x and y at Q are labelled (x_1, y_1).

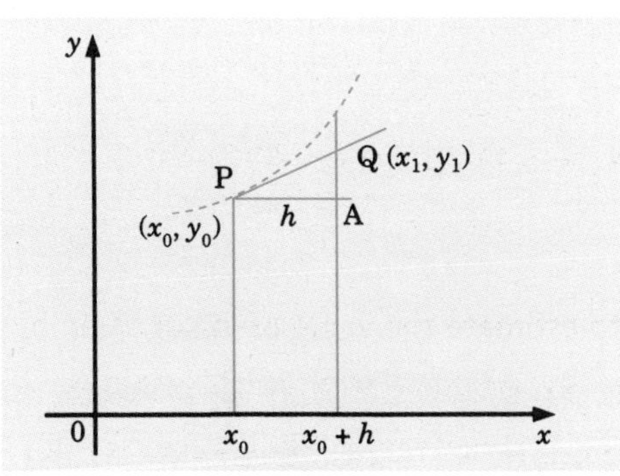

From triangle PAQ:

$$\frac{AQ}{AP} = \text{slope of tangent to solution curve at P} = f(x_0, y_0)$$

$$\Rightarrow \quad \frac{y_1 - y_0}{h} = f(x_0, y_0)$$

$$\Rightarrow \quad y_1 = y_0 + hf(x_0, y_0)$$

This formula can be used to form a sequence of estimates for y as x increases.

Exploration 28.6

A general formula for the nth step

- Show that after two steps $y_2 = y_1 + hf(x_1, y_1)$
- Show that after three steps $y_3 = y_2 + hf(x_2, y_2)$
- Formulate a general formula $y_{n+1} = \ldots$ to generate the sequence $y_1, y_2, y_3, \ldots$

Interpreting the results

The formula for the sequence generated by Euler's method is:
$y_{n+1} = y_n + hf(x_n, y_n)$.

Example 28.8

Solve the differential equation $\frac{dy}{dx} = x + y^3$ *with initial condition* $y = 1$ *when* $x = 0$.

Write down the sequence for Euler's numerical method applied to the above problem. Use the sequence with step length $h = 0.1$ *to estimate the solution to* $y(0.5)$ *giving the answer correct to four decimal places.*

Solution

The general formula for y_{n+1} *from Exploration 28.6 is:*

$y_{n+1} = y_n + h(x_n + y_n^3)$

Starting with $x_0 = 0$, $y_0 = 1$, *and* $h = 0.1$. *The following table shows the sequence for* y_1, y_2, …. *The layout of the table includes all the calculations as they occur.*

n	x_n	y_n	$x_n + y_n^3$	$h(x_n + y_n^3)$	$y_{n+1} = y_n + h(x_n + y_n^3)$
0	0	1	1	0.1	1.1
1	0.1	1.1	1.431	0.1431	1.2431
2	0.2	1.2431	2.120 95	0.212 095	1.455 19
3	0.3	1.455 19	3.381 47	0.338 147	1.793 33
4	0.4	1.793 33	6.167 40	0.616 74	2.410 06
5	0.5	2.410 06			

The estimate of y *when* $x = 0.5$ *is* $y(0.5) = 2.4101$ *correct to four decimal places.*

CALCULATOR ACTIVITY

Using Euler's method

The program generates the x-values and y-values using Euler's numerical method to estimate the solution of a first-order differential equation. Values for the initial values of x and y are stored as X and Y. The step size is stored as H and the final x-values as Z. Use the program to check estimates in Exploration 28.5. Before using the program you need to store the function f(x, y) as $\sqrt{x+y}$ within the program.

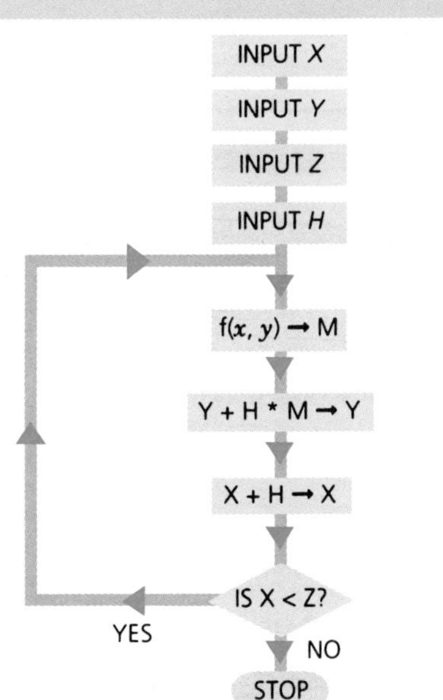

Example 28.9

Solve the differential equation $\frac{dy}{dx} = (1+x)y$ *with initial condition* $y = 1$ *when* $x = 2$.

a) *Use the program to estimate the solution to* $y(3)$ *correct to four decimal places with step sizes* $h = 0.1$, $h = 0.05$, $h = 0.01$, $h = 0.005$, $h = 0.001$.

b) *Plot a graph of these estimates to* $y(3)$ *against the step size* h *and use your graph to estimate the true value of* $y(3)$, *correct to two significant figures.*

c) *Find the analytical solution to the problem and compare the exact value of* $y(3)$ *with your estimate in part* **b)**.

Solution

a) *The program generates the following estimates to* $y(3)$.

Step size h	0.1	0.05	0.01	0.005	0.001
Estimate to $y(3)$	19.3300	24.5956	31.0290	32.0247	32.8960

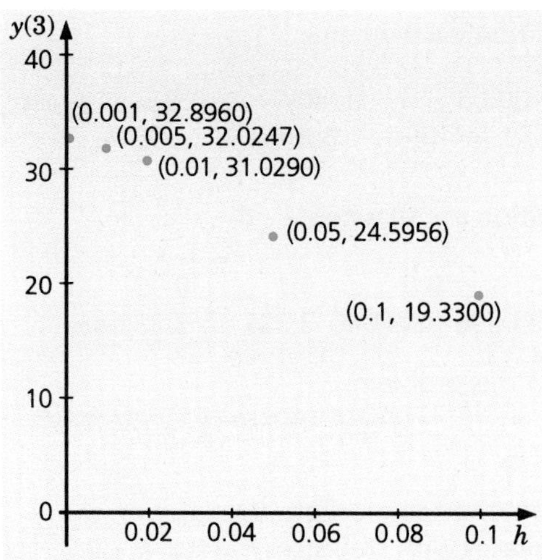

b) *The diagram shows the graph of* $y(3)$ *against* h. *As* $h \to 0$, *the graph suggests that* $y(3) \to 33$ *(correct to two significant figures). For small values of* h, *the relationship between* $y(3)$ *and* h *is approximately linear. However, the analysis of these ideas is beyond the scope of this book.*

c) *The differential equation* $\frac{dy}{dx} = (1+x)y$ *can be solved analytically using the method of separation of variables.*

$$\int \frac{1}{y}\,dy = \int (1+x)\,dx$$

$$\Rightarrow \ln y = x + \tfrac{1}{2}x^2 + c$$

The initial conditions $y = 1$ *when* $x = 2 \to c = -4$, *so:*

$$y = e^{x+\frac{1}{2}x^2-4}$$

When $x = 3$, *the analytical solution for* $y(3)$ *is:*

$y(3) = e^{3+\frac{1}{2}3^2-4} = e^{3.5} = 33.1155$, *correct to four decimal places.*
The percentage errors in the estimates from part b) are summarised in this table.

Step size h	0.1	0.05	0.01	0.005	0.001
Estimate to $y(3)$	19.3300	24.5956	31.0290	32.0247	32.8960
Percentage error	41.6	25.7	6.3	3.3	0.7

You can see that by reducing the step size considerably the percentage error in the estimate is also reduced to an acceptable level.

EXERCISES

28.3 A

For questions **1–3**, use Euler's method by hand to find the solution for each differential equation with the given initial condition and step length h.

1 $\dfrac{dy}{dx} = x^2 - y^2$ initial condition $y = 0$ when $x = 1$, and $h = 0.1$

Estimate $y(1.2)$.

2 $\dfrac{dy}{dx} = (x - y)^2$ initial condition $y = 0.5$ when $x = 1$, and $h = 0.1$

Estimate $y(0.5)$.

3 $v\dfrac{dv}{ds} = 10 - 3v^{\frac{3}{2}}$ initial condition $v = 10$ when $s = 0$, and $h = 0.1$

Estimate $v(0.5)$.

In questions **4** and **5** the tables show Euler's numerical solutions to the given differential equation. Use an appropriate graph to estimate the exact value correct to five significant figures.

4 $\dfrac{dy}{dx} = x^2 - y^2$ initial condition $y = 0$ when $x = 1$

Step size h	0.1	0.05	0.025	0.01
Estimate to $y(2)$	1.563 85	1.565 09	1.565 65	1.565 98

5 $v\dfrac{dv}{ds} = 10 - 3v^{\frac{3}{2}}$ initial condition $v = 10$ when $s = 0$

Step size h	1	0.5	0.2	0.1	0.05
Estimate to $v(5)$	10.6536	2.232 96	2.231 45	2.231 45	2.231 45

Explain what is happening in this problem.

6 Solve the differential equation $\dfrac{dy}{dx} = -x^2y^2$ with initial condition $y = 0.5$ when $x = 0$.

a) Use the program for Euler's method to estimate the solution to $y(2)$ correct to four decimal places with step sizes $h = 0.2$, $h = 0.1$, $h = 0.05$, $h = 0.02$.

b) Plot a graph of these estimates to $y(2)$ against step size h and use your graph to estimate the true value $y(2)$ correct to three decimal places.

c) Find the analytical solution to the problem and compare the exact value of $y(2)$ with your estimate in part **b)**.

EXERCISES

28.3 B

For questions **1–3**, use Euler's method by hand to find the solution for each differential equation with the given initial condition and step length h.

1 $\dfrac{dy}{dx} = 2x - y^2$ initial condition $y = 1$ when $x = 0$, and $h = 0.1$

Estimate $y(0.5)$.

2 $\dfrac{dy}{dx} = 1 + x\sin y$ initial condition $y = 0.5$ when $x = 0$, and $h = 0.1$

Estimate $y(0.5)$.

3 $\dfrac{dv}{dt} = 10 - 0.1v^2$ initial condition $v = 15$ when $t = 0$, and $h = 0.2$

Estimate $v(2)$.

In questions **4** and **5** the tables show Euler's numerical solutions to the given differential equation. Use an appropriate graph to estimate the exact value correct to five significant figures.

4 $\dfrac{dy}{dx} = 2x - y^2$ initial condition $y = 1$ when $x = 0$

Step size h	0.2	0.1	0.05	0.01
Estimate to $y(2)$	1.84774	1.84396	1.84264	1.84184

5 $\dfrac{dy}{dx} = 1 + x\sin y$ initial condition $y = 0.5$ when $x = 0$

Step size h	0.2	0.1	0.05	0.01
Estimate to $v(5)$	1.860 69	1.910 62	1.934 95	1.954 01

6 The graph of a function passes through the point (1, 0.5). The gradient function of the graph is $\sqrt{x} + \sqrt{y}$. Use the program for Euler's method to estimate the solution correct to one decimal place at $x = 2$, $x = 3$, $x = 4$, $x = 5$. Sketch the graph of the function for $1 \le x \le 5$ using your estimates.

You may need to choose different h values within each interval (1, 2), (2, 3), (3, 4), (4, 5) to give an accuracy of two decimal places.

MATHEMATICAL MODELLING ACTIVITY

Problem statement

1 → 2 → 3
6 ← 5 ← 4

Specify the real problem

Yogurt can be made at home by placing a small amount of commercially-made yogurt into warm milk and leaving it in a warm place. The bacteria in the yogurt thrive in the warm milk and multiply. By this process the milk is turned into yogurt.

- Formulate a model describing the growth of the population of yogurt bacteria. Use the model to find the time it takes to make yogurt.

Set up a model

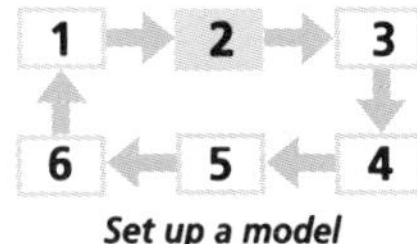

Set up a model

As described above, the problem is described in vague terms, so the first step is to find out about the biological process of turning milk into yogurt. Essentially what happens is that the *Lactobacillus bulgaris* bacteria present in a small batch of yogurt ferment in the warm milk producing lactic acid from the fermentation of the carbohydrates. Gradually, if the correct temperature of approximately 45°C is maintained, the milk becomes yogurt. Typically 10 grams of yogurt should be placed in 500 ml of warm milk. Experiments show that under ideal conditions the amount of yogurt doubles approximately every 30 minutes. To proceed we make the following assumptions.

- The mixture of milk and yogurt is kept in ideal conditions.
- The yogurt increases so that its growth rate is proportional to the amount present which leads to an exponential model.

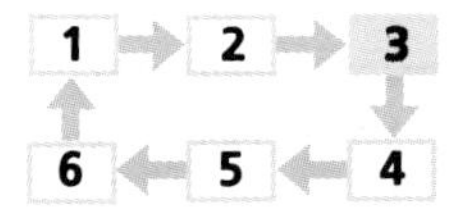

Formulate the mathematical problem

Mathematical model

The assumptions lead to the following relationship between the amount of yogurt present P, measured in grams, at time t, measured in hours.

$$\frac{dP}{dt} = kP \text{ or } P = P_0e^{kt} \qquad (1)$$

where P_0 is the initial amount of yogurt and k is a constant describing the growth rate. The initial problem is to find the value of k.

1 → 2 → 3
6 ← 5 ← 4
Solve the mathematical problem

Mathematical solution

Under ideal conditions the amount of yogurt doubles every half-hour. Hence $P = 2P_0$ when $t = 0.5$.

Substitution into (1) gives:

$$2P_0 = P_0e^{0.5k}$$

and solving for k:

$$k = 2\ln(2) \approx 1.39 \qquad \text{(correct to two decimal places)}$$

The model of the growth of yogurt is:

$$P = P_0e^{1.39t}$$

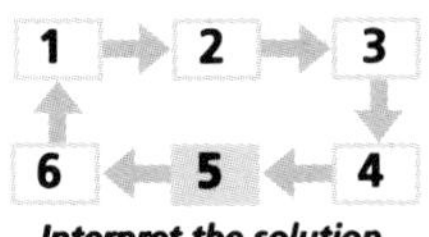

Interpret the solution

Interpretation

Suppose that we start with 10 grams of yogurt. Then the figure shows the growth of the yogurt in the milk.

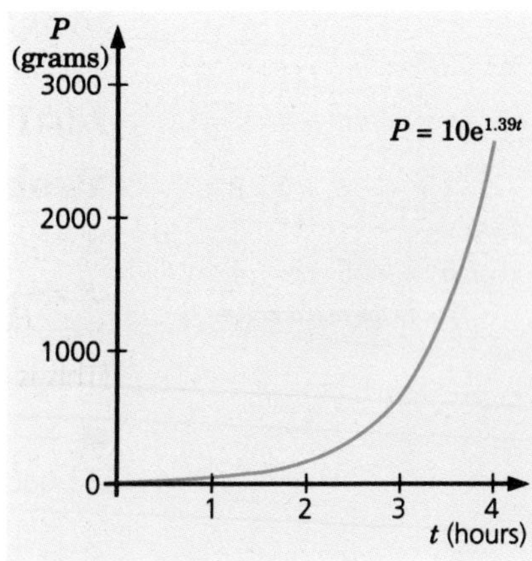

After four hours we have a lot of yogurt! If we assume that the density of yogurt is the same as that of water, then 2598 grams of yogurt is 2.598 litres (just over 4 pints). According to a recipe book, it takes between four and six hours for 10 grams of yogurt culture to turn 500 grams of milk into yogurt.

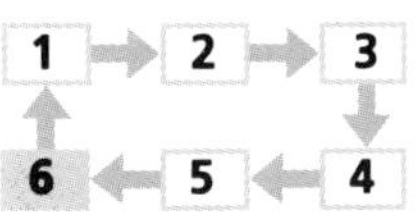

Compare with reality

Criticism of the model

The pitfall in our model is the assumption that the bacteria will continue to grow indefinitely according to the exponential model. The basic assumption underlying the exponential model is that the growth rate is proportional to the amount of yogurt present, i.e:

$$\frac{dP}{dt} = kP$$

The growth of the yogurt will be approximately exponential at the start but the growth rate will decline as the milk is being used up. If we start with 500 grams of milk then the limiting amount of yogurt is 500 grams.

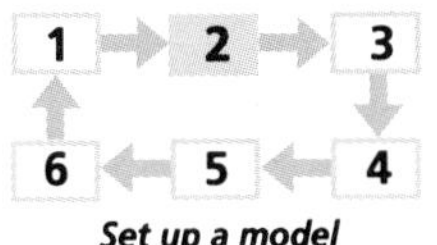

Set up a model

Refinement of the model

We shall still assume that the mixture of milk and yogurt is kept under ideal conditions of approximately 45°C, but we change the second assumption.

- The growth rate of the yogurt increases initially and then decreases to zero as the amount of yogurt increases.

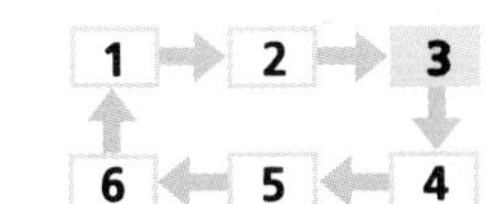

Formulate the mathematical problem

Set up an improved model

A quadratic function satisfies this new assumption. We have:

$$\frac{dP}{dt} = aP(M - P) \qquad (2)$$

where M is called the equilibrium population and a is a constant. In our problem M is the amount of milk at the start of the yogurt-making process. When $P = M$ there is no more milk to be turned into yogurt, and the bacteria stop growing.

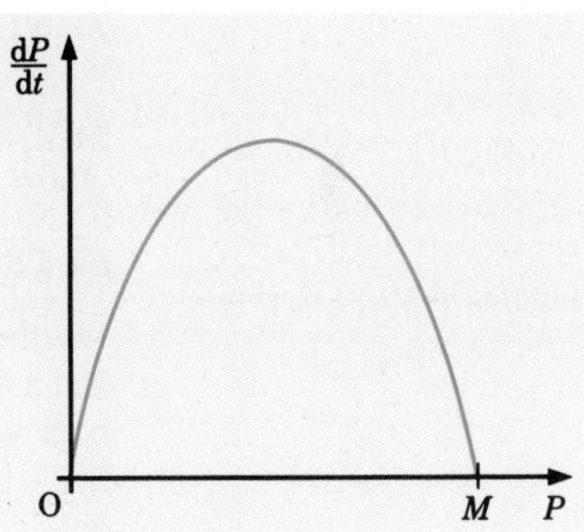

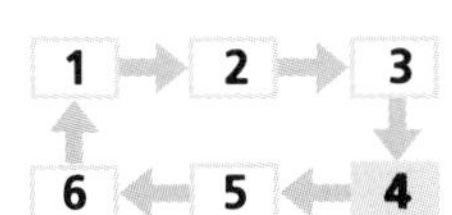

Solve the mathematical problem

Mathematical solution

Equation (2) is a differential equation with separable variables. To solve it, we rearrange it in the form:

$$\int \frac{1}{P(P - M)} dP = \int a \, dt$$

The left-hand side can be integrated by first using partial fractions. The solution for P is:

$$P = \frac{P_0 M}{P_0 + (M - P_0)e^{-aMt}} \qquad (3)$$

(This is left as an activity for you to do.)

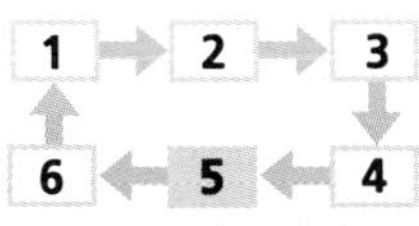

Interpret the solution

Interpretation

At first sight this looks a complicated formula for P. However if we choose some values of the constants and draw a graph of P against t then we can see that this model has all the ingredients of the real situation.

Let $P_0 = 10$, $M = 500$, and $aM = 1.39$.

Equation (3) becomes:

$$P = \frac{5000}{10 + 490e^{-1.39t}} \qquad (4)$$

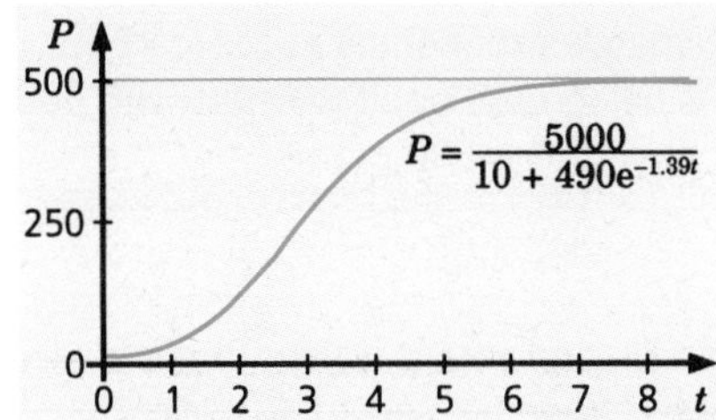

A graph of this function is shown in the diagram.

This is known as the **logistic curve**. The yogurt grows rapidly at the start with the same form as the exponential model. But as time increases the amount of yogurt levels off towards the maximum amount of 500 grams.

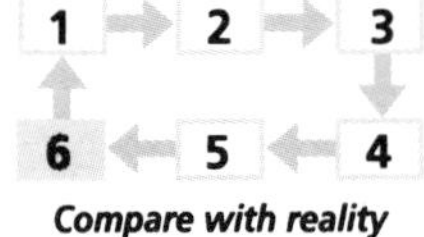

Compare with reality

Answer to the problem

The initial statement asked for the time required to make yogurt. According to equation (4) it would take an infinite time to turn 500 grams of milk into yogurt. However we would probably be satisfied with, say, 90 per cent of the milk becoming yogurt. Equation (4) can be used to find the value of t for $P = 0.9 \times 500 = 450$. We need to solve for t.

$$450 = \frac{5000}{10 + 490e^{-1.39t}}$$

$t = 4.38$

This model predicts that it takes almost $4\frac{1}{2}$ hours to make 450 grams (i.e. half a litre) of yogurt, using 10 grams of culture. This agrees with a cookery book which suggests removing the yogurt to a refrigerator after four to six hours so that the yogurt does not become too thick and the acid flavour does not become too strong.

Activities

- Show that equation (3) is the general solution to the differential equation (2).
- Explain why we chose $aM = k = 1.39$.

CONSOLIDATION EXERCISES FOR CHAPTER 28

1 **The body of a murder victim was discovered in the early hours of the morning at 2.00 a.m. The police doctor arrived at 2.30 a.m. and immediately took the temperature of the body, which was 34.8°C. One hour later the temperature of the body was 34.1°C. The room temperature was constant at 32.2°C. (The normal body temperature is 37°C.)**

a) **Formulate a differential equation model for the temperature of the body as a function of time. State your assumptions clearly.**

b) **Solve the differential equation and use the given information to find any unknown constants in the general solution.**

c) **Use your solution to estimate the time of death.**

2 **The slope of the tangent to a curve at the point (x, y) is given by $x + y^2$. The curve passes through the point (0, 1).**

a) **Set up a problem consisting of a differential equation and an initial condition to find the equation of the curve.**

b) Write down the sequence for Euler's numerical method applied to the above problem.

c) Use this sequence with a step length 0.2 to estimate the curve between $x = 0$ and $x = 1$.

d) Improve your solution by choosing a smaller step length.

3 Radium is a radioative substance. You can model its decay by the differential equation $\frac{dR}{dt} = -kR$ where t is the time in years, R is the amount of radium in grams present at time t, and k is a positive constant.

Suppose that when $t = 0$, 10 g of radium are present.

a) Solve the equation $\frac{dR}{dt} = -kR$ to find R in terms of t and k.

b) It is known that the amount of radium will have halved after about 1600 years. Use this information to show that $k = \frac{\ln 2}{1600}$.

c) According to this model, how many grams of radium will be left after 100 years?

(Oxford (Nuffield) Question 9, Specimen Paper 2, 1994)

4 Find:

a) $\int x \cos x \, dx$, **b)** $\int \cos^2 y \, dy$.

Hence find the general solution of the differential equation

$$\frac{dy}{dx} = x \cos x \sec^2 2y, \quad 0 < y < \frac{\pi}{4}.$$

(ULEAC Question 9, Specimen Paper 2, 1994)

5 **a)** Write $\frac{1}{y(3-y)}$ in partial fractions.

b) Find $\int \frac{1}{y(3-y)} dy$.

c) Solve the differential equation $x\frac{dy}{dx} = y(3-y)$ where $x = 2$ when $y = 2$, giving y as a function of x.

(MEI Question 4, Specimen Paper 3, 1994)

6 If left undisturbed, the population, P, of mice on a small island would increase at two per cent of its current value every day. However, it is estimated that the number of mice killed each day by predators is $0.8\sqrt{P}$. These facts can be summarised in a differential equation.

$$\frac{dP}{dt} = 0.02P - 0.8\sqrt{P}$$

[In this model, P and t are to be treated as continuous variables.]

a) If when $t = 0$ the population is 1000 mice, use a step-by-step method to estimate how it will change over the next 20 days. Use a step-length of $\delta t = 10$, and work to the nearest whole number of mice at each step.

b) Use the differential equation to show that, if P exceeds a certain critical value, then the population will increase, but that if P is below that value, the population will decrease. State the value of this critical number.

c) Explain why, if the population is larger than the critical number found in **b)**, then it will go on increasing indefinitely, but that if it is less than the critical value, it will go on decreasing (until it is wiped out).

(Oxford and Cambridge Question 1, Specimen Paper 1, 1994)

7 Water flows out of a tank through a pipe at the bottom, and at time t minutes the depth of the water in the tank is x metres. The rate at which the depth of water remaining in the tank is decreasing at any instant is proportional to the square root of the depth at that instant.

a) Explain how the above information leads to the differential equation:

$-\frac{dx}{dt} = k\sqrt{x}$ where k is a positive constant.

b) Find the general solution of the differential equation in part **a)**.

c) At time $t = 0$ the depth of the water in the tank is 2 m. After 10 minutes the depth is 1.5 m. Find, to the nearest minute, the time at which the depth is 1 m.

(UCLES (Linear) Question 14, Specimen Paper 1, 1994)

8 A particle P moves along the positive x-axis and at time $t = 0$ is at the origin O. At time t seconds, P is x metres from O and $\frac{dx}{dt} = 2t(x+4)$.

Solve this differential equation and use the condition that $x = 0$ at $t = 0$ to show that $x = 4(e^{t^2} - 1)$.

By expressing $\frac{dx}{dt}$ in terms of t, or otherwise, find the acceleration of P at $t = 2$.

(ULEAC Question 7, Paper 2, January 1992)

9 The current, I, in an electrical circuit containing resistance R and inductance L varies with time t according to the differential equation:

$L\frac{dI}{dt} + IR = E \ (t \geq 0)$

where L, R and E are constants.

a) Given that $I = 0$ when $t = 0$, show that:

$$I = \frac{E}{R}\left(1 - e^{\frac{-Rt}{L}}\right)$$

b) State the limiting value of I as $t \to \infty$, and find an expression for the time at which I is equal to half this limiting value.

(WJEC Question 9, Specimen Paper A1, 1994)

10 A canal lock is modelled as a tank with a rectangular base area of 50 m² which is filled by letting water run in. The height of the water surface above its lowest level is h m at time t, where t is the number of seconds after the lock starts to fill.

In the model, h satisfies the differential equation:

$\frac{dh}{dt} = k(2.5 - h)$

where k is constant.

Initially, when $t = 0$, water flows into the lock at 2 $m^3 s^{-1}$.

a) Show that initially $\frac{dh}{dt} = 0.04$ m s^{-1} and hence that $k = 0.016$.

b) Solve the original differential equation for t in terms of h.

The lock is full when $h = 2.5$.

c) What does your solution to the differential equation predict about the time required to fill the lock?

d) How long would it take for the water to rise to within 2 mm of the lock being full?
(In practice, the lock gates can be opened before the lock is completely full.)

(MEI Paper 9, Mechanics 3, January 1994)

Summary

- A differential equation is an equation involving derivatives.
- A first order differential equation only involves a first derivative. The general form of a first order differential equation is:
  $\frac{dy}{dx} = f(x, y)$.
- If $f(x, y)$ is a function of x only then the differential equation can be solved by direct integration.
- If $f(x, y)$ is in the form $u(x)v(y)$ then we say that the variables have separated and the method of separation of variables can be used to give:
  $\int \frac{1}{v(y)} dy = \int u(x) dx$
- Euler's numerical step-by-step method gives the following sequence for estimating a solution:
  $y_{n+1} = y_n + hf(x_n, y_n)$
- The general solution of a first-order differential equation contains one unknown constant and represents a 'family of solution curves'.
- A particular solution is one member of the family of solutions.

PURE MATHS

29 Mathematics of uncertainty

In this chapter we summarise core ideas in statistics and probability covering the following areas:

- *collecting and presenting data,*
- *averages and spread of data,*
- *probability: calculating the chance of single and combined events.*

PATTERNS IN DATA

There are two types of data, **discrete** and **continuous**. Discrete data can be counted, e.g. the number of children in a family, the score on a die when it is thrown. Continuous data can be measured, e.g. the height of a person, the time it takes to run 100 m.

Raw data may be arranged in a **frequency table** or **frequency distribution**, which lists the frequency or number of times each value occurs. The **cumulative frequency** is a running total of all the individual frequencies, and the cumulative frequency of a whole range of data is therefore the same as the total number of items of data in the range. A discrete frequency distribution may be represented by a **line diagram**. A continuous frequency distribution or a grouped discrete frequency distribution may be illustrated by a **stem-and-leaf** diagram, a **histogram** or **cumulative frequency curve**. These diagrams are illustrated in the examples that follow.

There are three popular measures used to locate a single value that represents a data set:

- **the mode:** the most frequently occurring value,
- **the median:** the middle value when the data are arranged in order: $\frac{1}{2}(n+1)$th value if n is odd, half of the $[\frac{1}{2}n + (\frac{1}{2}n + 1)]$th values if n is even,
- **the mean** ($\bar{x}$): the average of the data set.

For raw data: $\bar{x} = \dfrac{\sum x}{n}$

For a frequency distribution: $\bar{x} = \dfrac{\sum fx}{\sum f}$

where $\sum x$ or $\sum fx$ represents the sum of all the data items and n or $\sum f$ represents the number of data items (total frequency).

There are three popular measures used to represent the spread or dispersion of a data set:

- **range:** highest value – lowest value
- **inter-quartile range:** upper quartile – lower quartile = $Q_3 - Q_1$

  where $Q_1 = (\frac{1}{4}n + \frac{1}{2})$th value and $Q_3 = (\frac{3}{4}n + \frac{1}{2})$th value
- **standard deviation (s):** square root of the average of the squares of the deviations from the mean.

For raw data: $s = \sqrt{\dfrac{\Sigma(x-\bar{x})^2}{n}} \equiv \sqrt{\dfrac{\Sigma x^2}{n} - \bar{x}^2}$

For a frequency distribution: $s = \sqrt{\dfrac{\Sigma f(x-\bar{x})^2}{\Sigma f}} \equiv \sqrt{\dfrac{\Sigma fx^2}{\Sigma f} - \bar{x}^2}$

An outlier is a data item that lies more than two standard deviations from the mean, i.e. a value x such that $x < \bar{x} - 2s$ or $x > \bar{x} + 2s$.

The relationship between the median, range and inter-quartile range may be illustrated using a **box-and-whisker plot**, as in the diagram.

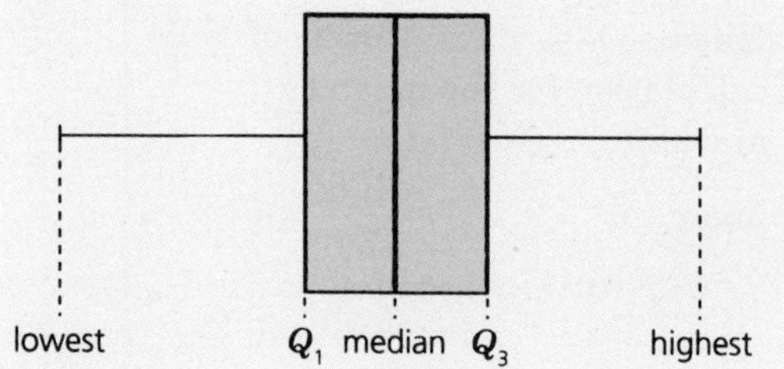

Representing discrete data

Example 29.1

*In 50 randomly chosen households the numbers of under-18s were recorded (**raw data**) and the results recorded as follows.*

22410 32235 11021 43722 04311
13342 23502 24113 43252 43230

a) *Find the mode, median and mean of the data set.*
b) *Find the range, inter-quartile range and standard deviation.*
c) *Identify any outliers.*
d) *Represent the data using a suitable diagram.*

Solution

The raw data may be ordered like this.

00000 11111 11112 22222 22222
22233 33333 33334 44444 45557

a) *Mode = 2* *most frequently occurring value*
Median = 2 *average of 25th and 26th values*
Mean ($\bar{x}$) = 2.4 $\bar{x} = \dfrac{\Sigma x}{n} = \dfrac{120}{50}$

b) *Range = 7* *highest value – lowest value = 7 – 0*

Inter-quartile range = 2 $Q_3 - Q_1$ = 38th value – 13th value = 3 – 1
Standard deviation

s = 1.50 $s = \sqrt{\dfrac{\Sigma(x-\bar{x})^2}{n}} \equiv \sqrt{\dfrac{\Sigma x^2}{\Sigma f} - \bar{x}^2} = \sqrt{\dfrac{400}{50} - 2.4^2}$

c) *An outlier is a data item which is more than two standard deviations from the mean.*
$\bar{x} + 2s = 2.4 + 2 \times 1.50 = 5.40 \Rightarrow x = 7$ *is an outlier.*

d) *The data may be arranged into a* ***frequency distribution*** *chart.*

x	0	1	2	3	4	5	6	7	
f	5	9	14	11	7	3	0	1	$\Sigma f = 50$
fx	0	9	28	33	28	15	0	7	$\Sigma fx = 120$
fx²	0	9	56	99	112	75	0	49	$\Sigma fx^2 = 400$

This may be illustrated by a ***line diagram.***

Note
When the data are arranged in a frequency distribution we may use alternative forms of calculation for the mean ($\bar{x}$) and standard deviation (s).
Mean ($\bar{x}$) = 2.4, $\bar{x} = \dfrac{\Sigma fx}{\Sigma f} = \dfrac{120}{50}$

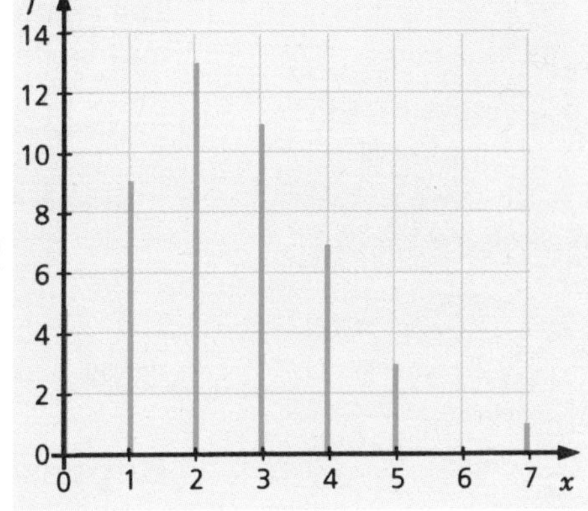

Standard deviation

$s = 1.50$ $s = \sqrt{\dfrac{\Sigma f(x-\bar{x})^2}{\Sigma f}} \equiv \sqrt{\dfrac{\Sigma f x^2}{\Sigma f} - \bar{x}^2} = \sqrt{\dfrac{400}{50} - 2.4}$

Representing continuous data

Example 29.2

The heights of 50 randomly chosen 11-year-olds were recorded, to the nearest cm, in a stem-and-leaf diagram, like this.

100	0 1 3
105	1 1 2 4
110	0 0 1 1 2 3 3 4
115	0 0 1 1 1 2 2 2 3 3 3 3 4 4
120	0 0 0 1 1 2 2 2 3 3 4 4
125	0 1 2 2 3 4 4
130	1 3 4

a) *Group the data into* ***classes*** *of width 5 cm, centred on 102 cm, 107 cm, etc.*

b) *Illustrate the data using a histogram and estimate the mean and standard deviation of the children's heights.*

c) Illustrate the data using a cumulative frequency curve. Use it to estimate the median and the inter-quartile range.

Solution

a)

Height, h cm	Frequency, f	Mid-interval height, x	Cumulative frequency, F
$99.5 \le h < 104.5$	3	102	3
$104.5 \le h < 109.5$	4	107	7
$109.5 \le h < 114.5$	8	112	15
$114.5 \le h < 119.5$	14	117	29
$119.5 \le h < 124.5$	11	122	40
$124.5 \le h < 129.5$	7	127	47
$129.5 \le h < 134.5$	3	132	50

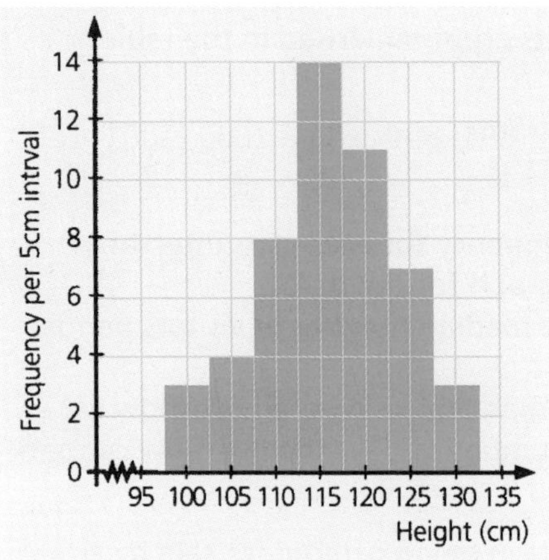

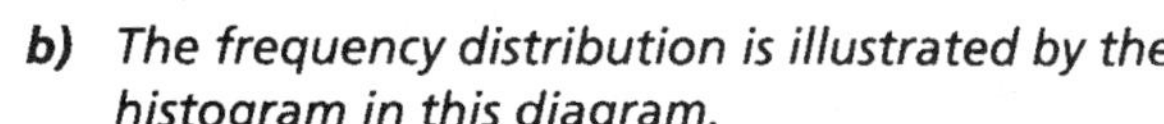

b) The frequency distribution is illustrated by the histogram in this diagram.

Estimated mean ($\bar{x}$) = 117.9 $\quad \bar{x} = \dfrac{\Sigma fx}{\Sigma f} = \dfrac{5895}{50}$

Estimated standard deviation

$s = 7.60$

$$s = \sqrt{\frac{\Sigma f(x-\bar{x})^2}{\Sigma f}} \equiv \sqrt{\frac{\Sigma f x^2}{\Sigma f} - \bar{x}^2} = \sqrt{\frac{697\,905}{50} - 117.9^2}$$

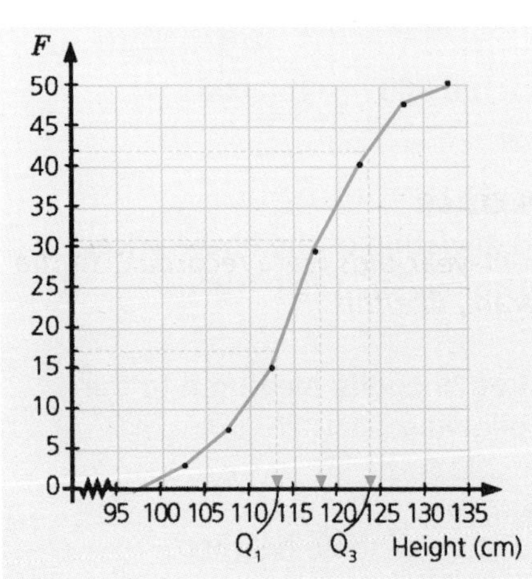

c) The cumulative frequency curve in this diagram is an alternative way of illustrating the data.

*The following **estimates** may be found from the cumulative frequency curve:*

- *the median is 118 (the 25th value),*
- *the inter-quartile range is* 10.

$(Q_3 - Q_1 = 37\tfrac{1}{2}\text{th value} - 12\tfrac{1}{2}\text{th value} = 123 - 113)$

The range, median and inter-quartile range may be illustrated by a box-and-whisker plot.

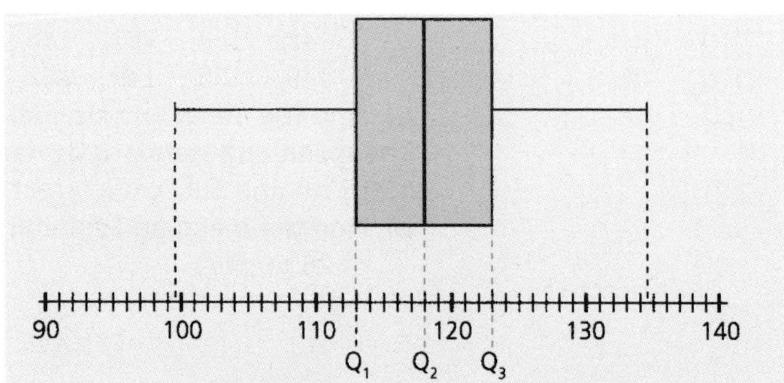

EXERCISES

29.1

1 The temperatures in 20 European capital cities at noon one day in May were recorded in degrees Celsius, and the results are shown in this list.

20	23	16	18	9	17	10	25	22	23
11	18	21	16	24	15	24	17	19	28

a) Represent the data on a stem-and-leaf diagram.
b) Find the median temperature and inter-quartile range.
c) Calculate the mean and standard deviation.
d) Equivalent temperatures at noon one day in August had a mean of 22.4 and a standard deviation of 4.1. Comment on any differences between the temperatures recorded in May and those recorded in August.

2 During 1994, a record was kept of the number of visitors per day to a National Trust property. The results are summarised in the table below.

Number of visitors per day	1–200	201–400	401–600	601–800	801–1000	1001–1500
Number of days	2	29	64	47	26	12

a) Write down the cumulative frequency for each class interval.
b) Draw a cumulative frequency graph for the data.
c) Use your graph to estimate the median number of visitors per day and the inter-quartile range.
d) On days when there are more than 900 visitors, an overflow car park is opened. Estimate the number of days the overflow car park was used in 1994.

3 The prices of houses in a certain town being offered for sale by a particular estate agent are summarised in the following table.

House price (£)	0–30 000	–50 000	–70 000	–100 000	–150 000
Number of houses	3	8	13	5	1

a) Represent the data by a histogram.
b) Estimate the mean and standard deviation of the house prices.
c) Which house price is likely to be an outlier? Explain your reasoning.

4 The heights of the 12 girls in a class of 26 pupils, measured to the nearest cm, were as shown in the following list.

142	153	161	148	163	158
149	155	149	152	156	150

a) Find the mean and standard deviation of these heights.
The mean and standard deviation of the boys' heights are respectively 156.5 cm and 5.37 cm (correct to three significant figures).
b) Find the mean and standard deviation for the heights of the class of 26 pupils.

5 A group of children were given a simple jigsaw to complete. The times taken (in seconds) were recorded and are summarised in the following table.

Time to complete jigsaw (seconds)	Number of children
10–29	4
30–39	6
40–44	7
45–49	6
50–59	9
60–79	8
80–109	5

a) Draw a histogram of these data.
b) Draw a cumulative frequency graph and estimate the median and inter-quartile range.
c) Calculate estimates of the mean and the standard deviation.
d) The same group of children were given the same jigsaw to complete 24 hours later. What might you expect for the distribution of times taken and the various summary measures?

PROBABILITY

Probability is a measure of the chance of a **random event** occurring, on a scale of 0 to 1.

A **probability of 0** means that an event cannot occur (impossible), e.g. scoring a 7 when an ordinary die is thrown.

A **probability of 1** means that an event must occur (certain), e.g. scoring less than 7 when an ordinary die is thrown.

The probability of an event which is neither impossible nor certain may be computed in one of two ways, described below.

Experimental probability (relative frequency)

Suppose a trial is repeated a number of times, e.g. throwing a drawing pin in the air. The probability of a particular event is estimated by comparing the number of times that event (e.g. pin lands on its head, pointing upwards) occurs, with the total number of events. This is called the **relative frequency**.

Example 29.3

A drawing pin is thrown in the air 50 times and lands on its head 30 times. Find the experimental probability of its landing on its head.

Solution

An estimate of the probability of the event 'pin lands on its head' is $\frac{30}{50} = 0.6$.

Theoretical probability (symmetry method)

For any experiment the number of possible (equally likely) **outcomes** is known. For a particular event the number of favourable outcomes is known. The probability of the event is calculated as the number of favourable outcomes divided by the number of possible outcomes.

Example 29.4

A fair coin is tossed three times. Find the probability of getting three heads or three tails.

Solution

Number of (equally likely) possible outcomes = 8:
HHH, HHT, HTH, HTT, THH, THT, TTH, TTT
Number of favourable outcomes = 2:
HHH, TTT
Probability of getting three heads or three tails = $\frac{2}{8}$ = 0.25

Rules of probability

Some formal rules of probability are given below and are illustrated in the examples which follow.

For an event A, we write the probability of A occurring as $P(A)$.

Probability
For any event A:

$$P(A) = \frac{n(A)}{n(E)}$$

where $n(E)$ represents the number of equally likely outcomes and
$n(A)$ represents the number of favourable outcomes.

Multiplication rule
For any two events A and B:
$P(A \text{ and } B) = P(A) \times P(B \mid A)$

where $P(B \mid A)$ is the probability of event B happening if event A has occurred.

Addition rule
For any two events A and B:
$P(A \text{ or } B) = P(A) + P(B) - P(A \text{ and } B)$

In the special case of **mutually exclusive** events (where if one occurs the other cannot occur):
$P(A \text{ or } B) = P(A) + P(B)$, since $P(A \text{ and } B) = 0$

Conditional probability: $P(B \mid A) = \dfrac{P(A \text{ and } B)}{P(A)}$

In the special case of **independent** events (where the occurrence of one event does not influence the occurrence of the other):
$P(A \text{ and } B) = P(A) \times P(B)$

since $P(B \mid A) = P(B)$

Example 29.5

To play a new board game, a blue die and a red die are thrown together. What is the probability that:

a) *the blue die scores a 6,*

b) *the total score is at least 9,*

c) *the scores differ by 2,*

d) *the total score is at least 9 and the scores differ by 2,*

e) *the total score is at least 9 or the scores differ by 2 (or both)?*

Solution

*There are 36 equally likely outcomes which can be illustrated by a **sample space diagram**.*

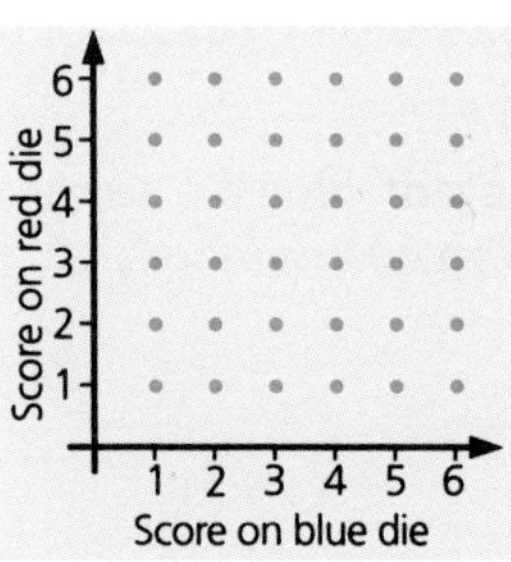

a) *There are six favourable outcomes (a 6 on the blue die and any score on the red die).*

P(blue die scores 6)

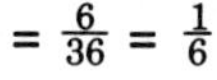

$= \frac{6}{36} = \frac{1}{6}$

b) *There are ten favourable outcomes: (6, 3), (5, 4), (4, 5), (3, 6), (6, 4), (5, 5), (4, 6), (6, 5), (5, 6), (6, 6)*

P(total score ≥ 9) $= \frac{10}{36} = \frac{5}{18}$

c) *There are eight favourable outcomes: (6, 4), (5, 3), (4, 2), (3, 1), (1, 3), (2, 4), (3, 5), (4, 6)*

P(scores differ by 2) $= \frac{8}{36} = \frac{2}{9}$

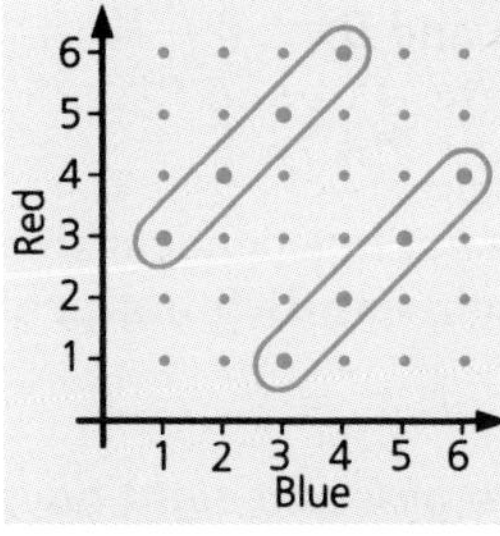

d) *There are two favourable outcomes: (6, 4), (4, 6)*

P(total score ≥ 9 and scores differ by 2) $= \frac{2}{36} = \frac{1}{9}$

e) *There are 16 favourable outcomes:*

P(total score ≥ 9 or scores differ by 2) $= \frac{16}{36} = \frac{4}{9}$

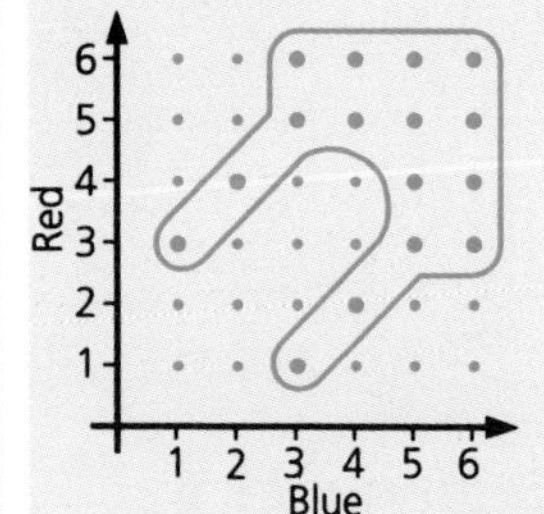

Extending the example

We can use the same conditions as laid down in the last example, to extend our study.

Let A represent the event 'total score is at least 9' and B represent the event 'scores differ by 2'.

Then part **e)** could be obtained using the addition rule:
$P(A) = \frac{5}{18}$ $\quad P(B) = \frac{2}{9}$ $\quad P(A \text{ and } B) = \frac{1}{18}$ $\quad P(A \text{ or } B) = \frac{4}{9}$

$\Rightarrow P(A \text{ or } B) = P(A) + P(B) - P(A \text{ and } B) = \frac{5}{18} + \frac{2}{9} - \frac{1}{18} = \frac{4}{9}$

If events A and B are mutually exclusive then $P(A \text{ and } B) = 0$, in which case the addition rule reduces to:
$P(A \text{ or } B) = P(A) + P(B)$

Now let A represent the event 'total score is at least 9' and B represent the event 'total score = 6'.

Since $P(A \text{ and } B) = 0$, $P(A \text{ or } B) = P(A) + P(B) = \frac{5}{18} + \frac{5}{36} = \frac{15}{36} = \frac{5}{12}$

Tree diagrams

When we are considering a problem which involves combining events, it can be useful to draw a **tree diagram** to represent all different combinations.

Let A represent the event 'blue die scores a 6' and B represent the event 'red die scores an odd number'.

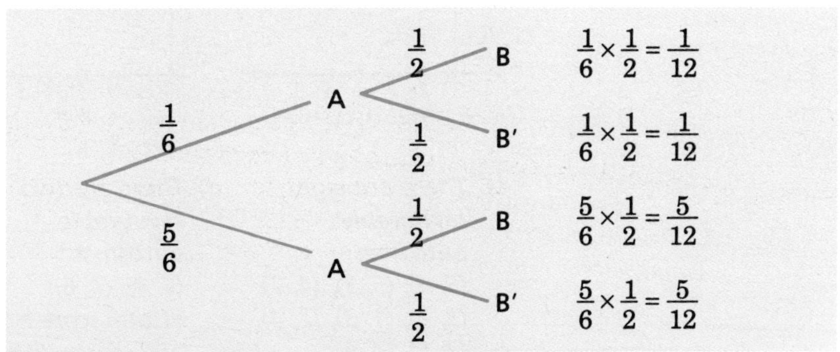

Since events A and B are independent:

$P(A \text{ and } B) = P(A) \times P(B) = \frac{1}{6} \times \frac{1}{2} = \frac{1}{12}$

Note
A' means 'not A'; $P(A) + P(A') = 1$

If events A and B are *not* independent then the second and subsequent sets of branches contain **conditional probabilities**.

Example 29.6

From a class of twelve girls and eight boys, two pupils are chosen at random as class representatives.

a) *What is the probability that both representatives are girls?*

b) *What is the probability of choosing one of either sex?*

Solution

Let A represent the event 'first student chosen is a girl' and B represent the event 'second student chosen is a girl'.
A tree diagram for these dependent events looks like this.

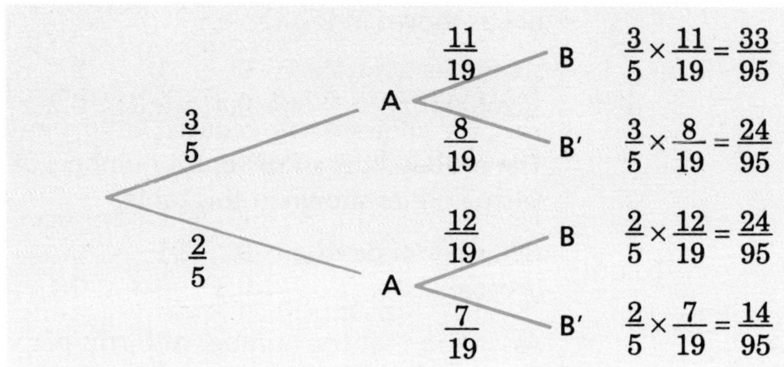

a) *P(both students girls) = P(A and B)* = $\frac{3}{5} \times \frac{11}{19} = \frac{33}{95}$

b) *P(one of either sex) = P(A and B′) + P(A′ and B)*

$= \frac{3}{5} \times \frac{8}{19} + \frac{2}{5} \times \frac{12}{19} = \frac{48}{95}$

EXERCISES

29.2

1 Jean tosses a fair coin five times. Find the probability that:
a) her first two tosses result in heads,
b) she gets just one head in the first three tosses,
c) she gets no heads in all five tosses,
d) she gets more heads than tails in all five tosses.

2 A player is dealt a hand of four cards from a pack of 52 ordinary playing cards which have been shuffled. Find the probability that the player receives:
a) four hearts,
b) four cards of the same suit,
c) four cards of different suits.

3 Sam takes a proficiency test. The probability that he passes is 0.6. If he fails, he takes the test again until he passes, up to a maximum of five attempts. Find the probability that:
a) he needs two attempts before he passes,
b) he passes in fewer than four attempts,
c) he fails to pass the test.

4 Suzie is an amateur weather forecaster. She has a simple system. If a day is fine, the probability that the next day is fine is $\frac{2}{3}$. If a day is wet, the probability that the next day is wet is $\frac{3}{4}$. She classifies a day as either fine or wet.
a) Monday is fine. Draw a tree diagram to show probabilities for Tuesday, Wednesday and Thursday.

b) For Tuesday, Wednesday and Thursday find the probabilities that:
i) all three days are wet, **ii)** just one day is wet.
c) Find the probability that Friday is fine.

5 The probability of different numbers of births per week in a village are as shown in this table.

Number of births	0	1	2	3
Probability	0.4	0.3	0.2	0.1

The probabilities of different numbers of deaths per week in the village are as shown in this table.

Number of deaths	0	1	2	3
Probability	0.1	0.3	0.4	0.2

Assuming that the number of births per week and the number of deaths per week are independent, find the probability that during any particular week:
a) there are exactly two births and two deaths,
b) the numbers of births and deaths are the same,
c) the population increases.

6 An *Instant Win* scratchcard has ten circles, with three smiling faces and seven sad faces hidden underneath them. You can only scratch off three circles. The cost of the card is £1.
If you scratch off three smiling faces you win £10. If you scratch off two smiling faces you win £2. Otherwise you lose.
a) Find the probability of winning: **i)** £10, **ii)** £2.
b) Estimate the amount you are likely to win in 120 goes.
What percentage of the takings would the promoter of the scheme expect to keep in the long run?

CONSOLIDATION EXERCISES FOR CHAPTER 29

1 In an engineering experiment, a student took ten measurements of the time taken by a trolley to run down an inclined plane. The following times, in seconds, were obtained.

8.2	8.7	8.4	9.0	8.4
8.5	8.3	8.7	8.4	8.8

Calculate the mean, the variance and the standard deviation of these times.

(ULEAC Question 1, Specimen Paper P2, 1994)

2 Newborn babies are routinely screened for a serious disease which affects only two per 1000 babies. The result of screening can be positive or negative. A positive result suggests that the baby has the disease, but the test is not perfect. If a baby has the disease, the probability that the result will be negative is 0.01. If the baby does not have the disease, the probability that the result will be positive is 0.02.
a) Find the probability that a baby has the disease, given that the result of the test is positive.

b) Comment on the value you obtain.

(ULEAC Question 3, Specimen Paper P2, 1994)

3 A golfer observes that, when playing a particular hole at his local course, he hits a straight drive on 80 per cent of the occasions when the weather is not windy but only on 30 per cent of the occasions when the weather is windy. Local records suggest that the weather is windy on 55 per cent of all days.

a) Show that the probability that, on a randomly chosen day, the golfer will hit a straight drive at the hole is 0.525.

b) Given that he fails to hit a straight drive at the hole, calculate the probability that the weather is windy.

(NEAB Question 7, Specimen Paper 1, 1994)

4 A player has two dice, which are indistinguishable in appearance. One die is fair, so that the probability of getting a six on any throw is $\frac{1}{6}$, and one is biased in such a way that the probability of getting a six on any throw is $\frac{1}{3}$.

i) The player chooses one of the dice at random and throws it once.

a) Find the probability that a six is thrown.

b) Show that the conditional probability that the die is the biased one, given that a six is thrown, is $\frac{2}{3}$.

ii) The player chooses one of the dice at random and throws it twice.

a) Show that the probability that two sixes are thrown is $\frac{5}{72}$.

b) Find the conditional probability that the die is the biased one, given that two sixes are thrown.

iii) The player chooses one of the dice at random and throws it n times. Show that the conditional probability that the die is the biased one, given that n sixes are thrown, is $\frac{2^n}{2^n+1}$.

(UCLES (Modular) Question 10, Specimen Paper 1, 1994)

5 The following information appears in the Annual Report 1992 of Eurotunnel PLC.

Size of shareholding	Number of shareholders
1–99	133 853
100–499	347 495
500–999	79 087
1000–1499	31 638
1500–2499	27 547
2500–4999	10 655
5000–9999	3842
10 000–49 999	2188
50 000–99 999	283
100 000–249 999	216
250 000–499 999	82
500 000–999 999	54
1 000 000 and over	50
Total	636 990

i) Explain briefly what difficulties would arise in attempting to:
   a) represent the data by means of an accurate and easily comprehensible diagram,
   b) find the mean size of shareholding per shareholder.

ii) Estimate the median size of shareholding.

(UCLES (Modular) Question 6, Specimen Paper 2, 1994)

6 A toy factory produces plastic elephants on a machine. It has been found by experience that five per cent of the output is defective.

a) Find the probability that, in a random sample of ten elephants selected for inspection:
   i) none is defective, ii) more than one is defective.

b) To meet the pre-Christmas rush the factory brings two old machines back into operation. The usual machine (known as machine A) produces 50 per cent of the output, machine B produces 30 per cent and machine C produces 20 per cent. The percentages of defective items in the output of B and C are 10 per cent and 15 per cent respectively.
   Draw a tree diagram to represent this information, marking the corresponding probabilities on each branch. Hence show that the probability that an elephant chosen at random from the whole output is defective is 0.085.

(Oxford and Cambridge Question 13, Specimen Paper 1, 1994)

7 Two events A and B are such that $P(A) = 0.4$, $P(B) = 0.7$, $P(A \text{ or } B) = 0.8$. Calculate:

a) $P(A \text{ and } B)$ b) the conditional probability $P(A|B)$.

(AEB Question 2, Specimen Paper 1, 1994)

8 a) The two box-plots in the figure summarise the percentage unemployment rates for towns in two regions of Britain: the North-west and the South-west.

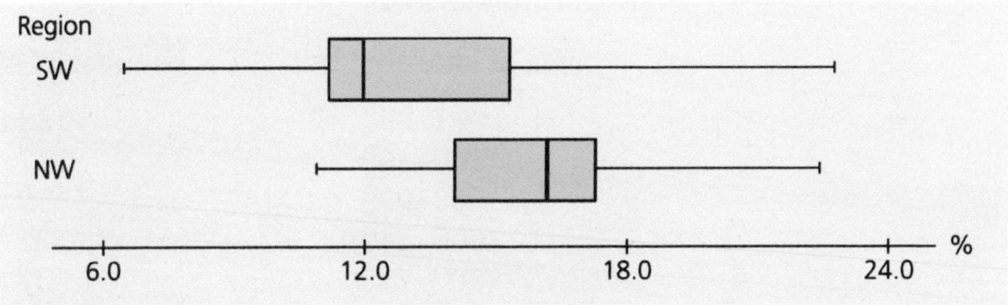

Use these box-plots to write two statements comparing the unemployment rates between the two regions.

b) Suppose that there are 1000 residents who live in your area. You have been asked to obtain a random sample of 20 of these residents from your area.
   Give two reasons why you would reject a method which picks the first 20 people in your area in the telephone book.

(Oxford (Nuffield) Question 1, Specimen Paper 2, 1994)

9 Doctors estimate that three people in every thousand of the population are infected by a particular virus. A test has been devised which is not perfect, but gives a positive result for 95 per cent of those who have the virus. It also gives a positive result for two per cent of those who do not have the virus.
Suppose that someone selected at random takes the test and that it gives a positive result. Calculate the probability that this person really has the virus.

(Oxford (Nuffield) Question 5, Specimen Paper 2, 1994)

10 A committee of three pupils is to be chosen from a class containing eight girls and twelve boys. Assuming that the three pupils are chosen at random, calculate the probability that they are all of the same sex.

(WJEC Question 1, Specimen Paper A1, 1994)

PURE MATHS

Answers

CHAPTER 1
Problem solving in mathematics

Exercises 1.1 *(p.5)*

1 **c) i)** 8.8 cm **ii)** 7 cm **iii)** 530 g
d) $l = 7 + 0.01m$

2 **c)** £178.50, 143
e) i) $d = 500 - 2p$
ii) $s = 1.5p - 125$

3 **b)** $y = -0.4x + 260$ **c) i)** 2000
ii) 2025

Exercises 1.2 *(p.9)*

1 **c) i)** £18 **ii)** side 3.3 ft
d) $c = 0.5l^2 + 6l$ **e) ii)** £23, side 2.7 ft **iii)** $c = 0.5l^2 + 6l + 5$

2 **c) i)** 240 ft **ii)** 28.7 mph
d) $d = v + 0.05v^2$
f) i) 420 ft **ii)** 21.9 mph

Exercises 1.3 *(p.14)*

1 **c)** $w = \dfrac{330}{f}$ **e)** 0.825 m
f) 220 cycles per second

2 **a)** $F = \dfrac{409600000}{d^2}$ **c)** 1.5 N
d) 750 km

3 **a)** 10 cm square base, 5 cm height
b) $A = 2x^2 + \dfrac{2000}{x}$, cube side 7.94 cm

Exercise 1.4 *(p.17)*

1 **c) i)** 14.2 **ii)** 4.96 years
d) $q = 10 \times 1.15^t$

2 **c) i)** 7.43 **ii)** 3.32 weeks
d) $r = 20 \times 0.5^t$

3 **i)** 70°C **ii)** 15.5 min
d) $T = 100 \times 0.8^{0.2t}$

Exercises 1.5 *(p.22)*

1 **b) i)** 11 hours **ii)** 9 November to 2 February
c) $h = 12 + 6\sin(30m°)$

2 **c) i)** 11.7 m **ii)** from 14 to 22 s
d) $h = 9 - 8\cos(10t°)$

CHAPTER 2
Basic algebra

Exercises 2.1A *(p.27)*

1 19 **2** 9.5 **3** 15 **4** 39 **5** $5\frac{1}{3}$
6 8 **7** 9 **8** −3 **9** 19 **10** 10
11 $\frac{1}{3}$ **12** −4 **13** 6 **14** 1 **15** 2
16 −8 **17** 3.5 **18** $\frac{2}{9}$ **19** $-\frac{5}{8}$ **20** 2.5

Exercises 2.1B *(p.27)*

1 83 **2** 0.3 **3** 2.8 **4** 28 **5** 6
6 8.75 **7** 174 **8** −3.5 **9** −50
10 $\frac{1}{3}$ **11** 2 **12** −0.2 **13** $\frac{2}{3}$ **14** 3
15 1.75 **16** −0.25 **17** $1\frac{5}{7}$
18 −0.4 **19** $3\frac{2}{3}$ **20** 3.5

Exercises 2.2A *(p.30)*

1 **a)** $3x + 12$ **b)** $12 - 3x$ **c)** $-4x - 20$
d) $20 - 4x$ **e)** $14p + 21q$
f) $14p - 21q$ **g)** $2x^2 - 6x$
h) $2x^2 + 6x$ **i)** $6pr - 3pq$
j) $-3pq - 6pr$ **k)** $5a^2 - 5ab + 5ac$
l) $5a^2 + 5ab - 5ac$
m) $6r^2 - 10rs + 2rt$
n) $-6r^2 + 10rs - 2rt$

2 **a)** $5x + 11$ **b)** $x - 5$ **c)** $25a - 14$
d) $-5a - 26$ **e)** $x - 8$ **f)** $16 - 17x$
g) $4x^2 + 7x$ **h)** $-9x$ **i)** $6x + 9$
j) $ab - ac$ **k)** $-3p - 5q$ **l)** $5x - 4y$

3 **a)** $5(2a - b)$ **b)** $7(x + 3)$
c) $3(4y + 1)$ **d)** $3x(4y - 1)$
e) $x(7x - 5)$ **f)** $7t(t - 5)$
g) $ab(a - b)$ **h)** $5d(3c + 4e)$
i) $5(r - 4s + 3t)$
j) $3x(2x + 10y - 5)$

Exercises 2.2B *(p.30)*

1 **a)** $18x + 12$ **b)** $-72x - 64$
c) $10 - 14x$ **d)** $3 - 4x$
e) $64a - 16b$ **f)** $24a - 36b$
g) $28m^2 + 24m$ **h)** $-x^2 - 9x$
i) $4x - 20x^2$ **j)** $-10t^2 - 25t$
k) $x^2 + xy - xz$ **l)** $-xy + y^2 + yz$
m) $-24a^2 + 24ab - 3ac$
n) $8a^2 - 12ab - 4a$

2 **a)** $15x + 105$ **b)** $8 - x$
c) $3x - 69$ **d)** $30x + 8$ **e)** $12t + 6$
f) $-31x - 28$ **g)** $67x^2 - 38x$
h) $25x^2 - 2x + 28$ **i)** x
j) $2xy + 4xz$ **k)** $20x - 63$
l) $5a - 47b$

3 **a)** $4(7x + 6)$ **b)** $7(6 - 5a)$
c) $3(1 - x)$ **d)** $2m(2m + 5)$
e) $y(8y - 3)$ **f)** $2a(a + 9)$
g) $xy(8x - 7y)$ **h)** $3a(2ad + b)$
i) $x(1 - x + 7y)$
j) $6t(4t - 7u^2 - 5)$

Exercises 2.3A *(p.32)*

1 3.5 **2** 5.4 **3** −2 **4** 5 **5** −5.8
6 2 **7** −2 **8** 0.25 **9** 0.8 **10** 4
11 −11 **12** 1.25

Exercises 2.3B *(p.32)*

1 3.75 **2** −10 **3** 1 **4** −16 **5** 3
6 −4 **7** −0.5 **8** $11\frac{2}{3}$ **9** $-2\frac{1}{9}$
10 −15.5 **11** $\frac{34}{53}$ **12** $\frac{17}{26}$

Exercises 2.4A *(p.34)*

1 10 **2** 6 **3** 0.3 **4** 6 **5** 15
6 −40 **7** 4 **8** 12 **9** 13 **10** −1
11 −10 **12** 3 **13** −3 **14** 1 **15** 5
16 8 **17** 1 **18** 13 **19** 7 **20** 2

Exercises 2.4B *(p.34)*

1 24 **2** 0.6 **3** $1\frac{1}{9}$ **4** −18 **5** 4.8
6 $-11\frac{9}{17}$ **7** $\frac{1}{74}$ **8** −7 **9** $\frac{2}{3}$ **10** −1
11 −53 **12** $1\frac{9}{67}$ **13** 3.1 **14** $-\frac{1}{6}$
15 $\frac{9}{37}$ **16** −3.6 **17** $3\frac{2}{7}$ **18** 0.25
19 $-1\frac{9}{14}$ **20** $-2\frac{5}{73}$

Exercises 2.5A *(p.38)*

1 $u = v - at$ **2** $a = \dfrac{v - u}{t}$
3 $h = \dfrac{2A}{a + b}$ **4** $a = \dfrac{2A}{h} - b$
5 $c = y - mx$ **6** $x = \dfrac{y - c}{m}$
7 $d = \dfrac{u - a}{n - 1}$ **8** $n = \dfrac{u - a - d}{d}$
9 $y = 5x - 12$ **10** $x = \dfrac{y + 12}{5}$
11 $y = \dfrac{20 - 4x}{5}$ **12** $x = \dfrac{20 - 5y}{4}$
13 $s = \dfrac{3r + 5}{8}$ **14** $r = \dfrac{8s - 5}{3}$
15 $r = \dfrac{5p - 2q}{w}$ **16** $p = \dfrac{2q + rw}{5}$
17 $q = \dfrac{5p - rw}{2}$ **18** $a = \dfrac{de}{b - c}$

19 $d = \frac{a(b-c)}{e}$ **20** $b = \frac{de}{a} + c$

21 $f = \frac{3e}{g} + 7$ **22** $e = \frac{g(f-7)}{3}$

23 $g = \frac{3e}{f-7}$ **24** $a = b$

25 $x = \frac{s-q}{p-r}$ **26** $m = \frac{-5n}{7}$

27 $m = \frac{5n}{3n-5}$ **28** $x = \frac{ab+2}{a+2}$

29 $x = \frac{a+2t}{t-1}$ **30** $x = \frac{bd}{a+b-d}$

31 $x = \frac{y+1}{y-1}$ **32** $x = \frac{2y+5}{2y+1}$

Exercises 2.5B *(p.39)*

1 $x = \frac{a-c}{b}$ **2** $s = \frac{v^2-u^2}{2a}$

3 $t = \frac{2s}{u+v}$ **4** $v = \frac{2s}{t} - u$

5 $m = \frac{y-y_1}{x-x_1}$ **6** $r = \frac{C}{2\pi}$

7 $w = \frac{gp}{f}$ **8** $h = \frac{A}{2\pi r} - r$

9 $x = \frac{7y+3}{8}$ **10** $y = \frac{6-9x}{2}$

11 $y = \frac{7x-3}{9}$ **12** $p = \frac{100i}{cn}$

13 $x = ag - 3y$ **14** $n = \frac{3x-7y}{2m}$

15 $\frac{3a-5m}{2}$ **16** $a = \frac{18x-2}{3}$

17 $x = \frac{tu}{p+4}$ **18** $y = mx - am + b$

19 $x = \frac{a-b}{6}$ **20** $x = -7y$

21 $q = \frac{8p-5pt}{t+7}$ **22** $x = \frac{3ay+14y}{2-3a}$

23 $y = \frac{nx-mn}{m+n}$ **24** $y = \frac{2x-3}{x+1}$

25 $y = \frac{mx-p}{2m-1}$ **26** $f = \frac{8x}{m+5}$

27 $x = \frac{ab+5bn}{b-10n}$ **28** $x = \frac{an+2a-b}{nu+2u+3}$

29 $x = \frac{m}{y-c}$ **30** $u = a - b$

31 $a = \frac{1}{c-b}$ **32** $v = \frac{u}{Fu-1}$

Exercises 2.6A *(p.43)*

1 $x > 2$ **2** $x \geq 4$ **3** $x \leq 4.5$
4 $x < 7$ **5** $x \leq -5$ **6** $x > 1$
7 $x > -3$ **8** $x > 2.125$ **9** $x > 2.2$
10 $x \leq -0.5$ **11** $2 \leq x \leq 5$
12 $-11 < x < -3$ **13** $x < 4, x > 10$
14 $x \leq -3, x > 2$ **15** $4 < x \leq 6$
16 $x > 2$ **17** $-5 \leq x \leq 2$
18 $x < 1\frac{2}{3}, x > 5$ **19** $x \leq -1.2, x \geq 2$
20 $-7 < x < 2.2$

Exercises 2.6B *(p.44)*

1 $x < 0.8$ **2** $x \leq 5$ **3** $x \geq -2.2$
4 $x < 0.875$ **5** $x > -8$ **6** $x \geq \frac{7}{9}$
7 $x < 2.4$ **8** $x \leq \frac{31}{47}$ **9** $x > -2$
10 $x > -\frac{5}{32}$ **11** $-2 < x < 11$
12 $1\frac{2}{3} \leq x \leq 4$ **13** $x < 1, x > 5$
14 $x \leq -6.5, x \geq -2$ **15** $-3 < x < \frac{4}{9}$
16 $x > 2$ **17** $5 \leq x \leq 11$
18 $x < -\frac{4}{7}, x > 2$
19 $x \leq -3.4, x \geq 6.6$
20 $-21 < m < 29$

Consolidation Exercises for Chapter 2 *(p.44)*

1 **a)** –1 **b)** $\frac{4}{9}$ **c)** 0.25
2 **a)** $30x - 35$ **b)** $-15p - 24p^2$
3 **a)** $29x^2 + 7x$ **b)** $-5x$
4 **a)** $3ab(3a + 2b)$
b) $5x(4y + 1 - 4x)$
5 **a)** 5 **b)** –0.6 **c)** –2
6 **a)** 0.7 **b)** $2\frac{5}{9}$ **c)** 2.4 **d)** 0.25
e) 1 **f)** 1
7 **a)** $k = \frac{m+n-fm}{2}$
b) $c = ax - am - bm$
c) $g = \frac{7b}{4a-1}$ **d)** $y = \frac{3x+1}{x+2}$
e) $x = \frac{40m-9}{24m+1}$
8 **a)** $x < 1$ **b)** $x > 1\frac{1}{3}$
c) $-2\frac{2}{3} \leq x \leq -1$
d) $x < -1$ **e)** $x \leq -1, x \geq 4\frac{2}{3}$

CHAPTER 3 Linear functions

Exercises 3.1A *(p.49)*

1 36 km (165, 38)
2 **a)** 6.7, (5, 2.5) **b)** 9.43, (–1,2.5)
c) 8.54, (–2.5, 1) **d)** 12.5, (1, 2.5)
3 **a)** XY = 5, YZ = 5, XZ = √50
b) isosceles, right-angled
c) M(2.5, 0.5) **d)** 12.5
4 **a)** M(7, 0), N(4, 4)
b) PR = 10, MN = 5
5 **a)** (–0.5, 2) **b)** S(0, –1)
6 **a)** E(–1, 3), F(–0.5, 1.5), G(3.5, 2.5), H(3, 4) **b)** EF = GH = √2.5, FG = HE = √17

Exercises 3.1B *(p.50)*

1 (199, 46), 52.9 km
2 **a)** 6 **b)** 10.3 **c)** 7.28 **d)** 11.7
e) 14.2 **f)** 11.2
3 **a)** (6, –1) **b)** (2, 4.5) **c)** (3.5, –2)
d) 1, –0.5) **e)** (2.5, –7)
f) (–1.5, –1.5)
4 **a)** 9.90; 13, 13
c) (3.5, 3), (–2.5, 5.5)
5 **a)** 13.2, 13.2, 13.2, 13.2
b) (3.5, 0.5), (3.5, 0.5)
c) PR = 21.2, QS = 15.6
6 3.354

Exercises 3.2A *(p.54)*

2 AF = $\frac{12}{7}$, HD = $-\frac{2}{5}$,
AD = $-\frac{7}{3}$, CH = $\frac{1}{12}$
3 Gradient DE = gradient EF = 0.5, common point E
4 Gradient AC = gradient DH = –0.4
5 BD is parallel to FH
6 Gradient BC = $-\frac{7}{2}$, gradient DG = $\frac{2}{7}$, product = –1
7 EG and GF, AC and CD
8 I (1, 0)
9 EG and GF perpendicular ⇒ right-angled triangle
10 J (3, 4)

Exercises 3.2B *(p.54)*

2 2, –0.1, 0.15
5 e.g. (–5, –3)
6 1, PN
7 (9, 1)
8 (13, –5), (7, –10) or (3, 7), (–3, 2)
9 (–10, 2)
10 (5, 2), (2, –4) (–7, –4), (–10, 2), (–7, 8), (2, 8)

Exercises 3.3A *(p.60)*

1 **a)** 2, –5 **b)** –3, 10 **c)** 0.2, –2.4
d) –1, 7 **e)** 1.5, 4 **f)** –0.8, 6
g) $\frac{1}{3}$, 1.5 **h)** $-\frac{4}{3}$, 4
2 **i)** (–4, 0), (0, 8) **ii)** (3, 0), (0, 12)
iii) (–6, 0), (0, 3) **iv)** (6, 0), (0, 4)
v) (10, 0), (0, 10) **vi)** (5, 0), (0, 10)
vii) (6, 0), (0, –4.5)
viii) (3, 0), (0, 2.5)
3 **a)** $y = 3x - 5$ **b)** $y = 7 - 2x$
c) $4y = 3x + 10$ **d)** $3x + 2y = 24$
e) $3y = 5x$ **f)** $y = 8$
g) $5y + 2x + 12 = 0$

h) $2y = 5x + 7$
4 a) $y = 4x - 21$ **b)** $y = 3 - 3x$
c) $5y = x + 19$ **d)** $3y + 2x - 3 = 0$
5 a) $5y - x + 2 = 0$
b) $2y + x - 11 = 0$
c) $3y - x - 1 = 0$
d) $5y + 13x - 14 = 0$
6 a) (4, 3) **b)** –1 **c)** $y = x - 1$
7 a) $y = x - 4$ **b)** $y = 7 - x$
d) $\mathrm{AB} = 3\sqrt{2} = 4.24$,
$\mathrm{CD} = \frac{7\sqrt{2}}{2} = 4.9$
Area = 10.5 units^2
8 a) $\mathrm{L} = (0,1.5)$, $\mathrm{M} = (2, -1)$, $\mathrm{N} = (4, 2.5)$,
b) LZ: $y = 1.5 - 0.25x$
MX: $x = 2$ NY: $4y = 3x - 2$
9 a) $3y - 4x + 3 = 0$

Exercises 3.3B *(p.61)*

1 a) 1, 2 **b)** –3, 2 **c)** 0.125, –0.5
d) –2, 0 **e)** 2, $-1\frac{2}{3}$ **f)** $-2\frac{1}{3}$, 3
g) –0.5, –3.5 **h)** –0.1, 4.8 **i)** 0, 7
2 a) $(-\frac{6}{7}, 0)$, (0, 6), 7
b) (1.5, 0), (0, 6), –4
c) $(-2\frac{2}{3}, 0)$, (0, 2), 0.75
d) $(\frac{2}{3}, 0)$, $(0, \frac{2}{9})$, $-\frac{1}{3}$
e) (5, 0), (0, –5), 1
f) $(1\frac{2}{7}, 0)$, (0, 1.8), –1.4
g) $(1\frac{2}{3}, 0)$, $(0, -1\frac{2}{3})$, 1
h) $(-\frac{2}{3}, 0)$, (0, –0.6), –0.9
i) $(-\frac{2}{3}, 0)$, $(0, \frac{4}{9})$, $\frac{2}{3}$
j) (0, –7.1), 0
3 a) $y = 7x + 4$ **b)** $27x + 3y = 2$
c) $12y = 16x + 3$ **d)** $4y = 18x - 11$
4 a) $y = 3x - 23$ **b)** $5x + y = 3$
c) $3y = 2x - 23$ **d)** $3x + 2y = 21$
5 a) $6y = 5x + 2$
b) $13x + 5y + 66 = 0$
c) $6x + 14y + 11 = 0$
d) $4x + 13y = 66.2$
6 a) $3x - y + 6 = 0$
b) $2x + y - 3 = 0$
c) $3x - 7y - 63 = 0$
d) $4x - y - 32 = 0$
e) $2x + 3y + 6 = 0$
f) $9x - y - 30 = 0$
g) $4x + y + 27 = 0$
h) $5x + 2y - 2 = 0$
7 a) (2, 2.5) **b)** 0.3 **c)** $6y + 20x = 55$
8 a) $2x + 5y = 10$ **b)** $2y = 5x + 62$
d) $\sqrt{1044}$ (32.31), $\sqrt{116}$ (10.77), 348
9 $4y = 2x + 7$

Exercises 3.4A *(p.65)*

1 b) $F = 1.8C + 32$ **c)** 32 **d)** –40
2 b) 300 g, 32.0 cm
c) $l = 0.04m + 20$ **d)** 20 cm
e) 250 g, **f)** ≫ 500 g
3 b) 6 years, 796 g
c) $m = 135a - 14$
d) 391 g **e)** 7.5 years
4 b) $R = 0.15T + 15$
d) i) 34.5 ohms, **ii)** $T < -33$°C,
iii) –100°C, no

Exercises 3.4B *(p.66)*

1 b) 4.84 m s^{-1}, 3.46 m s^{-1}
c) $v = 0.64u + 0.36$ **d)** 2.92 m s^{-1}
e) No, yes, poor prediction
2 b) $y = 0.06x + 0.03$
c) 0.15 mm, poor prediction
3 b) Yes, $y = 0.03x + 1.5$
c) 5.43 million **d)** –1.5 million!
4 b) $y = 310 - 2.2x$
c) i) 3: 55.2, **ii)** 3: 20

Exercises 3.5A *(p.71)*

1 $x = 6, y = 6$. **2** $x = 2\frac{4}{7}, y = \frac{2}{7}$.
3 $s = 2, t = 3$. **4** $p = 7, r = 2$.
5 $x = 5, y = 2$. **6** $x = -1.3, y = 3.4$.
7 $x = 1.5, y = 2.5$.
8 (10, 1), (6, 6), (1, 3).
9 All meet at (7,4)
10 a) Lines coincident, infinite number of solutions
b) Lines parallel, no solutions
11 Median equations: $x + 4y = 6$, $x = 2$, $4y = 3x - 2$ (2, 1)
12 a) PQ = PR = $\sqrt{50}$ **b)** $y = 8 - x$
c) $y = x$ **d)** (4, 4) **e)** 15
13 b) $t = 20, s = 50$
14 a) $5x + 3y = 105$, $2x + 4y = 70$
b) $x = 15, y = 10$
15 a) $s = 80x - 400$, $d = -20x + 3400$, x = cost per pair
b) $x = £38$, $s = d = 2640$ pairs

Exercises 3.5B *(p.72)*

1 $x = -\frac{5}{7}, y = 3\frac{1}{7}$
2 $x = \frac{20}{21}, y = -\frac{16}{21}$
3 $x = 4, y = 2$
4 $x = \frac{23}{39}, y = -\frac{11}{39}$
5 $x = \frac{13}{15}, y = -\frac{2}{5}$
6 $x = 1.5, y = 0.75$
7 (0.5, –0.5)
8 $8x + 12y + 9 = 0$ and $4x + 6y = 3$ are parallel (1.5, –1.75), (0., 0.5)
9 (6, –8)
10 a) ii) **b)** i) **c)** iii) **d)** (–12, –12)
11 $(-1\frac{15}{23}, -\frac{17}{23})$
12 a) $-\frac{7}{6}$ **b)** $7y = 6x - 46$
c) $6y = -7x + 28$
d) $(5\frac{47}{85}, -1\frac{69}{85})$, **e)** 15.5
13 42.5
14 40 first class, 10 second class
15 194 g beans, 1330 g rice

Consolidation Exercises for Chapter 3 *(p.74)*

1 a) $y = 0.5x + 3$ **b)** $3\sqrt{5}$
2 a) $\sqrt{32}$, $\sqrt{72}$, $\sqrt{104}$, 90° **b)** 1, –1
c) MA = MB = $\sqrt{26}$
3 $5x - 6y = 2$
4 $3x + 5y - 4 = 0$, $(1\frac{1}{3}, 0)$
5 a) $x + 2y = 8$ **b)** $x + 2y = 3$, (1, 1)
c) AP = BQ = PQ = $\sqrt{5}$
6 a) (–1, 5), (3, 7), (0, 7)
b) 36.9°, 26.6°, 116.6°
7 b) $v = 5.6t - 0.8$
c) 5.6 mph per second
8 c) $b = 0.06a + 25$
d) $b = 0.06a + 40$
e) 7100 g (2 s.f.)
9 a) $b + s = 20$, $30b + 20s = 480$
b) $b = 8$, $s = 12$
10 a) $4c + 3t = 48$, $200c + 90t = 1800$ **b)** 4 caravans, 10 tents

CHAPTER 4 Quadratic functions

Exercises 4.1A *(p.82)*

1 i) b) (0, 5) **c)** (1, 0), (5, 0)
d) $x = 3$ **e)** (3, –4)
ii) b) (0, 3) **c)** (–3, 0), (1, 0)
d) $x = -1$ **e)** (–1, 4)
iii) b) (0, 2.25) **c)** (1.5, 0)
d) $x = 1.5$ **e)** (1.5, 0)
iv) b) (0, –5) **c)** (–2.5, 0), (0.5, 0)
d) $x = -1$ **e)** (–1, –9)
v) b) (0, 25) **c)** (–2.5, 0), (5, 0)
d) $x = 1.25$ **e)** (1.25, 28.125)
vi) b) (0, –16) **c)** (–4, 0), (4, 0)
d) $x = 0$ **e)** (0, –16)
(vii) b) (0, 5) **c)** none **d)** $x = -1$
e) (–1, 4)

(viii) b) (0, 25) **c)** (–2.5, 0), (2.5, 0)
d) $x = 0$ **e)** (0, 25)
(ix) b) (0, 0) **c)** (0, 0), (4, 0)
d) $x = 2$ **e)** (2, –12)
(x) b) (0, –20) **c)** none **d)** $x = 4$
e) (4, –4)
(xi) b) (0, 0) **c)** (0, 0), (3, 0)
d) $x = 1.5$ **e)** (1.5, 4.5)
(xii) b) (0, –9) **c)** (–3, 0) **d)** $x = -3$
e) (–3, 0)

2 $x = 4.5$
3 (–1.75, 15.125)
4 a) 192 m **b)** 192 m **c)** $x = 96$
5 a) 23.6 min **b)** 41.6 min **c)** 10.5 min, 9.7

Exercises 4.1B *(p.83)*

1 a) (0, 4) **b)** (0, –3) **c)** (0, 7)
d) (0, 0) **e)** (0, 8) **f)** (0, 0)
g) (0, –7) **h)** (0, –3.5)
2 $x = 0.5$
3 (6.25, –15.125)
4 a) 0.818 sec **b)** 0.235 sec, 1.67 m
5 a) 6 cm **b)** 10 cm

Exercises 4.2A *(p.89)*

1 a) $x^2 + 13x + 40$ **b)** $x^2 - 3x - 40$
c) $x^2 + 3x - 40$ **d)** $x^2 - 13x + 40$
e) $2x^2 + 15x + 7$ **f)** $2x^2 + 17x - 9$
g) $5t^2 - 2t - 3$ **h)** $4y^2 - 11y + 6$
i) $15x^2 + 23x + 4$ **j)** $14u^2 - 9u + 1$
k) $4t^2 + 4t - 15$ **l)** $3a^2 - 8ab - 3b^2$
m) $16x^2 + 8x + 1$ **n)** $4x^2 - 12x + 9$
o) $16z^2 - 1$ **p)** $c^2 - 49$
q) $-3x^2 + 11x + 20$
r) $-2a^2 + ab + b^2$
s) $15x^2 + 40x - 15$
t) $12x^2 - 60xy + 75y^2$
2 a) $2x^2 - 6x - 29$ **b)** 1 **c)** $26x$
d) $x^2 + 17x - 21$
e) $13x^2 - 23x - 39$
f) $2x^2 + 17x - 9$
g) $-a^2 - 18ab - 6b^2$
h) $22x^2 - 5xy - 8y^2$
i) $7e^2 - 25ef - 3f^2$
j) $-20x^2 - 140x + 202$
3 a) $(x + 5)(x + 4)$ **b)** $(x - 5)(x - 4)$
c) $(x - 5)(x + 4)$ **d)** $(x + 5)(x - 4)$
e) $(x + 7)(x + 4)$ **f)** $(x - 7)(x - 5)$
g) $(t + 13)(t - 7)$ **h)** $(y - 9)(y + 7)$
i) $(x + 3)^2$ **j)** $(x - 7)^2$
k) $(2x + 1)(x + 1)$ **l)** $(3x + 5)(x + 4)$
m) $(3p - 4)(p + 2)$
n) $(2p - 1)(5p - 4)$
o) $(2x - y)(x + 3y)$ **p)** $(4a - 5b)^2$
q) $5(4x + 3)(x + 4)$
r) $(3 - 4z)(3z + 8)$ **s)** $3(2x + 5)^2$
(t) $x(2x - 7)(x + 3)$
4 a) $(x - 7)(x + 7)$ **b)** $(10 - t)(10 + t)$
c) $(p - q)(p + q)$ **d)** $(8s - 5)(8s + 5)$
e) $(3 - 4y)(3 + 4y)$
f) $(11t - 1)(11t + 1)$
g) $(6a - b)(6a + b)$
h) $(3e - 7f)(3e + 7f)$
i) $10(x - 2)(x + 2)$
j) $5(3 - 2x)(3 + 2x)$
k) $7(a + b)(a - b)$
l) $3(5p - 7q)(5p + 7q)$
m) $x(3 - 7x)(3 + 7x)$
n) $5(8x - 5)(8x + 5)$
o) $ab(a - b)(a + b)$

Exercises 4.2B *(p.90)*

1 a) $x^2 + 5x + 4$ **b)** $x^2 - 3x - 4$
c) $x^2 + 3x - 4$ **d)** $x^2 - 5x + 4$
e) $4x^2 + 31x + 42$ **f)** $6x^2 + x - 7$
g) $4y^2 + 11y - 45$
h) $2t^2 - 23t + 56$
i) $27x^2 + 57x + 28$
j) $41x^2 + x - 1$ **k)** $32x^2 - 92x + 63$
l) $2x^2 - 3xy - 2y^2$
m) $25x^2 + 30x + 9$
n) $36x^2 - 60x + 25$ **o)** $x^2 - 64$
p) $49u^2 - 4$ **q)** $-3x^2 + 2x + 21$
r) $-16a^2 + 6ab + b^2$
s) $216x^2 - 48x - 64$
t) $81x^3y + 108x^2y^2 + 36xy^3$
2 a) $2x^2 + 6x - 44$ **b)** $-6x$
c) $2x^2 + 54$ **d)** $4x^2 + 8x + 4$
e) $14x^2 + 30x + 27$ **f)** $x - 4$
g) $39a^2 + 37ab - 84b^2$
h) $-4xy - 8y^2$ **i)** $54ab - 10a^2$
j) $78x^2 + 293x - 273$
3 a) $(x + 2)(x + 5)$ **b)** $(x - 2)(x + 5)$
c) $(x - 5)(x + 2)$ **d)** $(x - 5)(x - 2)$
e) $(x + 8)(x + 4)$ **f)** $(8 - x)(x - 4)$
g) $(u - 14)(u + 6)$ **h)** $(t - 8)(t + 2)$
i) $(x - 5)^2$ **j)** $(x + 11)^2$
k) $(x + 1)(3x - 1)$ **l)** $(x - 5)(3x - 1)$
m) $(x - 3)(6x + 1)$ **n)** $(x + 6)(5x - 2)$
o) $(a + 2b)(7a - b)$ **p)** $(9x + 2y)^2$
q) $2(2x + 5)(3x - 4)$
r) $-2(3x - 5)(4x - 7)$ **s)** $8(5x - 7)^2$
t) $x(25x^2 - 195x - 84)$
4 a) $(x - 4)(x + 4)$ **b)** $(7 - u)(7 + u)$
c) $(a - 2b)(a + 2b)$
d) $-4(2x - 3)(2x + 3)$
e) $(5n - 3m)(5n + 3m)$
f) $(3x - 1)(3x + 1)$
g) $4(a - 3b)(a + 3b)$
h) $(8x - 9y)(8x + 9y)$
i) $5(3x - 2)(3x + 2)$
j) $9(a - b)(a + b)$
k) $3(9p - 4q)(9p + 4q)$
l) $5a(2a - b)(2a + b)$
m) $13(7x - 11)(7x + 11)$
n) $-t(4t - 9)(4t + 9)$
o) $4xy(3x - 5y)(3x + 5y)$

Exercises 4.3A *(p.93)*

1 a) translation 3 units // y-axis
b) translation –7 units // y-axis
c) translation 2 units // x-axis
d) translation –5 units // x-axis
e) stretch, factor 4 // y-axis
f) stretch, factor $\frac{1}{3}$ // y-axis
g) reflection in x-axis
h) translation –3 units // x–axis
2 a) $y = x^2 - 5$ **b)** $y = x^2 + 3.5$
c) $y = (x - 4)^2$ **d)** $y = (x + 7)^2$
e) $y = 3x^2$ **f)** $y = \frac{1}{3}x^2$
3 a) stretch, factor 2 // y-axis, translation 6 units // y-axis
b) translation 3 units // y-axis, stretch, factor 2 // y-axis
c) translation –5 units // y-axis, stretch, factor 0.2 // y-axis
d) stretch, factor 0.5 // y-axis, translation –1 units // y-axis
e) translation –4 units // y-axis, reflection in x-axis
f) reflection in x-axis, translation 4 units // y-axis
g) translation –4 units // x-axis, stretch, factor 3 // y-axis
h) translation 1 unit // x-axis, stretch, factor 5 // y-axis
i) translation 2 units // x-axis, translation 3 units // y-axis
j) translation –5 units // x-axis, translation –7 units // y-axis
k) translation –3 units // x-axis, stretch, factor 5 // y-axis, translation 5 units // y-axis
l) translation –4 units // x-axis, reflection in x-axis
m) translation –4 units // x-axis, reflection in x-axis, translation 7 units // y-axis
n) translation –6 units // x-axis, stretch, factor 0.4 // y-axis, translation –4.4 units // y-axis
o) translation 3 units // x-axis,

stretch, factor 2 // y-axis, reflection in x-axis, translation 10 units // y-axis
p) translation –2 units // x-axis, stretch, factor 5 // y-axis, translation 3 units // y-axis, reflection in x-axis

4 **a)** $y = 2x^2 - 5$ **b)** $y = -3x^2$
c) $y = 0.5(x - 3)^2$ **d)** $y = -(x + 5)^2$
e) $y = (x - 2)^2 + 5$ **f)** $y = 9 - x^2$
g) $y = -(x^2 + 5)$
h) $y = 3(x + 4)^2 - 7$
i) $y = 4 - (x - 6)^2$
j) $y = 5(x - 3)^2 - 12$

Exercises 4.3B *(p.94)*

2 **a)** $y = 6x^2$ **b)** $y = (x - 3)^2$
c) $y = x^2 + 3$ **d)** $y = (x + 0.5)^2$
e) $y = 0.2x^2$ **f)** $y = x^2 - 8$

3 **a)** $y = (x - 5)^2 + 5$ **b)** $y = 7x^2 - 1$
c) $y = -4x^2$ **d)** $y = 0.5(x^2 + 3)$
e) $y = -x^2 - 6$ **f)** $y = 2 - 3x^2$
g) $y = -0.7(x - 5)^2$
h) $y = 7(x + 2)^2 + 3$
i) $y = 2(x - 9)^2 - 3$
j) $y = 1 - 9(x - 4)^2$

Exercises 4.4A *(p.99)*

1 **a)** $y = (x + 2)^2 + 7$
b) $y = (x - 3)^2 - 4$
c) $y = (x - 2.5)^2 - 9.25$
d) $y = (x + 1.5)^2 - 10.25$
e) $y = (x + 0.5)^2 + 0.75$
f) $y = (x - 3.5)^2 - 12.25$

2 **a)** $y = 2(x - 2)^2 - 3$
b) $y = 3(x + 1)^2 - 10$
c) $y = 2(x + 1)^2 + 7$
d) $y = 5(x - 0.5)^2 + 6.75$
e) $s = 3(t - \frac{1}{6})^2 + 4\frac{11}{12}$
f) $d = 5(t + 0.2)^2 - 12.2$
g) $y = 9 - (x - 1)^2$
h) $y = 12.25 - (x + 1.5)^2$
i) $y = 15\frac{1}{3} - 3(x + 1\frac{2}{3})^2$
j) $y = 35.02 - 12(z - \frac{23}{24})^2$

3 **a)** $x = 1$, $(1, -4)$
b) $x = -2.5$, $(-2.5, 1.75)$
c) $x = 1.25$, $(1.25, 5.4375)$
d) $x = 1$, $(1, 7)$
e) $x = -\frac{5}{6}$, $(-\frac{5}{6}, -10\frac{1}{12})$
f) $x = 0.3$, $(0.3, -0.45)$
g) $x = 5$, $(5, 25)$
h) $x = -0.5$, $(-0.5, 12.25)$
i) $t = 0.3$, $(0.3, 15.45)$
j) $t = 0$, $(0, 4)$

4 **a)** **ii)** $x = 500$, **iii)** $(500, -222)$
b) 2.5 m

5 **b)** **ii)** 720 m^2, **iii)** 8 m, 7.5 m

Exercises 4.4B *(p.100)*

1 **a)** $y = (x - 5)^2 + 3$
b) $y = (x - 1.5)^2 + 0.75$
c) $y = (x + 6)^2 - 9$
d) $y = (x - 2.5)^2 - 5.75$
e) $y = (x + 3.5)^2 - 12.25$
f) $y = x^2 + 7$

2 **a)** $y = 2(x + 3)^2 - 11$
b) $y = 3(x - 6)^2 + 1$
c) $y = 8(x - 5)^2 - 95$
d) $s = 3(t - 2.5)^2 - 23.75$
e) $y = 8(x - 0.125)^2 + 8.875$
f) $y = 4 - (x - 7)^2$
g) $v = 9 - 2(t + 2)^2$
h) $y = 0.5 - 2(x - 0.5)^2$
i) $y = 3(x - \frac{1}{3})^2 + \frac{2}{3}$
j) $y = -1.2 - 5(x - 0.6)^2$

3 **a)** $(0, 7)$, $(2, 3)$ **b)** $(0, 8)$, $(0, 8)$
c) $(0, -4)$, $(-1, -5)$
d) $(0, -2)$, $(-2.5, -8.25)$
e) $(0, 7)$, $(-2.5, 0.75)$
f) $(0, 57)$, $(4, -7)$
g) $(0, 1)$, $(-0.5, 1.25)$
h) $(0, -5)$, $(-\frac{2}{3}, -7\frac{2}{3})$
i) $(0, 0)$, $(1\frac{1}{6}, 4\frac{1}{12})$
j) $(0, -23)$, $(0.9, -18.95)$

4 **b)** 225 feet **c)** 62 feet
d) 125 feet

5 **b)** 1.2 m by 2.4 m, 1.6 m by 1.6 m

Consolidation Exercises for Chapter 4 *(p.103)*

1 **a)** $x^2 + 5x - 36$ **b)** $6x^2 - 5x - 6$
c) $25 - 20x + 4x^2$
d) $7x^2 - 7x - 69$ **e)** $43x + 96$

2 **a)** $(x - 7)(x - 13)$
b) $(x + 11)(2x - 1)$
c) $2(2x + 5)(3x + 2)$
d) $4(5x^2 - 30x + 9)$
e) $5(7x - 1)(7x + 1)$

3 **a)** $(x - 10)^2 - 9$
b) $(x + 1.5)^2 + 7.75$
c) $2(x + 5)^2 - 15$
d) $8.25 - (x - 0.5)^2$
e) $4(x - 2)^2 - 9$

4 **a)** $4(x + 0.75)^2 + 4.75$
b) $(-0.75, 4.75)$, $x = -0.75$

5 **a)** $(2, 1)$, $x = 2$
b) $(-1, -3)$, $x = -1$
c) $(1, 5)$, $x = 1$

6 **a)** $a = -18$ **b)** $x = -2$, $(-2, 3)$

7 **a)** $p = 2$, $q = 9$ **b)** 9, –2

8 **b)** **i)** $y = (x + 5)^2$, **ii)** $y = 3x^2$,
iii) $y = x^2 - 2$ **c)** **i)** $y = 3(x + 5)^2 - 2$
ii) $y = 3(x + 5)^2 - 2$
iii) $y = 3(x + 5)^2 - 6$

9 $a\left(x + \frac{b}{2a}\right)^2 + \frac{4ac - b^2}{4a}$ **a)** $b^2 > 4ac$
b) $b^2 = 4ac$ **c)** $b^2 < 4ac$

10 **a)** $y = 2x^2 - 10$
b) $y = 0.1875x^2 + 0.75x + 8$
c) $y = x^2 - 4x$
d) $y = 16.25 - 4x - x^2$

CHAPTER 5
Quadratic equations and inequalities

Exercises 5.1A *(p.110)*

1 0, 7 **2** 4, 8
3 3 **4** –3, 4
5 –2, 4 **6** 4
7 0, 2.5 **8** –5, 5
9 0, 17 **10** 0, 2
11 –4, 0 **12** –7, 7
13 –5, 5 **14** –7, 7
15 –1.5, 1.5 **16** –2.5, 2.5
17 none **18** none
19 –9, 9 **20** none
21 –2, –3 **22** –2, –4
23 –6, –9 **24** 1, 3
25 –5, 8 **26** –3, 20
27 –1, –4 **28** –4, 1
29 –3, 4 **30** 1
31 0.5, 5 **32** $-2, 2\frac{1}{3}$
33 $-\frac{1}{3}, -\frac{1}{2}$ **34** $-2\frac{1}{2}, -\frac{2}{3}$
35 –7, 0.4 **36** $-1\frac{1}{2}, -\frac{1}{3}$
37 $2, 2\frac{1}{3}$ **38** $-1\frac{1}{3}, 2\frac{1}{2}$
39 $-\frac{2}{3}$ **40** –4
41 –3, 2 **42** –1.5, 4
43 –3, –2, 2, 3 **44** –2, 2
45 –3, –2 **46** 1.2 sec, 2 sec
47 **b)** 70 m by 45 m
48 **b)** 5, 12, 13
49 8 m by 15 m

Exercises 5.1B *(p.112)*

1 0, –2 **2** –3, 3
3 –3, 0.5 **4** $-\frac{1}{3}$
5 –2, 4 **6** $\frac{2}{5}, 1\frac{1}{2}$
7 –2.5, 0 **8** none
9 –4, 0 **10** 0, 3

11 $0, 1\frac{2}{3}$ **12** $-\sqrt{17}, \sqrt{17}$
13 $-\sqrt{26}, \sqrt{26}$ **14** $-\sqrt{10}, \sqrt{10}$
15 −4, 4 **16** $-3\frac{1}{3}, 3\frac{1}{3}$
17 none **18** none
19 −5, 0 **20** $0, 2\frac{3}{7}$
21 −3, −1 **22** −6, −4
23 −17, −3 **24** 4, 8
25 2, 5 **26** −6, 1
27 −1, 3 **28** 3
29 −2, 3 **30** 1
31 $-\frac{1}{7}, 1$ **32** −7, 2
33 $-3, -\frac{1}{3}$ **34** $-5, -2\frac{1}{3}$
35 $-1\frac{1}{2}, \frac{1}{3}$ **36** −0.25, 0.2
37 1.5, 2 **38** −2.5, 1.5
39 $-\frac{1}{3}$ **40** 3
41 $1\frac{1}{3}, 1\frac{1}{2}$ **42** 2, 6
43 −3, 6 **44** 4, 6
45 −1, 1 **46** **b)** 6, 8, 10
47 **a)** 27, 29 **b)** 35, 44
48 11

Exercises 5.2A *(p.117)*

1 2, 6 **2** −18, 1
3 −14, 6 **4** −23, 11
5 $-3, -\frac{1}{3}$ **6** $-2, \frac{5}{9}$
7 $-1, 1\frac{1}{11}$ **8** 0.438, 4.56
9 −2.12, 0.786
10 −0.773, −0.369
11 −4, 1 **12** $\frac{2}{3}, 3$
13 0.382, 2.62 **14** $-\frac{2}{3}, 5$
15 −0.75, 2.5 **16** −0.907, 2.57
17 −1.5, 0.8 **18** none
19 −136, 25 **20** $-12, 12\frac{1}{12}$
21 0.177, 2.82 **22** none
23 −1.45, 3.45 **24** 0.314, 3.19
25 −0.577, 0.577 **26** 1.62

Exercises 5.2B *(p.117)*

1 −10, −2 **2** −4, 5
3 −9, 20 **4** 5, 9
5 −1.13, 2.27 **6** −3.5, −1.25
7 0.551, 5.45 **8** −8.53, −0.469
9 −0.135, 2.47 **10** −1.44, 0.693
11 −7, −1 **12** −1.69, 1.19
13 −0.549, 1.22 **14** 0.4, 2
15 $-2\frac{1}{2}, \frac{2}{3}$ **16** none
17 $0, 1\frac{2}{7}$ **18** $\frac{3}{4}, 1\frac{1}{3}$
19 $-10\frac{2}{7}, 14$ **20** −75, 80
21 −1.18, 0.847 **22** −1.73, 1.73
23 4.27, 7.73 **24** −18.5, 30.5
25 −11.6, 0.603

Exercise 5.3A *(p.120)*

1 $-\frac{1}{3} < x < \frac{1}{3}$ **2** $x < -\frac{1}{3}, x > \frac{1}{3}$
3 all x values **4** no x values
5 $x < -3, x > 7$ **6** $x < 0, x > \frac{3}{2}$
7 $-2 < x < 4$ **8** $x < 2, x > 5$
9 $-1 < x < 7$ **10** all x values
11 $x < -3, x > 0.5$ **12** $-\frac{2}{3} < x < 2$
13 all x values except −2
14 all x values except 2
15 $x < -4.83, x > 0.828$
16 $x < -0.828, x > 4.83$

Exercises 5.3B *(p.121)*

1 $-2.646 < x < 2.647$
2 no x values
3 $x < -3.606, x > 3.606$
4 all x values
5 $-5 \le x \le 2$
6 $-4 \le x \le -1$
7 $x < 0, x > 2$
8 $x < 1, x > 4$
9 $x \le 0.586, x \ge 3.414$
10 $-0.449 \le x \le 0.449$
11 $-1.618 \le x \le 0.618$
12 $-6 < x < 1$
13 $-3.137 < x < 0.637$
14 $x < -0.291, x > 2.291$
15 $-0.646 < x < 4.646$
16 $-0.079 \le x \le 1.579$

Exercises 5.4A *(p.123)*

1 (−2, 0), (4, 6)
2 (3, 2)
3 (−2.5, 18.5), (0.5, 3.5)
4 (−10, 10), (3, 3.5)
5 none
6 $(-2, 2), (8, -\frac{2}{9})$
7 (−1, 1), (3, 4)
8 (−2, −11), (11, 2)
9 $(1\frac{2}{3}, 2\frac{2}{3}), (3, 2)$
10 (−9.52, 5.88), (7.65, 1.59)

Exercises 5.4B *(p.123)*

1 (−1, 4), (4, 19)
2 (0, 4), (−5, 9)
3 (−5.15, 10.1), (1.49, 5.68)
4 none
5 (2, 2.25), (3, 1.5)
6 (1, 5), (2.5, 8)
7 (−0.4, −2.8), (2, 2)
8 (4, 2), (−6, 12)
9 (10, 9)
10 (13, 10)

Consolidation Exercises for Chapter 5 *(p.124)*

1 $x = -0.5$ or 1.5 $y = -3.5$ or -1.5
2 $x = \frac{1}{3}$ or $x = 1$
3 $\{x: -5 < x < 1.75\}$
4 **a)** $(5x + 3)(x - 2)$
b) $\{x: x < -0.6 \text{ or } x > 2\}$
5 **a)** $-3 < x < 4$ **b)** $1 < x < 2$
c) $-1 < x < 1.5$ **d)** $x < -2, x > 3$
e) $x < 3, x > 5$
f) $x < -\frac{1}{3}, x > 3$
g) $x < 0.5, x > 2$
6 (2, −1)
7 **a)** $(3x - 7)(2x - 1)$
b) $x < \frac{1}{2}, x > 2\frac{1}{3}$
c) $(0, 7), (2\frac{2}{3}, 4\frac{1}{3})$
8 $-1\frac{15}{17}, 2$
9 (−2, −8), (5, 13)
10 (−1, 1), (3, 5)
11 $(-\frac{2}{3}, 9), (3, -2)$
12 (2, 3), (15, −10)

CHAPTER 6 Polynomial functions

Exercises 6.1A *(p.129)*

1 **b)** (−1.57, 13.1), (1.57, −1.13)
c) −3, 1, 2
2 **b)** none **c)** −0.246
3 **b)** (0, 0), (2.67, 9.48) **c)** 0, 4
4 **b)** (0, 0), (2.67, −9.48) **c)** 0, 4
5 **b)** (−0.881, 1.82), (3.21, 70.5)
c) 5.29
6 **b)** (0.451, 6.37), (2.22, 9.11)
c) 3.70
7 **b)** (−1.57, −4.14), (0.130, 10.2), (3.69, −94.0) **c)** −2, −1, 1, 5
8 **b)** (1.14, −7.07) **c)** −1.10, 2.05
9 **b)** (−1.5, 4.69) **c)** −2.26, 1
10 **b)** (−4.65, 187), (−0.904, −0.240), (2.56, 151)
c) −6.19, −1, −0.807, 4

Exercises 6.1B *(p.130)*

1 **b)** (−1.73, 10.4), (1.73, −10.4)
c) −3, 0, 3
2 **b)** none **c)** 1
3 **b)** (0, 0), (1.67, 4.63) **c)** 0, 2.5
4 **b)** (0, 0), (1.67, −4.63) **c)** 0; 2.5
5 **b)** (1, −3), (3.67, 6.48) **c)** 0.209, 2, 4.79
6 **b)** (−4, −14), (−0.667, 4.52)
c) −5.47, −1.74, 0.210
7 **b)** (−2.83, −64), (0, 0), (2.83, 64)

c) –4, 0, 4

8 b) (–1.94, –18.1), (0.558, 3.58), (1.38, 2.48) c) –2.81, –0.264

9 b) (–2.58, 341), (1.22, 8.60), (2.35, 21.1) c) –4, 3

10 b) (–0.573, 3.11) c) –1.08, 0.160

Exercises 6.2A *(p.133)*

1 a) $2x^2 + 4x - 3$, $6x - 13$
b) $x^3 + 10x + 2$, $x^3 - 4x^2 - 4x - 4$
c) $4x^2 - 5x + 24$, $4x^3 - 2x^2 - 3x - 6$
d) $3x^4 + x^3 + 5x^2 - 9x - 8$, $3x^4 - x^3 + 5x^2 + 9x - 8$
e) $x^5 + 4x^3 + x^2 - x + 5$, $-x^5 + 6x^3 - x^2 - x + 11$

2 a) $3x^3 + 5x^2 + 17x + 10$
b) $10x^3 - 17x^2 + 8x - 1$
c) $x^4 + 5x^3 + 19x^2 + 34x + 40$
d) $2x^4 + 13x^3 - 3x^2 + 73x - 30$
e) $2t^4 - 6t^3 - 21t^2 + 11t - 30$
f) $t^5 - 2t^4 - 18t^3 + 27t^2 - 22t + 20$
g) $10x^6 - 11x^5 - 24x^4 + 66x^3 + 82x^2 - 33x - 45$
h) $2x^5 + 7x^4 + 6x^3 + 59x^2 + 10x - 21$

3 a) $13x^3 - 13x^2 + 79x - 43$
b) $7t^3 - 5t^2 - t + 23$
c) $8x^4 - 5x^3 + 78x^2 - 5x + 29$
d) $-4x^3 + 90x^2 - 12x - 92$
e) $7t^4 - 17t^3 + 123t^2 - 285t + 118$

4 a) $x^3 + 6x^2 + 2x - 10$ i) 3 ii) 1
b) $-2x^2 + 6x + 29$ i) 2 ii) 0
c) $-2x^3 - 3x^2 + 20x + 30$ i) 3 ii) –2
d) $-x^5 - 5x^4 + 10x^3 + 53x^2 - 30$ i) 5 ii) 10
e) $-2x^6 - 13x^5 + 5x^4 + 136x^3 + 159x^2 - 60x - 90$ i) 6 ii) 136
f) $x^4 + 13x^3 + 35x^2 - 6x - 109$ i) 4 ii) 13
g) $10x^5 + 55x^4 - 6x^3 - 330x^2 + 1009$, i) 5 ii) –6
h) $16x^4 + 96x^3 + 216x^2 + 216x + 81$ i) 4 ii) 96

Exercises 6.2B *(p.134)*

1 a) $4x^2 + 3x - 4$, $3x + 4$
b) $x^3 - 2x^2 + x - 6$, $x^3 - 4x^2 - x + 6$
c) $3x^3 + x^2 - 6x$, $x^3 - x^2 - 6x + 8$
d) $x^4 - 2x^3 - 2x^2 + 10$, $x^4 - 4x^3 + 2x^2 + 2$
e) $x^5 - 2x^2 + 8$, $-x^5 + 4x^2 + 8x$

2 a) $x^4 - x^2 - 2$
b) $2x^3 - 10x^2 + 9x + 9$
c) $2x^4 - 9x^3 + 12x^2 - 11x + 6$
d) $-18x^5 + 9x^4 - 30x^3 + 24x^2 - 14x + 4$
e) $12x^5 - 27x^3 + 8x^2 + 15x - 10$
f) $x^8 - 3x^7 + 4x^6 + 2x^4 - 7x^3 + 11x^2 - 4x$
g) $18x^8 - 15x^7 + 6x^5 + 28x^4 - 9x^3 - 17x^2 + 18x + 6$
h) $4x^6 - 4x^5 - 33x^4 + 33x^3 + 62x^2 - 36x - 40$

3 a) $3x^3 - 2x^2 + 9x - 5$ b) $4x$
c) $2x^3 - 2x^2 - 4x - 12$
d) $x^4 - x^3 + 9x^2 - 3x + 31$

4 a) $x^3 + 4x^2 + 9$ b) $-x^2 + 3x - 11$
c) $3x^3 + 9x^2 - 6x + 4$
d) $3x^3 - 2x^2 + 27x - 18$
e) $2x^5 + 6x^4 + 18x^3 + 54x^2$
f) $6x^5 - 18x^4 - 243x^2 - 729$

Exercises 6.3A *(p.138)*

1 $(x - 4)(x - 1)(x + 3)$
2 $(x - 5)(x + 2)(2x - 1)$
3 $(x - 2)(x - 1)(x + 3)$
4 $(x - 2)^2(x + 1)$
5 $(x - 2)(x + 3)(3x - 1)$
6 $(x + 1)(2x - 3)(2x + 5)$
7 $(x - 2)(x^2 - 5x + 1)$
8 $(x - 3)(x - 1)(x + 2)(x + 4)$
9 $(x - 3)(x + 1)(2x - 1)(2x + 7)$
10 $(x - 2)^3(2x + 5)$
11 $(x - 1)(x + 3)(x^2 + x + 4)$
12 $(x + 4)(5x - 2)(x^2 - x - 5)$

Exercises 6.3B *(p.138)*

1 $(x - 3)(x - 2)(x - 1)$
2 $(x - 2)(x^2 - x + 3)$
3 $5(x - 2)(x - 1)(x + 1)$
4 $(2x - 1)(x^2 + 1)$
5 $(x + 3)(2x - 1)(2x + 1)$
6 $(x + 1)(x^2 + x + 5)$
7 $(x - 4)(x - 1)(x + 1)(x + 3)$
8 $(x + 1)^2(x^2 - x + 1)$
9 $(x^2 - 2)(3x^2 + 7x - 2)$
10 $(x - 3)(3x + 1)(x^2 + 4)$

Exercises 6.4A *(p.142)*

1 a) –2, 0.5, 5 c) $-2 < x < 0.5$ or $x > 5$, $x \le -2$ or $0.5 \le x \le 5$

2 a) –3.5, –1, 0.5, 3 c) $x < -3.5$ or $-1 < x < 0.5$ or $x > 3$, $-3.5 \le x \le -1$ or $0.5 \le x \le 3$

3 a) – 2.5, 2 c) $-2.5 < x < 2$ or $x > 2$, $x \le -2.5$ or $x = 2$

4 a) –4, 1, 2 c) $-4 - < x < 1$ or $x > 2$, $x \le -4$ or $1 \le x \le 2$

5 a) –1, 1.5, 3 c) $-1 < x < 1.5$ or $x > 3$, $x \le -1$ or $1.5 \le x \le 3$

6 a) –2, 3 c) $x \le 3$, $x > 3$

7 a) 0.6, 5 c) $0.6 < x < 5$ or $x > 5$, $x \le 0.6$ or $x = 5$

8 a) –2, 0.697, 4.30 c) $-2 < x < 0.697$ or $x > 4.30$, $x \le -2$ or $0.697 \le x \le 4.30$

9 a) 3 c) $x > 3$, $x \le 3$

10 a) –2, –1, 1,5 c) $x < -2$ or $-1 < x < 1$ or $x > 5$, $-2 \le x \le -1$ or $1 \le x \le 5$

11 a) – 6.19, –1, –0.807, 4 c) $x < -6.19$ or $-1 < x < -0.807$, $x > 4$, $-6.19 \le x \le -1$ or $-0.807 \le x \le 4$

12 a) –1, 1.5, 4 c) $x < -1$ or $x > 4$, $-1 \le x \le 4$

13 a) –4, 3 c) $x < -4$ or $x > 3$, $-4 \le x \le 3$

14 a) –3, –1, 1 c) $-1 < x < 1$ or $x > 1$, $x \le 1$ or $x = 1$

15 $f(x) = x^3 + x^2 - 6x$

16 $f(x) = x^3 + 2x^2 - 4x - 8$

17 $f(x) = 18 + 18x - 2x^2 - 2x^3$

18 $f(x) = x^3 + 0.5x^2 - 12.5x + 6$

Exercises 6.4B *(p.143)*

1 a) –3, –1.5, 2 c) $x < -3$ or $-1.5 < x < 2$, $-3 \le x \le -1.5$ or $x \ge 2$

2 a) –2.5, 1, 1.5 c) $x < -2.5$ or $1 < x < 1.5$, $-2.5 \le x \le 1$ or $x \ge 1.5$

3 a) –4, 0.382, 2.62 c) $x < -4$ or $0.382 < x < 2.62$, $-4 \le x \le 0.382$ or $x \ge 2.62$

4 a) –1, –0.457 or 1.457 c) $x < -1$ or $-0.457 < x < 1.457$, $-1 \le x \le 0.457$ or $x \ge 1.457$

5 a) 2 c) $x < 2$, $x \ge 2$

6 a) –1, 1.5 c) $x < -1$, $x \ge -1$

7 a) –4, –3, –2, –1 c) $-4 < x < -3$ or $-2 < x < 1$, $x < -4$ or $-3 < x < -2$, or $x > -1$

8 a) –3.30, –2, 0.303, 3 c) $-3.30 < x < 2$ or $0.303 < x < 3$, $x \le -3.30$ or $-3 \le x \le 0.303$ or $x \ge 3$

9 a) –1.85, –1, 1, 1.35 c) $-1.85 < x < -1$ or $1 < x < 1.35$, $x \le -1.85$ or $-1 \le x \le 1$ or $x \ge 1.35$

10 a) −3, −1.73, 1, 1.73 **c)** $x < -3$ or $-1.73 < x < 1$ or $x > 1.73$, $-3 \le x \le -1.73$ or $1 \le x \le 1.73$

11 a) −3, −2, −1, 1, 2, **c)** $x < -3$ or $-2 < x < -1$ or $1 < x < 2$, $-3 \le x \le -2$ or $-1 \le x \le 1$ or $x \ge 2$

12 a) −2.5, −1, 0, 2 **c)** $x < -2.5$ or $0 < x < 2$, $-2.5 < x \le 0$ or $x \ge 2$

13 a) 2.5, 5 **c)** $2.5 < x < 5$, $x \le 2.5$ or $x \ge 5$

14 a) 1, 3 **c)** $x > 3$, $x \le 3$

15 $f(x) = -x^3 + 6x^2 - 8x$

16 $f(x) = \frac{1}{4}x^3 - \frac{3}{4}x^2 - \frac{5}{2}x + 6$

17 $f(x) = -2x^3 - x^2 + 8x - 5$

18 $f(x) = 15x^3 - 21x^2 - 18x$

Consolidation exercises for Chapter 6 *(p.144)*

1 b) $f(x) = (x - 3)(x + 4)$ $(x^2 + 2x + 2)$, $x^2 + 2x + 2 = 0$ has no real roots

2 $f(x) = 6x^2 - 10x^2 - 5x + 10$

3 a) $t = -34$, **b)** − 2.5, 0.5, 1.5

4 b) ii) $p = 1$, $q = 3$, $a = 0.5$ **iv)** $h = 0.1$

5 b) $(x - 3)(x^2 - 2)$ **c)** $-\sqrt{2}$, $\sqrt{2}$, 3 **d)** $-\sqrt{2} - 1$, $\sqrt{2} - 1$, 2

6 $a = 1$, $b = -4$, $(x + 1)$

7 $x < 2$ or $3 < x < 4$

8 b) $b = 1$, $c = 2$, $a = 2$ **c)** $y = -2x^3 + 10x^2 14x + 6$

CHAPTER 7
Differentiation 1

Exercises 7.1A *(p.150)*

1 $2x$ **2** $10x$ **3** $6x^2$
4 $-6x^2$ **5** $4x^3$ **6** $12x^3$
7 $2x + 1$ **8** $2x + 5$ **9** $2x - 7$
10 $3x^2 - 3$

Exercises 7.1B *(p.150)*

1 1 **2** −5
3 $8t$ **4** $-6t$
5 $9t^2$ **6** $5t^4$
7 $2t - 2$ **8** $2t - 3$
9 $3t^2 - 12t + 12$ **10** $2t - 8$

Exercises 7.2A *(p.150)*

1 $20x$ **2** $-21x^2$
3 −5 **4** $3x^2 - 4x$
5 $21x^6 - 20x^4 + 6x^2$
6 $x + \frac{1}{3}$ **7** $9x^2 - 12$
8 $2x - 8$ **9** $4x + 1$
10 $3x^2 - 16x + 15$
11 $4x$, 20 **12** $2 - 3x^2$, 1.25
13 $6x - 21$, 0 **14** $15x^2 - 20x$, 35
15 $-4x + 3$, −5
16 $6x^2 - 10x + 1$, 1
17 $30x - 2$, −62
18 $3x^2 - 6x + 10$, 19
19 $2 - 8t$, −6
20 $45t^2 - 4t - 8$, 41
21 (1, 7)
22 (3, 2), (−3, 8)
23 (0, 0), (1.5, −1.69)
24 $(-\frac{1}{3}, -\frac{4}{27})$, (1, 0)
25 (1, −7), (−1, −23)
26 −1 **27** −7, 7
28 −1, 1.5, 3 **29** (−1, 11)
30 A$(-\frac{2}{3}$, 9.59), B(1.5, −4.25)

Exercises 7.2B *(p.154)*

1 $15x^2$ **2** $4x^3 + 3$
3 $-10 + 20x - 30x^2$
4 5 **5** 0
6 $1.24x^3 - 0.62x$ **7** $3u^2 - 2u + 5$
8 $98x^6 - 50x^4$ **9** $6t - 11$
10 $30t - 70$ **11** 0, 0
12 −5, −5 **13** $4x$, 4
14 $8x$, 0 **15** $-2x - 1.6$, −5.6
16 $6t^2 + 2t - 5$, 15
17 $4t^3 - 6t^2 + t - 1$, −1
18 $4t - 10$, 6
19 $3t^2 + 4t - 1$, 6
20 $36t^2 - 96t + 15$, −33
21 (−0.707, −0.25), (0, 0), (0.707, −0.25)
22 $(-1\frac{2}{9}, \frac{1}{9})$ **23** (2, −9)
24 (−1, 0), (2, −6)
25 (1.423, 0.385), (2.577, −0.385)
26 a) i) 11, **ii)** 5, **iii)** −7, 7 **b)** $(-\frac{5}{6}, -4\frac{1}{12})$
27 a) i) −4, **ii)** 21, −12, 28 **b)** (−3.694, 12.60), (0.361, −20.75)
28 a) 1 **b)** $y = x - 28$
29 (−0.5, −8.125), (1, −7) **30** b), d)

Exercises 7.3A *(p.157)*

1 a) 6 m s^{-1}, **b)** 6.05 m, 0.9 s, **c)** 11 m s^{-1}
2 a) 0, 89, 208, 405, 728 **b)** 90, 96, 150, 252, 402
3 a) $4\pi r^2$ **b)** 100π **c)** $333\frac{1}{3}$
4 a) 12 m **b)** $20t - 6t^2$ **d)** −4 m s^{-2} **e)** $3\frac{1}{3}$ sec
5 a) 4000 **b)** 5

Exercises 7.3B *(p.158)*

1 a) 0 m s^{-1}, **b)** 10 m s^{-1} **c)** −10 m s^{-2}
2 b) $f(t) = 5t^2$ **c)** $10t$
3 a) $24 - 6t$ **c)** 48 cm, s = 21, 36, 45, 21, v = 18, 12, −6, −18
4 b) 6π **c)** $16/\pi$
5 b) i) 55, **ii)** −40 **c)** $8\frac{2}{3}$

Exercises 7.4A *(p.161)*

1 B: (0, 10) max, C: (0, 0) inflexion
2 a) 13, −7 **b)** (0.25, 1.125)
3 a) 21, −6 **b)** (1.414, −1.657), (−1.414, 9.657)
4 (2.5, 20.25) max
5 (−1.5, 36.75) max
6 (2.5, 0) min
7 (0, 9) max $(2\frac{2}{3}, -\frac{13}{27})$ min
8 (−3, −17) min, $(2, 24\frac{2}{3})$ max
9 (1, 4) max, (3, 0) min
10 (2.5, 0) inflexion
11 $(\frac{1}{3}$, −1.19) min, (−1, 0) max
12 (0, 0) inflexion, (3, 27) max
13 (−2, 19) max, (0, 3) min, (2, 19) max
14 (0, −10) min, (2, 6) inflexion

Exercises 7.4B *(p.162)*

1 a) −3, 5 **b)** (−1.5, −6.25)
2 a) −24, 12, −24 **b)** (−1, −7), (2, 20)
3 a) −12, 0, 12 **b)** (0, 0), (0.5, 0.0625), (1, 0)
4 (−2.5, −0.25) min
5 (1.5, 51.25) max
6 (−5, 0) max
7 $(\frac{1}{3}, -\frac{37}{54})$ max, (2, −3) min
8 (−1.22, −3.25) min, (0, −1) max, (1.22, −3.25) min
9 (−2, 0) inflexion
10 (−1, −4) min, (1, 12) max, (3, −4) min
11 (3, −13.75) min
12 none
13 (0.606, 0.326) max
14 (0, −5) min, (2, 27) max, (3, 22) min

Exercises 7.5A *(p.164)*

1 a) $y = 30 - x$ **b)** $A = 30x - x^2$ **c)** 15, 225 m^2, square
2 Site A: **a)** $2x + y = 60$ **b)** $A = 60x - 2x^2$ **c)** 450 m^2, $x = 15$, $y = 30$

Site B: **a)** $2x + y = 68$
b) $A = 68x - 2x^2 - 40$
c) 538 m^2, $x = 17$, $y = 34$
3 b) $x = 9$, $y = 8432$ m^2
4 a) 12 miles from A **b)** B
5 a) $V = 4x^3 - 260x^2 + 4000x$
b) 10, 18 000 cm^3
6 b) $V = \pi h(400 - \frac{1}{4}h^2$
c) $h = 23.1$, $r = 16.3$, 19 350 cm^2
7 $x = 100$, $r = \frac{100}{\pi} \approx 31.8$

Exercises 7.5B *(p.166)*

1 $\frac{2}{3}$ m
2 75 m by 100 m
3 51 mph
4 A_1: 25 m square, A_2: 25 m by 12.5 m
5 c) $\frac{18}{\pi + 4} \approx 2.52$ m^2

Consolidation Exercises for Chapter 7

1 a) 4 **b)** 8 0, $2\frac{2}{3}$
2 a) $12x^3 + 12x^2$ **b), c)** (0, 0) inflexion, (–1, –1) min
d) $(-1\frac{1}{3}, 0)$, (0, 0)
3 a) $-2x + 2$ **b)** 4
4 a) $3x^2 - 1$ **b)** 6 **c)** 2 **d)** $y = 2x + 8$
e) (1, 6)
5 a) $15x^2 - 4x$ **b)** 0, $\frac{4}{15}$
6 a) $3t^2 - 18t + 24$, $t = 2, 4$ **b)** 4 m
c) –6 m s^{-2}, 6 m s^{-2} **d)** 18 m
7 a) $3x^2 + 6x + 4$
8 a) $6x^2 - 18x + 12$ **b)** (1, 1), (2, 0)
c) 0.5, 2
9 a) 0, 0 **b)** (1, 1)
c) i) (–1.414, –0.943), (1.414, 0.943)
ii) (–3.08, –0.520), (–1.59, –1.00), (1.59, 1.00), (3.08, 0.520)
d) i) –2.45, 0, 2.45 **ii)** 0
10 $a = 13500$, $b = -20$ £462.50, 4250, £898 125

CHAPTER 8 Trigonometry 1

Exercises 8.1A *(p.179)*

1 a) (90, 1), (–90, –1) (–180, –1), (180, –1), (0, 1) b) (–135, –0.7), (45, 0.7)
2 a) i) –315, –45, 315, **ii)** –225, –135, 135, 225 **b) i)** –315, –225, 135, **ii)** –135, –45, 225, 315 **c) i)** –315, –135, 225, **ii)** –225, –45, 135, 315
3 a) –330°, –150°, 210°
b) –300°, –120°, 240°
c) –210°, –30°, 150°, 330°
d) –240°, –60°, 120°, 300°
4 a) –90°, 90° **b)** (–180°, 0°), (0°, 0°), (180°, 0°)
5 a) 44.4°, 135.6° **b)** 14.5°, 165.5°
c) 203.6°, 336.4° **d)** 240°, 300°
e) 45.6°, 314.4° **f)** 75.5°, 284.5°
g) 113.6°, 246.4° **h)** 150°, 210°
i) 35°, 215° **j)** 68.2°, 248.2°
k) 158.2°, 338.2° **l)** 120°, 300°
6 a) –81.4°, 81.4° **b)** none
c) –45°, 135° **d)** –120°, 120°
e) –160.7°, –19.3°
f) –95.7°, 84.3° **g)** none
h) 60°, 120° **i)** –36.9°, 143.1°
j) –111.8°, 68.2° **k)** –150°, 30°
l) –76.0°, 104.0°

Exercises 8.1B *(p.180)*

3 (38.2, 0.786)
4 a) $\frac{\sqrt{3}}{2}, -\frac{1}{2}, -\sqrt{3}$
b) $\frac{1}{2}, -\frac{\sqrt{3}}{2}, -\frac{1}{\sqrt{3}}$
c) $-\frac{1}{2}, -\frac{\sqrt{3}}{2}, \frac{1}{\sqrt{3}}$
d) $-\frac{\sqrt{3}}{2}, \frac{1}{2}, -\sqrt{3}$
5 a) 36.9, 143.1 **b)** 5.7, 174.3
c) 228.6, 311.4 **d)** 197.5, 342.5
e) 53.1, 306.9 **f)** 84.3, 275.7
g) 138.6, 221.4 **h)** 107.5, 252.5
i) 31.0, 211.0 **j)** 5.7, 185.7
k) 143.1, 323.1 **l)** 163.3, 343.3
6 a) –126.9, –53.1 **b)** –110.5, 110.5
c) –91.1, 88.9 **d)** none
e) 1.15, 178.9 **f)** –21.8, 158.2
g) none **h)** –19.9, 19.9
i) –89.4, 90.6 **j)** 30, 150
k) –116.7, 116.7 **l)** –168.7, 11.3

Exercises 8.2A *(p.183)*

1 i) b) (0, 3), **c)** $2 \le y$ 4
ii) b) (0, 0), (180, 0), (360, 0),
c) $-1 < y < 1$
iii) b) (0, 0), (180, 0), (360, 0),
c) $-0.5 < y < 0.5$
iv) b) (0, 2), (90, 0), (270, 0),
c) $-2 < y < 2$
v) b) (0, 1), **c)** $1 < y < 3$
vi) b) (0, –3), (90, 0), (270, 0),
c) $-3 < y < 3$
vii) b) (0, 1), (30, 0), (90, 0), (150, 0), (210, 0), (270, 0), (330, 0), **c)** $-1 < y < 1$
viii) b) (0, 1), (180, 0),
c) $-1 < y < 1$
ix) b) (0, 0), (90, 0), (180, 0), (270, 0), (360, 0), **c)** $\infty < y < \infty$
x) b) (0, 1), (90, 0), (270, 0),
c) $-1 < y < 1$
xi) b) (0, 0.5), (150, 0), (330, 0),
c) $-1 < y < 1$
xii) b) (0, 1), (135, 0), (315, 0),
c) $\infty < y < \infty$
2 b) i) stretch, factor 0.5 // x-axis
ii) translation –1 unit // y-axis
iii) stretch, factor $\frac{1}{3}$ // x-axis
iv) translation –60 // x-axis
v) stretch, factor $\frac{1}{3}$ // x-axis, reflection in x-axis
vi) stretch, factor $\frac{1}{3}$ // x-axis, stretch, factor $\frac{1}{2}$// y-axis
vii) stretch, factor $\frac{1}{2}$ // y-axis, translation 3 units // y-axis
viii) stretch, factor 2 // x-axis
ix) stretch, factor 2 // x-axis, stretch, factor 5 // y-axis
x) translation –60 // x-axis, stretch, factor 3 // y-axis
xi) translation –60 // x-axis, translation –5 unit // y-axis
xii) translation –60 // x-axis, stretch, factor 3 // y-axis, translation –5 unit // y-axis.
3 i) b) $y = 4\cos x$, **c)** (0, 4), (180, –4), (360, 4) **ii) b)** $y = \cos x$ 3, **c)** (0, –2), (180, –4), (360, –2)
iii) b) $y = \cos(x + 90)$,
c) (90, –1), (270, 1)
iv) b) $y = \cos(0.4x)$, **c)** (0, 1)
v) b) $y = \cos 3x$ 5, **c)** (0, –4), (60, –6), (120, –4), (180, –6), (240, –4), (300, –6), (360, –4)
vi) b) $y = 4\cos(x + 90)$, **c)** (90, –4), (270, 4)
vii) b) $y = -\cos(x$ 120), **c)** (120, –1), (300, 1)
viii) b) $y = 0.5\cos(x + 60)$,
c) (120, –0.5), (300, 0.5)
ix) b) $y = 3\cos(x + 45) + 3$,
c) (145, 0), (315, 6)
b) $y = 1.5\cos(0.5x)$ 3.5,
c) (0, –2), (360, –5)

Exercises 8.2B (p.184)

1 **i) b)** (0, 3), **c)** $3 \le y \le 5$
ii) b) (0, 2), **c)** $1 \le y \le 3$
iii) (0, 3), (90, 0), (270, 0)
c) $-3 \le y \le 3$
iv) b) (0, 0), (180, 0), (360, 0)
c) $-2.5 \le y \le 2.5$
v) b) (0, 2), (90, 0), (270, 0),
c) $-2 \le y \le 2$
vi) b) (0, 0), (60, 0), (120, 0), (180, 0), (240, 0), (300, 0), (360, 0) **c)** $-\infty < y < \infty$
vii) b) (90, 0), (270, 0),
c) $-\infty < y < \infty$
viii) b) (0, 0.5), (150, 0), (330, 0), **c)** $-1 \le y \le 1$
ix) b) (0, –1), (30, 0), (150, 0), (270, 0), **c)** $-2 \le y \le 0$
x) b) (0, 2.863), **c)** $1 \le y \le 3$

2 **b) i)** stretch, factor $\frac{1}{3}$ // y-axis
ii) translation 4 units // y-axis
iii) stretch, factor $\frac{1}{6}$ // x-axis
iv) translation 90 units // x-axis
v) stretch, factor $\frac{1}{6}$ // x-axis, reflection in x–axis
vi) stretch factor $\frac{1}{2}$ // x-axis, stretch factor $\frac{1}{3}$ // y-axis
vii) stretch, factor $\frac{1}{3}$ // y-axis, translation 1 unit // y-axis
viii) stretch, factor 3 // x-axis
ix) stretch, factor 3 // x-axis, stretch, factor 6 // y-axis
x) stretch, factor 2 // x-axis, translation 2 units // y-axis

3 **b) i)** $y = 0.5\sin x$ **ii)** $y = \sin 4x$
iii) $y = \sin x - 2$ **iv)** $y = \sin(x - 60°)$
v) $y = 2\sin x + 1$ **vi)** $y = -\sin 2x$
vii) $y = \sin(x + 30°) - 2$
viii) $y = 3 - \sin x$
ix) $-2.5\sin(x - 90°)$
x) $y = -0.25\sin(0.5x)$

Exercises 8.3A (p.187)

1 **a)** 36.9°, 323.1° **b)** 22.6°, 157.4°
c) 60.3°, 240.3° **d)** 107.5°, 252.5°
e) 202.0°, 338.0°
f) 112.6°, 292.6° **g)** none
h) 106.3°, 286.3°

2 **a)** 26.6°, 63.4°, 206.6°, 243.4°
b) 33.7°, 146.3°, 213.7°, 326.3°
c) 18.9°, 108.9°, 198.9°, 288.9°
d) 268.9° **e)** none **f)** 236.1°
g) 24.6°, 95.4°, 144.6°, 215.4°, 264.6°, 335.4° **h)** 318.8°

3 **a)** 26.6°, 153.4°, 206.6°, 333.4°
b) 38.3°, 141.7°, 218.3°, 321.7°
c) 52.9°, 127.1°, 232.9°, 307.1°
d) 41.4°, 138.6°, 221.4°, 318.6°
e) 37.8°, 142.2°, 217.8°, 322.2°
f) 60°, 120°, 240°, 300°

4 **a)** –148.9°, –31.1°, 31.1°, 148.9°
b) none **c)** –149.0°, –31.0°, 31.0°, 149.0° **d)** –130°, 110° **e)** none
f) –150°, –120°, –60°, –30°, 30° 60°, 120°, 150° **g)** –60°, 60°
h) –131.8°, –120.2°, –59.8°, –48.2°, 12.2°, 23.8°, 84.2°, 95.8°, 156.2°, 167.8°
i) –93.7°, 86.3° **j)** none

Exercises 8.3B (p.187)

1 **a)** 44.4°, 135.6° **b)** 66.4°, 293.6°
c) 69.4°, 249.4° **d)** 204.6°, 335.4°
e) 104.5°, 255.5° **f)** 113.2°, 293.2° **g)** 216.9°, 323.1°
h) 152.7°, 207.3°

2 **a)** 22.8°, 157.2°, 202.8°, 337.2°
b) 11.8°, 78.2°, 191.8°, 258.2°
c) 16.3°, 76.3°, 136.3°, 196.3°, 256.3°, 316.3° **d)** 249.7°
e) none **f)** 18.4°, 71.6°, 198.4°, 251.6° **g)** 230.8° **h)** 136.4°

3 **a)** 48.2°, 131.8°, 228.2°, 311.8°
b) 39.2°, 140.8°, 219.2°, 320.8°
c) 42.4°, 137.6°, 222.4°, 317.6°
d) 49.8°, 130.2°, 229.8°, 310.2°
e) 37.8°, 142.2°, 217.8°, 322.2°
f) 61.9°, 118.1°, 241.9°, 298.1°

4 **a)** –166.0°, –104.0°, 14.0°, 76.0°
b) none **c)** –116.6°, –63.4°, 63.4°, 116.6° **d)** –178.5°, –29.5°
e) 41.8°, 138.2° **f)** none
g) –148.3°, –121.7°, –58.3°, –31.7°, 31.7°, 58.3°, 121.7°, 148.3° **h)** none **i)** –60°, 60°
j) –144.4, –54.4, 35.6, 125.6

Exercises 8.4A (p.192)

1 **b)** 03:00 and 15:00, 09:00 and 21:00 **c) i)** 4.2 m, **ii)** 04:20 and 16:20 **iii)** 1 h 20 min
d) i) 4.17 m **ii)** 04:23 and 16:23
iii) 1 h 26 min

2 **b) i)** 7 h 40 min **ii)** 22 Oct, 8 Feb
iii) 4 months 12 days

3 **a)** $a = 4$, $b = 30$, $c = 17$
b) i) 13.5, **ii)** 0, 1, 7, 8, 9, 10, 11, 12

4 **a)** $a = -0.5$, $c = 1.5$
c) i) after 0.5, 2.5, 3.5, 5.5 s, **ii)** 0.6 to 0.9 s, 2.1 to 2.4 s, 3.6 to 3.9 s, 5.1 to 5.4 s

5 **a)** $a = 1$, $b = 30$, $c = 1$
b) $a = 2$, $b = 30$, $c = 3$
c) $a = 1$, $b = 29$, $c = 1$

Exercises 8.4B (p.194)

1 **a)** $a = -0.25$, $b = 192$
c) $0.35 < t < 0.59$, $1.38 < t < 1.53$, $2.23 < t < 2.47$, $3.16 < t < 3.41$, $4.10 < t < 4.34$
d) i) 1.875 m, **ii)** 1.522 m

2 **a)** $h = \sin 12t$ **b)** $h = 5 + \sin 12t$
c) i) 5.87 m, **ii)** 4.13 m,
iii) 4.79 m

3 **b) i)** 2, **ii)** 264
c) $y = 5\sin(47520t)$

4 **a)** $y = 4\cos(22.5t)$
b) $y = 0.5\sin(5400t)$

5 **a)** $a = 2$, $b = 30$, $c = 3$
b) $a = 1$, $b = 29$, $c = 1$

Consolidation Exercises for Chapter 8 (p.196)

1 4, 2 (90, 4), (270, 4) (0, 2), (180, 2), (360, 2)

2 6, 2 (90, 6), (210, 6), (330, 6) (30, 2), (150, 2), (270, 2)

3 **i) b)** $y = 2\sin x$, **c)** (90, 2), (270, –2)
ii) b) $y = \sin x - 5$, **c)** (90, –4), (270, –6) **iii) b)** $y = \sin(x - 60°)$,
c) (150, 1), (330, –1)
iv) b) $y = -\sin(x + 120°)$,
c) (150, 1), (330, –1)

4 **b)** $y = 2.5\cos(0.5x) - 3$
c) (0, –0.5), (360, –5.5)

5 **c)** (60, 0), (180, 0), (300, 0), 2, –1.125

6 3, –1, 60°, 120°, 240°, 300°

7 30°, 150°, 210°, 330°

8 **a)** 7 m **b)** 13 m 29.03

9 **a)** 15 hr 6 min **b)** 29 June, 1 January or 27 December
c) $n = 12 - 6\cos(\frac{360}{365}x + 10)$

10 **a)** $d = 2\sin(92\,160t)$

CHAPTER 9
Integration 1

Exercises 9.1A (*p.204*)

1 **a)** 23.75 **b)** 15 **c)** 14.5 **d)** $42\frac{2}{3}$
2 **a)** 31.5 **b)** 66 **c)** 42.5 **d)** 0.16
3 **a)** (2, 0), 2 **b)** (–1, 0), 15 **c)** (6, 0), 20 **d)** (–4, 0), $\frac{1}{2}$
4 **a)** 40 m s^{-1} **b)** 20 s **d)** 400 m **e)** $A(t) = 40t - t^2$, 75 m
5 **b)** $A(t) = 0.4t^2 + t$ **d)** 7.6 million

Exercises 9.1B (*p.206*)

1 **a)** 41.25 **b)** $53\frac{1}{3}$ **c)** 8.5 **d)** 36.25
2 **a)** 8 **b)** 7.5 **c)** 2.5 **d)** $5\frac{5}{6}$
3 **a)** 0.5 **b)** 6.5 **c)** 0.25 **d)** 8.5
4 **a)** 24 m s^{-1} **b)** 4 s **d)** $24t - 3t^2$, 48 m **e)** $\frac{1}{12}$s
5 **b)** $\frac{t^2}{3500} + \frac{t}{175}$ **d)** 1 litre **e)** 33 s

Exercises 9.2A (*p.212*)

1 **a)** 9.13 **b)** 0.696 **c)** 2.19 **d)** 35
2 **b)** $y = 4 - x^2$ **c)** 3.00 **d)** 3.10, 3.14, 3.14
3 81.2
4 **a)** 4.725 m^2 **b)** 425 m^3

Exercises 9.2B (*p.213*)

1 **a)** 68 **b)** 0.34 **c)** 4.02 **d)** 31
2 40.6 m
3 **a)** $6 + \frac{\pi}{2}$ **b)** 7.57 **d)** 7.37 (3 sf) **e)** about 50
4 **b)** 7.59, 30.4 **c)** 7.16, 31.0 **d)** 3.3%, 1.2%

Exercises 9.3A (*p.218*)

1 **a)** $12\frac{1}{3}$ **b)** $29\frac{1}{3}$ **c)** $42\frac{2}{3}$ **d)** $20\frac{1}{4}$
2 **a)** 3.75 **b)** $\frac{1}{6}$ **c)** 12.8 **d)** 72
3 $6\frac{1}{3}$ **4** $57\frac{1}{6}$ **5** $21\frac{1}{3}$ **6** 24
7 $\frac{1}{6}$ **8** $10\frac{2}{3}$ **9** $6\frac{1}{6}$ **10** $58\frac{2}{3}$
11 31

Exercises 9.3B (*p.219*)

1 **a)** $85\frac{1}{3}$ **b)** 20 **c)** 41.25 **d)** 22.5
2 **a)** 18.6 **b)** 1.5 **c)** $10\frac{2}{3}$ **d)** 256
3 $21\frac{1}{3}$ **4** $\frac{2}{3}$ **5** $\frac{1}{18}$ **6** 0.5
7 0.5625 **8** $33\frac{1}{3}$
9 $2\frac{2}{3}$ **10** $\frac{1}{3}$ **11** $143\frac{5}{6}$

Exercises 9.4A (*p.222*)

1 **a)** $7x + c$ **b)** $x^4 + c$ **c)** $2x^3 + c$ **d)** $\frac{1}{3}x^3 - 2x^2 + 5x + c$
e) $\frac{1}{8}x^8 + c$ **f)** $x^3 - 3x^2 + c$
g) $\frac{1}{3}x^3 - \frac{1}{2}x^2 - 12x + c$
h) $x^4 - 1\frac{1}{3}x^3 + c$ **i)** $\frac{1}{2}x^2 - x + c$
j) $\frac{1}{4}(x + 2)^4$ or $\frac{1}{4}x^4 + 2x^3 + 6x^2 + 8x + c$
2 **a)** $2x^2 + 7x + x$ **b)** $\frac{1}{4}x^4 - \frac{1}{3}x^3 + c$
c) $\frac{2}{5}x^5 + 4x^2 + c$
d) $2\frac{1}{3}x^3 - 14x^2 + 28x + c$
e) $\frac{1}{3}x^3 + \frac{1}{2}x^2 - 2x + c$
f) $\frac{1}{8}(2x - 1)^4 + c$ or $2x^4 - 4x^3 + 3x^2 - x + c'$
3 **a)** $y = x^2 + 3$ **b)** $y = 2x^2 + 5x + 2$
c) $y = 10x - x^3 - 8$
d) $y = \frac{1}{4}x^4 + \frac{1}{3}x^3 + \frac{5}{12}$
e) $y = 2x^3 - \frac{3}{2}x^2 - 10$
f) $y = \frac{2}{3}x^3 - x^2 - 5$
g) $y = 3x^3 + 15x^2 + 25x - 43$
h) $y = \frac{3}{4}x^4 + 1\frac{2}{3}x^3 - 1\frac{1}{2}x^2 - 5x - 3$

Exercises 9.4B (*p.223*)

1 **a)** $-3x + c$ **b)** $x^5 + c$ **c)** $0.04x^5 + c$ **d)** $0.5x^7 + c$ **e)** $\frac{1}{3}x^3 + \frac{1}{2}x^2 + c$
f) $7\frac{1}{2}x^2 - 3\frac{1}{3}x^3 + c$
g) $\frac{1}{3}x^3 - \frac{1}{2}x^2 + 2x + c$
h) $1.2x^5 + 5.25x^4 + c$
i) $0.5x^2 + 2x + c$
j) $6.75x^4 + 9x^3 + 4.5x^2 + x + c$
2 **a)** $1.5x^2 - 4x + c$
b) $\frac{1}{6}t^6 - \frac{1}{4}t^4 + c$
c) $\frac{1}{12}t^6 + \frac{1}{6}t^3 + c$
d) $\frac{1}{6}x^3 - 2\frac{1}{2}x^2 + 12\frac{1}{2}x + c$
e) $0.25x^4 + x^3 + x^2 + c$
f) $\frac{1}{4}(3 + u)^4$
3 **a)** $y = 0.5x + 0.5$ **b)** $y = 4x^2 - 32$
c) $y = 11x - x^2 - 18$
d) $y = 11x - \frac{2}{3}x^3$
e) $y = x^3 + 3x^2 - 1$
f) $y = \frac{1}{4}x^4 + 1\frac{1}{3}x^3 - 137\frac{1}{4}$
g) $y = 1\frac{1}{3}x^3 - 6x^2 + 9x + 17\frac{1}{3}$
h) $y = \frac{1}{12}(3x - 2)$

Consolidation Exercises for Chapter 9 (*p.224*)

1 $\frac{1}{6}$ **2** $20\frac{5}{6}$
3 **a)** P(2, 20), Q(4, 16) **b)** Area L = 28; area R = 36
4 **a)** $19\frac{8}{15}$ **b)** $1049\frac{1}{3}$
5 **a)** A(4, 0) B(–4, 8) **c)** $13\frac{1}{3}$, $-2\frac{2}{3}$
d) $32 - (13\frac{1}{3} - 2\frac{2}{3}) = 21\frac{1}{3}$
6 **a)** A(–1, 4), B(2, 4) **b)** 12 – 5.25 = 6.75 **c)** AB = OC = 3
7 48 m.
8 **a)** 18.5 m s^{-1}
9 **a)** $\frac{2}{3}$ or 2 **b)** $3\frac{5}{12}$ m

CHAPTER 10
Trigonometry 2

Exercises 10.1A (*p.230*)

1 10.6 cm, 11.7 cm
2 **a)** 11.5 cm, 6.62 cm **b)** 9.51 cm, 8.75 cm
3 55.3 m and 63.4 m
4 394 m
5 56.4 m, 76.0 m 48.8 m
6 16.5 cm 35.4°, 39.6°
7 **a)** 36.2°, 91.8°, 50.7 cm **b)** either 71.0°, 57.0°, 42.6 cm or 109.0°, 19.0°, 16.5 cm
8 23.6 cm
9 Q by 0.42 s

Exercises 10.1B (*p.231*)

1 53.3°, 11.5 cm
2 146.8°, 13.2°, 3.33 cm
3 **a)** 45.6° or 134.4°
4 48.4 m
5 14.2 m
6 **a)** 177 m **b)** 156 m
7 2402 m
8 **a)** 2.1.3°, 112.7°, 100.5 cm **b)** 66.7°, 73.3°, 52.2 cm or 113.3°, 26.7°, 24.4 cm
9 32.1°, 616 cm^2

Exercises 10.2A *(p.236)*

1 11.3 cm, 40.1, 74.9
2 a) 8.1 cm, 96.5, 41.5
b) 31.4 cm, 39.4, 25.6
c) 42.8, 61.0, 76.2
d) 40.1, 122.6, 17.3
3 61.9, 90, 28.1
5 33.9 km
6 a) 16.7 km/h **b)** 131.7 **c)** 12.48
7 100 m

Exercises 10.2B *(p.237)*

1 11.9 cm 36.4°, 103.6°
2 90°
3 impossible
4 6.55 cm
5 3.69 km, 266°
6 b) 11.5 km **c)** 20.1 km/h **d)** 015.4°
7 9.4 km, 9.508 km, 1.1%, 9.217 km, –1.9%

Exercises 10.3A *(p.239)*

1 a) 31.5 cm² **b)** 41.1 m²
2 a) 13.4 m² **b)** 51.5 cm²
3 a) 10.8 m² **b)** 83.1 cm²
4 1810 m²
6 a) 259.8 cm² **b)** 82.7%
c) $3x^2\sin 60°$ or $2.598x^2$

Exercises 10.3B *(p.241)*

1 a) 14.9 cm² **b)** 47.1 cm²
c) 26.0 cm²
2 21.4 cm² **3** 7.21 km²
4 43.0 cm²
5 c) $0 < x < 10$
d) max area of 19.2 cm² when triangle is equilateral ($x = 6\frac{2}{3}$)

Exercises 10.4A *(p.245)*

1 a) 12.8 cm **b)** 13.7 cm **c)** 21.3°
2 a) 11.3 cm **b)** 55.6 **c)** 8.25 cm
3 a) 2.5 m **b)** 4.0 m **c)** 38.7
d) 8.76 m **e)** 116.8 m²
4 a) 57.7 m **b)** 124 m **c)** 135
d) 207 m **e)** 025.0
5 a) 9.09 **b)** 75 km
c) 40.4 km, 158.3 **d)** 15:39

Exercises 10.4B *(p.247)*

1 a) 8.94 cm, **b)** 12 cm,
c) 13.9 cm, **d)** 35.3
2 526 m
3 a) 1.19 m **b)** 1.15 m **c)** 3.38 m
d) 21.6 m²
4 a) 1370 m, 977 m **b)** 6 min 10 s from A, 600 m

Consolidation Exercises for Chapter 10 *(p.251)*

1 3 h 20 min, 048.5°
2 a) 264 m **b)** 101 m
3 440 m
4 a) 18.7 m **b)** 69.3 **c)** 75.0
5 a) 138.6 **b)** 992 m² **c)**
6 a) 3a **b)** 19.5 **c)** 35.3
9 b) 84

CHAPTER 11
Sequences and series 1

Exercises 11.1A *(p.257)*

1 a) $u_1 = 1, u_{k+1} = u_k + 4$, $u_k = 4k - 3$ **b)** $u_1 = 1$, $u_{k+1} = -2u_k$, $u_k = (-2)^{k-1}$
c) $u_1 = 16000$, $u_{k+1} = u_k\frac{1}{4}$, $u_k = 16000\,(0.25)^{k-1}$
d) $u_1 = 1$, $u_{k+1} = u_k + (3k^2 + 3k + 1)$ $u_k = k^3$
2 a) 55 **b)** 55 **c)** 55 **d)** 60 **e)** 60
f) 60
3 a) $\sum_{k=1}^{25} k^2$ **b)** $\sum_{k=1}^{15} k(k+1)$ **c)** $\sum_{k=1}^{100} \frac{1}{k}$
4 a) i) $4, 4\frac{2}{3}, 4\frac{8}{9}, 4\frac{26}{27}$,
ii) $u_k = 5 - \left(\frac{1}{3}\right)^{k-2}$, **iii)** 5
b) i) $2, \frac{2}{3}, \frac{6}{5}, \frac{10}{11}$, **ii)** $u_k =$, **iii)** 1

Exercises 11.1B *(p.258)*

1 a) $u_1 = 65, u_{k+1} = u_k - 9$, $u_k = 74 - 9k$
b) $u_1 = 1, u_{k+1} = 0.9u_k$, $u_k = (0.9)^{k-1}$
c) $u_1 = 1, u_{k+1} = u_k + (k+1)$, $u_k = \frac{1}{7}k(k+1)$
d) $u_1 = 1, u_{k+1} = u_k / (u_k+1)$, $u_k = \frac{1}{k}$
2 a) $\frac{341}{280}$ **b)** 63 **c)** 0 **d)** 85
e) 85 **f)** 85
3 a) $\sum_{k=1}^{10} 2k^3$ **b)** $\sum_{k=2}^{12} k(k+3)$
c) $\sum_{k=1}^{21} (-1)^{k+1}\frac{1}{k}$
4 a) $-\frac{1}{4}, -\frac{13}{16}, -\frac{61}{64}, -\frac{253}{256}$
b) $u_k = -1 + \frac{3}{4^{k-1}}$, $u_k \to -1$

Exercises 11.2A *(p.262)*

1 a) 11 **b)** 34, 26 **c)** 19.5, 27, 34.5
d) 13, 6, –1, –8, –15
e) $3\frac{1}{4}, 4\frac{3}{4}, 5\frac{1}{2}, 6\frac{1}{4}$
f) $3\frac{2}{3}, -\frac{1}{3}, -2\frac{1}{3}$
2 a) 10 **b)** 8 **c)** 19 **d)** 11
3 a) $u_k = 3k - 1, S_n = \frac{1}{2}n(3n+1)$, $u_{10} = 29, S_{10} = 155$
b) $u_k = 157 - 7k$, $S_n = \frac{1}{2}n(307 - 7n)$, $u_{10} = 87, S_{10} = 1185$
c) $u_k = 1.9k + 1.7, S_n = \frac{1}{2}$ $n(1.9n + 5.3), u_{10} = 20.7$, $S_{10} = 121.5$
d) $u_k = 3\frac{3}{4} - \frac{3}{4}k$, $S_n = \frac{1}{2}n(6\frac{3}{4} - \frac{3}{4}n), u_{10} = -3\frac{3}{4}$, $S_{10} = -3\frac{3}{4}$
e) $u_k = 3k - 16.5$, $S_n = \frac{1}{2}n(3n - 30), u_{10} = 13.5$, $S_{10} = 0$
f) $u_k = (k-1)x$, $S_n = \frac{1}{2}n((n-1)x), u_{10} = 9x$, $S_{10} = 45x$
g) $u_k = p + 2(k-1)q$, $S_n = n(p + (n-1)q)$
$S_n = \frac{1}{2}n(p + (n-1)q)$, $u_{10} = p + 18q, S_{10} = 10(p + 9q)$
h) $u_k = p + (2-k)q$, $S_n = \frac{1}{2}n(2p + (3-n)q)$, $u_{10} = p - 8q, S_{10} = 5(2p - 7q)$
4 $u_k = 1.5k + 2, u_8 = 14$
5 $S_n = n(4+n), S_8 = 96$
6 a) 279.5 **b)** 92 **c)** 93.5 **d)** 188.5
7 124.8
8 $u_k = 3.5k + 9.5, u_{10} = 44.5$
9 10
10 32
11 a) 74 m **b)** 12 s
12 a) $10n - 7$ **b)** $a = 3, d = 10$

Exercises 11.2B *(p.263)*

1 a) 20 **b)** 26, 31 **c)** 35, 33, 31
d) 17.5, 15, 12.5, 10, 7.5
e) $1, \frac{5}{3}$ **f)** $2\frac{1}{6}, 1\frac{1}{6}$
2 a) 11 **b)** 12 **c)** 13 **d)** 14
3 a) $u_k = 2k, S_n = n(n+1)$, $n_8 = 16, S_8 = 72$
b) $u_k = 2k + 1, S_n = n(n+1)$, $u_8 = 17, S_8 = 80$
c) $u_k = 21 - 4k, S_n = n(19 - 2n)$,

$u_8 = -11$, $S_8 = 24$
d) $u_k = \frac{2}{5}k - \frac{1}{5}$, $S_n = \frac{n^2}{5}$,
$u_8 = 3$, $S_8 = 12\frac{4}{5}$
e) $u_k = 10 - 5k$, $S_n = \frac{1}{2}5n(3 - n)$,
$u_8 = -30$, $S_8 = -100$
f) $u_k = 30 - (k - 1)x$,
$S_n = \frac{1}{2}n(60 - (n - 1)x)$,
$u_8 = 30 - 7x$, $S_8 = 240 - 28x$
g) $u_k = p - 4(k - 1)q$,
$S_n = n(p - 2(n - 1)q)$,
$u_8 = p - 28q$, $S_8 = 8(p - 14q)$
h) $u_k = p + q + (k - 1)(p + 2q)$,
$S_n = \frac{1}{2}n(2(p + q) + (n - 1)(p + 2q))$,
$u_8 = 8p + 15q$, $S_8 = 36p + 64q$

4 $u_k = 2k + 1$, $u_{11} = 23$
5 $S_n = \frac{1}{2}n(5n - 13)$, $S_{10} = 185$
6 **a)** 688 **b)** −18 **c)** 1500 **d)** 1550
7 2
8 6
9 **a)** $\frac{1}{2}n(n + 1)$ **i)** 5050, **ii)** 500500, **iii)** 15050, **iv)** 375250
b) $n(2n + 1)$, $\frac{1}{2}n(3n + 1)$
10 **a)** **i)** 1, **ii)** 4, **iii)** 9, 16, 25, ...
b) n^2
11 $n(n + 1)$
12 460 men, £705 600

Exercises 11.3A *(p.269)*

1 **a)** 5, 625 **b)** −9, 3 **c)** 10, 2.5
d) 1500, 2250, 5062.5
e) 900, 600, $266\frac{2}{3}$
f) 1, $\frac{9}{64}$, $\frac{27}{512}$ **g)** x^4, x^6, x^{10}
h) $p^{11}q^6, p^9q^5, p^7q^4, p^5q^3$
2 **a)** 10 **b)** 6 **c)** 10 **d)** 10 **e)** 8 **f)** 8
3 **a)** $u_k = 2^{k-1}$, $S_n = 2^{n-1}$ $u_8 = 128$, $S_8 = 255$
b) $u_k = 500(\frac{1}{2})^{k-1}$,
$S_n = 1000(1 - (\frac{1}{2})^n)$, $u_8 = 3\frac{29}{32}$,
$S_8 = 996\frac{3}{32}$
c) $u_k = 16(-\frac{1}{2})^{k-1}$,
$S_n = \frac{32}{3}\left(1-\left(-\frac{1}{2}\right)^n\right)$,
$u_8 = -\frac{1}{8}$, $S_8 = 10\frac{5}{8}$
d) $u_k = 1000(1.1)^{k-1}$,
$S_n = 10\,000(1.1^n - 1)$,
$u_8 = 1948.7171$,
$S_8 = 11\,435.8881$
e) $u_k = \frac{1}{81}(-3)^{k-1}$
$S_n = \frac{1}{324}\left(1-(-3)^n\right)$
$u_8 = -27$, $S_8 = -20\frac{20}{81}$
f) $u_k = (\frac{1}{x})^{k-1}$, $S_n = \dfrac{1-\left(\frac{1}{x}\right)^n}{1-\frac{1}{x}}$,
$u_8 = \frac{1}{x^7}$, $S_8 = \dfrac{1-\left(\frac{1}{x}\right)^8}{1-\frac{1}{x}}$
4 $u_k = 2700(\frac{4}{3})^{k-1}$, $u_5 = 8533\frac{1}{3}$
5 $u_k = 10 \times 3^{k-1}$, $u_9 = 65\,610$
6 $r = \pm 0.6$, $S_n = 25\,000(1 - (0.6)^n)$
$\Rightarrow S_5 = 23\,056$,
$S_n = 6250(1 - (0.6)^n)$, $S_5 = 6736$
7 **a)** $r = 1.5 \Rightarrow a = 32$,
$r = -1.5 \Rightarrow a = -32$
b) 32, 48, 72, 108, 162, 243, 364.5, 546.75 −32, 48, −72, 108, −162, 243, −364.5, 546.75
c) $S_n = 64(1.5^n - 1) \Rightarrow S_8 = 1576.25$, $S_n = 12.8(1 - (-1.5)^n)$
$\Rightarrow S_8 = -315.25$
8 **a)** $a = -2$, $r = -2$
b) $S_n = -\frac{2}{3}\left(1-(-2)^n\right)$, $S_{10} = 682$
9 7
10 10
11 **a)** $u_k = 20 \times 1.1^{k-1}$
b) 18th week **c)** 912 km (3 s.f.)

Exercises 11.3B *(p.270)*

1 **a)** $u_k = 2 \times 3^{k-1}$, $S_n = 3^n - 1$,
$u_{10} = 39\,366$, $S_{10} = 59\,048$
b) $u_k = 27(\frac{1}{3})^{k-1}$,
$S_n = \frac{81}{2}\left(1-\left(\frac{1}{3}\right)^n\right)$,
$u_{10} = 0.0014$, $S_{10} = 40.5$
c) $u_k = 12(-\frac{1}{2})^{k-1}$,
$S_n = 8(1 - (-\frac{1}{2})n)$,
$u_{10} = -0.0234$, $S_{10} = 7.99$
d) $u_k = 1000 \times (0.9)^{k-1}$,
$S_n = 10\,000 \times (1 - 0.9^n)$,
$u_{10} = 387.42$, $S_{10} = 6513.22$
e) $u_k = 1000 \times (1.2)^{k-1}$,
$S_n = 5000(1.2^k - 1)$,
$u_{10} = 5159.78$, $S_{10} = 25\,958.7$
f) $u_k = (\frac{\pi}{2})^k$, $S_n = \dfrac{\frac{\pi}{2}\left(\left(\frac{\pi}{2}\right)^n - 1\right)}{\frac{\pi}{2} - 1}$,
$u_{10} = 91.45$, $S_{10} = 248.9$
2 $u_k = 21 \times 2^{k-1}$, $u_9 = 5376$
3 $u_k = 18(\frac{1}{3})^{k-1}$,
$u_9 = \frac{2}{2187} \approx 0.000\,91$
4 **a)** $r = \pm\frac{2}{3} =$, $a = 2$
b) 2, $\frac{4}{3}, \frac{8}{9}, \frac{16}{27}, \frac{32}{81}, \frac{64}{243}$
or 2, $-\frac{4}{3}, \frac{8}{9}, -\frac{16}{27}, \frac{32}{81}, \frac{64}{243}$
c) $S_n = 6(1 - (\frac{2}{3})^n) \Rightarrow$
$S_{10} = 5.896$
or $S_n = \frac{6}{5}(1 - (-\frac{2}{3})^n) \Rightarrow$
$S_{10} = 1.179$
5 **a)** **i)** 0.2 mm **ii)** 0.4 mm
iii) 0.8 mm **b)** 0.1×2^n
c) **i)** 12.8 mm **ii)** 20 cm
iii) 209.7 m **d)** 20 folds
6 **a)** 210 by 297, 62 370 mm²
b) 148.5 by 210, 297 by 420, 420 by 594, 594 by 840, 840 by 1188
c) 31 185, 124 740, 249 480, 498 960, 997 920 mm² **d)** $\frac{1}{2}$
e) $\frac{1}{\sqrt{2}}$ **f)** 1 m², 210.224 by 297.302
g) A10: area 974.5 mm² (26.25 by 37.125), A(−2): area 4 m² (1680 by 2376)
7 18 weeks
8 **a)** 3 bounces **b)** 6 bounces
9 **a)** pays £10 000,
b) pays £10 485.75
10 1.84×10^{19} (3 s.f.) $= 2^{64} - 1$

Exercises 11.4A *(p.275)*

1 **a)** diverges **b)** 10 **c)** 448 **d)** 64
e) diverges **f)** 50
2 **a)** $\frac{2}{3}$ **b)** $\frac{3}{11}$ **c)** $\frac{1}{27}$ **d)** $\frac{1}{7}$ **e)** $\frac{5}{13}$ **f)** $\frac{1}{10}$
3 **a)** 160 **b)** $53\frac{1}{3}$
4 **a)** −54 or 27 **b)** $333\frac{1}{3}$ or $142\frac{6}{7}$
c) $31\frac{1}{4}$ or $20\frac{5}{6}$
5 36, 9, 2.25, 0.5625, 0.140 625
6 $r = 0.6$, $a = 60$
7 **a)** 1.08 m **c)** 20 m

Exercises 11.4B *(p.276)*

1 **a)** diverges **b)** $\frac{1}{3}$ **c)** 0.54
d) diverges **e)** 12.8
f) diverges
2 **a)** $\frac{7}{9}$ **b)** $\frac{16}{9}$ **c)** $\frac{4}{33}$ **d)** $\frac{41}{333}$ **e)** $\frac{97}{225}$
f) $\frac{227}{495}$
3 **a)** 1822.5 **b)** −911.25
4 $r = 0.75$; 20, 15, 11.25, 8.4375
5 $a = 2$; 2.00, 1.83, 1.68, 1.54, 1.41, 1.29 (2 d.p.)
6 **a)** 125 **b)** 18th term **c)** 19
7 **a)** 0.9 **b)** No **c)** Yes, for $-1 < r < 0$

Consolidation Exercises for Chapter 11 *(p.278)*

1 $a = -4, d = 7$
2 14.9
3 $a = 0, d = 4$
4 **a)** $n = 33, d = 3.5$ **b)** 1450
5 **a)** $p = 4.5$ **b)** 19 900
6 **a)** **i)** £70 620, **ii)** £1062
7 **a)** **i)** £3370, **ii)** £5679 **b)** £52 423
8 $u_0 = 50\,000$ $u_1 = 1.1 \times 50\,000\,R$, $u_2 = 1.1^2 \times 50\,000 - R(1.1 + 1)$, $u_3 = 1.1^3 \times 50\,000 - R(1.1^2 + 1.1 + 1)$, £5508.40
9 **a)** **i)** $14n + 11$ **ii)** $n(7n + 18)$ **b)** 880 cans
10 **a)** 9.22° **b)** 74.62° **c)** 81
11 40.2 km, 20 km
12 $0 < a < 2$ **a)** 4.91 **c)** 22
13 **a)** 1.08 **b)** £29 985 **d)** £1500
14 **a)** **i)** 15 cm **ii)** 41.5 m **b)** **i)** ar^9 **ii)** $a = 0.0301, r = 1.25$
15 **a)** 40 **b)** – 9.5, 1 **c)** $k - 10.5$
16 **a)** 2 **b)** 40 **c)** $\frac{1}{2}n(3n + 1)$ **d)** $\frac{1}{2}n\,(9n + 1)$
17 **a)** $a + 4d = 205, a + 18d = 373$ **b)** $d = 12, a = 157$ **c)** 28

CHAPTER 12
Indices

Exercises 12.1A *(p.288)*

1 **a)** 9 **b)** $\frac{1}{4}$ **c)** 1 **d)** $35x^8$ **e)** $8a^5b^5$ **f)** $60p^6q^4r^6$ **g)** $9r^6s^2$ **h)** $32a^5b^5$ **i)** $-64x^9y^6$ **j)** $\frac{x^2z^2}{2y^2}$ **k)** $3r^2s^2$ **l)** $\left(\frac{3ab^2}{4c^2}\right)^3$ **m)** $\frac{3p^3}{7q}$
2 **a)** $\frac{1}{8}$ **b)** 7 **c)** 343 **d)** $\frac{16}{81}$ **e)** $\frac{81}{16}$ **f)** $\frac{1}{5}$ **g)** $\frac{5}{2}$ **h)** $\frac{7}{2}$ **i)** $\frac{3}{2}$ **j)** $1\frac{1}{3}$
3 **a)** $n = -3$ **b)** $n = 2$ **c)** $n = 4$ **d)** $n = 5$ **e)** $n = \frac{8}{27}$ **f)** $n = 343$ **g)** $n = 10$ **h)** $n = 4$ **i)** $n = -4$ **j)** $n = \frac{1}{2}$ **k)** $n = \frac{1}{3}$ **l)** $n = -\frac{2}{3}$ **m)** $n = -\frac{1}{2}$

Exercises 12.1B *(p.288)*

1 **a)** 32 **b)** 25 **c)** 1 **d)** $15y^{11}$ **e)** $12c^6d^{12}$ **f)** $25x^8y^6$ **g)** $40a^8b^8c^6$ **h)** $11^9a^{18}b^9$ **i)** $-243m^{10}n^{15}p^{20}$ **j)** $\frac{5q^2r^2}{p}$ **k)** $\frac{a^2}{2b^2}$ **l)** $\frac{4x}{y^2}$ **m)** $\frac{25r^2t^2}{49s^4}$
2 **a)** $\frac{1}{32}$ **b)** 8 **c)** 512 **d)** $\frac{1}{8}$ **e)** $\frac{27}{64}$ **f)** $\frac{64}{27}$ **g)** $\frac{2}{5}$ **h)** $\frac{7}{4}$ **i)** $\frac{9}{2}$ **j)** $\frac{4}{3}$
3 **a)** $n = 3$ **b)** $n = -2$ **c)** $n = 2$ **d)** $n = \frac{1}{10}$ **e)** $n = \frac{64}{125}$ **f)** $n = 216$ **g)** $n = 6$ **h)** $n = 6$ **i)** $n = 0$ **j)** $n = -3$ **k)** $n = \frac{1}{2}$ **l)** $n = -2$

Exercises 12.2A *(p.291)*

1 $3\sqrt{2}$ **2** $4\sqrt{2}$ **3** $2\sqrt{7}$ **4** $5\sqrt{3}$
5 $3\sqrt{2}\sqrt{5}$ **6** $\sqrt{3}\sqrt{5}\sqrt{23}$ **7** $13\sqrt{3}$
8 $100\sqrt{2}\sqrt{3}$ **9** $\frac{1}{2}\sqrt{3}$ **10** $\frac{5}{3}\sqrt{2}$
11 $8\sqrt{3}$ **12** $2\sqrt{2}$ **13** $\sqrt{5}$ **14** $18\sqrt{2}$
15 $5\sqrt{2} - 2$ **16** $\sqrt{2}$ **17** $7\sqrt{2}$
18 $\sqrt{3} - 3$ **19** 20 **20** –2 **21** $\frac{5\sqrt{2}}{2}$
22 $\frac{\sqrt{3}}{3}$ **23** $\sqrt{7}$ **24** $\frac{\sqrt{5}\sqrt{6}}{2}$
25 $\frac{1}{6}\left(3\sqrt{2} - 2\sqrt{3}\right)$ **26** $\frac{15\sqrt{7}+14}{21}$
27 $\frac{1}{2}\left(\sqrt{3} - 1\right)$ **28** $7\sqrt{5} - 14$
29 $\frac{1}{7}\left(3 + \sqrt{2} - 3\sqrt{3} - \sqrt{2}\sqrt{3}\right)$
30 $4 - \sqrt{5}\sqrt{3}$

Exercise 12.2B *(p.292)*

1 $2\sqrt{11}$ **2** $5\sqrt{5}$ **3** $1000\sqrt{7}$
4 $8\sqrt{5}$ **5** $3\sqrt{29}$ **6** $4\sqrt{19}$
7 $4\sqrt{31}$ **8** $29\sqrt{5}$ **9** $\frac{4}{5}\sqrt{2}$
10 $\frac{3}{4}\sqrt{3}$ **11** $7\sqrt{7}$ **12** $6\sqrt{2}$
13 $16\sqrt{3}$ **14** $7\sqrt{3}$ **15** $7\sqrt{3} + 6$
16 0 **17** $14 + 2\sqrt{5}$ **18** 4
19 $-12\sqrt{13}$ **20** $x–y$ **21** $\frac{7\sqrt{2}}{2}$
22 $-\frac{4\sqrt{5}}{5}$ **23** $2\sqrt{7}$ **24** $\sqrt{13}$
25 $\frac{\sqrt{3}\sqrt{7}}{3}$ **26** $\frac{1}{15}\left(3\sqrt{5} + 5\sqrt{3}\right)$
27 $2 + \sqrt{3}$ **28** $\frac{1}{2}\left(5 - \sqrt{3}\right)$
29 $\frac{1}{3}\left(4 - \sqrt{7}\right)$
30 $2 + \sqrt{5}\sqrt{2} - \sqrt{5} - 2\sqrt{2}$

Exercises 12.3A *(p.296)*

1 **a)** $y = \frac{4}{x} - 3$, asymptotes $y = -3$ $x = 0$
b) $y = \frac{4}{x+3}$ asymptotes $y = 0$ $x = 3$
c) $y = \frac{12}{x}$ asymptotes $y = 0$ $x = 0$
d) $y = -\frac{4}{x} + 5$ asymptotes $y = -1\sqrt{x}, x \geq 0$ $y = 5$ $x = 0$
2 **a)** stretch, factor 3 // to y-axis, asymptotes $y = 0$ $x = 0$
b) translation through –2 units // x-axis, asymptotes $y = 0, x = -2$
c) translation through 5 units // y-axis, asymptotes $y = 5, x = 0$
d) reflection in x-axis and translation through 3 units // x-axis, asymptotes $x = 3, y = 0$.
3 **a)** $y = \sqrt{x+1}, x \geq -1$
b) $y = 2 + \sqrt{x}, x \geq 0$
c) $y = -\sqrt{x}, x \leq 0$
d) $y = -1\sqrt{x}, x \geq 0$
4 $x = 1$ $y = 2$ is a local minimum asymptotes $y = x, y = 0, x = -1$, $y = -2$ is a local maximum
5 x-intercept (2, 0), minimum point (–1.59, –2.52)
6 **b)** $I = \frac{12}{R}$ **c)** $E = 12$ volts
7 2.4 units

Exercises 12.3B *(p.297)*

1 **a)** $y = \frac{5}{x} + 2$, asymptotes $y = 2$, $x = 0$
b) $y = \frac{5}{x+2}$, asymptotes $x = -2$, $y = 0$
c) $y = \frac{10}{x}$, asymptotes $y = 0$, $x = 0$
d) $y = \frac{-5}{x-4}$, asymptotes $y = 0$, $x = 4$
2 **a)** reflection in x-axis, stretch, factor 8 // y-axis, asymptotes $x = 0, y = 0$
b) reflection in x-axis, translation through 3 units // x-axis, asymptotes $x = 3, y = 0$
c) reflection in x-axis, translation through 3 units // y-axis, asymptotes $x = 0, y = 3$
d) reflection in x-axis, translation through –2 units // x-axis, stretch, factor 5 // y-axis, asymptotes $x = -2, y = 0$
3 asymptotes $y = x, x = 0$, no stationary points
4 x-axis intercepts (2, 0), (–2, 0), asymptote $x = 0$

5 **a)** $p = \frac{540}{n} + 0.01$

c) 1103

6 $f = \frac{1}{T}$, $T = 0.0039$ s

7 **b)** $v = 4.4\ \sqrt{h}$

Exercises 12.4A *(p.302)*

1 **a)**

x	−3	−2	−1	0	1	2	3
y	0.037	0.111	0.333	1	3	9	27

c) $x = 2.7$

d)

x	−3	−2	−1	0	1	2	3
gradient	0.04	0.12	0.37	1.1	3.3	9.9	29.7

2 **a)** $x \approx 2.73$

3 **a)**

x	−3	−2	−1	0	1	2	3
y	40	20	10	5	2.5	1.25	0.625

c) $x = -1.58$

d)

x	−2	−1	0	1	2
gradient	−13.9	−6.9	−3.5	−1.7	−0.87

4 $x \approx -1.59$

5 **b)** reflection in y-axis.

c) reflection in y-axis followed by a translation $\frac{1}{2}$ unit // x-axis ***or*** reflection in y-axis followed by stretch, factor 2, // y-axis

7 **a)** $k = 3, a = 2$ **b)** $k = -2, a = 1.5$

8 **c)** $\alpha = 0.741, \beta = 5.47$

9 **c) i)** 4 million **ii)** 81.27 yrs

10 **b) i)** Drug B **ii)** Drug A

c) $t = 2.63$ h and $C = 1.238$ mg/ml

Exercises 12.4B *(p.297)*

1 **a)**

x	−3	−2	−1	0	1	2	3
y	0.008	0.04	0.2	1	5	25	125

c) $x \approx 1.7$ **d)** 24.8

2 **a)** $x = 1.68$

3 **a)**

x	−3	−2	−1	0	1	2	3
y	108	36	12	4	1.333	0.444	0.148

c) $x \approx -1.9$

4 **a)** $x = -1.89$

5 **b)** reflection in y-axis

c) translation through −1 // x-axis ***or*** a stretch, factor 5, // y-axis

d) translation through 1 unit // x-axis ***or*** stretch, factor $\frac{1}{5}$, // y-axis

7 **a)** $P = 4 \times 1.5t$

b) $P = 10 \times 0.5t$

8 **b)** $\alpha = 1.20$

9 **c)** 63 627 visitors

d) during the 16th year

10 $\frac{R}{R^E} = \frac{1}{10}(3.2^n + 4)$, $n = -2$

Consolidation Exercises for Chapter 12 *(p.305)*

1 **a)** $\frac{1}{49}$ **b)** $10a^5$ c) $6a^6b^7$

d) $16x^{28}y^2$ **e)** $16m^2n^2p^6$

f) $\frac{5x^3y}{2^2}$ **g)** $4p^2q^7$ **h)** $3r^4s^3t^2$

i) $\frac{81}{16}x^4y^4z^8$

2 **a)** $25a^2$ **b)** $1a^{-3}$ **c)** $a^{\frac{5}{2}}$

d) $16a^{-4}$ **e)** $4a^{\frac{1}{2}}$ **f)** $a^{-\frac{3}{2}}$

g) $3a^{\frac{1}{2}}$ **h)** $27a^{12}$

3 **a)** $x = -0.8$ **b)** $x = 7$ **c)** $a = -3$

d) $x = 2$ **e)** $\frac{1}{256}$ **f)** $x = 0$

g) $x = -6$ **h)** $x = 3$ **i)** $\frac{1}{2}$

4 **a)** $14\sqrt{3}$ **b)** $6\sqrt{2}$ **c)** $93\sqrt{5}$

d) $-16 + 3\sqrt{2}$ **e)** $-\frac{13}{22} + \frac{7}{22}\sqrt{3}$

f) $\frac{4}{3} + \frac{\sqrt{7}}{3}$ **g)** $\frac{15}{37} - \frac{\sqrt{3}}{37}$ **h)** $\frac{1}{2} + \frac{1}{2}\sqrt{5}$

5 **a)** $5\sqrt{5}$ **b)** $\frac{4\sqrt{2}}{5}$ **c)** $15 + 5\sqrt{7}$

d) $\frac{11}{4} - \frac{3}{4}\sqrt{13}$

6 **a)** $-2 - \sqrt{3}$

b) $\frac{1}{4}(5 - \sqrt{17})$

c) $\frac{1}{6}(-1 - \sqrt{61})$

d) $\frac{1}{8}(1 - \sqrt{113})$

7 translation through −2 units // x-axis; stretch, factor 4, // y-axis stretch, factor 4, // y-axis *followed by* translation through 2 units // x-axis

8 **a)** $y = \frac{10}{(x-3)^2}$, asymptotes $y = 0$, $x = 3$

b) $y = \frac{10}{x^2} - 3$, asymptotes $y = -3$, $x = 0$

c) $y = \frac{30}{x^2}$, asymptotes $y = 0$, $x = 0$

d) $y = \frac{10}{-x^2} + 3$, asymptote $y = 3$, $x = 0$

9 x intercept (−1.91, 0), $y = x$ and $x = 0$, local minimum at (2.41, 3.62)

10 **b) i)** local minimum at (2, 2)

ii) $x = -1.3$

c) $x = -1.3146$

11 $x = 0$ or $x = -1$

12 **a)** translation through 3 units // x-axis

b) reflection in x-axis then translation through 10 units // y-axis

c) reflection in x-axis *followed by* a stretch, factor 3, // y-axis

d) translation through $\frac{3}{2}$ // x-axis *followed by* stretch, factor $\sqrt{2}$ // y-axis

13 **a)**

x	−3	−2	−1	0	1	2	3
y	0.1875	0.375	0.75	1.5	3	6	12

c) $x = 2.3$ **d)** 0.13, 0.26, 0.52, 1.04, 2.08, 4.16, 8.32

14 **a)** $x = 2.32$

15 **b) i)** $d = 57$ cm

ii) $t = 22$ weeks

c) $t = 21.6$ weeks

16 *either* a translation through −1 unit // x-axis *or* a stretch, factor 2 // y-axis.

17 **b)** local maximum is (2.89, 1.13)

CHAPTER 13 Functions

Exercises 13.1A *(p.312)*

1 **a)** range {y: 16, 10, 7, −2}, one-to-one

b) range {y: 18, 2,}, many-to-one

c) range {y: 13, −3}, many-to-one

d) range {y: −2, −5, 5, 2}, one-to-one

e) range {y: 6, 2}, many-to-one

f) range {y: 0}, many-to-one

g) range {y: −54, 0, 54}, one-to-one

2 **a)** range {y: $2 \le y \le 9$}, one-to-one

b) range {y: $0 \le y \le 50$}, many-to-one

c) range {y: $0 \le y \le 108$}, one-to-one

d) range {y: $-2.5 \le y \le 18$}, many-to-one

e) range {y: $0 \le y \le 9$}, many-to-one

f) range {y: $-20 \le y \le 0$}, many-to-one

g) range $\{y: -2 \le y \le 20\}$, many-to-one
h) range $\{y: 0 \le y < \infty\}$, one-to-one
i) range $\{y: 0 \le y < \infty\}$, many-to-one
j) range $\{y: - < y \le 0\}$, one-to-one

Exercises 13.1B *(p.312)*

1 a) range $\{y: -4, 0, 4, 6, 8\}$, one-to-one
b) range $\{y: 15, 3, -9, -21\}$, one-to-one
c) range $\{y: 24, 6, 0\}$, many-to-one
d) range $\{y: 19, 1, -5,\}$, many-to-one
e) range $\{y: 4, 2, 1\frac{1}{3}, 1\}$, one-to-one
f) range $\{y: 6, 0\}$, many-to-one
g) range $\{y: -16, 0, 16\}$, one-to-one
h) range $\{y: 3, 0, -3\}$, one-to-one
2 a) range $\{y: -1 \le y \le 7\}$, one-to-one
b) range $\{y: -\frac{1}{2} \le y \le 6\frac{1}{2}\}$, one-to-one
c) range $\{y: 0 \le y \le 12\frac{1}{2}\}$, many-to-one
d) range $\{y: -1 \le y \le 15\}$, many-to-one
e) range $\{y: -0.125 \le y \le 45\}$, many-to-one
f) range $\{y: -45 \le y \le 0.125\}$, many-to-one
g) range $\{y: 0 \le y \le 12\}$, many-to-one
h) range $\{y: y \ge 0\}$, many-to-one
i) range $\{y: -4 \le y \le -0.4\}$, many-to-one
j) range $\{y: -39 \le y \le 39\}$, many-to-one

Exercises 13.2A *(p.315)*

1 a) i) $2(5 - x)$ **ii)** $5 - x^2$
iii) $25 - 20x + 4x^2$
b) i) 14 **ii)** 1 **iii)** 81
c) $x = 1, 4$
2 a) $5 - 4x$ **b)** x **c)** self-inverse
3 $x^3 - 3x^2 + 2x + 8$
4 x
5 $\frac{x}{1-2x}$
6 a) $f(g(x))$ **b)** $h(f(x))$ **c)** $g(h(x))$
d) $g(f(x))$ **e)** $f(h(x)$ **f)** $h(h(x))$
7 a) $p = 1, r = 46$
b) $4x^2 + 20x + 22$
8 b) $f(x) = x$, $g(x) = 25 - x^2$, $h(x) = x$

Exercises 13.2B *(p.316)*

1 a) $3x^3 - 4$ and $(3x - 4)^4$
b) i) 20 **ii)** 8 **c)** $x = 3$
2 a) i) $6x - 3$ **ii)** $12x + 1$
iii) $24x - 56$
b) i) -21 **ii)** -35 **iii)** -128
c) $6\frac{2}{3}$
3 i) x **ii)** x
4 $3(x^2 + 1)^2 + 2(x^2 + 1) - 4$
5 $\sqrt{4 - x^2}$, $2 - x$
6 $\frac{1}{2}(x^3 - 1)$
7 $\frac{x^2}{4(2x^2 - 1)}$, $-(4x^2 + 24x + 31)$
8 $f(x) = x^{\frac{2}{3}}$, $g(x) = 1 - x$, $h(x) = \sqrt{x}$

Exercises 13.3A *(p.320)*

1 $x + 4$ **2** $\frac{x}{3}$ **3** $\frac{x+4}{3}$ **4** $\sqrt{5x}$
5 $\sqrt[3]{x}$ **6** $\frac{7-x}{2}$ **7** $\frac{12}{x} - 1$ **8** $x^3 + 3$
9 $\sqrt{\frac{x-5}{2}}$ **10** $-\sqrt{x} - 2$
11 a) i) $k = 0$ **ii)** $\{f : 2 \le f \le \infty\}$
iii) $f^{-1}(x) = -\sqrt{x-2}$
b) i) $k = 3$ **ii)** $\{f : 0 \le f \le \infty\}$
iii) $f^{-1}(x) = -\sqrt{x+3}$
c) i) $k = -1$ **ii)** $\{f : f \le 5\}$
iii) $f^{-1}(x) = -1 - \sqrt{5-x}$
d) i) $k = 2$ **ii)** $\{f : f \ge -1\}$
iii) $f^{-1}(x) = 2 - \sqrt{1+x}$
e) i) $k = -3$ **ii)** $\{f : f \le 21\}$
iii) $f^{-1}(x) = -3 - \sqrt{21-x}$
12 a) i) $k = 5$ **ii)** $f^{-1}(x) = 5 - \frac{20}{x}$
iii) $\{f^{-1}: x \ne 0\}$, range $\{f : -\infty < f < \infty\}$
b) i) $k = 2$
ii) $f^{-1}(x) = \frac{5+2x}{x-1}$
iii) $\{f^{-1}: x \ne 1\}$, range $\{f: -\infty < f < \infty\}$
c) i) $k = -2\frac{1}{2}$
ii) $f^{-1}(x) = \frac{5x+2}{3-2x}$
iii) $\{f^{-1}: x \ne 1\frac{1}{2}\}$, range $\{f: -\infty < f < \infty\}$
13 b) $x \ne \frac{a}{b}$ **c)** $f^{-1}(x) = \frac{4-x}{4x+1}$
14 a) $\{f: f \ge -3\}$ **b)** $\{f^{-1}: x \ge -3\}$, range $\{f^{-1}: f^{-1} \ge 0\}$
e) $x = \frac{1}{2} - \sqrt{\frac{13}{4}}$

Exercises 13.3B *(p.321)*

1 $f^{-1}(x) = x + 3$
2 $f^{-1}(x) = \frac{1}{6}x$
3 $f^{-1}(x) = \frac{1}{2}(x - 3)$
4 $f^{-1}(x) = \sqrt[4]{x}$
5 $f^{-1}(x) = \sqrt[3]{3x}$
6 $f^{-1}(x) = \sqrt{2x+5}$
7 $f^{-1}(x) = 4x^2 - 6$
8 $f^{-1}(x) = \frac{4}{x} - 2$
9 $f^{-1}(x) = \sqrt{\frac{x+1}{3}}$
10 $f^{-1}(x) = \sqrt{\frac{3}{x} - 2}$
11 a) i) $k = 0$ **ii)** $\{y: y \le 3\}$
iii) $f^{-1}(x) = \sqrt{3-x}$
b) i) $k = -2$
ii) $\{y: y \ge -5\}$
iii) $f^{-1}(x) = \sqrt{x+5} - 2$
c) i) $k = 2.5$
ii) $\{y: y \ge -0.25\}$
iii) $f^{-1}(x) = \frac{5}{2} + \frac{1}{2}\sqrt{1+x}$
d) i) $k = -4$
ii) $\{y: y \le 7\}$
iii) $f^{-1}(x) = -4 + \sqrt{7-x}$
12 a) i) $k = -5$
ii) $f^{-1}(x) = \frac{2}{x} - 5$
iii) $\{x : x \ne 0\}$
b) i) $k = 5$
ii) $f^{-1}(x) = \frac{5x}{2+x}\{x: x \ne -2\}$
c) i) $k = 4$
ii) $f^{-1}(x) = \frac{3+4x}{x-2}$
iii) $\{x : x \ne 2\}$
13 a) $\frac{9}{5}$,32 **b) ii)** $f^{-1}(x) = \frac{5}{9}(x - 32)$
d) $x = -40$
14 a) $\{y: y \ge -6\}$
b) $g^{-1}(x) = \sqrt{6 + x} + 1$
$x = 4.193$

Exercises 13.4A *(p.325)*

1 a) i) a translation through -3 units // x-axis
b) i) a stretch, factor 2 // y-axis
c) i) stretch, factor 2 // x-axis
d) i) stretch, factor $\frac{1}{2}$ // y-axis *followed by* translation through -4 units // y-axis
e) i) translation through 2 units // x-axis *followed by* reflection in the x-axis
f) i) stretch, factor 2, // x-axis *followed by* translation through

2 units // y-axis.
g) i) translation through 5 units // x-axis *followed by* reflection // y-axis
h) i) stretch, factor $\frac{1}{2}$ // x-axis *followed by* translation through 1 unit // x-axis *followed by* stretch, factor 3 // y-axis

2 a) i) $\frac{1}{2}g(x)$
b) i) $g\left(\frac{x}{3}\right)$
c) i) $g(x) - 3$
d) i) $g(x - 3)$
e) i) $-g\left(\frac{1}{2}x\right)$
f) i) $g(-x) + 4$
g) i) $g(4 - x)$
h) i) $-g\left(\frac{1}{2}(x+3)\right)$

3 a) $2^x + 1$
b) $2^x + 1$
c) $f(x + 3)$
d) $\frac{1}{2^x}f(x)$

4 c) $\{f: x > 1\}$, $\{g : g(x) \geq 0\}$
$\{h: x > -2\}$, $\{h : h(x) \geq 0\}$

5 a) $h(x) = \sqrt{25 - x^2}$
b) stretch, factor $\frac{25^2}{4}$ // y-axis
stretch, factor $\frac{5}{2}$ // x-axis

6 a) odd **b)** even **c)** even
d) odd **e)** odd **f)** odd
g) even **h)** neither **i)** even
j) even

Exercises 13.4B *(p.327)*

1 a) i) translation through –1 units // x-axis
b) i) stretch, factor 3 // y-axis
c) i) stretch, factor $\frac{1}{2}$ // x-axis
d) i) stretch, factor $\frac{1}{2}$ // y-axis *followed by* translation through 2 units // y-axis
e) i) translation through 3 units // x-axis
f) i) translation through 3 units // x-axis *followed by* stretch, factor 2 // y-axis
g) i) translation through 4 units // x-axis *followed by* reflection in the y-axis
h) i) stretch, factor $\frac{1}{2}$ // x-axis *followed by* translation through 1 unit // x-axis *followed by* stretch, factor 3 // y-axis *followed by* translation through 4 units // y-axis.

2 a) i) $g(x - 3)$ **b) i)** $g(x) + 3$
c) i) $2g(x)$ **d) i)** $g(2x)$
e) i) $g(3x) + 4$ **f) i)** $g(-x) - 2$
g) i) $2g(x - 3)$ **h) i)** $g(2x) + 3$

3 a) i) $\sin(-x)$ **b) i)** $-\sin x$
c) $f(-x) = -f(x)$ **d) i)** $\cos x$
ii) $-\cos x$, $\cos x$ is an even function

4 b) reflection in the y-axis

5 a) translation through 2 units // x-axis
b) $\sqrt{9-(x-2)^2}$

6 a) even **b)** neither **c)** even
d) odd **e)** neither **f)** even
g) neither **h)** odd **i)** even
j) neither

Consolidation Exercises for Chapter 13 *(p.328)*

2 a) $\{y : y \geq 1\}$ **b)** 1, 0 **c)** $-2\frac{1}{3}$, 3
d) $-1\frac{1}{2}$

4 (1, 1), (3, 3)

5 (–1, 2)

6 a) $\{x : x \leq 3\}$ **b)** $0 < k < 4$
c) f one to one g many to one
d) $\{x: 0 \leq x \leq 4\}$, $\{y: 0 \leq y \leq 2\}$

8 a) $4x^2 + 4x + 4$
b) $\{y : y \geq 3\}$
c) $x = 3$

9 a) $-1 < x < 1$, $x > 3$
b) $f^{-1}(x) = \frac{3+x}{x-1} = f(x)$

10 b) $(gf)^{-1}: x \rightarrow \frac{2x}{x-2}$

11 a) i) $gf(2) = 3$
ii) $h(x) = x^2 - 1$
b) There are no limitations to x, f or g, $x \in \Re$, $f : f(x) \in \Re$, $g : g(x) \in \Re$
c) $g^{-1}(x) = x + 2$, $x \in \Re$
d) $x \geq 0$, $y \geq 1$

CHAPTER 14
Sequences and series 2: Binomial expansions

Exercises 14.1A *(p.335)*

1 a) $a^6 + 6a^5b + 15a^4b^2 + 20a^3b^3 + 15a^2b^4 + 6ab^5 + b^6$
b) $p^5 - 5p^4q + 10p^3q^2 - 10p^2q^3 + 5pq^4 - q^5$
c) $6561r^8 + 87\,480r^7 + 510\,300r^6 + 1\,701\,000r^5 + 3\,543\,750r^4 + 4\,725\,000r^3 + 3\,937\,500r^2 + 1\,875\,000r + 390\,625$

2 a) 495
b) $a^{12} + 12a^{11}b + 66a^{10}b^2 + 220a^9b^3$

3 a) $x^4 + 12x^3y + 54x^2y^2 + 108xy^3 + 81y^4$
b) $32f^5 + 240f^4g + 720f^3g^2 + 1080f^2g^3 + 810fg^4 + 243g^5$
c) $729a^6 - 1458a^5b + 1215a^4b^2 - 540a^3b^3 + 135a^2b^4 - 18ab^5 + b^6$
d) $p^5 + 2.5p^4q + 2.5p^3q^2 + 1.25p^2q^3 + 0.3125pq^4 + 0.03125q^5$
e) $6561x^8 - 34\,992x^7y + 81\,648x^6y^2 - 108\,864x^5y^3 + 90\,720x^4y^4 - 48\,384x^3y^5 + 16128x^2y^6 - 3072xy^7 + 256y^8$

4 a) $84a^6$
b) 180
c) $-448p^3$
d) $316\,800\,000f^5$
e) –77 520
f) $\binom{20}{r}.(-1)^r \,.\, x^r$

5 a) $78x^3$ **b)** $84x^3$
c) $32x^3$ **d)** $2\,427\,570x^3$

Exercises 14.1B *(p.335)*

1 a) $m^5 + 5m^4n + 10m^3n^2 + 10m^2n^3 + 5mn^4 + n^5$
b) $2187 - 5103a + 5103a^2 - 2835a^3 + 945a^4 - 189a^5 + 21a^6 - a^7$
c) $16x^4 + 96x^3 + 216x^2 + 216x + 81$

2 a) $-252x^5y^5$
b) $x^{10} - 10x^9y + 45x^8y^2 - 120x^7y^3$

3 a) $a^5 + 10a^4b + 40a^3b^2 + 80a^2b^3 + 80ab^4 + 32b^5$
b) $16x^4 + 160x^3b + 600x^2b^2 + 1000xb^3 + 625b^4$
c) $16\,384a^7 - 57\,344a^6b + 86\,016a^5b^2 - 71\,680a^4b^3 + 35\,840a^3b^4 - 10\,752a^2b^5 + 1792ab^6 - 128b^7$
d) $0.015\,625m^6 - 0.375m^5n + 3.75m^4n^2 - 20m^3n^3 + 60m^2n^4 - 96mn^5 + 64mn^6$
e) $3125 - 9375x + 11250x^2 - 6750x^3 + 2025x^4 - 243x^5$

4 a) $4a^3b$ **b)** $225\,173\,520x^4$

c) $1056p^{10}q^2$ d) $11\,160\,261a^5b^2$
e) $-192x^5$ f) $\binom{8}{n}2^{8-n}x^n$

5 a) $2000x^3$ b) $-17496x^3$
c) $66\,542x^3$ d) $20\,520\,960x^3$

Exercises 14.2A *(p.338)*

1 a) $1 + 10x + 45x^2 + 120x^3 + 210x^4$
b) $1 - 7x + 21x^2 - 35x^3 + 35x^4$
c) $1 + 12x + 54x^2 + 108x^3 + 81x^4$
d) $1 + 6x + 16.5x^2 + 27.5x^3 + 30.9375x^4$
e) $1 - 2x + 1.667x^2 - 0.741x^3 + 0.185x^4$
f) $1 + 8ax + 28a^2x^2 + 56a^3x^3 + 70a^4x^4$

2 a) $1 - 3x - 3x^2 + 25x^3$
b) $1 + 19x + 160x^2 + 780x^3$
c) $3 - 44x + 255x^2 - 720x^3$
d) $5 + 7x + 4x^2 + 1.12x^3$

3 $n = 5, a = 2$

4 $n = 10, r = 0.1$

5 a) $a = 4, b = -1$
b) $a = -\frac{44}{9}, b = \frac{1}{9}$

6 1.23 to 3 s.f.

7 0.817 to 3 s.f.

8 277 to 3 s.f.

9 a) 1.0121 to 5 s.f.
b) 60 240 to 5 s.f.
c) 16 298 to 5 s.f.

10 $h \approx 0.0111$

Exercises 14.2B *(p.339)*

1 a) $1 + 15x + 105x^2 + 455x^3 + 1365x^4$
b) $1 - 9x + 36x^2 - 84x^3 + 126x^4$
c) $1 + 35x + 525x^2 + 4375x^3 + 21\,875x^4$
d) $1 + 4x + 7\frac{1}{3}x^2 + 8.148148x^3 + 6\frac{1}{9}x^4$
e) $1 - 3\frac{1}{3}x + 4\frac{4}{9}x^2 - 2.963x^3 + 0.987x^4$
f) $1 - 10ax + 45a^2x^3 - 120a^3x^3 + 210a^4x^4$

2 a) $1 + 9x + 42x^2 + 77x^3$
b) $1 - 17x + 120x^2 - 460x^3$
c) $3 - 215x + 6732x^2 - 120\,204x^3$
d) $8 + 23x + 30x^2 + 23.375x^3$

3 $a = -6, n = 9$

4 $n = -3, r = -2$

5 $b = 2$ or $\frac{4}{7}$, $a = -3$ or $5\frac{4}{7}$

6 $a = 2$

7 $n = 10$

8 0.709 to 3 s.f.

9 525.98 to 5 s.f.

10 $\delta = 0.0061$

Exercises 14.3A *(p.344)*

1 a) $1 - x + x^2 - x^3 + x^4$
b) $1 + \frac{3}{2}x - \frac{9}{8}x^2 + \frac{27}{16}x^3 - \frac{405}{128}x^4$
c) $1 + \frac{2}{3}x + \frac{1}{3}x^2 + \frac{4}{27}x^3 + \frac{5}{81}x^4$
d) $2\left(1 + \frac{x}{24} - \frac{x^2}{576} + \frac{5x^3}{41472} - \frac{5x^4}{497664}\right)$
e) $1 - x + \frac{3}{x}x^2 - \frac{5}{2}x^3 + \frac{35}{8}x^4$
f) $1 + \frac{15}{4}x - \frac{75}{32}x^2 + \frac{625}{128}x^3 - \frac{28\,125}{2048}x^4$

2 a) $1 + x - \frac{1}{2}x^2 + \frac{1}{2}x^3$, 1.02
b) 9.80 to 3 s.f.
c) 2.03

3 $1 - 2x + 2x^2 - 2x^3$

4 $a^{\frac{1}{2}}b^{-\frac{1}{2}}$ 0.816

5 $a = \frac{1}{2}, P = -1, Q = \frac{3}{4}$

6 $n = -\frac{1}{5}, r = \frac{15}{2}, -0.06 \le x \le 0.07$

Exercises 14.3B *(p.345)*

1 a) $1 - 3x + 6x^2 - 10x^3$
b) $1 - 6x + 24x^2 - 80x^3$
c) $1 + 0.2x + 0.004x^2 + 0.008x^3$
d) $3(1 + 0.0123x - 0.000\,151x^2 + 3.14 \times 10^{-6}x^3)$
e) $1 + 3.5x - 6.125x^2 + 21.4375x^3$
f) $1 - 3.75x + 1.406\,25x^2 + 1.0547x^3$

2 a) $1 - 2.5x - 3.125x^2 - 7.8125x^3$, 0.975
b) 0.986 c) 2.0037

3 $1 + 1.5x + 0.875x^2 + 0.6875x^3 + 0.5859x^4$

4 $a = 4, b = -\frac{1}{32}$

5 $1 + 0.5x + 0.375x^2 - 0.1875x^3 + 0.023\,437x^4, -0.8 \le x \le 1.1$

6 $1 - \frac{2.667}{x} + \frac{7.111}{x^2} - \frac{31.61}{x^2}$

Consolidation Exercises for Chapter 14

1 $30z^3$

2 240

3 a) $1 + 0.5x^2 + 0.375x^4 + 0.3125x^6$
c) 1.4141

4 a) $d^2 = 25 - 24x$
b) $p = 5, q = -2.25, r = 0.506\,25$
c) $d = 4.988$

5 a) $k = \frac{3}{2}, p = 63, q = 189$ b) 126

6 a) $16(1 + 2x + 1.5x^2 + 0.5x^3 + 0.0625x^4)$
$16(1 - 2x + 1.5x^2 - 0.5x^3 + 0.0625x^4)$

7 a) $P = (\frac{1}{2}, -1)$
b) $8x^3 - 36x^2 + 54x - 27$
c) $I = \int_{\frac{3}{2}}^{\frac{5}{2}}\left(x^3 - \frac{9}{2}x^3 + \frac{27}{4}x - \frac{27}{8}\right)dx$
d) 0.25 e) 3

8 a) $16 - 32x + 24x^2 - 8x^3 + x^4$
b) $1 - 6x + 24x^2 - 80x^3$, converge for $-0.12 \le x \le 0.12$
c) $a = -128$ $b = 600$

9 a) $1 + 3x + 3x^2 + x^3$
b) $1 + 4x + 10x^2 + 20x^3$ valid for $x : -0.17 \le x \le 0.17$
c) $a = 25, b = 63$

10 a) $x = -\frac{1}{2} + \frac{1}{2}\sqrt{1-4p}$
or $x = -\frac{1}{2} - \frac{1}{2}\sqrt{1-4p}$
b) $A = -2$ $B = -2$ $C = -4$, valid for $p : -0.30 \le p \le 0.23$
c) $x = -(2p^3 + p^2 + p)$ (1)
$x = 2p^3 + p^2 + p - 1$ (2)
d) $x = -0.112, x = -0.888$

CHAPTER 15 Trigonometry 3

Exercises 15.1A *(p.352)*

1 a) $\frac{\pi}{6}$ b) $-\frac{3\pi}{4}$ c) 4π
d) $-\frac{7\pi}{6}$ e) $\frac{5\pi}{3}$

2 a) −120° b) 22.5° c) −330°
d) 75° e) 315°

3

Degrees°	Radians
0	0
30	$\frac{\pi}{6}$
60	$\frac{\pi}{3}$
90	$\frac{\pi}{2}$
120	$\frac{2\pi}{3}$
150	$\frac{5\pi}{6}$
180	π
210	$\frac{7\pi}{6}$
240	$\frac{4\pi}{3}$
270	$\frac{3\pi}{3}$
300	$\frac{5\pi}{3}$
330	$\frac{11\pi}{3}$
360	2π

4 a) 1.75 b) −0.506
c) 5.36 d) −3.49
e) 17.5

5 a) 114.6° **b)** 28.6°
c) 361.0° **d)** 99.2°
6 a) $\frac{\pi}{2}, \frac{3\pi}{2}$ **b)** $0, \pi, 2\pi$
c) $\frac{\pi}{3}, \frac{5\pi}{3}$ **d)** $\frac{2\pi}{3}, \frac{4\pi}{3}$ **e)** $\frac{\pi}{6}, \frac{11\pi}{6}$
f) $\frac{\pi}{4}, \frac{3\pi}{4}, \frac{5}{4}\pi, \frac{7\pi}{4}$
7 b) i) $0, \pi, 2\pi$ **ii)** $\frac{\pi}{2}, \frac{3\pi}{2}$
iii) $\frac{\pi}{6}, \frac{5\pi}{6}, \frac{7\pi}{6}, \frac{11\pi}{6}$
iv) $\frac{\pi}{4}, \frac{3\pi}{4}, \frac{5\pi}{4}, \frac{7\pi}{4}$

Exercises 15.1B *(p.353)*

1 a) $\frac{\pi}{3}$ **b)** $\frac{7\pi}{6}$ **c)** $\frac{-\pi}{4}$ **d)** 3π **e)** $-\frac{2\pi}{3}$
2 a) 45° **b)** –30° **c)** 300°
d) –105° **e)** 225°
3 a) 0.349 **b)** 1.26 **c)** 6.98
d) –2.44 **e)** 13.3
4 a) 85.9° **b)** 22.9 **c)** 171.9
d) 286.5 **e)** 412.5
5 b) i) $0, \pi, 2\pi$ **ii)** $\frac{\pi}{2}, \frac{3\pi}{2}$
iii) $\frac{\pi}{6}, \frac{5\pi}{6}$ **iv)** $\frac{7\pi}{6}, \frac{11\pi}{6}$ **v)** $\frac{\pi}{3}, \frac{2\pi}{3}$
vi) $\frac{\pi}{4}, \frac{3\pi}{4}, \frac{5\pi}{4}, \frac{7\pi}{4}$
6 c) i) $-\frac{\pi}{2}, \frac{\pi}{2}$ **ii)** $-\pi, 0, \pi$
iii) $-\frac{2\pi}{3}, -\frac{\pi}{3}, \frac{\pi}{3}, \frac{2\pi}{3}$
iv) $\frac{-3\pi}{4}, \frac{-\pi}{4}, \frac{\pi}{4}, \frac{3\pi}{4}$
7 c) i) $-2\pi, -\pi, 0, \pi, 2\pi$
ii) $-\frac{7\pi}{4}, -\frac{5\pi}{4}, -\frac{3\pi}{4}, -\frac{\pi}{4}$
iii) $-\frac{5\pi}{3}, -\frac{4\pi}{3}, -\frac{2\pi}{3}, -\frac{\pi}{3}$
iv) $-\frac{11\pi}{6}, -\frac{7\pi}{6}, -\frac{5\pi}{6}, -\frac{\pi}{6}$
v) $-\frac{3\pi}{2}, -\frac{\pi}{2}$

Exercises 15.2A *(p.355)*

1 a) PQ = 5.50 cm
POQ = 9.62 cm²
[shaded] = 3.50 cm²
b) PQ = 26.18 cm
POQ = 130.90 cm²
[shaded] = 105.90 cm²
2 $\theta = \frac{4}{3}$ radian = 76.4°
Area AOB = 150 cm²
3 θ = 28.65°, area = 36 cm²,
volume = 72 cm³
4 r = 6.25 cm
θ = 2.56 radians = 146.7°
5 perimeter = 50.83 cm,
area = 100π – 41.56 =
272.59 cm²
6 a) 0.199 m²
Water is travelling at 0.228 m s⁻¹
7 a) area of ΔPQR = 43.30 cm²
b) area S = 43.30 + 3(9.06) =
70.50 cm²
8 perimeter of cross-section =
19.13 m
9 θ = 1.28
10 $r = \frac{1}{4}l$, area $= \frac{1}{8}l^2$, $\theta = 2$

Exercises 15.2B *(p.356)*

1 a) i) 4.4 cm **ii)** 9.24 cm²
iii) 1.60 cm²
b) i) 10.47 cm **ii)** 26.18 cm²
iii) 15.36 cm²
2 θ = 264.44°, area = 24.375 cm
3 r = 5.76 cm, θ = 248.70
4 49.66°
5 a) 1.107 m²
b) speed of the water = 0.54 m s⁻¹
6 material needed = 10 708 m²
8 $\frac{1}{2}r^2(\alpha = \sin\alpha)$, $\alpha \approx \sin\alpha$
10 a) ∠CAB = 0.5857 radians
b) perimeter of R = 4.5142 cm
i) area ACX = 2.635 65 cm²
ii) area of shaded region
R = 0.296 cm²

Exercises 15.3A *(p.359)*

1 a) 2.000 **b)** 1.015
c) –0.364 **d)** 2.358
e) –6.000 **f)** 0
g) 2 **h)** 2 **i)** $\frac{4}{3}$
j) $\frac{4}{3}$, $1 + \tan^2 x = \sec^2 x$
$1 + \cot^2 x = \operatorname{cosec}^2 x$
2 a) 2 **b)** –1 **c)** –1
d) ∞ **e)** –1 **f)** –9.233
3 a) 60°, 300°
b) 191.54°, 348.46°
c) 63.44°, 243.44°
d) 9.74°, 80.26°
e) 25.66°, 154.34°
f) θ = 316.4
g) 45°, 135°, 225°, 315°
h) no solution
i) 30°, 60°, 120°, 150°, 210°,
240°, 300°, 330°
j) 53.13°, 306.87°
4 a) π **b)** 0.25, 2.89
c) 0.38, 3.52 **d)** 0.52, 1.05
e) 0.42, 2.72 **f)** 1.20, 1.94
g) no solution
h) 0.79, 2.36, 3.93, 5.50

Exercises 15.3B *(p.360)*

1 a) 1.06 **b)** –1.31 **c)** –5.67 **d)** 0
e) ∞ **f)** $\frac{4}{3}$ **g)** $\frac{4}{3}$ **h)** 1.19 **i)** 1.19
2 a) $\frac{1}{\sqrt{3}}$ **b)** $\sqrt{2}$ **c)** $\frac{2}{\sqrt{3}}$
d) ∞ **e)** ∞ **f)** 4.21
3 i) $\theta = 0, \pi, 2\pi$ **ii)** $\frac{\pi}{2}, \frac{3\pi}{2}$
iii) $0, \pi, 2\pi$
4 a) 0.52, 3.67 **b)** 3.67, 5.76
c) no solution **d)** 1.47, 3.05
e) 0.28, 1.81 **f)** no solution
g) 0.98, 2.35 **h)** no solution

Exercises 15.4A *(p.362)*

1 a) $\sec\theta$ **b)** $\sin\theta$ **c)** $\operatorname{cosec}\theta$
d) $\tan^4\theta$ **e)** $\sin\theta$ **f)** 1

Exercises 15.4B *(p.363)*

1 a) $\cos\theta$ **b)** $\sin^3\theta$ **c)** $\tan\theta$
d) $\frac{1}{2}\cos\theta$

Exercises 15.5A *(p.364)*

1 a) θ = 126.87°, 306.87°
b) θ = 90°, 270°
c) θ = 63.4°, 116.6°, 243.4°,
296.6°
d) θ = 30°, 150°
2 a) θ = 210°, 330°
b) θ = 60°, 120°, 240°, 300°
c) θ = 0°, 360°
θ = 131.8°, 228.2°
d) θ = 36.87°, 323.13°
θ = 180°
3 a) 30°, 150°, 138.2°, 318.2°
b) 51.8°, 308.2°
c) 14.5° or 165.5°, 199.5°
or 340.5°
d) 36.87°, 180°, 323.13°
4 a) –1.89 radians, 1.25 radians
–1.107 radians, 2.03 radians
b) –2.03 radians, 1.107 radians
–2.68 radians, 0.47 radians
c) 0.27, 2.87
d) –2.78, 0.37
–1.94, 1.21
e) –0.48, –2.67
5 i) –1.05, 1.05
ii) 1.13, 2.01
iii) –0.19, 0.91
6 a) $s = 1$, $s = \frac{1}{2}$, $s = -\frac{1}{2}$
b) 1.57, 0.52, 2.62, –0.52, –2.62

Exercises 15.5B *(p.365)*

1 a) 0°, 180°, 360°
b) 30°, 150°, 210°, 300°

c) 70.5°, 109.5, 250.5, 289.5°
d) 45°, 225°

2 a) 30°, 150°, 19.5°, 160.5°
b) –160.5°, –19.5°
c) –116.6, 63.4. –135, 45
d) $\theta =$ 53.1°, 1.26.9°
$\theta = -90°$

3 a) $\frac{\pi}{6}, \frac{5\pi}{6}, \frac{7\pi}{6}, \frac{11\pi}{6}$
b) 0.75 radians, 3.07 radians
c) 1.21, 4.35, 0.35, 3.51
d) 0.58, 2.56, 0.52, 2.62

4 –2.03 radians, 1.11 radians, –2.36 radians, 0.79 radians

5 3.48 radians, 5.94 radians, 0.52 radians, 2.62 radians

6 a) $(s + 1)(s + 2)(2s - 1)(2s + 1)$
b) $\frac{3\pi}{2}, \frac{\pi}{6}, \frac{5\pi}{6}, \frac{7\pi}{6}, \frac{11\pi}{6}$

Consolidation Exercises for Chapter 15 (*p.366*)

1 a) $k = -4$ **b)** 6 **c)** $\frac{5\pi}{4}, \frac{7\pi}{4}$

2 a) $\cos x = \frac{1}{3}$, $\cos x = -\frac{1}{3}$
b) 289.5°, 430.5°, 240, 480

3 a) at low tide y = 7 m
b) at high tide y = 13 m, $k = 0.51$

4 a) i) $l = r\theta$ **ii)** $A = \frac{1}{2}r^2\theta$
c) $r = 1$, $\theta = 2$ **d)** F

5 a) 0°, 180°, 360°
b) $\frac{\pi}{6} - 2n\pi$, $\frac{5\pi}{6} - 2n\pi$

6 b) $(3x + 2)(x + 2)(x - 1)$ ·
c) 222°, 318°, 90°

7 a) 10°, 50° **b)** 38.7°

9 30°, 150°, 194°, 346°

10 area of shaded region = r^2

CHAPTER 16
Logarithms

Exercises 16.1A (*p.371*)

1 a) $y = \log_{10}x$ is a reflection in $y = x$ of 10^x.

3 a) $8 = 2^3$ **b)** $100\,000 = 10^5$
c) $16 = 4^2$ **d)** $1 = 7^0$
e) $3 = a^b$ **f)** $x = a^5$
g) $1 = a^0$ **h)** $q = p^y$
i) $r = a^m$

4 a) $\log_3 9 = 2$ **b)** $\log_{11} 1 = 0$
c) $\log_{10} 10\,000 = 4$
d) $\log_{10} 0.001 = 10^{-3}$
e) $\log_8 2 = \frac{1}{3}$ **f)** $\log_n m = 2$
g) $\log_x 2 = y$ **h)** $10_{10}q = p$
i) $\log_u w = v$

5 a) 2.18 **b)** 0.70 **c)** 0.93
d) 1.20 **e)** 6.99 **f)** 1.28

6 a) 125.90 **b)** 15 848.93
c) 8.37 **d)** 632.46
e) 1000 **f)** 2.49
g) –0.34 **h)** 0.35
i) $x = 1.35$ **j)** $x = 4.11$
k) $x = 1.83$, $x = 0.69$

7 a) $\log_a \frac{63}{125}$ **b)** log2000
c) $\log_2 8pq^2$ **d)** $\log\left(\frac{r^2s^3}{100t^4}\right)$
e) $\log(x^{\frac{1}{2}}y^{\frac{1}{3}})$ **f)** $\log_a(x - 2)$
g) $\log_2(x + 1)^2$

8 a) $\log_2 u + \log_2 v$
b) $2\log_3 u + 3\log_3 v$
c) $a\log u + b\log v - \log w$
d) $\frac{1}{2}\log_a u + 2\log_a v + \frac{1}{3}\log_a w$
e) $\frac{3}{2}\log u + \frac{5}{2}\log u$
f) $3\log_{10} v + 2$ **g)** $-2\log w$
h) $\log(u - 1) + \log(v + 1) - \log(w - 2)$

Exercises 16.1B (*p.372*)

3 a) $128 = 2^7$ **b)** $1\,000\,000 = 10^6$
c) $3125 = 5^5$ **d)** $1 = 9^0$
e) $7 = a^b$ **f)** $x = a^{31}$ **g)** $1 = a^0$
h) $s = r^p$ **h)** $s = \frac{1}{2}q^n$

4 a) $\log_7 49 = 2$ **b)** $\log_{130} 1 = 0$
c) $\log_{16} 4 = \frac{1}{2}$ **d)** $\log_{10} 0.01 = -2$
e) $\log_{2401} 7 = \frac{1}{4}$ **f)** $\log_p r = 3$
g) $\log_x 2.5 = t$ **h)** $\log_{10} q = -p$
i) $\log_a 3c = 2b$

5 a) $x = 2.23$ **b)** $x = 0.52$
c) $x = 0.65$ **d)** $x = 0.28$
e) $x = 8.45$ **f)** $x = 8.45$

6 a) $x = 12.59$ **b)** $x = 316.23$
c) $x = 6.33$ **d)** x 0.099
e) $x = 50$ **f)** $x = 0$
g) $x = 36.31$ **h)** $x = \frac{2}{9}$
i) $x = 1.73$ **j)** $x = 4$

7 a) $\log_a\left(\frac{4}{3}\right)$ **b)** log 6075
c) $\log_3\left(\frac{rs^2}{9}\right)$ **d)** $\log\left(\frac{10xy^2}{t^5}\right)$
e) $\log\left(\frac{r^2}{s}\right)^{\frac{1}{8}}$ **f)** $\log_a(x + 3^2)$
g) $\log_5\left(\frac{x^3 - 1}{x - 1}\right)$

8 a) $\log_3 r + \log_3 s$
b) $4\log_4 s + \log_4 t$
c) $m\log r + n\log s + p\log t$
d) $\frac{1}{2}\log_a r - 3\log_a s$
e) $\frac{4}{3}\log s + \frac{5}{3}\log t$
f) $2\log_{10} r - 2$
g) $-(\log_a r + \log_a s + \log_a t)$
h) $\log(2r + 1) + \log(3s - 3) - \log(t + 3)$

Exercises 16.2A (*p.375*)

1 9.34×10^5 times greater

2 $E = 5.77 \times 10^{17}$ J

3 $\frac{I_C}{I_B} = 1 \times 10^5$

4 L = 63.01 dB

5 a) $\frac{S_1}{S_2} = 100$ **b)** 4.79×10^5
c) 5.75×10^{11}

Exercises 16.2B (*p.376*)

1 31 times stronger

2 $\frac{E_1}{E_2} = 11.086$

4 $r \geq 1.55$ m to hold a conversation.

5 a) 7.96, the solution is alkaline.
b) 1×10^{-5} moles per litre

Exercises 16.3A (*p.377*)

1 a) 0.6021 **b)** 2.1206
c) –1.8261 **d)** 0.7157
e) –0.9308 **f)** 11.4922
g) 4.0875 **h)** 0.1610
i) 0.4192 **j)** 1.0767
k) 1.3856 **l)** 0.0795

2 a) 2.7268 **b)** –1.5850

3 a) $x = 0.1260$ **b)** $x = 17.6549$
c) $x = 0.3562$, $x = 0$
d) $x = -1.5806$

4 $n \geq 65.5630$

Exercises 16.3B (*p.378*)

1 a) 3.2406 **b)** –0.6990
c) 0.1584 **d)** –0.2075
e) 0.3398 **f)** 4.0137
g) 1.8957 **h)** –0.2519
i) 7.9248 **j)** –4.1918
k) –0.5135 **l)** 3.1739

2 a) 1.6826 **b)** –1.8928

3 a) $x = 2$, $x = 1.5850$
b) $t = -1.3502$
c) 0.732 d) $x = -0.0534$

4 The least value of n is 9.

Exercises 16.4A (*p.382*)

1 $y = 4x^{\frac{1}{2}}$

2 $y = 5 \times \left(\frac{3}{2}\right)^x$

3 $w = 3x^{-2.5}$

4 $u = 2.5t^{1.6}$

5 $v = 1.7x^{-1.3}$

6 $s = 2.99 \times 0.02^r$

7 a) $\lambda = 1.4$, $C = 15\,971.81$
b) $P = 25.31$

8 $P = 17.1 \times 10^{(0.0133t)}$

9 a) $T_0 = 0.01$, $k = 1.03$

b) $T = 0.01 \times 1.03^x$ in the year 2081

10 $P = 108.73 \times 0.86^h$

Exercises 16.4B *(p.384)*

1 log – log $y = 3.4x^{2.7}$

2 log – ln $x = 0.3.4.1^t$

3 log – ln $p = 1.5 \times 2.72^t$

4 log – log $p = 1.5 \times t^{2.16}$

5 log – ln $v = 1.7 \times 5^u$

6 log – log $s = 4.9t^2$

7 $M = 0.04.108^l$

8 $\mu = 48.38T^{-2.06}$

9 $H = 16.16M^{0.93}$, $H = 603.68$

Consolidation Exercises for Chapter 16 *(p.385)*

1 a) $P = 7$ grams

b) $t > 6$ days 3 h 48 min

2 $t = 2.630$

3 $y = 3.(0.02)^x$, satisfies neither of the laws.

$s = 4.2.t^{1.7}$

4 Since linear graph is a log – log, $N = AT^B \Rightarrow N = 200T^{1.5}, A = 200$, $B = 1.5$

5 $yx^{0.50} = 56 \Rightarrow n = 0.50$, $k = 56$

6

$\log x$	0.477	0.778	1.000	1.176	1.301
$\log y$	1.017	1.468	1.801	2.065	2.253

b) Since it is a linear log log relationship, $y = Ax^n$.

c) $A = 2$, $n = 1.5$

7 $a = 3$

8 a) $1.5 < x < 2.0$, $x \approx 1.75$

b) $x = 1.76$ to 2 d.p.

9 b) $y = ab^x \Rightarrow \log y = \log(ab^x)$ $= \log a + x\log b$, $Y = mX + C$ where $Y = \log y$, $m = \log b$, $X = x$, $C = \log a$

c)

x	0.000	1.000	2.000	3.000	4.000
$\log y$	1.079	1.231	1.398	1.568	1.724

$y = 11.87 \times (1.45)^x$

10 a) $b = 1.32$, $a = 2.27$

b) $b = 10m$, $a = 10c$

CHAPTER 17 Exponentials and logarithms

Exercises 17.1A *(p.392)*

1 a) £2251.02 b) £225.65

c) £3187.70 d) £4065.59

2 a) 6.09% b) 8.24%

c) 12.68% d) 10.51%

3 a) £5986.09 b) £2983.65

c) £1822.12 d) £3664.21

4 a) 8.66 yrs b) 13.73 yrs.

5 a) 20.009 b) 1.396

c) 0.135 d) 1.948

6 a) $1.0835p$ b) $1.0811p$

therefore choose bank A

7 a) i) APR = 12.89%

ii) APR = 12.96%

b) A: £9167.68,

B: £9196.59, £28.91

Exercises 17.1B *(p.393)*

1 a) £3963.46 b) £3965.43

c) £6517.76 d) 6533.96

2 a) 5.10% APR b) 11.57% APR

c) 7.25% APR d) 9.42% APR

3 a) £1832.10 b) £1436.89

c) £7168.28 d) £94 877.36

4 a) 5 yrs 7 months

b) 8 yrs 9 months

c) 11 yrs 1 month

5 £20 702.63 in 2000, £143.88 in 1900

6 A: 5.875% APR

B: 5.654% APR

7 a) 7.389 non-oscillating

b) 0.368 oscillating

c) 1.649 non-oscillating

d) 0.607 oscillating

Exercises 17.2A *(p.395)*

2 a) i) at $y = 2$, $x = 1.897$

ii) at $x = 2$, $\frac{dy}{dx} = 0.815$

b) i) at $y = 4$, $x = 3.386$

ii) at $x = 1$, $\frac{dy}{dx} = 0.368$

c) i) at $y = 3.5$, $x = -0.560$

ii) at $x = 1$, $\frac{dy}{dx} = -0.736$

d) i) at $y = 5.3$, $x = 1.22$

ii) at $x = 1$, $\frac{dy}{dx} = 4.077$

e) i) at $y = -2$, $x = 4.605$

ii) at $x = 1$, $\frac{dy}{dx} = -0.824$

f) i) at $y = 0.9$, $x = -0.325$

ii) at $x = 1$, $\frac{dy}{dx} = -0.736$

3 a) stretch, factor 2 // y-axis *followed by* reflection in x-axis

b) reflection in y-axis *followed by* translation through –10 units // y-axis

c) stretch, factor 2 // x-axis

d) translation through –1 unit // x-axis *followed by* stretch, factor 0.5 // y-axis

e) stretch, factor $\frac{1}{2}$ // x-axis *followed by* translation through –5 units // y-axis

f) stretch, factor 5 // y-axis *followed by* reflection in x-axis *followed by* translation through 12 units // y-axis.

4 a) 13.86 yrs b) $\frac{dP}{dt} = 4121.80$

5 b) i) $t = 8$, $v = £2018.97$

ii) $t = 2.55$ yrs

c) $R = 20\%$

6 a) –0.405 c) 5473 books

d) $s = 6000$ books

Exercises 17.2B *(p.396)*

3 a) i) at $y = 0.75$, $x = 0.981$

ii) at $x = 1$ $\frac{dy}{dx} = -5.437$

b) i) at $y = 3$, $x = 0$

ii) at $x = -1$, $\frac{dy}{dx} = 0.812$

c) i) at $y = -18$, $x = 0.999$

ii) at $x = -1$, $\frac{dy}{dx} = -0.149$

d) i) at $y = 20$, $x = -1.004$

ii) at $x = -1$, $\frac{dy}{dx} = 20.086$

e) i) at $y = 4$, $x = -0.8326$

ii) at $x = -1$, $\frac{dy}{dx} = -10.873$

4 a) stretch, factor $\frac{1}{5}$ // y-axis

b) translation through –1 unit // y-axis

c) stretch, factor $\frac{1}{2}$, // x-axis *followed by* translation through 3 units // x-axis

d) stretch, factor $\frac{1}{2}$ // x-axis *followed by* reflection in y-axis *followed by* stretch, factor 2.5 // y-axis

e) reflection in the y-axis *followed by* translation through –1 unit // y-axis *followed by* reflection in x-axis *followed by* stretch, factor 4 // y-axis.

5 a) i) N will increase as $t \to \infty$

ii) N will decrease as $t \to \infty$

iii) $N = N_0$

b) i) $N = 5.18 \times 10^{25}$

ii) $N = 2.69 \times 10^{47}$

iii) $N > 10^{100}$
6 b) i) $Q = 3.679$ mg
ii) $t = 11$ h 30 mins
iii) $\frac{dQ}{dt} = -0.7358$ mg s^{-1}

Exercises 17.3A *(p.399)*

1 a) ln 21 **b)** ln 3.75 **c)** ln 3
d) ln 15 **e)** ln ba^n
2 a) 10 **b)** $\frac{1}{2}$ **c)** 5 **d)** x **e)** $\frac{4}{5}\tan x$
5 b) i) $2\ln x$ **ii)** $\frac{1}{2}\ln x$
c) $y^{-1} = e^{x/2}$, $y^{-1} = e^{2x}$
6 d) $y^{-1} = e^{-x}$

Exercises 17.3B *(p.399)*

1 a) ln 24 **b)** $\ln\frac{17}{3}$ **c)** ln 40
d) ln 6 **e)** ln 25
2 a) 1 **b)** 2 **c)** 25 **d)** 2 + ln 3
e) x^4 **f)** x^3
3 a) e^{12} **b)** $e^5 + e^7$ **c)** e^{-2}
6 c) $y^{-1} = e^{x/3}, x \in \Re$
$y^{-1} = e^{3x}, x \in \Re$

Exercises 17.4A *(p.402)*

1 a) $x = \ln 10 = 2.30$
b) $t = \frac{1}{5}\ln 15 = 0.54$
c) $t = \frac{1}{2}\ln\frac{5}{8} = -0.24$
d) $t = -\ln\frac{15}{4} = -1.32$
e) $t = -\frac{1}{3}\ln\frac{10}{7} = -0.12$
2 a) $x = e^3 = 20.09$
b) $x = e^{-\frac{5}{2}} = 0.08$
c) $x = \sqrt[2]{e^{2.1}} = 1.35$
d) $x = 8992.27$ **e)** $x = 3$
3 a) $x = 1.39$ **b)** $t = 0.3466$
c) $t = \frac{1}{5}\ln 5 = 0.32$
d) $x = 1.10$
e) $x = 1.67$
4 In the year 2005
5 a) $k = 0.000\,44$ **b)** 6.47 grams
c) 5281 yrs 10 months
6 a) 78%, 28.9%
b) 5589 yrs 11 months
7 2516.49
8 a) 90°C **b)** 5.61×10^{-3}
c) 2 min 31 s
9 a) when $k = 0.693$, $A = 20$
b) $17.5 \Rightarrow t = 4.32$
(no measurement)
10 b) 6 h

Exercises 17.4B *(p.403)*

1 a) $x = \ln 5 = 1.61$
b) $t = \frac{1}{4}\ln\frac{7}{3} = 0.21$
c) $t = -\ln\frac{3}{2} = -0.41$
d) $x = 0$
e) $x = \ln 1.5 + 2 = 2.41$
2 a) $x = e^7 = 1096.63$
b) $x = e^{\frac{3}{4}} = 2.12$
c) $x = 2$ **d)** $x = 0.28$ **e)** $x = 3$
3 a) $x = 1.61$ $x = 0.69$
b) $x = 1.95$ **c)** $x = 0.23$
d) Involves complex answers therefore no real solutions.
e) i) $x = -1.32$
ii) or same answer different method, $x = 1.32$, $x = -1.32$
f) no solutions
g) i) $x = 0.88$
ii) or same answer different method, $x = 0.88$
4 a) $M_0 = 80$ g **b)** $k = 0.069$
c) i) $M = 20$ g **ii)** $M = 2.5$ g
d) $t = 33$ days
5 $k = 4.33 \times 10^{-4}$
a) $t = 664$ yrs **b)** $t = 3200$ yrs
c) $t = 6915$ yrs
6 a) $k = 0.021$, $t = 32.89$ days
b) 13.51 yrs **c)** 1.51×10^{-3} s
7 a) 10 h 20 min **b)** $N = 2500$
c) 7 812 500 cells
8 after 10 h $N = 450\,000$
9 3.58×10^{-3}
a) $N = 61.5$ million
b) $N = 61.7$ million
c) $N = 73.6$ million
10 a) i) 933.90 mbar
ii) 887.43 mbar
iii) 862.56 mbar
b) $x = 205\,853$ feet

Consolidation Exercises for Chapter 17 *(p.406)*

1 a) APR = 10.47%
b) APR = 8.24%
c) APR = 13.00%
2 a) A = 85
$k = 0.064$
b) $t = 16.38$ seconds
3 a) B **b)** F **c)** A **d)** E **e)** C **f)** D
4 a) 783.71 years old **b)** no
5 a) $L = 70$, $D = 6$, $a = \ln\frac{2}{3}$
b) 3 days **d)** 70 mg
6 e) 20 000
7 a) $S_n = \frac{e^{nx} - 1}{e^x - 1}$
b) S_∞ exists for $x < 0$, $S_\infty = \frac{1}{1 - e^x}$
c) $T_n = eS_n$
8 a) $p = 3$, $q = 2$
b) $f^{-1}: x \rightarrow 2 + e - x, x \varepsilon \Re$
c) (0, 3), asymptote $y = 2$
9 a) 409 g **b)** 34.7 years
10 a) $a = 5$, $k = 0.462$ (3 s.f.)
b) $y = 5e^{0.462(x-3)}$
11 a) $f^{-1}(x) = -\sqrt{\ln x - 1}$
c) $g(x) = -\sqrt{x - 1}$, yes

CHAPTER 18 Differentiation 2

Exercises 18.1A *(p.413)*

1 a) $-\frac{7}{x^2}$ **b)** $\frac{24}{x^3}$ **c)** $5 + \frac{48}{x^4}$
d) $-\frac{6}{x^2} - 15$ **e)** $\frac{15}{2\sqrt{x}}$ **f)** $\frac{2}{x^{\frac{3}{2}}}$
g) $\frac{3}{2}\sqrt{x} - \frac{4}{\sqrt{x}}$
h) $8x + \frac{5}{2}x^{\frac{3}{2}} = 8x + \frac{5}{2}x\sqrt{x}$
2 a) (1, 2) is a local minimum, (–1, –2) is a local maximum
b) $(-3\frac{1}{3}, 3\frac{4}{3})$ is a local maximum
c) $(3, \frac{2}{9})$ is a local maximum, $(-3, -\frac{2}{9})$ is a local minimum
d) (4, 4) is a local maximum
e) (5.76, –19.9065) is a local maximum
f) $(36, -\frac{25}{6})$ is a local minimum
3 a) $h = \frac{500}{x^2}$ **b) i)** $A = x^2 + \frac{2000}{x}$
ii) $A = 2x^2 + \frac{2000}{x}$
c) i) $x = 10$, $h = 5$, $A = 300$ cm^2
ii) $x = \sqrt[3]{500}$, $h = \sqrt[3]{500}$,
$A = \frac{3000}{\sqrt[3]{500}}$ cm^2
4 a) $h = \frac{2}{\pi r^2}$ **b)** $A = \pi r^2 + \frac{4}{r}$
c) $r = \sqrt[3]{\frac{2}{\pi}}$; $h = r$ in this case
5 a) (0, 0), (12, 0) **b)** $f'(0) = \infty$, $f'(12) = \sqrt{3}$ **c)** (4, –16) is a local minimum
6 a) $2x + \frac{338}{x}$
b) $x = 13$ m so minimum length = 52 m

Exercises 18.1B *(p.414)*

1 a) 4 **b)** –125 **c)** 24 **d)** $6\frac{1}{16}$
e) $\frac{31}{9}$ **f)** 0 **g)** 13 **h)** $\frac{9}{4}$
2 a) $\left(-\frac{2}{2^{\frac{3}{2}}}, -\frac{3}{2^{\frac{1}{3}}}\right)$ is a local maximum
b) $\left(2, \frac{1}{4}\right)$ is a local maximum
c) $\left(\sqrt{15}, \frac{2\sqrt{15}}{45}\right)$ is a local maximum,

$\left(-\sqrt{15}, \frac{2\sqrt{15}}{45}\right)$ is a local minimum

d) $\left(\frac{25}{9}, \frac{125}{27}\right)$ is a local maximum

e) (3.2, 4.5795) is a local maximum

f) (100, –0.15) is a local minimum

3 a) $A = 24x^2 + \frac{1464}{x}$

b) $x = 7$ cm

c) $A = 3528$ cm^2 is a minimum value

4 a) $V = \pi r^2 h = 512\pi$, $S = 2\pi r^2 + 2\pi r$

b) $S = 2\pi r^2 + \frac{1024\pi}{r}$

c) $r = \sqrt[3]{256}$ gives a minimum value of $\frac{1536}{\sqrt[3]{256}}$

d) $\frac{d^2S}{dr^2} > 0$

e) $S = 192\ \sqrt[3]{2\pi}$; $h = 8\ \sqrt[3]{4\pi^2}$

5 a) $x = -3\ \sqrt[3]{10}$

b) f′(3) = 8 f′(9) = $\frac{268}{27}$ (= 9.9259)

c) $\left(\frac{3}{\sqrt[3]{5}}, 9\sqrt[3]{25}\right)$ is a local minimum

6 180 $\sqrt[3]{2}$ = 227 cm (to nearest cm)

7 i) $x = c$ **ii)** $v = \sqrt{2k}$

Exercises 18.2A *(p.420)*

1 a) $\frac{1}{2}e^{1.25}$ **b)** $10e^{-5}$ **c)** $-e^{0.3}$

d) $-40e^{-0.8}$ **e)** $5e^{1.5} - 6e^{-3}$

f) $-6e^{-2} - 12$ **g)** $5^{-1.2}\ln 5$

h) $\frac{3^{2.1x}\ln 3}{50}$ **i)** $4\ln\left(\frac{10}{3}\right)$ **j)** $\frac{5}{8}$

k) $-\frac{1}{5}$ **l)** 7.5 **m)** 2.5 – 0.5e

n) 2

2 a) $f'(x) = 2e^{2x} - 0.5$, $x = \frac{1}{2}\ln\left(\frac{11}{4}\right)$

b) $f'(x) = -0.2e^{-0.8x}$, $x = -1.25\ln 50$

c) $f'(x) = 0.2^{1-2x}\ln 2$, $x = 0.5$

d) $f'(x) = -\frac{2}{x}$, $x = \frac{8}{3}$

e) $f'(x) = -2e^{-x} - \frac{2}{x}$, $x = 2.428$ or $x = -0.679$

3 a) (ln 2, 2 – 2ln 2) is a local minimum

b) $\left(\frac{5}{6}\ln\frac{25}{6}, \frac{35}{6} + \frac{25}{6}\ln\frac{26}{6}\right)$ = (1.189, 11.79) is a local maximum

c) (1.5, 3 – 3ln 1.5) = (1.5, 1.784) is a local minimum

d) (2, 24ln 2 – 8) = (2, 8.636) is a local maximum

e) $\left(\frac{4}{7}, \frac{8}{7} - 7\ln\frac{4}{7}\right)$ = (0.5714, 6203) is a local minimum

4 a) $25\ln\frac{3}{2}$ = 10.14 years

b) $40\ \sqrt[5]{162}$ = 110.65 per year

5 a) $-3000e^{-0.6}$ (= –1646.43) £ per year

b) $\frac{10}{3}\ln 5$ = 5.36 years

6 a) $f(t) = 2500 \times 2^{\frac{1}{3}t}$

b) $f'(t) = \frac{2500\ln 2}{3}2^{\frac{1}{3}t}$

c) $t = 3 + \frac{3\ln 3}{\ln 2}$ = 7.75 hours

Exercises 18.2B *(p.421)*

1 a) 0.25e **b)** $21e^{14}$

c) e^6 **d)** –3.5

e) $6(e^6 + e^{-6})$ **f)** $-6e^{-4} + 84$

g) $7^5\ln 7$ **h)** –4000ln 5

i) $\frac{1}{15}\ln 10$ **j)** $\frac{7}{5}$

k) $\frac{16\,381}{8}$ = 2047.625

l) $\frac{1}{2}$ **m)** $-\frac{1}{9}$ **n)** $3\frac{2}{3}$

2 a) $f'(x) = \frac{1}{2}(e^x - e^{-x})$, $x = 0$

b) $f'(x) = 1 - \frac{1}{x}$, $x = 3$

c) $f'(x) = \frac{1}{2\sqrt{x}} - \frac{1}{2x}$, $x = 1$

d) $f'(x) = \frac{1}{2}(e^x + e^{-x})$, $x = \ln(\sqrt{5} + 2)$ or $x = \ln(\sqrt{5} - 2)$

e) $f'(x) = -\frac{10}{x}$, $x = -2.5$

3 a) (2ln 2, 2 – 2ln 2) is a local minimum

b) $\left(\frac{1}{2\sqrt{3}}, \frac{1}{4}(1 + \ln 12)\right)$ = (0.2887, 0.8712) is a local minimum

c) (ln 2, 8 – 12ln 2) is a local minimum

d) $(-1, -\frac{1}{4})$ is a local minimum

e) (1, –2) is a local maximum

4 b) $c'(t) = k(be^{-bt} - ae^{-at})$ rate of decay of drug in bloodstream

c) $\frac{1}{(a-b)}\ln\left(\frac{a}{b}\right)$

d) maximum value of $c(t)$ ($b > a$ for positive $c(t)$)

5 a) $t = \frac{1}{2}$ hour **b)** $P = 2000e^{2t\ln 2}$

c) $P'(1) = 16\,000\ln 2$, $P'(2) = 64\,000\ln 2$

6 a) $(2e^3)e^{2x} = 2e^{2x+3}$

b) $3e^{3x} - 2$, $-7e^{5-7x}$

c) ae^{ax+b} **d)** 0.74

Exercises 18.3A *(p.424)*

	tangent	normal
1	$y = 4x - 9$	$y = -\frac{1}{4}x - \frac{1}{2}$
2	$y = -6x$	$y = \frac{1}{6}x + \frac{37}{6}$
3	$y = -3x - 12$	$y = \frac{1}{3}x - \frac{16}{3}$
4	$y = \frac{5}{3}x + 15$	$y = -\frac{3}{5}x + \frac{177}{5}$
5	$y = -4t$	$s = \frac{1}{4}t + \frac{17}{4}$
6	$x = -2t + 2$	$x = \frac{1}{2}t + 2$
7	$y = 8x + 24$	$y = -8x + 40$ (1, 32)
8	$y = 4x - 11$	$y = 4x + 21$
9	$p = -0.6$	$q = 12.24$
10	$y = 3.5x - 3.75$	

Exercises 18.3B *(p.424)*

	tangent	normal
1	$y = -8x - 28$	$y = \frac{1}{8}x + 4.5$
2	$s = -12t + 30$	$s = \frac{1}{12}t + \frac{35}{6}$
3	$y = 3x - 4$	$y = -\frac{1}{3}x - \frac{2}{3}$
4	$y = 12x - 8$	$y = -\frac{1}{12}x + \frac{49}{12}$
5	$u = 6x - +12$	$y = -\frac{1}{6}x + 36\frac{2}{3}$
6	$y = -x + 3$	$y = x + 1$
7	$24y = 2x + 1$	$24y = -2x + 5$ (1, 32)
8	$y = -\frac{1}{3} + 36\frac{1}{3}$	$y = -\frac{1}{3}x + 3$

9 $y = -4x$, $y = 17x - 24$, $y = 12x + 36$

10 20.25

Exercises 18.4A *(p.426)*

1 a) $f''(x) = 6x$, $f''(2) = 12$

b) $f''(x) = 2 - 12x^2$, $f''(3) = -106$

c) $f''(x) = 2a$, $f''(-1) = 2a$

d) $f''(x) = 2 - \frac{6}{x^4}$, $f''(-1) = -4$

e) $f''(x) = e^{-x}$, $f''(0) = 1$

f) $f''(x) = -\frac{2}{x^2}$, $f''(1) = -2$

g) $f''(x) = \frac{(4 + 3x)}{4x^{\frac{3}{2}}}$, $f''(4) = -\frac{1}{2}$

h) $f''(x) = 2^x(\ln 2)^2$, $f''(3) = 8(\ln 2)^3$

2 a) (1, –2) is a local minimum: $f''(1) = 18 > 0$

(–2, 25) is a local maximum: $f''(-2) = 18 < 0$

b) (2, 3) is a local minimum: $f''(2) = \frac{3}{2} > 0$

c) $(\frac{1}{2}\ln 5, 5(\ln 5 - 1)$ is a local maximum: $f''(\frac{1}{2}\ln 5) = -20 < 0$

d) (0, 7) is a local minimum: $f''(0) = 4 > 0$

(1, 8) is a local maximum: $f''(1) = -8 < 0$

(–1, 8) is a local maximum: $f''(-1) = -8 < 0$

3 a) (1, 13) **b)** (0, 0), $\left(\frac{2}{3}, -\frac{16}{27}\right)$
c) (1, –10) (–1, 10)
d) (2.25, –8.4375)
e) (–ln 6, 3(ln 6)²)
(–1.79, –3.63))
f) (0.5, 0.5 – 3ln 2)

4 a) i) – 20m s^{-2} **ii)** 1.5 s
b) i) $-10e^{-2}$ m s^{-2} **ii)** ln 5 = 1.61 s
c) i) 5 m s^{-2} **ii)** 0.5 s

Exercises 18.4B *(p.427)*

1 a) $f''(x) = x + 1$, $f''(2) = 3$
b) $f''(x) = 5 - 3x$, $f''(2) = -1$
c) $f''(x) = 12x + \frac{6}{x^3}$, $f''(-1) = -18$
d) $f''(x) = e^{\frac{1}{2}x}$, $f''(-1) = e^{-0.5}$
e) $f''(x) = \frac{2}{x^2}$, $f''(3) = \frac{2}{9}$
f) $f''(x) = \frac{3}{2}(8 - 5\sqrt{x})$,
$f''(5) = \frac{3}{2}(8 - 5\sqrt{5}) \approx -4.7705$
g) $f''(x) = 5(\ln 3)^2 3^{-x}$,
$f''(0.5) = \frac{5\sqrt{3}}{3}(\ln 3)^2 \approx 3.4842$
h) $f''(x) = 3^x(\ln 3)^2 - 2^x(\ln 2)^3$,
$f''(-1) = \frac{1}{3}(\ln 3)^2 - \frac{1}{2}(\ln 2)^2 \approx$
0.1621

2 a) (0, 10) is a local minimum,
$f''(x) = 6 > 0$
(2, 14) is a local maximum,
$f''(x) = -6 < 0$
b) $\left(-\frac{1}{2}\sqrt[3]{10}, \frac{3}{2}\sqrt[3]{100}\right)$ is a local minimum,
$f''\left(-\frac{1}{2}\sqrt[3]{10}\right) = 12 > 0$
c) (5.76, 39.8131) is a local maximum, $f''(5.76) = -6 < 0$
d) (2, 10) is a local minimum,
$f''(2) = 6 > 0$
(4, 4) is a local maximum,
$f''(4) = -6 < 0$

3 a) $\left(\frac{2}{3}, -\frac{70}{27}\right)$
b) (0, 0), $\left(\frac{3}{4}, -\frac{81}{128}\right)$
c) no points of inflexion
d) $\left(\frac{5}{2}, 0\right)$
e) $(\frac{1}{2}\ln 5, 5(\frac{1}{2}(\ln 5)^2 - 1$
f) $\left(\frac{1}{10^{\frac{1}{5}}}, \frac{1}{10}(104\ln 10)\right)$ or
(0.6310, –0.8210)

4 a) i) $a = -3.6$ m s^{-2}
ii) $t = 1$ s or $t = 5$ s
b) i) $a = -\frac{15}{16}$ m s^{-2}
ii) $t = 3$ s
c) i) $a = 3e{-}0.4$ m s^{-2}
ii) $t = 2\ln(3)$ s

Consolidation Exercises for Chapter 18 *(p.429)*

1 b) $V = 2x^2\left(\frac{50}{x} - \frac{2x}{3}\right)$
c) $\frac{100}{3}$ cm³ (when $x = 5$ cm)
d) $V''(5) = -40 < 0$

2 a) $f'(x) = e^x - 5$ **b)** $x = 1.61$
d) $p = 25$

3 a = 0.90, $k = 0.039$, $\frac{dy}{dx} = -0.3$

4 4.58 hours

5 (–2, –4) is a (local) maximum,
(2, 4) is a (local) minimum,
(–∞, –2) and (2, ∞)

6 a) $x \geq 5$

7 a) 1.65, 65%
b) $k = 0.5008$ (to 4 d.p.)
(or $k = \ln 1.65$ exactly)
d) 0.5

8 a) $z = \frac{100}{x}$
b) $(x + 24)$ cm by $\left(12 + \frac{100}{x}\right)$
c) $1288 + 480\sqrt{5}$ cm² =
2361 cm² (to 4 s.f.)

9 a) $f'(x) = 6x^2 - 12x^3$, $x = 0$, $x = 5$,
(9, 4), $\left(\frac{1}{2}, \frac{65}{16}\right)$
b) $f''(x) = 12x - 36x^2$, $f''(0) = 0$,
$f''(0.5) = -3$
c) (0, 4) is a point of inflexion
d) $\left(\frac{1}{2}, \frac{65}{16}\right)$ is a (local) maximum

10 b) $40\sqrt{\pi}$

CHAPTER 19
Integration 2

Exercises 19.1A *(p.434)*

1 a) $-\frac{7}{x} + c$ **b)** $\frac{6}{x^2} + c$
c) $\frac{5}{2}x^2 + \frac{16}{3x^3} + c$
d) $-\frac{3}{2x}(4 + 5x^3) + c$
e) $10x\sqrt{x} + c$ **f)** $-8\sqrt{x} + c$
g) $\frac{2\sqrt{x}}{15}(3x^2 - 40x) + c$
h) $\frac{2x^3}{21}(14 + 3\sqrt{x}) + c$
i) $\frac{8\sqrt{x}}{5}(x^2 - 45) + c$
j) $-\frac{9}{x}(1 + 2\sqrt{x}) + c$

2 a) 3 **b)** 76 **c)** $\frac{31}{4} = 7.75$
d) 16
e) 7.0083 (to 4 d.p.)
$\left(= \frac{17}{5}\sqrt{6} - \frac{14}{15}\sqrt{2}\right)$
f) 3.5156 (to 4 d.p.)

3 a) 10 **b)** 7.5 **c)** $\frac{63}{6} = 12.6$
d) 16 **e)** $\frac{625}{6} = 104.167$
f) $\frac{48\sqrt{6}}{5} = 23.5151$ **g)** $-\frac{1}{12}$ **h)** $\frac{9}{32}$

4 a) 2 **b)** $\frac{155}{3} = 51.67$
c) $\frac{64}{3} = 21.33$ **d)** $\frac{20}{3} = 6.67$

5 a) $f(x) = 5x - 4\sqrt{x} - 3$, $f(4) = 9$
b) $f(x) = 4x + \frac{3}{x}$, $f(4) = \frac{67}{4} = 16.75$
c) $f(x) = 6x\sqrt{x} + 2$, $f(4) = 50$
d) $f(x) = \frac{1}{x^3} - \frac{1}{x^2} + 2$, $f(4) = \frac{125}{64} =$
1.9531

Exercises 19.1B *(p.435)*

1 a) $-\frac{1}{4x^2} + c$ **b)** $\frac{5}{8x^8} + c$
c) $x^3 + \frac{1}{x^3} + c$ **d)** $\frac{77}{3}x^3 + \frac{33}{4x^4} + c$
e) $\frac{2}{x}\sqrt{x}(x + 3) + c$
f) $\frac{2}{x}\sqrt{x}(x - 3) + c$
g) $\sqrt{x}(x - 1) + c$
h) $\frac{2}{5}x^{\frac{5}{2}} - \frac{7}{2}x^2 + c$
i) $-\frac{2}{3x^{\frac{3}{2}}} + c$ **j)** $\frac{2}{5}\sqrt{x}(x^2 - 20)$

2 a) $\frac{20}{9}$ **b)** $54.4\left(\frac{272}{5}\right)$ **c)** $22\sqrt{2} - 16$
d) $\frac{20}{9}$ **e)** $-\frac{8}{21}$ **f)** $\frac{1}{6}$

3 a) $\frac{3}{2}$ **b)** 0.2508 (4 d.p.) **c)** 1
d) $4 - \frac{8}{3}\sqrt{2} = 0.2288$ (to 4 d.p.)
e) 8.5
f) $\frac{3}{5}(6 + \sqrt{2}) = 4.4485$ (to 4 d.p.)
g) $\frac{1}{110} = 0.0091$ (to 4 d.p.)
h) $\frac{41}{3}$

4 a) $\frac{1}{3}$
b) $\frac{15}{5} - 8\ln 2 = 1.9548$ (to 4 d.p.)
c) $\frac{70}{3} - 8\sqrt{6} = 3.7341$ (to 4 d.p.)
d) $\frac{16}{3}$

5 a) $f(x) = 6\sqrt{x} - x - 3$,
$f(2) = 6\sqrt{2} - 5$
b) $f(x) = 3x + \frac{14}{x^2} - \frac{11}{2}$, $f(2) = \frac{11}{8}$
= 1.375
c) $f(x) = \frac{2\sqrt{x}}{3}(3 - x) + \frac{19}{x}$,
$f(2) = \frac{1}{3}(19 + 2\sqrt{2}) = 7.2761$
d) $f(x) = \frac{1}{x^4} - \frac{1}{x^3} + 2$, $f(2) = \frac{31}{16}$

Exercises 19.2A *(p.438)*

1 a) $4e^x + c$ **b)** $-e^{-x} + c$
c) $3e^{2x} + c$ **d)** $-\frac{2}{3}e^{-3x} + c$
e) $\frac{2^x}{\ln 2} + c$ **f)** $-\frac{(0.5)^{2x}}{2\ln 2} + c$
g) $-\frac{3^{-x}}{\ln 3} + c$ **h)** $-\frac{5^{-2x}}{2\ln 5} + c$

2 a) $4\ln x + c$ **b)** $-3\ln x + c$
c) $\frac{1}{2}x^2 - 9\ln x + c$
d) $2\ln x + \frac{5}{x} + c$

3 a) $\frac{1}{2}(e^2 - 1)$ **b)** 4 **c)** $\frac{8}{3\ln 3}$
d) 12ln6 **e)** $40 + e - e^3$
f) $\frac{137}{12} - e^{-5}$ **g)** 16 – 16ln3
h) $\frac{15}{2\ln 2}$
4 a) $34 - \frac{15}{\ln 2} = 12.3595$
b) $e^3 - e^{-1}$ **c)** $30 + 10e^{-4}$
d) 16ln2 **e)** $(e^1 - e^{-1})$
f) $e^1(2\ln 2 + 1) - 1$
5 a) $\frac{1}{2}e^{2x+3} + c$ **b)** $\frac{1}{4}e^{4x-7} + c$
c) $\frac{1}{a}e^{ax+b} + c$ **d) i)** $\frac{1}{5}(e^{1.5} - e^{-1})$
ii) $e - e^{-1}$ **iii)** $2(e - e^{-1})$
6 a) e – 1 **b)** 1
7 $e^2 - e - \ln 2$

Exercises 19.2B *(p.439)*

1 a) $\frac{7}{3}e^{3x} + c$ **b)** $-\frac{1}{2}e^{-2x} + c$
c) $-4e^{-3x} + c$
d) $-3e^{-3x} - 3e^{-5x} + c$ **e)** $\frac{5^x}{\ln 5} + c$
f) $\frac{(0.25)^{-x}}{\ln 4} + c$ or $\frac{2^{2x}}{\ln 4} + c$
g) $\frac{7^{3x}}{3\ln 7} + c$ **h)** $-\frac{8^{-4x}}{4\ln 8}$
2 a) $11\ln x + c$ **b)** $-17\ln x + c$
c) $13\ln x - \frac{1}{2}x^2 + c$
d) $7\ln x - \frac{15}{x} + \frac{1}{2x^2} + c$
3 a) $\frac{13}{10}(e^{1.5} - 1)$ **b)** $\frac{1}{4}$ **c)** $\frac{4\sqrt{5}}{5\ln 5}$
d) $\frac{1}{3}\ln 4$ **e)** $\frac{8}{\ln 2} - \frac{37}{3}$
f) $-\frac{e^{-15}}{3} - \frac{207}{4}$ **g)** $10 - 25\ln\left(\frac{2}{3}\right)$
h) $\frac{124}{25\ln 5}$
4 a) $\frac{8}{\ln 2} - \frac{3}{\ln 2}$ **b)** $\frac{15}{2\ln 2} - \frac{32}{4}$
c) $\ln 3 + e^{-1} - e^{-3}$ **d)** $\frac{3}{2}e^{-1} - e^{-2}$
e) $e + e^{-1} = 2$
f) $\frac{98\sqrt{7}}{5} - \frac{9}{5}\sqrt{\frac{3}{2}} + \frac{1}{\ln 1.5}\left(\frac{3}{2}\sqrt{\frac{3}{2}} - 1 - 1.5^7\right)$
(= 14.1085 to 4 d.p.)
5 a) $\frac{1}{3}e^{3x+5} + c$ **b)** $\frac{1}{4}e^{4x-11} + c$
c) $\frac{1}{a}e^{ax+b} + c$ **d) i)** $\frac{1}{2}(e^{-1} - e^{-3})$
ii) $e - e^{-1}$ **iii)** $2(e^{-1} - e^{-2})$
6 b) 1 – ln 2 **c)** ln 2
d) ln 2 **e)** $\ln(x + a)$
7 a) $\frac{14}{3} - 2\ln 2$

Exercises 19.3A *(p.442)*

1 a) 36π **b)** 32π **c)** $\frac{32\pi}{5}$ **d)** $\frac{24\pi}{\ln 2}$
e) $50\pi(4e^{-2} - e^{-4} + 1)$
2 a) $\frac{16^3\pi}{3}$ **b)** $\frac{49\sqrt{3}\pi}{5}$ **c)** 4π
3 a) $\frac{(36)^6\pi}{5}$ **b)** $\frac{7^5\pi}{30}$ **c)** $\frac{27\pi}{4}$
d) $\frac{\pi}{30}$
4 a) $\frac{19 \times 2^7\pi}{15}$ **b)** $\frac{3\pi}{10}$
5 a) 3 **b)** $\frac{21\pi}{2}$
6 a) A: (√3, 0), B: (1, 2)
b) $2\sqrt{3} - \frac{5}{3}$
c) $\left(\frac{24\sqrt{3}}{5} - \frac{88}{15}\right)\pi$
7 a) $y = \frac{a}{h}x$ **d)** $\frac{\pi a^2(h^3 - b^3)}{3h^2}$

Exercises 19.3B *(p.443)*

1 a) $\frac{6272\pi}{3}$ **b)** $\frac{9\pi}{2}$ **c)** $\frac{384\pi}{7}$
d) $\frac{40\pi}{\ln 3}$ **e)** $\frac{\pi}{2}(8\ln 2 - 5)$
2 a) $\frac{9\pi}{4}$ **b)** $\frac{3^5 2\pi}{5}$ **c)** 8π
3 a) $\frac{5^4\pi}{6}$ **b)** $\frac{5^6\pi}{21}$ **c)** $\frac{32\pi}{15}$ **d)** $\frac{64\pi}{21}$
4 a) $\frac{128\pi}{15}$ **b)** $\frac{2\pi}{35}$
5 a) 3ln 2 + 6 **b)** $\left(12\ln 2 + \frac{27}{2}\right)\pi$
6 a) A: (1, 1) **b)** $2\ln 2 + \frac{2}{3}$ **c)** $\frac{5\pi}{4}$
7 d) $\frac{\pi}{3}(a - b)^2(2a + b)$

Consolidation Exercises for Chapter 19 *(p.445)*

1 a) 117π **b)** 24π **c)** $\frac{\pi}{2}(e^5 - 1)$
d) 288π **e)** $\frac{243}{5}\pi$ **f)** 192πln2
g) $\frac{\pi}{2}(e^5 - 1)$ **h)** 44π **i)** $\frac{7\pi}{8}$
2 b) 1 **c)** ln 2
3 $\frac{4}{3}\pi ab^2$ If $a = b$ then volume $= \frac{4}{3}\pi b^3 \equiv$ volume of sphere
4 a) $\frac{13}{6}$ **b)** $\frac{59\pi}{15}$
5 a) 162π **b)** $\frac{567}{8}\pi$ **c)** $c = 1$
d) $c^3 - 75c - 88 = 0$, $c = -1.196$
7 a) $k = 8$ **b)** –4 **c)** 1.72
8 a) 39 **b)** 78 **c)** 10π
9 ii) $\int_0^{12} \pi(4 + y)dy$ **iii)** 120 cm³
10 a) $\int_0^k (\sqrt{x} - x^2)dx$, $\int_k^1 (\sqrt{x} - x^2)dx$
c) $k = 0.4410$ (to 4 d.p.)

CHAPTER 20
Differentiation 3

Exercises 20.1A *(p.457)*

1 a) $10(2x - 1)^4$ **b)** $8x(x^2 + 1)^3$
c) $-\frac{4}{(4x+1)^2}$ **d)** $-\frac{3x^2 - 2x}{(x^3 - x^2)^2}$
e) $\frac{1}{2\sqrt{x-2}}$ **f)** $\frac{2x}{(x^2 - 1)}$
g) $2(x + 1)e^{x^2 + 2x}$
h) $\frac{1}{2\sqrt{x}}e^{\sqrt{w}}$ **i)** $4(e^x + 1)(e^x + x)^3$
j) $\frac{12}{(4x - 1)}$
2 (0, 1)
3 a) at (0.5, 6.25): –15 decreasing, at (2, 6.25): 3.75 increasing
b) (1, 4) is a local minimum
4 a) –8 m s^{-2}
b) 0.88 m s^{-2}
c) 0.073 m s^{-2}
5 a) 0.000 16 $m^3 s^{-1}$
b) 0.016 $m^2 s^{-1}$
6 a) 4.8c
b) 0.01c (to 2 d.p.)
c) 0 (actually $6 \times 10^{-173}c$)
7 $\frac{5}{4}\sqrt{\frac{3}{\pi}} = 1.2215$ mm s^{-1}

Exercises 20.1B *(p.452)*

1 a) $12(3x + 2)^3$ **b)** $6t^2(t^3 - 1)$
c) $-\frac{6}{(s-7)^2}$ **d)** $\frac{-6}{(2t+1)^4}$
e) $\frac{-2x(6x+1)}{(4x^3 + x^2)^2}$ **f)** $x(3x^2 - 4)e^{x^2 - 2x^2}$
g) $\frac{-1}{\sqrt{x}}e^{-2\sqrt{x}}$ **h)** $\frac{(2x - e^x)}{2(e^x - x^2)^{\frac{3}{2}}}$
i) $\frac{8}{2s+3}$ **j)** $\frac{8x+3}{4x2+3x}$
2 a) 1.25 increasing
b) no stationary points
c) $x < -3$ and $x > 0$
d) ∞
3 a) $(0, e^{-1})$ is a (local) minimum
b) (0, 0) is a (local) minimum
c) (–1, 0) and (1, 0) are (local) minima
d) $\left(1, \frac{1}{\sqrt{3}}\right)$ is a (local) maximum
4 3π $cm^2 s^{-1}$
5 $\frac{1}{16}$m min^{-1}
6 $-240(5 + 2e^{-3t})e^{-3t}$
7 b) $\frac{90}{\pi^2 h^2}$cm s^{-1}

Exercises 20.2A *(p.454)*

1 a) $3x^2 - 2x + 1$ **b)** $\frac{3x+1}{2\sqrt{x}}$
c) $2(2x + 1)(4x^2 + x - 2)$
d) $-\frac{1}{(x-1)^2}$ **e)** $2(x - 1)^2(2x + 1)$
f) $t(5t^3 + 6t - 2)$ **g)** $\frac{(s-1)(5s+3)}{2\sqrt{s+1}}$
h) $(2u - 3u^2)e^{-3u}$
i) $(x - 1)(1 + x)e^x$
j) $(2 - 6x - 4x^2)e^{-x^2}$
k) $2(1 + \ln(2x + 1))$

l) $\left(\frac{1}{x}-\ln x\right)e^{-x}$

2 a) $y = 4x - 2$ increasing
b) $(-1, -e^2)$ is a local minimum
$(2, 2e^{-4})$ is a local maximum

3 at $(0, 0)$: $y = 0$ neither increasing nor decreasing
at $(1, e^1)$: $y = (5x - 4)e^1$ increasing,
at $(-1, -e^1)$: $y = (5x + 4)e^1$ increasing,
$(0, 0)$ is a point of inflexion

4 a) $\frac{(5-2t)}{4}e^{-0.25t}$
b) $t < 2.5$, the object is moving towards the origin
c) $t = 0$, $v = 1.25$ m s^{-1}
d) $x \to \infty$, $v \to 0$

5 b) 0.32 m^2

Exercises 20.2B (*p.455*)

1 a) $4x - 1$ **b)** $2(6t^2 + 6t - 5)$
c) $4s^3 + 9s^2 - 4s - 10$
d) $-(4x + 3)x^{-4}$
e) $\frac{2s^2+1}{\sqrt{s^2+1}}$ **f)** $(1 - 2u)e^{-u^2}$
g) $-2(3v^2 - 3v + 1)e^{-2v}$
h) $\frac{(5x+3)}{(1+x)^3}$ **i)** $\frac{3x-2}{4-x} - 3\ln(0.5x - 2)$
j) $(2t^2 - 1 + t^{-2})e^{t^2}$
k) $\left(3\ln(2x-1)+\frac{2}{(2x-1)}\right)e^{3x}$
l) $\frac{1}{3}(25t^2 + 32t - 23)(t^2 + 2t - 1)^{\frac{-2}{3}}$

2 a) $x(12x^4 + 5x^3 - 12^x - 3x + 2)$
b) $-2x(6x^2 - 6x - 1)e^{-2x}$
c) $2xe^{-3x} - 2(3x^2 + x - 1)e^{-3x}\ln(x + 1)$

3 $(0, 0)$ point of inflexion,
$(1, 0)$ local minimum,
$(-1, 0)$ local maximum,
$\left(-\sqrt{\frac{3}{7}}, \frac{48}{7^3}\sqrt{\frac{3}{7}}\right)$ local minimum

4 a) $y = -\frac{2}{3}x - \ln 3$, decreasing,
$y = \left(\frac{5}{12}+3\ln 6\right)x-\left(\frac{5}{8}+3.25\ln 6\right)$ increasing
b) $(-1, 0)$ is local maximum,
$(0.22, -1.18)$ is local maximum

5 a) $t = 2.857$ h, $r = 5.0850$ m
b) $t = 0$, $r = 1.1$ m and $t = \infty$, $r = 0.6$ m

Exercises 20.3A (*p.457*)

1 a) $\frac{(x+4)(x-4)}{x^2}$ **b)** $\frac{t(t+6)}{(t+3)^2}$
c) $\frac{(x-4)}{x^3}$ **d)** $\frac{t^{\frac{1}{4}}(4t-15)}{4(4t-3)^2}$
e) $-\frac{(10x^2+18x+5)}{(2x^2-1)^2}$ **f)** $-\frac{4t}{(t^2-1)^2}$
g) $\frac{-e^{-2x}(2x+3)}{(1+x)^2}$ **h)** $\frac{-2t\ln(t+1)-(1-t)}{(\ln(t+1))^2}$
i) $\frac{(1+2x-2x^2)e^{x^2}}{(x-1)^2}$ **j)** $\frac{2(1-\ln(2t-3))}{(2t-3)^2}$
k) $2x$ **l)** $\frac{(t^3-2)-3t^3\ln(t)}{t(t^3-2)^2}$

2 a) $(\sqrt{3} - 2, \sqrt{3} - 2)$ is a local minimum
b) $(-\sqrt{3} - 2, -\sqrt{3} - 2)$ is a local maximum

3 $(1, \frac{1}{2}e^1)$ is a point of inflexion

4 $v = 2\sqrt{\frac{10}{7}}$

Exercises 20.3B (*p.458*)

1 a) $-\frac{(x+2)(x+6)}{x^2}$ **b)** $\frac{5(2u+1)}{(u-1)^4}$
c) $\frac{2x(x-4)}{(x-2)^2}$ **d)** $\frac{s^{\frac{4}{3}}(12s-7)}{3(3s-1)^2}$
e) $\frac{3x^2-2x-3}{(x^2+1)^2}$ **f)** $\frac{(s+5)}{(1-s)^3}$
g) $\frac{2e^{4x}(7-4x)}{(3-2x)^2}$
h) $\frac{x(2x+1)\ln x-(x^2+x+1)}{x(\ln x)^2}$
i) $-e^{-2x^2}\frac{(12x^2+8x+3)}{(3x+2)^2}$
j) $\frac{1-\ln(x-2)}{(x-1)^2}$ **k)** $\frac{e^{-x}(e^{-2x}-1)}{(e^{-2x}+1)^2}$
l) $\frac{(s-4)}{2\sqrt{1-s}(s+2)^2}$

2 $(0, 0)$ is a (local) minimum

3 a) $(0.28, 8.82)$ is a local minimum, $(-1.78, 0.06)$ is a local maximum
b) no stationary points

4 a) $v=\frac{1-t^2}{(t^2+1)^2}$, $v = 0$ when $t = 1$
b) 4.5 **c)** 1

5 $x = e$

Exercises 20.4A (*p.459*)

1 $\frac{dy}{dx}=\frac{e^y}{(1-y)}$

2 $\frac{dy}{dx}=(1-y)$ or $\frac{dy}{dx}=e^x$

3 $\frac{dy}{dx}=\frac{2^x}{\ln 2}$

4 $\frac{dy}{dx}=\frac{1}{e^y+e^{-y}}$

5 $\frac{(3x+4)}{2\sqrt{x+2}}$

6 $24(2x^3 - 3)^3$

7 $\frac{x+2}{2(x+1)^{\frac{3}{2}}}$

8 $(8x + 1)e^{4x^2+x-1}$

9 $e^{x^2}(2x^3 - 6x^2 + 6x - 3)$

10 $\frac{1}{10}(8 - t)e^{-0.1t}$

11 $\frac{-8x^3}{(x^4+1)^3}$

12 $(t - 2)(3t - 4)$

13 $\frac{(2t-1)}{2\sqrt{t^2-t+1}}$

14 $\frac{e^t(t^3+t+1)}{(t^2+1)^2}$

Exercises 20.4B (*p.460*)

1 $\frac{dy}{dx}=\frac{2y+1}{2}$ or $\frac{dy}{dx}=\frac{1}{2}e^x$

2 $\frac{dy}{dx}=\frac{4(3^x)}{\ln 3}$ **3** $\frac{dy}{dx}=\frac{1}{2y-2}$

4 $\frac{dy}{dx}=\frac{1}{2(e^{2y}-e^{-2y})}$ **5** $-\frac{(x^2+4)}{(x^2-4)^2}$

6 $1 + \ln u$ **7** $4t + \frac{3}{2}\sqrt{t} - 1$

8 $(2t - 3)e^{t^2-3t+4}$

9 $2e^{2x-5}(x^2 - 3x - 3)$

10 $\frac{2t}{t^2+3}$ **11** $\frac{(x-1)e^x+(x+1)e^{-x}}{(e^x-e^{-x})^2}$

12 $\frac{10^x}{\ln 10}$ **13** $-\frac{7}{(x-2)^2}$ **14** $\frac{(2-s)^2(1-2s)}{\sqrt{s(2-s)^3}}$

Consolidation Exercises for Chapter 20 (*p.462*)

1 $\left(\frac{1}{2},\frac{3}{2}\right),\left(-\frac{1}{2},-\frac{3}{2}\right)$

2 a) $x(x^3 + 4)^2(22x^3 - 9x + 16)$
b) $\frac{(3x^2-1)}{2x^{\frac{3}{2}}}$ **c)** $6x(x^2 + 2)^2$
d) $e^x\ln x + \frac{1}{x}e^x$ **e)** $-\frac{1}{(2x-3)^2}$
f) $\frac{(1-x^2)}{(1+x^2)^2}$

3 $\left(\frac{1}{3}, \sqrt{10}\right)$ is a (local) maximum

4 a) $(c, 0)$ is a local minimum
$\left(\frac{1}{3}(c+2), \frac{4}{27}(c-1)^3\right)$ is a local maximum, when $c = 1$ then $(1, 0)$ is a point of inflexion

5 0.5π mm^2 s^{-1}

6 a) $r = \frac{\sqrt{6}}{3}$

7 a) $-\frac{f^2}{(f-u)^2}$
b) $\frac{3}{32}$ cm s^{-1} increasing

8 0.0134 m s^{-1} (to 4 d.p.)

9 a) $\frac{1}{3}$s **b)** 0.28 cm s^{-1}
c) $t \to \infty$, $r \to 4$ cm

10 b) £127 million, 109 km

CHAPTER 21
Numerical solution of equations

Exercises 21.1A *(p.468)*

1 **b)** [1, 2] **c)** 1.33
2 **b)** [–5, –4], [–2, –1], [1, 2]
c) –4.57, –1.64, 1.20
3 **b)** [–5, –4], [1, 2] **c)** –4.99, 1.94
4 **b)** [–1, 0], [1, 2], [4, 5]
c) –0.752, 1.16, 4.59
5 **b)** [0, 1], [1, 2] **c)** 0.793, 1.64
6 **b)** –1.73, –1.62
7 **a)** 0.739 **b)** (0.739, 0.739)
8 **b)** $2 - 2x - 4x^3$ **c)** (0.590, 8.71)
9 **b)** $\pm\sqrt{2}$
10 **b)** **i)** 2.7 (6 steps)
ii) 2.68 (9 steps)
iii) 2.677 (12 steps)
c) $\frac{1}{2}^{n-1}$

Exercises 21.1B *(p.469)*

1 **b)** [1, 2] **c)** 1.28
2 **b)** [–4, –3], [–2, –1], [1, 2]
c) –3.39, –1.77, 1.16
3 **b)** [–1, 1], [4, 5] **c)** 0, 4.58
4 **b)** [–1, 0], [1, 2]
c) –0.775, 1.93
5 **b)** [–1, 0] **c)** –0.927
6 **b)** 1.38. 1.62
7 **a)** 0.329 **b)** (0.329, 0.658)
8 **b)** $2x + e^x$ **c)** (–0.352, 0.827)
9 **b)** $\sqrt[3]{(-3)} \approx -1.44$
10 **b)** 0.318
c) infinite number of roots, all in the interval [–0.5, 0.5]

Exercise 21.2A *(p.474)*

1 **b)** 0.761
2 **b)** 0.297
3 **b)** 0.610
4 –1.88, 0.347, 1.53
5 **b)** converges to 0.703

Exercises 21.2B *(p.475)*

1 **b)** $-3 - 6x^2 < 0$ for all x **c)** 1.28
2 **c)** 0.259 **d)** 2.54
3 **b)** 1.39 **c)** –1.07, 4.68
4 **a)** (0.728, 5.63) **b)** minimum
5 **b)** ±4.47

Exercises 21.3A *(p.479)*

1 **b)** –1.59
2 **b)** –1.39, 0.145, 1.24
3 **b)** 0.178, 1.70
4 **b)** –2.21, 1.24
5 **b)** 3.35
6 **c)** –3.98, 1.75
7 $x_{n+1} = \dfrac{x_n(3a - x_n^3)}{2a}$, 3.16
8 –1, 0.177, 2.823
9 (–2.41, –16.1), (–0.265, 0.412), (1.17, –5.28)
10 **b)** **i)** 0.806 in 25 steps
ii) 0.806 in 4 steps

Exercises 21.3B *(p.480)*

1 **b)** 2.29
2 **b)** –3.88, –1.22, 2.11
3 **b)** –1.67, 0.195
4 **b)** –1.45, –0.747, 0.681
5 **b)** –1.19, 4
6 **b)** –1.31 **c)** **i)** division by zero
ii) initial sequence erratic
7 $x_{n+1} = \dfrac{x_n(4a - x_n^3)}{3a}$, 2.15
8 –2.39, 0.091, 2.30 (to 3 s.f.)
9 (–0.459, 0.729), (0.910, 1.73), (3.73, –10.2)
10 **b)** **i)** –0.726 in 22 steps
ii) –0.726 in 4 steps

Consolidation Exercises for Chapter 21 *(p.483)*

1 **a)** $a = 7$, $b = 3$
b) $x_1 = 1.183$, $x_2 = 1.294$, $x_3 = 1.277$, $x_4 = 1.279$, $x_5 = 1.28$ (2 d.p.)
2 1.935
3 **b)** $x_4 = 3.447$ **d)** converges
e) $x_6 = -0.562$ (3 s.f.)
4 **a)** 1.733 **b)** 1.620, 1.666
5 1.969
6 $x_0 = 0.5$, $x_1 = 0.630$, $x_2 = 0.615$, $x_3 = 0.619$, $x_4 = 0.618$, $x_5 = 0.618$, 0.62 (2 d.p.)
7 2.002, 2 iterations
8 2.218
9 13.80
10 No, 55 weeks
11 **a)** –4, 1, 0.268, 3.732 **c)** 0.16
12 **b)** 1.442

CHAPTER 22
Trigonometry 4

Exercises 22.1A *(p.491)*

1 **a)** $3\cos x$ **b)** $-4\sin x$
c) $\frac{1}{2}\sec^2 x$ **d)** $\frac{1}{4}\cos x$
e) $3\cos x - 4\sin x$ **f)** $-\sec^2 x$
g) $2x + 8\cos x$ **h)** $3e^{3x} + \sec^2 x$
2 **a)** $21\cos 7x$ **b)** $44\sin 11x$
c) $10\sec^2 4x$ **d)** $6\cos(2x - 5)$
e) $2\sec^2 \frac{1}{2}x - 2\cos\frac{1}{3}x$
f) $-3\sin\frac{1}{5}x - 3\cos\frac{1}{4}x$
g) $\frac{1}{2x} - \frac{9}{2}\sin(5x + 7)$
h) $4e^{4x} + 4^x \ln 4 + 3\cos(1 - 3x)$
3 **a)** $2\sin x \cos x$ **b)** $2x\cos x^2$
c) $-\frac{1}{x^2}\sin\frac{1}{x}$ **d)** $-\dfrac{\sin x}{2\sqrt{1+\cos x}}$
e) $6\tan^2 2x\sec^2 2x$
f) $\dfrac{2\cos 2x}{1+\sin 2x}$ **g)** $e^{\sin x}\cos x$
h) $e^x\cos e^x$ **i)** $-\cot x$
j) $-\dfrac{\sin \ln x}{x}$
4 **a)** $x\sec^2 x + \tan x$
b) $\cos x \cos 2x - 2\sin x \sin 2x$
c) $3x^2 \cos 2x - 2x^3 \sin 2x$
d) $e^x(\cos x - \sin x)$
e) $e^{-\frac{1}{2}x}(2\cos 2x - \frac{1}{2}\sin 2x)$
f) $\sqrt{x}\sec^2 x + \dfrac{1}{2\sqrt{x}}\cos x$
g) $\dfrac{\sin x}{x} + \cos x \ln x$
h) $e^{\sin x}(x\cos x + 1)$
5 **a)** $\dfrac{x\cos x - \sin x}{x^2}$
b) $\dfrac{\sin^3 x(4\cos^2 x - 3\sin^2 x)}{\cos^4 x}$
c) $\dfrac{e^x(\sin x - \cos x)}{\sin^2 x}$
d) $\dfrac{2x\cos x - \sin x}{2\sqrt{x}^3}$
e) $\dfrac{x(2 - x)\cos x + x(2 + x)\sin x}{(\sin x + \cos x)^2}$
f) $\dfrac{(2x + 1)\tan x - (x^2 + x + 1)\sec^2 x}{\tan^2 x}$
6 $\frac{d}{dx}(\operatorname{cosec} x) = -\operatorname{cosec} x \cot x$
and $\frac{d}{dx}(\cot x) = -\operatorname{cosec}^2 x$
$3\sec(3x - 1)\tan(3x - 1)$,
$-\frac{1}{2}\operatorname{cosec}\frac{1}{2}x\cot\frac{1}{2}x$,
$-\frac{3}{5}\operatorname{cosec}^2(\frac{3}{4}x - \frac{\pi}{4})$
7 Equation of normal is
$y = -\pi x + (\pi^2 + \ln 4x - 3)$
8 Equation of tangent is
$2y = -20x + 5(\pi - \sqrt{3})$
9 tangent: $2y = 4x - (\pi - 3)$,
normal: $8y = -2x + (12 + \pi)$
10 **a)** maximum $\frac{1}{2}$, minimum $-\frac{1}{2}$
b) maximum 1, minimum $-\sqrt{2}$
c) maximum $\frac{3\sqrt{3}}{16}$,

minimum $-\frac{3\sqrt{3}}{16}$

d) maximum 3, minimum −1

11 a) (0, 1) and $\left(\frac{2\pi}{3},1\right)$ max.

$\left(\frac{\pi}{3},-1\right)$ and $(\pi, -1)$ min

b) $\left(\frac{\pi}{6},\frac{\pi}{6}-\frac{\sqrt{3}}{2}\right)$ min

$\left(\frac{5\pi}{6},\frac{5\pi}{6}+\frac{\sqrt{3}}{2}\right)$ max

c) $\left(\frac{3\pi}{4},\frac{\sqrt{2}}{2}e^{\frac{3\pi}{4}}\right)$ max

d) No stationary points in this domain.

12 c) $\frac{1}{1+x^2}$ **e)** $4y = 2x + (\pi - 2)$

Exercises 22.1B *(p.493)*

1 a) $7\cos x$ **b)** $\frac{3}{4}\sin x$

c) $-11\sec^2 x$ **d)** $-\frac{\sin x}{5}$

e) $\frac{\cos x+\sin x}{10}$ **f)** $\frac{x^4}{3}-\frac{1}{3x}\frac{\sin x}{5}$

g) $3\sec^2 x+\frac{1}{x}$ **h)** $\frac{1}{4}e^{\frac{x}{5}}+\frac{1}{10}\sin x$

2 a) $\frac{25}{4}\cos 5x$ **b)** $-\frac{1}{3}\sec^2\frac{4x}{3}$

c) $-\frac{3}{2}\sin 2x$

d) $\frac{66}{7}\sin 11x+\frac{11}{4}\sec^2 11x$

e) $10\cos(4x-3)-2\sin(3x-4)$

f) $-5\cos(\frac{2\pi}{3}-5x)-6\cos(\frac{3\pi}{4}+6x)$

g) $\frac{11}{x}-\frac{5}{4}\sec^2\frac{x}{4}+\frac{1}{1\sqrt{x}}$

h) $\frac{\sqrt{x}^{-5}}{6}+6\sin(\frac{\pi}{9}-6x)+\frac{5}{x^2}$

3 a) $-3\cos^2 x\sin x$

b) $4\sin^3 5x\cos 5x$

c) $-\frac{\sin\sqrt{x}}{2\sqrt{x}}$ **d)** $\frac{1-\sin x}{2\sqrt{x+\cos x}}$

e) $-3\cos^5\frac{1}{2}x\sin\frac{1}{2}x$

f) $\frac{1+\cos x}{x+\sin x}$ **g)** $-e^{\cos x}\sin x$

h) $-e^x\sin e^x$ **i)** $\frac{3(\cos 3x-\sin 3x)}{\cos 3x+\sin 3x}$

j) $-\frac{2}{x^3}\sec^2\frac{1}{x^2}$

4 a) $2x(x\cos x+\sin x)\sin x$

b) $2\cos x(2\cos x\cos 4x-\sin x\sin 4x)$

c) $4\tan 3x\cos 4x+3\sec^2 x\, 3x\sin 4x$

d) $4x\cos 4x+\sin 4x$

e) $4\cos(4x+\frac{\pi}{3})\cos(2x-\frac{\pi}{4})-2\sin(4x+\frac{\pi}{3})\sin(2x-\frac{\pi}{4})$

f) $e^{-x}(\cos x-\sin x)$

g) $3(1+x)^2\cos 3x-3(1+x)^3\sin 3x$

h) $\frac{\sec x}{x}+\ln x\tan x\sec x$

5 a) $-\frac{x\sin x+\cos x}{x^2}$

b) $\frac{\cos x-\sin x-1}{(1-\cos x)^2}$

c) $\frac{(\sec^2 x-\tan x)}{e^x}$

d) $\frac{1}{x\tan x}-\frac{\ln x\sec^2 x}{\tan^2 x}$

e) $\frac{2\cos x}{(1-\sin x)^2}$

f) $\frac{2\sec x\tan x}{(1-\sin x)^2}$

6 $x° = \frac{\pi x}{180}$ radians, $-\frac{\pi}{180}\cos x°$, $-\frac{\pi}{180}\sin x°$, $\frac{\pi}{180}\sec^2 x°$, $\frac{\pi}{36}\cos 5x°$, $-\frac{\pi}{360}\sin\frac{1}{2}x°$, $\frac{\pi}{60}\sec^2(3x-45)°$

7 Equation of normal is $8y=\pi^3-2\pi^2x$

8 tangent : $y = 5x - 5(\pi - 1)$, normal : $5y = -x + (\pi + 25)$

9 tangent : $y = -16x - 4$, normal : $16y = -x - 64$

10 a) max: $\sqrt{5}$, min: −1

b) max: 3, min: −5

c) max: $-2\sqrt{2}$, min: $2\sqrt{2}$

d) max: 1, min: −1

11 a) $\left(\frac{\pi}{10},1\right)$, $\left(\frac{\pi}{2},1\right)$ and $\left(\frac{9\pi}{10},1\right)$ max, $\left(\frac{3\pi}{10},-1\right)$ and $\left(\frac{7\pi}{10},-1\right)$ min

b) $\left(\frac{3\pi}{4},\frac{3\pi}{2}+1\right)$ min, $\left(\frac{\pi}{4},\frac{\pi}{2}-1\right)$ max

c) $\left(\frac{3\pi}{4},-1\right)$ min

d) $\left(\frac{\pi}{2},8.5\right)$ max, $\left(\frac{\pi}{6},8\right)$ and $\left(\frac{5\pi}{6},8\right)$ min

12 b) i) $V = 2.26$, $a = 0.186$

ii) $V = -1.17$, $a = 0.234$

d) $t = 5.64$, $h = 9.41$

Exercises 22.2A *(p.496)*

1 All answers are $+ c$.

a) $-\frac{\cos 3x}{3}$ **b)** $\frac{\sin 6x}{6}$ **c)** $2\sin\frac{x}{2}$

d) $\frac{\sin x}{2}$ **e)** $\frac{2}{3}\sin\frac{3}{2}x$ **f)** $\frac{\tan 4x}{4}$

g) $4\tan x$ **h)** $-3\cos x-2\sin x$

i) $\frac{1}{2}(\sin x-\cos x)$

j) $-\frac{1}{2}\cos(2x+4)$

k) $-3\tan(\frac{\pi}{3}-x)$ **l)** $\frac{1}{2}\sin(2x+\frac{\pi}{4})$

2 a) 2 **b)** $\frac{1}{2}$ **c)** 2 **d)** $\frac{1}{2}$ **e)** π

f) −3 **g)** 2 **h)** $\frac{\pi}{9}$ **i)** 4

3 $\frac{2\sqrt{2}}{3}+\frac{1}{2} = 1.4428$ to 4 d.p.

4 a) $2-\sqrt{2} = 0.58578$

b) $\sqrt{2} = 0.41421$

c) $\sqrt{2} = 1.41421$

5 b) 0.739 085 133

c) 0.599 513 417

d) 0.400 486 583

6 a) $\frac{d}{dx}(\ln\sin x) = \cot x$, $\int\cot x\,dx = \ln\sin x + c$

b) $\frac{1}{5}\ln\sin x$, $12\ln\sin\frac{1}{4}x$, $\frac{1}{4}\ln\sin(4x+\pi)$

Exercises 22.2B *(p.497)*

1 All answers are $+ c$

a) $\frac{3}{4}\sin 4x$ **b)** $-15\cos\frac{1}{3}x$

c) $-\cos\frac{1}{5}x$ **d)** $-\frac{1}{5}\cos x$

e) $-5\cos\frac{1}{5}x$ **f)** $\frac{2}{3}\tan\frac{3}{2}x$

g) $49\tan\frac{1}{7}x$

h) $-\frac{4}{3}\cos 3x-\frac{5}{7}\sin 7x$

i) $\sin\frac{1}{2}x-\cos\frac{1}{2}x$

j) $-2\cos\left(\frac{x}{2}-1\right)$

k) $2\tan(x-\frac{\pi}{2})$

l) $-\frac{1}{3}\sin(\frac{\pi}{4}-3x)$

2 a) 1 **b)** −1 **c)** 0 **d)** $\frac{3}{4}$ **e)** 1

f) 6 **g)** 1.2071 **h)** π **i)** $\frac{1}{2}$

3 a) $\frac{\pi}{4}(9\pi+8) = 9.0686$ to 4 d.p.

4 a) $x = \frac{2\pi}{3}$ **b)** 0.526 67

c) 0.171 46

5 b) 0.588 532 744 and 3.096 363 932

c) 1.320 805 545

d) 0.276 614 887

6 a) $\frac{d}{dx}(\ln\cos x) = -\tan x$, $\int\tan x\,dx = -\ln\cos x + c$

b) $-\frac{1}{3}\ln\cos 3x$, $2\ln\cos\frac{1}{2}x$, $-\frac{1}{2}\ln\cos(\frac{\pi}{4}-2x)$

Exercises 22.3A *(p.500)*

1 a) $\frac{1}{2}x-\frac{1}{12}\sin 6x + c$

b) $-\frac{1}{2}x-\frac{1}{12}\sin 6x + c$

2 a) 3π **b)** 0.129 72

3 a) volume $= \frac{\pi}{4}(3\pi+8) = 4.3562\pi$

b) volume $= 3\pi$

c) volume $= \pi-\frac{\pi^2}{4}$

4 a) $3\sin^2 x\cos x$

b) $\sin x - \frac{1}{3}\sin^3 x + c$ **c)** $\frac{2\pi}{3}$

5 c) $\frac{3x}{8} - \frac{1}{4}\sin 2x + \frac{1}{32}\sin 4x + c$

d) $\frac{27\pi^2}{16}$

6 b) Both are equal to $2\sec^2 x \tan x$ since $1 + \tan^2 x = \sec^2 x$.

7 b) $\ln(\sec x + \tan x) + c$

c) 2.6399 and 1.7627

8 0.9014

Exercises 22.3B *(p.501)*

1 a) $\frac{1}{2}x + \frac{1}{2}\sin 10x + c$

b) $\frac{1}{4}x - \frac{1}{2}\sin\frac{1}{2}x + c$

2 a) $\frac{\pi}{12}$ **b)** π

3 a) $c - \frac{1}{4}\cos 2x$

b) $\frac{\pi}{4}(3\pi + 2) = 2.8562\pi$

4 b) $c - \cos x + \frac{1}{3}\cos^3 x$ **c)** 6

5 a) $2e^x \sin x$ and $2e^x \cos x$

b) $\frac{1}{2}(e^x \sin x - \cos x) + c$

and $\frac{1}{2}(e^x \sin x + \cos x) + c$

c) $\frac{1}{2}(e^{\pi} + 1) = 12.07$

d) $\frac{\pi}{2}(e^{\frac{\pi}{2}} + {}^{-\frac{\pi}{2}}) = 2.5092\pi$

6 b) Both are equal to $-2\operatorname{cosec}^2 x \cot x$ since $1 + \cot^2 x = \operatorname{cosec}^2 x$.

7 b) $c - \ln(\operatorname{cosec} x + \cot x)$

c) 1.7627 and 2.303

8 –0.6469

Exercises 22.4A *(p.505)*

1 a) 2.5 **b)** 2 **c)** $\frac{1}{x^2} - \frac{1}{2}$

d) $\frac{2}{3}$ **e)** 2 **f)** $-\frac{x}{2}$

g) $\sqrt{4 - x^2}$ **h)** $\frac{9}{16}$ **i)** $\frac{\sqrt{2}}{2}x - \frac{1}{2}x$

2 c) estimate is 0.571 44 to 5 d.p., both are 0.57 to 2 d.p.

d) cubic approximation is $2x - 3x^3$

e) estimated value = 0.617 to 3 d.p., true value = 0.625

3 a) quadratic approximation is $f(x) = 3x - 3x^2$ giving integral = 0.0140 exactly

b) cubic approximation is $f(x) = 3x - 3x^2 + 3x^3$ giving integral = 0.014 075 to 6 d.p. (The true value of the integral is 0.014 090 to 6 d.p.)

Exercises 22.4B *(p.505)*

1 d) $f'(x) = -2\cos x \sin 2x - \sin x \cos 2x = -0.94629$ when $x = 0.2$

e) cubic approximation to $f'(x) = 4x^3 - 5x = -0.968$ when $x = 0.2$

f) Integration involves higher powers of small numbers and is hence more accurate.

2 c) true value = 0.049 979 165 to 9 d.p., estimated value = 0.049 979 166 to 9 d.p.

e) estimated value = 0.025 041 666 to 9 d.p. which compares well with the true answer of 0.025 041 646 to 9 d.p.

3 a) cubic approximation is $x + 2x^3$

c) approximate value = 0.011 503 to 6 d.p.

e) The approximation for $\sin 4x$ is $4x$ and although the largest value of x used is 0.15, this gives $4x = 0.6$ radians, which is not particularly small.

Consolidation Exercises for Chapter 22 *(p.507)*

1 $y = 2e^{3x} - \frac{3}{2}\cos x + c$

2 a) 0.397 **b)** 0.740

3 b) 2.79

d) 0.146, 3 roots

4 0.035%

5 $e^{2x}(2\cos x - \sin x)$

b) $y = 2x + 1$

6 a) $e^{-x}(\cos x - \sin x)$

b) (–2.4, –7.5), (0.79, 0.32)

7 a) $2\cos 2x$

b) $1 - \frac{1}{2}\sin^2 2x$

8 a) 13 cm, $\frac{\pi}{2}$

b) 0.478

c) $v = -12\sin 4t$

$v(2) = -11.87$ m s^{-1} moving up

9 a) $2\cos x(\sin x + \cos x) = 0$

c) $1 + \frac{3\pi}{4}$

10 5.5102, 3.046 38π

CHAPTER 23
Rational functions

Exercises 23.1A *(p.519)*

1 a) $x = 0$ **b)** $x = 3$

c) $x = -2$ and $x = 1$

d) $x = 3$ **e)** none

f) $x = \frac{1}{2}$, $x = 2$ and $x = 3$

2 a) $y = 1$ **b)** $y = 0$

c) $y = 1$ **d)** $y = -1$

e) $y = \frac{1}{4}$ **f)** $y = 0$

3 a) yes **b)** no **c)** yes **d)** no

6 vertical asymptotes: $x = 2$ and $x = -2$, horizontal asymptotes: $y = 4$ and $y = -4$ domain: $-\infty < x -2, 2 < x < \infty$

Exercises 23.1B *(p.520)*

1 a) $x = 0$ **b)** $x = -4$ **c)** $x = 5$

d) $x = -\frac{1}{2}$ and $x = \frac{1}{2}$ **e)** none

f) $x = 1$, $x = 2$ and $x = 3$

2 a) $y = 1$ **b)** $y = 0$ **c)** $y = \frac{1}{3}$

d) $y = \frac{5}{7}$

3 a) yes **b)** no **c)** yes **d)** no

Exercises 23.2A *(p.524)*

1 a) $\frac{1}{x-1} - \frac{1}{x+1}$ **b)** $\frac{1}{x-2} - \frac{1}{x-1}$

c) $\frac{4}{x-1} - \frac{4}{x}$ **d)** $\frac{1}{2(x+3)} + \frac{1}{2(x-1)}$

e) $\frac{3}{(x-2)} - \frac{1}{(x-1)}$ **f)** $\frac{1}{3(x-1)} - \frac{1}{3(x+2)}$

g) $x - 1 + \frac{1}{3(x-1)} + \frac{8}{3(x+2)}$

h) $-\frac{3}{x-5} - \frac{2}{(x+5)}$ **i)** $\frac{1}{x} - \frac{x}{x^2+1}$

j) $\frac{2}{19(4-x)} - \frac{2x+8}{19(x^2+3)}$

k) $\frac{12}{(3x+2)} - \frac{1}{(x+1)^2} - \frac{4}{(x+1)}$

l) $\frac{2}{(x-2)^2} - \frac{1}{(x-2)} + \frac{1}{(x-1)}$

2 $\frac{5}{(x-3)} - \frac{2}{(3x+2)}$, (–4.90625) or $-\frac{157}{32}$

3 $y = \frac{1}{(x-3)} - \frac{1}{(x-1)}$

$\frac{dy}{dx} = -\frac{1}{(x-3)^2} + \frac{1}{(x-1)^2}$

$\frac{d^2y}{dx^2} = \frac{2}{(x-3)^3} - \frac{2}{(x-1)^3}$

(2, –2) is a local maximum

Exercises 23.2B *(p.525)*

1 a) $\frac{1}{3(x-3)} - \frac{1}{3x}$ **b)** $\frac{1}{(x+1)} - \frac{1}{(x+2)}$

c) $\frac{17}{10(x-10)} + \frac{3}{10x}$

d) $\frac{5}{3(2x-1)} - \frac{4}{3(x+1)}$

e) $\frac{1}{6(2x-1)} - \frac{1}{6(2x+3)}$

f) $\frac{1}{x+3}$ **g)** $\frac{2}{(x-1)} - \frac{1}{(x-1)^2}$

h) $\frac{9}{(3x-1)^2} - \frac{9}{(3x-1)} - \frac{3}{10x}$

i) $\frac{3x}{10(x^2+10)} - \frac{2}{(x^2+10)} - \frac{3}{10x}$

j) $\frac{1}{2(x-1)} - \frac{1}{2(x+1)} + 1$

k) $\frac{1}{(x+2)} + \frac{1}{(x-1)} + x - 1$

l) $-\frac{x}{(x^2+2)}-\frac{31}{(x^2+2)}+\frac{x}{(x^2+1)}+\frac{3}{(x^2+1)}$

2 $y=-\frac{4}{(x+6)^2}-\frac{1}{(x-2)}$,

$\frac{dy}{dx}=-\frac{4}{(x+6)^2}+\frac{1}{(x-2)^2}$,

$\frac{d^2y}{dx^2}=-\frac{84}{(x+6)^3}+\frac{2}{(x-2)^3}$

3 $y=1-\frac{5}{4(x-2)}+\frac{5}{4(x+2)}$,

$\frac{dy}{dx}=1-\frac{5}{4(x-2)^2}-\frac{5}{4(x+2)^2}$,

$\frac{d^2y}{dx^2}=-\frac{5}{2(x-6)^3}+\frac{5}{2(x+2)^3}$

Exercises 23.3A *(p.527)*

1 a) $2+3x+3x^2+3x^2$ $\quad |x|<1$
b) $\frac{1}{2}+\frac{x}{4}-\frac{x^2}{8}+\frac{x^3}{16}$ $\quad |x|<1$
c) $-\frac{1}{3}+\frac{x}{9}-\frac{8x^2}{27}-\frac{8x^3}{81}$ $\quad |x|<2$
d) $1-\frac{3x}{2}+\frac{11x^2}{8}-\frac{23x^2}{16}$ $\quad |x|<3$
e) $\frac{1}{2}-\frac{3x}{2}+\frac{13x^2}{8}-\frac{51x^3}{16}$ $\quad |x|<\frac{1}{2}$
f) $\frac{1}{2}+\frac{3x}{4}+\frac{7x^2}{8}+\frac{15x^3}{16}$ $\quad |x|<1$
g) $5+5x+35x^2+65x^3$ $\quad |x|<\frac{1}{3}$
h) $\frac{1}{3}+\frac{5x}{9}+\frac{13x^2}{27}+\frac{41x^3}{81}$ $\quad |x|<1$

2 $\frac{1}{2}+\frac{3x}{3}$ $\quad |x|<1$

3 $4+\frac{20x}{3}+\frac{88x^2}{9}$

Exercises 23.3B *(p.527)*

1 a) $-2+4x-4x^2+4x^3$ $\quad |x|<1$
b) $x+2x^2+4x^3$ $\quad |x|<$ _
c) $1+x+3x^2+3x^2$ $\quad |x|<1$
d) $\frac{1}{2}-\frac{x}{4}+\frac{3x^2}{8}-\frac{5x^3}{16}$ $\quad |x|<1$
e) $2+\frac{23x}{4}+\frac{1103x^2}{64}+\frac{2647x^3}{512}$ $\quad |x|<\frac{1}{3}$
f) $-\frac{1}{2}+\frac{x}{4}+\frac{5x^2}{8}+\frac{17x^3}{8}$ $\quad |x|<\frac{2}{3}$
g) $\frac{x}{2}-\frac{5x^2}{4}+\frac{17x^3}{8}$ $\quad |x|<1$
h) $-\frac{1}{4}-\frac{3x}{16}-\frac{29x^2}{64}-\frac{99x^2}{256}$ $\quad |x|<\frac{1}{3}$

2 $\frac{1}{2}+\frac{x}{4}-\frac{9x^2}{8}+\frac{9x^3}{16}$

3 $p_0=2$
$p_1=2+3x$
$p_2=2+3x+3x^2$
$p_3=2+3x+3x^2+3x^2$
$p_4=2+3x+3x^2+3x^3+3x^4$

Consolidation Exercises for Chapter 23 *(p.528)*

1 a) $x=-1$ and $x=5$ and $y=0$
b) $\frac{1}{6(x-5)}-\frac{1}{6(x+1)}$
c) $\left(2,-\frac{1}{9}\right)$ is a local maximum

2 a) $x=-2$ and $x=1$
b) $2x-1+\frac{1}{(x-1)}-\frac{3}{(x+2)}$
c) slant asymptote $y=2x-1$

3 $-\frac{8}{9(x-2)}-\frac{4}{3(x+1)^2}+\frac{8}{9(x+1)}$,
$y=-x+2$

4 $-\frac{4}{3(2x-1)}-\frac{1}{3(x+1)}$,
$1+3x+5x^2+11x^3$ for $|x|<\frac{1}{2}$

5 $f(x)\cong\frac{2}{(x-2)}-\frac{3}{(x-1)}$
$f(x)\cong 2+2.5x+2.75x^2+2.875x^3$

6 a) $A=6$, $B=2$, $C=5$
b) $9-12x+18x^2-42x^3$
c) $f'(0)=-12$

7 a) $\frac{14}{9(2x+1)}+\frac{5}{3(x-1)^2}-\frac{7}{9(x-1)}$
b) $4+x+12x^2-5x^3$

8 a) $A=3$, $B=1$, $C=-2$
b) $\frac{dy}{dx}=3$ when $x=0$

9 $f'(x)=-\frac{1}{(x+3)^2}<0$ for all x
Range of $f'(x)$: $y\in\Re$, $y\neq 2$,
$f'(x)=2+\frac{1}{x+3}$, translation -2 units // y-axis *followed by* translation 3 units // x-axis

10 $\frac{1}{(x-1)}+\frac{1}{(x-1)^2}-\frac{7}{(2x-1)}$
$1+3x+6x^2+11x^3$

11 $x=-1$ is vertical asymptote,
$y=x-1$ is a slant asymptote

CHAPTER 24
Integration 3

Exercises 24.1A *(p.535)*

1 a) $\frac{1}{5}(x+4)^5+c$
b) $\frac{1}{12}(2x-3)^6+c$
c) $\frac{14}{3}$ **d)** $\frac{2}{27}(3x^2+2)^{\frac{9}{4}}+c$
e) $\frac{1}{4}(e^7-e^3)$ **f)** e^4-1
g) $\frac{1}{2}(x^2+3x+7)^2+c$ **h)** 114

2 a) $\frac{1}{7}(x-5)^7+c$ **b)** 10
c) $\frac{2}{9}(3x-1)^{\frac{3}{2}}+c$
d) $\frac{1}{6}(x^2-1)^3+c$
e) $2(x^2-x+3)^{\frac{1}{2}}+c$
f) $\frac{1}{6}(6^6-4^6)=\frac{21280}{3}$
g) $\frac{1}{2}e^{2x+5}+c$ **h)** $e^{x^2-3x+1}+c$

3 $\frac{256}{5}$ **4** $\frac{1}{3}(e^2-e^{-1})$ **5** $\frac{64\pi}{3}$

6 $\frac{4\pi}{3}$ **7** $\frac{9\sqrt{30}-\sqrt{10}}{125}$ (= 0.369 062)

8 $\frac{80\sqrt{10}-13\sqrt{13}}{27}$ (= 7.633 70)

Exercises 24.1B *(p.536)*

1 a) $\frac{1}{12}(3x+4)^4+c$
b) $-\frac{2}{3}(x-3)^{-3}+c$
c) $\frac{1}{3}(2x-1)^{\frac{3}{2}}+c$
d) $\frac{125\sqrt{5}-1}{7}$ (= 39.7869)
e) $\frac{1}{5}$ **f)** $e^{x-1}+c$
g) $\frac{1}{2}(x^3-3)^5+c$ **h)** $e^{2x^2+x-2}+c$

2 a) $\frac{1}{5}(x+3)^5+c$ **b)** $\frac{247}{3}$
c) $\frac{1}{3}(2x+3)^{\frac{3}{2}}+c$
d) $\frac{1}{3}(x^2-2x+3)^{\frac{3}{2}}+c$
e) $\frac{-1}{8(4x+2)^2}+c$ **f)** $\frac{1}{3}e^{3x-1}+c$
g) $e^{x^2+x+4}+c$ **h)** $\frac{1-e^{-4}}{2}$

3 $\frac{8}{3}$ **4** $2(1-e^{-0.25})$

5 $\frac{8\pi}{3}$ **6** $\frac{2\pi}{5}$

7 $\bar{x}=\frac{185}{336}(\approx 0.55)$ **8** $\frac{1-e^{-0.25}}{2}$

Exercises 24.2A *(p.539)*

1 a) $\frac{3}{2}\ln|1+2x|+c$
b) $2\ln|2+3x|+c$
c) $\frac{3}{2}\ln\frac{11}{3}$
d) $\ln|1-2x+x^3|+c$
e) $\frac{3}{2}(\sqrt{6}-\sqrt{5})$
f) $\frac{1}{2}\ln|1-e^{-2x}|+c$

2 a) $-\frac{1}{3}\ln|2-3x|+c$
b) ln5 **c)** $\sqrt{x^2+2x+2}+c$
d) $\ln|\ln x|+c$ **e)** $\ln|e^x+e^{-x}|+c$
f) $\frac{1}{3}\ln|x^3+3x^2+1|+c$
g) $\frac{1}{2}\ln 13$ **h)** $\ln|xe^x-1|+c$

3 $10\ln 3-5\ln 5=5\ln\frac{9}{5}$

4 1

5 a) $s=-\frac{5}{4}\ln(10-0.4u^2)+\frac{5}{4}\ln(10)$
b) $\frac{5}{2}\ln\frac{5}{3}$
c) $u=5\sqrt{1-e^{-0.8t}}$
d) $u\to 5$ as $s\to\infty$

6 a) $t=-2\ln(X-40)+2\ln 40$
b) $2\ln 2\approx 1.386$ minutes
c) $X=40(1+e^{-\frac{1}{2}t})$
d) $X\to 40$°C as $t\to\infty$
The object cools to 40°C.

Exercises 24.2B *(p.540)*

1 a) $\frac{5}{4}\ln|1+4x|+c$
b) $-\frac{3}{2}\ln|0.5-2x|+c$
c) $\frac{1}{4}\ln\frac{5}{3}$

d) $-\ln 3$

e) $2\sqrt{t^4 - 3t} + c$

f) $\frac{5}{3}\ln|2 + e^{3x}| + c$

2 a) $\frac{1}{7}\ln|2 + 7x| + c$

b) $\ln 2 - \frac{1}{2}\ln 3 = \frac{1}{2}\ln\frac{4}{3}$

c) $\sqrt{v^2 + v - 1} + c$

d) 2

e) $\frac{1}{6}\ln|2x^3 - 3x^2 + 6x + 1| + c$

f) $1 - \frac{1}{\sqrt{22}}$

g) $\ln|e^{2x} - x| + c$

h) $\left(\frac{1}{1-e^5}\right) - \left(\frac{1}{1-e^1}\right) \approx 0.575193$

3 $\frac{\pi}{8}$ **4** $\frac{5}{2}\ln 3$

5 a) $s = -\frac{5}{2}\ln(10 + 0.2u^2) + \frac{5}{2}\ln 30$
$= \frac{5}{2}\ln\frac{30}{10 + 0.2u^2}$

b) maximum height $= \frac{5}{2}\ln 3$
= 2.7465 metres

6 a) $t = -\frac{10}{3}\ln(10 - 0.3u) + \frac{10}{3}\ln 10$

b) $u = \frac{100}{3}(1 - e^{-0.3t})$

c) $u \to \frac{100}{3}$ as $t \to \infty$
$u = \frac{100}{3}$ m s^{-1} is the terminal speed.

Exercises 24.3A *(p.544)*

1 a) $-\frac{1}{9}(1 + 3x)e^{-3x} + c$

b) $\frac{1}{4}(1 + e^2)$

c) $\frac{x^5}{25}(5\ln x - 1) + c$

d) $\frac{1}{27}(17e^{-3} - 50e^{-6})$

e) $(x + 2)e^x + c$

f) $\frac{1}{9}(\sin 3x - 3x\cos 3x) + c$

g) $\frac{\pi^2}{4} - 2$ **h)** $8\ln 2 - 3$

i) $\frac{1}{2}(2x + 3)\ln(2x + 3) - x + c$

j) $e^x(x^2 - x + 3) + c$

k) $\frac{2^{n+1}((n+1)\ln 2 - 1) + 1}{(n+1)^2}$

l) $\frac{1}{30}(x - 1)^5(5x + 1) + c$

m) $\frac{1663}{14}$

n) $\frac{1}{13}e^{2x}(2\sin 3x - 3\cos 3x) + c$

2 c) $2(\ln 2)^4 - 8(\ln 2)^3 + 24(\ln 2)^2 - 48\ln 2 + 24$

3 b) $\int x^2 e^{-x^2}\,dx = I_2 = -\frac{1}{2}xe^{-x^2} + \frac{1}{2}I_0$

$\int x^4 e^{-x^2}\,dx = I_4 = -\frac{1}{2}x^3 e^{-x^2} + \frac{3}{4}I_0$

$\int x^6 e^{-x^2}\,dx = I_6 = -\frac{1}{2}x^5 e^{-x^2} - \frac{5}{4}x^3 e^{-x^2} - \frac{15}{8}xe^{-x^2} + \frac{15}{8}I_0$

4 c) $I_0 = \sin x$

$\int x^4 \cos x\,dx = I_4 = x^4\sin x + 4x^3\cos x - 12x^2\sin x - 24x\cos x + 24\sin x$

$\int x^6\cos x\,dx = I_6$
$= x^6\sin x - 6x^5\cos x - 30I_4$
$= \sin x(x^6 - 30x^4 + 360x^2 - 720) + \cos(6x^5 - 120x^3 + 720x)$

Exercises 24.3B *(p.546)*

1 a) $\frac{1}{25}(5x - 1)e^{5x} + c$

b) $4 - 6e^{-0.5}$

c) $\frac{1}{4}(2x^2 - 2x + 1)e^{2x} + c$

d) $\frac{32}{3}\ln 2 - \frac{7}{4}$ **e)** $\frac{1}{660}$

f) $\frac{1}{2}(x + 1)^2\ln(x + 1) - \frac{1}{4}x(x + 2) + c$

g) $-(2x^2 + x + 2)e^{-x} + c$

h) $\frac{1}{16}(\sin 4x - 4x\cos 4x + c$

i) $\frac{\pi^3}{128} - \frac{3\pi}{16} + \frac{3}{8}$

j) $\frac{1}{2}(x^2 - 1)e^{x^2} + c$

k) $\frac{1}{27}(2e^3 - 101e^6)$

l) $\frac{448}{45}$

m) $\frac{2}{3}(x - 2)\sqrt{x + 1} + c$

n) $-\frac{1}{5}(2\cos 2x + \sin 2x)e^{-x}$

2 a) $I_n = \frac{x^{n+1}((n+1)\ln x - 1) + 1}{(n+1)^2}$

3 b) $\frac{5\pi}{32}$

4 $I_n = -x^n\cos x + nx^{n-1}\sin x - n(n - 1)I_{n-2}$
$I_1 = \sin x - x\cos x$
$I_5 = x^5\sin x\,dx$
$= (5x^4 - 60x^2 + 120)\sin x - (x^5 - 20x^3 + 120x)\cos x$

Exercises 24.4A *(p.549)*

1 a) $\frac{1}{2}\ln|x| - \frac{1}{2}\ln|x + 2| + c$

b) $2\ln 2 - \ln 3$

c) $\frac{1}{2}\ln|x + 2| - \frac{1}{2}\ln|x - 2| + c$

d) $\frac{1}{2}\ln|x - 1| + \frac{1}{2}\ln|x + 1| + c$

e) $5\ln 2 - 3\ln 3$

f) $\ln 3 - 2\ln 2 + \frac{1}{2}$

g) $\ln|x - 2| - \ln|x + 3| + c$

h) $\frac{3}{2}(\ln|x - 1| - \ln|x + 1|) + c$

i) $\frac{1}{5}(\ln|x + 3| + 4\ln|x - 2|) + c$

j) $\frac{1}{2}\ln 5 - \ln 2$

k) $x + 4\ln|x| - 3\ln|x - 1| + c$

l) $\frac{1}{2}x^2 - x + 2\ln|x| - \ln|x + 1| + c$

2 $2\ln 15 + 4$

3 $2\ln 2 - \ln 3$

4 $\frac{4}{5}\pi\ln 2$

5 a) $t = 10\ln|H| - 10\ln|5 - 2H| + 10\ln 23$

b) i) $10\ln\frac{23}{3}$ (≈ 20.37),
ii) $10\ln 46$ (≈ 38.29)

c) $H = \frac{5e^{0.1t}}{23 + 2e^{0.1t}}$

d) $H_{max} = 2.5$

6 a) $kt = \frac{1}{2a}\ln|a - x| - \ln|a + 3x|$

b) $x = \frac{a(e^{2akt} - 1)}{(3e^{2akt} - 1)}$

c) $x \to \frac{a}{3}$ as $t \to \infty$

7 $y = \frac{1}{2}\ln\left(\frac{x+1}{x+3}\right) + \frac{1}{2}\ln 3$

Exercises 24.4B *(p.530)*

1 a) $\ln|x| - \ln|x + 1| + c$

b) $\ln\frac{2}{3}$

c) $\frac{3}{5}(\ln|x - 2| - \ln|x + 2|) + c$

d) $2\ln|x - 2| - \ln|x - 1| + c$

e) $\frac{1}{3}\ln(x + 2) + \frac{2}{3}\ln|x - 1| + c$

f) $\frac{1}{2}(5\ln(x - 1) - \ln(x + 1) + c)$

g) $\frac{1}{10}(\ln|t - 5| - \ln|t + 5|) + c$

h) $\ln\frac{5}{3}$

i) $\frac{1}{6}(2\ln|x + 2| + \ln|x - 1| - 3\ln|x + 1|) + c$

j) $-2\ln|x| + \frac{1}{4}(5\ln|x + 2| + 3\ln|x - 2|) + c$

k) $\frac{1}{2}(9\ln 2 - 5\ln 5)$

l) $\ln 2 - \frac{1}{2}$

2 $20 - 20\ln 2$

3 $7\ln 3 - 14\ln 2 + \frac{14}{3}$

4 $\frac{\pi}{3}\ln 2$

5 a) $kt = \frac{1}{M}(\ln x - \ln(M - x))$

b) $x = \frac{Me^{KMt}}{1 + e^{KMt}}$

c) $x \to M$ as $t \to \infty$

6 a) $t = 2\ln\left(\frac{2(u + 5)}{3u}\right)$

b) $4\ln 2 - 2\ln 3$ (≈ 0.5754) s

c) $u = \frac{10}{3e^{0.5t} - 2}$

d) $u \to 0$ as $t \to \infty$

7 a) $t = \frac{\sqrt{2}}{4}(\ln(5\sqrt{2} + u) - \ln(5\sqrt{2} - u))$

b) $u = 5\sqrt{2}\,\frac{e^{2\sqrt{2}t} - 1}{e^{2\sqrt{2}t} + 1}$

c) $u \to 5\sqrt{2}$ as $t \to \infty$

$u = 5\sqrt{2}$ m s^{-1} is called the terminal speed.
The parachute will eventually travel with this constant speed.

Consolidation Exercises for Chapter 24 *(p.532)*

1 a) $2\ln(2) - \frac{1}{2}\ln(7)$

b) $-\frac{1}{4}(2x + 1)e^{-2x} + c$

c) $-\frac{1}{6}e^{-3x^2} + c$ **d)** $\ln\frac{3}{2}$

e) 2 **f)** $\frac{1}{3}\ln(t^2 + 7)^{\frac{3}{2}} + c$

g) $\frac{1}{9}(3x^2 \ln(x) - x^3 + c)$

h) $\frac{1}{2}(45\sqrt{5} - \sqrt{3})$

i) $\frac{1}{3}\ln|3x + 5| + c$

j) $-\frac{1}{12(t3+9)4} + c$

k) $\frac{1}{32}((1 - 8x^2)\cos 4x + 4x\sin x) + c$

l) $\frac{1}{13}(3\sin 3x - 2\cos x)e^{-2x} + c$

2 a) $T = 90 + \frac{40}{2t+1}$

b) $T \to 90$ as $t \to \infty$

3 $\frac{1}{243}(85\sqrt{85} - 8) \approx 3.192$

4 $24\pi\ln\frac{3}{2}$

5 a) $kt = -\ln(a + x) + \ln(a)$ $T = \frac{1}{k}\ln 2$

b) $kt = \frac{1}{(n-1)(a-x)^{n-1}} - \frac{1}{(n-1)a^{n-1}}$,

$T = \frac{(2^{n-1}-1)}{(n-1)a^{n-1}}$

6 a) $kt = \frac{1}{2}(2\ln(2 - x) - \ln(1 - x)$
$- \ln(3 - x)) + \frac{1}{2}\ln\frac{3}{4}$

b) $x = \frac{2\left(\sqrt{4e^{2kt} - 3} - \sqrt{e^{2kt}}\right)}{\sqrt{4e^{2kt} - 3}}$

7 a) $t = \ln(1 + 0.1u) - \ln(u) + \ln\frac{100}{11}$

b) $2\ln 2 - \ln\frac{11}{3}$

c) $u = \frac{100}{11e^t - 100}$

d) $u \to -10$ as $t \to \infty$

8 a) $A = 1, B = -1, C = 2$

9 a) A is point $(-\frac{1}{2}, 0)$, asymptote is $x = 1$

b) $y = 4x + 1$, tangent cuts curve at $\left(\frac{7}{4}, 8\right)$

10 a) $\frac{3x}{(x+1)} - \frac{1}{(x^2+1)} - \frac{4}{(x-2)}$

11 a) $\frac{3}{2}x^2(1 + x^3)^{-\frac{1}{2}}$ **b)** $\frac{4}{3}$

12 a) A is point (1, 0), gradient at A is 1

b) B is point $(e^{-0.5}, -\frac{1}{2}e^{-1})$

$\frac{d^2y}{dx^2} = 2$ at B

13 $v = \frac{1}{(1+t)} - 1$, 2.1, 2.13, 2.14, 2.145, 2.146, 2.146, 2.146

14 $\frac{\pi}{12}$

15 A is point (e^1, e^1), area $= \frac{1}{4}(e^2 - 1)$

16 0.91 $3\ln 3 - 2\ln 2 - 1$

17 $\frac{1}{2}\ln 3$

18 a) $8(3 + 2x)^3$ **b)** 68

19 a) $\frac{1}{2}\ln(x^2 + 4) + c$

c) $f'(0) = -\frac{17}{4}$, $f'(8) = -\frac{1}{1156}$

d) $\frac{8-x}{(x^2+4)(2x+1)}$

21 b) $\frac{25}{7}\ln 5 - 6$, trapezium rule gives 14.25

c) percentage error is 0.94%

22 a) i) $(x - 1)(x + 2)(2x - 3)$

ii) $-2 < x < 1$ and $x > 1.5$

b) $A = 1, B = 0, C = 1$

ii) $\frac{1}{2}\ln(2x + 3) + \frac{1}{2}\arctan\left(\frac{x}{2}\right)$

CHAPTER 25
Curves

Exercises 25.1A *(p.560)*

1 a) $x^2 + y^2 = 49$

b) $(x - 4)^2 + y^2 = 16$

c) $(x - 2)^2 + (y - 1)^2 = 25$

d) $(x + 3)^2 + (y - 2)^2 = 9$

e) $x^2 + (y + 5)^2 = 6.25$

f) $(x + 1)^2 + (y + 1)^2 = 36$

2 a) centre (0, 0), radius 11

b) centre (0, 5), radius 9

c) centre (3, 4), radius 5

d) centre (4, –3), radius 4

e) centre (–4, 0), radius 7

f) centre (0, 1), radius 6

g) centre (2, 0), radius 8

h) centre (–1.5, 2.5), radius $\sqrt{75}$

i) centre (1, 2), radius $2\sqrt{34}$

j) centre (3.5, 2), radius $3\sqrt{34}$

3 (a), (c), (f) are circles

4 a) $(x - 5)^2 + (y - 4)^2 = 25$

b) $(x + 0.5)^2 + (y - 1.5)^2 = 0.5$

c) $(x + 0.5)^2 + (y + 0.5)^2 = 8.5$

5 (3, 0), (27, 0)

6 $(1 - 2\sqrt{33}, 0)$, $(1 + 2\sqrt{33}, 0)$, $(0, 1.5 + \sqrt{133.25})$, $(0, 1.5 - \sqrt{133.25})$
or (–10.49, 0), (12.49, 0), (0, 13.04), (0, –10.04)

Exercises 25.1B *(p.561)*

1 a) $x^2 + y^2 = 36$

b) $(x - 3)^2 + y^2 = 1$

c) $x^2 + (y + 2)^2 = 0.25$

d) $(x - 3)^2 + (y + 2)^2 = 16$

e) $(x + 4)^2 + (y + 5)^2 = 2.25$

f) $(x - 0.5)^2 + (y - 0.25)^2 = 4$

2 a) centre (0, 0), radius 9

b) centre (1, 0), radius 7

c) centre (0, –2), radius 10

d) centre (1, –2), radius 2

e) centre (–3, 1), radius 4

f) centre (3, –1), radius 3

g) centre (6, 0), radius $2\sqrt{22}$

h) centre (0, 2.5), radius $\sqrt{6}$

i) centre (1.5, 2.5), radius $\frac{1}{2}\sqrt{34}$

j) centre (1, –2), radius $\sqrt{22}$

3 (b), (d), (e) are circles.

4 (0, 1), (0, 15), $(6 + \sqrt{21}, 0)$, $(6 - \sqrt{21}, 0)$
PQ is not a diameter.

5 a) $(x - 1)^2 + (y - 1)^2 = 2$

b) $(x - 0.5)^2 + (y - 1)^2 = 6.25$

c) $(x - 2.5)^2 + (y - 6)^2 = 42.25$

6 A: (–1, –3), B: (6, –3)

Exercises 25.2A *(p.566)*

2 a) $\frac{x^2}{4} + \frac{y^2}{9} = 1$

b) $\frac{1}{4}(x - 1)^2 + \frac{4}{9}(y - 2)^2 = 1$

or $\frac{9}{4}(x - 1)^2 + 4(y - 2)^2 = 9$

c) $\frac{4(x+2)^2}{25} + \frac{(y+1)^2}{4} = 1$

or $16(x + 2)^2 + 25(y + 1)^2 = 100$

3 a) $x^2 - y^2 = 9$

b) $(x - 1)^2 - y^2 = 9$

c) $y^2 - x^2 = 4$

4 a) $\frac{(x+1)^2}{5^2} + \frac{(y-3)^2}{\left(5/\sqrt{2}\right)^2} = 1$ ellipse,
centre (–1, 3), major axis 10, minor axis $\frac{10}{\sqrt{2}}$

b) $\frac{(x+2)^1}{3^2} + \frac{y^2}{\left(3/\sqrt{2}\right)^2} = 1$ hyperbola,
centre (–1, 0), with asymptotes $y = \pm\frac{1}{\sqrt{2}}(x + 1)$

c) $\frac{(x-1.5)^2}{4^2} - \frac{(y-2.5)^2}{4^2} = 1$
rectangular hyperbola, centre (1.5, 2.5), with asymptote $y - 2.5 = \pm(x - 1.5)$.

Exercises 25.2B *(p.567)*

3 a) $\frac{x^2}{4^2} + y^2 = 1$ or $x^2 + 16y^2 = 16$

b) $\frac{(x+1)^2}{\left(\frac{1}{2}\right)^2} + \frac{(y-3)^2}{\left(\frac{3}{2}\right)^2} = 1$

or $36(x + 1)^2 + 4(y - 3)^3 = 9$

c) $\frac{(x-0.5)^2}{\left(\frac{3}{2}\right)^2} + \frac{(y-2.5)^2}{2^2} = 1$

or $16(x - 0.5)^2 + 9(y - 2.5)^2 = 36$

4 a) $\frac{x^2}{16} - \frac{y^2}{64} = 1$ or $4x^2 - y^2 = 64$

b) $x^2 - y^2 = 1$

c) $\frac{x^2}{4} - \frac{y^2}{24} = 1$

Exercises 25.3A *(p.569)*

1 a) (0, –3), (3, 0)

b) $\left(\frac{\sqrt{19}-2}{5}, \frac{2\sqrt{19}-9}{5}\right)$

$\left(-\frac{\sqrt{19}-2}{5}, -\frac{2\sqrt{19}-9}{5}\right)$

c) (–1, 4), $\left(\frac{4}{5}, \frac{-7}{5}\right)$

d) (–1, 1), (6, –6)

e) (–2, 0), $\left(\frac{10}{13}, \frac{36}{13}\right)$

2 $m = -3$, (2, –6)

$m = \frac{13}{9}$, $\left(\frac{-18}{5}, -\frac{26}{5}\right)$

3 $m = 3$, (2, 7),

$m = \frac{1}{3}$, (6, 3)

4 a) $\sqrt{51}$ **b)** $2\sqrt{3}$ **c)** $4\sqrt{2}$

5 a) $4y = -3x + 25$ **b)** $3y = x + 3$

c) $3y = 4x - 34$

7 a) $c = \sqrt{5} - 2$ and $c = -\sqrt{5} - 2$

b) $-\sqrt{5} - 2 < c < \sqrt{5} - 2$

c) $c < -\sqrt{5} - 2$ or $c > \sqrt{5} - 2$

8 (1, 1), $(\sqrt{3} - 2, -\sqrt{3} - 2)$, $(-2\sqrt{2}, 1)$

9 $(2\sqrt{2}, 1)$, $(-2\sqrt{2}, 1)$

11 $7\pi + 4\sqrt{3}$

Exercises 25.3B *(p.571)*

1 a) (1, –1), $\left(-\frac{1}{5}, \frac{5}{7}\right)$

b) $(-1, 2)$, $\left(-\frac{1}{2}, \frac{1}{2}\right)$

c) no points of intersection

d) (–1, –5), (5, 1)

e) $\left(\frac{6}{\sqrt{37}}, \frac{3}{\sqrt{37}}\right)$, $\left(\frac{-6}{\sqrt{37}}, -\frac{2}{\sqrt{37}}\right)$

2 $m = 1$, (–1, –1)

$m = -\frac{1}{7}$, $\left(\frac{7}{5}, -\frac{1}{5}\right)$

3 $m = -\frac{12}{5}$, $\left(\frac{-10}{13}, -\frac{15}{13}\right)$

$m = 0$, (2, 3)

4 a) $m = -\sqrt{3}$ and $m = \sqrt{3}$

b) $m < -\sqrt{3}$ or $m > \sqrt{3}$

c) $-\sqrt{3} < m < \sqrt{3}$

6 a) 4 **b)** $2\sqrt{10}$ **c)** $\sqrt{23}$

7 For (–2, –4), $y = -\frac{3}{2}x - 7$

For (3, 1), $y = -\frac{2}{3}x + 3$

For (–2, 0), $y = \frac{2}{3}x + 3$

8 $(2\sqrt{2}, \sqrt{2})$, $(-2\sqrt{2}, -\sqrt{2})$

9 (1, 2), (1, –2)

10 $\frac{25\sqrt{3}}{2}$

11 $(x - 5)^2 + (y - 13)^2 = 9$

Exercises 25.4A *(p574)*

1 $\frac{dy}{dx} = -\frac{x}{y}$ **2** $\frac{dy}{dx} = \frac{x}{y}$

3 $\frac{dy}{dx} = \frac{1-x}{2+y}$ **4** $\frac{dy}{dx} = \frac{(x-1)}{2(y+1)}$

5 $\frac{dy}{dx} = \frac{3x^2+y}{x+2y}$ **6** $\frac{dy}{dx} = \frac{5y-2x}{8y-5x}$

7 $\frac{dy}{dx} = \frac{5\cos 5x}{2\sin 5x}$ **8** $\frac{dy}{dx} = \frac{\cos x}{\sin y}$

9 $\frac{dy}{dx} = \frac{-y\sin(xy)}{1+x\sin(xy)}$ **10** $\frac{dy}{dx} = \frac{2x-y}{x+2e^{3y}}$

11 –0.75 **12** –2

13 0.2 **14** 1.4

15 –1 **16** –2

17 0 **18** $-\frac{\sqrt{3}}{6}$

	tangent	normal
21	$y = -\frac{1}{\sqrt{3}}x + \frac{8}{\sqrt{3}}$	$y = \sqrt{3}x$
22	$y = -\frac{4}{3}x + \frac{23}{3}$	$y = \frac{3}{4}x - \frac{11}{4}$
23	$y = \frac{2}{3}x - \frac{4}{3}$	$y = -\frac{3}{2}x - \frac{7}{2}$
24	$y = -3x + 7$	$y = \frac{1}{3}x + \frac{11}{3}$
25	$y = \frac{1}{2}x + 2$	$y = -2x + 7$
26	$y = x - 2$	$y = -x + 4$

Exercises 25.4B *(p.575)*

1 $\frac{dy}{dx} = -\frac{x}{y}$ **2** $\frac{dy}{dx} = \frac{x}{y}$

3 $\frac{dy}{dx} = -\frac{2x}{3y}$ **4** $\frac{dy}{dx} = \frac{3}{2(y-2)}$

5 $\frac{dy}{dx} = \frac{4-y}{3+y}$ **6** $\frac{dy}{dx} = \frac{2x+3}{4y-5}$

7 $\frac{dy}{dx} = \frac{y\sin x + \sin y}{\cos x - x\cos y}$ **8** $\frac{dy}{dx} = \frac{2x - ye^y}{x(x+e^y)}$

9 $\frac{dy}{dx} = \frac{1}{\cos(x+y)} - 1$ **10** $\frac{dy}{dx} = \frac{y\cos xy}{1 - x\cos xy}$

11 $-\frac{1}{2}$ **12** $\frac{1}{6}$ **13** $-\frac{7}{9}$ **14** $-\frac{4}{15}$

15 2 **16** $\sqrt{\frac{6}{3}}$ **17** 0 **18** $\frac{9}{16}$

	tangent	normal
21	$y = -\sqrt{\frac{5}{5}}x + \frac{12}{5}\sqrt{5}$	$y = \sqrt{5}x$
22	$y = x - 3$	$y = -x - 1$
23	$y = -\frac{1}{4}x + \frac{11}{4}$	$y = 4x - 10$
24	$y = -8x + 21$	$y = \frac{1}{8}x + \frac{19}{4}$
25	$y = 2x + 4$	$y = -\frac{1}{2}x + \frac{3}{2}$
26	$y = -\frac{10}{3}x + \frac{43}{3}$	$y = \frac{30}{10}x - \frac{11}{5}$

Consolidation Exercises for Chapter 25

1 hyperbola, centre (0,0), asymptotes $y = \pm x$

2 circle, centre (2, –3), radius 10

3 ellipse, centre (2, 3), major axis $\frac{10}{\sqrt{2}}$, minor axis 10

4 parabola, $x \geq -2$, axis $y = 1$

5 tangent $y = \frac{1}{3}x + 1$, normal $y = -3x + 21$

6 a) $\frac{6-\sqrt{6}}{4} < m < \frac{6+\sqrt{6}}{4}$

b) $m = \frac{6 \pm \sqrt{6}}{4}$

c) $m < \frac{6-\sqrt{6}}{4}$ or $m > \frac{6+\sqrt{6}}{4}$

8 a) $3y = 4x - 150$

b) $x^2 + y^2 = 900$

c) (24, –18)

9 a) i) $7\sqrt{5}$ **ii)** $7\sqrt{5} - 13 = 133$ mm

b) i) 1

ii) AC has equation $y = -x + 10$

10 A: (1, 2)

P: (5, –1), Q : (6, 2)

CHAPTER 26 Trigonometry 5

Exercises 26.1A *(p.583)*

1 a) $\frac{\sqrt{6}}{4} + \frac{\sqrt{2}}{4}$ **b)** $\frac{3-\sqrt{3}}{3+\sqrt{3}}$

c) $\frac{\sqrt{6}}{4} + \frac{\sqrt{2}}{4}$ **d)** $\frac{\sqrt{3}+3}{3-\sqrt{3}}$

e) $\frac{\sqrt{2}}{4} - \frac{\sqrt{6}}{4}$ **f)** $\frac{\sqrt{6}}{4} - \frac{\sqrt{2}}{4}$

2 a) $\frac{1}{2}$ **b)** $\frac{1}{2}$ **c)** $\sqrt{3}$

3 a) $\frac{171}{221}$ **b)** $\frac{220}{221}$ **c)** $-\frac{171}{140}$

4 a) $\frac{44}{125}$ **b)** $\frac{3}{5}$ **c)** $\frac{44}{117}$

5 0 **7a)** $\frac{5}{2}$ **b)** 2 **c)** $\tan PQR = \frac{5}{12}$

8 $\tan\theta = \frac{5}{12}$ **9** $\tan\theta = \frac{3}{4}$

10 $\frac{\tan 60° + \tan 30°}{1 - \tan 60° \tan 30°}$, $\tan 90° = \infty$

11 18.43° **12** 45°, 225°

15 a) $\tan x = \frac{(1+k)}{(1-k)}\tan\alpha$

b) 60°, 240°
16 15°, 195° **17** –76.64°, 103.36°
18 156.2°, 336.2° P: (156.2°, 0.81), Q: (336.2°, –0.81)
19 (0.20), 0.60), (2.81, 0.99)
20 (–2.08, –3.56), (–0.27, 0.56), (1.06, –3.56), (2.87, 0.56)

Exercies 26.1B *(p.584)*

4 a) sin 104° **b)** sin 4° **c)** cos 40° **d)** –cos 60° **e)** sin $\frac{\pi}{3}$ **f)** sin $\frac{\pi}{2}$ **g)** sin x **h)** sin $3x$ **i)** cos $2x$ **j)** sin x
5 a) $\frac{\sqrt{2}}{4}(\sqrt{3}+1)$ **b)** $\frac{\sqrt{2}}{4}(\sqrt{3}-1)$
c) $\frac{3+\sqrt{3}}{3-\sqrt{3}}$
7 a) $\frac{4}{5}, \frac{4}{3}, \frac{5}{13}, \frac{5}{12}$
b) $\frac{63}{65}, \frac{33}{65}, \frac{16}{65}, \frac{56}{65}, \frac{63}{16}, \frac{33}{56}$
8 $\cos(A+B) = -\frac{63}{65}$, $\tan(A-B) = \frac{56}{33}$
9 $\frac{1}{4}, \frac{1}{4}, \frac{1}{4}, -\frac{1}{4}$
10 $\frac{1}{4}(\sqrt{6}-\sqrt{2}), \frac{1}{4}(\sqrt{6}+\sqrt{2}),$ $\frac{1}{4}(\sqrt{2}-\sqrt{6}), \frac{1}{4}(\sqrt{6}+\sqrt{2})$
12 a) $\beta = \alpha + \theta$
c) $\theta = \arctan \frac{1}{7} = 0.1419$rad or 8.13°
14 $\frac{325}{36}, \frac{204}{253}$
18 a) (80.2°, 0.34). (–99.9°, –0.34)
b) (80.1035°, 0.3437), (–99.8965, –0.3437)
19 (30°, 0), (210°, 0)
20 $a = -5$, $b = 8$

Exercises 26.2A *(p.586)*

1 a) $\frac{3}{5}$ **b)** $\frac{4}{5}$ **c)** $\frac{24}{25}$ **d)** $\frac{7}{25}$ **e)** $\frac{24}{7}$
2 a) –0.96 **b)** –0.28 **c)** 0.5376
3 a) $-\frac{527}{625}$ **b)** $-\frac{336}{625}$ **c)** $\frac{336}{527}$
7 a) $\frac{2\sqrt{s}}{25}$ **b)** $\frac{91}{125}$
8 $a = 3$, $b = -4$
a) $\frac{23}{27}$ **b)** $\frac{42\,471}{68\,921}$ (≈ 0.616)
9 a) $\frac{\pi}{2}, \frac{5\pi}{6}, \frac{3\pi}{2}, \frac{11\pi}{6}$
b) 0.4636, $\frac{\pi}{2}$, 3.6052, $\frac{3\pi}{2}$
c) 0, 1.8755, π, 4.4077, 2π

10 a) 60°, 180°, 300° **b)** 90°
c) 56.44°, 123.56°, 270°,
11 P: (30°, 0.5) Q: (150, 0.5)
12 (0, 0) $(\frac{\pi}{4}, 1)$, $(\frac{3\pi}{4}, -1)$, (π, 0), $(\frac{5\pi}{4}, 1)$, $(\frac{7\pi}{4}, -1)$, (2π, 0)

Exercises 26.2B *(p.589)*

1 A translation of 1 unit parallel to the y axis, $2\cos^2 x = \cos 2x + 1$.
4 a) $\frac{\sqrt{3}}{2}$ **b)** $\frac{\sqrt{2}}{2}$ **c)** $\frac{\sqrt{3}}{2}$ **d)** 0 **e)** $\sqrt{3}$
5 a) i) $\frac{\sqrt{7}}{4}$ **ii)** $\frac{3\sqrt{7}}{8}$ **iii)** $\frac{1}{8}$ **iv)** $3\sqrt{7}$
b) $\frac{3\sqrt{7}}{32}, -\frac{31}{32}, -\frac{3\sqrt{7}}{31}$
6 a) i) $\frac{2\sqrt{5}}{5}$ **ii)** $-\frac{4}{5}$ **iii)** $\frac{3}{5}$ **iv)** $-\frac{4}{3}$
b) $-\frac{24}{25}, -\frac{7}{25}, \frac{24}{7}$
7 $4\cos^3 x - 3\cos x$
8 b) $\frac{336}{625}$
9 b) $\pm\frac{3}{4}$ **c)** $\frac{1}{2}$ and 2
11 a) 90°, 210° **b)** 60°, 300°
c) 30°, 150°, 194.5°, 345.5°
12 $\tan\theta$

Exercises 26.3A *(p.594)*

1 a) translation of +36.87° x–axis, $3\cos x + 4\sin x = 5\sin(x + 36.87°)$
b) translation of –53.13° x–axis, $3\cos x + 4\sin x = 5\cos(x - 53.13°)$
2 a) $13\sin(x + 157.38°)$, $13\cos(x - 292.62°)$
b) $25\cos(x + 73.74°)$, $25\sin(x - 196.26°)$
3 a) $\sqrt{2}\cos(x - 45°)$ or $\sqrt{2}\cos(x - 0.7854)$
b) $5\cos(x - 36.87°)$ or $5\cos(x - 0.6435)$
4 a) $5\cos(x + 233.13°)$ or $5\cos(x + 4.0689)$
b) $13\cos(x + 22.62°)$ or $13\cos(x + 0.3948)$
5 a) $\sqrt{5}\sin(x - 26.57°)$ or $\sqrt{5}\sin(x - 0.4637)$
b) $\sqrt{12}\sin(x - 60°)$ or $\sqrt{12}\sin(x - \frac{\pi}{3})$
6 a) $\sqrt{41}\sin(x + 54.74°)$ or $\sqrt{41}\sin(x + 0.8961)$
b) $\sqrt{3}\sin(x + 54.74°)$ or $\sqrt{3}\sin(x + 0.9554)$
7 $17\sin(x - 61.93°)$

a) greatest value = 17, least value = –17
b) greatest value = ∞, least value = –∞
8 a) maximum value = 2, minimum value = –2, turning points (150°, –2), (330°, 20)
b) maximum value = 20, minimum value = –14, turning points (118.07°, –14), (298.07°, 20)
c) maximum value = ∞, minimum value = –14, turning points (45°, $\frac{1}{\sqrt{2}}$), (225°, $-\frac{1}{\sqrt{2}}$)
d) maximum value = 169, minimum value = 0, turning points (67.38°, 169), (157.38°, 0), (247.38°, 169), (337.38°, 0)

Exercises 26.3B *(p.595)*

1 a) translation of +67.38°// x-axis, $13\sin(x + 67.38°)$
b) translation of –22.67°// x-axis, $13\cos(x - 22.67°)$
2 a) $5\cos(x + 53.13°)$, $5\sin(x + 143.13°)$
b) $17\cos(x + 61.93°)$, $17\sin(x - 208.07°)$
3 a) $\sqrt{3}\cos(x - 35.26°)$ or $\sqrt{3}\cos(x - 0.6154)$
b) $\sqrt{10}\cos(x - 18.43°)$ or $\sqrt{10}\cos(x - 0.3217)$
4 a) $2\cos(x + 60°)$ or $2\cos(x + \frac{\pi}{3})$
b) $5\cos(x + 36.87°)$ or $5\cos(x + 0.6435)$
5 a) $5\sin(x - 53.13°)$ or $5\sin(x - 0.9273)$
b) $\sqrt{2}\sin(x - 45°)$ or $\sqrt{2}\sin(x - \frac{\pi}{4})$
6 a) $\sqrt{53}\sin(x + 74.05°)$ or $\sqrt{53}\sin(x + 1.2924°)$
b) $\sqrt{130}\sin(x + 15.26°)$ or $\sqrt{130}\sin(x + 0.2663)$
7 $25\cos(x - 73.74°)$
a) greatest value = 25, least value = –25 **b)** greatest value = ∞, least value = –∞
8 a) maximum value = $\sqrt{13}$ at x = 146.31, minimum value = $-\sqrt{13}$ at x = 326.31°
b) maximum value = $2 + \sqrt{58}$ at x = 293.2°, minimum value = $2 - \sqrt{58}$ at x = 113.2°
c) maximum value = ∞,

minimum value = $-\infty$, turning points $(144.74°, -\frac{1}{\sqrt{3}})$, $(324.74°, \frac{1}{\sqrt{3}})$,

d) maximum value 13 at $x = 56.31°$ and $236.31°$, minimum value 0 at $x = 146.31°$ and $326.31°$

Consolidation Exercises for Chapter 26 *(p.598)*

1 b) 61.32°, 241.32°
2 0.6155, 3.75707
3 150°, 330°
4 $(3n + 1)\frac{\pi}{3}$
5 $t = -2 \pm \sqrt{7}$ or $t = 0.6458$ and -4.6458, $x = 180n + 32.85°$ and $x = 180n - 77.85°$
6 0°, 60°, 180°, 300°, 360°
7 a) 30°, 150°, 270°
b) 51.33°, 128.67°
c) 0°, 180°, 210°, 330°, 360°
8 0.3747, 2.7669
9 a) 90°, 210°, 270°, 330°
b) 0°, 146.31°, 180°, 326.31°, 360°
c) 0°, 180°, 270°, 360°
d) 0°, 30°, 150°, 180°, 360°
e) 90°, 216.87°, 323.13°
f) 60°, 120°, 240°, 300°
10 $\cos 4x \equiv \cos^2 2x - \sin^2 2x \equiv 2\cos^2 2x - 1 \equiv 1 - 2\sin^2 2x$
a) 0°, 30°, 150°, 180°, 210°, 330°, 360°
b) 105°, 165°, 285°, 345°
11 $0, \frac{\pi}{4}, \pi, \frac{3\pi}{4}, \frac{5\pi}{4}, \frac{7\pi}{4}, 2\pi$
12 a) 0°, 180°, 360°
b) 39.2°, 140.8°
c) 35.3°, 144.7°, 215.3°, 324.7°
13 $R = 17$, $\alpha = 61.93°$, $-208.07°$, 151.93°
14 a) 103.3°, 330.5°
b) 50.5°, 256.3°
c) 0°, 90°, 360°
d) 106.27°, 180°
15 a) $7.4° + 360n$, $127.4° + 360n$
b) $360n$, $109.5° + 360n$
c) $45° + 360n$
d) $360n$; $360n - 143.2°$
16 a) $2n\pi + 1.38$, $2n\pi + 0.48$
b) $2n\pi + 0.1$, $2n\pi + 5.8$
c) $2n\pi - 0.13$, $2n\pi - 2.23$
d) $2n\pi$, $2n\pi - \frac{2\pi}{3}$
17 $360n \pm 60° + 73.74°$, 13.74°, 133.74° **a)** 27.48°, 267.48°
b) 6.87°, 66.87°, 186.87°, 246.87°
18 $\sqrt{5}\sin(x + 26.57°)$, 0°, 126.87°, 360° **a)** 0°, 253.74°
b) 0°, 42.29°, 120°, 162.29°, 240°, 282.29°, 360°
19 $R = \sqrt{2}$, $A = 45°$, or $\frac{\pi}{4}$
a) $2n\pi + \frac{\pi}{4}$
b) greatest value = $\sqrt{2}$
20 a) $\alpha = 30.96°$
b) 15.7°, 282.4°
c) 7.9°, 141.2°, 187.9°, 321.2°
21 $\frac{\pi}{12}, \frac{5\pi}{12}, \frac{13\pi}{12}, \frac{17\pi}{12}$
22 a) $5\sin(x + 36.97°)$
b) 5 **c)** 143.13°, 323.13°
23 15°, 75°, 195°, 255°
24 0°, 30°, 150°, 180°, 210°, 330°, 360°
25 a) $p = 4r\cos\theta + 2r\sin\theta$
b) maximum value of $a = r^2$ when $\theta = \frac{\pi}{4}$
c) p is maximum when $\tan^{-1}\frac{1}{2} = 26.57°$ $a = 0.8r^2$
26 a) $a = 2$, $b = \sqrt{3}$
b) $\sqrt{7}\sin(x + 40.89°)$
c) i) $x = 49.10°$ **ii)** $x = 0°$
27 a) $c = \sqrt{a^2 + b^2}$ **b)** $\tan\alpha = \frac{b}{a}$
c) $\alpha = 36.87°$
d) 103.29°, 330.45°
28 a) $\cos x \cos\alpha - \sin x \sin\alpha$
b) $\sqrt{29}\cos(x + 68.2°)$
c) maximum value is $\sqrt{29}$ at $x = 291.8°$ minimum value is $-\sqrt{29}$ at $x = 111.8°$
d) 235.65°, 347.95°
29 a) $R = 17$ **b)** $A = 61.9°$
c) maximum value = 17
d) 298.1°
30 b) 5.76 **c)** 0.40, 4.84
d) 3.4
31 a) $R = \sqrt{10}$, $\alpha = 18.43°$
b) 90°, 306.86°
c) $\sin x = -\frac{1}{2}$ and $\sin x = 1$

CHAPTER 27
Curves 2

Exercises 27.1A *(p.606)*

1 a) $x^2 + y^2 = 4$, circle, centre (0, 0), radius 2
b) $\frac{x^2}{16} + \frac{y^2}{9} = 1$
ellipse: centre (0,0), major axis 8, minor axis 6
c) $y = \frac{2}{x}$, rectangular hyperbola
d) $y = x^2 + 2x + 2$, parabola with $x = -1$ as axis of symmetry
e) $(x - 1)^2 + (y - 2)^2 = 9$, circle, centre (1, 2), radius 3
2 b) $\frac{x^2}{4} - \frac{y^2}{9} = 1$
c) hyperbola
3 $2y = x + 2$
a) $a = 2$
b) $T = \frac{3}{2}$
4 $y^2 = 1 - \frac{1}{5}x$, parabola with x–axis as axis symmetry of symmetry, $x \le 5$
7 (3, –1), (7, 5)
8 (4, 3), (7, –3)

Exercises 27.1B *(p.607)*

1 a) $x^2 + y^2 = 25$, circle; centre (0, 0), radius 5
b) $y = \frac{10}{x}$, rectangular hyperbola, asymptotes $y = 0$ and $x = 0$
c) $2(y - 1)^2 = 9(1+x)$, parabola with $y = 1$ as axis of symmetry
d) $\frac{(x+3)^2}{4} + \frac{(y-1)^2}{25} = 1$, ellipse, centre (–3, 1), major axis 4, minor axis 10
e) $y = x - 2$, straight line, slope 1, y–intercept (0, –2)
2 a) $x^2 - y^2 = 4$
b) hyperbola, asymptotes $y = x$ and $y = -x$
3 $(1 - y)^2 - (x - 1)^2 = 1$, hyperbola, asymptotes $y = 2 - x$, $y = x$
4 a) $2x + y = 5\cos\theta$, $x - 2y = 5\sin\theta$
b) $(2x + y)^2 + (x - 2y)^2 = 25$ or $x^2 + y^2 = 5$
c) circle, centre (0, 0), radius $\sqrt{5}$
7 (3, 1), (6, –5)
8 b) $y = 3$

Exercises 27.2A *(p.609)*

1 tangent $y = \frac{3}{2}x + 2$
normal $y = -\frac{2}{3}x + 2$

2 tangent $y = -\frac{1}{\sqrt{3}}x + \frac{7\sqrt{3}}{3} - 1$
normal $y = \sqrt{3}x - 1 - \frac{7\sqrt{3}}{2}$

3 tangent $y = -3x + 11$
normal $y = \frac{1}{3}x + \frac{13}{3}$

4 tangent $y = 2$, normal $x = 0$

5 tangent $y = -\frac{x}{\sqrt{2}} + 1$
normal $y = \sqrt{2}x + 1 - 3\sqrt{2}$

6 tangent $y = \frac{9}{4}x - 1$
normal $y = -\frac{4}{9}x + \frac{43}{54}$

7 (–2, 0) is a local minimum for $t = -1$

8 $(-\frac{1}{\sqrt{3}}, \frac{2\sqrt{3}}{9})$ is a local maximum for $t = -\frac{1}{\sqrt{3}}$
$(\frac{1}{\sqrt{3}}, -\frac{2\sqrt{3}}{9})$ is a local minimum for $t = \frac{1}{\sqrt{3}}$

Exercises 27.2B *(p.609)*

1 tangent $y = -2x - 2$
normal $y = \frac{1}{2}x - \frac{9}{2}$

2 tangent $y = -\frac{x}{\sqrt{3}} + 1 + 3\sqrt{3}$
normal $y = \sqrt{3}x + 1 + \sqrt{3}$

3 tangent $y = -\frac{2x}{3\sqrt{3}} + 3 - \frac{16\sqrt{3}}{9}$
normal $y = \frac{3\sqrt{3}x}{2} + 3 + \frac{17}{4}\sqrt{3}$

4 tangent $y = 2x - 1$
normal $y = -\frac{1}{2}x + \frac{3}{2}$

5 tangent $y = x - 1$
normal $y = -x + 1$

6 tangent $y = \left(\sqrt{2}-1\right)x + 1 + \frac{\pi}{4}\left(1-\sqrt{2}\right)$
normal $y = \frac{1}{1-\sqrt{2}}x - 1 + \frac{\pi}{4}\left(1+\sqrt{2}\right)$

7 $\frac{dy}{dx} = -4x$

8 a) $\frac{dy}{dx} = -\frac{1}{2}$ tangent $y = -\frac{1}{2}x + \frac{7}{2}$
normal $y = 2x + 1$
b) Tangent intersects again at point (16, –4.5). Normal does not intersect again.

Consolidation Exercises for Chapter 27 *(p.610)*

1 a) position $(1, \sqrt{3})$ speed 2
b) direction of line with slope $-\frac{1}{\sqrt{3}}$
c) $x^2 + y^2 = 4$

2 c) $y = -\frac{x}{t^2} + \frac{4}{t}$
d) area of triangle = 8

3 a) $(2(t_1+t_2), 2 + (\frac{1}{t_1} + \frac{1}{t_2})$
c) $y = 2x$
d) $(2\sqrt{2}, 4\sqrt{2})$, $(-2\sqrt{2}, -4\sqrt{2})$

4 c) $y = tx - 2t^2$
d) $(2(t_1 + t_2), 2t_1 t_2)$
e) $x = 4$

5 a) $t = 1$
c) $y = -x + 3$
e) (–8, –5)

7 a) $\frac{dy}{dx} = \frac{t^2+1}{t^2-1}$
c) $\frac{29}{12}\pi$

8 a) $y = -\frac{2}{3}x + \frac{5}{3}$ b) $\frac{23}{20}$

10 $p^2 = b - 2$ $s = -\frac{2+r^2}{r}$

CHAPTER 28
Differential Equations

Exercises 28.1A *(p.616)*

1 $\frac{dy}{dx} = x + y$, $y = 1$ when $x = 0$

2 $\frac{dx}{dt} = k\sqrt{1-x^2}$, k is constant,
$x = 0$ when $t = 0$

3 h is height in cm, t is time in days
$\frac{dh}{dt} = 0.25h$, $h = 2$ when $t = 0$

4 P is the population, t is time in years, $\frac{dP}{dt} = 0.02P$

5 $\frac{dT}{dt} = -0.4(T - 20)$,
$T = 90$ when $t = 0$

6 a) $\frac{dA}{dt} = kr$, k is constant
b) $\frac{dA}{dr} = 2\pi r$ c) $\frac{dr}{dt} = \frac{k}{2\pi}$

7 r is radius in cm, t is time in s
$\frac{dr}{dt} = -\frac{2.5}{144\pi}$

8 h is height in m, t is time in hours,
$\frac{dh}{dt} = -\frac{(1+2h)}{35}$

9 c is concentration in g cm^{-3},
t is time in s, $\frac{dc}{dt} = -0.2c$

10 h is height in m, t is time in s,
$\frac{dh}{dt} = -0.004 - 0.0008h$,
$h = 0.5$ when $t = 0$. Tank fills.

Exercises 28.1B *(p.617)*

1 m is mass and t is time.
$\frac{dm}{dt} = -km$, k is constant

2 P is number of rabbits, t is time in months
$\frac{dP}{dt} = 0.07P$ $P = 10$ when $t = 0$

3 $\frac{dy}{dx} = x^2 - y^2$ $y = 1$ when $x = 1$

4 T is temperature in °C, t is time in minutes
$\frac{dT}{dt} = -0.3(T - 20)$
$T = 2$ when $t = 0$

5 r is radius in cm, t is time in s
a) $\frac{dV}{dt} = \frac{2r^3}{675}$ b) $\frac{dr}{dt} = -\frac{r}{1350\pi}$
$r = 20$ when $t = 0$

6 r is radius in cm, t is time in s,
$\frac{dr}{dt} = \frac{5}{\pi r^2}$

7 m is mass of carbon–14 in grams, t is time in years,
$\frac{dm}{dt} = -km$, k is a constant
$m = 10^{-22}$ when $t = 0$

8 i is current in mA, t is time in s,
$\frac{di}{dt} = -\frac{1}{80}i$

9 h is depth of water in cm, t is time in s, $\frac{dh}{dt} = \frac{8}{\pi h^2}(60 - h)$
$h = 50$ when $t = 0$.
Cone initially fills. Depth is constant when $t = 60$.

10 $\frac{dP}{dt} = kP(1500 - P)$,
k is a constant

Exercises 28.2A *(p 622)*

1 a) $y = \frac{1}{3}x^3 + c$ b) $y = -\cos x + c$
c) $y = -\frac{1}{2}e^{-2x} + c$
d) $y = \ln\frac{x}{x+1} + c$ e) $y = \frac{2}{3}t^{\frac{3}{2}} + c$
f) $S = \frac{8}{5}t^{\frac{5}{2}} + c$ or $y = \frac{Ae^{x2} - 1}{Ae^{x2} + 1}$

2 a) $y = \sqrt{c + x^2}$ b) $y = -\frac{1}{(c + \frac{1}{2}x^2)}$
c) $y = \sqrt{c - 2\cos x}$
d) $y = \frac{1}{4}\ln(2x^2 + c)$
e) $y = e^{\sin x} + c$ or $y = Ae^{\sin x}$
f) $\ln\frac{1+y}{1-y} = x^2 + c$ or $y = \frac{Ae^{x^2} - 1}{Ae^{x^2} + 1}$

3 a) $y = \frac{1}{3}x^3 + x - 8$
b) $\ln y = (1 + x)^2 - 1$ or $y = e^{(1+x)^2 - 1}$

c) $y = \frac{1}{\cos x}$

d) $\ln \frac{x-1}{x+1} = 2t^2 - 2 + \ln\frac{1}{3}$

or $x = \frac{3e^{2t^2-2}+1}{1-3e^{2t^2-2}}$

e) $\ln \frac{P}{1-P} = 3t$ or $P = \frac{e^{3t}}{1+e^{3t}}$

f) $e^{2x} = \frac{5}{3} - \frac{2}{3}\cos 3t$

or $x = \frac{1}{2}(\frac{5}{3} - \frac{2}{3}\cos 3t)$

g) $y^2 = 4\frac{(1+x)^2}{(1-x)^4} - 1$

h) $y^2 = 1 + \frac{8}{3}(x^2-1)$

4 $y = \frac{1}{2} + \frac{3}{2}e^{2x}$

5 $y = \sqrt{\frac{x-1}{x+1}} - 1$

6 $y = 5 - \frac{4}{x}$

7 (3) $h = 2e^{0.25t}$ (4) $P = P_0e^{0.02t}$
(5) $T = 20 + 70\,e^{-0.4t}$
Time to reach 30°C is 4.9 s.
(6) $r = \frac{kt}{2\pi} + r_0$

(7) $r = -\frac{2.5t}{144\pi} + r_0$

(8) $h = Ae^{-\frac{2t}{35}} - \frac{1}{2}$

$h = \frac{9}{2}e^{-\frac{2t}{35}} - \frac{1}{2}$

Time to half empty is 10.3 hours.
(9) $c = Ae^{-0.2t}$
Time to decay to 25% of initial dose is 6.93 s.
(10) $h = 5 - 4.5\,e^{-0.0008t}$
Time to overflow is 147 s.

Exercises 28.2b *(p.623)*

1 a) $y = \frac{1}{4}x^4 + c$
b) $y = \sin x - x + c$
c) $y = \frac{1}{4}e^{4x} + c$
d) $s = \ln\frac{t+1}{t+2} + c$
e) $P = (t-1)\,e^t + c$
f) $v = -\frac{16}{9}s^{\frac{9}{4}} + c$

2 a) $y = Ae^{-\frac{1}{x}}$
b) $\ln\frac{P}{3-P} = 6t + c$

or $P = \frac{3Ae^{6t}}{1+Ae^{6t}}$

c) $y = \sqrt{(c+(x+2)^2)}$

d) $r^2 = \theta - \frac{1}{2}\sin 2\theta + c$

e) $\frac{1+y}{3+2y} = \frac{A}{1-x}$ or $y = \frac{1-x-3A}{2A-1+x}$

f) $v^2 = \frac{1}{2}(3 - Ae^{-4s})$

3 a) $y = x^3 - x^2 + x + 1$
b) $\ln y = 1 - \cos x$ or $y = e^{1-\cos x}$
c) $y = (\frac{5}{2} - \frac{1}{2}e^{-x})^2$
d) $y = \frac{1+x}{3-x}$ e) $T = 20 - 15e^{2t}$

f) $v + \frac{10}{3}\ln\frac{10-3v}{7} = 1 - 3s$

g) $s = \frac{1}{2}\ln\frac{5-2\cos 3t}{3}$

h) $v = \frac{40e^{-0.2t}}{41-40e^{-0.2t}}$

4 a) $h = \frac{1}{1+\frac{3}{4}t}$

b) Time to 10 cm is 12 minutes.

5 $y = 4x - 10$

6 $y = \frac{1}{4}(x+4)^2$

7 (2) $P = 10e^{0.07t}$
(4) $T = 20 - 18e^{-0.3t}$
Time to reach 10°C is 1.96 minutes (approximately 2 minutes).
(5) $r = 20e^{\frac{-t}{1350\pi}}$
Time to become 10 cm is 2940 seconds.
(6) $r = \sqrt[3]{\frac{15t}{\pi} + c}$

(7) $m = 10^{-22}\,e^{-kt}$
Time for mass to be 3.5×10^{-23} g is 1.05/*k* years.
(8) $i = Ae^{\frac{1}{80}t}$

(9) $h^2 + 120h + 7200\ln\frac{60-h}{10} = 8500 - \frac{16t}{\pi}$

(10) $P = \frac{1500Ae^{1500kt}}{1+Ae^{1500kt}}$

Exercises 28.3A *(p.630)*

1 $y(1.2) \approx 0.22$
2 $y(0.5) \approx 0.436\,071$
3 $v(0.5)$ 6.265 97
4 1.566 23
5 2.231 06 (neglect the value for $h = 1$)
6 a)

h	$y(2)$
0.02	0.214 452
0.05	0.214 688
0.1	0.215 041
0.20	215 555 5

b) 0.214 374
c) $y = \frac{1}{2+\frac{1}{3}x^3}$, 0.214 285

Exercises 28.3B *(p. 630)*

1 $y(0.5) \approx 0.818\,706$ (6 s.f.)
2 $y(0.5) \approx 1.072\,84$ (6 s.f.)
3 $v(2) \approx 10.0207$ (6 s.f.)
4 1.841 2
5 1.959 3
6

x	2	3	4	5
y	2.9	6.7	11.5	17.4

Consolidation Exercises for Chapter 28 *(p.634)*

1 a) $\frac{dT}{dt} = -k(T - 32.2)$
$T(0) = 34.8$, $T(1) = 34.1$
T is temperature in °C, *t* is time in hours after 2.30 a.m. and *k* is constant.
Assume that the rate of temperature change is proportional to the diffference between body temperature and room temperature.
Assume room temperature is constant.
b) $T = 32.2 + 2.6\,e^{-0.314t}$
c) Time of death was about 0.33 a.m. ($t = -117$ minutes)

2 a) $\frac{dy}{dx} = x + y^2$,
$y = 1$ when $x = 0$
b) $Y_{n+1} = Y_n + h(x_n + Y_n^2)$
$Y_0 = 1$

3 a) $R = 10e^{-kt}$
c) 9.58 grams

4 a) $\cos x + x\sin x + c$
b) $\frac{1}{2}y + \frac{1}{4}\sin 2y + c$
c) $\frac{1}{2}y + \frac{1}{8}\sin 4y = \cos x + x\sin x + c$

5 a) $\frac{1}{3}\left(\frac{1}{y} + \frac{1}{3-y}\right)$

b) $\frac{1}{3}\ln\frac{y}{3-y} + c$

c) $y = \frac{3x^3}{4+x^3}$

6 a) 947, 890
b) Critical value of *P* is 1600 mice.
c) $P > 1600$ then $\frac{dP}{dt} > 0$ so *P* increases.
$P<1600$ then $\frac{dP}{dt} < 0$ so *P* decreases.

7 b) $2\sqrt{x} = -kt + c$

c) 22 minutes

8 $\frac{dx}{dt} = 8tet^2$,
acceleration is 3931 m s^{-2}

9 b) $I \to \frac{E}{R}$, $0.693 \frac{L}{R}$

10 b) $t = -\frac{1}{k} \ln \frac{2.5-h}{2.5}$
c) an infinite time to fill
d) 7.43 min

CHAPTER 29 Mathematics of uncertainty

Exercises 29.1 *(p.623)*

1 b) 18.5, 7 **c)** 18.8, 5.01
d) Average temperature higher in August, temperatures less widely spread in August.
2 c) 580, 310 **d)** 23
3 b) 56 500, 23 280
4 a) 153, 5.70 **b)** 154.7, 9.65
5 b) 49, 20 **c)** 52.9, 20.2
d) on average lower, but spread wider

Exercises 29.2 *(p.623)*

1 a) 0.25 **b)** 0.375
c) 0.031 25 **d)** 0.5
2 a) 0.002 64 **b)** 0.0106 **c)** 0.105
3 a) 0.24 **b)** 0.936 **c)** 0.010 24
4 b) i) 0.1875 **ii)** 0.2593 **c)** 0.4458
5 a) 0.08 **b)** 0.23 **c)** 0.19
6 a) i) $\frac{1}{120}$ **ii)** $\frac{7}{40}$ **iii)** £52, 57%

Consolidation Exercises for Chapter 29 *(p.634)*

1 8.54, 0.0564, 0.237
2 a) 0.090
b) low probability → poor test
3 b) 0.811
4 i) a) $\frac{1}{4}$ **ii) b)** $\frac{4}{5}$
5 ii) 210
6 a) i) 0.599 **ii)** 0.0861
7 a) 0.3 **b)** $\frac{3}{7}$
9 0.125
10 0.242